ANNUAL REVIEW OF BIOCHEMISTRY

ANNUAL REVIEW OF BIOCHEMISTRY

ESMOND E. SNELL, *Editor*
University of Texas at Austin

PAUL D. BOYER, *Associate Editor*
University of California, Los Angeles

ALTON MEISTER, *Associate Editor*
Cornell University Medical College

CHARLES C. RICHARDSON, *Associate Editor*
Harvard Medical School

VOLUME 50

1981

ANNUAL REVIEWS INC.   4139 EL CAMINO WAY   PALO ALTO, CALIFORNIA 94306 USA

ANNUAL REVIEWS INC.
Palo Alto, California, USA

REPRINTS   The conspicuous number aligned in the margin with the title of each article in this volume is a key for use in ordering reprints. Available reprints are priced at the uniform rate of $2.00 each postpaid. The minimum acceptable reprint order is 5 reprints and/or $10.00 prepaid. A quantity discount is available.

International Standard Serial Number: 0066-4154
International Standard Book Number: 0-8243-0850-6
Library of Congress Catalog Card Number: 32-25093

PRINTED AND BOUND IN THE UNITED STATES OF AMERICA

PREFACE

This volume marks the 50th anniversary of the *Annual Review of Biochemistry*. Anniversaries call for a pause to reflect on where we are, how we got here, and where we may be going. The *Annual Review of Biochemistry*, and the other reviews published by Annual Reviews Inc., were created by the vision and organizational ability of J. Murray Luck, founder and for many years Editor of this *Review*. We are pleased that Murray accepted our request to furnish a prefatory chapter for this volume. His interesting account tells much about the early years of the *Annual Review* and its vital founding Editor.

Perusal of earlier prefaces to this series reveals a scarcity of comment on their quality and acceptance of the volumes, primarily because of the unpretentiousness and modesty of the founder, J. Murray Luck. An exception was the preface 15 years ago, when he retired from the Editorship after 35 years. The present occasion also warrants recognition of the status the series has achieved—indeed, its acceptance by the community of biochemists and molecular biologists is outstanding. One measure of this quality is the willingness of the best scientists to accept invitations to prepare chapters even though they receive no payment, and face the difficult challenge of preparing the expected scholarly appraisal of their area. Only those who have accomplished the task of assembling, sorting, evaluating, condensing, organizing, and interpreting the subject area under review can truly appreciate the time and effort demanded. Sometimes authors are rewarded by the new perspectives or insights they gain; other rewards come from peer group approval and acceptance. The vitality and accomplishment of the *Annual Review of Biochemistry* is only possible because it represents the efforts of skilled practitioners of a highly successful field of science.

Perhaps the most common theme underlying prefaces to past volumes or individual chapters is the immensity of the literature to be covered. Those writing 50 years ago had little concept of the expanded field with which we must now cope. But although faced with an ever increasing mass of scientific data, we do have the emerging generalizations and molecular explanations to help us build our accruing knowledge into a coherent edifice.

Some other indebtedness should be recorded here. One is to the many Members of Editorial Committees for the past 50 volumes. The close intellectual interplay that goes into planning a volume is a heartening experience. A list of Editorial Committee Members of the past (see p. vii) reads like a Who's Who of Biochemistry and Molecular Biology. To all, our thanks.

Another debt of thanks is due to the frequently overworked and often underappreciated production staff of Annual Reviews Inc. who must produce at minimum cost a timely volume of excellence out of the varying degrees of chaos often supplied by authors and editors. Without this staff and the continued fine management, Annual Reviews Inc. could not function.

Perhaps it may not be amiss to record here a pride felt about the nonprofit status of Annual Reviews Inc. The cost of the review is kept as low as possible. We contend that it is the best bargain in science. The value of the *Review* to emerging scholars is amply reflected in the response of graduate students who continue to use *ARB* throughout their careers.

The next fifty years is likely to see many changes in the reporting and review of science. The well written, individual treatise covering a selected portion of a field, and the keen appraisal that comes from a qualified mind, will remain as an essential and welcome contribution to science. A few members of the present *Annual Review* family and readers of this volume will have the pleasure of perusing the 100th Anniversary volume of the *Annual Review of Biochemistry*. From this perspective we hope that the next fifty years will be as glorious as the past fifty have been.

The privilege of writing this preface was extended to me because of my long association with Murray Luck, starting with my arrival in his laboratory at Stanford University in 1943 as a fresh PhD, and continuing later with many years of association at Annual Reviews Inc.

PAUL D. BOYER

EDITORS, ASSOCIATE EDITORS, AND EDITORIAL COMMITTEE MEMBERS

Annual Review of Biochemistry, 1932–1981

Editors

J. Murray Luck, 1932–1965
Paul D. Boyer, 1965–1968
Esmond E. Snell, 1963–1964 (Acting), 1969–present

Associate Editors

Carl R. Noller, 1938
James H. C. Smith, 1939–1946
Hubert S. Loring, 1946–1955
Gordon MacKinney, 1947–1965
Frank W. Allen, 1956–1963

Esmond E. Snell, 1965–1968
Paul D. Boyer, 1964, 1969–present
Alton Meister, 1965–present
Robert L. Sinsheimer, 1966–1972
Charles C. Richardson, 1973–present

Editorial Committee Members

J. M. Luck, 1932–present
C. L. Alsberg, 1932–1940
D. R. Hoagland, 1932–1949
C. L. A. Schmidt, 1932–1946
H. A. Spoehr, 1941–1953
H. J. Almquist, 1941–1955
H. J. Deuel, Jr., 1947–1956
A. K. Balls, 1950–1954
W. Z. Hassid, 1954–1958
B. L. Horecker, 1954–1959
T. H. Jukes, 1955–1957
F. S. Daft, 1956–1960
E. L. Smith, 1957–1961
E. Stotz, 1958–1962
H. E. Carter, 1959–1963
E. E. Snell, 1960–present
A. Meister, 1961–present
C. B. Anfinsen, 1962–1966
E. R. Stadtman, 1963–1967
M. D. Kamen, 1963–1968
E. Racker, 1965–1969
P. D. Boyer, 1965–present

D. S. Hogness, 1966–1970
R. L. Sinsheimer, 1966–1972
H. K. Schachman, 1966–1975
H. A. Sober, 1967–1971
E. P. Kennedy, 1968–1972
M. Nirenberg, 1969–1973
H. A. Lardy, 1970–1974
G. M. Tomkins, 1971–1975
J. Hurwitz, 1972–1976
M. J. Osborn, 1973–1977
C. C. Richardson, 1973–present
D. Nachmansohn, 1974–1978
P. R. Vagelos, 1975–1979
P. Berg, 1976–1980
G. Felsenfeld, 1976–1980
N. R. Cozzarelli, 1977–1981
P. W. Robbins, 1978–present
H. Tabor, 1979–present
M. D. Lane, 1980–present
J. E. Darnell, Jr., 1981–present
J. C. Wang, 1981–present

SOME RELATED ARTICLES IN OTHER *ANNUAL REVIEWS*

From the *Annual Review of Genetics*, Volume 14 (1980)

Transposable Elements in Drosophila and Other Diptera, M. M. Green

Molecular Arrangement and Evolution of Heterochromatic DNA, Douglas L. Brutlag

Phenylketonuria and Other Phenylalanine Hydroxylation Mutants in Man, Charles R. Scriver and Carol L. Clow

Biochemistry of the Gene Products From Murine MHC Mutants, R. Nairn, K. Yamaga, and S. G. Nathenson

The Genetics of Protein Degradation in Bacteria, David W. Mount

From the *Annual Review of Microbiology*, Volume 35 (1981)

Physiology and Biochemistry of Aerobic Hydrogen-Oxidizing Bacteria, Botho Bowien and Hans G. Schlegel

Genetics and Regulation of Nitrogen Fixation, Gary P. Roberts and Winston J. Brill

Translational Initiation in Prokaryotes, Larry Gold, David Pribnow, Tom Schneider, Sidney Shinedling, Britta S. Singer, and Gary Stormo

Evolutionary Significance of Accessory DNA Elements in Bacteria, Allan Campbell

From the *Annual Review of Physiology*, Volume 43 (1981)

Regulation of Smooth Muscle Actomyosin, David J. Hartshorne and Raymond F. Siemankowski

Recent Advances in the Metabolism of Vitamin D, H. F. DeLuca

Epidermal and Nerve Growth Factors in Mammalian Development, Denis Gospodarowicz

"Endorphins" in Pituitary and Other Tissues, Hiroo Imura and Yoshikatsu Nakai

From the *Annual Review of Medicine*, Volume 31 (1980)

Intestinal Absorption and Malabsorption of Folates, Charles H. Halsted

Folate-Binding Proteins, Neville Colman and Victor Herbert

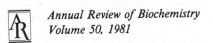 *Annual Review of Biochemistry*
Volume 50, 1981

CONTENTS

ix

ERRATA

ANNUAL REVIEW OF BIOCHEMISTRY, Volume 49 (1980)

In *Replication of Eukaryotic Chromosomes: A Close-Up of the Replication Fork,* by Melvin L. DePamphilis and Paul M. Wassarman

Due to a printer's error, the incorrect version of Figure 3 was printed on page 631. The correct figure is shown below:

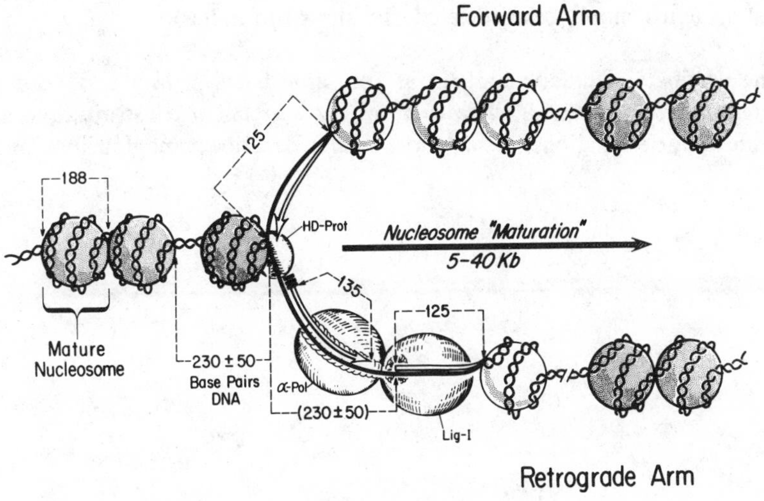

Figure 3 Replication fork of a eukaryotic chromosome. Only those components for which there is substantial experimental evidence are represented: mature (heavy shading) and imma- ture (light shading) nucleosomes, an RNA-primed Okazaki fragment (||| ⟹) on the retrograde arm, helix destabilizing (HD) protein, DNA polymerase α (α-pol), and DNA ligase I (lig-I). The sizes of the components are drawn to reflect their molecular weights. Numbers shown are for SV40 chromosomes since most of the available data comes from viral genomes. The size of the "Okazaki fragment initiation zone" (Figure 1) may be determined by the spacing of nucleosomes in front of the fork (see text). Nucleosomes are actually cylindrical in shape [57 Å × 110 Å (458)]. The direction of DNA synthesis is opposite to that of fork movement on the retrograde arm.

(continued next page)

In *Interconvertible Enzyme Cascades in Cellular Regulation,* by P. B. Chock, S. G. Rhee, and E. R. Stadtman

Due to a typesetter's error, the last line of text on page 825 was deleted. The last sentence should read (deleted line shown in italics):

The different serine residues on glycogen synthase that are phosphorylated by the cAMP-dependent kinase,[1] SPK, PK-3, and PK-4 to yield glucose-6-*P-dependent forms are referred to as sites 1, 2, 3, and 4, respectively (100).*

In *Structurally Distinct Collagen Types,* by Paul Bornstein and Helene Sage

Due to a typesetter's error, the last line of text on page 960 was deleted. The last sentence should read (deleted line shown in italics):

The products of different alleles at the same locus, if *they exist, can be referred to by superscript letters and, when the amino acid* substitutions are defined, by designations such as those used to describe hemoglobin variants.

Ann. Rev. Biochem. 1981. 50:1–22

CONFESSIONS OF A BIOCHEMIST

♦12068

J. Murray Luck

Professor of Chemistry Emeritus, Stanford University,
Stanford, California 94305

I regard it as an honor to be invited to write the Prefatory Chapter for the fiftieth volume of the *Annual Review of Biochemistry.* Anyone who was born in the nineteenth century can only be flattered when those about him entertain the belief that he is still able to write something that might be worth reading.

That 50 is more significant than 49 or 51 in the commemoration of events is lacking in any rational explanation. Yet I confess that I have been charmed and enslaved for years by a peculiar attachment to the number 50. On April 18, 1956 I found myself at the Golden Jubilee dinner of the American Society of Biological Chemists. As President of the Society, it was clear that I should address the members and guests who were present, if only in obedience to a circular letter to the Council in 1912 which would oblige future presidents to address the Society, in order to "show to the world at large that the President of our Society is capable of exerting a function." At this event, with fiftieth anniversaries in mind, I mentioned the *Biochemical Journal* and the *Biochemische Zeitschrift,* both of which were launched in 1906, and told of several other notable events in 1906. When I had concluded my remarks, Newton Richards, my next-chair neighbor, lost no time in reminding me that I had failed to mention the great San Francisco earthquake, possibly the worst calamity, apart from forest fires and floods, in the history of California. This earthquake had occurred exactly fifty years earlier, to the very day.

The Editorial Committee of the *Annual Review of Biochemistry,* concerned lest I might forget another fiftieth anniversary—the birth of the *Annual Review of Biochemistry* in 1931—issued a strict injunction in its invitation to me that I should discuss the origin and early years of this Review.

1

0066-4154/81/0701-0001$01.00

I suppose it all began with my coming to Stanford University in September 1926, as Acting Assistant Professor of Biochemistry. The settling-in process lasted for two or three years, during which I wrestled with the problem of fashioning lecture and laboratory courses in biochemistry, appropriate to the presumed needs of medical students. The nature of laboratory instruction in biochemistry in those far-off days can be inferred from a laboratory manual entitled "Quantitative Analysis of Blood and Urine" which, as author, and until I knew better, I inflicted upon the students. Meyer Bodansky' book *Physiological Chemistry* was the recommended textbook for the lecture course, supplemented by a miscellany of fact and theory from other sources. The pearls of wisdom that dropped from the lips of this youthful novice were avidly seized upon by a few, some of whom behaved as if they were inspired. This was so rewarding to the lecturer that he decided to give a course on current research in biochemistry to a group of ten or fifteen graduate students. *Chemical Abstracts* and three or four of the principal periodicals in biochemistry provided the material for the course. I soon found myself knee-deep in trouble, and this brings us at once to the conception of the *Annual Review of Biochemistry*.

The chemistry and metabolism of amino acids and proteins and two or three other areas in biochemistry were the only parts of the whole field wherein I was sufficiently knowledgeable to feel comfortable before those advanced students. Doubtless there were those gifted biochemists who could review with authority and confidence any area of biochemistry. But I was not among them. However, I was possibly not the only one who suffered from this painful ignorance: there must have been others and, among them, some kindred souls who sought to review for advanced students current research in biochemistry, only to find themselves as dismayed as I was by the immensity of the task. We must remember that even in 1930 *Chemical Abstracts* published about 6500 abstracts of papers on biochemistry, as it was then defined.[1]

In mid 1930, I inquired of about 50 well-known biochemists in the United States and abroad whether an annual volume of critical reviews on the research of the preceding year or two in biochemistry would be a useful addition to the biochemical literature. The volume, of course, would be international in scope and the 30 or so cooperating authors would have, so it was hoped, the necessary expertise to satisfy the expectations of their fellow biochemists the world over.

---

[1]The number of biochemical abstracts published by Chemical Abstracts in 1979 was over 148,000, the increase being attributable to a more comprehensive coverage of the literature, a more expansive definition of the subject, and an enormous increase, worldwide, in the number of research projects being pursued in biochemistry.

The responses to the inquiry were numerous and encouraging. A suggested list of topics received the benefit of very helpful advice by many old hands at the game who knew more about the structure of biochemistry than the young fellow who made the inquiry. Out of this emerged a list of 31 topics for the introductory volume and 14 more to be included with others in Volume II. The names of possible authors were also submitted. The invitations were extended and the declinations were surprisingly few.

Carl Alsberg (Stanford), Denis Hoagland, and Carl Schmidt (both at the University of California, Berkeley) agreed to serve as an Advisory Committee. And so we were off and running—almost. All that remained was to find a publisher. Six commercial publishers regarded the project as interesting but not the kind of a venture they would choose to undertake. Stanford University Press would publish the *Review* if it were adequately subsidized. Fortunately, the Chemical Foundation came to the rescue. We needed $10,000 to cover the deficits anticipated during the first three years. With the help of Carl Alsberg, a man of considerable influence, and with a few letters to the "right" people, the Foundation granted our request. Very few strings attached, other than annual reports of course—those were the days! We had the money, we had a publisher, we were on our way. Volume I, 724 pages, with authors from nine countries, came off the press in July 1932. All of the authors who promised to prepare the 30 reviews kept their word: there were no defaults! The volume sold for $5 as did succeeding volumes until 1948 when the list price was increased to $6.

The Advisory Committee, meanwhile, organized Volumes II and III, each with a lead time of two years. With myself as editor and chairman, the Committee functioned as an embryonic Board of Directors as well as an Editorial Committee for the *Annual Review of Biochemistry*. Except for occasional trivial spats between the publisher and the Advisory Committee (which had the money) in matters of policy regarding, inter alia, advertising and sales price, the relations between the two parties were most cordial and the Press served us well. However, on termination of our three-year contract, our relationship changed fundamentally. It had become more and more apparent that the Advisory Committee should assume a legal identity and accept the responsibilities of a publisher but with Stanford University Press continuing as our printer. On December 12, 1934, the Articles of Incorporation, signed by the four members of the predecessor committee, were filed with the California Secretary of State, and Annual Review of Biochemistry, Ltd., was formed (changed to Annual Reviews Inc. in 1937).

The company was organized as a nonprofit corporation which means, in effect, no shareholders, no dividends, no division of earnings among the members of the corporation, and, in the event of dissolution, no division of the assets among the members. Why did the incorporators not follow the usual route and become a commercial publisher? Briefly stated, "it was the only way to go." The enterprise was totally committed to the unstinted cooperation of all participants for the benefit of the worldwide community of scientists. From the beginning, the Directors and the Editorial Committee Members have donated their services, except for a modest expense allowance for meetings attended. The editors receive a token honorarium. The authors receive neither honoraria nor royalties. No less than 1974 of our colleagues have "done their bit" as authors during the first half century. Their only reward has been a short-term complimentary subscription to the Review, a number of reprints of their article, and the unsung gratitude of many fellow scientists. There is an implicit understanding that Annual Reviews Inc. is uniquely a service organization.

Life can be a grand adventure, rich in excitement and thrills of accomplishment, withal punctuated by many errors and narrow escapes. Annual Reviews Inc., now 50 years of age, has had a full life and has run the entire gamut of corporate hazards, perils, and modest achievements in the face of growing complexities.

At first, it was all delightfully simple. For 18 years, as editor of the *Annual Review of Biochemistry*, I could boastfully claim to know by name and close acquaintanceship all who had served on the Advisory Committee/Board of Directors and the Editorial Committee. It was easy. Apart from myself, only twelve were involved: Alsberg, Hoagland, and Schmidt —all three passed away in the 1940s—Herman Almquist (elected in 1940, retired in 1972), Herman Spoehr (1940–1953), H. Albert Barker (1946–1962), Harry Deuel (1946–1953), John Fulton (1946–1950), Henry Eyring (1948–1969), Ernest Hilgard (1948–1973), Andrew Ivy (1950–1953), and Douglas Whitaker (1950–1956). Until 1950 the Board of Directors and the Editorial Committee were identical in composition. For the next seven years the Editorial Committee of the *Annual Review of Biochemistry* consisted of two or three directors and several other biochemists (Arnold Balls, Floyd Daft, Zev Hassid, Bernie Horecker, Tom Jukes, and Emil Smith), each of whom served for five-year terms. The Board members were gradually rotated into retirement. By 1956 the Editorial Committee was entirely external to the Board of Directors, except for myself, who is still "hanging in there," though rather tenuously. It is with pleasure that we list, in the front of this volume, the names of all those who have been or still are members of the Editorial Committee.

During my 33 years in office, I received invaluable help from those who served as associate editors: Carl Noller in 1938, James H. C. Smith (1939–1946), Hubert Loring (1946–1956), Gordon Mackinney (1946–1965), Frank Allen (1947–1963), Robert Sinsheimer (1965–1972), and Alton Meister (1964–present). Paul Boyer and Esmond Snell have served with conspicuous ability and devotion as editors, jointly or on an alternating basis, since 1962.

There have been many corporate adventures, the first in 1938 when Annual Reviews and the American Physiological Society (APS) agreed to share in the parentage of an *Annual Review of Physiology* (*ARP*). The numerous discussions with officers of the Society and their Board of Publications Trustees are minutely documented in the files of Annual Reviews. From the first it was perfectly clear to both parties that, ideally, they should initiate the new *Review* under a plan of joint participation. But it soon emerged that the obstacles to be overcome were numerous and the initial resistance within the Society was formidable. The details need not be recited. The names of those who were singularly helpful deserve to be mentioned. Their wisdom, good sense, integrity, and simple honesty had earned for them the deepest respect of their fellow physiologists: A. J. Carlson, Walter J. Meek, Wallace Fenn, Frank C. Mann, Carl J. Wiggers, and Ralph W. Gerard. Participation of the Society in publication of the *Review* was formally approved and a Joint Board of Management (JBM) was agreed to: Walter Meek and Chauncey Leake as representatives of the APS and Carl Schmidt and Murray Luck on behalf of Annual Reviews. Volume I appeared in 1939. After a few years of JBM direction, and a subsequent series of gradual transitions, the *ARP* became a full-fledged member of the Annual Reviews family with its own Editorial Committee responsible to the Directors of Annual Reviews.

It was just as well. Quite innocently, the two parties had fashioned a working arrangement that seemed to make sense: Annual Reviews would be the business partner, the APS would elect the Editorial Committee of the new *Review,* and the two parties, year by year, would share equally in any profits or any losses. But it was quite illegal: Annual Reviews Inc., in charge of operations, was thereby enabled to jeopardize the assets of its partner and might indeed run the APS into bankruptcy! So said the attorneys. A new agreement provided that the two parties would divide equally any profits, but any losses would have to be borne by Annual Reviews. Finally, at the request of the Society, the annual operating surplus, if any, ceased to be divided.

Without any special fanfare or problems the *Annual Review of Microbiology* was initiated in 1947. Until his retirement in 1972 Charles Clifton (Stanford) served as Editor and Sidney Raffel (Stanford) loyally did his part as Associate Editor from 1947 to 1979.

The inception of these two new *Reviews* presented scheduling problems to our printers. Stanford University Press, responsible as it was for the timely appearance of the University's Bulletins, Announcements of Courses, Schedules, and many other items, indicated that the production of our *Reviews* on rigid and rather inflexible time schedules would no longer be possible. Hence it came about that Annual Reviews Inc. contracted with the George Banta Company in Menasha, Wisconsin to print the *Annual Review of Physiology* from Volume I forward, the *Annual Review of Microbiology* from its beginnings in 1947, and the *Annual Review of Biochemistry* from Volume 16 forward. The cooperation of the George Banta Company was unstinted: it was more than we could reasonably expect and perhaps more than we deserved.

An ancient Chinese philosopher is alleged to have said, "Since everything is full of something it is not possible to create anything new without pressing that which already exists into an uncomfortable position." In spite of this sage warning, the Company in 1950 plunged deeply into the unknown and, with the assurance of many knowledgeable colleagues in the sciences that the four proposed *Reviews* would satisfy a very real need, founded the *Annual Reviews of Medicine, Physical Chemistry, Plant Physiology,* and *Psychology.*

Something should be said about our relations with Stanford University. In 1931, the University generously made office space available to us to house the editorial and business operations of the enterprise. For 25 years we continued to occupy these premises on a rent-free basis with the University assuming almost all of the costs of utilities and general maintenance. In the meantime, the University became beset with serious occupancy problems. Pressures had developed for additional space for purely academic purposes. With great understanding, forbearance, and patience the Administration permitted us to continue occupancy of the very choice space that had been granted us, until we were able to find other appropriate quarters.

In 1956 we managed to construct a suitable building in nearby Palo Alto that served us well for the next 11 years. The county government, however, eventually cast its covetous and acquisitive eyes upon the building. By exercise of the right of eminent domain, the County forced us to move. Fortunately, they gave us plenty of time to acquire another building site and to construct a new and larger building which we moved into in 1968 and which should be adequate for quite a few more years. The company is now the publisher of 24 separate *Annual Review* series and a number of special publications.

During these first 50 years we have witnessed a progressive fragmentation of the sciences and a loss of any possibility to integrate the bits and pieces into a comprehensible whole. Biochemistry, almost by metastasis, has penetrated many other fields of science and may never again be reassembled into a neatly defined discipline with well-recognized boundaries. Nevertheless, the *Annual Review* series continue to provide a valuable overview of the important research being done in the various disciplines.

On rereading the text, I sense something nostalgic in much of the preceding—a "good old days" sort of flavor. In some respects, the good old days were good. In other respects, they were anything but good, as Otto Bettmann has pointed out in a book entitled *The Good Old Days: They were Terrible.* Let me first of all mention some of the unhappy qualities of the past. I would have to emphasize the lack of adequate information about the cause and treatment of disease, and the absence of many diagnostic aids that are now virtually indispensable. As a child, I suffered through the common diseases that were rampant among the young: measles, mumps, whooping cough, chicken pox, and perhaps one or two others. But so also did my playmates. Our parents believed that such diseases were inescapable; they were part of the business of growing-up. Among the adults, and perhaps also the young, diphtheria and typhoid fever were all too common. Respiratory diseases and epidemic influenza were experienced by many, especially in the winter months. As a freshman in the University of Toronto, I was hospitalized with many others in 1918 because of the terrible influenza plague. It was worldwide and more people died of the disease than were killed in World War I. Hospital space was insufficient in many towns. In Brantford, Canada, a huge circus tent was erected to take care of the hospital overflow.

Sanitary facilities—outdoor privies in abundance—were not good. Waterborn diseases were common. There was little protection against infectious diseases. From the biomedical point of view, the good old days left much to be desired. The statistical summary presented in Table 1 concerns the state of the public health in the past 100 years. The data pertain to Switzerland, only because I happen to have it conveniently at hand. I suspect that the corresponding data for the United States, Great Britain, and Western Europe would be qualitatively similar. The remarkable progress in public health in recent years and the virtual elimination of many diseases is largely due to biochemical and microbiological research.

The good old days were also good; the underlying reason in the early years of the century was the low population density in the towns and countryside. All of us lived close to nature. The radio, television, and movies had not yet infected our lives with a sham sort of entertainment. We did not experience life in the shadows of a ghetto. A crowded life, attended by the mounting pressures of the 1970s, was unknown in much of Canada and the United States. Our present frantic concern over our diminishing resources, our energy needs, and the heavy hand of government was nonexistent. But unrestrained growth of the human population, which many seem to regard with great satisfaction, is now "doing us in."

The dusty or muddy unpaved streets in many towns and villages of my boyhood years never seemed to be worrisome until the automobile came

Table 1 Health: The good old days were awful

A. Deaths from selected diseases in Switzerland[a, b] (numbers per 10,000 of population)

	1880	1930	1970	1978
Diptheria	6.12	0.55	0	0
Erysipelas[c]	0.42	—	—	—
Infantile enteritis	14.60	1.05	0	0
Measles	0.84	0.10	0	0
Puerperal fever	1.27	0.73	0.02	0
Poliomyelitis		0.05	0.005	0
Scarlet fever	2.66	0.10	0	0
Smallpox	0.61	0	0	0
Typhoid fever				
(including paratyphoid)	3.68	0.10	0	0
Whooping cough	2.05	0.32	0	0
Tabes (locomotor ataxia)[d]		0.32	0.03	0
Tuberculosis[e]	19.4	12.4	0.24	0.24
Apoplexy	8.01	5.02	3.40	3.40
Pneumonia and other				
respiratory diseases	24.0	15.2	8.45	5.14

B. Total deaths from all causes (numbers per 1,000 of population)

1880	1900	1930	1970	1965–1978 yearly average
21.9	19.3	11.6	9.1	8.7–9.3

C. Deaths of infants, 0 to 1 year of age (numbers per 100 live births)

1880	1900	1930	1970	1978
18.9	13.3	5.08	1.50	0.86

D. Life expectancies (in years)

| | 1881–1888 | | 1921–1930 | | 1968–1973 | |
At age	M	F	M	F	M	F
0	43.3	45.7	58.14	61.41	70.29	76.22
65	9.7	9.9	10.83	11.84	13.32	16.33

[a] Sources: *Statistisches Jahrbuch der Schweiz*, 1894, 1932, and 1979.

[b] The mortality tables in the *Yearbooks* are heavily footnoted. Readers interested in the definition of any disease or group of diseases should consult the *Yearbooks* for 1950 (p. 81) and 1979 (p. 69).

[c] Not listed in the mortality tables for 1930, 1970, and 1978.

[d] Not listed prior to 1921.

[e] In 1880 only pulmonary tuberculosis (phthisis) was listed. From 1930 on, other forms and other locations of tubercular lesions were included in the *Yearbooks*.

into use. At first the automobile was a novelty and, among the wealthy, it must have been pleasantly ostentatious. I remember well that Brantford was almost free of motor vehicles. Kerosene lamps were abundantly used in the homes and how overjoyed we were as a family when natural gas was introduced and tungsten-filament lighting illuminated several of the rooms. The well from which we pumped the drinking water and the cistern where rainwater was stored, continued in use for many years. But while I was still young, water lines were laid down and indoor faucets came into use. Sewage lines also were installed and the outdoor privy disappeared. The electric light followed the tungsten-filament lamp and shortly therafter we had a hand-cranked telephone. Especially in the rural areas, where six or more phones were commonly on the same line, all of the parties knew everything about each other. There were few secrets. Everyone for several miles around knew promptly when Mary Smith's baby was born, how much John Conway got for his corn, and that Henry Barber was trying to sell one of his old nags. One of my sisters was married to a farmer and rushed frequently to the phone when someone else on the line was being called. Listening in to other people's conversations was a favorite pastime and an excellent way to keep up with the local news.

And what about keeping up with the exciting news on the forefront of scientific research? The annual meetings of the professional scientific societies in the good old days were stimulating and delightful. A meeting was almost a family gathering! One knew almost all of the other participants and discussions were lively and exciting. Multiple contemporaneous sessions were not yet invented, thanks to the small number of papers. The huge meetings of the present are of dubious value and should be abolished. Our universities might be well advised to give to their research people a few hundred dollars for staying away from monstrous meetings rather than for attending.

My father was born in the good old days of the 1850s—the youngest of a family of sixteen children. He was a good man whose goodness had depth and substance; it was not a veneer. He seldom chastized or rebuked his own five children, even though the youngest was inclined toward waywardness and a questioning of the "eternal verities." Neighbourliness in the good old days was common and it was good. The family was accepted and recognized as the irreplaceable unit in the social structure. I believe that the bonds between the members of a family were stronger then than now—perhaps a dubious generalization because great differences can exist from family to family and from one social order to another.

My father was a blacksmith and, later, a workman in an implement factory. He never enjoyed much of an income. It was a hand-to-mouth existence for the family, but none of us suffered. Brantford is an industrial town seven miles from Paris, where I was born in October 1899, the youngest in the family. All but a few in the town knew only a very simple life. Incidentally, strikes were unknown. Doubtless there was exploitation of the working class, but industrial peace, however shaky its foundations might have been, managed to prevail.

In general, the schools in the days of my youth were better than they are now. Teachers were overworked but dedicated. Much was demanded of the pupils, discipline was severe, and substantial homework assignments could not be taken lightly.

After exposure to a rigorous and highly disciplined schooling through the high school level, I was fortunate in being awarded a four-year scholarship that gave me tuition-free undergraduate instruction in the University of Toronto. In the final year, as a chemistry major, I was introduced to biochemistry through a research project on the effect of sodium chloride on the growth and metabolism of yeast. Haldane Gee, a fellow student, shared in the project and in the exhilaration of discovery. Whatever we discovered was modest but for us a real thrill. H. B. Speakman was our mentor and overseer. He encouraged us to use our own wits in designing and pursuing the study. We floundered around but eventually the inevitable bungling of such novices in research faded into the past. The days and nights were all too short. The "work" was fun: for us a new kind of excitement and pleasure.

I spent the summer of 1922 in research at the Marine Biological Station of Canada in St. Andrews, New Brunswick (Director: Professor A. G. Huntsman). While there I learned to my surprise that I had won an 1851 Exhibition Scholarship—nominated by the University of Toronto—for several years of postgraduate study in Britain. I chose Cambridge University, to which I was duly admitted as a research student after acceptance by Gonville and Caius College. Biochemistry was to be my field of research.

What an interesting collection of biochemists occupied the laboratory! Each was different from the next, and each, in his own way, contributed to the exciting, entertaining, and brainy atmosphere of the place. I can now recall only Sir Gowland Hopkins (the lovable Hoppy), Rudolph Peters, J. B. S. Haldane, Marjorie Stephenson, The Hon. Mrs. Onslow, R. A. McCance, Dorothy Moyle (later Mrs. Needham), Joseph Needham, Tim Hele, Robin Hill, Malcolm Dixon, H. F. Holden, Bill Pirie, J. H. Quastel, Margaret Whetham, (later Mrs. Bruce Anderson), Vincent Wigglesworth, and my research partner for a time, Trilok Nath Seth. There were others, many others, each pursuing his or her research in glorious independence.

Hoppy, with much on his mind, seldom knew what each of us was up to but was always interested in hearing about our biochemical doings, and was ever affectionately concerned about our welfare. Haldane was a walking encyclopedia. He came to Cambridge in 1922 as Reader in Biochemistry and soon knew what everyone was doing. In his constant roamings about the laboratory, and in frequent chance encounters, he would discuss with remarkable insight the intricacies of one's research activities.

Hoppy suggested to Robin Hill and me that it might be interesting, as a starter, to collaborate with Haldane. It was indeed interesting! He strongly believed in "being one's own rabbit." As such, he swallowed in three days a 3.5-liter aqueous solution of 85 grams of calcium chloride to induce a good acidosis. Robin and I were responsible for analyzing the great man's urine. He developed an acidosis that was noteworthy. I recall swimming in the river Cam on a Sunday during the height of the 13-day experiment. Haldane was also there. Soon, a punt, bent on ascending the river, made its approach. Seated therein were Hoppy, who had been knighted but recently, Lady Hopkins, and two distinguished-looking guests. Haldane at once swam under and around the punt, describing in his booming voice his experiment on acidosis: "I am now excreting the most acid urine that has ever been excreted. . . ." "Yes, yes," replied Hoppy, rubbing his brow in characteristic fashion. Later it emerged that someone, somewhere, had reached a slightly lower pH but Haldane took his "defeat" in stride.

In my post-Cambridge years, I have felt greatly indebted to Haldane for the introduction in physiology he gave to four or five of us from the lab. We met in his rooms in Trinity on frequent occasions. Questions and more questions always preceded a very informal but informative "lecture" by Haldane. At the first session, I remember he started off with the query "How big do you think my liver is?" He weighed 100 kg. We answered with widely differing percentages of his body weight. "How much blood do you suppose I have?" Answers: A few pints up to a few gallons. "How may one determine the blood volume?" And so on. At the end of the evening, there was always an unorthodox and entertaining summary by Haldane.

Shortly after finishing the acidosis collaboration I investigated sources of the amide nitrogen of caseinogen. The phosphotungstate precipitate of the basic fraction was extracted with an amyl alcohol/ether solution. A remarkable discovery followed: not only ammonia but methylamine constituted the volatile components of the bases! Fortunately, a control run established that the methylamine originated as an impurity in Kahlbaum's C. P. (chemically pure) amyl alcohol. Strange as it may seem, the "discovery" thrilled me immensely—almost as much as if methylamine had been proven to be a product of caseinogen hydrolysis. Years later I had a somewhat similar experience. I have had to conclude that the joy of discovery in research is not necessarily limited to the discovery of truth but can also stem from the discovery of error.

Apart from research I had the never-to-be-forgotten and humiliating experience of "baptizing" the new laboratory (the Sir William Dunn Institute) into which Hoppy's "family" moved in 1924. Seth and I occupied a laboratory on the top floor. Late in the day I opened several faucets to be sure that water running through a condenser was cold. An hour or so later, I left for home and forgot about the wretched faucets. One of them misbehaved; during the night the water overflowed, reached electrical outlets in the floor, shorted the building's electrical circuits, and soaked the beautiful, plastered ceiling below. All of this happened only days before the building was to be dedicated and a great assembly of distinguished visitors would be present. Hoppy, though visibly dismayed, never even hinted, at least to me, that biochemistry was possibly not my "cup of tea."

In 1925 I emerged with a PhD degree and returned to the University of Toronto for a year as a Demonstrator in Biochemistry. Andrew Hunter, Hardolph Wasteneys, and Arthur Wynne constituted the top echelon. Speakman, by this time, headed a nearby institute for industrial biochemical research.

During the year in Toronto I was offered an assistant professorship by T. P. Nash in the Medical School of the University of Tennessee. We were both interested in the origin of blood ammonia. I promptly accepted the offer but, unexpectedly, because of a citizenship problem, the opportunity came to naught. Shortly thereafter, Robert Swain of Stanford University offered me an assistant professorship in his Chemistry Department with responsibility for the teaching of biochemistry. I went to Stanford in September 1926 and remained as an active member of the Department until retirement in 1965.

But there were many interruptions and discontinuities: life was not characterized by total immersion in research and teaching. Like other scientists I was drawn into many extramural committees and organizations during those 40 years: as an administrative officer of the Pacific Division, American Association for the Advancement of the Sciences (AAAS) for about 15 years; as a member of fund-granting committees of the American Cancer Society, the National Institutes of Health, the Medical Fellowship Board, one of the panels of the National Research Council, and the National Science Foundation. For several years I was head of the Section on Biological Chemistry of the International Union for Pure and Applied Chemistry —a more time-consuming diversion than I ever expected. But it was a memorable time because the groundwork was already being laid for the International Union of Biochemistry.

When I was a young fellow, even as old as 30, I was eager to see injustice corrected and the institutions of our country, if not of the world, reshaped into something that savored of heaven on earth. The severe depression of 1929–1933 was a tremendous incentive to the young to single out the misdeeds and mistakes of their elders. I became interested in the Mooney-Billings case which stemmed from a bomb explosion in San Francisco in July 1916. Ten persons were killed and forty wounded. Mooney was convicted and sentenced to death. Billings was sentenced to life imprisonment. Even before the remaining three defendants were tried, it was established that gross perjury had been committed by the principal witnesses for the prosecution. The three were acquitted but it was too late, because of legal technicalities, to release Mooney and Billings. They were still in prison in 1930. In March 1930, I joined a delegation that waited upon Governor Young who had the power, but not the inclination, to release the two men. Shortly thereafter, I visited Mooney in San Quentin and Billings in Folsom prison. Not until 1939 were the two men released. I wrote a report of the case (1).

At about the same time, I became interested in a nine-year old boy in the State of Washington whose parents were members of the Elijah Voice Society. The parents, in September 1925, withdrew the boy from school because of their objection to compulsory participation of children in flag-salute exercises in the schools. The father served eight days in jail for refusal to send the boy to school. After a succession of court orders, the boy was transferred by order of the court in January 1926 to an orphanage. In June 1926, all intercourse, direct or indirect, with the parents was prohibited. The parents refused to compromise their beliefs and, in June 1927, the court ordered that the boy be held for adoption. In November of that year, Judge Hardin, who succeeded Judge Brown in the County Superior Court, ordered that the boy be returned to his parents. I wrote a long account of the case but never submitted it for publication.[2]

In 1935, supported by a General Education Board Fellowship, I spent three months in Cambridge and three in Copenhagen in research on the liver proteins. For visiting biochemists, the Chemical Department of the Carlsberg Laboratory, headed by S. P. L. Sørensen, was a sort of Mecca. Sørensen, I suspect, was a very kind and considerate man of the old-school type. He made frequent rounds of the laboratory and derived, so it seemed to me, a special pleasure in finding everyone hard at work. I never felt able to confide in him, partly because he seemed to be very formal and I regarded him as up there somewhere among the Olympians of science, quite distant from the mere mortals in the lab. When I mentioned to him that I was planning a hurried pre-Christmas trip to Paris, Brussels, Göttingen, Berlin,

[2]*A History of the Case of Russell Tremain.* Photocopies are available from the author.

Munich, Zürich, Prague, and Budapest to meet certain biochemists, I felt that he considered such a departure from the laboratory as a wasteful fortnight.

Sørensen was 67 when I studied in the Carlsberg Laboratory. He died at the age of 71. "Time made him gentler, he became a delightful old man, calm and wise . . . just, unselfish, and warm-hearted. He expected a lot from his pupils, but that is no more than every good teacher ought to do" (2).

Sørensen's first assistant, later his successor, was Kai Linderstrøm-Lang. I am sure that everyone who knew Kai loved him. He was friendly, affable, helpful, and possessed of a delightful charm and good humour. Like Sørensen, he was a superb and imaginative scientist–at that time engaged in the development and use of ultramicro analytical techniques. I was especially interested because of his identification of the cells in pig gastric mucosa in which urease activity is localized.

Organization of the *Annual Review of Biochemistry* apparently caused me to be regarded by some as an authority, not only on reviews, but on abstracts, and on many other facets of the dissemination of science information. In 1948 I was sent to London as a US delegate to the Royal Society Conference on Science Information, and in 1949 to Paris as a US delegate to the UNESCO Conference on Science Abstracting.

At several conferences, beginning in 1946, on documentation, storage, and retrieval of information, it was quite the thing to point to the rising flood of science papers and to bemoan the gaps in our abstracting services. Crash programs to attain a total coverage in abstracting were proposed by some. I recall also a proposal at an International Conference, by an expert on snails, that we busy ourselves in abstracting papers published anywhere at any time in journals that have perished—the defunct literature of science.

Others have pointed to the "incredible waste in time and money" through duplication of effort. Hubert Humphrey, in 1961, as Chairman of a Senatorial subcommittee concerned with government operations, reported that an estimated 160,000 projects in research and development in the United States cost about $12 billion, of which the federal government's share approximated $8.1 billion. He stated that the various agency information systems were a "hodge-podge" . . . "overlapping, underplanned, undernourished and under-used."

I participated, with Chauncey Leake and Maurice Visscher, in a conference called to consider the Humphrey report and the role of government in the management of our information services. His subcommittee had recommended a total view of the information problem, commencing with the very conception of a project. Publication of the ideas and research plans of scientists prior to initiation of a contemplated investigation would be a beginning. Negative results from projects that turned sour should be published. Papers delivered orally at scientific meetings should be drawn within the compass of the overall government-managed information service—the argument, among others, being that "if they had not been registered, abstracted, and indexed on a prepublication basis, they may "disappear" except within the circle which will have heard the oral presentation." Also to be included would be the squeezings from progress reports and from projects cancelled without reaching formal publication; . . . "unless this is done, the chances are 'pretty high' that in later years scientists may needlessly and unwittingly repeat the earlier work of cancelled projects" (3).

I was horrified by the recommendations, which would generate, if implemented, a flood of information of dubious value and at an outrageous cost. Having little trust in the ability of government to manage anything, and believing that government should do nothing for people that people can do for themselves, I made a very negative contribution to the conference.

I have had a lasting interest in the science of nutrition since those unforgettable years in Cambridge. Hopkins, in his discovery of tryptophane and "accessory food factors," laid a firm foundation for much that followed, the world over, in our knowledge of essential amino acids and vitamins. About all that I myself ever published on nutrition appeared in two brief papers (4, 5), and as a book which dealt with human nutrition, poverty, and consumer cooperation (6). In case this leaves the impression that the book must have been a curious *Gemisch,* or catch all, I hasten to point out that nutritional disease is sometimes a result of poverty. Poverty, in turn, can find a partial solution in cooperation whereby groups of people, through a united attack on the economic problems that beset them, can do much through a great variety of cooperative enterprises to pull themselves out of the mud by tugging on their own bootstraps. Unfortunately, with the encouragement of politicians, we find ourselves beseeching our Washington uncle to provide salvation through a multitude of giveaway programs. And all such demands to government, at all levels, are compounded by the insistence of many that we be enabled to live in a risk-free environment. I see about me an almost universal hysteria over food additives and the products of our chemical and pharmaceutical industries, expressed in strident voices by many who should know better, that everything we eat, drink, breathe, wear, and look at be convincingly proven to be absolutely safe. The Environmental Protection Agency and the Food and Drug Administration, hedged in by a massive burden of legislative directives, are the unfortunate instruments of Congress in seeking attainment of such impossible goals.

In 1957, as President of the Pacific Division, AAAS, I gave an address on the population problem: *Man against his Environment: the Next Hundred Years* (7). It contained the sentence: "Abortion, at the request of the prospective mother, should not only be permitted but, in some instances, encouraged." To me, this was an innocent and reasonable proposal, but what a storm of protest it generated! About 90 letters of outrage were received in the weeks that followed, and 110 clippings of editorials and letters to the editor that appeared in the daily press arrived by mail. Protests and approvals were almost equally divided. One concerned citizen urged that I be fired from Stanford and another that I merely be castrated. The whole collection gives an illuminating picture of Americana in 1957.

My interest in the population problem has continued undiminished. In 1960, the following quotation expressed a widely shared concern:

> The dilemma that unfolds before us involves an explosively burgeoning population in a world that is threatened with the loss of many of its natural resources and increasing assaults upon its precious personal freedoms. The years before us may prove to be chastening in the extreme, and I would respectfully urge that the problems before us today be faced with all the courage, humanity and intelligence that we can summon to the task.[3]

In the last year or two preceding my retirement from Stanford, I conducted a senior-student discussion group on the population problem. We had several exciting sessions devoted to possible solutions. One suggestion that caught my attention, as if approaching the scaffold, called for the painless dispatch of all people on attainment of age 65: "On the average, they have nothing more to contribute to society." Like the aging Eskimo, we should wander out in the dark of night and disappear.

In 1960, as a member of an exchange mission, I went to the USSR. There were five of us in the group (Henry Sebrell Jr., Clifford Barborka, Floyd Daft, Currier McEwen, and myself). As guests of the Ministry of Health, we visited a number of institutes in Moscow, Leningrad, Kiev, Sukhumi, and Yerevan hoping to learn something about the status of research on arthritis and metabolic diseases. Although assured by our hosts that we could go anywhere in the USSR that we wished, practical considerations, such as shortage of time, limited us to European Russia.

Something like 15 million Russian men were reported to have lost their lives in the Second World War. Hence, we were not surprised that the directors and many staff members of most of the institutes we visited were women. Many other occupations, including heavy labor (e.g. hod carriers) in building construction, usually restricted to men, were at this time taken over by women. As an official policy designed to restore population losses, special encouragement and allocation of resources were then given to pediatrics. The research in progress, notably in biochemistry in the institutes

---

[3]Concluding paragraph of an address on "Hunger and Want or Population Control" at the Founding Conference of the *World Population Emergency Campaign,* Princeton, NJ, March 20, 1960.

visited, was quite pedestrian, partly because of wartime destruction of facilities, inadequate access to foreign periodicals, and constant bureaucratic political meddling. In 1961, I attended the biochemical congress in Moscow. I was accompanied by my wife who is of Russian parentage. After the congress we escorted a party of twelve American colleagues on a prearranged tour from Moscow to Leningrad, Kiev, and Yalta. All of us were Intourist "hostages." The heavy hand of the bureaucracy and the inefficiency of the system plagued us throughout the tour—from a near inability to learn which of the 12 were lodged in the Hotel Ukrainya (and to communicate by telephone with any in the same hotel) to the daily insistence by Moscow, from Leningrad on, that we were six in number even though only two had left at Leningrad for Helsinki. Of course, a head count always revealed a remainder of ten, each with the appropriate Intourist travel documents. In retrospect, the daily war of words for accommodations in hotels and on planes for a party of ten instead of six was highly entertaining.

The Luck family lived in Switzerland from March 1962 to the spring of 1964. A stranger in Washington—an official in the Department of State—phoned me in 1961 to inquire if I would be interested in serving for two years as the Science Attaché in the American Embassy in Bern. How could one fail to be interested in a two-year tour of duty in Switzerland? Nevertheless, months passed before the prospective appointment became a reality. Involved in all of this was also a cherished belief that the human organism needs a real jolt from time to time—a radical change of scene, a new type of intellectual or physical activity, a greatly altered human and physical environment. A two-year change seemed to be about right—long enough to be jolted out of those comfortable ruts that had become deeper and deeper in the preceding wonderful years at Stanford, but not long enough to require another severe jolt to effect a return to California and to Stanford.

As the two years came to an end, I learned that I would be expected to prepare a report on science in Switzerland and to indicate to the State Department what its Science Attaché had been up to during his tour of duty. I labored over that report for two years. It became so obese that the Department was obliged to approve its publication as a book (8): photocopies of the manuscript—100 or so for interoffice distribution—would have made unreasonable demands upon the filing space available to recipients.

I fell completely in love with Switzerland, her people, and their institutions, tempered though my affection was by recognizing that "all is not gold that glitters." I later found myself immersed in writing a general history of Switzerland which may never be finished, because I have been so fascinated with the subject and have loaded the manuscript with so much trivia and detail that any professional historian would probably be aghast in its presence. The history of preceding centuries was difficult enough to encompass, but the twentieth century posed problems that, to me, were insuperable. In short, a retrospective view of 50 or 100 years may be required to understand current events and discuss them intelligently. So what was the solution? In 1975 I discussed the problem with a number of friends in Switzerland. Out of our conversations emerged a volume *Modern Switzerland* in which 27 Swiss, possessed of the necessary expertise in different topical areas, cooperated in authorship (9).

I returned to Stanford in 1964, confident that my life as a science attaché had come to an end. But it was not to be so. The State Department sent me to the US Embassy in London for two months in 1967 and to our Embassy in Stockholm for a similar period to serve as the acting science attaché during temporary absences of the regular appointees. Since 1964 I have returned to Switzerland annually, for a month or so, to study in the National Library, where an enormous collection of works on Switzerland, wherever published, plus stacks of newspapers, journals, and reference materials are to be found. I owe a debt of gratitude to Dr. Raetus Luck (unrelated to me), the Vice-Director of the Library, for his interest and frequent help.

And now, deserting science and chronology, I return to the distant past. I was always a lover of books. As a boy I read everything by G. A. Henty and Horatio Alger that I could find. Some of the works of Fenimore Cooper and Mark Twain were delightful. And there were many others. As a freshman in the University of Toronto, I became interested in an opportunity to spend the summer in Western Canada as a book salesman. The book to be sold was a very large volume entitled *Better Farming* or *Farm Economy*— I forget which. It was not difficult to persuade a few fellow students to join in the venture. We took the required Knox Course in Salesmanship, then, youthfully confident that we could sell the book to anyone, we went to Saskatchewan, where each was assigned a huge block of the province as his exclusive territory. We almost felt as if we were mediaeval counts, each with vested property rights in a few hundred square miles of prairie. The *Hauptstadt* in my vast territory was the small town of Lanigan. Each of us travelled by bicycle over roads of a sort to reach our "tenants," some of whom farmed an entire square mile (640 acres) while others had a mere 160 acres. The distances between farmhouses were considerable and we spent the night in the last place we visited in the late afternoon. We were always received as welcome guests (lodging and meals freely provided) because social gatherings were infrequent and visitors were few in number in such sparsely populated regions.

The selling procedure was marvellous. In the distance, several hundred yards away, we might see the farmer at work on the land. We knew his name for it had been given to us by one of his neighbours. "Good morning, Mr.

Brown," we would say with a friendly handshake, "I am Dick Turpin (or whoever). Such wonderful weather for the crops—but also for smut, which of course goes after everybody's wheat at some time or other." "It sure does," says Brown. "Well, I'm here as a representative of the Better Farming Association to talk to everyone I meet about plant diseases." Taking two small vials from a pocket, I continue, "You will recognize, Mr. Brown, this one here is a specimen of wheat infested with 'loose smut', the other is 'stinking smut.' Which one do you usually have around here?" Before he can answer, I reach for a good-sized prospectus in my rucksack and continue "I just happen to have here some very good pictures of both." If lucky, I open the prospectus at the right page instead of one which pictures a cotton plantation or a watery field of rice. Then the real sales talk begins and I answer all possible objections against purchase—we had learned the answers in the course on salesmanship! Finally, opening the prospectus at the last page, where I had already inserted the names of a few of the prominent farmers in the area, and holding it so Brown could see the names, I give him the final thrust "I'm sure, Mr. Brown, you want one of these books like the rest of your neighbors." The reply that I always received was, "No, I don't want it." To this, as we had learned, there was no triumphant reply and I had to accept defeat.

As itinerant salesmen we were supposed to carry a provincial license, obtainable for a significant fee. If, however, one was selling hymnals, bibles, or religious publications, the license was not required. In consequence, we also had in the rucksack a second prospectus descriptive of religious literature. If we ever saw a uniformed member of the Royal Canadian Mounted Police approaching in the distance, we were instructed to switch promptly from one prospectus to the other and engage Brown in a serious conversation about the Bible. I never met an RCMP under such circumstances, though later I met one who served in the Lanigan area and I accompanied him on several occasions when he held court in neighboring towns as judge and jury to adjudicate various petty crimes.

As a salesman of books, I was a total failure and soon abandoned the job. Instead, I served as bookkeeper for the Ford/McLaughlin garage in Lanigan, wrote the tax notices for the town, "audited" the books of the Lanigan branch of the Saskatchewan Grain Growers' Association, and worked part-time in the local dairy determining the fat content of milk received daily from Lanigan-area farmers.

My love of books was not without its problems. In Paris, in 1949, I tried to purchase two books by my fellow Californian, Henry Miller. The bookstore, finding them temporarily out of stock, offered to send them by mail a few days later. Shortly thereafter, I received a polite but peremptory request from an officer of the US Customs: "A package of books by Henry Miller addressed to you is being held by the US Customs. Their importation into the United States is not permitted. Please be good enough to sign the enclosed form to authorize their immediate destruction" (or words to this effect). I replied, "Thank you for your kind letter of recent date. However, I do not choose to sign the enclosed form." Soon a second and similar request arrived. In the meantime I transferred title to the books to Stanford University Libraries, the Director being happy to add them to the library's collection of Americana. The US Customs disapproved of the pending transfer and again asked permission to consign the books to the flames. I refused to join in this contemplated burning of books—a shameful practice which for many centuries has impaired the study of history and literature. The American Civil Liberties Union, Western Division, then proposed to the judge of the appropriate court that a committee of 12 or so professors of English literature be appointed to pass upon the alleged obscenity of the books. The judge replied that a professor in one of our West Coast Catholic colleges had already convinced him that the books were obscene. The cost of a legal challenge to the judge's decision would have been prohibitive. This closed the case, but sometimes I doubt if the books were ever destroyed: I hope not.

During my student days in Toronto, Henri Lasserre, a Swiss professor of French, entertained a few of us in his home on frequent occasions to discuss consumer cooperatives, especially in a greatly expanded setting as total cooperative communities. The pleasure of these "bull" sessions was enhanced by the tea and cookies which Mme. Lasserre regularly provided. A few years later, as a student in Cambridge, I joined the local Consumers' Cooperative Society. It was in the local co-op food store that my wife and I bought our groceries. The society was clearly a partial realization of the expansive economic organization that Lasserre had talked about. The co-op appealed to me as a reasonable way to do business—the ultimate in free enterprise, and a valuable adjunct to the usual type of retail stores.

In the first few years at Stanford, I found myself invited to give talks—formal lectures or in discussion groups—on consumer cooperatives. In the late 1920s and early 1930s, during the severe economic depression, self-help enterprises of many types were organized throughout the country. At Stanford, several students urged me to take the initiative in organizing in Palo Alto a consumers' cooperative society. The Stanford community was fertile ground. The Founding Grant of the University contains the interesting injunction to the trustees "to have taught in the University the right and advantages of association and cooperation."

Despite many difficulties, the Palo Alto Consumers' Cooperative Society was founded in 1934 and by the end of the year had enough members to operate a small grocery store. In subsequent years, most of the initial difficulties vanished and the growth of the Society has been considerable.[4]

A few years later, nine of us, as members of the Palo Alto Co-op, organized the Palo Alto Credit Union, which now has about 5500 members and assets in excess of five million dollars.

Another adventure in consumer cooperation was formally initiated on April 17, 1944. I was invited by an officer of the Stanford YMCA/YWCA to attend a Christmas vacation conference (1942/1943) in Santa Cruz, California, at which many subjects of student interest were to be debated. I would be expected to preside over a discussion group on consumer cooperatives. Quite unexpectedly, I had the good fortune to meet Sumner Spaulding, a Los Angeles architect, who led a discussion group on city and community planning. Summer and I realized at once that our two discussion groups shared a common interest—cooperative housing. The land for a prospective housing development would be acquired by the collective action of the future occupants. Continuing to act collectively, the group would engage a planner to lay out the tract—roads, walkways, utility systems, etc. An architect, engaged by the group, would design the houses, which would be of ten to fifteen different plans according to size and interior and exterior design. Finally, advantage would be taken of the economies inherent in the coordinated construction of all the residences by a single firm of contractors.

In the months that followed, the Peninsula Housing Association (PHA) was incorporated (April 17, 1944). Many meetings of the members and prospective members were held and addressed by architects and community and regional planners. A news bulletin "The Four Hundred" was issued biweekly. When our resources permitted we purchased, at auction, for $155,000, a beautiful 253-acre tract of rolling land a few miles from Stanford's 8000 acres. Roads were laid out and utilities partially installed. However, the difficulties in carrying the plan to completion were insuperable: it was much too ambitious, and the Federal Housing Administration ceaselessly objected to this and that. Eventually PHA sold out to a Palo Alto firm of realtors who stripped the original plan of some of its attractive features in order to maximize the number of building lots. Members had each invested $2500 as a down payment on a stated building lot designated in the original PHA plan. In time, the new owners returned to the members the sums thus invested. Lots were sold and houses constructed according to conventional practices, and the community of Ladera, named by PHA, came into being—a mere shadow of the wonderful creature of which the founders had dreamed.

[4]The Palo Alto Society now has approximately 23,000 members and five supermarkets.

In 1976, I became involved with others in organizing a nonprofit tax-exempt enterprise—the Society for the Promotion of Science and Scholarship Inc. (SPOSS Inc.)—to encourage the publication of scholarly works that our university presses and commercial publishers, for financial reasons, are frequently prone to reject. The Society has published three scholarly works and is committed to several others.

The professional lives of many of my research students and associates have been centered in biochemistry or kindred sciences. Their encouragement in some of the peripheral activities in which I have indulged so heavily and, above all, their lasting friendships constitute precious fringe benefits that are rewarding beyond words. I can mention only a few of these colleagues: Paul Boyer, Emmett Chappelle, David Cornwell, Denis Fox, Clark Griffin, Erik Heegaard, Janet Ingalls, Venkataraman Jagannathan, Donald Kupke, Kenneth Murray, Lafayette Noda, Gordon Nordby, and Kazuo Satake. John Eudin, for years, was a most helpful research assistant. In certain specific projects, I enjoyed a happy association with John Marrack (London) and with David Bassett, Arthur Giese, Victor Hall, and C. V. Taylor, formerly active members of the Stanford University faculty.

It will be clear to the reader that much of my life has been given over to the organization of this and that. As one consequence, my contributions to biochemical research may have been meager. An enormous amount of time has been given to activities that were remote from science and far afield from biochemistry in particular. Yet it was my initial explorations into chemistry and biochemistry as a student that opened to me many of the doors and opportunities that have helped make my life as a biochemist, professor, and publisher a full and rewarding one. I have greatly enjoyed almost everything in which I have engaged and have had the lasting pleasure of association with many students and others in teaching, research, and other activities. Should one ask for more?

Literature Cited

1. Luck, J. M. 1930 *Friends Intelligencer.* Fourth Month 26, pp. 326–29; see also p. 324
2. Linderstrøm-Lang, K. 1938. S. P. L. Sorensen (1868–1939). *C. R. Trav. Lab. Carlsberg, Ser. Chim.* 23:27
3. Committee on Government Operations (US Senate) and its Subcommittee on Reorganization and International Organizations, Senator Hubert Humphrey, Chairman. Rep. dated Sept. 20, 1961, p. xix. See also Report #263, May 18, 1961, *Coordination of Information on Current Scientific Research and Development supported by the United States Government.* Washington DC: GPO
4. Luck, J. M. 1941. Fortification of foodstuffs. *Science* 94:31–33
5. Luck, J. M. 1945. Nutrition—retrospect and prospect. *Nutr. Rev.* 3:65–69
6. Luck, J. M. 1945. *The War on Malnutrition and Poverty–The Role of Consumer Cooperatives.* New York: Harper. 203 pp.
7. Luck, J. M. 1957. *Science* 126:903–8
8. Luck, J. M. 1967. *Science in Switzerland.* Irvington-on-Hudson, NY: Columbia Univ. Press. xvi + 419 pp.
9. Luck, J. M., ed. 1978. *Modern Switzerland.* xvi + 515 pp. Palo Alto, Calif: SPOSS (Soc. for the Promotion of Sci. and Scholarship.)

Ann. Rev. Biochem. 1981. 50:23–40
Copyright © 1981 by Annual Reviews Inc. All rights reserved

AMINO ACID DEGRADATION BY ANAEROBIC BACTERIA

♦12069

H. A. Barker

Department of Biochemistry, University of California,
Berkeley, California 94720

CONTENTS

PERSPECTIVES AND SUMMARY

The study of amino acid degradation by anaerobic bacteria has always been complicated by the fact that such organisms require complex media for growth. Until recently, satisfactory methods were not available for following the degradation of all the individual amino acids in such media. This difficulty was partially overcome by adding a high concentration of one amino acid to a medium containing low concentrations of other nutrients and determining the predominant products formed from the major substrate. In this way several investigators demonstrated that single amino acids, including aspartic acid, glutamic acid, histidine, lysine, glycine, alanine, γ-aminobutyrate, δ-aminovalerate, serine, threonine, and tyrosine, can serve as major energy sources for selected species of anaerobic bacteria. In 1934, Stickland, using washed suspensions of *Clostridium sporogenes*,

23

made the important discovery that this organism uses most single amino acids poorly, but readily degrades certain pairs of amino acids in coupled oxidation-reduction reactions ("Stickland reaction"). Subsequently the Stickland reaction was found to be used by many clostridia.

The abilities to ferment single amino acids and to use several amino acids either as oxidants or reductants, sometimes combined in one organism, imply that the degradation of the amino acids in a protein hydrolysate is a complex process. Even the determination of the relative rates of utilization or formation of each of the individual amino acids is difficult and has only recently been accomplished with a representative group of clostridia grown in a casein hydrolysate medium (1). Undoubtedly the relative rates of utilization of the various amino acids and the product yields will vary with the composition of the medium and other experimental conditions.

Amino acid degradations by anaerobic bacteria always involve oxidation and reduction reactions between one or more amino acids or non-nitrogenous compounds derived from amino acids. The oxidation reactions are usually similar or identical to corresponding reactions catalyzed by aerobic organisms, except for limitations imposed by the absence of molecular oxygen or other high potential oxidant. Thus oxidative deaminations, transaminations, and α-keto acid oxidations commonly occur in anaerobic amino acid degradation, but oxygenation reactions and fatty acid oxidations are excluded. The reduction reactions are more distinctive, since each organism must generate one or more electron acceptors of suitable potential from the amino acid(s) that it can metabolize. Electron acceptors used by amino acid–fermenting bacteria include amino acids, α- and β-keto acids, α, β unsaturated acids or their coenzyme A thiolesters, and protons. The ultimate reduction products include a variety of short-chain fatty acids, succinic acid, δ-amino valeric acid, and molecular hydrogen. Some of the more novel reactions catalyzed by amino acid–fermenting bacteria participate in the conversion of the substrate into oxidants and reductants, and in energy storage. An example of the former is the conversion of lysine to 3,5-diaminohexanoate and crotonyl·CoA; an example of the latter is ATP formation coupled with the reduction of glycine to acetate and ammonia.

During the past ten years, the period mainly covered by this review, the major interest in a few laboratories has been the determination of the metabolic pathways and enzymatic reactions involved in the degradation of selected amino acids by anaerobic bacteria. Several novel pathways and reactions, not previously observed in aerobic organisms, have been identified. A special stimulus for this type of investigation was the earlier demonstration that *Clostridium tetanomorphum* and *Peptococcus aerogenes* ferment glutamate to identical products, but use entirely different metabolic pathways. This emphasized again the fallacy of assuming that two organisms converting the same substrate to the same products necessarily contain the same enzymatic machinery.

Some reviews dealing with various aspects of this topic have appeared during the past few years (2–5).

AMINO ACID UTILIZATION BY VARIOUS CLOSTRIDIA

Mead (1) investigated the abilities of a number of amino acid–fermenting clostridia to utilize individual amino acids in a 3% casein hydrolysate medium supplemented with various growth factors. This is the first extensive systematic investigation of amino acid utilization by clostridia. Patterns of amino acid utilization or formation were determined semiquantitatively by a two-dimensional separation on paper by means of electrophoresis and chromatography, and quantitatively by use of an amino acid analyzer. The species and strains were divided into four groups and several subgroups on the basis of the patterns of amino acid utilization or formation.

Group I consists of organisms that carry out the Strickland reaction between certain pairs of amino acids. A consistent feature of the reaction is the reduction of proline, and generally arginine, to δ-aminovaleric acid, coupled to the oxidation of serine, phenylalanine, and possibly other amino acids. Some strains of *C. sporogenes* and *C. bifermentans* were shown to grow in the basal medium with only proline and phenylalanine added. Most of the organisms in Group I produce α-aminobutyric acid, which was shown to be derived from either threonine or methionine; they also utilize aspartate, glycine, alanine, leucine, tyrosine, and tryptophan in various degrees and form γ-aminobutyrate from glutamate. The organisms in Group I are *C. sordelli, C. sticklandii, C. botulinum* types A and B, *C. caloritolerans, C. sporogenes, C. difficile, C. putrificum,* and one strain of *C. cochlearium.*

The organisms in Groups II, III, and IV produce neither δ-aminovalerate, α-aminobutyrate nor γ-aminobutyrate from casein hydrolysate. All Group II organisms (*C. botulinum* type C, *C. histolyticum, C. subterminale,* and one strain of *C. cochlearium*) utilize glycine, glutamate, serine, aspartate, and arginine; *C. cochlearium* and *C. subterminale* also decompose lysine, proline, leucine, phenylalanine, and histidine. All Group III organisms (*C. tetani, C. tetanomorphum, C. microsporum,* and one strain of *C. cochlearium*) utilize glutamate, histidine, aspartate, and serine readily. Threonine, methionine, and tyrosine are degraded by all the organisms except *C. microsporum.* Acetate and butyrate are the major fatty acids formed by Group III organisms, along with a smaller amount of propionate derived from threonine. Group IV contains only one species, *C. putrefaci-*

ens, that uses threonine and serine readily, and glutamate, aspartate, arginine, lysine, glycine, and proline more slowly.

Elsden & Hilton (la) have developed a relatively simple and rapid method for qualitative identification of amino acids metabolized by anaerobic bacteria growing in a casein hydrolysate medium. The method involves the conversion of the amino acids to fluorescent dimethylaminonaphthalene sulfonyl derivatives and separation of the latter by two-dimensional chromatography on polyamide sheets. The method has been applied to determine amino acid utilization patterns of nine species of *Clostridium* not previously examined by Mead (1).

DEGRADATION OF SPECIFIC AMINO ACIDS

Valine, Leucine, Isoleucine, and Threonine

Volatile fatty acid formation by 27 *Clostridium* species from a trypticase-yeast extract medium and from the same medium supplemented with 20 mM threonine, valine, leucine, or isoleucine was investigated by Elsden & Hilton (6). The fatty acids were identified and quantitated by gas-liquid chromatography. The identity of branched-chain valeric and caproic acids in some samples was determined by mass spectrometry. This investigation confirmed and extended the results of earlier studies of the patterns of formation of C2 to C6 fatty acids by clostridia. The number of acids formed by a given species from a complex amino acid mixture was found to vary from one (acetic) to eight (acetic, propionic, *n*-butyric, isobutyric, *n*-valeric, 2-methylbutyric, 3-methylbutyric, and isocaproic); one species also formed small amounts of *n*-caproic acid.

By supplementing the complex medium with individual amino acids and observing the change in volatile acid yields, evidence for the origin of specific volatile acids was obtained. When the medium was supplemented with threonine, the majority of species gave a large increase in the yield of propionic acid and a smaller increase in the yield of *n*-butyric acid. A few species (*C. histolyticum, C. sporospheroides, C. sticklandii,* and *C. subterminale*) formed only acetic acid from threonine; two species (*C. difficile* and *C. scatologenes*) gave increased yields of *n*-valeric acid. Supplementation of the medium with valine, leucine, or isoleucine resulted in greatly increased yields of branched-chain fatty acids in 19 of 27 species studied. Organisms that oxidized valine to isobutyric acid also oxidized leucine to 3-methylbutyric acid and isoleucine to 2-methylbutyric acid. The isocaproic acid formed from leucine by several species was shown to be 4-methylvaleric acid, a reduction product. No comparable reduction of valine to isovaleric acid or of isoleucine to 3-methylvaleric acid was observed.

Poston (7) reported that isobutyric and acetic acids are major products of the fermentation of leucine by *C. sporogenes*. Cell-free extracts of this organism also formed unspecified amounts of ^{14}C-labeled isocaproic, α-ketoisocaproic, β-ketoisocaproic, and α-ketoisovaleric acids from L-α[U-^{14}C]leucine. The formation of isobutyric acid from leucine suggested that β-leucine (β-aminoisocaproic acid) may be involved in leucine degradation. This was confirmed by demonstrating that *C. sporogenes* extracts contain an enzyme, leucine 2,3-aminomutase, which interconverts leucine and β-leucine. This activity is stimulated by AdoCbl and inhibited by intrinsic factor. These observations indicate that isobutyric acid is formed from leucine via β-leucine and β-ketoisocaproic acid. Other products of leucine degradation are evidently formed by a pathway involving α-oxidation.

Proline

Proline has long been known to serve as a major oxidizing agent for a number of amino acid–degrading clostridia, and is reduced to δ-aminovaleric acid. Early studies (8, 9) established that extracts of *C. sticklandii* contain a very active proline racemase and a D-proline reductase that were partially purified and completely separated. The most effective electron donors with the reductase were found to be dithiols such as 1,3-mercaptopropanol and dithiothreitol. Pyridine nucleotides cannot replace dithiols with highly purified enzyme, although with crude extracts NADH may be more effective than a dithiol (10). The utilization of NADH requires two proteins in addition to D-proline reductase, an FAD-contained NADH dehydrogenase, and a presumed metalloprotein that is inhibited by EDTA and diethyl dithiocarbamate (11a, 12). The latter protein is thought to function as an electron carrier between the NADH dehydrogenase and proline reductase. It is also apparently required when leuco-methylene blue is used in place of NADH as a reducing agent.

Hodgins & Abeles (13) made a major contribution to the mode of action of D-proline reductase by showing that the enzyme contains a pyruvate moiety, probably linked to the enzyme by a peptide bond, that is essential for catalytic activity. More recent improvements in the method of isolating the reductase from the membrane of *C. sticklandii* (10) have permitted more detailed studies of the enzyme structure and the location of the pyruvate residue (11, 14). The reductase has been found to have a molecular weight of about 300,000 and to contain 10 probably identical subunits. Each subunit contains one pyruvate residue, which blocks the amino-terminal end of the protein. Mild alkaline hydrolysis of the enzyme liberates a pyruvate-containing polypeptide, which is also blocked at the amino terminus and has a molecular weight of about 4,600. An apparently identical polypeptide, as judged by electrophoretic mobility, COOH-terminal analysis, and amino acid composition, is released by treating the reductase with either lithium borohydride or hydroxylamine. The pyruvate-containing polypeptide has a serine at its carboxyl end. The hydroxyl group of the serine is evidently esterified with the γ-carboxyl group of a glutamate residue. The existence of this unusual ester bond is indicated by the mild conditions required for its hydrolysis and by the identification of α-amino-δ-hydroxyvalerate as a product of acid hydrolysis of the borohydride-cleaved enzyme. The pyruvate moiety is evidently derived from serine, since cultures provided with ^{14}C-serine form proline reductase containing ^{14}C-labeled pyruvate.

Glycine

When glycine serves as an oxidizing agent in the Stickland reaction it is reduced to acetate and ammonia in a complex enzymatic reaction that is only partially understood. Earlier studies by T. C. Stadtman and her associates (15) established that either NADH or a dimercaptan such as dithiothreitol can serve as the reducing agent in extracts of *C. sticklandii.* NADH is presumably the normal source of reducing power. The utilization of NADH for glycine reduction requires a complex electron transport system apparently containing at least three components: a flavoprotein dehydrogenase, ferredoxin, and a labile protein that has not been further characterized (16). When a dimercaptan replaces NADH, the need for these electron carriers is eliminated, but at least three other proteins or protein fractions, designated protein A, protein B, and fraction C, are required for glycine reduction (17). The dimercaptan-linked system also normally requires orthophosphate and ADP, and forms approximately one mole of high energy phosphate per mole of glycine reduced, although some enzyme preparations show little or no dependency on these compounds. Arsenate can replace phosphate and adenylate nucleotides.

Proteins A and B have been extensively purified, whereas fraction C is a heterogenous mixture of proteins (16–18). Protein A has a molecular weight of about 12,000 and is formed abundantly only when *C. sticklandii* is grown in a medium containing 1 μM selenite. When protein A was isolated from bacteria grown in the presence of $Na_2{}^{75}SeO_3$, it was found to contain 1 g atom of selenium per mole. When ^{75}Se-labeled protein A was reduced, treated with an alkylating reagent, and acid hydrolyzed, selenocysteine was isolated and characterized as its Se-carboxymethyl, Se-carboxyethyl, and Se-aminoethyl derivatives (19). The amino acid composition of protein A showed that it contains two sulfur-cysteine residues and one Se-cysteine residue. It is very easily oxidized by molecular oxygen with a distinctive change in the absorption spectrum between 240 and 280 nm, which is thought to be caused by the selenide anion. The oxidized form is readily reduced with KBH_4 to restore the original spectrum. The ease of oxidation-reduction and the low molecular weight led to the hypothesis that protein A functions as an electron carrier in glycine reduction.

Protein B and the proteins in fraction C have been shown by gel filtration to have molecular weights in excess of 200,000 (16, 17). Protein B is readily inactivated by hydroxylamine or borohydride, and therefore is thought to contain an essential carbonyl group. It is also inactivated by iodoacetamide in the absence, but not in the presence, of glycine, which suggests that an essential sulfhydryl group is located near the substrate-binding site. Protein B is also reported to catalyze a very slow exchange of tritium between [2-^{3}H]glycine and water. This activity seems to correlate with the catalytic activity of protein B in a reconstituted glycine reductase system. Tanaka & Stadtman (17) have postulated that the exchange reaction results from the interaction of the carbonyl function of protein B with the amino group of glycine.

C. sporogenes has also been shown to have an absolute requirement for selenium when using glycine as oxidant in a Stickland-type fermentation (20). No glycine reductase activity was detectable in cells grown without selenium. In a medium containing excess alanine and glycine, growth of *C. sporogenes* is proportional to selenite concentration in the range of 0.01–0.12 μM.

Addition of relatively high concentrations of proline and pyruvate, or a pyruvate precursor such as serine, to a growth medium containing a low concentration of glycine, suppresses the formation of protein A and at least one additional protein component of glycine reductase in *C. sporogenes* (21). The presence of proline in the growth medium also suppresses the ability of cells to take up glycine.

Ornithine

Ornithine is oxidized by resting cells of *C. sticklandii* mainly to acetate, alanine, and ammonia, and reduced to δ-aminovalerate (22). When equimolar amounts of ornithine and proline are provided, ornithine is oxidized to the same products plus carbon dioxide, while proline is reduced to δ-aminovalerate. By using appropriately labeled ^{14}C-ornithine, Dyer & Costilow (22) showed that ornithine is cleaved between carbon atoms 3 and 4 to form acetate from carbon atoms 4 and 5, and alanine from carbon atoms 1–3.

The first step in the pathway of L-ornithine oxidation is its conversion to the D isomer by a racemase. This is followed by an AdoCbl- and pyridoxal phosphate–dependent migration of the amino group at carbon atom 5 to carbon atom 4 to form D-*threo*-2,4-diaminopentanoate (23–25). The AdoCbl presumably functions as a hydrogen carrier between the two carbon atoms as in several other AdoCbl-dependent reactions. The specific function of pyridoxal phosphate in this reaction is unknown. The third step in the pathway is an NAD- or NADP-dependent oxidative deamination of 2,4-diaminopentanoate at carbon atom 4 to form 2-amino-4-ketopentanoate (24, 26). Both the aminomutase and the dehydrogenase have been extensively purified. In the fourth step, 2-amino-4-ketopentanoate undergoes a thiolytic cleavage by reaction with coenzyme A to form acetyl·CoA and alanine (27). The latter is presumably the D isomer, although this could not be demonstrated because the enzyme was contaminated with alanine racemase. The partially purified 2-amino-4-ketopentanoate thiolase is stimulated by addition of pyridoxal phosphate and can be made completely dependent on this cofactor by exposure to cysteine and subsequent dialysis against 1 mM dithiothreitol. Full activation of the apoenzyme requires exposure to 0.1 mM pyridoxal phosphate for 30 min at 25°C. The enzyme is not stimulated by pyridine nucleotides or α-keto acids, thus supporting the conclusion that no oxidative deamination or transamination is involved in the reaction.

In the presence of excess proline, part of the alanine formed by *C. sticklandii* from ornithine is oxidized to acetate, carbon dioxide, and ammonia (27, 28). It is not known whether the first step in alanine oxidation by this organism is a transamination or an oxidative deamination.

The utilization of ornithine as an enzymatic oxidizing agent involves the formation of proline as an intermediate. The conversion of ornithine to proline has been examined in dialyzed extracts of *C. botulinum* and *Clostridium* PA 3670, reported to be similar to *C. sporogenes*. The first study of this reaction led to the conclusion that ornithine is oxidized in an NAD-dependent reaction to glutamic-δ-semialdehyde, which spontaneously cyclizes to Δ'-pyrroline-5-carboxylate and is then enzymatically reduced to proline by NADH (29). This conclusion was based upon the following evidence. The conversion of ornithine to proline required a catalytic amount of NAD; Δ'-pyrroline-5-carboxylate was formed from either ornithine or proline in the presence of enzyme and NAD; and an active NAD-dependent Δ'-pyrroline-5-carboxylate reductase was demonstrated in extracts. This evidence seemed reasonably convincing, despite the fact that all the experiments were done with relatively crude enzyme preparations. However, when the same investigators undertook to purify the enzymes responsible for the conversion of ornithine to proline and ammonia they found that the Δ'-pyrroline-5-carboxylate reductase could be completely separated from an activity ("ornithine cyclase") catalyzing the overall reaction by either gel filtration, gel ionophoresis, or sucrose gradient centrifugation (30). Proline formation was not increased by recombining the two activities. The possibility that the cyclase contained a bound form of the reductase that cannot utilize free Δ'-pyrroline-5-carboxylate seems to have been eliminated by the observation that the cyclase has a much lower molecular weight (80,000) than the reductase (about 200,000). A later study established that highly purified ornithine cyclase contains approximately one mole of tightly bound NAD per mole of enzyme (31). Removal of bound NAD resulted in loss of activity that could be partially restored by adding NAD but not NADH. A role of NAD in the cyclase reaction was also indicated by a transient increase in A_{340}, lasting up to 2 min, when L-ornithine was added to a high concentration of holoenzyme (32). However, incorporation of tritium from exogenous NADT into proline during the cyclase reaction could not be detected, presumably because no exchange occurs between bound and free coenzyme.

By the use of [δ-^{15}N]ornithine it was shown that the α-amino nitrogen is converted to ammonia and the δ-amino nitrogen is retained in proline during the cyclase reaction. Consequently, the conversion of ornithine to proline apparently involves an oxidative deamination at the α position, with formation of 2-keto-5-aminopentanoic acid and Δ'-pyrroline-2-carboxylic acid as intermediates. The latter compound is presumably reduced to proline by reaction with the bound NADH formed in the oxidation step.

The NAD-dependent proline dehydrogenase and Δ'-pyrroline-5-carboxylic acid reductase activities of *C. sporogenes* are associated with a single protein or protein complex (33). The two activities are not separated by fractionation on DEAE-cellulose, hydroxylapatite, or Sephadex columns, and they have identical sedimentation coefficients and isoelectric points, and similar heat stabilities and inactivation rates in low ionic strength buffers. Unlike this enzyme, a Δ'-pyrroline-5-carboxylate reductase from *Escherichia coli* is unable to oxidize proline (34). The physiological role of proline dehydrogenase in *C. sporogenes* is not certain, though it may be involved in the conversion of proline to glutamate.

Tyrosine, Phenylalanine, and Tryptophan

Elsden et al (35) determined the identities and amounts of aromatic products formed by growing cultures of 23 Clostridium species from phenylalanine, tyrosine, and tryptophan. The aromatic products derived from phenylalanine are phenylacetic, phenylpropionic, and phenyllactic acids; those derived from tyrosine are hydroxyphenylacetic acid, hydroxyphenyllactic acid, hydroxyphenylpropionic acid, phenol, and *p*-cresol; those derived from tryptophan are indole, indoleacetic acid, and indolepropionic acid. The aromatic rings are apparently not modified by the clostridia. The species fall into five groups with respect to the amino acids used and products formed. One group, including *C. sporogenes,* uses all three aromatic amino acids as oxidizing agents and reduces them to propionic acid derivatives. A second group, including *C. bifermentans,* uses mainly phenylalanine, which is both oxidized to phenyllactic and phenylacetic acids and reduced to phenylpropionic acid; a little indole is also formed from tryptophan. A third group, including *C. sticklandii,* oxidizes all three amino acids to the corresponding acetic acid derivatives; in addition, one species, *C. difficile,* forms considerable *p*-cresol and a little phenylpropionic acid. A fourth group, including *C. tetani,* produces indole and phenol as the only aromatic products. A fifth group apparently does not utilize the aromatic amino acids.

The formation of phenol from tyrosine by *C. tetanomorphum* was investigated by Brot et al (36). They purified an enzyme, tyrosine phenol lyase, that catalyzes the conversion of L-tyrosine to phenol, pyruvate, and ammonia and requires pyridoxal phosphate and Mg^{2+} as cofactors. The enzyme is highly specific for L-tyrosine; neither D-tyrosine, phenylalanine, tryptophan, nor several tyrosine analogues are utilized. Crystalline tyrosine phenol lyases with similar properties, but lower substrate specificities, have been isolated from aerobic bacteria, *Escherichia intermedia* (37) and *Erwinia herbicola* (38).

The enzyme responsible for indole formation from tryptophan by clostridia has apparently not been characterized, although it is probable that tryptophanase performs this function as in aerobic and facultative bacteria. A comparable formation of benzene from phenylalanine has not been reported.

The oxidation of aromatic amino acids to the substituted acetic acids by anaerobic bacteria presumably involves formation of the α-keto acids by transamination, and either a direct oxidative decarboxylation of the keto acid or a decarboxylation followed by oxidation of the resulting aldehyde. The presence of at least two aromatic amino acid : α-ketoglutarate transaminases in *C. sporogenes* has been demonstrated (39). The oxidation of α-keto acids to acetic acid derivatives by extracts of *C. sporogenes* was shown to require NAD, CoA, and cocarboxylase (40), but it is not known whether the reaction occurs by an oxidative decarboxylation or by a decarboxylation and subsequent oxidation of the aldehyde, as has been demonstrated in the oxidation of phenylpyruvate by the aerobic *Achromobacter eurydice* (41).

Indoleacetic acid can be converted to skatole presumably by direct decarboxylation. This was shown by incubating ruminal microorganisms with [2-^{14}C]indoleacetic acid; 38% of the ^{14}C was recovered in skatole (42a). In a similar reaction, *p*-hydroxyphenylacetic acid had been shown earlier to be converted to *p*-cresol by an unidentified facultative organism under anaerobic conditions (42). No enzymatic studies of cresol or skatole formation have been reported.

The reductions of phenylalanine, tyrosine, and tryptophan to phenylpropionic, *p*-hydroxyphenylpropionic, and indolepropionic acids by clostridia apparently do not involve an initial desaturation by an "ammonia lyase" such as that found in many plants (43). Rather, there is considerable evidence that the pathway of amino acid reduction goes via the α-keto, α-hydroxy, and α, β-unsaturated acids. O'Neil & DeMoss (39) purified a pyridoxal phosphate–dependent tryptophan : α-ketoglutarate transaminase from C. sporogenes to a specific activity of 23.6 units per milligram. The enzyme also catalyzed transamination with tyrosine and phenylalanine. A second transaminase preferentially utilizing phenylalanine and α-ketoglutarate was present in the organism (see also 44). Jean & DeMoss (45) partially purified an NAD-specific indolelactate dehydrogenase from extracts of C. sporogenes. The enzyme was more active with phenyllactate and p-hydroxyphenyllactate than with indolelactate and was inactive with lactate. Presumably it participates in the metabolism of all three aromatic amino acids. Bühler et al (44) reported the presence of a relatively high level of an NADH-dependent phenylpyruvate reductase in crude extracts of C. sporogenes. The reduction product is presumably (R)-phenyllactate, since only this isomer is converted to phenylpropionate by resting cells of C. sporogenes under an atmosphere of hydrogen.

The conversion of aromatic α-hydroxy acids to α, β-unsaturated acids has not been demonstrated directly in amino acid–fermenting bacteria. Evidence for participation of an unsaturated acid in the reduction of aromatic amino acids is indicative but incomplete. Moss et al (46) demonstrated the formation of both cinnamic and phenylpropionic acids when washed cell suspensions of C. sporogenes were incubated with L-phenylalanine. The time course of cinnamic acid formation and utilization suggested that it was converted to phenylpropionic acid. Boezi & DeMoss (47) earlier reported that crude extracts of C. sporogenes can convert indoleacrylate as well as indolelactate to indolepropionate. Recently, several clostridia, including C. sporogenes, have been reported to contain a 2-enoate reductase that converts (E)[1]-cinnamate to phenylpropionate in an NADH-dependent reaction (44). Extracts of C. sporogenes also reduce 4-methyl-2-pentenoate, which could be derived from leucine, but not 3-methyl-2-pentenoate, which could be derived from isoleucine. Since these reductions are not inhibited by 0.2 M hydroxylamine, they evidently do not involve the participation of CoA thiolesters.

A 2-enoate reductase having a broader substrate specificity than the C. sporogenes enzyme has been extensively purified from Clostridium kluyveri, an organism that does not degrade amino acids readily (48). This enzyme has a molecular weight of about 450,000 and contains six apparently identi-

---

[1]The latest nomenclature for cis-trans isomers. E = entgegen = trans; Z = zusammen = cis.

cal subunits. Each subunit contains 1 mole of FAD and about four atoms each of iron and labile sulfur. A distinctive feature of the enzyme is the sterochemistry of double bond reduction, which is a mirror image of that catalyzed by butyryl-CoA dehydrogenase (49). The stereochemistry of double bond reduction by the 2-enoate reductases of *C. kluyveri* and *C. sporogenes* are reported to be the same (H. Simon, private communication).

In summary, the available evidence indicates that the reduction of both aromatic and nonaromatic amino acids by clostridia involves the sequence: amino acid, 2-keto acid, 2-hydroxy acid, 2,3-unsaturated acid, saturated acid. All of the component reactions have been demonstrated, except the conversion of the hydroxy acid to the unsaturated acid. This may involve the participation of the coenzyme A thiolesters of these compounds as in the conversion of 2-hydroxyglutarate to glutaconate by extracts of *Acidaminococcus fermentans* (50). Also lacking is a demonstration that the rates of the individual reactions are sufficient to account for the overall rate of reductive deamination.

Glutamate

Two pathways of glutamate degradation are utilized by anaerobic bacteria to form identical products: acetate, butyrate, carbon dioxide, ammonia, and hydrogen. *C. tetanomorphum* uses the methylaspartate pathway, in which the successive intermediates are 2S,3S-β-methylaspartate, mesaconate, S-citramalate, and pyruvate; the latter is further converted to the final products (51). *Peptococcus aerogenes* was later found to use the hydroxyglutarate pathway, in which the successive intermediates are α-ketoglutarate, R-α-hydroxyglutarate, and probably glutaconyl-CoA and crotonyl-CoA; the latter is presumably converted to acetate, butyrate, and hydrogen. The existence of this second pathway of glutamate degradation was deduced by Whitely (52), who found that extracts of *P. aerogenes* are unable to utilize mesaconate, citramalate, or β-methylaspartate. By fermenting ^{14}C-labeled glutamates with cell suspensions, Horler et al (53) found that the linear carbon chain of glutamate remained unchanged in butyrate, whereas with *C. tetanomorphum* the carbon chain is rearranged. Cell suspensions of *P. aerogenes* were shown to convert glutamate to glutaconate (54), and cell-free extracts converted glutamate to α-hydroxyglutarate (55). Finally, the DPN-dependent enzymes responsible for this conversion, glutamic dehydrogenase and α-hydroxyglutarate dehydrogenase, were shown to be present at high levels in extracts of *P. aerogenes* and *A. fermentans* (56–58). The fate of 2-hydroxyglutarate remained unclear for some years, although it was postulated to be converted to crotonyl-CoA via glutaconyl-CoA (54, 59). Buckel (50, 60) recently obtained evidence in a support of this postulate. He investigated the cofactor requirements for the conversion of (R)-2-hydroxyglutarate to glucaconate by extracts of *A. fermentans* and found that catalytic amounts of coenzyme A and acetylphosphate or acetyl-CoA are needed. Extracts were then shown to contain a CoA transferase that transfers CoA reversibly between acetate and (R)-2-hydroxyglutarate or (E)-glutaconate. This enzyme has been extensively purified and crystallized, although detailed data on its properties have not yet appeared.

The dehydratase catalyzing the conversion of (R)-2-hydroxyglutaryl-CoA to (E)-glutaconyl-CoA is a particulate enzyme that has not yet been purified. Buckel (50) has shown that this conversion in extracts can be conveniently assayed by the stereospecific release of the *pro*-3S-hydrogen as ^{3}HOH from (2R,3RS)-2-hydroxy[3-^{3}H]glutarate in the presence of the cofactors mentioned above. In addition, for the development of full activity, the enzyme requires a prior incubation with acetyl phosphate, NADH, $FeSO_4$, $MgCl_2$, and dithioerythritol under anaerobic conditions. The enzyme is inhibited by low concentrations of uncouplers of oxidative phosphorylation, such as 2,4-dinitrophenol, arsenate, and azide. Buckel concludes that a catalytic phosphorylation is probably involved in the action of the enzyme, but a more detailed analysis of the reaction mechanism must await its purification. The requirement of the dehydratase for NADH is thought to be analogous to the requirement of NADPH for the enzymatic elimination of phosphate from 3-*enol*pyruvoyl-shikimate 5-phosphate to give chorismate (61).

The decarboxylation of glutaconate, presumably as glutaconyl-CoA, by extracts of *A. fermentans* is stimulated by addition of biotin and inhibited by avidin, as expected (50). The decarboxylase, like the dehydratase, is membrane bound.

The two pathways of glutamate degradation can be distinguished either by determining the distribution of ^{14}C in butyrate derived from [4-^{14}C]glutamate or, more easily, by determining activities of 3-methylaspartase and 2-hydroxyglutarate dehydrogenase in extracts (59). In application of the [4-^{14}C]glutamate tracer method, organisms using the methylaspartate pathway form ^{14}C-butyrate containing half of the ^{14}C in the carboxyl group and half in the remaining three carbon atoms, whereas organisms using the 2-hydroxyglutamate pathway have virtually all the ^{14}C in carbon atoms 2–4. Extracts of organisms using the 2-hydroxyglutarate pathway contain high levels of the corresponding dehydrogenase and no β-methylaspartase, whereas the reverse is true with extracts of organisms using the methylaspartate pathway.

A third method of distinguishing between the two pathways of glutamate degradation utilizes L-[4-^{14}C, 3-^{3}H]glutamate as substrate (62). The resulting butyrate is isolated and converted chemically to propylamine. The ^{3}H/

[14]C ratio of propylamine divided by the same ratio of butyrate gives a clear identification of the pathway. The quotient is approximately 2.0 for the methylaspartate pathway and 1.0 for the hydroxyglutarate pathway.

A survey of 15 species of glutamate-fermenting clostridia and nonsporulating anaerobes by the above methods showed that nine out of eleven species of *Clostridium* use the methylaspartate pathway, whereas all four of the nonsporulation species of the genera *Peptococcus, Acidaminococcus,* and *Fusobacterium* use the hydroxyglutarate pathway. No evidence for a third pathway was obtained.

The citramalate lyase which catalyzes the conversion of L-citramalate to pyruvate and acetate, the third reaction in glutamate degradation by *C. tetanomorphum,* is very unstable in crude extracts (63). Buckel & Bobi (64) have found that this instability is the result of deacetylation; the enzyme can be fully reactivated by treatment with acetic anhydride or other acetylating agents following reduction by dithioerythritol. The properties of the enzyme, other than substrate specificity, are similar to those of citrate lyase, which is also inactivated and activated by deacetylation, and acetylation, respectively. The citramalate lyase contains pantothenate and cysteamine residues. Acetylation occurs at a cysteamine residue, which can also be carboxymethylated by iodoacetamide. The purified enzyme contains three polypeptide chains, α, β, and γ, in approximately equimolar amounts and has a hexameric structure. The pantothenate and cysteamine residues are exclusively present in the γ-chain, which is the acyl carrier protein of citramalate lyase. Acyl exchange between acetyl-S-enzyme and citramalate to form acetate and citramalyl-S-enzyme, and cleavage of the latter to pyruvate with regeneration of acetyl-S-enzyme are presumably catalyzed by the α and β subunits. Buckel (65) has presented evidence that a mixed citramalic-acetic anhydride participates in the acyl exchange reaction.

An earlier investigation of the glutamate decarboxylase from *C. perfringens* established that the enzyme is not activated by pyridoxal phosphate, although its spectral properties are similar to those of the *E. coli* glutamate decarboxylate and other pyridoxal phosphate–dependent enzymes (66). Treatment with sodium borohydride both inactivated the clostridial enzyme and modified its absorption spectrum. Reinvestigation of the purified clostridial enzyme demonstrated that it contains approximately 2 moles of pyridoxal phosphate per mole of molecular weight 290,000 (67). The coenzyme is released by diluting the enzyme in neutral or weakly alkaline solution. Although the apoenzyme is relatively labile, it is stabilized in the presence of glutathione and can be reconstituted by addition of the coenzyme.

Lysine

Lysine is actively used by *C. subterminale* and *C. sticklandii* as a major energy source and is converted to acetate, butyrate, and ammonia (68). Tracer experiments have indicated that at least three more or less distinct pathways participate in this process in intact bacteria. One pathway involves a cleavage of the six-carbon chain of a lysine metabolite between carbon atoms 2 and 3 to form acetate derived from carbon atoms 1 and 2 and butyrate from carbons atoms 3–6. A second pathway involves a cleavage between carbon atoms 4 and 5 to form butyrate from carbon atoms 1–4 and acetate from carbon atoms 5 and 6. A third pathway involves the formation of butyrate from two acetate or acetyl moieties. Only the first pathway has been studied in detail.

The first intermediates in the first pathway are β-lysine (3,6-diaminohexanoate), 3,5-diaminohexanoate, and 3-keto-5-aminohexanoate. The available information about the enzymes catalyzing the formation of these compounds has been summarized by T. C. Stadtman (68). More recent studies have been concerned with later steps in the fermentation.

The degradation of 3-keto-5-aminohexanoate was found to occur by a novel type of enzymatic reaction with acetyl·CoA which results in the formation of L-3-aminobutyryl·CoA and acetoacetate (69). The responsible enzyme, 3-keto-5-aminohexanoate cleavage enzyme, has been extensively purified, and shown to be activated by Co^{2+} or Mn^{2+}, and inhibited by orthophosphate and thiol reagents. The reaction it catalyzes is readily reversible, with the equilibrium slightly favoring the back reaction. Tracer experiments have shown that carbon atoms 1 and 2 of acetoacetate are derived from carbon atoms 1 and 2 of 3-keto-5-aminohexanoate, and carbon atoms 3 and 4 of acetoacetate are derived from the acetyl moiety of acetyl·CoA. The amino acid moiety of 3-aminobutyrl·CoA is derived entirely from carbon atoms 3–6 of the β-keto acid substrate. Neither CoA nor other intermediates could be detected by several possible group exchange reactions. Consequently the reaction probably occurs by a concerted exchange of CoA for the carboxymethyl moiety of 3-keto-5-aminohexanoate.

L-3-Aminobutyryl·CoA is deaminated reversibly to crotonyl·CoA by an active enzyme that has been extensively purified from *C. subterminale* (70). The enzyme cannot utilize the D-isomer. Crotonyl·CoA is presumably reduced mainly to butyryl·CoA, although this enzymatic reaction has not been directly demonstrated. Butyryl·CoA can then react with acetoacetate to form butyrate and acetoacetyl·CoA under the influence of a highly active and stable CoA transferase that has been purified from the same organism (71). Acetoacetyl·CoA is converted by a thiolase to two moles of acetyl·CoA, one of which participates in the 3-keto-5-aminohexanoate cleavage reaction, whereas the other is used to form ATP and acetate via the phosphotransace-tylase and acetate kinase reactions. An alternative pathway for crotonyl-CoA utilization involves its conversion by crotonase to L-3-hydroxybutyryl-CoA and the oxidation of the latter by 3-hydroxyacyl·CoA dehydrogenase to acetoacetyl·CoA. This sequence can provide both additional NADH and ATP that may be needed for biosynthetic reactions. All enzymes required for these reactions have been demonstrated in extracts of *C. subterminale.*

The second pathway of lysine degradation, forming butyrate from carbon atoms 1–4 and acetate from carbon atoms 5 and 6, has been postulated to start with the conversion of D-lysine to 2,5-diaminohexanoate (68). This product, which is formed by D-lysine 4,5-aminomutase, is a potential sub-strate for oxidation at carbon 5 and subsequent cleavage to yield C2 and C4 products. An enzyme catalyzing a pyridine nucleotide–dependent oxida-tive deamination at carbon 5 has been purified from *C. sticklandii,* but since it oxidizes carbon 4 of 2,4-diaminopentanoate, an ornithine metabolite, almost 1000 times faster than carbon 5 of 2,5-deaminohexanoate, it is doubtful that it functions in the second pathway of lysine degradation. Jeng et al (27) have suggested that 2,5-diaminohexanoate may be converted to 3,5-diaminohexanoate, presumably a 3D diasteromer, and this compound is oxidized to 3-amino-5-ketohexanoate. The latter compound may then be cleaved by a specific thiolase to acetyl·CoA and D-3-aminobutyrate. Direct evidence for such a reaction sequence is still lacking.

Aspartate

L-Aspartic acid is readily fermented by a *Campylobacter* sp. and by *Bacter-oides melaninogenicus.* In the degradation of one mol of aspartate by the *Campylobacter,* 0.3 mol is converted by an oxidative pathway to 0.6 mol of CO_2 0.1 mol of acetate, 0.1 mol of succinate, and 0.3 mol of ammonia; this is coupled with the reductive conversion of 0.7 mol of aspartate to succinate and ammonia via the fumarate reductase system, which is known to participate in electron transport phosphorylation (72–75). These results, combined with quantitative data on cell yield, led to the estimate that 0.66 mol of ATP equivalents is formed in the reduction of aspartate to succinate (76). The utilization of aspartate by the reductive pathway can also be coupled to the oxidation of hydrogen or formate. With excess formate, virtually all of the aspartate is reduced to succinate.

The initial step in aspartate utilization is catalyzed by aspartase, which is present in high levels in extracts (72). Fumarase, fumarate reductase, malic enzyme, malate dehydrogenase, isocitrate dehydrogenase, and hy-drogenase are also present in extracts. Most of these enzymes are presum-ably involved in fumarate oxidation or reduction. Cells grown anaerobically with aspartate contain cytochromes *b* and *c* and menaquinone, compounds associated with fumarate reductase systems in other organisms. In *Campylobacter* extracts, NADH-reduced cytochrome *b* is readily reoxidized by fumarate. The rate of reduction of cytochrome *b* by NADH is decreased by the addition of 2-heptyl-4-hydroxyquinoline N-oxide, a known inhibitor of electron transport between quinones and cytochromes. These and other observations suggest that electrons from NADH are transferred sequentially to menaquinone, cytochrome *b,* and fumarate in this organism.

B. melaninogenicus forms the same products as *Campylobacter* from L-aspartate, but the yields of carbon dioxide and acetate are higher and the yield of succinate is lower (77). These results indicate that succinate is formed only by the reductive pathway in this organism.

Literature Cited

1. Mead, G. C. 1971. *J. Gen. Microbiol.* 67:47–56
1a. Elsden, S. R., Hilton, M. G. 1979. *Arch. Microbiol.* 123:137–41
2. Kikuchi, G. 1973. *Mol. Cell. Biochem.* 1:169–87
3. Thauer, R. K., Jungermann, K., Decker, K. 1977. *Bacteriol. Rev.* 41:100–80
4. Gottschalk, G. 1979. *Bacterial Metabolism,* Ch. 8. New York: Springer. 281 pp.
5. Stadtman, T. C. 1980. *Ann. Rev. Biochem.* 49:93–110
6. Elsden, S. R., Hilton, M. G. 1978. *Arch. Microbiol.* 107:283–88
7. Poston, M. J. 1976. *J. Biol. Chem.* 251:1859–63
8. Stadtman, T. C. 1956. *Biochem. J.* 62:614–21
9. Stadtman, T. C., Elliott, P. 1957. *J. Biol. Chem.* 228:983–97
10. Seto, B., Stadtman, T. C. 1976. *J. Biol. Chem.* 251:2435–39
11. Seto, B. 1978. *J. Biol. Chem.* 253:4525–29
11a. Seto, B. 1978. *Fed. Proc.* 37:1521
12. Schwartz, A. C., Müller, W. 1979. *Arch. Microbiol.* 123:203–8
13. Hodgins, D. S., Abeles, R. H. 1969. *Arch. Biochem. Biophys.* 130:274–85
14. Seto, B. 1980. *J. Biol. Chem.* 255:5004–6
15. Stadtman, T. C., Elliott, P., Tiemann, L. 1958. *J. Biol. Chem.* 231:961–73
16. Turner, D. C., Stadtman, T. C. 1973. *Arch. Biochem. Biophys.* 154:366–81
17. Tanaka, H., Stadtman, T. C. 1979. *J. Biol. Chem.* 254:447–52
18. Cone, J. E., Del Rio, R. M., Stadtman, T. C. 1977. *J. Biol. Chem.* 252:5337–43
19. Cone, J. E., Del Rio, R. M., Davis, J.
N., Stadtman, T. C. 1976. *Proc. Natl. Acad. Sci. USA* 73:2659–63
20. Costilow, R. N. 1977. *J. Bacteriol.* 131:366–68
21. Venugopalan, V. 1980. *J. Bacteriol.* 141:386–88
22. Dyer, J. K., Costilow, R. N. 1968. *J. Bacteriol.* 96:1617–22
23. Dyer, J. K., Costilow, R. N. 1970. *J. Bacteriol.* 101:77–83
24. Tsuda, Y., Friedmann, H. C. 1970. *J. Biol. Chem.* 245:5914–26
25. Somack, R., Costilow, R. N. 1973. *Biochemistry* 12:2597–2604
26. Somack, R., Costilow, R. N. 1973. *J. Biol. Chem.* 247:385–88
27. Jeng, I. M., Somack, R., Barker, H. A. 1974. *Biochemistry* 13:2898–903
28. Schwartz, A. C., Schafer, R. 1973. *Arch. Mikrobiol.* 93:267–76
29. Costilow, R. N., Laycock, L. 1969. *J. Bacteriol.* 100:662–67
30. Costilow, R. N., Laycock, L. 1971. *J. Biol. Chem.* 246:6655–60
31. Muth, W. L., Costilow, R. N. 1974. *J. Biol. Chem.* 249:7457–62
32. Muth, W. L., Costilow, R. N. 1974. *J. Biol. Chem.* 249:7463–67
33. Costilow, R. N., Cooper, D. 1978. *J. Bacteriol.* 134:139–46
34. Rossi, J. J., Vender, J., Berg, C. M., Coleman, W. H. 1977. *J. Bacteriol.* 129:108–14
35. Elsden, S. R., Hilton, M. G., Waller, J. M. 1976. *Arch. Microbiol.* 107:283–88
36. Brot, N., Smit, Z., Weissbach, H. 1965. *Arch. Biochem. Biophys.* 112:1–6
37. Kumagai, H., Yamada, H., Matsui, H., Ohkishi, H., Ogata, K. 1970. *J. Biol. Chem.* 245:1767–72
38. Kumagai, H., Kashima, N., Torii, H., Yamada, H., Enei, H., Okumura, S. 1972. *Agric. Biol. Chem.* 36:472–82

39. O'Neil, S. R., DeMoss, R. D. 1968. *Arch. Biochem. Biophys.* 127:361–69
40. Nisman, B. 1954. *Bacteriol. Rev.* 18:16–42
41. Fujioka, M., Morino, Y., Wada, H. 1970. *Methods Enzymol.* 17A:585–96
42. Stone, R. W., Machamer, H. E., McAleer, W. J., Oakwood, T. S. 1949. *Arch. Biochem.* 21:217–23
42a. Yokoyama, M. T., Carlson, J. R. 1974. *Appl. Microbiol.* 27:540–48
43. Camm, E. L., Towers, G. H. N. 1973. *Phytochemistry* 12:961–73
44. Bühler, M., Giesel, H., Tischer, W., Simon, H. 1980. *FEBS Lett.* 109:244–46
45. Jean, M., DeMoss, R. D. 1968. *Can. J. Microbiol.* 14:429–35
46. Moss, C. W., Lambert, M. A., Goldsmith, D. J. 1970. *Applied Microbiol.* 19:375–78
47. Boezi, J. A., DeMoss, R. D. 1959. *Bacteriol. Proc.*, p. 124
48. Tischer, W., Bader, J., Simon, H. 1979. *Eur. J. Biochem.* 97:103–12
49. Hashimoto, H., Rambeck, B., Gunther, H., Mannschreck, A., Simon, H. 1975. *Hoppe-Seyler's Z. Physiol. Chem.* 356:1203–8
50. Buckel, W. 1980. *Eur. J. Biochem.* 106:439–47
51. Barker, H. A. 1961. Fermentations of nitrogenous compounds. In *The Bacteria,* ed. I. C. Gunsalus, R. Y. Stanier, 2:151–207. New York: Academic. 572 pp.
52. Whiteley, H. R. 1957. *J. Bacteriol.* 74:324–30
53. Horler, D. F., Westlake, D. W. S., McConnell, W. B. 1966. *Can. J. Microbiol.* 12:47–53
54. Horler, D. F., McConnell, W. B., Westlake, D. W. S. 1966. *Can. J. Microbiol.* 12:1247–52
55. Johnson, W. M., Westlake, D. W. S. 1969. *Can. J. Biochem.* 47:1103–7
56. Johnson, W. M., Westlake, D. W. S. 1972. *Can. J. Microbiol.* 18:881–92
57. Kew, O. M., WoolFolk, C. A. 1970. *Biochem. Biophys. Res. Commun.* 39:1126–33
58. Lerud, R. F., Whiteley, H. R. 1971. *J. Bacteriol.* 106:571–77
59. Buckel, W., Barker, H. A. 1974. *J. Bacteriol.* 117:1248–60
60. Buckel, W. 1979. Biochemie der buttersäuregärung. In *Viertes Symp. Techn. Mikrobiol.,* pp. 243–50. Berlin.
61. Onderka, D. K., Floss, H. G. 1969. *J. Am. Chem. Soc.* 91:5894–96
62. Buckel, W. 1981. *Arch. Microbiol.* In press
63. Barker, H. A. 1967. *Arch. Microbiol.* 59:4–12
64. Buckel, W., Bobi, A. 1976. *Eur. J. Biochem.* 64:255–62
65. Buckel, W. 1976. *Eur. J. Biochem.* 64:263–67
66. Cozzani, I. 1965. *Ital. J. Biochem.* 14:363–70
67. Cozzani, I., Misuri, A., Santani, C. 1970. *Biochem. J.* 118:135–41
68. Stadtman, T. C. 1973. *Adv. Enzymol.* 38:413–48
69. Yorifuji, T., Jeng, I. M., Barker, H. A. 1977. *J. Biol. Chem.* 252:20–31
70. Jeng, I. M., Barker, H. A. 1974. *J. Biol. Chem.* 249:6578–84
71. Barker, H. A., Jeng, I. M., Neff, N., Robertson, J. M., Tam, F. K., Hosaka, S. 1978. *J. Biol. Chem.* 253:1219–25
72. Langbroek, H. J., Lambers, J. T., DeVos, W. M., Veldkamp, H. 1978. *Arch. Microbiol.* 117:109–14
73. Kröger, A. 1977. Phosphorylative electron transport with fumarate and nitrate as terminal hydrogen acceptors. In *27th Symp. Soc. Gen. Microbiol. and Microb. Energetics,* ed. B. A. Haddock, W. A. Hamilton, pp. 61–93. London: Cambridge Univ. Press. 442 pp.
74. Gottschalk, G., Andreesen, J. R. 1979. Energy metabolism in anaerobes. In *Microb. Biochem., Int. Rev. Biochem.,* ed. J. R. Quayle, 21:85–115. Baltimore: Univ. Park Press. 381 pp.
75. Langbroek, H. J., Stal, L. J., Veldkamp, H. 1978. *Arch. Microbiol.* 119:99–102
76. Langbroek, H. J., Veldkamp, H. 1979. *Arch. Microbiol.* 120:47–51
77. Wong, J. C., Dyer, J. K., Tribble, J. L. 1977. *Appl. Environ. Microbiol.* 33:69–73

Ann. Rev. Biochem. 1981. 50:41–68

GENETIC TRANSFORMATION ❖12070

Hamilton O. Smith and David B. Danner

Department of Microbiology, The Johns Hopkins University School
of Medicine, Baltimore, Maryland 21205

Robert A. Deich

Cancer Biology Program, Frederick Cancer Research Center,
Frederick, Maryland 21701

CONTENTS

PERSPECTIVES AND SUMMARY

Genetic transformation is a process by which a cell takes up naked DNA
from the surrounding medium and incorporates it to acquire an altered
genotype that is heritable. Bacteria are the only organisms (with the possible
exception of yeast) known to transform in nature, although cells of higher

41

organisms, including those of mammals, are transformable by artificial techniques in the laboratory. Probably a minority of bacterial species are transformable, but for some, such as *Bacillus, Streptococcus,* and *Haemophilus* spp., transformation is a major means of genetic exchange, comparable in importance to conjugation and transduction (1–3). To accomplish this exchange, these organisms have evolved highly efficient, genetically determined mechanisms for release, uptake, and incorporation of DNA. In their natural environments, they grow in large populations and collectively carry a vast array of mutations. Under conditions conducive to transformation, some cells in the population, acting as donors, release DNA by autolysis (or perhaps by secretion), while other cells (the recipients) take up the DNA and incorporate it to produce new genetic combinations.

Transformation has been extensively studied in several different organisms since Griffith's discovery of the phenomenon in pneumococcus in 1928 (4). The overall process in these naturally transformable organisms may be divided into several common steps, each of which can be experimentally isolated and studied. These are: (*a*) development of competence (the ability to take up DNA), (*b*) binding of DNA, (*c*) uptake of DNA, (*d*) formation of a preintegration complex, and (*e*) integration of DNA into the recipient cell chromosome. Although both gram-positive and gram-negative bacteria follow these basic steps, they appear to use different mechanisms to perform them, probably because they differ fundamentally in their cell envelope structures (5). As a result, it is likely that nature had to evolve two different methods of transporting large DNA molecules through bacterial cell envelopes.

Typical gram-positive bacteria have a cytoplasmic membrane surrounded by a cell wall of thick, rigid peptidoglycan. Apparently, as these cells become competent, there are significant changes in this cell envelope; receptor proteins, DNA-binding proteins, and nucleases—all specific for the transformation process—make their appearance and are responsible for the binding, cleavage, and processing of DNA into single-stranded fragments during uptake. The single-stranded fragments then become coated with a competence-specific DNA-binding protein, forming an "eclipse complex" in which the DNA is protected from cellular nucleases. The mechanism of integration has been clarified by biochemical studies that reveal how the *rec*A protein works. The use of pure cloned DNA segments to transform cells may yield information at the nucleotide level concerning the initiation and termination of the recombination event.

Gram-negative bacteria have a three-layered cell envelope consisting of a cytoplasmic membrane, a thin peptidoglycan cell wall, and an outer membrane of unique composition. How DNA is able to penetrate these layers remains a puzzle. Recent work, focused on the specificity of the DNA uptake mechanism in these cells, has shown that *Haemophilus* and *Neisseria* spp. recognize their own DNA at the cell surface. In *Haemophilus*, an 11- base-pair (bp) sequence on the donor DNA identifies this DNA to the cell. Competence development involves the synthesis of several new membrane proteins that are presumably involved in recognition of the specific DNA sequence and in uptake of the DNA. Advances over the next decade will probably depend on a better understanding of membrane biochemistry.

An exciting new achievement in the study of transformation is the development of artificial methods for introducing DNA into nontransformable bacteria and into eukaryotic cells. Such systems have opened up the study of genetics in these organisms.

This review provides a broad view of the field; consequently, citations are selective. The main features of transformation in gram-positive and gram-negative bacteria are presented separately to highlight their differences. Another section deals with the special problems of transfection with viral and plasmid DNA, and a final section covers artificial transformation systems. For detailed treatments of particular systems, the reader is directed to other recent reviews (1, 2, 6–9).

TRANSFORMATION OF GRAM-POSITIVE BACTERIA

The best studied species among the transformable gram-positive bacteria are *Streptococcus pneumoniae, Streptococcus sanguis,* and *Bacillus subtilis.* Figure 1, a composite diagram that outlines the major steps in the transformation pathway, is included for illustration but should not be construed as completely representative of any particular organism.

Competence Development

Competence refers to the ability of a cell to take up DNA into a DNase-resistant form that is not readily removed by washing. Most transformable bacteria only become competent under certain conditions of growth; thus competence appears to be an induced state. For example, if a transformable strain of *Streptococcus* is grown in appropriate media, the entire culture rapidly becomes competent as a critical cell concentration is reached, in the range of 10^7–10^8 cells/ml (10, 11). This response is due to cellular secretion of a small protein called activator or competence factor (CF) (11, 12). CF appears to be released continuously in growing cultures, is autocatalytic, and reaches effective concentrations rapidly at higher cell densities (13); its action has been likened to that of a hormone (14). CF is readily recovered from supernatants of streptococcal cultures (15) as a small protein of 5000–10,000 daltons (16), and, when added to suspensions of noncompetent cells,

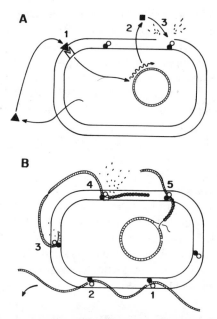

Figure 1 Some of the features of transformation in gram-positive bacteria. *A* Competence development: *1.* competence factor (activator, ▼) interacts with a cell-surface receptor (M); *2.* competence-specific proteins are induced, one of which is an autolysin (■); *3.* autolysis exposes membrane-associated DNA-binding protein (○) and nuclease (●). *B* Binding, uptake, and integration. A single long DNA molecule is shown bound to several sites on the cell surface. A different stage of binding and uptake is illustrated at each site: *1.* binding and nicking by DNA-binding protein; *2.* cleavage by nuclease; *3.* one strand is taken up while the other is degraded and nucleotides are released into the medium; *4.* the single-strand becomes associated with a competence-specific protein to form the eclipse complex; *5.* integration involves replacement of a recipient strand by the donor strand.

induces high levels of competence within 10–20 min (12, 17, 18). Several CF specificity types are found among the various streptococcal species (12, 19–21).

The mechanism of action of CF is obscure. It may trigger competence development by binding to a specific receptor on the cell surface. Recently, an "activation inhibitory protein" has been isolated from pneumococcal membrane preparations that could be the postulated receptor (22). It is an acidic protein of 68 kilodaltons and inactivates pneumococcal CF at concentrations as low as 1.4 ng/ml. It has been known for some time that protein synthesis is required for competence development (17), but only recently has it been shown that dramatic shifts in the level of gene expression take place (18, 23). In *S. sanguis,* transient synthesis of ten new polypeptides coincides with the onset of DNA binding ability and transformability (18). Eight species appear early and two appear after a lag of 5–10 min. One prominent species of 16 kilodaltons coelectrophoreses with eclipse complex protein (see below). Concomitant with the synthesis of these new polypeptides, there is a net decrease of up to 90% in the rate of protein synthesis (18). RNA synthesis is transiently depressed by about 90%, but the level of DNA synthesis remains normal. A new, pulse-labeled RNA species of high molecular weight, sufficient for the synthesis of ten induced polypeptides, has been detected by gel electrophoresis (18). Similar changes have been observed in pneumococcus, but without a net depression in protein synthesis (23). Competence factors may play as much of a role in causing donor cells to release DNA as they do in causing recipient cells to take it up, yet little evidence of this has been obtained. Pneumococcal CF has been observed to cause protoplast formation and leakage of intracellular proteins and metabolites (including nucleoside phosphates, DNase activity, and autolysin) into the medium, depending on postincubation conditions (24).

In *B. subtilis,* a precompetence phase begins 1.5–3 hr before the appearance of transformability, and is associated with decreases in the rate of DNA synthesis and in the ratio of RNA to protein synthesis (25). These alterations lead to a decrease in cell density and to the onset of competence in some 10–20% of the cell population; the remaining cells do not become competent. This heterogeneity has led to the use of Renografin density gradients for the separation of light competent cells (1.100 g/ml) from heavy noncompetent cells (1.311 g/ml) (26, 27). Once competence is reached, it is maintained for about 3 hr (28), during which time the cells neither multiply nor replicate their DNA (29). Competence factors have been reported for *Bacillus cereus* (30), *Bacillus stearothermophilus* (31), and *B. subtilis* (32, 33). The last appears to have autolysin activity.

Binding of DNA

Binding of DNA to the cell may be reversible or irreversible. Reversible binding, which implies that the DNA can easily be removed by elution without change, can occur in two ways: the DNA may interact with cell envelope components uninvolved in transformation, or it may bind to transformation-related components. With some bacteria the transformation-related binding may then become irreversible (non-elutable), even though the DNA still remains exposed and sensitive to DNase I (34).

In the presence of EDTA, competent pneumococcal cells bind DNA but do not convert it to a DNase-resistant form (34). In addition, *end* mutants of pneumococcus, deficient in a membrane-bound endonuclease, are blocked in uptake but bind DNA efficiently (35, 36). By either of these means it is possible to isolate the initial DNA-binding reaction from subsequent steps. This reaction proceeds rapidly in the first 5–10 min at 30°C and then more gradually for up to 30 min (34, 35). Bound DNA is sensitive to hydrodynamic shear (37) and pancreatic DNase treatment, but is resistant to repeated gentle washes and is not dissociated when excess carrier DNA is added (34, 38). Although bound DNA is not removed by such treatments, it does not appear to be covalently bound (39). Bound DNA remains duplex but contains nicks produced during binding. These are spaced about 6 kb apart, on the average, and are formed even in the presence of EDTA (40, 41). One possibility is that the binding receptor itself has the endonucleolytic activity.

A competent pneumococcal cell is estimated to carry 30–80 receptor sites on its surface for binding DNA (8, 42). A binding factor (BF) with the properties of a protein has been detected in both competent and noncompetent pneumococcal cell wall-membrane complexes (37); it can be washed off the complexes from competent cells (43). An autolysin leaks into the medium following treatment of pneumococcal cells with pure activator, and may unmask the BF of noncompetent cells (24). A cell surface BF has also been detected in $S.$ $sanguis$ (44, 45). Pneumococcal mutants called ntr are deficient in DNA binding and may lack BF (35). The mechanism and specificity of DNA binding in $B.$ $subtilis$ (7) and $S.$ $Sanguis$ (46) is very similar to that in pneumococcus. Only duplex DNA is bound; single-stranded DNA and RNA, and RNA : DNA hybrids are not bound (47–49) nor is glycosylated DNA (49–51). Heterologous DNAs are taken up readily so there is no obvious requirement for specific sequences on the DNA (48, 52–55). In $B.$ $subtilis,$ denatured DNA will bind to and transform competent cells only at low pH, and is optimal when EDTA is present to prevent DNA degradation (56–58). In pneumococcus, competent cells are transformed by denatured DNA at a 200-fold lower rate than that of comparably sized native DNA (59).

Uptake of DNA

DNA uptake (absorption) is operationally defined as the conversion of DNA into a DNase-resistant, nonelutable form. The conditions could be met by a tightly bound surface DNA, but more generally they imply removal of the DNA from the external environment into one of the cell compartments. This could mean movement into a vesicle, the periplasmic space, or the cytoplasm. The initiating event in pneumococcal DNA uptake is completion of a double-strand break in a bound DNA molecule, presumably by cleavage opposite the nick introduced during binding. The large membrane-bound endonuclease (endonuclease I or major endonuclease) responsible for this has been isolated and extensively studied (35, 36, 60, 61). Endonuclease-deficient end mutants appear to be in the structural gene (60); noz mutants occurring at the same genetic locus abolish the residual endonuclease activity of many of the leaky end mutants. Competent cells bind large amounts of DNA when uptake is blocked by the addition of EDTA. After Mg^{2+} (or Ca^{2+}) is added, much of the bound DNA is released into the medium as smaller double-stranded fragments, presumably by the initial action of endonuclease I (34, 38). That portion of the cleaved DNA that remains bound to the cell surface subsequently participates in uptake (34). Uptake of the DNA into a form inaccessible to pancreatic DNase is absolutely dependent on Ca^{2+} (38) and on endonuclease I activity (60). During uptake, bound duplex DNA fragments are converted to single-stranded DNA (62, 63) and an approximately equivalent amount of TCA-soluble oligonucleotides is released into the medium (39). Apparently, one strand is degraded as the other is taken up (8). Either strand can be taken up with equal probability (64, 65). In *S. sanguis* the single-stranded DNA is sequestered outside the cytoplasmic membrane, presumably in the periplasmic space, as an eclipse complex (see next section) with a DNA-binding protein (18).

The mechanism of binding and uptake in *B. subtilis* is similar to that in pneumococcus, with some minor exceptions. Binding occurs in the absence of divalent ions, but is somewhat inhibited by the presence of EDTA, probably because of a direct effect on the DNA receptors. Furthermore, the DNA-binding receptors remain associated with the cytoplasmic membrane during conversion to protoplasts and after isolation of membrane vesicles (66). Mg^{2+}, but not Ca^{2+}, is essential for DNA uptake. A Mn^{2+}- and Ca^{2+}-activated endonuclease localized on the cell surface produces duplex breaks (67, 68). Bound DNA of high molecular weight is broken to shorter duplex lengths, averaging 9×10^6 daltons, before uptake (69). Competent cells also contain an exonuclease, associated with the cell envelope, which is thought to be responsible for the digestion of one of the DNA strands during uptake (70). DNA is taken up in a single-stranded form (71) that averages 3 to 5×10^6 daltons in weight (72). Like *S. sanguis, B. subtilis* may take up DNA into a DNase-resistant single-stranded form and keep it outside the cytoplasmic membrane until integration.

Eclipse Complex

When the single-stranded intermediate form of DNA taken up by gram-positive cells is recovered and used to transform other cells, its transforming activity is temporarily lost; the DNA is therefore said to be in eclipse. This activity is restored when the DNA is returned to a duplex state by integration. In pneumococcus, the single-stranded DNA becomes associated noncovalently with a 19.5-kilodalton protein (73–75), which is made in large amounts during competence development (75). This eclipse complex is quite stable and can be isolated readily (76); the complex is insensitive to pancreatic ribonuclease (73). Susceptibility of the DNA in the eclipse complex to digestion by pancreatic DNase, micrococcal nuclease, *Neurospora* endonuclease, and nuclease P1 is 50- to 1000-fold less than that of naked single-stranded DNA (77). The complex is also particularly resistant to the nucleases in pneumococcal extracts, which suggests that one of its functions is the protection of donor DNA. Partial digestion with pancreatic DNase produces a series of fragment sizes, which suggests a repeating structure analogous to that in chromatin. More extensive digestion reduces the resistant DNA fraction to a major and a minor class centered at 70 and 107 nucleotides. Complete digestion with high levels of nuclease releases the eclipse protein in native form; its basic protein unit may consist of up to ten 19.5-kilodalton monomers (77).

The eclipse complex in *S. sanguis* (78–80) consists of single-stranded donor DNA and a single polypeptide species about 15.5 kilodaltons in size (18). Interestingly, the complex remains outside the cytoplasmic membrane: it is resistant to pancreatic DNase, but sensitive to micrococcal nuclease, and is readily released into the supernatant when the recipient cells are converted to spheroplasts. DNA appears to be integrated directly from the complex into the chromosome without a definite intracellular phase.

In *B. subtilis*, a single-stranded DNA-binding protein that appears specifically in competent cells has been reported (81), and an eclipse complex appears to be present (82, 83).

Three roles for an eclipse complex have been suggested. The decrease in free energy associated with its formation could promote transfer of a DNA strand into the cell (or into the periplasmic space), it could protect the DNA from cellular nucleases, and it could promote integration by holding the bases in an open position for pairing (73).

Integration

Donor DNA is integrated by a recombination event between the recipient chromosome and a single-stranded donor DNA fragment. The integration machinery appears to work very efficiently and nonselectively on any homologous DNA that is presented to it by the uptake mechanism. Classic studies with density-labeled donor DNA have shown that the single strand of the donor physically displaces an homologous strand of the recipient chromosome to form a heteroduplex segment (84–86). In *B. subtilis*, transformant heteroduplexes were directly visualized in the electron microscope by Fornili & Fox (87). They were able to see the regions of integration because the bromouracil-substituted donor strand contained in the heteroduplex caused it to melt at a lower temperature than the thymine-containing recipient DNA, which allowed precise denaturation mapping. They confirmed that donor DNA is inserted as a single-stranded segment;

no duplex donor segments were seen. The average size of integrated segments was estimated to be about 10 kb, but as much as 10% exceeded 30 kb, in agreement with previous estimates (69, 85). Insertion of more than one donor segment within a recipient DNA molecule was frequently observed and could not be accounted for by chance; such coupled events are presumably a result of the binding and uptake mechanism. Single large donor DNA molecules are presumably bound at several sites, processed into fragments of approximately the size that is eventually integrated, and then taken up (Figure 1). Because integration is very efficient [involving about 50% of the DNA taken up by the cell (88)] several pieces are likely to be integrated in the same region of the cell genome. Either strand of the donor duplex can be taken up and then integrated with equal probability (85).

The mechanism of integration is still not completely understood. A major question is whether an induced recombination pathway is used, rather than the normal constitutive pathways. It is possible, for example, that homologous pairing is achieved through a direct association of the eclipse complex with the recipient DNA duplex, the eclipse protein serving somehow to catalyze the process. On the other hand, a recA-like recombination pathway could be used (89–91). When cloned pneumococcal DNA segments flanked by phage lambda sequences are used as donor DNA, mutations are induced, at the high frequency of 2/1000 transformants, that are mostly deletions of recipient DNA and insertions of lambda DNA that extend outward at the extremities of the pneumococcal DNA segment (92). This may imply that the alignment between donor and recipient DNA during homologous pairing can extend into nonpaired regions, and supports the notion of an energetically driven process. Whatever the basis of recombination, it is both efficient and rapid, since donor-recipient DNA complexes begin to form within one minute (72) and reach completion in 10–12 min (93).

The integration process can be separated into at least four steps by the use of a combination of mutational analysis (94) and a variety of inhibitors (18, 95, 96). These steps are: 1. translocation of the eclipse complex into proximity with the chromosome, 2. formation of an unstable donor-recipient complex, 3. formation of a stable noncovalent complex, and 4. formation of a covalently closed (repaired) complex. Raina & Ravin (18) found that ethidium bromide prevents Step 1 in S. sanguis. The same authors also showed that coumermycin, a DNA gyrase inhibitor (97), allowed formation of donor-recipient complexes but decreased the yield of transformants; they therefore suggested that chromosomal superhelicity promotes recombination. Seto & Tomasz (95) found that in pneumococcus, ethidium bromide and a variety of other intercalating agents inhibited the formation of biologically active recombinant molecules, as if abnormal pairing and inadequate covalent closure had occurred. An unstable donor-recipient complex could be observed in *B. subtilis* by using cross-linking agents as fixatives (98). This labile complex may be the displacement loop postulated by Shibata et al (91). Buitenwerf & Venema (94, 96) found that donor-recipient complexes were formed, but arrested at an intermediate, S1 nuclease-sensitive stage, when *B. subtilis* was incubated at 17°C after allowing uptake at 30°C. A complete analysis of integration will depend on in vitro biochemical studies.

Mismatched-Base Correction

Donor genes that differ by a single base pair from recipient genes create a base-mismatch when integrated. In pneumococcus, these mismatches appear to be corrected at variable efficiencies (99, 100). Very high efficiency (VHE) markers transform at a relative efficiency of 1.0, high efficiency (HE) markers at 0.5, intermediate efficiency (IE) markers at 0.2, and low efficiency (LE) markers at 0.05 (100). The properties of this mechanism in pneumococcus can be summarized as follows (1): 1. Correction occurs after integration (101, 102). 2. LE markers appear to be of the transition ($A:T \leftrightarrow G:C$) type, and transversions account for the other classes (100). 3. Since either strand of a donor LE marker can be integrated, the cell must have some way to ensure that the correction mechanism specifically recognizes the donor strand of the heteroduplex (100, 103). 4. Those LE markers that survive correction become homozygous prior to replication (103, 104), which indicates that in this case the recipient strand can be corrected. 5. When one of two linked markers is corrected, the other marker is usually corrected as well, which suggests that corrections are made in a single sweep or scan along the DNA (100, 102, 105). 6. Single-step *hex* mutants exist that no longer show specific correction of donor markers; all classes become VHE (106). 7. The correction mechanism can be saturated by transforming DNA from a related strain containing many base-mismatches; the cell becomes functionally *hex* (65, 107). 8. Recent experiments with heteroduplex donor DNA molecules having a marker on one strand and its wild-type allele on the other allow one to follow the fate of each strand. With LE markers, either donor strand can be corrected. With HE markers, only one strand is correctable, and the strand chosen for correction varies with different mutations. In addition, surviving transformants produced by either HE or LE markers introduced on the correctable strand, become homozygous, as if in this case the *hex* function had acted to correct the recipient strand (65).

A model for the *hex* correction mechanism must account for (*a*) specific recognition of the donor strand, (*b*) differential affinity for the various types of mismatches, and (*c*) highly correlated correction of linked markers. Perhaps the *hex* mechanism is analogous to the methyl-directed mismatch repair system operating in phage λ–infected *E. coli* cells. In this system, the methylation enzyme responsible for directing the repair is a DNA adenine methylase encoded by the *dam* gene (108), acting at (5')G-A-T-C sites. Apparently, the unmethylated GATC sites on the newly replicated DNA strand act as entry points for the enzymatic repair mechanism, which then scans the DNA and corrects mismatches along the unmethylated strand. Although the *hex* system could possibly act in a similar fashion, the mechanism for strand recognition may differ. Advances may come from the sequence analysis of specific classes of mismatches generated by mutations in cloned genes, such as those recently isolated by Claverys et al (92), and from the development of an in vitro correction system. Regarding the latter, the *hex* mechanism may catalyze base removal at the mismatched site, thus feeding into the common apurinic-apyrimidinic (AP) site repair pathway (109).

TRANSFORMATION OF GRAM-NEGATIVE BACTERIA

Genetic transformation has been reported in a number of genera of gram-negative bacteria, including *Haemophilus, Neisseria, Moraxella,* and *Acinetobacter.* Of these, transformation in the genus *Haemophilus* (and in particular, *H. influenzae* and *H. parainfluenzae*) is by far the best studied. However, *Neisseria* strains are receiving increasing attention because of their clinical relevance and because they appear to have a transformation mechanism similar to that of *Haemophilus.* These gram-negative bacteria have evolved a system of DNA binding and uptake for genetic transformation that is markedly different from that used by the gram-positive organisms. The principle differences are their specific binding of homologous DNA via a highly repeated "uptake site" sequence (this last has not been verified directly in *Neisseria* to date), and their uptake of intact duplex DNA molecule into a DNAse-resistant form.

Competence Development

Competence development in gram-negative bacteria appears to be internally regulated: no "competence factors" have been reported, and competence development is insensitive to external proteases (R. A. Deich, unpublished observations). Conditions for competence development in gram-negative bacteria vary between species. In *H. influenzae,* competence develops when cell division is blocked under conditions that allow continued protein synthesis (110–112). By transferring exponentially growing cells into a defined nongrowth medium, it is possible to generate cultures that are essentially 100% competent for transformation. Low levels of competence may be produced in log phase *Haemophilus* cells by cyclic adenosine monophosphate (113). In *Neisseria gonorrhoeae,* only pilated forms are competent; competence in these lines is constitutive (114, 115). Transformable *Acinetobacter* strains become competent when entering stationary phase (116).

Competence development in *H. influenzae* is stimulated by inosine and lactate and inhibited by valine, NAD, and inhibitors of protein synthesis (111, 117), but once cells reach a competent state, these factors do not affect their ability to absorb DNA (118). If competent cells are returned to a rich growth medium, competence is rapidly lost when cell division resumes (after a lag of about 30 min) (119). Competence development is also inhibited by lysogeny with some temperate phages (120).

Induction of competence in *H. influenzae* involves a number of changes in the cell surface. Antiserum to noncompetent cells does not block DNA uptake, while antiserum to competent cells does (121). A number of specific inner and outer membrane proteins are induced during competence development, and the synthesis of these proteins closely parallels the kinetics of competence induction (122, 123). In addition, two normally cytoplasmic proteins become membrane- associated. Competence also causes a general stimulation of recombination activity; the recombination frequency between resident prophage recombination and superinfecting phage is increased by nearly an order of magnitude (124–126).

DNA Binding and Uptake

The amount of DNA taken up by competent cells can be affected by a number of factors. Generally, optimal pH and temperature for uptake are similar to the optimal growth conditions for the organism involved. Transformation systems require, optimally, 1 mM Ca^{2+} and 100 mM Na^+ and/or K^+ for DNA uptake. DNA uptake by competent cells also requires an active metabolism in the recipient cell (114, 116, 118). Normally, competent cells specifically bind and absorb only double-stranded DNA molecules. A small amount of uptake of single-stranded DNA has been attributed to rapid renaturation of DNA by natural "cross-links" (118, 127). RNA and other polycations bind to the cell surface but are not absorbed. Postel & Goodgal (128, 129) have described DNAse-resistant binding of single-stranded DNA by competent *H. influenzae* cells under low pH (4.5–4.8) conditions and in the presence of 1 mM EDTA. The exact form of this bound DNA is unknown, as much of it is released back into the medium after returning the cells to neutral pH. However, the efficiency of this system approaches that of double-stranded DNA transformation.

One of the most surprising observations in transformation of gram-negative bactera is the specificity of the DNA uptake system for homologous DNA. In *Haemophilus* (130, 131) and *Neisseria* (132–134) transformation, only DNA from the same or closely related species is absorbed. Foreign DNAs can bind to the competent cell surface in large amounts, but they are absorbed poorly and do not compete well with homologous DNA for uptake.

Fragments of *H. influenzae* or *H. parainfluenzae* DNA cloned into pBR322 and grown in *E. coli* are still recognized by the *Haemophilus* DNA uptake system (135, 136). Recognition regions (uptake sites) on these fragments have been localized by restriction mapping (135). Several small fragments containing uptake sites have been sequenced and shown to have a common 11-bp DNA sequence, 5'-AAGTGCGGTCA-3' (137). Ethylation of DNA phosphate groups in this sequence with ethylnitrosourea (138) interferes with fragment uptake, which confirms that this is the uptake site sequence (137).

Using a competition assay with cloned fragments of *Haemophilus* DNA of known length and number of uptake sites as standards, it has been possible to estimate that there are about 600 uptake sites in the *Haemophilus* genome, or one site per 4000 bp (135, 139). Non-*Haemophilus* DNAs (e.g. those from *E. coli*, phages λ and T7, salmon sperm, SV40, and pBR322) compete for uptake 25- to 100-fold less efficiently than *Haemophilus* DNA and are taken up at this lower level (139, 140). Even so, the binding and uptake of foreign DNAs are higher than predicted by chance. For example, *E. coli* DNA should contain about two copies of the uptake site per genome, based on the probability of occurrence of an 11-bp sequence in a random sequence of 4×10^6 bp. However, this DNA competes with *Haemophilus* DNA as if it had about 40 sites per genome. This could be due to the presence of degenerate versions of the uptake sequence in *E. coli* DNA, which are individually less efficient for uptake but compete at this level because of their greater number. Consistent with this hypothesis is the observation that the synthetic copolymer poly(dG)·poly(dC) competes 40-fold better than SV40 DNA, yet neither molecule bears a single "perfect" site (R. A. Deich, unpublished observation).

What selection pressure maintains the large number of uptake sites in the *Haemophilus* genome? One possibility is that the site is also involved in an important second function (such as promotion or termination), which ensures its maintenance. However, a one-to-one association of sites with individual genes is not obligatory; a 2.2-kb cloned fragment of *H. haemolyticus* DNA, known to contain three structural genes, does not have an uptake site (B. E. Hausler, personal communication). Possibly, the sites maintain themselves by autoselection during transformation (137). Opposing this spreading by autoselection will be three factors: the relative advantage of gaining a new site will decrease as the number of sites increases, the loss of sites through mutation will increase, and the sites may begin to interfere with cell function.

A specific cell-surface receptor may enable *Haemophilus* to recognize its own DNA (R. A. Deich, unpublished observations). Six polypeptides have been purified from competent *Haemophilus* cell membranes by chromatography on *Haemophilus* DNA-agarose. These polypeptides may be components of a membrane-bound complex that recognizes the DNA at the cell surface and transports it through the membrane. A soluble periplasmic DNA-binding protein has also been implicated in DNA uptake (141). Presumably, a similar system is involved in determining the specificity of DNA uptake in neisserial transformation, although this has not been directly verified to date.

DNA uptake in *Haemophilus* can be described by a simple kinetic model involving (*a*) reversible binding between a DNA uptake site and a cell surface receptor, (*b*) formation of an irreversible (non-exchangable) site-receptor complex, and (*c*) uptake of the entire DNA duplex into a DNAse-resistant, nonelutable form (see Figure 2). Estimates of the rate constants for these processes have been made (142). Under saturating DNA conditions, binding and uptake are complete within one minute, whereas absorption in pneumococcus requires up to 30 min.

The initial reversible receptor-DNA binding appears to be a simple bimolecular interaction between receptors and uptake sites. The probability of a given DNA molecule being bound to the cell is thus directly proportional to the number of sites it has. Conversely, the probability of any given site binding to a receptor is independent of the length or sequence of the rest of the DNA molecule on which it resides (139, 142). The number of DNA molecules that may be absorbed by competent *Haemophilus* cells is limited (142, 143); and DNA uptake in *Neisseria* is similarly limited (134). It has been suggested that each Haemophilus cell has 4–8 receptors each of which can be used only once to take up one molecule of DNA (142). The irreversible binding rate (see Figure 2) in *N. gonorrhoeae* is about one third the rate in *H. influenzae* (134). Other kinetic constants for *Neisseria* have not been determined.

The biological significance of the homospecificity for DNA uptake in gram-negative bacteria is not clear. It is undoubtedly more efficient for the cell to bind only DNA that it can subsequently use to become transformed, but the gram-positive bacteria do perfectly well without such discrimination.

Fate of Transforming DNA

Donor DNA for genetic transformation in *Haemophilus* and *Neisseria* is taken into competent cells as intact, duplex molecules (144, 145). There is no initial release of acid-soluble counts from labeled donor DNA (133, 140) and no "eclipse" of DNA-transforming activity (144–146). These molecules

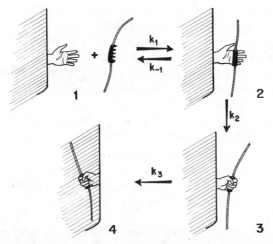

Figure 2 Possible mechanism of DNA uptake in *H. influenzae*: *1.* uptake site and receptor form a reversible complex; *2.* reversible complexes may either dissociate to *1* or form a nonexchanging complex, *3;* *3.* once a nonexchanging complex is formed, the entire DNA duplex is bound in a DNAse-resistant, nonelutable form, *4.*

are not attacked by cellular exonucleases or restriction endonucleases after uptake. They are, however, susceptible to a ligation activity that acts on both blunt and sticky ends [(135); D. B. Danner and H. O. Smith, unpublished]. It is not known whether this ligation activity plays any role in subsequent integration of the DNA.

The protection of transforming DNA from cellular nucleases suggests that it is either protected by a cellular DNA-binding protein similar to the pheumococcal eclipse complex (74, 77) or physically isolated from the bulk of the cytoplasm in some special cell structure. Recent studies on DNA uptake in *Haemophilus* suggest that both of these factors play a role in protecting donor DNA. A periplasmic DNA-binding protein required for DNA uptake in *Haemophilus* has been reported (141), which could correspond to the eclipse complex protein of pneumococcus. On the other hand, electron microscopic studies of competent *Haemophilus* suggest that DNA is initially bound in vesicle-like cell surface structures and transported into the cell inside these structures (147), a conclusion supported by the discovery of mutants of *H. influenzae* and *H. parainfluenzae* that shed DNA-binding vesicles after the normal competence induction procedure.

During integration, one strand of the donor DNA is incorporated into the host chromosome (133, 144, 148), while the other strand and the displaced host strand are degraded and released into the medium (149). The mechanism of the strand displacement is not known; however, it has been observed that unintegrated donor DNA in gram-negative bacteria does not lose transforming activity (144–146) and no single-stranded intermediates are formed (133, 148). A low-molecular-weight, biologically inactive DNA (species II DNA) can be recovered from competent *H. influenzae* cells after DNA uptake, but this form appears to be a byproduct of integration, probably an intermediate in the breakdown of the unintegrated donor strand (150).

The presence of single-stranded DNA regions in the host DNA of competent *H. influenzae* cells has been observed (151). These regions are implicated in donor integration because they are not present in non-transformable *Haemophilus rec2* mutants. Presumably, these regions are involved in the initial synapsis between host and donor DNAs, as donor DNA does not interact with the host chromosome in *rec2* cells (see next section).

Genetic studies on transformation-deficient mutants of *H. influenzae* that absorb DNA normally have shown three loci to be involved in the integration of donor DNA (152–156). The *rec1* phenotype suppresses transformation to a level of 10^{-6} relative to wild type. Donor DNA becomes associated with the host chromosome, but is then rapidly degraded. The *rec1* mutants also fail to show recombination between prophage and superinfecting phage DNAs. The *Haemophilus rec1* gene appears to be analogous to the *E. coli recA* gene: *rec1* cells are sensitive to UV and methylating agents and fail to induce prophages. The *rec2* phenotype suppresses transformation to a level of 10^{-7}. The *rec2* mutants also block recombination between prophage and superinfecting phage DNAs. Donor DNA does not become associated with the host genome and remains stably in the cell; biological activity can be recovered for at least an hour after absorption. The *rec2* mutants fail to generate single-stranded regions in host DNA during competence development (151). In a third type of mutant, strain KB6, transformation is suppressed to a level of 10^{-4}. Donor DNA does not become associated with the host genome and is maintained stably as with *rec2* cells. KB6 does not affect prophage/superinfecting phage DNA recombination.

The use of recombinant DNA technology has proven invaluable in determining the molecular mechanism of the homospecific DNA uptake in *Haemophilus*. It should also play a major role in solving many of the remaining problems in the integration of *Haemophilus* DNA. By cloning small, specific fragments of the genome and feeding pure populations of these fragments to competent cells, one should be able to follow the processing and integration of donor DNA at the nucleotide level. It should also be possible to clone and characterize all of the genes involved in genetic transformation. It may then prove feasible to transfer these highly efficient DNA-uptake systems into other organisms.

TRANSFECTION WITH VIRAL AND PLASMID DNA

Traditionally, the term transfection refers to the infection of a host cell with naked viral DNA by means of the cell's transformation machinery. Plasmid DNAs are similar in many respects to viral DNAs and seem to be handled by cells in the same fashion during transformation. Therefore we use transfection in this review to refer to transformation with plasmid and with viral DNAs.

Phage, plasmid, and chromosomal DNAs all seem to be taken up by the same mechanism in naturally competent cells. For example, in *S. pneumoniae,* phage MP2 enters cells as readily as chromosomal DNA, is bound to mutant cells (which are defective for uptake) as well as chromosomal DNA, and enters the cell in a single-stranded linear form, as does chromosomal DNA (41). In *H. influenzae,* the same DNA sequence required for efficient uptake of chromosomal DNA (137) is also required for efficient uptake of plasmid DNA (J. V. Israel and D. B. Danner, unpublished). In *B. subtilis,* competence for transformation by plasmid DNA develops with kinetics similar to those of competence for chromosomal DNA (157); plasmid and chromosomal DNAs compete for transformation (158), and both types of transformation are first-order processes (159).

On the other hand, phage and plasmid DNAs are typically inefficient at transfecting naturally competent cells. In gram-positive bacteria, this inefficiency is based on the conversion of incoming DNA into a single-stranded form that cannot replicate and usually does not have sufficient homology to be integrated into the recipient genome. In gram-negative bacteria, this inefficiency is probably due to a compartmentalization of the DNA, perhaps inside a membrane vesicle, which denies it access to the host-cell replication machinery. Thus in *B. subtilis,* transformation with monomeric plasmid is 10^3- to 10^4-fold less efficient than transformation with chromosomal markers (161). In *H. influenzae,* plasmid DNA transfers its antibiotic resistance markers at least 10^4-fold less frequently than chromosomal DNA transfers its auxotrophic markers (160). This low efficiency applies even to plasmids that bear the *Haemophilus* uptake sequence (J. V. Israel and D. B. Danner, unpublished). Although all competent cells in a nonlysogenized culture of *Haemophilus* will take up phage DNA, only about one in 10^5 will form an infective center; lysogeny of phage HP1 can only be established by DNA entering the cell through phage injection (126). In addition, the transformation efficiency into resident prophage markers is over 50-fold greater for phage DNA entering by the competence system than for DNA injected into cells by phage.

There are some exceptions to the typically low efficiency of transfection with plasmid DNA. In *B. subtilis,* for example, plasmids that are multimeric will transform at a high efficiency (161). Perhaps this is because the single-stranded fragments that enter the cells have enough sequence redundancy so that they can anneal into gapped circles and be repaired (159). Presumably the single-stranded plasmid DNA can only be propagated after conversion to a duplex circular form, since it does not ordinarily share sufficient sequence homology with the recipient genome to be integrated. However, if the recipient cells already contain a resident plasmid that shares homology with the incoming plasmid, then even monomeric plasmids can transform cells at high efficiency. To transform most efficiently, the incoming plasmid must be a linear piece of DNA, in which a region on each end is homologous to the resident plasmid (158). This also explains why plasmids that share sequence homology with the resident chromosome will transform effectively; in this case the plasmid is incorporated into the genome (162).

The ability to introduce plasmids into cells by transfection has proven to be extremely useful. Such plasmids have been used as cloning vectors to isolate, amplify, and analyze genes from many sources, and as convenient minichromosomes to facilitate the examination of DNA integration during transformation. The artificial transfection systems that have made some of these advances possible are discussed in the next section.

TRANSFORMATION BY ARTIFICIAL MEANS

To develop the "natural" transformation systems discussed above, bacteria have evolved elaborate mechanisms whose main purpose is to internalize and integrate exogenous DNA in order to supply the cell with new genetic information. On the other hand, "artificial" transformation refers to the introduction of DNA into a cell without the help of this special machinery, either because it is not present or because, in these artificial systems, the DNA enters by a different route after the cell envelope has been rendered permeable with some chemical or enzymatic treatment. Nearly all of the artificial transformation techniques have been developed because of their usefulness in genetic engineering and gene analysis.

Transformation of Prokaryotes

Ten years ago, a new and efficient transfection technique was developed for *E. coli*, based on soaking log-phase cells in a calcium chloride solution at 0°C (163). This technique was quickly applied to the transformation of *E. coli* cells with the DNA of antibiotic-resistant plasmids (164) and is now a cornerstone of recombinant DNA technology (165, 166). Since that time, a tremendous amount of work has been done with transformed *E. coli*, but almost nothing has been determined about the mechanism by which calcium promotes the competence of these cells. Many possible roles for calcium can be imagined. It may act as a chelating agent to neutralize and/or remove molecules at the cell surface that block DNA entry (167, 168), or it may bind to DNA and, in effect, precipitate it onto the surface of cells in a form favorable for uptake. It may activate an enzyme involved in uptake (168), protection, integration, or replication of incoming DNA, or decrease the activity of an enzyme antagonistic to one of these processes. Or it may produce some combination of these effects. What is known for certain? Clearly, calcium treatment does not activate an uptake system that is selective, since it allows the cell to internalize and propagate DNA from many sources (165, 166, 169). Nor is the incoming DNA converted to a single-stranded intermediate form during uptake (170), as is the case for the natural systems of *B. subtilis* and *S. pneumoniae* (8).

Much progress has been made in improving the convenience and efficiency of calcium-mediated *E. coli* transformation. Once competent, cells can be frozen in a calcium and glycerol solution so that they retain high levels of viability and transformability upon thawing (171). Defined conditions for competence in the strain $\chi1776$ have been determined (172). Recently, a simple modification of the calcium treatment method has brought the transformation rate per cell to within an order of magnitude of the maximum theoretical level (173). By extending to 24-hr the incubation time in calcium solution at $0°C$, it is possible to transform as many as 20% of remaining viable cells (about 4% of the viable cells in the original culture). As many as 2×10^7 transformants per μg of plasmid DNA (an efficiency of 10^{-4} per molecule) can be produced by this technique. Similar efficiencies can be obtained by treatment with a combination of rubidium and calcium chloride (174).

Calcium-mediated transformation has also been applied to other cells, such as *Staphylococcus aureus* (175), *Pseudomonas putida* (176), and *H. influenzae* (177). It thus appears to be a general method for bacteria. In addition, old methods have been improved (178) and new methods have been developed for the introduction of both plasmid and bacteriophage DNA into bacterial cells by artificial means. Freeze-thawing has been used successfully in *Agrobacterium tumefaciens* (179, 180). Growth under anaerobic conditions, in a broth supplemented with magnesium and/or calcium, has been used in *H. parainfluenzae* (181). This system appears to induce competence for both natural and artificial pathways simultaneously. A remarkably effective artificial system for plasmid transformation in *B. subtilis* has recently been developed; by removing the peptidoglycan layer and exposing the resulting protoplasts to plasmid DNA in the presence of polyethylene glycol, transformants can be produced at a rate of 4×10^7 per

μg of DNA, and as many as 80% of viable cells (or 20% of starting cells) can be transformed (182).

As discussed in the previous section, the efficiency of plasmid or viral transfection in natural systems is much lower than that for chromosomal DNA. In artificial systems the situation is reversed. For example, calcium-treated *E. coli* cells are transformed by plasmid DNA at a frequency far exceeding that of chromosomal markers (164, 169), and artificial transformation of *B. subtilis* can give rates of plasmid transformation approaching 100% of remaining viable cells (182). Presumably, plasmids are not inactivated or sequestered when they enter cells by these artificial pathways, and, once inside, find themselves in a better position than transforming DNA of chromosomal origin, both because the marker to be selected is present on all fragments that enter, rather than on just a few (183), and because the plasmids are replicons, which do not require an integration step in order to be propagated.

Transformation of Yeast

Natural transformation has been reported in yeasts (184–186) but at very low levels. Fortunately, yeasts can now be transformed at high efficiency by artificial means. This system involves exposing yeast spheroplasts (187) to DNA in the presence of polyethylene glycol (188). The DNA used in the first such experiment was a hybrid plasmid formed by ligating a yeast gene into a ColE1 plasmid vector and growing the recombinant in *E. coli* cells (189). Using the cloned DNA as a hybridization probe (190), it was shown that the whole plasmid became integrated into the recipient genome at several different locations, presumably those sharing some homology with the incoming yeast gene (191). However, this system still did not give optimal transformation frequencies, presumably because it depended on the relatively inefficient integration of donor DNA, since ColE1 DNA will not replicate autonomously in yeast. This problem was solved by linking the yeast genes to a yeast origin of replication so that integration was no longer necessary. Both a *Saccharmyces coelicolor* plasmid (191) and *Saccharomyces cerevisiae* plasmids ligated to *E. coli* plasmids (192, 193) have been used; these give transformation rates of $1/10^3$ of remaining viable cells and 10^4–10^5 transformants/μg DNA (192). Such hybrid plasmids can grow in either yeast or *E. coli* cells. Using one of these hybrid plasmids, any yeast chromosomal gene can be replaced by its cloned homologue (194); this is analogous to the replacement of *E. coli* genes following transduction with lambda phage (195). Such a technique could, in theory, be applied to other eukaryotic cells. Yeast plasmids have also been developed that bear a chromosomal replication origin (193).

Transformation of Mammalian Cells with Metaphase Chromosomes

In this system (see Figure 3), cells in tissue culture are exposed to purified chromosomes and, after several days growth, are exposed to a selective medium in which transformants survive but untransformed recipients do not. Since this system can produce both genuine transformants and spontaneous mutants to the transformed phenotype with a similar frequency (about $1/10^7$), it is necessary to prove that the new gene product is the same as that found in the donor cells. This has been done by DEAE-cellulose chromatography (196), gel electrophoresis (196), isoelectric focusing (197), and immunological techniques (198). The presence of donor DNA sequences can be detected by hybridization (199). Recessive genes, such as those for hypoxanthine-guanine phosphoribosyl transferase (196), thymidine kinase, and galactokinase (200) have been transferred, as well as codominant genes, such as anchorage-independence and the resistances to methotrexate, ouabain, and α-amanitin (201, 202). (The transfer of these codominant genes allows wild-type cells rather than mutants to be used as recipients.) Although some progress has been made in increasing the frequency of transformation through the use of a calcium phosphate precipitate of transforming chromosomes (202) and dimethyl-sulfoxide treatment of recipient cells (203), the highest frequencies reached are only $4/10^5$ (203).

The mechanism by which metaphase chromosomes enter the cell has not been determined; perhaps they are phagocytosed (197). In any case, they are not replicated in an intact form inside the cell (196). The size of the propagated chromosomal fragment appears to be variable. In some experiments, it has been visualized in a metaphase spread of the recipient chromosomes (204). Genetic markers 1% of a genome apart ($\sim$10 megabases) were rarely seen to co-transform (205), while markers about 0.2% apart co-transformed in two out of eight clones analyzed (200). Several transferred sequences have been shown by hybridization to be greater than 17 kb in size (199).

Both stable and unstable transformants are produced during chromosome-mediated gene transfer (196). Unstable transformants lose the transferred gene if selection pressure is removed. Unstable transformants fall into two different classes (204). Slow-loss transformants have a cytologically visible donor chromosome fragment with a centromere-like region and a low level of donor gene activity. Fast-loss transformants do not have a visible donor chromosome fragment but do have a high level of donor gene activity. The high level of gene activity in the latter class is probably due to an amplification of gene sequences (199). Both the high gene copy number (199) and the high level of gene activity (206) revert to normal when

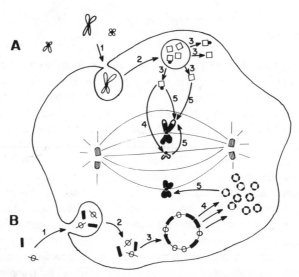

Figure 3 Possible steps in two genetic transformation systems for mammalian cells. *A* Transformation with metaphase chromosomes: *1.* donor chromosomes (𝕏) are taken up by phagocytosis; *2.* these are partially digested; *3.* they yield fragments with (⊡) and without (□) centromere-like regions; *4.* a small number of fragments with centromere-like regions will subsequently make use of the cells' mitotic apparatus and will then be lost slowly. (Those without centromere-like regions cannot do this and are lost quickly); *5.* all types of fragments can integrate at low efficiency into a resident chromosome (🟊) and become stably heritable. *B* Transformation with purified DNA; *1.* and *2.*—most of the cells can take up both a selectable marker (▬) and a piece of DNA bearing a replication origin (⊖); *3.* cellular enzymes act on the DNA and some large concatameric molecules are produced; *4.* in a very small fraction of cells, autonomously replicating concatemers will be found; *5.* these may recombine into a recipient chromosome to become stably heritable.

unstable transformants become stable, a rare event occurring in about $1/10^5$ cells/generation (207). Stability is achieved by apparently random integration into a host chromosome (204, 208, 209). Thus the transferred gene in unstable transformants is presumably propagated on an autonomously replicating element (205), analogous to a bacterial plasmid, which tends to be better controlled by the cell with respect to copy number and segregation if it has a centromere-like region (204, 209).

Transformation of Mammalian Cells with Purified DNA

As might be expected, the transfer of genes to mammalian cells with purified DNA is quite similar to gene transfer with metaphase chromosomes (see Figure 3). Ironically, however, the two systems were developed independently, based on different techniques. Although the idea for chromosome-mediated transformation was based on experience with mammalian cell fusion, the idea for transformation by naked DNA was based on experiences with transfection by oncogenic viral DNA and on infection by inactivated virions that bore a functional thymidine kinase gene.

In transformation with purified DNA, cells grown in tissue culture are exposed to a calcium phosphate precipitate (210) of the purified gene-bearing fragments, and then, some hours later, are exposed to a selective medium that allows only transformants or spontaneous mutants to survive. As in chromosome-mediated transformation, these two classes are then distinguished by biochemical or immunological characterization of the acquired gene product, or by DNA hybridization, to reveal acquired gene sequences. Such approaches have been used to transfer the gene for thymidine kinase (211–213), for methotrexate-resistance (214), and for adenine phosphoribosyl transferase (215), among others.

Much has been learned about the steps in transformation of mammalian cells by DNA (216). While it is likely that most cells in the culture take up and transiently express the transforming DNA (216), only about $1/10^5$–$1/10^6$) become transformed to the new phenotype; this is analogous to the abortive transformation seen after transfection with viral DNA (217). Both stable and unstable transformants are produced (218). Most of the cells in the same small group that are transformed with the selected gene will also be co-transformed with a nonselected gene if a mixture of the two DNAs is fed to the cells. Thus, only a small subpopulation of cells in a given culture is truly transformable, and transformability is probably induced rather than constitutive (218). The level of transformability varies widely between different cell types, so that only certain types are acceptable recipients. In addition, smaller heritable differences in transformability are observed between different subclones of a given cell type (C. M. Corsaro and M. L. Pearson, personal communication). Once inside the cell, co-transforming DNAs are probably ligated end-to-end, and the resulting large concatenates may then be integrated into a chromosome (216). The location of the integrated DNA is random, but constant for a given cell line (219). It is likely that if some of the transforming DNA bears a replication origin, then the gene of interest can be propagated on an autonomously replicating element (220).

Transformation of mammalian cells with purified DNA has already found applications in genetic engineering. A rabbit β-globin gene has been put into a mouse fibroblast and its expression examined in this heterologous system; intervening sequences are processed correctly but the transcripts have an aberrant 5'-terminus (221, 222). Using such systems it may be possible to determine how the expression of this gene is controlled. In addition, the nonselected sequences, such as pBR322, which co-integrate with a selected marker, have been used to locate that selected marker, after the transformed genome has been converted into a library of cloned fragments (216, 223).

Gene transfer in mammalian cells offers at least one advantage over molecular cloning in *E. coli* for the analysis of mammalian genes: the ability to examine gene expression in the cell of origin. On the other hand, certain improvements are needed that are analogous to the special advantages of the *E. coli* system. Specifically, an understanding of the mechanism underlying transformability in mammalian cells may permit a huge increase in the frequency of transformation (or vice versa). In addition, the development of an autonomously replicating vector for the transforming DNA, which could be induced to amplify under defined conditions (analogous to the chloramphenicol-induced amplification of pBR322 in *E. coli*), might facilitate the recovery of the transferred gene.

Acknowledgments

We would like to thank our colleagues who contributed preprints of their recent work. We would also like to thank Donald Morrison, John Scocca, Robert Yuan, and Howard Young for helpful criticism and advice. Special thanks go to JoAnn Olsen and Mildred Kahler for skillful preparation of the manuscript.

This work was supported by NIH grant number PO1-CA16519 to H. O. Smith, who is a Research Professor of the American Cancer Society. D. B. Danner was supported by NIH predoctoral training grant number 5-T32-GMO7445. R. A. Deich was supported by the National Cancer Institute under contract number NO1-CO-75380 with Litton Bionetics, Inc.

Literature Cited

1. Fox, M. S. 1978. *Ann. Rev. Genet.* 12:47–68
2. Low, K. B., Porter, D. D. 1978. *Ann. Rev. Genet.* 12:249–87
3. Graham, J. B., Istock, C. A. 1978. *Mol. Gen. Genet.* 166:287–90
4. Griffith, F. 1928. *J. Hyg.* 27:113–59
5. Costerton, J. W. 1977. *Microbiology-1977*, Washington DC: Am. Soc. Microbiol. pp. 151–57.
6. Notani, N. K., Setlow, J. K. 1974. *Prog. Nucleic Acid Res. Mol. Biol.* 14:39–100
7. Dubnau, D. 1976. *Microbiology-1976*, Washington DC: Am. Soc. Microbiol. pp. 14–27
8. Lacks, S. A. 1977. *Microbial Interactions, Receptors and Recognition*, ed. J. L. Reissig, pp. 179–232. London: Chapman & Hall
9. Hotchkiss, R. D., Gabor, M. 1970. *Ann. Rev. Genet.* 4:193–224
10. Hotchkiss, R. D. 1954. *Proc. Natl. Acad. Sci. USA* 40:49–55
11. Tomasz, A., Hotchkiss, R. D. 1964. *Proc. Natl. Acad. Sci. USA* 51:480–87
12. Pakula, R., Walczak, W. 1963. *J. Gen. Microbiol.* 31:125–33
13. Tomasz, A. 1966. *J. Bacteriol.* 91: 1050–61
14. Tomasz, A. 1965. *Nature* 208:155–59
15. Kohoutová, M., Malek, I. 1966. *The Physiology of Gene and Mutation Expression*, ed. M. Kohoutová, J. Hubacek, pp. 195–200. Prague: Academia
16. Leonard, C. G., Cole, R. M. 1972. *J. Bacteriol.* 110:273–280
17. Tomasz, A. 1970. *J. Bacteriol.* 101: 860–71
18. Raina, J. L., Ravin, A. W. 1981. *Proc. Natl. Acad. Sci. USA.* In press
19. Tomasz, A., Mosser, J. L. 1966. *Proc. Natl. Acad. Sci. USA* 55:58–66

20. Gaustad, P., Eriksen, J., Henriksen, S. D. 1979. *Acta Pathol. Microbiol. Scand. Sect. B* 87:117–22
21. Gaustad, P. 1979. *Acta Pathol. Microbiol. Scand. Sect. B* 87:123–28
22. Horne, D., Plotch, S., Tomasz, A. 1977. *Modern Trends in Bacterial Transformation and Transfection,* ed. A. Portolés, R. Lopez, M. Espinosa, pp. 11–34. New York/Amsterdam Elsevier/North Holland.
23. Morrison, D. A., Baker, M. F. 1979. *Nature* 282:215–17
24. Seto, H., Tomasz, A. 1975. *J. Bacteriol.* 121:344–53
25. Dooley, D. C., Hadden, C. T., Nester, E. W. 1971. *J. Bacteriol.* 108:668–79
26. Cahn, F. H., Fox, M. S. 1968. *J. Bacteriol.* 95:867–75
27. Hadden, C., Nester, E. W. 1968. *J. Bacteriol.* 95:876–85
28. Nester, E. W. 1964. *J. Bacteriol.* 87:867–75
29. McCarthy, C., Nester, E. W. 1967. *J. Bacteriol.* 94:131–40
30. Felkner, I. C., Wyss, O. 1964. *Biochem. Biophys. Res. Commun.* 16:94–102
31. Streips, U. N., Young, F. E. 1971. *J. Bacteriol.* 106:868–75
32. Akrigg, A., Ayad, S. R., Barker, G. R. 1967. *Biochem. Biophys. Res. Commun.* 28:1062–67
33. Akrigg, A., Ayad, S. R. 1970. *Biochem. J.* 117:397–403
34. Seto, H., Tomasz, A. 1974. *Proc. Natl. Acad. Sci. USA* 71:1493–98
35. Lacks, S., Greenberg, B., Neuberger, M. 1974. *Proc. Natl. Acad. Sci. USA* 71:2305–9
36. Lacks, S., Neuberger, M. 1975. *J. Bacteriol.* 124:1321–29
37. Seto, H., Lopez, R., Tomasz, A. 1975. *J. Bacteriol.* 122:1339–50
38. Seto, H., Tomasz, A. 1976. *J. Bacteriol.* 126:1113–18
39. Morrison, D. A., Guild, W. R. 1973. *Biochim. Biophys. Acta* 299:545–56
40. Lacks, S., Greenberg, B. 1976. *J. Mol. Biol.* 101:255–75
41. Lacks, S. 1979. *J. Bacteriol.* 138:404–9
42. Fox, M. S., Hotchkiss, R. D. 1957. *Nature* 179:1322–25
43. Seto, H., Tomasz, A. 1975. *J. Bacteriol.* 124:969–76
44. Ceglowski, P., Niedzielski, A. J., Dobrzanski, W. T. 1979. *Bull. Acad. Pol. Sci. Ser. Sci. Biol.* 27:99–103
45. Ceglowski, P., Kawczynski, M., Dobrzanski, W. T. 1980. *J. Bacteriol.* 141:1005–14
46. Ranhand, J. M. 1979. *Transformation, 1978,* ed. S. W. Glover, L. O. Butler, Oxford: Cotswold
47. Lerman, L. S., Tolmach, L. J. 1957. *Biochim. Biophys. Acta* 26:68–82
48. Tomasz, A. 1973. *Bacterial Transformation,* ed. L. J. Archer, pp. 81–88. New York: Academic
49. Lacks, S. 1977. See Ref. 22, pp. 35–44
50. Soltyk, A., Shugar, D., Piechowska, M. 1975. *J. Bacteriol.* 124:1429–38
51. Ceglowski, P., Fuchs, P. F., Soltyk, A. 1975. *J. Bacteriol.* 124:1621–23
52. Ciferri, O., Barlati, S., Lederberg, J. 1970. *J. Bacteriol.* 104:684–88
53. Dubnau, D., Cirigliano, C. 1972. *J. Mol. Biol.* 64:9–29
54. Piechowska, M., Soltyk, A., Shugar, D. 1975. *J. Bacteriol.* 122:610–22
55. Ranhand, J. M. 1980. *J. Bacteriol.* 142:568–80
56. Tevethia, M. J., Mandel, M. 1970. *J. Bacteriol.* 101:844–50
57. Tevethia, M. J., Mandel, M. 1971. *J. Bacteriol.* 106:802–7
58. Tevethia, M. J., Caudill, C. P. 1971. *J. Bacteriol.* 106:808–11
59. Miao, R., Guild, W. R. 1970. *J. Bacteriol.* 101:361–64
60. Lacks, S., Greenberg, B., Neuberger, M. 1975. *J. Bacteriol.* 123:222–32
61. Rosenthal, A. L., Lacks, S. A. 1981. *J. Mol. Biol.* In press
62. Lacks, S. 1962. *J. Mol. Biol.* 5:119–31
63. Morrison, D., Guild, W. R. 1972. *J. Bacteriol.* 112:1157–68
64. Guild, W. R., Robison, M. 1963. *Proc. Natl. Acad. Sci. USA* 50:106–12
65. Claverys, J. P., Roger, M., Sicard, A. M. 1980. *Mol. Gen. Genet.* 178:191–201
66. Garcia, E., Lopez, P., Perez Urena, M. T., Espinosa, M. 1978. *J. Bacteriol.* 135:731–40
67. Scher, B., Dubnau, D. 1973. *Biochem. Biophys. Res. Commun.* 55:595–602
68. Scher, B., Dubnau, D. 1976. *J. Bacteriol.* 126:429–38
69. Dubnau, D., Cirigliano, C. 1972. *J. Bacteriol.* 111:488–94
70. Joenje, H., Venema, G. 1975. *J. Bacteriol.* 122:125–33
71. Piechowska, M., Fox, M. S. 1971. *J. Bacteriol.* 108:680–89
72. Davidoff-Abelson, R., Dubnau, D. 1973. *J. Bacteriol.* 116:146–53
73. Morrison, D. A. 1977. *J. Bacteriol.* 132:576–83
74. Morrison, D. A., Baker, M., Mannarelli, B. 1979. See Ref. 46, pp. 120–26
75. Morrison, D. A., Baker, M. F. 1979. *Nature* 282:215–17
76. Morrison, D. A. 1978. *J. Bacteriol.* 136:548–57

77. Morrison, D. A., Mannarelli, B. 1979. *J. Bacteriol.* 140:655–65
78. Raina, J. L., Ravin, A. W. 1977. See Ref. 22, pp. 143–48
79. Raina, J. L., Ravin, A. W. 1978. *J. Bacteriol.* 133:1212–23
80. Raina, J. L., Metzer, E., Ravin, A. W. 1979. *Mol. Gen. Genet.* 170:249–59
81. Eisenstadt, E., Lange, R., Willecke, K. 1975. *Proc. Natl. Acad. Sci. USA* 72:323–27
82. Pieniazek, D., Piechowska, M., Venema, G. 1977. *Mol. Gen. Genet.* 156:251–61
83. Brietenwerf, J., Venema, G. 1979. *Mol. Gen. Genet.* 156:145–55
84. Fox, M. S., Allen, M. K. 1964. *Proc. Natl. Acad. Sci. USA* 52:412–19
85. Gurney, T. Jr., Fox, M. S. 1968. *J. Mol. Biol.* 32:83–100
86. Dubnau, D., Davidoff-Abelson, R. 1971. *J. Mol. Biol.* 56:209–21
87. Fornili, S. L., Fox, M. S. 1977. *J. Mol. Biol.* 113:181–91
88. Bodmer, W. F. 1966. *J. Gen. Physiol.* 49:233–58
89. Clark, A. J. 1973. *Ann. Rev. Genet.* 7:67–86
90. Radding, C. M. 1978. *Ann. Rev. Biochem.* 47:847–80
91. Shibata, T., Cunningham, R. P., Das-Gupta, C., Radding, C. M. 1979. *Proc. Natl. Acad. Sci. USA* 76:5100–4
92. Claverys, J. P., Lefevie, J. C., Sicard, A. M. 1980. *Proc. Natl. Acad. Sci. USA* 77:3534–38
93. Lacks, S., Greenberg, B., Carlson, K. 1967. *J. Mol. Biol.* 29:327–47
94. Buitenwerf, J., Venema, G. 1977. *Mol. Gen. Genet.* 156:145–55
95. Seto, H., Tomasz, A. 1977. *Proc. Natl. Acad. Sci. USA* 74:296–99
96. Buitenwerf, J., Venema, G. 1978. *Mol. Gen. Genet.* 160:67–75
97. Cozzarelli, N. R. 1980. *Science* 207: 953–60
98. Popowski, J., Venema, G. 1978. *Mol. Gen. Genet.* 166:119–26
99. Ephrussi-Taylor, H., Sicard, A. M., Kaman, R. 1965. *Genetics* 51:455–75
100. Lacks, S. 1966. *Genetics* 53:207–35
101. Shoemaker, N. B., Guild, W. R. 1974. *Mol. Gen. Genet.* 128:283–90
102. Roger, M. 1977. *J. Bacteriol.* 129:298–304
103. Ephrussi-Taylor, H. 1966. *Genetics* 54:11–22
104. Louarn, J.-M., Sicard, A. M. 1968. *Biochem. Biophys. Res. Commun.* 30: 683–89
105. Ephrussi-Taylor, H., Gray, T. C. 1966. *J. Gen. Physiol.* 49(6):P. 2, pp. 211–321
106. Lacks, S. 1970. *J. Bacteriol.* 101:373–83
107. Guild, W. R., Shoemaker, N. B. 1974. *Mol. Gen. Genet.* 128:291–300
108. Meselson, M., Pukkila, P., Rykowski, M., Peterson, J., Radman, M., Wagner, R., Herman, G., Modrich, P. 1980. *J. Supramol. Struct.,* Suppl. 4, p. 311
109. Friedberg, E. C., Bonura, T., Cone, R., Simmons, R., Anderson, C. 1978. *DNA Repair Mechanisms,* ed. P. C. Hanawalt, E. C. Friedberg, C. F. Fox, pp. 163–73. New York: Academic
110. Leidy, G., Jaffee, I., Alexander, H. E. 1962. *Proc. Soc. Exp. Biol. Med.* 111:725–31
111. Spencer, H. T., Herriott, R. M. 1965. *J. Bacteriol.* 90:911–20
112. Herriott, R. M., Meyer, E. M., Vogt, M. 1970. *J. Bacteriol.* 101:517–24
113. Wise, E. M., Alexander, S. P., Powers, M. 1973. *Proc. Natl. Acad. Sci. USA* 70:471–75
114. Sparling, P. F. 1966. *J. Bacteriol.* 92:1364–71
115. Biswas, G. D., Sox, T., Sparling, P. F. 1977. *J. Bacteriol.* 129:983–92
116. Juni, E. 1972. *J. Bacteriol.* 112:917–31
117. Ranhand, J. M., Lichstein, H. C. 1969. *J. Gen. Microbiol.* 55:37–43
118. Barnhart, B. J., Herriott, R. M. 1963. *Biochim. Biophys. Acta* 76:25–39
119. Scocca, J. J., Haberstat, M. 1978. *J. Bacteriol.* 135:961–67
120. Piekarowicz, A., Siwinska, M. 1977. *J. Bacteriol.* 129:22–29
121. Bingham, D. P., Barnhart, B. J. 1973. *J. Gen. Microbiol.* 75:249–58
122. Zoon, K. C., Scocca, J. J. 1975. *J. Bacteriol.* 123:666–67
123. Zoon, K. C., Haberstat, M., Scocca, J. J. 1976. *J. Bacteriol.* 127:545–54
124. Stuy, J. H., Hoffman, J. F. 1971. *J. Virol.* 7:127–36
125. Hoffman, J. F., Stuy, J. H. 1972. *Biochem. Biophys. Res. Commun.* 46: 1388–93
126. Boling, M. E., Setlow, J. K., Allison, D. P. 1972. *J. Mol. Biol.* 63:335–48
127. Mulder, C., Doty, P. 1968. *J. Mol. Biol.* 32:423–35
128. Postel, E. H., Goodgal, S. H. 1966. *J. Mol. Biol.* 16:317–27
129. Postel, E. H., Goodgal, S. H. 1967. *J. Mol. Biol.* 28:247–59
130. Schaeffer, P., Edgar, R. S., Rolfe, R. 1960. *C. R. Soc. Biol.* 154:1978–83
131. Scocca, J. J., Poland, R. L., Zoon, K. C. 1974. *J. Bacteriol.* 118:369–73
132. Jyssum, S., Jyssum, K. 1970. *Acta Pathol. Microbiol. Scand. Sect. B* 78: 140–48

133. Jyssum, K., Jyssum, S., Gunderser, W. B. 1971. *Acta Pathol. Microbiol. Scand. Sect. B.* 79:563–71
134. Dougherty, T. J., Asmus, A., Tomasz, A. 1979. *Biochem. Biophys. Res. Commun.* 86:97–104
135. Sisco, K. L., Smith, H. O. 1979. *Proc. Natl. Acad. Sci. USA* 76:972–76
136. Chung, B. C., Goodgal, S. H. 1979. *Biochem. Biophys. Res. Commun.* 88: 208–14
137. Danner, D. B., Deich, R. A., Sisco, K. L., Smith, H. O. 1980. *Gene.* 11:311–18
138. Siebenlist, U., Gilbert, W. 1980. *Proc. Natl. Acad. Sci. USA* 77:122–26
139. Deich, R. A., Smith, H. O. 1979. See Ref. 46, pp. 377–84
140. Stuy, J. H., vander Have, B. 1971. *J. Gen. Microbiol.* 65:147–52
141. Sutrina, S. L., Scocca, J. J. 1979. *J. Bacteriol.* 139:1021–27
142. Deich, R. A., Smith, H. O. 1980. *Mol. Gen. Genet.* 177:369–74
143. Stuy, J. H., Stern, D. 1964. *J. Gen. Microbiol.* 35:391–400
144. Voll, M. J., Goodgal, S. H. 1961. *Proc. Natl. Acad. Sci. USA* 47:505–12
145. Notani, N. K., Setlow, J. K. 1972. *J. Bacteriol.* 112:751–60
146. Stuy, J. H. 1965. *J. Mol. Biol.* 13:554–70
147. Kahn, M., Concino, M., Gromkova, R., Goodgal, S. H. 1979. *Biochem. Biophys. Res. Commun.* 87:764–72
148. Notani, N., Goodgal, S. H. 1966. *J. Gen. Physiol.* 49:197–209
149. Steinhart, W. L., Herriott, R. M. 1968. *J. Bacteriol.* 96:1725–31
150. Notani, N. K. 1971. *J. Mol. Biol.* 59:223–36
151. LeClerc, J. E., Setlow, J. K. 1975. *J. Bacteriol.* 122:1091–1102
152. Caster, J. H., Postel, E. H., Goodgal, S. H. 1970. *Nature* 227:515–17
153. Beattie, K. L., Setlow, J. K. 1971. *Nature New Biol.* 231:177–79
154. Setlow, J. K., Boling, M. E., Beattie, K. L., Kimball, R. F. 1972. *J. Mol. Biol.* 68:361–78
155. Notani, N. K., Setlow, J. K., Joshi, V. R., Allison, D. P. 1972. *J. Bacteriol.* 110:1171–81
156. Postel, E. H., Goodgal, S. H. 1972. *J. Bacteriol.* 109:292–97
157. Contente, S., Dubnau, D. 1979. *Mol. Gen. Genet.* 167:251–58
158. Contente, S., Dubnau, D. 1979. *Plasmid* 2:555–71
159. Dubnau, D. 1980. In *Proc. FEBS Symp. DNA, Prague, 1979,* ed. S. Zadrazil, J. Sponar. Oxford: Pergamon
160. Bendler, J. W. 1976. *J. Bacteriol.* 125:197–204
161. Canosi, U., Morelli, G., Trautner, T. A. 1978. *Mol. Gen. Genet.* 166:259–67
162. Duncan, C. H., Wilson, G. A., Young, F. E. 1978. *Proc. Natl. Acad. Sci. USA* 75:3664–68
163. Mandel, M., Higa, A. 1970. *J. Mol. Biol.* 53:159–62
164. Cohen, S. N., Chang, A. C. Y., Hsu, L. 1972. *Proc. Natl. Acad. Sci. USA* 69:2110–14
165. Cohen, S. N., Chang, A. C. Y. 1973. *Proc. Natl. Acad. Sci. USA* 70:1293–97
166. Sinsheimer, R. L. 1977. *Ann. Rev. Biochem.* 46:415–38
167. Leive, L. 1974. *Ann. NY Acad. Sci.* 235:109–129
168. Oishi, M., Irbe, R. M. 1977. See Ref. 22, pp. 121–34
169. Cosloy, S. D., Oishi, M. 1973. *Proc. Natl. Acad. Sci. USA* 70:84–87
170. Strike, P., Humphreys, G. O., Roberts, R. J. 1979. *J. Bacteriol.* 138:1033–35
171. Morrison, D. A. 1977. *J. Bacteriol.* 132:349–51
172. Norgard, M. V., Keen, K., Monahan, J. J. 1978. *Gene* 3:279–92
173. Dagert, M., Ehrlich, S. D. 1979. *Gene* 6:23–28
174. Kushner, S. R. 1978. *Genetic Engineering,* ed. H. B. Boyer, S. Nicosia, pp. 17–23. Amsterdam/New York Elsevier/North Holland
175. Lindberg, M., Sjöström, J. E., Johansson, T. 1972. *J. Bacteriol.* 109:844–47
176. Chakrabarty, A. M., Mylroie, J. R., Friello, D. A., Vacca, J. G. 1975. *Proc. Natl. Acad. Sci. USA* 72:3647–51
177. Stuy, J. H. 1979. *J. Bacteriol.* 139:520–29
178. Benzinger, R. 1978. *Microbiol. Rev.* 42:194–236
179. Dityatkin, S. Y., Lisovskaya, K. V., Panzhava, N. N., Il'yashenko, B. N. 1972. *Biochim. Biophys. Acta* 281: 319–23
180. Holsters, M., deWade, D., Depicker, A., Messens, E., van Montagu, M., Schell, J. 1978. *Mol. Gen. Genet.* 163:181–87
181. Gromkova, R., Goodgal, S. 1979. *Biochem. Biophys. Res. Commun.* 88: 1428–34
182. Chang, S., Cohen, S. N. 1979. *Mol. Gen. Genet.* 168:111–15
183. Oishi, M., Cosloy, S. D., Basu, S. R. 1974. *Mechanisms in Recombination,* ed. R. F. Grell, pp. 145–54. New York: Plenum
184. Fowell, R. R. 1969. *The Yeasts,* ed. A.

H. Rose, J. S. Harrison, pp. 303–83. London: Academic

185. Mishra, N. C., Tatum, E. L. 1973. *Proc. Natl. Acad. Sci. USA* 70:3875–79

186. Khan, N. C., Sen, S. P. 1974. *J. Gen. Microbiol.* 83:237–50

187. Hutchinson, H. T., Hartwell, L. H. 1967. *J. Bacteriol.* 94:1697–1705

188. Hinnen, A., Hicks, J. B., Fink, G. R. 1978. *Proc. Natl. Acad. Sci. USA* 75:1929–33

189. Ratzkin, B., Carbon, J. 1977. *Proc. Natl. Acad. Sci. USA* 74:487–91

190. Southern, E. M. 1975. *J. Mol. Biol.* 98:503–17

191. Bibb, M. J., Ward, J. M., Hopwood, D. A. 1978. *Nature* 274:398–400

192. Beggs, J. D. 1978. *Nature* 275:104–8

193. Struhl, K., Stinchcomb, D. T., Scherer, S., Davis, R. W. 1979. *Proc. Natl. Acad. Sci. USA* 76:1035–39

194. Scherer, S., Davis, R. W. 1979. *Proc. Natl. Acad. Sci. USA* 76:4951–55

195. Morse, M. L., Lederberg, E. M., Lederberg, J. 1956. *Genetics* 41:758–75

196. McBride, O. W., Ozer, H. L. 1973. *Proc. Natl. Acad. Sci. USA* 70:1258–62

197. Burch, J. W., McBride, O. D. 1975. *Proc. Natl. Acad. Sci. USA* 72:1797–1801

198. Shani, M., Huberman, E., Aloni, Y., Sachs, L. 1974. *Virology* 61:303–5

199. Scangos, G. A., Huttner, K. M., Silverstein, S., Ruddle, F. H. 1979. *Proc. Natl. Acad. Sci. USA* 76:3987–90

200. Willecke, K., Lange, R., Kruger, A., Reber, T. 1976. *Proc. Natl. Acad. Sci. USA* 73:1274–78

201. Spandidos, D. A., Siminovitch, L. 1977. *Cell* 12:675–82

202. Spandidos, D. A., Siminovitch, L. 1977. *Proc. Natl. Acad. Sci. USA* 74:3480–84

203. Miller, C. L., Ruddle, F. H. 1978. *Proc. Natl. Acad. Sci. USA* 75:3346–50

204. Klobutcher, L. A., Miller, C. L., Ruddle, F. H. 1980. *Proc. Natl. Acad. Sci. USA* 77:3610–14

205. Willecke, K., Ruddle, F. H. 1975. *Proc. Natl. Acad. Sci. USA* 72:1792–96

206. Degnen, G. E., Miller, I. L., Adelberg, E. A., Eisenstadt, J. M. 1977. *Proc. Natl. Acad. Sci. USA* 74:3956–59

207. Degnen, G. E., Miller, I. L., Eisenstadt, J. M., Adelberg, E. A. 1976. *Proc. Natl. Acad. Sci. USA* 73:2838–42

208. Athival, R. S., McBride, O. W. 1977. *Proc. Natl. Acad. Sci. USA* 74:2943–47

209. Klobutcher, L. A., Ruddle, F. H. 1979. *Nature* 280:657–60

210. Graham, F. L., van der Eb, A. J. 1973. *Virology* 52:456–67

211. Wigler, M., Silverstein, S., Lee, L. S., Pellicer, A., Cheng, Y. C., Axel, R. 1977. *Cell* 11:223–32

212. Maitland, N. J., McDougall, J. K. 1977. *Cell* 11:233–41

213. Bachetti, S., Graham, F. L. 1977. *Proc. Natl. Acad. Sci. USA* 74:1590–94

214. Wigler, M., Perucho, M., Kurtz, D., Dana, S., Pellicer, A., Axel, R., Silverstein, S. 1980. *Proc. Natl. Acad. Sci. USA* 77:3567–70

215. Wigler, M., Pellicer, A., Silverstein, S., Axel, R., Urlaub, G., Chasin, L. 1979. *Proc. Natl. Acad. Sci. USA* 76:1373–76

216. Pellicer, A., Robins, D., Wold, B., Sweet, R., Jackson, J., Lowy, I., Roberts, J. M., Sim, G. K., Silverstein, S., Axel, R. 1980. *Science* 209:1414–22

217. Goodfellow, P. N., Jones, E. A., van Heyningen, V., Solomon, E., Bobrow, M., Maggiano, V., Bodmer, W. F. 1975. *Nature* 254:267–69

218. Wigler, M., Sweet, R., Sim, G. K., Wold, B., Pellicer, A., Lacy, E., Maniatis, T., Silverstein, S., Axel, R. 1979. *Cell* 16:777–85

219. Pellicer, A., Wigler, M., Axel, R., Silverstein, S. 1978. *Cell* 14:133–41

220. Huttner, K. M., Scangos, G. A., Ruddle, F. H. 1979. *Proc. Natl. Acad. Sci. USA* 76:5820–24

221. Wold, B., Wigler, M., Lacy, E., Maniatis, T., Silverstein, S., Axel, R. 1979. *Proc. Natl. Acad. Sci. USA* 76:5684–88

222. Mantei, N., Boll, W., Weissmann, C. 1979. *Nature* 281:40–46

223. Perucho, M., Hanahan, D., Lipsich, L., Wigler, M. 1980. *Nature* 285:207–10

Ann. Rev. Biochem. 1981. 50:69–83
Copyright © 1981 by Annual Reviews Inc. All rights reserved

NMR STUDIES OF TISSUE   ♦12071
METABOLISM

David G. Gadian and George K. Radda

Department of Biochemistry, Oxford University, Oxford, United Kingdom

CONTENTS

Perspectives and Summary

During the past six or seven years, high resolution nuclear magnetic resonance (NMR) has been established as a powerful tool for the study of cellular metabolism in isolated cells and organs. The non-invasive nature of the measurement offers many advantages over conventional analytical techniques, but limitations on sensitivity restrict the method to the observation of molecules present in the cell at relatively high concentrations (in the order of 0.5–1 mM). Inevitably, many of the initial investigations tended to confirm existing knowledge, but this was important in establishing the validity of the NMR method. In the last few years, NMR has contributed new and interesting information on several aspects of metabolism and metabolic regulation. The majority of reports deal with ^{31}P, ^{13}C, and, to a lesser

69

0066-4154/81/0701-0069$01.00

extent, ^{1}H NMR. The particular areas of interest include measurements of metabolic fluxes in the steady state and transport processes, observations of cellular pH and H^+ ion distribution, and compartmentation and spatially resolved NMR of whole animals. The ability to directly observe the intracellular milieu is perhaps the major single advantage of NMR. A very large number of different systems can be studied, but it is important to keep in mind that NMR is simply an additional method for studying metabolism and must always be combined with carefully designed biochemical experiments.

Why Study Metabolism by NMR?

Historically, the study of metabolism is one of the oldest branches of biochemistry, and the well-known "metabolic charts" are unlikely to be changed by the introduction of the relatively recent technique of NMR. But an understanding of metabolism requires considerably more than a description of pathways. We are dealing with cellular dynamics in which the energy-producing and -consuming reactions are controlled and the different forms of energy are tightly coupled. In this article we describe how recent developments in NMR can be used to study some of the problems associated with control, energetics, and dynamics in whole tissue metabolism. We assume a basic knowledge of NMR as applied to cellular metabolism since there have been a number of recent review articles on the subject (1–5).

A Statement of the Problems

Developing a new method is interesting in its own right but the end value of the work must depend on the contributions the technique offers to the solution of specific questions. Some of the outstanding problems in cellular metabolism that can be tackled by NMR are enumerated below.

1. The concentrations of metabolites determined by techniques other than NMR generally rely on measurements of total amounts of metabolites within the tissues. It is essential to evaluate the "free" (i.e. metabolically available) concentrations, which may differ from measured values as a result of tight binding of metabolites to specific macromolecules.

2. The cellular environment has an important role in determining molecular structures, interactions, and functional activities. We need to increase our knowledge about the environment of molecules within the cytoplasm and the various cellular compartments.

3. Compartmentation of small molecules is in some cases known, and is invoked in others to explain deviations from expected behavior. We need to devise experiments to evaluate the distribution of molecules within the cell.

4. The knowledge of in vivo fluxes is a prerequisite for the definition of cellular dynamics. Both rate-limiting reactions and those operating close to equilibrium must be followed in vivo.

5. A central hypothesis of metabolic control is the presence of "regulatory enzymes." Their regulatory role is normally inferred by studying the properties of the isolated enzyme and by showing, from measurements of mass-action ratios, that the appropriate substrates are far from equilibrium in vivo. There are, however, several notable anomalies (e.g. pyruvate kinase) based on this approach. It is therefore important to measure interactions with regulatory ligands and enzyme activities in vivo.

6. Most tissues and organs are composed of a heterogeneous assembly of cells. In some cases, careful dissection of an organ (e.g. kidney) can be used to prepare different cell types. It is, however, necessary to determine to what extent cell heterogeneity has a role in the function of the whole intact organ.

The NMR Observation

High resolution NMR of tissues and cells essentially detects only mobile molecules, because highly immobilized molecules give signals that are too broad to observe. Thus signals are observed from metabolites, but not, in general, from macromolecular structures such as membrane phospholipids and DNA. However, special techniques are available for observing immobile molecules, as illustrated by some ^{31}P NMR studies of bone (6). The NMR spectrum is characterized by the position (frequency) of the absorption (which yields chemical identification), by the intensity of the signal (which gives quantitative analytical information), and by the two relaxation times (which give information about molecular motion and interactions).

The most suitable nuclei for high resolution studies of organelles, cells, and whole organs are ^1H, ^{13}C, and ^{31}P, and we restrict our discussion to these nuclei. We concentrate on studies of whole tissues, although some reference is made to work on isolated cells and subcellular organelles.

Identification and Quantification

^{31}P NMR spectra have been recorded from a variety of cells and perfused organs (e.g. 1, 5). In general, assignment of ^{31}P resonances is relatively simple, since the chemical shift range for biologically occurring phosphates is large ($\sim$40 ppm), and only components present in a mobile form in concentration larger than about 0.5 mM are detected. The major signals are from ATP (3 resonances), ADP (2 resonances, but generally obscured by ATP), phosphocreatine, P_i, sugar phosphates (mainly glucose 6-phosphate), 2,3-diphosphoglycerate (2 peaks in blood and red blood cells), and AMP (e.g. in kidney). In addition, hitherto unsuspected metabolites are detected by their signals in the phosphodiester frequency range in spectra from rabbit muscle, frog gastrocnemius (7–9), and dystrophic chicken pectoralis (10). On the basis of chromatographic separation and chemical analysis, these compounds were identified as glycero-3-phosphorylcholine in mammalian muscle (7) and as L-serine-ethanolamine phosphodiester in dystrophic chicken muscle (10). The presence of several such compounds in toad muscle (9, 11), kidney (12), brain (13), and rabbit heart (14) was also reported, though they have not yet been chemically identified. It was also shown that the concentration of glycero-3-phosphorylcholine in Duchenne dystrophic quadriceps was significantly lower than that in normal human quadriceps, whereas elevated levels were detected for muscles with Werd-nig-Hoffman syndrome (15). The search for the function of these compounds continues (16).

We note the very low levels of phosphocreatine in liver (17) and kidney (18), a fact that is used in the spatial selection of signals from in vivo organs (see below).

Quantification of the metabolites is based on the fact that the NMR signal area, or intensity, is proportional to the total amount of the detectable species. The derivation of an absolute concentration is, however, not simple and has been achieved only in a few instances [for a detailed discussion of the problems see (4)]. The concentrations of ATP and phosphocreatine determined by NMR and freeze extraction are in broad agreement for frog sartorii (11) and perfused rat heart (19). This implies that NMR detects the full intracellular content, within ~20%, of these compounds.

In contrast, the ^{31}P spectra from muscle and brain of live rats show that the analytically determined ADP and P_i levels are considerably higher than those measured by NMR (20). This is also true for P_i in perfused heart [see, for example, spectra in (21)]. The suggestion (20) that estimates of the phosphorylation potentials in intact tissues derived from freeze extraction experiments are considerably lower than the real values has important consequences in relation to the thermodynamic considerations of oxidative phosphorylation.

Measurement of the upper limit to the concentration of mobile ADP can be obtained directly from NMR spectra. However, even if the ADP signals are not detectable, the concentration of free ADP can sometimes be evaluated if the creatine kinase reaction is close to equilibrium within the tissue. This is because it is possible to estimate, primarily by NMR, the concentrations of all the substrates of this reaction other than ADP. In resting skeletal muscle, the creatine kinase reaction is close to equilibrium (see below), and ~20 μM is obtained for the free ADP concentration in anaerobic frog muscle at 4°C (22). This very low value has profound implications with regard to theories of metabolic control. Furthermore, if the adenylate kinase reaction is also shown to be at equilibrium, the level of free AMP determined from this reaction would be about 0.1 μM. This value is orders of magnitude lower than the value normally quoted for AMP, and again this would have most interesting effects on our ideas about metabolic control.

The discrepancy in the ADP concentrations measured by NMR and by other means is still unexplained. In muscle, the explanation is straightforward, for it has been assumed for some time that a large percentage of the ADP is tightly bound to the proteins of the myofilaments. Presumably this ADP is too immobilized to generate detectable signals. In brain, there may be unavoidable breakdown of high energy phosphates in the extraction procedures used; this would explain the relatively high levels of both ADP and P_i measured in this way. Alternatively, a significant fraction of the intracellular content of these compounds could be sequestered in such a way [e.g. in the mitochondria (23)] that it generates no detectable signal. Unfortunately, the NMR signals from metabolites within the mitochondria have yet to be fully characterized. Ogawa et al, in their studies of purified rat liver mitochondria, have observed signals that can be assigned to intramitochondrial ATP and P_i (24). However, their experiments were done at 0°C and the nature of the signals from the mitochondrial compartment requires further quantitative study.

The assignment and quantification of ^{13}C resonances is not as simple as for ^{31}P NMR spectra. Using ^{13}C-enriched substrates, Shulman and his collaborators have, however, identified a large number of resonances in isolated cells (3) [e.g. liver cells (25)] and in perfused liver (26). In general, only relative concentrations have been reported.

Because of the large number of overlapping peaks and underlying broad signals, observation of proton resonances from mobile components requires special pulse sequences (27, 28). Using these procedures in studies on red cells (27, 29) and a variety of gland storage systems (28, 30), signals from small molecules (e.g. glutathione, lactate in red cells, and adrenaline in the adrenal medulla), and from some parts of proteins (e.g. the histidine resonance of hemoglobin, chromogranin) have been assigned. The redox state of the red cells can be deduced from the nature of the gluthathione signals (27).

Cellular Environment

MOLECULAR INTERACTIONS   The NMR signals obtained from intracellular components depend on the cellular environment and on specific interactions within the cell. In general, the resonances are broader and the relaxation times shorter than in aqueous solutions of similar ionic strength to the intracellular milieu. An attempt has been made using HeLa cells (31), to analyze the numerous factors responsible, but only tentative conclusions could be drawn. Detailed relaxation studies in storage granules have yielded more information (see below). In an interesting series of papers, Fossel & Solomon (32, 33) presented results that were interpreted in terms of a network of enzymes involving the membrane of the red blood cell and phosphorylase kinase, phosphoglucomutase, and glyceraldehyde 3-phosphate dehydrogenase. The experiments were based on observations of very small chemical shifts in the ^{31}P spectra of intracellular 2,3-diphosphoglycerate under different conditions (e.g. on addition of oubain outside the cell). The validity of these experiments however, has been seriously questioned (34).

The fact that over 90% of cellular ATP is complexed with Mg^{2+} was deduced from the first study on muscle, as the three ^{31}P resonances of ATP are shifted in a characteristic manner (35). Similar observations have been made on other tissues (18, 36) and cells (37). Recently, measurements on the extent of ATP complexing with Mg^{2+} have been used to derive the free Mg^{2+} concentration in erythrocytes (38) and frog skeletal muscle (39). The value of 0.6 mM for the latter system is very low in comparison to the 4.4 mM (amended to 3mM), estimated from the effect of Mg^{2+} on the T_2 relaxation times of the phosphocreatine signal in frog gastrocnemius (40). Both of these values rely on assumptions that may not be totally valid. For example, the value of 0.6 mM was obtained by deducing from the chemical shifts of the ATP resonances that 93% of the intracellular ATP is complexed to Mg^{2+}. The latter figure relies on very precise measurements of chemical shifts and on the assumption that no intracellular features other than Mg^{2+}, pH, and ionic strength have any detectable effect on the ATP chemical shifts. This assumption, if proved, would be surprising in view of the very specific way in which (Mg)ATP binds to a variety of macromolecules. In our opinion, the chemical shifts observed in frog skeletal muscle suggest a lower limit of about 0.5 mM for the free Mg^{2+} concentration, but cannot provide an accurate upper limit.

Since the first report by Moon & Richards on erythrocytes (41), ^{31}P NMR has been widely used to measure the intracellular pH in a variety of systems (1–5). pH can be measured from the frequency of ^{31}P NMR signals of phosphate groups of pK values close to the pH. The two ionic species are in "fast exchange," and the frequency of the observed resonance is therefore determined by the relative amounts of the two species concerned.

The validity of the pH measurements requires some assumptions about intracellular environment (e.g. ionic strength and composition) and about the lack of effects (e.g. binding to protein) that might shift the resonances in an unpredictable way. Nevertheless, there is now general agreement that cytoplasmic pH in muscle (11), heart (42, 43), kidney (18), and other tissues can be reliably measured. Some recent important developments include the recognition that contributions from extracellular P_i must and can be excluded (42, 43), and that the use of more than one pH marker is an advantage (44). For example, the uptake of deoxyglucose into perfused rat hearts results in the accumulation of deoxyglucose 6-phosphate in the cytoplasm. From the resonances of this molecule, pH values were obtained identical to those measured from the P_i resonance. This confirms that both P_i and the sugar phosphate measure cytoplasmic pH, and the criticisms that the possible sequestering of P_i into the mitochondria distorts the pH measurement (45, 46) can be discounted. Further support for the validity of the pH measurement by NMR comes from a comparison the values derived by NMR and microelectrodes in barnacle muscle fibers [C. R. Bagshaw, R. D. Vaughan-Jones, D. G. Gadian, and G. K. Radda, unpublished observations; see also (4)].

In a study of acute renal acidosis it was demonstrated that the intracellular pH in perfused rat kidneys decreased by only 0.3 units when the extracellular pH is changed by 0.6 units. This fall in total intrarenal pH is insufficient to explain the changes observed in the metabolites of the glutamate dehydrogenase reaction in acute metabolic acidosis (47).

COMPARTMENTATION   It was noted above that the observed linewidth of the P_i signal from skeletal muscle is often considerably greater than that for phosphocreatine (8, 48). This was particularly noticeable during the metabolic rundown of anoxic muscle (8), which suggests that P_i experiences several pH environments (ΔpH $\sim$0.2) in the cell. Perturbations of the pH in muscle by acetate resulted in the detection of two P_i and sugar phosphate signals (48). Studies on frog muscle supported the tentative interpretation that this represents compartmentation of these molecules between the cytoplasm and sarcoplasmic reticulum. When the muscle was treated with 2,4-dinitro-1-fluorobenzene, a new "acidic" component appeared, which was not detected when the muscle was pretreated with glycerol (to disrupt the transverse tubules) (49). pH gradients were clearly seen between the cytoplasmic and mitochondrial compartments in isolated liver cells (50); such gradients have not yet been observed in perfused organs. In perfused kidneys the width of the signal for P_i corresponds to a pH heterogeneity of 0.4 units (47, 51), although this is most likely to be a result of cellular heterogeneity and not of intracellular distribution.

More convincing and readily identifiable compartmentation is seen in various storage systems. For example, the ^{31}P NMR spectrum of the adrenomedullary chromaffin granules shows ATP resonances at frequencies different from those of cytoplasmic ATP, as a result of the low pH (5.7) and interaction with catecholamine inside the vesicles (52–54). The internal pH drops by 0.4–0.5 units when external ATP is hydrolyzed by the membrane-

bound ATPase (52, 53). Cytoplasmic and intragranular ATP can be observed as distinct signals in isolated chromaffin cells, perfused adrenal gland
(A. Bevington, R. W. Briggs, G. K. Radda, and K. R. Thulborn, unpublished observations), and platelets (55, 56). Chromaffin granules, adrenal
medulla, and many other storage systems have been examined by ^1H, ^{13}C,
and ^{19}F NMR and important information about the nature of the "packaged" materials is emerging (28, 57–59). In a recent report, ^{31}P NMR
studies on maize root tips have shown that plant vacuoles also have a low
pH environment (60).

Reaction Rates

CHANGES IN METABOLIC STATE   In vivo reaction rates have been measured by NMR in several ways. Relatively slow metabolic transition can be
monitored by successive accumulations of the spectra, the time resolution
being determined by the time required for signal averaging (typically 0.5–1
min for ^{31}P spectra). Thus metabolic changes during global ischemia, and
recovery of perfused hearts (42), kidney (61), liver (62), and brain (63) have
been followed. In perfused rat hearts and kidney the decrease in pH (in
heart from 7.05 to 6.4) is considerably diminished by the use of buffers like
bis-tris-propane in the perfusion medium, and this appears to have a significant protective effect on the ischemic tissue (42, 64). Quantification of the
production of H^+ in rat hearts led to the conclusion that all the protons
produced during ischemia come from the anaerobic breakdown of glycogen
(42). The rate of glycogenolysis can therefore also be measured in a nondestructive way. Using this information, and the fact that accumulated deoxyglucose 6-phosphate inhibits phosphorylase b, the extent and rate of b to
a conversion and phosphorylase b activation in rat heart during global
ischemia have been estimated to show significant (~50%) phosphorylase
b activity after about 1 min (44).

Several of the ^{31}P resonances in the spectra of perfused and in vivo liver
have unusually short spin-lattice relaxation times [about 100 ms for liver
ATP (62) in contrast with about 1–2 s for ATP in other tissues]. This
considerably enhances the time resolution of kinetic studies of the liver, as
illustrated by the studies of McLaughlin et al (62) on oxygen supply and
Iles et al on fructose metabolism (65).

GATED NMR   In situations where rapid and repetitive changes occur,
time resolution can be greatly improved by synchronizing NMR data collection with different phases of the cycle. For example, electrical stimulation
of frog sartorious muscles has been used to gate ^{31}P NMR measurements

of the metabolic events associated with contraction (11). In more recent experiments on frog gastrocnemius muscles, synchronization of data collection with electrical stimulation has enabled detailed kinetic analysis to be performed of the reactions taking place during contraction (66) (see below). Experiments on the fluctuations of PCr and ATP concentrations during the contraction-relaxation cycle of perfused hearts have been reported (67).

SATURATION TRANSFER   The measurement of reaction rates in the steady-state or equilibrium conditions in functional organs can be achieved using the technique of saturation-transfer NMR. In this approach the spin magnetization of one chemical species is perturbed from its thermal equilibrium value, and the rate of appearance of the nonequilibrium spin magnetization of the second species is monitored to determine the reaction rate (68). The first in vivo application was for *Escherichia coli,* where the unimolecular exchange rates between P_i and ATP, catalyzed by the coupling ATPase, were measured (69). The role of the creatine kinase–catalyzed exchange between phosphocreatine and ATP in the energetics of frog muscle and perfused rat hearts was studied by ^{31}P saturation-transfer NMR (70).

In anaerobic frog muscle at rest at 4°C, the forward- (PCr to ATP) and reverse-reaction fluxes are about 1.6×10^{-3} M sec^{-1}, which shows that, as expected, the creatine kinase is at equilibrium (71). We note that the measured fluxes do show a small discrepancy (reverse "flux" is 1.2×10^{-3} M sec^{-1}) perhaps because the adenylate kinase reaction also contributes to the observed effects. When this reaction is taken into account, the two derived fluxes for the creatine kinase reaction become closer. By gating the saturation-transfer measurements with respect to 3-sec contractions, it was found that the forward-reaction flux remained at 1.6×10^{-3} M sec^{-1}, but the reverse flux decreased to 0.85×10^{-3} M sec^{-1}. The net rate of phosphocreatine breakdown during such a contraction was 0.75×10^{-3} M sec^{-1}. From these observations the following conclusions were reached (71):

1. During contraction, the creatine kinase reaction is no longer close to equilibrium; contrary to expectation, the creatine kinase reaction is not considerably faster than the rate of ATP utilization during contraction.
2. Nevertheless, the well-known observation that during contraction the ATP concentration remains constant can be accounted for by the measured rates.
3. That the forward flux does not increase during contraction and that the backward flux does decrease can be explained on the basis of competition of reactants for a limited amount of enzyme.

4. The formation of the non-productive enzyme-creatine-ADP complex could have an important regulatory role.

The situation in perfused beating hearts is more complex than in resting muscle, partly because of the relatively high rate of ATP utilization. Even though the system is in a steady state, the flux derived by NMR for the forward reaction of creatine kinase was measured to be larger than the reverse flux (61, 70, 72), yet, of course, they must be the same. This discrepancy may be a result either of competing reactions [e.g. ATP hydrolysis and snythesis or adenylate kinase activity (61, 70)] or of compartmentation (70). In spite of arguments in favor of compartmentation (72) we still believe that the alternative hypothesis cannot be discounted (73). There are indications that the adenylate kinase reaction can also be followed in this way (71).

ISOTOPE EXCHANGE AND INCORPORATION   The exchange of the C-2 and C-3 protons of lactate and pyruvate with solvent ^2H_2O has been followed by "spin-echo" proton NMR in red cells (27). From the exchange of the C-3 protons, the equilibrium rates across lactate dehydrogenase can be measured (29). The C-2 exchange of lactate depends on several reactions; the distinct differences between the exchange rates in normal blood and blood from patients with a pyruvate kinase deficiency may have metabolic importance.

Shulman and his collaborators have carried out extensive ^{13}C NMR studies of isolated cells (3 and references therein) examining the relative rates of utilization of the α- and β-anomers of glucose and the degree of disequilibrium of the reactions catalyzed by aldolase and triose phosphate isomerase (74, 75). They have also examined perfused mouse liver by ^{13}C NMR. Using isotopically enriched [^{13}C]alanine and [^{13}C]ethanol as substrates, they measured the enrichment of specific carbons of glucose, glutamate, glutamine, aspartate, acetate, acetoacetate, β-hydroxybutyrate, and lactate (26). The authors state that "the implications of the experimental results presented here are so extensive that only limited interpretations can be presented." Their significant findings, simply stated, are that isotopic scrambling, distribution, and incorporation can be used to follow individual enzyme activities, partitioning between different pathway fluxes, and substrate selection (26). Detailed interpretation and consolidation of this work are awaited with interest.

TRANSPORT   If the intracellular and extracellular signals for a given compound can be distinguished, or the uptake of the molecule results in a change of some NMR signal from a given cell, the rate of transport can be followed. The few reports available include measurements on the influx of alanine into red cells (76) and of deoxyglucose into perfused hearts (44), and a brief mention of the uptake of Mn^{2+} into the heart (77, 78).

Relationship Between the Biochemical and the Physiological States

An attractive feature of NMR is that metabolism, monitored continuously, can be related to physiological function. Dawson, Gadian & Wilkie (22, 79, 80) have studied the biochemical basis of fatigue in anaerobic frog gastrocnemius muscles at 4°C. In muscles subjected to various patterns of electrical stimulation, force development was found to be closely correlated with metabolite levels, rather than with changes in excitatory conduction. During fatigue, the force developed remained proportional to the rate of ATP hydrolysis during contraction. In addition, the decline in the rate of mechanical relaxation correlated with the change in the free energy for ATP hydrolysis in vivo. The latter change, which accompanies fatigue, may be responsible for a reduction in the rate of Ca^{2+} uptake into the sarcoplasmic reticulum, which in turn slows down the rate of mechanical relaxation.

Several brief reports on the relation between cardiac function (e.g. left ventricular pressure) and metabolism have appeared, including studies on acidosis (81), heart rate (61), and the effect of inotropic agents (82). Renal function (e.g. glomerular filtration rate, Na^+ reabsorption, etc) following ischemic periods, has also been measured in relation to NMR observations on blood-perfused rabbit kidney (61) and in vivo rat kidney (64). The evaluation of these important observations awaits the appearance of detailed reports.

Spatially Resolved NMR

Over the last ten years, there has been increasing interest in an NMR technique known as "spin-imaging" or "zeugmatography," which involves the use of magnetic field gradients to provide information about the spatial distribution of molecules within a sample. Various ingenious methods of spin-imaging have been devised. [For contributions from many laboratories see (83).] An underlying theme of the method is to obtain two- or three-dimensional images of the proton signals from water within animals and human beings. The images are a function not only of proton density, but also of the proton relaxation times, which reflect the average mobility of the water molecules. Consequently, high intensity in an image generally reflects a region containing a large concentration of mobile water molecules. Remarkably clear discrimination can be obtained between different tissues, as illustrated by the image of a human wrist obtained by Hinshaw et al, which has a resolution of about 0.4 mm (84).

There are indications that protons in tumorous regions have longer spin-lattice relaxation times than in the corresponding healthy tissues (85 and references therein). This raises the possibility that NMR can provide a way of screening for cancer. A wide range of other disorders associated with water concentration, diffusion, and flow may also be amenable to study. Clinical trials on this method are imminent.

For the biochemist, it would be of great interest to combine a spin-imaging method with high resolution ^{31}P NMR, in order to evaluate the metabolic state of different tissues within an intact animal. There are, however, severe problems in sensitivity associated with ^{31}P spin-imaging (86).

Fortunately, there are several simpler methods for obtaining high resolution information from a defined region of a live specimen. The simplest (but clearly restricted to animal experiments, as it requires surgery) is to place the radio frequency coil around the organ to be investigated, which has been done for rat heart (21) and kidney (47, 64). In rat heart the ratio of PCr to ATP in vivo is higher than in perfused heart (21), which relfects the efficiency of O_2 delivery by blood in comparison to that by conventional buffers.

In recent ^{31}P NMR experiments, use has been made of a simple, unconventional "surface coil" (20), a circular loop of copper wire. When placed, for example, against the leg of a rat, the copper loop will detect a signal from an approximately disc-shaped region of the muscle in front of the coil. The signal can be obtained very rapidly, and by placing the coil at various positions on the leg, one can map out variations of the metabolic state throughout different regions of the leg muscle. For example, ischemic regions can readily be distinguished from healthy regions. Similar studies on brain metabolism can be done by placing a surface coil against the head of a rat (20).

Investigations of internal organs by this approach are rather difficult and inefficient, and require the use of a further method of localization. One method, topical magnetic resonance (TMR), utilizes magnetic field gradients designed so that the field is only homogeneous over a selected central volume. It is only from this volume that high resolution signals are detected. This technique has been successfully used to investigate the metabolic state of the liver within anaesthetized intact rats (87), and more recently ^{31}P NMR spectra were recorded from a selected region of human arm muscle (88). In those experiments a surface coil was used in combination with the TMR method. The time course of changes in phosphocreatine and P_i following the application of a tourniquet were observed.

In summary, high resolution NMR has already proved to be a useful tool in metabolic studies and we expect that it may well have direct clinical applications in human investigations and diagnosis. The method seems particularly suitable for studying tissue energetics in normal, pathological, and stress conditions. Insufficiencies in oxygen delivery (resulting, for example, from vascular disease), metabolic recovery following therapy, and other diseased states that result in an imbalance in energy metabolism, should all be amenable to investigation. Before such uses further detailed studies on animal models and isolated organs are necessary.

Literature Cited

1. Radda, G. K., Seeley, P. J. 1979. *Ann. Rev. Physiol.* 41:749–69
2. Burt, C. T., Cohen, S. M., Bárány, M. 1979. *Ann. Rev. Biophys. Bioeng.* 8:1–25
3. Shulman, R. G., Brown, T. R., Ugurbil, K., Ogawa, S., Cohen, S. M., den Hollander, J. R. 1979. *Science* 205:160–66
4. Gadian, D. G., Radda, G. K., Richards, R. E., Seeley, P. J. 1979. *Biological Applications of Magnetic Resonance,* pp. 463–535, ed. R. G. Shulman. New York: Academic
5. Garlick, P. B., Radda, G. K. 1979. *Tech. Life Sci. Ser. B* 216:1–24
6. Herzfeld, J., Roufosse, A., Haberkorn, R. A., Griffin, R. G., Glimcher, M. J. 1980. *Philos. Trans. R. Soc. London Ser. B.* 289:459–69
7. Burt, C. T., Glonek, T., Bárány, M. 1976. *Biochemistry* 15:4850–53
8. Seeley, P. J., Busby, S. J. W., Gadian, D. G., Radda, G. K., Richards, R. E. 1976. *Biochem. Soc. Trans.* 4:62–64
9. Burt, C. T., Glonek, T., Bárány, M. 1976. *J. Biol. Chem.* 251:2584–91
10. Chalovich, J. M., Burt, C. T., Cohen, S. M., Glonek, T., Bárány, M. 1977. *Arch. Biochem. Biophys.* 182:683–89
11. Dawson, M. J., Gadian, D. G., Wilkie, D. R. 1977. *J. Physiol.* 267:703–35
12. Sehr, P. A., Radda, G. K., Bore, P. J., Sells, R. A. 1977. *Biochem. Biophys. Res. Commun.* 77:195–202
13. Chance, B., Nakase, Y., Bond, M., Leigh, J. S., McDonald, G. 1978. *Proc. Natl. Acad. Sci. USA* 75:4925–29
14. Hollis, D. P., Nunnally, R. L., Jacobus, W. E., Taylor, G. J. IV. 1977. *Biochem. Biophys. Res. Commun.* 75:1086–91
15. Chalovich, J. M., Burt, C. T., Danon, M. J., Glonek, T., Bárány, M. 1979. *Ann. NY Acad. Sci.* 317:649–69
16. Chalovich, J. M., Bárány, M. 1980. *Arch. Biochem. Biophys.* 199:615–25
17. Salhany, J. M., Stohs, S. J., Reinke, L. A., Pieper, G. M., Hassing, J. M. 1979.

Biochem. Biophys. Res. Commun. 86:1077–83
18. Sehr, P. A., Bore, P. J., Papatheofanis, J., Radda, G. K. 1979. *Br. J. Exp. Pathol.* 60:632–41
19. Garlick, P. B. 1979. *Molecular Studies of Cardiac Metabolism* D. Phil. thesis. Oxford Univ., UK. 117 pp.
20. Ackerman, J. J. H., Grove, T. H., Wong, G. G., Gadian, D. G., Radda, G. K. 1980. *Nature* 283:167–70
21. Grove, T. H., Ackerman, J. J. H., Radda, G. K., Bore, P. J. 1980. *Proc. Natl. Acad. Sci. USA* 77:299–302
22. Dawson, M. J., Gadian, D. G., Wilkie, D. R. 1978. *Nature* 274:861–66
23. Veech, R. L., Lawson, J. W. R., Cornell, N. W., Krebs, H. A. 1979. *J. Biol. Chem.* 254:6538–47
24. Ogawa, S., Rottenberg, H., Brown, T. R., Shulman, R. G., Castillo, C. L., Glynn, P. 1979. *Proc. Natl. Acad. Sci. USA* 75:1796–1800
25. Cohen, S. M., Ogawa, S., Shulman, R. G. 1979. *Proc. Natl. Acad. Sci. USA* 76:1603–7
26. Cohen, S. M., Shulman, R. G., McLaughlin, A. C. 1979. *Proc. Natl. Acad. Sci. USA* 76:4808–12
27. Brown, F. F., Campbell, I. D., Kuchel, P. W., Rabenstein, D. C. 1976. *FEBS Lett.* 82:12–16
28. Daniels, A., Williams, R. J. P., Wright, P. E. 1976. *Nature* 261:321–23
29. Brown, F. F., Campbell, I. D. 1980. *Philos. Trans. R. Soc. London Ser. B* 289:395–406
30. Williams, R. J. P. 1980. *Philos. Trans. R. Soc. London Ser. B* 289:381–94
31. Evans, F. E. 1979. *Arch. Biochem. Biophys.* 193:63–75
32. Fossel, E. T., Solomon, A. K. 1977. *Biochim. Biophys. Acta* 464:82–92
33. Fossel, E. T., Solomon, A. K. 1979. *Biochim. Biophys. Acta* 553:142–53

34. Momsen, G., Rose, Z. B., Gupta, R. K. 1979. *Biochem. Biophys. Res. Commun.* 91:651–57
35. Hoult, D. I., Busby, S. J. W., Gadian, D. G., Radda, G. K., Richards, R. E., Seeley, P. J. 1974. *Nature* 252:285–87
36. Garlick, P. B., Radda, G. K., Seeley, P. J. 1977. *Biochem. Biophys. Res. Commun.* 74:1256–62
37. Navon, G., Ogawa, S., Shulman, R. G., Yamane, T. 1977. *Proc. Natl. Acad. Sci. USA* 74:87–91
38. Gupta, R. J., Benovic, J. L., Rose, Z. B. 1978. *J. Biol. Chem.* 253:6165–71
39. Gupta, R. J., Moore, R. D. 1980. *J. Biol. Chem.* 255:3987–93
40. Cohen, S. M., Burt, C. T. 1977. *Proc. Natl. Acad. Sci. USA* 74:4271–75
41. Moon, R. B., Richards, J. H. 1973. *J. Biol. Chem.* 248:7276–78
42. Garlick, P. B., Radda, G. K., Seeley, P. J. 1979. *Biochem. J.* 184:547–54
43. Salhany, J. M., Pieper, G. M., Wu, S., Todd, G. L., Clayton, F. C., Eliot, R. S. 1979. *J. Mol. Cell. Cardiol.* 11:601–10
44. Bailey, I. A., Williams, S. R., Radda, G. K., Gadian, D. G. 1981. *Biochem. J.* In press
45. Poole-Wilson, P. A. 1978. *J. Mol. Cell. Cardiol.* 10:511–26
46. Gillies, R. J., Deamer, D. W. 1979. *Curr. Top. Bioenerg.* 9:63–87
47. Radda, G. K., Ackerman, J. J. H., Bore, P. J., Sehr, P. A., Wong, G. G., Ross, B. D., Green, Y., Bartlett, S., Lowry, M. 1980. *Int. J. Biochem.* 12:277–81
48. Busby, S. J. W., Gadian, D. G., Radda, G. K., Richards, R. E., Seeley, P. J. 1978. *Biochem. J.* 170:103–14
49. Yoshizaki, K., Nishikawa, H., Yamada, S., Morimoto, T., Watari, H. 1979. *Japn. J. Physiol.* 29:211–25
50. Cohen, S. M., Ogawa, S., Rottenberg, H., Glynn, P., Yamane, T., Brown, T. R., Shulman, R. G., Williamson, J. R. 1978. *Nature* 273:554–56
51. Deleted in proof
52. Casey, R. P., Njus, D., Radda, G. K., Sehr, P. A. 1977. *Biochemistry* 16:972–77
53. Njus, D., Sehr, P. A., Radda, G. K., Ritchie, G. A., Seeley, P. J. 1978. *Biochemistry* 17:4337–43
54. Pollard, H. B., Shindo, H., Creutz, C. E., Pazoles, C. J., Cohen, J. S. 1979. *J. Biol. Chem.* 254:1170–77
55. Johnson, R. G., Scarpa, A., Salganicoff, L. 1978. *J. Biol. Chem.* 253:7061–68
56. Ugurbil, K., Holmsen, H., Shulman, R. G. 1979. *Proc. Natl. Acad. Sci. USA* 76:2227–31
57. Sharp, R. R., Richards, E. P. 1977. *Biochim. Biophys. Acta* 497:260–71
58. Costa, J. L., Dobson, C. M., Kirk, K. L., Poulsen, F. M., Valeri, V. R., Vecchione, J. J. 1979. *FEBS Lett.* 99:141–46
59. Stadler, H., Fuldner, H. R. 1980. *Nature* 286:293–94
60. Roberts, J. K. M., Ray, P. M., Wade-Jardetzky, N., Jardetzky, O. 1980. *Nature* 283:870–72
61. Ackerman, J. J. H., Bore, P. J., Gadian, D. G., Grove, T. H., Radda, G. K. 1980. *Philos. Trans. R. Soc. London Ser. B* 289:425–36
62. McLaughlin, A. C., Takeda, H., Chance, B. 1979. *Proc. Natl. Acad. Sci. USA* 76:5445–49
63. Norwood, W. I., Norwood, C. R., Ingwall, J. S., Castaneda, A. R., Fossel, E. T. 1979. *J. Thorac. Cardiovasc. Surg.* 78:823–30
64. Bore, P. J., Chan, L., Sehr, P. A., Thulborn, K. R., Ross, B. D., Radda, G. K. 1980. *Europ. Surg. Res.* 12:20–21
65. Iles, R. A., Griffiths, J. R., Stevens, A. N., Gadian, D. G., Porteous, R. 1981. *Biochem. J.* 192:191–202
66. Brown, T. R., Chance, E. M., Dawson, M. J., Gadian, D. G., Radda, G. K., Wilkie, D. R. 1980. *J. Physiol.* 304:110–11P
67. Fossel, E. T., Morgan, H. E. L., Ingwall, J. S. 1980. *Proc. Natl. Acad. Sci. USA* 77:3654–58
68. Forsen, S., Hoffman, R. A. 1963. *J. Chem. Phys.* 39:2892–2901
69. Brown, T. R., Ugurbil, K., Shulman, R. G. 1977. *Proc. Natl. Acad. Sci. USA* 74:5551–53
70. Brown, T. R., Gadian, D. G., Garlick, P. B., Radda, G. K., Seeley, P. J., Styles, P. 1978. *Frontiers Biol. Energ.* 2:1341–49
71. Gadian, D. G., Radda, G. K., Brown, T. R., Chance, E. M., Dawson, M. J., Wilkie, D. R. 1981. *Biochem. J.* 194:215–28
72. Nunnally, R. L., Hollis, D. P. 1979. *Biochemistry* 18:3642–46
73. Radda, G. K. 1980. *Modern Structural Methods Proc. Welch Found.* Conf. on Chem. Res., 23rd, Ch. 7. In press
74. Ugurbil, K., Brown, T. R., den Hollander, J. A., Glynn, P., Shulman, R. G. 1978. *Proc. Natl. Acad. Sci. USA* 75:3742–46
75. den Hollander, J. A., Brown, T. R., Ugurbil, K., Shulman, R. G. 1979. *Proc. Natl. Acad. Sci. USA* 76:6096–6100
76. Brindle, K. M., Brown, F. F., Campbell, I. D., Grathwohl, C., Kuchel, P. W. 1979. *Biochem. J.* 180:37–44

77. Hollis, D. P., Bulkley, B. H., Nunnally, R. L., Jacobus, W. E., Weisfeldt, M. L. 1978. *Clin. Res.* 26:240A
78. Lauterbur, P. C., Dias, M. H. M., Rudin, A. M. 1978. *Frontiers Biol. Energ.* 1:752–59
79. Dawson, M. J., Gadian, D. G., Wilkie, D. R. 1980. *J. Physiol.* 299:465–84
80. Dawson, M. J., Gadian, D. G., Wilkie, D. R. 1980. *Philos. Trans. R. Soc. London Ser. B* 289:445–55
81. Jacobus, W. E., Pores, I. H., Taylor, G. J., Nunnally, R. L., Hollis, D. P., Weisfeldt, M. L. 1978. *J. Mol. Cell. Cardiol.* 10: Suppl. 1, p. 39
82. Williams, S. R., Matthews, P. M., Schwartz, A., Radda, G. K. 1980. *Fed. Proc.* 39:2113
83. 1980. *Philos. Trans. R. Soc. London Ser. B* 289:471–551 (several articles)
84. Hinshaw, W. S., Bottomley, P. A., Holland, G. N. 1977. *Nature* 270:722–23
85. Hollis, D. P. 1979. *Bull. Magn. Reson.* 1:27–37
86. Bendel, P., Lai, C., Lauterbur, P. C. 1980. *J. Magn. Reson.* 38:343–56
87. Gordon, R. E., Hanley, P., Shaw, D., Gadian, D. G., Radda, G. K., Styles, P., Chan, L. 1980. *Nature.* 287:736–38
88. Cresshull, I. D., Gordon, R. E., Hanley, P. E., Shaw, D., Gadian, D. G., Radda, G. K., Styles, P. 1981. *Bull. Magn. Reson.* In press

Ann. Rev. Biochem. 1981. 50:85–101

MEMBRANE RECYCLING BY COATED VESICLES

♦12072

Barbara M. F. Pearse[1] and Mark S. Bretscher

Medical Research Council Laboratory of Molecular Biology, Hills Road, Cambridge, England

CONTENTS

Perspectives and Summary

Eukaryotic cells specifically endocytose macromolecules by a process called adsorptive endocytosis. These molecules are generally transferred to lysosomes, where they may be degraded. In this process the macromolecules first bind to receptors that are localized in coated pits—indented sites on the plasma membrane that have a characteristically bristly cytoplasmic surface. The coated pit buds into the cytoplasm forming a coated vesicle containing the endocytosed macromolecules. Seconds later the vesicle sheds its coat, endocytic vesicles fuse with each other and eventually the receptors are returned to the cell surface; the contents are then transferred to lysosomes. Lysosomal and plasma membranes are presum-

[1]Supported by the Science Research Council.

85

0066-4154/81/0701-0085$01.00

ably different, and therefore, how this difference is maintained during endocytosis leads one to look more carefully at coated pits. These appear to act as molecular filters, selecting certain proteins (receptors) and excluding others. Those proteins selected to enter a coated pit presumably bind directly, or indirectly, to clathrin, the major scaffolding protein on the cytoplasmic surface of coated pits and vesicles.

Coated vesicles mediate other intracellular membrane translocations and, in particular, are implicated in the transfer of newly synthesized proteins from the endoplasmic reticulum to the Golgi and then on to the plasma membrane. It is during such transfers that the specificity of the different membranous organelles of a cell is established. Current ideas suggest that coated vesicles are the apparatus by which membrane (and secreted) proteins are sorted out to travel to their different destinations.

Introduction

Eukaryotic cells are distinguished from their prokaryotic counterparts by several features, not the least of which is their extensive membrane system. The larger features of this system include the plasma membrane, the Golgi apparatus, the endoplasmic reticulum (ER), and the nuclear and mitochondrial (or chloroplast) membranes (1). These can often be subdivided, as in the case of an epithelial cell where apical and basolateral plasma membranes have different functions and constituents. All these organelles are readily recognized in thin sections in the electron microscope. There is also a host of smaller vesicles that, *in toto,* must account for a sizeable fraction of cellular membrane: some of these are lysosomes, others no doubt include compartments of unknown function, and yet others may be vesicles in transit between the established organelles. Last, there are the coated vesicles, recognized by their bristly coats.

Each of these cellular compartments has its own character, which reflects its biological function and its molecular composition. Here we describe some of the features of coated vesicles and try to indicate why these organelles may be of central importance for establishing the specificity of different cellular compartments.

A discussion of these problems presupposes an understanding of membrane structure and membrane biosynthesis (2, 3), topics beyond the scope of this article. We assume that membranes always exist in a bilayer configuration, and that transfer between organelles occurs by vesicles budding from one membrane and fusing with another such that the membrane's molecular topology is never altered. In addition, as we mention the intermembrane transfer of newly synthesized proteins, it may be helpful to distinguish two separate levels at which proteins are compartmentalized. First, Nature has devised mechanisms such that the initial residence of a newly synthesized protein is the cytoplasm, the mitochondrion, the lumen of the endoplasmic reticulum, or the endoplasmic reticulum itself. The control of these processes is partly understood: the presence of a "signal" sequence in the protein may help determine its location (4–6), or the hydrophobic nature of the folded protein itself may lead to its solution into a membrane (2, 7). Second, those proteins intially inserted into the endoplasmic reticulum membrane, or into its lumen, are delivered later to their appointed sites in the cell. This process is known as the "sorting problem."

So far as is known, all transmembrane proteins of cell organelles (aside from those of the mitochondrion) are first inserted into the endoplasmic reticulum. Those that don't remain there seem to be sequestered into vesicles in which they are transferred to the Golgi apparatus. From there, some of these proteins are transferred to lysosomes, others to the plasma membrane, while some remain in the Golgi. In each of these transfers, those proteins that move on must be segregated from those that remain behind. A similar sorting occurs for noncytoplasmic soluble proteins, which are initially secreted into the lumen of the endoplasmic reticulum. Some are then moved to the Golgi—and perhaps a few stay there, some may go on to lysosomes, and yet others to secretory vesicles, to be released later from the cell. This process of segregation of soluble proteins into different compartments is, at least superficially, just the same as for the membrane proteins. That this sorting must be done, at least in part, on the noncytoplasmic side of the membrane is most evident for these soluble proteins, but presumably may hold true also for membrane proteins. In order to travel and arrive at a particular destination the proteins need a "ticket"—a region of the protein that specifies whether, or to where, the protein should be advanced. The nature of this address ticket is obscure. That secreted proteins are sorted out leads one to believe that the ticket lies on the noncytoplasmic region of the protein; the best candidate would then be the protein's oligosaccharide chain (8). That this is the case for some lysosomal enzymes, where phosphomannose residues are implicated (9, 10), seems reasonably clear. However, there is evidence that, at least for some plasma membrane proteins, oligosaccharide chains can be dispensed with, yet the proteins are properly sorted out. Some of this evidence rests on the effects of the drug tunicamycin [see, for example, (11)], which blocks the synthesis of N-linked oligosaccharide chains. For our discussion, we assume that normal sorting can be done by the recognition of oligosaccharide chains. However, if this turns out to be incorrect, the fact that some specific recognition has to occur may not invalidate our overall picture.

The existence of some type of sorting, irrespective of the mechanism, requires that molecular filters exist—structures that concentrate certain membrane proteins into a budding vesicle, while excluding others. Thus, newly synthesized plasma membrane proteins in the endoplasmic reticulum must be able to enter a forming vesicle so that they may be transferred to the Golgi, while regular endoplasmic reticulum proteins are excluded. Indeed, every time a vesicle buds from a membrane on its way to a biochemically distinct compartment, some kind of sorting must occur.

At the heart of this process is the molecular filter. Recent evidence indicates that the forming coated vesicle is a molecular filter, and that it may be responsible for the selectivity that exists in intercompartmental exchange.

Coated Vesicles

Coated vesicles with diameters in the range 50–150 nm have been isolated from several different cell types (Figure 1a) (12, 13). After removal of cell debris from homogenates, the coated vesicles are isolated by a centrifugation procedure that depends both on their sedimentation properties and density. They are located during purification by scanning a sample of each fraction in the electron microscope after negative staining, and by looking for their major protein, clathrin, on SDS polyacrylamide gels. Those fractions containing most of the coated vesicles, which also have a high proportion of the clathrin, are collected. However, some clathrin does occur in other fractions, perhaps attached to membrane fragments (14). As it seems

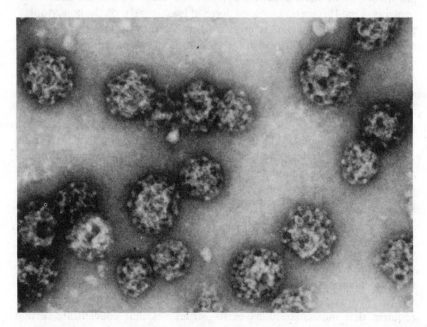

Figure 1a.   Electron micrograph of isolated coated vesicles from adrenal medulla (X 125,000). Sample was stained in 1% uranyl acetate.

likely that there are classes of coated vesicles with different functions in a given cell, those isolated may well be a subfraction of the coated vesicle population. This is a serious problem because in most cases the biological function of the isolated preparations is undetermined.

The major protein component of all coated vesicles examined is clathrin, a 180,000-dalton nonglycosylated protein (12). It resides on the outer (cytoplasmic) surface of the coated vesicles because (a) it is sensitive to added proteases and (b) it is easily dissociated from the vesicles by comparatively mild treatments—2 M urea (15) or low ionic strength (0.02 M) Tris chloride at pH 7.5 (16)—that would not be expected to disrupt a lipid bilayer. Little is known about the chemistry of clathrin, except that its sequence seems to be highly conserved between different species (13). Whether closely related clathrins of different sequence and function exist within a single cell is unknown.

Clathrin accounts for about 70% of the total protein of coated vesicles. About 5% of the protein is made up of polypeptides of about 35,000 daltons, while various other polypeptides—notably those of 100,000 and 50,000 daltons—are also seen (15, 17–19). So far the function of these smaller proteins is unknown; indeed, some may be contaminants or arise by proteolytic degradation during preparation. Only about half of these isolated coated vesicles (50–150 nm in diameter) seem to contain an enclosed vesicle (20).

Larger coated vesicles, with a diameter of 150–250 nm, have been isolated (in minute quantities) from chicken oocytes under isotonic conditions (Figure 1b) (21). These all contain an enclosed vesicle and are the same size as those observed in thin sections of chicken oocytes engaged in the uptake of yolk proteins (22). Again, these preparations contain clathrin as the major

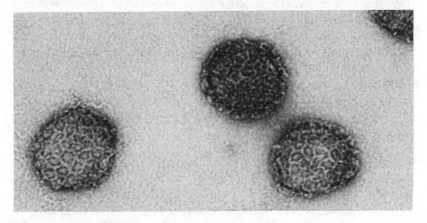

Figure 1b   Electron micrograph of isolated coated vesicles from chicken oocytes (X 125,000). Sample was stained in 1% uranyl acetate.

polypeptide, the other polypeptides previously mentioned, as well as some other components (some of which are probably yolk precursors).

The striking feature of coated vesicles is their remarkable surface lattice of hexagons and pentagons, first described by Kanaseki & Kadota (23). The smaller isolated coat structures are built from 12 pentagons (required by Euler's theorem for closed polyhedra) plus a variable number of hexagons, thus giving rise to the polymorphic range of particles observed (20). The same rule—12 pentagons plus many hexagons—presumably defines the structure of the larger coats. Chemically, it is likely that three clathrin molecules contribute to each polyhedral vertex and that two clathrin molecules contribute to each edge. This would mean that the smallest structures observed are built from 12 pentagons plus 4 or 8 hexagons, and these would contain 84 or 108 molecules of clathrin, respectively (Figure 2). A coated vesicle of about 200-nm diameter would have of the order of 1000 clathrin molecules. Thus, clathrin can form flexible lattices that act as scaffolds for the formation of vesicles during budding.

When coated vesicles are dissociated by low ionic strength (16), 2 M urea (15) or by 0.1% cholate, the solubilized clathrin exists in discrete complexes containing at least three clathrin molecules in association with the 35,000-dalton components (17). These are structural subunits of the coat presumably representing the vertices of the polygonal network. The components of coated vesicles disrupted in such conditions may be reconstituted to form coat-like structures (16, 18, 19).

Functions of Coated Vesicles

Coated vesicles have been observed by electron microscopy in the cytoplasm of many eukaryotic cells (24). It was first recognized by Roth & Porter in 1964 that they mediate the selective uptake of extracellular yolk protein in developing mosquito oocytes (25)—an example of adsorptive pinocytosis. A similar process occurs in chicken oocytes, where coated vesicles can pinocytose up to 1.5 g of yolk protein a day into a single cell (22, 26). These yolk proteins are subsequently sequestered into large nascent yolk vacuoles. Many different macromolecules are known to be taken up in this way. For example, maternally derived immunity is often conferred by the selective transport of maternal immunoglobulins across an epithelium to the young (27). In humans, this transfer occurs in the placenta, the antibody is selectively taken up into a coated vesicle from the mother's blood stream and transferred to the fetal blood system (28). On the other hand, rats obtain this immunity from their mother's milk; the baby rat selectively absorbs maternal antibodies from the milk through its intestinal epithelium, and again this uptake is mediated by coated vesicles (29, 30). And indeed, the developing chick oocyte takes up not only yolk proteins, but also maternal

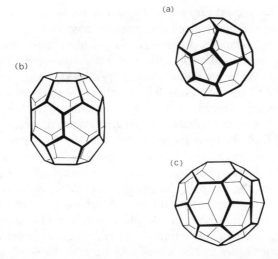

Figure 2.   Three structures identified among the smallest particles of a preparation of coated vesicles. Structure (*a*) contains 12 pentagons and four hexagons, the latter lying at the vertices of a tetrahedron; (*b*) has a barrel shape built of 12 pentagons and eight hexagons; structure (*c*) also has twelve pentagons and eight hexagons, but the latter are arranged in two arcs of four, related in the same way as the two parts of a tennis ball. Larger coats seem to be constructed on similar principles, with the addition of further hexagons.

immunoglobulins; here again, it is thought that these are taken up by coated vesicles, together with the yolk proteins.

In fibroblasts grown in culture, coated vesicles have been shown to mediate the uptake of low-density lipoprotein (LDL) (31), epidermal growth factor (EGF) (32), α_2-macroglobulin (33), and insulin (34). In rat hepatocytes, asialo orosomucoid and some other asialo proteins, which are recognized by their oligosaccharide chains, are also endocytosed by coated vesicles (35). Likewise, Semliki Forest Virus is taken up by coated vesicles into BHK cells (36). Most of these endocytosed molecules are ultimately transferred to lysosomes, where their degradation takes place or, in the case of the virus, where the infective process probably occurs.

Thus, excluding phagocytosis and the direct transport of molecules across the plasma membrane (different processes), coated vesicles are likely to provide the major system for the selective uptake of most macromolecules and peptide hormones that bind to cells.

However, the function of coated vesicles is not limited to budding from the plasma membrane in the process of adsorptive endocytosis. Evidence from electron microscopy suggests that coated vesicles are involved in secretory pathways and the delivery of newly synthesized molecules to their operational sites inside cells. Examples include the secretory cells of the pancreas and parotid salivary gland (37, 38) and the actively exporting cells of the lactating mammary gland (39), where coated vesicles appear to bud from the Golgi apparatus. In the rat vas deferens, large coated vesicles appear to be associated with uptake of extracellular molecules, while smaller coated vesicles are seen in the vicinity of the Golgi apparatus in the same cells (40). This is also true for fibroblasts (M. S. Bretscher, unpublished observation). In cells infected with vesicular stomatitis virus, biochemical analysis indicates that the newly synthesized viral coat (G) protein is transported from the endoplasmic reticulum to the plasma membrane (6). This occurs in two stages: first, from the endoplasmic reticulum to the Golgi apparatus, and second, after modification of the carbohydrate region of the G protein, from the Golgi to the plasma membrane. Rothman and colleagues (41, 42) have demonstrated that both forms of the G protein become transiently associated with clathrin, which indicates that coated vesicles mediate a step in each of these transfers. Their kinetic evidence shows that this association occurs in two waves: rapidly after its synthesis, for the endoplasmic reticulum to Golgi transfer, and more slowly (at a later time) for exit from the Golgi. An in vitro system for studying the first transfer has recently been developed (43); this holds great promise for opening up a direct biochemical route for unraveling the process of membrane transfer.

Obviously there is a considerable range in the known and suspected functions of coated vesicles. What seems clear is that coated areas of membrane are specialized for pinching off vesicles into the cytoplasm. Such vesicles may be derived not only from the plasma membrane, for which the evidence is conclusive, but also from the endoplasmic reticulum and the Golgi apparatus.

The Budding Process

The first critical step in the uptake of extracellular molecules is the binding of particular ligands to high affinity, cell surface receptors. These receptors collect into coated pits, and are therefore presumed to interact with components of the coated pit. These associations have been highlighted by Anderson, Brown & Goldstein in their study of LDL uptake in fibroblasts from patients with Familial Hypercholesterolemia (31, 44). Cells from these patients lack functional cell surface LDL receptors; these either do not bind LDL, or do bind LDL but fail to localize in coated pits (45). In both cases the cells cannot take up LDL by the usual process, which leads to aberrant cholesterol metabolism and results in an early condition of atherosclerosis.

It is believed that a given coated pit will contain receptors for several different ligands because most coated pits on fibroblasts contain LDL receptors (44), yet these cells take up a wide variety of other molecules. Whether a ligand binds first to its receptor, or whether the receptor first binds to the pit, or whether it makes any difference, is unknown. At least in some cases, budding seems to be a constitutive process as it occurs in the absence of particular external ligands that bind to coated pits (44).

Thus, coated pits are able to concentrate specific receptors from the surrounding plasma membrane. Equally important, they appear to exclude others. We have recently shown that θ antigen (Thy-1) and an 80,000-mol wt antigen (H63) on the surface of 3T3 fibroblasts are expressed in coated regions at a level of about 1% (or less) of that expected if they were randomly distributed throughout the whole membrane (46). The mechanism by which these proteins are excluded from the coated pit is unclear, but it may be that there is simply not enough room for them and they are squeezed out. It may be relevant here that, unlike the rest of the plasma membrane, coated pits often have a lot of fuzz on their noncytoplasmic surfaces when viewed in thin sections.

These observations of concentration and exclusion lead to the concept that the forming coated vesicle acts as a molecular filter, separating protein molecules required elsewhere in the cell from those necessary for the function and integrity of the parent membrane. This filtration may not be restricted to protein components, since cholesterol may be partly excluded from coated pits (47, 48).

The clathrin coat is thought to serve two principal functions: to select or exclude molecules from the coated pit, and to provide a structural scaffold for the invagination of a vesicle. Stages in this latter process have been vividly captured by Heuser (49) when he applied his deep etch technique to the cytoplasmic surface of the plasma membrane of fibroblasts (Figure 3). The coated regions so revealed show a smooth gradation of size and thus appear to form by the addition of coat subunits to the edge of the forming coated pit, rather than by the fusion of small coated vesicles to an edge. This is consistent with the existence of a pool of soluble clathrin in the cytoplasm.

These pictures of budding coated pits hint how the clathrin lattice may be used to form a vesicle. The flat membrane is presumably first coated with a planar array of hexagons. In order to curve the structure, pentagons must be introduced (23), and this can be done in at least two ways. Either (a) pentagons form at an edge and diffuse by a series of dislocations into the hexagonal array, which gives rise to a curved pit, or (b) dislocations could arise inside the hexagonal array, each giving rise to a pentagon and heptagon from two hexagons. In this latter case the heptagons would have to diffuse, by a series of dislocations, out of the hexagonal lattice to the edge, leaving behind pentagons and thereby inducing curvature. The pictures obtained by deep etching do indeed show the presence of neighboring pentagons and heptagons (49), which suggests that the second scheme is that which occurs in the cell. Whatever the mechanism, it is likely that a

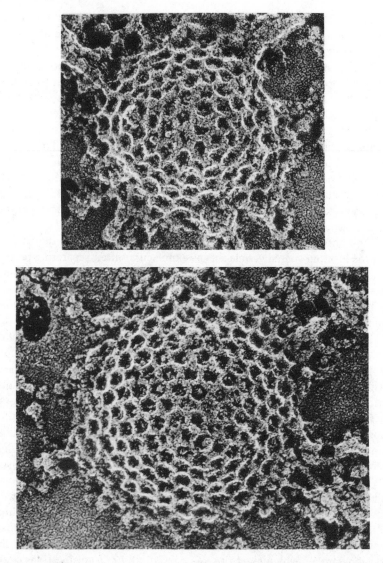

Figure 3.   Regions of coated membrane from fibroblasts at intermediate stages of the budding process, demonstrated by deep etching and rotary replication (kindly provided by J. E. Heuser).

source of energy is needed to promote the process: it stops when the cells are poisoned by metabolic inhibitors or are chilled to 0°C.

How the coated pit finally buds off to form a coated vesicle is not known. But there is a hint that the coated pit may be pulled into the cytoplasm (perhaps by components of the cytoskeleton) as it pinches off: coated pits quite deep in the cytoplasm (0.3 μm or more from the plasma membrane) are often seen in serial thin sections of fibroblasts and these are joined to the plasma membrane by a thin neck of uncoated membrane. Frequently, what appear to be coated vesicles in one section turn out to be connected to the cell surface in an adjacent section, and hence are actually deep coated pits (46). Furthermore, it appears that when cells enter mitosis, coated pits form on the plasma membrane but remain there without budding into the cell (50). As the cell's cytoskeleton is reorganized during this part of the cell cycle, it is possible that this reorganization could be responsible for the temporary suppression of invagination.

After Budding

Everyone seems agreed that once a plasma membrane–derived coated vesicle has formed, it sheds its coat very quickly. At about the same time, two other events occur. The contents of the vesicle appear, in thin sections, to become released from their receptors (i.e. they are no longer closely associated with the vesicle membrane) (25, 26, 51). In addition, these vesicles seem to fuse with each other or with a nearby vacuole. How these processes are controlled, or how the coat knows that the vesicle has pinched off, is a matter for conjecture. But it is worth noting that if there were, for example, a proton pump in the membrane of the coated vesicle, the pH inside the vesicle would be altered very quickly (because of the small volume of the vesicle) as soon as the connection to the outside world was severed.

As far as the receptors are concerned, once the vesicle has been uncoated the mystery begins. For example, once internalized, the LDL receptors on fibroblasts seem to be returned to the cell surface in no time at all. That is, even though nearly all the receptors are seen in coated pits at 0°C, a few minutes after a shift to 37°C most of the bound LDL is internalized, yet there is a full complement of receptors on the cell's surface (44). This might be explained if there were a large pool of free receptors inside the cell, but this is not the case (52). Nor are receptors being synthesized at a large rate —cycloheximide has no detectable effect on LDL receptor recycling (44). The clear conclusion to be drawn is that, once a coated vesicle is endocytosed, it is processed very quickly and the LDL receptors are returned to the surface. However, the contents of the vesicles are retained in the cell, and eventually are transferred to lysosomes. How receptors and contents are segregated from each other is unknown.

Those macromolecules that reach the lysosomes are usually degraded. However, this can be inhibited by certain drugs (such as chloroquine and amines) that are believed to raise the intralysosomal pH. It has been reported that amines inhibit the actual uptake of macromolecules (53), and that this inhibition may be due to the effect of the amines on transglutaminase, an enzyme hypothesized to cross-link receptors covalently into coated pits (54). It seems more probable that the amines, in elevating the lysosomal and perhaps other compartmental pHs, may have more general effects (55). This could include a slowdown in receptor recycling (56).

Tracing the fate of membrane internalized by coated vesicles has only been achieved in the presynaptic terminal of the frog neuromuscular junction. Artificial stimulation of the nerve leads to a massive fusion of synaptic vesicles with the presynaptic membrane, thereby releasing transmitter. Resorption of presynaptic membrane is achieved by coated vesicles, which deposit this membrane in reticula from which new synaptic vesicles arise, thereby completing the cycle (57).

Magnitude of Endocytosis

How fast is the uptake of extracellular macromolecules and does this fit with the turn-around time of specific receptors? Quantitative studies are hard to do and are in a rather primitive state. The best-studied case is the uptake of LDL by fibroblasts: cells are labeled in the cold with LDL conjugated to ferritin (LDL-Ft), brought to 37°C, and then the rate of internalization is examined by thin section electron microscopy (51). This approach has its drawbacks—it is not a steady-state study and what appears to be an internal coated vesicle in one section is often seen to be a deep coated pit in an adjacent section (46). Nevertheless, the results indicate that the LDL-Ft is internalized within a few minutes. As about 2% of the cell surface is made up of coated pits (44), a rate of uptake of the entire cell's surface must occur each 100 min or so. A different estimate of the rate of surface uptake may be obtained if one assumes that in the LDL-receptorless mutant cells, LDL uptake proceeds by fluid phase pinocytosis mediated by coated vesicles (there are no other obvious vesicles forming) (58); the figures give a lifetime for coated pits of rather less than a minute, which corresponds to an uptake of the cell's surface each 15 min or so.

A similar approach has been taken using Semliki Forest Virus which is endocytosed via coated vesicles bound to an unidentified receptor on BHK cells. Coated pits are labeled with virus at 0°C; 15 sec after warming to 37°C, viruses are seen inside cellular coated vesicles. Fifteen seconds later, these begin to shed their coats and the remaining vesicles fuse with each other (36). Studies of the rate of uptake of viruses by these cells (together with an estimate of the number of virus particles in each endocytic vesicle) indicate a lifetime of less than a minute for a coated pit, which corresponds to internalization of the cell's surface area each 30 min or so (59). Furthermore, the rate of coated vesicle internalization quantitatively accounts for the rate of fluid phase pinocytosis, which indicates that coated vesicles are the major route of endocytosis in these cells (59). This contrasts with fluid phase pinocytosis in macrophages, where the cell surface is taken up each 30 min (60); yet this endocytosis appears not to involve coated vesicles, and the internalized membrane is a sample of the plasma membrane (61), not selected components of it.

The conclusion of the studies on LDL and Semliki Forest Virus uptake is that coated pits have a short lifetime—probably about a minute or so— and that receptors are returned to the cell surface rather quickly.

It would be valuable to know whether the in vivo lifetime of a coated pit on an actively endocytosing cell is the same as for these tissue culture cells. Perhaps the best place to look at this is the developing chicken oocyte where a 2-cm diameter oocyte takes up about 1.5 g of very low density lipoprotein (VLDL) per day. To accommodate this huge uptake, the oocyte surface is increased enormously by deep infoldings that are studded with coated pits (22, 26). If the endocytosing coated area is taken as ten times the surface of a spherical oocyte, one can deduce a lifetime for a coated pit of about a minute. However, present knowledge of the extent of infolding and coating of the oocyte surface makes this an approximate guess.

Possible Implications

THE SORTING PROBLEM   At the outset, we posed the question of how different membrane proteins get to their homes, starting out from the endoplasmic reticulum after their synthesis. Since this type of sorting has to be done for soluble (lumenal or secreted) proteins as well as for membrane proteins, it seems likely that the same kinds of process are used for both sets of proteins. This sorting could most easily be achieved if each protein acquired a label to specify its next destination in the cell. One type of label might be an oligosaccharide chain. It is clear that oligosaccharides can be of prime importance in the steering of molecules to specific organelles. This has been established for the phosphomannose recognition system (molecules bearing this group are transferred to lysosomes) (9, 10); in addition, several other systems exist in which glycoproteins are recognized by their sugar chains, leading to their endocytosis (62). However, as mentioned earlier, in some cases N-linked oligosaccharides can be dispensed with, and yet sorting occurs. Thus oligosaccharide chains can act as labels, but they may not be the only type of label used by the cell in sorting processes. However, for what follows, we consider them as the address tickets.

There are several further points to bear in mind:

1. Coated vesicles are reasonable candidates for the selective transfer of pieces of membrane. The crucial observation by Rothman & Fine that

newly synthesized vesicular stomatitis virus (VSV) spike proteins become associated with coated vesicles during their transfer from the endoplasmic reticulum to the Golgi (and later also to the plasma membrane) leads to the conclusion that coated vesicles are involved in most membrane sorting processes (41), although this is far from established. But coated pits do seem to have the necessary requirements to act as sorters—they behave as molecular filters, selecting certain molecules and excluding others.

2. The organelle to which the coated vesicle is targeted in the cell (after it has shed its coat) must be determined by some signal on the vesicle's cytoplasmic surface. Yet the address ticket (if it is carbohydrate) that determines where the proteins should go is on the noncytoplasmic side. This problem is perhaps more clearly seen for secreted proteins, which are also sorted out (and which clearly do not span the bilayer).

3. A receptor, such as the LDL receptor, can shuttle between two compartments—the plasma membrane and some internal organelle. But originally it was inserted into the endoplasmic reticulum, and presumably was transported to the plasma membrane via the Golgi. This all suggests that if tickets that specify an address exist, they may be so designed that they are capable of easy modification. Chemically, oligosaccharides have the complexity needed for use as labels; in addition they can easily be modified.

If all plasma membrane proteins pass through a coated pit on their way from endoplasmic reticulum to Golgi (leaving behind endoplasmic reticulum–specific proteins), they must be actively recognized by the coated pit. All these proteins may well have cytoplasmic domains, but it seems improbable that they could all be recognized directly by clathrin, and thereby be drawn into the coated pit. Further, soluble proteins in the lumen could obviously not interact with clathrin. This leads one to believe that there is a family of adaptors that mediate between the address tickets and clathrin. We would then expect these adaptors (a) to interact with clathrin on the cytoplasmic side of the membrane, (b) to recognize specifically an address ticket (which, if it were carbohydrate, would be on the noncytoplasmic side of the membrane), and (c) perhaps to have some signal on their cytoplasmic sides indicating to which organelle they should go.

In this view, the function of the adaptors is to sort out those molecules that are to travel on from those that are to remain behind. It suggests that a protein selected to enter a coated pit does not interact directly with clathrin, but does so via an adaptor. Thus the cell surface LDL receptor would be a glycoprotein, and a portion of its extracellular oligosaccharide chain might be bound by an adaptor that is itself bound to clathrin on the cytoplasmic side of the coated pit. This adaptor might also recognize other plasma membrane receptors (such as the EGF or α_2-macroglobulin receptors). The coated vesicle then formed would thus contain clathrin, adaptors, receptors, and ligands (such as LDL).

The possible existence of adaptors raises other questions and problems. For example, do adaptors operate on adaptors? Are they allosteric, so that they can transfer information from one side of the bilayer to the other? Are they oligomeric? Speculating further may simply lead us into deeper pits; the only real hope would come from finding and studying them. Perhaps the oligosaccharide receptors (9, 10, 62) detected on so many different cell types are adaptors.

LIPID RECYCLING   One of the few membranes for which we have an idea of the rate of coated vesicle formation is the fibroblast plasma membrane. There, the whole cell surface is endocytosed every 10–100 min or so. This uptake occurs almost uniformly over the cell's surface, as shown by the beautiful pictures obtained by immunofluorescence using anticlathrin antibodies (63). But to where on the cell surface is it returned? The only hint we have comes from one observation that newly synthesized proteins (coming from the Golgi, presumably) are added at the cell's leading edge (64). If this is the site for return to the cell surface of all the lipid interiorized by coated vesicles, the bulk lipid phase in the cell surface would be flowing backwards from the leading edge. This bulk, rearward flow of lipid would tend to sweep proteins in the plasma membrane toward the cell's tail, but Brownian motion would tend to randomize them against this flow (47). Whether flow or diffusion is the dominant feature depends critically on the length of the cell, the rate of this lipid flow, and the diffusion coefficient of the protein. What seems clear is that a macroscopic object on the cell's membrane [such as an antibody-induced patch (65) or a carbon particle (66)] whose diffusion coefficient is rather small, would be unable to randomize against the flow and would therefore be cleared away from the front end of the cell. This is what is observed in the process of capping cross-linked surface antigens on fibroblasts (67, 68). ˙

The vesicles that bud from the endoplasmic reticulum transfer membrane to the Golgi and elsewhere in the cell. As this occurs, there is a net transfer of lipid away from the endoplasmic reticulum to the Golgi, plasma membrane, and other organelles. This transfer may be made up in part by lipid synthesis, which occurs in the endoplasmic reticulum. However, it seems unlikely that the rate of lipid synthesis is sufficiently high to drive the process on its own, and hence a mechanism for the return of lipid molecules to the endoplasmic reticulum from the Golgi and/or plasma membrane is indicated. Certainly, for other organelles where lipid synthesis does not occur, such a return cycle must operate. In this context it is worth noting that membranes such as the endoplasmic reticulum are cholesterol-poor (cholesterol : phospholipid molar ratio of about 0.1), whereas others, such as the plasma membrane, are cholesterol-rich (cholesterol : phospholipid is about 1). Other lipid differences also exist. In thinking about which routes are permissible to coated vesicles, it would help if we knew whether coated pits preferentially accommodate certain lipids above others. It was earlier suggested that they may tend to exclude cholesterol (47), and indeed there is an indication that this may be so (12, 48), but the evidence is not as firm as one would wish. If we knew, for example, that there is *no* selection for particular lipids in a coated pit, we could be reasonably sure that the lipid endocytosed by coated vesicles from the plasma membrane is not returned there via the endoplasmic reticulum.

It may be useful to think of such cycles of lipid transfer, mediated by the budding and fusion of vesicles, as providing selective conveyor belts that link particular membranous organelles. Certain proteins with special labels would join the flow and be carried to their destinations, while others would be excluded. The forming coated vesicle may be the molecular filter that permits entry at each stage to these cyclical conveyor systems.

Literature Cited

1. Palade, G. E. 1975. *Science* 189:347–58
2. Bretscher, M. S. 1973. *Science* 181: 622–29
3. Rothman, J. E., Lenard, J. 1977. *Science* 195:743–53
4. Milstein, C., Brownlee, G. G., Harrison, T. M., Mathews, M. B. 1972. *Nature New Biol.* 239:117–20
5. Blobel, G., Dobberstein, B. 1975. *J. Cell Biol.* 67:835–62
6. Rothman, J. E., Lodish, H. F. 1977. *Nature* 269:775–80
7. Wickner, W. 1979. *Ann. Rev. Biochem.* 48:23–45
8. Rodriguez Boulan, E., Sabatini, D. D. 1978. *Proc. Natl. Acad. Sci. USA* 75:5071–75
9. Neufeld, E. F., Lim, T. W., Shapiro, L. J. 1975. *Ann. Rev. Biochem.* 44:357–76
10. Kaplan, A., Achord, D. T., Sly, W. S. 1977. *Proc. Natl. Acad. Sci. USA* 74: 2026–30
11. Roth, M. G., Fitzpatrick, J. P., Compans, R. W. 1979. *Proc. Natl. Acad. Sci. USA* 76:6430–34
12. Pearse, B. M. F. 1975. *J. Mol. Biol.* 97:93–98
13. Pearse, B. M. F. 1976. *Proc. Natl. Acad. Sci. USA* 73:1255–59
14. Mello, R. J., Brown, M. S., Goldstein, J. L., Anderson, R. G. W. 1980. *Cell* 20:829–37
15. Blitz, A. L., Fine, R. E., Toselli, P. A. 1977. *J. Cell Biol.* 75:135–47
16. Schook, W., Puszkin, S., Bloom, W., Ores, C., Kochwa, S. 1979. *Proc. Natl. Acad. Sci. USA* 76:116–20
17. Pearse, B. M. F. 1978. *J. Mol. Biol.* 126:803–12
18. Woodward, M. P., Roth, T. F. 1978. *Proc. Natl. Acad. Sci. USA* 75:4394–98
19. Keen, J. H., Willingham, M. C., Pastan, I. H. 1979. *Cell* 16:303–12
20. Crowther, R. A., Finch, J. T., Pearse, B. M. F. 1976. *J. Mol. Biol.* 103:785–98
21. Pearse, B. M. F. 1980. *Trends Biochem. Sci* 5:131–34
22. Roth, T. F., Cutting, J. A., Atlas, S. B. 1976. *J. Supramol. Struct.* 4:527–48
23. Kanaseki, T., Kadota, K. 1969. *J. Cell Biol.* 42:202–20
24. Ockleford, C. D., Whyte, A. eds., 1980. *Coated Vesicles.* London & New York: Cambridge Univ. Press
25. Roth, T. F., Porter, K. R. 1964. *J. Cell Biol.* 20:313–32
26. Perry, M. M., Gilbert, A. B. 1979. *J. Cell Sci.* 39:257–72
27. Brambell, F. W. 1970. *Front. Biol.* Vol. 18

28. Ockleford, C. D. 1976. *J. Cell Sci.* 21:83–91
29. Rodewald, R. 1973. *J. Cell Biol.* 58:189–211
30. Rodewald, R. 1980. *J. Cell Biol.* 85:18–32
31. Anderson, R. G. W., Goldstein, J. L., Brown, M. S. 1976. *Proc. Natl. Acad. Sci. USA* 73:2434–38
32. Gorden, P., Carpentier, J., Cohen, S., Orci, L. 1978. *Proc. Natl. Acad. Sci. USA* 75:5025–29
33. Willingham, M. C., Maxfield, F. R., Pastan, I. H. 1979. *J. Cell Biol.* 82:614–25
34. Maxfield, F. R., Schlessinger, J., Shechter, Y., Pastan, I., Willingham, M. C. 1978. *Cell* 14:805–10
35. Wall, D. A., Wilson, G., Hubbard, A. L. 1980. *Cell* 21:79–93
36. Helenius, A., Kartenbeck, J., Simons, K., Fries, E. 1980. *J. Cell Biol.* 84:404–20
37. Jamieson, J. D., Palade, G. E. 1971. *J. Cell Biol.* 50:135–58
38. Castle, J. D., Jamieson, J. D., Palade, G. E. 1972. *J. Cell Biol.* 53:290–311
39. Franke, W. W., Luder, M. R., Kartenbeck, J., Zerban, H., Keenan, T. W. 1976. *J. Cell Biol.* 69:173–95
40. Friend, D. S., Farquhar, M. G. 1967. *J. Cell Biol.* 35:357–76
41. Rothman, J. E., Fine, R. E. 1980. *Proc. Natl. Acad. Sci. USA* 77:780–84
42. Rothman, J. E., Bursztyn-Pettegrew, H., Fine, R. E. 1980. *J. Cell Biol.* 86:162–71
43. Fries, E., Rothman, J. E. 1980. *Proc. Natl. Acad. Sci. USA* 77:3870–74
44. Goldstein, J. L., Anderson, R. G. W., Brown, M. S. 1979. *Nature* 279:679–85
45. Anderson, R. G. W., Goldstein, J. L., Brown, M. S. 1977. *Nature* 270:695–99
46. Bretscher, M. S., Thomson, J. N., Pearse, B. M. F. 1980. *Proc. Natl. Acad. Sci. USA* 77:4156–59
47. Bretscher, M. S. 1976. *Nature* 260:21–23
48. Montesano, R., Perrelet, A., Vassalli, P., Orci, L. 1979. *Proc. Natl. Acad. Sci. USA* 76:6391–95
49. Heuser, J. 1980. *J. Cell Biol.* 84:560–83
50. Fawcett, D. W. 1965. *J. Histochem. Cytochem.* 13:75–91
51. Anderson, R. G. W., Brown, M. S., Goldstein, J. L. 1977. *Cell* 10:351–64
52. Basu, S. K., Goldstein, J. L., Brown, M. S. 1978. *J. Biol. Chem.* 253:3852–56
53. Maxfield, F. R., Willingham, M. C., Davies, P. J. A., Pastan, I. 1979. *Nature* 277:661–63
54. Davies, P. J. A., Davies, D. R., Levitzki, A., Maxfield, F. R., Milhaud, P., Willingham, M. C., Pastan, I. H. 1980. *Nature* 283:162–67
55. King, A. C., Hernaez-Davis, L., Cuatrecasas, P. 1980. *Proc. Natl. Acad. Sci. USA* 77:3283–87
56. Gonzalez-Noriega, A., Grubb, J. H., Talkad, V., Sly, W. S. 1980. *J. Cell Biol.* 85:839–52
57. Heuser, J. E., Reese, T. S. 1973. *J. Cell Biol.* 57:315–44
58. Goldstein, J. L., Brown, M. S. 1974. *J. Biol. Chem.* 249:5153–62
59. Marsh, M., Helenius, A. 1980. *J. Mol. Biol.* 142:439–54
60. Steinman, R. M., Brodie, S. E., Cohn, Z. A. 1976. *J. Cell Biol.* 68:665–87
61. Mellman, I. S., Steinman, R. M., Unkeless, J. C., Cohn, Z. A. 1980. *J. Cell Biol.* 86:712–22
62. Ashwell, G., Morell, A. G. 1977. *Trends Biochem. Sci.* 2:76–78
63. Anderson, R. G. W., Vasile, E., Mello, R. J., Brown, M. S., Goldstein, J. L. 1978. *Cell* 15:919–33
64. Marcus, P. I. 1962. *Cold Spring Harbor Symp. Quant. Biol.* 27:351–65
65. Taylor, R. B., Duffus, W. P. H., Raff, M. C., de Petris, S. 1971. *Nature New Biol.* 233:225–29
66. Abercrombie, M., Heaysman, J. E. M., Pegrum, S. M. 1970. *Exp. Cell Res.* 62:389–98
67. Edidin, M., Weiss, A. 1972. *Proc. Natl. Acad. Sci. USA* 69:2456–59
68. Vasiliev, J. M., Gelfand, I. M., Domnina, L. V., Dorfman, N. A., Pletyushkina, O. Y. 1976. *Proc. Natl. Acad. Sci. USA* 73:4085–89

Ann. Rev. Biochem. 1981. 50:103–31
Copyright © 1981 by Annual Reviews Inc. All rights reserved

THE EXPRESSION OF ISOTOPE EFFECTS ON ENZYME-CATALYZED REACTIONS

♦12073

Dexter B. Northrop

School of Pharmacy, University of Wisconsin, Madison, Wisconsin 53706

CONTENTS

0066-4154/81/0701-0103$01.00

PERSPECTIVES AND SUMMARY

The title of this review emphasizes a profound distinction between enzymatic reactions and the nonenzymatic reactions familiar to organic chemistry. The latter tend to provide full expression of the isotope effect on measured rates of reaction, with the exceptions displaying less because of the presence of a second reactive step that is slower than the isotopically sensitive step. In either case, a single value is obtained and this value has been interpreted traditionally within transition-state theory (the first case) or as a means of identifying the rate-limiting step (the second case). In contrast, enzymatic reactions rarely provide full expression of the isotope effect, and often display a very broad range of values that are highly dependent upon reaction conditions and on the particular experimental design employed. Furthermore, a less than full expression may be due to reasons other than the presence of a slower step among a complex series of steps, and thus not necessarily correlated with rate-limiting steps.

The failure to come to terms with this distinction and instead apply traditional interpretations to enzymatic isotope effects has led to considerable conceptual error, poorly designed experiments, and faulty interpretations of results. This review illustrates these caveats and summarizes the recent advances that have brought about a "renaissance" in this field. The word is appropriate because, when the kinetic functions that govern the expression of isotope effects on enzyme-catalyzed reactions are identified and stripped away, then one can go back and readdress the enzymatic isotope effect in traditional terms. Significant items to this end are:

(a) Recognition that isotopic discrimination is a V/K isotope effect and often fails to encompass a complete enzymatic turnover;

(b) Discovery of a method for determining the intrinsic isotope effect, which is the only isotope effect on enzyme-catalyzed reactions that can be interpreted in terms of transition-state theory;

(c) Discovery of the equilibrium perturbation technique for detecting and quantitating small isotope effects, using instrumentation familiar to the enzymologist;

(d) Recognition of the importance of equilibrium isotope effects in the analyses of enzymatic data, and the parallel development of methodology for calculating and measuring equilibrium effects;

(e) Invention of a systematic nomenclature for understanding and distinguishing among the various expressions of isotope effects within the complexities of enzymatic mechanisms;

(f) Improvements in the experimental designs and methods of measurement that raise isotopic kinetic data from qualitative to quantitative significance;

(*g*) Definition of the kinetic functions that govern the expression of isotope effects, within general equations encompassing most enzymatic mechanisms.

This review excludes discussion of the origin, magnitude, or physical significance of isotope effects, and thus excludes some of the recent and most of the older literature. As a guide to the latter, consult the bibliography complete with identifying commentary compiled by O'Leary in the proceedings of a symposium on both theoretical and experimental topics of isotope effects relevant to enzymology (1), plus the thorough review and references therein by Klinman (2).

INTRODUCTION AND NOMENCLATURE

To dramatize the multiplicity of isotope effects on enzyme-catalyzed reactions and to demonstrate the need for coming to terms with different levels of expression, a family of deuterium isotope effects on an ordered reaction mechanism with two substrates and two products is illustrated in Scheme 1. The nomenclature in the scheme and below is that introduced by Northrop (3) in which an effect and the isotope responsible are indicated by a leading superscript. The conventional expression for an isotope effect, k_H/k_D, is here written ^Dk; similarly, the carbon-13 isotope effect on V_{max}/K_m is written $^{13}V/K$, and so forth.

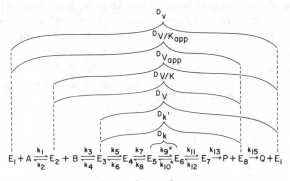

Scheme 1

What one measures directly in a comparison of isotopic and nonisotopic rates is an "observed" isotope effect, ^Dv, which depicts a single data point for one set of reaction conditions, and as such is dependent upon the entire steady-state distribution of enzyme forms. $^DV/K_{app}$ and $^DV_{app}$ are "apparent" isotope effects estimated by extrapolation of observed effects to very low and very high substrate concentrations, respectively; they are termed apparent because the co-substrate (or activator, in this case substrate B) is fixed at an arbitrary level. $^DV/K$ and ^DV are "limiting" isotope effects at low and high co-substrate concentrations, respectively, and represent the true kinetic constants describing the steady-state expression of the isotope effect. The limiting isotope effects share a portion of the reaction sequence in common. The "internal" isotope effect, $^Dk'$, is the effect on this common portion, representing all steps of the catalytic sequence following substrate binding and leading up to and including the first irreversible step (in this case, the release of the first product). Finally, ^Dk is the "intrinsic" isotope effect representing the full effect imposed on the isotopically sensitive step and is analogous to isotope effects in nonenzymatic reactions. Intrinsic isotope effects may be different in the forward and reverse reactions, thus giving rise to an isotope effect on the equilibrium constant, $^DK_{eq} = {}^Dk_f/{}^Dk_r$, which is often easier to calculate than to measure (4, 5).

The expression of isotope effects on V/K is governed by kinetic functions contained in equations of the following form:

$$^DV/K = \frac{^Dk + C_f + C_r \cdot {}^DK_{eq}}{1 + C_f + C_r}$$ 1.

C_f is the "forward commitment to catalysis" and represents the tendency of the enzyme complex poised for catalysis to continue forward as opposed to its tendency to partition back to free enzyme and unbound substrate. Similarly, C_r is the "reverse commitment to catalysis" and represents the tendency for the first enzyme complex following the isotopically sensitive step to undergo reverse catalysis as opposed to partitioning forward through the first irreversible step leading to unbound product.

The expression of isotope effects on the maximal velocity is governed by equations of similar form (6) as follows:

$$^DV = \frac{^Dk + R_f/E_f + C_r \cdot {}^DK_{eq}}{1 + R_f/E_f + C_r}$$ 2.

R_f is the forward "ratio of catalysis" and consists of the sum of the ratios of the rate constant for the isotopically sensitive step to the net rate constants for each of the other forward steps. E_f is the forward "equilibration preceding catalysis" and consists of a geometric series of the reciprocal of the equilibrium constants for each of the steps preceding isotopic transfer but following substrate binding. Mastery of these equations and the origins of their kinetic functions enables one to write appropriate equations for the expression of isotope effects by simple inspection of reaction mechanisms, and to think about changes in the expression of isotope effects without getting lost in a lot of algebra.

THEORY

Effect of Substrate Concentrations

Changes in the expression of isotope effects with changing concentrations of substrates are mediated by commitment factors in a predictable manner. For the ordered mechanism in Scheme 1, the apparent isotope effect at low concentrations of A and fixed levels of B is governed by the following commitments to catalysis (3):

$$C_{fa} = \frac{k_9}{k_8} \left\{ 1 + \frac{k_7}{k_6} \left[1 + \frac{k_5}{k_4} \left(1 + \frac{k_3 [B]}{k_2} \right) \right] \right\} \qquad 3.$$

$$C_r = \frac{k_{10}}{k_{11}} \left(1 + \frac{k_{12}}{k_{13}} \right) \qquad 4.$$

Substitution of C_{fa} for C_f in Equation 1 reveals the apparent isotope effect on V/K_a to be dependent upon the concentration of B. At very high B, the commitment of A becomes large and the isotope effect is abolished. At very low B, the expression approaches the limiting isotope effect on V/K where the forward commitment factor reduces to

$$C_{fb} = \frac{k_9}{k_8} \left[1 + \frac{k_7}{k_6} \left(1 + \frac{k_5}{k_4} \right) \right] \qquad 5.$$

For random mechanisms, expressions for the commitment factors are more complex due to the branch points: the immediate rate constant forward through the common sequence is divided by the net rate constant away through the separate branches. For example, in the random segment of Scheme 2:

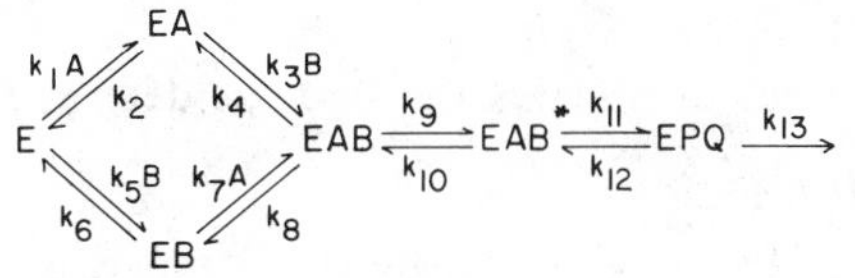

Scheme 2

the forward commitment to catalysis at low A is:

$$C_{fa} = \frac{k_{11}}{k_{10}} \left(1 + \frac{k_9}{k_8 + \dfrac{k_2 k_4}{k_2 + k_3 [B]}}\right) \qquad 6.$$

Similarly, the forward commitment to catalysis at low B is:

$$C_{fb} = \frac{k_{11}}{k_{10}} \left(1 + \frac{k_9}{k_4 + \dfrac{k_6 k_8}{k_6 + k_7 [A]}}\right) \qquad 7.$$

Equations 6 and 7 reveal that the isotope effects at low concentrations of one substrate should be determined at low concentrations of the other, in which case both equations reduce to:

$$C_{fab} = \frac{k_{11}}{k_{10}} \left(1 + \frac{k_9}{k_4 + k_8}\right) \qquad 8.$$

At high concentrations of all substrates, the expression of isotope effects is governed by three kinetic functions (6). For the mechanism in Scheme 1, Equation 2 applies in which the forward ratio of catalysis is:

$$R_f = \frac{k_9}{k_5''} + \frac{k_9}{k_7} + \frac{k_9}{k_{11}'} + \frac{k_9}{k_{13}} + \frac{k_9}{k_{15}} \qquad 9.$$

with net rate constants (indicated by primes) defined as:

$$k_{11}' = k_{11} k_{13}/(k_{12} + k_{13}) \qquad 10.$$

$$k_5'' = k_5 k_7/(k_6 + k_7) \qquad 11.$$

The forward equilibration proceding catalysis is:

$$E_f = 1 + k_8/k_7 + k_8 k_6/k_7 k_5 \qquad 12.$$

The reverse commitment remains the same as defined in Equation 4.

Isotopic Discrimination

The measurement of isotope effects by the competitive method in which substrate B of Scheme 1 is trace-labeled yields the limiting isotope effect on V/K directly, independent of the concentration of A or B. If A carries a label such as tritium, then $^{T}V/K_{app}$ is obtained, which varies as a function of the concentration of B, and this concentration must be extrapolated to zero. In an ordered mechanism, the extrapolation is linear on a reciprocal plot, according to the following equation (3):

$$\frac{1}{^{T}(V/K_{app})-1} = \frac{1}{^{T}(V/K_b)-1} + \frac{K_a[B]/K_{ia}K_b}{^{T}(V/K_b)-1} \qquad 13.$$

In a random mechanism the extrapolation is not linear.

Isotope effects are obtained from competitive experiments by measuring isotopic enrichment in substrate or product during the initial course of reaction, and extrapolating to zero time. For an irreversible reaction, tritium enrichment is related to the isotope effect by the equation (3, 7):

$$^{T}V/K = \frac{\log(1-f)}{\log[1-f \cdot (SA)_p/(SA)_o]} = \left| \frac{(SA)_o}{(SA)_p} \right|_{f \to 0} \qquad 14.$$

where f is the fractional conversion of substrate to product, and $(SA)_o$ and $(SA)_p$ are the specific radioactivities (or isotopic mole fractions) of the initial substrate and isolated product, respectively.

Intramolecular Isotope Effects

Isotopic discrimination between two positions on the same molecule resembles a V/K isotope effect. For the mechanism in Scheme 3, where ES_H and ES_D are the enzyme-substrate complexes for two possible conformers of an asymmetrically deuterated substrate,

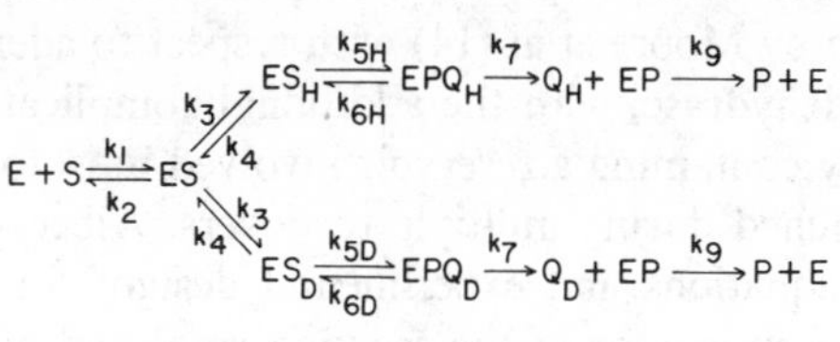

Scheme 3

the intramolecular isotope effect obeys the following equation (8):

$$Q_H/Q_D = \frac{^{D}k_5 + k_5/k_4 + k_6 \cdot {^{D}K_{eq}}/k_7}{1 + k_5/k_4 + k_6/k_7} = \frac{^{D}k + C_{if} + C_r \cdot {^{D}K_{eq}}}{1 + C_{if} + C_r} \qquad 15.$$

where C_{if} is an internal forward commitment to catalysis.

Intrinsic Isotope Effects

Swain et al (9) calculated a fixed relationship between intrinsic isotope effects of deuterium and tritium, which led Northrop (3, 10) to derive a relationship between deuterium and tritium isotope effects on V/K.

$$\frac{^D V/K - 1}{^T V/K - 1} = \frac{^D k - 1 + C_r(^D K_{eq} - 1)}{^D k^{1.44} - 1 + C_r(^D K_{eq}^{1.44} - 1)} = \left| \frac{^D k - 1}{^D k^{1.44} - 1} \right|_{^D K_{eq} = 1} \qquad 16.$$

By comparing deuterium and tritium isotope effects on V/K it is possible to calculate intrinsic isotope effects in both the forward ($^D k_f$) and reverse ($^D k_r$) directions of reaction. Given an equilibrium isotope effect, an exact answer is not calculable: the true $^D k_f$ lies between the calculated $^D k_f$ and $^D K_{eq} \cdot ^D k_r$ (3, 9).

Limitations

At present, general theory and equations for describing the steady-state kinetic functions that govern the expression of isotope effects on enzyme-catalyzed reactions are possible only for reaction mechanisms in which a single step is isotopically sensitive. When more than one step is affected (for example if the isotope is passed first from a donor substrate to a group on the enzyme and then from the enzyme to an acceptor substrate, with each passage subject to an isotope effect) unusual isotope effects are possible, such as tritium isotope effects of 250 and 160 accompanied by normal deuterium effects, which do not fit Equation 16, observed on adenosylcobalamine-dependent enzymes (12, 13).

The problem has been addressed with partial consideration of steady-state suppression by Moore et al (14) with respect to adenosylcobalamine-dependent diol dehydrase, with the additional complicating problem of a reaction pathway containing a reservoir involved in isotopic transfers that can become enriched during multiple turnovers. Albery & Knowles (15) have presented equations and experimental designs for consideration of isotope effects on every step of the reaction catalyzed by triosephosphate isomerase. Unfortunately, their symbolism is abstract, their equations are foreign to enzyme kinetics, and their logic connecting different experimental designs is not at all clear, so their system does not appear to have a general utility. An obvious second isotopically sensitive step might be the binding of substrate to enzyme. Several direct measurements have failed to detect any significant isotope effects on noncovalent Michaelis complexes, including NADH-liver alcohol dehydrogenase (16), propanediol-diol dehydratase (17, 18), and benzylamine-benzylamine oxidase (19).

THE UNCERTAINTY OF DETECTING
A RATE-LIMITING STEP

Isotopic Discrimination

The purpose of measuring isotope effects on enzyme-catalyzed reactions has generally been to determine whether or not the isotopically sensitive step is rate determining (20). That measurements of isotopic discrimination cannot give this information was noted by Abeles et al (21), elaborated by Simon & Palm (22), and reiterated by Northrop (10); yet failure to grasp this fact remains the most frequent error in the literature. Because isotopic discrimination measures effects only on V/K, it is independent of events following the first irreversible step, and thus often fails to detect a rate-limiting step by failing to encompass a complete enzymatic turnover. But even for a reaction mechanism with a single irreversible step, Northrop (6) has argued that discrimination fails because V/K represents an apparent rate constant at zero substrate concentration, conditions under which there is no rate, and that ground states do not contribute to the kinetic functions governing the expression of isotope effects on V/K. This was illustrated by the reaction coordinate diagram shown in Figure 1. Stabilization of the Michaelis complex (that is, lowering the free energy level of E_2, perhaps through the use of a series of alternate substrates) will have no effect on the

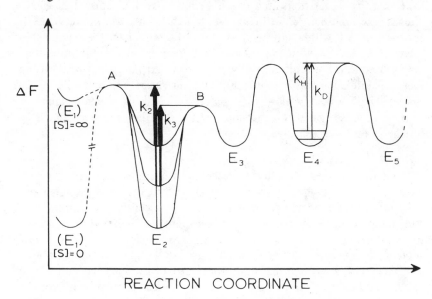

REACTION COORDINATE

Figure 1 A reaction coordinate diagram illustrating the stabilization of the Michaelis complex (E_2). Lowering the energy level of E_2 may either increase or decrease $^D V$, but will have no effect on $^D V/K$.

expression of an isotope effect on V/K, because the relative tendency of E_2 to cross energy barrier A versus B will not have changed. Both k_3 and k_2 will be decreased, but not their ratio, and this ratio is the commitment factor for E_2. Even if k_3 becomes totally "rate limiting" and equal to V/E_t, the mechanism will still express the same isotope effect on V/K.

Isotope Effects on Maximal Velocities

Northrop (6) also argues that the kinetic and thermodynamic properties (that is, the barrier heights and energy levels of intermediates) are unequally distributed among the three kinetic functions that govern the expression of isotope effects on maximal velocities, and this unequal distribution renders the singular concept of a rate-limiting step an inadequate basis for interpreting isotope effects. To dramatize this inadequacy, one can factor out k_5 from R_f/E_f governing the mechanism of Scheme 1 and obtain:

$$\frac{R_f}{E_f} = \frac{a(k_5) + b}{c(k_5) + d} \qquad 17.$$

where coefficients a–d are collections of nonisotopic rate constants. If $a/c < b/d$, the relationship between k_5 and $^D V$ will be normal: as k_5 becomes rate limiting, $^D V$ will decrease, but then only to a finite value. If $a/c = b/d$, $^D V$ will be independent of k_5. If $a/c > b/d$, the relationship between k_5 and $^D V$ is incompatible with the concept of a rate-limiting step: as k_5 becomes rate limiting, $^D V$ will *increase*. An uncertain relationship between rate constants comprising coefficient ratios a/c and b/d determines whether fast or slow steps preceding catalysis either enhance or suppress the isotope effect, and not how rate-limiting that particular step or the isotopically sensitive step may be.

Internal Isotope Effects

The catalytic sequence is often represented by a single, apparent rate constant (i.e. $E_3 \xrightarrow{\ k_{cat}\ } E_7$ in Scheme 1), because neither the number and sequence of catalytic and conformational events occurring within the central complex, nor their relationship to one another, is altered or expressed in traditional steady-state kinetics where reaction velocities vary as a function of the concentrations of substrates, products, and dead-end inhibitors. Northrop (10) suggested that in a comparison of isotope effects on V and V/K, the larger of the two represented a minimal possible value for the isotope effects as expressed on this single, apparent rate constant. Differences between $^D V$ and $^D V/K$ could then be used to set limits on the relative rates of the catalytic sequence and the release of either substrate or product.

This reasoning has subsequently been applied to isotopic interpretations of a number of enzymes, but fails to come to terms with the nontraditional character of the isotope effect. Because of the uncertainty of Equation 17 and lack of ground states in $^D V/K$, the internal isotope effect itself need not be the same when contributing to $^D V$ as to $^D V/K$ (cf Equations 1 and 2 when k_4 and k_{15} in Scheme 1 are infinitely fast). Therefore, differences between $^D V$ and $^D V/K$ cannot be interpreted solely in terms of the contributions of the rates of release of substrates or products.

Minimal Mechanisms

In order to come to terms with these uncertainties and to ensure that the kinetic and thermodynamic properties governing the expression of isotope effects are fully represented, possible reaction mechanisms and related rate equations must include at least two precatalytic and one postcatalytic reversible steps, in addition to substrate binding and product release steps (6). Two steps before catalysis are necessary because the full uncertainty expressed in Equation 17 requires at least two steps. An additional step after catalysis is needed to allow for variety in the reverse commitment to catalysis independent of the rate of the reverse catalytic step. Finally, the isotopically sensitive step itself must be included, and considered reversible, even if only microscopically so. An example of a minimal mechanism satisfying these requirements is shown in Scheme 1. Based upon transient kinetic and deuterium isotope effect studies, Dubrow & Pizer (23) have proposed an equivalent minimal mechanism for phosphoglycerate dehydrogenase.

LIMITS OF EXPERIMENTAL METHODS

Purity of Labeled Reactants

Isotopic and nonisotopic velocities are measured separately in the noncompetitive method, and any differences in purity appear as artifactual isotope effects. Northrop (3), Cleland (11), and others have assumed that isotope effects measured by a noncompetitive method in the presence of inhibitory contaminants will still be accurate provided the impurities are equally present in the isotopic and nonisotopic reactant. However, Grimshaw & Cleland (24) have examined this assumption and found it to be untrue for determinations of $^D V$ and, when the substrate not carrying the isotopic label is being varied, also untrue for $^D V/K$. The observed $^D V$ as a function of labeled and unlabeled inhibitor is given by the equation (24):

$$^D V_{obs} = \frac{(K_{is}/K_H)^D V + ([I_D]/[S_D])^D (V/K)}{(K_{is}/K_H) + [I_H]/[S_H]}$$

18.

where K_H is the Michaelis constant for S_H, and K_{is} is the slope inhibition constant for either competitive or noncompetitive inhibitors that have no isotope effect on binding. Equation 18 shows that under the conditions of the initial assumption, the observed $^D V$ will be at a value between the true $^D V$ and $^D V/K;$ when the isotopic reactant is more contaminated, the observed value may exceed either $^D V$ or $^D V/K;$ and when the nonisotopic reactant is more contaminated, large inverse isotope effects are possible. Determinations of $^D V/K$ by the competitive method, or by the noncompetitive method when the labeled substrate is varied, remain free from artifacts due to binding inhibitors.

Competitive Methods

In addition to the independence from inhibitors, isotopic discrimination measured by the competitive method is also independent of the concentration of the labeled substrate [which must be precisely determined for noncompetitive measurements (22)] and capable of great precision [particularly in the case of heavy-atom isotope effects, where precision of the order of ± 0.000002 is possible (25)]. Deuterium isotope effects are normally determined by the noncompetitive method (which provides more information) but need not be.

Dahlquist et al (26) introduced an elegant double label technique for competitive determination of deuterium isotope effects and achieved sufficient precision to apply it to the detection of small secondary isotope effects. In this design, the reaction progress of the deuterated substrate is followed with a carbon-14 trace label and compared to the reaction progress of the non-deuterated substrate, followed with a tritium trace label, to take advantage of the precision available in $T/^{14}C$ channels ratio scintillation counting. The technique has also been extended to heavy-atom isotope effects (27). More recently, both Goiten et al (28) and Bull et al (29) have refined the technique and achieved a higher level of precision, limited primarily by the precision of $T/^{14}C$ counting ratios, which are of the order of $\pm 0.5\%$. Perhaps the most important result in both studies was the documentation of the decrease in observed isotope effects as a function of the percent of the reaction. It has been frequently assumed that negligible error ensues from allowing isotopic discrimination to proceed for 5–10% of the reaction, but these more precise studies show considerable error even at low levels of product formation. In order to obtain accurate quantitative data, several early determinations must be obtained and extrapolated back to 0% of reaction. Equation 14 will correctly compensate only for isotopic enrichment of substrate, and then only in an irreversible reaction. It does not account for the enrichment of product that occurs with reaction reversibility and its second isotope effect, nor does it compensate for a decreasing co-substrate or increasing product, which may influence the apparent V/K isotope effect underlying the isotopic discrimination being measured.

Failure to compensate isotopic discrimination for the effects of isotopic enrichment that accompany reaction progress, or to appreciate the independence of discrimination from the concentration of labeled substrate or inhibitor, can lead to serious distortions of the values assigned to kinetic isotope effects. For example, Wang et al (30) observed a large and inverse ^{18}O discrimination in transesterification catalyzed by chymotrypsin. The initial pool of ethyl ester formed from acylenzyme and $^{16}O/^{18}O$-ethanol was depleted in ^{18}O, which indicates a normal isotope effect. But as transesterification continued, the pool became enriched in ^{18}O, supposedly because of an inverse isotope effect. The possibility of the ethyl ester undertaking a second catalytic turnover subject to a second isotope effect via a hydrolytic reaction was dismissed because of its high K_m and low concentration relative to the substrate ester. As noted above, however, isotopic discrimination is independent of the concentration of labeled reactant relative either to its K_m or concentration of competitor (and in this case, the labeled ester originated already bound to the enzyme). Initially, formation of the ethyl ester exceeds its disappearance, and its isotopic ratio should show ^{18}O depletion as observed. As the pool approaches steady state between transesterification and hydrolysis, its isotopic ratio should approach isotopic equilibrium, which ought to be inverse (5). And as the pool decreases, its isotopic ratio should show ^{18}O enrichment, which will become excessively high relative to the true kinetic isotope effect during the last few turnovers (7).

Deviation from the Swain-Schaad Relationship

Calculation of intrinsic isotope effects by use of Equation 16 is based on measurements of deuterium and tritium isotope effects on V/K, and depends on how a comparison of these effects deviates from the Swain-Schaad relationship, the deviation being a measure of the steady-state suppression of the intrinsic isotope effect. Albery & Knowles (31) examined this deviation and found it reached a maximum near observed $^D V/K$ values of about two [corresponding to a value for the sum of commitment factors of eight or nine and a deviation of 31% (2)]. At either side of the maximum, they found the deviation to be similar to the experimental error frequently found on kinetic data for most enzymes, and concluded that the experimental error on the kinetic data therefore needs to be less than 3%.

To achieve this goal, improvements in experimental design and methods are needed. Regarding design, it makes little sense to measure $^T(V/K)_{app}$ by the competitive method and $^D(V/K)_{app}$ by the non-competitive method when accuracy and precision are at a premium, because both must then be extrapolated to limiting values. Regarding methods, the recent improvements in the double label technique for competitive methods show what is possible once the need for quantitative isotopic data is perceived. Combining these, C. J. Newton and D. B. Northrop (unpublished results) designed a "triple competitive" method for determining primary intrinsic isotope effects using Equation 16. To illustrate the method, consider the following sequence of reactions of lactate dehydrogenase:

$$^{14}C\text{-NADD} + {}^{14}C\text{-pyruvate} \rightarrow {}^{14}C/D\text{-lactate} + {}^{14}C\text{-NAD} \qquad 19.$$

$$T\text{-NADH} + {}^{14}C\text{-pyruvate} \rightarrow {}^{14}C/H\text{-lactate} + T\text{-NAD} \qquad 20.$$

$$\text{NADT} + {}^{14}C\text{-pyruvate} \rightarrow {}^{14}C/T\text{-lactate} + \text{NAD} \qquad 21.$$

Reaction 19 represents deuteride transfer accompanied by the production of NAD labeled in the adenine ring with ^{14}C. Reaction 20 represents hydride transfer accompanied by the production of tritiated NAD, also labeled in the adenine ring. Consequently, the deuterium isotope effect is expressed in the $T/^{14}C$ ratio of NAD. Reaction 21 represents tritiide transfer, which produces tritiated lactate. All three reactions are accompanied by the production of ^{14}C-labeled lactate, which will reflect hydride transfer *providing* NADH is trace-labeled in the transferable position. Consequently, the tritium isotope effect is expressed in the $^{14}C/T$ ratio of lactate. The net result is a determination of the intrinsic isotope effect from a single incubation containing all four labeled reactants, which should provide both increased accuracy and precision because: identical reaction conditions are assured for both isotope effects, errors of the non-competitive method are avoided, advantage is taken of the precision of double label counting methods, and an extrapolation step is avoided. In addition, the reduction of labor makes it practical to perform multiple determinations for the purpose of establishing confidence limits and to extrapolate to 0% of reaction.

Equilibrium Perturbation

The most significant advance in experimental methods was the discovery of the equilibrium perturbation technique by Schimerlik et al (32). In this technique, reactions are initiated at chemical equilibrium, but with only the substrate carrying an isotope. The forward reaction proceeds more slowly than the reverse because of the isotope effect, and a transitory increase in substrate and decrease in product is observed. The transient displacement between chemical and isotopic equilibrium is very sensitive to small isotope effects (i.e. the precision is ± 0.001) because it represents a product of many molecular turnovers sensitive to the isotope effect, versus a sum of molecular turnovers produced by the usual methods. It is therefore suitable for secondary and heavy-atom isotope effects, using an ordinary spectrophotometer if a substrate or product absorbs light. The perturbation represents essentially a V/K isotope effect, but involves all the steps of a reaction mechanism. Definitions, equations, experimental methods, and applications have recently been summarized by Cleland (5). Equally as important as the sensitivity provided by the method is the interest it triggered in equilibrium isotope effects. Although they are usually small, they do contribute to observed kinetic isotope effects (see Equations 1, 2, 15, and 16) and need to be known. Available data relevant to enzyme kinetic isotope effects have also been tabulated by Cleland (5).

ISOTOPIC DIAGNOSTICS OF KINETIC MECHANISMS

Primary Isotope Effects

How the expression of isotope effects changes with changing concentrations of reactants can be useful in distinguishing between kinetic mechanisms. These distinctions have in practice been restricted to primary isotope effects because detection of significant changes requires a large and precisely determined isotope effect.

ORDERED ADDITION OF SUBSTRATES   Northrop (3, 10) proposed that a distinction between ordered and random addition of substrates can be established by examining an apparent V/K isotope effect as a function of the concentration of the second substrate. In an ordered mechanism, saturation of the enzyme by high concentrations of the second substrate drives the forward commitment to catalysis of the first substrate to infinity and abolishes its V/K isotope effect (see Equation 3). In a random mechanism, the commitment becomes finite, and the V/K effect is reduced but not abolished (see Equation 6). Blanchard & Cleland (33) observed the former behavior in the ordered kinetic mechanism of yeast formate dehydrogenase, in which $^D V/K_{app}$ of NAD was abolished by saturating formate. In contrast, Rife & Cleland (34) observed finite values for both V/K isotope effects at saturating concentrations of co-substrates in a study of glutamate dehydrogenase consistent with the random mechanism proposed earlier.

RAPID EQUILIBRIUM BINDING   Rapid equilibrium conditions are frequently assumed in steady-state kinetic analysis, often without justification, primarily because the reciprocal rate equations become linear and easier to analyze. Isotope effect data can be used to test these conditions. When substrates are in rapid equilibrium with their binary enzyme complexes, equivalent V/K isotope effects obtain at all levels of the other substrate (i.e. the condition is that k_2 in Scheme 1 and both k_2 and k_6 in Scheme 2 are much greater than V/E_t, with the result that Equation 3 reduces to Equation 5, and Equations 6 and 7 reduce to Equation 8). Alternatively, if either substrate is rapidly released from the ternary complex in a random mechanism, equivalent V/K isotope effects obtain that are independent of cosubstrate concentrations, because they are governed by internal commitment factors only (i.e. $C_{fab} = k_{11}/k_{10}$ replaces Equation 8). The earlier assertion by Northrop (3) and Cleland (11) that in a fully rapid equilibrium mechanism, $^D V/K_a$ equals $^D V$ as well as $^D V/K_b$, is dependent upon equivalent expression of the internal isotope effect on V and V/K, and can no longer be assumed because of the uncertainty expressed in Equation 17.

ORDERED RELEASE OF PRODUCTS   Extension of these ideas invites a proposal for determining the order of release of products. Because V/K terminates in the first irreversible step, which is the release of the first product under initial velocity conditions for a reversible reaction, addition of the first product will extend V/K to encompass one more step, which will be incorporated into the reverse commitment to catalysis. Moreover, increasing concentrations of the first product will eventually abolish the kinetic isotope effect and cause the equilibrium isotope effect to be expressed. For the mechanism in Scheme 1, the reverse commitment to catalysis becomes (cf Equation 4):

$$C_r = \frac{k_{10}}{k_{11}} \left[1 + \frac{k_{12}}{k_{13}} \left(1 + \frac{k_{14}[P]}{k_{15}} \right) \right] \qquad 22.$$

Thus:

$$^D V/K = \frac{^D k + C_f + C_r \cdot {}^D K_{eq}}{1 + C_f + C_r} = \left. \frac{C_r \cdot {}^D K_{eq}}{C_r} \right|_{[P] \to \infty} = {}^D K_{eq} \qquad 23.$$

In contrast, no change in $^D V/K$ will follow addition of the second product to be released. However, care must be taken to use labeled product if the first product released contains the isotope originating from the substrates. In a random mechanism the addition of either product causes $^D V/K$ to be suppressed, but only to a finite value. The logic of this diagnostic can easily be extended to the order of product release in terreactant mechanisms (which can be difficult to ascertain by steady-state kinetics alone) by adding appropriate pairs of product inhibitors.

Intramolecular Isotope Effects

Abdel-Monem (35) reported deuterium intramolecular isotope effects on N-demethylation of tertiary amines by liver homogenates of 1.31 (rats) and 1.45 (mice) and concluded that cleavage of the C–H bond is a rate-limiting step in N-demethylation by rodent microsomal enzymes. Cheung & Walsh (36) interpreted intramolecular effects as equivalent to k_{cat} effects and employed comparisons between inter- and intramolecular isotope effects in pyruvate carboxylase catalysis as a means of estimating relative values of various rate constants. Hjelmeland et al (37) obtained a large intramolecular isotope effect in hydroxylations by cytochrome P-450 ($^Dk=11$) and interpreted it as equivalent to an intrinsic effect, whereas Miwa et al (8) obtained small intramolecular values for N-demethylation by the same enzyme ($^Dk < 2$) and argued that, like an intermolecular effect, it was subject to steady-state suppression. Each assertion may be true in its particular context, but caution is advised in these comparisons; differences between inter- and intramolecular effects do not necessarily reveal the full level of suppression of the intrinsic isotope effect on the overall reaction, because both may be subject to partial suppression, and similarities do not guarantee an absence of suppression because both are equally subject to the reverse commitment to catalysis (cf Equations 1 and 15). Indeed, the major diagnostic value of intramolecular isotope effects may be in determining values for C_f and C_r by comparisons to intermolecular and intrinsic effects, when it is reasonable to assume that the internal forward commitment to catalysis must be zero (for example, in the rapid equilibrium positioning of a monodeuterated methyl group), which leaves only the reverse commitment responsible for suppression of the intramolecular isotope effect.

Enzyme-Mediated Solvent Isotope Effects

Solvent isotope effects are usually obtained by measuring enzymatic reaction rates with an excess of substrate in H_2O and in D_2O as a function of pH(D). A proton inventory in the form of a variation of the atom fraction of deuterium determines if the effect observed is of single or multiple origins (38). Rarely is the concentration of substrate varied to obtain the V/K isotope effect. Consequently, with only the single value of ^DV available, interpretations of the kinetic functions governing the expression of solvent isotope effects are severely limited.

But what about changes in the pK_a? Normally, the apparent pK_a in D_2O shifts to a higher value by 0.5–0.6 pH units, because of the difference in acidity of D_3O^+ versus H_3O^+ (38). A "normal" shift is interpreted as reflecting no isotope effect at equivalent pH(D). This concept of normal is, in effect, a balancing of an equilibrium isotope effect by an adjustment in reactant concentration, which is acceptable for the measurement of equilibrium isotope effects, but it does not apply to kinetic isotope effects, particularly when proton concentrations are far from the pK_a values. Instead, solvent proton concentrations must be treated the same as other reactant concentrations if their effect is to be interpreted within a similar context.

When acting as a varied substrate-like reactant, high concentrations of protons yield a V isotope effect, and the protonation and deprotonation rate constant are absent; low concentrations of protons give a V/K effect, with the isotope effects on protonation versus deprotonation unequally represented. Therefore adjustments based on ΔpK_a are inappropriate. We are left with the problem of more than one isotopically sensitive step (see section on limitations), but then only at low proton concentrations. When acting as a product-like inhibitor, the proton concentration is multiplied by the equilibrium isotope effect on catalysis, cancelling out the need for an adjustment (cf Equations 3 and 28). [Note, the situation is different yet again if the proton acts as a dead-end inhibitor or as an activator (see section on pH effects)].

To illustrate this approach, consider the "abnormal" behavior reported by Welsh et al (39) in their steady-state kinetic studies of yeast alcohol dehydrogenase. An inverse solvent isotope effect of 0.5 ± 0.05 on the maximal velocity and a much less than normal shift in pK_a of 0.02 were observed in the reduction of p-methoxybenzaldehyde. In the oxidation of p-methoxybenzyl alcohol, a small but normal solvent $^D V$ of 1.20 ± 0.09 and less than normal ΔpK_a of 0.21 were observed. Significantly, no change in the solvent isotope effect was observed when NADH was replaced by NADD (which by itself caused an isotope effect of 2.9), which indicates that proton transfer and hydride transfer are not concerted. Scheme 4 portrays the catalytic sequence for uncoupled proton and hydride transfer in alcohol dehydrogenase.

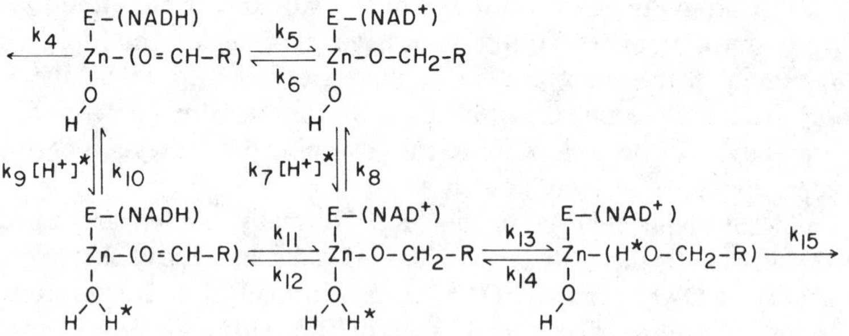

Scheme 4

Treating H^+ as a substrate during aldehyde reduction, then at high $[H^+]$:

$$^D V_f = \frac{^D k + R_f/E_f + C_r \cdot {}^D K_{eq}}{1 + R_f/E_f + C_r}$$

$$= \frac{^D k_{13} + \dfrac{[k_{13}/k_{11} + k_{13}/k_{15}]}{[1 + k_{12}/k_{11}]} + [k_{14} \cdot {}^D K_{eq}/k_{15}]}{1 + \dfrac{[[k_{13}/k_{11} + k_{13}/k_{15}] + [k_{14}/k_{15}]}{[1 + k_{12}/k_{11}]} \quad f} \qquad 24.$$

Because α-secondary kinetic isotope effects of tritiated alcohols equaled equilibrium values (39), hydride transfer favors the aldehyde, k_{12} exceeds k_{11}, and E_f is large, which in turn means R_f/E_f must be small. The inverse isotope effect on the maximal velocity is then seen as being predominantly an expression of the equilibrium isotope effect because of a large reverse commitment to catalysis.

$$^D V_f \cong \frac{C_r \cdot {}^D K_{eq}}{C_r} = {}^D K_{eq} \qquad 25.$$

The equilibrium isotope effect on proton transfer between H_3O^+ and R-OH should be (38):

$$^D K_{eq} = \frac{^D([H_3O^+]/[H_2O])^3}{^D([ROH]/[H_2O])} = \frac{(0.69)^3}{1} = 0.33 \qquad 26.$$

A large reverse commitment to catalysis should also cause an inverse effect to be expressed on $^D V/K_{H^+}$, because it also contains C_r. We must consider first, however, the magnitude of C_f at low $[H^+]$:

$$^D V/K_{H^+} = \frac{^D k + C_f + C_r \cdot {}^D K_{eq}}{1 + C_f + C_r}$$

$$= \frac{^D k_{13} + \{k_{13}/[k_8 + k_{10}k_{12}/(k_{11} + k_{10})]\} + [k_{14} \cdot {}^D K_{eq}/k_{15}]}{1 + \{k_{13}/[k_8 + k_{10}k_{12}/(k_{11} + k_{10})]\} + [k_{14}/k_{15}]} \qquad 27.$$

An earlier report of $^D V/K_b = 3.6$ for the oxidation of deuterated benzyl alcohol (40) requires that deprotonation (k_8 and k_{10}) must exceed hydride transfer (k_{11} and k_{12}); thus C_f of Equation 27 approaches k_{13}/k_8, which is likely to be small. Consistent with prediction, the very small ΔpK_a value means that $^D K_H^+ \cong 1$, and thus that $^D V/K_H^+ \cong 0.5/1 = 0.5$. In the reverse reaction, H^+ acts as a product inhibitor (Equation 28):

$$
^D V_r = \frac{^D k + R_r/E_r + C_f/^D K_{eq}}{1 + R_r/E_r + C_f}
$$

$$
= \frac{^D k_{14} + R_r/E_r + (k_{13}/k_8)(1 + k_7[H^+]/k_6)/^D K_{eq}}{1 + R_r/E_r + (k_{13}/k_8)(1 + k_7[H^+]/k_6)} \qquad 28.
$$

Expression of either kinetic or equilibrium isotope effects should be normal for alcohol oxidation, as is $^D V_r$ of 1.20 plus or minus 0.09 at low $[H^+]$. From the ΔpK_a, a value of 1.62 can be calculated for the isotope effect on the product (i.e. proton) inhibition constant, which is multiplied by $^D V_r$ to find the limit isotope effect at very high $[H^+]$:

$$
^D V_{r[H^+] \to \infty} = 1.62 \times 1.20 = 1.94 \cong 1/^D K_{eq} \qquad 29.
$$

The limit isotope effect by product inhibition agrees remarkably well with the reciprocal of substrate activation—both expressions of the equilibrium isotope effect. Thus, neither the "inverse" isotope effect nor the "small" ΔpK_a's reveal anything abnormal about yeast alcohol dehydrogenase. A similar analysis may account for the small normal (1.03–1.45) and inverse (0.63–0.74) solvent isotope effects observed with various substrates during hydrolysis by papain, which may not be inconsistent with general base catalysis supposedly eliminated by these data (41).

Abandoning adjustments for shifts in pK_a's, and letting the reverse commitment to catalysis take care of equilibrium isotope effects, invites a radical reinterpretation of solvent isotope effects on carbonic anhydrase. Assuming that H_2O is a substrate and H^+ a product, and using a "normal" pK_a shift of 0.5 units in D_2O, Steiner et al (42a) have calculated solvent isotope effects of 3.8 on the maximal velocities, and negligible effects on V/K in both forward and reverse directions of catalysis by human carbonic anhydrase. Similar results were later obtained by Pocker & Bjorkquist (42b) using bovine carbonic anhydrase. Although results showing that $^D V/K$ is much less than $^D V$ have been interpreted to indicate a slow addition of substrate to enzyme (2), this cannot be true because the rate constant for substrate addition always cancels out of $^D V/K$. Steiner et al (42a) proposed a rate-limiting and isotopically sensitive, intraenzymatic proton transfer step following the release of the first product to account for the absence of an effect on V/K, because V/K is insensitive to steps following the first irreversible step. However, this mechanism requires that the hydration and dehydration reactions must be even faster than the turnover rates, by several orders of magnitude, for an enzyme with one of the highest turnover rates known. Alternatively, calculating from "real" pH(D) values with CO_2 varied, an impressive $^DV = 15.8$ and $^DV/K = 8.8$ are obtained. With increasing hydroxide ion concentrations, $^DV/K$ is abolished and ^DV is suppressed to the pH-independent value of 3.8 With varying HCO_3^-, ^DV and $^DV/K$ are both normal, but both become inverse with increasing hydroxide ion concentrations.

Examined in this way, these isotopic data argue for an ordered addition of CO_2 and OH^- as substrates in the "hydration" reaction, with OH^- acting as a product inhibitor in the "dehydration" reaction. Consistent with this mechanism is the sensitivity to anion inhibition [especially SH^- (43)], and a pH-independent inverse solvent isotope effect on V/K of 0.64–0.74 for ester hydrolysis by carbonic anhydrase (42a, 42b). This inverse effect suggests that an alkoxide anion is the first hydrolytic product [which accounts for the lack of transesterification to methanol (44)] and that it can be replaced by the hydroxide anion, which binds as an alternate product to the acylenzyme and drives a proton transfer step back toward equilibrium. The hydroxide ion failed to inhibit the overall reaction at the pH of the measurements because, in the much slower esterase activity (i.e. $\times 10^{-4}$), deacylation and not proton transfer is rate limiting. The common inverse isotope effects at alkaline pH shows that the catalytic mechanism for esterase activity more closely resembles the dehydration rather than the hydration reaction.

Solvent isotope effects may also arise from direct transfer between solvent and substrate, instead of mediation by the enzyme. Yamada & O'Leary (45) have argued that a distinction between the two can sometimes be made by performing competitive experiments in 50:50 H_2O:D_2O. Specifically, they argue that isotopic discrimination will not be observed if the proton transfer occurs through the mediation of a monoprotic catalytic group that is shielded from exchange with the solvent. The reason for restricting the mediation to monoprotic groups is an expectation of a large discrimination between the mixed population of isotopic hydrogens on groups such as a lysine ammonium group. However, Northrop (46) determined that alternate reaction pathways of di- and triprotic mechanisms compensate for the isotope effect and dilute the expression of the intramolecular component. The upper limits for isotopic discrimination were established as follows:

$$(P_H/P_D)_{diprotic} = (3\,^Dk + 1)/(^Dk + 3) \tag{30.}$$

$$(P_H/P_D)_{triprotic} = [7\,(^Dk)^2 + 10\,^Dk + 1]/[(^Dk)^2 + 10\,^Dk + 7] \tag{31.}$$

Given a normal intrinsic isotope effect of $^Dk = 7$, maximal isotopic discrimination in 50:50 H_2O:D_2O is therefore 2.2 and 3.3, respectively, versus 1.0

for a monoprotic group. An experimental protocol was proposed for interpreting intermediate values of isotope discrimination with respect to distinguishing enzyme-mediated catalytic mechanisms from those of direct transfer between solvent and substrate, and for identifying mediating groups. The protocol entails comparisons of isotopic discrimination at high and low concentrations of substrates and references to intrinsic and intermolecular isotope effects (46). Substrate dependence of isotopic discrimination indicates shielding and therefore enzyme-mediated proton transfer. Maximal values of shielded isotopic discrimination are calculated from Equations 30 and 31 after first determining the intrinsic isotope effect from Equation 16. Because the commitment factors governing the V/K isotope effect were shown to be at least as great as those governing intramolecular isotopic discrimination, minimal values are calculated by substituting the V/K effect for the intrinsic in Equations 30 and 31. If isotopic discrimination is less than both minima, the group is monoprotic; if less than the triprotic minimum, the group is either a shielded diprotic or a partially shielded monoprotic; if less than the triprotic maximum, all three groups are candidates at various levels of shielding; and if greater than the triprotic maximum, then proton transfer must be mediated by an unknown group that cannot be fully shielded from exchange with solvent.

Secondary Isotope Effects

Because secondary isotope effects are small, they are usually measured by the competitive method and therefore subject to the limitations of V/K. Also, they are rarely examined as functions of the concentration of cosubstrate or other co-factor concentrations. Therefore, the level of expression of the secondary effect on the apparent V/K measured is usually unknown, making it difficult to interpret secondary effects that are smaller than those of model reactions. Goiten et al (28) neatly sidestepped the problem by exploiting a primary ^{14}C isotope effect as an internal reference to the level of expression. In this way, they were able to distinguish between isotope effects near unity arising either from expressed S_N2-like or suppressed carbocation-like transition states, arguing for the latter as the mechanism of several phosphoribosyltransferases. Similarly, by reference to large primary effects and an equivalent equilibrium secondary isotope effect, Welsh et al (39) were able to interpret a secondary kinetic isotope effect ($^T V/K = 1.3$) on the oxidation of aromatic alcohols by yeast alcohol dehydrogenase as implicating a transition-state structure that resembles aldehyde with regard to bond hybridization.

But the notable exception to both deficiencies is the determination of V and V/K secondary deuterium isotope effects on catechol O-methyltransferase activity by Hegazi et al (47). Within a moderate experimental error, no V/K effect was detected, but a significant and inverse effect of 0.83 ± 0.03 was found on V. Unlike primary isotope effects, inverse secondary kinetic isotope effects are possible without involving a large reverse commitment to catalysis and subsequent expression of $^D K_{eq}$. To distinguish between the two causes, Hegazi et al (47) also determined the primary carbon-13 isotope effect arising from the same catalytic step, using the same methods. Again, the V/K effect was difficult to distinguish from unity but the ^{13}V was a normal 1.09 ± 0.04, which rules out a large reverse commitment and instead supports a highly expressed S_N2-like transition state.

Heavy-Atom Isotope Effects

Because heavy-atom isotope effects are normally measured by the competitive method (or more recently by equilibrium perturbation), the results obtained are V/K effects only, and subject to the limitations of that parameter (25). Until recently, there have been few attempts to alter or manipulate the steady-state functions that govern the expression of heavy-atom isotope effects, hence most results reflect a fixed but unknown level of expressions. The rest reflect the influence of a perturbant whose mechanism of altering the level of expression is unknown. For example, $^{13}V/K_{app}$ of yeast pyruvate decarboxylase varies from 1.002 at pH 7.5 to 1.011 at pH 5.0, and the latter value is temperature independent (48).

A significant exception, however, is the attempt by O'Leary & Limburg (49) to shift the rate-limiting step of isocitrate dehydrogenase from nucleotide release to catalysis, by substituting nickel for magnesium as the activating ion, because the former causes a tenfold reduction in the rate of the overall reaction (50). Although no change was detected on $^D V/K$, the $^{13}V/K$ shifted from 0.9989 ± 0.0004 with magnesium activation to 1.0051 ± 0.0012 with nickel-ion activation. Being inverse, the former suggests the expression of an equilibrium isotope effect arising from a large reverse commitment to catalysis, and indeed, later experiments by O'Leary & Yapp (51) established that $^{13}K_{eq} = 0.9963$. The shift from inverse to normal is indicative of a large reduction in the reverse commitment to catalysis, which together with the decrease in the maximal velocity and no change in $^D V/K$, can only arise from a large reduction in the catalytic rate constants and not from the effects of metal on substrate and product binding and dissociation as suggested. Thus, despite the small magnitude of these isotope effects, they do reveal that the primary contribution of metal ions to catalysis is in the carbon-carbon bond-breaking step.

But the most significant exception is the recent attempt by Blanchard & Cleland (38) to perturb the ^{13}C isotope effect on yeast formate dehydrogenase by deuteration of the ^{13}C-formate. Because both the ^{13}C isotope effect ($^{13}V/K = 1.043$) and the deuterium isotope effect ($^D V/K = 2.8$) originate by cleavage of the same bond, both obey the same equation for an isotope effect on V/K, and the level of perturbation of the ^{13}C isotope effect by deuterated formate should reveal the magnitude of the commitment factors. No change in $^{13}V/K$ was observed; thus the commitment factors equal zero and both isotope effects have intrinsic values. (A $^TV/K$ of 4.8 was in agreement with prediction from the Swain-Schaad relationship, within experimental error.) This is the only heavy-atom isotope effect on an enzymatic reaction for which one can confidently attempt a quantitative analysis of transition-state structure.

Transient-State Kinetics

Investigators of transient-state kinetics have been somewhat slower than those of steady-state kinetics to realize that the isotope effects they observe on enzyme-catalyzed reactions are not necessarily full, intrinsic isotope effects imposed on catalysis. But the pH dependence of a substrate deuterium isotope effect on the first transient (52), and inverse solvent deuterium isotope effect on the second transient (53) (indicative of an equilibrium isotope effect) of the reaction catalyzed by liver alcohol dehydrogenase, clearly reveals the presence of kinetic functions that control the expression of the full effect. These two papers also illustrate the primary utility of isotope effects in transient-state kinetics, which is to determine the number, identity, and order of the reactive steps comprising the catalytic sequence. In this case, they demonstrate that hydride transfer preceeds proton transfer, which clearly rules out a concerted reaction mechanism, proposed earlier by Dworschack & Plapp (54) on the basis of very interesting enhanced isotope effects obtained by chemical modification of the enzyme.

Schopfer & Massey (55) exploited an eightfold transient deuterium isotope effect to show that a charge-transfer complex observed in the oxidation of reduced 3-acetylpyridine adenine dinucleotide by melilotate hydroxylase is on the catalytic pathway. Combining pH dependence, allosteric inhibition, and transient kinetic isotope effects, Dubrow & Pizer (23) identified two conformational equilibrations preceding hydride transfer and one following it, in the catalytic sequence of phosphoglycerate dehydrogenase. Combining solvent and substrate deuterium isotope effects in stopped-flow measurements, Jenkins & Harruff (56) detected a proton transfer from substrate to a shielded diprotic group on pig heart aspartate aminotransferase as an early but not rate-limiting step in the first half-reaction of pyridoxal-dependent transamination. Combining stopped-flow spectrophotometry and the proton inventory technique of characterizing solvent isotope effects, Hankapillar et al (57) identified concerted transfer of two protons in the transition state leading to the rate-limiting breakdown of the tetrahedral intermediate of serine proteases. In an unusual application of the isotope effect on transient-state kinetics, Dunn et al (58) used deuterated NADH to slow down by threefold the first of two transients in the catalysis of liver alcohol dehydrogenase, in order to obtain a more accurate determination of both the rates and amplitudes of the two steps in an investigation of half-site reactivity.

Apparent Isotope Effects on Binding

Isotope effects on binding need not be a nuisance (see section on limitations) but may reveal useful mechanistic information. A secondary tritium isotope effect of 1.23 on the slow dissociation of FdUMP from thymidylate synthetase was interpreted, along with supporting data, as evidence for a covalently bonded complex (59). Lewis & Wolfenden (60) tabulated solvent and secondary deuterium isotope effects on the hydration of a series of antiproteolytic aldehydes and ketones, then compared these data to secondary isotope effects on aldehyde binding to papain; they concluded that the inhibitory aldehydes bind to the active site of this sulfur protease by thiohemiacetal formation and not as either free aldehyde or covalent hydrate (61).

Extending the inverse isotope effect on aldehyde hydration to indirect effects on binding and catalysis, Viola & Cleland (62) reported that chitose-6-P deuterated at C-1 yields a normal $^D V/K$ of 1.23 but has no effect on V for chitose-induced ATPase activity of phosphofructokinase, which indicates activation by free aldehyde and eliminates unstable phosphorylation of the hydrate as the mechanism of activation. Stein et al (63) measured secondary kinetic isotope effects on phosphorolysis catalyzed by purine nucleoside phosphorylase from calf spleen. They observed values of 1.047 ± 0.017 with adenosine and 1.043 ± 0.004 with inosine as substrates deuterated in the C-$1'$ position. These small values suggested that binding of the nucleoside to the active site of the enzyme occurs with significant change in the purine geometry, because the measurements were made with levels of the phosphate co-substrate at 100–400 times its K_m, and earlier kinetic data supported an ordered kinetic mechanism with nucleoside binding to the enzyme before phosphate (64). However, the isotope effect diagnostic (see section on ordered addition of substrates) is more sensitive to a partially random order of binding than are the steady-state kinetic patterns, which suggest as an alternative interpretation of the observed secondary isotope effects that the reaction mechanism is not strictly ordered.

Effect of pH

The joining of isotope effects to pH effects promises powerful diagnostics to kinetic mechanisms of enzymes. Both are complex and their union (which is still in progress) exceeds the limits of this review. Nevertheless, Figure 2 illustrates an example of these diagnostics. Four variants of the reaction mechanism in Scheme 5 are symbolically represented with the predicted pH profiles of isotope effects. All four mechanistic variants give a unique pair of pH profiles and can therefore be distinguished. In contrast, only the first variant gives a unique pH profile on initial velocity data alone (11). As of this writing, P. F. Cook and W. W. Cleland are preparing an extensive and significant analysis of pH-isotope effects on alcohol dehydrogenases, preliminary reports of which have appeared (65–67).

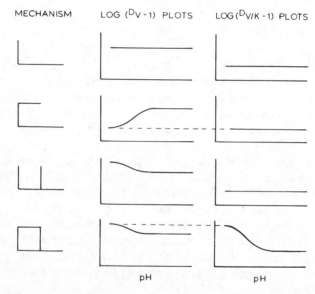

Figure 2   The pH dependence of the isotope effects minus one on V and V/K. The symbols for mechanisms represent variants of the mechanism in Scheme 5, which signify the presence or absence of steps governed by k_5, k_6, k_7, and k_8.

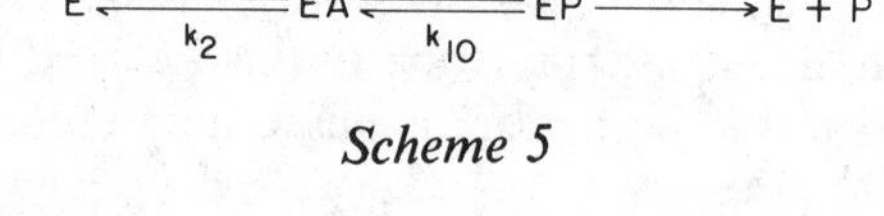

Scheme 5

Apart from diagnostics, changes in pH can be exploited to enhance small apparent isotope effects, despite the presence of other, rate-limiting steps. For example, the conversion of EAH to EA in the mechanism of Scheme 5 attains an unfavorable equilibrium at low pH, which lowers the maximal velocity but raises the expression of an isotope effect. One expects the isotope effect to be abolished rather than enhanced by a shift in the rate-limiting step to something other than catalysis, but the unfavorable equilibrium suppresses the ratio of catalysis. For the mechanism in Scheme 5 at low pH:

$$R_f/E_f = \frac{k_9/k_7 + k_9/k_{11}}{1 + k_8 [H^+]/k_7} \qquad\qquad 32.$$

Therefore

$$^DV = \frac{^Dk + R_f/E_f + C_r \cdot {}^DK_{eq}}{1 + R_f/E_f + C_r} = \left. \frac{^Dk + C_r \cdot {}^DK_{eq}}{1 + C_r} \right|_{pH\, <\, pK_a} \qquad 33.$$

That pH enhancement of ^DV arises from an unfavorable equilibrium preceding catalysis contradicts the earlier assertion that in order for isotope effects to be pH dependent, alternate reaction pathways leading up to, but not necessarily including, the catalytic step must exist whose relative contribution to the overall reaction is pH dependent (3, 24).

CONCLUSION

The purpose of applying the isotope effect to enzyme-catalyzed reactions has shifted dramatically from detecting the elusive "rate-limiting" step, to performing subtle distinctions between kinetic mechanisms. In retrospect, it seems naive to expect an entity so simple as this single, dimensionless number to reveal very much about something as complex as enzymatic catalysis. Many numbers are needed, which means isotope effects must be determined and examined as functions of something else in order to define and to strip away the many kinetic layers that prevent the full expression of the effect on enzymatic turnovers. When this is done, the isotope effect emerges as a very sensitive and very specific tool that is unique in enzyme kinetics.

Steady-state kinetics tell us about the comings and goings of reactants during enzymatic turnover, but reveal little about events within the catalytic sequence. Transient-state kinetics are sensitive to catalytic events, but often the identity of the event detected is uncertain and reference points on which to base quantitative analysis are lacking. Non-kinetic methods such as X-ray crystallography can reveal structural information about active sites of enzymes but fail to accurately portray the structures during catalysis because of their dependence upon static analogues. What makes the isotope effect unique is that its immediate perturbation is exercised, in most cases, on a single and identifiable catalytic step; the magnitude of the immediate perturbation provides clues to the structure of the transition state; and differences between isotope effects at various levels of expression on measurable transient- and steady-state kinetic parameters portray the relative kinetic and thermodynamic contributions of all other events comprising an enzymatic turnover. Consequently the isotope effect provides a needed link between steady-state, transient-state, and transition-state processes.

ACKNOWLEDGMENTS

Support during the preparation of this review was provided by a Career Development Award from the National Institutes of Health. Enthusiasm for enzymology during this career was provided long ago by the example of Professor Harland G. Wood, to whom this work is dedicated, in honor of the golden anniversary shared by his active research in biochemisty and the *Annual Review of Biochemistry*.

Literature Cited

1. Cleland, W. W., O'Leary, M. H., Northrop, D. B. 1977. *Isotope Effects on Enzyme-Catalyzed Reactions.* Baltimore: Univ. Park Press
2. Klinman, J. P. 1978. *Adv. Enzymol.* 46:415–94
3. Northrop, D. B. 1977. *See Ref. 1, pp. 122–52*
4. Buddenbaum, W. E., Shiner, V. J. 1977. See Ref. 1, pp. 1–36
5. Cleland, W. W. 1980. *Methods Enzymol.* 64:104–25
6. Northrop, D. B. 1981. *Biochemistry.* In press
7. Bigeleisen, J., Wolfsberg, M. 1958. *Adv. Chem. Phys.* 1:15–76
8. Miwa, G. T., Garland, W. A., Hodshon, B. J., Lu, A. Y. H., Northrop, D. B. 1980. *J. Biol. Chem.* 255:6049–54
9. Swain, C. G., Stivers, E. C., Reuwer, J. F. Jr., Schaad, L. J. 1958. *J. Am. Chem. Soc.* 80:5885–93
10. Northrop, D. B. 1975. *Biochemistry* 14:2644–51
11. Cleland, W. W. 1978. *Adv. Enzymol.* 45:273–387
12. Essenberg, M. K., Frey, P. A., Abeles, R. H. 1971. *J. Am. Chem. Soc.* 93: 1242–51
13. Weisblat, D. A., Babior, B. M. 1971. *J. Biol. Chem.* 246:6064–71
14. Moore, K. W., Bachovchin, W. W., Gunter, J. B., Richards, J. H. 1979. *Biochemistry* 18:2776–82
15. Albery, W. J., Knowles, J. R. 1976. *Biochemistry* 15:5588–600
16. deJuan, E., Taylor, K. B. 1976. *Biochemistry* 15:2523–27
17. Bachovchin, W. W., Moore, K. W., Richards, J. H. 1978. *Biochemistry* 17:2218–24
18. Eagar, R. G. Jr., Bachovchin, W. W., Richards, J. H. 1975. *Biochemistry* 14: 5523–28
19. Olsson, B., Olsson, J., Pettersson, G. 1976. *Eur. J. Biochem.* 64:327–31
20. Walsh, C. 1979. *Enzymatic Reaction Mechanisms*, p. 110. San Francisco: Freeman
21. Abeles, R. H., Frisell, W. R. Mackenzie, C. G. 1960. *J. Biol. Chem.* 235:853–56. (Corrected 235:1544)
22. Simon, H., Palm, D. 1966. *Angew. Chem. Int. Ed. Engl.* 5:920–33
23. Dubrow, R., Pizer, L. I. 1977. *J. Biol. Chem.* 252:1539–51
24. Grimshaw, C. E., Cleland, W. W. 1980. *Biochemistry* 19:3153–57
25. O'Leary, M. H. 1978. *Transition States of Biochemical Processes*, ed. R. D. Grandour, R. L. Schowen, pp. 285–316. New York: Plenum
26. Dahlquist, F. W., Rand-Meir, T., Raftery, M. A. 1969. *Biochemistry* 8: 4214–21
27. Stark, G. R. 1971. *J. Biol. Chem.* 246:3064–68
28. Goiten, R. K., Chelsky, D., Parson, S. M. 1978. *J. Biol. Chem.* 253:2963–71
29. Bull, H. G., Ferraz, J. P., Chordes, E. H., Ribbi, A., Apitz-Castro, R. 1978. *J. Biol. Chem.* 253:5186–92

30. Wang, C. A., Trout, C. M., Calvo, K. C., Klapper, M., Wong, L. K. 1980. *J. Am. Chem. Soc.* 102:1221–23
31. Albery, W. J., Knowles, J. R. 1977. *J. Am. Chem. Soc.* 99:637–38
32. Schimerlik, M. I., Rife, J. E., Cleland, W. W. 1975. *Biochemistry* 14:5347–54
33. Blanchard, J. S., Cleland, W. W. 1980. *Biochemistry* 19:3543–50
34. Rife, J. E., Cleland, W. W. 1980. *Biochemistry* 19:2321–28
35. Abdel-Monem, M. M. 1975. *J. Med. Chem.* 18:427–30
36. Cheung, Y., Walsh, C. 1976. *Biochemistry* 15:3749–54
37. Hjelmeland, L. M., Aronow, L., Trudell, J. R. 1977. *Biochem. Biophys. Res. Commun.* 76:541–49
38. Schowen, R. L. 1977. See Ref. 1, pp. 64–99
39. Welsh, K. M., Creighton, D. J., Klinman, J. P. 1980. *Biochemistry* 19:2005–16
40. Klinman, J. P. 1976. *Biochemistry* 15:2018–26
41. Polgar, L. 1979. *Eur. J. Biochem.* 98:369–74
42a. Steiner, H., Jonsson, B., Lindskog, S. 1975. *Eur. J. Biochem.* 59:253–59
42b. Pocker, Y., Bjorkquist, D. W. 1977. *Biochemistry* 16:5698–5707
43. Thorslund, A., Lindskog, S. 1967. *Eur. J. Biochem.* 3:117–23
44. Verpoorte, J. A., Mehta, S., Edsall, J. T. 1967. *J. Biol. Chem.* 242:4221–29
45. Yamada, H., O'Leary, M. H. 1977. *J. Am. Chem. Soc.* 99:1660–61
46. Northrop, D. B. 1981. *J. Am. Chem. Soc.* In press
47. Hegazi, M. F., Borchardt, R. T., Schowen, R. L. 1979. *J. Am. Chem. Soc.* 101:4359–65
48. Jordon, F., Kuo, D. J., Monse, E. V. 1978. *J. Am. Chem. Soc.* 100:2872–78
49. O'Leary, M. H., Limburg, J. A. 1977. *Biochemistry* 16:1129–35
50. Northrop, D. B., Cleland, W. W. 1970. *Fed. Proc.* 19:408
51. O'Leary, M. H., Yapp, C. J. 1978. *Biochem. Biophys. Res. Commun.* 80:155–60
52. Morris, R. G., Saliman, G., Dunn, M. F. 1980. *Biochemistry* 19:725–31
53. Schmidt, J., Chen, J., DeTroglia, M., Minkel, D., McFarland, J. T. 1979. *J. Am. Chem. Soc.* 101:3634–40
54. Dworschack, R. T., Plapp, B. V. 1977. *Biochemistry* 16:2716–25
55. Schopfer, L. M., Massey, V. 1979. *J. Biol. Chem.* 254:10634–43
56. Jenkins, W. T., Harruff, R. C. 1979. *Arch. Biochem. Biophys.* 192:421–29
57. Hunkapiller, M. W., Forgac, M. D., Richards, J. H. 1976. *Biochemistry* 15:5581–88
58. Dunn, M. F., Bernard, S. A., Anderson, D., Copeland, A., Morris, R. G., Rogue, J. 1979. *Biochemistry* 18:2346–54
59. Santi, D. V., McHenry, C. S., Sommer, H. 1974. *Biochemistry* 13:471–81
60. Lewis, C. A. Jr., Wolfenden, R. 1977. *Biochemistry* 16:4886–90
61. Lewis, C. A. Jr., Wolfenden, R. 1977. *Biochemistry* 16:4890–95
62. Viola, R. E., Cleland, W. W. 1980. *Biochemistry* 19:1861–66
63. Stein, R. L., Romero, R., Bull, H. G., Cordes, E. H. 1978. *J. Am. Chem. Soc.* 100:6249–51
64. Krenitsky, T. A. 1967. *Mol. Pharmacol.* 3:526–36
65. Cook, P. F. 1978. *Fed. Proc.* 37:1605
66. Cook, P. F. 1979. *Fed. Proc.* 38:646
67. Cook, P. F. 1980. *Fed. Proc.* 39:1641

Ann. Rev. Biochem. 1981. 50:133–57

METABOLIC PATHWAYS IN ❖12074
PEROXISOMES AND GLYOXYSOMES

N. E. Tolbert

Department of Biochemistry, Michigan State University, East Lansing, Michigan 48824

CONTENTS

PERSPECTIVES AND SUMMARY

Peroxisomes and glyoxysomes or microbodies from plants have been reviewed in the *Annual Review of Plant Physiology* (1, 2), but in the *Annual Review of Biochemistry* their general metabolic properties have never been reported. Microbodies are present in most eucaryotic cells and have many properties in common regardless of the tissue. The reviews by De Duve (3) and De Duve & Baudhhin (4) in 1966 and 1969 about the properties of microbodies from mammalian liver and kidney describe the belated discov-

133

ery of a major subcellular organelle, and should be examined by all interested in peroxisomes. In addition, other historical reviews are now available (5–8). Summaries are available on microbodies from liver (6, 9), plants (1, 2, 10–13), germinating seed endosperm (2, 14), algae (15), protozoa (16), and yeast (17, 18) and other fungi (19).

Microbodies are a subcellular respiratory organelle for catabolic pathways usually characterized by the presence of flavin oxidases and catalase. They are thought to contain no energy-coupled electron transport system or nucleic acid in spite of some reports to the contrary; rather they are formed by budding from the smooth endoplasmic reticulum (ER) (20). The metabolic pathways in microbodies are variable depending on tissue, substrate availability, and stage of development. It is estimated that they contain 1–1.5% of the cellular protein in liver (3) and in leaves of C_3 plants (1), and that there are about 1000 microbodies per hepatocyte (21). Microbodies are ~ 0.5–1.5 μm in diameter and have a single, boundary lipid bilayer membrane surrounding a granular matrix of soluble proteins. They have no inner membranes but may have a dense core or crystalloid structure of protein, and appear to be permeable to most small substrates, although shuttles into and out of the organelle have been proposed.

Various laboratories use the terms microbodies, peroxisomes, or glyoxysomes for the same organelle (8). Microbody is used when describing morphological appearance and as a general term, particularly if the biochemical properties of the tissue-specific organelle are unknown. Peroxisome is usually used if the metabolic pathway or the flavin oxidase and catalase content have been characterized, as in liver or leaf peroxisomes. When the microbody contains any part of the glyoxylate cycle, it is called a glyoxysome by some laboratories. Microbody-like organelles, such as glycerol phosphate oxidase bodies and hydrogenosomes in protozoa (17), are also being investigated.

Although microbody respiration can be a significant part of total respiration, and is different from mitochondrial or ER respiration, in mammals there is no physiological terminology for it. In plants, photorespiration, which occurs partly in leaf peroxisomes, is a manifestation of peroxisomal metabolism. During mitochondrial respiration part of the energy is conserved in ATP synthesis, whereas in the peroxisomes the energy released at the flavin oxidase plus catalase step is lost as heat. Energy from other oxidative reactions catalyzed by dehydrogenases in the peroxisomes is conserved through shuttles linked to the reduced nucleotide.

Characterization of Peroxisomes

MORPHOLOGY   Many morphological descriptions of microbodies from various tissues have been published (3–5, 9, 22, 23) and their distribution catalogued (5). They are probably present in all eucaryotic cells and are as widespread as catalase. Since microbodies are characterized by catalase, their cytochemical detection is based on peroxidation of diaminobenzidine (DAB) by added H_2O_2 to its conjugated unsaturated structure, which binds osmium tetraoxide (5, 22, 23). There is also a cytochemical test for peroxisomal α-hydroxy acid oxidase (24). The core or nucleoid in leaf peroxisomes gives a dense DAB reaction for catalase, but catalase is also in the matrix. The core in rat liver and kidney peroxisomes is enriched in urate oxidase (4, 5, 9). Otherwise microbodies usually appear structureless, although branching filaments and plates have been observed in some.

Small microbodies, called microperoxisomes ($\sim$0.1 μm diameter) are found in a variety of animal cell types (25–27). They give the cytochemical test with DAB for catalase, do not have cores, and show frequent continuities of their delimiting single membrane with that of the ER. Since they are abundant in tissues engaged in lipid metabolism, e.g. heart (25), it has been speculated that the microperoxisomes, like the hepatic peroxisomes, are involved in β oxidation.

ISOLATION AND ASSAY   Microbodies were first studied by De Duve's group as a component of the "light-mitochondrial" fraction. The development of buoyant density centrifugation in zonal rotors or large tubes made possible procedures for isolating the microbody fraction in sucrose gradients with minimal contamination by other organelles, and has permitted their enzymatic characterization (28–32). In general, catalase and urate oxidase are used as markers for microbodies; glutamate dehydrogenase, succinate dehydrogenase, or cytochrome c oxidase are used for mitochondria; acid phosphatase is used for lysosomes; and NADH:cytochrome c reductase is used for the ER. Microbodies equilibrate at $\sim$1.24 g/cm^3 (a relatively high density) on sucrose gradients because of their high protein content, their low lipid content from the single outer membrane, and their rapid loss of water to the sucrose gradient. The specific activities of microbody enzymes are increased 50- to 100-fold by isolation of the organelle from the homogenate.

The isolated microbodies may be broken by osmotic shock or by overnight incubation in pyrophosphate buffer (28). Upon recentrifugation in a sucrose gradient the components may be grossly separated into the soluble matrix, which contains most of the enzymes [e.g. catalase and carnitine acyltransferases (33)]; the membrane fraction, which contains NADH:cytochrome c reductase (34); and the core, which contains urate oxidase.

Development of Microbodies

Microbodies are formed by budding from the smooth ER (5, 20), to which they may remain attached (35), and as a consequence the enzymatic composition of new and old microbodies may interchange via the ER channels (20). The half-life of hepatic peroxisomal catalase is only 1.5 days (3). Microbodies are formed during tissue development and differentiation (9), and their appearance and activity are substrate dependent. For example, they are nearly absent in poorly differentiated hepatomas such as Morris 3683. During postnatal development in mouse or rat liver, catalase and peroxisomal enzymes for β oxidation develop to maximum activity during the first two weeks, while urate oxidase increases more slowly over a 4-week period (36). Glyoxysomes develop during seed germination and leaf peroxisomes develop during greening of the etiolated leaf in the light (1, 2). The enzymatic content of the microbody may change during a developmental sequence, such as when an etiolated cotyledon with glyoxysomal enzymes is exposed to light and develops into a cotyledonary leaf with enzymes for glycolate oxidation.

Hepatic catalase has been extensively investigated in developmental studies (9, 20, 37). According to De Duve (20) and De Duve et al (37) the apomonomer is formed with a half-life of 14 min at a ribosomal site; addition of heme and tetramerization then occur in the peroxisomes. After synthesis of a microbody enzyme (catalase) on the ribosome, the protein may move through the ER channel or through the cytoplasm to the microbody. However Goldman & Blobel (38) report that hepatic catalase and uricase arise from translation products directed by free polysomes, and they predict a post-translational transfer from the cytoplasm into the peroxisomes. In mice, five isoenzymes of catalase that differ in their content of sialic acid are the products of a single structural gene (9). The peroxisomal catalase is the most highly sialated.

The hypolipidemic drug, ethyl p-chlorophenoxyisobutyrate (Clofibrate), or its many structural analogues, induce hepatomegaly and two- to three-fold proliferation of peroxisomes per liver volume (39–41). Di-(2-ethylhexyl)-phthalate (DEHP), a mildly hypolipidemic plasticizer, also induces hepatic peroxisomes (42). In relatively large dosage (2% wt/wt in the feed daily for one to two weeks) these agents cause a tenfold increase in the total activity, as well as an increase in the specific activity of peroxisomal enzymes associated with β oxidation in livers from male rats (33, 42–46), including carnitine transferases (43, 47) and NAD : glycerol-P dehydrogenase (R. Gee, unpublished). Although Clofibrate treatment increases total catalase, its specific activity does not increase, while that of urate oxidase decreases (43). However, numerous effects of Clofibrate on metabolism have been reported and its effect on lipid turnover is not restricted to the peroxisomes (48). There is no direct evidence for hormonal regulation of peroxisomes, yet there are differences in the amount of peroxisomal enzymes in male and female rats and in the effect of Clofibrate on such enzymes.

GENERAL METABOLIC PROPERTIES

Enzymes in Peroxisomes

Table 1 lists 40 enzymes that have been reported in microbodies from different tissues. Microbodies from a single tissue never contain all of these activities. Peroxisomes are characterized by having a flavin oxidase that forms H_2O_2 and excess catalase to remove it. The universal presence of catalase is the basis of the cytochemical stain for microbodies, but one exception is a glyoxysome in which only the glyoxylate cycle is present, without an oxidase or catalase (19). Recent reviews on catalase emphasize its role in peroxisomal metabolism (9, 20, 37, 49, 50). The various terminal oxidases have as their prosthetic group either FMN (e.g. glycolate oxidase (19, 51, 52) or FAD (e.g. fatty acyl-CoA oxidase) (45, 53–55). Microbody enzymes, particularly the oxidases, have pH optima around 8.5, in comparison with lower pH profiles in other subcellular compartments. The combination of oxidase and catalase may direct carbon metabolism toward catabolism because of the negative ΔG of the reaction. There is little evidence that the oxidase activities are regulated, but since they are of low activity they may be rate limiting. These flavin oxidases have K_m values of $\sim 10^{-4}$ M for the substrate, and the fatty acyl-CoA oxidase has a high affinity for O_2. Glycolate oxidase has a low affinity for O_2, but this does not seem limiting since glycine rather than glycolate accumulates during photorespiration.

The flavin oxidases and the associated metabolic pathway in microbodies are variable and tissue dependent, in contrast to the relative constancy of mitochondrial metabolism. Thus liver peroxisomes have some enzymes that are different from kidney peroxisomes, and the metabolic pathway in leaf peroxisomes is different from that in seed glyoxysomes. In spite of these significant variations, the physical appearance of microbodies from all sources is similar.

Variable enzymatic composition of microbodies depends not only on the tissue, but also on substrate, stage of development, and chemical agents. Examples of substrate induction of microbodies are numerous. In yeast grown on a sugar, microbodies are repressed and respiration is mainly mitochondrial. Yeast grown on a percursor substrate for glyoxylate (e.g. malate) develop glyoxysomes, while those grown on methanol develop gigantic microbodies that fill the cell and contain an alcohol oxidase (17, 18). Yeast (17), *Euglena* (56), and fungi (19) grown on fatty acids or alkanes contain numerous microbodies with the β oxidation pathway.

Great variability in enzyme composition of microbodies indicate that they will eventually have to be characterized from each tissue. Only hepatic peroxisomes have been examined in detail in animals. The assumption that

Table 1 Enzymes reported in microbodies from five tissues[a]

	Liver	Kidney	Leaves	Germinating seeds	Yeast
Catalase	9, 20, 37, 49, 50[a]	3, 4, 9	1	2	17, 19
Oxidases					
Glycolate (α-hydroxy acid)	4, 5, 9, 52	3, 4, 9, 52	1, 108	1, 2	17–19, 99
Fatty acyl-CoA	45, 54, 69, 82	[b]	none	53	72
Urate	3, 4, 9	3, 4, 9	124	124	
D-Amino acid	3, 4, 9	3, 4, 9	trace	trace	
L-Amino acid	none[c]	9			
Alcohol (methanol)					17, 18, 130–132
Amine					129
Polyamine	128				
Dehydrogenases					
NAD: L-3-hydroxy fatty acyl CoA	84, 85	none	none	86	17–19
NAD: malate	none	none	1, 57, 58	2, 60	
NAD: Glycerate	none	none	1, 123	1, 2	
NAD: glycerol-P	61, 62	none	none	none	64
NADP: isocitrate	28				
Xanthine	125				
Other enzymes for fatty acid oxidation					
Fatty acyl CoA synthetase	78, 79		none	77	80
Acetyl CoA synthetase			none	77	
Enoylhydratase	84, 85	none	none	86	
Thiolase	88	none	none	53	17, 18
Acyl transferases					
Carnitine acetyl CoA	33, 43, 75	none	none	none	98
Carnitine octanoyl CoA	43, 76	none	none	none	
Acyl CoA: dihydroxyacetone-P	100				
Aminotransferases					
Glutamate-glyoxylate			118	2	
Serine-glyoxylate			118	2	
Glutamate-oxalacetate			118		
Alanine-glyoxylate	120				
Serine-pyruvate	120				
Leucine-glyoxylate	122				
Other enzymes of glyoxylate cycle					
Isocitrate lyase			none	2, 53	17–19
Malate synthase			none	2, 53	17–19
Citrate synthase			none	2, 103	
Aconitase			none	2, 53	
Membrane					
NADH: cytochrome c reductase	34		34	2, 34	
Lipase			none	81	
Other enzymes of ureide metabolism					
Allantoinase	125, 126				
Allantoicase	125, 126				

Other enzymes in protozoa microbodies (16) are glycerol-P oxidase, hydrogenase, and pyruvate synthetase. The following enzymes have been reported in microbodies but the results are either not confirmed or refuted: polyphenol oxidase, phenylalanine ammonia lyase, lactate dehydrogenase, neutral protease, and nitrate reductase.

[a] Numbers refer to references.
[b] No reference means enzyme not reported in this tissue.
[c] "None" means enzyme has been cited as absent.

β oxidation will be in peroxisomes from other tissues such as kidney, intestinal mucosa, or heart may be incorrect. β oxidation is present in the seed glyoxysomes but not in the leaf peroxisomes of the same plant. Leaf peroxisomes from all C_3 plants have similar enzymes for the dominant glycolate pathway, but peroxisomes in the mesophyll cells of C_4 plants appear to be different. Glyoxysomes from various tissues may contain only isocitrate lyase and malate synthase or all the enzymes of the glyoxylate cycle. In many tissues the enzymes and function of microbodies remain unknown, since only the presence of catalase has been noted by the DAB cytological test. In various plant tissues (tubers, roots), the partially purified microbody fraction contains catalase and a trace of glycolate oxidase, but other known enzymes of microbodies have not been reported.

The activities of microbody enzymes are often duplicated in other cellular compartments. To prove the presence of a specific peroxisomal enzyme it is necessary to first isolate it from the organelle fraction, and then to show that it is either isoenzymic with or different from the enzyme that shows the same activity elsewhere in the cell. In leaves there are three isoenzymic NAD:malic dehydrogenases (57, 58); the peroxisomal and mitochondrial isoenzymes are of about equal activity, while the cytoplasmic form is of lower total activity. In addition there is an NADP:malic dehydrogenase in the chloroplast. In liver peroxisomes, NAD:glycerol-P dehydrogenase is isoenzymic with the more abundant cytosolic form, whereas the enzymes for β oxidation in the hepatic peroxisomes are different from, rather than isoenzymic with, those in the mitochondria. It is not known in such cases whether the same genetic components are involved in formation of similar activities in different compartments.

Microbody enzymes from different tissues may also differ in substrate specificity. For example, α-hydroxy acid oxidase from hepatic or leaf peroxisomes oxidizes best glycolate and α-hydroxy isocaproate, whereas that from kidney peroxisomes oxidizes only the long-chain, α-hydroxy acids (9, 52).

Most metabolic pathways in microbodies are alternates to those elsewhere in the cell. For example, the tricarboxylic acid cycle in mitochondria, and the glyoxylate cycle partially in glyoxysomes, represent alternate pathways for acetyl-CoA utilization. The β oxidation pathway of mitochondria and peroxisomes both form acetyl-CoA. Ureide metabolism in microbodies represents only one of the pathways for nitrogen metabolism. Serine can be synthesized in leaf peroxisomes from glycerate, or in other plant tissues from P-glycerate, by a pathway not reported in the microbodies. Another possible example of duplication of metabolism by the microbodies is provided by the possibility of peroxidative oxidation of such substrates as ethanol, methanol, and formate by peroxisomal catalase when H_2O_2 is present. Perfusion experiments with liver show that addition of peroxisomal substrates, such as glycolate or urate, to generate H_2O_2 stimulates turnover of the catalase·H_2O_2 complex (49, 59), but such experiments do not indicate whether this complex is being used to oxidize another substrate.

The reaction catalyzed by a microbody oxidase is sometimes catalyzed by a dehydrogenase elsewhere in the cell, so that part of the energy may be conserved as a reduced pyridine nucleotide. For example, lactate may be oxidized by lactate dehydrogenase in the cytoplasm or by α-hydroxy acid oxidase in the microbody. The relative contribution of each site to the overall oxidation has not been determined, but the activity of LDH exceeds that of the peroxisomal oxidase by many fold. Although a glycolate oxidase is in peroxisomes from all green plants and multicellular algae, a glycolate dehydrogenase, probably linked to a cytochrome b, is in the mitochondria or chloroplasts of unicellular green or blue-green algae (15). In the latter there are few if any peroxisomes and little catalase.

Most reactions involving glyoxylate are located in the microbody. This is true for the two reactions of the glyoxylate cycle, for the glycolate pathway in photorespiration, for microbody aminotransferases from all sources, and for glyoxylate formation during ureide metabolism. Possible reasons for such a compartmentation are that any glyoxylate in the cytoplasm would be instantaneously converted to glycolate and oxalate by excess LDH or oxidized nonenzymatically by a trace of H_2O_2 to CO_2 and formate.

Latency, Shuttles, and Outer Membrane

No translocase in the microbody membrane has so far been reported. The ER membrane, the outer membrane of the nucleus and mitochondria, and the single microbody membrane all may have a common origin and are relatively permeable to small molecules. Therefore no latency might be expected during assays with isolated microbodies, but assays involving larger factors such as NAD and palmitate have exhibited latency. Experiments with intact microbodies are difficult, since they may be broken during isolation or upon dilution for assaying. Assays in dense sucrose to retain intact particles are diffusion limited. Because microbodies contain enzymes associated with organelle shuttles, and microbody activity is greater than expected for diffusion of NAD or other cofactors, exchange of organic and amino acids have been proposed between the inside of the microbody and the cytoplasm (60). The transport of these carrier acids is considered to be by passive diffusion. Such shuttles for export of reducing equivalents are needed to oxidize NADH formed in microbodies, as NADH oxidase has not been detected.

Hepatic peroxisomes from rat or mice contain a small part (5–10%) of the total NAD:glycerol-P dehydrogenase, which is isoenzymic with the large cytoplasmic pool of this enzyme [(61, 67) R. Gee, unpublished]. This dehydrogenase contains an unidentified peptide factor (with covalently bound FAD) which can be removed from the enzyme by heat and assayed by its stimulation of the peroxisomal glycerol-P dehydrogenase (63). A peroxisomal glycerol-P shuttle linked to NAD reduction is proposed in Figure 1 to shuttle the reducing equivalent out of the peroxisomes as glycerol-P. This shuttle may be continuous with the glycerol-P shuttle between the cytoplasm and the mitochondria, which functions to reoxidize the glycerol-P. A glycerol-P dehydrogenase present in yeast peroxisomes (64) changes in activity in proportion to the β oxidation activity (36).

In plant microbodies the malate-aspartate shuttle is present (60, 60a), and the glycerol-P shuttle is absent. We have found no evidence for a malate-aspartate shuttle with liver or kidney peroxisomes. Leaf peroxisomes and seed glyoxysomes contain large amounts of a microbody-specific NAD: malate dehydrogenase isoenzyme (1, 2, 57, 58). The activity of malate dehydrogenase in leaf peroxisomes (50–100 μmol per minute per milligram of protein) makes this the most active enzyme in the organelle excepting catalase. Low activities of three isoenzymes of aspartate: α-ketoglutarate aminotransferase are also found in leaf peroxisomes (65). The proposed malate shuttle in Figure 3 is modeled after similar malate-aspartate shuttles from the mitochondria. Malate dehydrogenase and aspartate: α-ketoglutarate aminotransferase are present in glyoxysomes from germinating seeds (60a). A malate-aspartate shuttle (Figure 2) has been proposed to oxidize NADH formed during β oxidation (60a).

The mechanism of initiating fatty acid oxidation and exiting of carnitine acyl products are discussed under β oxidation. The carnitine acetyl- and octanyltransferases are in the peroxisomal matrix and the carnitine derivatives are presumed to diffuse from the peroxisome.

The single membrane of the microbody membrane is similar in composition to the ER, from which it arises by budding during development of microbodies. It contains phosphatidylcholine, phosphatidylethanolamine, phosphatidylinositol (2), and antimycin A–insensitive NADH: cytochrome c reductase (34). Some enzymes apparently associate with microbody ghosts by ionic forces after osmotic shock, but are not thought to be intrinsic membrane components (2). Studies of the small available amounts of microbody membrane are limited by a significant amount of ER contamination relative to the small amount of microbody membrane.

METABOLIC PATHWAYS

Fatty Acid β Oxidation

The β oxidation pathway for fatty acyl-CoA was first described in glyoxysomes from germinating castor bean endosperm (53), then in peroxisomes from *Euglena* (56) and *Tetrahymena* (66, 67), and later in rat liver (45, 68–70), mouse liver (62), yeast (71, 72), and human liver (73). A peroxisomal fatty acid β oxidation pathway thus seems well established as an important general function of microbodies. However, β oxidation has not been reported in microbodies from all tissues (i.e. kidney or leaf peroxisomes). Based on maximum enzyme activity, the hepatic peroxisomal β oxidation capacity in mice is about the same as that in the mitochondrial system (62), but in rats only about 25–33% of the total activity is peroxisomal (74). All of the β oxidation is in the glyoxysomes from germinating seed endosperm (2) or *Euglena,* and in yeast grown on a fatty acid. The same intermediates are formed during β oxidation in both the peroxisomal and mitochondrial pathway, but the enzymes are different. There seem to be two β oxidation systems: in mitochondria and *Clostridium* individual enzymes are composed of multisubunits; whereas in peroxisomes and *E. coli* multienzyme complexes with some bifunctional proteins may exist.

Hepatic peroxisomal and mitochondrial fatty acid β oxidation differ in many ways including location, enzymes, end products, and control (Figure 1):

(*a*) The two pathways are in different organelles. Equilibrium sucrose density gradients of homogenates clearly show nearly complete separation of peroxisomes and mitochondria, and fatty acid β oxidation is present in

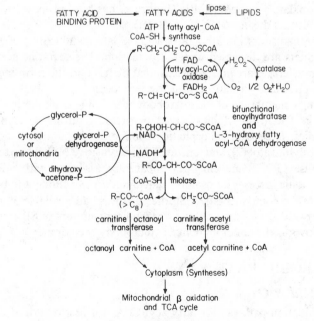

Figure 1   Fatty acid β oxidation in liver peroxisomes.

both fractions. Each enzyme of the β oxidation complex has been partially purified from isolated peroxisomes from liver, castor bean cotyledons, and yeast.

(b) The initiation of liver peroxisomal β oxidation is carnitine independent, whereas mitochondrial β oxidation may be carnitine dependent for long-chain acids. Long-chain fatty acyl-CoA derivatives (palmitoyl and oleoyl) are substrates for the fatty acyl-CoA oxidase, but palmitoylcarnitine is not utilized by isolated peroxisomes, and in contrast to mitochondria, little if any carnitine palmitoyltransferase has been detected in peroxisomes (33, 75). A protein-palmitate complex reportedly sustains peroxisomal β oxidation with ATP (76). Microbodies from seed endosperm (77), liver (78, 79), and yeast (80) contain an active fatty acyl-CoA synthetase that forms the substrate for the oxidase. Initiation of fatty acid metabolism in germinating seeds is by a lipase in the glyoxysomal membrane (81).

(c) Instead of the FAD-linked acyl-CoA dehydrogenase of mitochondria, liver (45, 54, 69, 82) and yeast peroxisomes (18, 83), and seed glyoxysomes (53) contain an FAD-linked acyl-CoA oxidase that transfers the electrons to O_2 to form H_2O_2, which is decomposed in turn by catalase. In the presence of inhibitors of mitochondrial electron transport, one atom of O_2 per turn of the cycle is taken up during peroxisomal β oxidation, as measured by the initial rate of O_2 uptake or NAD reduction upon adding palmitoyl-CoA. This rate of O_2 uptake is increased (53, 54) and the amount doubled if enough cyanide is added to completely inhibit catalase. In the presence of a second substrate that can act as a hydrogen donor for the catalase$\cdot H_2O_2$ complex I, the oxygen uptake is two atoms per turn of the cycle. This situation exists when ethanol is present in the peroxisomal preparation to protect the catalase (36) or if the peroxidation of [^{14}C]-HCOOH to CO_2 or [^{14}C]CH_3OH to nonvolatile products is used as an assay procedure (28).

(d) In rat liver peroxisomes (84, 85) or seed glyoxysomes (86) there is a bifunctional enzyme with enoylhydratase and L-3-hydroxy fatty acyl-CoA dehydrogenase activities that differs from the two separate mitochondrial enzymes. An increase in the amount of this protein, of 80,000 daltons, has been observed during Clofibrate treatment (87). The peroxisomal dehydrogenase is linked to NAD reduction and can be measured in the presence of mitochondrial inhibitors such as CN, rotenone, or antimycin A, which prevent reoxidation of NADH by contaminating mitochondria.

(e) Peroxisomal thiolase is different from the two thiolases in mitochondria and the one in cytoplasm (88), and it is readily separated from them because it is not adsorbed to phosphocellulose. It requires the presence of 1 mM dithiothreitol (DTT) during its assay, whereas no DTT is required for assay of the mitochondrial activity (J. Krahling and N. E. Tolbert, in press). Peroxisomal thiolase does not cleave β-ketobutyrate.

(*f*) Peroxisomal β oxidation is most active with long-chain (C_{10}–C_{22}) fatty acyl CoA substrates (42, 45, 46, 54, 89); maximum activity occurs with the C_{14} acyl-CoA. Peroxisomes oxidize mono-unsaturated fatty acids, in particular erucic acid [C_{22}:1 (9)]. Partially purified peroxisomal acyl-CoA synthetase from seed (53) or liver (78, 79), fatty acyl-CoA oxidase (75), and thiolase [(88), J. Krahling and N. E. Tolbert, unpublished] require a chain length of C_8 or longer, and have only minimal activity with shorter-chain substrates. It appears that hepatic peroxisomal β oxidation is specific for long-chain fatty acyl CoAs, whereas mitochondrial β oxidation utilizes both long- and short-chain substrates. Consistent with the long-chain specificity and substrate induction of peroxisomal enzymes, feeding male rats high fat diets increased hepatic peroxisomal β oxidation (90, 91) and feeding partially hydrogenated marine oil increased erucic acid metabolism and catalase in perfused heart (92, 93) and hepatocytes (94).

(*g*) Isolated rat hepatic peroxisomes contain two carnitine acyltransferases in the organelle matrix (33, 43, 75, 95), carnitine acetyltransferase and a transferase for medium-length chains such as octanyl carnitine. The specific activities of these enzymes in the peroxisomes exceed those in the mitochondria. The peroxisomal and mitochondrial transferases are similar, but whether they are isoenzymic is not known. Their substrate specificity suggests that the C_2 and C_8 acyl-CoA products of hepatic peroxisomal β oxidation are converted to the corresponding carnitine derivatives and then passively diffuse out of the peroxisomes into the cytoplasm. Thus peroxisomal β oxidation may provide acetyl units for other cellular syntheses, or the acyl carnitine products may be further oxidized in the mitochondria. Peroxisomal β oxidation can not be said to be carnitine independent, since the end products appear to be acetylcarnitine or octanoylcarnitine. There is no acyl-CoA hydrolase in the peroxisome.

Regulation of peroxisomal β oxidation and the division of β oxidation between mitochondria and peroxisomes require further investigations. Malonyl-CoA may regulate mitochondrial β oxidation by inhibiting palmitoylcarnitine acyl-CoA transferase (96). Mitochondrial β oxidation in vitro is increased by added albumin, but peroxisomal β oxidation is maximum in the absence of albumin (97). That thiolase activity in isolated peroxisomes is nearly undetectable in the absence of 1 mM DTT suggests regulation by a sulfhydryl requirement. The fatty acyl-CoA oxidase from yeast also requires sulfhydryl activation. There is little evidence for regulation of any microbody flavin oxidase other than by substrate availability, although the oxidase catalyzes an irreversible step.

During β oxidation by hepatic peroxisomes the energy from fatty acyl-CoA oxidase is lost, but would be conserved as 2 ATP if the oxidation occurred in the mitochondria. If the products of palmitate oxidation in peroxisomes were 1 octanoyl-CoA, 4 acetyl-CoA, and 4 NADH and these were further oxidized in the mitochondria, 122 ATP would be generated as compared to 130 ATP if the complete oxidation to CO_2 had occurred solely in mitochondria. Thus the initiation of palmitate oxidation in peroxisomes may result in the loss of about 6% of the energy expected from total mitochondrial β oxidation. This modest loss is presumably the price paid for the presently unknown advantage of peroxisomal β oxidation. Though small, this loss could impose a considerable change in the overall metabolic balance regulating body weight. When this was considered experimentally, hepatic peroxisomes from obese mice were found to have the capacity for about three times more β oxidation of palmitoyl-CoA per milligram of liver protein, than did peroxisomes from their lean litter mates (62). The obesity was certainly not due to the absence of peroxisomal β oxidation, and the increase in activity was viewed as an effort to dispose of the excess fatty acids. Peroxisomal β oxidation provides a pathway not limited by electron transport for production of acyl-CoA, especially acetyl-CoA, and reducing equivalents for use elsewhere in the cell.

The yeast, *Candida tropicalis*, when transferred for 8 hr to an *n*-alkane medium or when grown on a fatty acid, develops microbodies with catalase (18, 99), palmitoyl-CoA oxidase (72), and other enzymes associated with β oxidation such as carnitine acetyltransferase (98). The initial steps of alkane oxidation occur in the microsomes (99) and β oxidation occurs in the peroxisomes. The substrate induction of the β oxidation system is characteristic of microbodies. The yeast peroxisomal fatty acyl-CoA oxidase has been crystallized and characterized (72). It contains FAD and forms H_2O_2. Thus, these yeast cells contain a microbody with a β oxidation system similar to that in rat liver peroxisomes; in addition the yeast peroxisomes contain isocitrate lyase and malate synthetase similar to microbodies from germinating seed, *Euglena*, and *Tetrahymena*.

The significance of sn-glycerol-3-P acyltransferase or dihydroxyacetone-P acyltransferase in guinea pig liver peroxisomes (100) is not clear. When yeast were grown on oleic acid, microbodies capable of β oxidation developed, but the glycerol-P acyltransferase was in the microsomes and mitochondria (80). This transferase is thought to be involved in lipid synthesis.

β Oxidation in glyoxysomes from endosperm or cotyledons of germinating fatty seeds differs from that in liver peroxisomes as to the mechanism of the entry of fatty acids and the utilization of the acetyl-CoA formed (Figure 2). The glyoxysomes in the plant cells are closely associated with the lipid bodies (22); they develop during germination, reach maximum activity on the third to fifth day, and disappear as the lipid bodies are used up (14, 101). During these times there is no detectable mitochondrial β oxidation. The bounding outer glyoxysomal membrane contains lipase activity (81), and the fatty acyl-CoA synthetase is located in the matrix. The cofactor for the fatty acyl-CoA oxidase is FAD [(53), J. Uhlig and N. E. Tolbert, unpublished]. A bifunctional enzyme contains the enoylhydratase and the hydroxy fatty acyl-CoA dehydrogenase activity (86). The thiolase cleaves β-keto fatty acyl-CoA substrates of all chain lengths. The sole products of glyoxysomal β oxidation appear to be acetyl-CoA and NADH, and the fatty acyl-CoA oxidase system oxidizes all chain lengths as well as ricinoleic acid (101). There is no carnitine dependency either to initiate or terminate β oxidation in the glyoxysomes. Acetyl-CoA from β oxidation is used directly by the glyoxylate cycle in the same glyoxysomes to form succinate. Each turn of the glyoxylate cycle generates one extra C_4 acid, which is converted to P-enolpyruvate for gluconeogenesis elsewhere in the cell. This tight coupling between β oxidation and the glyoxylate cycle in the glyoxysome accounts for a very high conservation of carbon in converting lipid into carbohydrate during seed germination (14).

Glyoxylate Cycle

In his initial studies of glyoxylate cycle activity in microbodies ("peroxisomes") of *Tetrahymena,* Hogg (102) found the two unique glyoxylate bypass enzymes, isocitrate lyase and malate synthase. The other enzymes common to both the TCA and glyoxylate cycle were present only in the mitochondria. Microbodies (glyoxysomes) from germinating castor beans contain in addition citrate synthase (103) and a small amount of aconitase (53), needed for a glyoxylate pathway to produce succinate from 2 acetyl-CoA (Figure 2).

The location of the enzymes common to both the TCA and the glyoxylate cycle has not been extensively studied in other germinating seeds. Glyoxysomes or peroxisomes from yeast (17–19), *Tetrahymena,* (7) *Euglena* (56), and various fungi (19) have been reported to contain isocitrate lyase and malate synthase, but not the other enzymes of the glyoxylate cycle. The scheme in Figure 2 after Mettler & Beevers (60a) for seed glyoxysomes and Hogg (102) for Tetrahymena peroxisomes indicates that succinate produced by isocitrate lyase is metabolized in the mitochondria. The malate formed by malate synthase from glyoxylate and acetyl CoA also diffuses from the glyoxysomes to the malate pools in the cytoplasm and mitochondria. Since these reactions and pools are distributed among the glyoxysomes, mitochondria, and cytosol, the term glyoxylate pathway is used in Figure 2 rather than glyoxylate cycle, which might imply a complete cycle within one organelle. Biological systems utilizing acetyl units for growth or gluconeogenesis generally have microbodies with isocitrate lyase and malate synthetase. *Turbatrix aceti,* in which these two enzymes are present in

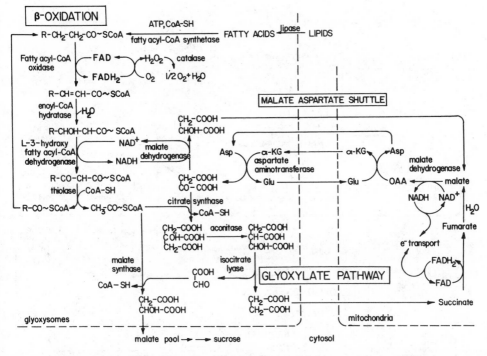

Figure 2 Fatty acid β oxidation and the glyoxylate pathway in germinating castor bean endosperm.

the mitochondria (104), is an exception for which separate acetyl-CoA pools are proposed (105). This worm, which belongs to an early limited evolutionary branch, does have catalase-containing microbodies. Animals are thought not to have the glyoxylate cycle in their peroxisomes, but a recent report (106) notes the possibility of this in fetal guinea pig liver.

Glycolate Pathway

Leaf peroxisomal metabolism plays a major role in the conversion of glycolate to glycine in the oxidative photosynthetic carbon cycle of photorespiration (1, 11, 107) (Figure 3). Photorespiration is the uptake of O_2 and the formation of CO_2 in light resulting from glycolate biosynthesis in chloroplasts and subsequent glycolate metabolism in peroxisomes and mitochondria. In C_3 plants, photorespiration may occur at 25–50% of the rate of photosynthesis. The total cyclic process for photorespiration is thought of as an oxidative photosynthetic carbon cycle to emphasize that it is composed of several energy-consuming portions: (*a*) ribulose-P_2 oxidation by ribulose-P_2 carboxylase/oxygenase to P-glycolate and P-glycerate, fol-

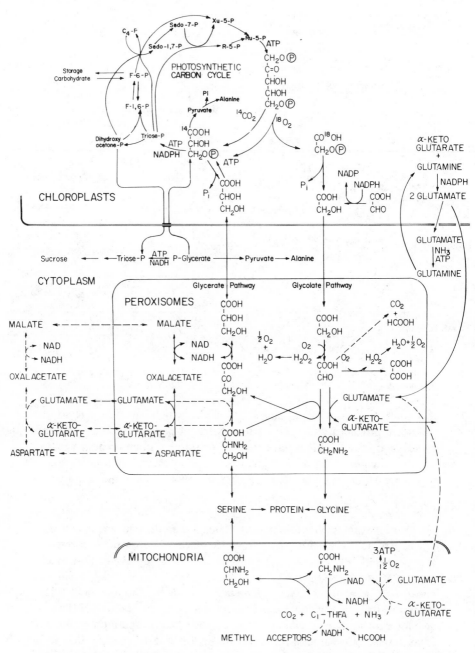

Figure 3   The oxidative photosynthetic carbon cycle of photorespiration associated with leaf peroxisomes.

lowed by P-glycolate hydrolysis by a specific P-glycolate phosphatase in the chloroplast; (*b*) glycolate oxidation to glyoxylate and conversion to glycine in the peroxisomes; (*c*) glycine oxidation to CO_2 and a C_1-tetrahydrofolic acid (C_1-THFA) derivative in the mitochondria followed by serine formation from this C_1 with another glycine; (*d*) serine conversion to glycerate in the peroxisomes; and (*e*) glycerate phosphorylation combined with the photosynthetic carbon cycle in the chloroplast to regenerate the ribulose-P_2. This cycle accomplishes the oxidation of newly formed sugar phosphates from photosynthesis or from starch degradation in the chloroplasts, to CO_2.

The source of glycolate for leaf peroxisomal oxidation has been established by [^{18}O]-O_2 and [^{14}C]CO_2 labeling in whole plant experiments. Carbon atoms C_1 and C_2 of ribulose-P_2 (Figure 3) are oxidized by ribulose-P_2 carboxylase/oxygenase to P-glycolate. The dual activity of this major chloroplast protein (50% of the soluble protein) establishes a competition between the two gaseous substrates, CO_2 and O_2, as to whether ribulose-P_2 is carboxylated in the reductive photosynthetic carbon cycle or oxygenated in the oxidative photosynthetic carbon cycle. In all photosynthetic tisues this enzyme has both activities, and it represents a branch point between photorespiration and photosynthesis for which there has been no confirmed proof for differential regulation other than by substrate competition. The P-glycolate phosphatase of chloroplasts is very active in all photosynthetic tissues examined, and the glycolate is excreted from the chloroplasts or the algae (15).

The two consecutive reactions, glycolate oxidation to glyoxylate and glyoxylate conversion to glycine, are present in leaf and liver peroxisomes, but not in all microbodies (e.g. kidney peroxisomes). Leaf peroxisomes from C_3 plants have a catalytic activity of about 1 μmol glycolate oxidase, 8000 μmol catalase, and 40 μmol of malate dehydrogenase per minute per milligram of peroxisomal protein (108, 109). In C_4 plants, glycolate oxidase is primarily in the peroxisomes of the bundle sheath cells (110), and only a few peroxisomes are in the mesophyll cells (22). This FMN-linked oxidase has been called glycolate oxidase, particularly in plants where there is a large flux through the glycolate pathway, or α-hydroxy acid oxidase in animals, as its substrate in vivo is uncertain. α-Hydroxy acid oxidase in mammalian peroxisomes exists in two forms (9, 52). In rat liver peroxisomes, it has a specific activity of 40–80 nmol per minute per milligram of protein, and like the leaf enzyme it oxidizes glycolate and lactate, is much less active with C_4 and C_5 α-hydroxy acids, but is as active with C_6 (α-hydroxyisocaproate) as with glycolate (9, 52). An oxidase in kidney peroxisomes oxidizes only the long-chain α-hydroxy acids, (9, 52); its function is unknown. The kidney, but not the liver enzyme also oxidizes L-amino acids, including L-tryptophan (111).

Glycolate oxidase from liver or leaf peroxisomes also oxidizes glyoxylate to oxalate (K_m 10^{-4} M), since glyoxylate in its hydrated form is structurally similar to glycolate. Thus peroxisomal activity may be a major source of oxalate, an undesirable end product of metabolism. Glycolate is not oxidized by LDH, but LDH dismutates two glyoxylates to one oxalate and one glycolate (112). The relative amounts of oxalate formed in plants depend both on the species and on the nitrogen status. In the presence of abundant nitrogen most of the glyoxylate is transaminated to glycine.

Perfusion of liver with glycolate results in the formation of the catalase·H_2O_2 complex (59). There is normally about 0.5 μg of glycolate per milliliter of human plasma and 10–20 μg glycolate per milliliter of urine (J. Vrbanac and C. Sweeley, unpublished). The glycolate in animals can not be from plant material, where it is very low due to its oxidation by leaf peroxisomes. Feeding [^{14}C]xylose or [^{14}C]xylitol to rats produces [^{14}C]glycolate (113). This result suggests that glycolate may be formed by oxidation of the C_2-thiamine pyrophosphate complex generated by transketolase. Perhaps related to glycolate oxidation in peroxisomes is P-glycolate phosphatase in human erythrocytes (114) and other tissues (115), in which P-glycolate seems to have a regulatory role on 2,3,-P-glycerate phosphatase (116). Pyruvate kinase has some glycolate kinase activity (117).

Peroxisomes from leaves and liver contain unique aminotransferases that preferentially form glycine by utilizing glyoxylate as the amino acceptor from specific amino donors. The peroxisomal aminotransferases using glyoxylate are physiologically irreversible but can be reversed by 0.1 M glycine (Y. Nakamura and N. E. Tolbert, unpublished). Leaf peroxisomes contain two different glyoxylate aminotransferases, with activities of about 1–2 μmol per minute per milligram peroxisomal protein (118). One is a glutamate:glyoxylate aminotransferase. In the glycolate pathway, during the conversion of two glycolates via two glycine to one serine, CO_2, and NH_3, the toxic ammonia must be refixed by glutamine and glutamate synthetases or by glutamate dehydrogenase (Figure 3). This nitrogen balance provides one glutamate for the transamination of one glyoxylate to glycine. The serine formed by the glycolate pathway is further converted to hydroxypyruvate by an irreversible serine:glyoxylate aminotransferase in the peroxisomes. The net yield from both transaminations is the conversion of two glyoxylates to one hydroxypyruvate and one CO_2, which is the balance depicted in Figure 3 for the total cycle. The glutamate:glyoxylate aminotransferase is equally as active when alanine is the amino donor, although there is little activity with other amino acids (Y. Nakamura and N. E. Tolbert, unpublished). The role of alanine in photorespiration is uncertain. The tryptophan aminotransferase activity in leaf peroxisomes (119) may be due to these same two enzymes.

Aminotransferases in mammalian peroxisomes are similar to those in plants in that glyoxylate is the best amino acceptor. Isolated human liver peroxisomes have an alanine:glyoxylate (120–121) and a serine:pyruvate aminotransferase, and rat liver peroxisomes a phenylalanine or leucine: glyoxylate aminotransferase (122). The significance in liver peroxisomes of the oxidation of glycolate to glyoxylate followed by transamination to glycine is unknown.

Glycerate Pathway

Leaf peroxisomes catalyze both the energy-wasting conversion of glycolate to glycine and serine and the reversible NAD-linked interconversion between glycerate and serine. During photorespiration, carbon flows from glycolate to glycine and from serine to glycerate in the peroxisomes, before reentering the photosynthetic carbon cycle (Figure 3). Reducing equivalents, ultimately from photosynthesis, must be shuttled to the peroxisomes through malate to provide NADH for hydroxypyruvate reduction. This conversion of serine to glycerate is a gluconeogenic function of leaf peroxisomes, and is not linked to a flavin oxidase and catalase. In addition, the glycerate pathway in leaf peroxisomes seems essential for serine synthesis from glycerate when the pathway from glycolate is blocked, as in the dark.

In leaf peroxisomes two aminotransferases are needed for the interconversion of hydroxypyruvate and serine. One is the irreversible serine: glyoxylate aminotransferase, needed to form hydroxypyruvate whenever glyoxylate is available during photorespiration. To allow for serine formation from glycerate at other times, another reversible aminotransferase is necessary, which perhaps uses glutamate or alanine.

The NAD-linked glycerate dehydrogenase is also an NADH:hydroxypyruvate reductase (K_m 2×10^{-4} M for hydroxypyruvate) with a V of 50 μmol per minute per milligram of peroxisomal protein. Its K_m (glyoxylate) of 2×10^{-2} M is apparently too large for it to be functional in the peroxisomes (123), which instead convert glyoxylate to glycine by aminotransferases with K_m values of 10^{-4} M. There is no evidence for the participation of this enzyme in vivo as a glyoxylate reductase, which if present, would set up a glycolate-glyoxylate terminal oxidase system.

Interconversion of glycerate and serine is also assumed to occur in mammalian cells, but a low level of NAD-glycerate dehydrogenase in rat liver has been found only in the cytoplasm (52). In animals, hydroxypyruvate is an excellent substrate for LDH and the product is L-glycerate rather than D-glycerate. Since reactions involving glyoxylate are postulated as being protected from LDH by being in the peroxisomes, it is not apparent why the same protection would not occur for hydroxypyruvate.

In summary, the leaf peroxisomal system has two pathways for glycine and serine synthesis. The glycolate pathway is typical of peroxisomes in that it dissipates energy by a flavin-catalyzed oxidase and catalase. The glycerate pathway is reversible and linked to NAD-malic dehydrogenases, and the only apparent reason for its presence in peroxisomes is to permit coupling to the glyoxylate aminotransferase reaction and to spatially separate hydroxypyruvate from lactate dehydrogenase.

UREIDE METABOLISM   Uricase is a microbody oxidase that forms H_2O_2 (4, 9). In ureotilic animals it is a significant hepatic and renal peroxisomal protein that is a major component of the nucleoid, perhaps because of its insolubility. A trace of uricase activity is present in plant microbodies (124). The other enzymes for the ureide pathway—allantoinase and allantoicase —have been reported in fish liver (125), and allantoinase was found in frog hepatic peroxisomes (126), but neither investigation has been confirmed. Peroxisomes from fish (125) and chicken (126) livers were reported also to contain xanthine dehydrogenase, but in general, degradation of purine bases to uric acid occurs in the cytosol. In the root nodule of plants that fix N_2 and transport it to the leaf as allantoin and allantoic acid, uricase and catalase are peroxisomal, xanthine dehydrogenase is in the cytosol, but allantoinase is in the ER (127). Thus there seems to be considerable variation in the location of enzymes for ureide metabolism. As expected, uricase is always in the peroxisomes, since it generates H_2O_2. Since allantoicase forms glyoxylate, it too, if present, is likely to be peroxisomal.

OTHER OXIDASES   A polyamine oxidase has been purified from rat liver (128) and shown to be located in the peroxisomes. Its cofactor is FAD, and H_2O_2 is formed during the oxidation of spermidine and spermine to 3-aminopropionaldehyde. A peroxisomal amine oxidase develops in yeast grown on glucose with methylamine as the nitrogen source (129).

Hepatic peroxisomes contain D-amino acid oxidases that are flavoproteins and form H_2O_2. Like other peroxisomal enzymes, D-amino acid oxidase is apparently substrate inducible, for in the absence of D-amino acids in bacteria-free mice, D-amino acid oxidase is absent.

Unique microbodies are present in protozoa (19). Microbodies in aerobic *Kinetoplastida,* have an β-glycerol phosphate oxidase. The anaerobic protozoa contain microbodies with a hydrogenase and pyruvate synthetase.

Alcohol Oxidation and Peroxidation

Oxidation of methanol and perhaps ethanol can be a major metabolic function of microbodies. When yeast are grown on methanol, gigantic peroxisomes are formed that contain an FMN-linked alcohol oxidase and catalase (17, 18, 83, 130–132). The purified methanol oxidase with FMN has been characterized as an H_2O_2 generating system typical of microbodies. Two methanols are oxidized by O_2 to two formaldehydes. The H_2O_2 formed in the oxidation of the first methanol by alcohol oxidase is used by the catalase·H_2O_2 complex I to oxidize the second methanol. The initial loss of energy during methanol oxidation to initiate methanol metabolism is typical of a microbody pathway.

Whether methanol and ethanol are peroxidatively metabolized in other peroxisomes in vivo remains uncertain (6, 59, 133). Isolated hepatic peroxisomes rapidly and quantitatively oxidize ethanol, if an H_2O_2-generating substrate such as palmitoyl-CoA, glycolate, or urate are also present. The catalase·H_2O_2 complex I is the oxidant. This complex has an absorption maximum at 660 nm which disappears following addition of a hydrogen donor such as methanol or ethanol (49, 59). During liver perfusion with a substrate for a peroxisomal oxidase, the catalase·H_2O_2 complex forms in vivo. Inhibition of alcohol dehydrogenase accelerates the rate of its decomposition. Myocardial catalase activity and peroxisomal number increase when rats are fed ethanol (134). That methanol detoxification in rats, but not monkeys, is peroxidative, involving hepatic catalase, has been demonstrated with the catalase inhibitor, aminotriazole (135). Because alcohol dehydrogenase has a low K_m of 10^{-4}, it has been argued that the peroxisomal oxidation of ethanol could not be significant. As mentioned in the introduction, peroxisomal metabolism generally represents an alternate pathway, and in this case, the potential for peroxidation of ethanol may have significance at higher alcohol concentrations, and when linked to metabolism of other peroxisomal substrates that generate H_2O_2.

Because the peroxidation of an organic substrate by the catalase·H_2O_2 complex is faster than its reaction with another H_2O_2 to form 2 H_2O plus O_2, several investigators (34, 49, 61) have considered that catalase acts in vivo almost entirely as a peroxidase. However, except for the possible peroxidation of alcohol, few data exist for a general peroxidative mechanism of oxidation by catalase in peroxisomes. Such a process does not seem to function when leaf peroxisomes oxidize glycolate to glyoxylate plus H_2O_2 and transaminate the glyoxylate to glycine. The CO_2 formed from photorespiration arises from glycine oxidation in the mitochondria. Glyoxylate oxidation by H_2O_2 or by catalase plus H_2O_2 in vitro is extremely rapid, yet this does not occur to any significant extent in the leaf peroxisomes. Excess oxidation of glyoxylate to CO_2 and formate would prevent reclamation of part of the carbon in the glyoxylate, during photorespiration (Figure 3). It is as if the peroxisomal complex protects against peroxidation of glyoxylate.

CONCLUDING DISCUSSION

Peroxisomes are found in most eucaryotic cells and contain various metabolic pathways depending on substrate induction and tissue function. Most flavin oxidases that form H_2O_2 are or can be suspected of being in peroxisomes along with the catalase. Isocitrate lyase and malate synthase of the glyoxylate cycle are in the organelle, which may then be called a glyoxysome. Enzymes for β oxidation of fatty acids in peroxisomes are different from those in the mitochondria. Peroxisomal β oxidation is a significant pathway or the main one in some tissues, and β oxidation seems established as a major function of peroxisomes. Glycolate oxidation to glyoxylate and conversion of the latter to glycine by a specific aminotransferase occurs in all peroxisomes. Other peroxisomal metabolic pathways that occur in some tissues include a glycerate pathway to serine, ureide metabolism of xanthine and urate, and various amine and amino acid oxidases. Methanol oxidation by an oxidase in yeast peroxisomes has been extensively investigated. The peroxidative role of the catalase·H_2O_2 complex of peroxisomes remains uncertain.

The essentiality of peroxisomes is confused by their variable composition and the apparent duplication of catabolic pathways present elsewhere in the cell. The fatal cerebrohepatorenal syndrome of infants is associated with the absence of peroxisomes in the liver and renal proximal tubules (136). Metabolic diseases, such as oxaluria and glycinuria, may be related to the peroxisomal metabolism. The near absence of peroxisomal enzymes in rapidly developing tissue or in unicellular algae, and the development of peroxisomes in mature tissue or in complex cellular associations, suggest that the development of alternate metabolic pathways in peroxisomes may be related physiologically to energy balances in the whole tissue. Such evidence hints that peroxisomes are associated with higher forms of development rather than being a primitive respiratory organelle.

Literature Cited

1. Tolbert, N. E. 1971. *Ann. Rev. Plant Physiol.* 22:45–74
2. Beevers, H. 1979. *Ann. Rev. Plant Physiol.* 30:159–93
3. de Duve, C. 1969. *Proc. R. Soc. London Ser. B* 173:71–83
4. de Duve, C., Baudhuin, P. 1966. *Physiol. Rev.* 46:323–57
5. Hruban, Z., Rechcigl, M. 1969. *Microbodies and Related Particles.* New York: Academic. 296 pp.
6. Masters, C. J., Holmes, R. S. 1977. *J. Biochem.* 8:549–53
7. Hogg, J. F. 1969. *Ann. NY Acad. Sci.* 168:209–381
8. Tolbert, N. E., Essner, E. 1980. *Discovery in Cell Biology,* ed. P. Siekevitz. New York: Rockefeller Univ. Press. In press
9. Masters, C., Holmes, R. 1977. *Physiol. Rev.* 57:816–77
10. Tolbert, N. E. 1980. *Plant Biochemistry,* Vol. 1, ed. N. E. Tolbert, pp. 359–88. New York: Academic
11. Tolbert, N. E., Yamazaki, R. K. 1969. *Ann. NY Acad. Sci.* 168:325–41
12. Gerhardt, B. 1978. *Microbodies/Peroxisomes Pflanzlicher Zellen.* Berlin: Springer. 283 pp.

13. Vigil, E. L. 1973. *Sub-Cell. Biochem.* 2:237–85
14. Beevers, H. 1969. *Ann. NY Acad. Sci.* 168:313–24
15. Tolbert, N. E. 1972. *Algal Physiology and Biochemistry* ed. W. D. P. Steward pp. 474–504. Oxford: Blackwell
16. Müller, M. 1975. *Ann. Rev. Microbiol.* 29:467–83
17. Fukui, S., Tanada, A. 1979. *J. Appl. Biochem. Sci.* 4:246–49
18. Fukui, S., Tanaka, A. 1979. *Trends Biochem. Sci.* 4:246–49
19. Maxwell, D. P., Armentrout, V. N., Graves, L. B. Jr. 1977. *Ann. Rev. Phytopathol.* 15:119–34
20. de Duve, C. 1973. *J. Histochem. Cytochem.* 21:941–48
21. Sternlieb, I. 1979. *Prog. Liver Dis.* 11:85–104
22. Frederick, S. E., Gruber, P. J., Newcomb, E. N. 1975. *Protoplasma* 84:1–29
23. Fahimi, H. D. 1975. *Tech. Biochem. Biophys. Morphol.* 2:197–245
24. Thomas, J., Trelease, R. N. 1980. *Plant Physiol.* 65:Suppl. 65
25. Herzog, V., Fahimi, H. D. 1975. *J. Mol. Cell Cardiol.* 8:271–81
26. Navikoff, P. M., Novikoff, A. B. 1972. *J. Cell. Biol.* 53:532–60
27. Novikoff, A. B., Novikoff, P. M., Davis, C., Quintana, N. 1973. *J. Histochem. Cytochem.* 21:737–55
28. Leighton, F., Poole, B., Beaufay, H., Baudhuin, P., Coffey, J. W., Fowler, S., deDuve, C. 1968. *J. Cell Biol.* 37:482–513
29. Tolbert, N. E. 1971 *Methods Enzymol.* 23: 665–82.
30. Tolbert, N. E. 1974. *Methods Enzymol.* 31: 734–46
31. Tolbert, N. E. 1978. *Methods Enzymol.* 52:493–505
32. Beevers, H., Breidenbach, R. W. 1974. *Methods Enzymol.* 31: 565–71
33. Markwell, M. A. K., McGroarty, E. J., Bieber, L. L., Tolbert, N. E. 1973. *J. Biol. Chem.* 248:3426–32
34. Donaldson, R. P., Tolbert, N. E., Schnarrenberger, C. 1972. *Arch. Biochem. Biophys.* 152:199–215
35. Novikoff, A. B., Shin, W-Y. 1964. *J. Microsc.* 3:187–206
36. Krahling, J. B., Gee, R., Gauger, J. A., Tolbert, N. E. 1979. *J. Cell. Physiol.* 101:375–90
37. de Duve, C., Lazarow, P. B., Poole, B. 1974. *Adv. Cytopharmacol.* 2:219–23
38. Goldman, B. M., Blobel, G. 1978. *Proc. Natl. Acad. Sci. USA* 75:5066–70
39. Svoboda, D., Azarnoff, D. 1966. *J. Cell. Biol.* 30:442–50
40. Leighton, F., Coloma, L., Koenig, C. 1975. *J. Cell Biol.* 67:281–309
41. Reddy, J. K. 1973. *J. Histochem. Cytochem.* 21:967–71
42. Osumi, T., Hashimoto, T. 1979. *J. Biochem.* 85:131–39
43. Markwell, M. A. K., Bieber, L. L., Tolbert, N. E. 1977. *Biochem. Pharmacol.* 26:1697–1702
44. Lazarow, P. B. 1977. *Science* 197: 580–81
45. Lazarow, P. B. 1978. *J. Biol. Chem.* 253:1522–28
46. Osmundsen, H., Neat, C. E., Norum, K. R. 1979. *FEBS Lett.* 99:292–96
47. Moody, D. E., Reddy, J. K. 1971. *Res. Commun. Chem. Pathol. Pharmacol.* 9:501–10
48. Crane, D., Holmes, R., Masters, C. 1980. *Biochem. Biophys. Res. Commun.* 93:258–63
49. Sies, H. 1974. *Angew. Chem. Int. Ed. Engl.* 13:706–18
50. Ruis, H. 1979. *Can. J. Biochem.* 57:1122–30
51. Cromartie, T. H., Walsh, C. T. 1975. *Biochemistry* 14:2588–96
52. McGroarty, E., Hsieh, B., Wied, D., Tolbert, N. E. 1974. *Arch. Biochem. Biophys.* 161:194–210
53. Cooper, T. G., Beevers, H. 1969. *J. Biol. Chem.* 244:3507–13 and 3514–20
54. Inestrosa, N. C., Bronfman, M., Leighton, F. 1979. *Biochem. J.* 182:779–88
55. Inestrosa, N. C., Bronfman, M., Leighton, F. 1979. *Life Sci.* 25:1127–36
56. Graves, L. B., Becker, W. M. 1974. *J. Protozool.* 21:771–74
57. Yamazaki, R. K., Tolbert N. E. 1969. *Biochim. Biophys. Acta.* 178:11–20
58. Ting, I. P., Fuhr, I., Curry, R., Zschoche, W. C. 1975. in *Enzymes* 2:369–86
59. Oshimi, N., Jamieson, D., Sugano, T., Chance, B. 1975. *Biochem. J.* 146:67–77
60. Tolbert, N. E. 1973. *Symp. Soc. Exp. Biol.* 27:215–39
60a. Mettler, I. J., Beevers, H. 1980. *Plant Physiol.* 66:555–60
61. Gee, R., McGroarty, E., Hsieh, B., Wied, D. M., Tolbert, N. E. 1974. *Arch. Biochem. Biophys.* 161:187–93
62. Murphy, P. A., Krahling, J. B., Gee, R., Kirk, J. R., Tolbert, N. E. 1979. *Arch. Biochem. Biophys.* 193:179–85
63. Gee, R., Hasnian, S. H., Tolbert, N. E. 1975. *Fed. Proc.* 34:599
64. Kawamoto, S., Yamada, T., Tanaka, A., Fukui, S. 1979. *FEBS Lett.* 97:253–56

65. Rehfeld, D. W., Tolbert, N. E. 1972. *J. Biol. Chem.* 247:4803–11
66. Blum, J. J. 1973. *J. Protozool.* 20: 688–92
67. Hryb, D. J., Hogg, J. F. 1976. *Fed. Proc.* 35:1501
68. Lazarow, P., de Duve, C. 1976. *Proc. Natl. Acad. Sci. USA* 73:2043–46
69. Osumi, T., Hashimoto, T. 1978. *Biochem. Biophys. Res. Commun.* 83:479–85
70. Osumi, T., Hashimoto, T. 1978. *J. Biochem.* 83:1361–63
71. Kawamoto, S., Nozaki, C., Tanaka, A., Fukui, S. 1978. *Eur. J. Biochem.* 83:609–13
72. Shimizu, S., Yasui, K., Tani, Y., Yamada, H. 1979. *Biochem. Biophys. Res. Commun.* 91:108–13
73. Bronfman, M., Inestrosa, N. C., Leighton, F. 1979. *Biochem. Biophys. Res. Commun.* 88:1030–36
74. Krahling, J. B., Gee, R., Murphy, P. A., Kirk, J. R., Tolbert, N. E. 1978. *Biochem. Biophys. Res. Commun.* 82: 136–41
75. Markwell, M. A. K., Tolbert, N. E., Bieber, L. L. 1976. *Arch. Biochem. Biophys.* 176, 479–88
76. Appelkvist, E. L., Dallner, G. 1980. *Biochim. Biophys. Acta* 617:156–60
77. Cooper, T. G. 1971. *J. Biol. Chem.* 246:3451–55
78. Shindo, Y., Hashimoto, T. 1978. *J. Biochem.* 84:1177–81
79. Kriasans, S. K., Mortensen, R. M., Lazarow, P. B. 1980. *J. Biol. Chem.* In press
80. Mishina, M., Kamiryo, T., Tashiro, S., Hagihara, T., Tanaka, A., Fukui, S., Osumi, M., Numa, S. 1978. *Eur. J. Biochem.* 89:321–28
81. Muto, S., Beevers, H. 1974. *Plant Physiol.* 54:23–28
82. Osumi, T., Hashimoto, T., Ui, N. 1980. *J. Biochem* 87:1735–46
83. Osumi, M., Fukuzumi, F., Teranishi, Y., Tanaka, A., Fukui, S. 1975. *Arch. Microbiol.* 103:1–11
84. Osumi, T., Hashimoto, T. 1979. *Biochem. Biophys. Res. Commun.* 89: 580–84
85. Osumi, T., Hashimoto, T. 1980. *Arch. Biochem. Biophys.* 203:372–83
86. Frevert, J., Kindl, H. 1980. *Eur. J. Biochem.* 107:79–122
87. Reddy, M. J., Hollenberg, P. F., Reddy, J. F. 1980. *Biochem. J.* 188:731–40
88. Miyazawa, S., Osumi, T., Hashimoto, T. 1980. *Eur. J. Biochem.* 103:589–96
89. Hryb, D. J., Hogg, J. F. 1979. *Biochem. Biophys. Res. Commun.* 87:1200–6
90. Leighton, F., Inestrosa, N., Gonzalez, O., Bronfman, M. 1978. *XI Int. Congr. Nutr.* p. 103
91. Neat, C. E., Thomassen, M., Osmundsen, H. 1980. *Biochem. J.* 186: 369–71
92. Norseth, J. 1979. *Biochim. Biophys. Acta* 575:1–9
93. Norseth, J. 1980. *Biochim. Biophys. Acta* 617:183–91
94. Christiansen, R. Z., Christiansen, E. N., Bremer, J. 1979. *Biochim. Biophys. Acta* 473:417–29
95. Moody, D. E., Reddy, J. K. 1974. *Res. Commun. Chem. Pathol. Pharmacol.* 9:501–10
96. McGarry, J. D., Leatherman, G. F., Foster, D. W. 1978. *J. Biol. Chem.* 253:4129–36
97. Mannaerts, G. P., Debeir, L. J., Thomas, J., Schepper, P. J. 1979. *J. Biol. Chem.* 254:4585–95
98. Kawamoto, S., Ueda, M., Nozaki, C., Yamamura, M., Tanaka, A., Fukui, S. 1978. *FEBS. lett.* 96:37–40
99. Yamada, T., Nawa, H., Kawamoto, S., Tanaka, A., Fukui, S. 1980. *Arch. Microbiol.* In press
100. Jones, C. L., Hajra, A. K. 1977. *Biochem. Biophys. Res. Commun.* 76: 1138–43
101. Beevers, H. 1975. *Recent Advances in Chemistry and Biochemistry of Plant Lipids*, eds. T. Galleard, E. I. Mercer, pp. 287–99. New York: Academic
102. Hogg, J. F. 1969. *Ann. NY Acad. Sci.* 168:281–91
103. Breidenbach, R. W., Beevers, H. 1967. *Biochem. Biophys. Res. Commun.* 27:462–69
104. McKinley, M. P., Trelease, R. N. 1980. *Protoplasma* 94:249–61
105. McKinley, M. P., Trelease, R. N. 1980. *Comp. Biochem. Physiol.* 67B:17–32
106. Jones, C. T. 1980. *Biochem. Biophys. Res. Commun.* 95:849–56
107. Tolbert, N. E. 1980. *Plant Biochem.* 2:487–523
108. Tolbert, N. E., Oeser, A., Kisaki, T., Hageman, R. H., Yamazaki, R. K. 1968. *J. Biol. Chem.* 243:5170–84
109. Tolbert, N. E., Oeser, A., Yamazaki, R. K., Hageman, R. H., Kisada, T. 1969. *Plant Physiol.* 44:135–47
110. Rehfeld, D. W., Randall, D. D., Tolbert, N. E. 1970. *Can. J. Bot.* 48:1219–26
111. Nakano, M., Ushijima, Y., Saga, M., Tsutsumi, Y., Asami, H. 1967. *Biochim. Biophys. Acta* 139:40–48
112. Warren, W. A. 1970. *J. Biol. Chem.* 245:1675–81

113. Hauscheldt, S., Chalmers, R. A., Lawson, A. M., Schultis, K., Watts, R. W. E. 1976. *Am. J. Clin. Nutr.* 29:258–73
114. Badwey, J. A. 1977. *J. Biol. Chem.* 252:2441–443
115. Barker, R. F., Hopkinson, D. A. 1978. *Ann. Hum. Genet. Land.* 42:143–51
116. Rose, Z. B. 1980. *Adv. Enzymol.* 51:211–53
117. Kayne, F. J. 1974. *Biochem. Biophys. Res. Commun.* 59:8–13
118. Rehfeld, D. W., Tolbert, N. E. 1972. *J. Biol. Chem.* 247:4803–11
119. Noguchi, T., Hayashi, S. 1980. *J. Biol. Chem.* 255:2267–69
120. Noguchi, T., Takada, Y. 1979. *Arch. Biochem. Biophys.* 196:645–47
121. Noguchi, T., Takada, Y. 1978. *J. Biol. Chem.* 253:7598–600
122. Hsieh, B., Tolbert, N. E. 1976. *J. Biol. Chem.* 251:4408–15
123. Tolbert, N. E., Yamazaki, R. K., Oeser, A. 1970. *J. Biol. Chem.* 245:5129–36
124. Huang, A. H. C., Beevers, H. 1971. *Plant Physiol.* 48:637–41
125. Noguchi, T., Takada, Y., Fujiwara, S. 1979. *J. Biol. Chem.* 254:5272–75
126. Scott, P. J., Visentin, L. P., Allan, J. M.

127. Hanks, J. F., Schubert, K. R., Tolbert, N. E. 1980. *Plant Physiol.* 65:Suppl. 111
128. Holtta, E. 1977. *Biochemistry* 16:91–100
129. Zwart, K., Veenhuis, M., Van Dijken, J. P., Harder, W. 1980. *Arch. Microbiol.* 126:117–27
130. Sahm, H., Hinkelmann, W., Roggenkamp, R., Wagner, F. 1975. *J. Gen. Microbiol.* 88:218–22
131. Van Dijken, J. P., Veenhuis, M., Kreger-van Rij, N. J. W., Harder, W. 1975. *Arch. Microbiol.* 102:41–44
132. Zwart, K., Veenhuis, M., Van Dijken, J. P., Harder, W. 1980. *Arch. Microbiol.* 126:117–26
133. Thurman, R. G. 1977. *Fed. Proc.* 36:1640–46
134. Fahimi, H. D., Kino, M., Hicks, L., Thorp, K. A., Abelman, W. H. 1979. *Am. J. Pathol.* 96:373–90
135. Mannering, G. J., Van Harken, D. R., Makar, A. B., Tephly, T. R., Watkins, M. S., Goodman, J. I. 1969. *Ann. NY Acad. Sci.* 168:265–80
136. Goldfischer, S. 1979. *J. Histochem. Cytochem.* 27:1371–73

1969. *Ann. NY Acad. Sci.* 168:244–63

Ann. Rev. Biochem. 1981. 50:159–92
Copyright © 1981 by Annual Reviews Inc. All rights reserved

DNA MODIFICATION AND CANCER[1]

♦12075

Michael J. Waring

Cancer Research Laboratory, School of Medicine, Auckland, New Zealand

CONTENTS

PERSPECTIVES AND SUMMARY

The cancer problem is full of paradoxes. One of them concerns the strangely reciprocal behavior of the drugs used for its treatment. All too often, given

[1]The preparation of this review and research in the author's laboratory were supported by grants from the Auckland Division of the Cancer Society of New Zealand Inc., the Medical Research Councils of Great Britain and New Zealand, The Wellcome Trust, The Royal Society, and the Cancer Research Campaign.

159

the appropriate circumstances these drugs are themselves capable of producing cancer, and there is justified disquiet that those circumstances may include the conditions under which antitumor drugs are routinely employed in the clinic. The interest of the biochemist is aroused by the finding that these drugs are, more often than not, inhibitors of nucleic acid synthesis (Figure 1). Indeed, most of them actually interact physically with DNA so as to distort its structure and function.

It can fairly be argued that most aspects of the total cancer problem revolve around the behavior of DNA. As the biochemical repository of the genetic inheritance of cells its dominant influence over the processes of cell growth, division, and differentiation is taken for granted by the vast majority of biochemists and oncologists. Abundant evidence exists to support the assertion that interference with DNA, in almost any conceivable way, is likely to serve as a prelude to the induction of cancer. Indeed, to put it another way, one is hard pushed to point to any instance of a recognizable oncogenic event that has *not* involved DNA modification, one way or another. Conversely, insofar as DNA replication is a necessary precondition for cell division it constitutes a vulnerable target for the action of cytotoxic drugs intended to block the multiplication of disease-producing cells. Thus the biochemical basis for the reciprocal property of anticancer drugs to act also as potential carcinogens (we should really say oncogens) is plausibly related to their capacity to interact with DNA.

For these reasons the subject of drug-DNA interactions has become a focus for the attention of a rapidly expanding portion of the scientific community. In recent years all aspects of the problem have come under increasingly close scrutiny, and the advent of new techniques (not to mention genetic engineering) has provided added impetus to the pace of research. The relevant literature has burgeoned. Fortunately the topic of carcinogenesis by chemicals has been extraordinarily well served with reviews and critical assessments (2–9) so that further treatment is unnecessary here; accordingly this review concentrates on drug-DNA interactions from the standpoint of therapeutic effects that appear to be mediated at this level.

Any consideration of DNA modification by ligand binding must begin by acknowledging the preeminence of the concept of intercalation, first explicitly formulated by Lerman (10) in 1961 to explain the binding of aminoacridines to DNA (and thence their remarkable effectiveness as frameshift mutagens). Subsequent work has revealed that the property of intercalation is shared by a variety of DNA-binding drugs, including several important antitumor agents as well as a few carcinogens. Early doubts about the validity of claims that chemotherapeutic drugs might intercalate into DNA have been largely dispelled by the application of better, more direct methods for assessing the character of drug-DNA interactions, perhaps most importantly the circular DNA–binding test (11). Indeed, a whole new area of

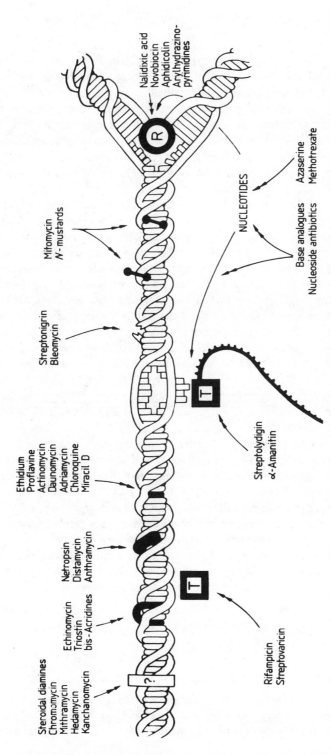

Figure 1 Sites for inhibition of nucleic acid synthesis by antibiotics and drugs. R represents the replicating enzyme complex, T the transcribing enzyme (RNA polymerase). Actions of inhibitors are represented by double-headed arrows and are purely diagrammatic; it is not intended to imply that the sites of action represented here account for the whole mode of action of any drug. From (1).

biochemical investigation based on the circularity of DNA has grown out of the original discovery that intercalation affects the supercoiling of circular DNA (12, 13); developments in this field have been ably reviewed by Bauer (14). While confirming the intercalative potential of some drugs, the sophisticated new methods have ruled it out as a molecular explanation for the action of other DNA binders; here the state of knowledge is developing comparatively slowly. Only recently have tentative ideas as to how drugs may lodge in one or other of the grooves of the helix been placed on a firmer footing, and there remains a long way to go.

The nature of the damage to DNA occasioned by ligand binding has also been clarified by the development and application of novel approaches and experimental techniques. It is clear that strand breakage is a more common and widespread response to assault on DNA than once was supposed, and the nature of the breaks and their ontogeny have been illuminated by several penetrating studies. In one or two instances the involvement of powerful oxidizing fragments has been established. Attention has also been paid to the chain of consequences resulting from DNA damage, and how the effect(s) of such damage may be propagated or modified by the operation of repair mechanisms. A hotly debated issue concerns the relationship between mutagenic potency and the potential for carcinogenesis, in particular whether the anticipated correlation between them is real or illusory. Less attention has been given to biochemical mechanisms underlying blockade of nucleic acid synthesis, and to the exact sequence of events leading ultimately to cell death.

It is generally acknowledged that the pressing need in cancer medicine is for drugs endowed with better specificity and selectivity of action (15). Of the various rational approaches to this problem the study of bifunctional or polyfunctional DNA-binding ligands has yielded promising results. Synthetic as well as naturally occurring agents that interact with DNA in this fashion have been identified and characterized; they display very tight binding constants and have proved active against a variety of experimental tumor models. There are grounds for cautious optimism that future advances in our understanding of the molecular basis of drug-DNA interaction may lead to significant improvements in cancer treatment.

MECHANISMS OF PHYSICAL INTERACTION BETWEEN DRUGS AND DNA

Reversible Interactions: the Classical Intercalation Concept

Modification of DNA by binding of dyestuffs suspected to have possible chemotherapeutic utility was noted as early as 1947 in the pioneering work of Michaelis (16). Gradually over the ensuing decades the potential importance of DNA as a target for drug action became realized, but it was early work on the actinomycin-DNA complex (17–19), and the formulation of the intercalation model for aminoacridine-DNA binding (10, 20) at the beginning of the 1960s that set the scene for future developments. The character of the intercalation process was initially established with reference to aminoacridines (e.g. proflavine, 9-aminoacridine) and the trypano-cidal phenanthridine ethidium (Figure 2), substances known to be active against tumors (21, 22). These drugs still stand as the type-specific examples of intercalating agents whose properties and behavior are regarded as classical. But not surprisingly, the elegance and simplicity of the basic intercalation postulate that the flat polycyclic aromatic ring system of a drug molecule might slip between the stacked base pairs of the double helix led to rapid extension of the concept to explain the binding of other molecules to DNA, especially those such as antitumor antibiotics whose mode of action seemed to involve interference with nucleic acid synthesis.

Initial evidence for the intercalation model was necessarily somewhat indirect, and rested largely on inferences from X-ray diffraction of oriented fibers and two supposedly characteristic changes in hydrodynamic behavior caused by drug binding: an increase in viscosity and a decrease in sedimentation coefficient (10, 23). Later evidence from diverse sources that involved the application of varied physical techniques [summarized in (1)] consistently supported the model. However, the most generally accepted criteria for intercalation employed today come from what might be termed direct techniques that measure three fundamental features based on an unashamedly simple-minded view of intercalation: the helix must be (a) extended, and (b) locally unwound by the binding reaction, and (c) the plane of the aromatic chromophore of the bound drug must be parallel to that of the base pairs (ostensibly more or less perpendicular to the helix axis). The required extension has been verified by autoradiographic (24) and electron microscopic (25–27) measurements of the contour length of DNA molecules in the presence of drugs, but the most readily applicable technique is simply to measure the enhancement of viscosity of sonicated rod-like DNA fragments, which can be done in a simple capillary viscometer and related to the changes in molecular length by a cube root relation (28, 29). The local unwinding is evidenced by monitoring the effects of drug binding on the superhelical state of circular DNA; as the level of intercalative drug binding rises the negatively (right-handed) supercoiled state of a naturally occurring circular DNA is progressively diminished, lost, and reversed in a very characteristic fashion that can be observed by following the sedimentation coefficient (11–13), the viscosity (30, 31), or the mobility in gels (32). The orientation of the drug chromophore with respect to the long axis of DNA can be investigated using dichroic techniques in which partial alignment of the molecules along a known vector is achieved by flow (33) or electric

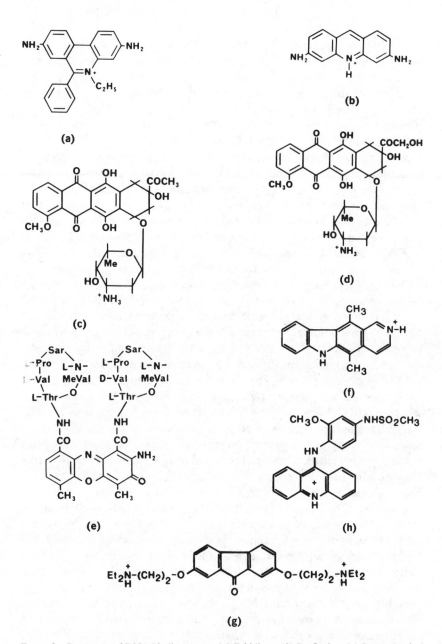

Figure 2 Structures of DNA-binding drugs. (*a*) Ethidium, (*b*) Proflavine, (*c*) Daunomycin, (*d*) Adriamycin, (*e*) Actinomycin D, (*f*) Ellipticine, (*g*) Tilorone, (*h*) *m*-AMSA, (*i*) Mithramycin, (*j*) Netropsin, (*k*) Irehdiamine A, (*l*) Anthramycin, (*m*) Streptonigrin, (*n*) Bleomycin. (R represents a terminal amine moiety. Phleomycin antibiotics are identical to bleomycins but for the arrowed double bond, which is missing in the phleomycin series.)

(i)

(j)

(k)

(l)

(m)

(n)

discharge (34, 35); the absorption at wavelengths characteristic of the drug and of the base pairs is then monitored in two mutually perpendicular directions.

No one of these lines of evidence is sufficient by itself to constitute a firm diagnosis of intercalation. For example, positive evidence of helix extension has been recorded for netropsin and distamycin (36, 37), and unwinding of circular DNA has been seen with steroidal diamines (31, 38), yet neither of these groups of substances is believed to intercalate. In applying the circular DNA–unwinding test the experimenter must demonstrate removal *and reversal* of the supercoiling if he wishes to establish that helix unwinding is a necessary concomitant of drug binding (as required by the intercalation hypothesis) and not just a nonspecific effect reflecting generalized loosening of the helical structure. It is easy to imagine drug-induced loss of supercoiling due to nonintercalative "weakening" of the helix aided by release of superhelical free energy (11,14). So far as the goal of proving intercalation is concerned, it remains true that confidence in the validity of the conclusion increases in proportion to the number of tests applied, and that consonance between the results of independent experimental approaches (such as helix extension and unwinding measurements) does more to enhance credibility than repetition of one test with variations.

An interesting and perhaps unexpected finding that has emerged from application of these tests is that all three characteristics of intercalation vary considerably in magnitude between different drugs. The measured helix extension may fall anywhere in the range 1.8–4.5 Å, and relatively few compounds (not even the classical ones like proflavine and ethidium) attain the "theoretical" value of 3.4 Å (28, 29, 35, 39, 40). Likewise, the apparent helix-unwinding angle per bound drug molecule may vary more than twofold between different ligands (Table 1), though, discounting the *bis*-intercalating quinoxaline antibiotics and diacridines, no drugs provoke a larger unwinding effect than the standard reference intercalator, ethidium. Also the orientation of the bound chromophore with respect to the long axis of DNA is not invariant but generally indicates a significant angle of tilt, apparently by as much as 28°, from perpendicularity to the helix axis (34,35). These results speak for the occurrence of more far-reaching perturbations of the regularity of the DNA helix than implied by the minimally distorted models originally proposed, and lend caution to any attempt to interpret DNA-binding data in strict conformity to existing models. There seems little justification for reiterating the suggestion that different unwinding angles may reflect the persistence of nonintercalated "outside-bound" drug molecules at equilibrium (11).

CRYSTALLINE DRUG-DINUCLEOTIDE COMPLEXES Recent years have seen elegant confirmation of the intercalation concept with the deter-

Table 1 Unwinding of the DNA helix by intercalating antibiotics and drugs[a]

Compound	Remarks	Apparent unwinding angle (degrees)	Ref.
Ethidium	Trypanocidal phenanthridine	26	1, 14
Propidium	Phenanthridine; biochemical tool	26	11
Proflavine	Antibacterial acridine; mutagen	17	41
9-Aminoacridine	Antibacterial agent; mutagen	17	42
m-AMSA	Antitumor acridine	21	43
Daunomycin	Anthracycline antibiotic	11	11
Adriamycin	Anthracycline antibiotic	11	44
Nogalamycin	Anthracycline antibiotic	18	11
Hycanthone	Schistosomicidal drug	15	11
Tilorone	Antiviral agent	~13	42
Ellipticine	Antitumor alkaloid	17	45
Actinomycin D	Antitumor antibiotic	26	11, 46
C6 diacridine	Antitumor drug	33	47
Spermidine diacridine	Antitumor drug	38	48
Echinomycin	Quinoxaline antibiotic	48	49
Triostin A	Quinoxaline antibiotic	47	50
TANDEM	Quinoxaline peptide	46	51
Benzo(a)pyrene diol epoxide	Carcinogen	~26	52
N-Acetoxy-2-acetyl aminofluorene	Carcinogen	~26	52

[a] The list is far from exhaustive; entries have been selected on the basis of established antitumor activity and/or unwinding angle. Values quoted for the apparent unwinding angle have been corrected, where necessary, to refer to an assumed unwinding angle of 26° for ethidium.

mination of crystal structures for model complexes between drugs and dinucleotides. Table 2 lists relevant properties of eleven such complexes whose structures have been solved at atomic resolution. With two exceptions they all contain equimolar proportions of drug and dinucleotide, which no doubt reflects the requirement for exact neutralization of electric charges to permit crystal formation; (in exceptional instances an additional inorganic ion is included in the unit cell). The most common 2:2 complexes contain one drug molecule intercalated between the base pairs and another stacked between successive drug-dinucleotide units so as to enable the formation of infinite columns of alternating base pairs and drug molecules. Recurrent features common to nearly all of the structures include the following: imperfect twofold rotational symmetry, an alternating pattern of sugar puckering on individual dinucleotide strands, more or less symmetrical intercalation of the aromatic chromophore of the drug molecule, and a resultant base-turn angle between upper and lower base pairs of about 10°. (None of these generalizations is perfect, however, and a careful study of

Table 2 Crystalline complexes containing Watson-Crick base-paired dinucleoside phosphates with intercalated drug molecules

Complex	Stoichi-ometry	Base-turn angle[a] (degrees)	Sugar pucker	Ref.
Ethidium:5-iodo-CpG	2:2	8	C3' endo-(3',5')-C2' endo	53
Ethidium:5-iodo-UpA	2:2	8	C3' endo-(3',5')-C2' endo	54
Proflavine:5-iodo-CpG	2:2	36	C3' endo-(3',5')-C3' endo	55
Proflavine:CpG	3:2	33	C3' endo-(3',5')-C3' endo	56
Proflavine:dCpG	2:2	17	C3' endo-(3',5')-C2' endo C3' endo-(3',5')-C3' endo	57
Acridine orange:CpG	1:2	10	C3' endo-(3',5')-C2' endo	58
Acridine orange:5-iodo-CpG	2:2	10	C3' endo-(3',5')-C2' endo	55
9-Aminoacridine: 5-iodo-CpG	4:4	10	C3' endo-(3',5')-C2' endo	59
Ellipticine:5-iodo-CpG	2:2	10–12	C3' endo-(3',5')-C2' endo	60
3,5,6,8-Tetramethyl-N-methyl-phenanthrolinium:5-iodo-CpG	2:2	10–12	C3' endo-(3',5')-C2' endo	60
Platinum terpyridine:dCpG	2:2	13	C3' endo-(3',5')-C2' endo	61

[a] Defined as the angle between the vectors connecting C1' atoms of each base pair when viewed approximately perpendicular to the average plane in which the base pairs lie.

the exceptions may prove instructive. For example all the proflavine complexes display a much larger base-turn angle, that in the ribo structures appears as large as that of a ten- or elevenfold DNA or RNA helix.) Not infrequently, hydrogen bonds occur connecting the exocyclic amino groups of drug molecules to substituents in the sugar-phosphate backbones.

The availability of detailed structural information on these model complexes has stimulated the building of detailed polynucleotide intercalation models more or less firmly based on the structural parameters of drug-dinucleotide complexes [(62,63) and references cited therein]. Confidence in the validity of this approach is reinforced by the evident pyrimidine-(3',5')-purine sequence selectivity in the complexes listed in Table 2, which is consistent with that observed in solution (64–66), and by the fact that the characteristic C3' endo (3',5') C2' endo alternating pattern of sugar pucker can apparently be generated in solution by interaction of dinucleotides with an intercalating drug (67–69). Probably the best-known model is that proposed for ethidium-DNA binding, by Sobell et al (70, 71), derived from the coordinates of the ethidium: iodo-CpG and ethidium:iodo-UpA complexes. In this model the drug molecule is intercalated into a pyrimidine-(3',5')-purine sequence with its phenyl and ethyl groups located in the narrow groove; the preference for this particular type of nucleotide sequence is explained on the basis of optimizing stacking interactions between the phenanthridine chromophore and the base pairs. The continuity of the helix axis is disrupted at the intercalation site: it is dislocated about 1 Å by sideways displacement of the base pairs and it is kinked about 8°, a residuum of a much larger kink that is postulated as an essential intermediate on the way to establishment of the intercalated state. The effective helix-unwinding angle is 26°. A most important feature of this model lies in its requirement for strictly alternating C3' *endo* (3',5') C2' *endo* sugar puckering. Because all the sugars in the Watson-Crick tenfold helical B form of DNA are puckered C2' *endo*, this necessitates a change of pucker in *one* deoxyribose ring of each strand at the intercalation site, and automatically imposes a condition of neighbor exclusion (72, 73) if the next sugar ring in the 3' direction is constrained to remain C2' *endo*. Not all authors agree that this is a legitimate explanation for the occurrence of neighbor exclusion (63), and it is true that alternating patterns of puckering are not invariably seen with drug-dinucleotide complexes (Table 2). However, the concept is attractive and the exceptions noted in the table could be explained on the basis of the extra degrees of freedom available to dinucleotides in isolation. The initial formation of a kink may also be related to the postulated change in sugar ring pucker: it arises naturally in a DNA sequence when the pucker of sugar rings appropriately located on the complementary strands is altered to C3' *endo*, which provokes the appearance of a wedge-shaped entry notch accessible from the narrow groove. The helix axis suffers an abrupt bend of about 40°, a sideways dislocation of -1.0 Å (opposite to that eventually established after the ethidium slips in), and an angular unwinding of about 10° (70, 71).

It is prudent to emphasize that these proposals have their origins in observations on model complexes, for models can occasionally be misleading. For instance, there are at least two reported instances of crystalline complexes forming between aminoacridines and noncomplementary dinucleotides (74,75), and in another case the base pairing that occurs is non-Watson-Crick despite the nominal self-complementarity of the dinucleotide involved, ApU (76). Even more striking is the manifestly nonintercalative binding of ethidium seen in a crystalline ethidium-tRNA molecular complex (77), notwithstanding the expectation that the drug would be found intercalated between A and U base pairs of the amino acid acceptor stem (78). These oddities, and various other unexpected features noted in crystalline model complexes, may prove relevant to interaction between intercalating drugs and single-stranded regions in polynucleotides (59, 63, 74, 76).

KINETICS   A new dimension has been added to our understanding of the intercalation reaction as a result of kinetic studies employing stopped-flow and temperature-jump methods. With aminoacridines such as proflavine there is an initial fast, concentration-dependent binding that is attributable to external attachment of the drug to the DNA helix, mediated primarily by electrostatic forces. This is followed by one or more slow first-order processes attributed to the insertion reaction proceeding from the outside-bound state to the intercalated state (79–82). By contrast, the interaction of ethidium with DNA is characterized by the occurrence of two distinct intercalated species that can rapidly interconvert at equilibrium without passing through a free-solution state (83; D. Ryan and D. M. Crothers, personal communication). This process of direct ligand transfer, envisaged as the direct passage of an ethidium molecule, bound to one DNA molecule, to an empty binding site on another DNA molecule, provides a highly efficient means for the ligand to equilibrate over its available binding sites, and may be important in the binding of natural regulatory substances to DNA. Direct ligand transfer seems to be characteristic of phenanthridinium drugs and has also been observed with one aminoacridine (82). The nature of the two complexed forms remains mysterious: Wakelin & Waring (82) have suggested that they may represent drug molecules intercalated via the two different grooves of the DNA helix.

Intercalation as a Property of Antitumor Agents

ANTHRACYCLINES The most important anticancer drugs whose mode of action seems to depend on intercalative binding to DNA are the antibiotics of the anthracycline group and actinomycin D. Daunomycin (daunorubicin) and adriamycin (Figure 2) are the best-known representatives of the anthracycline group, which also includes nogalamycin, cinerubin, rhodomycins, and rubomycins. All are characterized by a tetrahydrotetracenequinone chromophore that contains three flat, coplanar, six-membered rings (cf proflavine and ethidium, Figure 2). The phenoxazone chromophore of actinomycin is in essential respects comparable in size and shape. By all the usual criteria these antibiotics are identified as intercalating agents (1, 11, 69, 85), and reasonably detailed molecular models for their binding to DNA have been formulated. The daunomycin-DNA model of Pigram et al (86) places the aromatic portion of the chromophore between successive base pairs in fairly classical intercalative fashion, while the puckered fourth (cyclohexene) ring with its attached daunosamine (amino sugar) substituent projects into the wide groove of the helix. Additional stabilization comes from strong electrostatic interaction between the charged amino group and a nearby DNA phosphate residue. The gross unwinding angle compared to normal B-DNA is 12° (cf Table 1). This model is consistent with the known requirement for the aminosugar (85) and with the crystal structures of the unmodified antibiotic, its N-bromoacetyl derivative, and related substances (87–89), which point to a surprising lack of flexibility around the glycosidic linkage. Various proposals have been made for hydrogen bond formation that might stabilize the conformation of the antibiotic itself and its complex with DNA; they include one that involves the additional C (14) hydroxyl group peculiar to adriamycin (88). All in all, the agreement between the basic postulates of the intercalation model and the fairly extensive data on properties of derivatives (89) is good, though this may be partly due to the lack of detailed information at the atomic level such as is available for other intercalators (cf Table 2).

Daunomycin and adriamycin are extremely valuable agents in the treatment of cancer, and no discussion of their action would be complete without mention of the discovery of a derivative, AD32 (N-trifluoracetyladriamycin-14-valerate), which is therapeutically effective but does not appear to bind to DNA, does not become localized in cell nuclei, and is not significantly metabolized to adriamycin (90–92). Its mode of action may be quite different from that of the parent antibiotics. Other biochemical sites of action are also suggested by the existence of a dose-limiting cardiotoxic side effect of anthracycline antibiotics, for which several diverse explanations have been put forward; most have nothing to do with DNA-binding (see 93).

ACTINOMYCIN   With this antibiotic there is general agreement that its antitumor activity is directly attributable to its ability to bind to DNA (19,94). A specific requirement for deoxyguanosine-containing nucleotides in the binding reaction was early established and later narrowed down to an apparent dependence on the presence of a purine 2-amino group (19, 95, 96). This feature has been considered of prime importance in the development of molecular models for the actinomycin-DNA complex, though recent studies have suggested that the need for a 2-amino substituent may not be so much a primary requirement as an indication of the presence of a conformational environment suitable for antibiotic binding (96–98). Be that as it may, the successive proposal of four distant molecular models (with added refinements) over a period of 15 years indicates something of the interest in this problem, and the saga of their development [recounted in (1)] contains lessons for everyone.

That actinomycin binds to DNA by intercalation can now be taken as established (11, 35, 39, 46, 66, 68, 71). The latest and best-defined model for the complex is that of Sobell et al (71), which has its origins in the X-ray structure of a crystalline complex containing one molecule of antibiotic and two molecules of deoxyguanosine (99). Interestingly, although no one has yet reported the structure of an intercalated dinucleotide complex akin to those listed in Table 2, there is evidence that an actinomycin:2dGpC complex forms quite readily in solution (68,69,100). The model has the phenoxazone chromophore intercalated more or less symmetrically between two

GC base pairs, with H bonds connecting the 2-amino groups of the guanine bases to the CO residues of the threonine residues in the peptide lactone rings, which pack neatly into the narrow groove of the DNA helix. In the refined form of the model (71) the usual C3' *endo* (3',5') C2' *endo* pattern of sugar puckering is retained, together with a total unwinding angle of 28° and residual kinking, but there is nonideal stacking of base pairs surrounding the chromophore, and significant departure from exact twofold symmetry such that the critical H bonds become unequal, differing in length as well as linearity. These modifications to the original form of the model (99) proved necessary to account for the inclination of the chromophore at an angle to the normal to the helix axis (35, 101) as well as the unwinding angle (11, 46), and to explain the occurrence of tight binding to sites containing only a single GC base pair (71, 96).

ELLIPTICINE, TILORONE, AND *m*-AMSA   Of these three small molecules (Figure 2), each of which has reported antitumor activity and binds to DNA by intercalation, one is a plant alkaloid [ellipticine; (102)], another is an antiviral agent [tilorone; (103)] and the third is a synthetic drug showing great promise in clinical trials against a variety of tumors, principally diffuse neoplasms [4'-(9-acridinylamino) methanesulfon-*m*-anisidide (*m*-AMSA) (104, 105)].

Ellipticine is characterized as an intercalator by the usual three criteria (106), and its helix-unwinding angle is indistinguishable from that of the simple aminoacridines (Table 1). Its mode of binding is confirmed by the structure of a crystalline intercalative complex with a dinucleotide (Table 2). In a careful study of naturally occurring and synthetic derivatives of ellipticine, Le Pecq et al (107) have probed the possibility of rational design of antitumor drugs based on the behavior of this series. They concluded that reactivity toward DNA may be a necessary but not sufficient condition for antitumor activity.

The mode of action of tilorone is not entirely clear; its antiviral action is thought to involve induction of interferon synthesis as well as direct interference with the functional expression of viruses in infected cells (103). Either or both of these actions may provide a basis for its antineoplastic activity. Its complex with DNA is clearly of the intercalative type, characterized by a rather low unwinding angle (42), but is unusual in respect of a strong preference for AT-rich regions of duplex DNA (103, 108). Müller and his colleagues (109–111) have pointed out that whenever an intercalating agent displays specificity it is practically always in favor of GC-rich sequences; they found preference for AT-rich DNAs to be restricted to nonintercalating ligands.

The synthesis of *m*-AMSA represents the culmination of many years of effort to enhance and exploit the antitumor activity of basic drugs and dyestuffs, including the acridines (112). It is an effective inhibitor of nucleic acid synthesis in cultured cells, though its binding to DNA in vitro is not particularly strong. What may be significant is its marked preference for double-helical DNA as opposed to heat-denatured DNA or RNA, and its helix-unwinding angle (Table 1), which is unusually high for an aminoacridine (43, 113, 114). The latter is likely to be a consequence of the resemblance between the anilino 9-substituent of m-AMSA and the out-of-plane 6-phenyl group of ethidium, which might provoke additional distortion of the intercalation site over and above that necessary to accommodate a simple polycyclic aromatic chromophore as present in the other acridines. A vast array of information is available concerning structure-activity relations for antitumor activity, mutagenic potency, and DNA-binding of drugs in the AMSA series (115). It is tentatively interpreted as indicating the outcome of an interplay between constraints dictated by the steric and electronic demands of intercalation sites affecting primarily the acridine portion of the molecule, and requirements for contact with a third (macromolecular) species affecting the acceptability of substituents on the anilino ring (113, 115).

Nonintercalative DNA binding

MITHRAMYCIN, CHROMOMYCIN, AND OLIVOMYCIN   Chiefly as a result of applying the newer techniques for characterizing intercalative drug binding to DNA, it is now possible to be reasonably confident that certain groups of ligands do *not* bind by that mechanism, despite the presence in some instances of structural features that appear suitable for intercalation. A case in point is the mithramycin group of antibiotics, which have a similar three-ring chromophoric moiety. [That of chromomycin A_3 is identical to the chromophore of mithramycin (Figure 2) with different sugar substituents (116).] All three have found clinical application for treatment of various unusual tumors, and seem to share a common mechanism of action that involves inhibition of nucleic acid synthesis (116). They are also of value in experimental cytochemistry as chromosome stains that produce specific fluorescent banding patterns (117, 118). Despite the nominal planarity of at least two thirds of their polycyclic chromophore, however, the evidence argues consistently against an intercalative mode of DNA binding (11, 119, 120). It does seem that they are specific for double-helical DNA, and there is an evident requirement for guanine residues that, as was found for actinomycin, seems to relate to the presence of the purine 2-amino group (95, 119, 120). The most characteristic feature of the binding of these antibiotics to DNA is their requirement for a stoichiometrically equivalent amount of Mg^{2+} (116, 119–121). In the absence of a divalent cation only very limited interaction with DNA, if any, is detectable. One might speculate that the role of the ion is merely to effect a neutralization of charges, but the affinity of these antibiotics for divalent cations is real, and doubtless is not without relevance to their pharmacological and pharmacokinetic properties in the treatment of cancer patients (see 93).

NETROPSIN AND DISTAMYCIN   These closely related basic oligopeptide antibiotics (Figure 2) exemplify a large class of DNA-binding ligands that can be thought of as more or less linear, flexible molecules possessing cationic charges at one (or more usually both) ends, and having a variety of functional substituents such as H-bond donors/acceptors spaced in between. Some are quite well known, such as the aromatic diamidines (1), whereas others are purely of interest to the experimentalist [see, for example (122)]. Yet activity against tumors is a frequent property in these molecules. Characteristically they display AT-specificity in binding to DNA, sometimes to a marked degree, consonant with the "outside-binding" hypothesis of Müller & Gautier (111).

Netropsin and distamycin are probably better known for their antiviral effects than their antitumor activity (123,124). Both antibiotics bind avidly to DNA and form highly ordered complexes whose properties show a strong dependence on the AT content, which is attributable to preferential interaction with AT-rich regions or clusters estimated to be 3–5 base pairs long (125–129). Neither netropsin nor distamycin perceptibly affect the supercoiling of circular DNA at levels of binding up to saturation (125,126), so that intercalation may be ruled out as a possible mode of interaction, and any alteration in the average helical winding of DNA associated with the binding reaction must be strictly limited. Magnetic circular dichroism similarly excludes the occurrence of large structural distortions in the region(s) to which the antibiotics bind (130,131). Nevertheless, some small changes in the conformational parameters of the helix do seem to occur as a result of antibiotic binding: the hydrodynamic behavior of DNA in viscometric experiments is perceptibly altered, but the exact nature of changes in contour and/or persistence length seems to depend upon the level of binding as well as on the source of the DNA (36,37).

Molecular model building for the netropsin- and distamycin-DNA complexes has been greatly facilitated by the availability of crystallographically determined structures for the antibiotics or their analogues (132,133). Both molecules adopt the form of gently curved, flattened rods that appear ideally shaped to lodge in the narrow groove of a B-form DNA helix in close contact with the base pairs. The amide nitrogen atoms are well disposed to engage in hydrogen-bonding interactions with the O(2) atoms of the thymine bases and/or the N(3) atoms of the adenine bases within one strand of the helix, and the amidine or guanidine substituents at the end(s) of the antibiotic molecule could form salt links with nearby phosphate residues. The AT-specificity of binding is explicable as resulting from the precise disposition of the potential H-bond acceptor substituents on the base pairs, perhaps augmented by some steric interference from the 2-amino group of guanine residues (125,126,132,134). Support for these suggestions may be gleaned from several sources. First, the placement of the ligands in the narrow groove of the helix agrees with their ability to interfere with the methylation of narrow-groove substituents, but not wide-groove substituents, by dimethyl sulfate (135). Second, flow dichroism studies have revealed that the transition moment of DNA-bound distamycin molecules is strongly inclined to the helix axis, in contrast to the situation found with intercalating drugs (136,137). Third, an NMR study of a 1:1 complex between netropsin and a self-complementary octanucleotide duplex having the sequence dG-dG-dA-dA-dT-dT-dC-dC has yielded evidence of H-bonding interactions, close proximity of the pyrrole rings and the edges of the base pairs in the narrow groove of the helix, and shifts of internucleotide phosphate resonances that may reflect the occurrence of electrostatic interactions with the charged terminal groups of the antibiotic (138). Although these models go a long way toward explaining the properties and sequence-specificity of netropsin- and distamycin-DNA complexes they fall short of a complete description, particularly as regards the cooperative or "allosteric" character of interaction observed between distamycin and certain polymers (37,134). With calf thymus DNA, the binding of small quantities of distamycin seems to provoke a long-range cooperative transition to a new form having enhanced affinity for the antibiotic and altered structural properties (37). Why a similar transition is not observed with other DNAs, even another mammalian DNA, remains obscure.

STEROIDAL DIAMINES   A particularly intriguing mode of DNA-binding that has recently come to light is seen with steroidal diamines such as irehdiamine A (Figure 2). The first hint of something unusual arose with the discovery that irehdiamine A and its *bis*-quaternary saturated analogue, malouetine, very efficiently remove and reverse the supercoiling of closed circular duplex DNA (11,38); yet in other respects their actions on DNA are unremarkable and distinct from those of low-molecular-weight intercalating drugs (139–141). The unwinding effect also occurs with dipyrandium and other isomeric synthetic diamino androstanes (31). It seems to be a common property of steroidal diamines having two cationic centers spaced 10Å or more apart. Measured unwinding angles per bound ligand molecule fall within the range 6°–13° [corrected to the standard 26° reference for ethidium; (31,38)], i.e. generally below the values determined for intercalating drugs (cf Table 1). Structure-activity relations effectively eliminate any straightforward intercalation-type model involving the bulky, nonaromatic steroid nucleus as a basis for the unwinding effect (31), but Sobell et al (71) have put forward the suggestion that steroidal diamines may bind to and stabilize the kink that they envisage as an intermediate state on the way to intercalation (and a normal mode of dynamic structural fluctuation in DNA). Evidence that binding of irehdiamine and dipyrandium is indeed associated with kinking of DNA and partial exposure of the base pairs to the solvent as a result of local unstacking has recently been adduced from hydrodynamic and dichroic experiments (142) and NMR studies on a complex formed with poly(dA-dT) (143).

Covalent Binding to DNA

BIFUNCTIONAL ALKYLATING DRUGS   Probably the best-known anticancer drugs, universally employed in clinical treatment of the disease and in experimental work, are the alkylating agents. These include nitrogen and sulfur mustards, nitrosoureas, triazenes, and ethyleneimines (93,144–146). They are characteristically reactive compounds that will readily form covalent bonds to nucleophilic centers in biological materials, especially proteins and nucleic acids, and it is natural to look to the latter for explanation of the cytotoxic (or indeed mutagenic/carcinogenic) consequences of exposure to alkylating agents. In DNA the most susceptible (but not sole) site of reaction seems to be the N (7) position of guanine residues. With nitrogen mustards it has long been established that antitumor activity requires bifunctional character so that the molecule can become linked to two different DNA bases forming a cross-link (144, 145, 147, 148). More recently, the same has been established for 2-haloethylnitrosoureas (146, 149, 150) and for cis-dichlorodiammineplatinum (II), the chief representative of an important new class of antitumor drugs (151). Cross-linking of bases in DNA is not a unique property of antitumor agents, however; it has been reported to occur with certain carcinogens (152) and seems to form the basis for the skin-photosensitizing action of plant furocoumarins (psoralens) (153). Understandably, attention has been concentrated upon the ability of all these agents to cause the formation of interstrand cross-links, readily detectable via the appearance of reversible DNA (i.e. DNA whose complementary strands cannot be physically separated and that consequently resists irreversible denaturation). On closer inspection it has become evident that only a fraction of the cross-linked bases can be accounted for by the number of true interstrand cross-links; the remainder must presumably originate from intrastrand bifunctional alkylation reactions. Moreover, it is also clear that a substantial amount of DNA-protein cross-linking can occur. With all these different reactions taking place, not to mention the formation of a large amount of monofunctionally alkylated product, which seems to be the predominant fate of even the most avid bifunctional alkylators, it is not surprising that there is confusion as to the nature of the critical lesion(s) responsible for effective antitumor activity (144–154).

Whereas the biochemistry associated with the action of synthetic alkylating agents used to treat cancer has been well documented (see reviews cited above), rather less has been written about two groups of antibiotics that react covalently with DNA and whose actions are significantly different.

MITOMYCIN Diverse aspects of the biochemistry, pharmacology, and efficacy of this antibiotic in the treatment of cancer have recently been summarized (155). It is a very effective cross-linker of DNA both in vivo and in vitro, but this effect demands prior activation of the molecule, a property that distinguishes mitomycin from the majority of alkylating agents and gives it the edge in clinical treatment of certain tumors. Activation is accomplished by reduction of its quinone moiety (see Figure 3), mediated in vivo by NADPH-dependent enzyme systems and in vitro by a variety of reducing agents such as dithionite or H_2/Pd (120, 156, 157). Once activated, the molecule appears every bit as reactive as typical alkylating agents, and produces the usual spectrum of products among which interstrand DNA cross-links figure only as a minor (though perhaps crucial) component (157–160). The two reaction centers that are believed to provide the bifunctional alkylating character of the molecule are the strained three-membered aziridine ring and the methylurethane side chain at position 10. At one time it was considered that the aromatic NH_2 group at position 7 might be involved (it is known to be important for biological activity), but since analogues that lack this substituent as well as the aziridine ring are still capable of alkylating DNA, the second reactive site must lie elsewhere, i.e. the carbamate at position 10 by a process of elimination (156, 159–161).

After the initial reduction the pathway of activation could proceed along several routes; Moore (156) has considered possibilities of bioreductive activation in general, and the scheme suggested in Figure 3 derives from his ideas. An important feature is the postulated formation of an essential semiquinone intermediate (b), for recent evidence implicates such a species in preliminary noncovalent binding to DNA, quite possibly by intercalation, as a prerequisite for cross-link formation with specific sites in the polymer (159, 162). Several lines of evidence establish that those sites are associated with guanine residues: the extent of cross-linking of natural DNAs rises with increasing GC content, depurination releases the drug

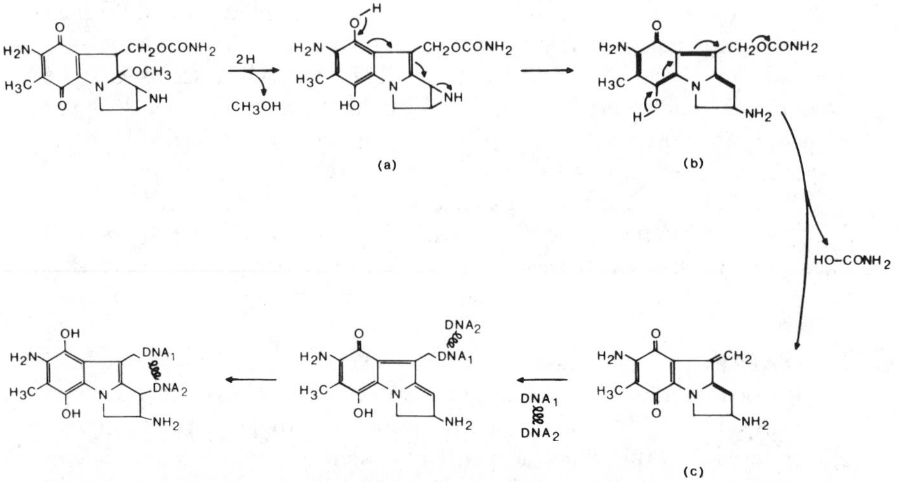

Figure 3 A proposed reaction scheme for the reductive activation of mitomycin C [after Moore (156)]. The key step is the conversion of the fully reduced hydroquinone (*a*) to the semiquinone (*b*) powered by release of the steric strain energy of the aziridine ring; this triggers elimination of the carbamate side chain to yield the bifunctional alkylating agent (*c*) which is the proposed biologically active form. Subsequent reaction with two nucleophilic centers in DNA leads to the establishment of a cross-link, as indicated.

together with normal amounts of free guanine and adenine; apurinic acid fails to react detectably with the antibiotic; poly(dG) binds mitomycin well, whereas binding to other homopolymers is negligible; and alkylation of poly(dG-dC) or poly(dI-dC) occurs readily, though not to the same extent as with poly(dG) alone (159, 160). It seems likely that the preferred, if not sole, sites of alkylation are the O(6) groups of guanine residues; N(7) and C(8) are ruled out by the failure of reaction with mitomycin to labilize ^{3}H attached to C(8) (163), and attachment to O(6) would account for the relative stability of mitomycin cross-links compared to those formed by nitrogen mustards (157, 159, 163). A long time ago Szybalski & Iyer (157) decided that if some local distortion of the helix could be tolerated, the best fit with molecular models was found by connecting a mitomycin molecule between O(6) groups of the nearest guanines on opposite strands. There is, in fact, evidence from CD spectra that alkylation of DNA by mitomycin can provoke conformational changes (164), and it has recently been speculated that those changes may be of the same character as the transition to formation of Z-DNA (165, 166). It would indeed be an exciting discovery if binding of mitomycin were found to provoke a local change in handedness of the DNA helix.

In addition to forming cross-links, when mitomycin is incubated with DNA in vitro in the presence of a reducing agent and molecular O_2 it causes the appearance of single-strand breaks that may represent the first stage of

DNA degradation as promoted by the antibiotic in vivo. This cleavage of DNA in vitro has been shown to proceed by a mechanism precisely analogous to that postulated to explain DNA strand breakage by streptonigrin, outlined below (160, 162).

ANTHRAMYCIN AND RELATED ANTIBIOTICS   Anthramycin (Figure 2) is the best-known member of the pyrrolo-(1,4)-benzodiazepine family of antitumor antibiotics that also includes sibiromycin and tomaymycin. Their modes of action appear to be essentially the same, and involve inhibition of nucleic acid synthesis as a consequence of DNA binding (167–170). Although their interaction with DNA seems to involve formation of a covalent bond it is quite different from the classical alkylation processes noted above, first in respect of the remarkable slowness of reaction, and second in respect of its specificity: neither anthramycin, sibiromycin, nor tomaymycin react significantly with RNA, protein, or simple nucleotides, and heat-denatured DNA binds less antibiotic than native DNA (169–175). Pertinent observations from these studies, which support the idea that a labile covalent bond is formed with DNA, are as follows: (a) The complex resists dissociation by treatments such as dialysis, gel filtration, alcohol precipitation, or addition of detergents. It can, however, be denatured by alkali, and the anthramycin remains bound to the separated strands (under conditions that would not permit significant direct reaction between anthramycin and single-stranded DNA). Evidently cross-links are not formed; (b) DNA-bound anthramycin is protected from decomposition by alkali or heat, and it can be recovered in active, unchanged form upon acid treatment of the complex; (c) Neither strand breaks nor unwinding of the DNA helix can be detected using sensitive assays. Evidently intercalation is not involved; and (d) The interaction with synthetic polynucleotides reveals a clear requirement for guanine residues; moreover poly(dG)·poly(dC) binds high levels of anthramycin, whereas poly(dI)·poly(dC) does not, which suggests that the 2-amino group of guanine may be directly implicated in the interaction. Reaction with N(7) or C(8) is ruled out by the failure to labilize ^{3}H attached to the C(8) position, and by the lack of increased susceptibility to depurination of anthramycin-modified [8-^{3}H]guanine-labeled DNA.

With the publication of a crystal structure for anthramycin (176) sufficient information was available to permit Hurley & Petrusek (177) to propose a tentative molecular model for its complex with DNA. They suggested formation of an aminal linkage between C(11) of the carbinolamine function of the antibiotic and the 2-amino group of guanine, locating the antibiotic in the narrow groove of the DNA helix with its chromophore inclined at 45° to the helix axis, in tolerable agreement with the angle reported from measurements of electric dichroism (173). Further stabilization could arise from H bonding involving the phenolic hydroxyl group. The additional amino sugar substituent of sibiromycin could engage in electrostatic interaction with the deoxyribose-phosphate backbone (169, 177). Like any good molecular model this one makes predictions that can be tested.

Strand-Breaking Interactions

A common characteristic of many DNA-binding drugs, including a substantial number of cytotoxic and antitumor agents, is their capacity to provoke breakage of DNA strands in vivo. For two groups of antibiotics this seems to form the basis of their primary effect on cells.

STREPTONIGRIN This antibiotic contains the same o-aminoquinone moiety present in mitomycin (cf Figures 2 and 3), but it is not a cross-linker. However, like mitomycin it selectively affects DNA synthesis and promotes extensive DNA degradation in vivo; its lethal effect is quite well correlated with the extent of DNA damage, and for maximum lethal effect both molecular O_2 and a reducing agent are required (157, 178–180). In vitro it will attack DNA to produce single-strand breaks, provided O_2 and an electron source are present, and the reaction is inhibited by addition of free radical scavengers such as KI and also by the enzymes catalase and superoxide dismutase, whose effect appears to be synergistic (178,181). Cone et al (181) have proposed that the hydroxyl radical OH· is the ultimate reactive species that initiates attack on DNA, most likely via a series of reactions commencing with the abstraction of a hydrogen atom from position 4 of the deoxyribose ring (182). They put forward the following scheme to explain the involvement of reduced intermediates, where SN represents the antibiotic molecule in its starting (quinone) form:

$$SN + NADH + H^+ \longrightarrow SNH_2 + NAD^+ \qquad\qquad 1.$$

$$SNH_2 + O_2 \longrightarrow SNH^· + HO_2^· \qquad\qquad 2.$$

$$HO_2^· \rightleftharpoons O_2^{\bar{·}} + H^+ \qquad\qquad 3.$$

$$2\,O_2^{\bar{·}} + 2H^+ \xrightarrow{\text{superoxide dismutase}} H_2O_2 + O_2 \qquad\qquad 4.$$

$$2\,H_2O_2 \xrightarrow{\quad\text{catalase}\quad} 2\,H_2O + O_2 \qquad\qquad 5.$$

$$O_2^{\bar{·}} + H_2O_2 \longrightarrow OH^· + OH^- + O_2 \qquad\qquad 6.$$

It appears probable that divalent metal ions, most likely Fe^{2+}, play an important role in the process by complexing with the reduced quinolinequinone in the autoxidation step (Equation 2) and/or facilitating the generation of OH· radicals [Equation 6; (181)]. Lown (162) has suggested that the effect of streptonigrin on the integrity of the cell's DNA is more pronounced than that of mitomycin because streptonigrin inactivates superoxide dismutase.

BLEOMYCIN AND PHLEOMYCIN   Of these two closely related antibiotics (Figure 2) bleomycin is much the better known; it is in the front line of chemotherapeutic attack on cancer (182) and has been the subject of intense research, excellently reviewed (183–188). Both bleomycin and phleomycin promote massive degradation of DNA in vivo and in vitro under conditions that include the presence of molecular O_2, a metal ion, and a reducing agent such as mercaptoethanol or dithiothreitol. The nature of the bleomycin-induced damage has been investigated in several laboratories: both single-strand and double-strand cleavages occur, the latter apparently as a result of staggered single-strand breaks formed in the complementary strands about two base pairs apart (189–192). Careful characterization of the sites of cleavage has revealed several remarkable features: (*a*) For each direct single-strand break there appears approximately one additional alkali-labile site that seems to correspond to the removal of a base moiety (193–195); (*b*) The excised base is most frequently thymine. Cytosine, adenine, and guanine are released in progressively lesser amounts (184, 195–198); and (*c*) There is considerable specificity as regards the site(s) of cleavage: in natural, double-helical DNA, GpT and GpC sequences are especially susceptible to attack (189,196).

It appears that a 1:1 complex between bleomycin and Fe(II) is formed that acts, in the presence of O_2, as a "quasi-enzyme" to mediate catalytic release of DNA bases by a cyclic oxidation-reduction process during which the iron, which probably remains chelated to the antibiotic, is transiently converted to the Fe(III) state (195–199). It is estimated that only one base is released per strand break; that the oxidised Fe(III) bleomycin complex can dissociate from DNA, be reduced to Fe(II) bleomycin, bind back to DNA and again oxidise, eventually releasing up to three DNA bases by successive oxidation-reduction cycles; and that on average 0.36 base is released per oxidation of an Fe(II) bleomycin complex (195). The latter estimate is significant and indicative of a certain lack of specificity in the action of bleomycin; it would be consistent with a reaction mechanism involving the generation of an intermediate free radical, produced at a specific site in the iron·bleomycin·DNA complex, but having finite probabilities for attacking deoxyribose, bleomycin itself (leading to eventual inactivation), or escaping into solution (195). The exact chemistry of events subsequent to the oxidation of the iron-antibiotic complex is presently obscure. The ultimate reactive species may be superoxide anion, or OH⁻ as postulated for streptonigrin, but reports of inhibition (or lack of) by enzymes such as superoxide dismutase or free-radical scavengers have been conflicting (195, 198, 200).

Physical binding of bleomycin to DNA is primarily dependent upon the portion of the molecule containing the bithiazole rings and the cationic side chain (R in Figure 2), but there is uncertainty as to the precise nature of the interaction. Povirk et al (201) reported classical symptoms of intercalation, including unwinding of circular DNA and lengthening of the helix, whereas Chien et al (202) concluded that intercalation must be considered unlikely because they were unable to detect much increase in the viscosity of sonicated DNA fragments. Phleomycin also yielded an ambiguous result when added to closed-circular duplex PM2 DNA: it was found to remove, but not reverse, the supercoiling (M. J. Waring, unpublished observations). The notion that physical binding to DNA involving some change in the winding of the helix is relevant to bleomycin-induced DNA breakage is supported by two further findings: that supercoiled Col E1 DNA is nicked at a rate of 1.5 times greater than relaxed-circular DNA (201), and that certain sites in PM2 DNA appear relatively more reactive when the DNA contains superhelical turns (191). Perhaps intercalation of the bithiazole rings between DNA bases at the iron (II)·bleomycin binding site distorts the helix in such a fashion as to make the deoxyribose more accessible to radical attack (195). This is the latest of a series of ingenious suggestions advanced to account for the selectivity and mechanism of attack on DNA that results from interaction with bleomycin (184, 190, 203, 204).

Another antitumor agent that causes DNA strand breakage both in vivo and in vitro by mechanisms that appear similar, though not identical, to those described for bleomycin, is neocarzinostatin (205,206). It is a single-chain, acidic protein of molecular weight 10,700 that contains a prosthetic chromophore; the latter appears to be entirely responsible for the cleavage activity (207). Single-strand DNA breaks produced by neocarzinostatin in vitro occur almost exclusively at thymidylate and adenylate residues (189, 206).

CELLULAR RESPONSES TO DNA MODIFICATION

One obvious consequence that might reasonably be expected to result from any physical or chemical assault upon DNA in vivo is the production of mutations. The study of mutagenesis is now a highly developed science, and in several instances the physical basis for generation of particular types of mutations (e.g. frameshifts) as a result of interaction with small molecules has been quite well defined (1, 8, 208, 209). Perhaps because of the ease and extreme sensitivity of mutagenicity testing it has been possible to establish that the great majority of known carcinogens and not a few antitumor drugs are mutagens (210, 211), so that mutagenicity assays have become adopted as indispensable short-term tests in screening programs aimed at detecting potential carcinogens (212). It is not, however, generally believed that such effects may be held accountable for the cytotoxic, therapeutic actions of anticancer drugs, and indeed a profitable line of current research in drug development consists of efforts to dissociate the two, i.e. to come up with effective cytostatic/cytotoxic agents that have zero or negligible mutagenic (and hopefully, by extension, carcinogenic) potential.

What, then, does kill cells? It is easy enough in the case of cross-linkers like nitrogen mustards and mitomycin to point to the cross-links as strong interstrand joints that would physically prevent separation of the parental DNA strands and their incorporation into different daughter duplexes during semiconservative replication, but there is a dearth of hard evidence that this really is the lethal lesion inside cells (145–151, 154, 157). Attractive though that postulate may be, there is little evidence that favors it over alternative explanations [e.g. that DNA-protein cross-links are of critical significance (151)]. For those drugs that bind reversibly to DNA, the origins of their lethal effect(s) are still harder to trace. Many of them manifest their presence through selective inhibition of DNA synthesis, and it may be that the mechanism of their killing action is much the same as that of the alkylating agents. Others, however, are clearly selective inhibitors of RNA synthesis, so that unless we are to attribute their action to prevention of synthesis of RNA primers required for DNA replication, alternative mechanisms must be sought. To explain the impressive potency of actinomycin D as an inhibitor of DNA-directed RNA synthesis, Müller & Crothers (39) originally suggested that its effectiveness might be related to the slow dissociation rate constant of the antibiotic-DNA complex. According to this view actinomycin would inhibit transcription by providing more or less long-lived blocks to the progression of RNA polymerase along its DNA template. This concept has stood up well in the light of subsequent work and has been extended to the action of other antitumor agents: in the chromomycin (119), daunomycin (213, 214), and ethidium (82, 215) series, correlations have been found between efficacy in vivo, inhibitory activity towards RNA polymerase, and the rate constants for dissociation of the ligand-DNA complex. There still remains a gap, however, between understanding the molecular basis of inhibition of RNA synthesis and relating that biochemical blockade to cell death, if indeed the two truly are causally related.

Some light has been shed on likely mechanisms of lethality, or at least on important contributory factors, from the results of careful studies on the repair of DNA modified by cytotoxic, mutagenic, and carcinogenic chemicals (216). An unexpected finding, doubtless related at least in some cases to the operation of excision-repair enzymes, was the discovery that DNA strand breakage is a common phenomenon in cells exposed to intercalating or other DNA-binding drugs (217–222). This raises the interesting possibility that the cytotoxic action of such drugs may be better related to consequential damage sustained by the DNA as a result of the action of nucleases that "recognize" distortion of the helical structure than to the physical presence of the bound drug itself. In any event, the idea that DNA strand breakage is a critical step on the path to cell death is strengthened by this evidence. It receives further support from the recent finding that treatment of human lymphoid cells with methotrexate leads to substantial misincorporation of dUMP residues into DNA, with consequent accumulation of DNA damage due to the concerted action of excision-repair enzymes and Ura-DNA glycosylase (223). Only a year or two ago few people would have attached credence to any suggestion that antimetabolites might kill cells by a mechanism other than simple inhibition of macromolecular biosynthesis due to starvation for essential substrates. It can be predicted with confidence that the pursuit of these findings, and the role of repair mechanisms in healthy and compromised cells, will increasingly engage the attention of biochemists over the coming decade.

RATIONAL APPROACHES AND THE QUEST FOR SPECIFICITY

Future Directions

Without wishing to decry the achievements of classical medicinal chemistry as a route to efficacious drug development (and several outstanding examples of its success have been described above), there is a widespread feeling that the time has come for radically new directions of attack upon the cancer problem, in particular via the application of so-called rational approaches to chemotherapy. One such direction that commends itself naturally to medicinal chemists lies in the development of quantitative structure-activity relationships [QSAR (115, 122, 224–226)]. Others, which similarly depend upon computer technology for their prosecution, include quantum-mechanical calculations and sophisticated molecular model building. It would be invidious to try to select representative examples of each of these approaches. However, since so much is expected from molecular biology as the future servant of medicine, and since this article is specifically concerned with DNA modification, I discuss a group of related efforts to exploit the specificity inherent in the structure of DNA by means of molecular interactions that we can claim to understand, at least in principle.

Bifunctional (Bis) Intercalative Binding to DNA

In the mid-1970s several laboratories independently recognized the possibility of synthesizing bifunctional intercalating drug molecules as a means of enhancing the binding constant and perhaps attaining selective binding to preferred sites in DNA. They hoped that this would be a step toward the goal of attacking defined, biologically meaningful nucleotide sequences that might provide the effective discrimination between cell types and/or function so sadly lacking in existing anticancer drugs. Paradoxically, the first such compound to be characterized was naturally occurring: a quinoxaline antibiotic already known to possess antitumor activity, echinomycin (1, 227–229) (Figure 4).

Echinomycin was shown to remove and reverse the supercoiling of closed-circular duplex DNA with an apparent unwinding angle almost twice as large as that of ethidium (Table 1), and to extend the DNA helix by 6.3 Å per bound antibiotic molecule (49, 232), again twice as large as the value reported for ethidium and only slightly short of the theoretical value for two intercalation events per bound ligand molecule. Measured by a specially developed solvent-partition method (233), the binding was shown to be specific for double-helical DNA (not RNA) and to vary considerably in strength depending upon the source of the polymer: a broad preference for GC-rich DNA sequences was inferred (232). Closely related quinoxaline antibiotics were also shown to be efficient bifunctional intercalators; these include triostin A, which is the exact homologue of echino-

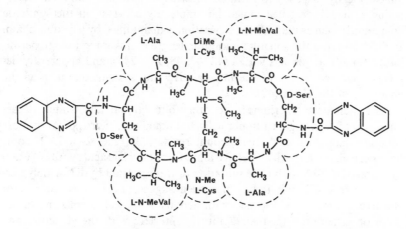

Figure 4 Structure of echinomycin (quinomycin A) (230, 231).

mycin having a disulfide cross-bridge spanning the octapeptide lactone ring (as opposed to a thioacetal; cf Figure 4) (50). The same mechanism of binding was established for a synthetic analogue, des-N-tetramethyltriostin A (TANDEM) (51). Most significantly, however, the patterns of sequence-preference for DNA binding by these agents were found to be quite different (50, 51, 232). Detailed investigation of structure-activity relations for DNA binding has confirmed that small, defined changes in either the chromophores or the peptide ring structure of these antibiotics can radically alter their sequence-recognition properties (228, 229, 234, 235). A molecular model, based on the findings of NMR experiments (236, 237) and energy calculations (238), proposes that quinoxaline antibiotics form a *bis*-intercalated complex with DNA that has two base pairs sandwiched between the chromophores. Specific sequence-recognition is attributed to H-bonding interactions between substituents on the peptide ring and functional groups of the base pairs exposed in one of the grooves of the helix, most probably the narrow groove (1, 229, 236).

The first synthetic *bis*-intercalating compounds were dimeric molecules containing two moieties of 2-methoxy-6-chloro-9-aminoacridine (the chromophore of the well-known antimalarial drug, quinacrine) linked by spermine or spermidine chains (48). Subsequently, bifunctional intercalators containing a variety of chromophores have been reported (42, 47, 239–243), though not all putative bifunctional molecules have proved to *bis*-intercalate on careful examination (244). One factor that evidently affects the capacity of a dimer to *bis* intercalate is the length and character of the "linker" chain. In echinomycin and its congeners, the chromophores are separated by a relatively rigid peptide ring that constrains them to lie almost ideally disposed to form a sandwich of two base pairs, effectively acting as a well-designed molecular staple. By contrast, in the synthetic compounds such as diacridines, separation is maintained by a flexible chain. In different series the apparent minimum length of linker required to permit bifunctional reaction has varied (42, 47, 48, 240–245), and apparently depends upon the nature of the intercalative ring systems as much as the character of the chain (42).

As anticipated, diacridines and other bifunctional DNA-intercalators have proved active against tumor cells, but not always more potent or more selective than their monomeric counterparts (239, 240, 243, 246). Some have been characterized as effective inhibitors of nucleic acid synthesis (239) that preferentially inhibit the initiation of transcription by RNA polymerase in vitro (247), but it has also been suggested that their antitumor effectiveness may be associated with actions at the cell membrane rather than, or in addition to, their ability to interfere with the structure and function of DNA (246, 248). Be that as it may, the search for "magic bullets" to fulfil a human need demands that every avenue which shows promise be explored, and if by tailoring trial molecules to fit DNA we arrive at useful drugs which act by additional unexpected mechanisms that can only be regarded as a bonus. The urgency of the need is not in dispute.

Literature Cited

1. Gale, E. F., Cundliffe, E., Reynolds, P. E., Richmond, M. H., Waring, M. J. 1981. *The Molecular Basis of Antibiotic Action,* London: Wiley. 2nd ed.
2. Searle, C. E., ed. 1976. *ACS Monogr. 173: Chemical Carcinogens.* Washington DC: Am. Chem. Soc. 788 pp.
3. Pegg, A. E. 1977. *Adv. Cancer Res.* 25:195–269
4. Miller, J. A., Miller, E. C. 1977. *Origins of Human Cancer,* ed. H. H. Hiatt, J. D. Watson, J. A. Winsten, pp. 605–27. Cold Spring Harbor, NY: Cold Spring Harbor Lab.
5. Cairns, J. 1978. *Cancer, Science and Society.* San Francisco: Freeman. 199 pp.
6. Gelboin, H. V., Ts'o, P. O. P., eds. 1978. *Polycyclic Hydrocarbons and Cancer,* Vols. 1, 2. New York: Academic. 432 pp., 480 pp.
7. Grunberger, D., Weinstein, I. B. 1979. *Prog. Nucleic Acids Res. Mol. Biol.* 23:105–49
8. Grover, P. L., ed. 1979. *Chemical Carcinogens and DNA,* Vols. 1, 2. Boca Raton, Fla: CRC. 236 pp., 210 pp.
9. Brookes, P., ed. 1980. *Br. Med. Bull.* 36:No. 1
10. Lerman, L. S. 1961. *J. Mol. Biol.* 3:18–30
11. Waring, M. J. 1970. *J. Mol. Biol.* 54:247–79
12. Crawford, L. V., Waring, M. J. 1967. *J. Mol. Biol.* 25:23–30
13. Bauer, W. R., Vinograd, J. 1968. *J. Mol. Biol.* 33:141–71
14. Bauer, W. R. 1978. *Ann. Rev. Biophys. & Bioeng.* 7:287–313
15. Tattersall, M. H. N. 1980. *Cancer: Causes and Control,* ed. D. Metcalfe. Canberra: Australian Acad. Sci.
16. Michaelis, L. 1947. *Cold Spring Harb. Symp. Quant. Biol.* 12:131–142
17. Kirk, J. M. 1960. *Biochim. Biophys. Acta* 42:167–69
18. Rauen, H. M., Kersten, H., Kersten, W. 1960. *Z. Physiol. Chem.* 321:139–47
19. Reich, E., Goldberg, I. H. 1964. *Prog. Nucleic Acid Res. Mol. Biol.* 3:183–234
20. Peacocke, A. R., Skerrett, J. N. H. 1956. *Trans. Faraday Soc.* 52:261–79
21. Acheson, R. M., ed. 1973. *Acridines.* London: Wiley. 878 pp. 2nd ed.
22. Waring, M. J. 1975. *Antibiotics. III. Mechanism of Action of Antimicrobial and Antitumor Agents,* eds. J. W. Corcoran, F. E. Hahn, pp. 141–65. Berlin/ Heidelberg/New York: Springer
23. Fuller, W., Waring, M. J. 1964. *Ber. Bunsenges. Physik. Chem.* 68:805–8
24. Cairns, J. 1962. *Cold Spring Harbor Symp. Quant. Biol.* 27:311–18
25. Lang, D. 1971. *Philos. Trans. R. Soc. London Ser. B* 261:151–58
26. Freifelder, D. 1971. *J. Mol. Biol.* 60:401–3
27. Butour, J. L., Delain, E., Coulaud, D., Le Pecq, J. B., Barbet, J., Roques, B. P. 1978. *Biopolymers* 17:873–86
28. Cohen, G., Eisenberg, H. 1969. *Biopolymers* 8:45–55
29. Reinert, K. E. 1973. *Biochim. Biophys. Acta* 319:135–39
30. Revet, B. M. J., Schmir, M., Vinograd, J. 1971. *Nature New Biol.* 229:10–13
31. Waring, M. J., Henley, S. M. 1975. *Nucleic Acids Res.* 2:567–86
32. Espejo, R. T., Lebowitz, J. 1976. *Anal. Biochem.* 72:95–103
33. Lerman, L. S. 1963. *Proc. Natl. Acad. Sci. USA* 49:94–102
34. Houssier, C., Hardy, B., Fredericq, E. 1974. *Biopolymers* 13:1141–60
35. Hogan, M., Dattagupta, N., Crothers, D. M. 1979. *Biochemistry* 18:280–88
36. Reinert, K. E. 1972. *J. Mol. Biol.* 72:593–607
37. Hogan, M., Dattagupta, N., Crothers, D. M. 1979. *Nature* 278:521–24
38. Waring, M. J., Chisholm, J. W. 1972. *Biochim. Biophys. Acta* 262:18–23
39. Müller, W., Crothers, D. M. 1968. *J. Mol. Biol.* 35:251–90
40. Waring, M. J., González, A., Jiménez, A., Vázquez, D. 1979. *Nucleic Acids Res.* 7:217–30
41. Müller, W., Crothers, D. M., Waring, M. J. 1973. *Eur. J. Biochem.* 39:223–34
42. Wright, R. G. Mc. R., Wakelin, L. P. G., Fieldes, A., Acheson, R. M., Waring, M. J. 1980. *Biochemistry* 19: 5825–36
43. Waring, M. J. 1976. *Eur. J. Cancer* 12:995–1001
44. Waring, M. J. 1975. *Topics in Infectious Diseases, Vol. 1, Drug-Receptor Interac-*

tions in Antimicrobial Chemotherapy, ed. J. Drews, F. E. Hahn, pp. 77–90. Vienna: Springer

45. Kohn, K. W., Waring, M. J., Glaubiger, D., Friedman, C. A. 1975. *Cancer Res.* 35:71–76

46. Wang, J. C. 1971. *Biochim. Biophys. Acta* 232:246–51

47. Wakelin, L. P. G., Romanos, M., Chen, T. K., Glaubiger, D., Canellakis, E. S., Waring, M. J. 1978. *Biochemistry* 17:5057–63

48. Le Pecq, J. B., Le Bret, M., Barbet, J., Roques, B. 1975. *Proc. Natl. Acad. Sci USA* 72:2915–19

49. Waring, M. J., Wakelin, L. P. G. 1974. *Nature* 252:653–57

50. Lee, J. S., Waring, M. J. 1978. *Biochem. J.* 173:115–28

51. Lee, J. S., Waring, M. J. 1978. *Biochem. J.* 173:129–44

52. Drinkwater, N. R., Miller, J. A., Miller, E. C., Yang, N. C. 1978. *Cancer Res.* 38:3247–55

53. Jain, S. C., Tsai, C. C., Sobell, H. M. 1977. *J. Mol. Biol.* 114:317–31

54. Tsai, C. C., Jain, S. C., Sobell, H. M. 1977. *J. Mol. Biol.* 114:301–15

55. Reddy, B. S., Seshadri, T. P., Sakore, T. D., Sobell, H. M. 1979. *J. Mol. Biol.* 135:787–812

56. Neidle, S., Achari, A., Taylor, G. L., Berman, H. M., Carrell, H. L., Glusker, J. P., Stallings, W. C. 1977. *Nature* 269:304–7

57. Shieh, H. S., Berman, H. M., Dabrow, M., Neidle, S. 1980. *Nucleic Acids Res.* 8:85–97

58. Wang, A. H. J., Quigley, G. J., Rich, A. 1979. *Nucleic Acids Res.* 6:3879–90

59. Sakore, T. D., Reddy, B. S., Sobell, H. M. 1979. *J. Mol. Biol.* 135:763–86

60. Jain, S. C., Bhandary, K. K., Sobell, H. M. 1979. *J. Mol. Biol.* 135:813–40

61. Wang, A. H. J., Nathans, J., Van der Marel, G., Van Boom, J. H., Rich, A. 1978. *Nature* 276:471–74

62. Sobell, H. M., Jain, S. C., Sakore, T. D., Reddy, B. S., Bhandary, K. K., Seshadri, T. P. 1978. *Int. Symp. Biomol. Struct. Conform. Funct. Evol.,* Madras, India

63. Neidle, S., ed. 1981. *Topics in Nucleic Acid Structure,* pp. 177–96. London: Macmillan

64. Krugh, T. R., Wittlin, F. N., Cramer, S. P. 1975. *Biopolymers* 14:197–210

65. Reinhardt, C. G., Krugh, T. R. 1978. *Biochemistry* 17:4845–54

66. Patel, D. J. 1979. *Acc. Chem. Res.* 12:118–25

67. Patel, D. J., Shen, C. 1978. *Proc. Natl. Acad. Sci. USA* 75:2553–57

68. Krugh, T. R., Nuss, M. E. 1979. *Biological Applications of Magnetic Resonance,* ed. R. G. Shulman, pp. 113–75. New York/London: Academic

69. Patel, D. J. 1980. In *Nucleic Acid Geometry and Dynamics,* ed. R. H. Sarma, pp. 185–231. Oxford: Pergamon

70. Sobell, H. M., Tsai, C. C., Gilbert, S. G., Jain, S. C., Sakore, T. D. 1976. *Proc. Natl. Acad. Sci. USA* 73:3068–72

71. Sobell, H. M., Tsai, C. C., Jain, S. C., Gilbert, S. G. 1977. *J. Mol. Biol.* 114:333–65

72. Crothers, D. M. 1968. *Biopolymers* 6:575–84

73. McGhee, J. D., Von Hippel, P. H. 1974. *J. Mol. Biol.* 86:469–89

74. Neidle, S., Taylor, G., Sanderson, M., Shieh, H. S., Berman, H. M. 1978. *Nucleic Acids Res.* 5:4417–22

75. Westhof, E., Sundaralingam, M. 1980. *Proc. Natl. Acad. Sci. USA* 77:1852–56

76. Seeman, N. C., Day, R. O., Rich, A. 1975. *Nature* 253:324–26

77. Liebman, M., Rubin, J., Sundaralingam, M. 1977. *Proc. Natl. Acad. Sci. USA* 74:4821–25

78. Jones, C. R., Kearns, D. R. 1975. *Biochemistry* 14:2660–65

79. Li, H. J., Crothers, D. M. 1969. *J. Mol. Biol.* 39:461–77

80. Schmechel, D. E. V., Crothers, D. M. 1971. *Biopolymers* 10:465–80

81. Ramstein, J., Dourlent, M., Leng, M. 1972. *Biochem. Biophys. Res. Commun.* 47:874–82

82. Wakelin, L. P. G., Waring, M. J. 1980. *J. Mol. Biol.* 144:183–214

83. Bresloff, J. L., Crothers, D. M. 1975. *J. Mol. Biol.* 95:103–23

84. Deleted in proof

85. Di Marco, A., Arcamone, F., Zunino, F. 1975. See Ref. 22, pp. 101–28

86. Pigram, W. J., Fuller, W., Hamilton, L. D. 1972. *Nature* 235:17–19

87. Angiuli, R., Foresti, E., Riva di Sanseverino, L., Isaacs, N. W., Kennard, O., Motherwell, W. D. S., Wampler, D. L., Arcamone, F. 1971. *Nature New Biol.* 234:78–80

88. Neidle, S., Taylor, G. 1977. *Biochim. Biophys. Acta* 479:450–59

89. Neidle, S. 1979. *Prog. Med. Chem.* 16:151–221

90. Sengupta, S. K., Seshadri, R., Modest, E. J., Israel, M. 1976. *Proc. Am. Assoc. Cancer Res.* 17:109

91. Krishan, A., Israel, M., Modest, E. J., Frei, E. 1976. *Cancer Res.* 36:2114–16

92. Israel, M., Pegg, W. J., Wilkinson, P. M. 1978. *J. Pharm. Exp. Ther.* 204:696–701
93. Pratt, W. B., Ruddon, R. W. 1979. *The Anticancer Drugs.* New York/Oxford: Oxford Univ. Press. 323 pp.
94. Meienhofer, J., Atherton, E. 1977. *Structure-Activity Relationships among the Semisynthetic Antibiotics,* ed. D. Perlman, pp. 427–529. New York: Academic
95. Cerami, A., Reich, E., Ward, D. C., Goldberg, I. H. 1967. *Proc. Natl. Acad. Sci. USA* 57:1036–42
96. Wells, R. D., Larson, J. E. 1970. *J. Mol. Biol.* 49:319–42
97. Krugh, T. R., Young, M. A. 1977. *Nature* 269:627–28
98. Fasy, T. M., Kallos, J., Bick, M. D. 1979. *J. Biol. Chem.* 254:4492–98
99. Sobell, H. M., Jain, S. C., Sakore, T. D., Nordman, C. E. 1971. *Nature New Biol.* 231:200–5
100. Davanloo, P., Crothers, D. M. 1976. *Biochemistry* 15:5299–305
101. Gellert, M., Smith, C. E., Neville, D., Felsenfeld, G. 1965. *J. Mol. Biol.* 11:445–57
102. Kohn, K. W., Ross, W. E., Glaubiger, D. 1979. *Antibiotics. V, Part 2, Mechanism of Action of Antieukaryotic and Antiviral Compounds,* ed. F. E. Hahn, pp. 195–213. Berlin/Heidelberg/New York: Springer
103. Chandra, P., Woltersdorf, M., Wright, G. J. 1979. See Ref. 102, pp. 385–413
104. Cain, B. F., Atwell, G. J. 1974. *Eur. J. Cancer* 10:539–49
105. Legha, S. S., Gutterman, J. U., Hall, S. W., Benjamin, R. S., Burgess, M. A., Valdivieso, M., Bodey, G. P. 1978. *Cancer Res.* 38:3712–16
106. Kohn, K. W., Waring, M. J., Glaubiger, D., Friedman, C. A. 1975. *Cancer Res.* 35:71–76
107. Le Pecq, J. B., Nguyen Dat Xuong, Gosse, C., Paoletti, C. 1974. *Proc. Natl. Acad. Sci. USA* 71:5078–82
108. Chandra, P., Woltersdorf, M. 1976. *Biochem. Pharmacol.* 25:877–80
109. Müller, W., Crothers, D. M. 1975. *Eur. J. Biochem.* 54:267–77
110. Müller, W., Bünemann, H., Dattagupta, N. 1975. *Eur. J. Biochem.* 54:279–91
111. Müller, W., Gautier, F. 1975. *Eur. J. Biochem.* 54:385–94
112. Cain, B. F., Seelye, R. N., Atwell, G. J. 1974. *J. Med. Chem.* 17:922–30
113. Wilson, W. R., Baguley, B. C., Wakelin, L. P. G., Waring, M. J. 1981. *Mol. Pharmacol.* In press
114. Gormley, P. E., Sethi, V. S., Cysyk, R. L. 1978. *Cancer Res.* 38:1300–6
115. Baguley, B. C., Denny, W. A., Atwell, G. J., Cain, B. F. 1981. *J. Med. Chem.* In press
116. Gause, G. F. 1975. See Ref. 22, pp. 197–202
117. Schweizer, D. 1976. *Chromosoma* 58:307–24
118. Van de Sande, J. H., Lin, C. C., Jorgenson, K. F. 1977. *Science* 195:400–2
119. Behr, W., Honikel, K., Hartmann, G. 1969. *Eur. J. Biochem.* 9:82–92
120. Kersten, H., Kersten, W. 1974. *Inhibitors of Nucleic Acid Synthesis. Biophysical and Biochemical Aspects. Molecular Biology, Biochemistry and Biophysics* Vol. 18. Berlin/Heidelberg/New York: Springer. 184 pp.
121. Ward, D., Reich, E., Goldberg, I. H. 1965. *Science* 149:1259–63
122. Denny, W. A., Atwell, G. J., Baguley, B. C., Cain, B. F. 1979. *J. Med. Chem.* 22:134–50
123. Hahn, F. E., 1975. See Ref. 22, pp. 79–100
124. Zimmer, C. 1975. *Progr. Nucleic Acid Res. Mol. Biol.* 15:285–318
125. Luck, G., Triebel, H., Waring, M., Zimmer, C. 1974. *Nucleic Acids Res.* 1:503–30
126. Wartell, R. M., Larson, J. E., Wells, R. D. 1974. *J. Biol. Chem.* 249:6719–31
127. Patel, D. J., Canuel, L. L. 1977. *Proc. Natl. Acad. Sci. USA* 74:5207–11
128. Nosikov, V. V., Sain, B. 1977. *Nucleic Acids Res.* 4:2263–73
129. Reinert, K. E., Stutter, E., Schweiss, H. 1979. *Nucleic Acids Res.* 7;1375–92
130. Sutherland, J. C., Duval, J. F., Griffin, K. P. 1978. *Biochemistry* 17:5088–91
131. Zimmer, C., Marck, C., Schneider, C., Thiele, D., Luck, G., Guschlbauer, W. 1980. *Biochim. Biophys. Acta* 607:232–46
132. Berman, H. M., Neidle, S., Zimmer, C., Thrum, H. 1979. *Biochim. Biophys. Acta* 561:124–31
133. Gurskaya, G. V., Grokhovsky, S. L., Zhuze, A. L., Gottikh, B. P. 1979. *Biochim. Biophys. Acta* 563:336–42
134. Krylov, A. S., Grokhovsky, S. L., Zasedatelev, A. S., Zhuze, A. L., Gursky, G. V., Gottikh, B. P. 1979. *Nucleic Acids Res.* 6:289–304
135. Kolchinsky, A. M., Mirzabekov, A. D., Zasedatelev, A. S., Gursky, G. V., Grokhovsky, S. L., Zhuze, A. L., Gottikh, B. P. 1975. *Mol. Biol.* 9:19–27
136. Krey, A. K., Hahn, F. E. 1970. *FEBS Lett.* 10:175–78

137. Sveshnikov, P. G., Grokhovsky, S. L., Zhuze, A. L., Kondratyeva, N. O., Makarov, V. L., Poletayev, A. I. 1978. *Mol. Biol.* 12:557–64
138. Patel, D. J. 1979. *Eur. J. Biochem.* 99:369–78
139. Mahler, H. R., Green, G., Goutarel, R., Khuong-Huu-Qui 1968. *Biochemistry* 7:1568–82
140. Gabbay, E. J., Glaser, R. 1971. *Biochemistry* 10:1665–74
141. Silver, S., Wendt, L., Bhattacharyya, P. 1975. See Ref. 22, pp. 614–22
142. Dattagupta, N., Hogan, M., Crothers, D. M. 1978. *Proc. Natl. Acad. Sci. USA* 75:4286–90
143. Patel, D. J., Canuel, L. L. 1979. *Proc. Natl. Acad. Sci. USA* 76:24–28
144. Ludlum, D. B. 1975. *Handb. Exp. Pharm.* 38:Pt. 2, pp. 6–17
145. Connors, T. A., 1975. See Ref. 144, pp. 18–34
146. Wheeler, G. P. 1975. See Ref. 144, pp. 65–84
147. Kohn, K. W., Spears, C. L., Doty, P. 1966. *J. Mol. Biol.* 19:266–88
148. Lawley, P. D., Brookes, P. 1967. *J. Mol. Biol.* 25:143–59
149. Kohn, K. W., 1977. *Cancer Res.* 37:1450–54
150. Lown, J. W., McLaughlin, L. W. 1979. *Biochem. Pharmacol.* 28:2123–28
151. Roberts, J. J., Thomson, A. J. 1979. *Progr. Nucl. Acid Res. Mol. Biol.* 22:71–133
152. Maher, V. M., Miller, E. C., Miller, J. A., Szybalski, W. 1968. *Mol. Pharmacol.* 4:411–26
153. Song, P. S., Tapley, K. J. 1979. *Photochem. Photobiol.* 29:1177–97
154. Kohn, K. W. 1979. *Methods in Cancer Res.* 16:291–345
155. Carter, S. K., Crooke, S. T., eds. 1979. *Mitomycin C. Current Status and New Developments.* New York: Academic. 254 pp.
156. Moore, H. W. 1977. *Science* 197:527–32
157. Szybalski, W., Iyer, V. N. 1967. *Antibiotics, I, Mechanism of Action,* ed. D. Gottlieb, P. D. Shaw, pp. 211–45. Berlin, Heidelberg & New York: Springer
158. Weissbach, A., Lisio, A. 1965. *Biochemistry* 4:196–200
159. Tomasz, M., Mercado, C. M., Olson, J., Chatterjie, N. 1974. *Biochemistry* 13:4878–87
160. Lown, J. W., Begleiter, A., Johnson, D., Morgan, A. R. 1976. *Can. J. Biochem.* 54:110–19
161. Lown, J. W., Weir, G. 1978. *Can. J. Biochem.* 56:296–304
162. Lown, J. W. 1979. See Ref. 155, pp. 5–26
163. Tomasz, M. 1970. *Biochim. Biophys. Acta* 213:288–95
164. Mercado, C. M., Tomasz, M. 1977. *Biochemistry* 16:2040–46
165. Wang, A. H. J., Quigley, G. J., Kolpak, F. J., Crawford, J. L., Van Boom, J. H., Van der Marel, G., Rich, A. 1979. *Nature* 282:680–86
166. Arnott, S., Chandrasekaran, R., Birdsall, D. L., Leslie, A. G. W., Ratliff, R. L. 1980. *Nature* 283:743–45
167. Horwitz, S. B. 1971. *Progr. Mol. Subcell. Biol.* 2:40–47
168. Kohn, K. W. 1975. See Ref. 22, pp. 3–11
169. Gause, G. F. 1975. See Ref. 22, pp. 269–73
170. Hurley, L. H. 1977. *J. Antibiot.* 30:349–70
171. Kohn, K. W., Spears, C. L. 1970. *J. Mol. Biol.* 51:551–72
172. Kohn, K. W., Glaubiger, D., Spears, C. L. 1974. *Biochim. Biophys. Acta* 361:288–302
173. Glaubiger, D., Kohn, K. W., Charney, E. 1974. *Biochim. Biophys. Acta* 361:303–11
174. Kozmyan, L. I., Gauze, G. G., Galkin, V. I., Dudnik, Yu. V. 1978. *Antibiotiki* 23:771–75
175. Lown, J. W., Joshua, A. V. 1979. *Biochem. Pharmacol.* 28:2017–26
176. Mostad, A., Romming, C., Storm, B. 1978. *Acta. Chim. Scand.* B32:639–45
177. Hurley, L. H., Petrusek, R. 1979. *Nature* 282:529–31
178. White, H. L., White, J. R. 1966. *Biochem. Biophys. Acta* 123:648–51
179. White, H. L., White, J. R. 1968. *Mol. Pharmacol.* 4:549–65
180. Mizuno, N. S. 1979. See Ref. 102, pp. 372–84
181. Cone, R., Hasan, S. K., Lown, J. W., Morgan, A. R. 1976. *Can. J. Biochem.* 54:219–23
182. Carter, S. K., Crooke, S. T., Umezawa, H., eds. 1978. *Bleomycin: Current Status and New Developments.* New York: Academic. 365 pp.
183. Umezawa, H. 1975. See Ref. 22, pp. 21–33
184. Müller, W. E. G., Zahn, R. K. 1977. *Prog. Nucleic Acid Res. Mol. Biol.* 20:21–57
185. Haidle, C. W., Lloyd, R. S. 1979. See Ref. 102, pp. 124–54
186. Hecht, S. M., ed. 1979. *Bleomycin: Chemical, Biochemical and Biological Aspects.* Berlin/Heidelberg/New York: Springer. 351 pp.

187. Hori, M. See Ref. 186, pp. 195–206
188. Earhart, C. F. 1979. See Ref. 102, pp. 298–312
189. D'Andrea, A. D., Haseltine, W. A. 1978. *Proc. Natl. Acad. Sci. USA* 75:3608–12
190. Haidle, C. W., Lloyd, R. S. 1979. See Ref. 102, pp. 124–54
191. Lloyd, R. S., Haidle, C. W., Robberson, D. L. 1979. *Gene* 7:289–302
192. Lloyd, R. S., Haidle, C. W., Robberson, D. L. 1979. *Gene* 7:303–16
193. Lloyd, R. S., Haidle, C. W., Hewitt, R. R. 1978. *Cancer Res.* 38:3191–96
194. Ross, S. L., Moses, R. E. 1978. *Biochemistry* 17:581–86
195. Povirk, L. F. 1979. *Biochemistry* 18:3989–95
196. Takeshita, M., Grollman, A. P., Ohtsubo, E., Ohtsubo, H. 1978. *Proc. Natl. Acad. Sci. USA* 75:5983–87
197. Sausville, E. A., Peisach, J., Horwitz, S. B. 1978. *Biochemistry* 17:2740–46
198. Sausville, E. A., Stein, R. W., Peisach, J., Horwitz, S. B. 1978. *Biochemistry* 17:2746–54
199. Oppenheimer, N. J., Rodriguez, L. O., Hecht, S. M. 1979. *Proc. Natl. Acad. Sci. USA* 76:5616–20
200. Lown, J. W., Sim, S. K. 1977. *Biochem. Biophys. Res. Commun.* 77:1150–57
201. Povirk, L. F., Hogan, M., Dattagupta, N. 1979. *Biochemistry* 18:96–101
202. Chien, M., Grollman, A. P., Horwitz, S. B. 1977. *Biochemistry* 16:3641–47
203. Murakami, H., Mori, H., Taira, S. 1976. *J. Theor. Biol.* 59:1–23
204. Takeshita, M., Grollman, A. P. 1979. See Ref. 186, pp. 207–21
205. Sim, S. K., Lown, J. W. 1978. *Biochem. Biophys. Res. Commun.* 81:99–105
206. Goldberg, I. H. 1979. See Ref. 102, pp. 262–74
207. Kappen, L. S., Napier, M. A., Goldberg, I. H. 1980. *Proc. Natl. Acad. Sci. USA* 77:1970–74
208. Roth, J. R. 1974. *Ann. Rev. Genet.* 8:319–46
209. Drake, J. W., Baltz, R. H. 1976. *Ann. Rev. Biochem.* 45:11–37
210. Ames, B. N., Sims, P., Grover, P. L. 1972. *Science* 176:47–49
211. McCann, J., Ames, B. N. 1976. *Proc. Natl. Acad. Sci. USA* 73:950–54
212. Bridges, B. A. 1976. *Nature* 261:195–200
213. Gabbay, E. J., Grier, D., Fingerle, R. E., Reimer, R., Levy, R., Pearce, S. W., Wilson, W. D. 1976. *Biochemistry* 15:2062–70
214. Wilson, D. W., Grier, D., Reimer, R., Bauman, J. D., Preston, J. F., Gabbay,

E. J. 1976. *J. Med. Chem.*, 19:381–84
215. Aktipis, S., Panayotatos, N. 1977. *Mol. Pharmacol.* 13:706–18
216. Roberts, J. J. 1978. *Adv. Radiat. Biol.* 7:211–436
217. Elkind, M. M. 1971. *Biophys. J.* 11:502–20
218. Schwartz, H. S. 1975. *Res. Commun. Chem. Path. Pharmacol.* 10:51–64
219. Schwartz, H. S. 1976. *J. Med. Basel* 7:33–46
220. Bases, R., Leifer, A., Rozycki, H., Blake, C., Neubort, S. 1977. *Cancer Res.* 37:2177–81
221. Ross, W. E., Glaubiger, D. L., Kohn, K. W. 1978. *Biochim. Biophys. Acta* 519:23–30
222. Ross, W. E., Glaubiger, D., Kohn, K. W. 1979. *Biochim. Biophys. Acta* 562:41–50
223. Goulian, M., Bleile, B., Tseng, B. Y. 1980. *Proc. Natl. Acad. Sci. USA* 77:1956–60
224. Gould, R. F., ed. 1972. *Biological Correlations: The Hansch Approach* (Adv. Chem. Ser. 114). Washington, DC: Am. Chem. Soc. 304 pp.
225. Purcell, W. P., Bass, G. E., Clayton, J. M. 1973. *Strategy of Drug Design: A Guide to Biological Activity.* New York: Wiley. 193 pp.
226. Martin, Y. C. 1978. *Quantitative Drug Design: A Critical Introduction* (Med. Res. Ser., Vol. 8). New York: Dekker. 425 pp.
227. Katagiri, K., Yoshida, T., Sato, K. 1975. See Ref. 22, pp. 234–51
228. Waring, M. J. 1977. *Drug Action at the Molecular Level,* ed. G. C. K. Roberts, pp. 167–89. London: Macmillan
229. Waring, M. J. 1979. See Ref. 102, pp. 173–94
230. Dell, A., Williams, D. H., Morris, H. R., Smith, G. A., Feeney, J., Roberts, G. C. K. 1975. *J. Am. Chem. Soc.* 97:2497–2502
231. Martin, D. G., Mizsak, S. A., Biles, C., Stewart, J. C., Baczynskyj, L., Meulman, P. A. 1975. *J. Antibiot.* 28:332–36
232. Wakelin, L. P. G., Waring, M. J. 1976. *Biochem. J.,* 157:721–40
233. Waring, M. J., Wakelin, L. P. G., Lee, J. S. 1975. *Biochim. Biophys. Acta* 407:200–12
234. Fox, K. R., Olsen, R. K., Waring, M. J. 1980. *Br. J. Pharmacol.* 70:25–40
235. Fox, K. R., Gauvreau, D., Goodwin, D. C., Waring, M. J. 1980. *Biochem. J.* 191:729–42
236. Cheung, H. T., Feeney, J., Roberts, G. C. K., Williams, D. H., Ughetto, G.,

Waring, M. J. 1978. *J. Am. Chem. Soc.* 100:46–54

237. Kalman, J. R., Blake, T. J., Williams, D. H., Feeney, J., Roberts, G. C. K. 1979. *J. Chem. Soc. Perkin Trans. 1,* 1313–21

238. Ughetto, G., Waring, M. J. 1977. *Mol. Pharmacol.* 13:579–84

239. Canellakis, E. S., Fico, R. M., Sarris, A. H., Shaw, Y. H. 1976. *Biochem. Pharmacol.* 25:231–36

240. Cain, B. F., Baguley, B. C., Denny, W. A. 1978. *J. Med. Chem.* 21:658–68

241. Kuhlmann, K. F., Charbeneau, N. J., Mosher, C. W. 1978. *Nucleic Acids Res.* 5:2629–41

242. Dervan, P. B., Becker, M. M. 1978. *J. Am. Chem. Soc.* 100:1968–70

243. Roques, B. P., Pelaprat, D., Le Guen, I., Porcher, G., Gosse, C., Le Pecq, J. B. 1979. *Biochem. Pharmacol.* 28:1811

244. Gaugain, B., Barbet, J., Capelle, N., Roques, B. P., Le Pecq, J. B. 1978. *Biochemistry* 17:5078–88

245. Wakelin, L. P. G., Creasy, T. S., Waring, M. J. 1979. *FEBS Lett.* 104:261–65

246. Fico, R. M., Chen, T. K., Canellakis, E. S. 1977. *Science* 198:53–56

247. Sarris, A. H., Niles, E. G., Canellakis, E. S. 1977. *Biochim. Biophys. Acta* 474:268–78

248. Chen, T. K., Fico, R., Canellakis, E. S. 1978. *J. Med. Chem.* 21:868–74

Ann. Rev. Biochem. 1981. 50:193–206

PROTEOLIPIDS ❖12076

Milton J. Schlesinger

Department of Microbiology and Immunology, Washington University School of Medicine, St. Louis, Missouri 63110

CONTENTS

Perspectives and Summary (A Matter of Definition)

It is now thirty years since Folch & Lees (1) described a protein fraction present in brain myelin that could be extracted into organic solvents under conditions that were used to isolate cell lipids. They named this kind of protein a proteolipid—an appropriate term since the proteins were, operationally, lipid-like. Subsequently, Folch-Pi and his colleagues as well as other investigators described nonmyelin proteins extractable from membranes into organic solvents. These various proteins were also classified as proteolipids. The terminology, therefore, has been one based on an experimental procedure, applicable to proteins on the basis of their solubility in lipophilic solvents.

In their recent comprehensive review of these proteolipids, Lees et al (2) point out the difficulty in using an experimental procedure to classify proteins, but they contend that "in the absence of a functional criteria, the term proteolipid is useful for describing certain hydrophobic membrane proteins." In this review, I propose an alternative definition for proteolipid— namely a protein that contains a lipid moiety as part of its *primary* structure. Proteolipids defined in this way would be analogous to glycoproteins, phosphoproteins, metalloproteins, hemeproteins, and nucleoproteins. Un-

193

0066-4154/81/0701-0193$01.00

fortunately, it is not feasible to call these newly defined proteolipids "lipo-proteins;" since that term is well-entrenched in the literature to identify a water-soluble complex composed of specific proteins and lipids. No data about lipoproteins are included in this article; the topic has been extensively reviewed in recent publications (3–4).

Some of the classical proteolipids do contain covalently bound lipid in the form of esterified fatty acids, and thus meet the new criterion proposed here for this terminology. Others probably do not, although such information is not available for many of these proteins. However, a number of enveloped virus membrane glycoproteins have recently been shown to contain cova-lently bound fatty acids, and preliminary data indicate that a substantial number (∼50) of cell membrane proteins exist with bound lipid (see below). The virus glycoproteins and most of the latter cell membrane proteins are not extractable into organic solvents; thus, they are not classical proteolip-ids. They are described in some detail in this review.

Lipid covalently bound to protein was first described for a small protein localized to the outer membrane of gram-negative bacteria (5). The lipid-protein linkage has been well characterized, and the protein's complete primary sequence determined (6). This subject has also been thoroughly reviewed recently (7) but I describe some new information regarding the biosynthesis of this kind of proteolipid. Much of the current interest in this bacterial system is concerned with how these proteins are transported from their site of biosynthesis to their final destination at the surface membrane of the cell. Intracellular transport of membrane proteins, an area of active research is not described in depth here but I refer to this subject in discuss-ing possible functions of covalently bound lipid.

One of the major interests of investigators currently studying proteolipids concerns how these proteins interact with membrane lipid and what their biological function may be. The brain myelin proteolipid is one of the most extensively studied for it can be purified to homogeneity in milligram amounts and reinserted into lipid vesicles. I describe some of the recently published information about this protein, mainly its primary sequence and its possible role in myelin membrane structure. Two reviews (2, 8) cover information about proteolipids up to 1979.

A related topic of considerable interest and research activity concerns membrane ATPase systems associated with ion transport. A proton trans-port complex has been isolated from a variety of organisms and its protein composition characterized. In all of these a small hydrophobic protein (mol wt ∼8,000) of highly conserved structure seems to be involved with the translocation of protons across the membrane. Whether a related type of protein functions as a type of ionophore in Ca^{2+} and Na^+ transport is under current study. Most investigators are attempting to purify components of these systems and to reconstitute them into membranes so that they function efficiently. In microbial systems, mutants that have alterations in various protein subunits of the complex provide an important and powerful tool to probe the function and interaction of members of the complex.

The recent discovery of fatty acyl transmembranal glycoproteins of animal viruses (discussed in the latter part of this review) may provide an important system for determining the role of covalently bound lipid in the synthesis, structure, and function of proteolipids. It is likely that proteolipids will prove to be an increasingly important topic as investigators probe more deeply into the molecular basis of lipid-protein interactions and the metabolic activities of cell membranes.

Myelin Proteolipids

The archetypal proteolipid is the major membrane protein of brain myelin. In the purified, delipidated state it is termed proteolipid apoprotein, or lipophilin. This protein was the major species in the organic soluble fraction described by Folch & Lees (1) and accounts for almost half the total myelin protein (9). It has been purified to homogeneity; the considerable information about its structure and its interaction with various types of membrane vesicles published up to 1979 has been well described in two recent reviews (2, 8) and is not presented in detail here. Virtually nothing is known about its function in myelin.

The primary structure of lipophilin is about 50% completed. The protein is very hydrophobic: 66% of the amino acids are either apolar or hydrophobic (10) and the protein contains about 2% esterified fatty acid ($\sim$2 mol/mole polypeptide chain), putatively linked directly to amino acids in the polypeptide chain (11). Lipophilin is thus part of the newly proposed category of proteolipids, since lipid is part of its primary structure.

There is now convincing evidence that rat brain myelin lipophilin has a subunit mol wt of approximately 23,500. Three cyanogen bromide fragments were isolated by Biogel P100 gel filtration, and the largest fragment further separated into two species after performic acid oxidation of disulfide bonds (12). The largest fragment (CNBr1, $M_r = 10,000$) corresponds to the N-terminal segment of the intact protein on the basis of amino-terminal sequence data. CNBr1 is disulfide-linked to CNBr2. CNBr3 has 13 residues and is adjacent to a 6-residue CNBr piece that contains the carboxylterminal phenylalanyl residue. Thus, these 4 fragments can be ordered in the manner shown below, although only the CNBr3–4 overlap has been confirmed. Of the sequences described thus far, the N terminus is very hydrophobic whereas the C terminus is highly positive charged. As of this review, about 120 of the estimated 200 residues have been sequenced from the rat and bovine brain myelin proteolipid apoprotein:

(N terminus) Gly-Leu-Leu-Glu-Cys-Cys-Ala-Arg-Cys-Leu-Val-Gly-Ala-
Pro-Phe-Ala-Ser-Leu-Val-Ala-Thr-Gly-Leu— . . . .

   Met-Tyr-Gly-Val-Leu-Pro-Trp-Asn-Ala-Phe-Pro-Gly-Lys-Val-X-Gly-
Ser-Asn-Leu— . . . . *Met*

   Met-Ile-Ala-Ala-Thr-Tyr-Asn-Phe-Ala-Leu-Lys-Leu-*Met*-Gly-Arg-Gly-
Thr-Lys-Phe (C terminus)

Jollès et al (13) had earlier separated and sequenced 11 tryptic peptides.
Six of these are Arg-peptides and account for all the arginine in the protein.
The largest of these has 13 amino acids and contains a fatty acid molecule.
Five of the 10 lysine residues are in sequenced peptides, and one of these,
a hexapeptide, contains fatty acid. The sequences of the fatty acid–con-
taining peptides are:

Thr-Thr-Ile-Cys-Gly-Lys; Thr-Ser-Ala-Ser-Ile-Gly-Ser-Leu-Cys-Ala-
Asp-Ala-Arg.

The precise amino acid residues to which fatty acids are esterified are
not known, but from the tryptic peptide analyses noted above they appear
to be either threonine or serine. Recently, the metabolic incorporation of
fatty acid into rat brain myelin proteolipid was demonstrated by intra-
cranial inoculation of ^{3}H-palmitic acid into young rats. The labeled myelin
proteolipid apoprotein could be separated from bulk lipid and the other
myelin proteins by SDS-polyacrylamide gel electrophoresis. The radioactive material bound to proteolipid was released by mild alkaline hydrolysis
in methanol, and identified as methyl palmitate (M. F. G. Schmidt, M. J.
Schlesinger, and H. Agrawal unpublished experiments).
   The secondary structure and state of subunit aggregation of the purified
proteolipid are highly dependent on the method of isolation. Water-soluble
preparations could be obtained that contained from 16 to 70% α-helix (14),
and the helical content was further increased when these forms were introduced into lipid-micelles. Forms soluble in organic solvents had even higher
helical contents; one preparation in 2-chloroethanol showed almost 100%
α-helix (15). Apparent molecular weights ranged from values of 24,000 to
28,000 for the minimal subunit form in 0.5% SDS, to aggregates of $M_r =$
500,000. Stable oligomeric species have been obtained with material soluble
in 2-chloroethanol. Sedimentation velocity measurements gave an $M_r =$
86,000 for one preparation (15), but light-scattering experiments gave a
value of 165,000 for a similar sample (16). Lavialle et al (16) have reported
recently that protein soluble in 1.0% acetic acid was not monodisperse and
was asymmetric in shape, with an M_r range of 64,000–80,000. An oligom-

eric species of about six subunits, obtained from 2-chloroethanol, appeared to consist of α-helical rod-like subunits with flexible junctions, although possibly restricted by intrachain disulfide bonds. It has been suggested that the protein can change reversibly from a worm-like shape in organic solvents to an ellipsoid form in aqueous solutions (16).

The protein contains a relatively high proportion of cystine and cysteine, and Cockle et al (17) have investigated the extent to which the cysteinyl residues in the human brain myelin proteolipid exist in the reduced sulfhydryl state in various forms of the molecule. They used 5,5'-dithiobis(2-nitrobenzoic acid) to titrate SH groups in hydrolyzed and intact protein samples that were soluble in both an aqueous medium and in 2-chloroethanol. A total of 11 cysteines were measured in hydrolyzed preparations of the protein: of these, at least 6 were present as 3 disulfide bridges in the intact protein. The reduced cysteine residues in the aqueous protein samples appeared to be in hydrophobic cores, since they were accessible to -SH reagents only after acid denaturation. In the organic-soluble state of the protein, however, two cysteines could be reacted with DTNB or alkylated with iodoacetamide. Thus, the methods of preparation of this protein from tissue can influence its structure.

The possibility that the esterified fatty acids could be in thiolester linkage to cysteines was eliminated (17), since treatment with hydroxylamine at pH 7 and pH 11 released the same amount (2 equivalents) of hydroxamic acid both before and after reductive treatment sufficient to give a protein that had 10 of the 11 cysteines reduced. The fatty acid containing tryptic peptides (see above) were derived from protein in which cysteines were converted to cysteic acid—a further indication that fatty acids are in alcohol ester linkages.

The strong hydrophobic character of the protein, its disulfide linked regions, and its relatively high number of SH groups endow it with many unusual properties. For example, the protein aggregates when heated with SDS or with thiol reagents, and is not unfolded by 8 M urea or 6 M guanidium chloride (18).

The orientation of proteolipid apoprotein in membranes has been examined by recombining the purified, delipidated protein with lipid vesicles, lipid bilayers, or liposomes. Results to 1978 have been summarized elsewhere (8). Again, the experimental procedures and the forms of the lipophilin used are critical to the kinds of lipid-protein interactions obtained. Single layered membranes could be prepared with the lipophilin oriented so that it spanned the bilayer, and this kind of orientation has been proposed for the native protein in the myelin membrane (8).

The proteolipid appears to be more strongly associated with acidic phospholipids, mostly phosphatidylserine, on the basis of both delipidation experiments (19, 20) and results obtained by use of differential scanning calorimetry to measure phase transitions of mixture of lipids in the presence of lipophilin (21). The latter measurements permitted the investigators to propose a "boundary" lipid matrix containing about 15 molecules of phosphatidyl-serine surrounding a molecule of lipophilin, in which the acidic lipid head groups are electrostatically bound to positively charged regions of the protein. Since the latter regions appear to be asymmetrically oriented in the protein, the acidic boundary lipids may also be asymmetrically arranged in the membrane bilayer.

Data based on electron spin resonance (ESR) and Raman spectra measurements (reviewed in 8) have been interpreted to indicate a pronounced effect of proteolipid on the "ordering" of lipid in the membrane. In contrast to this, however, are recent data from NMR measurements with lipophilin and proteolipid apoprotein inserted into membranes composed of ^{2}H-labeled phospholipid that show no effect or even a slight "disordering" of the membrane lipid (22). NMR measurements with ^{31}P-phospholipid showed that lipid head groups do interact with protein via electrostatic charges, and at high protein-lipid ratios the protein could "entrap" lipid around the protein. At lower protein-lipid ratios ($\sim$20% protein) the general picture seems to be one of disorder of the fatty acid chains of lipid that surround the regions of protein embedded in the bilayer. There also appears to be a preferential partitioning of the unsaturated fatty acids into the lipid boundary gradient surrounding the protein (23).

Few studies have been reported that bear on the role of proteolipid in the formation, structure, and function of the myelin membrane. Ting-Beall et al (24) have inserted a water-soluble form of the Folch-Lees proteolipid apoprotein prepared from bovine brain white matter into black lipid membranes, and studied both the structure of the protein in the membrane and effects of the protein on voltage-dependent conductances across the membrane. Addition of the proteolipid to either side of the membrane induced an increase in conductance across the bilayer when 0.1 M NaCl was added to the chambers. Orientation of the proteolipid in the bilayer was such that its ability to induce the conductance could not be affected by addition of pronase to the chambers.

It has been suggested (8) that defective forms of lipophilin or alterations in its biosynthesis could be responsible for abnormal structures of myelin, thereby leading to certain types of demyelinating diseases. A report by Greenfield et al (25) has shown that brain proteolipid is synthesized normally in the Quaking mouse but is not incorporated effectively into myelin. Such genetically altered strains of mice may prove to be important tools for studying the role of proteolipid in myelin function.

Nonmyelin Proteolipids

The application of organic solvent extraction procedures to other animal tissues showed that proteolipids probably are present in all cellular material (26). Much of this organic-soluble material has been found in extracts of mitochondria (27). The mitochondria-derived hydrophobic proteins differ importantly from the myelin proteolipid in that they do not appear to contain covalently bound fatty acid, and they have few or no disulfide bonds. They range in molecular weights from 2300 to 35,000. It is likely that many of these proteins are the hydrophobic, membrane-embedded subunits of various membrane-associated enzyme complexes and membrane-transport structures.

The major organic-soluble proteins (classical proteolipids) associated with the energy-transducing (proton-pump) ATPase complex have been purified from a variety of sources including *Escherichia coli* and bovine heart mitochondria (reviewed in 28–30). The ATPase complex from all species examined thus far contains a hydrophobic protein subunit with M_r = 8000. The structural features of these proteins have recently been summarized (28). Complete sequences of the *E. coli, Neurospora,* and yeast proteins are known, and amino acid substitutions have been determined for mutationally altered forms of the bacterial and yeast proteins (31). There is a striking sequence homology among the proteins, particularly between those of yeast and *Neurospora.* These data show several common features that include an N-terminal polar region and two very hydrophobic sequences of about 25 amino acids separated by another polar loop of amino acids. In the middle of the second hydrophobic sequence (around residues 61–65) there is a single acidic residue: aspartic acid in the *E. coli* protein and glutamic acid in the yeast and *Neurospora* proteins (32). It is this residue that is bound by the mitochondrial ATPase inhibitor, dicyclohexylcarbodiimide (DCCD); this feature has led to the assignment of these proteins as DCCD-binding proteins. This same protein also appears to be the site of oligomycin inhibition of the ATPase (33).

The DCCD-binding protein accounts for about 10% of the protein in the energy-transducing ATPase complex. It is a subunit of that portion of the complex, noted as F_o, that is an intrinsic part of the membrane. The protein is believed to be present as a hexamer in the complex (34, 35) and is postulated to form a pore through the membrane that facilitates proton transport (36). In the ATPase complex from bovine heart mitochondria, the DCCD-binding protein is only one of seven chloroform/methanol-soluble species separable by high pressure liquid chromatography (37).

The DCCD-binding protein of yeast is coded by the mitochrondrial

DNA, but in *Neurospora* the protein is synthesized on cytoplasmic polyribosomes as a precursor with about 35 extra residues (38). Reconstitution of an active proton-translocating F_o complex (from *E. coli*) in phospholipid liposomes has been achieved, and the activity is blocked by DCCD (39). A similar kind of reconstitution was reported for a butanol extract of bovine heart mitochondria with phospholipid liposomes (40). Preparations of spinach chloroplasts also can synthesize the DCCD-binding protein (41).

The Ca^{2+}-dependent ATPase complex of sarcoplasmic reticulum has been reported to contain a proteolipid subunit ($M_r = 6000$) with 1 mol esterified fatty acid per mol of subunit (42), and there is evidence that the proteolipid contributes to the efficient channeling of Ca^{2+} through the membrane (43, 44). The Na,K-ATPase may also have a proteolipid subunit as part of its biologically active complex (45). When purified from the medulla of mammalian kidney, the complex contains a large subunit of $M_r = 100,000$, a glycoprotein of $M_r = 45,000$, and a chloroform-methanol soluble protein of $M_r = 13,500$. Hydrophobic proteins of about this size were also labeled by a photoaffinity probe that was used to identify subunits of purified preparations of the pig kidney medulla Na,K-ATPase (46) and the Na,K-ATPase from *Electrophorus electricins* (47). The affinity label bound to the oubain-sensitive sites of the ATPase complex and was distributed equally between the large and small polypeptides of the mammalian complex.

With the exception of the proteolipid of the mammalian Ca-ATPase, there is no evidence thus far that the other organic-soluble proteins associated with these membrane ATPase–coupled transport systems contain fatty acyl groups in their primary structure. In fact, current published data argue against the presence of fatty acid in the mitochondrial membrane proteins (26).

Bacterial Outer Membrane Proteolipid

The first and, so far, only protein that has been rigorously shown to contain covalently attached lipid is the small polypeptide (58 amino acids) found in very large amounts bound to the peptidoglycan matrix associated with the outer membrane of several gram-negative bacteria (reviewed in 48). The amino acid sequence of the protein, isolated from *E. coli*, is known (6) as well as the nucleotide sequence of its mRNA (49) and its DNA genome (50). The linkage between protein and lipid consists of one fatty acid in amide linkage to the amino-terminal cysteinyl residue of the polypeptide. Two additional fatty acids are esterified in the 2,3 positions of glycerol, which is attached by a thioether bond to the cysteinyl sulfhydryl group (51).

Recent investigations of this protein have focused on the biosynthesis of the lipid-protein bonds and on the intracellular transport of the protein from its translation site to the site of membrane attachment. The glycerol portion of the lipid appears to be donated to the protein by a nonacylated glycerol moiety of phosphatidylglycerol, to form the thioether linkage between the cysteine sulfhydryl group and the carbon 1 position of glycerol (52, 53). The fatty acids in the lipoprotein appear to come from preexisting fatty acids in phospholipids, which suggests a biosynthetic pathway whereby the apolipoprotein combines initially with phosphatidylglycerol and then with phospholipid to form the diglyceride and amide-linked fatty acid (54). The protein portion of this bacterial proteolipid is initially synthesized as a higher-molecular-weight polypeptide that contains 20 additional amino acids at the amino-terminal position (55). This sequence is postulated to act as a signal for transfer of the protein to the bacterial outer membrane and, in fact, a mutation that alters the sequence in the amino-terminal preprotein does inhibit transport of the polypeptide (56, 57). Proteolytic processing of the precursor occurs at the cytoplasmic membrane, and assembly into the outer membrane is postulated to occur at specialized regions of the cell where the two membranes are closely linked (58). The glyceride moiety with acylated fatty acids may be bound before the precursor is processed and transported to the outer membrane (59).

There are reports of additional bacterial proteins with covalently attached lipid, possibly with linkages analogous to the *E. coli* outer membrane protein described above, but distinct from the latter in amino acid composition. Mizuno (60) described two proteins with covalently linked lipid; one of these is from the outer membrane of *Pseudomonas aeruginosa,* the other is from *E. coli.* Both have an $M_r = 21,000$. This kind of protein appears to be common to a number of gram-negative bacteria, but its function is unknown.

Virus Glycoprotein Proteolipids

The proteolipid nature of transmembranal virus glycoproteins was first detected in the course of studies on the replication in tissue culture cells of a small enveloped animal virus called Sindbis virus (61). We made the serendipitous observation that the two glycoproteins (called E1 and E2, $M_r \simeq 50,000$) that form part of the envelope of Sindbis virus could be radiolabeled when virus was grown in cells given ^3H or ^{14}C palmitic acid. The labeled protein-bound fatty acid could be removed from the glycoproteins, which had been purified free of any phospholipid or sphingolipid, by mild alkaline hydrolysis in methanol, and subsequently was identified by gas chromatography as methyl palmitate. Indications that this fatty acid moiety was covalently bound directly to the protein followed from the observation that fatty acid was retained on protein after its treatment with denaturants (urea and guanidine), detergents (sodium dodecyl sulfate and triton), or extraction with organic solvent. Subsequently, chemically derived or proteolytic fragments of the protein were isolated and shown to contain esterified fatty acid (61, 62). Amino acid analysis of some of the smaller fragments, which were purified by chromatography on thin layer plates or on hydrophobic sizing gel columns, indicate that the fatty acid is linked to a serine residue on that portion of the glycoprotein that is embedded in the lipid bilayer of the virus (A. I. Magee, M. J. Schlesinger, unpublished experiments).

Sindbis virions have equal amounts of E2 and E1, which form spikes protruding from the virus lipid envelope (Figure 1). The E2 glycoprotein had 3 times more fatty acid than the E1 glycoprotein, and the latter had 1–2 mol of fatty acid per mole of protein. There are 240 spikes per virion, thus there would be 1000–2000 fatty acids in the bilayer covalently bound to protein. There are estimated to be 15,000 molecules of phospholipid in the virion bilayer; thus the protein-bound fatty acid accounts for about 3–6% of the total lipid fatty acid. In fact, a quantitative analysis of the distribution of radioactive label between virus glycoprotein and membrane lipid showed that the glycoproteins contain 5% of the total lipid label (61).

The transmembranal glycoproteins of several other enveloped animal viruses have been shown to contain bound fatty acid. These include the G protein of vesicular stomatitis virus (VSV) (62), the HA protein of influenza and New Castle disease viruses, the F protein of Sendai virus, the E1 and E2 proteins of Semliki forest virus, the glycoprotein of Coronavirus (M. F.

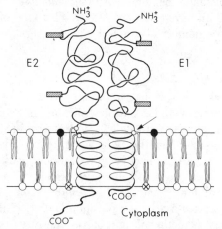

Figure 1   The Sindbis virus envelope spike. The transmembranal regions are shown with two fatty acids on E2 and one fatty acid attached to E1 (arrow) at the extracellular side of the lipid bilayer. The positioning of the fatty acid for the E1 protein is based on amino acid sequence data for this protein from Semliki forest virus (67). O, ●, ⊗ indicate different phospholipids; ▧ indicates oligosaccharide portions of the protein.

G. schmidt, unpublished experiments). The VSV G protein has been analyzed in considerable detail and contains one to two fatty acids per mole of protein (62). This fatty acid appears also to be linked to a serine residue on that portion of the polypeptide that spans the lipid bilayer.

Schmidt and Schlesinger were able to detect fatty acid labeling of the VSV G and Sindbis virus glycoproteins in virus-infected tissue culture cells given very short pulses of ^{3}H-palmitic acid. By employing a pulse-chase labeling experiment, and the judicious choice of metabolic inhibitors that affect glycoprotein synthesis and intracellular transport, they showed that fatty acid is added to protein at a distinct stage in its post-translational processing, i.e., approximately 10–15 min after polypeptide chain synthesis had been completed, and about 5–10 min before the glycoproteins reached the surface membrane of the cell (63). For VSV G protein, fatty acid addition preceded by about 1–2 min the "trimming" of the high mannose oligosaccharide of the G protein. There is evidence that the latter reaction occurs in the Golgi apparatus of the cell (64, 65); thus, acylation with fatty acids may also occur in this organelle.

Membrane Proteolipids of Cells in Tissue Culture

On the basis of the studies described above with virus transmembranal glycoproteins, there are four criteria to identify a putative proteolipid (defined here as protein with covalently esterified fatty acid):

1. The fatty acid must remain bound to protein after denaturation and purification, for example by electrophoresis in SDS-polyacrylamide gel electrophoresis;
2. The fatty acid must be removable from the protein by alkaline hydrolysis in methanol or by mild treatment with neutral or alkaline hydroxylamine, and recovered as the fatty acid ester or hydroxamate;
3. The protein must be localized in a membrane fraction of the cell;
4. The in vivo labeling of the protein by fatty acid must be blocked in cells treated with inhibitors of protein synthesis such as cycloheximide.

On the basis of these criteria we have detected a substantial number of putative proteolipids in several kinds of tissue culture cells (66). Figure 2 shows a pattern of those proteins labeled when ^{3}H-palmitate was added to primary cultures of chick embryo fibroblasts or to an established mouse plasmacytoma cell line. About 20–50 discrete protein bands ranging in molecular weight from about 10,000 to 100,000 were detected. A protein of 20,000 mol wt in both cell types was labeled to a relatively greater extent than other proteins. It will be of considerable interest to discover what functions these proteins serve and where they are located in the cell membranes.

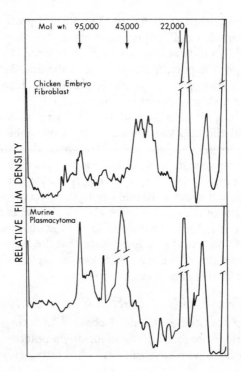

Figure 2 Densitometric traces of an autoradiogram prepared from SDS-polyacrylamide gel electropherograms that contained extracts of chicken embryo fibroblast cells (upper panel) and mouse plasmacytomal cells (lower panel) that had been given ^{3}H-palmitic acid. Experimental details are in (66). The radioactivity in the protein bands shown was released by hydrolysis with hydroxylamine

Structure and Function of Proteolipids

Clearly, the proteolipids include a variety of proteins, and discussion of structure or function of the class as a whole is not useful. On the other hand, one feature is common to them all, regardless of whether they are defined on the basis of an experimental procedure or on the basis of a chemical structure: all have segments of their polypeptide chain embedded in a lipid bilayer, and there may be some commonality of the structures that provides for protein-lipid contacts. Some are subunits of more complex structures involved in ion transport across membranes, and some of the proteins noted in Figure 2 probably play a similar role in transport of nutrients and metabolites. The larger transmembranal virus glycoproteins that have been listed here as proteolipids serve an essential role in the release of virus from infected cells and in attachment of the virus to a host cell membrane. Thus, many of the glycoproteins that serve as cell surface receptors and contain transmembranal segments may eventually prove to contain covalently bound fatty acid.

What role the fatty acids play in the structure of these proteins is not at all clear. The observation that fatty acid attachment occurs midway in intracellular transport of the virus glycoprotein to the cell surface suggests that the fatty acids could serve as signals for membrane targeting. Or they may serve to select, from among the various lipids in the membranes, specific lipids for binding to the protein (see Figure 1). This selectivity could lead to formation of unique regions of the membrane oriented around the protein, and facilitate certain cell surface activities such as transmembrane signaling and membrane fusions that occur during endocytosis and exocytosis. It will be of considerable interest to follow the studies of proteolipids in the coming years to determine how they contribute to membrane structure and function.

Literature Cited

1. Folch-Pi, J., Lees, M. B. 1951. *J. Biol. Chem.* 191:807–17
2. Lees, M. B., Sakura, J. D., Sapirstein, V. S., Curatolo, W. 1979. *Biochim. Biophys. Acta* 559:209–30
3. Scanu, A. M., Landsberger, F. R., ed. 1980. *Ann. NY Acad. Sci.* 348:
4. Smith, L. C., Pownall, H. S., Gotto, A. M. Jr. 1978. *Ann. Rev. Biochem.* 47:751–77
5. Braun, V., Rehn, K. 1969. *Eur. J. Biochem.* 10:426–38
6. Braun, V., Bosch, V. 1972. *Eur. J. Biochem.* 28:51–69
7. Inouye, M. 1979. *Biomembranes* 10:141–208.
8. Boggs, J. M., Moscarello, M. A. 1978. *Biochim. Biophys. Acta* 515:1–21
9. Eng, L. F., Chao, R. C., Gerstl, B., Pratt, D., Tavaststjerna, M. G. 1968. *Biochemistry* 7:4455–65
10. Gagnon, J., Finch, P. R., Wood, D. D., Moscarello, M. A. 1971. *Biochemistry* 10:4756–63
11. Sherman, G., Folch-Pi, J. 1971. *Biochem. Biophys. Res. Commun.* 44:157–61
12. Jollés, J., Schoentgen, F., Jollés, P., Vacher, M., Nicot, C., Alfsen, A. 1979. *Biochem. Biophys. Res. Commun.* 87:619–26
13. Jollés, J., Nussbaum, J-L., Schoentgen, F., Mandel, P., Jollés, P. 1977. *FEBS Lett.* 74:190–93
14. Cockle, S. A., Epand, R. M., Boggs, J. M., Moscarello, M. A. 1978. *Biochemistry* 17:624–29
15. Moscarello, M. A., Gagnon, J., Wood, D. D., Anthony, J., Epand, R. M. 1973. *Biochemistry* 12:3402–6
16. Lavialle, F., Foresta, B. de, Vacher, M., Nicot, C., Alfsen, A. 1979. *Eur. J. Biochem.* 95:561–67
17. Cockle, S. A., Epand, R. M., Stollery, J. G., Moscarello, M. A. 1980. *J. Biol. Chem.* 255:9182–88
18. Cockle, S. A., Epand, R. M., Moscarello, M. A. 1978. *J. Biol. Chem.* 253:8019–26
19. Folch-Pi, J., Stoffyn, P. J. 1972. *Ann. NY Acad. Sci.* 195:86–107
20. Uda, Y., Nakazawa, Y. 1973. *J. Biochem. Tokyo* 74:545–49
21. Boggs, J. M., Wood, D. D., Moscarello, M. A., Papahadjopoulos, D. 1977. *Biochemistry* 16:2325–29
22. Rice, D. M., Meadows, M. D., Scheinman, A. O., Goñi, F. M., Gomez-Fernandez, J. C., Moscarello, M. A., Chapman, D., Oldfield, E. 1979. *Biochemistry* 18:5893–903
23. Verma, S. P., Wallach, D. F. H., Sakura, J. D. 1980. *Biochemistry* 19:574–79
24. Ting-Beall, H. P., Lees, M. B., Robertson, J. D. 1979. *J. Membr. Biol.* 51:33–46
25. Greenfield, S., Williams, N. I., White, M., Brostoff, S. W., Hogan, E. L. 1979. *J. Neurochem.* 32:1647–51
26. Folch-Pi, J., Sakura, J. A. 1976. *Biochim. Biophys. Acta* 427:410–27
27. Murakami, M., Ozawa, Y., Funahashi, S. 1963. *J. Biochem. Tokyo* 54:166–72

28. Sebald, W., Wachter, E. 1978. *Energy Conservation in Biological Membranes,* ed. G. Schafer, J. Kingenberg, pp. 228–36. Berlin: Springer
29. Tzagoloff, A., Macino, G., Sebald, W. 1979. *Ann. Rev. Biochem.* 48:419–41
30. Downie, J. A., Gibson, F., Cox, G. B. 1979. *Ann. Rev. Biochem.* 48:103–31
31. Hoppe, J., Schairer, H. U., Sebald, W. 1980. *FEBS Lett.* 109:107–11
32. Sebald, W., Machleidt, W., Wachter, E. 1980. *Proc. Natl. Acad. Sci. USA* 77:785–89
33. Criddle, R. S., Arulanadan, C., Edwards, T., Johnston, R., Scharf, S., Enns, R. 1977. *Genetics and Biogenesis of Chloroplast and Mitochondria,* ed. Th. Bucher et al, pp. 151–57. Amsterdam: North-Holland
34. Sigrist-Nelson, K., Azzi, A. 1979. *J. Biol. Chem.* 254:4470–74
35. Sebald, W., Graf, Th., Lukins, H. B. 1979. *Eur. J. Biochem.* 93:587–99
36. Criddle, R. S., Packer, L., Shieh, P. 1977. *Proc. Natl. Acad. Sci. USA* 74:4306–10
37. Blondin, G. A. 1979. *Biochem. Biophys. Res. Commun.* 90:355–61
38. Michel, R., Wachter, E., Sebald, W. 1979. *FEBS Lett.* 101:373–76
39. Negrin, R. S., Foster, D. L., Fillingame, R. H. 1980. *J. Biol. Chem.* 255:5643–48
40. Celis, H. 1980. *Biochem. Biophys. Res. Commun.* 47:26–31
41. Nelson, N., Nelson, H., Schatz, G. 1980. *Proc. Natl. Acad. Sci. USA* 77:1361–64
42. MacLennan, D. H. 1975. *Can. J. Biochem.* 53:251–61
43. Racker, E., Eytan, E. 1975. *J. Biol. Chem.* 250:7533–34
44. Knowles, A., Zimniak, P., Alfonzo, M., Zimniak, A., Racker, E. 1981. *J. Membr. Biol.* 56: In press
45. Schwartz, A., Adams, R. J., Ball, W. J. Jr., Collins, J. H., Gupte, S., Lane, L. K., Reeves, A. S., Wallick, E. T. 1980. *Int. J. Biochem.* 12:287–93
46. Forbush, B. III, Kaplan, J. H., Hoffman, J. F. 1978. *Biochemistry* 17:3667–75
47. Rogers, T. B., Lazdunski, M. 1979. *FEBS Lett.* 98:373–76
48. DiRienzo, J. M., Nakamura, K., Inouye, M. 1978. *Ann Rev. Biochem.* 47:481–532
49. Nakamura, K., Pirtle, R. M., Pirtle, I. L., Takeishi, K., Inouye, M. 1980. *J. Biol. Chem.* 255:210–16
50. Nakamura, K., Inouye, M. 1979. *Cell* 18:1109–17
51. Hantke, K., Braun, V. 1973. *Eur. J. Biochem.* 34:284–96
52. Chattopadhyay, P. K., Wu, H. C. 1977. *Proc. Natl. Acad. Sci. USA* 74:5318–22
53. Chattopadhyay, P. K., Lai, J.-S., Wu, H. C. 1979. *J. Bacteriol.* 137:309–12
54. Lai, J.-S., Philbrick, W. M., Wu, H. C. 1980. *J. Biol. Chem.* 255:5384–87
55. Inouye, S., Wang, S., Sekizawa, J., Halegoua, S., Inouye, M. 1977. *Proc. Natl. Acad. Sci. USA* 74:1004–8
56. Lin, J. J. C., Kanazawa, H., Ozols, J., Wu, H. C. 1978. *Proc. Natl. Acad. Sci. USA* 75:4891–95
57. Lin, J. J. C., Lai, J.-S., Wu, H. C. 1980. *FEBS Lett.* 109:50–54
58. Wu, H. C., Lin, J. J. C., Chattopadhyay, P. K., Kanazawa, H. 1980. *Ann. NY Acad. Sci.* 343:368–83
59. Hussain, M., Ichihara, S., Mizushima, S. 1980. *J. Biol. Chem.* 255:3707–12
60. Mizuno, T. 1979. *J. Biochem. Tokyo* 86:991–1000
61. Schmidt, M. F. G., Bracha, M., Schlesinger, M. J. 1979. *Proc. Natl. Acad. Sci. USA* 76:1687–91
62. Schmidt, M. F. G., Schlesinger, M. J. 1979. *Cell* 17:813–19
63. Schmidt, M. F. G., Schlesinger, M. J. 1980. *J. Biol. Chem.* 255:3334–39
64. Tabas, I., Schlesinger, S., Kornfeld, S. 1978. *J. Biol. Chem.* 253:716–22
65. Hubbard, S. C., Ivatt, R. J. 1981. *Ann. Rev. Biochem.* 50: In press
66. Schlesinger, M. J., Magee, A. I., Schmidt, M. F. G. 1980. *J. Biol. Chem.* 255:10021–24
67. Garoff, H., Frischauf, A. M., Simon, K., Lehrach, H., Delius, H. 1980. *Nature* 288:236–41

Ann. Rev. Biochem. 1981. 50:207–31

LECTINS: THEIR MULTIPLE ENDOGENOUS CELLULAR FUNCTIONS[1]

❖12077

S. H. Barondes[2]

Department of Psychiatry, University of California, San Diego, La Jolla,
California 92093, and Veterans Administration Medical Center, San Diego,
California 92112

CONTENTS

PERSPECTIVES AND SUMMARY

Lectins are divalent or multivalent carbohydrate-binding proteins that are grouped together because they agglutinate cells or other materials that display more than one saccharide of sufficient complementarity. They were discovered in plants, but are also found in all other categories of living

---

[1]The US Government has the right to retain a non-exclusive royalty-free license in and to any copyright covering this paper.

[2]This research was supported by grants from the McKnight Foundation, the US Public Health Service, and the Veterans Administration Medical Center.

things. To more sharply define this category it has recently been proposed that a number of related proteins be excluded (1). Among these are monovalent carbohydrate-binding toxins such as ricin (because of their valence); carbohydrate-binding immunoglobulins (presumably because they can be included in another structural and functional category); and carbohydrate-binding enzymes (because they are generally not agglutinins). Polyvalent enzymes such as a glycosidase (2) that can agglutinate cells under certain conditions are considered to "act as a lectin" (1), which reflects a conflict between strictly operational and functional considerations.

Failure to emphasize function as a defining characteristic is due to the fact that the endogenous roles of lectins are not yet understood very well. This is now changing, and the conclusion that "nothing is known about their role in nature," which appeared appropriate in the first review about lectins in this series (3), merits updating.

This paper evaluates current knowledge about the endogenous biological functions of lectins. To this end I will discuss approaches to an analysis of this problem, then illustrate them with some well-developed examples from three biological systems—plants, cellular slime molds, and animal tissues. It appears that these proteins have been adapted for a variety of specific roles, but a unifying hypothesis is that at some point they all involve interaction with components of the cell surface and/or its extracellular environment. Thus, studies with a number of systems indicate that: (a) lectins, although predominantly intracellular, are also detectable on cell surfaces or extracellularly; and (b) these lectin molecules appear destined to interact with complementary saccharides on other cell surfaces, in extracellular materials, or in solution. Work with chicken lectins also suggests that the same protein may have different functions in different tissues. This may be a consequence of the nature and localization of the complementary molecules to which it binds.

GENERAL CONSIDERATIONS

Some Properties of Lectins that Suggest Biological Functions

It is reasonable to expect that the known biochemical properties of lectins dictate their endogenous biological function. Some of these properties and the functions that might be inferred from them are summarized in Table 1. Two recent reviews of properties of lectins may be consulted for details (4, 5).

One major property of lectins is their specific saccharide-binding sites. In the many lectins that are multimers of an identical subunit these binding sites are the same. In contrast, some lectins are composed of subunits with different binding sites. These include the lectin from the red kidney bean,

Table 1 Some common properties of lectins that suggest biological functions

Property	Function suggested
1. Specific binding sites: a. All of one kind b. Of different kinds	Recognition of complementary oligosaccharide receptors (range of specificities)
2. More than one carbohydrate-binding site	a. Cross-linking glycoproteins or glycolipids in membranes and/or solution b. High affinity (multisite) binding to molecules or a cell surface with multiple receptors
3. Agglutinin	Binding cells together: a. Like cells (promoting adhesion, fusion, etc) b. Unlike cells (promoting symbiosis, infection, phagocytosis, etc)
4. Abundant	Structural rather than catalytic function
5. Generally not integrated in membranes	Relative freedom of movement in or between cellular compartments

Phaseolus vulgaris. It is composed of two different subunits combined into five different forms of noncovalently bound tetramers (6). Since the subunits have markedly different specificities for cell surface receptors, each combination could be envisioned to have a different function. For example the homotetramer of one subunit might agglutinate cells with an appropriate receptor, whereas a tetramer that contains only one of these subunits per molecule might inhibit such agglutination. Similar mixed tetramers of two functionally distinct subunits have been observed in a plant seed lectin from *Bandeiraea simplicifolia* (7) and in a slime mold lectin, purpurin (D. N. Cooper and S. H. Barondes, unpublished). In the latter two cases the seeds of *B. simplicifolia* and the cells of *Dictyostelium purpureum* contain still other lectins besides the five forms possible from tetrameric association of the two different subunits.

The common finding of more than one lectin in seeds, slime molds, and even vertebrate tissues (8, 9) raises the possibility of concerted specific reactions due to concurrent display of these proteins on a structure like a cell surface. A highly specific interaction directed by the binding properties of more than one type of lectin molecule could result.

The specificity of the binding sites of the lectins suggests that there are endogenous saccharide receptors in the tissues from which they are derived or on other cells or glycoconjugates with which the lectin is specialized to interact. Unfortunately, no endogenous receptor for a lectin has yet been unambiguously identified, despite the fact that their carbohydrate-binding sites may be specialized for association with highly specific complex oligosaccharides (10). One difficulty in searching for a specific receptor is that lectins bind with measurable affinity to a wide range of simple and complex saccharides. Although the criterion of high affinity binding is often used to identify a "true" receptor, and the term receptor has been applied to oligosaccharides that bind to lectins with high affinity, regardless of their source (11), it remains possible that the soluble form of a biologically meaningful receptor for a given lectin may not bind with high affinity. Instead, high affinity binding may result from the polyvalent interactions between the multiple binding sites of a lectin and a cluster of receptor molecules at a cellular site such as a membrane. Furthermore, biologically significant lectin-receptor interactions may be determined not by the very restrictive complementarity normally associated with ligand-receptor interactions, but, rather by what is available to the lectin in the cell or cells with which it is associated. Uncertainty about the nature of a true lectin receptor is a major impediment to our understanding of lectin function.

The other properties listed in Table 1 require much less discussion. The fact that lectins have more than one carbohydrate-binding site suggests that they could act to cross-link glycoproteins and glycolipids in membranes of the same cell for various organizational purposes. Agglutination activity suggests functions in binding like or different cells for a variety of purposes ranging from morphogenesis in embryos to phagocytosis of one cell by another. The marked abundance of lectins suggests that they play a structural role rather than an enzymatic or catalytic function. The fact that many lectins are readily isolated as water-soluble materials suggests that if they play a role in membrane function it is by association with oligosaccharides on membranes without the constraints imposed by being integrated within the membrane bilayer.

It of course remains possible that some functions of lectins may be mediated by properties that have not yet been discovered. Nevertheless, the types of considerations listed in Table 1 have suggested a number of hypotheses about lectin function, which have led to fruitful investigation.

Experimental Approaches to Identification of Lectin Function

Elucidation of the function of lectins makes use of information gathered in a variety of ways (Table 2). This multitude of approaches is necessary since each, although valuable, is subject to misinterpretation. For example, localization may indicate where the bulk of lectin protein is but not where it is utilized for a specific function. Thus, localization of a lectin in intracellular vesicles could indicate either that it functions in the vesicle or that it is in the vesicle in a storage form and only exerts its biological effect upon secretion. Likewise, studies of mutants with absent or abnormal lectin, although potentially very informative, can also be misleading. For example,

Table 2 Experimental approaches to identification of lectin function

1. Localization:
   A. Histochemical (light and electron microscope):
      a. With labeled antibodies
      b. With labeled haptens
   B. Subcellular fractionation
      a. Immunological assays
      b. Functional assays
2. Developmental studies:
   A. Overall concentrations
   B. Localization
3. Biological effects of mutations:
   A. Lectin absent
   B. Lectin functionally altered
   C. Replacement studies
4. Biological effects of reagents that interact with lectin (haptens, antibodies) or of excess lectin:
   A. In vitro
   B. In vivo
5. Endogenous receptors(s):
   A. Identification and purification
   B. Application of approaches 1–4 to the receptor

the correlation of the absence of a lectin with the failure of a specific biological function may indicate a casual relationship; but an alternative explanation is that both abnormalities are pleiotropic expressions of a mutation in a gene that controls a specific developmental pathway.

In the following sections, I consider the various experimental approaches that have been used in an attempt to identify lectin function in several important biological systems. The value and limitations of an individual approach and the importance of evaluating the problem in different ways should become apparent. Since little is presently known about biologically significant lectin receptors, the strong corroborative evidence that could be marshalled from simultaneous studies of these interacting molecules has only rarely been brought to bear on a given problem.

LECTINS IN PLANTS

Binding Nitrogen-Fixing Bacteria to Legume Roots

Among the possible functions of lectins in plants, their participation in binding nitrogen-fixing bacteria to legume roots has received the most attention. In this reaction bacteria of the genus *Rhizobium* adhere to the surface of differentiated root cells and are then internalized into the root hair to form nitrogen-fixing nodules. This symbiosis is specific in that certain species of *Rhizobium* can only associate with a particular species of legume.

The possible involvement of lectins in this symbiotic interaction was first suggested by Hamblin & Kent (12) who showed that bacteria that nodulate beans are agglutinated by bean extracts, presumably by interaction with the lectin in these extracts, and that erythrocytes that could be agglutinated by the bean lectin bound to root hairs. Based on these preliminary studies they proposed the possibility of binding of bacteria to the root hairs by a lectin-bacterial interaction. Much subsequent work has been done with the soybean–*Rhizobium japonicum* symbiosis and with the clover–*Rhizobium trifolii* symbiosis.

Evidence for lectin involvement in directing the binding of specific strains of *Rhizobium* to soybean root hairs was first provided by Bohlool & Schmidt (13). They prepared a fluorescent derivative of soybean lectin and found that it bound to 22 of the 25 strains of *R. japonicum* that nodulate soybean plants. In contrast, 23 other strains of *Rhizobium* that do not nodulate soybeans did not bind the fluorescent lectin. Of the 22 *R. japonicum* strains tested, 15 bound fluorescent or tritiated soybean lectin and the binding was inhibitable by specific haptens (14). All nine strains of *Rhizobium* that do not nodulate soybean did not bind the lectin. The failure of non-nodulating strains to bind soybean lectin, and the general tendency of nodulating strains to bind the lectin were all consistent with the hypothesis under discussion. However, the failure of some nodulating strains to bind soybean lectin seemed puzzling. The possibility that these were false negatives due to culture conditions was raised by subsequent findings (15). Eleven strains of *R. japonicum* were cultured either in synthetic media or in association with root exudate of soybean. In synthetic media only five strains developed receptors for soybean lectin, whereas in root exudate all eleven strains developed these receptors. Several other *Rhizobium* strains that do not nodulate soybean developed soybean lectin receptors when grown in the exudate medium. The authors conclude that all strains of *Rhizobium* capable of nodulating soybean have the capacity to synthesize receptors for soybean lectin. The fact that some other strains that are unable to nodulate soybean may also synthesize such receptors is not inconsistent with this hypothesis, since a factor other than lectin-receptor interaction could limit infectivity.

Morphological evidence in support of lectin involvement in *Rhizobium* binding has included the observation that some strains of *R. japonicum* bind soybean lectin only at one cell pole (16, 17), which suggests that this might be correlated with the polar attachment of some strains of *Rhizobium* to roots. However, the receptor is not confined to one pole of *R. japonicum*, since labeled soybean lectin receptor sites were distributed throughout the capsular material of this bacterium (18). Localization with the electron microscope suggested that the receptors were in capsular polysaccharides rather than lipopolysaccharides. Related polysaccharides that bind soybean lectin are secreted into the medium of cultured *R. japonicum* (19).

Evidence against the hypothesis under discussion comes from studies with soybean lines that apparently lack the soybean lectin, but can be nodulated by *R. japonicum*. Pull et al (20) screened 102 lines of soybean seeds by powdering them, defatting with petroleum ether, and extracting the seed meal with phosphate-buffered saline containing tracer amounts of tritium-labeled soybean lectin of known specific activity. Each soluble extract was then fractionated by affinity chromatography. After specific elution of the lectin fractions, which were detected by liquid scintillation counting of the tracer lectin, the amount of extracted lectin from the seed was determined from a protein determination and a calculation based on isotope dilution. Of the 102 lines studied, five showed no detectable lectin. In these five lines recovery of tritium-labeled tracer lectin was good, which indicates that the lectin had not been degraded or lost during the purification procedure. Extracts of these five lines also showed no detectable hemagglutination activity. In a follow-up study of these five lines with a radioimmunoassay no antigen that reacted with antibody to soybean lectin was detected in extracts of seeds or of parts of soybean plants of various ages (21). This assay would have picked up extractable lectin that might have been masked in the previous experiments by high concentrations of an endogenous inhibitor. In the tissue studies, the roots, hypocotyls, stems, leaves, and cotyledons were dissected at various ages, and were frozen, lyophilized, ground to fine powders, and extracted in phosphate-buffered saline containing D-galactose and other materials. Unfortunately, no studies of the possible presence of unextracted antigen in the particulate material were reported.

A further argument against the participation of soybean lectin in the specific interaction with *R. japonicum* comes from studies of the distribution of soybean lectin in plant tissues derived from normal seeds (22). The tissue samples were ground to a fine powder at dry ice temperatures, and filtered extracts made with phosphate-buffered saline containing galactose were studied with a radioimmunoassay. Lectin was highly concentrated in the cotyledons of young plants that had several milligrams per gram fresh weight for several days after planting, with a marked decline thereafter. Levels in roots were 10–50 μg per gram fresh weight four days after planting, but had fallen to undetectable levels about two weeks after planting. Since the highest rate of nodulation occurs in soybean plants infected two to four weeks after sowing, a time when soybean lectin levels in roots were not detectable, the participation of the lectin in nodulation seemed difficult to support. However, the authors raised the possibility of low levels of lectin in the tips of the growing root hairs that may be sufficient to mediate *Rhizobium* attachment yet be below the level of detection of the assay. They also mentioned that root lectin may not have been extracted by their procedure.

Indeed, failure to demonstrate lectin in the roots of soybean plants at the normal time of nodulation must now be reevaluated in light of new evidence presented by Bowles et al (23). They also studied lectin distribution in soybean plant tissues of different ages but used a different extraction procedure. Roots and other tissues were removed, homogenized at 4°C in buffer containing 15% sucrose, and centrifuged at 100,000 X g for 90 min. The membrane pellets were then sequentially extracted by: (*a*) suspension in 0.1 M D-galactose; (*b*) sonication in 0.5 M NaCl; (*c*) sonication in 0.5% Triton X-100. They found that under these conditions, sonication in the detergent was required to extract lectin activity. For example, roots of two-week-old soybean plants showed no detectable lectin activity either in the initial soluble extract or in the extract after 0.1 M galactose extraction. Upon sonication in 0.5 M NaCl, 512 hemagglutination units per 100 microliters were found, and after sonication in the detergent, 8192 were found. Similar levels of lectin activity were found in the Triton extracts of roots of five- and seven-week-old plants.

Another important result of these studies was the difference in inhibitory potency of oligosaccharides and monosaccharides when tested with purified soybean lectin derived from seeds, compared with the lectin activity in the detergent-solubilized membrane extracts. Whereas N-acetyl-D-galactosamine and galactose were potent inhibitors of the seed lectin, they had no detectable effect on the detergent-solubilized membrane extracts. In contrast, the inhibitory potency of a number of glycoproteins, including asialofetuin, were identical with the two materials. It is notable also that attempts to purify the lectin in the detergent-solubilized materials by affinity chromatography on immobilized glycoprotein columns was not successful. It would appear then that failure to detect lectin in previous studies was due to inadequate extraction, and that the root lectin may be different from the seed lectin. Further biochemical, immunochemical, and immunohistochemical studies are obviously needed to evaluate this problem. For now the role of soybean lectin or a related material in binding specific *Rhizobium* species to root hairs has much support and no decisive negative evidence.

Added support for the role of lectins in specific root-bacterial symbiosis comes from a series of studies with the clover–*Rhizobium trifolii* system. Much of this work has recently been reviewed (24), and only some highlights are considered here.

The first fairly direct evidence for lectin involvement in the clover-bacterial symbiosis system came from the finding that 2-deoxyglucose inhibited the attachment of *R. trifolii* to clover root hairs, whereas 2-deoxygalactose or D-glucose were ineffective (25). The 2-deoxyglucose was presumably inhibiting the binding of a receptor on the surface of *R. trifolii* to a lectin on the clover root epidermal cells, since Dazzo & Hubbell (26) had previously shown that the capsular polysaccharide of *R. trifolii* contains 2-deoxyglucose, and that the agglutination of *R. trifolii* by a crude lectin preparation from clover seed is inhibited by 2-deoxyglucose. Other sugars in the capsular polysaccharide, including galactose, glucose, and glucuronic acid, had no detectable effect on this bacterial agglutination reaction, whereas N-acetyl-D-galactosamine, the only other active inhibitor, was not a constituent of the polysaccharide. The capsular polysaccharide that contained 2-deoxyglucose was inferred to be specific to infective strains of *R. trifolii,* since it bound an antiserum raised against infective strains and adsorbed with noninfective strains (26). A plant antigen that cross-reacted with this absorbed antiserum was localized by immunofluorescence to the exposed surface of the root epidermal cells of clover. The results, when taken together, suggested the preferential adsorption of infective versus noninfective cells of *R. trifolii* on the surface of clover roots by a cross-bridging of the related polysaccharides on their surfaces by a multivalent clover lectin.

Further work supported this general conclusion. The purified lectin from clover seeds, named trifoliin, had the same subunit molecular weight and appeared to be immunologically identical with a protein eluted from clover root hairs with 2-deoxyglucose and further purified by preparative gel electrophoresis (27). It was also shown that the lectin purified from root tissue agglutinated *R. trifolii,* and that this reaction was inhibited specifically by 2-deoxyglucose. Immunohistochemical studies with antiserum raised against trifoliin indicated that the lectin is concentrated at the growing root hair tip, the site of attachment of *R. trifolii* and of its capsular polysaccharides (28). It is notable that trifoliin receptors only appear on *R. trifolii* under certain growth conditions (29), which is reminiscent of findings with *R. japonicum* noted above; and that immunologically detectable trifoliin in the root hair region of clover seedlings was markedly inhibited by high NO_3^- concentrations, as was binding of *R. trifolii* to the root hairs (29). Thus, the evidence that root hair trifoliin mediates the binding of infective *R. trifolii* seems compelling.

Relationship of Seed Lectins to Related Material in Other Tissues

Among the functional studies of plant lectins, those that concern lectins in tissues other than seeds merit special consideration. Evidence that soybean lectins in plant tissues differ from those in seeds (23) has already been considered.

A detailed comparison of the seed lectin from *Dolichos bifluorus* with a related but distinct material from stems and leaves of six-week-old *D. bifluorus* plants has been reported (30, 31). Using antiserum raised against purified *D. bifluorus* seed lectin, immunologically cross-reactive material was demonstrated in buffer extracts of powdered leaves and stems, with a radioimmunoassay. Levels in these tissues were low and constant between two and eight weeks after germination, then increased severalfold in the next ten weeks. At this time extracts of stems and leaves contain several nanograms of lectin-like material per microgram of nitrogen, whereas mature seeds have about 1000 ng lectin per microgram. It is notable that in developing seeds the amount of lectin rises very abruptly; it is undetectable during the first 26 days after flowering but reaching a maximal level by day 28. It is also notable that no immunologically cross-reactive material was detectable in extracts of root tissue.

Immunologically cross-reactive material from extracts of stems and leaves of five- to six-week-old *D. bifluorus* plants did not bind to the affinity column used for purification of the seed lectin. However, this material could be purified by alternative methods, and yielded about 9 mg of immunologically cross-reactive material from 1 kg of plant tissue. The purified material did not agglutinate erythrocytes that could be agglutinated by the seed lectin, and also did not inhibit this agglutination. It behaved as a dimer with a molecular weight of 68,000, in contrast to the tetrameric seed lectin, which has a molecular weight of 110,000. Upon polyacrylamide gel electrophoresis in the presence of 8 M urea and 0.1% sodium dodecy sulfate, the plant tissue material showed two bands, one of which migrated exactly with subunit IA of the seed lectin. The other had a higher apparent molecular weight than that of the subunits of the seed lectin, and no other bands corresponding to those of the seed lectin were detected. The amino acid sequence of the first 13 amino-terminal residues of the two subunits of the cross-reactive material was identical. The only difference from the comparable sequences of subunits IA and IIA of the seed lectin was that the latter contain an asparagine rather than an aspartic acid at position 2.

Similarities of amino-terminal sequences raised the possibility that the subunits of the cross-reactive material and the seed lectin may represent different stages of completion or proteolytic modification of the carboxyl terminus of a common polypeptide chain. One possibility (31) is that seed lectin is formed by several modifications including proteolytic conversion of the heavy subunit of the cross-reactive material to subunit IIA, and transport of the precursor may account for the very rapid increase of lectin during maturation of the seed. In this scheme the related materials may be functional only at one site or at both.

Recently, Etzler and colleagues (32, 33; M. Etzler, personal communication) found that the immunologically cross-reactive material is in the cell walls of cells from the leaves and stems of the *D. bifluorus* plant, and that fungal infection or wounding of the plant seems to cause an increase in the amount of this material, as determined by fluorescence microscopy. In addition, it appears that this material is capable of binding to blood group substances at low ionic strength, but that it has a somewhat different specificity than the seed lectin (M. Etzler, personal communication). Localization in cell walls of the mature plant contrasts with the absence of another plant lectin from seed cell walls, and its distribution intracellularly and on the plasma membrane, as determined by immunohistochemistry (34). Evidence for the role of lectins in defense against infection has been reviewed recently (35) but is not considered here. The work with *D. bifluorus* suggests that lectin function in plants is followed by its transport and further utilization in the seed; this merits detailed evaluation.

LECTINS IN CELLULAR SLIME MOLDS

In contrast with plants, lectins in slime molds were discovered in the course of studies of cellular function. These eukaryotes are popular because of their value for studies of cellular differentiation and morphogenesis, since they change from discrete amoebae to aggregating colonies with 9–12 hr of starvation. In the ensuing 12 hr, this aggregate differentiates into a fruiting body consisting of stalk cells and spore cells.

Because slime mold amoebae can easily be made to differentiate from a vegetative to an aggregating and adhesive form they lend themselves to studies of the cellular changes that correlate with the development of cell-cell adhesiveness. One major change correlated with the development of this property is the synthesis of lectins. For example, in *Dictyostelium discoideum,* these lectins, which are absent in the vegetative state, may comprise more than 1% of the cellular protein of the aggregating amoebae (36–38). Lectin synthesis is regulated by the synthesis of mRNA for these particular proteins (39). In *D. discoideum* two lectins, discoidin I and discoidin II, have been identified. Each is a tetrameric protein with molecular weight of approximately 100,000. The discoidin II peptide is a product of a different gene than discoidin I (38, 39). Synthesis of the discoidin II subunit and the mRNA that directs its synthesis is relatively more advanced earlier in differentiation than synthesis of discoidin I. In *D. discoideum* in the aggregating stage, 90–95% of the lectin is discoidin I and the remainder is discoidin II. There is some evidence that discoidin I (40) and the mRNA that directs its synthesis (41) may be heterogeneous. In the case of purpurin (42), the lectin from a related species, *Dictyostelium purpureum,* seven forms of the lectin have been identified (D. N. Cooper and S. H. Barondes, unpublished). Five forms are tetramers made up of different combinations of two subunits, whereas the remaining two forms are homotetramers of still other subunits. All seven forms are found in *D. purpureum* populations cloned from a single cell. Each slime mold species examined contains lectins in the aggregating stage (43), and all those examined in some detail, including *Polysphondylium pallidum* (44) and *Dictyostelium mucuroides* (42), contain more than one lectin subunit.

Because slime mold lectins are synthesized concurrently with the development of cell-cell adhesiveness, it seemed possible that these processes were related. This hypothesis is supported by many lines of evidence, much of which has been reviewed recently (45); the work is of interest not only in evaluating the function of lectins in slime molds, but as an example of the types of biological approaches to determination of lectin function and the importance of the use of many experimental strategies to evaluate this problem.

One important property of slime mold lectins that suggests that they could play a role in cell-cell adhesion is that they are detectable on the cell surface. This was originally shown by the fact that aggregating but not vegetative slime mold cells could form rosettes with erythrocytes that were agglutinated by the slime mold lectins (36, 46). Participation of the lectins in rosette formation was shown by the fact that specific saccharides that inhibit the active site of the lectin blocked rosette formation, whereas others did not. The location of slime mold lectins on the cell surface has also been shown by immunohistochemical techniques using immunofluorescence and ferritin labeling (46, 47) and by surface iodination techniques (48, 49). The lectin appears to be diffusely distributed on the cell surface (46, 47). Immunohistochemical studies indicate that most of the lectin is intracellular (46, 47), but adequate evaluation of the intracellular localization of the lectin has not yet been satisfactorily accomplished. Examination with the electron microscope showed lectin associated with the cytoplasmic surface of endoplasmic reticulum (46), but its possible localization within the endoplasmic reticulum or within vesicles could not be evaluated with the technique used.

Of the total cellular lectin in aggregating *D. purpureum* or *D. discoideum* cells, only about 2% (i.e. 1×10^5 molecules per cell) is detectable on the cell surface by immunological techniques (49). The remainder of the approximately 5×10^6 molecules per cell is intracellular. All the intracellular and cell surface lectin can be solubilized by disrupting the membranes in the presence of a saccharide that reacts with the carbohydrate-binding site (49), which suggests that the lectin may be normally bound to a receptor and indicates that it is not integrated into the membrane bilayer. The cell surface lectin eluted by incubation with an appropriate saccharide is replaced within 45 min (49). By iodinating cell surface lectin before elution, and evaluation of eluted and noneluted lectin by immunoprecipitation and gel electrophoresis, it was shown that all cell surface lectin accessible to iodinating reagents was eluted with low concentrations of specific saccharide (49).

Whereas only a small fraction is normally present on the cell surface, considerable additional cell surface lectin can apparently be elicited to appear on the cell surface by appropriate treatment. This was first discovered in the course of studies that sought to quantitate cell-surface lectin by measuring binding of univalent fragments or divalent antibody directed against this protein. Univalent antibody fragments in saturating concentrations bound less than 10^5 molecules per cell, whereas divalent antibody molecules bound at least 10^6 molecules per cell (49). The divalent antibody binding was specific, as shown by both adsorption with lectin and by analysis of specific binding using a radioimmunoassay. The large amount of cell surface lectin detected by divalent antibody binding could either be due to cell breakage by the antibody or to a reaction by which it elicits appearance of intracellular lectin on the cell surface. Since direct observation with fluorescent antibody showed no evidence of cell breakage, it was concluded that the immunoglobulin elicited the appearance of additional cell surface lectin. Incubation of *D. purpureum* cells with polyvalent glycoproteins that bind to purpurin also elicited the appearance of additional cell surface lectin (W. R. Springer and S. H. Barondes, unpublished). Since both the glycoproteins and antibody directed against purpurin could cross-link cell surface lectin, elicitation of additional cell surface purpurin could be a consequence of this cross-linking. However, the participation of purpurin is not essential since cross-linking other surface molecules with concanavalin A also elicits the appearance of some cell surface purpurin.

How the additional cell surface purpurin is transferred from the cell interior is presently unclear. One possibility is that it is secreted from intracellular vesicles into the extracellular medium, then becomes bound to unoccupied cell surface receptors (49). In support of this idea, there is evidence that lectins in chicken intestinal goblet cells are contained in secretory granules and are secreted onto the intestinal mucosal surface (E. C. Beyer and S. H. Barondes, unpublished), as discussed below. However, no intracellular slime mold lectin has yet been detected in vesicles (46). Evidence for the secretion of soluble cytoplasmic proteins across the cell membrane (50) raises the possibility of a similar process here. Whatever the mechanism for the elicitation of additional cell surface purpurin, the results suggest that one function of the large amount of intracellular lectin is as a reservoir that may be tapped to mediate cell surface reactions.

In addition to its location on the cell surface, there is other evidence that supports the possible role of lectins in cell-cell adhesion in slime molds. Complementary oligosaccharides on the surface of fixed (51, 52) and living (49) differentiated slime mold cells have been detected, and the agglutinability of *D. discoideum* cells by purified discoidin increases markedly as the cells differentiate (51). Unfortunately, the nature of the cell surface receptors has not been determined. It is notable that upon polyacrylamide gel electrophoresis of *D. discoideum* glycoproteins and reaction of the gel with labeled lectins there is no detectable binding of discoidin, whereas concanavalin A and wheat germ agglutinin label many glycoproteins intensely (53). This suggests that the discoidin receptors are not glycoproteins or that they are very scarce. It remains possible that the labeled discoidin is a poor reagent for such studies and failed to detect the receptor molecules.

Discovery of a *D. discoideum* mutant that does not differentiate beyond the loose aggregate stage and that appears to have a specific defect in discoidin I, the major lectin of *D. discoideum,* also supports the role of this lectin in cell-cell adhesion (54). Extracts of aggregating cells of this mutant show very little hemagglutination activity, but are rich in a protein that is immunologically cross-reactive with discoidin I and has the same subunit molecular weight on SDS gels. Revertants that recover hemagglutination activity and that show complete differentiation have also recently been identified (55). The inference is that the mutant synthesizes an abnormal discoidin I subunit that is defective in carbohydrate binding. Failure of this mutant to develop beyond the loose aggregate stage supports a role for discoidin I in the differentiative process that leads to cell-cell adhesion. Whether the lectin binds the cells together or acts by eliciting some other molecule that directly mediates cell-cell adhesion cannot yet be determined. Failure of univalent (56, 57) or divalent (58) immunoglobulin directed against slime mold lectins to block cell-cell adhesion, except under unusual in vitro assay conditions (56), argues against, but does not exclude a direct role for the lectins in binding the cells together.

Like much of the work with plants, studies with slime molds suggest that lectins, although predominantly intracellular, play a role in some critical cell surface reactions. In both cases the lectins appear to mediate specific cell-cell associations, but the evidence is not yet conclusive.

LECTINS IN ANIMALS

The existence of lectins in animals has been known since the discovery of an agglutinin in the hemolymph of the horseshoe crab, early in the century. Because it is a circulating protein that could agglutinate pathogens, a protective function was considered (59) and has not yet been thoroughly evaluated. Many recent studies of animal lectins have been guided by two hypotheses. One proposes a role for cell surface lectins in receptor-mediated pinocytosis of partially degraded glycoproteins, as a prelude to their catabolism. The other, stimulated by studies of developmentally regulated lectins in slime molds, proposes that lectins in developing vertebrates play a role in intercellular interactions during tissue differentiation. Both hypotheses have led to discoveries that will probably have implications beyond what was envisioned initially.

Hepatic Lectins in Receptor-Mediated Pinocytosis

In contrast with the studies of plant lectins, this line of investigation began with the search for the molecular basis of a functional property—the pinocytosis of circulating asialoglycoproteins by rabbit liver cells. The history and many details of this work have been reviewed (60, 61). The binding protein from liver that was presumed to be responsible for this reaction was initially assayed by its formation of a precipitable complex with radioactive asialoorosomucoid. This protein was subsequently purified from detergent extracts of an acetone powder of rabbit liver by affinity chromatography on Sepharose derivatized with asialoorosomucoid (62), and was isolated as a highly aggregated water-soluble material containing 10% sugar by weight. The requirement for detergent in the extraction suggests, but does not prove, that this material is an integral membrane protein. In Triton X-100, aggregation of the purified material was reversed, with the formation of a single component with an estimated molecular weight of 2.5×10^5 and comprised of subunits with apparent molecular weights of 48,000 and 40,-000 (63). The purified carbohydrate-binding protein agglutinates human or rabbit erythrocytes, which leads to its explicit designation as a lectin (64). One other property that it shares with some plant lectins is induction of mitosis in peripheral lymphocytes (65). Like many other lectins it has a fairly broad range of saccharide-binding specificities. Although originally identified as a galactose-binding protein, methyl-N-acetyl-α-D-galactosaminide has a higher affinity (66). Among a group of synthetic neoglycoproteins the glucosyl derivative binds at least as well as the galactosyl derivative (67).

The finding of a purified lectin in liver that shared some of the properties that would be expected of the galactose-binding receptor, suggested, but did not prove, that the lectin mediated the clearance of partially degraded glycoproteins. The most direct supporting evidence was recently provided by Stockert et al (68) who infused antilectin IgG or control IgG into the portal vein of an isolated perfused rat liver prior to injection of labeled asialoorosomucoid or bilirubin. The antilectin IgG reduced the rate of influx of asialoorosomucoid into the liver by 80%, but did not affect the influx of bilirubin. This tends to directly implicate the cell surface lectin in uptake of asialoorosomucoid, although the possibility of a nonspecific effect secondary to the binding of antibody to this hepatic surface protein could not be excluded. It is also notable that uptake of asialoorosomucoid remained blocked for at least 90 min following antibody infusion, which suggests that there is no significant restoration of functional lectin to the hepatocyte surface during this period.

Although this liver lectin appears to exert its biological function at the plasma membrane, only about 5% (7 X 10^4 molecules/hepatocyte) is found at this site (69). About 95% of the asialoorosomucoid-binding protein of rat hepatocytes, as assayed in these studies, is localized intracellularly (69) in association with the Golgi complex, smooth microsomes, and lysosomes (70). One especially surprising finding is that most of the lectin activity associated with lysosomes appears to be oriented on the cytoplasmic surface of the membrane (70), in contrast with the other organelles in which it appears to be localized at the luminal surface. This localization was inferred from the fact that intact lysosomes bound labeled ligand without detergent disruption, whereas exposure of lectin in the other organelles required detergent treatment. The nature and significance of this apparent localization is not presently known.

The rat liver lectin is a fairly stable protein that apparently is not consumed in the degradation of serum glycoproteins. Tanabe et al (70) determined its half-life as approximately 88 hr, in the range of turnover of overall membrane proteins, and orders of magnitude slower than the degradation rate of serum glycoproteins taken up by the liver for destruction. Furthermore, the rate of degradation of the lectin was not increased by administration of a large exogenous load of asialoorosomucoid for liver clearance. This and other studies (69) indicate that the plasma membrane lectin is used catalytically and is regenerated in an active form by an unknown mechanism. Since the internalized asialoglycoproteins become rapidly associated with the lysosomes of hepatocytes (71), where they are presumed to be degraded, the possibility of a mechanism for protecting the lectin from degradation by eversion onto the cytoplasmic surface of the lysosomal membrane was considered (70). However, it remains possible that only the lectin of the plasma membrane is involved in the pinocytosis reaction, and that it may remain there during pinocytosis. The structural and functional relationship between the plasma membrane and intracellular lectin remains to be determined.

Added support for the role of the hepatic lectin in receptor-mediated pinocytosis comes from the finding that avian liver contains a carbohydrate-binding protein specific for N-acetylglucosamine–terminated glycoproteins, but is devoid of one specific for galactose-terminated proteins (72). This correlates with the fact that partially degraded glycoproteins with exposed

N-acetylglucosamine residues are cleared by avian liver. The estimated molecular weight of the subunit of the purified protein from chicken liver is 26,000 (73). A protein with subunit molecular weight of 31,000 that is presumed to participate in the clearance of mannose and N-acetylglucosamine–terminated glycoproteins has been purified from rabbit liver (74), and a hepatic fucosyl glycoprotein clearance system has been inferred (75). The material on human fibroblasts that directs pinocytosis of glycoproteins containing mannose 6-phosphate (61, 76) is also apparently related and could have both intracellular and cell surface functions. A possible pinocytotic role for the three other chicken lectins found in liver and other tissues (8, 9) that are discussed in the next section, remains to be evaluated.

An additional line of investigation provides a different perspective on the liver proteins that are presumed to mediate pinocytosis of glycoproteins. These studies showed that rat hepatocytes, which have a galactose-binding cell-surface lectin, specifically adhere to polyacrylamide gels containing covalently linked galactosides, and that chicken hepatocytes specifically bind to N-acetylglucosaminyl gels (77). Therefore, in these assays the lectins, or as yet unidentified molecules with similar specificities, behave as specific cell-adhesion molecules that bind the whole cell to a surface containing complementary saccharides. Whereas these lectins appear to function in glycoprotein pinocytosis, it is notable that in a different cellular environment they can function quite differently. This underscores the point that lectin function may be determined not only by the characteristics of its binding site but also by factors controlling its cellular localization and concentration.

Developmentally Regulated Vertebrate Lectins

Since cellular differentiation is currently of great interest, evidence for a possible mechanistic role of lectins in the development of vertebrate tissues has stimulated their identification and functional evaluation. In this section I describe what is known about three vertebrate lectins or classes of lectins, all of which may be designated "developmentally regulated" in that they have been shown to be especially prominent at a specific stage in the development of some tissues. The inference is that they play a specific role in the differentiative process that may diminish in importance or become totally unnecessary when the tissue has matured. One finding of this work is that, although all three lectins under discussion appear to be developmentally regulated in some tissues, they are especially prominent in others in adulthood, which suggests that they have also been adapted for other functions. This is one indication that the same lectin may be utilized in different ways.

Much of the work with developmentally regulated lectins has focused on muscle, since this system, which is so popular for in vitro studies of vertebrate cellular differentiation, contains a lectin that undergoes striking changes with development (78, 79). Thus extracts of pectoral muscle from 8-day-old chick embryos show very low levels of this lectin, but it rises to a level 1 to 2 orders of magnitude higher by 16 days of development (78), and then declines to very low levels in adult muscle. At its maximum, this lectin, which interacts with lactose and thiodigalactoside, constitutes about 0.1% of the protein in 16-day-old chicken embryo muscle extracts. It has been purified by affinity chromatography and is a dimer with subunit molecular weight of approximately 15,000 (80, 81). Antibodies raised to it have been used to demonstrate that it is predominantly intracellular, although some is detectable on the surface of cultured myoblasts (81, 82).

Because the striking increase in muscle cell lectin levels correlates temporally with fusion of these cells to form polynucleated myotubes, there has been considerable interest in the possibility that the lectin plays a role in this process. The finding of some lectin on the cell surface is considered to be consistent with this possibility. Direct support for this hypothesis was first provided by the finding that thiodigalactoside, a potent inhibitor of the hemagglutination activity of this lectin, blocks the fusion of cultured myoblasts from an established rat cell line (83). However, the same saccharide did not affect the fusion of chick myoblasts in primary cultures (79). Recent studies (84) showed that continued exposure of cultured L_8 myoblasts to 25 mM thiodigalactoside, beginning 2 days after plating, markedly inhibited fusion, which normally begins at day 5 and progresses thereafter. In contrast, maintenance of 25 mM thiodigalactoside beginning at day 4 after plating had no effect on fusion. These findings are consistent with both the previous positive and negative reports, and indicate that precise experimental conditions are decisive. The authors interpret their results as challenging the participation of cell surface lectin in myoblast fusion, since exposure beginning 24 hr before detectable fusion would be expected to be inhibitory; and because they confirm, by immunohistochemistry, that most of the muscle lectin is intracellular.

In contrast, other recent studies showing inhibition of myoblast fusion by repeated additions of purified chicken muscle lectin to the culture medium of primary chick myoblasts have been interpreted to support a role for this lectin in myoblast fusion (85). However, interpretation of this result is, in turn, difficult, since plant lectins with different specificities have also been shown to impair fusion; but higher concentrations were required. Unfortunately, simultaneous comparison of the effects of the muscle lectin and the other lectins on myoblast fusion has not yet been presented. Nevertheless, this finding has been taken as evidence that cell surface muscle lectin molecules may mediate fusion, perhaps by cross-linking available receptors on adjacent cells. Presumably the excess exogenous lectin blocks this bridging reaction by allowing each site to be occupied by its own lectin molecule; but alternative explanations are possible. Despite many suggestive findings, a specific role for this lectin in myoblast fusion is not yet established.

Because of its striking developmental regulation in embryonic muscle, the presence of this lectin has also been sought in other embryonic chick tissues at varying stages of development. Significant changes in a lectin that is probably identical have been observed in developing chick nervous system (86–88) and may be correlated with neuronal maturation (89). A similar or identical lectin activity has been identified as early as the pregastrula stage in chick embryos (90). In contrast with skeletal muscle, however, where virtually no lectin is found in the adult, other adult chicken tissues including liver and pancreas are very rich in this lectin (91). In adult liver the levels far exceed those found in embryonic liver, and the lectin is predominantly localized within the cells lining hepatic sinusoids, presumably Kupffer cells (91). The lectin purified from adult chicken liver by affinity chromatography appears indistinguishable from that purified from embryonic skeletal muscle, as evaluated by a variety of techniques including peptide mapping (8). Since the same lactose-inhibitable lectin is detectable in many chicken tissues, but is distinct from another lactose-inhibitable lectin described below, it has been operationally designated chicken lactose lectin-I (CLL-I). In contrast with its predominantly intracellular localization in cultured muscle cells and in sections of adult liver, an apparently identical lectin is highly concentrated in extracellular spaces in adult chicken pancreas (91), as demonstrated by immunohistochemistry. This striking difference in localization suggests a different but yet unidentified function of CLL-I in pancreatic organization.

A second lactose-inhibitable chicken lectin, chicken lactose lectin-II, (CLL-II) also shows striking developmental changes in several chick tissues. This lectin, which is scarce in embryonic muscle and brain, is very prominent in embryonic liver and kidney and then falls to much lower levels in adulthood (E. C. Beyer and S. H. Barondes, unpublished). It has, however, been studied most extensively in intestine, where it is found at low levels in the embryo but at extremely high levels in the adult. Indeed, adult chicken intestine, which contains a lectin apparently identical with CLL-I, contains about 200 times as much CLL-II.

CLL-II has been purified by affinity chromatography from chicken intestine and has a subunit molecular weight of approximately 14,000 and a peptide map and isoelectric point that are strikingly different from CLL-I (8). It is isolated predominantly as a monomer, and is a much less potent agglutinin than CLL-I, perhaps because it exists in reversible equilibrium with a dimeric form that may actually be the agglutinin (8). CLL-I and CLL-II are easily distinguishable immunochemically since, in radioimmunoassays with antibodies to each protein, the other shows no cross-reactivity even at concentrations three orders of magnitude higher (E. C. Beyer and S. H. Barondes, unpublished). However, hapten inhibitors of hemagglutination activity of both lectins are very similar (8). Whereas nothing is yet known about the possible function of CLL-II in development, a possible role in secretory functions of intestinal goblet cells is indicated. This is suggested by the concentration of CLL-II in the secretory granules of the intestinal goblet cells, as shown by immunohistochemical methods (E. C. Beyer and S. H. Barondes, unpublished). Similar techniques previously localized CLL-I in intestine to the same site (91); this is probably not due to immunological cross-reactivity, since antibody against CLL-I reacts so poorly with CLL-II. The mucin that is concentrated in secretory granules of the goblet cells is a potent inhibitor of CLL-II and CLL-I (E. C. Beyer and S. H. Barondes, unpublished) and is presumably associated with these lectins in the secretory granules of the goblet cells. Both lectin and mucin can be eluted from the intestinal mucosal surface. This series of findings suggest that CLL-II is somehow involved in the secretion of mucin from goblet cells into the intestinal lumen.

Demonstration of related lectins in many different vertebrate species has further focused attention on them. Thus far, the only lactose-binding lectins found in vertebrates other than the chicken behave as dimers and share other properties with the lectin referred to as CLL-I. The initial observation of a lectin of this type in vertebrate tissues, ranging from the electric organ of the electric eel to muscle extracts, was made by Teichberg et al (92) who suggested the name electrolectin. A similar lectir has since been identified in a number of adult bovine tissues including heart and lung (93) thymus, liver, and spleen (94), and no differences among them have been observed. Whereas bovine and human (95) muscle lectins show considerable immunological cross-reactivity, CLL-I is immunologically distinct (94). A related lectin that has been shown to be developmentally regulated has been described in rat lung (96).

As with many other lectins, little is presently known about the endogenous receptors for these lactose-binding lectins. As indicated above, CLL-II appears to be associated with mucins within the secretory granules of intestinal goblet cells. CLL-I is also found at this site, but it has different localizations in other tissues, including the extracellular space surrounding pancreatic lobules. Recent studies indicate that purified CLL-I added to intestinal sections binds to material in the basement membrane and in the intercellular spaces of intestinal epithelial cells, although it is not normally localized at these sites (97). However, it is difficult to interpret this, since purpurin, a lactose-binding lectin from *D. purpurem*, also localizes at these sites. This supports the presence of appropriate complementary glycoconjugates, but indicates that the capacity to bind does not necessarily indicate a functional relationship.

The only other developmentally regulated chicken lectin that has been characterized was initially identified as an agglutinin of trypsinized, glutaraldehyde-fixed rabbit erythrocytes that had been modified either by aging (98) or by heat or alcohol treatment (99). Hemagglutination activity is best inhibited by heparin, although dermatan sulfate and heparan sulfate also have activity (9, 99). Of the simple saccharides, N-acetyl-D-galactosamine is the most potent inhibitor (9, 98, 99). Although this lectin is developmentally regulated in embryonic chick muscle (98), it is extremely abundant in adult chicken liver and has recently been purified from both of these sources by gel filtration followed by affinity chromatography on heparin-Sepharose (9). The lectin exists as a very large aggregate that voids on Sepharose CL-2B, but can be dissociated into subunits with molecular weights of 13,000 and 16,000, upon polyacrylamide gel electrophoresis in SDS. Studies with cultured muscle cells indicate that the lectin is intracellular before fusion, but becomes detectable extracellularly after fusion (100). A potent inhibitor of the lectin has been identified in substrate-attached material from primary chick muscle cultures and appears to be glycosaminylglycan, with characteristics that suggest that it could be related to dermatan sulfate (100). When taken together these findings imply that this lectin may play some role in organization of extracellular glycosaminylglycans and/or the association of these substances with the cell surface. A lectin with very similar properties has recently been purified from young rat lung (M. M. Roberson, H. Ceri and S. H. Barondes, unpublished).

OTHER RELEVANT WORK

The studies summarized thus far illustrate the general approaches and types of experimental systems that are being used to elucidate lectin functions. There are many other related studies that will be mentioned briefly.

One major role of lectins, typified by the bacteria-legume symbiosis, appears to be to bind together cells of two different species. There is evidence that lectins acting in this way participate in both the prevention of plant infection (35, 101), by binding to saccharides on bacteria or fungi, and in the promotion of bacterial infection of vertebrate cells (102–104). In the latter case, bacterial lectins apparently mediate the adhesion of these microorganisms to oligosaccharides on animal cells, which could be a prelude to infection. Thus, *Escherichia coli* contains a lectin that binds D-mannose and its α-glycosides (102) and that presumably mediates bacterial attachment to cells (103). Evidence for binding of *Vibrio cholera* to intestinal cell surfaces by a reaction inhibited by L-fucose has been presented (104).

Support for a role of lectins in cell-cell interactions in lower organisms, conceptually similar to work with cellular slime molds, comes from several sources. For example, the bacterium *Myxococcus xanthus,* which has a complex life cycle that includes cellular aggregation and formation of a fruiting body, synthesizes a galactose-binding lectin with an apparent subunit molecular weight of 30,000, as it differentiates to the aggregating stage (105). Evaluation of the proposal that this lectin plays a role in cellular interactions will presumably exploit the background of genetic studies with this organism. Evidence that a sponge lectin, also specific for β-galactose residues, plays a role in sponge cell aggregation has been presented (106).

Animal lectins continue to be discovered and will undoubtedly stimulate much additional functional investigation. For example, multiple lectin activities have been identified in lymphocyte membrane preparations (107), and studies with fluoresceinyl derivatives of glycosylated cytochemical markers suggest that some of this lectin activity is displayed on the lymphocyte cell surface (108). Participation of a lectin-like receptor on the surface of rat macrophages in phagocytosis of senescent erthrocytes has been suggested (109). Evidence for a cell surface carbohydrate-binding component in teratocarcinoma stem cells that bind oligomannosyl residues has been presented (110), and its role in intercellular adhesion has been considered. A similar substance has been detected in frog embryos (111), where it presumably plays a developmental role.

Materials with binding properties similar to the chicken heparin-lectin have also been observed. Thus fibronectin, an adhesive glycoprotein that plays a role in cellular interactions, binds glycosaminoglycans, including heparin (112). It has also been shown to have hemagglutination activity that can be inhibited by hexosamines (113), which permits its classification as a lectin. It is notable, however, that the N-acetylated derivatives of the hexosamines, the form in which they are generally found in macromolecules, are not hapten inhibitors of this hemagglutination reaction, whereas nonsaccharide amines are inhibitory. These results suggest that the hemagglutination activity of fibronectin may not be due to its carbohydrate-binding properties. Similar hemagglutination activities that are inhibited by hexosamines but not by N-acetylhexosamines have been observed at a specific stage in differentiation of rat brain (114) as well as in human platelets (115). Their identity and function remain to be determined.

CONCLUSION

Although a specific cellular function cannot yet be unequivocably assigned to any lectin, a large body of evidence indicates that lectins have been adapted for a variety of cell surface and intercellular functions in which the specific carbohydrate-binding site of the lectin binds a complementary saccharide-containing substance as a prelude to one of a number of biological actions. The actions apparently include specific binding of cells of the same (slime molds) or different (*Rhizobium*–root cell) species; binding of soluble glycoproteins to the cell surface as a prelude to their pinocytosis; and organization of extracellular materials. Work with chickens suggests that the same lectin may be utilized for different functions in different tissues, and that the same lectin may play a developmental role in embryonic tissue and a different role in an adult tissue of the same organism.

Some lectins, such as those in root hairs, apparently bind complementary saccharides in a fairly discriminating way, which leads to highly specific symbiosis. In this case the lectin is apparently playing a highly refined recognition function, although its binding site is not so exclusive as to reject receptors on test erythrocytes. In contrast, evidence for different localizations and functions of the same lectin in different animal cells suggests that the specificity of lectin function is not dictated solely by the precise nature of its binding sites, but also by opportunistic factors (being in the right place at the right time) that determine which of many receptors of adequate complementarity are available. As with neurotransmitters that can interact with more than one receptor, the specificity of some lectin functions appears to be determined by structural features not only at the molecular but also at higher (cellular and developmental) levels of biological organization.

Literature Cited

1. Goldstein, I. J., Hughes, R. C., Monsigny, M., Osawa, T., Sharon, N. 1980. *Nature* 285:66
2. Hankins, C. N., Shannon, L. M. 1978. *J. Biol. Chem.* 253:7791–97
3. Lis, H., Sharon, N. 1973. *Ann. Rev. Biochem.* 42:541–74
4. Goldstein, I. J., Hayes, C. E. 1978. *Adv. Carbohydr. Chem. Biochem.* 35:127–340
5. Lis, N., Sharon, N. 1980. *The Biochemistry of Plants: A Comprehensive Treatise,* Vol. 6, ed. A. Marcus. New York: Academic. In press
6. Miller, J. B., Hsu, R., Heinrikson, R., Yachnin, S. 1975. *Proc. Natl. Acad. Sci. USA* 72:1388–91
7. Murphy, L. A., Goldstein, I. J. 1977. *J. Biol. Chem.* 252:4739–42
8. Beyer, E. C., Zweig, S. E., Barondes, S. H. 1980. *J. Biol. Chem.* 255:4236–39
9. Ceri, H., Kobiler, D., Barondes, S. H. 1981. *J. Biol. Chem.* 256:390–94
10. Pereira, M. E. A., Kabat, E. A. 1979. *Crit. Rev. Immunol.* 1:1–78

11. Kornfeld, R., Kornfeld, S. 1970. *J. Biol. Chem.* 245:2536–45
12. Hamblin, J., Kent, S. P. 1973. *Nature New Biol.* 245:28–30
13. Bohlool, B. B., Schmidt, E. L. 1974. *Science* 185:269–71
14. Bhuvaneswari, T. V., Pueppke, S. G., Bauer, W. D. 1977. *Plant Physiol.* 60:486–91
15. Bhuvaneswari, T. V., Bauer, W. D. 1978. *Plant Physiol.* 62:71–74
16. Bohlool, B. B., Schmidt, E. L. 1976. *J. Bacteriol.* 125:1188–94
17. Tsien, H. C., Schmidt, E. L. 1977. *Can. J. Microbiol.* 23:1274–84
18. Calvert, H. E., Lalonde, M., Bhuvaneswari, T. V., Bauer, W. D. 1978. *Can. J. Microbiol.* 24:785–93
19. Tsien, H. C., Schmidt, E. L. 1980. *Appl. Environ. Microbiol.* 39:1100–4
20. Pull, S. P., Pueppke, S. G., Hymowitz, T., Ord, J. H. 1978. *Science* 200:1277–79
21. Su, L. C., Pueppke, S. G., Friedman, H.

P. 1980. *Biochim. Biophys. Acta* 629:292–304

22. Pueppke, S. G., Keegstra, K., Ferguson, A. L., Bauer, W. D. 1978. *Plant Physiol.* 61:779–84

23. Bowles, D., Lis, H., Sharon, N. 1979. *Planta* 145:193–98

24. Dazzo, F. B. 1980. *The Cell Surface: Mediator of Developmental Processes,* ed. S. Substelny, N. K. Wessels, pp. 277–304. New York: Academic

25. Dazzo, F. B., Napoli, C. A., Hubbell, D. H. 1976. *Appl. Environ. Microbiol.* 32:168–71

26. Dazzo, F. B., Hubbell, D. H. 1975. *Appl. Microbiol.* 30:1017–33

27. Dazzo, F. B., Yanke, W. E., Brill, W. J. 1978. *Biochim. Biophys. Acta* 539:276–86

28. Dazzo, F. B., Urbano, M. R., Brill, W. J. 1979. *Curr Microbiol.* 2:15–20

29. Dazzo, F. B., Brill, W. J. 1978. *Plant Physiol.* 62:18–21

30. Talbot, C. F., Etzler, M. E. 1978. *Plant Physiol.* 61:847–50

31. Talbot, C. F., Etzler, M. E. 1978. *Biochemistry* 17:1474–79

32. Etzler, M., Gibson, D., Scates, S. 1979. *Plant Physiol.* 63:134 (Abstr.)

33. Gibson, D., Etzler, M. 1979. *Plant Physiol.* 63:134 (Abstr.)

34. Kilpatrick, D. C., Yeoman, M. M., Gould, A. R. 1979. *Biochem. J.* 184:215–19

35. Sequeira, L. 1978. *Ann. Rev. Phytopathol.* 16:453–81

36. Rosen, S. D., Kafka, J. A., Simpson, D. L., Barondes, S. H. 1973. *Proc. Natl. Acad. Sci. USA* 70:2554–57

37. Simpson, D. L., Rosen, S. D., Barondes, S. H. 1974. *Biochemistry* 13:3487–93

38. Frazier, W. A., Rosen, S. D., Reitherman, R. W., Barondes, S. H. 1975. *J. Biol. Chem.* 250:7714–21

39. Ma, G. C. L., Firtel, R. A. 1978. *J. Biol. Chem.* 253:3924–32

40. Ishiguro, A., Weeks, G. 1978. *J. Biol. Chem.* 253:7585–87

41. Roewekamp, W., Poole, S., Firtel, R. A. 1980. *Cell* 20:495–505

42. Barondes, S. H., Haywood, P. L. 1979. *Biochim. Biophys. Acta* 550:297–308

43. Rosen, S. D., Reitherman, R. W., Barondes, S. H. 1975. *Exp. Cell Res.* 95:159–66

44. Rosen, S. D., Kaur, J., Clark, D. L., Pardos, B. T., Frazier, W. A. 1979. *J. Biol. Chem.* 254:9408–15

45. Barondes, S. H. 1980. *Cell Adhesion and Motility,* ed. A. S. G. Curtis, J. Pitts, pp. 309–28. Cambridge: Cambridge Univ. Press

46. Chang, C.-M., Rosen, S. D., Barondes, S. H. 1977. *Exp. Cell Res.* 104:101–9

47. Chang, C.-M., Reitherman, R. W., Rosen, S. D., Barondes, S. H. 1975. *Exp. Cell Res.* 95:136–59

48. Siu, C.-H., Lerner, R. A., Ma, G. C. L., Firtel, R. A., Loomis, W. F. 1976. *J. Mol. Biol.* 100:157–78

49. Springer, W. R., Haywood, P. L., Barondes, S. H. 1980. *J. Cell Biol.* 87:682–90

50. Wickner, W. 1980. *Science* 210:861–68

51. Reitherman, R. W., Rosen, S. D., Frazier, W. A., Barondes, S. H. 1975. *Proc. Natl. Acad. Sci. USA* 72:3541–45

52. Bartles, J. R., Frazier, W. A. 1980. *J. Biol. Chem.* 255:30–38

53. Burridge, K., Jordan, L. 1979. *Exp. Cell Res.* 124:31–38

54. Ray, J., Shinnick, T., Lerner, R. A. 1979. *Nature* 279:215–21

55. Shinnick, T. M., Lerner, R. A. 1980. *Proc. Natl. Acad. Sci. USA* 77:4788–92

56. Rosen, S. D., Chang, C.-M., Barondes, S. H. 1977. *Dev. Biol.* 61:203–13

57. Bozzaro, S., Gerisch, G. 1978. *J. Mol. Biol.* 120:265–79

58. Springer, W. R., Barondes, S. H. 1980. *J. Cell Biol.* 87:703–7

59. Finstad, C. L., Good, R. A., Litman, G. W. 1974. *Ann. NY Acad. Sci.* 234:170–80

60. Ashwell, G., Morell, A. G. 1974. *Adv. Enzymol.* 41:99–128

61. Neufeld, E. F., Ashwell, G. 1980. *The Biochemistry of Glycoproteins and Proteoglycans,* ed. W. Lennarz, pp. 241–66. New York: Plenum

62. Hudgin, R. L., Pricer, W. E. Jr., Ashwell, G., Stockert, R. J., Morell, A. G. 1974. *J. Biol. Chem.* 249:5536–43

63. Kawasaki, T., Ashwell, G. 1976. *J. Biol. Chem.* 251:1296–1302

64. Stockert, R. J., Morell, A. G., Scheinberg, I. H. 1974. *Science* 186:365–66

65. Novogrodsky, A., Ashwell, G. 1977. *Proc. Natl. Acad. Sci. USA* 74:676–78

66. Sarkar, M., Liao, J., Kabat, E. A., Tanabe, T., Ashwell, G. 1979. *J. Biol. Chem.* 254:3170–74

67. Stowell, C. P., Lee, Y. C. 1978. *J. Biol. Chem.* 253:6107–10

68. Stockert, R. J., Gartner, U., Morell, A. G., Wolkoff, A. W. 1980. *J. Biol. Chem.* 255:3830–31

69. Steer, C. J., Ashwell, G. 1980. *J. Biol. Chem.* 255:3008–13

70. Tanabe, T., Pricer, W. E. Jr., Ashwell, G. 1979. *J. Biol. Chem.* 254:1038–43

71. Hubbard, A. L., Stukenbrok, H. 1979. *J. Cell Biol.* 83:65–81

72. Lunney, J., Ashwell, G. 1976. *Proc. Natl. Acad. Sci. USA* 73:341–43
73. Kawasaki, T., Ashwell, G. 1977. *J. Biol. Chem.* 252:6536–43
74. Kawasaki, T., Etoh, R., Yamashina, I. 1978. *Biochem. Biophys. Res. Commun.* 81:1018–24
75. Prieels, J. P., Pizzo, S. V., Glascow, L. R., Paulson, J. C., Hill, R. L. 1978. *Proc. Natl. Acad. Sci. USA* 75:2215–19
76. Sly, W. S., Stahl, P. 1979. *Transport of Molecules in Cellular Systems,* Dahlem Conferenzen, Berlin, ed. S. Silverstein, pp. 229–44
77. Weigel, P. H., Schnaar, R. L., Kuhlenschmidt, M. S., Schmell, E., Lee, R. T., Lee, Y. C., Roseman, S. 1979. *J. Biol. Chem.* 254:10830–38
78. Nowak, T. P., Haywood, P. L., Barondes, S. H. 1976. *Biochem. Biophys. Res. Commun.* 68:650–57
79. Den, H., Malinzak, D. A., Rosenberg, A. 1976. *Biochem. Biophys. Res. Commun.* 69:621–27
80. Den, H., Malinzak, D. A. 1977. *J. Biol. Chem.* 252:5444–48
81. Nowak, T. P., Kobiler, D., Roel, L. E., Barondes, S. H. 1977. *J. Biol. Chem.* 252:6026–30
82. Podleski, T. R., Greenberg, I. 1980. *Proc. Natl. Acad. Sci. USA* 77:1054–58
83. Gartner, T. K., Podleski, T. R. 1975. *Biochem. Biophys. Res. Commun.* 67:972–78
84. Kaufman, S. J., Lawless, M. L. 1980. *Differentiation* 16:41–48
85. MacBride, R. G., Przybylski, R. J. 1980. *J. Cell. Biol.* 85:617–25
86. Kobiler, D., Barondes, S. H. 1977. *Dev. Biol.* 60:326–30
87. Kobiler, D., Beyer, E. C., Barondes, S. H. 1978. *Dev. Biol.* 64:265–72
88. Eisenbarth, G. S., Ruffolo, R. R. Jr., Walsh, F. S., Nirenberg, M. W. 1978. *Biochem. Biophys. Res. Commun.* 83:1246–52
89. Gremo, F., Kobiler, D., Barondes, S. H. 1978. *J. Cell Biol.* 78:491–99
90. Cook, G. M. W., Zalik, S. E., Milos, N., Scott, V. 1979. *J. Cell Sci.* 38:293–304
91. Beyer, E. C., Tokuyasu, K., Barondes, S. H. 1979. *J. Cell Biol.* 82:565–71
92. Teichberg, V., Silman, I., Beitsch, D. D., Reshoff, G. 1975. *Proc. Natl. Acad. Sci. USA* 72:1383–87
93. DeWaard, A., Hickman, S., Kornfeld, S. 1976. *J. Biol. Chem.* 251:7581–87
94. Briles, E. B., Gregory, W., Fletcher, P., Kornfeld, S. 1979. *J. Cell Biol.* 81:528–37
95. Childs, R. A., Feizi, T. 1979. *Biochem. J.* 183:755–58
96. Powell, J. T., Whitney, P. L. 1980. *Biochem. J.* 188:1–8
97. Beyer, E. C., Barondes, S. H. 1980. *J. Supramol. Struct.* 13:219–27
98. Mir-Lechaire, F. J., Barondes, S. H. 1978. *Nature* 272:256–58
99. Kobiler, D., Barondes, S. H. 1979. *FEBS Lett.* 101:257–61
100. Ceri, H., Shadle, P. J., Kobiler, D., Barondes, S. H. 1979. *J. Supramol. Struct.* 11:61–67
101. Sharon, N. 1979. *Glycoconjugate Research,* Vol. 1, Proc. Int. Symp. Glycoconjugates, 4th, pp. 459–91. New York: Academic
102. Eshdat, Y., Ofek, I., Yashouv-Gan, Y., Sharon, N., Mirelman, D. 1978. *Biochem. Biophys. Res. Commun.* 85:1551–59
103. Ofek, I., Mirelman, D., Sharon, N. 1977. *Nature* 265:623–25
104. Jones, G. W., Freter, R. 1976. *Infect. Immunol.* 14:240–45
105. Cumsky, M., Zusman, D. R. 1979. *Proc. Natl. Acad. Sci. USA* 76:5505–9
106. Muller, W. E. G., Kurelec, B., Zahn, R. K., Muller, I., Vaith, P., Uhlenbruck, G. 1979. *J. Biol. Chem.* 254:7479–81
107. Kieda, C. M. T., Bowles, D. J., Ravid, A., Sharon, N. 1978. *FEBS Lett.* 94:391–96
108. Kieda, C. M. T., Roche, A.-C., Delmotte, F., Monsigny, M. 1979. *FEBS Lett.* 99:329–32
109. Kolb, H., Kolb-Bachofen, V. 1978. *Biochem. Biophys. Res. Commun.* 85:678–83
110. Grabel, L. B., Rosen, S. D., Martin, G. R. 1979. *Cell* 17:477–84
111. Roberson, M. M., Armstrong, P. B. 1980. *Proc. Natl. Acad. Sci. USA* 77:3460–63
112. Yamada, K. M., Olden, K. 1978. *Nature* 275:179–84
113. Yamada, K. M., Yamada, S. S., Pastan, I. 1975. *Proc. Natl. Acad. Sci. USA* 72:3158–62
114. Simpson, D. L., Thorne, D. R., Loh, H. H. 1977. *Nature* 266:367–69
115. Gartner, T. K., Williams, D. C., Minion, F. C., Phillips, D. R. 1978. *Science* 200:1281–83

Ann. Rev. Biochem. 1981. 50:233–60

PROTEINS CONTROLLING THE HELICAL STRUCTURE OF DNA

♦12078

Klaus Geider and Hartmut Hoffmann-Berling[1]

Max-Planck-Institut für Medizinische Forschung, Abteilung Molekulare Biologie, D 6900 Heidelberg, West Germany

CONTENTS

SUMMARY AND PERSPECTIVES

Cells produce a variety of proteins for manipulating the helical structure of DNA and for shaping the physical form of the DNA molecule. Enzymatically nonactive species of these proteins such as the histones and the nonhistone DNA binding proteins serve to organize the genetic material for storage. They combine with DNA irrespective of the nucleotide sequence, ordering the message in such a way that it can be read by RNA polymerase and regulatory proteins. Other types of DNA-binding proteins interact

[1]We thank colleagues for sending information prior to publication. We also acknowledge the help of Brunhilde Werner, Fizz Marvin, and Dr. Mo-Quen Klinkert in preparing the manuscript.

233

preferentially with the separated DNA strands. They serve to modulate the accessibility of the DNA bases for DNA polymerase action and for complementary pairing with the bases of a homologous DNA.

Enzymes involved in shaping the conformation of DNA can be divided into those using the energy of ATP (or some other nucleoside triphosphate) to drive the transition of the DNA conformation and those that are independent of the supply of chemical energy. Enzymes in the first category are DNA-stimulated ATPases; in the presence of Mg^{2+} they hydrolyze ATP to ADP and P_i. This group includes the various types of DNA unwinding enzymes, the *recA* protein of *Escherichia coli* (a DNA annealing enzyme), and the topoisomerases of type II (represented by bacterial gyrase). Topoisomerases of type I are independent of a nucleotide cofactor. They catalyze the conversion of superhelical DNA to the energetically favored relaxed form of the DNA.

Our article reviews a topic discussed in the 1978 volume of this series (1), and as far as possible we describe only results obtained since then. Histones, recently the subject of a separate review (2), are not considered here. Additional reviews are available on the physical and chemical properties of DNA binding proteins (3), DNA unwinding enzymes (4, 4a) and topoisomerases (5, 6).

DNA BINDING PROTEINS

DNA binding proteins are nonenzymatic proteins with affinity for DNA. They lack nucleotide specificity and bind to DNA in stoichiometric amounts. Binding proteins specific for single-stranded DNA can be distinguished from those with affinity also for double-stranded DNA. Proteins of the first type have been named DNA unwinding proteins (7), DNA extending proteins (8), DNA melting proteins (9), single strand binding proteins (10), and helix destabilizing proteins (11). The last two terms are common in recent publications.

Purification and Characterization of DNA Binding Proteins

DNA has been removed from the supernatant of a cell homogenate by nuclease digestion (12) or precipitation with polyethylene imine (13, 14) or polyethylene glycol 6000 (12). *E. coli* extracts low in DNA can be obtained by treating the cell suspension with lysozyme followed by brief heating at 37°C (15) or freezing in liquid nitrogen (16) and subsequent centrifugation. DNA binding proteins can be purified on a column to which DNA has been adsorbed (12) or cross-linked by UV-treatment (17). Other chromatographic procedures make use of phosphocellulose (18), heparin-Sepharose (19), blue dextran-Sepharose (20), or Bio-Rex 70 (14, 21). Since the DNA binding proteins I and II of *E. coli* are heat stable, other proteins in the solution can be denatured by boiling, and after centrifugation the binding proteins are found in the supernatant (18, 22). Antibodies against DNA binding protein immobilized on a matrix have been used to prepare binding protein in a single chromatographic step (23). If required, contaminants in an apparently homogenous preparation of DNA binding protein have to be removed by additional purification steps (24).

DNA binding proteins can be identified by gel electrophoresis. After transferring the protein bands to nitrocellulose sheets the proteins can be assayed for DNA binding (25). Complexes between binding protein and DNA can be identified by adsorbing the protein to nitrocellulose filters (18, 26), velocity sedimentation (18, 27), or photometry (28–30).

Many single strand binding proteins bind cooperatively to DNA. Cooperative binding can be assayed by electron microscopy, by studying the melting of double-stranded DNA, or by sedimentation of single-stranded DNA in the presence of the binding protein (7, 9, 22, 28, 31). The dissociation of the protein-DNA complexes can be followed by velocity sedimentation (32), by eluting the protein at constant salt from DNA immobilized on cellulose (33), or by transferring the protein from labeled to unlabeled DNA strands (18).

Single Strand Binding Proteins

E. COLI DNA BINDING PROTEIN I   This protein is a tetramer of 20,000-dalton subunits. The isoelectric point is at pH 6.0. The protein binds cooperatively to DNA, covering eight nucleotides per monomer (7, 18). Fast binding to DNA is followed by slow release. In solutions of low ionic strength the protein destabilizes the helix (7); at physiological salt concentrations it catalyzes the renaturation of the separated strands. At pH 6–7 renaturation is enhanced 5000-fold when either spermidine or spermine is also present (34). The protein may function in general recombination (35). DNA binding protein I protects single-stranded DNA against degradation by nucleases (27, 36) and interferes with uptake of the DNA by *E. coli* spheroplasts (37).

About 1000 monomeric protein copies of DNA binding protein I are found in a cell. This amount is probably sufficient to cover all transient single-stranded areas of the cellular DNA. The protein hinders uncontrolled single strand transcription by RNA polymerase (38, 39) and guides the initiation of DNA synthesis to specific sites, as shown for filamentous phages (fd, M13, f1) (40). DNA binding protein I covers fd single strands, leaving a hairpin structure for RNA polymerase to bind. Transcription of about 30 nucleotides of the hairpin at a specific site of the phage genome results in a stable RNA/DNA hybrid that replaces the incompletely base–paired DNA region (40). DNA binding protein I binds subsequently to the nontranscribed part of the hairpin and, on collision with the transcribing RNA polymerase, causes termination of the RNA. This occurs within a range of eight nucleotides, because the binding protein apparently does not always bind in the same frame. The hairpin structure also causes termination of DNA synthesis. A gap in the phage replicative form (RF) synthesized in vitro (41) is found near the site where *E. coli* RNA polymerase binds to the DNA.

Similarly, DNA binding protein I causes specific initiation of replication on the single strands of phage G4. The initiating *E. coli dnaG* protein forms the primer RNA only in the presence of the binding protein (42). In the initiation on phage øX174 single strands, an *E. coli* protein, named n' protein, enters the DNA at a specific site (43) and forms, with other *E. coli* proteins, e.g. n' protein, n" protein, i protein, *dnaB* protein, and *dnaG* protein, a preinitiation complex (44) that apparently migrates along the DNA (45). Mediated by the initiation complex, appropriate signals on the DNA enable *dnaG* protein to start RNA synthesis (46). Without the binding protein, *dnaB* protein and *dnaG* protein alone can prime øX174 DNA and other single-stranded DNAs, but initiation occurs at multiple and unspecific sites (47).

DNA binding protein I enhances chain elongation by DNA polymerase III holoenzyme (48) and *E. coli* DNA polymerase II (7). The latter polymerase forms a complex with the binding protein (49). *E. coli* DNA polymerase I or phage T4 DNA polymerase are not affected by *E. coli* protein I.

E. coli DNA binding protein I is also required for the replication of double-stranded DNA of bacteriophages øX174 and fd (50, 51). During separation of the strands by *rep* protein (see below) the binding protein forms complexes with the single-stranded DNA regions, thus preventing their renaturation [(52); T. F. Meyer and K. Geider, manuscript in preparation]. Furthermore, the protein supports DNA polymerase III holoenzyme during synthesis of the viral strand. When viral and complementary DNA strands are replicated in vitro simultaneously, *E. coli* DNA binding protein I is needed at only 4% the level required for separate replication of the two strands (53).

An *E. coli* mutant that produces thermolabile DNA binding protein I has been isolated recently (20). The mutation (*ssb*-1, allelic wit *lexC*), which maps at 90.8 min, renders the cells temperature-sensitive for DNA replication and inefficient both for UV-induced repair and for recombination (54). The growth of a øX174-like phage is also affected by the *ssb* mutation (10). Cells carrying the *ssb* gene on a plasmid (55) produce about ten times more

DNA binding protein I than wild-type cells (56). The protein is active in vitro in phage fd replication (V. Berthold-Schmidt and K. Geider, manuscript in preparation).

BACTERIOPHAGE T4 GENE 32 PROTEIN   The first DNA binding protein discovered was the gene 32 product of phage T4. Mutations in this gene interfere with replication and recombination of the phage DNA. Some have a mutator effect for the phage (57) and others affect the activities of phage gene products like DNA polymerase, DNA ligase, or nucleases (58).

The level of gene 32 protein in infected cells is adjusted according to the demand for the protein in viral growth. Amber mutations in the gene cause the production of large quantities of nonfunctional amber fragment (59). Suppression of these mutants can lead to some restoration of gene 32 protein activity, while leaving a defect in the self-regulation of its production (60). The amount of gene 32 protein in the cell can be correlated with the level of single-stranded DNA (59). An increase of this level by UV-induced lesions stimulates gene 32 protein synthesis (61). On the other hand, gene 32 protein inhibits its own expression. The protein forms a complex with the gene 32 mRNA thereby inhibiting further synthesis (62, 63).

In cell extracts, some of the T4 DNA binding protein is found as a fragment of 27,000 daltons. This has a higher affinity to DNA than the 35,000-dalton protein (64). Controlled proteolysis of T4 gene 32 protein results in the formation of fragments with either the N terminus or the C terminus removed (65). The fragment lacking the N terminus has a weak affinity for single-stranded DNA. Calorimetric and spectrometric studies suggest that this terminus is essential for cooperative DNA binding of the protein (29, 66). In contrast, protein-protein interaction is not impaired by removing 60 amino acids from the C terminus. Not only has this cleavage product an increased affinity to single-stranded DNA (about five times greater than that of gene 32 protein), it also binds to double-stranded DNA. Binding to T4 DNA polymerase found for wild-type binding protein (67) is not observed with this fragment. Moreover, the fragment inhibits DNA synthesis by T4 DNA polymerase (68). This is, however, not the case when T4 DNA polymerase accessory proteins (gene 44/62 and 45 proteins) are also present (see below). The fragment also inhibits the synthesis of RNA primers in the presence of T4 gene 41/61 proteins and diminishes the utilization of the RNA primers. Whether processing of T4 DNA binding protein plays a role in phage growth remains to be shown. Phage-DNA-protein complexes from T4-infected cells contain T4 gene 32 protein (69). The 35,000-dalton protein is also a tool for enhancing single-stranded regions of DNA in the electron microscope (70).

DNA BINDING PROTEINS OF BACTERIOPHAGES T7 AND T5   A DNA binding protein coded by phage T7 stimulates the viral DNA polymerase. It can be replaced by *E. coli* DNA binding protein I in vitro and in vivo (32, 71). A gene for the T7 protein has only recently been assigned on the phage genome (R. Hayward, personal information; W. Studier, personal information). T7 DNA can be cleaved at specific sites in a reaction dependent on T7 DNA polymerase, T7 gene 4 protein, and a 33,000-dalton host protein T7. In this reaction T7 gene 4 protein can be replaced by T7 DNA binding protein and less efficiently by *E. coli* DNA binding protein I (72).

Gene D5 of bacteriophage T5 is required for viral DNA synthesis and transcription (73). The gene codes for a DNA binding protein that is produced in large quantities (2% of total cell protein) (74, 75). T5 DNA binding protein is a 29,000-dalton monomer in solution. It prefers cooperative binding to double-stranded DNA to noncooperative binding to single strands. Conditions favouring viral double strand synthesis increase the level of T5 DNA binding in the cell.

GENE 5 PROTEIN OF FILAMENTOUS BACTERIOPHAGES   Filamentous phages like fd, M13, and f1 produce single-stranded DNA by asymmetric replication of the RF circles (76). Conversion of the displaced single strand to a double strand is prevented by phage gene 5 protein (77). It displaces *E. coli* DNA binding protein I from the single strand (27) and inhibits complementary DNA synthesis by *E. coli* DNA polymerase III holoenzyme (48). Phage fd gene 5 protein is synthesized in at least 100,000 copies per cell (78, 79); the monomer has a molecular weight of 9,688 as deduced from the amino acid sequence (80, 81). The carboxyl terminus of the protein can be cross-linked to the viral DNA (82). Spectroscopic studies suggest that tyrosine residues stack to the DNA bases in the protein-DNA complex. They are not involved in protein-protein interaction with the gene 5 protein dimer (30). X-ray studies show that a tyrosine residue is found in the β-sheet structure of gene 5 protein. This structure is probably involved in DNA binding (83). A tyrosine residue binding to DNA has also been demonstrated by nuclear magnetic resonance studies (84).

ADENOVIRUS DNA BINDING PROTEIN   A DNA binding protein of M_r 72,000, most abundant among the early proteins synthesized by adenovirus, is coded by a region extending from 62.4 to 67.9% on the standard map (85). The protein binds cooperatively to single-stranded DNA, one monomer covering seven nucleotides (86). In 1 M salt it exists as trimers, and it aggregates more extensively in low ionic strength (87). The protein may be involved in the initiation of DNA replication and chain elongation (88) and in the control of its own level (89). The protein, originally believed to bind only to single-stranded DNA (90), can also bind to the termini of double strands (91). Complexes of adenovirus DNA and binding protein extracted from virus-infected cells appear as thick filaments that represent protein-covered single strands, and thinner double-stranded DNA (92). Terminal binding of the DNA binding protein renders the termini of the DNA accessible to single strand specific nucleases (91) and stabilizes the double-stranded part against melting (92). Chymotrypsin cleaves the adenovirus protein into an N-terminal 26,000-dalton fragment and a C-terminal 44,000-dalton fragment (93, 94). Phosphorylation occurs at the N terminus of the protein. Since binding to DNA is mediated by the C-terminal domain, the phosphate groups only weakly affect the interaction with DNA (93). A mutant producing a thermolabile protein is temperature-sensitive in viral DNA synthesis both in vivo and in nuclear extracts (95). Normal replication is obtained by adding wild-type adenovirus DNA binding protein (96) or the 44,000-dalton protein fragment to the extracts (97).

In contrast to wild-type infection, most of the viral genome integrates into host DNA in multiple copies when mutant infection occurs at a nonpermissive temperature (98). Furthermore, in immunological studies, the adenovirus DNA binding protein cross-reacts with T antigen prepared from virus-transformed cells (99). The antigen is expressed only in those transformed cells that contain the DNA fragment specific for adenovirus DNA binding protein (100).

OTHER EUKARYOTIC SINGLE STRAND BINDING PROTEINS   Several eukaryotic nonhistone DNA binding proteins alter their affinity with the degree of phosphorylation, and others stimulate cellular DNA polymerases when bound to single-stranded DNA. There has recently been a survey on the properties of these proteins (101).

Double Strand Binding Proteins

In prokaryotic chromosomes DNA is associated with polyamines, RNA, and protein. Electron microscopy of *E. coli* cells disrupted on the film shows regularly condensed chromatin-like fibers (102). Proteins of 17,000, 9,000 (103), 30,000, and 15,000 daltons (104) are associated with the chromosomal DNA.

E. COLI DNA BINDING PROTEIN II   DNA binding protein II (previously named protein HU, HD, and NS) was found on the basis of its ability to enhance the transcription of phage λ DNA (105) and to inhibit DNA synthesis with DNA polymerase III holoenzyme (22). The protein was also detected in preparations of the 30S subunit of ribosomes (106). The proteins in these preparations are identical on the basis of molecular weight, immunological properties, and electrophoretic behavior in two-dimensional gels.

DNA binding protein II is heat stable. It binds to both single-stranded and double-stranded DNA and to RNA (22). The molecular weight is 9,500, as confirmed by the amino acid sequence. At low pH the protein can be separated into two 90-amino acid long polypeptides that differ in 30% of the amino acid residues (107, 108). In solution the protein is a dimer or tetramer; the two components are present in a 1:1 ratio (109). DNA binding protein II is present in about 100,000 copies per cell. It forms nucleosome-like structures with double-stranded DNA, as shown by electron microscopy and by nucleolytic degradation (110). SV40 DNA is condensed twofold by a saturating amount of the protein, and a deficiency of about 18 superhelical turns is introduced into circular SV40 DNA. The complexed DNA molecules show an average of 14 beads in the electron microscope. In *E. coli* the protein is associated with the nucleoid (111).

Protein II, added to complexes of single-stranded DNA and *E. coli* DNA binding protein I, can displace about 50% of the DNA binding protein I molecules. In the presence of the two proteins and single-stranded and double-stranded DNA, protein II preferentially binds to the double strands (27). Complexes of proteins I or II with DNA can be distinguished in the electron microscope (112). The complex made by protein I, protein II, and single-stranded DNA is inert in DNA replication (22).

Double-stranded DNA associated with protein II appears in the electron microscope to be thickened (105) and to form flower-like structures (112). As suggested by velocity sedimentation studies, double-stranded DNA can be bound almost in a 1:1 ratio of nucleotides and protein monomers (22). On the other hand, one monomer of the protein binds four nucleotides of single-stranded DNA. In cells, DNA binding protein II is sufficient to bind about 1/10 of the chromosomal DNA. A 10,000-dalton DNA binding protein from blue-green bacteria shows an immunological cross-reaction with *E. coli* DNA binding protein II (113).

OTHER PROKARYOTIC DOUBLE STRAND BINDING PROTEINS   A histone-like protein of *E. coli,* H protein, with a molecular weight of 28,000, has been isolated (114). This protein binds 75 bases per dimer; the 30,000 molecules in the cell may thus complex 20% of the bacterial chromosome. H protein inhibits: transcription and DNA synthesis with isolated polymerases; DNA-dependent ATPase activity of *rep* protein, *dnaB* protein, and n' protein; and eukaryotic DNA topoisomerase activity. It protects DNA against nucleases and supports the reannealing of denatured DNA. Its amino acid composition resembles that of histone H2A. Histone H2A antibodies neutralize H protein, but H protein antibodies do not interfere with reannealing catalyzed by histone H2A.

A 17,000-dalton protein, HLP I (histone-like protein), interacts with *E. coli* RNA polymerase (115). It is positively charged at pH 8.7 and binds to DNA. HLP I is coded by the *firA* gene, which maps at 4 min on the *E. coli* map.

About 300 dimers of H1 protein can be bound per molecule of phage λ DNA. This amount covers only a part of the phage genome, but it is sufficient to stimulate transcription of λ DNA in vitro (116). The molecular weight of H1 protein is 15,000 in the presence of a reducing agent and 30,000 in its absence (117).

EUKARYOTIC NONHISTONE PROTEINS   The order of the nucleosomes in the eukaryotic chromosome is thought to be controlled by nonhistone DNA binding proteins. Proteins with molecular weights below 30,000, called high mobility group proteins (HMG proteins), occur in large quantities in the nucleus (about 10^6 molecules per cell). These proteins, which belong to four major species (HMG 1, 2, 14, and 17), seem to associate with the nucleosome (118). Although their amino acid composition has some similarity to *E. coli* DNA binding protein II (22, 105, 106, 119), the sequence data of HMG 14 and 17 (120, 121) show a relationship to histones. The HMG proteins bind to double-stranded DNA (122), which causes local unwinding of the helix (123) and formation of beads similar in appearance to nucleosomes (124). In addition HMG proteins form complexes with histone H1 (125). Like histones they are acetylated in a postsynthetic process (126). Furthermore, they inhibit histone deacetylase enzymes (127).

Actively transcribed eukaryotic genes lose their sensitivity to pancreatic DNase when the nuclear proteins HMG 14 and HMG 17 are removed from the chromatin by salt elution (128). The sensitivity to DNase is restored by adding HMG proteins to the HMG-depleted chromatin (129). The switch of transcription from embryonic to adult β-globin genes is accompanied by a loss of nuclease sensitivity of the embryonic genes (130). The reason for this altered sensitivity to nuclease may be that the embryonic genes release their HMG proteins in the course of differentiation.

Two further DNA binding proteins, D1 and D2, have been eluted from the chromatin of *Drosophila melanogaster* cells. D1 elutes at low salt and is present at a limited number of loci in polytene chromosomes (131). D2 elutes at 0.5 M salt together with histone H2A (132). Peptide mapping shows that protein D2 is not a histone. A third DNA binding protein from *D. melanogaster* binds preferentially to *Drosophila* satellite DNA (133). Some properties of DNA binding proteins are listed in Table 1.

ENZYMES DEPENDENT ON CHEMICAL ENERGY

DNA Unwinding Enzymes

Two groups of DNA unwinding enzymes can be distinguished: DNA polymerase accessory proteins and DNA helicases. Enzymes of the former type,

Table 1 Properties of specific DNA binding proteins[a]

DNA-binding protein	Gene	Molecular weight of subunit	Copies per cell	Binds	Cooperativity	Interaction with
E. coli I	ssb	20,000	1,000	ssDNA	yes	DNA polymerase II, DNA polymerase III holoenzyme
E. coli II	?	9,500	100,000	dsDNA	—	E. coli DNA binding protein I
				ssDNA	no	
				RNA	—	
E. coli H	?	28,000	30,000	dsDNA	—	
E. coli HLPI	firA	17,000		dsDNA	—	
E. coli H1	?	15,500		dsDNA	—	
phage T4	32	35,000	10,000	ssDNA	yes	T4 DNA polymerase
				RNA	—	
phage T5	D5	29,000	500,000	dsDNA	yes	
				ssDNA	no	
phage T7	2.5	30,000	5,000	ssDNA	low	T7 DNA polymerase
phage fd	5	9,688	100,000	ssDNA	yes	E. coli DNA binding protein I
adenovirus	E72K	72,000	10^7	ssDNA	yes	

[a] Sources for these data are cited in the text.

made by phages T4 and T7, unwind DNA only in conjunction with the DNA polymerase. DNA helicases, in contrast, are independent of the activity of other enzymes. They are found in *E. coli* and, together with DNA polymerase accessory proteins, in T4-infected cells. Two DNA unwinding enzymes identified in eukaryotic cells unwind limited portions of a DNA on their own, but at least one of these enzymes is stimulated by DNA polymerase action. The *recBC*-type nucleases of bacteria, not discussed here, also are able to unwind DNA at the expense of ATP (134).

DNA POLYMERASE ACCESSORY PROTEINS   In vitro replication studies (135, 136) have shown that the genes 44, 62, and 45 of phage T4 code for DNA polymerase accessory proteins (137, 138). The proteins 44 and 62 (molecular weights 34,000 and 20,000, respectively) copurify as a tight 4:2 molar complex with single-stranded DNA–dependent ATPase and dATPase activity. Gene 45 protein, a dimer of 27,000-dalton subunits, enhances this activity (135). In the presence of ATP, catalytic amounts of these proteins, together with sufficient viral single strand binding protein, enable the viral DNA polymerase to copy single-stranded DNA processively at a high rate and to perform strand displacement synthesis starting at nicks. Adenosine-5'-(γ-thio) triphosphate (ATP-γS) does not substitute for ATP. However, since less than one ATP is hydrolyzed per ten nucleotides polymerized, energy for helix unwinding (1.2–5.0 kcal/mol base pairs (bp) destablized) seems to be drawn not only from ATP but also from the dephosphorylation of the DNA precursors or the binding of the single strand binding protein to the DNA (139–141). ATP dependence of the system has not been reported by all investigators (142).

In the presence of the viral gene 41 and 61 products the system synthesizes DNA at a rate almost as high as in vivo (500 nucleotides/sec). The seven proteins together are further capable of synthesizing oligoribonucleotide primers and thus of converting displaced single strands to double strands (136, 141–145).

A DNA polymerase accessory protein made by T7 is coded by the viral gene 4 (146–148). The protein, a polypeptide of 66,000 and/or 58,000-dalton subunits, is a single stranded DNA–dependent NTPase which, in contrast to the other enzymes discussed here, prefers dATP and dTTP to ATP. Together with a single strand binding protein (from phage or host) gene 4 protein enables the viral DNA polymerase to copy nicked DNA by displacement synthesis (149–153). This synthesis is stimulated by ATP but occurs in the presence of dNTP only. In this case the polymerization of one mole of deoxynucleotides is coupled to the hydrolysis of about four moles of dNTP to dNDP and P_i. Apparently, the dNTP provides not only the DNA precursors but, in a separate reaction, also the energy needed for shifting the fork (153). Gene 4 protein can also act as a primase (154). Like the T4 seven-protein system the T7 three-protein system is thus capable of replicating the displaced DNA (151, 152).

DNA HELICASES Enzymes in this category are, in *E. coli,* the helicases I (8, 155, 156), II (157, 158), III (159, 160), and the *rep* gene product (52, 161–163). A helicase found in T4-infected cells (164) is specified by the viral gene *dda.* Helicase II (165) and the T4 helicase (166, 167) are identical to DNA-dependent ATPases described previously. The peptide molecular weights of the enzymes are: helicase I, 180,000; helicase II, 75,000; helicase III, 20,000; *rep* protein, 65,000; and T4 helicase, 56,000. Helicase I is a fibrous molecule with a tendency to form aggregates. The other enzymes are globular proteins. All enzymes are single-subunit molecules, except for helicase III, which is probably a dimer. Helicase I is present in approximately 600 copies per cell (8, 168), helicase II, 6,000 (168), helicase III, 20 (159), and *rep* protein, 50 (161). Based on ATPase activity, T4-infected cells contain as many copies of T4 helicase as helicase II (166). Cells overproducing *rep* have been obtained by cloning the gene in plasmid (161) or phage vectors (169).

Helicase I, II, III, and T4 helicase were identified by their DNA-dependent ATPase activity. In search of a mechanical function of the ATPase, helicase I (156) and later the other ATPases were shown to unwind DNA. The *rep* protein, needed to propagate phages øX174, fd (M13) (170), and P2 (171), was purified from overproducing cells as a factor required for the in vitro replication of øX174 RF (172–174). The protein unwinds the RF ahead of replication (52).

Helicase action can be measured as the conversion of double-stranded DNA to material degradable by a single-stranded specific nuclease (156, 175, 176), by velocity sedimentation of the DNA or electron microscopy. The effect of *rep* protein has been followed, in a coupled assay, as the incorporation of dNTP label into RF (52, 161, 176–178).

The helicases bind specifically to single-stranded DNA, both in the presence and absence of ATP. Single-stranded DNA is, further, the effector of the ATPase. The ATP turnover number is high, between 4,000 and 10,000/min (155, 157, 159, 162, 164). The consumption of ATP, measured for the *rep* protein, is two ATP per base pair opened for replication (162, 179).

A region of single-stranded DNA is required to initiate unwinding. Helicase I requires approximately 200 nucleotides, helicase II and T4 helicase require at least 12 (175, 180), while *rep* protein, in conjunction with an auxiliary protein (see below), can initiate unwinding at a nick (52, 173, 174, 176). Unwinding is unidirectional; helicase I, II, III, and T4 helicase act in the 5' to 3' direction at the initiating strand (160, 175, 180), while *rep* protein acts in the 3' to 5' direction (179). The rate of unwinding observed for the former three enzymes is in the range of 1,000 bp/sec (180), comparable to the rate of chromosomal fork movement in *E. coli*. These enzymes can completely dissociate a duplex of 119,000 bp (175, 180), while *rep* protein can unwind at least 6,400 contiguous bp (52, 162, 177).

Three different mechanisms of helix unwinding can be defined (4). Unwinding of the DNA by helicase I appears to be processive. Between 70 and 80 enzyme molecules are required to adsorb to each single-stranded DNA site before the start of unwinding. The observation that the enzyme can form aggregates led to the assumption that these molecules migrate along the DNA as a unit (8, 156, 175, 180).

Helicase II and T4 helicase seem to unwind DNA nonprocessively. Stoichiometric amounts of the enzymes are needed for unwinding. One helicase can unwind approximately five bp and one T4 helicase approximately two. Enzyme molecules continue to adsorb to the DNA after the start of unwinding, presumably until the initiating strand is saturated. The role of ATP hydrolysis is not clear (158, 164, 175, 180).

The *rep* enzyme and helicase III, like helicase I, appear to be processive in their action, except for the fact that DNA binding protein I in an amount stoichiometric with the unwound DNA is also required. This protein presumably serves to keep the unwound strands apart (52, 161, 162, 173, 176, 179).

Little is known concerning the biological function of the helicases. Mutants are available for *rep* (181), the T4 *dda* gene (182), and, recently, helicase II (183). None of these mutations affects the viability of the organism or, as far as tested, recombination and repair. A *rep* mutant appears to produce a residual amount of active *rep* protein (163). Besides inability to propagate øX174, fd, and P2 the mutant shows a reduced rate of chromosomal fork movement and a compensatory increase in the number of forks per nucleoid (184). Three helicase II mutants found by screening cell extracts for ATPase activity yield a temperature-labile helicase II, but they continue to synthesize chromosomal and phage λ DNA at elevated temperature [(183); M. Kohiyama, unpublished information]. At T4 *dda* mutant grows in wild-type bacteria and those defective for *dna, uvr, polA,* and *recBC* functions (182).

The RF molecules of øX174 and fd replicate as rolling circles initiated by site-specific endonucleases, the products of øX174 gene A and fd gene 2. After nicking, gene A product remains covalently attached to the 5′ terminus of the interrupted viral RF strand, thus conserving energy for the recircularization of the strand after its replicative displacement. The presence of DNA binding protein I and gene A product attached to the RF allows *rep* protein to initiate its unwinding (52, 172, 173). Gene 2 protein does not form a stable complex with the nicked RF, but it, too, enables the *rep* enzyme to initiate strand separation (51, 178).

Like the *E. coli* chromosome, the DNA of λ and plasmid ColE1 can replicate in *rep* mutant cells. The replication of these three molecules is inhibited by antibody against helicase II, whereas the *rep*-dependent replication of fd RF is not inhibited by the antibody. None of these replication processes is significantly inhibited by antibody against helicase I. The possibility thus exists that helicase II is important in the replication of chromosomes initiating as closed circles (i.e. the DNA of E. coli, λ, and ColE1), and the *rep* protein plays a role in replication of nicked-circular chromosome (168). Assuming that only *rep* protein can initiate unwinding at a nick the function of this enzyme in RF replication is clear. Difficult to interpret, however, is the fact that the replication of *E. coli* DNA and λ DNA requires the function of helicase II in vitro but, according to the behaviour of helicase II mutants, not in vivo.

Helicase II is the only helicase that promotes in vitro replication of forked DNA by DNA polymerase III holoenzyme, the bacterial replicase (in the presence of DNA binding protein I). Interestingly, DNA synthesis is enhanced when helicase III or *rep* protein are also present (B. Kuhn, unpublished information). Helicase II and III are expected to act together on the lagging side of the replication fork, while *rep* protein should act on the leading side.

EUKARYOTIC DNA UNWINDING ENZYMES   Single-stranded DNA-dependent ATPases with native molecular weights around 130,000 and 110,000 have been purified from *Lilium* meiotic cells (185) and a human cell line (186), respectively. The *Lilium* enzyme (ATP turnover number approximately 200/min) binds tightly to the 3'-OH termini of DNA duplexes and opens 400 to 500 bp per duplex end and 50 bp per nick in the presence of ATP. No stimulation of DNA polymerase is reported. The enzyme may play a role in genetic recombination (185).

The DNA-denaturing influence of the human enzyme can be assayed with DNA polymerase α on forked templates [also on poly(dAT), which is probably fork-like in structure], but not on DNA containing gaps or nicks. It is believed that the ATPase unwinds DNA in the 5' to 3' direction. It therefore stimulates the polymerase when it is bound to the opposite strand at a fork (186).

The recA Protein

The *recA* gene of *E. coli* is required for general genetic recombination, postreplication repair, phage induction, mutagenesis, and cell division (for reviews see 187, 188). Expression of the gene is induced by DNA damaging treatment, arrest of DNA replication, mutation of a control gene [*lexA* (189)], and mutation at a site called *tif* (190, 191). The *tif* mutation and a further mutation affecting *recA* expression—*lexB* (192)—are located within *recA,* which suggests that the gene is involved in controlling its expression.

Amplification of *recA* in phage (193) and plasmid vectors (194, 195) has enabled the purification of the gene product. The protein (194, 196, 197), a peptide of 37,842 daltons according to the sequence of the cloned gene (198), shows weak ATPase activity that is stimulated by single-stranded DNA and by superhelical DNA at high enzyme to DNA ratio. The ATP turnover number is approximately 30/min (196, 199). There is evidence for cooperative binding of the enzyme to DNA (199, 200).

The purified *recA* protein has two discrete functions. Most prominent is its ability to catalyze, in the presence of ATP, homologous base pairing between DNA molecules. The result is the renaturation of denatured DNA (194) or strand assimilation, i.e. the pairing of single-stranded DNA to a double strand (196). The product is a triple strand consisting of a heteroduplex and a displaced strand. The triple strand can be retained through its single strand portions on a nitrocellulose filter (196, 199).

Open double strand (linear or circular) and negatively superhelical DNA both can provide the recipient duplex. Superhelical DNA is especially effective, in accordance with results obtained previously in DNA annealing studies (201, 202). The assimilated DNA can derive from a linear single strand or a linear or open-circular duplex containing a single strand region (196, 203, 204); *recA* protein initiates DNA binding when at least 12 unpaired nucleotides are available (200). Stable pairing between two circular strands has not been observed, probably for topological reasons. The *recA* protein has no topoisomerase activity (205).

The *recA* protein acts as a tool in the search for DNA homology. In the presence of ATP the enzyme binds to single-stranded DNA and, in addition, to homologous or heterologous double-stranded DNA, at the same time partially unwinding it (205). With ATP hydrolysis the ternary complex dissociates, leaving either a base-paired heteroduplex (the triple strand described above) or, in the case of nonhomology of the complexed DNA sequences, two physically separated DNA molecules. ATP-γS replaces ATP in DNA binding and unwinding, but it does not allow base-pairing or dissociation of the ternary complex to take place (199, 206).

The requirement for *recA* protein is determined by the amount of single-stranded DNA (206) and the absence or presence of DNA binding protein. For the same amount of single-stranded DNA in the assay mixture, more *recA* is required to initiate homologous pairing without binding protein (one per 5 to 10 nucleotides) than with binding protein (1 per 50 nucleotides). The binding protein seems to cut down single strand binding sites available to *recA* protein, but apparently has no function in the search for homology (35, 207).

Heteroduplex molecules formed by *recA* protein can undergo branch migration. Open-circular duplexes thereby give rise to various forms of figure-8 molecules (203) that resemble products generated in vivo. [Duplex circles can fuse at a region of homology also, under the influence of a "synaptase," a 33,000 dalton-protein recently purified from *E. coli;* this protein requires Mg^{2+} but no nucleotide cofactor (208).] Activity of *recA* is also a prerequisite to "cutting in *trans,*" the scission of a DNA duplex in cells and cell extracts under the influence of an abortively repaired homologous duplex (209–211).

Besides homologous base pairing, *recA* protein also functions as an ATP-dependent protease. In the presence of single-stranded DNA, the enzyme cleaves and inactivates λ repressor protein (212–215) and, as recently shown, also the *lexA* product (216). These effects seem to account for the role of the gene in the induction of λ lysogens and the self-regulatory character of *recA* expression.

DNA Topoisomerases of Type II

Topoisomerases (217) is the now generally accepted name for enzymes that catalyze the concerted breakage and rejoining of DNA backbone bonds. By altering the linking number (218) of closed-circular DNA, the enzymes allow the interconversion of the superhelical and the relaxed forms of the DNA. Topoisomerases seem to store the energy of the interrupted internucleotide bond in an enzyme–DNA bond (219) such that rejoining of the DNA backbone is independent of a nucleotide cofactor.

Treatment of topoisomerase-DNA complexes with a protein-denaturing agent such as alkali, sodium dodecylsulfate, or a protease traps most of the topoisomerases in a covalent complex with the DNA while leaving the DNA broken. Analysis of such DNA has led to the present distinction between topoisomerases of type I and type II (6, 220).

Type II topoisomerases include the bacterial gyrase (221), a topoisomerase induced by T4 infection (222, 223), and activities found in eukaryotes (6, 220, 224, 225). Gyrase yields linear DNA in the aborted reaction, which suggests that the enzyme cuts DNA in both strands. Transverse scission of the helix enables the enzyme to catenate and decatenate homologous and heterologous duplex circles, to introduce topological knots into such circles, or to remove the knots. The same effects are shown by the other enzymes in this group, which suggests that they interrupt DNA in the same way as gyrase. Type II topoisomerases require ATP for activity but only gyrase is known to catalyze an endergonic reaction. This enzyme introduces negative superhelical turns into topologically closed DNA. Gyrase and the T4 topoisomerase are multisubunit proteins.

Type I topoisomerases [previously called untwisting (226), relaxing (227), or nicking-closing (228) enzymes or swivelases (229)] include the bacterial ω proteins (219) and topoisomerases found in the nuclei of eukaryotic cells (226). Type I topoisomerases yield nicked DNA in the aborted reaction, which suggests that they cut only one DNA strand. Although unable to supercoil DNA, type I topoisomerases may be involved in generating the underwound state of the DNA of chromatin. Wrapping of DNA around the histone core of nucleosomes is thought to generate compensatory domains of overwound DNA that are relaxed by topoisomerase action (1). The resulting net deficit in double-helical turns is similar to that generated in bacteria by gyrase—about 1 double-helical turn in 15 (5).

Schemes proposed previously for classifying topoisomerases have used the numerals I and II specifically for the ω protein and the gyrase of a bacterium (217), or to indicate specificity of topoisomerase action with respect to the handedness of the superhelical turns removed (5).

Topoisomerase activity can be assayed electrophoretically by separating the superhelical and the relaxed DNAs in an agarose gel (230) or by banding the DNA at equilibrium in CsCl in the presence of an intercalating dye (5, 231). Other assay techniques have also been described (5). Covalent topoisomerase-DNA complexes can be identified by equilibrium banding (232) or by adsorption of the denatured protein to a glass fiber filter (233).

DNA GYRASE   DNA gyrase has been purified from *E. coli* (221, 234–236), *Micrococcus luteus* (237), *Pseudomonas aeruginosa* (238), and *Bacillus subtilis* (239). The *E. coli* enzyme (234, 236) [probably also the *M. luteus* enzyme (237)] is an A_2B_2 tetramer where A and B have molecular weights around 110,000 and 90,000, respectively. Active gyrase, including chimeric *E. coli-M. luteus* molecules (240), can be obtained by assembling separately purified subunits (237, 240).

In the presence of ATP, gyrase acts catalytically and processively (234, 240–243), introducing approximately 100 negative superhelical turns/min (236) into a closed DNA circle. The density of the supertwist attained (approximately 1.5 that of natural DNA) (6, 221) is limited by the low affinity of the enzyme for the negatively superhelical DNA. Both relaxed DNA and positive supercoils are negatively supercoiled (242, 243).

The ATPase activity of the enzyme is stimulated by relaxed DNA and, ten times less, by single-stranded DNA (234, 240). The ATP turnover number (242) is one or two orders of magnitude lower than that of a helicase.

In the absence of ATP, *E. coli* gyrase relaxes negative supercoils (235, 244) but, according to recent evidence, not positive supercoils (240). The enzymes of *M. luteus* (237) and *P. aeruginosa* (238) fail to show topoisomerase activity in the absence of ATP.

Gyrase appears to supercoil its substrate by modulating the DNA conformation, in a reaction dependent on ATP, and to act subsequently as a topoisomerase. The A and B subunits contribute different functions to these activities. As shown by reconstitution studies, A is specified by a gene *gyrA* (previously *nalA*), which determines resistance to nalidixic acid and oxolinic acid. These antibiotics interfere preferentially with the topoisomerase activity of gyrase. B is specified by a gene *gyrB* (previously *cou*), which determines resistance to coumermycin A1 and novobiocin. They interfere preferentially with the energy-transducing ATPase activity, competing with ATP (234–237, 244, 245).

All activities, including binding to DNA, require the presence of both types of subunits. B can be replaced, however, for some reactions, by a 50,000-dalton fragment of the B subunit, which is found in a tenfold excess to B in *E. coli* extracts, which compensates for a deficiency in B relative to A. The B fragment, generated by a yet unidentified reaction, complements A to yield a topoisomerase that relaxes both negative and positive superhelical turns but lacks ATPase and supercoiling activity (240, 246). Topoisomerase activity described previously for purified gyrase A subunit (235) can be ascribed to a contamination with B fragment.

Oxolinic acid traps gyrase in a complex with its DNA substrate. Protein denaturing treatment of this complex (235, 236, 242–244, 247) causes a double strand break and leaves the A subunits connected by phosphotyrosinyl bonds (248) at the newly created 5'-phosphoryl DNA termini. These termini protrude as four nucleotides. The cleavage pattern is nonrandom (236, 241–244) but, except for flanking G and T residues in one strand, no nucleotide specificity on either side of the break is seen (247). The cleavage pattern is shifted by ATP or adenosine-5'-(β,γ-imido)triphosphate (App(NH)p) (236, 241–244), which suggests that enzyme and DNA are arranged differently relative to each other depending on whether or not the enzyme has bound one of these nucleotides.

Early hypotheses concerning the mode of gyrase action (237, 242, 249) have recently been replaced by a concept (250) that the enzyme binds to the DNA at two sites, thereby inducing the formation of a positively superhelical loop. Subsequent passage of one of the duplex arms of this loop through a transient double strand break introduced into the other arm would invert the sign of the loop, reducing, for topological reasons, the linking number by two. To conserve existing supercoil, the enzyme has to prevent rotation of the interrupted strands around each other. Reversal of this scheme would relax the DNA (6, 250, 251).

This interpretation accounts for the double strand character of the DNA breaks caused by the enzyme in the aborted reaction, i.e. the fact that it alters the linking number in steps of two rather than one as observed for a type I topoisomerase (250, 251), and its ability to catenate or decatenate duplex circles (depending on the conditions) or to knot and unknot such circles (251, 252).

M. luteus gyrase protects 143 ± 3 bp against degradation by staphylococcal nuclease and holds this DNA, in the absence of ATP, positively superhelical (253). With substrate levels of enzyme, App(NH)p, although not hydrolyzed, induces negative supercoiling (proportional to the amount of enzyme). Therefore ATP and App(NH)p are probably effectors of a conformational change in gyrase that drives the enzyme through one round of supercoiling (241, 243). Their influence on the pattern of DNA breaks induced by gyrase also raised the possibility of a conformational change in gyrase. ATP hydrolysis would be required only at the end of a supercoiling cycle, to allow further cycles (241, 243).

The negative superhelicity of prokaryotic DNA is required for many functions including the int-mediated recombination of λ DNA (254) [the reaction through which gyrase was detected (221)], the recognition of some but not all promoter signals by RNA polymerase (255–257), the recognition of the origin of RF replication by the initiating endonucleases of øX174 and fd (258, 177), and the uptake of a single strand into a closed-circular duplex (201). Coumermycin reduces the superhelicity of the folded E. coli chromosome (259) and prevents superinfecting λ DNA from being supercoiled (245). Supercoiling in E. coli is thus probably controlled by gyrase. Coumermycin also lowers the superhelicity of mitochondrial DNA (260).

Gyrase inhibitors (261–265) or temperature shift-up of a thermosensitive *gyrA* mutant (266) arrest the elongation synthesis of *E. coli* DNA. Gyrase may therefore act as the swivel (267) needed to separate the chromosomal strands for replication. The ω-protein of bacteria acts poorly on overwound DNA; it is therefore unlikely to provide the replication swivel.

BACTERIOPHAGE T4 DNA TOPOISOMERASE   Mutation in genes 39, 52, or 60 of T4 causes delayed initiation of viral replication forks without impeding fork movement (268, 269). Unlike the wild-type phage these mutants cannot grow in the presence of coumermycin or novobiocin, which suggests that their defect is compensated by a host gyrase function (270).

The gene 39 and 52 proteins (molecular weights 64,000 and 51,000, respectively) have recently been found to be constituents of an ATP-dependent topoisomerase (222, 223). The enzyme fraction contains, additionally, a 16,000-dalton peptide, presumably the viral gene 60 product (222). Instead of this small peptide, other investigators have found a 110,000-dalton peptide of nonviral origin, conceivably the bacterial gyrase A subunit or ω protein (223).

T4 topoisomerase catalyzes the stepwise relaxation of negative and positive supercoils by hydrolyzing one or two ATP for each superhelical turn removed. Despite the need for ATP, substrates as used in gyrase assays are not supercoiled (222, 223). ATP-γS inhibits the enzyme in the presence of ATP; however, with substrate levels of enzyme, ATP-γS induces relaxation, which suggests that by binding this nucleotide the enzyme is driven through one cycle of activity (222), which is essentially the same as observed with gyrase. The T4 topoisomerase activity is slightly sensitive to oxolinic acid (222, 223); it is not inhibited by antibody against ω protein (222). Its function is therefore probably independent of a host topoisomerase protein (220, 222).

The ability to catenate-decatenate (knot-unknot) duplex circles suggests that the T4 enzyme relaxes DNA by a strand passing reaction as proposed for gyrase. To explain the requirement for ATP and to account for the role of the enzyme in the initiation of T4 DNA replication it has been proposed that T4 topoisomerase acts normally as a supercoiling enzyme specific for the origin of T4 DNA replication. Using a nonphysiological substrate only the uncoupled relaxing activity of the enzyme would be seen (220, 222).

EUKARYOTIC TYPE II TOPOISOMERASES   Enzymes able to catenate-decatenate duplexes in the presence of ATP have now been identified in many eukaryotic organisms including *Xenopus laevis* (6, 224), Drosophila (220), and HeLa cells (225). The HeLa enzyme has been purified but a detailed description of its properties has not yet appeared.

ENZYMES INDEPENDENT OF CHEMICAL ENERGY

DNA Topoisomerases of Type I

Type I topoisomerases have been found in bacteria (219), the nuclei (226) and mitochondria (271) of eukaryotic cells, and vaccinia virus particles (272). Lists of sources are given in (1) and (5).

THE ω-PROTEIN OF BACTERIA   This protein has been prepared from $E.$ $coli$ (219), $M.$ $luteus$ (273, 274), and other bacterial sources (275–277). It relaxes negative supercoils while acting poorly on positive supercoils (219). Also, a weakly negative superhelical DNA is relaxed only slowly. Processive action of the Mg^{2+}-dependent enzyme is favored by high negative superhelicity and low salt (217, 274).

In the absence of added Mg^{2+} the enzyme forms a stable complex with its DNA substrate without relaxing it (232). Alkali treatment breaks the DNA at one strand, and leaves the enzyme protein at the newly created 5' DNA terminus (278) via a phosphotyrosinyl linkage. The cleavage pattern is nonrandom, but as with gyrase there is no nucleotide specificity on either side of the break (248). "Alkali-cleavable" complexes are also formed with single-stranded DNA but not with relaxed double-stranded DNA, although the enzyme can bind to such DNA. Conceivably, ω-topoisomerase action requires the separation of the DNA strands. This may be facilitated by negative superhelicity of the DNA (217, 232).

The ω protein introduces topological knots into single strand circles and it catalyzes the formation of completely base-paired closed-circular duplexes from complementary single strand circles (217, 279). This effect was first demonstrated for a eukaryotic type I topoisomerase (280).

It is not clear how ω protein is prevented from annihilating the effect of gyrase in cells. Mutational loss of $> 90\%$ of the ω activity causes no detectable biological defect (281).

EUKARYOTIC TYPE I TOPOISOMERASES   The supercoil relaxing activity of eukaryotic cells is found almost entirely in the chromatin fraction (282). Purification of nuclear type I topoisomerases from HeLa cells (225), *Drosophila melanogaster* (283), and calf thymus (H. P. Vosberg, unpublished information) under conditions of rigorously controlled proteolysis has shown that the molecular weights of these enzymes are around 110,000 rather than around 70,000 (282, 284) as reported previously for eukaryotic type I topoisomerases. The enzyme of vaccinia virus core particles has M_r 35,000 (272).

The 70,000-dalton topoisomerase proteins relax both negative and positive supercoils. They are most active in 0.2 M NaCl and under these conditions are independent of Mg^{2+} (5, 282, 284, 285). The 110,000-dalton HeLa enzyme, however, is strongly activated by Mg^{2+} (225). A reinvestigation of the nuclear type I topoisomerases is therefore needed.

Protein denaturing treatment of a complex between a 70,000 topoisomerase protein and double-stranded DNA causes a single strand break (286) accompanied by fixation of the protein to the resulting 3'-OH DNA terminus (233, 284) (rather than the 5' terminus, as observed for all other topoisomerases). Calf thymus topoisomerase also cleaves single-stranded DNA when protein denaturing treatment is omitted. The 5'-terminal nucleotide of these fragments is predominantly A or T (233).

Rat liver topoisomerase catalyzes in vitro assembly of chromatin-like material from core histones and relaxed closed-circular DNA. After deproteinisation the DNA is negatively supercoiled, thus providing evidence for the mechanism of chromatin assembly outlined above (287, 288). However, since chromatin may be assembled in vivo from newly replicated, still discontinuous DNA, a physiological role for topoisomerases in this reaction is not established. The nuclear topoisomerases may further provide the replication swivel. Their high concentration within a nucleus is consistent with a crucial role; approximately one topoisomerase copy is found per ten nucleosomes (1, 225).

OTHER TYPE I TOPOISOMERASES   Nicking of RF by fd gene 2 protein or øX174 gene A protein is occasionally followed by immediate closure of the nick. The resulting conversion of superhelical RF to the relaxed topoisomer is favored by high Mg^{2+} with gene 2 protein (40% closure of the nicked RF) (51) or by Mn^{2+} with gene A protein (20% closure) (289). Closing of nicks by gene 2 protein is inhibited by DNA binding protein II, which suggests that the effect plays no role in vivo.

In the absence of Mg^{2+} the λ *int* product also (290) can act as a topoisomerase (291). By altering the linking number in steps of one the enzyme relaxes positive and negative supercoils (283). Type I topoisomerase activity of the *int* protein is consistent with the mechanism proposed for the integrative recombination of λ DNA. Strand exchange is thought to be initiated by single strand interruptions of the recombining DNA molecules (292).

CONCLUDING REMARKS

As we show in this survey, single strand binding proteins are intimately connected with DNA replication. Double strand binding proteins play a role in organizing the bacterial chromosome; in eukaryotic cells their role in the regulation of gene expression and cell differentiation is still poorly understood. Enzymes involved in the control of DNA conformation are to a large extent chemomechanical devices capable of either distorting the helix or separating its strands by drawing their energy from ATP. Various other processes are known where DNA function is coupled to mechanical activity. Examples are the filling of phage heads, the sexual transfer of DNA between bacteria and, in particular, mitosis. For an organism, the physical manipulation of DNA is apparently as much a task as the synthesis of the nucleotide chains and the maintenance of their integrity.

Literature Cited

1. Champoux, J. J. 1978. *Ann. Rev. Biochem.* 47:449–79
2. Isenberg, I. 1979. *Ann. Rev. Biochem.* 48:159–91
3. Coleman, J. E., Oakley, J. L. 1980. *CRC Crit. Rev. Biochem.* 7:247–89
4. Abdel-Monem, M., Hoffmann-Berling, H. 1980. *Trends Biochem. Sci.* 5:128–30
4a. Falaschi, A., Cobianchi, F., Riva, S. 1980. *Trends Biochem. Sci.* 5:154–57
5. Bauer, W. R. 1978. *Ann. Rev. Biophys. Bioeng.* 7:287–313
6. Cozzarelli, N. R. 1980. *Science* 207: 953–60
6a. Gellert, M. 1981. *Ann. Rev. Biochem.* 50:879–910
7. Sigal, N., Delius, H., Kornberg, T., Gefter, M. L., Alberts, B. 1972. *Proc. Natl. Acad. Sci. USA* 69:3537–41
8. Abdel-Monem, M., Lauppe, H.-F., Kartenbeck, J., Dürwald, H., Hoffmann-Berling, H. 1977. *J. Mol. Biol.* 110:667–85
9. Jensen, D. E., von Hippel, P. H. 1976. *J. Biol. Chem.* 251:7198–7214
10. Meyer, R. R., Glassberg, J., Kornberg, A. 1979. *Proc. Natl. Acad. Sci. USA* 76:1702–5
11. Alberts, B., Sternglanz, R. 1977. *Nature* 269:655–61
12. Alberts, B., Herrick, G. 1971. *Methods Enzymol.* 21:198–217
13. Zillig, W., Zechel, K., Halbwachs, H.-J. 1970. *Hoppe-Seylers Z. Physiol. Chem.* 351:221–24
14. Burgess, R. R., Jendrisak, J. J. 1975. *Biochemistry* 14:4634–38
15. Wickner, W., Brutlag, D., Schekman, R., Kornberg, A. 1972. *Proc. Natl. Acad. Sci. USA* 69:965–69
16. Staudenbauer, W. L. 1976. *Mol. Gen. Genet.* 145:273–80
17. Litman, R. M. 1968. *J. Biol. Chem.* 243:6222–33
18. Weiner, J. H., Bertsch, L. L., Kornberg, A. 1975. *J. Biol. Chem.* 250:1972–80
19. Sternbach, H., Engelhardt, R., Lezius, A. G. 1975. *Eur. J. Biochem.* 60:51–55
20. Meyer, R. R., Glassberg, J., Scott, J. V., Kornberg, A. 1980. *J. Biol. Chem.* 255:2897–2901
21. McHenry, C., Kornberg, A. 1977. *J. Biol. Chem.* 252:6478–84
22. Berthold, V., Geider, K. 1976. *Eur. J. Biochem.* 71:443–49
23. Laine, B., Kerckaert, J. P., Sautiere, P., Biserte, G. 1978. *FEBS Lett.* 96:291–94
24. Bittner, M., Burke, R. L., Alberts, B. M. 1979. *J. Biol. Chem.* 254:9565–72
25. Bowen, B., Steinberg, J., Laemmli, U. K., Weintraub, H. 1980. *Nucleic Acids Res.* 8:1–20
26. Coombs, D. H., Pearson, G. D. 1978. *Proc. Natl. Acad. Sci. USA* 75:5291–95
27. Geider, K. 1978. *Eur. J. Biochem.* 87:617–22
28. Kelly, R. C., von Hippel, P. H. 1976. *J. Biol. Chem.* 251:7229–39
29. Spicer, E. K., Williams, K. R., Konigsberg, W. H. 1979. *J. Biol. Chem.* 254:6433–36
30. Pretorius, H. T., Klein, M., Day, L. A. 1975. *J. Biol. Chem.* 250:9262–69
31. Alberts, B. M., Frey, L. 1970. *Nature* 227:1313–18
32. Reuben, R., Gefter, M. L. 1974. *J. Biol. Chem.* 249:3843–50
33. deHaseth, P. L., Gross, C. A., Burgess, R. R., Record, M. T. Jr. 1977. *Biochemistry* 16:4777–83
34. Christiansen, C., Baldwin, R. L. 1977. *J. Mol. Biol.* 115:441–54
35. McEntee, K., Weinstock, G. M., Lehman, I. R. 1980. *Proc. Natl. Acad. Sci. USA* 77:857–61
36. Molineux, I. J., Gefter, M. L. 1975. *J. Mol. Biol.* 98:811–25
37. Uhlmann, A., Geider, K. 1977. *Biochim. Biophys. Acta* 474:639–45
38. Molineux, I. J., Friedman, S., Gefter, M. L. 1974. *J. Biol. Chem.* 249:6090–98
39. Niyogi, S. K., Ratrie, H. III, Datta, A. K. 1977. *Biochem. Biophys. Res. Commun.* 78:343–49
40. Geider, K., Beck, E., Schaller, H. 1978. *Proc. Natl. Acad. Sci. USA* 75:645–49

41. Tabak, H. F., Griffith, J., Geider, K., Schaller, H., Kornberg, A. 1974. *J. Biol. Chem.* 249:3049–54
42. Bouché, J. P., Zechel, K., Kornberg, A. 1975. *J. Biol. Chem.* 250:5995–6001
43. Shlomai, J., Kornberg, A. 1980. *Proc. Natl. Acad. Sci. USA* 77:799–803
44. Weiner, J. H., McMacken, R., Kornberg, A. 1976. *Proc. Natl. Acad. Sci. USA* 73:752–56
45. McMacken, R., Ueda, K., Kornberg, A. 1977. *Proc. Natl. Acad. Sci. USA* 74:4190–94
46. McMacken, R., Kornberg, A. 1978. *J. Biol. Chem.* 253:3313–19
47. Arai, K.-I., Kornberg, A. 1979. *Proc. Natl. Acad. Sci. USA* 76:4308–12
48. Geider, K., Kornberg, A. 1974. *J. Biol. Chem.* 249:3999–4005
49. Molineux, I. J., Gefter, M. L. 1974. *Proc. Natl. Acad. Sci. USA* 71:3858–62
50. Eisenberg, S., Scott, J. F., Kornberg, A. 1976. *Proc. Natl. Acad. Sci. USA* 73:3151–55
51. Geider, K., Meyer, T. F. 1978. *Cold Spring Harbor Symp. Quant. Biol.* 43:59–62
52. Scott, J. F., Eisenberg, S., Bertsch, L. L., Kornberg, A. 1977. *Proc. Natl. Acad. Sci. USA* 74:193–97
53. Arai, K.-I., Arai, N., Shlomai, J., Kornberg, A. 1980. *Proc. Natl. Acad. Sci. USA* 77:3322–26
54. Glassberg, J., Meyer, R. R., Kornberg, A. 1980. *J. Bacteriol.* 140:14–19
55. Sancar, A., Rupp, W. D. 1979. *Biochem. Biophys. Res. Commun.* 90:123–29
56. Chase, J. W., Whittier, R. F., Auerbach, J., Sancar, A., Rupp, W. D. 1980. *Nucleic Acids Res.* 8:3215–27
57. Koch, R. E., McGaw, M. K., Drake, J. W. 1976. *J. Virol.* 19:490–94
58. Mosig, G., Luder, A., Garcia, G., Dannenberg, R., Bock, S. 1978. *Cold Spring Harbor Symp. Quant. Biol.* 43:501–15
59. Gold, L., O'Farrell, P. Z., Russel, M. 1976. *J. Biol. Chem.* 251:7251–62
60. Wood, W. J. Jr., Bernstein, H. 1977. *J. Virol.* 21:619–25
61. Krisch, H. M., van Houwe, G. 1976. *J. Mol. Biol.* 108:67–81
62. Russel, M., Gold, L., Morrissett, H., O'Farrell, P. Z. 1976. *J. Biol. Chem.* 251:7263–70
63. Krisch, H. M., van Houwe, G., Belin, D., Gibbs, W., Epstein, R. H. 1977. *Virology* 78:87–98
64. Hosoda, J., Takacs, B., Brack, C. 1974. *FEBS Lett.* 47:338–42
65. Hosoda, J., Burke, R. L., Moise, H., Kubota, I., Tsugita, A. 1980. *Mechanis-tic Studies of DNA Replication and Genetic Recombination.* Presented at ICN-UCLA Symp. Mol. Cellul. Biol. New York: Academic. In press
66. Williams, K. R., Sillerud, L. O., Schafer, D. E., Konigsberg, W. H. 1979. *J. Biol. Chem.* 254:6426–32
67. Huberman, J. A., Kornberg, A., Alberts, B. M. 1971. *J. Mol. Biol.* 62:39–52
68. Burke, R. L., Alberts, B. M., Hosoda, J. 1981. *J. Biol. Chem.* In press
69. Manoil, C., Sinha, N., Alberts, B. 1977. *J. Biol. Chem.* 252:2734–41
70. Wu, M., Davidson, N. 1975. *Proc. Natl. Acad. Sci. USA* 72:4506–10
71. Scherzinger, E., Litfin, F., Jost, E. 1973. *Mol. Gen. Genet.* 123:247–62
72. Richardson, C. C., Romano, L. J., Kolodner, R., LeClerc, J. E., Tamanoi, F., Engler, M. J., Dean, F. B., Richardson, D. S. 1978. *Cold Spring Harbor Symp. Quant. Biol.* 43:427–40
73. Chinnadurai, G., McCorquodale, D. J. 1974. *Nature* 247:554–56
74. McCorquodale, D. J., Gossling, J., Benzinger, R., Chesney, R., Lawhorne, L., Moyer, R. W. 1979. *J. Virol.* 29:322–27
75. Rice, A. C., Ficht, T. A., Holladay, L. A., Moyer, R. W. 1979. *J. Biol. Chem.* 254:8042–51
76. Meyer, T. F., Geider, K. 1980. See Ref. 65. In press
77. Mazur, B. J., Zinder, N. D. 1975. *Virology* 68:490–502
78. Oey, J. L., Knippers, R. 1972. *J. Mol. Biol.* 68:125–38
79. Alberts, B., Frey, L., Delius, H. 1972. *J. Mol. Biol.* 68:139–52
80. Cuypers, T., van der Ouderaa, F. J., de Jong, W. W. 1974. *Biochem. Biophys. Res. Commun.* 59:557–63
81. Nakashima, Y., Dunker, A. K., Marvin, D. A., Konigsberg, W. 1974. *FEBS Lett.* 43:125
82. Lica, L., Ray, D. S. 1977. *J. Mol. Biol.* 115:45–59
83. McPherson, A., Jurnak, F. A., Wang, A. H. J., Molineux, I., Rich, A. 1979. *J. Mol. Biol.* 134:379–400
84. Garssen, G. J., Tesser, G. I., Schoenmakers, J. G. G., Hilbers, C. W. 1980. *Biochim. Biophys. Acta* 607:361–71
85. Chow, L. T., Roberts, J. M., Lewis, J. B., Broker, T. R. 1977. *Cell* 11:819–36
86. van der Vliet, P. C., Keegstra, W., Jansz, H. S. 1978. *Eur. J. Biochem.* 86:389–98
87. Schechter, N. M., Davies, W., Anderson, C. W. 1980. *Biochemistry* 19:2802–10

88. van der Vliet, P. C., Zandberg, J., Jansz, H. S. 1977. *Virology* 80:98–110
89. Carter, T. H., Blanton, R. A. 1978. *J. Virol.* 25:664–74
90. Sugawara, K., Gilead, Z., Green, M. 1977. *J. Virol.* 21:338–46
91. Fowlkes, D. M., Lord, S. T., Linné, T., Pettersson, U., Philipson, L. 1979. *J. Mol. Biol.* 132:163–80
92. Kedinger, C., Brison, O., Perrin, F., Wilhelm, J. 1978. *J. Virol.* 26:364–79
93. Klein, H., Maltzman, W., Levine, A. J. 1979. *J. Biol. Chem.* 254:11051–60
94. Linné, T., Philipson, L. 1980. *Eur. J. Biochem.* 103:259–70
95. van der Vliet, P. C., Levine, A. J., Ensinger, M. J., Ginsberg, H. S. 1975. *J. Virol.* 15:348–54
96. Horwitz, M. 1978. *Proc. Natl. Acad. Sci. USA* 75:4291–95
97. Ariga, H., Klein, H., Levine, A., Horwitz, M. S. 1980. *Virology* 101:307–10
98. Mayer, A. J., Ginsberg, H. S. 1977. *Proc. Natl. Acad. Sci. USA* 74:785–88
99. Gilead, Z., Arens, M. Q., Bhaduri, S., Shanmugam, G., Green, M. 1975. *Nature* 254:533–35
100. Levinson, A., Levine, A. J., Anderson, S., Osborn, M., Rosenwirth, B., Weber, K. 1976. *Cell* 7:575–84
101. Falaschi, A., Cobianchi, F., Riva, S. 1980. *Trends Biol. Sci.* 5:154–57
102. Griffith, J. D. 1976. *Proc. Natl. Acad. Sci. USA* 73:563–67
103. Varshavsky, A. J., Nedospasov, S. A., Bakayev, V. V., Bakayeva, T. G., Georgiev, G. P. 1977. *Nucleic Acids Res.* 4:2725–45
104. Busby, S., Kolb, A., Buc, H. 1979. *Eur. J. Biochem.* 99:105–11
105. Rouvière-Yaniv, J., Gros, F. 1975. *Proc. Natl. Acad. Sci. USA* 72:3428–32
106. Suryanarayana, T., Subramanian, A. R. 1978. *Biochim. Biophys. Acta* 520:342–57
107. Mende, L., Timm, B., Subramanian, A. R. 1978. *FEBS Lett.* 96:395–98
108. Laine, B., Kmiecik, D., Sautiere, P. Biserte, G., Cohen-Solal, M. 1980. *Eur. J. Biochem.* 103:447–61
109. Rouvière-Yaniv, J., Kjeldgaard, N. O. 1979. *FEBS Lett.* 106:297–300
110. Rouvière-Yaniv, J., Yaniv, M., Germond, J.-E. 1979. *Cell* 17:265–74
111. Rouvière-Yaniv, J. 1977. *Cold Spring Harbor Symp. Quant. Biol.* 42:439–47
112. Zentgraf, H., Berthold, V., Geider, K. 1977. *Biochim. Biophys. Acta* 474:629–38
113. Haselkorn, R., Rouvière-Yaniv, J. 1976. *Proc. Natl. Acad. Sci. USA* 73:1917–20

114. Hübscher, U., Lutz, H., Kornberg, A. 1980. *Proc. Natl. Acad. Sci. USA* 77:5097–5101
115. Lathe, R., Buc, H., Lecocq, J.-P., Bautz, E. K. F. 1980. *Proc. Natl. Acad. Sci. USA* 77:3548–52
116. Cukier-Kahn, R., Jacquet, M., Gros, F. 1972. *Proc. Natl. Acad. Sci. USA* 69:3643–47
117. Spassky, A., Buc, H. C. 1977. *Eur. J. Biochem.* 81:79–90
118. Goodwin, G. H., Woodhead, L., Johns, E. W. 1977. *FEBS Lett.* 73:85–88
119. Rabbani, A., Goodwin, G. H., Johns, E. W. 1978. *Biochem. Biophys. Res. Commun.* 81:351–58
120. Walker, J. M., Goodwin, G. H., Johns, E. W. 1979. *FEBS Lett.* 100:394–98
121. Walker, J. M., Hastings, J. R. B., Johns, E. W. 1977. *Eur. J. Biochem.* 76:461–68
122. Shooter, K. V., Goodwin, G. H., Johns, E. W. 1974. *Eur. J. Biochem.* 47:263–70
123. Javaherian, K., Liu, L. F., Wang, J. C. 1978. *Science* 199:1345–46
124. Mathis, D. J., Kindelis, A., Spadafora, C. 1980. *Nucleic Acids Res.* 8:2577–90
125. Cary, P. D., Shooter, K. V., Goodwin, G. H., Johns, E. W., Olayemi, J. Y., Hartman, P. G., Bradbury, E. M. 1979. *Biochem. J.* 183:657–62
126. Sterner, R., Vidali, G., Allfrey, V. G. 1979. *J. Biol. Chem.* 254:11577–83
127. Reeves, R., Candido, E. P. M. 1980. *Nucleic Acids Res.* 8:1947–63
128. Weisbrod, S., Weintraub, H. 1979. *Proc. Natl. Acad. Sci. USA* 76:630–34
129. Weisbrod, S., Groudine, M., Weintraub, H. 1980. *Cell* 19:289–301
130. Stalder, J., Groudine, M., Dogson, J. B., Engel, J. D., Weintraub, H. 1980. *Cell* 19:973–80
131. Alfageme, C. R., Rudkin, G. T., Cohen, L. H. 1976. *Proc. Natl. Acad. Sci. USA* 73:2038–42
132. Palmer, D., Snyder, L. A., Blumenfeld, M. 1980. *Proc. Natl. Acad. Sci. USA* 77:2671–75
133. Hsieh, T.-S., Brutlag, D. L. 1979. *Proc. Natl. Acad. Sci. USA* 76:726–30
134. MacKay, V., Linn, S. 1976. *J. Biol. Chem.* 251:3716–19
135. Morris, C. F., Sinha, N. K., Alberts, B. M. 1975. *Proc. Natl. Acad. Sci. USA* 72:4800–4
136. Alberts, B. M., Morris, C. F., Mace, D., Sinha, N., Bittner, M., Moran, L. 1975. *DNA Synthesis and its Regulation,* ed. M. Goulian et al, p. 241. Menlo Park, Calif: Benjamin
137. Morris, C. F., Hama-Inaba, H., Mace, D., Sinha, N. K., Alberts, B. M. 1979. *J. Biol. Chem.* 254:6787–96

138. Nossal, N. G. 1979. *J. Biol. Chem.* 254:6026–31
139. Piperno, J. R., Alberts, B. M. 1978. *J. Biol. Chem.* 253:5174–79
140. Piperno, J. R., Kallen, R. G., Alberts, B. M. 1978. *J. Biol. Chem.* 253:5180–85
141. Liu, C. C., Burke, R. L., Hibner, U., Barry, J., Alberts, B. M. 1979. *Cold Spring Harbor Symp. Quant. Biol.* 43:469–85
142. Nossal, N. G., Petterlin, B. M. 1979. *J. Biol. Chem.* 254:6032–37
143. Sinha, N. K., Morris, C. F., Alberts, B. M. 1980. *J. Biol. Chem.* 255:4290–303
144. Silver, L. L., Nossal, N. G. 1979. *Cold Spring Harbor Symp. Quant. Biol.* 43:489–94
145. Nossal, N. G. 1980. *J. Biol. Chem.* 255:2176–82
146. Strätling, W., Knippers, R. 1973. *Nature* 245:195–97
147. Hinkle, D. C., Richardson, C. C. 1975. *J. Biol. Chem.* 250:5523–29
148. Scherzinger, E., Seiffert, D. 1975. *Mol. Gen. Genet.* 141:213–32
149. Scherzinger, E., Lanka, E., Morelli, G., Seiffert, D., Yuki, A. 1977. *Eur. J. Biochem.* 72:543–58
150. Kolodner, R., Masamune, Y., LeClerc, J. E., Richardson, C. C. 1978. *J. Biol. Chem.* 253:566–73
151. Richardson, C. C., Romano, L. J., Kolodner, R., LeClerc, J. E., Tamanoi, F., Engler, M. J., Dean, F. B., Richardson, D. S. 1979. *Cold Spring Harbor Symp. Quant. Biol.* 43:427–40
152. Hillenbrand, G., Morelli, G., Lanka, E., Scherzinger, E. 1979. *Cold Spring Harbor Symp. Quant. Biol.* 43:449–59
153. Kolodner, R., Richardson, C. C. 1977. *Proc. Natl. Acad. Sci. USA* 74:1525–29
154. Scherzinger, E., Klotz, G. 1975. *Mol. Gen. Genet.* 141:233–49
155. Abdel-Monem, M., Hoffmann-Berling, H. 1976. *Eur. J. Biochem.* 65:431–40
156. Abdel-Monem, M., Dürwald, H., Hoffmann-Berling, H. 1976. *Eur. J. Biochem.* 65:441–49
157. Abdel-Monem, M., Chanal, M. C., Hoffmann-Berling, H. 1977. *Eur. J. Biochem.* 79:33–38
158. Abdel-Monem, M., Dürwald, H., Hoffmann-Berling, H. 1977. *Eur. J. Biochem.* 79:39–45
159. Yarranton, G. T., Das, R. H., Gefter, M. L. 1979. *J. Biol. Chem.* 254:11997–12001
160. Yarranton, G. T., Das, R. H., Gefter, M. L. 1979. *J. Biol. Chem.* 254:12002–6
161. Scott, J. F., Kornberg, A. 1978. *J. Biol. Chem.* 253:3292–97
162. Kornberg, A., Scott, J. F., Bertsch, L. L. 1978. *J. Biol. Chem.* 253:3298–304
163. Takahashi, S., Hours, C., Chu, A., Denhardt, D. T. 1979. *Can. J. Biochem.* 57:855–66
164. Krell, H., Dürwald, H., Hoffmann-Berling, H. 1979. *Eur. J. Biochem.* 94:387–95
165. Richet, E., Kohiyama, M. 1976. *J. Biol. Chem.* 251:808–12
166. Ebisuzaki, K., Behme, M. T., Senior, C., Shannon, D., Dunn, D. 1972. *Proc. Natl. Acad. Sci. USA* 69:515–19
167. Purkey, R. M., Ebisuzaki, K. 1977. *Eur. J. Biochem.* 75:303–10
168. Klinkert, M. Q., Klein, A., Abdel-Monem, M. 1980. *J. Biol. Chem.* 255:9746–52
169. Takahashi, S., Hours, C., Denhardt, D. T. 1977. *FEMS Microbiol. Lett.* 2:279–83
170. Denhardt, D. T. 1977. *Comprehensive Virology,* ed. H. Fraenkel-Conrat, R. R. Wagner, 7:1–104. London & New York: Plenum
171. Calendar, R., Lindqvist, B., Sironi, G., Clark, A. J. 1970. *Virology* 40:72–83
172. Eisenberg, S., Scott, J. F., Kornberg, A. 1976. *Proc. Natl. Acad. Sci. USA* 73:1594–97
173. Eisenberg, S., Scott, J. F., Kornberg, A. 1979. *Cold Spring Harbor Symp. Quant. Biol.* 43:295–302
174. Sumida-Yasumoto, C., Ikeda, J. E., Benz, E., Marians, K. J., Vicuna, R., Sugrue, S., Zipursky, S. L., Hurwitz, J. 1979. *Cold Spring Harbor Symp. Quant. Biol.* 43:311–29
175. Kuhn, B., Abdel-Monem, M., Hoffmann-Berling, H. 1979. *Cold Spring Harbor Symp. Quant. Biol.* 43:63–67
176. Duguet, M., Yarranton, G. T., Gefter, M. L. 1979. *Cold Spring Harbor Symp. Quant. Biol.* 43:335–43
177. Meyer, T. F., Geider, K. 1979. *J. Biol. Chem.* 254:12636–41
178. Meyer, T. F., Geider, K. 1979. *J. Biol. Chem.* 254:12642–46
179. Yarranton, G. T., Gefter, M. L. 1979. *Proc. Natl. Acad. Sci. USA* 76:1658–62
180. Kuhn, B., Abdel-Monem, M., Krell, H., Hoffmann-Berling, H. 1979. *J. Biol. Chem.* 254:11343–50
181. Denhardt, D. T., Dressler, D. H., Hathaway, A. 1967. *Proc. Natl. Acad. Sci. USA* 57:813–20
182. Behme, M. T., Ebisuzaki, K. 1975. *J. Virol.* 15:50–54
183. Richet, E., Kern, R., Kohiyama, M., Hirota, Y. 1980. See Ref. 65. In press
184. Lane, H. E. D., Denhardt, D. T. 1975. *J. Mol. Biol.* 97:99–112

185. Hotta, Y., Stern, H. 1978. *Biochemistry* 17:1872–80
186. Cobianchi, F., Riva, S., Mastromei, G., Spadari, S., Pedrali-Noy, G., Falaschi, A. 1979. *Cold Spring Harbor Symp. Quant. Biol.* 43:639–47
187. Witkin, E. M. 1976. *Bacteriol. Rev.* 40:869–907
188. Radding, C. M. 1978. *Ann. Rev. Biochem.* 47:847–80
189. Mount, D. W. 1977. *Proc. Natl. Acad. Sci. USA* 74:300–4
190. Castellazi, M., George, J., Buttin, G. 1972. *Mol. Gen. Genet.* 119:139–52
191. McEntee, K. 1977. *Proc. Natl. Acad. Sci. USA* 74:5275–79
192. Morand, P., Goze, A., Devoret, R. 1977. *Mol. Gen. Genet.* 157:69–82
193. McEntee, K., Epstein, W. 1977. *Virology* 77:306–18
194. Weinstock, G. M., McEntee, K., Lehman, I. R. 1979. *Proc. Natl. Acad. Sci. USA* 76:126–29
195. Sancar, A., Rupp, W. D. 1979. *Proc. Natl. Acad. Sci. USA* 76:3144–48
196. Shibata, T., DasGupta, C., Cunningham, R. P., Radding, C. M. 1979. *Proc. Natl. Acad. Sci. USA* 76:1638–42
197. McEntee, K., Hesse, J. E., Epstein, W. 1976. *Proc. Natl. Acad. Sci. USA* 73:3979–83
198. Sancar, A., Stachelek, C., Konigsberg, W., Rupp, W. D. 1980. *Proc. Natl. Acad. Sci. USA* 77:2611–15
199. McEntee, K., Weinstock, G. M., Lehman, I. R. 1979. *Proc. Natl. Acad. Sci. USA* 76:2615–19
200. West, S. C., Cassuto, E., Mursalim, J., Howard-Flanders, P. 1980. *Proc. Natl. Acad. Sci. USA* 77:2569–73
201. Holloman, W. K., Wiegand, R., Hoessli, C., Radding, C. M. 1975. *Proc. Natl. Acad. Sci. USA* 72:2394–98
202. Holloman, W. K., Radding, C. M. 1976. *Proc. Natl. Acad. Sci. USA* 73:3910–14
203. Cunningham, R. P., DasGupta, C., Shibata, T., Radding, C. M. 1980. *Cell* 20:223–35
204. DasGupta, C., Shibata, T., Cunningham, R. P., Radding, C. M. 1980. *Fed. Proc.* 39:1808
205. Cunningham, R. P., Shibata, T., DasGupta, C., Radding, C. M. 1979. *Nature* 281:191–95
206. Shibata, T., Cunningham, R. P., DasGupta, C., Radding, C. M. 1979. *Proc. Natl. Acad. Sci. USA* 76:5100–4
207. Shibata, T., DasGupta, C., Cunningham, R. P., Radding, C. M. 1980. *Proc. Natl. Acad. Sci. USA* 77:2606–10
208. Potter, H., Dressler, D. 1980. *Proc. Natl. Acad. Sci. USA* 77:2390–94
209. Ross, P., Howard-Flanders, P. 1977. *J. Mol. Biol.* 117:137–58
210. Ross, P., Howard-Flanders, P. 1977. *J. Mol. Biol.* 117:159–74
211. Cassuto, E., Mursalim, J., Howard-Flanders, P. *Proc. Natl. Acad. Sci. USA* 75:620–24
212. Roberts, J. W., Roberts, C. W. 1975. *Proc. Natl. Acad. Sci. USA* 72:147–51
213. Ogawa, T., Wabiko, H., Tsurimoto, T., Horii, T., Masukata, H., Ogawa, H. 1978. *Cold Spring Harbor Symp. Quant. Biol.* 43:909–15
214. Roberts, J. W., Roberts, C. W., Craig, N. L., Phizicky, E. M. 1978. *Cold Spring Harbor Symp. Quant. Biol.* 43:917–20
215. Craig, N. L., Roberts, J. W. 1980. *Nature* 283:26–30
216. Little, J. W., Edmiston, S. H., Pacelli, L. Z., Mount, D. W. 1980. *Proc. Natl. Acad. Sci. USA* 77:3225–29
217. Wang, J. C., Liu, L. F. 1979. *Molecular Genetics,* J. H. Taylor ed., Pt. III, pp. 65–88. New York: Academic
218. Crick, F. H. C. 1976. *Proc. Natl. Acad. Sci. USA* 73:2639–43
219. Wang, J. C. 1971. *J. Mol. Biol.* 55:523–33
220. Liu, L. F., Liu, C. C., Alberts, B. M. 1980. *Cell* 19:697–707
221. Gellert, M., Mizuuchi, K., O'Dea, M. H., Nash, H. A. 1976. *Proc. Natl. Acad. Sci. USA* 73:3872–76
222. Liu, L. F., Liu, C. C., Alberts, B. M. 1979. *Nature* 281:456–61
223. Stettler, G. L., King, G. J., Huang, W. M. 1979. *Proc. Natl. Acad. Sci. USA* 76:3737–41
224. Attardi, D. G., Martini, G., Mattoccia, E., Tocchini-Valentini, G. P. 1976. *Proc. Natl. Acad. Sci. USA* 73:554–58
225. Liu, L. F. 1980. See Ref. 65. In press
226. Champoux, J. J., Dulbecco, R. 1972. *Proc. Natl. Acad. Sci. USA* 69:143–46
227. Keller, W. 1975. *Proc. Natl. Acad. Sci. USA* 72:4876–80
228. Vosberg, H.-P., Vinograd, J. 1976. *DNA Synthesis and its Regulation,* ed. Goulian, M., Hanawalt, P. pp. 94–120. Menlo Park, Calif: Benjamin
229. Wang, J. C. 1973. *DNA Synthesis in vitro,* ed. R. D. Wells, R. B. Inman, pp. 163–74. Baltimore, Md: Univ. Park Press
230. Keller, W. 1975. *Proc. Natl. Acad. Sci. USA* 72:2550–54
231. Radloff, R., Bauer, W., Vinograd, J. 1967. *Proc. Natl. Acad. Sci. USA* 57:1514–21

232. Depew, R. E., Liu, L. F., Wang, J. C. 1978. *J. Biol. Chem.* 253:511–18
233. Prell, B., Vosberg, H.-P. 1980. *Eur. J. Biochem.* 108:389–98
234. Mizuuchi, K., O'Dea, M. H., Gellert, M. 1978. *Proc. Natl. Acad. Sci. USA* 75:5960–63
235. Sugino, A., Peebles, C. L., Kreuzer, K. N., Cozzarelli, N. R. 1977. *Proc. Natl. Acad. Sci. USA* 74:4767–71
236. Higgins, N. P., Peebles, C. L., Sugino, A., Cozzarelli, N. R. 1978. *Proc. Natl. Acad. Sci. USA* 75:1773–77
237. Liu, L. F., Wang, J. C. 1978. *Proc. Natl. Acad. Sci. USA* 75:2098–102
238. Scurlock, T. R., Miller, R. V. 1980. Manuscript submitted
239. Sugino, A., Bott, K. F. 1980. *J. Bacteriol.* 141:1331–39
240. Brown, P. O., Peebles, C. L., Cozzarelli, N. R. 1979. *Proc. Natl. Acad. Sci. USA* 76:6110–14
241. Sugino, A., Higgins, N. P., Brown, P. O., Peebles, C. L., Cozzarelli, N. R. 1978. *Proc. Natl. Acad. Sci. USA* 75:4838–42
242. Gellert, M., Mizuuchi, K., O'Dea, M. H., Ohmori, H., Tomizawa, J. 1979. *Cold Spring Harbor Symp. Quant. Biol.* 43:35–40
243. Peebles, C. L., Higgins, N. P., Kreuzer, K. N., Morrison, A., Brown, P. O., Sugino, A., Cozzarelli, N. R. 1979. *Cold Spring Harbor Symp. Quant. Biol.* 43:41–52
244. Gellert, M., Mizuuchi, K., O'Dea, M. H., Itoh, T., Tomizawa, J. 1977. *Proc. Natl. Acad. Sci. USA* 74:4772–76
245. Gellert, M., O'Dea, M. H., Itoh, T., Tomizawa, J. 1976. *Proc. Natl. Acad. Sci. USA* 73:4474–78
246. Gellert, M., Fisher, M. L., O'Dea, M. H. 1979. *Proc. Natl. Acad. Sci. USA* 76:6289–93
247. Morrison, A., Cozzarelli, N. R. 1979. *Cell* 17:175–184
248. Tse, Y., Kirkegaard, K., Wang, J. C. 1980. *J. Biol. Chem.* 255:5560–65
249. Forterre, P. 1980. *J. Theor. Biol.* 82:255–69
250. Brown, P. O., Cozzarelli, N. R. 1979. *Science* 206:1081–83
251. Mizuuchi, K., Fisher, L. M., O'Dea, M. H., Gellert, M. 1980. *Proc. Natl. Acad. Sci. USA* 77:1847–51
252. Kreuzer, K. N., Cozzarelli, N. R. In press
253. Liu, L. F., Wang, J. C. 1978. *Cell* 15:979–84
254. Nash, H. A., Mizuuchi, K., Weisberg, R. A., Kikuchi, Y., Gellert, M. 1977. *DNA Insertion Elements, Plasmids and Episomes,* ed. A. Bukhari, J. A. Shapiro, S. L. Adhya, pp. 363–70. Cold Spring Harbor, NY: Cold Spring Harbor Lab.
255. Smith, C. L., Kubo, M., Imamoto, F. 1978. *Nature* 275:420–23
256. Sanzey, B. 1979. *J. Bacteriol.* 138:40–47
257. Kubo, M., Kano, Y., Nakamura, H., Nagata, A., Imamoto, F. 1979. *Gene* 7:153–71
258. Marians, K. J., Ikeda, J., Schlagman, S., Hurwitz, J. 1977. *Proc. Natl. Acad. Sci. USA* 74:1965–68
259. Drlica, K., Snyder, M. 1978. *J. Mol. Biol.* 120:145–54
260. Castora, F. J., Simpson, M. V. 1979. *J. Biol. Chem.* 254:1193–95
261. Goss, W. A., Deitz, W. H., Cook, T. M. 1965. *J. Bacteriol.* 89:1068–74
262. Ryan, M. J. 1976. *Biochemistry* 15:3769–77
263. Smith, D. H., Davis, B. D. 1965. *Biochem. Biophys. Res. Commun.* 18:796–800
264. Staudenbauer, W. L. 1975. *J. Mol. Biol.* 96:201–05
265. Staudenbauer, W. L. 1975. *Eur. J. Biochem.* 62:491–97
266. Kreuzer, K. N., Cozzarelli, N. R. 1979. *J. Bacteriol.* 140:424–35
267. Cairns, J. 1963. *J. Mol. Biol.* 6:208–13
268. Yegian, C. D., Mueller, M., Selzer, M., Russo, V., Stahl, F. W. 1971. *Virology* 46:900–19
269. McCarthy, D., Minner, C., Bernstein, H., Bernstein, C. 1976. *J. Mol. Biol.* 106:963–81
270. McCarthy, D. 1979. *J. Mol. Biol.* 127:265–83
271. Fairfield, F. R., Bauer, W. R., Simpson, M. V. 1979. *J. Biol. Chem.* 19:9352–54
272. Bauer, W. R., Ressner, E. C., Kates, J., Patzke, J. V. 1977. *Proc. Natl. Acad. Sci. USA* 74:1841–45
273. Hecht, R., Thielmann, H. W. 1977. *Nucleic Acids Res.* 4:4235–47
274. Kung, V. T., Wang, J. C. 1977. *J. Biol. Chem.* 252:5398–402
275. Burrington, M. G., Morgan, A. R. 1978. *Can. J. Biol.* 56:123–28
276. LeBon, J. A., Kado, C. I., Rosenthal, L. J., Chirikjian, J. G. 1978. *Proc. Natl. Acad. Sci. USA* 75:4097–4101
277. Shishido, K., Tadahiko, A. 1979. *Biochim. Biophys. Acta* 563:261–65
278. Liu, L. F., Wang, J. C. 1979. *J. Biol. Chem.* 254:11082–88
279. Liu, L. F., Depew, R. E., Wang, J. C. 1976. *J. Mol. Biol.* 106:439–52
280. Champoux, J. 1977. *Proc. Natl. Acad. Sci. USA* 74:5328–32

281. Sternglanz, R., DiNardo, S., Wang, J. C., Nishimura, Y., Hirota, Y. 1980. See Ref. 65. In press

282. Vosberg, H.-P., Grossman, L. I., Vinograd, J. 1975. *Eur. J. Biochem.* 55: 79–93

283. Wang, J. C., Gumport, R. I., Javaherian, K., Kirkegaard, K., Klevan, L., Kotewicz, M. L., Tse, Y. C. 1980. See Ref. 65. In press

284. Champoux, J. J., McConaughy, B. L. 1976. *Biochemistry* 15:4638–42

285. Keller, W., Wendel, I. 1974. *Cold Spring Harbor Symp. Quant. Biol.* 39:199–208

286. Champoux, J. J. 1976. *Proc. Natl. Acad. Sci. USA* 73:3488–91

287. Germond, J. E., Hirt, B., Oudet, P., Gross-Bellard, M., Chambon, P. 1975. *Proc. Natl. Acad. Sci. USA* 72:1843–47

288. Germond, J. E., Rouvière-Yaniv, J., Yaniv, M., Brutlag, D. 1979. *Proc. Natl. Acad. Sci. USA* 76:3779–83

289. Langeveld, S. A., van Arkel, G. A., Weisbeek, P. J. 1980. *FEBS Lett.* 114:269–72

290. Nash, H. 1977. *Curr. Top. Microbiol. Immunol.* 1978:171–99

291. Kikuchi, Y., Nash, H. 1979. *Proc. Natl. Acad. Sci. USA* 76:3760–64

292. Enquist, L. W., Nash, H., Weisberg, R. A. 1979. *Proc. Natl. Acad. Sci. USA* 76:1363–67

Ann. Rev. Biochem. 1981. 50:261–84

ADVANCES IN PROTEIN SEQUENCING

◆12079

Kenneth A. Walsh, Lowell H. Ericsson, David C. Parmelee, and Koiti Titani

Department of Biochemistry, University of Washington, Seattle, Washington 98195

CONTENTS

PERSPECTIVES

Our present understanding of the detailed relationship of the structure and function of proteins has grown from discoveries in the nineteen fifties that these macromolecules have unique amino acid sequences that fit unique three-dimensional structures. Since that time, technological advances have facilitated the determination of the details of the amino acid sequences of a wide variety of proteins, and established: (*a*) that their complex biological functions can be described in terms of simple chemical concepts; (*b*) that

261

0066-4154/81/0701-0261$01.00

many proteins comprise domains and subunits that specify discrete subfunctions; (c) that proteins can be grouped into homologous, evolutionarily related families; and (d) that the structure of a protein can be directly related to its encoding gene.

In 1976, Edman (1) estimated that the data base of established sequences consisted of 80,000 amino acid residues. Three years later, this number had doubled, and about 160,000 residues had been placed in about 1100 completed protein sequences [W. C. Barker, personal communication from an analysis of the current data base of the *Atlas of Protein Sequence and Structure* (2)]. Ongoing research is providing a continuous stream of data, including sequences of proteins of larger molecular weight and analyses of increasing sensitivity and reliability.

Despite a recent view that "the decline and fall of protein chemistry" is at hand (3), it appears that detailed analyses of protein sequences will continue to provide a chemical base for understanding proteins and their role in biology. While it is true that recent advances in DNA sequencing technology provide more rapid methods, even analysis of cDNA (which minimizes the problem of untranslated information) fails to identify the sites of disulfide bonds, of attachment of prosthetic groups, or of removal of peptide segments, any of which may influence expression of function. Procedures to obtain pure nucleic acid starting material are rapidly improving, and one can confidently predict that reams of new sequence data will be translated from these analyses. In fact, DNA sequence technology will probably provide the simplest and fastest source of raw polypeptide backbone structure. Nevertheless, it is vital that sequences inferred in this way be correlated with those of the mature protein products, to avoid frame shift errors, to correct for intron excisions, and to identify the myriad of processing events (4) that modulate both structure and function. Ideally, one would like to describe the complete process of maturation of the protein from the control of gene expression through the control of its ultimate function, and this requires a cooperative blend of studies by molecular biologists, protein chemists, and X-ray crystallographers. Such cooperative descriptions are beginning to emerge, but there are very few examples to date where a combination of nucleic acid analysis, amino acid sequence analysis, and X-ray crystallography have focused on a single protein.

The increasing rate of accumulation of amino acid sequence data has resulted less from a proportional increase in the number of analysts, than from a striking increase in efficiency (Figure 1). Whereas in the fifties and sixties a small research group might place, on the average, only a single amino acid residue per week, by analyzing many small overlapping peptides isolated from a large protein, the availability of automated sequential degradative technology in the seventies has changed the strategy (5–9) and increased the efficiency by perhaps two orders of magnitude. Nevertheless, more than 95% of completed sequence analyses describe proteins smaller than 40,000 daltons (Figure 2), and the efficiency of their analysis falls off markedly with increasing chain length (Figure 1). Larger proteins require more sophisticated strategies than smaller ones, and call for the development of yet more efficient and more sensitive tactics.

In this review we focus on recent advances in tactics that have led to improvements in strategy requiring smaller amounts of proteins and being applicable to proteins of larger molecular weight. General reviews of methodology (10–13) and specific reviews of technology (14–16) have been published during the last five years. Shorter discussions of strategy have also

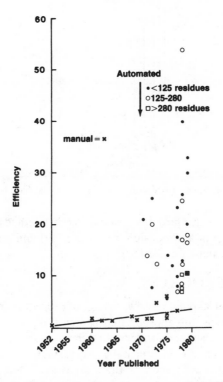

Figure 1 Progress in the efficiency of amino acid sequence analysis. The efficiency is calculated as the number of residues finally placed in sequence divided by the number of fragments isolated to prove that sequence. Of the 1100 proteins sequenced by 1978 (Figure 2), only 45 were randomly chosen for this illustration. Those analyzed by manual degradation procedures (x) are distinguished from the three symbols designating automated Edman degradative procedures on proteins of three arbitrarily different size classes. Sequences completed by automated techniques were first reported in 1971. The diagram demonstrates both the increased efficiency of automated analyses and a generally inverse relationship between chain length and efficiency.

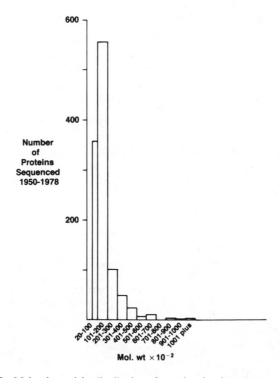

Figure 2 Molecular weight distribution of proteins that have been sequenced.

appeared (17–18). Periodical discussion of recent methods appear bienially in the proceedings of a symposium series (e.g. 19) and more frequently in a manufacturer's news service (20).

TACTICAL PROGRESS

From a technological point of view, a range of procedures are now available for cleaving proteins into appropriate fragments, for separating these fragments from each other, and for degrading the pure fragments to phenylthiohydantoins, which are identified by increasingly sensitive techniques. Recent improvements in each of these areas are continuing to provide broader strategies, more miniaturized separation systems, and more sensitive degradative analyses.

Cleavage Techniques

Table 1 is a partial list of specific chemical and enzymatic cleavage techniques that have been effective in sequence analysis. The list does not include the classical enzymatic and partial acid cleavage techniques that were at one time the only known cleavage methods. Although these are still used as primary tools in some analyses, particularly in conjunction with dansyl[1] -Edman procedures, the more restrictive cleavage techniques listed are more productive in concert with automated Edman degradations. In principle, it is more efficient to separate and analyze a small number of large fragments than a larger number of smaller ones. Three of the more parsimonious methods are finding wider usage, namely limited proteolysis of native proteins (17) and chemical cleavage of Asn-Gly or Asp-Pro bonds (12). The rarity of these dipeptides augments their value in subdividing a protein into large segments. Limited proteolysis of a native protein, although unpredictable in its site of attack, often separates domains at their "hinges" or nicks the protein on its "fringes" (31).

Among the chemical cleavage agents, iodosobenzoate shows promise for cleaving tryptophanyl bonds, providing p-cresol is present to minimize a side reaction with tyrosine (27). New enzymes have expanded the cleavage repertoire by severing bonds at the amino side of lysyl residues (28, 29) and at the carboxyl side of glutamyl residues (23). Other reports describe substrate modifications that redirect tryptic cleavage to aspartyl residues (32, 32a) and the *Armillaria* protease to the amino side of cysteinyl residues (33).

Table 1 Summary of useful specific cleavage techniques[a]

Bond cleaved	Cleavage agent
Met–X	Cyanogen bromide
Asn–Gly	Hydroxylamine
Asp–Pro	pH 2.5
Arg–X	Trypsin (N-acylated substrate), Clostripain, or mouse submaxillary protease (21)
Lys–X	Trypsin [block arginine with cyclohexanedione or with malonic dialdehyde (22)]
Glu–X	*Staphylococcal* protease (23)
Various	Limited proteolysis of *native* substrate (17)
X–Cys	2-nitro-5-thiocyanobenzoic acid (24, 24a) followed by Raney nickel (25)
Trp–X	Skatole derivative (26) or iodosobenzoate (27)
X–Lys	Myxobacter (28) or *Armillaria* (29) proteases

[a] General references (12, 30) describe techniques that are not specifically cited.

[1]Dansyl, 5-dimethylamino-1-naphthalenesulfonyl.

Separation Techniques

GENERAL METHODS Isolation of pure fragments of suitable size has been a major operational problem in sequence analysis. Classical methods such as paper chromatography, paper electrophoresis, and Dowex column chromatography are still applied to mixtures of small peptides in many studies, with some modifications. The former two techniques have been replaced by thin-layer plate techniques in some cases (34, 34a), and the latter by improved resins or ion exchange cellulose in microbore columns with automated apparati (35, 36). However, these methods are not generally applicable to mixtures of large, denatured fragments, which tend to aggregate and precipitate. By careful choice of denaturing solvents, such fragments have been separated by gel permeation chromatography and gel electrophoresis, and more recently by high performance liquid chromatography.

Gel filtration has been applied to large fragments in denaturing solvents such as high concentrations of formic, acetic, or propionic acid for basic and neutral fragments, and ammonium hydroxide or ammonium bicarbonate for acidic fragments. Certain chemical modifications, such as succinylation (37) or citraconylation (38) of lysine residues, and esterification (39) or amidation (40) of acidic residues, result in striking changes in the solubility of fragments and permit the separation of large fragments without using urea or detergents. Hydrophobic fragments have been separated on Sephadex LH columns with organic solvents (41) or by partitioning between butanol and dilute acid (42). Urea, guanidine hydrochloride, or detergents provide alternative solvents (17), but urea can lead to carbamylation of amino groups at high pH, guanidine hydrochloride and ionic detergents preclude ion exchange chromatography, and other detergents may be so strongly bound to apolar fragments as to perturb further separation. Recently, several methods have been reported for removal of SDS from proteins or fragments (43, 44).

Polystyrene-based ion exchange columns (e.g. Dowex columns) provide high resolution of small peptides but are not applicable to large fragments because of strong hydrophobic interactions and the small pore size of the resins. Various grades of hydrophilic cation or anion exchange columns are more useful for large fragments (35), provided that urea is present to solubilize the fragments. Examples of high resolution in urea solutions include the separation of ribosomal proteins on phosphocellulose or DEAE-cellulose (45), of β-galactosidase fragments on CM-cellulose (42, 46), and of phosphorylase fragments on SP-Sephadex (47).

Selective separation techniques have been described for methionine-, cysteine-, and tryptophan-containing peptides on special matrices (48–50).

Novel de facto separations can be achieved by selective attachment of fragments through C-terminal homoserine or lysine residues in preparation for solid phase sequencing (51).

The ninhydrin method has been routinely applied to detect small peptides, whereas absorbance measurements at 200–280 nm are suitable for large peptides. Fluorescamine- or o-phthalaldehyde–based methods have been developed for more sensitive detection of peptides (52, 53).

SDS-GEL SEPARATIONS   Polyacrylamide gel electrophoresis (PAGE) is a powerful analytical tool for assessing the purity of proteins and large peptides before subjecting them to sequence analysis. The usefulness of this technique is due largely to its high resolving capability. For example, a two-dimensional electrophoresis system has resolved 1100 different components from *Escherichia coli* and has the potential of separating 5000 proteins (54). This procedure utilizes isoelectric focusing in the first dimension and SDS electrophoresis in the second. The results of isoelectric focusing may be evaluated prior to the second electrophoresis by fixing and staining with Coomassie brilliant blue (55). This procedure does not alter the protein pattern, but does cause sharper spots. The order of iso-electric focusing and SDS-electrophoresis may be reversed (56) to yield material free from SDS. Such two-dimensional systems have been used analytically and preparatively on complex mixtures of proteins or peptides (57).

Difficult separations have been achieved by varying the experimental conditions. For example, the use of SDS-gradient gels selectively improves resolution and provides reliable molecular weight determinations (58); substitution of SDS by a cationic detergent, cetyltrimethyl ammonium bromide, yields better data with certain proteins (59).

As microsequencing techniques become more extensively used, sensitive analytical gels are needed to avoid waste. Recently, a cupric-silver stain has been reported (60) to be 100 times more sensitive than the conventional Coomassie blue stain (e.g. it detects 0.33 ng/mm^2 of serum albumin). These results were comparable to an autoradiogram of ^{14}C-methylated proteins following a five-day exposure. Another rapid staining procedure, ferrous bathophenanthroline sulfonate (61), has a detection limit of 1 μg for bovine serum albumin. The sensitivity of fluorescent labeling techniques allows continuous monitoring during electrophoresis. For example, a combination of microgel electrophoresis and fluorescent labeling with dansyl chloride produced distinct bands with 50 pmol of material (62). The fluorogenic reagent, 2-methoxy-2,4-diphenyl-3(2H)-furanone detected as little as 10 ng of peptide in a CNBr digest (63), whereas fluorescein-labeled lectin probes detected less than 100 ng of hexose in carbohydrate-containing proteins (64).

SDS-polyacrylamide gel electrophoresis probably will be used more routinely to purify sufficient quantities of material for microsequence analysis. A small portion of the sample can be dansylated as a marker of the underivatized material (65), but with caution, because the mobility of low-molecular-weight peptides may be altered by the label. Alternatively, phosphorescence (66), alkylbenzyldimethyl-ammonium chloride (67), or precipitation with sodium acetate (68) may be employed to locate the components of interest. After the materials have been separated they can be electrophoretically eluted, with recoveries of more than 90% (69). If necessary, residual SDS can be quantitatively removed by a micro method utilizing ion-pair extraction with triethylammonium cations (44). The resulting small amounts of proteins or peptides can then be subjected to amino acid microanalysis using o-phthalaldehyde (70). At low protein-to-gel ratios it may be necessary to apply a correction for possible background contamination (71). The ultimate goal of the various separation techniques has been achieved by subjecting the eluted materials (100 pmol) to microsequence analysis (e.g. 72, 73).

HIGH PERFORMANCE LIQUID CHROMATOGRAPHY (HPLC)   This technique is potentially the most powerful procedure currently available for both analytical and preparative isolation of peptides and proteins. The most efficient column materials are silica-based microporous particles (5–10 μm) that have coatings covalently linked to their surface silanol moieties (74–76). Several commercial HPLC packings are currently available that have been utilized for molecular exclusion (77) or high-capacity ion exchange chromatography (78). These materials have only recently been introduced and subsequent improvements will, it is hoped, result in products superior to the classical alternatives. The microporous particles with bonded nonpolar moieties (74–76) have been responsible for the growing interest in HPLC to fractionate peptides and proteins. These column packings are used in "reversed-phase" chromatography in which the eluent is more polar than the stationary phase. This is reversed from the conventionally more polar stationary packing material.

Reversed-phase HPLC has advantages of speed, sensitivity (79–82), and high resolving power (83, 84) when compared with other separation techniques. These features are only beginning to be utilized in primary structural studies. Although this application is far from routine, and experiments must be closely monitored to ensure that all components applied to the various columns are actually eluted, the practicality of reversed-phase HPLC for separation and recovery of biologically active peptides and proteins has been confirmed by many investigators. For example, one species of human leukocyte interferon (mol wt 18,000) has been purified to homogeneity (85). Thirty-two hormonal polypeptides and six proteins, ranging from five to 584 amino acid residues in length, have been successfully chromatographed (86). However, lactalbumin, elastase, and ovalbumin were apparently irreversibly bound to octadecyl silica (86). In many cases components fail to elute because of the use of inappropriate buffers and organic eluents.

Several buffer systems are routinely used. These include 0.1% phosphoric acid (87), 0.1% trifluoroacetic acid (88), pyridine (1 M) – acetic acid (0.5 M) (89), and 0.25 N phosphoric acid titrated to pH 3–3.5 with triethylamine (90). Other ion-pairing reagents, e.g. alkylsulfonates (80) or alkylammonium salts (91), have also been added to these buffers to improve the resolution of various peptides (92). Water-miscible organic solvents, such as methanol, n-propanol, or acetonitrile are common eluent components for peptides and proteins (83, 84, 89, 90). The most frequently used reversed-phase supports are cyanopropyl-, octadecyl-, octyl-, and phenyl-silica (74, 83, 84, 90). Recommendations have been made by several investigators (41, 83, 84, 90) for optimizing the various experimental conditions, such as flow rate, ionic strength, pH, temperature, capacity, gradient elution, and sample preparation.

Many have utilized reversed-phase HPLC to analyze small peptides produced by enzymatic digests of various proteins (41, 81, 82, 89). This is essentially an analytical peptide mapping technique that provides the means to characterize subnanomol quantities, to observe the extent of digestion, and to prepare sufficient material for further structural studies. For example, the total amino acid composition of bovine intestinal calcium-binding protein (75 residues) was accounted for on the basis of two tryptic maps that used a total of 4 μg of protein (81). Differences in the HPLC peptide maps of the tryptic digests of normal human hemoglobins and 25 variants were used to purify the specific abnormal peptides. Amino acid analysis of those peptides established the nature of the alterations (93). Many of these variant peptides showed differences in elution position that correlated well with that expected from the changed amino acid compositions. Methods are now available for the prediction of the relative retention times of small peptides ($<$ 20 residues) solely on the basis of their amino acid compositions (94). However, the absolute elution times vary with the chromatographic conditions.

The ultimate goal in using HPLC for primary structural studies is to rapidly purify relatively large, denatured peptides to homogeneity for sequence analysis or subdigestion and refractionation. For example, two chymotryptic fragments (19,000 and 6,900 daltons) of bacteriorhodopsin have been separated (41). These were further cleaved with CNBr, and two of the resulting peptides were sequenced after reversed-phase HPLC purification. Similarly, all three CNBr fragments (32, 44, and 66 residues) from the S-aminoethylated α-chain of human hemoglobin were recovered in yields of 59, 70, and 102%, respectively (82). Two of these were sequenced using 4-N,N-dimethylaminoazobenzene 4'-isothiocyanate (DABITC) with solid phase and liquid phase instruments, respectively (95, 96). Another investigation demonstrated that all but two of the CNBr peptides from human hemoglobin α-, β-, and γ-chains (13–91 residues in length) could be purified and recovered in yields of 71–95% (88). These various peptides were not sequenced, but their amino acid compositions agreed well with expected values.

These reports indicate that reversed-phase HPLC is an extremely powerful analytical and preparative alternative to more conventional techniques. The excellent resolution and high capacity may provide single-step purification of many peptides needed in primary structural studies. Unresolved fragments may be collected and rechromatographed on other reversed-phase columns or with different eluents. At present, the most useful HPLC separations utilize reversed-phase column packings. As improved alternative supports become available, analogous purifications may well be made in conjunction with molecular exclusion HPLC or ion-exchange HPLC.

Edman Degradation

The principal chemical method to determine the amino acid sequences of fragments of a protein is the stepwise degradation initiated with phenylisothiocyanate and completed with acid (1), as devised by Edman three decades ago. This "Edman degradation" has been used in a "subtractive" mode whereby the composition of the shortened peptide establishes the missing amino terminus (97), in the dansyl-Edman mode whereby the new amino-terminal residue is identified by end-group analysis (98), and in direct modes whereby the phenylthiohydantoin (PTH) of the released amino acid is identified directly by various chromatographic techniques or as the free amino acid after hydrolysis (99). A direct manual method was well described in 1977 (100). Other direct methods, as automated in spinning cups in 1967 (5) and on solid matrices in 1971 (9) offer controlled reaction conditions and reproducibility that permit extended degradation of fragments. Improvements in the spinning cup methodology (16) have focused recently on the chemistry, the instrumentation (101–103), and techniques of product identification. Together, these advances have led in 13 years to increases by 3 to 6 orders of magnitude in sensitivity, which has now approached 10 pmol in one laboratory (103) and 0.1 pmol in special cases (see section on microsequencing).

CHEMISTRY OF THE DEGRADATION   To achieve degradation yields of 94–98%, side reactions must be minimized by paying close attention to the purity of the sample, reagents, solvents and gaseous atmosphere, and to the cleanliness of the surrounding mechanical structures.

The nature and the purity of the subject peptide must be carefully considered. Prior to sequence analysis, impurities and salts can be removed by gel filtration, dialysis, or electrophoresis (104), but contaminating SDS may actually be advantageous during subsequent analysis (105). Blocking groups at the amino terminus are particularly troublesome. Enzymatic methods to remove amino-terminal pyrollidone carboxylic acid (106) have met with varying success (107). Formyl groups have been removed from aminoterminal methionine (108). Although no method is known to remove the acetyl groups that block the amino termini of many proteins, a technique is described (109) to prevent its introduction during in vitro syntheses.

Other covalently modified residues (4) present difficulties during both degradation and product identification. For example, glycosyl groups complicate estimations of molecular weight, isolation of peptides, and the Edman degradation. The PTH derivatives of glycosylated amino acids have been extracted during solid phase degradation, but not from the spinning cup (110). However, a general method for partial deglycosylation with HF has been reported (111).

General procedures for purifying reagents and solvents have been described (5, 112), but the strategy of continuously scavenging contaminants in the reagent bottles provides a simple alternative (113) that yields impressive results (114) and minimizes problems of reagent instability. The most widely used coupling buffer, Quadrol, is difficult to purify, but Begg recommends its replacement by tetrahydroxyethylethylenediamine (THEED), and claims very high degradative efficiency (115). Significant advantages in peptide retention are achieved by reducing the Quadrol concentration from 1 M to 0.1–0.18 M (116, 117), although many laboratories have now adopted intermediate concentrations of 0.25–0.33 M, provided that protecting carriers are included. Alternatives to phenylisothiocyanate have been employed, usually in order to achieve greater sensitivity of product detection. These are reviewed in the section on microsequencing.

An important operational improvement is the inclusion of a polyionic carrier to improve retention of small peptides and small quantities of peptide or protein in the spinning cup. The most widely used is Polybrene, a polycationic carrier introduced in 1978 (118, 119), whereas a polyanionic alternative should be better with basic peptides containing arginine or histidine (120).

The chemistry of the degradation can be optimized by appropriate control of the automation program. These programs are continually being modified in various laboratories, but minor changes tend not to be published until enough of them accumulate to merit a detailed description (e.g. 102,

116, 117, 121, 122). Improvements include schemes to avoid a postdelivery drip from the reagent delivery tubes during vacuum steps (117), to control film height by cup speed changes (102), to reduce partial acid cleavage by minimizing N to O shifts (122), to achieve azeotropic removal of acid (116, 121), and to apply gentler conditions for drying the film and extracting the product with minimal loss of peptide (116, 117, 121). More complex programs define shorter cleavage times, double coupling, and/or double cleavage. These can provide residue-specific experimental conditions at preselected degradation cycles (102, 103, 123). It should be noted that one impact of the widespread use of the Polybrene carrier has been to reduce the rate of evolution of programs designed to minimize peptide washout.

MODIFICATIONS OF THE INSTRUMENTATION   Effective improvement of the sensitivity and efficiency of automated Edman degradation requires a combination of engineering and chemical skills. Several laboratories have introduced minor mechanical changes, (e.g. 124, 125), and two have described major redesigns of the spinning cup instrument (101, 103). Unfortunately, the major modifications are not yet commercially available, and other laboratories have only introduced a few changes consistent with their engineering capabilities.

The vacuum system in the conventional instrument has many problems. The pneumatic valve in the fine vacuum system has a significant failure rate, but can be replaced by a more reliable alternative (125). A more complex vacuum control manifold has been designed (101, 126) and adopted in a few other laboratories. Edman's original simple design (5) may yet be readopted. The reliability of the vacuum system is reported to be improved by the interposition of a cold trap between the pump and the cup (101, 102, 124, 126a) or by recycling the pump oil through an activated carbon filter (127).

Several redesign features serve to remove potential sources of contamination of the intracup atmosphere. In-line purified argon has been substituted for nitrogen (102), delivery tubes have been shortened to minimize penetration of teflon by oxygen, and reagent delivery valves have been redesigned to eliminate dead volumes (101, 103, 126). One seemingly simple and quite important modification is to eliminate as far as possible contaminants and leaks from the o-rings that are throughout the reservoir bottles, the delivery line connectors, and the spinning cup chamber itself. Although their total elimination requires significant redesign (101, 103), replacement of some of the present o-rings with more resistant versions should reduce one source of contamination. It has been suggested that the size of the cup should be related to the amount of material being degraded (127), and that a straightwalled cup (as originally designed by Edman) minimizes the inefficiency of laminar flow extraction characteristic of undercut cups (103).

It is well known that the intermediate thiazolinone product of the Edman degradation is less stable than the ultimate phenylthiohydantoin. To minimize the period of instability, automatic conversion units have been designed, and fitted between the cup exit and the fraction collector. One, designed commercially for the solid-phase instrument, has been installed in liquid-phase sequenators (e.g. 127); another has been described in detail (128) and adapted to other instruments (103). Each offers a rapid and reproducible conversion that improves the identification of the end products, saves operator time, and avoids human error.

PRODUCT IDENTIFICATION   Major improvement in phenylthiohydantoin identification has resulted from the introduction of reversed-phase HPLC systems. Although thin layer chromatography, gas-liquid chromatography, and back hydrolysis systems are still used, HPLC offers the advantages of reasonable quantitation, nondestructive analysis (with the potential for product recovery), and excellent sensitivity. A variety of HPLC systems have been described using either isocratic or gradient elution with methanol (129, 130–130b) or acetonitrile (131–136) in acidic buffers from C_{18}-bonded microporous silica columns. Recent reports describe the use of cyanopropylsilane (137) and phenylalkylsilane (138) columns.

Mass spectrometric identification of phenylthiohydantoins have been reported (139, 140). Alternatively, the thiazolinones have been converted to methylamides before separation and detection (141).

Interpretation of the patterns of phenylthiohydantoin release tend to be judgmental rather than quantitative (47), although a more stringent approach is possible with data from HPLC, gas/liquid chromatography, or back hydrolysis than from thin layer chromatography. Even with the best HPLC data, it is necessary to compare the yield of products from a given cycle with those of the preceding and following cycles in order to recognize the rise and fall of legitimate product against a rising background from occasional acidic cleavage of peptide bonds. Interpretation is further complicated by the variable instability of the products, the decreasing yield of true product, and special problems with occasional resistant peptide sequences that place subsequent cycles partly out of register (6, 103, 127). Objective interpretations showing quantitative data for all amino acids at each step have been reported, for example by Frank and co-workers (114), but such documentation is expensive and could consume excessive journal space. Such considerations have led to diverse editorial responses to communicated proofs of sequence. In some cases, extensive and well-organized documentation has appeared, e.g. the six papers by Kühn and his colleagues (142). In other cases, only the final conclusion is published and the supporting proof accumulates somewhere between the laboratory and the editorial office.

Recently reference has been made to a computer-controlled procedure to subtract background and enhance the pattern of HPLC data obtained with five-picomole samples (103). Such procedures may enhance both the sensitivity and the reliability of future analyses.

MICROSEQUENCING   There tends to be an inverse correlation between the availability of starting material and the interest in a particular amino acid sequence. Such pressures have prompted continuing refinement of all phases of sequencing tactics, with particular emphasis on the degradative techniques. Whereas the spinning cup sequenator was designed for 100 nmol, is it now used routinely at the 10–50 nmol level in many laboratories and with much smaller quantities (0.1 pmol–5 nmol) in a few microsequencing adaptations. Many of the preceding improvements in the chemistry and instrumentation were designed to increase the sensitivity of the analyses, as in the recent determination of the amino-terminal 13–20 residues of 20–500 pmol (0.4–27 μg) of human interferon (143, 144). In this case, the protein was isolated from SDS-gels, degraded in the presence of a polybrene carrier in a totally redesigned sequenator with ultrapure reagents, and the products identified by HPLC with computerized enhancement programs (103).

There appear to be three general approaches to microsequence analysis. The first comprises variations of such instrument redesign (e.g. 101, 102, 124, 127), but it is not clear whether partial improvement in sensitivity would result from incorporation of part but not all of the suggested modifications. A second general approach uses radioactive, colored, or fluorescent isothiocyanates to generate more easily detected thiohydantoins. The third general approach involves radio-labeling the protein prior to degradation, to enhance the detection sensitivity.

Radiolabeled isothiocyanates have been used since 1961, but were applied to automated procedures in the solid phase in 1969 (145) and in the spinning cup in 1975 (146) using thin layer techniques for the final identifications. Microsequencing procedures were reviewed in 1977 for liquid (147) and solid phase (18, 148) modes. The amino-terminal 20-residue sequence of 70 pmol of lysozyme was determined with ^{35}S-phenylisothiocyanate (72). The more recent availability of HPLC systems and flow-through scintillation counters may facilitate application of such procedures, but the cost of equipment and radiolabeled reagent is a significant drawback. Alternative schemes using fluoresceinisothiocyanate (149) have been applied at the 6-nmol level (150). More recently the colored DABITC procedure (151, 152) has found application at the 1–10-nmol level in a manual mode, and in both solid phase (95) and liquid phase (96) modes using HPLC for sensitive identification of the products (153). A detailed study has been reported of the side reactions that destroy serine and threonine during this degradation (154).

The third general approach to microsequencing relies on the radioactivity of the amino acids themselves, either intrinsically labeled during synthesis or extrinsically labeled after synthesis. When suitable procedures have been available for isolation of traces of a pure labeled protein, the highest analytical sensitivity (~ 0.1 pmol) has been achieved. Several methods have been used to introduce labeled amino acids during synthesis in cell culture (e.g. 155–159) or by cell-free in vitro translation of mRNA (e.g. 160–166). In most cases immunological techniques were used to separate a single radiolabeled protein from the mixture. This inclusion of unlabeled protein (antibody) served as a convenient carrier during subsequent sequence analysis. It is particularly efficient to degrade a protein labeled with several different amino acids, then to separate, by HPLC, the labeled phenylthiohydantoins, together with unlabeled internal standards (164). The cell-free translation methods have detected precursor forms of proteins and identified them as primary translation products by demonstration of methionine incorporation from initiator tRNA (108, 161, 164). Such data facilitate the correlation of gene sequencing data with protein sequence data (158, 159, 167).

Proteins have also been labeled extrinsically by group-specific modification, for example of cysteinyl residues with ^{14}C-iodoacetate, of lysyl residues by acylation or methylation, and of acidic residues by esterification or amidation. Other procedures have radiolabeled histidyl, tyrosyl (168), or methionyl (169) residues. Attempts to label proteins uniformly by exchanging side chain hydrogens for microwave-excited tritium atoms have damaged a protein in ways that have precluded sequence analysis (170).

Other Degradation Schemes

Although the Edman degradation has clearly been the major process applied to determine the sequence of the proteins in Figure 2, other procedures have contributed, and still others may become more important in the future.

Exopeptidase digestions are useful to identify amidated residues and chemically modified residues, but the kinetics of residue release may be difficult to interpret. Stepwise chemical procedures have been described (171, 172) in which thioacylating reagents have been substituted for phenylisothiocyanate. The thioacyl product cyclizes to form a stable thiazolinone with the amino-terminal residue, which is then extracted and hydrolyzed to the free amino acid for identification.

Schemes to degrade the carboxyl terminus in a stepwise fashion have been reported (173–175), but have not yet found wide application in sequence analysis.

At one time there was a wave of optimism that refinements in techniques of X-ray crystallography would lead to direct description of both primary and tertiary structure. Kannan (176) concluded that this goal is unrealistic, but that cooperation between a sequence chemist with partial structures and a crystallographer with electron density maps can save much time and effort for each investigator. He cites examples of such productive cooperation, to which could be added the recent interaction of our laboratory with two crystallography groups during studies of glycogen phosphorylase (177). Approximately 190 three-dimensional structures have been described (E. E. Abola, personal communication), of which 126 are available from the Protein Data Bank (178). In several instances the primary sequence data are lacking in an otherwise complete three-dimensional structure. In many other cases, the primary structure is among the 1100 described (2), but the tertiary structure is lacking.

Applications of mass spectrometry to sequence determination have been reviewed recently (179). This approach has proven particularly useful in identifying the nature of blocked or chemically modified peptides. A case in point is the identification of γ-carboxyglutamyl residues in a blood coagulation factor (180). These residues are acid-labile and had proven to be refractory to conventional Edman degradation. Although mass spectrometric methods have been largely limited to analysis of peptides with less than 15 residues, a recent report (181) of a total amino acid sequence determination by mass spectrometry (dihydrofolate reductase) indicates that this methodology holds promise for the future. A new technique uses two tandem mass spectrometers (182), one for separation of peptides and the second for their identification. The potential for speed and sensitivity (10–100 pmol for derivatization) makes this approach attractive.

Problems of Covalently Modified Residues

Reference has been made to the possibility of removing blocked amino-terminal groups or glycosyl groups so that sequence analysis can proceed unhindered. However, it is important to identify the sites of these and many other covalent modifications or proteolytic processing events, which may be clues to biological function or control. Some covalent modifications are known to exist even before a structural analysis begins (e.g. heme attachment in cytochrome), but others have been discovered only by careful work duing the sequence analyses. Since Uy & Wold (4) list 140 different modified amino acid residues, this review needs only to emphasize the problems and tactics in a general way. But it is a safe prediction that the majority of proteins are covalently modified in some way during, or following, translation. If true, this represents a potential shortcoming of predictions of mature structure from gene or cDNA sequence data. Of course, one should also question how many modified amino acids have been missed in conventionally proven and published sequences.

A covalently modified amino acid residue may not be apparent during amino acid analysis of a protein either because it is marginally detectable, because it reverts to a conventional amino acid during acid hydrolysis [e.g. phospho serine or γ-carboxyglutamic acid (182a)], or because it coelutes with another amino acid during chromatographic analysis (183). Furthermore, it can be missed during the Edman degradation because it is acid-labile and progressively lost as cycles proceed (184), because it is too polar to be extracted from the spinning cup in conventional solvents (185), or because the product phenylthiohydantoin co-migrates with another conventional one (186). These problems point to the importance of seemingly redundant identifications of each residue during sequence analysis. Many laboratories adopt as a routine the principle that each residue should be identified by two techniques based on different recognition features (e.g. by HPLC, gas/liquid or thin-layer chromatography, or back-hydrolysis of PTH-amino acids; or by comparison of electrophoretic behavior with the sequence of peptides).

It would seem that a radio-labeled prosthetic group would be easy to identify with its attached residue, but one experience in our own laboratory (187) illustrates that problems can arise even in this simple case. In essence, incomplete cleavage of the second methionine in a sequence -Met-Lys*-Met-Glu-Thr-Glu (where Lys* is [^{14}C] biotinyl-lysine) yielded a mixture of a major Glu-Thr-Glu——peptide contaminated with a very minor Lys*-Met-Glu——peptide. The first cycle of Edman degradation of this mixture released the label, but a tentative conclusion of Lys*-Thr-Glu was later proven to be incorrect.

It should be noted that modified residues with polar character may be more easily recovered after Edman degradation by the solid phase procedure, where diverse solvents can be explored. However, as a general observation, mass spectrometric analysis of modified peptides or residues may be the most flexible and definitive technique for ultimate identification.

STRATEGIES

From a broader point of view, the organization of tactics used in a given sequence problem defines the strategy of attack. The availability of starting material influences the degree to which microanalytical considerations must be adopted. The feasibility of specific experimental methods in a given laboratory circumscribes the choice of tactics. But the complexity of the necessary strategy increases markedly with the chain length of the protein or subunit. Specific clues guiding the rational development of a strategy are given by an examination of the amino acid composition for potential sites of limited group-specific cleavage (e.g. at Arg, Trp, or Met), but there is no guarantee that their distributions will lend themselves to productive cleavage. The composition alone gives only statistical data about potentially cleavable dipeptide sequences (Table 1), and no clues as to sites susceptible to limited proteolysis.

At this point there is a great tendency to simply start an analysis with fragments generated by any of the more classical cleavage techniques, thus compiling a random collection of fragmentary sequence data, and then to consider an appropriate strategy to finish the molecule. With many proteins this has been successful, but experience with large proteins suggests that more exploratory experiments and planning lead to a more efficient attack. For example, analyses of seven different proteins (17) were found to be facilitated by exploiting their susceptibility to limited proteolysis, thereby dividing the original problems into several smaller and more tractable ones. The classical subdivisions of γ-globulin into F_{ab} and F_c fragments (188) and of ribonuclease into S-protein and S-peptide (189) are well-known examples. In our own work with six different proteins (8, 47, 190–193), limited proteolysis or autolysis greatly facilitated the development of a productive strategy, as best exemplified with phosphorylase, where an 841-residue problem was reduced to two much smaller ones (47). Although limited proteolysis is an empirical and unpredictable procedure, the potential advantage does justify exploration of proteolytic susceptibility in the presence and absence of ligands, or in supramolecular aggregates (194).

Another set of advisable exploratory experiments is finding wider usage. This comprises a search for chemical cleavage at the rare sites Asn-Gly and Asp-Pro (Table 1). If the 20 amino acids were present in equimolar amounts and randomly distributed, one would expect an average of one such dipeptide per 400 residues. In fact, each appears to occur with 2–3 times that frequency, which makes them ideal loci for specific cleavage. An exploratory search among the mixed reaction products by SDS-gels or by Edman degradation gives a quick guide to the number and size of fragments (17). Another approach is to artificially limit susceptible cleavage sites, for example by selective oxidative protection of exposed methionyl residues in a native protein from cleavage with CNBr (195).

It is expedient to minimize the complexity of fragment isolation schemes. It is desirable but not essential to recover all peptides from a single fragmentation mixture. Sometimes simple mixtures have been successfully analyzed (e.g. 184, 196, 197). Sometimes a pure peptide can be recovered in low yield by judicious pooling of a shoulder fraction of a chromatographic peak (8). One mixture was simplified by selectively blocking an internal glutamine of a contaminant during Edman degradation of the mixture (17). An elegant purification strategy utilizes residue-specific attachment to solid phase supports (51), where the act of purification is a part of the coupling preparation for solid phase degradation. At later stages of sequence analysis, purification schemes may be directed specifically toward particular peptides of suspected sequence, for overlap purposes. Hartley (198) has reviewed a variety of selective diagonal procedures that are simple and effective, for example, with methionyl peptides (199). Alternatively, methionyl peptides can be specifically radiolabeled for ready localization (200) or aligned after pulse-labeling during in vitro synthesis (201).

Strategic considerations can even be applied during specific Edman degradations. It is useful to extend each degradation into the region where only a few residues are being identified, so that overlaps can be established or good subsequent cleavage sites can be located. The extremely efficient analysis of the α-chain of C-phycocyanin illustrates this point dramatically (114). To the analyst, the most interesting part of any Edman degradation is the last few productive cycles, where one hopes to find a unique bond for subsequent cleavage and extension of the structure. An early example was an analysis of amyloid protein (202), where all methionines were placed during degradation of the starting material, and the last lysine was placed during analysis of the carboxyl-terminal fragment from CNBr degradation. An analogous strategy proved to be extremely effective with β-thromboglobulin (203), where arginine residue 65, which was placed in the initial analysis, served as the single cleavage point necessary to complete the 81-residue chain.

A related strategy involves blocking the amino terminus of a peptide, then cleaving a unique internal bond to generate a mixture of a blocked peptide and an unblocked peptide. This mixture can be subjected directly to sequence analysis to yield a single, internal sequence (e.g. 17, 47). Interesting variations of this approach were performed directly in the spinning cup by interrupting Edman degradations, blocking the exposed α-amino groups with an isocyanate, cleaving at a single remaining methionine or arginine, and continuing each degradation into a single new region (204).

Fowler & Zabin (205), in their heroic analysis of the 1021-residue chain of β-galactosidase, used both a complementation method and a termination mutant to place certain fragments in the amino-terminal region. In the same study, immunological aids were used to monitor the purification of overlapping peptides (206), and correlations with DNA sequence data provided confirmatory data (205).

One would like to be able to summarize general approaches to the development of strategies of sequence analysis. But a cursory look at the record, particularly with larger proteins, indicates that more efficient strategies evolve as new information accumulates. Just as a general must be prepared to refine his strategy in response to both current tactical strengths and perceived oppositional weaknesses, a sequence analyst should continually redefine more practical sets of tactics as salient features of the molecule become evident.

SUMMARY

Although this review cannot possibly cover the myriad of new ideas reported in the analyses of 1100 proteins, it does attempt to highlight procedures and strategies of interest, admittedly with a bias toward examples most familiar to us. It is clear that the scope, efficiency, and sensitivity of the analyses have increased and will continue to do so. Literally any protein sequence can now be solved. The diversity of tactics and strategies adds to the thrill of the chase.

If the next few years continue to show the growing interplay between sequence chemists, X-ray crystallographers, and nucleic acid sequencing groups, we may truly begin to understand the meaning underlying the 20-letter hieroglyphics of protein chemistry.

Literature Cited

1. Edman, P. 1977. *Carlsberg Res. Commun.* 42:1–9
2. Dayhoff, M. O., ed. 1978. *Atlas of Protein Sequence and Structure,* Vol. 5, Suppl. 3. Washington DC: Natl. Biomed. Res. Found. 414 pp.
3. Malcolm, A. D. B. 1978. *Nature* 275: 90–91
4. Uy, R., Wold, F. 1977. *Science* 198: 890–96
5. Edman, P., Begg, G. 1967. *Eur. J. Biochem.* 1:80–91
6. Hermodson, M. A., Ericsson, L. H., Titani, K., Neurath, H., Walsh, K. A. 1972. *Biochemistry* 11:4493–4502
7. Niall, H. D., Hogan, M. L., Sauer, R., Rosenblum, I. Y., Greenwood, F. C. 1971. *Proc. Natl. Acad. Sci. USA* 68:866–69
8. Hermodson, M. A., Ericsson, L. H., Neurath, H., Walsh, K. A. 1973. *Biochemistry* 12:3146–53
9. Laursen, R. A. 1971. *Eur. J. Biochem.* 20:89–102
10. Needleman, S. B., ed. 1975. *Protein Sequence Determination.* New York: Springer. 393 pp.
11. Needleman, S. B., ed. 1977. *Advanced Methods in Protein Sequence Determination.* New York: Springer. 189 pp.
12. Hirs, C. H. W., Timasheff, S. N., eds. 1977. *Methods Enzymol.* Vol. 47. 668 pp.
13. Konigsberg, W. H., Steinman, H. M. 1977. *The Proteins,* Vol. 3, ed. H. Neurath, R. L. Hill, pp. 1–178. New York: Academic. 663 pp. 3rd ed.
14. Perham, R. N., ed. 1975. *Instrumentation in Amino Acid Sequence Analysis.* London: Academic. 197 pp.
15. Niall, H. D. 1977. See Ref. 13, pp. 179–238
16. Niall, H. D. 1973. *Methods Enzymol.* 27:942–1010
17. Walsh, K. A., Ericsson, L. H., Titani, K. 1978. *Versatility of Proteins,* ed. C. H. Li, pp. 39–58. New York: Academic. 465 pp.
18. Bridgen, J. 1977. *Science Tools, The LKB Instrument Journal* 24(1):1–6
19. Birr, C. H., ed. 1980. *Methods in Peptide and Protein Sequence Analysis.* Amsterdam: Elsevier. 531 pp.
20. *In Sequence.* A news service published by the Spinco Division of Beckman Instruments, Palo Alto, Calif.
21. Schenkein, I., Levy, M., Franklin, E. C., Frangione, B. 1977 *Arch. Biochem. Biophys.* 182:64–70
22. Muranov, A. V., Modyanov, N. N. 1979. *Bioorg. Chem.* 5:210–16
23. Drapeau, G. R. 1977. See Ref. 12, pp. 189–91
24. Jacobson, G. R., Schaffer, M. H., Stark, G. R., Vanaman, T. C. 1973. *J. Biol. Chem.* 248:6583–91
24a. Degani, Y., Patchornik, A. 1974. *Biochemistry* 13:1–11
25. Otieno, S. 1978. *Biochemistry* 17: 5468–74
26. Omenn, G. S., Fontana, A., Anfinsen, C. B. 1970. *J. Biol. Chem.* 245:1895–1902
27. Mahoney, W. C., Hermodson, M. A. 1979. *Biochemistry* 18:3810–14; 1980. See Ref. 19. pp. 323–28
28. Wingard, M., Matsueda, G., Wolfe, R. S. 1972. *J. Bacteriol.* 112:940–49
29. Doonan, S., Doonan, H. J., Hanford, R., Vernon, C. A., Walker, J. M.,

Airoldi, L. P. daS., Bossa, F., Barra, D., Carloni, M., Fasella, P., Riva, F. 1975. *Biochem. J.* 149:497–506

30. Hirs, C. H. W., ed. 1967. *Methods Enzymol.* Vol. 11. 988 pp.
31. Neurath, H. 1981. In *Symp. Protein Folding.* ed. R. Jaenicke. London: Plenum. In press
32. Wang, T., Young, N. M. 1978. *Anal. Biochem.* 91:696–99
32a. Debons, F. E., Loudon, G. M. 1980. *Biochem. Biophys. Res. Commun.* 92:606–9
33. Doonan, S., Fahmy, H. M. A. 1975. *Eur. J. Biochem.* 56:421–26
34. Hitz, H., Schäfer, D., Wittmann-Liebold, B. 1977. *Eur. J. Biochem.* 75:497–512
34a. Fishbein, J. C., Place, A. R., Ropson, I. J., Powers, D. A., Sofer, W. 1980. *Anal. Biochem.* 108:193–201
35. Chin, C. C. Q., Wold, F. 1977. See Ref. 12, pp. 204–10
36. Machleidt, W., Otto, J., Wachter, E. 1977. See Ref. 12, pp. 210–20
37. Klotz, I. M. 1967. See Ref. 30, pp. 576–80
38. Atassi, M. Z., Habeeb, A. F. S. A. 1972. *Methods Enzymol.* 25:546–53
39. Wilcox, P. E. 1972. See Ref. 38, pp. 596–615
40. Carraway, K. L., Koshland, D. E. Jr. 1972. See Ref. 38, pp. 616–23
41. Gerber, G. E., Anderegg, R. J., Herlihy, W. C., Gray, C. P., Biemann, K., Khorana, H. G. 1979. *Proc. Natl. Acad. Sci. USA* 76:227–31
42. Fowler, A. V., Brake, A. J., Zabin, I. 1978. *J. Biol. Chem.* 253:5490–98
43. Kapp, O. H., Vinogradov, S. N. 1979. *Anal. Biochem.* 91:230–35
44. Henderson, L. E., Oroszlan, S., Konigsberg, W. 1979. *Anal. Biochem.* 93:153–57
45. Held, W. A., Mizushima, S., Nomura, M. 1973. *J. Biol. Chem.* 248:5720–30
46. Fowler, A. V. 1978. *J. Biol. Chem.* 253:5499–504
47. Koide, A., Titani, K., Ericsson, L. H., Kumar, S., Neurath, H., Walsh, K. A. 1978. *Biochemistry* 17:5657–72
48. Shechter, Y., Rubinstein, M., Patchornik, A. 1977. *Biochemistry* 16:1424–30
49. Bucklehurst, K., Carlsson, J., Kierstan, M. P. J., Crook, E. M. 1974. *Methods Enzymol.* 34:531–44
50. Rubinstein, M., Shechter, Y., Patchornik, A. 1976. *Biochem. Biophys. Res. Commun.* 70:1257–63
51. Laursen, R. A., Horn, M. J. 1977. See Ref. 11, pp. 21–37
52. Lai, C. Y. 1977. See Ref. 12, pp. 236–43

53. Udenfriend, S. 1978. See Ref. 17, pp. 23–37
54. O'Farrell, P. H. 1975. *J. Biol. Chem.* 250:4007–21
55. Jäckle, H. 1979. *Anal. Biochem.* 98:81–84
56. Tuszynski, G. P., Buck, C. A., Warren, L. 1979. *Anal. Biochem.* 93:329–38
57. Epstein, H. F., Wolff, J. A. 1976. *Anal. Biochem.* 76:157–69
58. Lambin, P. 1978. *Anal. Biochem.* 85:114–25
59. Eley, M. H., Burns, P. C., Kannapell, C. C., Campbell, P. S. 1979. *Anal. Biochem.* 92:411–19
60. Switzer, R. C. III, Merril, C. R., Shifrin, S. 1979. *Anal. Biochem.* 98:231–37
61. Graham, G., Nairn, R. S., Bates, G. W. 1978. *Anal. Biochem.* 88:434–41
62. Chan, W. Y., Seale, T., Rennert, O. M. 1980. *J. Chromatogr.* 181:259–65
63. Chen-Kiang, S., Stein, S., Udenfriend, S. 1979. *Anal. Biochem.* 95:122–26
64. Furlan, M., Perret, B. A., Beck, E. A. 1979. *Anal. Biochem.* 96:208–14
65. Stephens, R. E. 1978. *Anal. Biochem.* 65:369–79
66. Isenberg, I., Smerdon, M. J., Cardenas, J., Miller, J., Schaup, H. W., Bruce, J. 1975. *Anal. Biochem.* 69:531–35
67. Takagi, T., Kubo, K., Isemura, T. 1977. *Anal. Biochem.* 79:104–9
68. Higgins, R. C., Dahmus, M. E. 1979. *Anal. Biochem.* 93:257–60
69. Nelson, J. W., Cordry, A. L., McCullough, G. A., Meakin, G. 1978. *Anal. Biochem.* 85:188–96
70. Drescher, D. G., Lee, K. S. 1978. *Anal. Biochem.* 84:559–69
71. Brown, W. E., Howard, G. C. 1980. *Anal. Biochem.* 101:294–98
72. Bridgen, J. 1976. *Biochemistry* 15:3600–4
73. Bridgen, J., Snary, D., Crumpton, M. J., Barnstable, C., Goodfellow, P., Bodmer, W. F. 1976. *Nature* 261:200–5
74. Horvath, C., Melander, W., Molnar, I. 1976. *J. Chromatogr.* 125:129–56
75. Horvath, C., Melander, W. 1977. *J. Chromatogr. Sci.* 15:393–404
76. Colin, H., Guiochon, G. 1977. *J. Chromatogr.* 141:289–312
77. Ui, N. 1979. *Anal. Biochem.* 97:65–71
78. Chang, S. H., Gooding, K. M., Regnier, F. E. 1976. *J. Chromatogr.* 125:103–14
79. Frei, R. W., Michel, L., Santi, W. 1976. *J. Chromatogr.* 126:665–77
80. Hancock, W. S., Bishop, C. A., Meyer, L. J., Harding, D. R. K., Hearn, M. T. W. 1978. *J. Chromatogr.* 161:291–98
81. Fullmer, C. S., Wasserman, R. H. 1979. *J. Biol. Chem.* 254:7208–12

82. Hughes, G. J., Winterhalter, K. H., Wilson, K. J. 1979. *FEBS Lett.* 108: 81–86
83. Rubinstein, M. 1979. *Anal. Biochem.* 98:1–7
84. Brown, P. R., Krstulovic, A. M. 1979. *Anal. Biochem.* 99:1–21
85. Rubinstein, M., Rubinstein, S., Familletti, P. C., Miller, R. S., Waldman, A. A., Pestka, S. 1979. *Proc. Natl. Acad. Sci. USA* 76:640–44
86. O'Hare, M. J., Nice, E. C. 1979. *J. Chromatogr.* 171:209–26
87. Hancock, W. S., Bishop, C. A., Prestidge, R. L., Harding, D. R. K., Hearn, M. T. W. 1978. *J. Chromatogr.* 153: 391–98
88. Mahoney, W. C., Hermodson, M. A. 1980. *J. Biol. Chem.* 255:11199–11203
89. Rubinstein, M., Chen-Kiang, S., Stein, S., Udenfriend, S. 1979. *Anal. Biochem.* 95:117–21
90. Rivier, J. E. 1978. *J. Liq. Chromatogr.* 1:343–66
91. Hancock, W. S., Bishop, C. A., Battersby, J. E., Harding, D. R. K., Hearn, M. T. W. 1979. *J. Chromatogr.* 168: 377–84
92. Tomlinson, E., Jefferies, T. M., Riley, C. M. 1978. *J. Chromatogr.* 159:315–58
93. Wilson, J. B., Lam, H., Pravatmuang, P., Huisman, T. H. J. 1979. *J. Chromatogr.* 179:271–90
94. Meek, J. L., 1980. *Proc. Natl. Acad. Sci. USA* 77:1632–36
95. Hughes, G. J., Winterhalter, K. H., Lutz, H., Wilson, K. J. 1979. *FEBS Lett.* 108:92–97
96. Wilson, K. J., Hunziker, P., Hughes, G. J. 1979. *FEBS Lett.* 108:98–102
97. Konigsberg, W. 1972. See Ref. 38, pp. 326–32
98. Gray, W. R. 1972. See Ref. 38, pp. 333–44
99. Lai, C. Y. 1972. See Ref. 38, pp. 369–73
100. Tarr, G. E. 1977. See Ref. 12, pp. 335–57
101. Wittmann-Liebold, B. 1980. In *Polypeptide Hormones,* ed. R. F. Beers Jr., E. G. Bassett, pp. 87–120. New York: Raven
102. Hunkapiller, M. W., Hood, L. E. 1978. *Biochemistry* 17:2124–33
103. Hunkapiller, M. W., Hood, L. E. 1980. *Science* 207:523–25
104. Bhown, A. S., Mole, J. E., Hunter, F., Bennett, J. C. 1980. See Ref. 19, pp. 517–20
105. Bailey, G. S., Gillett, D., Hill, D. F., Petersen, G. B. 1977. *J. Biol. Chem.* 252:2218–25
106. Doolittle, R. F. 1972. See Ref. 38, pp. 231–44
107. Podell, D. N., Abraham, G. N. 1978. *Biochem. Biophys. Res. Commun.* 81 :176–85
108. Thibodeau, S. N., Palmiter, R. D., Walsh, K. A. 1978. *J. Biol. Chem.* 253: 9018–23
109. Palmiter, R. D. 1977. *J. Biol. Chem.* 252:8781–83
110. Katayama, K., Ericsson, L. H., Enfield, D. L., Walsh, K. A., Neurath, H., Davie, E. W., Titani, K. 1979. *Proc. Natl. Acad. Sci. USA* 76:4990–94
111. Mort, J. A., Lamport, D. T. A. 1977. *Anal. Biochem.* 82:289–309
112. Edman, P., Henschen, A. 1975. See Ref. 10, pp. 232–79
113. Frank, G. 1979. *Hoppe-Seylers Z. Physiol. Chem.* 360:997–99
114. Frank, G. Sidler, W., Widmer, H., Zuber, H. 1978. *Hoppe-Seylers Z. Physiol. Chem.* 359:1491–1507
115. Begg, G. S. 1980. See Ref. 19, pp. 153–56
116. Crewther, W. G., Inglis, A. S. 1975. *Anal. Biochem.* 68:572–85
117. Brauer, A. W., Margolies, M. N., Haber, E. 1975. *Biochemistry* 14:3029–35
118. Tarr, G. E., Beecher, J. F., Bell, M., McKean, D. J. 1978. *Anal. Biochem.* 84:622–27
119. Klapper, D. G., Wilde, C. E. III, Capra, J. D. 1978. *Anal. Biochem.* 85:126–31
120. Silver, J., Hood, L. E. 1974. *Anal. Biochem.* 60:285–92
121. Hermodson, M. A., Schmer, G., Kurachi, K. 1977. *J. Biol. Chem.* 252:6276–79
122. Thomsen, J., Bucher, D., Brunfeldt, K., Nexo, E., Olesen, H. 1976. *Eur. J. Biochem.* 69:87–96
123. Begg, G. S. 1980. See Ref. 19, pp. 485–90
124. McCumber, L. J., Qadeer, M., Capra, J. D. 1980. See Ref. 19, pp. 165–72
125. Bhown, A. S., Cornelius, T. W., Mole, J. E., Lynn, J. D., Tidwell, W. A., Bennett, J. C. 1980. *Anal. Biochem.* 102:35–38
126. Wittmann-Liebold, B. 1973. *Hoppe-Seylers Z. Physiol. Chem.* 354:1415–31
126a. Kumar, A. A., Blankenship, D. T., Kaufman, B. T., Freisheim, J. H. 1980. *Biochemistry* 19:667–78
127. Henschen-Edman, A., Lottspeich, F. 1980. See Ref. 19, pp. 105–14
128. Wittmann-Liebold, B., Graffunder, H., Kohls, H. 1976. *Anal. Biochem.* 75:621–33
129. Bridgen, P. J., Cross, G. A. M., Bridgen, J. 1976. *Nature* 263:613–14

130. Bhown, A. S., Mole, J. E., Weissinger, A., Bennett, J. C. 1978. *J. Chromatogr.* 148:532–35

130a. Rose, S. M., Schwartz, B. D. 1980. *Anal. Biochem.* 107:206–13

130b. Tarr, G. E. 1981. *Anal. Biochem.* In press

131. Downing, M. R., Mann, K. G. 1976. *Anal. Biochem.* 74:298–319

132. Zimmerman, C. L., Appella, E., Pisano, J. J. 1977. *Anal. Biochem.* 77:569–73

133. Ericsson, L. H., Wade, R. D., Gagnon, J., McDonald, R. M., Granberg, R., Walsh, K. A. 1977. In *Solid Phase Methods in Protein Sequence Analysis,* ed. A Previero, M. A. Coletti-Previero, pp. 137–42. Amsterdam: Elsevier. 298 pp.

134. Margolies, M. N., Brauer, A. 1978. *J. Chromatogr.* 148:429–39

135. Moser, P. W., Rickli, E. E. 1979. *J. Chromatogr.* 176:451–55

136. Van Beeumen, J., Van Damme, J., Tempst, P., De Ley, J. 1980. See Ref. 19, pp. 503–6

137. Johnson, N. D., Hunkapiller, M. W., Hood, L. E. 1979. *Anal. Biochem.* 100:335–38

138. Henderson, L. E., Copeland, T. D., Oroszlan, S. 1980. *Anal. Biochem.* 102:1–7

139. Fales, H. M., Nagai, Y., Milne, W. A., Brewer, H. B. Jr., Bronzert, T. J., Pisano, J. J. 1971. *Anal. Biochem.* 43:288–99

140. Nasimov, I. V., Levina, N. B., Shemyakin, V. V., Rosynov, B. V., Bogdanova, I. A., Merimson, V. G. 1980. See Ref. 19, pp. 475–83

141. Appella, E., Inman, J. K., Dubois, G. C. 1977. See Ref. 133. pp. 121–33

142. Fietzek, P. P., Allmann, H., Rauterberg, J., Henkel, W., Wachter, E., Kühn, K. 1979. *Hoppe-Seylers Z. Physiol. Chem.* 360:809–20

143. Knight, E. Jr., Hunkapiller, M. W., Korant, B. D., Hardy, R. W. F., Hood, L. E. 1980. *Science* 207:525–26

144. Zoon, K. C., Smith, M. E., Bridgen, P. J., Anfinsen, C. B., Hunkapiller, M. W., Hood, L. E. 1980. *Science* 207:527–28

145. Laursen, R. A. 1969. *Biochem. Biophys. Res. Commun.* 37:663–67

146. Jacobs, J. W., Niall, H. D. 1975. *J. Biol. Chem.* 250:3629–36

147. Silver, J. 1977. See Ref. 12, pp. 247–60

148. Bridgen, J. 1977. See. Ref. 12, pp. 321–35

149. Maeda, H., Ishida, N., Kawauchi, H., Tuzimura, K. 1969. *J. Biochem.* 65:777–83

150. Muramoto, K., Kawauchi, H., Tuzimura, K. 1978. *Agric. Biol. Chem.* 42:1559–63

151. Chang, J. Y., Brauer, D., Wittmann-Liebold, B. 1978. *FEBS Lett.* 93:205–14

152. Chang, J. Y. 1980. See. Ref. 19, pp. 115–22

153. Wilson, K. J., Rodger, K., Hughes, G. J. 1979. *FEBS Lett.* 108:87–91

154. Chang, J. Y. 1979. *Biochim. Biophys. Acta* 578:175–87

155. Jacobs, J. W., Kemper, B., Niall, H. D., Habener, J. F., Potts, J. T. Jr. 1974. *Nature* 249:155–57

156. Silver, J., Hood, L. 1975. *Nature* 256:63–64

157. Ballou, B., McKean, D. J., Freedlender, E. F., Smithies, O. 1976. *Proc. Natl. Acad. Sci. USA* 73:4487–91

158. Walker, J. E., Shaw, D. C., Northrop, F. D., Horsnell, T. 1977. See Ref. 133, pp. 277–85

159. Schwyzer, M., Weil, R., Frank, G., Zuber, H. 1980. *J. Biol. Chem.* 255:5627–34

160. Devillers-Thiery, A., Kindt, T., Scheele, G., Blobel, G. 1975. *Proc. Natl. Acad. Sci. USA* 72:5016–20

161. Kemper, B., Habener, J. F., Ernst, M. D., Potts, J. T. Jr., Rich, A. 1976. *Biochemistry* 15:15–19

162. Burstein, Y., Schechter, I. 1976. *Biochem. J.* 157:145–51

163. Strauss, A. W., Bennett, C. D., Donohue, A. M., Rodkey, J. A., Alberts, A. W. 1977. *J. Biol. Chem.* 252:6846–55

164. Palmiter, R. D., Gagnon, J., Ericsson, L. H., Walsh, K. A. 1977. *J. Biol. Chem.* 252:6386–93

165. McKean, D. J., Maurer, R. A. 1978. *Biochemistry* 17:5215–19

166. Atger, M., Mercier, J. C., Haze, G., Fridlansky, F., Milgrom, E. 1979. *Biochem. J.* 177:985–88

167. Gagnon, J., Palmiter, R. D., Walsh, K. A. 1978. *J. Biol. Chem.* 253:7464–68

168. Segre, G. V., Niall, H. D., Sauer, R. T., Potts, J. T. Jr. 1977. *Biochemistry* 16:2417–27

169. Link, T. P., Stark, G. R. 1967. *J. Biol. Chem.* 243:1082–88

170. Wessels, B. W., McKean, D. J., Lien, N. C., Shinnick, C., DeLuca, P. M., Smithies, O. 1978. *Radiat. Res.* 74:35–50

171. Doolittle, L. R., Mross, G. A., Fothergill, L. A., Doolittle, R. F. 1977. *Anal. Biochem.* 78:491–505

172. Cavadore, J. C. 1978. *Anal. Biochem.* 91:236–40

173. Rangarajan, M., Darbre, A. 1976. *Biochem. J.* 157:307–16

174. Parham, M. E., Loudon, G. M. 1978. *Biochem. Biophys. Res. Commun.* 80: 1–6
175. Parham, M. E., Loudon, G. M. 1978. *Biochem. Biophys. Res. Commun.* 80: 7–13
176. Kannan, K. K. 1977. See Ref. 11, pp. 75–122
177. Titani, K., Koide, A., Ericsson, L. H., Kumar, S., Hermann, J., Wade, R. D., Walsh, K. A., Neurath, H., Fischer, E. H., Sprang, S., Fletterick, R. J., Jenkins, J. A., Johnson, L. N., Wilson, K. S. 1978. *Biochemistry* 17:5680–95
178. Bernstein, F. C., Koetzle, T. F., Williams, G. J. B., Meyer, E. F. Jr., Brice, M. D., Rodgers, J. R., Kennard, O., Shimanouchi, T., Tasumi, M. 1977. *J. Mol. Biol.* 112:535–42
179. Morris, H. R. 1980. *Nature* 286:447–52
180. Morris, H. R., Dell, A., Petersen, T. E., Sottrup-Jensen, L., Magnusson, S. 1976. *Biochem. J.* 153:663–79
181. Morris, H. R. 1979. *Phil. Trans. R. Soc. London Ser. A* 293:39–51
182. Hunt, D. F., Sisak, M., Buko, A., Ballard, J., Giordani, A., Shabanowitz, J. 1980. *Am. Soc. Mass Spectrom. Ann. Conf. Mass Spectrom. & Allied Topics, 28th, New York*, pp. 390–91
182a. Hauschka, P. V., Henson, E. B., Gallop, P. M. 1980. *Anal. Biochem.* 108: 57–63
183. DeLange, R. J., Fambrough, D. M., Smith, E. L., Bonner, J. 1968. *J. Biol. Chem.* 243:5906–13
184. Enfield, D. L., Ericsson, L. H., Fujikawa, K., Walsh, K. A., Neurath, H., Titani, K. 1980. *Biochemistry* 19:659–67
185. Proud, C. G., Rylatt, D. B., Yeaman, S. J., Cohen, P. 1977. *FEBS Lett.* 80: 435–42
186. Hermodson, M. A., Chen, K. C. S., Buchanan, T. M. 1978. *Biochemistry* 17: 442–45
187. Maloy, W. L., Bowien, B. U., Zwolinski, G. K., Kumar, K. G., Wood, H. G.,
Ericsson, L. H., Walsh, K. A. 1979. *J. Biol. Chem.* 254:11615–22
188. Porter, R. R. 1959. *Biochem. J.* 73:119–26
189. Richards, F. M., Vithayathil, P. J. 1959. *J. Biol. Chem.* 234:1459–65
190. Titani, K., Ericsson, L. H., Walsh, K. A., Neurath, H. 1975. *Proc. Natl. Acad. Sci. USA* 72:166–70
191. Levy, P. L., Pangburn, M. K., Burstein, Y., Ericsson, L. H., Neurath, H., Walsh, K. A. 1975. *Proc. Natl. Acad. Sci. USA* 72:4341–45
192. Takio, K., Walsh, K. A., Neurath, H., Smith, S. B., Krebs, E. G., Titani, K. 1980. *FEBS Lett.* 114:83–88
193. Bloxham, D. P., Ericsson, L. H., Titani, K., Walsh, K. A., Neurath, H. 1980. *Biochemistry* 19:3979–85
194. Boosman, A. 1978. *J. Biol. Chem.* 253:7981–84
195. Schechter, Y., Burstein, Y., Patchornik, A. 1975. *Biochemistry* 14:4497–4503
196. Gray, W. R. 1968. *Nature* 220:1300–4
197. Elgin, S. C. R., Schilling, J., Hood, L. E. 1979. *Biochemistry* 18:5679–85
198. Hartley, B. S. 1970. *Biochem. J.* 119:805–22
199. Tang, J., Hartley, B. S. 1967. *Biochem. J.* 102:593–99
200. Link, T. P., Stark, G. R. 1968. *J. Biol. Chem.* 243:1082–88
201. Cozzone, P., Marchis-Mouren, G. 1972. *Biochim. Biophys. Acta* 257: 222–29
202. Hermodson, M. A., Kuhn, R. W., Walsh, K. A., Neurath, H., Eriksen, N., Benditt, E. P. 1972. *Biochemistry* 11:2934–38
203. Begg, G. S., Pepper, D. C., Chesterman, C. N., Morgan, J. 1978. *Biochemistry* 17:1739–44
204. Boosman, A. 1980. See Ref. 19, pp. 513–16
205. Fowler, A. V., Zabin, I. 1978. *J. Biol. Chem.* 253:5521–25
206. Brake, A. J., Celada, F., Fowler, A. V., Zabin, I. 1977. *Anal. Biochem.* 80: 108–15

Ann. Rev. Biochem. 1981. 50:285–315

STRUCTURE AND MECHANISM OF MULTIFUNCTIONAL RESTRICTION ENDONUCLEASES[1]

◆12080

Robert Yuan

Cancer Biology Program, Frederick Cancer Research Center, Frederick, Maryland 21701

CONTENTS

PERSPECTIVES AND SUMMARY

Restriction endonucleases are strain-specific enzymes, which enable bacteria to recognize and rapidly destroy foreign DNA and which cause double-stranded scissions at a limited number of sites on the DNA. In addition to having a restriction activity, each of these bacterial strains possesses a specific DNA methylase that transfers methyl groups from S-adenosylmethionine (AdoMet) to specific adenine or cytosine residues in the DNA. In this way, the DNA of the cell is protected against its own restriction enzyme.

[1]The US Government has the right to retain a nonexclusive royalty-free license in and to any copyright covering this paper.

285

The restriction enzymes were originally classified into two types, based on the complexity of their structure and their cofactor requirements (1). As our knowledge has increased, we have established additional criteria to differentiate among various restriction and modification systems and, as a result, a third class of restriction enzymes has been identified. The properties of these three types of restriction enzymes are listed in Table 1.

The Type I enzymes, such as those from *Escherichia coli* B and *E. coli* K, and the Type III enzymes, such as those coded by the phage P1 and the plasmid P15, are multifunctional proteins able to both cleave and methylate unmodified DNA. They are composed of three (Type I) or two different subunits (Type III) and require ATP and Mg^{2+} for endonuclease activity. Although AdoMet is required for restriction by Type I enzymes, it only stimulates the reaction by Type III enzymes. Type II enzymes, which are best represented by *Eco*RI,[2] appear to have simpler protein structures (e.g.

Table 1 Characteristics of restriction endonucleases

	Type I	Type II	Type III
1. Restriction and modification activities	Single multifunctional enzyme	Separate endo-nuclease and methylase	Single multifunctional enzyme
2. Protein structure	3 different subunits	Simple	2 different subunits
3. Requirements for restriction	AdoMet, ATP, Mg^{2+}	Mg^{2+}	ATP, Mg^{2+} (AdoMet)
4. Binding of enzyme-DNA complexes to filters	Yes	No	No
5. Sequence of host specificity sites	sB: T–G–A–N8–T–G–C–T sK: A–A–C–N6–G–T–G–C	Twofold symmetry	sP1: A–G–A–C–C s15: C–A–G–C–A–G
6. Cleavage sites	Possibly random, at least 1,000 bp from host specificity site	Generally, at host specificity site	24–26 bp to 3′ of host specificity site
7. Enzymatic turnover	No	Yes	Yes
8. DNA translocation	Yes	No	No
9. Requirements for methylation	AdoMet (ATP, Mg^{2+})	AdoMet	AdoMet (ATP, Mg^{2+})
10. Restriction vs methylation	Mutually exclusive	Separate reactions	Simultaneous
11. Site of methylation	Host specificity site	Host specificity site	Host specificity site

[2]The nomenclature in this paper follows the rules proposed by Smith & Nathans (2). *Eco*K refers to the restriction enzyme from *E. coli* K; *Eco*P15 is the enzyme coded by the plasmid P15. λ.O is phage replicated on a nonmodifying host, λ.K is replicated on *E. coli* K, and λ.O:K and λ.K:O are heteroduplex phages. The host specificity sites are named *s* followed by the name of the restriction and modification system, e.g. sB is the recognition site for *Eco* B. sB⁰ refers to a mutant site that is not a substrate for either restriction or modification.

*Eco*RI is a dimer) and only require Mg^{2+} for activity. DNA modification is catalyzed by a separate methylase activity. Under restriction conditions, only Type I enzymes form complexes with unmodified DNA in a form that is retained on nitrocellulose filters. In addition, no enzymatic turnover has been observed in the Type I restriction reaction, which is in sharp contrast to the other two systems. The susceptibility of a particular DNA to a restriction and modification system depends on the presence of certain nucleotide sequences (host specificity sites). In general, the sequences recognized by Type II enzymes consist of 4 to 6 bases and have twofold symmetry. However, sequence information strongly suggests that Type II systems may encompass more than one class of enzymes. Several enzymes are able to recognize a specific sequence but cleave at a site 5 to 9 bases away. In another instance, the sequence being recognized has a hyphenated structure. Both endonucleolytic cleavage and methylation take place within these sequences. In contrast, the Type I and III sites are asymmetric. In Type III, the sequence consists of 5 to 6 bases; in Type I, the sequences are hyphenated with two constant domains of 3 and 4 bases separated by a nonspecified spacer of 6 or 8 bases. Although recognition and methylation take place at these sites, DNA cleavage takes place elsewhere—24 to 27 (Type III) or several thousand (Type I) bases away. Methylation by the Type I and III systems is stimulated by ATP and Mg^{2+}, whereas no such effect is observed in Type II systems. In Type I systems, the protein can catalyze either restriction or modification, but the two activities are mutually exclusive; on the other hand, Type III enzymes are able to express both activities simultaneously.

The use of the Type II enzymes has resulted in revolutionary advances in biological research. In particular, the construction of recombinant genomes has attracted considerable publicity and widespread debate. Although it has not been as obvious, these systems also provide an excellent subject for the study of protein-DNA interactions. The detailed study of Type II systems has focused primarily on the *Eco*RI enzyme and an excellent review has appeared (3). In this article, I confine my discussion to the mechanism studies on the multifunctional restriction endonucleases (i.e. Types I and III). New developments in this area have resulted from experiments on the genetic structure of the restriction and modification locus and its transcription patterns, the nucleotide sequence of the host specificity sites, and the multiple steps involved in the interaction between the enzymes and the DNA. Certain basic questions have evolved: What conditions determine the activity that will be expressed by a multifunctional protein? What is the role of ATP and AdoMet in these reactions? How does an enzyme recognize a specific DNA sequence and then act at a site? In some instances, the answers have been unexpected. The modulation of enzymatic activity can be caused by the relative concentrations of ATP and AdoMet (Type III) or by the interaction of the protein with the host specificity sites (Type I). In the latter case, the site acts as an allosteric effector of the enzyme activity. ATP can act as an allosteric effector, enabling the enzyme to cut a short distance away (Type III). Alternatively, ATP hydrolysis is coupled to a novel mechanism of DNA translocation that brings the enzyme into contact with potential cleavage sites far removed from the recognition sites (Type I).

An understanding of the reaction mechanisms of these interesting enzymes is not only important in the area of restriction and modification but is also increasingly relevant to other important biological reactions such as genetic recombination, DNA splicing, and transposition of genetic elements.

TYPE I RESTRICTION ENDONUCLEASES

Genetics

The most extensively studied Type I restriction and modification systems are those of *E. coli* B and *E. coli* K. These two systems are allelic and are closely related to the A system of *E. coli* 15 (4) and to that of some strains of *Salmonella* (5, 6). A classical series of genetic experiments indicate that three genes are involved in restriction and modification: a restriction gene (*hsd*R), a modification gene (*hsd*M), and a gene for recognition of the host specificity site (*hsd*S). A *hsd*S mutation leads to the loss of site recognition and results in a restriction- and modification-deficient phenotype (r^-m^-). This three-gene model is supported by the experiments described below.

Mutants in restriction arise spontaneously or can be generated by mutagenesis. The r^- mutants isolated are divided into two classes: half with normal modification (r^-m^+) and the other half modification-deficient (r^-m^-) (7, 8). It is not likely that the high proportion of r^-m^- mutants can be attributed to true double mutants, so it was proposed that they arise from a mutation in a third gene essential for both restriction and modification. The r^+m^- phenotype is presumably lethal, so it is not surprising that none of these phenotypes was found in these early studies.

The r^-m^+ and r^-m^- mutants are characterized functionally by genetic complementation with F' merodiploids (9, 10). The r^-m^+ mutants complement r^-m^- strains in diploids, but independent r^-m^+ mutants do not complement each other. The same lack of complementation was observed with independent r^-m^- mutants. Experiments with mutants from two different strains, *E. coli* K and *E. coli* B, defined the genetic determinant for host specificity. The $r_K^-m_K^+/r_B^-m_B^-$ diploid had a $r_K^+m_K^+$ phenotype, whereas the $r_B^-m_B^+/r_K^-m_K^-$ diploid had a $r_B^+m_B^+$ phenotype. This was taken as evidence for a gene, hsdS, that codes for a polypeptide that recognizes the host specificity site. A mutation in the hsdS gene results in the r^-m^- phenotype; mutation in the hsdR gene leads to loss of restriction but not modification. The hsdR gene of one strain complements the hsdS gene of a different strain, but the host specificity is always determined by the functional hsdS gene.

The characterization of the hsdM gene proved to be more difficult because r^+m^- mutants are presumably lethal. This problem was solved by mutagenizing a r^-m^+ mutant and isolating a second step r^-m^- mutant. Because a two-step $r_B^- m_B^-$ mutant complements the wild-type $r_K^+ m_K^+$ and results in a $r_{KB}^+ m_{KB}^+$ phenotype (with both K and B specificities expressed), the second mutation must be in the hsdM gene. In addition, studies of two temperature-sensitive mutants with a $r_K^- r_K^{ts}$ phenotype showed that hsdM is required for normal restriction (11). The $r_K^- m_K^{ts}$ strain complements a $r_B^- m_B^+$ strain to yield a $r_{KB}^+ m_{KB}^+$ phenotype at both low and high temperatures. This result is consistent with the temperature-sensitive mutation being in hsdM, with both hsdR and hsdS genes normal. An unusual class of r^-m^- mutants was also isolated, which is *trans* dominant in diploids with a wild-type strain (9). This *trans* effect has been attributed either to a possible fourth gene or, more likely to negative complementation.

The role of the hsdS gene in determining host specificity suggests that mutation or recombination within this gene can result in a new host specificity. No such mutants or recombinants had been isolated until recombination experiments were carried out between *Salmonella typhimurium* and *S. postdam* (12). A new host specificity system (called SQ) was obtained that is different from the SB system of *S. typhimurium* and the SP system of *S. postdam*. All three systems are presumed to be Type I inasmuch as they are related to those from *E. coli* as shown by functional complementation.

The hsd locus has been mapped close to serB in the *E. coli* map (13). On the basis of experiments with *E. coli* B, it was proposed originally that the gene order is hsdM, hadS, hsdR, serB, thr (14). More recently, the hsd locus of *E. coli* K was cloned in λ, and complementation tests were used to show that the gene order is hsdR, hsdM, hsdS (15). Although transcription of all three genes is in the same direction, two different promoters have been identified. One promoter is responsible for transcription of hsdM and hsdS and the other for transcription of hsdR. This transcriptional pattern enables the cell to selectively synthesize a methylase composed of the hsdM and hsdS gene products without the concomitant production of the restriction endonuclease. Finally, probes carrying most of the hsdR gene and the complete hsdM and hsdS genes of *E. coli* K have been shown to have extensive homology with *E. coli* B DNA but no detectable homology with the DNA of *E. coli* C (a strain that naturally lacks restriction and modification).

As is apparent in the following sections, the genetic studies provided the theoretical framework for research into the structure and function of these restriction enzymes and the nucleotide sequences they recognize.

Enzyme Structure

Two enzyme structures were proposed for EcoK and EcoB on the basis of the genetic studies. It was postulated that one of these structures has a restriction endonuclease with three different subunits (the hsdR, hsdS, and hsdM gene products) and a modification methylase with two different subunits (the hsdM and hsdS gene products). Alternatively, a protein with all three subunits might catalyze both restriction and modification. It is evident that these genetic models agree remarkably well with the results from the purified enzymes.

EcoK has three enzymatic activities: a restriction endonuclease that cleaves unmodified DNA in the presence of ATP, AdoMet, and Mg^{2+} (16), a modification methylase that requires AdoMet (17), and an ATP hydrolase that is coupled to restriction but is not stoichiometric to the number of DNA cleavage events (18). Gel filtration and sedimentation on glycerol gradients showed that the enzyme has a molecular weight of approximately 400,000 (19). Three different subunits with molecular weights of 135,000, 62,000, and 55,000 (in molar ratios of 2:2:1) were identified by gel electrophoresis in the presence of sodium dodecyl sulfate. When UV-irradiated cells were infected with λ carrying the cloned hsd genes, the hsdR gene coded for the 135,00 polypeptide, the hsdM gene coded for the 62,000 polypeptide and the hsdS gene coded for the 55,000 polypeptide (15).

EcoB has the same enzymatic activities as EcoK (20–22). Purified EcoB is heterogeneous (with S values of 11 to 18), and two different enzyme species have been observed under nondenaturing conditions (23). Both of these enzymes have three subunits with molecular weights of 135,000, 60,000, and 55,000, but they differ in having molar ratios of 1:2:1 and 1:1:1, respectively. A separate modification methylase with a molecular weight of 120,000 has been purified from E. coli B and shown to have two subunits with molecular weights of 60,000 and 55,000 in molar ratios of 1:1 (24). No such methylase has been reported for E. coli K. On the basis of the enzymology, it is not possible to conclude whether the endonuclease or the methylase is responsible for modification in vivo. It is possible that under ordinary circumstances both reactions can be catalyzed by the endonuclease, but the existence of two separate promoters for the hsd locus should allow the methylase to be synthesized by itself under certain physiological conditions.

The identification of the gene products does not reveal which biochemical reaction is catalyzed by each subunit. The direct approach to this problem involves the purification and characterization of each subunit. Unfortunately, we have been unable to use this approach because of inactivation of the purified subunits. An indirect and more laborious approach requires the characterization of mutant enzymes. It had been observed that complementation between $r_B^- m_B^+$ and $r_B^- m_B^-$ strains could be detected in crude extracts (25). Using an in vitro complementation assay, we have purified three mutant enzymes. EcoK-18 was isolated from a $r_K^- m_K^+$ ($hsdR_K^-$) strain, EcoK-19 from a $r_K^- m_K^-$ ($hsdS_K^-$) strain and EcoK-46 from a $r_K^- m_K^-$ ($hsdM_K^-$) strain (26, 27). EcoK-18 lacks endonuclease activity, is unable to bind unmodified DNA to filters, and has a low level of DNA-independent ATPase activity that required AdoMet. It does, however, have normal methylase activity. EcoK-19 lacks the endonuclease, DNA-binding, and methylase activities but has an AdoMet-dependent ATPase. None of these four activities has been detected with EcoK-46. However, EcoK-18 complements EcoK-19 or EcoK-46, resulting in wild-type activitiy as measured by DNA cleavage, DNA binding to filters, restriction-dependent ATPase, and methylation.

The nature of each of the mutations was studied by following the interaction of these enzymes with ATP, AdoMet, and DNA. EcoK-18 and EcoK-19 form stable complexes with AdoMet just like wild-type EcoK (26), whereas EcoK-46 does not bind AdoMet (27). The isolated Eco K-18-AdoMet complex is complemented by EcoK-19 without any further addition of AdoMet (26). However, the reverse experiment with the isolated EcoK-19-AdoMet complex and EcoK-18 shows no complementation in the absence of AdoMet. These results indicate that EcoK-19 can bind AdoMet but is unable to recognize the host specificity sites. EcoK-18 does bind AdoMet and forms stable complexes with the DNA; however, the EcoK-18-DNA complex differs from that formed with the wild-type enzyme. It is sensitive to heparin unless ATP is added (the wild-type complex is heparin-insensitive) and the kinetics of formation of the filter-binding complex following ATP (and EcoK-46) addition is slower than that of the wild-type enzyme.

Several conclusions can be drawn from these experiments. A mutation in the hsdM subunit results in the loss of AdoMet binding. A mutation in the hsdS subunit allows normal AdoMet binding but the transition to the activated form of the enzyme is blocked and DNA sequence recognition is lost. A mutation in the hsdR subunit allows normal recognition and methylation, but the ATP-induced conformational change is lost. These results agree with the predictions from the genetic experiments and support the reaction mechanism described in a later section. Also, the ease of complementation and the inability to isolate the wild-type activity from a mixture of mutant proteins supports sequential catalysis of different steps in a complex reaction pathway.

The Nature of DNA Recognition and Cleavage Sites

The presence of host specificity sites on a given DNA molecule makes that DNA a substrate for a given restriction and modification system. The existence of such sites was first suggested by the isolation of phage fd mutants that were no longer restricted by *E. coli* B (28). It is also possible to generate host specificity sites by mutation (29). Modification-related methylation requires sB sites both in vivo (30) and in vitro (31). These sites are also required for specific binding of the enzyme to DNA (32). Several methods have been used to map these sites. Genetic procedures have served to localize the sB sites on f1 (33, 34) and the sK sites on the DNA of λ (35), øX174 (29), and G4 (36). DNA methylation with radioactive AdoMet, followed by restriction analysis (36) and electron microscopy of enzyme-DNA complexes, (32) has yielded more precise mapping data.

Although the Type I enzymes recognize and methylate at the genetically mapped host specificity sites, DNA cleavage takes place at other sites that are presumably random in nature (37). Therefore, in sharp contrast to Type II enzymes, the Type I proteins have to interact with DNA at two different sites: recognition sites (i.e. host specificity sites) and cleavage sites. The routine method for sequencing restriction sites involves labeling the 5'-end of cleaved DNA with ^{32}P-γ-ATP and polynucleotide kinase. This method cannot be used with *Eco*K and *Eco*B, because cleavage does not take place at the recognition sites. A different approach is required using small genomes that have been sequenced fully and for which a collection of recognition site mutants is available. The sequences in the vicinity of the recognition sites were then compared. The sequences of the wild-type sB1 site in f1 and nine independent sB1[0] mutants were determined, along with those of the sB2 site and four sB2[0] mutants (38). The same was done with the sB site of øXam3cs70 (a øX174 strain that is not restricted by *E. coli* B) and three sB[0] mutants of øXsB1 plus the unique sB site on SV40 (39). The wild-type sB sequence was shown to be:

5' T G A N$_8$ T G C T 3'

a hyphenated structure with two constant domains of 3 and 4 bases separated by an 8-base spacer of variable sequence. Base changes in the T and G of the trinucleotide and in the G, C, and T of the tetranucleotide result in a sB[0] phenotype. Computer analysis of a øX174 strain that is not restricted by *E. coli* B identified sequences that differ from that of sB at each of the seven constant positions. Computer analysis also revealed sequences in which the constant domains are the same as for the sB site, but the spacers vary from 5 to 15 bases. Therefore, the sB site is defined by two constant domains of T-G-A and T-G-C-T separated by a spacer of 8 bases with a variable sequence.

*Eco*B methylation results in the incorporation of two methyl groups per sB site, and the product of the reaction is N^6-methyladenine (30, 31). Methyl groups can be transferred to either DNA strand as demonstrated by the fact that *Eco*B methylates both forms of heteroduplex DNA (one strand modified and one unmodified) in vitro (40). Labeled oligonucleotides have been sequenced following in vitro methylation of fd DNA (41) and SV40 DNA (42) with ^{3}H-AdoMet and *Eco*B. These results have been interpreted to mean that the adenine in the T-G-A sequence and the first adenine in the A-G-C-A sequence that is complementary to the second domain are the methylated bases (38). Corroborating evidence was obtained by taking the methylated sequence and separating the two constant domains of the sB1 site in f1 DNA with *Hpa*II. Both restriction fragments were methylated.

A similar approach has been used for determining the sequences of the three sK sites on øXsK1, øXsK2, and G4 and the two sK sites on pBR322 (36). On the basis of sequence comparisons, it has been proposed that the sK sequence is

$$5' \text{ A A C } N_6 \text{ G T G C } 3',$$

a hyphenated structure with two constant domains of 3 and 4 bases separated by a 6-base spacer of variable sequence. A change in the C of the A-A-C sequence results in a sK^0 phenotype, whereas a sK^+ phenotype results from a change to T in the G-T-G-C sequence. Sequences that differ from the sK sequence at each of the seven constant positions were found by computer analysis of the four genomes but they do not appear to be functional sites. Because a large number of sK^0 mutants are lacking, the evidence for the sK sequence is considerably weaker than that for the sB site. It is known that the product of *Eco*K methylation is N^6-methyladenine (17) but no information is available on the methylated sequence.

The structures of the sK and sB sites are very similar. Both structures have a hyphenated sequence with two constant domains of 3 and 4 bases separated by a spacer of unspecified sequence with 8 bases for sB sites and 6 bases for sK sites. No twofold symmetry is present in either case. One direct result of the spacer is that both constant domains are on the same side of the DNA helix (38). Given these facts, it is likely that the enzyme interacts with one side of the DNA in an asymmetric fashion. The bases methylated by *Eco*K can be tentatively identified by lining up the sK and sB sequences in which 4 of the 7 conserved bases are the same:

```
sB 5'  T-G- | A* | -N-N-N-N-N-N-N-N- | T-G-C | -T 3'
       A-C- | T  | -N-N-N-N-N-N-N-N- | A*-C-G | -A


sK 5'     A- | A | -C-N-N-N-N-N-G- | T-G-C |    3'
          T- | T | -G-N-N-N-N-N-C- | A-C-G |
```

Assuming that the same positions are always methylated, it seems plausible that the second A in the A-A-C sequence and the A in the C-A-C-G sequence are the sites of *Eco*K methylation.

The characterization of the cleavage sites for *Eco*K and *Eco*B has turned out to be a singularly intractable problem. None of the cleavage sites can be labeled with polynucleotide kinase (23, 43) even though experiments with λ exonuclease indicate that the *Eco*B sites are double-stranded and have a 5' phosphate. A different aspect of the problem revolves around the presumed randomness of the *Eco*K and *Eco*B cleavage sites. The first indication that *E. coli* K restriction might not be random came from in vivo experiments with λ*trp* (44). A λ*trp* phage with one of three different sK sites was used to infect a *rec*BC⁻*trp*⁻ strain of *E. coli* K12, and both restriction of the infecting phage and expression of the *trp* genes were measured. Restriction of the phage is very efficient in all three cases but expression of the *trp* operon is unaffected, regardless of whether the sK site is far from, close to, or actually in the *trp* operon. Therefore, the region of the *trp* operon is relatively insensitive to restriction by *E. coli* K.

Five sK sites have been mapped on λ DNA (35). *Eco*K digestion of unmodified λ DNA resulted in a heterogeneous product consisting of a population of fragments approximately 25% and 35% of λ length and a variety of smaller fragments. The *Eco*K cleavage regions in λ were further defined by treating the *Eco*K digests with various Type II restriction endonucleases (*Eco*RI, *Hind*III, *Bam*HI, *Bgl*II, *Pvu*I, and *Pst*I) (R. Musso and R. Yuan, unpublished data). Each reaction was then analyzed for the selective disappearance of the Type II restriction fragments.

Three *Eco*K cleavage regions were identified:

(*a*) a region of approximately 3 kb about 25 map units from the left end of λ;
(*b*) a region of approximately 4 kb in the region from 45 to 54 map units;
(*c*) a region of approximately 3 kb in the region from 81 to 90 map units.

The region from 91 to 100 map units was cleaved at low efficiency, whereas two-thirds of the λ genome seemed to be relatively insensitive to EcoK restriction. Our conclusion is that, although EcoK cleavage is not sequence-specific, neither is it totally random. The localized recognition that we have observed during restriction is probably a reflection of the EcoK reaction mechanism and is discussed in the next section.

The Mechanism of the Restriction Reaction

The reaction mechanism of the Type I enzymes has some important implications for the study of protein–nucleic acid interactions. Three important problems are posed by these systems. One is the way in which the interaction between the enzyme and the recognition site determines whether the protein will express its methylation or endonuclease activity. Another is how the enzyme binds specifically to a recognition site and then cuts at sites that may be several thousand bases away. The nature of localized recognition and the role of ATP in all of these reactions constitute other important aspects of these systems.

The EcoK and EcoB experiments permit us to postulate the reaction pathway shown in Figure 1. The pathway consists of the following steps:

ADOMET BINDING   It has been shown that EcoK binds AdoMet in a noncovalent fashion with an apparent K_m of 3×10^{-7} M (45). AdoMet acts as an allosteric effector, and the Hill coefficient was determined to be 2.4, indicating that there is a minimum of three AdoMet-binding sites. The mutation in the modification subunit of EcoK (EcoK-46) resulted in the loss of AdoMet binding and, as a consequence, the inability to either restrict or modify DNA (27).

ENZYME ACTIVATION   Following AdoMet binding, EcoK undergoes a slow transition to an activated form called EcoK*. This is a first-order reaction with a half-time of 54 sec. EcoK* then proceeds to cleave DNA without added AdoMet. The mutation in the hsdS subunit of EcoK (EcoK-19) allows AdoMet binding but prevents enzyme activation (26).

FORMATION OF AN INITIAL COMPLEX EcoK* interacts with any DNA regardless of the presence or absence of sK sites. Indirect evidence for this was provided by the stabilization of EcoK* with mutant DNA. The half-life of EcoK* alone was shown to be 130 sec, versus 6 min in the presence of λ sK1⁰ sK2⁰ DNA (46). Electron microscopy proved the existence of initial complexes formed between EcoK* and restriction frag-

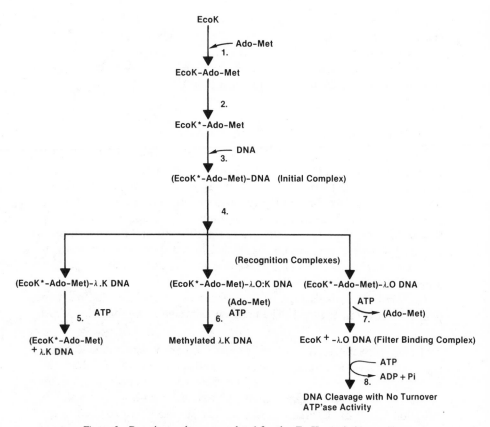

Figure 1   Reaction pathway postulated for the *Eco*K restriction enzyme.

ments lacking sK sites (32). In such complexes, the protein molecules were located at random sites.

FORMATION OF RECOGNITION COMPLEXES   Having formed initial complexes, *Eco*K* seeks and binds to sK sites. These recognition complexes are formed at both modified and unmodified sK sites (46) but can be differentiated on the basis of their stability (a half-life of 6 min for complexes with modified DNA versus 22.5 min for those with unmodified DNA). It is this enzyme-DNA interaction that determines the next step in the reaction mechanism.

FORMATION OF FILTER-BINDING COMPLEXES   The next step involves the reaction of recognition complexes with ATP. Gel filtration and electron microscopy experiments showed that $EcoK^*$ dissociated from modified DNA in the presence of ATP (47). When the sK sites were unmodified, ATP triggered a transformation of $EcoK^*$ into a new form, $EcoK^+$. $EcoK^+$ has two important characteristics: it loses the enzyme-bound AdoMet and forms a complex that is retained on filters, a property that was absent from both initial and recognition complexes. $EcoK^+$ also appears to have a smaller diameter, but it is not possible to determine whether this is because of a shape change or the loss of one or more subunits. The absence of enzymatic turnover in the endonuclease reaction is consistent with the latter possibility.

A related question is whether the ATP-induced conformational change to $EcoK^+$ is sufficient to allow DNA cleavage, or whether ATP hydrolysis is actually required. This could be tested by replacing ATP with adenosine-5'-(β, γ)-imidotriphosphate, a nonhydrolyzable analog of ATP. This analog also induces the conformational change to $EcoK^+$ but neither single- nor double-stranded DNA scission has been detected. This allows us to separate the role of ATP into two separate steps: first, as an allosteric effector that enables the enzyme to discriminate between the different forms of the sK site (modified, unmodified, or heteroduplex), and, second, as an energy source required for the actual endonucleolytic reaction.

The mutation in the hsdR subunit ($EcoK$-18) allows formation of recognition complexes and normal methylation, but no filter-binding complexes are formed upon ATP addition (27). Therefore, the hsdR lesion probably blocks the ATP-induced conformational change from $EcoK^*$ to $EcoK^+$.

DNA TRANSLOCATION   One of the most unusual features of these enzymes is the way in which they bind specifically to a recognition site and yet cleave the DNA elsewhere. The ability of the enzyme to specifically cleave unmodified DNA when incubated with a mixture of modified and unmodified DNA (16) makes it unlikely that it proceeds by a random walk mechanism in which the enzyme dissociates from the sK site and then goes on to cleave at a random site either on the same or a different molecule. The localization of $EcoK^+$ at sK sites following DNA cleavage (47) argues against the enzyme moving along the DNA in a directional walk mode. The actual mechanism appears to incorporate features from two other models. Both have $EcoK^+$ remaining at the sK site; but, in one, the enzyme makes a random second contact with another portion of the same DNA molecule, while in the other, the enzyme winds the DNA past it until it comes in contact with a cleavage region.

Our experiments have led us to propose the following mechanism of DNA translocation (48). In this model, $EcoK$ has two DNA-binding sites, one of which is specific for sK sites, whereas the second one becomes accessible only after the transition to $EcoK^+$ interacting with DNA randomly. The DNA on one side of the sK site can diffuse back in one of two possible orientations and is then bound by this second DNA-binding site (Figure 2). This second contact can take place on either side of the sK site as shown by restriction experiments using three different linear forms of a pBR322 DNA with a single sK site (with the sK site either in the middle or close to one of the two ends). All three DNAs were cleaved by EcoK. The size of such loop intermediates is probably only limited by the ability of DNA to bend back on itself. $EcoK^+$ then begins to wind the DNA past itself in a process that is coupled to ATP hydrolysis, resulting in twisted loops that have a higher helical density than that of the original pBR322 DNA. Such intermediates have been detected by electron microscopy. Whether these structures are overwound or underwound is not known. Similar structures have also been observed with relaxed covalent circles but not with various linear forms of pBR322 DNA. This has led us to speculate that circular DNA imposes certain constraints on the rate of translocation that are absent in the case of linear DNAs. Support for this view comes from experiments with longer linear DNA that resulted in the visualization of the same type of twisted loops.

Indirect evidence for DNA translocation comes from a number of independent observations. It has been shown that, although linear DNAs are poor substrates for EcoK or EcoB cleavage, small circular DNAs are readily cut (37, 49). ATP hydrolysis has been demonstrated to be a function of the size of unmodified linear DNA (50) but can also be detected in the absence of DNA cleavage (51). ATP hydrolysis presumably drives DNA translocation and, with short linear DNA, the DNA must be moved right to the end before DNA scission can take place.

Electron microscopy experiments of EcoB digests of single-site fd DNA showed regular loop structures formed by binding of the enzyme to the sB site and a cleavage site (51). This led the authors to propose a model for unidirectional "tracking" and DNA cleavage. The differences between their results and conclusions with EcoB and ours with EcoK appear to be caused by the experimental design and techniques. The argument for undirectionality is based on the fact that a HindII-treated fd DNA with the sB site located 959 bases from the closest end is not cleaved by EcoB. However, as a positive control, a substrate with the sB site located close to the opposite end was never tested. The EcoK reactions were achieved by adding ATP to recognition complexes (an intermediate) and taking samples at intervals ranging between 5 and 60 sec. The EcoB experiments involved adding enzymes to a complete reaction mixture and taking samples after 3 to 5 min, a period that allowed the DNA to be cleaved one or more times. Therefore,

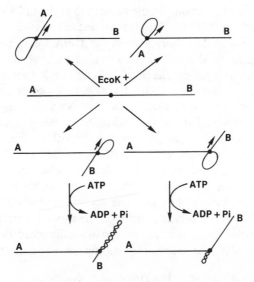

Figure 2 Model of the mechanism of DNA translocation. *Eco*K* binds at an unmodified sK site and forms a recognition complex. Addition of ATP leads to a conformational alteration of the enzyme. This makes a second DNA-binding site available. The DNA folds back in either of two possible ways to make contact with this second site and forms a loop. The arrow indicates the direction of DNA translocation. *Eco*K winds the DNA past itself while ATP is hydrolyzed. Depending on the orientation of the DNA at the second site, the loop will increase or decrease in size until cleavage occurs.

what was probably observed were the products of the *Eco*B reaction rather than intermediates. It is also likely that the sample preparation technique used for electron microscopy would not have detected the twisted loop intermediates, even under different incubation conditions.

DNA CLEAVAGE   Although endonuclease activity was the first aspect of these enzymes to be studied, it remains the least understood, primarily because of the phenomenon of localized recognition. What is known is that both *Eco*K and *Eco*B make a single-stranded break on one DNA strand followed soon afterwards by a second cut on the opposite strand very close to the original one (16, 20). That DNA cleavage is a stoichiometric event is supported by the following experiments. Neither *Eco*K nor *Eco*B turn over in the restriction reaction (23). High DNA-to-enzyme ratios lead to the accumulation of nicked circular DNA (49). These observations are consistent with a requirement for two endonuclease molecules for each double-stranded cut.

The possibility that the endonuclease action is accompanied by exonucleolytic degradation is based on the observation that *Eco*B digestion results in the production of small oligonucleotides in the ratio of 75 nucleotides per double-stranded cut (52). This was interpreted in terms of a mechanism by which the enzyme makes short gaps in the DNA following the first cut, which may result in short 3' single-stranded extensions that effectively prevent the action of polynucleotide kinase at the 5' termini.

Taking into consideration DNA translocation and the stoichiometric action of the enzyme, which carries out a two-step cleavage of the DNA, the most likely mechanism for the endonuclease step is one in which the enzyme translocates DNA at a constant rate with a given probability of cleavage at each phosphodiester bond. A slowing down of DNA translocation (possibly due to localized structural features of the DNA) increases the probability of DNA cleavage and may be responsible for the localized recognition. The enzyme then first cleaves one strand and then perhaps creates a gap by exonucleolytic digestion. The gapped structure could stop DNA translocation and a second enzyme molecule could then bind to the first one to complete the reaction by cutting the opposite strand.

The Mechanism of the Modification Reaction

The major difference between the modification and restriction reactions is the nature of the substrate DNA. Heteroduplex DNA is methylated rapidly by $EcoB$ (22) but is not susceptible to either single- or double-stranded scission by the restriction activity (16, 53). The unmodified DNA is also methylated by $EcoK$ or $EcoB$ in the absence of ATP but only following an extremely lengthy incubation (17, 22). The methylation of heteroduplex DNA requires wild-type recognition sites on both strands since heteroduplexes formed with a wild-type sequence and a mutant one are not methylated in vitro (40). The reaction only requires AdoMet but is stimulated by the presence of ATP and Mg^{2+} (22, 54).

Based on recent experiments in our laboratory, the mechanisms of the modification reaction can be divided into a number of steps, the first three of which are common also to the restriction mechanism:

ADOMET BINDING $EcoK$ binds AdoMet in a noncovalent fashion. The apparent K_m's for AdoMet in the methylation reaction are 2×10^{-7} M (with ATP) and 2.2×10^{-7} M (without ATP), values similar to those obtained for the restriction reaction. The minimum number of AdoMet-binding sites is estimated as five. The results from these experiments indicate that AdoMet behaves initially as an allosteric effector but that the enzyme-bound AdoMet is not transferred directly to the substrate DNA (54).

ENZYME ACTIVATION   Having once bound AdoMet, the enzyme undergoes the slow transition to an activated form, presumably $EcoK^*$. Attempts to measure enzyme activation directly by techniques similar to those used for restriction have been unsuccessful because AdoMet is required for both early and late steps in the methylation reaction. The AdoMet analog, Sinefungin, was used because it inhibits the EcoK methylase activity but does not block the reaction once the heteroduplex recognition complex has been formed. The enzyme activation rate is a first order reaction with a half-time of 57 sec, a value similar to that found in the restriction reaction (45).

FORMATION OF ENZYME-DNA COMPLEXES   The activated EcoK* proceeds to form initial complexes first and then recognition complexes. EcoK* was shown to bind to the heteroduplex sK sites (D. Hamilton and R. Yuan, unpublished data), and it is presumably this interaction between EcoK* and the sK sequence that determines the transition of the enzyme to its modification mode. If this is indeed the case, the heteroduplex recognition complexes would likely differ from those formed between the enzyme and modified or unmodified DNA. One major difference is the rate of methylation in the absence of ATP (the reaction was completed in 30 min with heteroduplex DNA versus 12–16 hr with unmodified DNA). The other differences become discernible upon interaction of the recognition complexes with ATP.

INTERACTION OF ATP WITH RECOGNITION COMPLEXES   Though ATP is not an absolute requirement, it does act as an allosteric effector in the methylase reaction with an apparent K_m 250 times lower than for restriction (54). The specific nature of the heteroduplex recognition complexes becomes more apparent in the presence of ATP. The addition of ATP does not lead to the formation of filter-binding complexes. The β, γ-imido analog of ATP sharply stimulates methylation but does not sustain restriction, and addition of heparin or Actinomycin D followed by ATP inhibits methylation but not restriction. Studies with isolated heteroduplex complexes have demonstrated that the enzyme does turn over in the methylation reaction. Electron microscopy has shown that ATP does not induce the conformational change to EcoK+, nor does it lead to the formation of twisted loop intermediates, but ATP does release EcoK* from the DNA during a time interval commensurate with the methylation reaction.

The conclusion drawn from these studies is that the early steps of the modification reaction (AdoMet binding, enzyme activation, and formation of recognition complexes) are the same as for restriction. The enzyme probably has at least two binding sites for AdoMet: an effector site and a methyl transfer site. It is postulated that this second site becomes accessible only following interaction with a heteroduplex site and catalyzes transfer of the methyl group to the unmodified strand. Alternatively, ATP or its analog may induce a conformational change to a form that methylates at a faster rate. In neither case is DNA translocation required, and the enzyme is released from the DNA in an active form once the methylation is completed.

TYPE III RESTRICTION ENDONUCLEASES

Genetics

The three Type III restriction and modification systems that have been investigated are those from phage P1, the related plasmid P15, and the bacterial strain *Haemophilus influenzae* Rf. *E. coli* 15T⁻ has been shown to carry two host specificity systems (4), one of which (*hsd*A) is linked to the *thr* locus on the bacterial chromosome. This A system is genetically related to the Type I systems found in *E. coli* K and B. The genetic information for the second system is carried on the plasmid P15, which is approximately 90% homologous with P1 (55). The restriction and modification functions of these two extrachromosomal elements seem to be closely related, and genetic recombination between $r_{P15}^- m_{P15}^+$ and a P1 $r_{P1}^+ m_{P1}^+$ phage results in plasmids with the r_{P15}^+ phenotype (4). The strain *H. influenzae* Rf has been shown to have two restriction and modification systems, FIII and FIV (56), neither of which codes for a Type II system (57). Both in vivo and in vitro studies have focused on the FIII system. The chromosomal marker of Novobiocin resistance appears linked to the FIII system in transformation experiments, which suggests that it is present on the chromosome (57).

The first indication that the P1 and P15 systems belong to a new class came from experiments with a mutant strain of *E. coli* requiring methionine (58). When this strain was grown in the presence of ethionine, a methionine analog, it degraded about 50% of its DNA after a round of replication. This effect was caused by the restriction and modification system since r⁻ mutants did not degrade their DNA under these conditions. However, a marked difference was observed, depending on which restriction system was present in the cell. The Type I systems (*E. coli* K, *E. coli* B, and *E. coli* A) did not degrade their DNA following growth in ethionine, whereas the Type III systems (P1 and P15) did. The conclusion drawn is that Type I systems require AdoMet for endonucleolytic cleavage, whereas the Type III systems do not.

Two classes of restrictionless mutants (r⁻m⁺ and r⁻m⁻) have been isolated in P1 (59), P15 (4), and *H. influenzae* Rf (56). Because many of these mutants appear spontaneously, it was assumed that the same three-gene model that had been postulated for the Type I systems also applies in these cases. The first evidence against this hypothesis came from studies on the c2 and c3 clear plaque mutants of P1 that are unable to plate efficiently on P1 cryptic lysogens (60). This P1 cryptic strain lacks immunity but has wild-type restriction and modification. The c2 and c3 mutants were shown to have a $r^+_{P1}m^-_{P1}$ phenotype, which results in their susceptibility to restriction by the host. When these same clear P1 mutants were used to infect nonlysogens, the unmodified bacterial DNA was destroyed following the appearance of the restriction activity after lysogenization. The c2 and c3 mutants complement each other to express P1 modification but neither complements a $r^-_{P1}m^-_{P1}$ mutant. It was concluded that c2 and c3 are two cistrons required for P1 modification, raising the possibility that four genes are involved in the P1 system. Studies with P1 and P15 showed that r^-m^+ and r^-m^- mutants complement each other to give the wild-type phenotype with the host specificity of the r^-m^+ mutant (61). Such results are similar to those for the Type I systems and are consistent with a *hsd* locus with three genes. However, P1 and P15 also showed certain major differences, such as the high proportion of mutants with partial modification ($r^-m^\pm$) and the existence of a class of r^-m^+ mutants that is complemented by both P1 r^-m^+ and r^-m^- mutants. This was taken as further support for the existence of four *hsd* genes.

The solution to this problem has come from recent experiments on the *hsd* regions of P1 and P15 that had been cloned into pBR322 (S. Jida, J. Meyer, and W. Arber, personal communication). The *hsd* genes are present in a 5 kb region on the *Bam*HI-4 fragments of both P1 and P15. Mutations were obtained by the insertion of transposons (Tn9 or Tn10), and they were mapped by restriction and heteroduplex analyses. r^-m^+ mutants have insertions in a 3 kb region, whereas r^-m^- mutants have insertions in an adjacent stretch of 2 kb to its right. The *hsd* locus of P1 and P15 consists of two cistrons: one responsible for restriction (*hsd*R) and one responsible for both modification and site recognition (*hsd*MS). Mutations in the *hsd*R gene affect restriction but not modification. Mutations in the *hsd*MS gene have different effects depending where the lesion maps. If the lesion is on the recognition part of the gene, the resulting phenotype is r^-m^-, whereas if it is on the modification part, the resulting phenotype is r^+m^-. The complementation between c2 and c3 mutants could have been the result of intracistronic complementation in the *hsd*MS gene. The P1 and P15 modification genes show regions of homology at both ends, with a central 1.2 kb region of nonhomology. It is this latter region that is presumably responsible for host specificity. In contrast, the restriction genes show a large degree of homology flanked by areas of partial homology. These genes are large enough to code for a modification polypeptide with a molecular weight of 73,000 and a restriction subunit with a molecular weight of 106,000.

R-loop analysis of in vitro transcription products has shown that two promoters are present in the P1 *hsd* locus (J. Meyer, personal communication). Both generate transcription leftwards, with one located at the right end of the modification gene and the other at the right end of the restriction gene. In some cases, transcripts originating at the modification promoter read through into the restriction gene. These results exclude the possibility that r^-m^- mutants are caused by lesions on a single promoter for the whole *hsd* locus. However, a mutation in the promoter for *hsd*MS would result in a r^-m^- phenotype caused by the loss of the recognition function. Furthermore, the existence of the two promoters might explain how modification occurs temporally before restriction following P1 infection.

Enzyme Structure

The *Eco*P1 (16, 62; T. A. Bickle, S. Jida, and B. Bächi personal communication), *Eco*P15 (63; T. A. Bickle, S. Jida, and S. Hadi personal communication), and *Hinf*III (57) enzymes have all been purified extensively and shown to have similar properties. Like the Type I enzymes, they restrict unmodified DNA in the presence of ATP and Mg^{2+} and methylate unmodified DNA when incubated with AdoMet. A closer examination shows that they are significantly different from the Type I enzymes. Type I enzymes require AdoMet for endonuclease activity; Type III enzymes are only stimulated by it. Digestion by Type I enzymes generates a heterogeneous population of DNA fragments, whereas Type III reactions produce distinct fragments, even though the digestion is not complete (i.e. scission does not occur at every possible cleavage site). Type I enzymes have an ATP hydrolase activity, whereas only a minimal activity is observed with Type III enzymes. Type I enzymes form a complex with unmodified DNA, which is retained on nitrocellulose filters. No such complexes are formed by Type III enzymes. Type III enzymes methylated unmodified DNA in both the presence and absence of ATP, although the reaction is more efficient in its presence. Therefore, these enzymes are able to modify and cleave DNA simultaneously. The larger fragments produced during digestion in the presence of AdoMet are presumably caused by partial modification of the substrate DNA. This is in sharp contrast to the Type I enzymes, which either methylate or restrict but do not do both simultaneously. These differences are not trivial, and they reflect basic differences in the structure and reaction mechanism of these enzymes.

*Hinf*III has a molecular weight of 200,000 and is composed of subunits with molecular weights of 110,000 and 80,000 (57). There is some uncertainty about the structure of the *Eco*P1 and *Eco*P15 enzymes. Earlier reports indicated that *Eco*P1 has four subunits with molecular weights of 100,000, 78,000, 63,000, and 39,000, whereas *Eco*P15 has four subunits with molecular weights of 100,000, 80,000, 27,500, and 24,500 (63). These results are somewhat disconcerting inasmuch as only two of the four subunits are of similar size. More recent studies have shown that EcoP15 is composed of two subunits with molecular weights of 106,000 and 73,000 (T. A. Bickle, personal communication). Given a molecular weight of 318,000, it was proposed that the enzyme has one large and three small subunits. EcoP1 is also composed of two subunits, one with a molecular weight of 106,000 and the other slightly larger than the small EcoP15 subunit. Based on the coding capacity of the hsd genes and the size of the subunits, the larger subunit is the restriction polypeptide, and the smaller is the modification one.

The functions of the two subunits were identified in experiments using an antibody against EcoP15 (T. A. Bickle, personal communication). A crude extract from a r⁻m⁺ mutant (with a large deletion in the hsdR gene) was subjected to gel electrophoresis in the presence of sodium dodecyl sulfate. When the gel was reacted with the EcoP15 antibody, only the small polypeptide was detected. A similar experiment with EcoP15 clearly showed the presence of both subunits. The EcoP15 antibody also reacted with both subunits of EcoP1 and with the smaller subunit from a r⁻m⁺ P1 mutant. Therefore, the presence of the hsdR gene is not required for modification activity, and the cross-reactivity between EcoP1 and EcoP15 is in clear agreement with the homology between the P1 and P15 hsd genes. The modification subunit has been purified from both P1 and P15. Both of them specifically methylate unmodified DNA, and restriction analysis of the product indicates that the modification takes place at the appropriate recognition sites. However, the modification subunit methylates less efficiently than the restriction enzyme, as indicated by a higher apparent K_m for AdoMet and a requirement for higher enzyme concentration. It also requires Mg^{2+} but is not stimulated by ATP. The relatively low efficiency of the modification subunit suggests that both EcoP1 and EcoP15 are responsible for normal modification in vivo but that under some circumstances (e.g. following phage infection) the reaction may be catalyzed by the modification subunit.

The sequence of the sP1 and sP15 sites (see below) only allows one strand to be methylated. This means that following replication one of the daughter chromosomes would be susceptible to the endogenous restriction activity. To circumvent this difficulty, it has been proposed that the modification subunit is part of the replication complex protecting the newly replicated DNA. To test this theory, cells carrying the wild-type cloned genes were fractionated into nuclear and cytoplasmic fractions. Analysis of these fractions showed that the nuclear fraction contains the restriction enzyme, whereas the cytoplasmic fraction has a large pool of modification subunits.

The apparent compartmentalization of the two enzyme species is the reverse of what had been expected, and its biological significance remains to be elucidated.

The Nature of DNA Recognition and Cleavage Sites

The Type III enzymes cleave unmodified DNA in both the presence and absence of AdoMet. In the presence of AdoMet, the DNA fragments produced are larger because of the competing methylase activity; in its absence, the fragments are smaller, but the digestion is still incomplete. Supercoiled SV40 DNA was cleaved once by EcoP1 at only one of four possible sites (64) as shown by the formation of circular DNA molecules when EcoP1-generated linears were denatured and renatured. The positions of the EcoP1 cleavage sites were then mapped by denaturation mapping and restriction analysis. The frequency of cleavage varies from site to site. In vitro reactions with EcoP1 showed that methylation occurs at, or in the vicinity of, the cleavage sites.

The sP1 sites on SV40 DNA (65), the sP15 sites on pBR322 DNA (66), and the sHinfIII sites on fd DNA (A. Piekarowicz, personal communication) have been sequenced by methylating the DNA in vitro and subjecting the product to restriction analysis. The sequences were then identified by a computer search for similar sequences in the methylated regions of the DNA. It was then necessary to prove that such sequences do not appear in those portions of the DNA that are not cut by the given enzyme. The sequences are

sP1   5' A G A C C 3'
sP15   5' C A G C A G 3'
sHinfIII 5' C G A A T 3'

which are composed of 5 or 6 bases with no symmetry. Once these recognition sequences were identified, it was necessary to determine whether they also represent the methylation sites. All three enzymes methylate adenine residues. Small restriction fragments have been isolated from EcoP1-methylated SV40 DNA and two of them have been sequenced (65). These limited results are consistent with methylation of the adenine at the third position of the A-G-A-C-C sequence. No sequencing has been done on the sites methylated by EcoP15 or HinfIII, but it can be assumed that the recognition sequence is the one methylated. A casual inspection of the three sequences indicates that, in two, there are no adenines that can be methylated on the opposite strand. Therefore, methylation can only take place on one strand. To test this, the two strands of restriction fragments containing the sP1, sP15, and sHinfIII sites have been separated by gel electrophoresis. In all cases, only one of the strands has been methylated.

DNA cleavage must occur at the recognition sequence or in its vicinity. The *Eco*P1 cleavage sites on SV40 DNA were analyzed by labeling the 5' termini with polynucleotide kinase (65). The termini were not unique but roughly 50% contained a thymidine. Three unique termini were isolated by cutting SV40 DNA under conditions that would produce linear molecules. These were then labeled with polynucleotide kinase, cut with *Hpa*II, isolated by gel electrophoresis, and sequenced. Cleavage was shown to have taken place by one cut 24 to 26 bases in the 3' direction from the A-G-A-C-C sequence and a second cut 27 to 29 bases away. This represents a stagger of 2–4 bases between the two cuts. A similar experiment has been done with *Eco*P15 in which the termini have been labeled by repair synthesis with DNA polymerase I. *Eco*P15 also cuts 25–26 bases to the 3' side of the C-A-G-C-A-G sequence with a stagger of two bases between the two cuts (66). As in the case of P1, there appears to be a predominance of thymidines at the cleavage ends. The results with *Hinf*III have been much less clear because of a low level of exonuclease contamination in the *Hinf*III preparation (A. Piekarowicz, personal communication). Nevertheless, it appears that *Hinf*III cuts 20–30 bases to the 3' side of the C-G-A-A-T sequence.

Our electron microscopy experiments with *Eco*P15-DNA complexes have shown that the enzyme binds to the sP15 sites (66a). Therefore, the Type III enzymes interact with two sites on the DNA, a recognition site and a cleavage site. One strand of the recognition sequence is also the site for the methylation reaction. DNA cleavage is not sequence-specific but occurs at a distance of 24–26 bases to the 3' side of the recognition sequence. The variability in the distance between recognition and cleavage sites and the preference for thymidines at the cleavage site have led to the proposal that these enzymes prefer to cleave on the 5' side of a thymidine within a certain distance from the recognition sequence (65). The relative efficiency of cleavage at any given site should reflect how closely these requirements are fulfilled.

The Mechanism of the Restriction and Modification Reactions

Our attempts to understand the reaction mechanisms of the Type III enzymes have been complicated by the simultaneous expression of both restriction and modification activities, the incompleteness of the restriction reaction, and the differences in cleavage efficiencies at different sites. Furthermore, although the reaction pathways of *Eco*P1 and *Eco*P15 are similar, they differ in certain aspects from that of *Hinf*III. These differences could be real or they could be caused by variations in the purification procedure and/or the age of the enzyme preparations used. Given the similarity in enzyme structures and nucleotide sequences, the latter is more likely. The various steps in the reaction pathway of *Eco*P15 described here also apply to *Eco*P1 and *Hinf*III, except where otherwise indicated.

ADOMET BINDING   *Eco*P15 has been shown to bind AdoMet and to form a different enzyme species called *Eco*P15* (67). The dependence of DNA cleavage on AdoMet concentration has been measured and shows that it behaves as an allosteric effector with a minimum of two binding sites on the enzyme (67). It has been proposed that *Eco*P15, as purified from the cell, has AdoMet bound to it. This would explain the incomplete DNA restriction reaction in the absence of AdoMet. In fact, the enzyme isolated from *H. influenzae* Rf was shown to be *Hinf*III* and had AdoMet bound to it (68). *Hinf*III* is converted to *Hinf*III following prolonged storage; however, *Hinf*III still does not cleave the DNA completely in the absence of AdoMet. It is also probable that the enzyme can exist in more than one multimeric structure, as *Eco*P1* at low or high enzyme concentrations restricted poorly (62).

FORMATION OF RECOGNITION COMPLEXES   One important aspect of the mechanism of Type III enzymes has been the existence of two active forms of the protein, e.g. *Eco*P15 and *Eco*P15*. Though there is no direct evidence, it can be reasonably assumed that both forms can interact with DNA randomly to form initial complexes. The enzymes would then proceed to bind at sP15 sites to form recognition complexes. *Eco*P15 was shown to bind to the sP15 sites but the efficiency of complex formation varies from site to site. The same is true for *Eco*P15*, but the pattern of binding varies from that of *Eco*P15 (66a). Though the kinetics of complex formation are the same for both enzyme forms, the *Eco*P15* recognition complexes are considerably more stable (67).

INTERACTION OF ATP WITH RECOGNITION COMPLEXES   In the case of *Hinf*III, ATP is required for the formation of recognition complexes. ATP acts as an allosteric effector in the *Eco*P15 cleavage reaction, and it has been estimated that the enzyme has a minimum of three ATP-binding sites (67). Unlike the situation with Type I enzymes, ATP does not induce an apparent size change in the enzyme nor does it lead to the formation of complexes that bind to nitrocellulose filters. ATP hydrolysis does not appear to be required for DNA cleavage since a limited reaction was observed when ATP was replaced by its imido analog. (No restriction reaction was observed when the imido analog was used with *Hinf*III.) The *Eco*P15* present in the recognition complexes interacts with ATP much more efficiently, as shown by its ability to cleave DNA at lower ATP concentrations and by its faster rate of endonuclease activity.

One interpretation of these experiments is that both $EcoP15$ and $EcoP15^*$ are sufficiently large to be able to interact with both the recognition and cleavage sites. ATP induces a conformational change in these enzymes that allows them to cleave at sites to the 3' side of the recognition sequences. This allosteric change is considerably faster for $EcoP15^*$ than for $EcoP15$. ATP hydrolysis per se may not be required for this effect but would be needed for it to act catalytically by releasing the enzyme from the cleaved DNA. ATP also strongly stimulates methylation, which competes with the endonuclease activity. The interaction between the Type I enzymes and their recognition sites determines which activity is expressed. No such stringent control exists with Type III enzymes. How then can the activities of such enzymes be modulated? It has been shown that the apparent K_m value for AdoMet in the endonuclease reaction is 10^{-8} M, whereas that for the methylase reaction is 4×10^{-7} M (63). It is likely that this 40-fold difference in K_m values serves to modulate the two reactions.

DNA CLEAVAGE   ATP triggers DNA cleavage by either $EcoP15$ or $EcoP15^*$. It had been assumed that the limited reaction with $EcoP15^*$ is caused by the competing methylation. That this is not the case was shown in an experiment with $Hinf$III in which only one of the six possible sites were methylated but 90% of the supercoiled DNA molecules were linearized (68). $Hinf$III also preferentially cleaved supercoiled DNA over linear DNA, whereas the rate of the methylation reaction was the same for both forms of the substrate. DNA cleavage in the absence of AdoMet also showed a preference for supercoiled DNA. These results agree with kinetic experiments using $EcoP15$, which showed an accumulation of linear recognition complexes during cleavage of supercoiled SV40 DNA that are not true intermediates inasmuch as they were not cut any further. Under such conditions, the first cleavage event on a supercoiled DNA would be very fast, but the rate of DNA scission would then slow down and be superseded by the methylase reaction.

One other problem concerns differences in cutting efficiency at various sites. Experiments with $EcoP15$ have allowed us to identify sites that are efficiently cut in pBR322 DNA in both the presence and absence of AdoMet. Most of our results have been negative in that inefficient DNA cleavage does not always correlate with methylation at a given site. There is no necessary correlation between efficient binding to a site and efficient cleavage. A comparison of the sequences at five sites that are frequently cut shows no significant homologies and no common structural features such as hairpin structures, palindromes, or AT-rich stretches. Though the sequence data are not inconsistent with the model based on a preference for a thymidine residue at the cleavage site, they are too limited to prove it. It is also likely that other sequence features are required for effective DNA cleavage. One possibility is that different DNA sequences can generate similar structures that could then be recognized by the enzyme in its endonucleolytic mode.

SOME IMPLICATIONS

Our present knowledge of restriction and modification systems enables us to fit them into a reasonable evolutionary scheme. Type II systems are the simplest. They have two genes, one coding for a restriction endonuclease and the other for a modification methylase. Each enzyme has an active site that recognizes the same nucleotide sequence. These active sites can be identical, arising by duplication of that portion of a gene, or can be partly homologous, arising by convergence. In either case, recognition of different sequences by the restriction and modification enzymes within the same cell would be lethal. Therefore there is a selection for enzymes that recognize the same nucleotide sequence and cleave or methylate it. This recognition sequence is a palindrome. The restriction reaction does not require AdoMet or ATP.

The next step in the evolutionary ladder is represented by a subclass of Type II enzymes (Type IIA?) such as *Hph* (69), *Mbo*II (70), and *Hga*I (71). Although these enzymes recognize a specific sequence, cleavage occurs 5 bases away in the case of *Hga*I and 8–9 bases away with *Hph* and *Mbo*II. The recognition sequences consist of 5 bases and are asymmetric. There is no requirement for AdoMet or ATP in the restriction reaction, and the efficiency of cleavage varies for each individual site. *Bgl*I also represents an exception to Type II enzymes since it recognizes a sequence with two constant domains of 3 bases separated by a 5-base spacer (72–74).

Type III systems have two genes: one codes for the restriction polypeptide and the other for the modification polypeptide. The modification subunit has two domains: one responsible for modification and one responsible for sequence recognition. A mutation in the former results in a lethal r^+m^- phenotype, and a lesion in the latter results in a r^-m^- phenotype. The enzyme from a r^+m^- strain continues to restrict normally because there is no AdoMet requirement for this reaction. The restriction endonuclease consists of both subunits and can catalyze both restriction and modification, though the presence of two promoters in the *hsd* region allows for independent synthesis of the endonuclease and the modification methylase. Under restriction conditions, the enzyme can also methylate so that there is a competition between the two reactions. The restriction enzyme recognizes and methylates at specific sites but cleavage occurs at a distance of 24–27 bases. The sequence itself consists of 5–6 bases and is asymmetric (very much like Type IIA) and can only be methylated on one strand. Therefore,

Type III sites can have two configurations: modified (with a methylated base on one strand) or unmodified (without methylated bases on either strand).

ATP is required for DNA cleavage and stimulates methylation. ATP appears to induce a conformational change that enables the protein to bind to the recognition and cleavage sites simultaneously, and the low level of ATP hydrolase seems to be involved in catalytic turnover of the endonucleotypic reaction. The variability in the distance between the two sites appears to be a limited case of localized recognition involving distance and probably some other structural features. The expression of restriction or methylase activities by such proteins appears to be modulated by the AdoMet concentration.

Type I systems have three genes: one codes for a specialized recognition subunit and the other two code for restriction and modification subunits. Mutations in either the *hsd*S or *hsd*M subunits result in r⁻m⁻ phenotypes, but no lethal r⁺m⁻ phenotypes are likely to be generated in this system. The restriction endonuclease is composed of all three subunits and is able to catalyze both restriction and modification reactions. However, in sharp contrast to the Type III enzymes, both activities are not expressed simultaneously but are modulated by the interaction between the protein and the host specificity site on the DNA. These sites can have three configurations: unmodified (neither strand methylated), modified (both strands methylated), and heteroduplex (one strand methylated). The enzyme recognizes and methylates at the host specificity sites, but DNA cleavage occurs at sites that may be several thousand bases away. In order to carry out this reaction at distal sites, these systems have developed a system of DNA translocation that is coupled to ATP hydrolysis. Furthermore, the conformational change in the protein following its first interaction with ATP prevents it from cutting more than once. Cleavage within certain regions that may consist of two- to three thousand bases appears to be an expansion of the localized recognition that we first encountered with the Type III systems. However, DNA scission is probably triggered by structural factors that slow down the rate of DNA translocation.

The symmetrical nature of the Type II sequences has led to the hypothesis that they form hairpin structures that are then recognized by the enzymes (19) (Figure 3). An interesting alternative to this model is one in which the duplex DNA retains its hydrogen bonding while forming additional intrastrand hydrogen bonds (Figure 3) (75). This cage structure would not require denaturation of the DNA and could be maintained with weaker forces than the hairpin. Methylation of adenine or cytosine residues within the sequence would prevent transition to the cage conformation by interference with the intrastrand hydrogen bonds. The cage model predicts that Type II enzymes recognize this four-stranded DNA structure. Type III and Type I enzymes also are presumed to have intermediates involving four-stranded DNA structures (which have been visualized in the latter case); in Type III, the separation between the two recognition and cleavage sites is roughly 2.5 turns, and in Type I it can be several hundred turns. The presence of one or two methylated bases at the recognition site would then interfere with the formation of such four-stranded structures.

The crucial question is what purpose would be served by four-stranded DNA intermediates in the TYPE I and III reactions. One attractive possibility is a search for limited homology between the recognition site and the cleavage sites. In Type I systems, the variable sequence spacer would be used to match up with the DNA that was being translocated past the enzyme. This type of model has been tested with both Type I and Type III enzymes, but no evidence in its favor has been found (R. Yuan, unpublished results). Nevertheless, such a model is extremely appealing in dealing with genetic recombination (e.g. the χ sites in λ and *E. coli*) (76). These are sites that promote a high rate of recombination in the neighborhood of the sites. The χ sites can be shown to act directionally at a distance.

Two features of Type I enzymes are not unique to such systems: DNA translocation and the role of nucleotide sequences as allosteric effectors. For example, it has been reported recently that the *rec*A protein cleaves the λ and P22 repressors (77). ATP and single-stranded DNA (or oligonucleotides) are required for this reaction. The ATP (or its thioanalog) is required for binding to the polynucleotide, which then induces a conformational change in the *rec*A protein, enabling it to cleave the repressors.

Finally, it is interesting to reconsider the biological role of Type I enzymes. Various phages such as T3 (78), T7 (79), μ (80), λ (81), and P1 (W. Arber, personal communication) synthesize proteins that block restriction

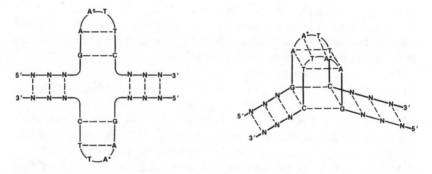

Figure 3   Possible structures of the *Eco*RI recognition sequence: the hairpin structure (*left*) and the cage structure (*right*).

by either inhibiting the endonuclease or stimulating methylation. All of these systems are specific for Type I enzymes and, in at least three of them, the amount of antirestriction protein made is vastly in excess of any restriction enzyme that they could possibly encounter. Such observations, taken in conjunction with the baroque complexity of the restriction mechanisms, have led to speculations that they may be involved in other reactions. Two kinds of experiments have been conducted that tend to discourage these speculations. In vivo, it has been shown that *E. coli* lacks the *hsd* locus without any apparent biological effects (15). In vitro, no additional enzymatic activities have been observed with *Eco*K by itself or when inhibited by the 0.3 antirestriction protein from T7 (R. Yuan, unpublished observations). On the other hand, it has been shown from an examination of many *Salmonella* strains that these can be classified into three classes, depending on which restriction systems are present (82). This classification correlates well with that obtained by the standard taxonomical methods. This study supports the role of Type I restriction and modification systems in controlling gene flow between different populations.

Acknowledgments

The author is grateful to Dr. Mark Pearson and Dr. Robert Deich for their critical reading of the manuscript and for useful discussions. He also would like to thank the researchers who made available to him their results prior to publication. Research sponsored by the National Cancer Institute under Contract No. NO1-CO-75380 with Litton Bionetics, Inc.

Literature Cited

1. Boyer, H. W. 1971. *Ann. Rev. Microbiol.* 25:153–76
2. Smith, H. O., Nathans, D. 1973. *J. Mol. Biol.* 81:419–23
3. Modrich, P. 1979. *Q. Rev. Biophys.* 12:315–69
4. Arber, W., Wauters-Willems, D. 1970. *Mol. Gen. Genet.* 108:203–17
5. Colson, C., Colson, A. M. 1971. *J. Gen. Microbiol.* 69:345–51
6. Colson, A. M., Colson, C. 1972. *J. Gen. Microbiol.* 70:123–28
7. Glover, S. W., Schell, J., Symonds, N., Stacey, K. A. 1963. *Genet. Res.* 4:480–82
8. Wood, W. B. 1966. *J. Mol. Biol.* 16:118–33
9. Boyer, H. W., Roulland-Dussoix, D. 1969. *J. Mol. Biol.* 41:459–72
10. Glover, S. W. 1970. *Genet. Res.* 15:237–50
11. Hubacek, J., Glover, S. W. 1970. *J. Mol. Biol.* 50:111–27
12. Bullas, L. R., Colson, C., Van Pel, A. 1976. *J. Gen. Microbiol.* 95:166–72
13. Boyer, H. W. 1964. *J. Bacteriol.* 88:1652–60
14. Bulkacz, J. 1972. *Genetic analysis of the cistrons controlling K-specific and B-specific restriction and modification in E. coli.* PhD thesis. Univ. Calif., San Francisco
15. Sain, B., Murray, N. 1980. *Mol. Gen. Genet.* 180:35–46
16. Meselson, M., Yuan, R. 1968. *Nature* 217:1111–14
17. Haberman, A., Heywood, J., Meselson, M. 1972. *Proc. Natl. Acad. Sci. USA* 69:3138–41
18. Yuan, R., Heywood, J., Meselson, M. 1972. *Nature New Biol.* 240:42–43
19. Meselson, M., Yuan, R., Heywood, J. 1972. *Ann. Rev. Biochem.* 41:447–66
20. Roulland-Dussoix, D., Boyer, H. 1969. *Biochim. Biophys. Acta* 195:219–29

21. Eskin, B., Linn, S. 1972. *J. Biol. Chem.* 247:6192–96
22. Vovis, G. F., Horiuchi, K., Zinder, N. D. 1974. *Proc. Natl. Acad. Sci. USA* 71:3810–13
23. Eskin, B., Linn, S. 1972. *J. Biol. Chem.* 247:6183–91
24. Lautenberger, J. A., Linn, S. 1972. *J. Biol. Chem.* 247:6176–82
25. Linn, S., Arber, W. 1968. *Proc. Natl. Acad. Sci. USA* 59:1300–6
26. Hadi, S. M., Yuan, R. 1974. *J. Biol. Chem.* 249:4580–86
27. Bühler, R., Yuan, R. 1978. *J. Biol. Chem.* 253:6756–60
28. Arber, W., Kühnlein, U. 1967. *Pathol. Microbiol.* 30:946–52
29. Sclair, M., Edgell, M. H., Hutchison, C. A. III. 1973. *J. Virol.* 11:278–85
30. Smith, J., Arber, W., Kühnlein, U. 1972. *J. Mol. Biol.* 63:1–8
31. Kühnlein, U., Arber, W. 1972. *J. Mol. Biol.* 63:9–19
32. Brack, C., Eberle, H., Bickle, T. A., Yuan, R. 1976. *J. Mol. Biol.* 108: 583–93
33. Lyons, L. B., Zinder, N. D. 1972. *Virology* 49:45–60
34. Horiuchi, K., Vovis, G. F., Enea, V., Zinder, N. D. 1975. *J. Mol. Biol.* 95:147–65
35. Murray, N. E., Manduca de Ritis, P., Foster, L. 1973. *Mol. Gen. Genet.* 120:261–281
36. Kan, N. C., Lautenberger, J. A., Edgell, M. H., Hutchison, C. A. III. 1979. *J. Mol. Biol.* 130:191–209
37. Horiuchi, K., Zinder, N. D. 1972. *Proc. Natl. Acad. Sci. USA* 69:3220–24
38. Ravetch, J. V., Horiuchi, K., Zinder, N. D. 1978. *Proc. Natl. Acad. Sci. USA* 75:2266–70
39. Lautenberger, J. A., Kan, N. C., Lackey, D., Linn, S., Edgell, M. H., Hutchison, C. A. III. 1978. *Proc. Natl. Acad. Sci. USA* 75:2271–75
40. Vovis, G. F., Zinder, N. D. 1975. *J. Mol. Biol.* 95:557–68
41. van Ormondt, H., Lautenberger, J. A., Linn, S., de Waard, A. 1973. *FEBS Lett.* 33:177–80
42. Dugaiczyk, A., Kimball, M., Linn, S., Goodman, H. M. 1974. *Biochem. Biophys. Res. Commun.* 61:1133–40
43. Murray, N. E., Batten, P. L., Murray, K. 1973. *J. Mol. Biol.* 81:395–407
44. Brammar, W. J., Murray, N. E., Winton, S. 1974. *J. Mol. Biol.* 90:633–47
45. Hadi, S. M., Bickle, T. A., Yuan, R. 1975. *J. Biol. Chem.* 250:4159–64
46. Yuan, R., Bickle, T. A., Ebbers, W., Brack, C. 1975. *Nature* 256:556–60
47. Bickle, T. A., Brack, C., Yuan, R. 1978. *Proc. Natl. Acad. Sci. USA* 75:3099–3103
48. Yuan, R., Hamilton, D. L., Burckhardt, J. 1980. *Cell* 20:237–44
49. Adler, S. P., Nathans, D. 1973. *Biochim. Biophys. Acta* 299:177–88
50. Horiuchi, K., Vovis, G. F., Zinder, N. D. 1974. *J. Biol. Chem.* 249:543–52
51. Rosamond, J., Endlich, B., Linn, S. 1979. *J. Mol. Biol.* 129:619–35
52. Kimball, M., Linn, S. 1976. *Biochem. Biophys. Res. Commun.* 68:585–90
53. Vovis, G. F., Horiuchi, K., Hartman, N., Zinder, N. D. 1973. *Nature New Biol.* 246:13–16
54. Burckhardt, J., Weisemann, J., Yuan, R. 1981. *J. Biol. Chem.* In press
55. Ikeda, H., Inuzuka, M., Tomizawa, J. 1970. *J. Mol. Biol.* 50:457–70
56. Piekarowicz, A., Kauc, L., Glover, S. W. 1974. *J. Gen. Microbiol.* 81:391–403
57. Kauc, L., Piekarowicz, A. 1978. *Eur. J. Biochem.* 92:417–26
58. Lark, C., Arber, W. 1970. *J. Mol. Biol.* 52:337–48
59. Glover, S. W., Schell, J., Symonds, N., Stacey, K. A. 1963. *Genet. Res.* 4:480–82
60. Rosner, J. L. 1973. *Virology* 52:213–22
61. Reiser, J. 1975. *The P1- and P15-specific restriction endonucleases: A comparative study* PhD thesis. Univ. Basel, Switzerland, pp. 136–46
62. Haberman, A. 1974. *J. Mol. Biol.* 89:545–63
63. Reiser, J., Yuan, R. 1977. *J. Biol. Chem.* 252:451–56
64. Risser, R., Hopkins, N., Davis, R. W., Delius, D., Mulder, C. 1974. *J. Mol. Biol.* 89:517–44
65. Bächi, B., Reiser, J., Pirrotta, V. 1979. *J. Mol. Biol.* 128:143–63
66. Hadi, S. M., Bächi, B., Shepherd, J. C. W., Yuan, R., Ineichen, K., Bickle, T. A. 1979. *J. Mol. Biol.* 134:655–66
66a. Yuan, R., Hamilton, D. L., Hadi, S. M., Bickle, T. A. 1981. *J. Mol. Biol.* In press
67. Yuan, R., Reiser, J. 1978. *J. Mol. Biol.* 122:433–45
68. Piekarowicz, A., Brzezinski, R. 1981. *J. Mol. Biol.* In press
69. Kleid, D., Humayun, Z., Jeffrey, A., Ptashne, M. 1976. *Proc. Natl. Acad. Sci. USA* 73:293–97
70. Gelinas, R. E., Myers, P. A., Roberts, R. J. 1977. *J. Mol. Biol.* 114:169–79
71. Brown, N. L., Smith, M. 1977. *Proc. Natl. Acad. Sci. USA* 74:3213–16
72. van Heuverswyn, H., Fiers, W. 1980. *Gene* 9:195–203

73. Bickle, T. A., Ineichen, K. 1980. *Gene* 9:205–12
74. Lautenberger, J. A., White, C. T., Haigwood, N. L., Edgell, M. H., Hutchison, C. A. III. 1980. *Gene* 9:213–31
75. Stasiak, A., Klopotowski, T. 1979. *J. Theor. Biol.* 80:65–82
76. Stahl, F. W. 1979. *Ann. Rev. Genet.* 13:7–24
77. Craig, N. L., Roberts, J. W. 1980. *Nature* 283:26–30

78. Studier, F. W., Movva, N. R. 1976. *J. Virol.* 19:136–45
79. Studier, F. W. 1975. *J. Mol. Biol.* 94:283–95
80. Toussaint, A. 1976. *Virology* 70:17–27
81. Zabeau, M., Friedman, S., Van Montagu, M., Schell, J. 1980. *Mol. Gen. Genet.* 179:63–73
82. Bullas, L. R., Colson, C., Neufeld, B. 1980. *J. Bacteriol.* 141:275–92

Ann. Rev. Biochem. 1981. 50:317–48

TRANSFER OF PROTEINS ACROSS MEMBRANES

◆12081

Günther Kreil

Institute of Molecular Biology, Austrian Academy of Sciences, Billrothstrasse 11, A-5020 Salzburg, Austria

CONTENTS

PERSPECTIVES AND SUMMARY

The signal hypothesis, formulated some twenty years after the pioneering studies of Palade and Siekevitz on protein secretion, made the following simple predictions: (*a*) secretory polypeptides contain an amino-terminal signal peptide that binds to the endoplasmic reticulum and initiates vec-

317

0066-4154/81/0701-0317$01.00

torial discharge of the growing polypeptide chain; (*b*) this transport must start early, before the chain reaches a certain length, and be strictly coupled to protein synthesis; (*c*) signal peptides are transient entities that are cleaved before chain completion.

To date, these basic predictions have been corroborated for numerous proteins of diverse function that have only one feature in common, namely that they are exported from cells. In addition, the biosynthesis of a group of transmembrane proteins and probably lysosomal enzymes proceeds, in its early steps, by the same mechanism.

The concept of specific signal sequences that guide polypeptides to different cellular compartments has proved very fruitful. Besides the cotranslational transport of secretory polypeptides across microsomal membranes, a second, post-translational route was soon established. It was shown that the small subunit of ribulose-bis-phosphate carboxylase, a protein made in the cytosol and subsequently translocated to the chloroplast stroma, was also derived from a precursor with an amino-terminal extension. Uptake into chloroplasts and cleavage of the precursor were shown to take place after completion of the polypeptide chain and to be energy-requiring processes. The selective import of proteins into mitochondria has now been shown to proceed by the same basic mechanism. Precursors for several proteins have been detected that are transported and cleaved independent of ongoing protein synthesis.

Yet this simple picture of two transport mechanisms, a secretory-type and an organelle-type, and transient signal sequences characteristic for export or for uptake into mitochondria or into chloroplasts, proved incomplete. Ovalbumin, the main secretory product of chick oviduct cells and a few membrane proteins synthesized on the endoplasmic reticulum were found not to be synthesized as precursors. Mitochondrial proteins, like cytochrome *c* and the ADP/ATP carrier protein, are similarly imported without partial processing. So there may be "uncleaved" signals and possibly even "internal" ones, as suggested for ovalbumin.

Experiments on protein transport across the plasma membrane of *Escherichia coli* yielded even more complex results. While on the one hand, the crucial role of terminal signal peptides for translocation could be defined more thoroughly than with animal cells, it was on the other hand found that in certain instances a signal peptide is not sufficient for secretion. Work with mutants producing hybrid proteins that contain various parts of a secreted or outer membrane protein fused to the cytoplasmic enzyme β-galactosidase, has shown that other regions in the protein, besides the amino-terminal signal peptide, are essential for determining the cellular location. The post-translational transport and processing of proteins located in one of the two membranes or the periplasm has also been demonstrated. At the inner membrane of *E. coli* more than one type of polypeptide transport appears to operate.

"The signal hypothesis" has now burst at the seams as it can no longer accommodate all these diverse findings. The definition of a signal has become vague and may indeed be meaningless in cases where regions widely separated in the primary structure of a polypeptide must cooperate in the translocation process. While there clearly are groups of proteins that follow a simple, signal-type pathway, it is nevertheless apparent that the mechanisms of protein transport across membranes are more diverse and complex than could earlier be envisaged.

INTRODUCTION

With the basic principles and many of the details of protein biosynthesis now long established, attention in recent years has increasingly been directed to the question of how polypeptides are distributed to different cellular compartments. Save for the limited biosynthetic activity found in mitochondria and chloroplasts, the translation of mRNAs occurs exclusively in the cytosol. Mechanisms must therefore exist through which proteins find their place in the complex edifice of subcellular structures.

Our present understanding of these mechanisms owes a great deal to studies on the "export compartment" of cells, which is responsible for secretion of a specific class of polypeptides. As certain cells show a high degree of specialization for protein export, this process was particularly amenable to ultrastructural and biochemical analysis. The decisive steps in the unraveling of the secretory pathway have been summarized by Palade in his Nobel lecture (1). Using mainly pancreas and liver cells the following key observations were made: (*a*) secretory proteins are produced on ribosomes bound to the endoplasmic reticulum, (*b*) newly synthesized proteins do not appear in the cytosol but are transferred directly into the membrane-bounded compartment, which is the lumen of the endoplasmic reticulum in vivo and of microsomes in vitro (2, 3). This latter point was further stressed by the important experiments of Redman & Sabatini (4), who demonstrated that even incomplete polypeptide chains, obtained by the addition of puromycin as chain-terminating inhibitor, were not located on the cytoplasmic side of the membrane. It was then shown (5) that microsomes protect nascent chains of secretory but not of cytoplasmic proteins from proteolytic degradation. These investigations led to the concept that the growing polypeptide chain is translocated through the membrane of the endoplasmic reticulum during polypeptide synthesis. This is now recognized as one type of mechanism of polypeptide transfer across membranes and has been termed cotranslational.

Later studies on the biosynthesis of cytochrome c (6, 7), other mitochondrial proteins (8), and chloroplast (9) and peroxisomal (10) proteins, led to the conclusion that transfer across membranes can also occur after completion of polypeptide chains by post-translational mechanisms. Several examples for this type of processing are now known, and the energy requirements and other factors are beginning to be unraveled.

Many polypeptides are not completely transported through a bilayer but remain partly inserted in a membrane. In their final disposition these proteins of different cellular membranes show just about every conceivable orientation with respect to the cytoplasm, where their synthesis is initiated and frequently also completed. Only a few clear examples on the biosynthesis of membrane proteins are currently available, but certain rules are beginning to emerge (11, 12).

The transport of polypeptides into the nucleus represents a special case that is not to be reviewed here. It has been demonstrated that both histones and nonhistone proteins injected into the cytosol of cells migrate into the nucleoplasm (13, 14). This selective uptake through pores in the nuclear envelope and not through a phospholipid bilayer is unrelated to size or net charge of the proteins, but depends on some unknown feature that distinguishes nuclear from cytoplasmic proteins.

THE MAIN THEME: PROTEIN SECRETION

Numerous experiments performed during the sixties, some of which are briefly mentioned in the preceding paragraph, led to the general acceptance of the concept that ribosomes bound to the endoplasmic reticulum synthesize the nascent polypeptide chain and discharge it vectorially through the membrane. Different reasons have been proposed to explain the presence of a particular class of mRNAs, mainly those for secretory proteins, in membrane-bound polysomes. The selection for this class of mRNAs was suggested to occur via: (*a*) binding of incomplete initiation complexes composed of mRNA and 40 S ribosomal subunits to membrane-bound 60 S subunits (15); (*b*) formation of a specific link between the mRNA and the membrane (16); and (*c*) attachment of bound polysomes via the amino-terminal region of the growing polypeptide chain (17).

Evidence for the last alternative was first presented by Milstein et al (18) in their studies on the translation of immunoglobulin light chain mRNA in cell-free systems. These authors showed that a larger precursor containing extra amino acids at the NH_2 terminus was synthesized in the reticulocyte lysate but not in cell-free systems still containing microsomal membranes. This finding was corroborated by others (19) and extended by Schechter (20) who presented the first data on the amino acid sequence of this NH$_2$-terminal segment. Similar precursors only detected in vitro were subsequently described for parathyroid hormone (21), placental lactogen (22), and melittin (23). Milstein et al (18) had already used the term signal for this region and they speculated about its possible role in binding polysomes to the endoplasmic reticulum shortly after the onset of protein synthesis.

These assumptions were raised to the status of a widely accepted scientific theory through the work of Blobel & Dobberstein (24) who presented the decisive experimental data that then led to the formulation of the signal hypothesis. Working also with light chain mRNA they showed that (a) translation in the presence of membranes yields the authentic product located inside the microsomal vesicles; (b) in the absence of microsomes, the precursor is synthesized; (c) upon addition of membranes after completion of the polypeptide, chain processing of the preprotein and translocation do not take place. According to the signal hypothesis, based on these and other data, initiation of proteins generally starts in the cytosol, and membrane attachment is guided by the first stretch of amino acids emerging from the larger subunit of the ribosome. If this constitutes a "signal sequence," it interacts with the endoplasmic reticulum, which leads to the subsequent binding of the ribosome through its large subunit, and the formation of a tunnel through which the growing polypeptide chain is threaded into the lumen. The cleavage of the signal peptide is also suggested to occur during polypeptide synthesis, even though cases where it is retained in the finished product would also be compatible with this scheme.

The signal hypothesis had a major impact, and in quick succession the existence of precursors was demonstrated for a wide variety of secretory polypeptides. A large number of signal peptides has by now been sequenced either by analysis of cell-free translation products or, with increasing frequency, through sequencing of cDNA or genomic clones. A partial list of secretory proteins possessing a well-characterized signal peptide is presented in Table 1. While this aspect of the signal hypothesis has been amply confirmed, progress toward a more detailed understanding of the molecular interactions required for vectorial transport has been slow.

Structure of Signal Peptides: Comparative Aspects

At the beginning, most researchers trying to determine the amino acid sequence of signal peptides probably expected to find a highly conserved structure responsible for the interaction of this region with a membrane receptor. This turned out not to be the case, as signal peptides were found to vary considerably in both length and structure. This part of different secretory proteins is in most cases not a homologous stretch of amino acids bearing the marks of a common evolutionary origin. A comparison of signal

Table 1 Secretory proteins with signal peptide

Protein (species)	Size of signal peptide	Ref.
Serum proteins		
Immunoglobulin (mouse)		
L chain, κ (several examples)	20–22	25, 26
L chain, λ_I, λ_{II}	19	25, 27
H chain, γ	18	28
Serum albumin (rat and bovine)	18	29, 30
Prothrombin (bovine)	22	31
Hormones		
Proparathyroid hormone (bovine)	25	32
Proinsulin (rat I, rat II, and human)	23	33–35
Prolactin (rat)	29	36
Growth hormone (rat and human)	26	37, 38
Pro-opiomelanocortin (rat)	26	39, 40
Placental lactogen (human)	25	41, 55
Chorionic gonadotropin α-chain (human)	24	42, 43
Milk proteins		
α-Lactalbumin (sheep)	19	44
β-Lactoglobulin (sheep)	18	45
Caseins α_{s1}, α_{s2}, and β (sheep)	15	45, 46
Casein κ (sheep)	21	45, 46
Egg proteins		
Lysozyme (chicken)	18	47
Conalbumin (chicken)	19	48
Ovomucoid (chicken)	23	49
Other proteins		
Interferons (human)	23	49a
Uteroglobin (rabbit)	21	50
Promelittin (honeybee)	21	51

sequences of different immunoglobulin light chains has demonstrated that this region could be classified as a variable region within the immunoglobulin structure (25). Also, the comparison of rat and human preproinsulin reveals that in these hormone precursors the signal peptides show almost as little evolutionary conservation as the highly variable C peptides (34, 35, 52).

Several attempts have been made to look for other types of similarities between different signal peptides. These principally involved an analysis of the distribution of polar and apolar amino acids, and predictions about secondary structures. In looking for common features in the primary structure, it is striking that all signal peptides contain a central region of hydrophobic residues with a minimal length of about nine residues. A decrease

in hydrophobicity in this region, achieved by replacing leucine with
β-hydroxyleucine in the in vitro translation, results in an inhibition of
translocation and processing of the growing polypeptide chain (55). With
bacterial mutants, the crucial role of this apolar part has been documented
more extensively (see below). Most signal peptides also contain, besides the
alpha amino group, at least one more charged residue near the amino end,
frequently an arginine or lysine. Toward the carboxyl end, the hydro-
philicity mostly increases again and the sequences invariably terminate with
an amino acid with a short side chain. Signal peptides also differ in overall
size; the smallest is 15 residues long.

Attempts have also been made to look for common elements in the
secondary structure of signal peptides, as predicted from rules established
for water-soluble, globular proteins (52–54). A certain degree of similarity
emerged from these studies that indicated a tendency for alpha-helix formation in the hydrophobic part and a β-turn near the COOH terminus. The
possibility that all signal sequences do contain a common element of secondary structure is intriguing, but the conformational analysis of signal peptides remains ambiguous.

An alternative approach has recently been chosen (56) that uses a synthetic peptide encompassing the pre and pro part of preproparathyroid
hormone to measure its circular dichroism spectra in both aqueous and
nonpolar solvents. Upon transfer from an aqueous to an organic environment, the structure of this peptide changed from a predominantly β structure to one rich in alpha helix. Rosenblatt and co-workers propose that the
signal part is entirely alpha-helical except for a β-turn near the carboxyl
end. This increase in helicity in an apolar environment has also been observed for peptides like melittin (57) and glucagon (58), which spontaneously interact with phospholipid bilayers. The physicochemical investigation of synthetic signal peptides opens a way toward an understanding
of the essential features currently hidden behind differing primary sequences.

Interaction between Membrane and Signal Peptide

In principle, the original interaction between the signal peptide emerging
from the ribosome and the endoplasmic reticulum could occur via specific
protein receptors or through contact with lipids. Both possibilities have
attractive features and inherent difficulties. In view of the observed difference between signal peptides, the usual type of ligand-receptor interaction,
with its exquisite specificity, does not appear to be a good model for these
considerations. For example, polypeptide hormones that interact with a
receptor present in the plasma membrane show a much higher degree of
sequence conservation than signal peptides. Their contact with a membrane

protein would then have to be more flexible so that a short, hydrophobic segment would be recognized without large contraints as to the nature of the side chains. On the other hand, such a hydrophobic structure would probably have a tendency to interact with the membrane lipids, a process that may be facilitated by ionic interactions. Common experience with hydrophobic peptides and the data of Rosenblatt et al (56) mentioned above would tend to favor such a mechanism. However, the interaction ought to be specific for the membrane of the endoplasmic reticulum and should not occur to any significant extent with the phospholipids of other cellular membranes.

The experimental evidence that would allow a distinction between the two possible types of primary interaction is scarce. Prehn et al (59) have shown that preproinsulin, but not proinsulin, binds to stripped membranes of the rough endoplasmic reticulum, and that this binding can be abolished by pretreatment of the membranes with protease. It has been demonstrated (60, 61) that protein translocation is highly susceptible to proteases and/or high salt washes, and the protein involved may also function as a receptor for signal peptides. Furthermore, competition experiments have shown that different secretory proteins compete for the translocation machinery of endoplasmic reticulum (62), which implies a limited number of sites for signal peptides, ribosomes, or some other limiting factor.

The direct binding of signal peptides to phospholipids has not been investigated. Comparative studies on the interaction of preproteins and synthetic signal peptides with microsomes (see 32) and with liposomes are needed for a better understanding of this important first step. Judging from the published schemes, several investigators currently favor the possibility of signal peptide-phospholipid interaction (52). The "loop model" proposed by Inouye et al (63) for the translocation of prolipoprotein across the cytoplasmic membrane of *E. coli* is also based on peptide-phospholipid interaction.

Generally, it seems that this first contact between a signal peptide and the endoplasmic reticulum need not be of very high precision or affinity, for two reasons: First, a few molecules of preprotein wrongly located in the cytosol may not have adverse effects. In fact, preproteins injected into oocytes were shown to have an unusually short half-life (64), which led to the suggestion that signal peptides, apart from their function in protein translocation across the endoplasmic reticulum, may exert a labilizing effect due to wrong folding if they are not cleaved in time. This has recently been demonstrated for the precursor of a periplasmic protein of *E. coli*, which was found to be cleaved in vitro by proteases under conditions where the mature protein was stable (65). Small amounts of preproteins have been actually detected in intact cells (52, 66, 67), but it is not yet known whether these represent true intermediates or simply mislocalized proteins. Second, once vectorial discharge of a polypeptide is under way, the corresponding polysome will constantly remain membrane-bound, and newly emerging signal peptides will be in a favorable position to interact with an adjacent membrane protein or phospholipids.

Ribosome Binding and Translocation of Polypeptide Chains

The binding of ribosomes to the membrane of the endoplasmic reticulum through their large subunits is well documented (1, 68). As depicted in the signal hypothesis, ribosome binding would occur shortly after interaction of the membrane with the signal peptide. Two proteins present only in the rough but not in the smooth endoplasmic reticulum have been invoked as being responsible for ribosome binding (69, 70) and these were termed ribophorins I and II. The interaction of ribosomes with these proteins is presumably ionic, since ribosomes can effectively be removed by a combination of puromycin (to terminate growing chains) and high ionic strength. These ribophorins are integral membrane glycoproteins, and cross-linking to ribsomes has been demonstrated.

The chemical nature of the pore or tunnel through which the growing polypeptide is threaded is again a matter of conjecture. Blobel and his collaborators clearly favor a tunnel formed through the assembly of membrane proteins. They envisage that binding of signal peptide and ribosome somehow favors interaction among membrane proteins, which leads to the formation of a hole (24). At least one protein is essential for the translocation activity of microsomes as has been demonstrated by Warren and Dobberstein and by Walter et al (60, 61). Both groups now agree that through the action of proteases and treatment with high salt a large polypeptide is released from the cytoplasmic side of the endoplasmic reticulum membrane. This cleavage requires only minute amounts of protease and mild incubation conditions, which first led to the conclusion that high ionic strength alone would also detach this fragment (61). The fragment released by elastase is highly basic and has a molecular weight of about 60,000 (160). As first shown by Warren & Dobberstein (60), addition of this fragment to inactive microsomes restores their translocation activity. This activity also requires a cytoplasmically exposed sulfhydryl group (71, 72), shown to be located in the polypeptide released by trypsin and high concentrations of salt (72).

An alternative view has, on purely theoretical grounds, been presented by von Heijne (73–75). In his "direct transfer model" the physicochemical parameters for a transmembrane transport of a nascent chain through the lipophilic core of a membrane are evaluated. Starting from estimated free energy differences for transferring amino acids from a random coil conformation in water to an alpha helix in an apolar environment, a computerized procedure has been presented that makes correct predictions about a number of secreted and membrane proteins. The putative protein tunnel could also be rendered superfluous by the assumption that signal peptides locally act as "lytic" agents and cause the collapse of the bilayer structure. However, the only experimental evidence available at present is the observation that part of a membrane protein protruding on the cytoplasmic side, which is easily detached by proteases and contains an essential sulfhydryl group, is required for protein translocation.

Signal Peptidase

The signal peptide is cleaved prior to polypeptide chain completion in a cotranslational reaction taking place on the growing polypeptide chain (24, 76–78). Palmiter et al, moreover, estimated that the cleavage of prelysozyme to lysozyme occurs at a chain length of only about 60 residues (49). This would indicate that binding of the signal peptide and the ribosome to the membrane, formation of the pore, and cleavage of the signal peptide all take place very early in the synthesis of a secretory polypeptide. Comparison of the cleavage sites in different preproteins is as disappointing as the comparison of signal peptide sequences. The "specificity" of the signal peptidase that emerges from this can be summarized as follows: a bond X–Y is hydrolyzed, where Y can be virtually any amino acid and X is an amino acid with a small side chain (generally glycine or alanine, sometimes serine, cysteine, or threonine). As for other parts of the secretory apparatus, the signal peptidase is also not species-specific and correctly processes presecretory polypeptides of diverse origin. Again, we have to invoke considerations about possible secondary structures to explain the observed specificity. A further level of complexity is introduced by the possibility that the hydrophobic parts of the signal peptide may stay in the membrane and that consequently the protease cleaves at the first suitable peptide bond that appears on the luminal side of the endoplasmic reticulum. It should, however, be mentioned that endoplasmic reticulum membranes also contain aminopeptidase activities (79), and the observed cleavage sites may not in all cases represent the primary action of signal peptidase but be the result of subsequent "nibbling" by such enzymes.

Several laboratories have reported on the post-translational cleavage of preproteins (71, 80–86), and this should eventually lead to a more thorough characterization of the intracellular protease involved. In the presence of fairly high amounts of detergent, such as 0.2% deoxycholate, partial but correct processing by microsomes has been observed. Signal peptidase is an integral membrane protein that can be solubilized in detergent and then reconstituted into vesicles (85). However, under all circumstances tested so far, it remains a cryptic enzyme requiring added detergent for its unmasking. Save for one report on the inhibition of signal peptidase by high amounts of chymostatin (83), other inhibitors of proteases were found to have no effect on this enzyme (86). Signal peptidase may in fact represent a new type of proteolytic enzyme hitherto not found in other biological systems.

The detection of signal peptides after their cleavage from the polypeptide chain has been attempted, but no clear results could be obtained (81, 87). In an heroic effort, Habener et al (88) devised experimental conditions for the detection of minute quantities of signal peptide, but it could not be found. We are thus left with the fact that cleaved signal peptides must be very short lived in vivo and in vitro and rapidly hydrolyzed via an unknown route. This rapid degradation could in fact be important, since the hydrophobic portions of signal peptides might otherwise have an adverse influence on the properties of the membrane.

The Ovalbumin Case: A Variation

Synthesis of egg white proteins in the oviduct of hormonally stimulated chicken has been studied in great detail. Three of these proteins, conalbumin, ovomucoid, and lysozyme are synthesized via preproteins, and the signal peptides of the latter two show some sequence homology (47–49). It was therefore surprising that ovalbumin, the main secretory product of oviduct gland cells, which accounts for about half of the total protein synthesis, does not contain a transitory signal sequence. The initiating methione is cleaved from the primary translation product of ovalbumin mRNA, and the newly exposed NH_2-terminal glycine is then acetylated (49, 89). It has been estimated that both reactions are completed before the growing chain is 50 amino acids long (49). N-acetyl-glycine is also the terminus of mature, secreted ovalbumin, which shows that cleavage of a signal peptide is not essential for secretion, and also raises the possibility of two different mechanisms of protein secretion. This view was apparently strengthened by the fact that the amino-terminal sequence of ovalbumin does not quite resemble a signal peptide.

It was, however, subsequently shown (62) that ovalbumin competes with prolactin, another secretory protein, for at least one essential constituent of the translocation system of microsomes. Competition was observed only with nascent ovalbumin chains, but not with the completed, folded polypeptide. In an attempt to locate the putative signal peptide of ovalbumin, Blobel and his colleagues checked whether its proteolytic fragments could inhibit translocation of preprolactin into dog pancreas membranes (90). It was found that a tryptic digest of ovalbumin inhibited this translocation, i.e. increased the amount of preprolactin recovered from the assays, and further analysis has shown that this inhibitory activity resides in a fragment containing residues 229–276 of the ovalbumin sequence. A part of this fragment shows some sequence homology to the signal sequences of two other oviduct products, ovomucoid and lysozyme. These and other data led Lingappa et al (90) to the conclusion that ovalbumin contains an internal signal sequence located around residue 240 in a chain of 385 amino acids. As expected from this data, membrane addition for half-maximal translocation can occur later in the case of ovalbumin than for other secretory proteins (about 5.5 vs 2 min). This type of experiment will, however, only determine the latest, not the earliest time when the signal peptide becomes available. In the case of ovalbumin, the putative signal sequence may simply remain exposed much longer. Also, the amount of fragment required for the inhibition of preprolactin translocation and processing were quite high, in the order of one milligram of peptide per milligram of membrane. This is certainly far in excess of any putative receptor for signal peptides present in the endoplasmic reticulum.

Contrasting, but still preliminary data on the ovalbumin example have been presented. From kinetic experiments to determine the minimum length of nascent chains needed to observe binding to dog pancreas microsomes, it was found that ovalbumin and ovomucoid can bind equally early with a chain length of less than 100 residues (91). These data suggest a signal sequence at or near the NH_2 end of ovalbumin. These latter experiments, however, only measure binding and not actual translocation.

Generally, the concept of an internal signal sequence poses conceptual difficulties. It is hard to visualize how the same tunnel through which a growing polypeptide chain is threaded from the NH_2 to the COOH end could accommodate a second chain running in the opposite direction, and transport a hairpin-like structure. At this time I consider the location of the signal peptide in ovalbumin an open question.

Lysosomal Enzymes

Lysosomal enzymes represent a group of intracellular glycoproteins that are frequently derived from proenzymes. It has generally been assumed that these proteins are synthesized on membrane-bound polysomes and therefore share a common route with secretory polypeptides. Separation of secretory and lysosomal enzymes is considered to take place in the Golgi apparatus or a specialized region of the smooth endoplasmic reticulum termed GERL (for review see 92). The close parallel between the biosynthesis of secretory and lysosomal enzymes is exemplified by the observation that normal fibroblasts secrete lysosomal enzymes. Export of lysosomal enzymes is prominent in certain hereditary storage diseases, particularly

I-cell disease (93). Fibroblasts from patients with this disease are deficient in several lysosomal enzymes, which instead are secreted into the extracellular medium. These findings have led to the formulation of the Secretion-Recapture Hypothesis for the packaging of lysosomal enzymes (94). This conjecture has played an important role in stimulating further research, but reuptake is now considered to play a minor part in the packaging of lysosomal enzymes. It is now generally assumed that the major portion of lysosomal enzymes of a given cell also originate from the same cell.

Irrespective of where the pathways for the biosynthesis of lysosomal and secretory polypeptides branch off, the primary events should be similar. This led to the prediction that lysosomal enzymes are also synthesized with NH_2-terminal signal peptides equivalent to those of secretory polypeptides. Direct evidence for this prediction is currently lacking but, as an important first step in this direction, the in vitro synthesis and translocation into microsomal vesicles of a lysosomal enzyme has recently been demonstrated. Using an antibody against cathepsin D, it was shown (95) that different products are precipitated after translation of total spleen mRNA in the presence or absence of dog pancreas membranes. The larger, presumably glycosylated polypeptide obtained in the former case was found to be insensitive to proteolysis, which indicates its segregation into the microsomal vesicles.

TRANSFER OF PROTEINS ACROSS BACTERIAL MEMBRANES

In *E. coli* and other gram-negative bacteria there exist specific sets of proteins located in the inner membrane, the outer membrane, and the periplasmic space. These latter two groups have to be secreted through the inner membrane, while some of the proteins of the inner membrane may be viewed as partially secreted proteins. Soon after the formulation of the signal hypothesis, data on a similar secretory mechanism for procaryotes were presented by several laboratories. Different aspects of protein localization in *E. coli* have recently been reviewed (96–99).

Procaryotes do not possess an endoplasmic reticulum. Its functional equivalent for protein secretion is the inner membrane. Polysome fractions bound to this membrane have been isolated from *E. coli* and these were shown to be rich in mRNAs for periplasmic and outer membrane proteins (100, 101). If one extends the concept of cotranslational secretion to procaryotes, it can be predicted that the growing polypeptide chain should be accessible from the external medium. To check this possibility, protoplasts have been treated with radioactive reagents that covalently modify polypeptide chains and that cannot penetrate a membrane. It could be demonstrated (97, 102) that such reagents label incomplete polypeptide chains still attached to polysomes and bound to tRNA. Using this approach, the co-translational secretion has been demonstrated for several bacteria other than *E. coli* (97, 103). One difference between the procaryotic and the eucaryotic systems appears to be the ribosome-membrane interaction. While both puromycin and high ionic strength are required to detach ribosomes from the endoplasmic reticulum, a weaker ribosome-membrane attachment susceptible to puromycin alone was found in bacteria (97, 103). This suggested that ribosomes are linked to the bacterial plasma membrane solely through their nascent chains, and led to interesting speculations about the energy requirement of polypeptide secretion (97).

Evidence for the existence of precursors was simultaneously presented for an outer membrane protein, a periplasmic enzyme, and a phage coat protein that accumulates in the inner membrane of *E. coli*. First, Inouye and colleagues showed that the mRNA for the lipoprotein of the outer membrane codes for a precursor with an additional twenty amino acids at the NH_2 terminus (104). This precursor as well as precursors for other outer membrane proteins were also detected in toluene-treated cells (105). The terminal sequence of pro- or prelipoprotein has the characteristics of a signal peptide in that it starts with a positively charged part followed by a cluster of hydrophobic residues. Second, H. Inouye & Beckwith (106) demonstrated that alkaline phosphatase of *E. coli*, a periplasmic enzyme, is synthesized as a larger precursor that could slowly be converted to the mature enzyme by incubation with an outer membrane fraction. The latter was found to be less hydrophobic than the precursor, which indicates the cleavage of a peptide rich in apolar amino acids during maturation. Finally, sequence analysis of the mRNA for the major coat protein of phage fd (107) demonstrated that it codes for a polypeptide starting with a segment of 23 amino acids not found in the final coat protein. This established the existence of transient precursors for proteins located in different compartments of *E. coli* and, in conjunction with the experiments of Smith et al (102), a secretory mechanism similar to the one operating in animal cells was strongly indicated. Signal sequences have since been identified for a variety of periplasmic and outer membrane proteins (see Table 2). While the co-translational transport and processing of these precursors is suggested by the in vivo studies discussed above (97, 103), slow cleavage of completed precursors has been observed in vitro (65, 106). Moreover, evidence for post-translational transport and processing in vivo has been presented for an outer membrane protein (123) and two phage proteins, β-lactamase (115) and coat protein (see below). Contrary to the nomenclature adopted for eucaryotic systems, the bacterial precursors have frequently been called proproteins. This is perhaps unfortunate, yet the above shows that the distinction between preproteins. i.e. precursors processed during chain growth, and proproteins, which are cleaved only after chain completion, is becoming somewhat ill-defined for certain bacterial proteins.

A comparison of the NH_2-terminal extensions found in bacterial proproteins with signal peptides of secretory polypeptides of eucaryotic cells reveals no principal differences. They are similar in length, distribution of polar and apolar amino acids, and in the nature of the COOH-terminal residue. The structure of these extensions is similar, irrespective of whether the mature protein is located in one of the two membranes or in the periplasmic space which suggests that the NH_2-terminal region functions as a true signal sequence required solely for vectorial discharge through the inner membrane, and that the subsequent location is determined by other factors. The experimental evidence indicates that this may be true for some but certainly not for all extracytoplasmic proteins of *E. coli.*

Protein Localization Mutants

Significant progress in the study of protein localization in *E. coli* has been possible through the construction of mutants that produce hybrid proteins. This genetic approach has already yielded significant results in the attempt to unravel the essential features of signal peptides and protein transport, and it holds great promise for future research.

A method was described several years ago that allows the fusion of genes widely separated on the *E. coli* chromosome (116). Using this technique,

Table 2 Bacterial proteins with known signal peptides

Protein[a]	Length of signal peptide	Ref.
Periplasmic proteins		
Maltose-binding protein	26	108
Leucine-binding protein	23	65
Arabinose-binding protein	24	109
β-Lactamase (plasmid pBR 322)	23	110
Outer membrane proteins		
Lipoprotein	20	104
Lipoprotein (*Serratia marcescens*)	19	111
LamB gene product (λ receptor)	21	112
ompA protein	21	113
Inner membrane protein		
Phage M13 coat protein (fd)	23	107
Phage minor coat protein (fd)	18	114

[a] Unless indicated otherwise, all proteins are from *E. coli.*

Bassford, Silhavy & Beckwith (96, 117) have isolated mutant strains carrying a fusion between the gene for the maltose-binding protein, normally present in the periplasm, and β-galactosidase, a cytoplasmic enzyme. The mutant strains constructed in these studies form large hybrid proteins with amino-terminal portions of different length of the maltose-binding protein, and almost the entire β-galactosidase. Synthesis of these proteins is induced by maltose. It was shown that the amino-terminal portion of the maltose-binding protein, including its signal sequence, generally suffices to partially initiate secretion. However, for some unknown reason, export remains incomplete and the hybrid proteins are largely localized in the cytoplasmic membrane. Accumulation of these proteins in the inner membrane has deleterious effects, since the export of other proteins to the periplasm and the outer membrane becomes blocked. These strains then show a maltose-sensitive phenotype in that the presence of maltose in the growth medium induces synthesis of hybrid proteins that eventually results in cell death. Mutations in the signal sequence of the maltose-binding protein can now lead to a reversion from the maltose-sensitive to the maltose-resistant phenotype (96, 117). The hybrid proteins formed in these revertants are localized largely in the cytoplasm of the cell, and its export apparatus is thus no longer blocked. The sequence of the signal sequence of wild-type maltose-binding protein and of several export-defective revertants was determined (108). In four of these, an apolar amino acid in the center region of the signal was replaced by an acidic or basic one; in a fifth mutant, a leucine residue in the same part was replaced by proline. These results clearly demonstrate that the central hydrophobic region of signal peptides is essential and the introduction of a single charge can block its ability to initiate export. The importance of secondary structure in this region, possibly alpha-helix formation, is emphasized by the mutant which contains proline instead of leucine.

Similar results have been obtained with *E. coli* strains in which the gene for β-galactosidase was fused to *lamB*, the gene for an outer membrane protein also involved in maltose uptake (118). Different classes of mutants could be discerned, depending on the amount of *lamB* DNA linked to the β-galactosidase gene, that produce hybrid proteins of different size. In one type, termed class III, a hybrid protein containing about half the *lamB* gene product was synthesized, which again renders these strains maltose-sensitive due to blockade of the export machinery of the cell. Several revertants were characterized in which the hybrid protein remains in the cytoplasm due to a mutation in the signal peptide of the *lamB* gene product (118). Replacement of an apolar by a charged amino acid or a deletion of four amino acids from the signal peptide were found to prevent export of the hybrid protein.

Other strains carrying *lamB*-β-galactosidase gene fusions produced hybrid proteins with surprising properties. One contains the first 39 amino acids of the outer membrane protein including its complete signal sequence (119). This protein remains in the cytoplasm, which demonstrates that in this case a signal sequence is not sufficient for export. Even more puzzling have been the characteristics of strains with the largest hybrid proteins, which contain more than half of the *lamB* gene linked to the β-galactosidase. Contrary to what would have been expected from the properties of class III mutants (see above) these were only weakly maltose-sensitive, and the bulk of the hybrid protein was found in the outer membrane (99). Clearly, a region in the second half of the *lamB* gene product must play a crucial role in directing this protein to the outer membrane.

Another example that a signal sequence may not be sufficient for export has been presented (115). Synthesis and secretion of β-lactamase was studied in *Salmonella* infected with phage P22. Several chain-termination mutants were investigated that produced β-lactamase lacking all or only parts of the COOH-terminal half of the protein. Both wild-type and mutant proteins were shown to be originally synthesized as precursors from which the signal sequence was subsequently cleaved. Surprisingly, only the wild-type protein was exported to the periplasm, while mutant proteins lacking as few as 21 amino acids from the carboxyl end remained in the cytoplasm. The post-translational processing of the precursor and the importance of carboxyl-terminal sequences for export are clearly at variance with the signal hypothesis. Secretion of β-lactamase may occur by a mechanism similar to the import of polypeptides into organelles (see later) or may depend on conformation of the complete protein as suggested by the "membrane trigger hypothesis" (132).

A different case has been described for an *E. coli* mutant with an altered precursor of the outer membrane lipoprotein (120). The mutant protein contained an aspartic acid instead of glycine in position 14 of the signal peptide. Contrary to what has been observed for maltose-binding protein and phage lambda receptor, this change in the hydrophobic segment of the signal sequence only partly impaired its export, and about 60% of the protein still reached the outer membrane. However, the bacterial signal peptidase could not process the mutant precursor, and this leads to a defect in the assembly of the outer membrane. Export without proteolytic processing has also been observed for several outer and inner membrane proteins encoded by F sex factor (137).

This genetic approach to the mechanism of protein localization in *E. coli* should conceivably also yield information beyond a more precise definition of the structural requirements for signal sequences and other regions of exported polypeptides. It is hoped that pleiotropic mutants defective in protein secretion in general will be found, that will help unravel the components of the secretory apparatus of bacteria. One such mutant with reduced amounts of a portion of the periplasmic and outer membrane proteins has already been described (121).

Conditions have also been found where precursors of outer membrane and periplasmic proteins accumulate. Treatment of *E. coli* cells with the local anesthetic procaine results in an inhibition of secretion of periplasmic proteins, and their precursors accumulate in the cytoplasmic membrane (122). The precursors for outer membrane proteins persist under a variety of conditions, such as cooling the cells below the transition point of their membrane lipids or exposure to phenylethyl alcohol (123, 124). Interestingly, these precursors were largely found in the outer membrane and they could be "chased" into mature protein upon returning to normal conditions. It thus appears that for some outer membrane proteins, processing of the signal peptide is not essential for localization but may instead be required for correct assembly, a characteristic reminiscent of eucaryotic pro- rather than preproteins. One explanation for this notion may be that outer membrane proteins can apparently reach their destination through Bayer's patches, regions where the inner and outer membrane are in contact, and thus do not have to cross the aqueous periplasm (98).

The Case of the Phage Coat Protein

An interesting controversy has arisen over the biosynthesis of a phage coat protein localized in the inner membrane of *E. coli.* Both co- and post-translational insertion and processing has been suggested for the precursor of phage M13 (also called fd or f1) coat protein, a polypeptide of 73 amino acids. In its mature form, phage coat protein spans the inner membrane with about 20 residues exposed to the periplasm and 11 on the cytoplasmic side.

Incubation of phage DNA in a coupled transcription-translation system yields as one of the major products the coat protein precursor that has been termed procoat (125, 126). Chang et al (125, 127) reported that in the presence of a fraction of inverted vesicles from *E. coli,* a partial conversion to coat protein was observed. In their experiments, vesicles had to be present during translation, while their subsequent addition did not lead to processing. More extensive processing was observed in the presence of detergent to activate the largely latent bacterial signal peptidase. This cotranslational insertion was subsequently shown to be in the correct orientation, i.e. only the COOH-terminus remained exposed to the medium and was sensitive to protease treatment. Completed procoat bound strongly to inverted vesicles, but it did not assume the correct orientation and could be entirely degraded by proteases (127).

Opposite findings were reported by Wickner and co-workers (126, 128). Using the same experimental approach, complete agreement was found on the cotranslational aspects of processing by added membranes and its enhancement in the presence of detergents (128). However, these authors reported that even after chain completion procoat could assemble correctly into membranes as well as liposomes (126). Conversion of procoat to coat in the absence of detergent was also observed (128). These conflicting results led to a series of in vivo studies (129–131) with *E. coli* cells infected with normal and mutant phages. The results show that procoat is synthesized on free polysomes in a soluble form and then post-translationally inserted into the cytoplasmic membrane and converted to coat protein. This insertion in vivo was inhibited by uncouplers but not by arsenate, i.e. ATP depletion. On the basis of these findings, the "membrane trigger hypothesis" was proposed (132), which predicts that certain membrane proteins can be formed as soluble precursors and subsequently be integrated into a membrane with or without cleavage of a terminal fragment. Insertion without proteolytic processing has been observed for several inner and outer membrane proteins of *E. coli* (137). This has in particular been demonstrated for lactose permease by sequence analysis of the gene and its product (138, 139). It is, however, not known whether these proteins are synthesized on free or on membrane-bound polysomes.

Bacterial Signal Peptidase

An approximately 6000-fold purification of bacterial signal peptidase has recently been reported (133) and the post-translational cleavage of procoat by this enzyme in the presence of detergent has been demonstrated. Interestingly, the protease was found to be sensitive to dinitrophenol. In their report on the cotranslational cleavage of procoat by inverted vesicles prepared from the inner membrane of *E. coli,* Chang et al (125) detected not only coat protein, but also varying amounts of the cleaved signal peptide. This has not yet been achieved for any eucaryotic system. Possibly, further hydrolysis of bacterial signal peptides by "signal peptide hydrolase" proceeds at a slower rate. In this context, a recent suggestion (134) should be mentioned that the bacterial factors, which stimulate neurophils during an inflammatory response, may be derived from signal peptides released during synthesis of extracytoplasmic polypeptides.

Comparative Aspects

In view of the overall similarity between the secretory apparatus of bacteria and animal cells, it would be of interest to know whether a eucaryotic preprotein can be translocated and processed by a procaryotic membrane and vice versa. Such information has been obtained by cloning genes of animal secretory polypeptides in *E. coli* under conditions where formation of a polypeptide product was possible. Two groups have achieved the synthesis of polypeptides closely resembling ovalbumin in *E. coli* (135, 136). From the strategy used for construction of the recombinant plasmids, the product made in *E. coli* should in both cases start with the NH_2-terminal sequence of β-galactosidase, a cytoplasmic polypeptide. While the actual amino termini of these ovalbumin-like polypeptides has in neither case been determined, the formation of the amino terminus of natural ovalbumin appears most unlikely. Fraser & Bruce (135) reported that about half of the ovalbumin synthesized in their system was recovered from the periplasmic space of *E. coli*, while Mercereau-Puijalon et al (136) found no trace of ovalbumin in the supernatant of bacterial cultures. One has to postulate the recognition of an internal signal sequence by the *E. coli* membrane to explain the former result.

More recently, Talmadge and colleagues (140) demonstrated that a eucaryotic signal sequence is functional in *E. coli*. Different plasmids were constructed that contained part or all of the signal sequence of the plasmid penicillinase gene fused to different locations of the rat preproinsulin gene. Most interesting for the present considerations, a hybrid with only the first four residues of the penicillinase signal, a linker tetrapeptide, and 21 of the 24 residues of the prepart of preproinsulin is efficiently transported into the periplasmic space. Moreover, it was shown by the same group (141) that the secreted penicillinase-preproinsulin hybrid is also correctly processed to proinsulin by bacterial signal peptidase. This demonstrates in a convincing way that the secretory apparatus has been as much conserved in evolution as protein synthesis itself. Complementary experiments on the processing of bacterial precursors with microsomal membranes have not yet been reported.

BIOSYNTHESIS OF MEMBRANE PROTEINS: MORE VARIATIONS AND NEW TUNES

Even though detailed studies on the biosynthesis of only a few membrane proteins are so far available, several different mechanisms of insertion are already apparent. This is not surprising in view of the fact that just about every conceivable orientation of polypeptide chains in the phospholipid bilayer has been observed. These include insertion, via amino- or carboxyl-terminal segments with exposure on either side of the membrane, of terminal and/or internal regions of the protein. This section discusses a few prototypes of membrane proteins for which the orientation and certain aspects of their biosynthesis have been established. More comprehensive reviews have been presented (11, 12). Proteins of mitochondrial and chloro-plast membranes are reviewed separately.

Incompletely Secreted Polypeptides

There exists a group of transmembrane proteins synthesized on membrane-bound polysomes, whose biosynthesis closely follows the route established for secretory polypeptides. These proteins have their NH_2 terminus and the bulk of the polypeptide chain on the luminal side of the endoplasmic reticulum and ultimately exposed at the cell surface. Close to the carboxyl end they contain a hydrophobic region of 20–25 amino acids that remains embedded in the membrane. This sequence has been called a stop-tranfer sequence—a signal where secretion comes to a halt and only a few residues (up to 39) remain exposed on the cytoplasmic side. To this group belong several viral glycoproteins, such as the G protein of vesicular stomatitis virus (VSV) and the hemagglutinin of influenza virus, as well as cellular glycoproteins, e.g. the heavy chain of the histocompatibility antigens and glycophorin. The biosynthesis of VSV G protein has been studied in great detail and thus serves as the prototype for this group of proteins (142–145).

In vitro studies on the biosynthesis of G protein have demonstrated that it is initiated on free polysomes with the synthesis first of a typical signal sequence containing 16 amino acids (144). Soon after initiation, microsomal membranes must be added to obtain the correct insertion into the mem-brane, i.e. vectorial discharge of the growing polypeptide chain. Rothman & Lodish (143) have estimated, that the "window" to initiate translocation across the microsomal membrane is open only until the polypeptide chain reaches a length of 80–100 residues. Cotranslational cleavage of the signal peptide, as well as core-glycosylation have been clearly established. Translo-cation of G protein then stops at a hydrophobic region close to the COOH terminus, and in its final orientation only about 5% of the polypeptide chain (29 residues) remain exposed on the cytoplasmic side (146). Experiments where mRNAs for G protein and the secretory protein prolactin were present simultaneously demonstrated, furthermore, that these mRNAs competed for the same "channel" or some other component of the secretory apparatus (144). The further transport of G protein via Golgi vesicles to the plasma membrane, and the concomitant changes in the carbohydrate struc-ture, are currently subjects of intense investigations in several laboratories (12, 147, 148).

Cotranslational insertion and cleavage of a signal peptide have also been demonstrated for the heavy chain of histocompatibility antigens (149, 150) and for the hemagglutinin of influenza virus (151–153). Glycophorin, a glycoprotein of the erythrocyte membrane, which carries MN blood group specificity, has the same type of orientation (154) and is probably synthesized along the same route. The cytoplasmic stubs of these transmembrane proteins appear to be generally shorter than the peptide embedded in the large subunit of the ribosome, which has been estimated to accommodate around 40 amino acids (5). This indicates that ribosomes may be able to detach from the endoplasmic reticulum only after chain completion but not during polypeptide synthesis.

An interesting variation has been found for one of the transmembrane glycoproteins of Sindbis virus. This virus and the closely related Semliki forest virus encode a large polyprotein starting from one initiation site on a 26S mRNA that is processed during translation (155–158). The complete nucleotide sequence of Semliki forest virus mRNA has now been determined (159) and has been shown to code, starting at the 5' end, for a cytoplasmic core protein and the two transmembrane glycoproteins p62 and E_1. (The former is subsequently cleaved to yield E_3 and E_2.) In the case of Sindbis, the identical gene order 5'-core-pE_2-E_1-3' has been established, where pE_2, just prior to virus maturation, is converted to E_2. In both cases synthesis of the polyprotein is initiated in the cytoplasm. Right after completion of the core protein, this is cleaved from the growing polypeptide chain and a new NH_2 terminus is exposed (157). This functions as a signal sequence that binds to the endoplasmic reticulum and initiates chain translocation across the membrane. It has recently been demonstrated that in the case of Sindbis virus this signal is not cleaved in vivo as well as in vitro (161). This emphasizes the earlier finding on ovalbumin (89) that signal sequences may be retained in the final product.

A more complex example appears to be opsin, the apoprotein of rhodopsin, which is synthesized on bound polysomes and subsequently transported via Golgi vesicles to the membraneous discs of outer rod segments (162). It has been shown that the amino-terminal sequences of opsin synthesized in vitro and of mature rhodopsin are identical save for the fact that the initiating methionine is acetylated in the latter (163). The exact orientation of opsin in the membrane is presently not known (164) but two asparagines close to the amino end, residues 2 and 15, are glycosylated (165) and are thus probably located on the extracytoplasmic side of the membrane.

An interesting group of hydrolytic enzymes is present in microvillar membranes of intestinal and kidney cells (166, 167). These include the sucrase-isomaltase complex (168), aminopeptidase N and A (169, 170), alkaline phosphatase (171), and dipeptidyl peptidase IV (172). These enzymes have the bulk of their polypeptide chains, including the active centers, located in the extracellular space, and can be detached from the membrane by mild proteolysis. However, in contrast to the viral and cellular glycoproteins discussed above, this group of proteins is bound to the membrane by a hydrophobic segment at or near the amino end. These anchors may contain as few as 40 residues (170, 172). It is presently not known whether the NH_2-terminal amino acids of these enzymes are located outside the cell or in the cytoplasm, although in case of the aminopeptidase, evidence for the latter has been presented (173).

The sucrase-isomaltase complex has been studied most extensively. It contains two polypeptide chains derived from a common precursor (174) that are tightly bound to each other by noncovalent interactions and that contain the respective enzymatic activites. The complex is anchored to the luminal surface of epithelial cells in the small intestine through the amino end of the isomaltase subunit. The amino acid sequence of this NH_2-terminal part has been determined (175); it contains a carbohydrate side chain at residue 11 immediately followed by an extremely hydrophobic segment composed of at least 20 amino acids.

The biosynthetic route by which these enzymes acquire their orientation in the cell membrane is currently not known. Theoretically, their amino ends could represent uncleaved signal peptides that are membrane bound at all times, or they could be secreted completely into the lumen of the endoplasmic reticulum and subsequently reinserted. The presence of a carbohydrate chain near the amino end of the isomaltase makes the latter possibility more likely.

Other Membrane Proteins Synthesized on Bound Polysomes

Several membrane proteins present in rough and smooth endoplasmic reticulum have been shown to be synthesized on bound polysomes. These include cytochrome P-450 (176), epoxide hydratase (177), and probably NADPH-cytochrome P-450 reductase. The latter enzyme is anchored to the membrane solely through an amino-terminal segment, while the rest of the protein is exposed to the cytoplasm (178). The others are more extensively buried in the membrane, but it is not known whether portions are exposed on the extracytoplasmic side.

Amino-terminal sequences of different species of cytochrome P-450 have been determined and were all found to be very rich in apolar amino acids (179, 180). The NH_2-terminal sequence of cytochrome P-450 synthesized in vitro was found to be identical to that of the mature protein (176). Translation of epoxide hydratase mRNA in a cell-free system also yielded a product identical to the native enzyme (177). Cleavage of terminal signal peptides is not required for the proper insertion of these proteins into membranes.

Two other proteins present in microsomal and several other cellular membranes—cytochrome b_5 and NADH cytochrome b_5 reductase—have been shown to be inserted at their carboxyl-terminal ends (181, 182). Apparently these are both synthesized on free polysomes (183, 184) and their binding to membranes is considered to occur upon exposure of the completed COOH-terminal domain.

SIGNAL SEQUENCES AS "LEITMOTIFS" FOR ORGANELLE PROTEINS

Even though mitochondria and chloroplasts contain genetic information and synthesize proteins, they both depend to a large extent on the import of polypeptides manufactured in the cytoplasm. The selective uptake of different classes of proteins into these organelles and their distribution to different compartments has been an active area of investigation in the past few years (185, 186). The general conclusion now is that protein import into chloroplasts and mitochondria is a post-translational process not coupled to protein synthesis. As reviewed elsewhere in this volume in the chapter by Tolbert, this principle also holds for the proteins found in peroxisomes and glyoxisomes.

Mitochondrial Protein Import

Mitochondria have an outer membrane that is permeable to molecules up to a molecular weight of several thousand, and an inner membrane. Specific sets of proteins are localized in these two membranes and in the intermembrane space and the mitochondrial matrix (187). Examples for the uptake of polypeptides into all of these compartments, except for the outer membrane, have been presented.

The mitochondrial protein investigated most extensively is cytochrome c, and accordingly, early attempts to study the biosynthesis of mitochondrial proteins have focused mainly on this component. There were several reports on the synthesis of cytochrome c on bound polysomes (reviewed in 188), until much later synthesis on free polysomes was strongly indicated (7, 189). The genetic studies on yeast cytochrome c (190) and finally the elucidation of the structure of this gene (191) established that, at least in this organism, apocytochrome c is the primary translation product. In vitro translation of cytochrome c mRNA also demonstrated that there exists no precursor for this protein (192, 193).

Cytochrome c has to traverse the outer mitochondrial membrane to reach its location at the cytoplasmic side of the inner membrane, where it is bound by ionic interactions. Neupert and colleagues (7, 193, 194) have demonstrated that apocytochrome c is taken up by mitochondria in vitro where it is converted to holocytochrome c. This import of the apoprotein is inhibited by carbonyl cyanide m-chlorophenylhydrazone, an uncoupler of oxidative phosphorylation (195). Preliminary data indicated that the recognition site for import resides in the carboxyl-terminal third of apocytochrome c (196). As the complete hemoprotein is not taken up into mitochondria, this site is apparently not exposed in the mature protein.

The studies on the biosynthesis of mitochondrial proteins entered a new phase when, as the result of a transatlantic collaboration, it was shown (8) that three subunits of the F_1-ATPase complex of yeast mitochondria were derived from larger precursors made in the cytoplasm. The post-translational uptake of these precursors into mitochondria and·their cleavage to the mature proteins could also be demonstrated.

In quick succession, similar findings were reported for a number of other mitochondrial proteins of cytoplasmic origin. These included subunit V of the cytochrome bc_1 complex of the inner membrane (197), cytochrome c peroxidase, located between the two membranes (198), and the matrix enzymes carbamyl phosphate synthetase (199, 200), ornithine transcarbamoylase (201), citrate synthetase (202), and aspartate aminotransferase (203). All these precursors of mitochondrial proteins are larger by 2000–6000 daltons than the mature proteins.

However, mitochondrial proteins imported from the cytosol are not always derived from precursors, as already mentioned for cytochrome c. This has also been shown for the ADP/ATP carrier of *Neurospora crassa,* an integral protein of the inner mitochondrial membrane. The in vitro product, synthesized in the presence of formylated initiatory methionyl-tRNA to exclude processing at the amino end, has the same size as the mature carrier protein (204). Precursors could also not be detected for one of the subunits of cytochrome oxidase (see below) and for carbamyl phosphate synthetase from tadpole liver (205).

Several laboratories have investigated the biosynthesis of cytochrome oxidase from yeast mitochondria. This enzyme is composed of seven subunits, three of which are made in mitochondria, while the four smaller ones are imported from the cytoplasm. It was originally suggested (206) that the four subunits of cytoplasmic origin are derived from a polyprotein through successive proteolytic cleavages. However, recent experiments from four laboratories are clearly at variance with the "polyprotein concept" and convincing evidence for the synthesis of each of the four subunits as individual polypeptides has been presented (207–210). Using in vivo labeling of yeast cells in the presence of uncouplers (207, 208) as well as in vitro translation, precursors for subunits IV, V, and VI could be detected. In the case of subunit VII, however, the primary translation product was found to be identical to the mature polypeptide (208). The in vitro translation products could be labeled with formylated initiator (^{35}S) methionyl tRNA, which ruled out the possibility that they could be derived from a polyprotein. As noted by Lewin at al (207), some of the antibody preparations directed against individual subunits did precipitate a protein of the size reported (206) for the polyprotein precursor. However, this probably represents a highly antigenic, unidentified polypeptide of yeast mitochondria that is not related to cytochrome oxidase.

As originally shown (195) import of cytochrome c and other proteins into mitochondria in vitro is inhibited by uncouplers. Depletion of the mitochondrial matrix ATP in intact yeast spheroplasts by various inhibitors prevents processing (211) and presumably also uptake of several precursors of inner membrane proteins. However, cleavage of the precursor of cytochrome c peroxidase, an enzyme located in the mitochondrial intermembrane space, was not impaired by these inhibitors (211). Transfer of the ADP/ATP carrier into *Neurospora* mitochondria in vitro was also found to be blocked by an uncoupler (212).

Transport of Proteins into Chloroplasts

The uptake of cytoplasmic polypeptides into chloroplasts resembles that observed in mitochondria. The first precursor of an organelle protein was actually discovered in studies on the biosynthesis of a prominent chloroplast protein, the small subunit of ribulose-bis-phosphate carboxylase. Working with *Chlamydomonas reinhardtii,* it could be demonstrated (9) that a precursor for this chloroplast protein is synthesized on free polysomes, which can be processed after chain completion by an endoprotease present in these algae. This precursor was subsequently shown to contain an amino-terminal extension of 44 amino acids that presumably represent the signal for uptake into the chloroplast stroma (213). This extension also has an excess of hydrophobic amino acids and is strikingly rich in alanine (15 residues). However, as opposed to signal peptides for secretory proteins, no stretch of apolar residues containing more than eight amino acids is present (213). A similar precursor was subsequently discovered for the small subunit of ribulose-bis-phosphate carboxylase from higher plants, and researchers have established its post-translational uptake into chloroplasts (214, 215). It was then demonstrated that this uptake is an ATP-dependent process (216).

There is also evidence for the existence of precursors of two components of the light-harvesting chlorophyll a/b complex that are imported, processed, and assembled into thylakoid membranes post-translationally (216,

217). In the case of ferredoxin, preliminary evidence indicates that it is taken up as the apoprotein without further processing (218).

General Aspects

Details about the selective import of completed polypeptide chains into mitochondria, chloroplasts, or other cell organelles are currently not known. The analogy between these processes and the uptake of toxins like diphtheria toxin (219), ricin, or abrin (220) through the plasma membrane of cells is obvious. The intact toxins could be likened to precursors, of which one part is taken up through the membrane, but it is not clear at present how far-reaching this analogy is.

The post-translational uptake of polypeptides into organelles may prove to be more amenable to experimental analysis, since translation and transport are not coupled. It has thus been possible to establish the energy requirement of this process, something that will be difficult to unravel for the secretory system. In the precursor of the ribulose-bis-phosphate subunit of chloroplasts, the extra amino acids are located at the NH_2 terminus (213) and the same is probably true for some precursors of mitochondrial polypeptides (207). However, the interaction of a completed, as opposed to a nascent, chain with the membrane could in principle occur with any part of the molecule. It need not even be a continuous stretch of amino acids— an assumption inherent in the current use of the term signal (221)—but may depend on a three-dimensional structure made up of different regions of the polypeptide chain (132). Nevertheless, the strategy employed in the attempts to locate the signal peptide in ovalbumin, where proteolytic fragments were used to compete with nascent chains for membrane interaction (90), could also be tried to study the mechanism of protein import into organelles. A report on the isolation of a fragment of cytochrome c that blocks the uptake of this and other mitochondrial proteins (196), raises the intriguing possibility that a common recognition sequence for the outer mitochondrial membrane exists. However, the experience gained with the concept of signal sequences has taught us not to be so optimistic as to expect a general mechanism.

ACKNOWLEDGMENTS

I thank Dr. Annelore Vaerst for her valuable help in the preparation of this manuscript, Dr. J. V. Small for patiently correcting my English, and Christa Mollay for many helpful discussions. I am indebted to B. Dobberstein, A. Frischauf-Lehrach, H. Lodish, and W. Wickner for providing manuscripts prior to publication.

Literature Cited

1. Palade, G. 1975. *Science* 189:347–58
2. Redman, C. M., Siekevitz, P., Palade, G. E. 1966. *J. Biol. Chem.* 241:1150–58
3. Chefurka, W., Hayashi, Y. 1966. *Biochem. Biophys. Res. Commun.* 24:633–38
4. Redman, C. M., Sabatini, D. D. 1966. *Proc. Natl. Acad. Sci. USA* 56:608–15
5. Sabatini, D. D., Blobel, G. 1970. *J. Cell Biol.* 45:146–57
6. Kadenbach, B. 1970. *Eur. J. Biochem.* 12:392–98
7. Korb, H., Neupert, W. 1978. *Eur. J. Biochem.* 91:609–20
8. Maccecchini, M. L., Rudin, Y., Blobel, G., Schatz, G. 1979. *Proc. Natl. Acad. Sci. USA* 76:4365–69
9. Dobberstein, B., Blobel, G., Chua, N.-H. 1977. *Proc. Natl. Acad. Sci. USA* 74:1082–85
10. Robbi, M., Lazarow, P. B. 1978. *Proc. Natl. Acad. Sci. USA* 75:4344–48
11. Wickner, W. 1980. *Science.* 210:861–68
12. Lodish, H. F., Braell, W. A., Schwartz, A. L., Strous, G. J. A. M., Zilberstein, A. 1980. *Int. Rev. Cytol.* In press
13. De Robertis, E. M., Longthorne, R. F., Gurdon, J. B. 1978. *Nature* 272:254–57
14. Rechsteiner, M., Kuehl, L. 1979. *Cell* 16:901–8
15. Baglioni, C., Bleiberg, I., Zauderer, M. 1971. *Nature New Biol.* 232:8–12
16. Mechler, B., Vassalli, P. 1975. *J. Cell Biol.* 67:25–37
17. Blobel, G., Sabatini, D. D. 1971. *Biomembranes* 2:193–95
18. Milstein, C., Brownlee, G. G., Harrison, T. M., Matthews, M. B. 1972. *Nature New Biol.* 239:117–20
19. Mach, B., Faust, C., Vassalli, P. 1973. *Proc. Natl. Acad. USA* 72:224–28
20. Schechter, I., 1973. *Proc. Natl. Acad. Sci. USA* 70:2256–60
21. Kemper, B., Habener, J. F., Mulligan, R. C., Potts, J. T. Jr., Rich, A. 1974. *Proc. Natl. Acad. Sci. USA* 71:3731–35
22. Boime, I., Boguslawski, S., Caine, J. 1975. *Biochem. Biophys. Res. Commun.* 62:103–9
23. Suchanek, G., Kindas-Mügge, I., Kreil, G., Schreier, M. H. 1975. *Eur. J. Biochem.* 60:309–15
24. Blobel, G., Dobberstein, B. 1975. *J. Cell Biol.* 67:835–51
25. Burstein, Y., Schechter, I. 1977. *Proc. Natl. Acad. Sci. USA* 74:716–20
26. Smith, G. P. 1978. *Biochem. J.* 171:337–47
27. Tonegawa, S., Maxam, A. M., Tizard, R., Bernard, O., Gilbert, W. 1978. *Proc. Natl. Acad. Sci. USA* 75:1485–89

28. Jilka, R. L., Pestka, S. 1977. *Proc. Natl. Acad. Sci. USA* 74:5692–96
29. Strauss, A. W., Bennett, C. D., Donohue, A. M., Rodkey, J. A., Alberts, A. W. 1977. *J. Biol. Chem.* 252:6846–55
30. MacGillivray, R. T. A., Chung, D. W., Davie, E. W. 1979. *Eur. J. Biochem.* 98:477–85
31. Chung, D. W., MacGillivray, R. T. A., Davie, E. W. 1980. *Ann. NY Acad. Sci.* 343:210–17
32. Habener, J. F., Rosenblatt, M., Kemper, B., Kronenberg, H. M., Rich, A., Potts, J. T. Jr. 1978. *Proc. Natl. Acad. Sci. USA* 75:2616–20
33. Chan, S. J., Keim, P., Steiner, D. F. 1976. *Proc. Natl. Acad. Sci. USA* 73:1964–68
34. Villa-Komaroff, L., Efstratiadis, A., Broome, S., Lomedico, P., Tizard, R., Naber, S. P., Chick, W. L., Gilbert, W. 1978. *Proc. Natl. Acad. Sci. USA* 75:3727–31
35. Ullrich, A., Dull, T. J., Gray, A., Brosius, J., Sures, I. 1980. *Science* 209:612–15
36. McKean, D. J., Maurer, R. A. 1978. *Biochemistry* 17:5215–19
37. Seeburg, P. H., Shine, J., Martial, J. A., Baxter, J. D., Goodman, H. M. 1977. *Nature* 270:486–94
38. Martial, J. A., Hallewell, R. A., Baxter, J. D., Goodman, H. M. 1979. *Science* 205:602–6
39. Nakanishi, S., Inoue, A., Kita, T., Nakamura, M., Chang, A. C. Y., Cohen, S. N., Numa, S. 1979. *Nature* 278:423–7
40. Gossard, F., Seidah, N. G., Crine, P., Roithier, R., Chrétien, M. 1980. *Biochem. Biophys. Res. Commun.* 92:1042–51
41. Birken, S., Smith, D. L., Canfield, R. E., Boime, I. 1977. *Biochem. Biophys. Res. Commun.* 74:106–12
42. Fiddes, J. C., Goodman, H. M. 1979. *Nature* 281:351–56
43. Birken, S., Fetherston, J., Desmond, J., Canfield, R., Boime, I. 1978. *Biochem. Biophys. Res. Commun.* 85:1247–53
44. Mercier, J.-C., Hazé, G., Gaye, P., Petrissant, G., Hue, D., Boisnard, M. 1978. *Biochem. Biophys. Res. Commun.* 85:662–70
45. Mercier, J.-C., Hazé, G., Gaye, P., Hue, D. 1978. *Biochem. Biophys. Res. Commun.* 82:1236–45
46. Gaye, P., Gautron, J.-P., Mercier, J.-C., Hazé, G. 1977. *Biochem. Biophys. Res. Commun.* 79:903–11

47. Palmiter, R. D., Gagnon, J., Ericsson, L. H., Walsh, K. A. 1977. *J. Biol. Chem.* 252:6386–93
48. Thibodeau, S. N., Lee, D. C., Palmiter, R. D. 1978. *J. Biol. Chem.* 253:3771–74
49. Palmiter, R. D., Thibodeau, S. N., Gagnon, J., Walsh, K. A. 1977. *FEBS Proc. Meet. 11th,* 47:89–101
49a. Taniguchi, T., Mantei, N., Schwarzstein, M., Nagata, S., Muramatsu, M., Weissmann, C. 1980. *Nature* 285:547–49
50. Atger, M., Mercier, J.-C., Hazé, G., Fridlansky, F., Milgrom, E. 1979. *Biochem. J.* 1977:985–88
51. Suchanek, G., Kreil, G., Hermodson, M. A. 1978. *Proc. Natl. Acad. Sci. USA* 75:701–5
52. Chan, S. J., Patzelt, C., Duguid, J. R., Quinn, P., Labrecque, A., Noyes, B., Keim, P., Heinrikson, R. L., Steiner, D. F. 1979. *Miami Winter Symp.* 16:361–78
53. Austen, B. M. 1979. *FEBS Lett.* 103:308–12
54. Garnier, J., Gaye, P., Mercier, J.-C., Robson, B. 1980. *Biochimie* 62:231–39
55. Hortin, G., Boime, I. 1980. *Proc. Natl. Acad. Sci. USA* 77:1356–60
56. Rosenblatt, M., Beaudette, N. V., Fasman, G. D. 1980. *Proc. Natl. Acad. Sci. USA* 77:3983–87
57. Lauterwein, J., Bösch, C., Brown, L. R., Wüthrich, K. 1979. *Biochim. Biophys. Acta* 556:244–64
58. Schneider, A. B., Edelhoch, H. 1972. *J. Biol. Chem.* 247:4986–91
59. Prehn, S., Tsamaloukas, A., Rapoport, T. A. 1980. *Eur. J. Biochem.* 107:185–95
60. Warren, G., Dobberstein, B. 1978. *Nature* 273:569–71
61. Walter, P., Jackson, R. C., Marcus, M. M., Lingappa, V. R., Blobel, G. 1979. *Proc. Natl. Acad. Sci. USA* 76:1759–1799
62. Lingappa, V. R., Shields, D., Woo, S. L. C., Blobel, G. 1978. *J. Cell Biol.* 79:567–72
63. DiRienzo, J. M., Nakamura, K., Inouye, M. 1978. *Ann. Rev. Biochem.* 47:481–532
64. Lane, C., Shannon, S., Craig, R. 1979. *Eur. J. Biochem.* 101:485–95
65. Oxender, D. L., Anderson, J. J., Daniels, C. J., Landick, R., Gunsalus, R. P., Zurawski, G., Yanofsky, C. 1980. *Proc. Natl. Acad. Sci. USA* 77:2005–9
66. Maurer, R. A., McKean, D. J. 1978. *J. Biol. Chem.* 253:6315–18
67. Habener, J. F., Potts, J. T. Jr., Rich, A. 1976. *J. Biol. Chem.* 251:3893–99
68. Unwin, P. N. T. 1979. *J. Mol. Biol.* 132:69–84
69. Kreibich, G., Ulrich, B. L., Sabatini, D. D. 1978. *J. Cell Biol.* 77:464–87
70. Kreibich, G., Czako-Graham, M., Grebenau, R., Mok, W., Rodriguez-Boulan, E., Sabatini, D. D. 1978. *J. Supramol. Struct.* 8:279–302
71. Thibodeau, S. N., Walsh, K. A. 1980. *Ann. NY Acad. Sci.* 343:180–91
72. Jackson, R. C., Walter, P., Blobel, G. 1980. *Nature* 286:174–76
73. von Heijne, G., Blomberg, C. 1979. *Eur. J. Biochem.* 97:175–81
74. von Heijne, G. 1980. *Eur. J. Biochem.* 103:431–38
75. von Heijne, G. 1980. *Biochem. Biophys. Res. Commun.* 93:82–86
76. Harrison, T. M., Brownlee, G. G., Milstein, C. 1974. *Eur. J. Biochem.* 47:613–20
77. Habener, J. F., Kemper, B., Potts, J. T. Jr., Rich, A. 1975. *Biochem. Biophys. Res. Commun.* 67:1114–21
78. Szczesna, E., Boime, I. 1976. *Proc. Natl. Acad. Sci. USA* 73:1179–83
79. Wachsmuth, E. D. 1967. *Biochem. Z.* 346:446–455
80. Kreil, G., Suchanek, G., Kaschnitz, R., Kindas-Mügge, I. 1977. *FEBS Proc. 11th,* 47:79–88
81. Jackson, R. C., Blobel, G. 1977. *Proc. Natl. Acad. Sci. USA* 74:5598–5602
82. Kaschnitz, R., Kreil, G. 1978. *Biochem. Biophys. Res. Commun.* 83:901–7
83. Strauss, A. W., Zimmermann, M., Boime, I., Ashe, B., Mumford, R. A., Alberts, A. W. 1979. *Proc. Natl. Acad. Sci. USA* 76:4225–29
84. Strauss, A. W., Zimmermann, M., Mumford, R. A., Alberts, A. W. 1980. *Ann. NY Acad. Sci.* 343:168–79
85. Kreil, G., Mollay, G., Kaschnitz, R., Haiml, L., Vilas, U. 1980. *Ann. NY Acad. Sci.* 343:338–46
86. Jackson, R. G., Blobel, G. 1980. *Ann. NY Acad. Sci.* 343:391–404
87. Patzelt, C., Labrecque, A. D., Duguid, J. R., Carroll, R. C., Keim, P. S., Heinrikson, R. L., Steiner, D. F. 1978. *Proc. Natl. Acad. Sci.* 75:1260–64
88. Habener, J. F., Rosenblatt, M., Dee, P. C., Potts, J. T. Jr. 1979. *J. Biol. Chem.* 254:10596–99
89. Palmiter, R. D., Gagnon, J., Walsh, K. A. 1978. *Proc. Natl. Acad. Sci. USA* 76:94–98
90. Lingappa, V. R., Lingappa, J. R., Blobel, G. 1979. *Nature* 281:117–21
91. Meek, R. L., Walsh, K. A., Palmiter, R. D. 1980. *Fed. Proc.* 39:1867 (Abstr.)

92. Farquhar, M. G. 1978. *Transp. Macromol. Cell. Syst.* 11:351–62
93. Neufeld, E. F., Lim, T. W., Shapiro, L. J. 1975. *Ann. Rev. Biochem.* 44:357–76
94. Neufeld, E. F., Sando, G. N., Garvin, J. A., Rome, L. H. 1977. *J. Supramol. Struct.* 6:95–101
95. Erickson, A. H., Blobel, G. 1979. *J. Biol. Chem.* 254:11771–74
96. Silhavy, T. J., Bassford, P. J. Jr., Beckwith, J. R. 1979. *Bacterial Outer Membrane: Biogenesis and Function,* ed. M. Inouye. New York: Wiley
97. Davis, B. D., Tai, P.-C. 1980. *Nature* 283:433–38
98. Inouye, M., Halegoua, S. 1980. *Crit. Rev. Biochem.* 7:339–71
99. Emr, S. D., Hall, M. N., Silhavy, T. J. 1980. *J. Cell Biol.* 86:701–11
100. Randall, L. L., Hardy, S. J. S. 1977. *Eur. J. Biochem.* 75:43–53
101. Randall, L. L., Hardy, S. J. S., Josefson, L.-G. 1978. *Proc. Natl. Acad. Sci. USA* 75:1209–12
102. Smith, W. P., Tai, P.-C., Thompson, R. C., Davis, B. D. 1977. *Proc. Natl. Acad. Sci. USA* 74:2830–34
103. Smith, W. P., Tai, P.-C., Davis, B. D. 1979. *Biochemistry* 18:198–202
104. Inouye, S., Wang, S., Sekizawa, J., Halegoua, S., Inouye, M. 1977. *Proc. Natl. Acad. Sci. USA* 74:1004–8
105. Sekizawa, J., Inouye, S., Halegoua, S., Inouye, M. 1977. *Biochem. Biophys. Res. Commun.* 77:1126–33
106. Inouye, H., Beckwith, J. 1977. *Proc. Natl. Acad. Sci. USA* 74:1440–44
107. Sugimoto, K., Sugisaki, T., Okamoto, T., Takanami, M. 1977. *J. Mol. Biol.* 111:487–507
108. Bedouelle, H., Bassford, P. J. Jr., Fowler, A. V., Zabin, I., Beckwith, J., Hofnung, M. 1980. *Nature* 285:78–81
109. Wilson, V. G., Hogg, R. W., 1980. *J. Biol. Chem.* 255:6745–50
110. Sutcliffe, J. G. 1978. *Proc. Natl. Acad. Sci. USA* 75:3737–41
111. Nakamura, K., Inouye, M. 1980. *Proc. Natl. Acad. Sci. USA* 77:1369–73
112. Hedgpeth, J., Clement J.-M., Marchal, G., Perrin, D., Hofnung, M. 1980. *Proc. Natl. Acad. Sci. USA* 77:2621–25
113. Movva, N. R., Nakamura, K., Inouye, M. 1980. *J. Biol. Chem.* 255:27–29
114. Schaller, H., Beck, R., Takanami, M. 1978. *The Single-Stranded DNA Phages,* ed. D. Denhardt, D. Dressler, D. Ray, pp. 139–53. Cold Spring Harbor, NY: Cold Spring Harbor Lab.
115. Koshland, D., Botstein, D. 1980. *Cell* 20:749–60
116. Casadaban, M. J. 1976. *J. Mol. Biol.* 104:541–55
117. Bassford, P. J. Jr., Silhavy, T. J., Beckwith, J. R. 1979. *J. Bacteriol.* 139:19–31
118. Emr, S. D. Hedgpeth, J., Clement, J.-M., Silhavy, T. J., Hofnung, M. 1980. *Nature* 285:82–85
119. Moreno, F., Fowler, A. V., Hall, M., Silhavy, T. J., Zabin, I., Schwartz, M. 1980. *Nature* 286:356–59
120. Lin, J. J. C., Kanazawa, H., Ozols, J., Wu, H. C. 1978. *Proc. Natl. Acad. Sci. USA* 75:4891–95
121. Wanner, B. L., Sarthy, A., Beckwith, J. 1979. *J. Bacteriol.* 140:229–39
122. Lazdunski, C., Baty, D., Pagès J.-M. 1979. *Eur. J. Biochem.* 96:49–57
123. Di Rienzo, J. M., Inouye, M. 1979. *Cell* 17:155–61
124. Halegoua, S., Inouye, M. 1979. *J. Mol. Biol.* 130:39–61
125. Chang, C. N., Blobel, G., Model, P. 1978. *Proc. Natl. Acad. USA* 75:361–65
126. Wickner, W., Mandel, G., Zwizinski, C., Bates, M., Killick, T. 1978. *Proc. Natl. Acad. Sci. USA* 75:1754–58
127. Chang, C. N., Model, P., Blobel, G. 1979. *Proc. Natl. Acad. Sci. USA* 76:1251–55
128. Mandel, G., Wickner, W. 1979. *Proc. Natl. Acad. Sci. USA* 76:236–40
129. Ito, K., Mandel, G., Wickner, W. 1979. *Proc. Natl. Acad. Sci. USA* 76:1199–1203
130. Ito, K., Date, T., Wickner, W. 1980. *J. Biol. Chem.* 255:2123–30
131. Date, T., Zwizinski, C., Ludmerer, S., Wickner, W. 1980. *Proc. Natl. Acad. Sci. USA* 77:827–31
132. Wickner, W. 1979. *Ann. Rev. Biochem.* 48:23–46
133. Zwizinski, C., Wickner, W. 1980. *J. Biol. Chem.* 255:7973–77
134. Bennett, J. P., Hirth, K. P., Fuchs, E., Sarvas, M., Warren, G. B. 1980. *FEBS Lett.* 116:57–61
135. Fraser, T. H., Bruce, B. J. 1978. *Proc. Natl. Acad. Sci. USA* 75:5936–40
136. Mercereau-Puijalon, O., Royal, A. Cami, B., Garapin, A., Krust, A., Gannon, F., Kourilsky, P. 1978. *Nature* 275:505–10
137. Achtmann, M., Manning, P. A., Edelbluth, C., Herrlich, P. 1979. *Proc. Natl. Acad. Sci.* 76:4837–41
138. Ehring, R., Beyreuther, K., Wright, J. K., Overath, P. 1980. *Nature* 283:537–40
139. Büchel, D. E., Gronenborn, B., Müller-Hill, B. 1980. *Nature* 283:541–45
140. Talmadge K., Stahl, S., Gilbert, W.

1980. *Proc. Natl. Acad. Sci. USA* 77:3369–73

141. Talmadge, K., Kaufman, J., Gilbert, W. 1980. *Proc. Natl. Acad. Sci. USA* 77:3988–92

142. Katz, F. N., Rothman, J. E., Lingappa, V. R., Blobel, G., Lodish, H. F., 1977. *Proc. Natl. Acad. Sci. USA* 74:3278–82

143. Rothman, J. E., Lodish, H. F. 1977. *Nature* 269:775–80

144. Lingappa, V. R., Katz, F. N., Lodish, H. F., Blobel, G. 1978. *J. Biol. Chem.* 253:8667–70

145. Toneguzzo, F., Ghosh, H. P. 1978. *Proc. Natl. Acad. Sci. USA* 75:715–19

146. Rose, J. K., Welch, W. J., Sefton, B. M., Esch, F. S., Ling, N. C. 1980. *Proc. Natl. Acad. Sci. USA* 77:3884–88

147. Schmidt, M. F. G., Schlesinger, M. J. 1979. *Cell* 17:813–19

148. Fries, E., Rothman, J. E. 1980. *Proc. Natl. Acad. Sci. USA* 77:3870–74

149. Dobberstein, B., Garoff, H., Warren, G., Robinson, P. J. 1979. *Cell* 17:759–69

150. Ploegh, H. L., Cannon, L. E., Strominger, J. L. 1979. *Proc. Natl. Acad. Sci. USA* 76:2273–77

151. Elder, K. T., Bye, J. M., Skehel, J. J., Waterfield, M. D., Smith, A. E. 1979. *Virology* 95:343–50

152. Air, G. 1979. *Virology* 97:468–72

153. Porter, A. G., Barber, C., Carey, N. H., Hallewell, R. A., Threlfall, G., Emtage, J. S. 1979. *Nature* 282:471–77

154. Tomita, M., Furthmayr, H., Marchesi, V. 1978. *Biochemistry* 17:4756–70

155. Wirth, D. F., Katz, F. N., Small, B., Lodish, H. F. 1977. *Cell* 10:253–63

156. Garoff, H., Söderlund, H. 1978. *J. Mol. Biol.* 124:535–49

157. Garoff, H., Simons, K., Dobberstein, B. 1978. *J. Mol. Biol.* 124:587–600

158. Bonatti, S., Cancedda, R., Blobel, G. 1979. *J. Cell Biol.* 80:219–24

159. Garoff, H., Frischauf, A.-M., Simons, K., Lehrach, H., Delius, H. 1980. *Nature.* 288:236–41

160. Meyer, D., Dobberstein, B. 1980. *J. Cell Biol.* 87:498–502

161. Bonatti, S., Blobel, G. 1979. *J. Biol. Chem.* 254:12261–64

162. Papermaster, D. S., Schneider, B. G., Zorn, M. A., Kraehenbuhl, J. P. 1978. *J. Cell Biol.* 77:196–210

163. Schechter, I., Burstein, Y., Zemell, R., Ziv, E., Kantor, F., Papermaster, D. S. 1979. *Proc. Natl. Acad. Sci. USA* 76:2654–58

164. Albert, A. D., Litman, B. J. 1978. *Biochemistry* 17:3893–900

165. Hargrave, P. A. 1977. *Biochim. Biophys. Acta* 492:83–94

166. Kenny, A. J., Booth, A. G. 1978. *Essays Biochem.* 14:1–44

167. Desnuelle, P. 1979. *Eur. J. Biochem.* 101:1–11

168. Brunner, J., Hauser, H., Braun, H., Wilson, K. J., Wacker, H., O'Neill, B., Semenza, G. 1979. *J. Biol. Chem.* 254:1821–28

169. Maroux, S., Louvard, D. 1976. *Biochim. Biophys. Acta* 419:189–95

170. Benajiba, A., Maroux, S. 1980. *Eur. J. Biochem.* 107:381–88

171. Colbeau, A., Maroux, S. 1978. *Biochim. Biophys. Acta* 511:39–51

172. MacNair, R. D. C., Kenny, A. J. 1979. *Biochem. J.* 179:379–95

173. Louvard, D., Semeriva, M., Maroux, S. 1976. *J. Mol. Biol.* 106:1023–35

174. Hauri, H.-P., Quaroni, A., Isselbacher, K. J. 1979. *Proc. Natl. Acad. Sci. USA* 76:5183–86

175. Frank, G., Brunner, J., Hauser, H., Wacker, H., Semenza, G., Zuber, H., 1978. *FEBS Lett.* 96:183–88

176. Bar-Nun, S., Kreibich, G., Adesnik, M., Alterman, L., Negishi, M., Sabatini, D. D. 1980. *Proc. Natl. Acad. Sci. USA* 77:965–69

177. Gonzalez, F. J., Kasper, C. B. 1980. *Biochem. Biophys. Res. Commun.* 93:1254–58

178. Black, S. D., French, J. S., William, C. H. Jr., Coon, M. J. 1979. *Biochem. Biophys. Res. Commun.* 91:1528–35

179. Haugen, D. A., Armes, L. G., Yasunobu, K. T., Coon, M. J. 1977. *Biochem. Biophys. Res. Commun.* 77:967–73

180. Botelho, L. H., Ryan, D. E., Levin, W. 1979. *J. Biol. Chem.* 254:5635–40

181. Ozols, J., Gerard, C. 1977. *Proc. Natl. Acad. Sci. USA* 74:3725–29

182. Mihara, K., Sato, R., Sakakibara, R., Wada, H. 1978. *Biochemistry* 17:2829–34

183. Rachubinski, R. A., Verma, D. P. S., Bergeron, J. J. M. 1978. *J. Cell Biol.* 79:362a (Abstr.)

184. Borgese, N., Gaetani, S. 1980. *FEBS Lett.* 103:216–20

185. Chua, C. H.-H., Schmidt, G. W. 1979. *J. Cell Biol.* 81:461–83

186. Schatz, G. 1979. *FEBS Lett.* 103:203–11

187. DePierre, J. W., Ernster, L. 1977. *Ann. Rev. Biochem.* 46:201–62

188. Schatz, G., Mason, T. L. 1974. *Ann. Rev. Biochem.* 43:51–87

189. Robbi, M., Berthet, J., Trouet, A.,

Beaufay, H. 1978. *Eur. J. Biochem.* 84:333–40

190. Sherman, F., Stewart, J. W. 1971. *Ann. Rev. Genet.* 5:257–96
191. Smith, M., Keung, D. W., Gillam, S., Astell, C. R., Montgomery, D. L., Hall, B. D. 1979. *Cell* 16:753–61
192. Zitomer, R. S., Hall, B. D. 1976. *J. Biol. Chem.* 251:6320–26
193. Zimmermann, R., Paluch, U., Neupert, W. 1979. *FEBS Lett.* 108:141–46
194. Harmey, M. A., Hallermayer, G., Korb, H., Neupert, W. 1977. *Eur. J. Biochem.* 81:533–44
195. Harmey, M. A., Hallermayer, G., Neupert, W. 1976. *Genetics and Biogenesis of Chloroplasts and Mitochondria.* ed. T. Bücher et al, pp. 813–18. Amsterdam: Elsevier/North Holland Biomed. Press
196. Arpin, M., Matsuura, S., Margoliash, E., Sabatini, D. D., Morimoto, T. 1980. *Eur. J. Cell Biol.* 22:152 (Abstr. M451)
197. Coté, C., Solioz, M., Schatz, G. 1979. *J. Biol. Chem.* 254:1437–39
198. Maccecchini, M.-L., Rudin, Y., Schatz, G. 1979. *J. Biol. Chem.* 254:7468–71
199. Raymond, Y., Shore, G. C. 1979. *J. Biol. Chem.* 254:9335–38
200. Mori, M., Miura, S., Tatibana, M., Cohen, P. P. 1979. *Proc. Natl. Acad. Sci. USA* 76:5071–75
201. Conboy, J. G., Kalousek, F., Rosenberg, L. E. 1979. *Proc. Natl. Acad. Sci. USA* 76:5724–27
202. Harmey, M. A., Neupert, W. 1979. *FEBS Lett.* 108:385–89
203. Sonderegger, P., Jaussi, R., Christen, P. 1980. *Biochem. Biophys. Res. Commun.* 94:1256–60
204. Zimmermann, R., Paluch, U., Sprinzl,

M., Neupert, W. 1979. *Eur. J. Biochem.* 99:247–52
205. Mori, M., Morris, S. M. Jr., Cohen, P. P. 1979. *Proc. Natl. Acad. Sci. USA* 76:3179–83
206. Poyton, R. O., McKemmie, E. 1979. *J. Biol. Chem.* 254:6763–71
207. Lewin, A. S., Gregor, I., Mason, T. L., Nelson, N., Schatz, G. 1980. *Proc. Natl. Acad. Sci. USA* 77:3998–4002
208. Mihara, K., Blobel, G. 1980. *Proc. Natl. Acad. Sci. USA* 77:4160–64
209. Parimoo, S., Padmanaban, G. 1980. *Biochem. Biophys. Res. Commun.* 95:1673–79
210. Schmelzer, E., Heinrich, P. C. 1980. *J. Biol. Chem.* 255:7503–6
211. Nelson, N., Schatz, G. 1979. *Proc. Natl. Acad. Sci. USA* 76:4365–4369
212. Zimmermann, R., Neupert, W. 1980. *Eur. J. Biochem.* 109:217–29
213. Schmidt, G. W., Devillers-Thiery, A., Desruisseaux, H., Blobel, G., Chua, N.-H. 1979. *J. Cell Biol.* 83:615–22
214. Highfield, P. E., Ellis, R. J. 1978. *Nature* 271:420–24
215. Chua, N.-H., Schmidt, G. W. 1978. *Proc. Natl. Acad. Sci. USA* 75:6110–14
216. Grossman, A., Bartlett, S., Chua, N.-H. 1980. *Nature* 285:625–30
217. Apel, K., Kloppstech, K. 1978. *Eur. J. Biochem.* 85:581–88
218. Chua, N.-H., Grossman, A., Bartlett, S., Schmidt, G. 1980. *Eur. J. Cell Biol.* 22:152 (Abstr. M452)
219. Pappenheimer, A. M. Jr. 1977. *Ann. Rev. Biochem.* 46:69–94
220. Olsnes, S. 1978. *Transp. Macromol. Cell. Syst.* Dahlem Konferenzen, 11:103–16
221. Blobel, G. 1980. *Proc. Natl. Acad. Sci. USA* 77:1496–1500

Ann. Rev. Biochem. 1981. 50:349–83

ORGANIZATION AND EXPRESSION OF EUCARYOTIC SPLIT GENES CODING FOR PROTEINS

♦12082

Richard Breathnach and Pierre Chambon[1]

Laboratoire de Génétique Moléculaire des Eucaryotes du CNRS, Unité 184 de Biologie Moléculaire et de Génie Génétique de L'INSERM, Institut de Chimie Biologique, Faculté de Médecine, Strasbourg, France

CONTENTS

[1]We are grateful to all those who sent us preprints of their work, and to our colleagues for comments on the manuscript. Work in our laboratory has been supported by grants from the INSERM, the CNRS, the DGRST, and the Foundation pour la Recherche Médicale Française.

PERSPECTIVES AND SUMMARY

Since the field was reviewed by Abelson in 1979 (1), a large number of eucaryotic protein-coding genes have been cloned and shown to be split, and the terms exon (for the sequences coding for messenger RNA) and intron or intervening sequence (for the sequences separating the exons) are now a part of the literature. The split gene has become a commonplace, to be found in organisms as diverse as yeasts, insects, chickens, and humans; the detailed molecular anatomy of several genes is now known. Ultimately, this should lead to the elucidation of the molecular mechanisms by which the expression of genes such as the globin and ovalbumin genes is controlled during development and by hormones, respectively. However, answers about evolution are more easily obtained from the molecular anatomy of genes than are answers about control mechanisms. In fact studies of cloned genes have already indicated the importance of duplication events in the evolution of individual genes and gene families. Investigations of control mechanisms have concentrated on the important preliminary work of defining the primary transcript and determining how it is processed, and on a search for the RNA polymerase B (or II) promoter.

Split genes are transcribed into colinear precursors from which the intron transcripts are removed by splicing. The sequencing of many intron-exon junctions has confirmed that they are all related in sequence, and that it is possible to derive consensus sequences describing the eight or nine nucleotides around the junctions. It is not clear what other sequences of the primary transcript play a role in the splicing process, though it seems that in some cases sequences distant from the junctions have an influence. Apparently, splicing is linked to transport of transcripts from nucleus to cytoplasm and the accumulation of stable messengers.

The sequences upstream from the presumed start of transcription of many genes transcribed by RNA polymerase B (or II) have been determined; an AT-rich region of homology, the Goldberg-Hogness or "TATA" box, has been observed about 30 base pairs (bp) upstream from this point. Techniques have been developed for accurate in vitro transcription of cloned genes, and have been used for testing the functional importance of this sequence homology. At least in vitro, the Goldberg-Hogness box plays a role in promotion of transcription by RNA polymerase B, and the limits of the in vitro promoter sequence have been determined. However, in vivo studies indicate that there may be more to a promoter region for RNA polymerase B than just the TATA box and its surrounding sequences.

Repeated genes (2), histone genes (3), immunoglobulin genes (4) and adenovirus and papovavirus genomes (5) have been reviewed recently. Split yeast mitochondrial protein-coding genes may not be closely related to nuclear genes. We therefore do not attempt a comprehensive review of any of these subjects, but mention them when appropriate for the sake of comparison.

ORGANIZATION OF PROTEIN-CODING NUCLEAR SPLIT GENES

Molecular Anatomy of the Split Genes

GLOBIN GENES Hemoglobin is composed of heme residues and two chains each of α-like and β-like globin polypeptides. Each type of chain is encoded by a family of genes under differential, but coordinate control during development (reviewed in 6, 6b). The split structure of globin genes was first noted for rabbit (7) and mouse (8) β-globins, and a comparison of the structure of mouse α-, β^{major}-, and β^{minor}-globin genes led to the hypothesis that all functional globin genes contain two introns interrupting the protein-coding sequences at homologous positions (9). Studies of the structure of globin genes from mouse (10–13a), rabbit (14–16), human (reviewed in 6, 6a), chicken (17–19), sheep (20), *Xenopus laevis* (21, 22), and goat (23) have confirmed this prediction, and suggest that introns were present before the duplication that gave rise to separate α- and β-globin genes, an event estimated to have occurred 500 million years ago (24).

Blotting studies (25) on genomic DNA from normal and thalassemic individuals have shown that the activity of a particular globin gene may be affected by deletions occurring some distance from it (reviewed in 6). Considerable effort has therefore gone into the cloning of extensive sequences surrounding globin genes. Analysis of such clones has shown that the human α- and β-like genes are present in two separate clusters, which confirms the results of blotting experiments. The analysis has also shown the clustering of the chicken α- and β-like globin genes (17, 18) and the rabbit (15, 16) and mouse (26) β-like globin genes.

The α- and β-like globin gene families are clustered separately on different chromosomes (27–33); coordinate expression of these two gene families is thus not dependent on their being clustered together. In a deviation from the norm the adult α- and β-globin genes of *X. laevis* are linked; they are separated by 7.7 kb in the genome (21, 22).

Comparison of the arrangement of globin clusters between species where data are available has indicated the following common features (34):

1. Globin genes in clusters are arranged in the order of their expression during development, and in the same orientation of transcription. Thus the human and chicken α-like globin genes are arranged in the order 5'-embryonic-adult-3' (18, 34), and the human [for references see (6)] and rabbit β-

like globin genes in the order 5'-embryonic-foetal-adult-3' (15, 16). The chicken β-like globin gene cluster may prove an exception to this rule, since a gene tentatively identified as being embryonic has been mapped 3' to the adult gene (17).

2. Genes expressed at the same stage of development are often arranged together in pairs. This is the case for the human $\hat{\delta}$ and β, $^G\gamma$ and $^A\gamma$, and $\alpha1$ and $\alpha2$ genes [for references see (6)]; the rabbit $\beta3$ and $\beta4$ genes (16); and the chicken α^A and α^D (18) genes. A comparison of the mouse β^{major} and β^{minor} gene pair at the nucleotide sequence level has shown that the sequences of their exons are much less divergent than the sequences of their flanking and intervening sequences. The former seem to have diverged largely by single-base substitutions, while the latter have diverged by insertion, deletion, and duplication events (11, 13, 35). This appears to be the case for many of the above-mentioned globin gene pairs. However, the human adult α-globin (and the γ-globin) gene pair differs little in sequence over the genes' coding, intervening, or flanking sequences (34, 36) . This result suggests that in these cases, mechanisms for gene matching have operated during evolution, possibly intrachromosomal gene conversion by homologous but unequal crossing-over. For the γ-globin gene pair, a "recombinational hot-spot" in the second intervening sequence may be responsible for the gene conversion (36).

3. Globin gene clusters are often associated with repeated sequence elements. Such elements have been mapped in the human (reviewed in 6, 37), rabbit (38, 39), and chicken (17) β-like globin clusters. The rabbit β-like globin cluster shows a particularly complex pattern of repeated sequences. One of the sequences present in the rabbit cluster is related to the Alu family (see below) and is also represented in the human α-like and β-like globin gene clusters [for references see (6)].

4. Globin gene clusters often contain sequences complementary to globin probes that cannot, however, encode functional globin polypeptides because the translational reading frame is changed by small deletions or insertions. Changes at the splice junctions may also occur. An example is a goat β-globin pseudogene (40) in which a frameshift mutation would lead to premature termination; the histidines linked to heme binding have been substituted, the termination codon has been changed, the GT (see below) at the 5'-border of the first intron has been changed to GC, and the Goldberg-Hogness box (see below) has been mutated. Other examples are a rabbit β-pseudogene (15, 40a) and a human α-pseudogene (34, 40b). A particularly interesting example is a mouse α-pseudogene (41, 42), in which the two intron sequences have been removed precisely. Nishioka et al (41) have suggested a removal mechanism involving heteroduplex formation between the original gene and its mature mRNA or a cDNA copy, and have noted

that a similar mechanism could explain the precise deletion of one of the introns of a rat insulin gene (see below). Pseudogenes appear not to give rise to stable RNA transcripts (15, 42). Their widespread occurrence may indicate that they have some role in the control of globin gene expression (42).

A large number of gene polymorphisms have been mapped in and around the human globin gene clusters (reviewed in 6).

Analysis of the human and other globin genes is proceeding rapidly, but there are still no clues as to the mechanisms used for the developmental regulation of these genes, although some possible mechanisms have been discussed (43). However, it seems likely that sequences immediately up-stream from the start of transcription, around the putative promoter sites, are not responsible for this regulation; a comparison of three goat β-like globin genes (44) expressed at different developmental stages shows very little sequence difference over the first 200 nucleotides upstream of the cap site.

GENES EXPRESSED IN THE CHICKEN OVIDUCT   The chicken ovalbu-min, conalbumin (ovotransferrin), lysozyme, and ovomucoid genes are un-der hormonal control in the chicken oviduct (45). Their structure has been elucidated with the aim of understanding the molecular mechanism of this control.

The split organization of the ovalbumin gene was first demonstrated by genomic blotting using ovalbumin cDNA plasmids as hybridization probes [(46–49); for additional references see (50)]. A complete ovalbumin gene was subsequently cloned in lambda phages (51, 52). A comparison of the sequence of extensive regions of the gene (53–56) with that of the messenger (57, 58) has definitively shown that the gene is composed of 8 exons and 7 introns. The first exon encodes the bulk of the 5'-untranslated region of the messenger (53); there are no interruptions in the sequences encoding the 650-nucleotide long 3'-untranslated region (57). Gene polymorphisms have been observed in and around the ovalbumin gene (48, 57, 59, 60).

Cloning of extensive sequences upstream from the ovalbumin gene using cosmids has led to the discovery of two related genes (61–63), X and Y. Both the X and Y genes are also under hormonal control although they are expressed at a much lower rate that the ovalbumin gene; they share a common ancestor with the ovalbumin gene. All three genes of the ovalbu-min family have 8 exons, and analogous exons of the ovalbumin, X, and Y genes are of similar length (except for the last exon, which encodes the C terminus of the protein and the 3'-untranslated region) and show varying degrees of sequence homology [(62); J. L. Mandel and R. Heilig, personal communication]. Comparison of selected regions of these genes indicates that while the protein-coding sequences and the locations and sequences of the splice junctions have been well conserved, the introns and sequences coding for the 3'-untranslated regions of the mRNAs have diverged much more rapidly (62).

The X, Y, and ovalbumin genes are oriented in the same direction for transcription (in the order 5'-X-Y-ovalbumin-3'), and messenger and precursor RNAs for genes X and Y have been identified (61–64). Two cytoplasmic polysomal RNAs are encoded by gene X (62, 63), and probably also by gene Y (63). These RNAs differ in the size of their 3'-untranslated regions, which suggests that partial read-through across a first poly(A) addition site is occurring. In vitro translation of hybridization-selected X and Y RNAs have allowed a tentative identification of protein products of these genes (63).

Four regions of apparently unrelated repetitive sequences are found in the X gene region; two of these lie within introns and the others between the X and Y genes. They have apparently no counterpart in the ovalbumin and Y genes (62).

The chicken conalbumin (ovotransferrin) gene has been cloned and shown to contain at least 16 introns (65, 66). The first exon of the gene encodes the bulk of the hydrophobic signal peptide. Amino acid sequence studies (for review see 67) have suggested that doubling of an ancestral structural gene occurred during evolution of transferrins, and that the conalbumin gene may thus have evolved from a split gene of seven or eight exons. To investigate this possibility exons 6 and 14 of the conalbumin gene have been sequenced and shown to encode protein sequences of significant homology. When these exons differ it is largely by addition or deletion of whole codons (J. M. Jeltsch, personal communication). Sequences of both high and middle repetitive frequency are found upstream of the gene and within two introns (66). The high repetitive frequency sequence in intron C is apparently related to the Alu family and the middle frequency repeat present in intron B is repeated in inverse orientation upstream from exon 1 (L. Maroteaux, personal communication).

The chicken ovomucoid and lysozyme genes have been cloned, and contain seven (68–73) and three (71, 74–76) introns, respectively. Sequence studies of the ovomucoid protein and gene suggest that the latter has been derived by triplication of an ancestral gene containing one intron (72). Unfortunately, a comparison of the structure of the six genes mentioned above does not suggest a mechanism for their hormonal control. Apart from their split organization, there are no common structural features. A comparison of the sequences of the 300 base pairs (bp) located upstream of the transcription initiation site of the ovalbumin (51, 56), X (J. L. Mandel and R. Heilig, personal communication), Y (J. L. Mandel and R. Muraskowski, personal communication), conalbumin [(66); L. Maroteaux, personal communication], ovomucoid [(69); P. Gerlinger and D. Dupret, personal communication], and lysozyme(A. E. Sippel, personal communication) genes has shown no consistently present sequence feature besides the putative promoter TATA box, to date.

VITELLOGENIN GENES   Studies of cloned *X. laevis* cDNAs have shown that vitellogenin is encoded in a small family of at least four genes (77). The genes fall into two pairs; members of a pair are about 95% homologous in sequence and about 80% homologous to members of the other pair. The genes coding for both members of one of these pairs have been cloned and show marked similarities (78). Both genes are spread over 21 Kilobase pairs (kb), for a messenger length of 6000 nucleotides, and contain at least 33 introns. The introns interrupt the coding sequences at homologous points in both genes, but the sequence and length of analogous introns has not been conserved.

INSULIN GENES   Chicken (79) and human (80) have a single preproinsulin gene. These genes have been cloned and sequenced; each contains two introns, similarly located, one within the sequences coding for the messenger's 5'-untranslated region and the other interrupting the C-peptide coding region. Rat, however, has two nonallelic insulin genes, both of which have been cloned and sequenced (81, 82). One of these genes (with two introns) is similar in structure to human and chicken genes; the second gene lacks the intron interrupting the C-peptide coding region. Precise excision of this intron must have occurred during the evolution of rat insulin genes.

COLLAGEN GENES   Type I collagen is composed of two α-1 subunits and one α-2 subunit. Each subunit contains a long helical portion of sequence $(gly-x-y)_n$ and a nonhelical C peptide. Portions of the sheep α-2 collagen gene have been cloned and contain many small exons (83). The chick α-2 collagen gene has been cloned in its entirety (84–84c) and has a length of about 40 kb, for a messenger size of 5000 nucleotides. It is made up of at least 51 exons. Exons specifying the helical portion are similar in size and smaller than those encoding the C peptide. Several separate exons from the helical region have been sequenced (84–84 c). All begin with a glycine codon and end before a glycine codon, and are multiples of nine base pairs in size. Many of the exons sequenced were 54 bp long, with a range in size from 45 bp to 108 bp. This has led to the suggestion that the collagen gene arose by amplification of a single polypeptide coding unit of 54 bp and additions (or deletions) of units of 9 bp or multiples of 9 bp encoding the basic gly-x-y repeat.

ACTIN GENES; SPLIT GENES IN INSECTS AND LOWER EUCARYOTES
The single actin gene of yeast has been cloned and sequenced (85, 86). It
contains a 309-bp intron in the protein-coding region, which separates the
DNA encoding the first three amino acids of the protein from the rest of
the coding sequence. The exon-intron junction sequences show redundancy
and are similar to those of other eucaryotic protein-coding split genes. The
actin gene is the only protein-coding nuclear gene of yeast thus far shown
to be split, though several mitochondrial genes contain introns. Several
yeast genes are unlikely to contain introns in their protein-coding regions,
as they may be expressed in *Escherichia coli* (for review see 87). Sequence
analysis of a cloned yeast iso-1-cytochrome *c* gene has indeed shown that
it contains no introns in the protein-coding sequences (88).

By coincidence, the first protein-coding gene of *Drosophila melanogaster*
shown to contain an intron was also the actin gene (89). There are at least
six actin genes par haploid genome, mapping at six widely dispersed sites
(89, 90), and at least two of them have an intron at a location close to that
of the yeast one. Two sea urchin actin genes have been cloned [(90a); W.
R. Crain, personal communication]; one of these (but not the other) contains an intron beginning immediately after amino acid 121 in the protein
sequence. A chick actin gene has at least three introns (90b). This variation
in intron position for a given gene between species is in marked contrast to
the globin and insulin genes (see above). Interestingly, there is no analogous
intron in the gene coding for the slime mold *Dictyostelium* actin (R. Firtel,
personal communication), reputedly a more highly evolved protein than
yeast actin [K. Weber, unpublished; quoted in (85)]. The only other examples of split, protein-coding genes in insects are the silk fibroin gene of
Bombyx mori [which has been extensively sequenced (91, 92)], the chorion
protein genes of *Antheraea polyphemus* (93), and the *Drosophila* alcohol
dehydrogenase gene (93a).

OTHER GENES   Other split genes that have been cloned or analyzed recently include the two nonallelic chicken δ-crystallin genes (94, 95); rat
albumin gene (96, 97); mouse α-fetoprotein (98) and metallothionein I
genes (99); human (100) and rat (101, 101a) growth hormone genes; rat
prolactin gene (101a); several rat amylase genes (101b); mouse dihydrofolate reductase gene (102); and hamster CAD gene, which codes for the
multifunctional protein involved in de novo UMP biosynthesis (103). The
last two cases, so-called house-keeping genes, show that it is not only genes
coding for highly specialized gene products expressed in terminally differentiated cells that contain introns. The gene coding for human interferon
α1 has been cloned (104). It appears to be one of not less than eight distinct
human interferon genes, at least some of which are clustered in the genome.

The interferon gene cloned does not contain introns (104). The only other protein-coding genes yet shown definitively not to contain introns are the histone genes (3) and the adenovirus polypeptide IX gene (105).

Repeated Sequences

Short repeated sequences are interspersed with longer single copy sequences in the genomes of higher eucaryotes. Such sequences have been implicated in models of regulation of gene expression (106). Repeated sequences have been mapped and analyzed in and around several cloned genes, such as human and rabbit globin gene clusters and chicken conalbumin, and X and Y genes (see above). The location of these sequences does not fit a simple pattern: the X gene region contains four regions of repeated sequences (62), while in a 20 kb region surrounding the ovalbumin gene, there are no repeated sequences (J. L. Mandel, personal communication). So far, repeated sequences have not been mapped in exons, only in introns and flanking sequences.

Recently, it has been shown that many of the 300-nucleotide interspersed repeats in human DNA belong to one family of 300,000 members—the Alu family (107–109). The chinese hamster genome appears to contain a similar family (110). Some of the repeated sequences mapping in human globin gene clusters belong to this family [for references see (6)] and a repeated sequence in the chicken conalbumin gene shows homology to the Alu family. The Alu family shows sequence homology with the repetitive double-stranded regions of HeLa cell hnRNA and with a low-molecular-weight RNA found hydrogen-bonded to CHO cell hnRNA (107). A sequence homology between the Alu family and some papovavirus origins of replication has also been noted, which led to the hypothesis that the Alu family may be cellular origins of replication (107).

Some interspersed repeated DNA sequences of the human β-globin gene cluster may be transcribed in vitro by RNA polymerase C (or III) to give discrete products (111). One of these repeats has been sequenced and belongs to the Alu family (107).

Rearrangements

Blotting of restriction enzyme digests of DNA extracted from various tissues and probing for rabbit (7) and chicken β-globin genes (112), chicken ovalbumin (46, 48), chicken conalbumin (66, 113), chicken lysozyme (114), and silk fibroin (115) genes have shown that the environment of these genes is, at least grossly, the same in tissues where the genes are actively expressed and where they are not. Comparison of the sequences around the putative promoter sites of fibroin (91) and ovalbumin (116) genes cloned from DNA of producer and nonproducer tissues has shown no significant difference. The number and variety of genes tested is however still limited. That the eucaryotic genome is not a rigidly unchangeable structure is evidenced by the immunoglobulin gene rearrangements (4), the mating-type rearrangements in yeast [for references see (117)] and the existence of elements such as copia in *D. melanogaster* or the Ty elements of yeast [for references see (118)], which are believed to be capable of transposition analogous to the transposable elements of bacteria. The analogy is strengthened by the fact that both copia and Ty generate 5-bp duplications of the host DNA on insertion (188–120a). Mechanisms thus evidently exist for genome rearrangements, and the possibility of rearrangements being used to activate e.g. regulatory genes remains open.

Generalization and Molecular Evolution

The first conclusion that can be drawn from the exponentially increasing number of eucaryotic gene structure studies is that all of the sophisticated methodology involved is surprisingly faithful. It is only occasionally that cloning or enzymatic errors are encountered (see, for example, 34, 57, 121).

The widespread occurrence of introns and the location of neighboring genes at distances comparable to or larger than their lengths, accounts in part for the C-value paradox of higher eucaryotic genomes; (in the 40-kb ovalbumin gene family region apparently not more than 9% of the DNA codes for protein.) From one gene to another and for the same length of mature mRNA, the number and lengths of introns and exons are variable, and there is no apparent rule. Total intron length can represent more than 95% of a gene [the dihydrofolate reductase gene (102)].

Introns are mostly localized in amino acid coding regions. They have been found in the 5'-untranslated regions of some genes [eg ovalbumin (53) and insulin genes (79–82)], but not yet in any 3'-untranslated region, which suggests that splicing the 3'-end of an mRNA to bring it closer to the termination codon is of no major selective advantage. There is no evidence as yet that any intron or exon from a given gene could be found as such elsewhere in the genome, if one excludes the genes that belong to the same family. However, introns (but not exons) can contain highly or middle repetitive sequences, as shown in the case of the X (62) and conalbumin (66) genes. These intronic repeated sequences appear to evolve rapidly, since, for instance, the repeated sequences that are present in the X gene are not found in the two other members of the ovalbumin gene family (see above).

Although one would predict from theoretical considerations that in the absence of a specific excision machinery the exact removal of an intron would be a very rare event, there are now several examples that show that precise intron deletion can occur in the course of evolution [some globin pseudogenes (41, 42), one rat insulin gene (81, 82) and probably the *Dictyostelium* actin gene (R. Firtel, personal communication)]. This suggests the existence of an enzyme system for intron excision and raises the question of why all introns have not been removed. There are basically two possibilities: the enzymatic system could be very inefficient, or introns could have some functions (see below); (it is interesting that the globin pseudogenes, which have lost both globin introns, are not expressed.)

Genetic polymorphisms, without effect on the protein product and due to allelic variants, are widespread both in exonic and intronic sequences and can be used as genetic markers (for review see 6). The comparison of members of a given gene family and of a given gene in different species has led to the conclusion that protein-coding sequences evolve slowly, mainly by point mutations. Noncoding sequences (introns and very often 3'-untranslated regions) evolve much more rapidly [except when a gene matching mechanism, like gene conversion, is operating (see above)], by deletion and insertion events that are most probably selected against in the coding regions. Therefore split genes appear to be made up of rather stable protein-coding elements embedded in a rapidly changing environment.

The problems of the origins of introns, and the possible evolutionary advantage of the split gene organization have been the subject of intense speculation (11, 122–129). Although we have very little chance ever to answer the basic question of whether the split gene organization is the most primitive one and whether present-day bacteria are "streamlined" cells, there are a number of observations that are relevant. Clearly, gene duplication by unequal crossing-over, a major mechanism in evolution (127, 130), is responsible for the existence of many present-day split genes. First, from the study of the various split gene families, including the immunoglobulin genes, it appears that many split genes have arisen by "discrete" duplications of ancestral genes that were already split in identical or very similar patterns. Secondly, gene duplication by unequal crossing-over can sometimes explain how the split ancestral genes have arisen in the course of evolution. Indeed, it has been postulated by Tonegawa and his colleagues that the split immunoglobulin light- (131) and heavy-chain (132) genes have evolved by multiple duplications of an ancestral unsplit DNA segment coding for a polypeptide chain with a size similar to that of the present-day homologous domains of the heavy chains. To account for the introns that separate these present-day domains, it has been postulated that some flanking segments of the ancestral domain-encoding DNA segment were duplicated also and that splicing signals have subsequently arisen at the ends of the domain-encoding DNA segments. The evolutionary advantage of the splicing mechanism is clear: it allows the creation of new genes by "fusion" of the duplicated domain-encoding DNA segments into one transcription unit and subsequent splicing of the RNA to yield a new mRNA, without waiting for an exact deletion of the spacer DNA that separates the duplicated DNA segments. [A similar mechanism for gene expansion by internal duplications of some exons of an already split ancestral gene has been proposed by Rogers et al (133).] Similarly, it is possible that the α-2 collagen's α-helical portion evolved by successive duplication of an ancestral DNA segment of 54 bp, the duplicated exons diverging by addition or deletion of multiples of 9 bp and by point mutations (84a, 84c). Obviously, the same mechanism could apply to the generation of a new split gene by fused duplication of an ancestral split gene. The chicken ovomucoid gene may have evolved by triplication of an ancestral protein-coding DNA segment split by one intron (72), and there is evidence that the conalbumin gene evolved by duplication from an ancestral gene with seven or eight exons [(66); J. M. Jeltsch, personal communication]. The lack of homology between the various exons of some genes, like the ovalbumin gene, definitely does not exclude the possibility that they originated through multiple ancient duplication events that may be difficult or impossible to recognize due to sequence drift during the course of evolution.

Obviously, the generation of exons and introns by fused duplication of an ancestral gene accounts for the observation that, for some split genes, like the immunoglobulin genes (131, 132, 134–143) or the ovomucoid gene (72), there is a close correspondence between exons and structural or functional domains of the protein. However, it is remarkable that the sequences coding for the several peptides with distinct biological activities that are encoded for by the human pro-opio-melano-corticotropin gene are not separated by introns (144). In the bovine gene, the signal peptide is, however, encoded by a separate exon and there is an intron within the sequences encoding the 5'-untranslated region of the mRNA (144a). In a similar vein, a single exon of the chick collagen gene codes for the last 15 amino acids of the triple helical coding region, the telopeptide, and the first 54 amino acids of the C-terminal propeptide (84c).

The correspondence between exons and functional protein domains (126) extends to some exons that apparently have not arisen in situ by fused duplication of an ancestral gene. For instance, the hydrophobic signal peptides of conalbumin (66), ovomucoid (72), immunoglobulins (134–137), and chorion proteins (93) [but not insulin (79–82)] are encoded in separate exons. Another possible example is the exons coding for the hydrophobic tail of the membrane-bound form of IgM (133, 145). Analysis of the globin gene has shown that the product of the central exon can bind heme tightly (146), and that the third exon encodes most of the residues of the α_1-β_1 contact involved in cooperativity (147). A relationship between exons and functional units of the chicken lysozyme has also been suggested (76). It is therefore possible that in some cases the origin of the introns may be found in the shuffling and assembly of exons initially scattered throughout the genome, as originally proposed by Gilbert (126). However, it is yet to be shown whether the origin of any split gene is related to such a random shuffling of exons. Exon shuffling between duplicated genes by unequal crossing-over events within intron sequences does, however, most likely occur, as indicated by the remarkable homology observed over the CH_1 domain and the first half of the first intervening sequence of γ_1 and γ_{2b} heavy-chain genes (148).

THE TRANSCRIPTION UNIT

A large body of evidence suggests that split genes are transcribed into colinear primary transcripts from which intron sequences are removed by splicing. Addition of poly(A) apparently precedes splicing (73, 149, 150–153).

The sizes of the largest transcripts observed for the globin genes (see, for example, 154), the ovalbumin gene (73, 151), the X and Y genes (63, 64), the ovomucoid gene (152, 153), and the conalbumin gene [(66); M. LeMeur and A. Krust, personal communication] are consistent with transcription initiating at the 5' end of the first exon and terminating at the 3' end of the last exon [see (154–158) for discussion of alternate views]. For a correlation of gene structure to expression it is vital to know, to the nucleotide, where this transcription initiates and terminates.

Transcription initiation may occur at the nucleotide corresponding to the first, capped nucleotide of the mature mRNA (cap site); alternatively it may occur upstream from this position, and the 5' end of the messenger would then be generated by nucleolytic processing. Analysis of precursor molecules by the Berk and Sharp procedure (159) [as modified by Weaver & Weissmann (160)] for chicken ovalbumin (161, 162), silk fibroin (163), and mouse β-globin (160) genes has not provided evidence for any precursor molecules with 5' termini other than those corresponding to the first, capped nucleotide of the mature messenger. These results suggest that initiation occurs at the cap site and are in agreement with an earlier report on the transcription initiation site of the Adenovirus-2 major late transcription unit (164). However, very rapid processing of a primary transcript that initiates upstream from the cap site can still not be ruled out entirely on the basis of this data, as is witnessed by the continuing controversy over the size of the primary transcript of globin genes (154–158). Nonetheless, in vitro transcription studies of cloned genes strongly point to the cap site being the transcription initiation site (see below).

In the case of the Adenovirus-2 major late transcription unit (165) and the units for early regions 2 and 4 (166) and the SV40 late region (167), transcription terminates beyond the poly(A) addition sites; the mature 3' ends of the messengers are generated by endonucleolytic cleavage at this site, and polyadenylation. Corresponding data for cellular genes are lacking except for ovalbumin, where termination of transcription seems to occur at (or very close to) the poly(A) addition site (161, 168).

Whether transcription normally stops at or beyond the polyadenylation site for cellular genes, it is clear that read-through across polyadenylation sites can occur, as evidenced by the two size classes of messengers for genes X and Y [see above and (62, 63)], and probably for the dihydrofolate reductase gene (102), and the read-through across a globin site in an SV40-globin recombinant (169). A particularly interesting example of read-through is that used to generate the mRNA for the membrane-bound form of IgM (IgM_m) (133, 145). Read-through across the polyadenylation site used to generate the messenger for the secreted form of IgM (IgM_s), and polyadenylation at a site downstream produces a longer primary transcript containing transcripts of two additional exons (encoding the hydrophobic C-terminus of IgM_m). These two exon transcripts are then spliced to a donor site lying upstream of the termination codon of IgM_s, which produces the messenger for the larger protein IgM_m. As IgM_s and IgM_m are produced at different stages of development of an immunoglobulin producer, it has been suggested that there may be developmental stage–specific recognition of different polyadenylation sites (145). Partial read-through of polyadenylation sites in a search for downstream exons to be introduced into the transcription unit has evident evolutionary advantage.

How is a polyadenylation site recognized? Despite the availability of many sequences of genomic clones surrounding polyadenylation sites (56 and references therein), there are few consistently present features. Exceptions are the ubiquitous sequence AATAAA (170) or a close relative [e.g. ATTAAA in the chicken lysozyme gene (76) and the mouse pancreas α-amylase gene (171)]; also, almost always the first one or more As of the polyA tail can be added transcriptionally, i.e. it is impossible to define the polyadenylation site to the nucleotide (56).

MECHANISM OF RNA SPLICING

Pre-mRNA Processing Intermediates

Putative pre-mRNA processing intermediates containing linked intron and exon transcripts have been identified for the genes of: chicken ovalbumin (73, 151, 152), X (63, 64), Y (63, 64), ovomucoid (152, 153), conalbumin (M. LeMeur and A. Krust, personal communication), various globins (154–158, 172–175), *Xenopus* vitellogenin (176), chicken α-2 collagen (176a), heavy- and light-chain immunoglobulin (177–181), many adenoviruses (for review see 5, 165), and yeast mitochondria (182–185). Pulse-chase experiments have in certain cases shown the existence of genuine precursors (see, for example, 152, 155, 177, 179). Candidates for excised intron transcripts have been put forward for the chicken ovalbumin gene (73) and the yeast mitochondrial cob-box gene (185, 186); in the latter case some of these excised transcripts are circular.

Individual intron transcripts may be excised in several steps. Many intermediates have been identified in the splicing pathway used to link the Adenovirus-2 major late tripartite leader to the main bodies of the late messengers (reviewed 5, 165). Similarly, the mouse β-globin mRNA precursor apparently requires at least two splicing events to excise the large intron transcript (156). Three distinct stepwise pathways have been identified for the removal of an intron transcript in the chick α-2 collagen mRNA precursor (176a). Multiple steps may occur during intron-transcript excision in yeast mitochondrial genes (185) and the chicken ovalbumin gene (M. Le-Meur, personal communication). In mouse β-globin mRNA precursor, the first of the two splicing events needed to excise the large intron transcript may be rate determining (156); if generally true, this would fit in with the fact that processing intermediates involving partial loss of intron transcripts were not seen in electron microscope analyses of heteroduplexes between vitellogenin (176) or ovalbumin mRNA precursors (73) and their respective cloned genes. Species with partially removed intron-transcripts have not been shown unequivocally to be genuine splicing intermediates as opposed to products of abortive events, however.

Studies on the vitellogenin (176), ovalbumin (73), and ovomucoid (152, 153) putative mRNA precursors suggest that there is not a rigid order of removal of intron transcripts from these molecules: thus in some cases processing intermediates that retain a particular intron transcript but that have lost another intron transcript and vice versa have been visualized in the electron microscope. Preferred orders of excision may exist, however, as reported for the ovomucoid precursor (152). The 5'-terminal splice has been reported to occur before others for the maturation of the Adenovirus-2 major late leader (187) and some adenovirus early messengers (149). This could be a kinetic effect or could reflect a requirement that must be fulfilled before further splicing events. That such requirements can exist is illustrated by the yeast mitochondrial cob-box mRNA precursor. Individual intron mutations in this gene block processing of the precursor at defined stages of its maturation and lead to accumulation of intermediates containing full intron transcripts (183–185).

The Splicing Signals

The specificity and accuracy of splicing (in the chick α-2 collagen gene, 51 exon transcripts must be linked precisely) must ultimately reside in the nucleotide sequence of the primary transcript. Recombinant DNA techniques have been used to define what sequences are required for splicing. Table 1 shows the results of the analysis of 90 donor- and 85 acceptor-site sequences available for protein-coding genes [donor site = 5'-splice junction, acceptor site = 3'-splice junction (188)]. We deduce a consensus sequence for the donor site of $5'\text{-}{}^{A}_{C}\text{AGGT}{}^{A}_{G}\text{AGT-3}'$, and for the acceptor site of $5'\text{-PyPyPyPyPyPyXCAGG}{}^{G}_{T}\text{-3}'$. These are similar to consensus sequences derived previously on the basis of fewer sites and include both exon and intron nucleotides (53, 54, 188, 197, 198). It is apparent from known junction sequences that if the second and third nucleotides of a particular donor sequence (measured relative to the consensus sequence

Table 1 Analysis of sequences around splice junctions[a]

DONOR SITE ← ———————— EXON ———————— → ← ———————— INTRON ———————— →

BASE FREQUENCY																				
A	21	36	22	20	13	26	22	38	53	9	0	0	50	61	9	14	30	24	30	16
G	23	19	22	14	22	11	20	11	12	68	90	0	30	12	77	9	19	17	17	18
C	25	19	26	29	27	22	25	37	13	4	0	0	4	8	1	11	19	26	23	23
T	15	16	20	28	28	31	23	4	12	9	0	90	6	9	3	56	21	21	18	26

CONSENSUS SEQUENCE:

$$A_{\underline{38}}/90 \qquad\qquad A_{\underline{50}}/90$$
$$A_{\underline{53}}\ G_{\underline{68}}\ G_{90}\ T_{90}\ \big|\ A_{\underline{61}}\ G_{\underline{77}}\ T_{\underline{56}}$$
$$C_{\underline{37}}/90 \quad 90 \quad 90 \quad 90 \qquad G_{\underline{30}}/90 \quad 90 \quad 90 \quad 90$$

ACCEPTOR SITE ← ———————— INTRON ———————— → ← ———————— EXON ———————— →

BASE FREQUENCY																				
A	11	6	13	7	4	6	20	3	85	0	19	11	21	16	23	20	19	10	29	27
G	5	8	9	5	7	5	19	1	0	85	40	27	19	25	13	21	28	38	19	15
C	28	27	24	36	31	27	25	63	0	0	17	15	26	25	26	22	21	22	23	22
T	35	44	39	37	43	47	21	18	0	0	9	32	19	19	23	22	17	15	24	16

CONSENSUS SEQUENCE:

$$\text{PY}_{\underline{63}}\ \text{PY}_{\underline{71}}\ \text{PY}_{\underline{63}}\ \text{PY}_{\underline{73}}\ \text{PY}_{\underline{74}}\ \text{PY}_{\underline{74}}\ X \quad C_{\underline{63}}/85 \quad A_{\underline{85}}\ G_{\underline{85}}\ G_{\underline{40}}/85 \quad G_{\underline{27}}/85$$
$$85 \quad 85 \quad 85 \quad 85 \quad 85 \quad 85 \qquad T_{\underline{18}}/85 \qquad 85 \quad 85 \quad 85 \quad T_{\underline{32}}/85$$

[a] 90 donor sites and 85 acceptor sites were analyzed; the table shows the number of times a given nucleotide occurred at a given position. Consensus sequences are shown with frequency of occurrence of the given nucleotide. The "GT–AG rule" splice point is marked with a vertical arrow. The direct repeats in the consensus sequences are underlined. Py = pyrimidine, Pu = purine, X = A, G, C, or T. The genes used are: globins of goat (23), human (36), mouse (10, 12, 13), rabbit (14), and X-laevis (22); yeast actin (85, 86); human growth hormone (100); insulins of chicken, rat, and human (79–82); mouse dihydrofolate reductase (102); gene X (62); ovalbumin (51, 53, 54); lysozyme (76); conalbumin (J. M. Jeltsch, personal communication); B. mori fibroin (91, 92); adenovirus E_1 A (189); chicken ovomucoid (72); collagen (B. de Crombrugghe, personal communication); adenovirus DNA–binding protein (190); SV40 (191, 192); polyoma (193); adenovirus tripartite leader (194, 195); γ_1, γ_2, and μ heavy chains (132, 138–141, 145); mouse heavy- and light-chain J segments (131, 136, 137, 196); and immunoglobulin signal peptides (134–137, 142).

above) are not AG, nucleotides six and eight almost without exception are purine and G, respectively. The 3' end of introns is pyrimidine rich, even upstream of the consensus sequence given (53, 54, 191).

While in most cases the intron-exon junction cannot be defined exactly because one to five nucleotides are repeated at the donor and acceptor sites, the junction may always be drawn such that the intron begins with the dinucleotide GT at its 5' end and ends with the dinucleotide AG at its 3' end (53). In the few cases (all belonging to the ovomucoid gene) where a splice site is defined exactly by the sequence (72), it falls in such a way as to conform to this "GT-AG" rule. The dinucleotides GT and AG are invariably present at donor and acceptor sites, respectively, of genes transcribed by RNA polymerase B. All previously reported exceptions have now been corrected. The existence of a direct repetition (underlined in Table 1) in the consensus sequence of both donor and acceptor sites (53, 54) is confirmed in this more extensive comparison of junction sequences. It may be of evolutionary relevance that such short direct repetitions are found at the extremities of both procaryotic and eucaryotic transposable elements (118).

The above observations do not apply for split ribosomal (199–202) or tRNA genes [for references see (1)], whose transcripts are probably spliced by different enzymes. For protein-coding genes they hold for organisms as diverse as yeasts, insects, chickens, and humans. This is reflected in the fact that chicken and rabbit gene transcripts can be correctly spliced in mouse cells (203–205), and mouse gene transcripts in monkey cells (169, 208). The GT-AG rule may not apply strictly to the intermediate splicing events that may be involved in the stepwise removal of sequences within a given intron transcript (176a).

The consensus sequences are clearly very simple. Can they alone be sufficient to define splice points or are other features, such as secondary structures, necessary? Extensive secondary structures involving intron transcripts do not seem to be necessary for correct splicing, as viable deletion mutants of SV40 lacking large segments of the large T intron have been obtained (206, 207). Some of these deletions extend to 13 and 12 bp from the upstream and downstream splice junctions, respectively, which suggests that almost all the intron sequences may be dispensed with. Extensive sequences of the exons may also be removed without stopping the splicing process: an SV40-mouse-β-globin recombinant containing only 18 bp of exon at the 3' end of the second globin exon, an intact large intron, and an intact third exon, all under the control of the SV40 late promoter, produced RNA where the globin splice had occurred correctly (169). This and another experiment have shown that correct excision of either globin intron transcript is not dependent on the presence of the other intron transcript or of transcripts of distant exons (169, 208). Splicing out of one of the two introns of a rat insulin gene is not dependent on the presence of the other, as a second rat insulin gene that has only one of the introns is correctly expressed (81). Comparison of the sequences of several different families of genes linked by duplication processes (see, for example, 14, 62) has shown that while the sequences of analogous introns can be widely divergent, the sequences at the intron-exon junctions are much better conserved. Taken together, these facts suggest that sufficient information to define a functional splice site may be encoded in a 20 to 30-nucleotide stretch, part of which is related to the consensus sequence.

Evidently, sequences related to the consensus sequences occur also within genes at sites not normally used for splicing. Splice sites that are not normally used probably do exist within genes, and can be "activated" if one of the normal splice junctions is somehow interfered with. A viable SV40 mutant from which the small t donor splice has been deleted still makes a truncated small t protein (209). An alternate splice site is almost certainly used, as splicing is needed for production of stable RNAs from SV40. An SV40 mutant with a deletion in the late region makes use of a "new" acceptor site at 0.86 map unit and not the normal 19S splice site junction (P. Gruss and G. Khoury, unpublished). An adenovirus-SV40 hybrid virus produces a hybrid mRNA that has spliced together a transcript from the adenovirus region and a transcript from the SV40 region; the use of the SV40 acceptor splice site involved has not been detected in SV40 infections or transformations (210).

The examples listed above strongly suggest that alternative splice junctions not normally used do exist within genes. Other interesting observations are that splicing can occur between globin gene donors and SV40 acceptors (G. Chu and P. Sharp, personal communication; P. Berg, personal communication) and that splicing can occur between the donor site of the hydrophobic leader exon and the acceptor site of a constant region of an abortively rearranged κ light-chain gene (211–213). In a mutant γ_1 heavy-chain gene with a deleted CH1 domain exon, splicing occurs between a V-J region donor site and the hinge acceptor site (214). Possibly, under suitable conditions, any donor site may be able to be spliced to any acceptor site, i.e. donor and acceptor sites may be separate entities that need not be considered as pairs. If this is the case, it is possible that sites capable of acting as splice sites may be defined by fairly simple consensus sequence rules, such as those defined above. The question then is what mechanisms make a particular site active, while keeping others dormant and so ensure that a given donor site does in fact become linked to its normal acceptor site, and not to the acceptor site of, for example, a later downstream exon? (No such events have yet been observed except for viral genes, and no intermolecular splicing has ever been observed.)

It is possible that sequences distant from the splice sites themselves may play a role in determining whether a particular splice site will be used and to what it will be linked. The relative frequencies of the splicing events used to generate the SV40 T and t mRNAs from their common precursor is heavily affected by deletions in the large T intron in a complex manner (209). Thus a virus bearing a relatively large deletion in the T intron close to the t splice junction produces much more truncated t mRNA for the same amount of T mRNA than a second virus bearing a smaller deletion further away from the splice junction. An unrearranged κ-light chain gene yields a transcript that is not detectably processed, in spite of the fact that both C_κ and J_κ splice junctions are contained in it (181); the corresponding J-C splice does of course occur during the maturation of the transcript of the productively rearranged gene. However, it only occurs for that J segment that has been joined to V, even though the precursor may harbor additional J segment transcripts with intact splice sites closer to the C transcript (181). While these findings indicate the influence that sequences distant from a splice site may have on splicing, the mechanism involved remains to be elucidated.

Further insight into the splicing mechanism should come from studies of the interaction between purified splicing enzymes and their substrates. This has been partially achieved for tRNA genes (215, 216), and the recent demonstration of splicing in vitro using isolated nuclei and cytoplasmic "extracts" (217–219) suggests that it may become possible for mRNA coding genes. A question that could rapidly be resolved in this way is whether the purified enzyme requires molecules other than the substrate to carry out the splicing reaction. It should be remembered that pre-mRNA molecules are associated in vivo (possibly specifically) with proteins (220 and references therein), and these RNP complexes also contain some small nuclear RNAs (snRNAs). It has been suggested that "splicer" RNAs could hybridize near the splice points and "guide" the splicing enzyme (221, 222). Several investigators have suggested that some small nuclear RNAs may be involved in splicing [for references see (197, 198, 221, 222)]. These RNAs are strictly conserved across the higher eucaryotic species possessing interchangeable processing systems (197), and the nucleotide sequence at the 5' end of one of them, U1 RNA, exhibits complementarity to splice junctions (197, 198). It could function by forming base pairs with a number of residues lying within the two ends of an intron, thus bringing the exon sequences into proximity. A similar role in the maturation of adenovirus late messengers has been proposed for the virus-associated RNAs (VA RNAs) (221, 222). VA RNA binds to unfractionated late virus mRNA and to a cloned cDNA copy of a single late mRNA species (223) but not to unrelated viral DNA sequences such as the unspliced polypeptide IX coding sequence (105). The observation that polyadenylation may precede splicing, and that all spliced molecules are polyadenylated has led to the suggestion that polyA could align splicing sites by formation of triple-stranded structures with sequences around both donor and acceptor sites (224).

The role of snRNAs or polyA as guides in splicing remains speculative. However, yeast mitochondrial genes may indeed use RNA guide molecules for splicing. The two genes, cob-box and oxi-3, produce a series of precursor processing intermediates involving intron-transcript excision (183, 185). Pleiotropic mutations exist that block synthesis of both gene products, and these mutations interfere with the normal splicing pattern of both genes at specific steps. Some of these mutants, the box-3 and box-7 mutants, map in introns. Functional *trans*-acting RNAs affected by the mutations may serve guide functions for the splicing enzyme and must be present for some intron excision step(s) in both cob-box and oxi-3 precursors (183). However, some studies show that the box-3 mutation may be located within a poly-peptide chain that may be involved in the splicing machinery (225).

Splicing and RNA Transcript Stability

Experiments using SV40 have linked a productive splicing event to the formation of a stable messenger. Precise deletion of SV40 late gene introns (226, 227) prevents the expression of VP1. Though the late region is appar-ently transcribed normally, the RNA product is rapidly degraded in the nucleus. Introduction of a pair of splice sites in the sense orientation from the mouse β-globin gene into the "intronless" SV40 16S RNA gene results in stable, spliced RNA being formed (208). D. A. Hamer, P. Leder, and their colleagues have constructed a series of SV40-transducing viruses that carry various combinations of splice junctions derived from the viral genome and a mouse globin gene. All of the viruses that possess a functional pair of splice junctions encoded stable hybrid RNAs, while a virus from which all the splice junctions had been removed failed to produce any detectable stable RNA (169, 228, 229). Mutants of SV40 early genes where the acceptor splice junction has been removed also fail in many cases to produce stable RNAs or proteins (207–209). When they do, it is probably because an alternative acceptor-site is used to replace the deleted one (209). These results suggest that a splicing event is required for the accumulation of stable RNA, and that splicing may be linked to transport of the RNA from the nucleus to the cytoplasm (207, 229, 221). The splicing enzyme could, for example, be bound to the nuclear membrane and make splicing and transport interdependent. However, it is clear that some messengers are unspliced; they include the nonpolyadenylated histone messengers (3), the polyadenylated messengers for adenovirus polypeptide IX (105), and interferon α-1 (104). These messengers, nonetheless, are transported to the cytoplasm, and accumulate as stable mRNAs; one must therefore postulate different transport pathways for these messengers in the splicing-transport model. It is an interesting contrast that precise removal of an intron from a tyrosine transfer RNA gene does not abolish its function (229a).

Splicing and Control of Gene Expression

It is tempting to speculate that splicing, with its apparent link to the production of stable mRNA, may be used as a control mechanism. There is very little experimental data along these lines. However, infection of undifferentiated F-9 murine teratocarcinoma cells by SV40 does not lead to production of early mRNAs; the block probably lies at the splicing level (230). After in vitro–induced differentiation, however, the block is lifted and stable early mRNAs are produced (231). A related observation is that during abortive infection of monkey cells by human Adenovirus type 2, a large proportion of cytoplasmic fiber gene RNA contains long sequences between the tripartite leader and the main body that have not been spliced out, while other late RNAs are correctly processed (232). This effect may be relieved by mutations that map in the Adenovirus 72 K DNA binding protein gene (233, 234). Late in adenovirus infection of HeLa cells, poly(A)-containing hnRNA is transcribed from the cellular genome, but no mature mRNA reaches the cytoplasm. This could be due to a lack of splicing of the cellular messengers; the splicing machinery is taken over by their viral counterparts (235). Evidently, despite the fact that all splice sites seem to fit a common consensus sequence and that transcripts can be spliced in heterologous systems, mechanisms do exist for modulating splicing activity.

SEQUENCES INVOLVED IN THE CONTROL OF TRANSCRIPTION

Procaryotic promoters have been characterized as regions of the DNA that are indispensable for specific initiation of transcription to which RNA polymerase binds. They contain a sequence of homology related to 5'-TATAATG-3' (the "Pribnow" box), which is located about 10 bp upstream from the mRNA start site (236, 237 and references therein). A second sequence of homology, the "recognition region" has also been noted in some promoters in a region centered about 35 bp upstream from the mRNA start site (236, 237).

The comparison of several cellular and viral protein-coding genes in a search for possible eucaryotic counterparts of these sequences has revealed the existence of an AT-rich region of homology centered about 25–30 bp upstream from the mRNA start sites (51 and references therein). This sequence, known as the TATA box, was first noticed by Goldberg and

Hogness (238), and has been found upstream from the known or putative mRNA start sites of all as yet sequenced eucaryotic mRNA-coding genes transcribed by RNA polymerase B (or II), with the exception of the papovavirus late genes and the Adenovirus early region 2 (190 and references therein). The consensus TATA box sequence shown in Table 2 has been compiled from sequences at the 5' end of 60 eucaryotic genes and shows a striking similarity to the bacterial Pribnow box. The exact location of the TATA box varies from gene to gene with the T in position (1) falling between positions -34 to -26 from the mRNA start site. However, for 54 out of the 60 genes this T falls within 2 bases of position -31. There are no obvious additional homologies within 10 bp around the TATA box, with the possible exception of a preference for G in the antisense strand. A second region of homology, 5'-GG^C_TCAATCT-3', has been noticed at positions -70 to -80 of several cellular and viral protein-coding genes (6b, 56), but its generality remains to be demonstrated. A third region of homology exists around the mRNA start site. Rather than having a particular sequence, as suggested previously by comparison of β-globin and Adenovirus major late genes (12, 164), the mRNA start site appears to consist of an A residue (in the antisense strand), surrounded by pyrimidines (Table 2). (However, this consensus is certainly biased, since only the 5' ends of 22 mRNAs, all but one of which start with an A, have yet been accurately mapped.) The TATA box is situated at roughly 3 turns of the DNA helix from the mRNA start site, whereas the procaryotic Pribnow box is located at about one helix turn from this site.

Recently, accurate initiation of transcription of protein-coding eucaryotic genes has been obtained in vitro with purified RNA polymerase B (or II). Initially, Weil et al (256) demonstrated that, in the presence of a cell-free extract, accurate initiation of transcription can occur in vitro on the cloned DNA of the Adenovirus-2 major late transcription unit. Using the same system, accurate in vitro initiation of transcription on chicken conalbumin and ovalbumin genes (162) and on a mouse β-globin gene (257) has been demonstrated. A second cell-free system promoting accurate in vitro initiation of transcription of the adenovirus major late genes (258) and human globin genes (6a) has been described.

Using deletion mutants of the conalbumin and of the Adenovirus-2 major late genes, constructed by in vitro genetic techniques, it has been shown that sequences located upstream from the mRNA start points are required for the initiation of specific transcription in vitro (162, 259, 259a). These sequences are located between positions -12 and -32 [the first T (position (1) in Table 2) of the TATA box of these two genes is located at position -31]. A deletion mutant lacking the sequence upstream from position -29 does not promote specific transcription (259). In addition, it has been found that the -12 to -32 Adenovirus-2 major late gene fragment alone, when cloned in the plasmid pBR322, can direct specific in vitro initiation of transcription, starting about 25 bp downstream from the TATA box (260). In vitro transcription of SV40 early gene mutants bearing various deletions located downstream from the TATA box also initiates about 25 bp downstream from the TATA box (261). These results indicate that sufficient information for specific initiation of transcription in vitro is contained within a 20-bp region including the TATA box, and that transcription initiates about 25 bp downstream from the box. However, in discord with this simple picture, initiation of transcription still occurs on the SV40 early gene region in the absence of the TATA region, but at multiple specific sites (261), which suggests the existence of functional TATA box–related substitute elements.

Results obtained with deletion mutants should however be interpreted with caution. Deletion cannot simply be considered as the absence of the deleted sequence, since new DNA sequences are fused at the deletion endpoint, which makes it impossible to rule out that the observed effects are due to an effect of the replacing sequences, rather than to direct alteration of the promoter region. This may lead to a misinterpretation in the identification of the location of the promoter sequences. However, a single basepair transversion (T to G) at position (3) (Table 2) of the conalbumin TATA box drastically decreases the efficiency of specific in vitro transcription of conalbumin DNA (262). This down-mutation, which is analogous to some promoter down-mutations in the procaryotic Pribnow box (236, 237), demonstrates that the TATA box fulfills at least one criterion that has been used to define procaryotic promoters; it is a DNA sequence required for initiation of specific in vitro transcription. However, it cannot be definitely concluded at the present time that the TATA box region belongs to a eucaryotic promoter region (at least for in vitro transcription), since it has not yet been shown that RNA polymerase binds to it. Interestingly, the mouse pseudo α-globin gene (41, 42), which is probably not transcribed in vivo (42), exhibits the same $T \rightarrow G$ transversion at position (3) (Table 2) of its TATA box when compared to the normal α-globin gene. There is no evidence in the above studies that the consensus region located around positions -70 to -80 is affecting the specificity or the efficiency of in vitro transcription. On the other hand, it appears that the nature of the sequences located around the mRNA startpoints influences the efficiency of specific in vitro transcription (259, 259a).

The sequence requirements for specific in vitro initiation of transcription by RNA polymerase B is in marked contrast with what has been reported for genes (5S RNA, tRNAs) transcribed by RNA polymerase C (or III), where an essential sequence component directing specific initiation of tran-

Table 2 Analysis of "sequences around and upstream from the mRNA start site[a]

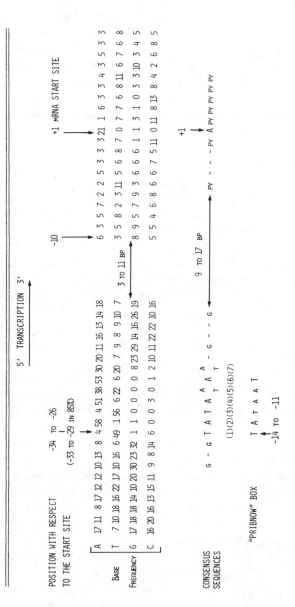

a The sequence of the antisense strand of 60 genes has been analyzed. The sequences around the mRNA start site are aligned with respect to the base coding for the first nucleotide of the mRNA (position +1, arrow). The "upstream" sequences (the numbering is negative upstream from the mRNA start point; numbers in parentheses below the consensus sequence refer to the position of the individual bases in this sequence) are aligned with respect to the almost invariant A, which occurs at position (2) of the upstream consensus sequence. The distance between the T at position (1) in the upstream consensus sequence and the mRNA start site varies from 26 to 34 bp, depending on the gene. Moreover 85% of these Ts are located at distances between −29 and −33 bp from the start site. Highly conserved positions are indicated by large uppercase letters; other conserved positions (> 50%) are indicated by smaller uppercase letters (PY = pyrimidine). The base frequency around the mRNA start site was established from the following 22 genes for which the start site is known unambiguously: major late Adenovirus-2 (164), Adenovirus-2 polypeptide IX (105), Adenovirus-5 E1A (239), Adenovirus-5 E1B (240), Adenovirus-2 E3 (241), Adenovirus-2 E4 (242; Baker and Ziff, personal communication), Moloney murine leukemia virus (243), chicken conalbumin (66), chicken ovalbumin (51), chicken ovomucoid (69; D. Dupret and P. Gerlinger, personal communication), chicken lysozyme (M. Grez, G. Schütz, A. Jung, and A. E. Sippel, personal communication), chicken preproinsulin (79), *Bombyx mori* fibroin (92), *Antheraea chorion* 292 protein (93), human fetal Gγ globin (36), human preproinsulin (80), mouse α-globin (10), mouse β-major globin (13), rabbit β-globin (35), rat preproinsulin I (81), goat β^A-globin (44), and *Drosophila* heat-shock protein 70 (244). The base frequency of the upstream sequence of homology was derived from both the preceding and the following genes: Adenovirus-2 IVa2 (194), Adenovirus-12 E1A (left and right) (245), Adenovirus-7 E1A and E1B (246), Adenovirus-5 E4 (242), SV40 T antigens (247, 248), BK virus T antigens (249), Moloney sarcoma virus (250), spleen necrosis virus (251), chicken X and Y (R. Heilig and J. L. Mandel, personal communications), yeast iso-1-cytochrome c (88), *Antheraea chorion* 10, 40, and 18 proteins (93), human fetal Aγ globin (36), human interferon α-1 (104), mouse β-minor globin (13), mouse IgG λII light chain (134), rat preproinsulin II (81), goat β^C and γ-globin (44), and 14 sea urchin histone genes (252–255). The sequence of the procaryotic Pribnow box of homology is taken from 236. (List compiled with help of J. Cozden.)

scription is intragenic (263–269). Furthermore, the homologous sequences that are found upstream from the mRNA start site are not found upstream from the start sites of genes transcribed by RNA polymerase A (or I) (270) or C (263, 264), which indicates that the specific transcription of different genes by the distinct classes of RNA polymerase (reviewed in 271, 272) could be due to the specific recognition of sequences characteristic of a class of gene.

What is the in vivo relevance of the in vitro transcription studies? Does the TATA box belong to the promoter region in vivo, and is there more to the RNA polymerase B promoter than the TATA box and the mRNA start site? In agreement with the in vitro results, studies with sea urchin histone H2A (273) and SV40 early genes (274, 275) deleted downstream from the TATA box have shown that the 5' end of the novel mRNAs synthesized in *Xenopus* oocytes (273) or cells in culture (274, 275) always map about 25 nucleotides downstream from the TATA box.

However, it has been demonstrated in several systems [sea urchin histone H2A (273), SV40 early genes (274, 276), and polyoma early genes (277)] that deletions of regions located around the mRNA start site, but including the TATA box, do not eliminate transcription in vivo. The new RNA species have heterogenous 5' ends (273, 274) that, in the case of SV40, have been shown to correspond to the 5' ends of the RNAs synthesized in vitro from the same deletion mutants (261; see above). In addition, there is evidence that sequences situated far upstream from the mRNA start site are important for in vivo transcription of the histone H2A and SV40 early genes. Mutants of SV40 early genes (274) and H2A (273, 278) genes with intact sequences around the mRNA start sites and TATA boxes, but with deletions mapping more than 150 bp upstream, have been constructed in vitro and introduced into cells in culture (274) or *Xenopus* oocytes (273, 278). There is very little, if any, in vivo transcription of these deletion mutants. Obviously, the interpretation of the results obtained in vivo with deletion mutants suffers from the same limitation as the corresponding in vitro transcription experiments, since a deletion is more than just the absence of a particular sequence. In vivo experiments with base-substitution mutants constructed in vitro by site-directed mutagenesis are required to confirm the interpretation of the results obtained with the deletion mutants. However, if one assumes that none of the new sequences fused at the deletion endpoints of the various mutants affects positively or negatively the functional properties of the remaining wild-type sequence, it appears that the DNA sequences involved in vivo in the promotion of specific transcription of eukaryotic protein-coding genes could correspond to larger regions than their procaryotic counterparts.

Taking together the limited amount of information available at the present time, it appears that at least two elements could be required in vivo for specific initiation of transcription at the mRNA start site. One element, which corresponds to the TATA box region, would be involved in the mechanism directing RNA polymerase B to initiate transcription about 25 bp downstream at the position corresponding to the 5' end of the mRNA (the efficiency of initiation is influenced also by the actual DNA sequence around the mRNA start site,) and may correspond to a strictu sensu promoter site. In its absence, mRNAs with 5'-terminal sequence heterogeneity would be synthesized. A second element would be located in a region located more than 100 bp upstream and its absence would preclude transcription. It could be visualized as a region required to make the downstream sequences accessible to RNA polymerase. It may correspond to binding sites for proteins involved in positive control of transcription. It is also possible that this region is involved in the generation of "open", "active," chromatin structure (279 and references therein), since there is evidence that chromatin structure of both viral (280, 281, and references therein) and cellular (282, 283, and references therein) genes is modified in regions corresponding to their 5' ends. Such a region would not be required for specific transcription in the in vitro cell-free systems, where there is no evidence that a chromatin structure is reconstituted (261). Alternatively (or in addition) the promoter site could consist of both the TATA box and the indispensable sequence situated further upstream, to both of which the RNA polymerase would bind. In this alternative a eukaryotic promoter region for RNA polymerase B should be viewed as a folded chromatin structure, since it is unlikely that RNA polymerase B could cover a 150-bp linear stretch of DNA. In any case, since the TATA box is absent from some genes (see above) there should be other sequences that can dictate where transcription should start. In this respect it is worth recalling that in the absence of the TATA box initiation occurs on SV40 early regions both in vivo and in vitro at multiple sites scattered over a broad area. It has been proposed by Grosschedl & Birnstiel (273, 278) that the TATA box is not the direct functional equivalent of a bacterial Pribnow box, but rather acts as a "selector," with the indispensable upstream region acting as a "modulator."

Our knowledge concerning the sequences specifying termination of transcription is still very limited. Although some sequence homologies have been recognized around the poly(A) addition site (56 and references therein) in addition to the AATAAA sequence (170), no site-directed mutagenesis experiments or the development of a faithful in vitro system permitting the study of the sequences involved in transcription termination and poly(A) addition, have been reported.

CONCLUSION AND PROSPECTS

Recombinant DNA and ancillary technologies have led to remarkable advances in our appreciation of the molecular anatomy of a number of genes and the general organization of eucaryotic genomes. Some long-standing paradoxes have been resolved at least partially, and there is a feeling that we have grasped the general evolutionary significance of the split gene organization. The case for the fundamental importance of duplication events in evolution has been strengthened, and a remarkable evolutionary plasticity of the eucaryotic genome (apart from the protein-coding sequences) has been discovered. "Dead" genes (pseudogenes) have been found and their discovery raises the interesting possibility that the origin of at least some of the intergenic DNA could be related to ancient gene duplication events no longer recognizable. Molecular evolutionists are having a field day.

Progress has been made in the identification of primary transcripts and in the characterization of the splicing pathways that lead to the appearance of the mature mRNAs. DNA sequence comparisons and studies (carried out mainly with mutants made in vitro) have given some insight into the nature of the splicing signals, but cell-free splicing systems would accelerate progress. The observation that there may be a relationship between the formation of stable polysomal mRNA and the presence of introns is a major advance toward an understanding of the functional importance of RNA splicing. However, the molecular mechanisms underlying this relationship and their possible role in the control of gene expression during differentiation and in differentiated cells are still mysterious, and here too, the establishment of cell-free splicing systems would be of distinct advantage.

With the development of cell-free systems capable of accurate in vitro initiation of transcription, it has become possible to study the functional importance of the regions of homology that have been recognized by comparison of the DNA sequences of various genes. Such in vitro studies have led to the characterization of a possible promoter region for RNA polymerase B. However, in vivo transcription studies performed by introduction of deletion mutants made in vitro into cells in culture or into *Xenopus* oocytes, indicate that the promoter region could be much larger in eucaryotic cells than in procaryotes. It remains to be seen to what extent this is related to the eucaryotic chromatin organization and whether the promoter regions in eucaryotes should be viewed as structures of higher order.

Gene structure comparisons in a search for DNA sequences that may be necessary for tissue-specific regulation of gene expression have been unsuccessful to date. Insight into the developmentally regulated control of gene expression should be gained from in vitro and in vivo studies of transcription of the cloned genes and mutants thereof. However, no such in vivo or in vitro transcription systems mimicking the differentiated states are available at the present time. Their development represents a most challenging problem for the coming years.

Literature Cited

1. Abelson, J. 1979. *Ann. Rev. Biochem.* 48:1035–69
2. Long, E. O., Dawid, I. B. 1980. *Ann. Rev. Biochem.* 49:727–64
3. Kedes, L. H. 1979. *Ann. Rev. Biochem.* 48:837–70
4. Molgaard, H. V. 1980. *Nature* 286:657–59
5. Tooze, J., ed. 1980. *Molecular Biology of Tumor Viruses,* 2nd Ed. Cold Spring Harbor, NY: Cold Spring Harbor Lab.
6. Maniatis, T., Fritsch, E. F., Lauer, J., Lawn, R. M. 1980. *Ann. Rev. Genet.* 14:145–78
6a. Proudfoot, N. J., Shander, M. H. M., Manley, J. L., Gefter, M. L., Maniatis, T. 1980. *Science* 209:1329–36
6b. Efstratiadis, A., Posakony, J. W., Maniatis, T., Lawn, R. M., O'Connell, C., Spritz, R. A., DeRiel, J. K., Forget, B. G., Weissman, S. M., Slightom, J. L., Blechl, A. E., Smithies, O., Baralle, F. E., Shoulders, C. C., Proudfoot, N. J. 1980. *Cell* 21:653–668
7. Jeffreys, A. J., Flavell, R. A. 1977. *Cell* 12:1097–1108
8. Tilghman, S. M., Tiemeier, D. C., Seidman, J. G., Peterlin, B. M., Sullivan, M., Maizel, J. V., Leder, P. 1978. *Proc. Natl. Acad. Sci. USA* 75:725–29
9. Leder, A., Miller, H. I., Hamer, D. H., Seidman, J. G., Norman, B., Sullivan, M., Leder, P. 1978. *Proc. Natl. Acad. Sci. USA* 75:6187–91
10. Nishioka, Y., Leder, P. 1979. *Cell* 18:875–82
11. Tiemeier, D. C., Tilghman, S. M., Polsky, F. I., Seidman, J. G., Leder, A., Edgell, M. H., Leder, P. 1978. *Cell* 14:237–45
12. Konkel, D. A., Tilghman, S. M., Leder, P. 1978. *Cell* 15:1125–32
13. Konkel, D. A., Maizel, J. V., Leder, P. 1979. *Cell* 18:865–73
13a. Leder, P., Hansen, J. N., Konkel, D. A., Leder, A., Nishioka, Y., Talkington, C. 1980. *Science* 209:1336–42
14. Van den Berg, J., Van Ooyen, A., Mantei, N., Schambock, A., Grosveld, G., Flavell, R. A., Weissmann, C. 1978. *Nature* 276:37–44
15. Hardison, R. C., Butler, E. T. III, Lacy, E., Maniatis, T., Rosenthal, N., Efstratiadis, A. 1979. *Cell* 18:1285–97
16. Lacy, E., Hardison, R. C., Quon, D., Maniatis, T. 1979. *Cell* 18:1273–83
17. Dodgson, J. B., Strommer, J., Engel, J. D. 1979. *Cell* 17:879–87
18. Engel, J. D., Dodgson, J. B. 1980. *Proc. Natl. Acad. Sci. USA* 77:2596–2600
19. Ginder, G. D., Wood, W. I., Felsenfeld, G. 1979. *J. Biol. Chem.* 254:8099–8102
20. Kretschmer, P. J., Kaufman, R. E., Coon, H. C., Chen, M-J., Geist, C. E., Nienhuis, A. W. 1980. *J. Biol. Chem.* 255: 3204–11
21. Jeffreys, A. J., Wilson, V., Wood, D., Simons, J. P., Kay, R. M., Williams, J. G. 1980. *Cell* 21:555–64
22. Patient, R. K., Elkington, J. A., Kay, R. M., Williams, J. G. 1980. *Cell* 21: 565–73
23. Haynes, J. R., Rosteck, P. Jr., Schon, E. A., Gallagher, P. M., Burks, D. J., Smith, K., Lingrel, J. B. 1980. *J. Biol. Chem.* 255:6355–67
24. Dayhoff, M. O., Hunt, L. T., McLaughlin, P. J., Jones, D. D. 1972. *Atlas of Protein Sequence and Structure,* ed. M. O. Dayhoff, pp. 17–30. Washington, DC: Natl. Biomed. Res. Found.
25. Southern, E. M. 1975. *J. Mol. Biol.* 98:503–17
26. Jahn C. L., Hutchison, C. A. III, Phillips, S. J., Weaver, S., Haigwood, N. L., Voliva, C. F., Edgell, M. H. 1980. *Cell* 21:159–68
27. Deisseroth, A., Nienhuis, A., Lawrence, J., Giles, R., Turner, P., Ruddle, F. H. 1978. *Proc. Natl. Acad. Sci. USA* 75:1456–60
28. Deisseroth, A., Nienhuis, A., Turner, P., Velez, R., Anderson, W. F., Ruddle, F., Lawrence, J., Creagan, R., Kucherlapati, R. 1977. *Cell* 12:205–18
29. Popp, R. A. 1979. *Inbred and Genetically Defined Strains of Laboratory Animals,* ed. P. L. Altman, D. D. Katz, Part I, p. 105. Bethesda: Fed. Am. Soc. Exp. Biol.
30. Hughes, S. H., Stubblefield, E., Payvar, F., Engel, J. D., Dodgson, J. B., Spector, D., Cordell, B., Schimke, R. T., Varmus, H. E. 1979. *Proc. Natl. Acad. Sci. USA* 76:1348–52

31. Lebo, R. V., Carrano, A. V., Burkhart-Schultz, K., Dozy, A. M., Yu, L-C., Kan, Y. W. 1979. *Proc. Natl. Acad. Sci. USA* 76:5804–8
32. Gusella, J., Varsanyi-Breiner, A., Kao, F-T., Jones, C., Puck, T. T., Keys, C., Orkin, S., Housman, D. 1979. *Proc. Natl. Acad. Sci. USA* 76:5239–43
33. Jeffreys, A. J., Craig, I. W., Francke, U. 1979. *Nature* 281:606–8
34. Lauer, J., Shen, C-K. J., Maniatis, T. 1980. *Cell* 20:119–30
35. Van Ooyen, A., Van den Berg, J., Mantei, N., Weissmann, C. 1979. *Science* 206:337–44
36. Slightom, J. L., Blechl, A. E., Smithies, O. 1980. *Cell* 21:627–38
37. Fritsch, E. F., Lawn, R. M., Maniatis, T. 1980. *Cell* 19:959–72
38. Shen, C-K. J., Maniatis, T. 1980. *Cell* 19:379–91
39. Hoeijmakers-van Dommelen, H. A. M., Grosveld, G. C., De Boer, E., Flavell, R. A., Varley, J. M., Jeffreys, A. J. 1980. *J. Mol. Biol.* 140:531–47
40. Cleary, M. L., Haynes, J. R., Schon, E. A., Lingrel, J. B. 1980. *Nucleic Acids Res.* 8:4791–4802
40a. Lacy, E., Maniatis, T. 1980. *Cell* 21:545–53
40b. Proudfoot, N. J., Maniatis, T. 1980. *Cell* 21:537–44
41. Nishioka, Y., Leder, A., Leder, P. 1980. *Proc. Natl. Acad. Sci. USA* 77:2806–9
42. Vanin, E. F., Goldberg, G. I., Tucker, P. W., Smithies, O. 1980. *Nature* 286:222–26
43. Bernards, R., Flavell, R. A. 1980. *Nucleic Acids Res.* 8:1521–34
44. Haynes, J. R., Rosteck, P. Jr., Lingrel, J. B. 1981. *Proc. Natl. Acad. Sci. USA* 77: In press
45. Palmiter, R. D. 1975. *Cell* 4:189–97
46. Breathnach, R., Mandel, J-L., Chambon, P. 1977. *Nature* 270:314–19
47. Doel, M. T., Houghton, M., Cook, E. A., Carey, N. H. 1977. *Nucleic Acids Res.* 4:3701–13
48. Weinstock, R., Sweet, R., Weiss, M., Cedar, H., Axel, R. 1978. *Proc. Natl. Acad. Sci.* 75:1299–1303
49. Lai, E. C., Woo, S. L. C., Dugaiczyk, A., Catterall, J. F., O'Malley, B. W. 1978. *Proc. Natl. Acad. Sci. USA* 75:2205–9
50. Kourilsky, P., Chambon, P. 1978. *Trends Biochem. Sci.* 3:244–47
51. Gannon, F., O'Hare, K., Perrin, F., LePennec, J. P., Benoist, C., Cochet, M., Breathnach, R., Royal, A., Garapin, A., Cami, B., Chambon, P. 1979. *Nature* 278:428–34

52. Dugaiczyk, A., Woo, S. L. C., Colbert, D. A., Lai, E. C., Mace, M. L. Jr., O'Malley, B. W. 1979. *Proc. Natl. Acad. Sci. USA* 76:2253–57
53. Breathnach, R., Benoist, C., O'Hare, K., Gannon, F., Chambon, P. 1978. *Proc. Natl. Acad. Sci. USA* 75:4853–57
54. Catterall, J. F., O'Malley, B. W., Robertson, M. A., Staden, R., Tanaka, Y., Brownlee, G. G. 1978. *Nature* 275:510–13
55. Robertson, M. A., Staden, R., Tanaka, Y., Catterall, J. F., O'Malley, B. W., Brownlee, G. G. 1979. *Nature* 278:370–72
56. Benoist, C., O'Hare, K., Breathnach, R., Chambon, P. 1980. *Nucleic Acids Res.* 8:127–42
57. O'Hare, K., Breathnach, R., Benoist, C., Chambon, P. 1979. *Nucleic Acids Res.* 7:321–34
58. McReynolds, L., O'Malley, B. W., Nisbet, A. D., Fothergill, J. E., Givol, D., Fields, S., Robertson, M., Brownlee, G. G. 1978. *Nature* 273:723–28
59. Lai, E. C., Woo, S. L. C., Dugaiczyk, A., O'Malley, B. W. 1979. *Cell* 16:201–11
60. Mandel, J. L., Chambon, P. 1979. *Nucleic Acids Res.* 7:2081–2103
61. Royal, A., Garapin, A., Cami, B., Perrin, F., Mandel, J. L., LeMeur, M., Brégégègre, F., Gannon, F., LePennec, J. P., Chambon, P., Kourilsky, P. 1979. *Nature* 279:125–32
62. Heilig, R., Perrin, F., Gannon, F., Mandel, J. L., Chambon, P. 1980. *Cell* 20:625–37
63. LeMeur, M., Glanville, N., Mandel, J. L., Gerlinger, P., Palmiter, R., Chambon, P. 1981. *Cell.* In press
64. Colbert, D. A., Knoll, B. J., Woo, S. L. C., Mace, M. L., Tsai, M-J., O'Malley, B. W. 1981. *Biochemistry* In press
65. Perrin, F., Cochet, M., Gerlinger, P., Cami, B., LePennec, J. P., Chambon, P. 1979. *Nucleic Acids Res.* 6:2731–48
66. Cochet, M., Gannon, F., Hen, R., Maroteaux, L., Perrin, F., Chambon, P. 1979. *Nature* 282:567–74
67. Aisen, P., Listowsky, I. 1980. *Ann. Rev. Biochem.* 49:357–93
68. Catterall, J. F., Stein, J. P., Lai, E. C., Woo, S. L. C., Dugaiczyk, A., Mace, M. L., Means, A. R., O'Malley, B. W. 1979. *Nature* 278:323–27
69. Lai, E. C., Stein, J. P., Catterall, J. F., Woo, S. L. C., Mace, M. L., Means, A. R., O'Malley, B. W. 1979. *Cell* 18:829–42
70. Lindenmaier, W., Nguyen-Huu, M. C., Lurz, R., Blin, N., Stratmann, M.,

Land, H., Jeep, S., Sippel, A. E., Schütz, G. 1979. *Nucleic Acids Res.* 7:1221–32

71. Lindenmaier, W., Nguyen-Huu, M. C., Lurz, R., Stratmann, M., Blin, N., Wurtz, T., Hauser, H. J., Giesecke, K., Land, H., Jeep, S., Grez, M., Sippel, A. E., Schütz, G. 1980. *J. Steroid Biochem.* 12:211–18

72. Stein, J. P., Catterall, J. F., Kristo, P., Means, A. R., O'Malley, B. W. 1980. *Cell* 21:681–87

73. Chambon, P., Benoist, C., Breathnach, R., Cochet, M., Gannon, F., Gerlinger, P., Krust, A., LeMeur, M., LePennec, J. P., Mandel, J. L., O'Hare, K., Perrin, F. 1979. *From Gene to Protein: Information Transfer in Normal and Abnormal Cells,* ed. T. R. Russell, K. Brew, H. Faber, J. Schultz, pp. 55–81. New York: Academic

74. Baldacci, P., Royal, A., Cami, B., Perrin, F., Krust, A., Garapin, A., Kourilsky, P. 1979. *Nucleic Acids Res.* 6:2667–81

75. Lindenmaier, W., Nguyen-Huu, M. C., Lurz, R., Stratmann, M., Blin, N., Wurtz, T., Hauser, H. J., Sippel, A. E., Schütz, G. 1979. *Proc. Natl. Acad. Sci. USA* 76:6196–6200

76. Jung, A., Sippel, A. E., Grez, M., Schütz, G. 1980. *Proc. Natl. Acad. Sci. USA* 77:5759–63

77. Wahli, W., Dawid, I. B., Wyler, T., Jaggi, R. B., Weber, R., Ryffel, G. U. 1979. *Cell* 16:535–49

78. Wahli, W., Dawid, I. B., Wyler, T., Weber, R., Ryffel, G. U. 1980. *Cell* 20:107–17

79. Perler, F., Efstratiadis, A., Lomedico, P., Gilbert, W., Kolodner, R., Dodgson, J. 1980. *Cell* 20:555–66

80. Bell, G. I., Pictet, R. L., Rutter, W. J., Cordell, B., Tischer, E., Goodman, H. M. 1980. *Nature* 284:26–32

81. Lomedico, P., Rosenthal, N., Efstratiadis, A., Gilbert, W., Kolodner, R., Tizard, R. 1979. *Cell* 18:545–58

82. Cordell, B., Bell, G., Tischer, E., DeNoto, F. M., Ullrich, A., Pictet, R., Rutter, W. J., Goodman, H. M. 1979. *Cell* 18:533–43

83. Schafer, M. P., Boyd, C. D., Tolstoshev, P., Crystal, R. G. 1980. *Nucleic Acids. Res.* 8:2241–53

84. Vogeli, G., Avvedimento, E. V., Sullivan, M., Maizel, J. V. Jr., Lozano, G., Adams, S. L., Pastan, I., de Crombrugghe, B. 1980. *Nucleic Acids Res.* 8:1823–37

84a. Yamada, Y., Avvedimento, V. E., Mudryj, M., Ohkubo, H., Vogeli, G., Irani, M., Pastan, I., de Crombrugghe, B. 1980. *Cell* 22:887–92

84b. Ohkubo, H., Vogeli, G., Mudryj, M., Avvedimento, V. E., Sullivan, M., Pastan, I., de Crombrugghe, B. 1980 *Proc. Natl. Acad. Sci. USA.* In press

84c. Wozney, J., Hanahan, D., Morimoto, R., Boedtker, H., Doty, P. 1980. *Proc. Natl. Acad. Sci. USA.* In press

85. Gallwitz, D., Sures, I. 1980. *Proc. Natl. Acad. Sci. USA* 77:2546–50

86. Ng, R., Abelson, J. 1980. *Proc. Natl. Acad. Sci. USA* 77:3912–16

87. Petes, T. D. 1980. *Ann. Rev. Biochem.* 49:845–76

88. Smith, M., Leung, D. W., Gillam, S., Astell, C. R., Montgomery, D. L., Hall, B. D. 1979. *Cell* 16:753–61

89. Fyrberg, E. A., Kindle, K. L., Davidson, N., Sodja, A. 1980. *Cell* 19:365–78

90. Tobin, S. L., Zulauf, E., Sanchez, F., Craig, E. A., McCarthy, B. J. 1980. *Cell* 19:121–31

90a. Durica, D. S., Schloss, J. A., Crain, W. R. Jr., *Proc. Natl. Acad. Sci. USA* 77:5683–87

90b. Ordahl, C. P., Tilghman, S. M., Ovitt, C., Fornwald, J., Largen, M. T. 1980. *Nucleic Acids Res.* 8:4989–5005

91. Tsujimoto, Y., Suzuki, Y. 1979. *Cell* 16:425–36

92. Tsujimoto, Y., Suzuki, Y. 1979. *Cell* 18:591–600

93. Jones, C. W., Kafatos, F. C. 1980. *Cell* 22:855–67

93a. Goldberg, D. A. 1980. *Proc. Natl. Acad. Sci. USA* 77:5794–98

94. Bhat, S. P., Jones, R. E., Sullivan, M. A., Piatigorsky, J. 1980. *Nature* 284:234–38

95. Jones, R. E., Bhat, S. P., Sullivan, M. A., Piatigorsky, J. 1980. *Proc. Natl. Acad. Sci. USA* 77:5879–83

96. Kioussis, D., Hamilton, R., Hanson, R. W., Tilghman, S. M., Taylor, J. M. 1979. *Proc. Natl. Acad. Sci. USA* 76:4370–74

97. Sargent, T. D., Wu, J-R., Sala-Trepat, J. M., Wallace, R. B., Reyes, A. A., Bonner, J. 1979. *Proc. Natl. Acad. Sci. USA* 76:3256–60

98. Gorin, M. B., Tilghman, S. M. 1980. *Proc. Natl. Acad. Sci. USA* 77:1351–55

99. Durnam, D. M., Perrin, F., Gannon, F., Palmiter, R. D. 1981. *Proc. Natl. Acad. Sci. USA.* In press

100. Fiddes, J. C., Seeburg, P. H., DeNoto, F. M., Hallewell, R. A., Baxter, J. D., Goodman, H. M. 1979. *Proc. Natl. Acad. Sci. USA* 76:4294–98

101. Soreq, H., Harpold, M., Evans, R., Dar-

nell, J. E. Jr., Bancroft, F. C. 1979. *Nucleic Acids Res.* 6:2471–82

101a. Chien, Y-H., Thompson, E. B. 1980. *Proc. Natl. Acad. Sci. USA* 77:4583–87

101b. MacDonald, R. J., Crerar, M. M., Swain, W. F., Pictet, R. L., Thomas, G., Rutter, W. J. 1980. *Nature* 287:117–22

102. Nunberg, J. H., Kaufman, R. J., Chang, A. C. Y., Cohen, S. N., Schimke, R. T. 1980. *Cell* 19:355–64

103. Wahl, G. M., Padgett, R. A., Stark, G. R. 1979. *J. Biol. Chem.* 254:8679–89

104. Nagata, S., Mantei, N., Weissmann, C. 1980. *Nature* 287:401–8

105. Aleström, P., Akusjärvi, G., Perricaudet, M., Mathews, M. B., Klessig, D. F., Pettersson, U. 1980. *Cell* 19:671–81

106. Davidson, E. H., Britten, R. J. 1979. *Science* 204:1052–59

107. Jelinek, W. R., Toomey, T. P., Leinwand, L., Duncan, C. H., Biro, P. A., Choudary, P. V., Weissman, S., Rubin, C. M., Houck, C. M., Deininger, P. L., Schmid, C. W. 1980. *Proc. Natl. Acad. Sci. USA* 77:1398–1402

108. Rubin, C. M., Houck, C. M., Deininger, P. L., Friedmann, T., Schmid, C. W. 1980. *Nature* 284:372–74

109. Jelinek, W. R. 1978. *Proc. Natl. Acad. Sci. USA* 75:2679–83

110. Houck, C. M., Rinehart, F. P., Schmid, C. W. 1979. *J. Mol. Biol.* 132:289–306

111. Duncan, C., Biro, P. A., Choudary, P. V., Elder, J. T., Wang, R. R. C., Forget, B. G., DeRiel, J. K., Weissman, S. M. 1979. *Proc. Natl. Acad. Sci. USA* 76:5095–99

112. Engel, J. D., Dodgson, J. B. 1978. *J. Biol. Chem.* 253:8239–46

113. Lee, D. C., McKnight, G. S., Palmiter, R. D. 1980. *J. Biol. Chem.* 255:1442–50

114. Nguyen-Huu, M. C., Stratmann, M., Groner, B., Wurtz, T., Land, H., Giesecke, K., Sippel, A. E., Schütz, G. 1979. *Proc. Natl. Acad. Sci. USA* 76:76–80

115. Manning, R. F., Gage, L. P. 1978. *J. Biol. Chem.* 253:2044–52

116. Gannon, F., Jeltsch, J. M., Perrin, F. 1980. *Nucleic Acids Res.* 8:4405–21

117. Leupold, U. 1980. *Nature* 283:811–12

118. Calos, M. P., Miller, J. H. 1980. *Cell* 20:579–95

119. Farabaugh, P. J., Fink, G. R. 1980. *Nature* 286:352–56

120. Gafner, J., Philippsen, P. 1980. *Nature* 286:414–18

120a. Dunsmuir, P., Brorein, W. J. Jr., Simon, M. A., Rubin, G. M. 1980. *Cell* 21:575–79

121. Fagan, J. B., Pastan, I., De Crombrugghe, B. 1980. *Nucleic Acids Res.* 8:3055–64

122. Doolittle, W. F. 1978. *Nature* 272:581–82

123. Darnell, J. E. Jr., 1978. *Science* 202:1257–60

124. Blake, C. C. F. 1979. *Nature* 277:598

125. Crick, F. 1979. *Science* 204:264–71

126. Gilbert, W. 1978. *Nature* 271:501

127. Doolittle, R. F. 1979. *The Proteins*, ed. H. Neurath, R. L. Hill, C-L Boeder, pp. 2–118. New York: Academic

128. Gilbert, W. 1979. *Eucaryotic Gene Regulation*, ed. R. Axel, T. Maniatis, C. F. Fox, pp. 1–12. New York: Academic

129. Reanney, D. 1979. *Nature* 277:598–600

130. Ohno, S. 1970. *Evolution by Gene Duplication.* Berlin & New York: Springer

131. Sakano, H., Hüppi, K., Heinrich, G., Tonegawa, S. 1979. *Nature* 280:288–94

132. Sakano, H., Rogers, J. H., Hüppi, K., Brack, C., Traunecker, A., Maki, R., Wall, R., Tonegawa, S. 1979. *Nature* 277:627–33

133. Rogers, J., Early, P., Carter, C., Calame, K., Bond, M., Hood, L., Wall, R. 1980. *Cell* 20:303–12

134. Tonegawa, S., Maxam, A. M., Tizard, R., Bernard, O., Gilbert, W. 1978. *Proc. Natl. Acad. Sci. USA* 75:1485–89

135. Seidman, J. G., Max, E. E., Leder, P. 1979. *Nature* 280:370–75

136. Early, P., Huang, H., Davis, M., Calame, K., Hood, L. 1980. *Cell* 19:981–92

137. Sakano, H., Maki, R., Kurosawa, Y., Roeder, W., Tonegawa, S. 1980. *Nature* 286:676–83

138. Tucker, P. W., Marcu, K. B., Newell, N., Richards, J., Blattner, F. R. 1979. *Science* 206:1303–6

139. Kawakami, T., Takahashi, N., Honjo, T. 1980. *Nucleic Acids Res.* 8:3933–45

140. Yamawaki-Kataoka, Y., Kataoka, T., Takahashi, N., Obata, M., Honjo, T. 1980. *Nature* 283:786–89

141. Honjo, T., Obata, M., Yamawaki-Kataoka, Y., Kataoka, T., Kawakami, T., Takahashi, N., Mano, Y. 1979. *Cell* 18:559–68

142. Bernard, O., Hozumi, N., Tonegawa, S. 1978. *Cell* 15:1133–44

143. Gough, N. M., Kemp, D. J., Tyler, B. M., Adams, J. M., Cory, S. 1980. *Proc. Natl. Acad. Sci. USA* 77:554–58

144. Chang, A. C. Y., Cochet, M., Cohen, S. N. 1980. *Proc. Natl. Acad. Sci. USA* 77:4890–94

144a. Nakanishi, S., Teranishi, Y., Noda, M., Notake, M., Watanabe, Y., Kakidani, H., Jingami, H., Numa, S. 1980. *Nature* 287:752–55

145. Early, P., Rogers, J., Davis, M., Calame, K., Bond, M., Wall, R., Hood, L. 1980. *Cell* 20:313–19
146. Craik, C. S., Buchman, S. R., Beychok, S. 1980. *Proc. Natl. Acad. Sci. USA* 77:1384–88
147. Eaton, W. A. 1980. *Nature* 284:183–85
148. Miyata, T., Yasunaga, T., Yamawaki-Kataoka, Y., Obata, M., Honjo, T. 1980. *Proc. Natl. Acad. Sci. USA* 77:2143–47
149. Weber, J., Blanchard, J-M., Ginsberg, H., Darnell, J. E. Jr. 1980. *J. Virol.* 33:286–91
150. Nevins, J. R., Darnell, J. E. Jr. 1980. *Cell* 15:1477–93
151. Roop, D. R., Nordstrom, J. L., Tsai, S. Y., Tsai, M-J., O'Malley, B. W. 1978. *Cell* 15:671–85
152. Tsai, M-J., Ting, A. C., Nordstrom, J. L., Zimmer, W., O'Malley, B. W. 1980. *Cell* 22:219–30
153. Nordstrom, J. L., Roop, D. R., Tsai, M-J., O'Malley, B. W. 1979. *Nature* 278:328–31
154. Tilghman, S. M., Curtis, P. J., Tiemeier, D. C., Leder, P., Weissmann, C. 1978. *Proc. Natl. Acad. Sci. USA* 75:1309–13
155. Curtis, P. J., Mantei, N., Van den Berg, J., Weissmann, C. 1977. *Proc. Natl. Acad. Sci. USA* 74:3184–88
156. Kinniburgh, A. J., Ross, J. 1979. *Cell* 17:915–21
157. Haynes, J. R., Kalb, V. F. Jr., Rosteck, P. Jr., Lingrel, J. B. 1978. *FEBS Lett.* 91:173–77
158. Bastos, R. N., Aviv, H. 1977. *Cell* 11:641–50
159. Berk, A. J., Sharp, P. A. 1978. *Proc. Natl. Acad. Sci. USA* 75:1274–78
160. Weaver, R. F., Weissmann, C. 1979. *Nucleic Acids Res.* 7:1175–93
161. Roop, D. R., Tsai, M-J., O'Malley, B. W. 1980. *Cell* 19:63–68
162. Wasylyk, B., Kédinger, C., Corden, J., Brison, O., Chambon, P. 1980. *Nature* 285:367–73
163. Tsuda, M., Ohshima, Y., Suzuki, Y. 1979. *Proc. Natl. Acad. Sci. USA* 76:4872–76
164. Ziff, E. B., Evans, R. M. 1978. *Cell* 15:1463–75
165. Ziff, E. B. 1980. *Nature* 287:491–99
166. Nevins, J. R., Blanchard, J-M., Darnell, J. E. Jr. 1980. *J. Mol. Biol.* 144:377–86
167. Ford, J. P., Hsu, M-T. 1978. *J. Virol.* 28:795–801
168. Tsai, S. Y., Roop, D. R., Stumph, W. E., Tsai, M-J., O'Malley, B. W. 1980. *Biochemistry* 19:1755–61
169. Hamer, D. H., Leder, P. 1979. *Cell* 17:737–47
170. Proudfoot, N. J., Brownlee, G. G. 1976. *Nature* 263:211–14
171. Hagenbuchle, O., Bovey, R., Young, R. A. 1980. *Cell* 21:179–87
172. Knöchel, W., Grundmann, U. 1979. *Biochim. Biophys. Acta* 563:143–49
173. Smith, K., Lingrel, J. B. 1978. *Nucleic Acids Res.* 5:3295–3301
174. Courtney, M., Williamson, R. 1979. *Nucleic Acids Res.* 7:1121–30
175. Kinniburgh, A. J., Mertz, J. E., Ross, J. 1978. *Cell* 14:681–93
176. Ryffel, G. U., Wyler, T., Muellener, D. B., Weber, R. 1980. *Cell* 19:53–61
176a. Avvedimento, V. E., Vogeli, G., Yamada, Y., Maizel, J. V. Jr., Pastan, I., de Crombrugghe, B. 1980. *Cell* 21:689–96
177. Gilmore-Hebert, M., Wall, R. 1979. *J. Mol. Biol.* 135:879–91
178. Gilmore-Hebert, M., Hercules, K., Komaromy, M., Wall, R. 1978. *Proc. Natl. Acad. Sci. USA* 75:6044–48
179. Gilmore-Hebert, M., Wall, R. 1978. *Proc. Natl. Acad. Sci. USA* 75:342–45
180. Schibler, U., Marcu, K. B., Perry, R. P. 1978. *Cell* 15:1495–1509
181. Perry, R. P., Kelley, D. E., Coleclough, C., Seidman, J. G., Leder, P., Tonegawa, S., Matthyssens, G., Weigert, M. 1980. *Proc. Natl. Acad. Sci. USA* 77:1937–41
182. Van Ommen, G-J. B., Groot, G. S. P., Grivell, L. A. 1979. *Cell* 18:511–23
183. Church, G. M., Slonimski, P. P., Gilbert, W. 1979. *Cell* 18:1209–15
184. Van Ommen, G-J. B., Boer, P. H., Groot, G. S. P., De Haan, M., Roosendaal, E., Grivell, L. A., Haid, A., Schweyen, R. J. 1980. *Cell* 20:173–83
185. Halbreich, A., Pajot, P., Foucher, M., Grandchamp, C., Slonimski, P. 1980. *Cell* 19:321–29
186. Arnberg, A. C., Van Ommen, G-J. B., Grivell, L. A., Van Bruggen, E. F. J., Borst, P. 1980. *Cell* 19:313–19
187. Berget, S. M., Sharp, P. A. 1979. *J. Mol. Biol.* 129:547–65
188. Seif, I., Khoury, G., Dhar, R. 1979. *Nucleic Acids Res.* 6:3387–97
189. Perricaudet, M., Akusjärvi, G., Virtanen, A., Pettersson, U. 1979. *Nature* 281:694–96
190. Baker, C. C., Herisse, J., Courtois, G., Galibert, F., Ziff, E. 1979. *Cell* 18:569–80
191. Reddy, V. B., Ghosh, P. K., Lebowitz, P., Piatak, M., Weissman, S. M. 1979. *J. Virol.* 30:279–96
192. Ghosh, P. K., Reddy, V. B., Swinscoe, J., Lebowitz, P., Weissman, S. M. 1978. *J. Mol. Biol.* 126:813–46

193. Kamen, R., Favalor, J., Parker, J., Treisman, R., Lania, L., Fried, M., Mellor, A. 1979. *Cold Spring Harbor Symp. Quant. Biol.* 44:63–75
194. Akusjärvi, G., Pettersson, U. 1979. *J. Mol. Biol.* 134:143–58
195. Zain, S., Gingeras, T. R., Bullock, P., Wong, G., Gelinas, R. E. 1979. *J. Mol. Biol.* 135:413–33
196. Max, E. E., Seidman, J. G., Leder, P. 1979. *Proc. Natl. Acad. Sci. USA* 76:3450–54
197. Lerner, M. R., Boyle, J. A., Mount, S. M., Wolin, S. L., Steitz, J. A. 1980. *Nature* 283:220–24
198. Rogers, J., Wall, R. 1980. *Proc. Natl. Acad. Sci. USA* 77:1877–79
199. Bos, J. L., Osinga, K. A., Van der Horst, G., Hecht, N. B., Tabak, H. F., Van Ommen, G-J. B., Borst, P. 1980. *Cell* 20:207–14
200. Wild, M. A., Sommer, R. 1980. *Nature* 283:693–94
201. Dujon, B. 1980. *Cell* 20:185–97
202. Allet, B., Rochaix, J-D. 1979. *Cell* 18:55–60
203. Breathnach, R., Mantei, N., Chambon, P. 1980. *Proc. Natl. Acad. Sci. USA* 77:740–44
204. Mantei, N., Boll, W., Weissmann, C. 1979. *Nature* 281:40–46
205. Wold, B., Wigler, M., Lacy, E., Maniatis, T., Silverstein, S., Axel, R. 1979. *Proc. Natl. Acad. Sci. USA* 76:5684–88
206. Thimmappaya, B., Shenk, T. 1979. *J. Virol.* 30:668–73
207. Volckaert, G., Feunteun, J., Crawford, L. V., Berg, P., Fiers, W. 1979. *J. Virol.* 30:674–82
208. Gruss, P., Khoury, G. 1980. *Nature* 286:634–37
209. Khoury, G., Gruss, P., Dhar, R., Lai, C-J. 1979. *Cell* 18:85–92
210. Khoury, G., Alwine, J., Goldman, N., Gruss, P., Jay, G. 1980. *J. Virol.* 36:143–51
211. Choi, E., Kuehl, M. Wall, R. 1980. *Nature* 286:776–79
212. Schnell, H., Steinmetz, M., Zachau, H. G. 1980. *Nature* 286:170–73
213. Seidman, J. G., Leder, P. 1980. *Nature* 286:779–83
214. Dunnick, W., Rabbitts, T. H., Milstein, C. 1980. *Nature* 286:669–75
215. Knapp, G., Ogden, R. C., Peebles, C. L., Abelson, J. 1979. *Cell* 18:37–45
216. Peebles, C. L., Ogden, R. C., Knapp, G., Abelson, J. 1979. *Cell* 18:27–35
217. Blanchard, J-M., Weber, J., Jelinek, W., Darnell, J. E. Jr. 1978. *Proc. Natl. Acad. Sci. USA* 75:5344–48
218. Hamada, H., Igarashi, T., Muramatsu, M. 1980. *Nucleic Acids Res.* 8:587–99
219. Goldenberg, C. J., Raskas, H. J. 1980. *Biochemistry* 19:2719–23
220. Beyer, A. L., Miller, O. L. Jr., McKnight, S. L. 1980. *Cell* 20:75–84
221. Murray, V., Holliday, R. 1979. *Genet. Res.* 34:173–88
222. Murray, V., Holliday, R. 1979. *FEBS Lett.* 106:5–7
223. Mathews, M. B. 1980. *Nature* 285:575–77
224. Bina, M., Feldmann, R. J., Deeley, R. G. 1980. *Proc. Natl. Acad. Sci. USA* 77:1278–82
225. Jacq, C., Lazowska, J., Slonimski, P. P. 1980. *C. R. Acad. Sci. Paris* 290:89–92
226. Gruss, P., Lai, C-J., Dhar, R., Khoury, G. 1979. *Proc. Natl. Acad. Sci. USA* 76:4317–21
227. Lai, C-J., Khoury, G. 1979. *Proc. Natl. Acad. Sci. USA* 76:71–75
228. Hamer, D. H., Smith, K. D., Boyer, S. H., Leder, P. 1979. *Cell* 17:725–35
229. Hamer, D. H., Leder, P. 1979. *Cell* 18:1299–302
229a. Wallace, R. B., Johnson, P. F., Tanaka, S., Schold, M., Itakura, K., Abelson, J. 1980. *Science* 209:1396–1400
230. Segal, S., Levine, A. J., Khoury, G. 1979. *Nature* 280:335–38
231. Segal, S., Khoury, G. 1979. *Proc. Natl. Acad. Sci. USA* 76:5611–15
232. Klessig, D. F., Chow, L. T. 1980. *J. Mol. Biol.* 139:221–42
233. Klessig, D. F., Grodzicker, J. 1979. *Cell* 19:957–66
234. Klessig, D. F., Chow, L. T. 1981. *J. Mol. Biol.* In press
235. Beltz, G. A., Flint, S. J. 1979. *J. Mol. Biol.* 131:353–73
236. Rosenberg, M., Court, D. 1979. *Ann. Rev. Genet.* 13:319–53
237. Siebenlist, U., Simpson, R. B., Gilbert, W. 1980. *Cell* 20:269–81
238. Goldberg, M. L. 1979. PhD thesis. Stanford Univ.
239. Van Ormondt, H., Maat, J., De Waard, A., Van der Eb, A. J. 1978. *Gene* 4:309–28
240. Maat, J., Van Ormondt, H. 1979. *Gene* 6:75–90
241. Hérissé, J., Courtois, G., Galibert, F. 1980. *Nucleic Acids Res.* 8:2173–92
242. Shinagawa, M., Padmanabhan, R. V., Padmanabhan, R. 1980. *Gene* 9:99–114
243. Van Beveren, C., Goddard, J. G., Berns, A., Verma, I. M. 1980. *Proc. Natl. Acad. Sci. USA* 77:3307–11
244. Török, I., Karch, F. 1980. *Nucleic Acids Res.* 8:3105–23

245. Sugisaki, H., Sugimoto, K., Takanami, M., Shiroki, K., Saito, I., Shimojo, H., Sawada, Y., Uemizu, Y., Uesugi, S-I., Fujinaga, K. 1980. *Cell* 20:777–86
246. Dijkema, R., Dekker, B. M. M., Van Ormondt, H. 1980. *Gene* 9:141–56
247. Fiers, W., Contreras, R., Haegeman, G., Rogiers, R., Van de Voorde, A., Van Heuverswyn, H., Van Herreweghe, J., Volckaert, G., Ysebaert, M. 1978. *Nature* 273:113–20
248. Reddy, V. B., Thimmappaya, B., Dhar, R., Subramanian, K. N., Zain, B. S., Pan, J., Ghosh, P. K., Celma, M. L., Weissman, S. M. 1978. *Science* 200:494–502
249. Seif, I., Khoury, G., Dhar, R. 1979. *Cell* 18:963–77
250. Dhar, R., McClements, W. L., Enquist, L. W., Vande Woude, G. F. 1980. *Proc. Natl. Acad. Sci. USA* 77:3937–41
251. Shimotohno, K., Mizutani, S., Temin, H. M. 1980. *Nature* 285:550–54
252. Busslinger, M., Portmann, R., Irminger, J. C., Birnstiel, M. L. 1980. *Nucleic Acids Res.* 8:957–77
253. Hentschel, C., Irminger, J. C., Bucher, P., Birnstiel, M. L. 1980. *Nature* 285:147–51
254. Sures, I., Levy, S., Kedes, L. H. 1980. *Proc. Natl. Acad. Sci. USA* 77:1265–69
255. Levy, S., Sures, I., Kedes, L. H. 1979. *Nature* 279:737–39
256. Weil, P. A., Luse, D. S., Segall, J., Roeder, R. G. 1979. *Cell* 18:469–84
257. Luse, D. S., Roeder, R. G. 1980. *Cell* 20:691–99
258. Manley, J. L., Fire, A., Cano, A., Sharp, P. A., Gefter, M. L. 1980. *Proc. Natl. Acad. Sci. USA* 77:3855–59
259. Corden, J., Wasylyk, B., Buchwalder, A., Sassone-Corsi, P., Kedinger, C., Chambon, P. 1980. *Science* 209:1406–14
259a. Hu, S-L., Manley, J. L. 1980. *Proc. Natl. Acad. Sci. USA.* In Press
260. Sassone-Corsi, P., Buchwalder, A., Corden, J., Chambon, P. 1981. *Nucleic Acids Res.* In press
261. Mathis, D., Chambon, P. 1981. *Nature.* In press
262. Wasylyk, B., Derbyshire, R., Guy, A., Molko, D., Roget, A., Téoule, R., Chambon, P. 1980. *Proc. Natl. Acad. Sci. USA.* In press
263. Kressmann, A., Hofstetter, H., Di Capua, E., Grosschedl, R., Birnstiel, M. L. 1979. *Nucleic Acids Res.* 7:1749–63
264. Sakonju, S., Bogenhagen, D. F., Brown, D. D. 1980. *Cell* 19:13–25
265. Bogenhagen, D. F., Sakonju, S., Brown, D. D. 1980. *Cell* 19:27–35
266. Birkenmeier, E. H., Brown, D., Jordan, E. 1978. *Cell* 15:1077–86
267. Engelke, D. R., Ng, S-Y., Shastry, B. S., Roeder, R. G. 1980. *Cell* 19:717–28
268. Pelham, H. R. B., Brown, D. D. 1980. *Proc. Natl. Acad. Sci. USA* 77:4170–74
269. DeFranco, D., Schmidt, O., Söll, D. 1980. *Proc. Natl. Acad. Sci. USA* 77:3365–68
270. Sollner-Webb, B., Reeder, R. H. 1979. *Cell* 18:485–99
271. Chambon, P. 1975. *Ann. Rev. Biochem.* 44:613–38
272. Roeder, R. G. 1976. *RNA Polymerase*, pp. 285–329. Cold Spring Harbor, NY: Cold Spring Harbor Lab.
273. Grosschedl, R., Birnstiel, M. L. 1980. *Proc. Natl. Acad. Sci. USA* 77:1432–36
274. Benoist, C., Chambon, P. 1981. *Nature.* In press
275. Gluzman, Y., Sambrook, J. F., Frisque, R. J. 1980. *Proc. Natl. Acad. Sci. USA* 77:3898–3902
276. Benoist, C., Chambon, P. 1980. *Proc. Natl. Acad. Sci. USA* 77:3865–69
277. Bendig, M. M., Thomas, T., Folk, W. R. 1980. *Cell* 20:401–09
278. Grosschedl, R., Birnstiel, M. L. 1980. *Proc. Natl. Acad. Sci. USA.* In press
279. Mathis, D., Oudet, P., Chambon, P. 1980. *Prog. Nucleic Acid Res. and Mol. Biol.* 24:1–54
280. Wigmore, D. J., Eaton, R. W., Scott, W. A. 1980. *Virology* 104:462–73
281. Saragosti, S., Mayne, G., Yaniv, M. 1980. *Cell* 20:65–73
282. Wu, C. 1980. *Nature* 286:854–60
283. Stalder, J., Larsen, A., Engel, J. D., Dolan, M., Groudine, M., Weintraub, H. 1980. *Cell* 20:451–60

Ann. Rev. Biochem. 1981. 50:385-432

THE BIOCHEMISTRY OF THE COMPLICATIONS OF DIABETES MELLITUS

♦12083

M. Brownlee and A. Cerami

Laboratory of Medical Biochemistry, The Rockefeller University, New York, New York 10021

CONTENTS

PERSPECTIVES AND SUMMARY

Diabetes mellitus, the clinical expression of absolute or relative insulin deficiency, afflicts 20 to 30 million people throughout the world (1), including 5.5 million in the United States (2). Although insulin deficiency can be ameliorated by diet, insulin injection, or oral hypoglycemic agent therapy,

385

standard treatment has not prevented the development of chronic complications affecting the eyes, kidneys, nerves, and arteries.

In the eye, retinal capillary damage leading to edema, new vessel formation, and hemorrhage makes blindness 25 times more common among diabetics. Cataracts also occur more frequently in patients with diabetes. Capillary damage in the glomerulus associated with basement membrane thickening makes chronic renal failure with proteinuria 17 times more common. In the diabetic peripheral nerve, axonal dwindling and segmental demyelination are associated with a very high prevalence of motor, sensory, and autonomic dysfunction, including impotence, which affects 40% of diabetic males. Increased atheromata in medium and large arteries make coronary artery disease and stroke twice as common among diabetics, and symptomatic peripheral arterial disease three to four times more common. Gangrene leading to amputation is at least five times more frequent among diabetics. The average life expectancy of diabetic patients is only two thirds that of the general population (2).

Until relatively recently, knowledge about diabetic complications was in large part limited to clinical description and morphological characterization. Information from physiological and biochemical investigations of diabetic complications has been accumulating rapidly, however, and this review summarizes and critically examines much of that current information.

Relative tissue hypoxia may play some part in the development of several diabetic complications. Hematological abnormalities in diabetes, which could lead to decreased tissue oxygenation are therefore considered first. Among these are: increased erythrocyte aggregation with increased microviscosity and decreased deformability; increased levels of nonenzymatically glycosylated hemoglobins, which have altered oxygen affinities; decreased effective levels of 2,3 diphosphoglycerate (2,3-DPG); abnormalities in platelet function, including increased adhesiveness, increased sensitivity to several aggregating agents, and accelerated production of thrombogenic prostaglandin derivatives; plasma protein abnormalities, which lead to increased blood viscosity; accelerated fibrinogen consumption; decreased levels of antithrombin III; increased levels of von Willebrand factor (vWF); and decreased fibrinolysis.

Since the description and characterization of glycosylated hemoglobin, glucose has been found also to modify a variety of other proteins in vivo. Increased formation of such adducts in insulin-independent tissues could play a role in the pathogenesis of some diabetic complications.

Microvascular disease in the diabetic kidney has been the focus of considerable investigation. Early changes in renal function and structure have been described that may be involved in the pathogenesis of diabetic renal disease. The discussion of these includes: increased filtration rate and filtration fraction; increased excretion of urinary albumin and other proteins; increased mesangial actomyosin-like material; persistence of certain extravasated serum proteins within glomerular basement membrane; diminished localized mesangial macromolecular clearing mechanisms; and increased renal size, glomerular tuft volume, capillary luminal volume, and capillary filtration surface area.

The relationship between these early changes and the renal disease associated with diabetes of long duration remains to be elucidated. Ultimately, continual basement membrane thickening over many years leads to progressive glomerular capillary occlusion and chronic renal failure. The chemical characterization of glomerular basement membrane is reviewed, compositional changes that occur in diabetes are described, and biosynthetic studies are discussed. Increased synthesis has been demonstrated to be a major factor in the excessive accumulation of glomerular basement membrane in the diabetic kidney.

Less information is available concerning diabetic retinal microvascular disease. Regional ischemia appears to be a central process in the development of diabetic retinopathy, exacerbated by an early, sustained increase in retinal capillary permeability, changes in endothelial and mural cell morphology, and perhaps basement membrane thickening.

Abnormal growth hormone secretion has been observed in some diabetic patients, and it has been proposed that this may contribute in various ways to the development of microangiopathy. The evidence supporting this hypothesis is examined, and conflicting data are considered.

Diabetic neuropathy is characterized by a variety of morphological changes associated with decreased sensory and motor conduction velocities. The discussion of these includes: decreased number of intramembranous particles on the myelin surface; endoneurial edema with resultant shrinkage of axons and Schwann cells; increased permeability of nodal gap substance; basement membrane thickening in the intra- and perineural vessels; axonal degeneration; and segmental demyelination. Particular emphasis is given to the following biochemical alterations in the nerve: changes in the composition and synthetic rate of various myelin lipid and protein components; increased activity of the polyol pathway; decreased concentration and synthetic rate of myoinositol and phosphatidylinositol; and decreased axoplasmic transport of choline acetylase, acetylcholinesterase, norepinephrine, and several glycoproteins.

Accelerated large vessel disease in diabetes may be due to synergistic pathological mechanisms involving hyperlipidemia, altered platelet behavior, and abnormalities in arterial wall function. Elevated levels of very low density lipoprotein (VLDL) triglyceride and low density lipoprotein (LDL) cholesterol have been observed in all major classes of diabetic patients. High density lipoprotein (HDL) cholesterol may be decreased, unchanged, or elevated. A removal defect, in part due to lipoprotein lipase (LPL) deficiency, is largely responsible for the hypertriglyceridemia of severe diabetes, while increased hepatic VLDL production in response to elevated plasma-free fatty acids makes a greater contribution to hypertriglyceridemia when permissive levels of insulin are present. Hypercholesterolemia may be due to increased intestinal VLDL and LDL production, and perhaps also to an LDL removal defect. Alterations in high density lipoprotein may reflect changes in LPL-mediated degradation of triglyceride-rich lipoproteins.

Hyperglycemia-induced endothelial cell dysfunction, increased platelet interaction with subendothelial structures, increased serum growth factor–mediated smooth muscle cell proliferation, increased secretion of altered collagen molecules, alterations in proteoglycans, and impaired intracellular degradation of LDL may all contribute to the development of macroangiopathy.

Finally, the relationship between diabetic complications and the degree of clinical metabolic dysfunction is discussed. The organizing concept that emerges from this review is that insulin deficiency and its metabolic consequences are the primary etiologic events in the pathogenesis of diabetic complications.

HEMATOLOGIC ABNORMALITIES

Introduction

Dilation of retinal veins that is reversible with good metabolic control is observed in diabetes (3). Since dilation of these vessels is thought to occur by autoregulation in response to local hypoxia, the hypothesis is that relative tissue hypoxia plays some part in the development of several diabetic complications (4). This concept, originally based on reports of some diabetic-like lesions occurring in such clinical entities as macroglobulinemia, sickle cell anemia, severe cyanotic heart disease, and cystic fibrosis, has gained credibility from more recent laboratory observations confirming abnormalities predisposing to decreased tissue oxygenation in diabetes. These include increased erythrocyte aggregation and decreased deformability, increased levels of glycohemoglobin, decreased effective levels of 2,3-DPG, increased platelet aggregation, increased blood viscosity, accelerated fibrinogen consumption, decreased levels of antithrombin III, increased levels of von Willebrand factor, and decreased fibrinolysis.

Red Blood Cells

CELL MEMBRANE   Increased erythrocyte aggregation has been observed in blood from diabetic patients, and it has been argued that this phenomenon may contribute to obliterative microvascular changes in the retina (5–7). Since washed red blood cells from both diabetics and normals aggregate in diabetic plasma but fail to aggregate in plasma from normal patients, this process appears to be more a consequence of increased plasma protein levels than of changes in the erythrocytes themselves (6, 7). Although the abnormal red cell aggregation of diabetes does correlate with the degree of glycohemoglobin elevation (8), at present there are no data to support the contention that this plays a causal role in the pathogenesis of microvascular sequelae.

Erythrocyte deformability, as measured by their ability to pass through five micron pores, is significantly lowered in diabetes (9). This has recently been confirmed using a different technique in which single cells are introduced into micropipettes and moved in a cyclic standard velocity pattern while pressure is measured (10, 11). The ratio of mean pressure, diabetic to normal, averaged 1.51. Since erythrocytes must often traverse capillaries considerably smaller than their own diameter, this decreased deformability may seriously impair rapid and homogeneous perfusion within the microcirculation. The basis for the reduced deformability is not known, but its correlation with degree of diabetic control (9) suggests a metabolic etiology. The higher levels of erythrocyte membrane microviscosity observed in diabetic patients with fasting blood glucose > 140 mg/dl may be one factor (12), perhaps as a result of increased membrane protein glycosylation (13). The decreased sialic acid and cholesterol content observed in red blood cells of chronically diabetic rats (14) and humans (15), and the membrane-bound hemoglobin molecules reported in human diabetic erythrocyte membranes (16) may also adversely affect the deformability characteristics of these cells. The possible role of other cell surface changes is discussed below.

Studies of circulating leukocytes in diabetic patients have demonstrated numerous functional abnormalities both in polymorphonuclear leukocytes and in lymphocytes (17, 18). Polymorphonuclear leukocytes from poorly controlled diabetic patients exhibit defective granulocytic adherence, chemotaxis, phagocytosis, and intracellular bactericidal activity (17, 19). All of these defects in host defense are reversed by more effective diabetic control. Lymphocytes from poorly controlled juvenile-onset diabetics show a decreased proliferative response to mitogen stimulation and a reduction in T- and B-cell surface membrane markers (18). Optimal blood glucose control for five days using an external artificial pancreas leads to a rapid normalization of mitogen responsiveness and membrane marker values in lymphocytes from previously poorly controlled subjects.

Although the clinical significance of these observed leukocyte cell surface abnormalities is unclear, their implications for other diabetic tissues could be very important. Current concepts of cell surface organization suggest that changes in the mobility and distribution of plasma membrane components, as well as changes in their chemical composition, may significantly alter cell function (20, 21). Alterations in a variety of cellular properties occur in diabetes that may play a role in the development of complications, and these in turn may result in part from changes in the cell surface. This is a new and relatively unexplored area, but one of great potential value for the further elucidation of the pathophysiology of diabetic complications.

Studies of lectin binding to liver cell plasma membranes of diabetic rats have demonstrated significant changes in membrane receptors for these molecules. The binding of concanavalin A and ricin were reduced 20–25% (22), which reflects an apparent reduction in membrane glycoprotein carbohydrate content (23). The binding of desialylated thyroxine-binding globulin, an interaction dependent on membrane sialic acid residues, was reduced 50% (22). Since this change represents a much greater reduction than that reported in sialic acid in these membranes (14), it suggests either a selective reduction in the sialic acid content of this specific receptor, or changes in the structure or mobility of receptor components on the membrane surface. Extensive changes in cell surface glycoproteins in this animal model of diabetes are also suggested by changes in lectin binding to membrane components of various molecular weights separated by SDS polyacrylamide gel electrophoresis (24). Such changes could conceivably influence a variety of cellular properties including red cell aggregation and deformability, as well as permeability, electrical resistance, and binding properties of mitogen, hormone, and lipoprotein receptors.

The chemical basis for the reported cell surface changes is not clearly understood. However, both insulin deficiency and secondary changes in metabolites and hormones could affect macromolecular synthesis in certain tissues. Postsynthetic chemical modification resulting from abnormally elevated substrates may also play a role, as suggested by the discussion of glycosylated hemoglobins in the section below.

HEMOGLOBIN GLYCOSYLATION AND OXYGENATION   It has been appreciated for some time that normal human hemolysates contain several minor hemoglobin components in addition to the single major hemoglobin, Hb A_o. Together, these minor hemoglobin species account for five to ten percent of the total found in normal adult erythrocytes. Two- to threefold elevations of three of these components—Hb A_{1a}, Hb A_{1b}, and Hb A_{1c}— are consistently found in diabetic patients (25). Recently, HbA$_{1a}$ has been resolved into two separate components, Hb A_{1a1} and Hb A_{1a2} (26). All four of these negatively charged minor hemoglobins have been proposed to be glycosylated, with carbohydrate present on the β-chain (27). For the most abundant of these glycosylated hemoglobins, Hb A_{1c}, it has been demonstrated that glucose reacts with the NH_2 terminal of the β-chain by means of a Schiff base and an Amadori rearrangement to form a 1-amino-1-deoxyfructose derivative (28). The carbohydrate moieties of the other glycohemoglobins are presumed to be attached to the protein through analogous ketoamine linkages. Recently a significant portion of Hb A_o has also been shown to be glycosylated (29). Glucose ketoamine linkages have been identified on the N terminus of the α-chain, and on several lysine residues on both the α- and β-chain. Results from diabetic subjects show increases comparable to the increases of Hb A_{1c} (29, 30).

In addition to carbohydrate, Hb A_{1a1} and Hb A_{1a2} also contain phosphate (27). Since synthetic glycohemoglobin formed by incubation of Hb A_o with glucose 6-phosphate cochromatographs with Hb A_{1a2}, it is presumed that phosphorylated sugars are the source of this phosphate in vivo. The assignment of structure for minor hemoglobins other than Hb A_{1c} has yet to be reported.

These glycosylated hemoglobins have unique functional properties that differ significantly from those of Hb A_o. Compared to Hb A_o, Hb A_{1a1} and Hb A_{1a2} have low oxygen affinities, Hb A_{1b} has high affinity, and Hb A_{1c} has moderately high affinity (31). Removal of organic phosphate increases the oxygen affinities of Hb A_{1b}, Hb A_{1c}, and Hb A_o, while the affinities of Hb A_{1a1} and Hb A_{1a2} remain low. Addition of organic phosphate substantially decreases the oxygen affinity of Hb A_{1b} and Hb A_{1c}, although to a much smaller degree than that of Hb A_o. Hb A_{1a1} and Hb A_{1a2} are essentially unaffected by addition of organic phosphate. These altered oxygen binding properties of the glycohemoglobins may be explained in part by the location of the carbohydrate moieties, which interferes with the binding of organic phosphates such as 2,3-DPG. The presence of covalently bound organic phosphate in the Hb A_{1a1} and Hb A_{1a2} molecules could account for their persistently low oxygen affinities in the absence of added phosphate.

In vivo, the organic phosphate 2,3-DPG (derived from red cell glucose metabolism) is believed to play an important regulatory role in oxygen exchange by hemoglobin. Increased levels of 2,3-DPG lower hemoglobin oxygen affinity, thus increasing oxygen release to tissues. Even in the presence of high concentrations of organic phosphate, however, all of the glycohemoglobins are still 50% saturated with oxygen at partial pressures where Hb A_o has already given up most of its oxygen (31). In diabetics, where a relative deficiency of 2,3-DPG may coexist with elevated glycohemoglobin levels during periods of changing blood glucose concentration (32, 33), decreased oxygen delivery to critical tissues may result. Blood from ambulatory diabetic children does in fact exhibit increased hemoglobin oxygen affinity (34), but the clinical significance of this phenomenon has not been established. Pronounced hyperlipoproteinemia has been reported to accentuate this defect (35).

Glycohemoglobin biosynthesis takes place in vivo throughout the life of the erythrocyte via postsynthetic, nonenzymatic glycosylation of Hb A_o. The rate of formation of this component is proportional to the integrated concentration of blood glucose (28, 36–38). It has been shown that the level of glycohemoglobin in diabetic patients is linearly correlated with the peak or area under the curve of the glucose tolerance test (39), and that concentrations of glycohemoglobin return toward normal after improvement of glucose regulation (40). The fall in glycohemoglobin levels reflects normal red cell destruction and the very slow reversibility of the reaction. In tissue components having longer half-lives, however, glycosylation would reflect the true chemical equilibrium. Covalent modification of such proteins by glucose could play a role in the pathogenesis of diabetic complications (41). Direct support of this hypothesis comes from recent studies of lens crystallins, where high glucose concentrations promoted nonenzymatic glycosylation of this protein's lysine residues both in vitro and in vivo. Crystallins thus modified by glucose exhibited increased susceptibility to sulfhydryl oxidation, which produced high-molecular-weight aggregates linked by disulfide bonds. The opalescence of these aggregates in solution strongly resembles that seen in diabetic cataracts, which suggests that analogous mechanisms may participate in their clinical formation (41–42a).

Increased nonenzymatic glycosylation has also been reported to occur in renal glomerular proteins isolated from diabetic animals (43), thus there is also a possible involvement of this mechanism in the development of diabetic nephropathy. Nonenzymatic glycosylation of erythrocyte membrane proteins occurs to a greater extent in diabetics, and may relate to the changes in deformability discussed above (13). The nonenzymatic glycosylation of human albumin both in vitro and in vivo has been described (44), with the level in diabetic patients averaging over twice that of normal controls (45–47). Similarly, other classes of serum protein undergo nonenzymatic glycosylation in vitro and in vivo, both in animals and in man (48, 49). The level of these adducts is increased in diabetes, and is proportional to the degree of hyperglycemia. Several other tissues in which diabetic complications occur are currently being evaluated for the presence and significance of similar glycosylated derivatives.

Platelets

Various functional and biochemical abnormalities have been reported in platelets obtained from diabetic subjects. These include alterations in platelet adhesion, increased sensitivity to several aggregating agents, and accelerated production of thrombogenic prostaglandin derivatives. Observations such as these have prompted speculation that abnormal platelet behavior may contribute to the development of obliterative microvascular disease in diabetes, although proliferative retinopathy has occurred in patients with defects of platelet function such as von Willebrand's disease (50). The composition of platelet-rich plasma prepared for such in vitro assays may be changed by other blood factors that affect the centrifugation equilibrium, however, which perhaps results in a diabetic preparation enriched with a subpopulation of more reactive platelets (51). The nonphysiologic laboratory assay conditions also preclude direct extrapolation of these data to the intact organism. As yet there has been no convincing evidence for in vivo platelet dysfunction (52).

Platelet participation in clot formation begins with adhesion to exposed subendothelial components of the vessel wall. Although a number of investigators have found increased platelet adhesiveness in diabetic patients (53), adhesion assays utilizing a rotating glass bulb or glass beads in a column are notoriously difficult and imprecise (54). Since fibrinogen and other plasma proteins known to be elevated in diabetes affect the result of these assays, proper interpretation of published data is even more difficult. In this regard it is interesting to note that the platelet adhesion of both normal and diabetic platelets increases when glucose is added in vitro (55). Adhesion experiments utilizing a physiologic substrate such as basement membrane might provide more relevant information in this important area of platelet physiology (56).

After platelets adhere to the subendothelium, they undergo aggregation. This process has both a surface and an intracellular component. The surface reaction occurs in response to low concentrations of such stimuli as ADP and epinephrine. This so-called primary wave of aggregation is reversible. At higher concentrations of these stimuli, and in the presence of collagen, irreversible aggregation occurs in conjunction with the release of stored intracellular thrombogenic substances from the platelets. This irreversible process is the secondary wave of aggregation. In diabetes, increased second phase aggregation has been observed in response to ADP, epinephrine, arachidonic acid, and collagen (53, 57–62). Increased release of platelet factors 3 and 4, two accelerators of the coagulation process, also occurs during this process in various subgroups of diabetic patients (54). The results obtained from individual diabetic patients range from normal to extremely abnormal, with no clear-cut association between this value and the class of diabetes. Good correlation with level of hyperglycemia may well exist, however, since the phenomenon of increased second phase aggregation can be corrected by strict diabetic control, as confirmed by blood glucose levels near mealtimes, by glycosuria, and by Hb A_{1c} assays (63). Plasma factors that enhance ADP-induced platelet aggregation have been reported to occur in 50% of unselected male diabetics (64). The initial characterization of this activity has recently been accomplished (65). Failure of other groups (57, 66) to confirm such activity has been attributed to methodological differences (64). Plasma β thromboglobulin, a platelet-specific protein released during aggregation, is also elevated in diabetic patients (67).

Hyperaggregation of diabetic platelets may be influenced by altered activities of thromboxane synthetase in platelets and prostacyclin synthetase in blood vessel endothelial cells. The platelet enzyme converts the cyclic endoperoxide prostaglandins G_2 and H_2 (PGG_2 and PGH_2) to thromboxanes A_2 and B_2, both potent stimulators of both platelet clumping and arterial constriction. Prostacyclin synthetase, located in blood vessel endothelial cells, converts these same endoperoxides to prostacyclin, a compound that inhibits the effect of the thromboxanes (68). Since these two prostaglandins have exactly opposing effects, it is thought that the balance between the two activities may determine platelet plug formation. Pharmacologic inhibitors of prostaglandin synthesis have been shown to reverse the accentuated second phase aggregation of diabetic platelets (59). Furthermore, platelets from diabetic patients demonstrate increased synthesis of prostaglandin-E-like material in response to several aggregating agents, and metabolize arachidonic acid to prostaglandin-E-like material at a significantly greater rate and extent than nondiabetic controls (69). Biosynthesis of thromboxane A_2 in response to ADP is also increased in diabetic platelets (70). Net synthesis of prostaglandins by whole blood was significantly higher in samples obtained from diabetic children (71). Some preliminary studies suggest that thromboxane B_2 synthesis is directly proportional to the plasma glucose at the time of the study (64). However, others (72) show no difference between diabetics and normals. Together, these observations have suggested that increased synthesis of prostaglandins may partially mediate the abnormal aggregation response of diabetic platelets. In addition, markedly decreased prostacyclin activity has been reported in tissues of diabetic animals and man (73–75). In both aorta and renal cortex of diabetic experimental animals, levels were reduced to less than one half of normal (73, 74). These values were restored to normal by insulin treatment (76). Decreases of similar magnitude occur in venous tissue from juvenile-onset diabetic patients (75). Plasma concentrations of the prostocyclin metabolite, 6-keto-$PGF_{1\alpha}$, remain unchanged (72, 77).

Increased amounts of lipid phosphorus have been reported in diabetic platelets, along with changes in the fatty acid pattern of the various phospholipids (78). Endogenous platelet protein phosphorylation also is higher in some diabetic patients (79). Increased myosin-ATPase activity is found in diabetic platelets (80). The significance of these observations is not clear at the present time.

Plasma Proteins

A variety of plasma protein abnormalities have been reported to occur in diabetes. Particular attention has been given to those proteins that contribute to plasma viscosity and to those that participate in the formation and dissolution of blood clots. Although it has been suggested that alterations in such factors may play a role in the pathogenesis of diabetic vascular disease, there is no direct evidence to support this hypothesis at present. Nevertheless, a significant association has been observed between the diabetic state and several factors that could interfere with microcirculatory flow.

Elevated levels of the blood glycoproteins, fibrinogen, haptoglobin, β lipoprotein, caeruloplasmin and α_2 macroglobulin have been found in diabetic patients (81–83). These protein changes, particularly increased fibrinogen and haptoglobin, raise the plasma viscosity at low shear-rates by as much as 16%, thus increasing the flow resistance of whole blood (83–86). In addition, increased levels of fibrinogen and α_2 globulins favor enhanced red cell aggregation (6, 7). Although the most striking increases in plasma viscosity and erythrocyte aggregation are reported to occur in patients having either proliferative retinopathy or nephropathy (6, 7, 83–85), this association most probably reflects a greater degree of hyperglycemia in these patients. Diminished elevation of these blood glycoprotein levels correlates with the use of regular insulin (81) and with blood glucose values in the normal range (82). A most significant correlation is found with the presence or absence of recent hypoglycemia symptoms; the level of all the glycoproteins studied is significantly lower in those patients reporting hypoglycemia. The association of near-normal values with recent symptoms of hypoglycemia (an indicator of short-term control) rather than with levels of hemoglobin A_{1c} (an integrated measure of control over several weeks) (66) suggests that these changes in plasma glycoproteins occur fairly rapidly.

Coagulation factors VIII and V are also reported to be increased in diabetes (87), which prompts speculation that a hypercoagulable state may be involved in the evolution of vascular complications. This concept has received strong support from recent studies of fibrinogen survival and turnover in diabetic patients (88). These studies demonstrated that the survival of autologous radioiodinated fibrinogen is reduced in adult-onset diabetics, which indicates that increased consumption of fibrinogen occurs at levels of hyperglycemia commonly observed in outpatients. This abnormality was rapidly corrected toward normal as euglycemia was achieved. While antiplatelet agents given during the hyperglycemic period were not effective in restoring fibrinogen survival, heparin infusion normalized the fibrinogen kinetics of hyperglycemic patients. This observation suggests the possible involvement of thrombin, Factor XI, IX, X, and/or antithrombin III.

The α globulin designated antithrombin III adsorbs thrombin released during clot formation and blocks the effect of this thrombin on fibrinogen, thus playing an important regulatory role in hemostasis. Deficiency of this protein has been reported in maturity-onset diabetes, and may contribute to a state of increased coagulability (89–91). Increased levels were observed in juvenile diabetics, however (92). No correlations have yet been published between levels of antithrombin III and glycemia, nor is turnover data to distinguish the relative roles of synthetic versus degradative defects currently available. The effect of nonenzymatic glycosylation on bioactivity is also unknown.

Abnormalities of platelet function in patients with diabetes have been discussed in a previous section. The glycoprotein plasma cofactor of platelet function, vWF, may play a significant role in these disturbances. Levels of this factor, thought to play a role in both platelet adhesion and platelet aggregation, have been reported to be elevated in patients with diabetes, which correlates with the degree of platelet aggregation in at least one study (57, 62, 93–96). This increased vWF activity in turn appears to be influenced by growth hormone (59, 97), and may be suppressed by insulin (98). Indirect evidence that vWF contributes to the development of vascular disease comes from studies of spontaneous arteriosclerosis in pigs and that induced by a high cholesterol diet. Normal animals developed significant arteriosclerotic plaques in 1–3 yr, while no animals with homozygous von Willebrand's disease (i.e. lacking vWF) showed these changes. Similar results were obtained when the animals were fed high cholesterol diets (99).

In addition to increased viscosity and hypercoagulability, changes in the fibrinolytic system may also be involved in the pathogenesis of diabetic vascular disease. The fluidity of blood is maintained not only by carefully regulated coagulation factor activity, but also by finely coordinated humoral factors that remove fibrin deposited in blood vessels. Plasminogen, the inactive circulating precursor of the enzyme plasmin that digests fibrin, requires the release of an activator substance from vascular endothelium to accomplish this conversion. The fibrinolytic activity of diabetic blood is significantly lower than normal (100), and the fibrinolytic response to venous occlusion is also decreased, possibly because of a decreased plasminogen activator release (101). These differences were most marked in patients with diabetic retinopathy, although no systematic attempt was made to correlate these findings with levels of hyperglycemia. The elevated levels of serum protease inhibitors such as α_2 macroglobulin, which occur in diabetes, may also contribute to decreased degradation of fibrin and other proteins (102). Recent histological studies suggest that fibrin deposition could play a part in the small vessel damage of diabetes (103, 104).

The changes in plasma protein levels and/or activities that are associated with hyperglycemia may constitute additional examples of glucose overutilization in diabetes. Increased availability of sugar-nucleotide precursors for glycoprotein synthesis may preferentially stimulate their production (105). In isolated, perfused livers from diabetic and nondiabetic rats, the rate of protein secretion in the diabetic liver is only one third that of the nondiabetic, while glycoprotein production is double (106). This may be due either to de novo synthesis of glycoproteins or to an excess glycosylation of existing proteins.

Circulating plasma proteins may also be subject to further glycosylation by a nonenzymatic process analogous to that described for glucosylation of hemoglobin A to hemoglobin A_{1c}. Incubation of human serum with glucose results in glucosylation of each of the major molecular weight classes of serum protein, and glucosylated albumin has been isolated and quantitated from normal and diabetic human serum (44–47). Higher levels of these derivatives are found in diabetic plasma, since their formation is a function of the ambient glucose concentration (48, 49, 107). Since the carbohydrate portion of vWF is the major determinant in its interaction with platelets (108), and the carbohydrate portion of clotting factor V is essential for its coagulation activity (109), excess glycosylation of such proteins could conceivably enhance their biological activity. Changes in carbohydration may also help explain the altered pattern of plasma protein degradation seen in diabetes. Ordinarily, a general correlation exists between isoelectric points of serum proteins and their degradative rates. In diabetic animals, this relationship is abolished (110). Elucidation of the biochemical mechanisms by which hyperglycemia produces plasma protein abnormalities awaits further investigation.

CATARACTS

Opacity of the ocular lens, cataract, occurs earlier and more frequently in patients with diabetes. In a number of experimental diabetic animals, and occasionally in severe juvenile-onset diabetics, the development of cataracts is particularly rapid. The biochemical factors involved in the formation of nondiabetic cataracts have been investigated in both humans and experimental animals. Dische & Zil (111) first noted that the proteins of human cataractous lenses contained a greater number of disulfide bonds than did normal lenses, which reflects markedly increased sulfhydryl oxidation. The association of increased disulfide bonds with cataract formation has been confirmed in humans and subsequently demonstrated in animals, by a number of other workers (112–116). Since disulfide bonds can function as intermolecular cross-links, it has been proposed that these bonds participate in the polymerization of lens proteins to form high-molecular-weight aggregates. Benedek (117) has proposed from theoretical considerations that cataracts could occur from the scattering of light produced by protein aggregates with molecular weights greater than 50×10^6 daltons. The aggregates found in human cataracts and in aged normal lenses are in the range of $50–200 \times 10^6$ daltons (118, 119).

Diabetic and galactosemic rat lenses contain similar disulfide-bonded high-molecular-weight aggregates that are capable of scattering light (120). It has been suggested that the formation of these high-molecular-weight aggregates in diabetes is facilitated by an increased susceptibility of lens proteins to sulfhydryl oxidation, which results from nonenzymatic glycosylation (41, 42). The lens, like other body tissues that develop diabetic complications, does not require insulin for glucose transport. Therefore, hyperglycemia would increase glucose concentration within the lens.

Elevated levels of aldoses and ketoses would result in more frequent interactions with amino groups of lens proteins, and lead to greater formation of stable Amadori or Heyns rearrangement products similar to that observed with hemoglobin A_{1c}. These modified lens proteins form disulfide bonds more easily than do normal lens proteins. Since the major lens proteins, α, β, and γ crystallins, undergo little or no replacement throughout the life of the individual, the total amount of glycosylated protein in the lens would increase continually over time. It has also been proposed that the yellow-brown pigments associated with lens proteins from human cataracts represent further rearrangements of the sugar-protein adducts, analogous to the well-studied nonenzymatic browning that occurs in stored foods (121). Accumulation of browning products would also be expected to occur in other long-lived proteins such as collagen.

Another mechanism that may participate in the development of diabetic cataracts has been investigated (122–124). In the lens, the enzyme aldose reductase reduces aldoses such as glucose and galactose to their corresponding sugar alcohols. Elevated amounts of glucose in the diabetic lens would result in an intracellular accumulation of sorbitol, since polyols do not readily diffuse across cell membranes. It is proposed that intracellular polyol accumulation causes osmotic swelling and eventual disruption of cell architecture. An associated abnormality in electrolyte transport leads to an influx of sodium ions, which is believed to increase swelling and thus accelerate opacification of the lens. The strongest evidence in favor of this hypothesis is the finding that several drugs that inhibit aldose reductase significantly retard cataract formation in diabetic and galactosemic rats (125–127).

MICROANGIOPATHY

Kidney

FUNCTIONAL AND MORPHOLOGICAL CHANGES   The renal disease associated with diabetes of long duration is characterized clinically by continuous proteinuria and a decreasing glomerular filtration rate (GFR). The severity of these clinical features correlates with the extent of glomerular basement membrane (GBM) thickening. At the onset of diabetes, human GBM is normal thickness. Increased thickness develops after several years of sustained metabolic derangement (128, 129). Progressive glomerular capillary occlusion due to continual thickening over many years ultimately results in chronic renal failure. The accumulation of basement membrane material that has altered filtration properties thus represents the ultimate structural and functional basis of diabetic renal disease. Although much information about the biochemistry of GBM has been obtained (see below), the sequence of events leading from abnormal glucose homeostasis to the late manifestations of diabetic renal pathology remains to be elucidated.

In recent years, early changes in renal function and structure have been described in diabetes, some of which may play a role in the pathogenesis of diabetic renal disease. Glomerular filtration rate and filtration fraction (GFR/renal plasma flow) are elevated in the first few weeks of diabetes, and return toward normal after intensive insulin treatment (130). Very early in the course of diabetes, excretion of urinary proteins spanning a molecular weight range of 44,000–150,000 is also significantly increased (131). Newly diagnosed juvenile diabetics in poor metabolic control have elevated urinary albumin excretion rates that are normalized by strict control of blood glucose (130). Similar findings have been reported in diabetic rats (132, 133) and in longer-duration insulin-dependent human diabetics with no clinical evidence of renal disease (133a). Abnormally large amounts of albumin are also excreted by such patients in response to exercise (130), when a higher glomerular filtration pressure operates. The amount excreted rises progressively with increasing duration of diabetes (134). Neutral dextran polymer clearance is similarly increased in newly diagnosed juvenile diabetics and returns to normal after several weeks of effective insulin treatment (130, 135). Since a pronounced increase in high molecular weight dextran clearance is not observed, it is unlikely that these findings represent alterations in the size-selective properties of the glomerular filtration barrier. Rather, increased filtration pressure across the glomerulus appears to be primarily responsible for these alterations in renal function (130, 135, 136). The striking increase of actomyosin-like material in the mesangium, specific for diabetes (137, 138), may contribute to this, since contractile proteins in the mesangium could play an important role in the regulation of glomerular blood flow and ultrafiltration (139).

In normal rats, interruption of blood flow for short periods of time results in a marked increase in glomerular permeability (140). If hemodynamic factors do influence permeability significantly, the hematological abnormalities favoring microvascular stasis in diabetes (discussed above) may induce intermittent permeability changes in a nonhomogeneous pattern consistent with the character of diabetic glomerular disease. Since glomerular permeability to anionic serum proteins is also markedly influenced by fixed negative charges in the glomerular capillary wall (141), decreased sialic acid in diabetic glomeruli (142) could also contribute to increased passage of appropriately charged serum components. Whatever the mechanism of increased glomerular permeability in diabetes, immunohistochemical studies have documented the presence of several serum proteins in glomerular capillary, muscle capillary, and muscle basement membranes of diabetic patients. The immunofluorescent pattern for IgG, albumin, and fibrin was identical with that observed using heterologous antibasement membrane antiserum, which indicates that these proteins are present within the basement membranes (143). A rapid accumulation of such extravasated serum proteins could help explain the remarkable rate of morphological changes observed very early in the course of human and experimental diabetes. Within a few weeks of the acute onset of the disease, significant increases are found in renal size, the glomerular tuft volume, the capillary luminal volume, and capillary filtration surface area (134, 144–146). Many of these findings are reversible with appropriate insulinization. Isotransplantation of kidneys from six-month-old diabetic rats into normal recipients results in the disappearance of IgG, IgM, and C3 from the mesangium and the arrest of glomerular mesangial thickening (147). Glomerular basement membrane thickness remains the same, however (148), which reflects the slow rates of GBM turnover (149).

Persistence of serum proteins or their degradation products deposited in the glomerulus as a result of increased permeability, along with excess fibrin due to decreased fibrinolysis (discussed above), may contribute to the capillary occlusion and progressive glomerular dropout in long term diabetes. Nonenzymatically glycosylated plasma proteins (discussed above) may be particularly important in this process, particularly if increased glycosylation reduces the susceptibility of these proteins to proteolysis (150). A link between plasma protein glucosylation and diabetic nephropathy is suggested by a recent report of GBM thickening induced in nondiabetic animals by repeated intravenous injection of glucosylated plasma proteins (151).

Expansion of "basement membrane–like" mesangial matrix material appears to diminish localized mesangial macromolecular clearing mechanisms (152, 153). In nondiabetic animals, such mesangial dysfunction leads to an accelerated thickening of the glomerular basement membrane (154). Although GBM thickening precedes mesangial change in juvenile diabetics (128), resultant mesangial damage may ultimately contribute to the process of long-term protein accumulation in the diabetic glomerulus.

BASEMENT MEMBRANE CHEMISTRY AND METABOLISM

Chemical characterization of GBM   Over the past decade, much information about the chemical structure of the normal GBM has been acquired. Although much of the basic structural work was done on bovine material, compositional studies performed in several mammalian species have yielded similar results (149, 155, 156). The basement membrane is composed of collagen-like glycoprotein material that is particularly rich in the amino acids glycine, hydroxyproline, and hydroxylysine. However, it clearly differs from fibrillar collagen, since it contains three times as much hydroxylysine, large amounts of half-cystine, and has a significant carbohydrate content (105). Recent analysis of the Clq component of complement has shown a very similar composition for this soluble serum protein (157). The sugar constituents of the basement membrane are distributed between two distinct types of carbohydrate units (158). One is a complex asparagine-linked heteropolysaccharide containing galactose, mannose, fucose, sialic acid, and glucosamine that occurs on relatively more polar, less collagen-like portions of the peptide chain. These heteropolysaccharide units are not particularly characteristic of basement membrane, and have been described in a variety of other glycoproteins (159).

The second type of carbohydrate unit consists of a glucose-galactose disaccharide unit attached to hydroxylysine. The two carbohydrate units are found in a ratio of 1 : 10 by number, but by total weight are about equal. Hydroxylysine is formed by the enzymatic hydroxylation of lysine residues that are already incorporated in nascent or released polypeptides. Eighty percent of this hydroxylysine is then linked to galactose, and, subsequently, to glucose by sequential sugar transferase enzymes, to form the disaccharide-hydroxylysine complex (160).

After reduction of its disulfide cross-links, the basement membrane can be solubilized in SDS or urea and subsequently fractionated into a large number of components whose molecular weights range from 25,000 to over

200,000 (161–163). These subunits vary greatly in their amino acid and carbohydrate composition. Some are rich in the collagen-like components, while others contain more polar amino acids such as half-cystine, tyrosine, and aspartic acid. Most subunits contain both types of carbohydrate units, but in quite different proportions. Even material of the same molecular size exhibits this compositional heterogeneity. Although the basis for this subunit polydispersity is unknown, it has been shown that these various polypeptides do not arise by proteolytic cleavage of a high-molecular-weight component during membrane isolation (164, 165). Rather, they may be the result of a limited physiologic in vivo proteolytic digestion of a smaller number of large basement membrane subunits (105). It has been suggested that basement membrane is composed primarily of three α1 collagen chains, and should thus be designated "Type IV" collagen. Since such material has been shown to represent < 2% of purified GBM after pepsin digestion, this proposed structure cannot be correct (166).

Compositional changes in diabetic GBM   Chemical composition studies of human GBM from normal and diabetic kidneys have been published by several groups of investigators (167–170). Beisswenger & Spiro first demonstrated an increased amount of basement membrane protein in glomeruli from long-term diabetics with histologic evidence of nephropathy. In addition, it was found that this material had a distinctly different composition from normal (155, 171). Compared to the normal, there was increased hydroxylysine with a proportional decrease in lysine, so that the sum of the two remained constant. Glucose and galactose were likewise elevated in the diabetic membrane, which corresponds to an increase in hydroxylysine-linked disaccharide residues. Smaller, but still significant increases of hydroxyproline and glycine were also found, as were decreases in valine and tyrosine. These compositional data suggest that there is an increased presence of hyroxylysine-rich subunits with collagen-like amino acid composition in the diabetic membrane. These findings have subsequently been confirmed by Canivet et al (23).

In two other studies (168, 169), these differences were not observed. However, a significant decrease in the half-cystine and sialic acid content of GBM from diabetic kidneys was noted. Decreased half-cystine and sialic acid content were also observed by Canivet et al (23). Important differences, both in technique of basement membrane preparation and in the diabetic kidneys selected for study, may explain these apparent discrepancies.

Glomeruli are isolated and purified from kidney cortex by a series of sieving steps through different mesh sizes. Selection of sieve mesh size is extremely important, since glomerular fractionation experiments (172) have demonstrated that glomeruli isolated from human diabetics are distributed among four major groups according to size, while normal glomeruli are equally distributed between two groups. Electron microscopic analysis of glomerular size and structure in long-term diabetes has shown hypertrophied, enlarged glomeruli with nonoccluded capillaries coexisting with another subpopulation of severely diseased glomeruli with capillary occlusion (173). Thus improper choice of sieve pore size could exclude the most severely involved diabetic glomeruli from analysis.

Differences in the clinical and pathological characteristics of diabetic kidneys selected for study could also account for some variation in compositional changes. Since basement membrane turnover is very slow, membrane isolated from diabetics includes material that was synthesized both before and after the onset of the metabolic derangement (105, 149). Thus, changes that may occur in the composition of the diabetic membrane would be minimized in these analyses by the presence of varying amounts of normal basement membrane material. A very substantial derangement in the hydroxylysine-rich, collagen-like subunit could well result in only a slight alteration of the whole membrane composition.

Because total compositional studies of GBM are an insensitive index of disease-induced synthetic changes, it is imperative that glomeruli with confirmed pathological alterations be included in the analyses of diabetic tissue. Analyses of diabetic kidney material without morphologic evidence of glomerular disease have not demonstrated chemical abnormalities (171). In addition to using kidney material with diabetic glomerular changes present histologically, it may also be important to compare diabetic and normal individuals who are age-matched. Extrapolating from rat data (174), it would appear that the GBM hydroxylysine/lysine ratio increases with age. Failure to take this into account may obscure some real compositional differences between normal and diabetic samples.

Biosynthesis and degradation of GBM   The compositional abnormalities discussed in the previous section may be partially responsible for the altered filtration properties of the GBM in diabetes that ultimately result in clinically significant continuous, nonselective proteinuria. It is speculated that increased glycosylation may interfere with polypeptide packing and/or hydroxylysine-derived cross-link formation. This may result in an increased effective pore size of the basement membrane.

While the proteinuric component of diabetic nephropathy almost certainly reflects changes in the structure of the basement membrane and perhaps other components of the glomerular filtration barrier, the progressive fall in GFR with chronic renal failure is a consequence of the abnormal accumulation of excessive amounts of basement membrane material with resultant glomerular capillary occlusion. This excessive accumulation of

GBM in diabetes could represent increased synthesis, decreased degradation, or a combination of the two processes. Increases in basement membrane synthesis could affect all subunits equally. Alternatively, there could be a relative overproduction of selected basement membrane components. Similarly, degradation may be uniform and nonspecific, or it may affect some portions of the basement membrane preferentially. Such differential subunit effects, if they do occur, could account for the qualitative, as well as quantitative alterations in diabetic GBM. Compositional analysis, however, would be a relatively insensitive means of detecting these changes.

Metabolic studies of GBM have, by necessity, been limited to animal systems. Although the appropriateness of the rat model has been questioned by some investigators (175), recent work has established that experimental diabetes results in increased kidney protein content and protein/DNA ratio as well as significant glomerular capillary basement membrane thickening (176–180). A highly significant, positive correlation exists between these parameters and the degree of hyperglycemia. Kidney ribosome preparations from rats made diabetic with alloxan, streptozotocin, or antiinsulin serum incorporate amino acids into protein at a greater rate than do ribosomes prepared from normal rat kidney (181), which suggests that diabetes leads to increased kidney protein synthesis.

Biosynthetic enzyme activities   Indirect evidence for accelerated GBM biosynthesis in diabetic kidneys comes from studies of enzymes involved in the assembly of its hydroxylysine-linked disaccharide units. These units are synthesized by a series of specific enzymatic, postribosomal steps. The first involves the hydroxylation of lysine residues that are already incorporated in the newly synthesized peptide chains (160). Evidence has been presented that suggests that lysyl hydroxylase activity is increased in the soluble fraction of kidney homogenates from diabetic animals, and is restored to normal by prior insulin treatment (182, 183). However, substrate was prepared from chick embryo tibia, rather than from basement membrane.

After the hydroxylation of lysine, the sequential participation of two very specific glycosyltransferases completes the synthetic process. One is a galactosyltransferase that functions only to transfer galactose from the activated sugar UDP-galactose to hydroxylysine in high-molecular-weight components (184). The other is a glucosyltransferase that only transfers glucose from UDP-glucose to hydroxylysine-linked galactose (185). In contrast to the hydroxylase, both of these enzymes have been intensively investigated in rat renal cortex using "glucose-free" and "disaccharide-free" GBM as the enzyme substrates. Purified hydroxylysine-linked disaccharide units synthesized in the assay system were isolated by ion-exchange and paper chromatography prior to determination of enzymatic specific activities. The activity of the specific basement membrane glucosyltransferase varies markedly with age; it reaches a peak at nine days and then falls rapidly to lower levels as the animal matures (185). In kidney cortex from rats, diabetic one to five months, the activity of this enzyme is significantly elevated over that of age-matched normal controls, and this difference increases with the duration of the disease. Simultaneous measurements of glucosyltransferase activity in several other tisuses of the diabetic rats do not show any differences from normal (186). Moreover, measurements of another kidney glycosyltransferase enzyme, one involved in the synthesis of the heteropolysaccharide unit of the basement membrane, also shows no difference from normal. The observed elevation of renal glucosyltransferase activity in the diabetic animals most likely reflects a specific increase in GBM synthetic activity.

This finding has been questioned by another group (187) who reported no changes in glycosyltransferase activity after either four weeks or twelve weeks of streptozotocin diabetes, although only three animals were analyzed in the twelve-week diabetic group. Others (188) have since confirmed that glucosyltransferase activity is increased in kidney cortex from diabetic rats.

Amino acid incorporation   While studies of glycosyltransferase activities provide indirect evidence suggesting accelerated GBM synthesis in diabetes, more direct evidence comes from investigations of amino acid incorporation into glomerular proteins. Incubations of isolated rat glomeruli have been utilized in several studies of basement membrane biosynthesis both in normal and in diabetic animals (189–193). Caution must be exercised in the interpretation of such studies, however, because of serious problems with the metabolic properties of these preparations. Although these glomeruli are viable as determined with trypan blue, there is a serious but apparently overlooked problem with the normal basement membrane synthetic process in such systems (149). Basement membrane hydroxylysine and hydroxyproline are formed by enzymatic hydroxylation of lysine and proline, respectively, after they are incorporated into the nascent peptide chain. Therefore, the ratios of the specific activity of hydroxylysine to lysine and of hydroxyproline to proline should both approximate unity. While this information is lacking in most published reports, a useful estimation can be made from the ratios of hydroxylysine cpm to lysine plus hydroxylysine cpm (or similarly for hydroxyproline and proline), which can be calculated from the published data. Since rat renal basement membrane contains 20.6 residues of hydroxylysine, 23.8 residues of lysine, 69.5 residues of hydroxyproline, and 73.5 residues of proline per 1000 amino acid residues, these calculated ratios should approach 0.46 and 0.49. In the work of Grant et al (192), this ratio for proline varies between 0.02 and 0.05. Although these authors have subsequently reported an increased synthesis of GBM in streptozotocin diabetes (193), the physiological meaning of this is unclear.

Similarly, the calculated values for hydroxylysine and lysine cpm reported by Beisswenger (189) are also in the range of 0.02. In contrast to Grant et al (192), recovery of label in hydroxylysine of diabetic GBM did not differ from controls. Although Cohen & Vogt (191) obtained a better hydroxylysine ratio in their glomerular culture system, it was still less than half the expected value. Therefore, although they also observed an increased basement membrane synthesis in diabetic glomeruli that was partially reversible by prior insulin treatment (190, 194), the same caveats mentioned previously also apply here. Even if isolated glomerular systems did function normally, it could still be argued that interpretation and extrapolation of such data would be difficult, due to the artificial, nonphysiologic incubation milieu. Consequently, intact animal studies have been undertaken in an effort to circumvent these difficulties.

In vivo studies of normal GBM turnover have been performed by Price & Spiro (149), using injected tracer doses of tritiated amino acids. The rat GBM in this study was well characterized by amino acid composition and by chemical assessment of cellular contamination. In these studies, the ratio of proline to hydroxyproline specific actvity approached unity, which indicates normal synthesis and hydroxylation. Basement membrane turnover rates were determined from the decay in specific activity of a number of incorporated amino acids. Data from the specific activity of GBM proline and hydroxyproline indicate that the turnover rate of GBM is extremely slow in comparison to other glomerular proteins, and approaches that of fibrillar collagen. Additional data from the specific activity of other amino acids suggest that GBM polypeptide components of dissimilar composition may turn over at different rates.

Utilizing injected tracer doses of tritiated proline, Brownlee & Spiro (195) have extended these analyses to include comparative data from diabetic and normal rats. Preliminary investigations revealed a significant difference in the specific activity of the precursor proline pool between the two groups of animals. This finding of substantially decreased proline specific activity in the serum of diabetic animals may represent dilution by an increased peripheral output of the unlabeled amino acid to meet the demands of augmented gluconeogenesis. A valid comparison of in vivo basement membrane synthesis between normal and diabetic animals could not be made without taking into account these differences in the specific activity of the injected radiolabeled amino acid in the precursor pools.

In this in vivo animal model, the incorporation of radioactivity into the basement membrane of both normals and diabetics was slow, and a period of almost constant specific activity of proline and hydroxyproline was observed after maximum incorporation was reached. In the diabetic basement membrane, however, the specific activity of both these amino acids was nearly twice that attained in normal basement membrane. These results are consistent with accelerated rates of GBM polypeptide synthesis and proline hydroxylation in diabetes, and constitute the most direct evidence presently available in support of this hypothesis.

Pathogenesis of basement membrane changes in diabetes mellitus   A conceptual framework for understanding the alterations in basement membrane synthesis that occur in diabetes is provided by the following considerations. Only certain tissues require insulin for the intracellular transport of glucose (196). Other tissues, including kidney, do not, and thus equilibrate with the extracellular glucose concentration. Hyperglycemia would therefore produce an elevated glucose concentration within the cells of such tissues. This effect would be magnified in diabetes by both a shunting of glucose from insulin-dependent to insulin-independent tissues, and by an increased glucose release from the liver. The resulting high glucose concentration in glomerular cells could significantly affect basement membrane synthesis in two ways. First, because glycosylation of hydroxylysine in the peptide chain is a postribosomal synthetic step, availability of sugar substrate may well be an important rate-limiting factor in the regulation of basement membrane disaccharide synthesis (197). Increased levels of the activated sugar UDP-glucose in diabetic rat kidney cortex are consistent with this (198). In addition to increasing the synthesis of disaccharide units, high glucose concentrations may also have a direct effect on the synthesis and secretion of the peptide portion of basement membrane. The data derived from diabetic rats discussed above is consistent with this notion. Additional support for this effect of hyperglycemia comes from observations of collagen secretion by human skin fibroblasts in vitro (199). Cells growing in a high glucose medium secreted more collagen than those in glucose at a physiologic concentration. Amino acid analysis of the extracted collagen demonstrated twice as much hydroxyproline in each sample of the high glucose medium. Isotope incorporation studies confirmed that incubation of fibroblasts in a medium equivalent to a plasma glucose concentration of 300 mg/dl does result in increased collagen synthesis.

From the previous discussion, it can be concluded that increased synthesis plays a central role in the excessive accumulation of GBM in diabetic kidney. While decreased degradation may also occur in diabetes, direct evidence for this is currently lacking. The findings of decreased glomerular β glycosidase activities (200, 201) and urinary glomerular basement membrane–like protein (202) in diabetic rats, however, have prompted speculation that decreased glycoprotein catabolism may in fact play a role in basement membrane accumulation. Since increased glycosylation may also reduce the susceptibility of glycoproteins to proteolysis (150), increased synthesis of disaccharide units or nonenzymatic glycosylation may also influence the degradation of basement membrane peptides. The elevation of serum protease inhibitors associated with diabetes may also be a contributing factor in this process (102).

Glomerular proteoglycans   Investigations of proteoglycans in diabetic nephropathy are of great interest and importance. These heteropolysaccharide polymers may help organize the quaternary structure of collagen, influence permeability, exert an antithrombotic effect, and perhaps bind selected macromolecules such as circulating immune complexes (203). Their recent localization at the interface between cell surface membranes and the basement membrane of the glomerulus (204, 205) suggests that quantitative or qualitative alterations in these molecules could be involved in the pathological processes that initiate diabetic nephropathy. Glomerular permeability to anionic serum proteins is markedly influenced by fixed negative charges in the glomerular capillary wall (141). Since proteoglycans are highly charged polyanions, diabetes-induced alterations in these molecules could well lead to the development of the proteinuric component of diabetic renal disease. It is also conceivable that increased production and deposition of proteoglycans adjacent to or within the GBM could contribute to the glomerulosclerosis associated with progressive renal failure. Analysis of proteoglycans isolated from cortex of a heterogenous group of human diabetic kidneys showed slight increases in the percentage of heparan sulfate in most samples, and variable increases in hyaluronic acid, depending on the degree of histological abnormality (206). Studies of proteoglycans extracted from glomerular and/or GBM preparations may be more meaningfully interpreted than those performed on kidney cortex homogenates, however, since the locus of diabetic nephropathy is known to be the glomerulus. Significant glomerular changes may otherwise be masked by dilution from other locations, or artifactually suggested by proteoglycan changes in nonglomerular extracellular matrix.

Retina: Morphologic, Functional, and Biochemical Changes

Regional ischemia appears to be a central process in the development of diabetic retinopathy (207–209). Initially, a compensatory increase in both volume and segmental retinal blood flow occurs, with autoregulatory dilatation of retinal vessels (208, 210). As background retinopathy progresses, there is a corresponding progression of increased retinal blood flow (210), although this mechanism may ultimately exacerbate regional hypoperfusion by the shunting of red blood cells through some capillaries, and only plasma through others ("plasma skimming"). Eventually, the autoregulatory response becomes limited by gradual, focal occlusion of capillaries. Retinal hypoxia inadequately compensated by increased blood flow results at first in the formation of microaneurysms. As retinal hypoxia becomes more severe, new vessel formation follows (210–212).

In early diabetes, before significant capillary occlusion has developed, the autoregulatory dilatation of retinal vessels can be rapidly reversed by adequate insulin treatment (3). This early, correctable ischemia in the diabetic retina probably reflects two major pathophysiologic processes. Within days of the onset of diabetes, a breakdown of the blood-retinal barrier occurs that is detectable clinically by vitreous fluorophotometry (213–216). In diabetic patients, the degree of this breakdown is correlated with the degree of metabolic control, as well as with the duration of diabetes (214). Early in the course of diabetes, normalization of blood glucose levels with insulin significantly reduces this increased capillary permeability (215, 216).

The morphologic basis for these permeability changes appears to be a diabetes-induced opening in the endothelial cell tight junctions (217). Decreased active transport of osmotically active substances from the extravascular to the intravascular space may also contribute to the breakdown of the blood-retinal barrier, since diabetes results in a decreased number of micropinocytotic vesicles in capillary endothelial cells (218). These abnormalities in endothelial cell function may in turn be accentuated by selective loss of accompanying pericytes, perhaps due to polyol-induced toxicity (discussed under neuropathy below) (219).

Eventually, increased retinal capillary permeability results in clinically apparent intraretinal edema and hard exudate formation. Before these changes become clinically detectable, however, fluid accumulation in the retina may significantly reduce oxygen diffusion. Increased polyol pathway activity (see below), which in the lens is associated with fluid accumulation, may further contribute to intraretinal edema formation. Formation of intraretinal edema and increased capillary leakage may also be affected by a simple Starling mechanism, since the intraocular pressure is decreased with elevated blood glucose (220), and asymmetric diabetic retinopathy has been associated with asymmetry in intraocular tensions (221). Removal of edema fluid from the retina may also be impeded, since thickening of Bruch's membrane (222) presents an increased diffusion barrier at the pigment epithelial cell layer.

Along with increased capillary permeability and the resultant intraretinal fluid accumulation, reversible diabetic hematological abnormalities also could contribute to early ischemia in the diabetic retina. These hematological abnormalities, discussed in detail above, would tend to decrease oxygen delivery to retinal tissue per unit of blood flow. Whether due to hemoglobin

A_{1c} or to some other red cell abnormality, the oxygen affinity of the diabetic red cell is increased. This may lead to a state of hypoxia at the venous end of the capillary bed, even when the arterial oxygen tension is normal. Elevated levels of plasma proteins such as fibrinogen with resulting hyperviscosity could cause microvascular "sludging," which would reduce local blood flow and increase intracapillary pressure, thus favoring leakage of serum contents into the tissue space. The increased platelet aggregability, red cell aggregability, decreased erythrocyte deformability, enhanced coagulability, and decreased fibrinolysis discussed above would all have qualitatively similar effects, some additive, some possibly synergistic.

Although both the increased capillary permeability and the hematological abnormalities are relatively reversible by insulin treatment, they may contribute to the gradual development of permanent capillary occlusion. Extravasated growth-promoting substances from plasma may play a role in the production of retinal capillary basement membrane thickening (223, 224). Endothelial cell proliferation and swelling may arise in response to local hypoxia (208, 209, 225). Both of these structural alterations would tend to decrease capillary lumen size, further reducing local blood flow and exacerbating any preexisting regional ischemia. In addition, the increased capillary basement membrane thickening would restrict the autoregulatory dilatation of retinal vessels in response to intermittent hypoxia. Progressive capillary occlusion and intensifying regional ischemia finally result in the appearance of microaneurysms, cotton wool spots, and neovascularization.

Panretinal photocoagulation has been shown to have a beneficial effect on proliferative diabetic retinopathy (226). It is speculated that this partial destruction of the retina may decrease the oxygen demands of that tissue, thereby reducing the degree of ischemia present.

Role of Growth Hormone

Abnormal growth hormone secretion has been observed in some diabetic patients, and it has been proposed that this may contribute in various ways to the development of microangiopathy (227). In ambulatory juvenile diabetics, mean 24-hr growth hormone levels have been reported to average three times those of normal subjects (228, 229). However, subsequent investigations revealed no differences in the 24-hr serum growth hormone level between obese, maturity-onset diabetics, and obese normals (230), or between clinically stable normal weight diabetics and their weight-matched control group (231). Diabetics with proliferative retinopathy were reported to have a markedly greater growth hormone response to insulin-induced hypoglycemia (232) and exercise (233) than did diabetics without retinopathy, and 60–80% of patients with proliferative retinopathy were said to stabilize after hypophysectomy (234, 235). It is probable that this merely reflects differing levels of diabetic control among these groups, since the degree of growth hormone elevation correlates with the level of hyperglycemia (236). In a group of glucose-intolerant, growth hormone–deficient dwarfs matched according to sex, age, and duration of diabetes with nongrowth hormone–deficient diabetics, no retinopathy was observed over a ten-year follow-up period (237–239). Although this suggests that increased growth hormone may be a contributing factor in the development of diabetic microangiopathy, the report of a growth hormone–deficient diabetic patient with severe nephropathy, neuropathy, and retinopathy tends to invalidate this hypothesis (240). Growth hormone excess without diabetes (acromegaly) does not result in diabetic tissue changes.

Somatomedins, the mediators of growth hormone's growth-promoting activity, are decreased in the serum of both diabetic animals and diabetic patients (241–244). An inverse correlation exists between levels of hemoglobin A_{1c} and serum somatomedin activity (244). Increased levels of a somatomedin inhibitor may also occur in diabetes (245–247). Since insulin facilitates growth hormone–mediated somatomedin release by the liver (248), the increased growth hormone levels in some diabetics may simply reflect compensatory adjustments to alterations in somatomedin levels.

Experimentally, the characteristic renal mesangial cell dysfunction observed in diabetic rats can be duplicated by administration of growth hormone to normal animals (249), and a two point five fold greater increase in GBM thickness can be produced in the kidneys of diabetic rats (177). In cell culture, supplementation of normal serum with growth hormone results in a significant stimulation of aortic smooth muscle cell proliferation (250). The elucidation of such processes awaits further investigation.

NEUROPATHY

Morphological and Functional Changes

The most frequent symptoms of diabetic peripheral neuropathy are numbness and paraesthesias. These are thought to be the clinical expression of changes in nerve electrophysiology, and reflect altered structural and biochemical properties of the axon and the myelin sheath (251). These changes in nerve electrophysiology can be detected in a majority of diabetic animals and patients, even when clinically apparent symptoms are absent. Untreated, newly diagnosed diabetics exhibit decreased sensory and motor nerve conduction velocities (252, 253), and animals rendered diabetic rapidly develop similar abnormalities (254). Nerve function improves in both animals and man after glucose intolerance is corrected by treatment (252, 254–256). In myelinated neurons, this decreased conduction velocity is thought to result from diminished internodal resistance, which would reduce current density at the nodes and delay excitation (257, 258). Changes in myelin composition could result in such alterations of internodal resistance, as discussed in the following section. Paranodal separation of myelin lamellae (259), perhaps resulting from interlammelar fluid accumulation (260), could also decrease internodal resistance and nerve conduction velocity. Alterations in axon structure and/or metabolism may produce additional electrophysiologic and clinical abnormalities.

Structural studies of peripheral nerves from diabetic patients characteristically show segmental demyelination (261). In rats, there is a direct correlation between the extent and degree of this pathological finding and the severity and duration of hyperglycemia (262, 263). Segmental demyelination may represent primary metabolic impairment of Schwann cells. Alternatively, it may occur secondarily in response to preceding axonal dysfunction (261, 263). In either case, loss of myelin would contribute to the slowing of the rate of nerve conduction.

Long before segmental demyelination can be demonstrated histologically in diabetic animals, alterations in the internodal myelin membrane structure are detectable by freeze-fracture electron microscopy (264). Significant decreases in the number of intramembranous particles on both the inner and outer myelin surface of diabetic nerve are observed at the time when impaired nerve conduction velocity first develops. Both the myelin structural alteration and the decreased conduction velocity are prevented by insulin treatment (264), which suggests that a morphologic basis for this ubiquitous functional abnormality may have been identified.

Another very early morphologic alteration that occurs in the peripheral nerves of diabetic rats is endoneurial edema with resultant shrinkage of axons and Schwann cells (265). Increased water content in the endoneurial space of diabetic nerve results in a 30% reduction in Schwann cell volume. Although the effects of this fluid accumulation on endoneurial capillary blood flow could be deleterious, such adverse consequences have not yet been documented. It has been claimed that increased endoneurial fluid in the diabetic nerve results from a greater permeability of endoneurial vessels in these animals (266). This claim has not been confirmed by subsequent investigations, however, and it thus seems unlikely that extravasation of protein into the endoneurial space plays a role in the pathogenesis of diabetic neuropathy (267, 268). The permeability of nodal gap substance is also said to be increased in diabetic nerve (269). Since this material prevents the diffusion of potassium ions away from the node, alterations in membrane resting potential or ion conductances could result. If this in turn led to excessive generation of action potentials in sensory nerves, paraesthesias might be expected to occur.

Recent quantitative electromyographic studies have demonstrated that extensive axonal dysfunction also occurs in diabetic patients (270). Axonal degeneration, the morphologic equivalent of this functional abnormality, is found in both diabetic humans and animals (263, 265, 271–273). In patients with diabetic neuropathy, loss of large and small myelinated and un-myelinated fibers is evident early in the disease. Although this suggests that axon fiber loss may be important in the evolution of peripheral nerve dysfunction, it has been calculated from fiber diameter that conduction slowing of 20–30% is primarily due to causes other than fiber loss (271). Axonal degeneration and decreased diameter of myelinated axons have also been described in diabetic rats (263, 265, 274, 275). Insulin-treated diabetic animals show fewer structural alterations of this type than do untreated animals.

In addition to the myelin and axon abnormalities described thus far, changes in neuronal vasculature also occur in diabetes. Marked thickening of basement membranes surrounding intra- and perineural vessels has been described in nerves from diabetic patients and from diabetic rats (263, 271, 273, 276). Hyperplasia of endothelial cells in small vessels also occurs frequently, and in some cases this completely occludes the vessel lumen (276). Fibrin deposition in the micro-vasculature was observed in a series of sural nerve biopsies performed in diabetic patients with clinical and electrophysiological evidence of diabetic neuropathy (104). It is possible that such abnormalities in the microvasculature of peripheral nerves may reduce blood flow through small vessels, which would produce regional ischemia or microinfarcts that could contribute to neuronal dysfunction and degeneration.

Biochemical Alterations

MYELIN COMPOSITION Myelin accounts for 75% of peripheral nerve by dry weight. Its composition and rate of synthesis in diabetes have there-fore attracted particular attention in the search for etiologies of peripheral neuropathy. Myelin is a complex lipid containing triglyceride, sphingomye-lin, cholesterol, cerebrosides, and proteins. Abnormalities in nearly all of these constituents have been reported in diabetes (277, 278), in association with decreased nerve conduction velocity. Decreased triglyceride is found in diabetic sciatic nerve (277), which results perhaps from defective incorpo-ration of acetate into fatty acids (279). A reversible 70% reduction in the activity of the enzyme acetic thiokinase may be responsible for these altera-tions (279). Cerebroside synthesis is decreased, and in addition there is a change in the type of cerebrosides, with a pronounced decrease in the incorporation of saturated fatty acids (280).

In myelin protein no compositional changes have been reported with diabetes, but there is a significant decrease in the total amount of protein in peripheral nerve myelin from diabetic patients (281). This may reflect alterations in either synthesis or degradation. A decreased rate of amino acid incorporation into some myelin proteins has been found in diabetic animals (282), while activities of two peripheral nerve proteinases appear to be increased in diabetes (281). The significance of these observations is unclear, however, since peripheral nerve myelin from experimental diabetic animals shows decreased susceptibility to proteolytic digestion (283). Since glycosylation apparently decreases the rate of proteolysis of ribonuclease B compared to the nonglycosylated ribonuclease A (150), there may be an alteration in the carbohydrate of myelin in diabetes that affects its degradation. The glycoproteins of myelin have been localized to the external surface of the myelin sheath, and it is speculated that such a localization enhances the potential role of these glycoproteins in specific interactions in the process of myelination of myelin maintenance (284). Specific analysis of the carbohydrates and glycopeptides of diabetic peripheral nerve myelin have not yet been reported, however.

Although the relative contributions of these individual abnormalities is unknown, together they produce major physico-chemical changes that result in abnormal peripheral myelin sedimentation and electrophoretic patterns in hyman diabetic nerve (285). The relationship between these changes and altered nerve conduction properties in diabetes remains to be elucidated.

SORBITOL PATHWAY   The Schwann cell sheath and certain other mammalian tissues have the unique ability to synthesize free (nonphosphorylated) fructose for various metabolic purposes (286). This biosynthetic pathway, termed the sorbitol pathway, consists of two reactions. Glucose is first reduced to its corresponding sugar alcohol, sorbitol, by the enzyme aldose reductase, with NADPH as the electron-donating coenzyme. The sorbitol molecule is then oxidized to fructose by the enzyme sorbitol dehydrogenase and NAD^+. Significantly elevated concentrations of both sorbitol and fructose have been reported in several tissues of diabetic animals and man, including lens, retina, arterial wall, erythrocytes, and Schwann cell sheath (219, 286–290). Activity of this pathway is regulated by the ambient glucose concentration in those tissues not requiring insulin for glucose transport, such as nerve (291), since the K_m for aldose reductase lies in the range of physiologic concentrations of glucose (292). Increased sorbitol pathway activity has been implicated in the formation of galactosemic and acute diabetic cataracts. It may also play a role in the pathogenesis of diabetic large vessel disease and peripheral neuropathy (287, 292, 293). In diabetic animals, hyperglycemia and the resultant increased sorbitol pathway activity in peripheral nerve is associated with decreased oxygen uptake in this tissue (294, 295). Such impairment of tissue oxygen delivery could conceivably contribute to the development of neurological dysfunction. The increased activity of the sorbitol pathway in nerve may also contribute to the endoneurial fluid accumulation described in the pervious section. Insulin treatment restores sorbitol pathway activity toward normal (296) and prevents the development of diminished nerve conduction velocity in diabetic animals (254, 287). The administration of an aldose reductase inhibitor can also prevent sorbitol accumulation in sciatic nerve of diabetic animals (297). However, compounds of this type have not yet been shown to have beneficial effects on nerve conduction velocity in diabetes.

MYOINOSITOL   One of the cyclic isomers of sorbitol, myoinositol, may play an important role in such nerve cell functions as electrolyte transport, amino acid transport, and nerve impulse transmission. Recent evidence suggests that alterations in the metabolism of this axon phospholipid precursor may be responsible for some of the functional changes of diabetic peripheral neuropathy (254, 261, 298). In severely diabetic animals, sciatic nerve myoinositol content is decreased (254, 299). Reductions in the content and rate of synthesis of phosphatatidylinositol from myoinositol also occur (299–301), and activities of CDP-diglyceride-inositol transferase and phosphatidylinositol 4-phosphate kinase are decreased in diabetic sciatic nerves (302, 303). When nerve myoinositol content is restored to normal by dietary supplementation, conduction velocity improves (254, 299, 304). The exact relevance of these observations to human diabetic neuropathy is unclear, however, since they have not been confirmed in animals having diabetes of less severe degree.

In uncontrolled human diabetes, urinary myoinositol excretion is markedly increased, and there is an apparent decrease in intracellular transport of myoinositol (305). These abnormalities are corrected toward normal with careful insulin treatment (254, 305). Dietary myoinositol supplementation did not alter the decreased motor conduction velocity or vibratory perception threshold in symptomatic or asymptomatic diabetic patients (306–308). Slight improvement in sensory nerve conduction velocity may have occurred, however (306, 307). The clinical importance of these observations remains to be determined.

AXOPLASMIC TRANSPORT   Many enzymes and substrates utilized for neurotransmitter synthesis in the nerve axon actually originate in the neuronal cell body. From here they are transported to distal locations by a process of axoplasmic flow. It has recently been suggested that reductions in axoplasmic transport of certain critical proteins may result in peripheral nerve dysfunction in diabetes (309, 310). In diabetic rats, both fast and slow axonal flow are markedly decreased. Cholinergic neurons showed a 32% reduction in the rate of acetyl-cholinesterase transport and a 41% reduction in the rate of choline acetylase transport (309). In noradrenergic neurons, axoplasmic transport of noradrenaline is reduced to a similar degree (310). Decreased retrograde axonal flux of transported glycoproteins also occurs in early diabetes (311). Such changes could adversely affect action potential regeneration at the synapse, with resulting decreases in propagated conduction velocity. All of the observed decreases in axoplasmic treatment are effectively reversed by insulin administration.

MACROANGIOPATHY

Accelerated large vessel disease in diabetes may be due in part to abnormalities in plasma lipids, and perhaps also to changes in the composition and metabolism of the arterial wall. Epidemiologic data appear to show no excess risk from diabetes independent of hypertension, cigarette smoking, and hyperlipidemia for a large proportion of the specific age-sex groups (312). However, relatively inaccurate and insensitive correlates of the integrated concentration of blood glucose were used to assess severity of diabetes. Insulin deficiency may well influence the progression of atherosclerosis through synergistic pathological mechanisms involving hyperlipidemia, altered platelet behavior, and abnormalities in arterial function.

Plasma Lipids and Lipoproteins

Abnormalities in lipid and lipoprotein metabolism are associated with atherosclerosis in nondiabetics, and the presence of even modest elevations in plasma lipids over many years could be a major contributory factor to the development of diabetic macroangiopathy. Fasting plasma lipid and lipoprotein levels may be relatively insensitive indicators of disordered lipid homeostasis, due to diurnal variation in levels and extensive heterogeneity within each lipoprotein class (313, 314). Nevertheless, fasting hyperlipidemia and hyperlipoproteinemia are observed in a significant percentage of ambulatory diabetic patients. Various studies report a 24–55% prevalence; hypertriglyceridemia is more frequent than elevated plasma cholesterol (315–324). Elevated levels of VLDL triglyceride have been found in all major classes of diabetic patients, usually associated with inadequate insulinization (319, 320, 325, 326). Similarly, elevated levels of LDL cholesterol have been consistently observed in both juvenile-onset and maturity-onset diabetic groups (318, 319, 322–324, 326). HDL cholesterol, which varies inversely with cardiovascular disease risk in nondiabetics, has been reported to be decreased (318, 319, 327, 328) or unchanged (323) in a number of studies of maturity-onset diabetics. Newly diagnosed juvenile diabetic patients exhibit the same abnormality (326), while chronic insulin-treated juvenile-onset patients may have either decreased (321, 329) or elevated HDL levels (320, 323). Among a heterogeneous group of diabetics, increased VLDL triglyceride, increased LDL cholesterol, and decreased HDL cholesterol each correlated with the presence of arteriosclerosis obliterans (330) in different subgroups. In another study, the frequency of hypertriglyceridemia among diabetics without atherosclerosis was no different than controls, while 50% of diabetics with atherosclerosis had triglyceride levels above 150 mg/dl (331).

The biochemical derangements involved in the production of diabetic lipid abnormalities are incompletely understood at present. Hypertriglyceridemia of similar degree in two different types of diabetic rat with comparable levels of hyperglycemia appears to arise by different mechanisms depending on the severity of insulin deficiency. In animals with absolute insulin deficiency, hepatic and intestinal VLDL production is decreased, despite high circulating levels of free fatty acids (FFA) resulting from unrestrained lipolysis (332–334). Perfusion of livers from these animals with increasing amounts of FFA did not result in additional VLDL production. In these animals, elevated very low density lipoprotein triglyceride (VLDL-TG) reflects a severe removal defect, due in part to a 20–30% decrease in acetone-ether–extracted adipose tissue LPL (332). In animals with relative insulin deficiency associated with obesity and peripheral insulin resistance, hepatic VLDL production is increased, and intestinal VLDL production is doubled (333, 334). Perfusion of livers from these animals with FFA results in a linear increase in VLDL-TG production (333). In these animals, elevated VLDL-TG is due primarily to increased production responding to elevated FFA in the presence of permissive levels of insulin.

Studies in man are for the most part consistent with these findings, although the possibility of familial hyperlipoproteinemia coexisting with diabetes makes interpretation more complex. In untreated juvenile-onset diabetics, there is little evidence for increased VLDL-TG production (325, 335, 336). The LPL component of plasma postheparin lipolytic activity is reduced 40%, however, and both adipose tissue and muscle heparin–releasable LPL are decreased by similar amounts (326, 335). Insulin treatment tends to lower plasma TG and normalize LPL activity. Hypertriglyceridemia in human diabetics with absolute insulin deficiency thus appears to result, as in animals, from a defect in removal of TG-rich lipoproteins.

In maturity-onset diabetics with significant fasting hyperglycemia and increased triglycerides, VLDL-TG production is markedly increased (325, 335). These patients may also have a mild removal defect, since plasma postheparin lipolytic activity is reduced 20%, and adipose tissue LPL is also decreased (335, 337, 338). In treated patients with elevated TG, these findings persist. The decreased fractional turnover of exogenous TG in untreated patients (317) is associated with a lower maximal removal capacity for plasma TG and an increased K_m for endogenous plasma TG-LPL interaction (336). These parameters do normalize with treatment.

The causes of hypercholesterolemia in diabetes have been less extensively investigated. In some patients, mildly elevated plasma cholesterol is primarily a reflection of the increased VLDL levels discussed above. In others, it represents a separate derangement of plasma LDL. LDL is formed by delipidation of TG-rich particles, and is degraded by various tissues after receptor-mediated endocytosis (339). High levels of LDL could result from an increase in the absolute rate of synthesis, a decrease in the absolute rate of degradation, or a diminished fractional catabolic rate (339). Turnover studies of ^{125}I-labeled LDL in diabetic plasma have not yet been reported, however.

Isotope incorporation studies in ketotic diabetic rats suggest that sterol synthesis is depressed in liver, intestine, and carcass. In nonketotic, hyperglycemic rats, sterol synthesis is unchanged in liver and carcass, but increased twofold in intestine (K. R. Feingold, personal communication). Activity of 3-hydroxy-3-methylglutaryl coenzyme A reductase (HMG CoA reductase), the rate-limiting enzyme in cholesterol synthesis, is also significantly increased in intestine from diabetic rats (340). Since liver and intestine are the only known sources of the predominant protein of plasma LDL, apolipoprotein B, increased intestinal production of LDL may be occurring in diabetes. However, these findings are equally consistent with an increased VLDL secretion by the intestine. Cholesterol absorption and sterol balance studies in diabetic patients have suggested that normolipemic diabetics have a mild increase in cholesterol synthesis (328, 341) and a slight decrease in cholesterol degradation (341). Others have reported no effect of diabetes on sterol balance (342).

LDL-receptor activity and subsequent postbinding intracellular LDL degradation in cultured human fibroblasts from diabetics do not differ from normal. Physiological concentrations of insulin result in significant stimulation of both these parameters, however, which suggests that insulin deficiency in vivo may lead to an LDL removal defect, analagous to that observed in patients with familial hypercholesterolemia (339, 343).

HDL is synthesized by liver and intestine. In addition, a portion of plasma HDL may be generated in the circulation by the action of LPL on TG-rich lipoproteins (343a). If VLDL degradation is responsible for some HDL production, then the decreased LPL activity occurring in untreated juvenile-onset and maturity-onset diabetic patients may partially account for the lower HDL levels observed in these groups (318, 319, 326, 327, 328, 335, 337, 338). The elevated levels of HDL reported to occur in some juvenile-onset diabetics (323), like LDL, could reflect increased synthetic rate, decreased rate of degradation, or a diminished fractional catabolic rate.

Arterial Wall

Sequential observations of arterial wall pathology in developing experimental atherosclerosis have provided a conceptual model for the pathogenesis of macroangiopathy, the response-to-injury hypothesis (344–346). In this model, the initial event is injury to the arterial intimal endothelium. The resultant increase in permeability of the endothelial barrier allows blood components to enter the subendothelial space, which creates an area of focal desquamation. This focal desquamation allows circulating platelets to contact the subendothelium. Factors are then released by these adherent platelets that stimulate a focal proliferation of smooth muscle cells. Excessive lipid deposition and formation of connective tissue ultimately result in a mature atheromatous plaque. In normal individuals, the process is probably reversible at an early stage. In hyperlipidemia and perhaps in diabetes, however, regression of the early lesions is prevented by continuous or repeated injury.

Each of the four elements in this proposed pathogenetic sequence may be accentuated in diabetes. Endothelial cell injury with increased permeability is described in previous sections of this review. Speculative mechanisms for this hyperglycemia-induced damage include excessive nonenzymatic glycosylation, altered production or composition of cell surface components, and lipoprotein abnormalities, all discussed in earlier sections.

Platelet interactions with subendothelial structures may be enhanced by the increased adhesiveness, hyperaggregability, and accelerated production of thrombogenic prostaglandin derivatives reported to occur with diabetic platelets. Decreased prostacyclin activity in diabetic arterial wall may be a further contributing factor (see above). Cell culture experiments suggest that smooth muscle cell proliferation may be promoted by increased levels of growth hormone (347) or other growth factors from diabetic serum (348), as well as by released platelet mitogens.

Increased collagen secretion is observed when human skin fibroblasts are grown in a high glucose medium (199), thus hyperglycemia in vivo may stimulate arterial wall collagen production. Decreased susceptibility to degradation by collagenase may also contribute to arterial wall collagen accumulation in diabetes (349, 350), perhaps resulting from diabetes-induced structural alterations. In diabetic rat aorta, increased nonenzymatic glycosylation of collagen has been observed (351), and diabetic fibroblasts appear to secrete increased amounts of a procollagen-like high-molecular-weight peptide that contains abnormal reducible cross-links and lacks normal non-reducible cross-links (199, 352). Granulation tissue collagen produced in the diabetic milieu also shows evidence of increased cross-links (350).

Arterial wall proteoglycans may help organize the quaternary structure of collagen, influence permeability, exert an antithrombotic effect, and bind LDL (203). Variable qualitative and quantitative alterations in these molecules have been described in diabetic animals and man, that could conceivably contribute to the development of accelerated large vessel disease (353–357). Decreased chondroitin sulfate and heparan sulfate have been reported in aorta of diabetic rats (353, 355) and dogs (354), with hyaluronic acid either increased (355) or unchanged (353, 354). A 20% increase in the amount of dermatan sulfate was found in the coronary arteries of diabetic dogs, while increased heparan sulfate, increased chondroitin sulfate, and decreased hyaluronic acid were found in various other arteries (357). Skin fibroblasts from diabetic patients were found to secrete a higher percentage of heparan sulfate relative to other sulfated proteoglycans (356), and similar changes have been reported to occur in human diabetic kidney cortex (206).

Excessive lipid accumulation in arterial wall cells of poorly controlled diabetics may be caused in part by impaired intracellular degradation of LDL. In aortic smooth muscle cells of diabetic rats, activity of acid cholesteryl esterase, the lysosomal enzyme that hydrolyzes cholesterol esters, is reduced 15–25% (358). Similar findings have been reported in circulating mononuclear cells from untreated noninsulin dependent diabetic patients (359). Mononuclear cell degradation of the protein moiety of LDL is also reduced. In both animals and man, insulin treatment increases the activity of all measured parameters.

DIABETIC COMPLICATIONS AND THE DEGREE OF METABOLIC DYSFUNCTION

The previous sections review evidence from a variety of morphologic, physiological, and biochemical studies that strongly suggest a central role for insulin deficiency and its metabolic consequences in the pathogenesis of diabetic complications. It is difficult, however, to derive clinically valid conclusions about the efficacy of diabetic control from such data alone. Experimentally diabetic dogs develop microaneurysms, hemorrhages, and exudates in the retina, and diffuse glomerulosclerosis with typical nodular lesions in the kidney (360, 361). The development of these lesions is inihibited by good control of hyperglycemia with insulin (362–364). Spontaneously diabetic *Macaca nigra* develop atherosclerosis in proportion to the severity of their hyperglycemia and insulin deficiency (365). However, the degree of relevance to the human situation is unclear, since the diabetic-like lesions are similar, but not precisely identical to those seen in man.

Implicit in the foregoing discussions has been the conclusion that perfect metabolic regulation of diabetes, although presently impossible to achieve, would prevent the development of diabetic sequelae. Do levels of excellent diabetic control that are attainable by current therapies approximate normal physiology enough to favorably alter the clinical course of the disease? Formidable methodological problems make conclusive clinical studies difficult if not impossible to design and execute (366). Nevertheless, the efficacy of excellent control is strongly supported by the available clinical data.

Knowles, reviewing the extant world literature in 1964 (367) found over 300 publications on this topic, representing data from 85 studies. Of these, 51 concluded that a correlation exists between degree of diabetic control and complications, 26 favored the opposite conclusion, and 8 were inconclusive. However, in 40 there was not enough data to support the impressions of the authors. The remaining 45 all lacked random allotment of treatment and utilized unsatisfactory measures of diabetic control. In 39 of the 45, there was no satisfactory description of material for interpreting the results (368). Thus, a critical final appraisal of a consensus opinion was not possible at that time.

Most of these clinical studies have been retrospective in nature. One of the largest and most frequently quoted studies, representative of those supporting a relationship between good control and fewer complications, is of 451 patients from the Joslin Clinic with onset of diabetes under the age of 20 and duration from 10 to 36 years (369). Among the 101 patients with nephropathy, only one came from the excellent/good control group.

In patients having diabetes for 20 years or more, nephropathy was present in 3% of those with excellent or good control, and in 32% of those with fair or poor control. Similarly, a 3% incidence of grade 3 retinopathy was found among the patients with excellent or good control who had had diabetes for 20 or more years, compared to a 31% incidence among the fair and poor control patients of like duration. No grade 4 (proliferative) retinopathy occurred in the excellent and good control groups. Other retrospective studies have shown that motor nerve conduction velocity is significantly diminished in poorly controlled diabetics (255). The degree of this conduction impairment is closely related to the level of fasting plasma glucose and to the level of glycosylated hemoglobin (253). Normal thickness of muscle capillary basement membrane (MCBM) has been observed after 5–20 yr of extremely well-controlled juvenile diabetes, while increased MCBM thickness has been observed after less than 5 yr of poorly controlled diabetes (370).

Some retrospective studies have not demonstrated a relationship between diabetic control and the incidence of complications (371, 372). In these studies, the criteria for "good control" more closely resemble those for the "fair control" groups in other studies, thus perhaps obscuring the real difference between excellent/good and fair/poor control groups. Where comparisons have been attempted between institutions advocating different therapeutic philosophies, failure to distinguish between clinic patients in the aggregate and the subgroup of "excellent/good" control patients has seriously undermined the stated conclusions.

It has been suggested that good control may be easier to achieve in one type of diabetes than in another, and thus that any observed correlation between degree of control and incidence of complications may actually reflect heterogeneity of the disease, rather than differences in control (373). On the other hand, any failure to observe such correlations may simply reflect the insensitivity of the chosen measure of diabetic control, since most of these do not adequately reflect the degree of minute to minute metabolic regulation. Studies in which "degree of control" is replaced by "therapeutic regimen" as an independent variable might avoid both these difficulties.

One large retrospective study that bears directly on this question compares all diagnosed diabetics in Malmö, Sweden from 1922 to 1935 with those diagnosed during 1936 to 1945 (374). The former group was treated with multiple daily injections of short-acting insulin and a strict dietary regimen. The latter group was treated primarily with longer-acting insulin in a single daily injection, along with a less exact dietary program. All of the patients examined from both groups had had diabetes for more than 15 yr at the time of the study. Despite the fact that the 1945 group had diabetes 8.6 yr less than the 1922–1935 group, the prevalence of grades 3 and 4 retinopathy in the single injection group was over 6 times, and the prevalence of nephropathy in this group nearly 2 times that observed in the multiple injection group. Retrospective analyses of diabetic patients surviving 20–40 (375) and 40 yr or more with their disease (376, 377) confirm that the great majority of these individuals have likewise received multiple insulin injections for most of their diabetic lives.

All of the retrospective studies discussed thus far support the thesis that more adequate insulinization reduces the occurrence of diabetic complications. The data on which these conclusions have been based, however, characterize available survivors of an undefined initial population. More valid information would result if analysis of previous deaths and possible transfers from one initial cohort to another were included in such studies. In one large retrospective study that does address itself to these considerations, relative survival was significantly better in relatively well-controlled juvenile diabetics than in moderately or poorly controlled patients (378). In fact, survival among the well-controlled group did not differ significantly from the general population.

Prospective clinical studies can avoid many of the design defects inherent in retrospective analyses. In recent years, important new information has thus become available from prospective investigations of diabetic control and complications. In one 20-year prospective mortality study, mortality in cases with poor control was 2.5 times that of cases with better control (379). Kohner et al (380) performed a prospective study on the progression of established diabetic retinopathy as related to the degree of diabetic control achieved during the study period. Only "very good" control seemed to protect patients from worsening of microaneurysms, hemorrhages, and new vessels. Similar findings were reported by Miki et al (381), where progression of established retinopathy was 3 times more common among the poor- that among the good-control groups over a 6 yr period.

One randomized, prospective study comparing the effect of single and multiple dose daily insulin regimens on the progression of retinopathy in insulin-dependent diabetics has provoked considerable controversy. Patients receiving multiple insulin injections were reported to have a slower rate of microaneurysm increase than patients receiving single insulin injections, despite minimal differences in mean urinary glucose excretion and fasting blood glucose levels between the two treatment groups (382). Questions about the statistical significance of these conclusions were raised, however, by subsequent analysis of the published data (383). Now it appears that recent follow-up data from the original study have reconfirmed the previously reported results (384). In other prospective studies, very good diabetic control has been associated with decreases in the frequency, degree, and progression with time of proteinuria in diabetic patients (385, 386). Only 9% of the good control group had worsening of proteinuria over a 6-yr period, compared to 22% of the poor control group.

The largest and most comprehensive clinical investigation of the relationship between diabetic control and long-term complications has been a prospective study of 4400 patients observed between 1947 and 1973 (387–390). Nearly 2800 of these patients were followed continuously from the time of diagnosis, which allowed the incidence as well as the prevalence of defined complications to be ascertained. This study is unique in fulfilling virtually all of the requirements that are necessary in order to derive valid conclusions from the data obtained (389). Although the incidence and prevalence of complications were found to be strongly correlated with the known duration of diabetes, there was no association with the inherent severity of the disease. Rather, a higher prevalence and incidence of neuropathy, retinopathy, and nephropathy were linked to poor control, assessed cumulatively over the years. In addition, the annual incidence of these complications was clearly and separately related to the glycemic control achieved during the year preceding the annual examination, whatever the prior degree of control.

This study and the others reviewed in this section provide compelling clinical arguments for maintaining blood glucose levels of diabetic patients as close to normal as possible. Perhaps the most effective means of accomplishing this at the present time is the use of divided dose daily insulin injections in conjunction with home self-monitoring of blood glucose (391). Attempts to develop more physiologic insulin delivery systems are currently in progress (392–395). The development of effective new pharmacologic agents for the prevention and treatment of diabetic complications awaits greater understanding of the biochemical mechanisms involved in their pathogenesis.

ACKNOWLEDGMENTS

We would like to thank Mrs. Hilary McCaleb for skilled secretarial assistance. Dr. Brownlee is the recipient of a Public Health Service Special Emphasis Research Career Award (1-K01-AM00589-01 SRC) from the National Heart, Lung and Blood Institute and the National Institute of Arthritis, Metabolism and Digestive Diseases.

Literature Cited

1. Marks, H. H., Krall, L. P., White, P. 1971. *Joslin's Diabetes Mellitus,* ed. A. Marble, P. White, R. F. Bradley, L. P. Krall, pp. 10–34. Philadelphia: Lea & Febiger. 11th ed.
2. Crofford, O. 1975. *Report of the National Commission on Diabetes.* DHEW Publ. No. (NIH) 76-1018
3. Ditzel, J. 1968. *Diabetes* 17:388–97
4. Ditzel, J., Standl, E. 1975. *Acta Med. Scand.* 578:49–58. (Suppl.)
5. Little, H. L., Sacks, A. H., Krupp, M., Johnson, P., Basso, L., Tichner, G., Vassiliadis, A., Zweng, H. C. 1973. *Congr. Int. Diabetes Fed., 8th,* p. 176
6. Little, H. 1977. *Trans. Am. Ophthalmol. Soc.* 75:397–426
7. Little, H., Sacks, A. H. 1977. *Trans. Am. Acad. Ophthalmol. Otolaryngol.* 83:522–34
8. Paulsen, E. 1979. *Metabolism* 28:Suppl. 1, p. 407
9. Schmid-Schonbein, H., Volger, E. 1976. *Diabetes* 25:Suppl. 2, pp. 897–902
10. McMillan, D. E., Utterback, N. G., LaPuma, J. 1977. *Diabetes* 26:Suppl. 1, p. 369
11. McMillan, D. E., Utterback, N. G., LaPuma, J. 1978. *Diabetes* 27:895–901
12. Baba, Y., Kai, M., Kamada, T., Setoyama, S., Otsuji, S. 1979. *Diabetes* 28:1138–40
13. Miller, J. A., Gravalles, E., Bunn, H. F. 1980. *J. Clin. Invest.* 65:896–901
14. Chandramouli, V., Carter, J. R. Jr. 1975. *Diabetes* 24:257–62
15. Baba, Y., Kai, M., Setoyama, S., Otsuji, S. 1978. *Clin. Chim. Acta* 84:247–49
16. Paulsen, E. P., Koury, M. 1976. *Diabetes* 25:Suppl. 2, pp. 890–96
17. Bagdade, J. D., Stewart, M., Walters, E. 1978. *Diabetes* 27:677–81
18. Selam, J. L., Clot, J., Andary, M., Mirouze, J. 1979. *Diabetologia* 16:35–40
19. Nolan, C. M., Beaty, H. N., Bagdade, J. D. 1978. *Diabetes* 27:889–94
20. Edelman, G. M. 1976. *Science* 192:218–26
21. Nicolson, G. L., Poste, G. 1976. *N. Engl. J. Med.* 295:197–203
22. Chandramouli, V., Williams, S., Marshall, J. S., Carter, J. R. Jr. 1977. *Biochim. Biophys. Acta* 465:19–33
23. Canivet, J., Cruz, A., Moreau-Lalande, H. 1979. *Metabolism* 28:1206–10
24. Carter, J. R. Jr., Chandramouli, V. 1977. *Diabetes* 26:Suppl. 1, p. 366
25. Trivelli, L. A., Ranney, H. M., Lai, H. T. 1971. *N. Engl. J. Med.* 284:353–57

26. McDonald, M. J., Shapiro, R., Bleichman, M. 1978. *J. Biol. Chem.* 253:2327–332
27. Shapiro, R., McManus, M., Garrick, L., McDonald, M. J., Bunn, H. F. 1979. *Metabolism* 28:427–30
28. Koenig, R. J., Blobstein, S. H., Cerami, A. 1977. *J. Biol. Chem.* 252:2992–97
29. Bunn, H. F., Shapiro, R., McManus, M., Garrick, L., McDonald, M. J., Gallop, P. M., Gabbay, K. H. 1979. *J. Biol. Chem.* 254:3892–98
30. Gabbay, K. H., Sosenko, J. M., Banuchi, G. A., Mininsohn, M. J., Flückiger, R. 1979. *Diabetes* 28:337–40
31. McDonald, M. J. 1979. *J. Biol. Chem.* 254:702–7
32. Standl, E., Kolb, H. J. 1973. *Diabetologia* 9:461–66
33. Ditzel, J., Jaeger, P., Standl, E. 1978. *Metabolism* 27:927–34
34. Ditzel, J. 1974. *Diabetologia* 10:363
35. Ditzel, J., Dyerberg, J. 1977. *Metabolism* 26:141–50
36. Bunn, H. F., Gabbay, K. H., Gallop, P. M. 1978. *Science* 200:21–27
37. Bunn, H. F., Haney, D. N., Gabbay, K. H., Gallop, P. M. 1975. *Biochem. Biophys. Res. Commun.* 67:103–9
38. Stevens, V. J., Vlassara, H., Abati, R., Cerami, A. 1977. *J. Biol. Chem.* 252:2998–3002
39. Koenig, R. J., Peterson, C. M., Kilo, C., Cerami, A., Williamson, J. R. 1976. *Diabetes* 25:230–32
40. Koenig, R. J., Peterson, C. M., Jones, R. L., Saudek, C., Lehrman, M., Cerami, A. 1976. *N. Engl. J. Med.* 295:417–20
41. Cerami, A., Stevens, V. J., Monnier, V. M. 1979. *Metabolism* 28: Suppl. 1, pp. 431–37
42. Stevens, V. J., Rouzer, C. A., Monnier, V. M., Cerami, A. 1978. *Proc. Natl. Acad. Sci. USA* 75:2918–22
42a. Monnier, V. M., Stevens, V. J., Cerami, A. 1979. *J. Exp. Med.* 150:1098–1107
43. Chang, A. Y., Noble, R. E. 1979. *Diabetes* 28:408
44. Day, M. F., Thorpe, S. R., Baynes, J. W. 1979. *J. Biol. Chem.* 254:595–97
45. Dolhofer, R., Wieland, O. H. 1979. *FEBS Lett.* 103:282–86
46. Gundersen, H. J. G., Mogensen, C. E., Seyer-Hansen, K., Osterby, R., Lundbaek, K. 1979. *Adv. Nephrol.* 8:43–62
47. Dolhofer, R., Wieland, O. H. 1980. *Diabetes* 29:417–22
48. McFarland, K. F., Catalano, E. W., Day, J. F., Thorpe, S. R., Baynes, J. W. 1979. *Diabetes* 28:1011–14
49. Yue, D. K., Morris, D., McLennan, S., Turtle, J. R. 1980. *Diabetes* 29:296–300
50. Harrison, H. E., Reece, A. H., Johnson, M. 1980. *Diabetologia* 18:65–68
51. Bern, M. 1979. *Metabolism* 28: Suppl. 1, p. 408
52. Abrahamsen, A. 1968. *Scand. J. Hematol.* 1: Suppl. 3, pp. 9–11
53. Colwell, J. A., Halushka, P. V., Sarji, K. E. 1978. *Med. Clin. N. Am.* 62:753–66
54. Bern, M. 1978. *Diabetes* 27:342–50
55. Bridges, J. M. 1965. *Lancet* 1:75–77
56. Freytag, J. W., Dalrymple, P. N., Maguire, M. H., Strickland, D. K., Carraway, K. L., Hudson, B. G. 1978. *J. Biol. Chem.* 253:9069–74
57. Bensoussan, D., Levy-Toledano, S., Passa, J., Caen, J., Canivet, J. 1975. *Diabetologia* 11:307–12
58. Dobbie, J. G., Kwaan, H. C., Colwell, J., Suwanwela, N. 1974. *Arch. Ophthalmol.* 91:107–9
59. Colwell, J. A., Halushka, P. V., Sarji, K., Levine, J., Sagel, J., Nair, R. M. G. 1976. *Diabetes* 25: Suppl. 2, pp. 826–31
60. Heath, H., Brigden, W. D., Canever, J. V., Pollock, J., Hunter, P. R., Kelsey, J., Bloom, A. 1971. *Diabetologia* 7:308–15
61. Sagel, J., Colwell, J. A., Crook, L., Laimins, M. 1975. *Ann. Intern. Med.* 82:733–38
62. Sarji, K. E., Nair, R. M. G., Chambers, A. L., Laimins, M., Smith, T., Colwell, J. A. 1975. *Diabetes* 24: Suppl. 1, p. 398
63. Peterson, C. M., Jones, R. L., Koenig, R. J., Melvin, E. T., Lehrman, M. L. 1977. *Ann. Intern. Med.* 86:425–29
64. Colwell, J. A., Nair, R. M. G., Halushka, P. V., Rogers, C., Whetsell, A., Sagel, J. 1979. *Metabolism* 28: Suppl. 1–4, pp. 394–400
65. Nair, R. M. G., Van Zile, J., Johnson, J., Laimins, M., Whetsell, A., Sagel, J., Colwell, J. A. 1979. *Diabetes* 28:385
66. Coller, B. S., Frank, R. N., Milton, R. C., Gralnick, H. R. 1978. *Ann. Intern. Med.* 88:311–16
67. Borsey, D. Q., Dawes, J., Fraser, D. M., Prowse, C. V., Elton, R. A., Clarke, B. F. 1980. *Diabetologia* 18:353–57
68. Marx, J. L. 1977. *Science* 196:1072–75
69. Halushka, P. V., Lurie, D., Colwell, J. A. 1977. *N. Engl. J. Med.* 297:1306–10
70. Ziboh, V. A., Maruta, H., Lord, J., Cagle, W. D., Lucky, W. 1979. *Eur. J. Clin. Invest.* 9:223–28
71. Chase, H. P., Williams, R. L., Dupont, J. 1979. *J. Pediat.* 94:185–89
72. German, G. A., Stampfer, H. G. 1979. *Lancet* 2:789

73. Harrison, H. E., Reece, A. H., Johnson, M. 1978. *Diabetologia* 15:237 (Abstr.
74. Harrison, H. E., Reece, A. H., Johnson, M. 1978. *Life Sci.* 23:351–56
75. Silberbauer, K., Schernthaner, G., Sinzinger, H., Piza-Katzer, H., Winter, M. 1979. *N. Engl. J. Med.* 300:366–67
76. Haft, D. E., Reddi, A. S. 1979. *Biochim. Biophys. Acta* 584:1–10
77. Davis, T. M. E., Mitchell, M. D., Dornan, T. L., Turner, R. C. 1980. *Lancet* 1:373
78. Nordoy, A., Rodset, J. M. 1970. *Diabetes* 19:698–702
79. Kranias, G., Kranias, E., Dobbie, G. J., Jungmann, R. A. 1978. *FEBS Lett.* 92:357–60
80. Muhlrad, A., Eldor, A., Bar-On, H., Kahane, I. 1979. *Thromb. Res.* 14:621–29
81. Boston Collaborative Drug Surveillance Center. 1974. *Diabetes* 23:151–53
82. Jonsson, A., Wales, J. K. 1976. *Diabetologia* 12:245–50
83. Barnes, A. J. 1977. *Lancet* 2:789–91
84. McMillan, D. E. 1974. *J. Clin. Invest.* 53:1071–79
85. McMillan, D. E. 1975. *Diabetes* 24: Suppl. 2, p. 438
86. Lowe, G. D. O., Lowe, J. M., Drummond, M. M., Reith, S., Belch, J. J. F., Kesson, C. M., Wylie, A., Foulds, W. S., Forbes, C. D., MacCuish, A. C., Manderson, W. G. 1980. *Diabetologia* 18:359–63
87. Egebert, O. 1963. *J. Clin. Lab. Med.* 15:833
88. Jones, R. L., Peterson, C. M. 1979. *J. Clin. Invest.* 63:485–93
89. Banerjee, R. N., Sahni, A. L., Kumar, V., Arya, M. 1974. *Thromb. Diath. Haemorrh.* 31:339–45
90. Monnier, L., Follea, G., Mirouze, J. 1978. *Horm. Metab. Res.* 10:470–73
91. Gandolfo, G. M., de Angelis, A., Torresi, M. V. 1980. *Haemostasis* 9:15–19
92. Corbella, E., Miragliotta, G., Masperi, R., Villa, S., Bini, A., de Gaetano, G., Chiumello, G. 1979. *Haemostasis* 8:30–37
93. Almer, L.-O., Pandolfi, M., Osterlin, S. 1975. *Ophthalmologica* 170:353–61
94. Pandolfi, M., Almer, L.-O., Holmberg, L. 1974. *Acta Ophthalmol.* 52:823–28
95. Lufkin, E. G. 1979. *Metabolism* 28: 63–66
96. Gensini, G. F., Abbate, R., Favilla, S., Neri Serneri, G. G. 1979. *Thromb. Haemostasis* 42:983–93
97. Sarji, K. E., Levine, J. H., Nair, R. M. G., Sagel, J., Colwell, J. A. 1977. *J. Clin. Endocrinol. Metab.* 45:853–56
98. Graves, J. M., Colwell, J. A., Nair, R. M. G., Sarji, K. E. 1977. *Clin. Endocrinol.* 6:437–42
99. Fuster, V., Bowie, E. J. W., Lewis, J. C., Fass, D. N., Owen, C. A. Jr., Brown, A. L. 1978. *J. Clin. Invest.* 61:722–30
100. Almer, L.-O., Nilsson, I. M. 1975. *Acta Med. Scand.* 198:101–6
101. Almer, L.-O., Pandolfi, M. 1976. *Diabetes* 25: Suppl. 2, pp. 807–10
102. Brownlee, M. 1976. *Lancet* 1:779–80
103. Cameron, J. S., Ireland, J. T., Watkins, P. J. 1975. *Complications of Diabetes,* ed. H. Keen, J. Jarret, p. 120. London: Edward Arnold
104. Timperly, W. R., Ward, J. D., Preston, F. E., Duckworth, T., O'Malley, B. C. 1976. *Diabetologia* 12:237–43
105. Spiro, R. G. 1976. *Diabetologia* 12:1–14
106. Bar-On, H., Berry, E., Ziv, E. 1977. *Diabetologia* 13:380
107. Day, J. F., Ingebretsen, C. G., Ingebretsen, W. R. Jr., Baynes, J. W., Thorpe, S. R. 1980. *Diabetes* 29:524–27
108. Gralnick, H. R., Coller, B. S., Sultan, Y. 1976. *Science* 192:56–59
109. Saraswathi, S., Colman, R. W. 1975. *J. Biol. Chem.* 250:8111–18
110. Dice, J. F., Walker, C. D., Byrne, B., Cardiel, A. 1978. *Proc. Natl. Acad. Sci. USA* 75:2093–97
111. Dische, Z., Zil, H. 1951. *Am. J. Ophthalmol.* 34:104
112. Pirie, A. 1968. *Invest. Ophthalmol.* 7:634
113. Harding, J. J. 1973. *Exp. Eye Res.* 17:377
114. Dilley, K. J. 1975. *Exp. Eye Res.* 20:73
115. Truscott, R. J. W., Augusteyn, R. C. 1977. *Exp. Eye Res.* 25:139
116. Takemoto, L. J., Azari, P. 1977. *Exp. Eye Res.* 24:63
117. Benedek, G. B. 1971. *Appl. Opt.* 10:459
118. Jedziniak, J. A., Kinoshita, J. H., Yates, E. M., Hocker, L. O., Benedek, G. B. 1973. *Exp. Eye Res.* 15:185
119. Jedziniak, J. A., Kinoshita, J. H., Yates, E. M., Benedek, G. B. 1975. *Exp. Eye Res.* 20:367–74
120. Rasch, R. 1979. *Diabetologia* 17: 243–48
121. Reynolds, T. M. 1963. *Adv. Food Res.* 14:167–277
122. Van Heyningen, R. 1962. *Exp. Eye Res.* 1:396–403
123. Kinoshita, J. H., Merola, L. O., Dikmak, E. 1962. *Exp. Eye Res.* 1:405–10
124. Obazawa, H., Merola, L. O., Kinoshita, J. H. 1974. *Ophthalmol.* 13:204–9
125. Kinoshita, J. H. 1974. *Invest. Ophthalmol.* 13:713–24

126. Varma, S. D., Mikuni, I., Kinoshita, J. H. 1975. *Science* 195:1215–16
127. Dvornik, D., Gabbay, K. H., Kinoshita, J. H. 1973. *Science* 182:1146–48
128. Osterby, R. 1971. *Proc. Congr. Int. Diabetes Fed., 7th,* pp. 793–803
129. Osterby, R. 1973. *Vascular and Neurological Changes in Early Diabetes,* ed. R. A. Camerini-Davalos, H. S. Cole, pp. 323–40, New York: Academic
130. Mogensen, C. E. 1976. *Diabetes* 25:872–79
131. Schnider, S., Aronoff, S. L., Tchou, P., Miller, M., Bennett, P. H. 1977. *Diabetes* 26: Suppl. 1, p. 362
132. Mauer, S. M. 1978. *Diabetes* 27:959–64
133. Rasch, R. 1980. *Diabetologia* 18:413–16
133a. Viberti, G. C., Pickup, J. C., Jarret, J., Keen, H. 1979. *N. Engl. J. Med.* 300:638–41
134. Mogensen, C. E., Osterby, R., Gundersen, H. J. G. 1979. *Diabetologia* 17:71–76
135. Parving, H.-H., Rutili, F., Granath, K., Noer, I., Deckert, T., Lyngsoe, J., Lassen, N. A. 1979. *Diabetologia* 17:157–60
136. Hostetter, T. H., Fray, J. L., Brenner, B. M., Green, A. A. 1978. *Kidney Int.* 14:725
137. Scheinman, J. I., Fish, A. J., Michael, A. F. 1974. *J. Clin. Invest.* 54:1144–54
138. Scheinman, J. I., Steffes, M. W., Brown, D. M., Mauer, S. M. 1978. *Diabetes* 27:632–37
139. Deen, W. M., Robertson, C. R., Brenner, B. M. 1974. *Fed. Proc.* 33:14–20
140. Ryan, G. B., Karnovsky, M. J. 1976. *Kidney Int.* 9:36–45
141. Bennett, C. M., Glassock, R. J., Chang, R. L. S., Deen, W. M., Robertson, C. R., Brenner, B. M. 1976. *J. Clin. Invest.* 57:1287–94
142. Westberg, N. G. 1976. *Diabetes* 25: Suppl. 2, pp. 920–24
143. Cohn, R. A., Mauer, S. M., Barbosa, J., Michael, A. F. 1978. *Lab. Invest.* 39:13–16
144. Kroustrup, J. P., Gundersen, H. J. G., Surface, R. 1977. *Diabetologia* 13:207–10
145. Rasch, R. 1979. *Diabetologia* 16:125–28
146. Osterby, R., Gundersen, H. J. G. 1980. *Diabetologia* 18:493–500
147. Lee, C. S., Mauer, S. M., Brown, D. M., Sutherland, D. E. R., Michael, A. F., Najarian, J. S. 1974. *J. Exp. Med.* 139:793–800
148. Steffes, M. W., Brown, D. M., Basgen, J. M., Matas, A. J., Mauer, S. M. 1979. *Lab. Invest.* 41:116–18
149. Price, R. G., Spiro, R. G. 1977. *J. Biol. Chem.* 252:8597–602
150. Birkeland, A. J., Christensen, T. B. 1975. *J. Carbohydr.-Nucleosides-Nucleotides* 2:83–90
151. McVerry, B. A., Hopp, A., Fisher, C., Huehns, E. R. 1980. *Lancet* 1:738–40
152. Mauer, S. M., Steffes, M. W., Michael, A. F., Brown, D. M. 1976. *Diabetes* 25: Suppl. 2, pp. 850–57
153. Mauer, S. M., Steffes, M. W., Chern, M., Brown, D. M. 1979. *Lab. Invest.* 41:401–6
154. Romen, W., Morath, R. 1979. *Virchows Arch. B* 31:205–10
155. Beisswenger, P. J., Spiro, R. G. 1973. *Diabetes* 22:180–93
156. Sato, T., Spiro, R. G. 1976. *J. Biol. Chem.* 251:4062–70
157. Reid, K. B. M. 1974. *Biochem. J.* 141:189–203
158. Spiro, R. G. 1967b. *J. Biol. Chem.* 242:1923–32
159. Spiro, R. G. 1973. *Adv. Protein Chem.* 27:349–467
160. Spiro, R. G. 1973. *N. Engl. J. Med.* 288:1337–42
161. Hudson, B. G., Spiro, R. G. 1972. *J. Biol. Chem.* 247:4229–38
162. Hudson, B. G., Spiro, R. G. 1972. *J. Biol. Chem.* 247:4239–47
163. Cohen, M. P., Klein, C. V. 1979. *J. Exp. Med.* 149:623–31
164. Cohen, M. P., Klein, C. V. 1977. *Biochem. Biophys. Res. Commun.* 77:1326–1332
165. Freytag, J. W., Ohno, M., Hudson, B. G. 1976. *Biochem. Biophys. Res. Commun.* 72:796–802
166. Spiro, R. G. 1978. *Ann. NY Acad. Sci.* 312:106–21
167. Beisswenger, P. J., Spiro, R. G. 1970. *Science* 168:596–98
168. Kefalides, N. A. 1974. *J. Clin. Invest.* 53:403–7
169. Westberg, N. G., Michael, A. F. 1973. *Acta Med. Scand.* 194:39–47
170. Canivet, J., Cruz, A., Moreau-Lalande, H. 1979. *Metabolism* 28:1206–10
171. Beisswenger, P. J. 1973. *Diabetes* 22:744–50
172. Butcher, D., Kikkawa, R., Klein, L., Miller, M. 1977. *J. Lab. Clin. Med.* 89:544–53
173. Gundersen, H. J. G., Osterby, R. 1977. *Diabetologia* 13:43–48
174. Hoyer, J. R., Spiro, R. G. 1978. *Arch. Biochem. Biophys.* 185:496–503

175. Klein, L., Yoshida, M., Miller, M. 1977. *Diabetes* 26:361
176. Fox, C. J., Darby, S. C., Ireland, J. T., Sonksen, P. H. 1977. *Br. Med. J.* 2:605–7
177. Osterby, R., Seyer-Hansen, K., Gundersen, H. J. G., Lundbaek, K. 1978. *Diabetologia* 15:487–89
178. Seyer-Hansen, K. 1977. *Diabetologia* 13:141–43
179. Yagihashi, S., Goto, Y., Kakizaki, M., Kaseda, N. 1978. *Diabetologia* 15: 309–12
180. Rasch, R. 1979. *Diabetologia* 16: 319–24
181. Peterson, D. T., Greene, W. C., Reaven, G. M. 1971. *Diabetes* 20:649–54
182. Khalifa, A., Cohen, M. P. 1975. *Biochim. Biophys. Acta* 386:332–39
183. Cohen, M. P., Khalifa, A. 1977. *Biochim. Biophys. Acta* 496:88–94
184. Spiro, R. G., Spiro, M. J. 1971. *J. Biol. Chem.* 246:4910–18
185. Spiro, R. G., Spiro, M. J. 1971. *J. Biol. Chem.* 246:4899–909
186. Spiro, R. G., Spiro, M. J. 1971. *Diabetes* 20:641–48
187. Risteli, J., Koivisto, V. A., Akerblom, H. K., Kivirikko, K. 1976. *Diabetes* 25:1066–70
188. Guzdek, A., Sarnecka-Keller, M., Dubin, A. 1979. *Horm. Metab. Res.* 11:107–11
189. Beisswenger, P. J. 1976. *J. Clin. Invest.* 58:844–52
190. Cohen, M. P., Vogt, C. 1972. *Biochem. Biophys. Res. Commun.* 49:1542–46
191. Cohen, M. P., Vogt, C. A. 1975a. *Biochim. Biophys. Acta* 383:78–87
192. Grant, M. E., Harwood, R., Williams, I. F. 1975. *Eur. J. Biochem.* 54:531–40
193. Grant, M. E., Harwood, R., Williams, I. F. 1976. *J. Physiol.* 257:56–57
194. Cohen, M. P., Khalifa, A. 1977. *Biochim. Biophys. Acta* 500:395–404
195. Brownlee, M., Spiro, R. G. 1979. *Diabetes* 28:121–25
196. Spiro, R. G. 1971. *Joslin's Diabetes Mellitus* ed. A. Marble, P. White, R. F. Bradley, pp. 146–156. Philadelphia: Lea & Febiger
197. Rasio, E., Bendayan, M. 1978. *Diabete Metab.* 4:57
198. Sochor, M., Baquer, N. Z., McLean, P. 1979. *Biochem. Biophys. Res. Commun.* 86:32–39
199. Villee, D. B., Powers, M. L. 1977. *Nature* 268:156–58
200. Chang, A. Y. 1978. *Biochim. Biophys. Acta* 522:491–502
201. Fushimi, H., Tarui, S. 1976. *J. Biochem.* 79:265–70
202. Weil, R. III, Nozawa, M., Koss, M., Weber, C., Reemtsma, K., McIntosh, R. 1976. *Arch. Pathol. Lab. Med.* 100:37–49
203. Lindahl, U., Höök, M. 1978. *Ann. Rev. Biochem.* 47:385–417
204. Kanwar, Y. S., Farquhar, M. G. 1979. *Proc. Natl. Acad. Sci. USA* 76:1303–7
205. Kanwar, Y. S., Farquhar, M. G. 1979. *Proc. Natl. Acad. Sci. USA* 76:4493–97
206. Berenson, G. S., Ruiz, H., Dalheres, E. R., Dugan, F. A. 1970. *Diabetes* 19:161
207. Goldberg, M. F. 1974. *Diabetic Retinopathy*, ed. J. R. Lynn, W. B. Snyder, A. Vaiser, pp. 47–63. New York: Grune & Stratton
208. Kohner, E. M. 1976. *Diabetes* 25: Suppl. 2, pp. 839–44
209. Kohner, E. M., Oakley, N. W. 1975. *Metabolism* 24:1085–1102
210. Cunha-Vaz, J. G., Fonseca, J. R. 1978. *Arch. Ophthalmol.* 96:809–11
211. Cunha-Vaz, J. G. 1978. *Br. J. Ophthalmol.* 62:351–55
212. Kohner, E. M. 1977. *Clin. Endocrinol. Metab.* 6:345–75
213. Cunha-Vaz, J. G., Faria de Abreau, J. R., Campos, A. J. Figo, G. M. 1975. *Br. J. Ophthalmol.* 59:649–56
214. Cunha-Vaz, J. G., Fonseca, J. R., Abreu, J. F., Ruas, M. A. 1979. *Diabetes* 28:16–19
215. Waltman, S., Krupin, T., Hanish, S., Oestrich, C., Becker, B. 1978. *Arch. Ophthalmol.* 96:878–79
216. Scharp, D., Krupin, T., Waltman, S., Oestrich, C., Feldman, S., Ballinger, W., Becker, B. 1978. *Diabetes* 27:435
217. Wallow, I. H. L., Engerman, R. L. 1977. *Invest. Ophthalmol. Vision Sci.* 16:447–61
218. Osterby, R., Gundersen, H. J. G., Christensen, N. J. 1978. *Diabetes* 27:745–49
219. Buzney, S. M., Frank, R. N., Varma, S. D., Tanishima, T., Gabbay, K. H. 1977. *Invest. Ophthalmol. Vision Sci.* 16: 392–95
220. Raman, P. G., Jain, S. C. 1976. *Congr. Int. Diabetes Fed.* 9:124 (Abstr.)
221. Duane, T. D., Thomas, B., Field, R. A. 1969. *Symp. Treat. Diabetic Retinopathy* USPHS Publ. No. 1890, pp. 657–63. Arlington, Va: GPO
222. Steffes, M. W., Brown, D. M., Mauer, S. M. 1976. *Clin. Res.* 24:584A
223. Papachristodoulou, D., Heath, H. 1977. *Exp. Eye Res.* 25:371–84
224. Sosula, L. 1974. *Microvasc. Res.* 7:274–76
225. Sosula, L., Beaumont, P., Hollows, F. 1972. *Invest. Ophthalmol* 11:926–35

226. Diabetic Retinopathy Study Research Group. 1976. *Am. J. Ophthalmol.* 81:383–95

227. Navalesi, R., Pilo, A., Vigneri, R. 1975. *Diabetics* 24:317–27

228. Johansen, K., Hansen, A. P. 1969. *Br. Med. J.* 2:356–57

229. Lundbaek, K., Jensen, V. A., Olsen, T. S., Orskov, H., Christensen, N. J., Johansen, K., Hansen, A. P., Osterby, R. 1970. *Lancet* 2:131–33

230. Kjeldsen, H., Hansen, A. P., Lundbaek, K. 1975. *Diabetes* 24:977–82

231. Merimee, T. J. 1979. *Diabetes* 28: 308–12

232. Beaumont, P., Schofield, P. J., Hollows, F. C., Williams, J. F. 1971. *Lancet* 1:579–81

233. Passa, P., Gauville, C., Canivet, J. 1974. *Lancet* 2:72–74

234. Bradley, R. F., Ramos, E. 1971. See Ref. 196, pp. 498–508

235. Goldberg, M. F., Fine, S. L., eds. 1969. See Ref. 221, 913 pp.

236. Vigneri, R., Squatrito, S., Pezzino, V., Filetti, S., Branca, S., Polosa, P. 1976. *Diabetes* 25:167–72

237. Merimee, T. J., Hall, J. G., Rimoin, D. L., Fineberg, S. E., McKusick, V. A. 1969. *J. Clin. Invest.* 48:58a

238. Merimee, T. J., Siperstein, M. D., Hall, J. D., Fineberg, S. E. 1970. *J. Clin. Invest.* 49:2161–64

239. Merimee, T. J. 1978. *N. Engl. J. Med.* 298:1217–22

240. Rabin, D., Bloomgarden, Z., Feman, S., Davis, T. 1979. *Diabetes* 28:412

241. Phillips, L. S., Young, H. S. 1976. *Diabetes* 25:516–27

242. Yde, H. 1969. *Acta Med. Scand.* 186:293–97

243. Baxter, R. C., Brown, A. S., Turtle, J. R. 1979. *Horm. Metab. Res.* 11:216–20

244. Winter, R. J., Phillips, L. S., Klein, M. N., Traisman, H. S., Green, O. C. 1979. *Diabetes* 28:952–54

245. Phillips, L. S., Belosky, D. C., Young, H. S., Reichard, L. A. 1979. *Endocrinology* 104:1519–24

246. Phillips, L. S., Vassilopoulou-Sellin, R., Reichard, L. A. 1979. *Diabetes* 28: 919–24

247. Phillips, L. S., Belosky, D. C., Reichard, L. A. 1979. *Endocrinology* 104: 1513–18

248. Daughaday, W. H., Phillips, L. S., Mueller, M. C. 1976. *Endocrinology* 98:1214–19

249. Wardle, E. N. 1975. *Biomedicine* 23:299–302

250. Ledet, T. 1976. *Diabetes* 25:1011–17

251. Moorhouse, J. A. 1976. *Diabetes Mellitus,* ed. S. S. Fajans, pp. 243–55. Bethesda, Md: Nat. Inst. Health

252. Ward, J. D., Fisher, D. J., Barnes, C. G., Jessop, J. D. 1971. *Lancet* 1:428–30

253. Graf, R. J., Halter, J. B., Halar, E., Porte, D. 1979. *Ann. Int. Med.* 90:298–303

254. Greene, D. A., DeJesus, P. V. Jr., Winegrad, A. I. 1975. *J. Clin. Invest.* 55:1326–36

255. Gregersen, G. 1967. *Neurology* 17: 972–80

256. Graf, R., Halter, J., Pfeifer, M., Halar, E. 1979. *Diabetes* 28:387 (Abstr.)

257. Eliasson, S. G. 1969. *J. Neurol. Neurosurg. Psychiatr.* 32:525–29

258. McDonald, W. I. 1963. *Brain* 86: 501–24

259. Babel, J., Bischoff, A., Spoendlin, H. 1970. *Atlas of Normal and Pathologic Anatomy.* St. Louis, Mo: C. V. Mosby

260. Thomas, K. P. 1971. *Proc. R. Soc. Med.* 64:295–98

261. Clements, R. S. Jr. 1979. *Diabetes* 28:604–11

262. Chopra, J. S., Sawhney, B. B., Chakravorty, R. N. 1977. *J. Neur. Sci.* 32:53–67

263. Yagihashi, S., Kudo, K., Nishihira, M. 1979. *Tohoku J. Exp. Med.* 127:35–44

264. Fukuma, M., Carpentier, J. L., Orci, L., Greene, D. A., Winegrad, A. I. 1978. *Diabetologia* 15:65–72

265. Jakobsen, J. 1978. *Diabetologia* 14: 113–19

266. Seneviratne, K. N. 1972. *J. Neurol. Neurosurg. Psych.* 35:156–62

267. Jakobsen, J., Malmgren, L., Olsson, Y. 1978. *Exp. Neurol.* 60:277–85

268. Sima, A. A. F., Robertson, D. M. 1978. *Acta Neuropathologica* 44:189

269. Seneviratne, K. N., Weerasuriya, A. 1974. *J. Neurol. Neurosurg. Psychiatr.* 37:502–13

270. Hansen, S., Ballantyne, J. P. 1977. *J. Neurol. Neurosurg. Psychiatr.* 40: 555–64

271. Behse, F., Buchthal, F., Carlsen, F. 1977. *J. Neurol. Neurosurg. Psychiatr.* 40:1072–82

272. Sima, A. A. F., Robertson, D. M. 1979. *Lab. Invest.* 40:627–32

273. Yagihashi, S., Matsunaga, M. 1979. *Tohoku J. Exp. Med.* 129:357–66

274. Jakobsen, J. 1976. *Diabetologia* 12: 539–46

275. Jakobsen, J. 1976. *Diabetologia* 12: 547–53

276. Timperly, W. R., Williams, E., Ward, J. D., Preston, F. E. 1977. *Diabetologia* 13:437

277. Pratt, J. H., Berry, J. F., Kaye, B., Goetz, F. C. 1969. *Diabetes* 18:556–61
278. Spritz, N., Singh, H., Marinan, B. 1975. *Diabetes* 24:680–83
279. Field, R. A. 1966. *Diabetes* 15:696–98
280. Eliasson, S. G. 1966. *Lipids* 1:237–40
281. Palo, J., Reske-Nielsen, E., Riekkinen, P. 1977. *J. Neurol. Sci.* 33:171–78
282. Spritz, N., Singh, H., Marinan, B. 1975. *J. Clin. Invest.* 55:1049–56
283. Spritz, N., Singh, H., Silberlicht, I. 1977. *Diabetes* 26: Suppl. 1, p. 362
284. Poduslo, J. F., Quarles, R. H., Brady, R. O. 1976. *J. Biol. Chem.* 251:153–58
285. Palo, J., Savolainen, H., Haltia, M. 1972. *J. Neurol. Sci.* 16:193–99
286. Ludvigson, M. A., Sorenson, R. L. 1978. *Diabetes* 27:463
287. Gabbay, K. H. 1973. *N. Engl. J. Med.* 288:831–36
288. Heath, H., Hamlett, Y. C. 1976. *Diabetologia* 12:43–46
289. Winegrad, A. I., Morrison, A. D., Clements, R. S. 1973. *Vascular and Neurological Changes in Early Diabetes,* ed. R. A. Camerini-Davalos, H. S. Cole, pp. 117–24. New York: Academic
290. Malone, J. I., Knox, G., Benford, S. 1979. *Diabetes* 28:386
291. Gonzalez, A. M., Sochor, M., Hothersall, J. S., McLean, P. 1978. *Biochem. Biophys. Res. Commun* 84:858–64
292. Winegrad, A. I., Clements, R. S., Morrison, A. D. 1971. *Diabetes Mellitus: Diagnosis and Treatment,* ed. S. S. Fajans, K. E. Sussman, 3:269–73. New York: Am. Diabetes Assoc.
293. Ludvigson, M. A., Sorenson, R. L. 1980. *Diabetes* 29:438–449
294. Winegrad, A. I., Morrison, A. D., Clements, R. S. 1974. *Proc. VIII Congr. Int. Diabetes Fed., 8th,* pp. 387–95
295. Greene, D., Winegrad, A. 1979. *Diabetes* 28:388 (Abstr.)
296. Gabbay, K. H. 1975. *Ann. Rev. Med.* 26:521–36
297. Peterson, M. J., Sarges, R., Aldinger, C. E. 1979. *Diabetes* 28:367
298. Clements, R. S. Jr., Reynertson, R. 1977. *Diabetes* 26:215–21
299. Palmano, K. P., Whiting, P. H., Hawthorne, J. N. 1977. *Biochem. J.* 167:229–235
300. Clements, R. S. Jr., Estes, T., Stockard, R. 1978. *Clin. Res.* 26:789A
301. Ho, P. C., Feman, S. S., Stein, R. S., McKee, I. C. 1979. *Am. J. Ophthalmol.* 88:37–39
302. Whiting, P. H., Palmano, K. P., Hawthorne, J. N. 1979. *Biochem. J.* 179:549–53
303. Clements, R. S. Jr., Stockard, C. R. 1980. *Diabetes* 29:227–35
304. Jeffrys, J. G. R., Palmano, K. P., Sharma, A. K., Thomas, P. K. 1978. *J. Neurol. Neurosurg. Psychiatr.* 41:333–39
305. Clements, R. S. Jr., Reynertson, R. 1977. *Diabetes* 26:215–21
306. Clements, R. S. Jr., Vouranti, B., Juba, T., Oh, S. J., Darnell, B. 1979. *Metabolism* 28: Suppl. 1, pp. 477–83
307. Clements, R. S. Jr., Vourganti, B., Darnell, B., Oh, S. 1978. *Diabetes* 27:436 (Abstr.)
308. Gregersen, G., Borsting, H., Theil, P., Servo, C. 1978. *Acta Neurol. Scand.* 58:241–48
309. Schmidt, R. E., Matschinsky, F. M., Godfrey, D. A., Williams, A. D., McDougal, D. B. Jr. 1975. *Diabetes* 24:1081–85
310. Giachetti, A. 1979. *Diabetologia* 16:191–94
311. Jakobsen, J., Sidenius, P. 1979. *J. Neurochem.* 33:1055–60
312. Gordon, T., Garcia-Palmieri, M. R., Kagan, A., Kannel, W. B., Schiffman, J. 1974. *J. Chronic Dis.* 27:329–44
313. Simpson, R. W., Carter, R. D., Moore, R. A., Penfold, W. A. F. 1980. *Diabetologia* 18:35–40
314. Eisenberg, S. 1979. *Progr. Biochem. Pharmacol.* 15:139–65
315. New, M. I., Roberts, T. N., Bierman, E. L., Reader, G. G. 1963. *Diabetes* 12:208–12
316. Wolff, O. H., Salt, H. B. 1958. *Lancet* 1:707–10
317. Lewis, B., Mancini, M., Mattock, M. 1972. *Eur. J. Clin. Invest.* 2:445–53
318. Schonfeld, G., Birge, C., Miller, J. P., Kessler, G., Santiago, J. 1974. *Diabetes* 23:827–34
319. Howard, B. V., Savage, P. J., Bennion, L. J., Bennett, P. H. 1978. *Atherosclerosis* 30:153–62
320. Nikkila, E. A., Hormila, P. 1978. *Diabetes* 27:1078–86
321. Chase, H. P., Glasgow, A. M. 1976. *Am. J. Dis. Child.* 130:1113–17
322. Kaufmann, R. L., Assal, J. P., Soeldnev, J. S. 1975. *Diabetes* 24:672–79
323. Mattock, M. B., Fuller, J. H., Maude, P. S., Keen, H. 1979. *Atherosclerosis* 34:437–49
324. Wilson, D. E., Schreibman, P. M., Dug, V. C., Arky, R. A. 1970. *J. Chronic Dis.* 23:501–6
325. Nikkila, E. A., Kekki, M. 1973. *Metabolism* 22:1–22
326. Taskinen, M.-R., Nikkila, E. A. 1979. *Diabetologia* 17:351–56

327. Lopes-Virella, M. F. L., Stone, P. G., Colwell, J. A. 1977. *Diabetologia* 13:285–91
328. Bennion, L. J., Grundy, S. M. 1977. *N. Engl. J. Med.* 296:1365–71
329. Lopes-Virella, M. F. L., Wohltmann, H. J. 1979. *Diabetes* 28:348
330. Beach, K. W., Brunzell, J. D., Conquest, L. L., Strandness, D. E. 1979. *Diabetes* 28:836–40
331. Santen, R. J., Willis, P. W. J., Fajans, S. S. 1972. *Arch. Intern. Med.* 130:833–40
332. Chen, Y.-D. I., Kisser, T. R., Cully, M., Reaven, G. M. 1979. *Diabetes* 28: 893–98
333. Weiland, D., Mondon, C. E., Reaven, G. M. 1980. *Diabetologia* 18:335–40
334. Rissen, T. R., Reaven, G. M., Reaven, E. P. 1978. *Diabetes* 27:902–8
335. Nikkila, E. A., Huttunen, J. K., Ehnholm, C. 1977. *Diabetes* 26:11–21
336. Brunzell, J. D., Porte, D. Jr., Bierman, E. L. 1979. *Metabolism* 28:901–7
337. Taylor, K. G., Galton, D. J., Holdsworth, G. 1979. *Diabetologia* 16:313–17
338. Pykalisto, O. J., Smith, P. H., Brunjell, J. D. 1975. *J. Clin. Invest.* 56:1108–17
339. Goldstein, J. L., Brown, M. S. 1977. *Ann. Rev. Biochem.* 46:897–930
340. Nakayamu, H., Nakagawa, S. 1972. *Diabetes* 26:439
341. Palumbo, P. J., Zimmerman, B. R., Wellik, D. 1978. *Diabetes* 27:448
342. Saudek, C. D., Brach, E. L. 1978. *Diabetes* 27:1059–64
343. Chait, A., Bierman, E. L., Albens, J. J. 1979. *Diabetes* 28:914–18
343a. Nilsson-Ehle, P., Garfinkel, A. S., Schotz, M. C. 1980. *Ann. Rev. Biochem.* 49:667–93
344. Ross, R., Glomset, J. A. 1976. *N. Engl. J. Med.* 295:369–77
345. Ross, R., Glomset, J. A. 1976. *N. Engl. J. Med.* 295:420–25
346. Ross, R., Harker, L. 1976. *Science* 193:1094–1100
347. Ledet, T. 1977. *Diabetes* 26:798–803
348. Koschinsky, T., Bunting, C. E., Schwippert, B., Gries, F. A. 1979. *Atherosclerosis* 33:245–52
349. Hamlin, C. R., Kohn, R. R., Luschin, J. H. 1975. *Diabetes* 24:902–4
350. Williamson, J. R., Chang, K., Uitto, J., Rowold, E. A., Grant, G. A., Kilo, C. 1980. *Diabetes* 29: Suppl. 2, p. 12A (Abstr.)
351. Rosenberg, H., Modrak, J. B., Hassing, J. M., Al-Turk, W. A., Stohs, S. J. 1979. *Biochem. Biophys. Res. Commun.* 91:498–501
352. Kohn, R. R., Hense, S. 1977. *Biochem. Biophys. Res. Commun.* 76:765–71
353. Cohen, M. P., Foglia, V. G. 1970. *Diabetes* 19:639–43
354. Brosnan, M. E., Sirek, O. V., Sirek, A., Przybylska, K. 1973. *Diabetes* 22:397–402
355. Malathy, K., Kurup, P. A. 1972. *Diabetes* 21:1162–67
356. Silbert, C. K., Kleinman, H. K. 1979. *Diabetes* 28:61–64
357. Sirek, O. V., Sirek, A., Cukerman, E. 1980. *Diabetes* 29: Suppl. 2, p. 12A (Abstr.)
358. Wolinsky, H., Goldfischer, S., Capron, L., Capron, F., Coltoff-Schiller, B., Kasak, L. 1978. *Circ. Res.* 42:821–31
359. Henze, K., Chait, A. 1980. *Diabetes* 29: Suppl. 2, p. 11A (Abstr.)
360. Bloodworth, J. M. B., Engerman, R. L., Davis, M. D. 1971. *Proc. Congr. Int. Diabetes Fed., 7th,* pp. 804–19
361. Engerman, R. L., Davis, M. D., Bloodworth, J. M. B. 1971. *Proc. Congr. Int. Diabetes Fed., 7th,* 261–67
362. Bloodworth, J. M. B., Engerman, R. L. 1973. *Diabetes* 22: Suppl. 1, p. 290
363. Engerman, R. L., Bloodworth, J. M. B. 1973. *Congr. Int. Diabetes Fed., 8th,* p. 188 (Abstr.)
364. Engerman, R., Bloodworth, J. M. B. Jr., Nelson, S. 1977. *Diabetes* 26:760–69
365. Howard, C. F. Jr. 1979. *Atherosclerosis* 33:479–93
366. Kaplan, M. H., Feinstein, A. R. 1973. *Diabetes* 22:160–74
367. Knowles, H. C. 1964. *Trans. Am. Clin. Climatol. Assoc.* 76:142–47
368. Ricketts, H. T. 1965. *On the Nature and Treatment of Diabetes,* ed. B. S. Leibel, G. A. Wrenshall, pp. 588–600. Amsterdam: Excerpta Medica
369. Keiding, N. R., Root, H. F., Marble, A. 1952. *J. Am. Med. Assoc.* 150:964–69
370. Jackson, R., Guthrie, R., Esterly, J., Bilginturan, N., James, R., Yeast, J., Saathoff, J., Guthrie, D. 1975. *Diabetes* 24: Suppl. 1, p. 400
371. Downie, E. 1959. *Diabetes* 8:383–87
372. Knowles, H., Guest, G. M., Lampe, J. 1965. *Diabetes* 14:239–73
373. Bondy, P. K., Felig, P. 1971. *Med. Clin. N. Am.* 55:889–97
374. Johnsson, S. 1960. *Diabetes* 9:1–8
375. Ryan, J. R., Balodimos, M. C., Chazan, B. I., Root, H. F., Marble, A., White, P., Joslin, A. P. 1970. *Metabolism* 19:493–501
376. Oakley, W. G., Pyke, D. A., Tattersall, R. B., Watkins, P. J. 1974. *Q. J. Med.* 43:145–156
377. Deckert, T., Poulsen, J. E., Larsen, M. 1978. *Diabetologia* 14:363–70

378. Deckert, T., Poulsen, J. E., Larsen, M. 1978. *Diabetologia* 14:371–77
379. Goodkin, G. 1975. *J. Occup. Med.* 17:716–21
380. Kohner, E. M., Fraser, T. R., Joplin, G. F., Oakley, N. W. 1968. *Symp. Treat. Diabetes Retinopathy* pp. 119–128
381. Miki, E., Fukuda, M., Kuzuya, T., Kosaka, K., Nakao, K. 1969. *Diabetes* 18:773–80
382. Job, D., Eschwege, E., Guyot-Argenton, C., Aubry, J.-P., Tchobroutsky, G. 1976. *Diabetes* 25:463–69
383. Ashikaga, T., Borodic, G., Sims, E. A. H. 1978. *Diabetes* 27:592–96
384. Eschwege, E., Job, D., Guyot-Argenton, C., Aubry, J. P., Tchobroutsky, G. 1979. *Diabetologia* 16:13–15
385. Miki, E., Kuzuya, T., Ide, T., Nakao, K. 1972. *Lancet* 1:922–24
386. Takazakura, E., Nakamoto, Y., Hayakawa, H., Kawai, K., Muramoto, S., Yoshida, K., Shimizu, M., Shinoda, A., Takeuchi, J. 1975. *Diabetes* 24:1–9
387. Lauvaux, J. P., Vassart, G., Pirart, J. 1973. *Congr. Int. Diabetes Fed., 8th,* pp. 189–90 (Abstr.)
388. Pirart, J., Lauvaux, J. P., Vassart, G. 1973. *Congr. Int. Diabetes Fed., 8th,* p. 185 (Abstr.)
389. Pirart, J. 1978. *Diabetes Care* 1:168–88
390. Pirart, J. 1978. *Diabetes Care* 1:252–63
391. Skyler, J. S. 1980. *Diabetes Care* 3:57–186
392. Raskin, P. 1979. *Metabolism* 28:780–96
393. Santiago, J. V., Clemens, A. H., Clarke, W. L., Kipnes, D. M. 1979. *Diabetes* 28:71–84
394. Skyler, J. S. 1980. *Diabetes Care* 3:253–370
395. Brown, J. 1980. *Diabetes* 29: Suppl. 1, pp. 1–128

Ann. Rev. Biochem. 1981. 50:433–64

THE PROTEOLYTIC ACTIVATION SYSTEMS OF COMPLEMENT[1]

◆12084

Kenneth B. M. Reid and Rodney R. Porter

Medical Research Council Immunochemistry Unit, Department of
Biochemistry, University of Oxford, Oxford OX1 3QU, United Kingdom

CONTENTS

[1]Abbreviations used are: $\overline{C3b,Bb}$ and $\overline{C(3b)_n,Bb}$, C3 and C5 convertase of the alternative
pathway, which contain the major cleavage fragments C3b and Bb of components C3 and
factor B, respectively; $\overline{C4b,2a}$ and C4b,2a,3b, C3 and C5 convertase of the classical pathway,
which contain the major cleavage fragments C4b, C2a, and C3b of components C4, C2, and
C3, respectively; C3bINA, C3b inactivator; C1INH, C1 inhibitor; DFP, diisopropyl phosphofluoridate; E, erythrocyte; IgG, immunoglobulin G; NPGB, nitrophenyl guanidine benzoate; P, properdin; SDS, sodium dodecyl sulfate.

0066-4154/81/0701-0433$01.00

INTRODUCTION

The principal biological role of the complement system, found in all vertebrate blood, appears to be as an effector mechanism in the immune defense against infection by microorganisms. It is a complete mechanism in which activation products of the complement components cause lysis of cellular antigens, attract phagocytic cells to the place of activation, and facilitate uptake and destruction by the phagocytes.

The activation of complement is initiated by two pathways and is dependent primarily, though not entirely, on the sequential activation of proteolytic zymogens to proteases. In the classical pathway, initiation of activation of complement in blood is caused mainly by formation of antibody-antigen aggregates or by antibody bound to cellular or particulate antigens. This pathway may also be activated in vitro, independently of antibody, by RNA tumor viruses (1) such as murine leukemia virus (2), some gram-negative bacteria (3, 4), and a variety of acidic polymers such as polyinosinic acid (5), dextran sulfate (6), cellulose sulfate (7), and double-stranded DNA (8). In the leukemia virus a surface protein was identified as the complement receptor (9).

In the alternative pathway, activation may also be initiated by aggregated antibody; this can be shown with F(ab')$_2$-Ag complexes which do not activate the classical pathway (10, 11), and by antibody of immunoglobulin classes, such as IgA (12) and guinea pig γ_1 (10, 13), which fail to activate the classical pathway. However, the major role of the alternative pathway in vivo is probably due to its ability to be activated by polysaccharides such as are present on bacterial and yeast cell walls and hence cause their destruction. A simplified view is that the alternative pathway of complement activation forms a first line of defense against infection prior to an immune response and that the classical pathway becomes quantitatively more important after antibody production has been stimulated, though both pathways may be activated simultaneously by antibody aggregates (14) and the true position is likely to be more complex. There is, for example, indirect evidence that activation products have a role in the lymphocyte interactions essential to the immune response. The presence of genes coding for some components of both pathways in the major histocompatibility locus suggests that they may also have a role on cell surfaces, and there is now evidence to support this. Much remains to be resolved of the relative roles and the biological significance of the two pathways of activation of complement.

Figure 1 is a diagram of the activation scheme of both pathways of complement. In the classical pathway the essential features are three complex proteases:

1. $C\bar{1}$, formed from C1q, $C\bar{1}r$, and $C\bar{1}s$; the latter two are typical serine proteases formed from zymogens C1r and C1s on activation;
2. C3 convertase, a complex of equimolar amounts of $C\bar{4}b$ and $C\bar{2}a$ formed after activation of C4 and C2 by $C\bar{1}$;
3. C5 convertase, formed from $C\bar{4}b$ and $C\bar{2}a$ together with the product of the C3 activation, $C\bar{3}b$.

In the alternative pathway there is still some uncertainty as to the mechanism of initiation or stimulation of the activation, as discussed later. It is clear that $\bar{D}$ is a serine-type protease and catalyzes the conversion of B together with C3b to form a C3 convertase $\overline{C3b,Bb}$. The product of C3 convertase, additional C3-b, then associates with $\overline{C3b,Bb}$ to form $(\overline{C3b})_n\overline{Bb}$, a C5 convertase. Thus in each of the two pathways there are two complex proteases, C3 convertase and C5 convertase, with equivalent specificity and function but different composition.

In both pathways the activation of C5 is the last proteolytic step, and the association of $C\bar{5}$ with C6, C7, C8, and C9 to form a lytic complex appears to be spontaneous. There is no evidence of proteolysis of the last four complement components on activation or during the lytic step itself.

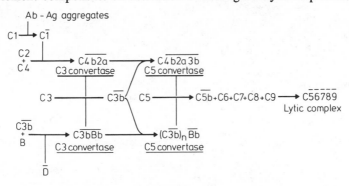

Figure 1   A simplified scheme of complement activation showing present views on the formation of the C3 and C5 convertases formed by both pathways of activation and the subsequent formation of the lytic complex. The method of initiation of the activation of the alternative pathway is omitted as the details are still not clear (19).

Not shown in the diagram are the proteins which control the complement pathway. Identified so far are the C$\overline{1}$ inhibitor (C1INH), which displaces and inhibits C$\overline{1}$r and C$\overline{1}$s in the C$\overline{1}$ complex, and C3b inactivator (C3bINA), a protease which in association with another serum protein, β1H, will hydrolyze a bond in C3b. C3bINA will also hydrolyze two bonds in C4b in association with a further protein, C4b-binding protein (C4bp). Another control protease is the anaphylatoxin inactivator, which acts similarly to carboxypeptidase B in removing the C-terminal arginine from the activation peptides C3a and C5a, thus destroying their biological activity. These peptides are released from the N-terminal end of the α chains of C3 and C5, respectively, when they are activated by their appropriate convertase (15).

Thus all the proteases in the complement system appear to be complex, requiring the association of two or more proteins for activity, with the exception of the anaphylotoxin inactivator. The proteases of the ongoing pathway are serine proteases, and their chain structures are compared in Figure 2. This article is concerned primarily with recent evidence on the nature of the proteolytic enzymes of the complement system, the structure

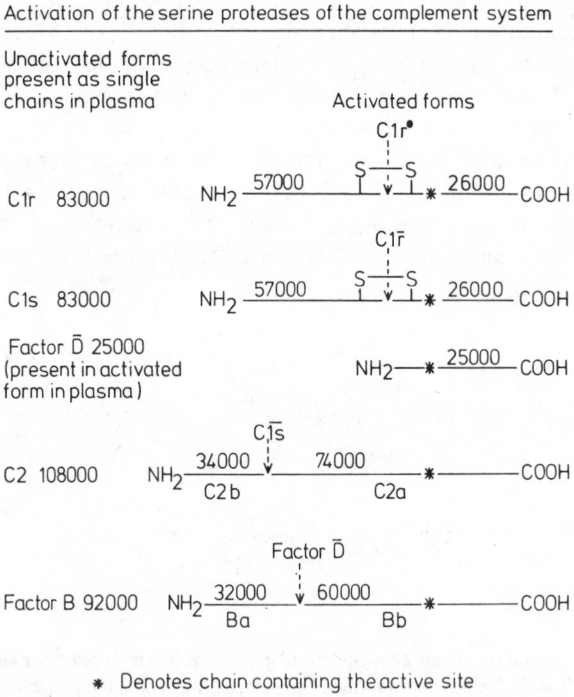

Figure 2 Activation of the serine proteases of the complement system. C1r* denotes the single-chain proenzyme form of C1r which has enzymatic activity (51).

of the component proteins, and the mechanism of their activation. Perhaps the most unexpected recent finding in the activation scheme is the rapidly accumulating evidence that C3 and C4 when activated form covalent bonds with adjacent polysaccharides or proteins.

Of general biological interest is the occurrence in the complement system of three proteins with structures not, so far, found elsewhere. These are C1q, which has a mixed half collagen fibril – half globular structure, and C2 and factor B, which appear to be novel types of serine protease with the typical catalytic site but an additional N-terminal 300 residues on the catalytic peptide chain and probably a different activation mechanism. C3b inactivator is another protease which does not appear to conform to known structural patterns. These apparently novel structures suggest that the complement system may have had an unusual evolutionary history.

Some aspects of the complex proteases of complement have been described recently (16) and more general reviews have also appeared (17–21).

CLASSICAL PATHWAY OF ACTIVATION

The pathway up to activation of C5 is shown in Figure 3 and includes seven proteins with C5.

Structure of Components

Evidence on the structure of these proteins has been reviewed recently (20) and is summarized only briefly here.

C1 (C1q, C1r, AND C1s)

Subcomponent C1q   C1q can be isolated by a variety of methods, but direct binding to IgG coupled to Sepharose, or to immune aggregates and subsequent elution at high salt concentrations (22, 23) appear to be the most

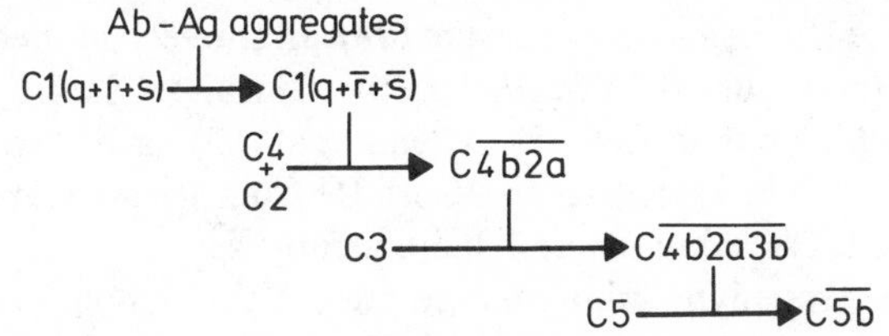

Figure 3   Activation of the early components of complement. The first component of complement (C1) binds to antibody-antigen (Ab-Ag) aggregates or antibody bound to cells and is activated through its subcomponents C1q, C1r, and C1s. Activated subcomponent C̅1̅s is a proteolytic enzyme which hydrolyzes components C2 and C4, yielding the enzyme complex of C4b,C2a comprising the major cleavage fragments of each component. The C̅4̅b̅,̅2̅a̅ complex, via its active site in the C2a portion, converts C3 into an activated form that associates with C̅4̅b̅,̅2̅a̅ to give a C̅4̅b̅,̅2̅a̅,̅3̅b̅ complex that will activate C5.

effective. It has a molecular weight of about 400,000 and is formed from 18 peptide chains of three types, 6A, 6B, and 6C, chains all of which are similar in size and structure. A unique feature is that each chain has about 80 residues near the N terminus which are of collagen sequence with the repeating triplet X–Y–Gly and with the disaccharide glucosyl-galactosyl substituted on many of the hydroxylysine residues in this sequence (24). In two chains the repeating triplet is broken at about the midpoint of the collagen sequence, by the change of an alanine for a glycine residue in the C chain, insertion of an additional threonine in the A chain; there is an additional triplet in the B chain. The presence of these collagen-like sequences led to the suggestion that the 18 peptide chains were associated in 6 collagen-like fibrils of 3 chains in a helix formation but with a bend about halfway corresponding to the breaks in the triplet sequences. The C-terminal 110 residues, which follow the collagen-like section in each chain, show no unusual features though they are very similar in sequence in the A, B, and C chains. The polysaccharide in this section is bound through an asparagine residue (25). The C-terminal sections are assumed to take up a globular-like structure. The suggested structure of C1q (20) is in agreement with the size and shape of the molecule seen in the electron microscope. C1q has also been studied by neutron diffraction where there is less likelihood of distortion. It was concluded that C1q molecules in solution have an open structure, with the angle of inclination of the stalks at least 60°, and it appeared that all molecules have a unique conformation (26).

Subcomponents C1r and C1s   As with C1q, two methods have been used to purify from human serum C1r, C1s, and the active enzymes $\overline{C1r}$ and $\overline{C1s}$. The first method is to precipitate C1q, C1r, and C1s as euglobulin from diluted or dialyzed serum and subsequently separate the components by ion exchange chromatography on DEAE cellulose, as first described by Lepow et al (27) and later modified in a number of laboratories. These methods are similarly successful using bovine serum (28). In the second method, the C1 is bound to IgG coupled to Sephadex; the C1r and C1s are eluted with EDTA containing buffers (29–35). Arlaud et al (23) have used ovalbumin-rabbit antiovalbumin aggregates to absorb C1 from serum; C1r and C1s are eluted with EDTA solutions and subsequently separated on DEAE cellulose. Unless proteolytic inhibitors are present throughout, activation occurs and the activated enzymes $\overline{C1r}$ and $\overline{C1s}$ are obtained. If DFP or NPGB are added there is little activation and the proenzymes are obtained.

C1r and C1s are both single polypeptide chain proteins of about 85,000 molecular weight that on activation give two disulfide-bonded chains of about 27,000 and 56,000 molecular weight (Figure 2). They appear to be typical serine-type protease with the active site in the smaller b chain.

Radioactive DFP reacts with the b chain at one molecule per mole of enzyme, and the N-terminal sequence of the b chain in both C̄1r and C̄1s shows strong homology to those of other serine proteases (36). Much more structural evidence will be necessary to explain the marked difference in specificity of C̄1r and C̄1s. The only effective substrate of the former appears to be C1s, but C1s will hydrolyze bonds in C4, C2, and a variety of peptide substrates showing a tryptic-like specificity.

COMPONENT C2   Only limited structural data is available on C2, which is present at about 15 mg/liter of serum and is particularly vulnerable to proteolysis during isolation. It can however be isolated in 15–20% yield with appropriate precautions (37, 38). It is a single-chain glycoprotein of molecular weight 100,000–110,000 (39, 40), which on activation by C̄1s is split into nondisulfide-bonded chains of about 70,000 (C2a) and 30,000 (C2b) molecular weight (Figure 2). Short N-terminal sequences reported for C2a and C2b show that C2b is derived from the N-terminal section and C2a from the C-terminal section of C2, respectively, but no homology with serine proteases is apparent (38). There has been conflicting evidence as to whether C2 can be inactivated by DFP (39–41). However, recent sequence data on factor B (42) suggest that this protein, which has the equivalent role to C2 in the alternative pathway C3 and C5 convertases and is homologous to C2 in gross structure, is a novel type of serine protease with a typical catalytic site but different activation mechanism. It seems likely that C2 will be similar to factor B with respect to its activation and catalytic site.

COMPONENT C4   Component C4 of approximately 200,000 molecular weight is composed of three disulfide-linked, polypeptide chains α, β, and γ with apparent molecular weights of 93,000, 78,000, and 30,000, respectively (43). The intact molecule has been isolated in good yield (44, 45) and separation of the chains after reduction and alkylation of disulfide bonds has been achieved by gel filtration or chromatography on calcium phosphate columns in the presence of SDS (44, 45). The N-terminal amino acid sequence of the α chain of C4 shows some similarity to the N-terminal sequences of the α chains of C3 and C5, which may reflect a common origin (for review see 20). The N-terminal regions of the β and γ chains show heterogeneity at various positions that may be related to the C4 variants found by electrophoretic techniques (46).

Biosynthetic studies have indicated that C4 is synthesized as a single chain which must be hydrolyzed subsequently to give the three-chain structure (47). The finding of a single-chain form of C4 in plasma at about 1–3% of the level of the three-chain form (48, 49) is consistent with the biosynthetic studies, but it is not clear where the conversion from the single-chain to multichain form takes place. The α, β, and γ chains of C4 appear to be aligned in the order β–α–γ in the single-chain pro-C4 molecule (50, 52, 53) as judged by comparison of the N-terminal amino acid sequences of the intact pro-C4, the β chain of C4, and the cleavage of the pro-C4 with $\overline{\text{C1s}}$.

In the formation of the classical pathway $\overline{\text{C14bC2a}}$ complex, C4 is split by the $\overline{\text{C1s}}$, in the $\overline{\text{C1}}$ complex, at probably only one point in its α chain, which releases the small C4a fragment which is probably composed of the N-terminal 66 residues of the α chain (54), by analogy with the release of C3a and C5a during C3 and C5 activation. The remaining large C4b fragment can be bound strongly, probably covalently, via its α' chain to cell surfaces or aggregated IgG antibody, as is discussed below. The attachment sites on cell surfaces have not been characterized, but the site on the IgG molecule has been shown to be in the Fd fragment—the N-terminal half of the heavy chain (55).

COMPONENTS C3 AND C5   These proteins have very similar gross structures of $\sim$ 185,000 molecular weight, and are formed from two disulfide-bonded chains—α, and β, of 115,000 and 70,000 molecular weight, respectively. As they have similar physical properties they are isolated together until the final stages of the preparation; the most successful method is that of Tack & Prahl (56, 57).

When C3 and C5 are activated by C3 and C5 convertases, respectively, each loses N-terminal peptides C3a and C5a from the α chain (about 70 residues) with obvious homology of sequence. C5a however has a large polysaccharide content (58) while C3a has none. The C3a and C5a peptides have been sequenced fully from several species (15) and the N- and C-terminal sequences reported for the C3α' and β chains and the C5α chain (57, 59). This work has been reviewed recently by Porter & Reid (20).

C3, as C4, is synthesized as a single peptide chain (60) with the β chain identified as N-terminal in the precursor molecule and followed by the α chain sequence (51).

CONTROL PROTEINS ($\overline{\text{C1}}$ INHIBITOR, C4b BINDING PROTEIN) CONCERNED WITH REGULATION OF $\overline{\text{C1}}$ AND $\overline{\text{C4B}}$ At least three proteins, $\overline{\text{C1}}$ inhibitor, C4b-binding protein, and C3bINA are concerned with the control of the activated protease complexes of the classical pathway. $\overline{\text{C1}}$ inhibitor binds very tightly to the activated forms of C1r and C1s, while C4b binding protein acts as a cofactor in the inactivation of C4b by the enzyme C3bINA. The characteristics and function of C3bINA are more fully described in the section concerned with the control of the alternative pathway.

CĪ inhibitor   CĪ inhibitor is a single-chain glycoprotein of ~ 100,000 molecular weight that has an unusually high content (40%) of carbohydrate (61, 62). It forms a very tight stoichiometric complex with CĪs, CĪr, and plasmin (62–64), which can be demonstrated on polyacrylamide gels after treatment with urea-SDS. No splitting of peptide bonds takes place in the CĪs, CĪr, or CĪ inhibitor on complex formation and in all cases the CĪ inhibitor is bound to the catalytic, light chain of the enzyme. Studies on the interaction of CĪ inhibitor with the CĪ complex bound to immune aggregates have shown that the inhibitor binds rapidly to CĪs and more slowly to CĪr. Once the interaction between CĪr and CĪ inhibitor has taken place there is a rapid dissociation of the CĪr-CĪs complex from the immune aggregates containing CĪ, and a complex of the composition CĪ inhibitor$_2$-CĪr-CĪs can be observed in solution (63–65). Although the CĪr-CĪs complex can be removed almost completely from immune aggregates containing bound CĪ, most of the C1q remains bound to the aggregates and therefore may be controlled by a separate protein, such as the C1q inhibitor described by Conradie et al (66), in order to prevent it from causing the further consumption of C1r and C1s.

C4b-binding protein   Human C4b-binding protein is a 10.7 S glycoprotein that serves as a cofactor in the splitting of C4b by C3bINA (67, 68). The C4b-binding protein appears functionally identical to the high-molecular-weight C3b-4b INA cofactor, described by Stroud and co-workers (69, 70). Human C4b-binding protein has an apparent molecular weight of 540,000–590,000 on SDS polyacrylamide gel electrophoresis in nonreducing conditions, and yields chains of 70,000–80,000 apparent molecular weight on the reduction of the disulfide bonds. Two forms of human C4b-binding protein which differ slightly in net charge and apparent molecular weight have been described by Fujita & Nussenzweig (68). Both these forms were found to behave as cofactors in the splitting of C4b or C3b, by C3bINA in solution. Thus C4b-binding protein is considered to act in a similar manner to β1H with respect to C3b inactivation by C3bINA. However, the cofactor activity of β1H is approximately 20 times greater than that of C4b-binding protein, on a weight basis, with respect to C3b inactivation by C3bINA in solution, and when the inactivation of cell-bound C3b was examined, the difference between the abilities of C4b-binding protein and β1H to serve as cofactors of the inactivation of C3b by C3bINA was more marked (68). Therefore it is probable that C4b-binding protein is utilized under physiological conditions primarily as a cofactor of C4b inactivation, while β1H is the cofactor utilized during C3b inactivation by C3bINA. Thus C4b-binding protein would be concerned with the control of the $C\overline{4b,2a}$ and $C\overline{4b,2a,3b}$ complex proteases of the classical pathway, while $\beta 1H$ could be involved with the control of proteases from either pathway which contains C3b.

Complex Proteases

C3 CONVERTASE C4b,2a   C3 convertase is a complex of activated $C\overline{4}$ and $C\overline{2}$ with the proteolytic site in the C2a portion of the activated $C\overline{2}$. The C4 when activated by $C\overline{1}$ forms a strong, probably covalent bond with antibody molecules or cell surfaces, and C2 when activated also by $C\overline{1}$ on or in close proximity to the bound C4b forms the C3 convertase $C\overline{4b,2a}$ in the presence of Mg^{2+}. The activated C2a dissociates spontaneously with loss of the C3 convertase activity. The convertase may be formed in solution by adding $C\overline{1}s$ to a mixture of C4 and C2 (71). C4 may be replaced by C4b but C2 cannot be replaced by C2 previously activated by $C\overline{1}s$. The C2 fragments formed—C2a containing the proteolytic site, 70,000 molecular weight, and C2b, 30,000 molecular weight—dissociate from the C4b and from each other and cannot reassociate. It is probable that whole C2 associates with C4b and is activated by $C\overline{1}s$ in the complex (72); the split products have a lower affinity for C4b than the whole C2 molecule. As only the complex is active, it is probable that the substrate C3 interacts with both components; or possibly a conformational change in C2 on activation, which causes exposure of the active site in C2a, is labile, and further change leads to dissociation from C4b and loss of affinity for C3.

C2 from human serum gives a stable C3 convertase on activation with C4 and $C\overline{1}s$ if it is first treated with weak iodine solution (73), possibly by oxidizing SH to S–S groups (39). With such modified C2 an apparent molecular weight of the complex was found by sedimentation in sucrose gradient, which suggests that it was formed from an equimolar complex of C4b and C2a (39). The presence of an equimolar complex was shown clearly using gel filtration (74).

Kerr (74) has reinvestigated the structure of the $C\overline{4b,2a}$ complex and has shown that it has a decay constant of 2.0 per minute at 37°C. Because of this, maximum C3 convertase activity is only achieved at 22°C with equimolar amounts of $C\overline{1}s$, C4b, and C2 in the presence of excess C3. Under these conditions, the rate of formation of C3 convertase is at a maximum and decay over short time intervals is minimal, and it was confirmed kinetically that the convertase is formed from equimolar amounts of C4b and C2. Using oxidized C2, gel filtration experiments showed that C2a must be the catalytic unit, as C2b dissociates from the active enzyme without loss of activity. The iodine oxidation presumably stabilized groups in C2a interacting with C4b, though there was some evidence that the initial interaction between C2 and C4b before activation of C2 was between the C2b part of C2 and C4b (74).

As discussed in section on component C2; (see also 42), the catalytic site of the C3 convertase in C2a appears by analogy with factor B to be that of a serine protease. The activation mechanism of C2 and B must differ from that of other serine protease zymogens, as C2 and B have an additional 300 amino acid residues that are N-terminal compared to the catalytic chains of all other serine proteases. It is presumably this section of C2a that is interacting with C4b and giving the complex its unique property as a C3 convertase, but little is known about this interaction or about the essential role of Mg^{2+} in the formation of the complex.

C5 CONVERTASE $\overline{\text{C4b,2a,3b}}$ The binding of one or more activated C3 molecules to, or adjacent to, the activating enzyme $\overline{\text{C4b2a}}$ changes its specificity from a C3 convertase to a C5 convertase (75). The change in specificity is apparently not great, as C3 and C5 have very similar structures, and in each case an N-terminal peptide with C-terminal arginine is split off from the α chain.

The mechanism of the change of specificity from C3 to C5 hydrolysis by interaction of the $\overline{\text{C4b,2a}}$ enzyme with its product C3b is far from clear. Using a mixture of $\overline{\text{C4b,2a}}^{\text{oxy}}$ with C3, C5 convertase activity in solution has been reported briefly for the classical pathway enzyme (76). Isenman et al (77) showed that in the presence of high concentrations (3.8×10^{-5} M) of C3b, C5 has a weak affinity for C3b in isotonic solutions. Effective C5 convertase activity could be found for the convertase of either pathway only when C3b was bound to a cell surface, or when other receptors such as agarose was present in addition to the C3 convertase (78). The C3 convertase enzyme could be either in solution or bound to a cell surface. It was suggested that C5 becomes susceptible to the $\overline{\text{C4b,2a}}$ (or $\overline{\text{C3b,Bb}}$) proteolytic activity only after interacting with the bound C3b. This interaction appears to be inhibited specifically by propamidine (79). C3b in solution could not replace bound C3b, possibly because the two molecules differ in their conformation. These results suggest that the C3 and C5 convertases are identical, i.e. both are $\overline{\text{C4b,2a}}$, and that there are two substrates, either C3 alone or C5 after interaction with bound C3b. The most efficient interaction between the bound C3b and C5 occurs when both are from human serum. Heterologous proteins are less effective (80). This model has some support in that addition of C5 reduced the rate of C3 cleavage by $\overline{\text{C4b,2a}}$ either cell bound or in solution. C5 itself was not hydrolyzed by a convertase in solution but was a competitive inhibitor (81), which suggests that it has an affinity for the $\overline{\text{C4b2a}}$ enzyme but is not in the correct conformation to permit hydrolysis.

The difficulties of investigating a complex of three large proteins and an equally large substrate are obvious, but if the C5 convertase could be assembled on aggregated antibody, the three-dimensional structure of which is available, it may be possible to get more information about the mechanism of the proteolysis.

Covalent Bond Formation by Complement Components

Müller-Eberhard and colleagues found that activation of C3 (82) and of C4 (83) by C3 convertase and C$\overline{1}$, respectively, led also to their inactivation. They postulated that these proteases convert C3 and C4 into activated forms able to bind to a red cell surface, and in the case of C4 also to IgG (83, 84). In both cases the activated forms had only short half-lives, and molecules not reacting with cell surfaces or IgG became inactive; these molecules are now named C3b and C4b.

The chemical nature of activated C3 and C4 and the mechanism of their binding remained unclear until Law & Levine (85) reported that C3 when activated by EAC$\overline{14b2a}$ forms a stable, probably covalent, bond with components on the red cell surface. Similarly, if the alternative pathway is activated by the addition of zymosan to serum, C3 is bound to the zymosan particles again, probably covalently. The evidence for covalent binding was that, using ^{125}I-labeled C3b bound to the red cell surface, radioactive bands moving more slowly than the C3 α' and β chains could be seen on acrylamide gels after reduction, alkylation, and electrophoresis in SDS containing buffers. This suggested that the C3 was forming a bond, stable to reduction and denaturing conditions, with components of the cell membrane. This was confirmed by double labeling, of the cell membrane with ^{125}I and the C3 with ^{131}I. The C3 bound to zymosan particles could be removed in part by SDS containing buffers but in full only be a combination of hydroxylamine and SDS extraction. Hydroxylamine also eluted most of the C3 from the red cell membrane. It was suggested that C3 binds by hydrophobic interaction and also by a covalent bond split by hydroxylamine. Further investigation of the stability of zymosan-bound C3b to hydroxylamine under neutral and alkaline conditions suggested that the covalent bond is probably an ester bond, not a thioester, tyrosyl ester, or acylimidazole (86).

The reactive group in C3 was localized to the C3d fragment produced by treatment of the cell-bound C3, with C3INA and β1H followed by trypsin. C3d, which is an ~ 30,000-molecular weight portion of the α chain, remained bound to the cell surface, while the other major fragment, C3c, could be eluted. C3d could be eluted after incubation with hydroxylamine (87) and was found to contain 0.66 mol of hydroxymate per mole of C3d, which suggests that the acyl group forming an ester bond is in C3d and that the hydroxyl group in the cell surface reactant is probably in a sugar residue. This conclusion is supported by experiments that showed strong binding of C3 to Sepharose (88). A model system was developed by coupling trypsin to Sepharose 4B after activation by CNBr. Sepharose alone would not bind either C3 or C3b and therefore strong binding to trypsin-Sepharose is likely to be by the C3 activated by trypsin in a manner probably similar to activation by C3 convertase. The binding could be inhibited by sugars and by IgG but in the latter case the inhibition was proportional to the hexose content and the inhibitory activity survived pronase digestion and partial purification of the glycopeptide. The results are in agreement with the conclusion that the activated C3 reacts with a sugar residue.

The trypsin-Sepharose system has been used to investigate further the characteristics of the reactive group formed on C3 during activation (89). A maximum of 26–28 mol of C3 were bound per mole of trypsin, and a half-life for the reactive group was calculated to be ~ 60 μsec. Hydroxamic acids, hydrazines, and hydroxylamine caused 50% inhibition of binding of C3 at about 1mM or lower concentration, while about 50-fold higher concentrations of amines and sugars were needed. Using radiolabeled compounds, efficient incorporation into C3 during activation was demonstrated for glycerol, glucosamine, methylamine, and phenylhydrazine. In a more extended study of inhibitors of C3 binding, Twose et al (90) showed a good correlation between the inhibitory potency of a range of compounds in the trypsin-Sepharose system and in hemolytic assays of whole complement, which emphasizes that the primary effect of the inhibitor is on the C3 binding step and hence confirms the validity of the model system being used. Inhibitory power is proportional to nucleophilicity, and among the most effective inhibitors is a range of hydroxamic acids. Hydroxamic acids in general appear to be specific in that they have no inhibitory effect on blood clotting or on transglutaminase.

C4 when activated on EAC1 cells also forms a strong but hydroxylamine-labile bond with the red cell surface, which suggests that when C4 is activated by C$\overline{1}$ a reactive group appears which similarly forms an ester bond with a saccharide on the cell surface (91). C4 can on activation dimerize and interact with IgG (83). From a comparison of the binding of C4 and development of the C3 convertase activity in EAC1 cells and on antibody antigen aggregates to which C$\overline{1}$ was bound, Goers & Porter (92) found that for each 1000 molecules of antibody some 20-fold more C4b bound to EAC$\overline{1}$ cells than to aggregates. This is as expected, as most of the activated C4 binds to the cell surface, probably by ester bonds (see above). However, the C3 convertase activities developed in excess C2 were similar with cells and aggregates, which probably means that only C4 bound to antibody can interact with activated C$\overline{2}$ to form the C3 convertase. The activating enzyme $C\bar{1}$ is fixed to the Fc portion of the antibody, and as C2 has a short half-life it may be able to form C3 convertase only after prior binding to C4b molecules close to the site of activation. Indirect evidence suggests that C4b is bound to the Fab part of the antibody. Further work showed that some of the C4b is bound covalently to the Fd section of the heavy chain (55). The reactive group is in the C4d section of the α chain of C4 (R. D. Campbell, unpublished), which is comparable to the C3d fragment in the α chain of C3. Important differences are that the C4 binding to the antibody aggregates is inhibited by diamines and by salicyl hydroxamates which inhibit C3 binding. The bond formed between C4b and IgG also seems significantly more stable to hydroxylamine and high pH than the C3b-polysaccharide bond, and it was suggested that it may be an amide bond with an ϵNH_2 group of lysine. Such a bond between a $\gamma COOH$ group of glutamic acid and an ϵNH_2 group of lysine has been shown to form in fibrin clots catalyzed by transglutaminase (93). A similar reaction cannot be occurring between C4b and IgG, as serum transglutaminase is a thiol-dependent enzyme and the C4b-IgG interaction is not inhibited by 20 mM iodoacetamide (55). More probably a reactive acyl group is released when C4 is activated, as suggested by Law et al (91), but in this case it is reacting with an amino acid residue, perhaps lysine, in the N-terminal half of the heavy chain of IgG and not with a sugar residue, as is suggested in experiments with red cell membrane. An apparently similar covalent interaction between C3 and Fab occurs when $F(ab')_2$-antigen aggregates activate the alternative pathway either in serum or with purified components (94). Judging by reactivity with low-molecular-weight compounds with activated C3, both amide and ester bonds would be expected to form (89). Covalent interactions between several proteins and cell surfaces have been suggested recently, e.g. epidermal growth factor (95), thrombin (96), and α_2-macroglobulin (97). In the latter case there was evidence to suggest that the formation of the covalent bond is catalyzed by transglutaminase. Other work suggests that there may be a potentially reactive acyl group in α_2-macroglobulin, as methylamine can bind covalently to a glutamyl or glutaminyl residue (98), and α_2-macroglobulin can form covalent bonds with proteases (99, 100).

ORIGIN OF THE POSTULATED REACTIVE ACYL GROUP FORMED ON ACTIVATION OF C3, C4, AND α_2-MACROGLOBULIN   The hemolytic activity of C3 and C4 and the interaction of α_2-macroglobulin with proteases are unusually sensitive to amines and ammonium salts and to strong nucleophiles such as hydrazine and hydroxylamine. C3 (101–103), C4 (103), and α_2-macroglobulin (104, 105) but not C5 (103) show the unusual property of peptide bond breakage on denaturation by SDS, guanidine, or heat. There is convincing evidence (101–105) that this is not due to contamination with proteases but that it is an autolytic process arising from an unusual structure in the α chain of C3 and C4 and in the single peptide chain of α₂-macroglobulin. Tack & colleagues (101, 102) have also shown that there is a simultaneous appearance of a reactive thiol group when C3 and C4 are denatured and when they are converted enzymically to C3b and C4b, respectively. When α₂-macroglobulin is inactivated by methylamine, some methylamine becomes incorporated into a residue in the peptide chain to give γ glutamylmethylamide, and the adjacent sequence has been established (Figure 4). When the α₂-macroglobulin is split by heat denaturation the split occurs between a glutamic acid residue and the residue reacting with methylamine, which shows that the peptide bond cleavage by denaturation and the release of a reactive acyl group arise from the same structure (105). When the peptide containing the thiol group released on denaturation of C3 was sequenced it proved to be identical over seven residues with that from the reactive acyl group in α₂-macroglobulin (106), thus there can be little doubt that the same unusual structure is present in all three proteins, though no sequence data on this section of the C4 α chain is available at present. Methylamine reacted with the equivalent glutamyl residue in C3 as in α₂-macroglobulin and the reactive SH group was on the cysteinyl residue, three positions N-terminal to the reactive glutamyl residue (Figure 4). Possibly there is a thiol ester bond between the thiol group and the γ carbonyl group of the glutamic acid residue which is stabilized in the native C3 molecule (90, 106). Denaturation or reaction with amines leads to breakage of the thiol ester and either release of a reactive acyl group or chain cleavage with a free thiol that is released in both cases.

Alternatively, it has been suggested that in α₂-macroglobulin there may be an internal pyroglutamic acid residue with γ carboxyl bound to a peptide bond α amino group (105). In this case the thiol would be unreactive in the native molecule because of its inacessibility, not because it is in a thioester bond. This was suggested because the new N-terminal residue appearing on denaturation is pyroglutamic acid. However, reduction of C3, C4, and α₂-macroglobulin before heating greatly decreased the fragmentation and could be interpreted as breaking a thiol ester bond.

Whatever the structure in the native proteins which is responsible for

$$\text{Gly–Cys–Gly–Glu} \overset{\downarrow}{\text{–}} \text{X–Asn–Met–Val–Leu–Phe–Ala–Pro–Asn–Ile}$$

Figure 4   Amino acid sequence around the site of methylamine incorporation into α₂-macroglobulin. X is a residue which reacts with ¹⁴C Methylamine and which on breakage of the peptide chain by heat denaturation (↓) appears as pyroglutamic acid (105, 186).

these unusual properties, it is clear that similar and profound conformational changes occur when C3, C4, and probably α_2-macroglobulin are:

1. Denatured by detergents, guanidine, or heat, or for C3, incubation at pH 4 at 37°C (101, 102). C3 and C4 also convert to an inactive form without chain splitting on freezing, thawing, or on long standing in the cold;
2. Inactivated by amines or chaotropic reagents such as 1.5 M KBr (107);
3. Reduced by thiols (103, 105);
4. Activated catalytically, C3 to C3b and C4 to C4b.
   The changes can be recognized by:

1. Loss of hemolytic activity in C3 and C4;
2. Appearance of a free sulfydryl group in C3 and C4;
3. Change in iodination pattern of α, β, and γ chain in C4 (108);
4. Resistance to hydrolysis of C3 and C4 by the activating enzymes C3 convertase and C$\overline{1}$s respectively;
5. Susceptibility of C3 and C4 to hydrolysis by C3bINA together with β 1H for C3 and C4 binding protein for C4;
6. Characteristic changes in antigenic pattern from native to denatured, amine-treated or activated form (107);
7. The capacity of C3 and C4 to form C3 convertase with B and $\overline{D}$ or C2 and C$\overline{1}$s, respectively, in solution, though chemically inactive forms are less effective than activated forms (107).

It seems clear that C3 and C4 (but not C5) and α_2-macroglobulin contain, in their native form, a structure that on denaturation, chemical treatment, or proteolytic activation is disrupted, which leads to exposure of a thiol group and a reactive acyl group. This novel structure could be a thiol ester or perhaps less likely an internal pyroglutamic acid residue, the disruption of which leads to exposure of a previously unreactive SH group, but proof is lacking (as indeed is proof that this structure is identical in all three proteins). It appears to be essential for the biological activities of these proteins and causes covalent bonding to associated polysaccharides or proteins. The same structure is also responsible for the nonproteolytic splitting of adjacent peptide bonds under certain types of denaturation.

ALTERNATIVE PATHWAY OF ACTIVATION

Factor $\overline{D}$ and its Role in the Formation of the $\overline{C3b,Bb}$ Complex

Factor $\overline{D}$ is a serine proteinase, of 24,000 molecular weight, present in low concentration ($\sim$ 1.5 mg/1) in plasma (109, 110). Most of the factor $\overline{D}$ in plasma is considered to be in an activated form (110) although there is one unconfirmed report that a small amount of proenzyme factor $\overline{D}$ is present in serum (111). This proenzyme form was resistant to treatment with diisopropyl phosphofluoridate (DFP) and could be activated by trypsin (111) but has not been further characterized. Despite its low plasma concentration, the activated form of factor $\overline{D}$ has been isolated in $\sim$ 20% yield, by conventional protein purification procedures, in several laboratories (109, 112–114) for structural and functional studies. Extensive N-terminal sequence analysis of the whole molecule (114, 115), and sequence analysis of the cyanogen bromide–derived (114) and tryptic (116) peptides has provided data which show a strong degree of homology between factor $\overline{D}$ and other serine proteases (Figures 5 and 6). Surprisingly, factor $\overline{D}$ shows greater homology with the serine proteases such as "group-specific protease," elastase, and trypsin than with $C\overline{1}s$ and $C\overline{1}r$, when the catalytic chains of these enzymes are compared with the N-terminal 20 residues of factor $\overline{D}$ (114). The comparison is limited to the first 20 residues since that is all the data available for the catalytic chains of C1s and C1r (36). The 60% identity over the first 16 residues of factor $\overline{D}$ with group-specific protease from rat small intestine is of interest, since it is considered that this enzyme may also have no zymogen form (117).

The active form of factor $\overline{D}$ is irreversibly inactivated by DFP (110, 111) and the amino acid sequence around the active site serine residue which reacts with DFP has been determined (114). The amino acid sequence data show that there is an aspartic acid residue at position 189 in factor $\overline{D}$ (on alignment with other serine esterases and using the chymotrypsin numbering), which indicates that factor $\overline{D}$ should have a tryptic-like specificity since all the trypsin-like enzymes examined so far have Asp_{189} present in the substrate-binding pocket. Factor $\overline{D}$ is a highly specific enzyme, since it splits the proenzyme form of factor $\overline{B}$ at a single Arg–Lys bond, in the

	190	200
Factor $\overline{D}$ (human)	-Asp-Ser-Cys-Lys-Gly-Asp-Ser*-Gly-Gly-Pro-Leu-Val-Cys	
Factor B (human)	-Asn-Thr-Cys-Arg-Gly-Asp-Ser*-Gly-Gly-Pro-Leu-Ile-Val-	
Trypsin (cow)	-Asp-Ser-Cys-Glu-Gly-Asp-Ser*-Gly-Gly-Pro-Val-Val-Cys-	
Plasmin (human)	-Asp-Ser-Cys-Glu-Gly-Asp-Ser*-Gly-Gly-Pro-Leu-Val-Cys-	
Chymotrypsin A (cow)	-Ser-Ser-Cys-Met-Gly-Asp-Ser*-Gly-Gly-Pro-Leu-Val-Cys-	
Elastase (pig)	-Ser-Gly-Cys-Glu-Gly-Asp-Ser*-Gly-Gly-Pro-Leu-His-Cys-	
Conserved residues	Cys Gly Asp Ser* Gly Gly Pro	

Figure 5   Amino acid sequences around the active site serine residue in factors D and B.
The asterisk denotes the active site serine residue. The residue numbering is that of the cow chymotrypsin A sequence. Sequence data for factor D is taken from Johnson et al (114), and that for factor B from Christie et al (42). Other sequences are taken from Young et al (187), and the conserved residues shown are invariant in these known sequences.

```
            _   15      20        30          40        50
Factor D       I L G G R E A E A H A R P Y M A S V Q L - - - - N X A E L C G G V L V A E Q W

Factor B       VV E HR KGT D Y HKQ PWQAKI SVIRPSK GXESCMGAV VSEY F


Conserved    ↓
sequence     R I V G G    A    G S   P W Q V S L    S G    H F C G G  L I    W
```

Figure 6 Comparison of portions of the amino acid sequences of factors D (114, 115) and B (42) with the conserved residues found, in other serine proteases, in the region C-terminal to the peptide bond, which is cleaved during activation of these other serine esterases (187). The residues underlined are invariant in the known sequences. The arrow denotes the peptide bond cleaved during the activation of other serine esterases. Factor B is not split at this point in its catalytic chain, and factor D is considered to be present in plasma in its activated form. The dashes denote that a gap was left to give maximum homology on alignment. Residues are given in the single letter code: A, Ala; B, Asx; C, Cys; D, Asp; E, Glu; F, Phe; G, Gly; H, His; I, Ile; K, Lys; L, Leu; M, Met; N, Asn; P, Pro; Q, Glu; R, Arg; S, Ser; T, Thr; V, Val; W, Tryp; X, unknown; Y, Tyr. The residue numbering is that of the cow chymotrypsin A sequence. The portion of factor D sequence shown is the N-terminal sequence of the intact enzyme. The portion of factor B sequence shown is located within the Bb chain of factor B.

sequence –Gln–Lys–Arg–Lys–Ile–Val–, and it does this only after the factor B has formed a complex with C3b in the presence of Mg^{+2} (38, 112, 118). Further evidence of factor $\overline{D}$'s specificity is shown by the finding that synthetic peptides containing the amino acid sequence around the –Arg–Lys– bond are not split by factor D but are split by trypsin, thrombin, and C1-s (119). These synthetic peptides may however, be bound, since they inhibit the splitting of factor B in a mixture of purified factor B, factor $\overline{D}$, C3b, and Mg^{+2} (119). Only one synthetic substrate, a nitroanilide, has been shown to be split by highly purified factor $\overline{D}$ preparations (112, 113). This high degree of specificity could explain why active form factor $\overline{D}$ is found in blood in the presence of the relatively high concentration of protease inhibitors.

After factor B has been split into its Ba (30,000 molecular weight) and Bb (70,000 molecular weight) fragments by factor $\overline{D}$, in the presence of C3b and Mg^{+2} (38, 112), the enzymatically active $\overline{C3b,Bb}$ complex is formed that can, via the active site in B, split C3 to yield more C3b and C3a (40, 120). The formation of a $\overline{C3b,B}$ complex on sheep erythrocytes, in the presence of properdin, requires Mg^{+2}, but once this proenzymic complex has been formed its activation by factor $\overline{D}$ can proceed in the presence of EDTA (110). There is some evidence that factor $\overline{D}$ binds Ca^{+2} or $Mg^{+,}$ in serum, and that upon addition of chelating agents it may bind to an as yet unidentified protein (113, 121). However, factor $\overline{D}$ apparently is not incorporated into the $\overline{C3b,Bb}$ complex as judged by the absence of binding of factor D hemolytic activity, or of radiolabeled factor $\overline{D}$, to surfaces coated with C3b, factor B, and properdin in the presence of Mg^{+2} (110). Furthermore, Lesavre & Müller-Eberhard (110) found that plasma or serum that had been treated with zymosan, thus causing about 90% utilization of C3 via the alternative pathway, had identical factor $\overline{D}$ activity when compared with untreated plasma or serum. On the other hand, Gadd & Reid (94) found that incubation of immune aggregates with human serum consumed up to 50% of the factor D activity in conjunction with C3 and factor B utilization. Fujita et al (122) have shown that C3 convertase activity can be bound to $F(ab')_2$ immune aggregates when the aggregates are incubated in serum and washed with low ionic strength buffers. The C3 convertase activity on the aggregates could be blocked by either anti-C3 or antifactor B; after allowing the activity to "decay" at 37°C it could be regenerated only by the addition of both factor D and factor B. These results would be consistent with a $\overline{C3b,Bb}$ complex being formed on the immune aggregates but provide no evidence for the binding of factor $\overline{D}$. Using zymosan particles which had been incubated with human serum, washed, and allowed to decay at 37°C for 2 hr, Brade et al (123) observed binding of factor $\overline{D}$, but this binding was very weak at physiological pH and low ionic strength.

Factor B and its Role in the $\overline{C3b,Bb}$ and $\overline{C(3b)_nBb}$ Complexes

Factor B is an unusual serine protease, since the serine at the active site is located on the Bb fragment of 60,000 molecular weight rather than on the smaller Ba fragment of 30,000 molecular weight (42). All other serine proteases have a catalytic chain of 24–28,000 molecular weight and activation peptides which can vary from a few residues up to a polypeptide of 60,000 molecular weight. Medicus et al (120) reported that, contrary to other reports (74), both the C3 and C5 convertases of the alternative pathway could be partially inhibited by DFP and that the DFP was incorporated into the Bb fragment in these enzyme complexes. Clear evidence of the serine protease nature of factor B has come from extensive amino acid sequence studies of the Bb fragment, which have shown that the highly conserved active site residues histidine, aspartic acid, and serine are all present in factor B in positions homologous to those found in typical serine proteases (42) when comparison is made from the C-terminal ends. The N-terminal section of Bb, which is approximately 300 residues longer than the catalytic chains of other serine proteases examined so far (with the exception of C2), does not contain the characteristic N-terminal sequence found in the catalytic chains of other serine proteases (42, 124). The activation mechanism of factor B must be different from that of other serine proteases (42).

It has been shown that a stable complex can be formed between C3b and factor B in the presence of Mg^{+2} and that this complex contains equimolar amounts of both proteins (125, 126). Some evidence indicates that a mixture of C3b and proenzyme factor B contains C3-splitting activity in the absence of factor $\overline{D}$, even after pretreatment of the C3b and proenzyme factor B with DFP (126). In view of the relatively high concentrations of DFP required to completely inactivate factor $\overline{D}$ it is possible that this is due to the presence of a low level of factor D. However, others (127–129) have found that factor B may be active in its unsplit form.

Once the proenzyme factor B has been activated, the activation peptide, Ba, dissociates (130). The $\overline{C3b,Bb}$ complex is labile and decays rapidly at 37°C. Once the Bb fragment is dissociated from C3b its ability to participate in the formation of C3 convertase is lost, but C3b can be reused to generate more $\overline{C3b,Bb}$ in the presence of factor B, factor $\overline{D}$, and Mg^{+2}. The Bb fragment still retains its ability to hydrolyze synthetic esters such as acetylglycyl-lysine methyl ester, after the decay of the $\overline{C3b,Bb}$ complex (126, 131).

Control Proteins Concerned with the Regulation of the $\overline{C3b,Bb}$ and $\overline{C(3b)_nBb}$ Complexes

The C3 convertase $\overline{C3b,Bb}$ is unusual in that the substrate, C3, once it is split, yields the C3b fragment which can be used to generate more $\overline{C3b,Bb}$. All the C3 present in the serum is not consumed by such an uncontrolled "feed-back" mechanism because the level of C3b in the serum is finely controlled by β1H and C3bINA, as is illustrated by the findings that small increases in the β1H level (132) or C3bINA level (133) can lead to marked suppression of alternative pathway activation in serum. The central role that C3b plays in the activation of complement system along with its biological functions, such as in immune adherence reactions, makes the control of C3b levels of β1H and C3bINA a vital point in the system. Lack of control of the C3b level can have a markedly deleterious effect in vivo as is shown in those cases where the genetic deficiency of C3bINA leads to depletion of factor B and functionally active C3 and consequently a lowering of resistance to bacterial infections (134, 135).

β1H β1H is a single-chain glycoprotein of 150,000 molecular weight required as a cofactor for the inactivation of C3b by the enzyme C3bINA (136, 137). β1H binds to C3b, probably in an equimolar ratio, thus preventing the interaction of C3b with other proteins, such as factor B, or even disrupting preexisting interactions between C3b and other proteins. This can be seen by the acceleration of the rate of dissociation of Bb from $\overline{EC3b,Bb}$ and $\overline{Ec3b,Bb,P}$ (136–138). There is an absolute requirement for β1H to allow the splitting of C3b by C3bINA to take place in solution (139, 140). Recently it has been shown that β1H competes with C5 for a binding site on cell-bound C3b; thus as well as being involved in control of C3 convertase activity, β1H appears to be directly involved in the control of C5 convertase activity (77).

C3bINA   C3bINA is an enzyme composed of two covalently linked polypeptide chains, of molecular weights 50,000 and 38,000, which is present in serum in its activated form in a concentration of 34 μg/ml. Highly purified C3bINA does not split cell-bound C3b or C3b in solution unless β1H is present (87, 139–143). The α' chain of the C3b is split at only one major site by C3bINA, which results in the destruction of the ability of C3b to form the $\overline{C3b,Bb}$ complex and to participate in other biological roles, such as immune adherence. C3bINA has no effect on native C3 even in the presence of β1H, but C3bINA and β1H will bring about the splitting of C3 that has been inactivated by freezing, thawing, or by treatment with amines (140). The inactive C3 is considered to have taken up similar conformation to C3b and thus is able to interact with β1H. C3bINA shows a small but significant amount of binding to cell-bound C3b in the absence of β1H and this binding is increased 30-fold in the presence of β1H. Factor B appears to compete with β1H for the interaction with cell-bound C3b and thereby limit the action of C3bINA on the C3b (144). The control protein, properdin, is considered to enhance the interaction between C3b and factor B when it binds to C3b, but appears to have little effect on β1H binding or the action of C3bINA.

PROPERDIN   Properdin is a glycoprotein, of approximately 220,000 molecular weight, composed of four, probably identical, noncovalently linked polypeptide chains (145–147). The amino acid composition of properdin is unusual in that four amino acids—glycine, proline, glutamic acid, and cysteine, account for approximately 46% of the total composition (146, 148). However, extensive amino acid sequence analysis of the peptides produced by cyanogen bromide treatment indicates that there are no regions in the molecule in which some type of repeating unit involving these four amino acids is found and that these four amino acids are evenly distributed throughout the molecule (148). Properdin is considered to be present in serum in a native form which is functionally quite distinct from an activated form which is obtained with most of the isolation procedures reported, or if the native properdin binds to an activating particle containing $\overline{C3b,Bb}$. Care has to be taken to prevent conversion of native properdin to the activated form during purification, and a rapid two-step procedure, which yields 24% has been described for native properdin (147). The two forms of properdin can be distinguished only by functional assays, since chemically and antigenically they appear to be identical (147, 149). Activated properdin can be distinguished from native properdin by the fact that it causes consumption of the alternative pathway components C3 and factor B to take place when it is added to normal serum, and binds to surfaces coated with C3b, in the absence of factors B and D, while native properdin does not (147). Both native properdin and activated properdin can interact with and stabilize the labile C3 convertase $E\overline{C3b,Bb}$ (147, 150). Since native properdin can bind to $E\overline{C3b,Bb}$, but not to particles coated with C3b (147), it presumably recognizes some portion of the C3b molecule which is exposed only when C3b is bound to Bb, or some portion of the complex to which both C3b and Bb contribute. The conversion of native properdin to activated properdin by freeze-thawing, and the surprising conversion of activated properdin to native properdin by treatment with guanidine HCl (147) emphasizes that the activation of properdin may be due to some small change in conformation of the molecule induced on binding to the $\overline{C3b,Bb}$ complex.

The Complex Proteases

C3 CONVERTASE $\overline{C3b,Bb}$ A mixture of the purified components C3, factors B and $\overline{D}$, the control proteins $\beta1H$ and C3bINA, and Mg^{+2} is stable for long periods at 37°C (151–153). For example, Schreiber et al (151) could detect no C3 or factor B consumption over periods of up to 8 hr at 37°C when the three components and two control proteins were incubated together at physiological concentrations. When the control proteins are omitted from the incubation mixture, C3 convertase activity is generated and C3 and factor B are activated (151, 153). The C3 used by Schreiber et al (151) was considered to be completely free of C3b, since it had been treated with $\beta1H$ and C3bINA and then passed down an immunoadsorbent column that removed the $\beta1H$ and C3bINA. Therefore the consumption of C3 and factor B, in the absence of the control proteins, is possibly due to the formation of a complex between C3 and factor B which, if activated by factor $\overline{D}$, yields a C3 convertase $\overline{C3b,Bb}$. Parkes et al (153) prepared highly purified C3 and C3b using a sulphated-Sepharose column and showed that C3b is, initially, several orders of magnitude more efficient than C3 (Figure 7) at supporting factor B splitting, in the presence of factor $\overline{D}$. They considered that any splitting of factor B by factor D in the presence of C3 could be attributed to the trace contamination of the C3 with C3b. Although it is of interest to know whether factor B splitting, which takes place in the presence of factor $\overline{D}$ and C3, is due entirely to its interaction with C3, or to its interaction with a trace of C3b in the C3 preparation, this fluid phase system is incomplete without the control proteins $\beta1H$ and C3bINA (see Figure 8). Thus any information obtained in the absence of control proteins may not be entirely relevant to the mechanism of activation that takes place under physiological conditions.

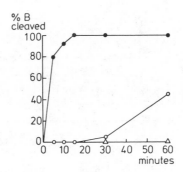

Figure 7 Reaction of C3 species with factors B and D. 25 µg of C3, degraded C3, or C3b was incubated with 1.5 µg of factor B and 1.5 µg of factor D in 1.5 ml of VB++ at 37°C. Samples were removed at various times, and after reduction run on a 10% sodium dodecyl sulphate polyacrylamide gel. The percentage of factor B cleaved was determined by densitometer scanning of the gel. ●, C3b; ○, C3; △, degraded C3 and also C3b or C3 in the absence of factor D. This Figure is taken from Parkes et al (153).

When a suitable activator, such as rabbit erythrocytes, is added to a mixture of the purified components C3, factors B and D̄, β1H, and C3bINA, all present at physiological concentrations, activation of the alternative pathway proceeds to a degree equivalent to that obtained with human serum depleted of C4, C5, and C6 (151). The deposition of C3b on the activator, and the hemolytic activity of the bound C3b, showed a very similar pattern in both the mixture and the C4, 5, and 6 depleted serum. The importance of the role of C3bINA and β1H in the control of the initiation of C3 splitting and utilization in the presence of an activator such as rabbit erythrocytes is well-illustrated from the results of Schreiber et al (151). At physiological concentrations of the two control proteins, the C3 consumption was 4% of that available and the C3b deposition was approximately 1×10^5 molecules/cell, which was 0.3% of the total C3 in the system. At zero concentration of the control proteins the C3 consumption was 100% but C3b deposition was less than 5×10^2 molecules/cell. Addition of native properdin, at physiological concentration, gave an eightfold increase in overall C3 consumption and a 50% increase in the level of C3b deposition (151). These results show that uncontrolled fluid phase activation of C3, in the presence of an activator, does not result in efficient deposition of C3b on the activating particle, while efficient deposition takes place in the presence of the control proteins. Similar results have been obtained by Fearon & Austen (154) using zymosan as an activating particle and purified alternative pathway proteins.

The exact role that any activating cell or particle plays in the initiation of the alternative pathway is not entirely clear (see below) but it seems

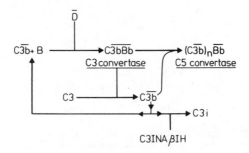

Figure 8   Activation and control of alternative pathway (19).

well-established that controlled C3b deposition on such an activator results in the efficient generation of a $\overline{\text{C3b},\text{Bb}},\text{P}$ complex that is protected from the combined action of β1H and C3bINA.

C5 CONVERTASE ($\overline{\text{C3b}}$)ₙBb,P   The C3 convertase of the alternative pathway in solution or in a cell-bound form is considered to be composed of equimolar amounts of C3b and the Bb fragment of factor B (126). The cell-bound alternative pathway, C3 convertase $\overline{\text{C3b},\text{Bb}},\text{P}$, can be converted to a C5 convertase by the deposition of further C3b on the surface (155, 120). This C5 convertase activity can then be selectively decreased by treatment with C3bINA to yield a surface with predominantly C3 convertase activity. This suggests the existence of two different classes of C3b molecules on the surface, one which is relatively protected in the $\overline{\text{C3b},\text{Bb}},\text{P}$ complex and one which is exposed to attack by C3bINA. It has been found by Vogt et al (78) that, for efficient development of C5 convertase activity, C3b fixed to a solid surface has to be present in addition to C3b and factors B and $\overline{\text{D}}$, and that the C3b which is fixed to the solid surface must be generated by a surface-fixed enzyme. Thus, in agreement with the work of Daha et al (155) and Medicus et al (120), it was proposed that at least two molecules of C3b are involved in the generation of C5 convertase; one is incorporated into the enzyme complex and the other binds C5, thus preparing it for activation (78). The C5 binding takes place only when the C3b is free of Bb and properdin, but it has not been established how many molecules of C3b are involved in binding one molecule of C5. Vogt et al (78) concluded that the C5 convertase probably has the same composition of $\overline{\text{C3b},\text{Bb}},\text{P}$ as the C3 convertase and that the close proximity of surface-bound C3b, generated by surface-bound $\overline{\text{C3b},\text{Bb}}$, allows the binding of C5 in such a conformation that it can be readily split. These results would help to explain why a completely fluid phase mixture of components often is not efficient at splitting C5. However, Isenman et al (77) have shown that under suitable conditions weak interaction between C3b and C5 can be detected in free solution; the association constant for the interaction was estimated to be 2×10^6 M^{-1} at 0.06 ionic strength. This interaction, which was fivefold weaker at physiological ionic strength, could be inhibited by β1H and did not occur with native C3 in place of C3b. The $\overline{C5}$ which was bound to C3b in free solution was split by fluid phase $\overline{C3b,Bb}$ that was stabilized by nephritic factor. Thus under certain conditions the ability of C5 to bind to C3b does not appear to be completely dependent upon the attachment of C3b to a surface. The apparent difference between the results of Vogt et al (78) and those of Isenman et al (77) may be explained by the different concentration of reagents. However, if, for example, C5 were bivalent for C3b, the surface-bound clusters of C3b would bind C5 much more efficiently than C3b in free solution or C3b randomly distributed over a surface, and then the apparent differences between the two sets of results could be explained.

Activation of the Alternative Pathway

Lachmann and co-workers (21, 133, 156) have proposed that in vivo C3b formation takes place continuously at a low level (the "tick-over mechanism") and that activators of the alternative pathway can be viewed as being able to accelerate the formation of C3b or slow down its breakdown. This view of alternative pathway activation has proved accurate considering that if no C3b, or modified native C3 (i.e. native C3 modified in such a way that it has properties similar to C3b), is generated then the pathway cannot function.

NONIMMUNOGLOBULIN ACTIVATORS   A large number of widely different particles and cells can activate the alternative pathway by a mechanism which does not appear to involve antibody. These antibody-independent activators include: particles of microbial origin, such as zymosan (157, 158); lipopolysaccharides from gram-negative bacteria (159); cell-wall teichoic acid from gram-positive bacteria (160); whole microorganisms, for example, certain strains of *Escherichia coli* (161); trypanosomes (162) and other parasites; mammalian cells, such as rabbit erythrocytes in human serum (163, 158); neuraminidase-treated sheep erythrocytes (164); measles virus–infected HeLa cells (165, 166); lymphoblastoid (Rajii) cells (167, 168); and a variety of other substances which are polyanionic in character, including dextran sulfate, polyvinyl sulfate (169), and DNP-albumin (170).

Faced with this large number of very different types of activators of the alternative pathway it is difficult to envisage what could be the essential common feature which allows them to efficiently generate and amplify the C3 convertase complex $\overline{C3b,Bb}$ in the presence of the two control proteins, β1H and C3bINA. Since most of these activators are polymeric in nature, being composed of or containing on their surface some type of repeating structure, this feature may be of importance in the provision of closely spaced acceptors for freshly activated C3b, so that any $\overline{C3b,Bb}$ complex formed is protected to some extent from the combined inactivating effect of β1H and C3bINA. Alternatively, rather than being bound directly to these repeating units, the C3b could be bound closely to the units in such a way that the repeating structure interacts with a site on the bound C3b, which prevents control by β1H (171).

The distinguishing feature which decides whether cells or certain particles will or will not activate the alternative pathway of human complement may be the amount of sialic acid (144, 172–174) or sulfated muccopolysaccharides (175) present on the surface. Particles which have no, or low levels of, sialic acid are considered to be potential activators of the alternative pathway, e.g. zymosan and rabbit erythrocytes. However, a low sialic acid content alone is unlikely to be a decisive feature of activators (166, 171, 173, 176) and it seems likely that other mechanisms which do not involve only the sialic acid content must also be important since, for example, human erythrocytes which had been 80% desialated did not activate the alternative pathway in human serum (171, 176). Other examples of the apparent noninvolvement of sialic acid in alternative pathway activation have been given by Pangburn et al (171) and Sissons et al (166).

Bacterial activation of the alternative pathway without the use of antibody may play a significant role in natural immunity (161). The lysis of measles virus–infected cells via the alternative pathway only proceeds with the aid of antibody, although antibody-independent activation of the alternative pathway can take place (166). Although the precise role of immunoglobulin in alternative pathway activation is not always clear it is of interest to consider how immune aggregates behave as activators since the immunoglobulin molecule is well-characterized and can provide clearer understanding of the initiation of the pathway.

IMMUNOGLOBULIN ACTIVATORS   Immunoglobulins that cause activation of the alternative pathway do so only after binding of antigen to form large complexes or on aggregation by chemical procedures. This suggests that some factor with several binding sites, e.g. C1q in activation of the classical pathway by immunoglobulins (20), may be required to recognize and bind to the aggregated antibody, thus initiating the pathway. However, unless C3b is envisaged as playing this role, there appears to be no factor in the alternative pathway that is directly comparable with the C1q subcomponent of the classical pathway. Reports for the evidence of an alternative pathway "initiating factor" (177) have now been retracted (178).

There are several reports that show that IgG antibodies or myeloma proteins, after aggregation, can activate the alternative pathway. For example, immune aggregates containing guinea pig IgG_1 or ruminant IgG_2 antibodies are unable to activate complement via the classical pathway but can cause activation via the alternative pathway (10, 179). Treatment of two endotoxin-free human IgG myeloma proteins yielded aggregates that caused alternative pathway activation (180). Also, Ferrone et al (181) found that some anti-Human Leukocyte Antigen (HLA) antisera, containing IgG-type antibodies, could activate complement by either the alternative pathway alone, the classical pathway alone, or both pathways when added to cultured human lymphoid cells. It has been shown that human IgG antibodies can enhance the lysis of rabbit erythrocytes by the alternative pathway of human complement (182), which suggests that some site in the human IgG molecule bound to a particle may be involved in the generation of the C3 and C5 convertases of the alternative pathway. There is no clear study of the ability of fragments of human IgG to activate the alternative pathway, but relevant data are available from studies using rabbit and guinea pig antibodies.

Using immune aggregates prepared from rabbit antibodies, it has been shown that rabbit IgG can activate the alternative pathway of human and rabbit complement under conditions permitting only alternative pathway activation (94). The site on the rabbit or guinea-pig IgG antibody that is involved in the activation of the alternative pathway is clearly located in the $F(ab')_2$ portion of the molecule (10, 11, 94, 122). The $F(ab')_2$ region of human secretory IgA may also be involved in the activation of the alternative pathway as judged by studies using interfacially aggregated preparations of the IgA (12). It has been shown (122) that C3/C5 convertase that survives two washing steps is formed on $F(ab')_2$ immune aggregates on their incubation with human serum. The activity gradually decayed at 37°C but could be regenerated on addition of purified factors B and $\overline{D}$. These results indicate that C3b is bound to the $F(ab')_2$ region in the aggregates. Mild reduction of the rabbit IgG prior to immune aggregate formation completely blocked the ability of the aggregates to cause alternative pathway activation when tested under conditions allowing only alternative pathway activation (94). Reduction of IgG antibodies under similar conditions also considerably reduced activation of complement by the classical pathway (183). This blocking of alternative pathway activation appears to depend principally upon the splitting of the single inter-heavy-chain disulfide bond in the rabbit IgG molecule (94). The splitting of this bond probably affects the conformation taken up by the IgG molecules in the immune aggregates (184) and therefore could interfere with C3 binding to the $F(ab')_2$ region, since it is known that C3b, like C4b (55), is bound strongly, probably covalently, to the $F(ab')_2$ region of the intact IgG molecule during complement activation. This interference could thus prevent the formation of the C3 convertase complex and consequently block activation. The contribution that the site on the $F(ab')_2$ region makes during complement activation may be small in the case of IgG antibodies which can utilize complement via the Fc classical pathway site, but could be important in the case of antibodies such as guinea pig IgG_1, which do not activate the classical pathway.

CONCLUSIONS

Rapid progress is being made in solving the structure of many of the complement components. The rewarding results reveal features apparently confined to the complement system, though no doubt similar structures will be found elsewhere. The ability of C3 and C4, when activated, to form covalent bonds with proteins and carbohydrates is almost certainly due to an unusual structure which is shared with α_2-macroglobulin; it may also be found in some ligand-receptor interactions at cell surfaces. Subcomponent C1q has an apparently unique structure, but there have been suggestions that acetylcholinesterase is another protein which may contain collagen-like fibrils and globular sections (185). There is no evidence so far that the other activation systems of the blood, nor indeed any other proteolytic system, contain the new class of serine proteases typified by factor B and presumably C2.

Also unusual is the necessity for two or three proteins to associate to hydrolyze peptide bonds in a single protein molecule. Many multienzyme systems are known where enzymes of related specificities are associated, but the coming together on a surface of several distinct soluble proteins to perform one catalytic function is uncommon. A similar mechanism may occur among proteins in a cell membrane, a context that is less easily studied. In the C3–C5 convertases, the catalytic site is in C2 or factor B but it is not clear whether association with C4 and C3 facilitates binding of the C3 or C5 substrates or whether it leads to conformational changes which expose either the catalytic site or the vulnerable bond in the substrate. Detailed physical studies and probably the crystal structures of the different proteins may be required before the mechanisms of these reactions are understood. Though not discussed in this article, the necessity for the association of the last five proteins of the complement sequence to form a complex which can lyse cells is even more surprising.

Major outstanding problems are the mechanism of activation of the alternative pathway and the respective roles of the two pathways in vivo. It is likely that the alternative pathway comes into play whenever the classical pathway is activated, which leads to the formation of C3b to stimulate the amplification cycle of the alternative pathway. The short half-lives of the activated forms of C4, C3, C2, and factor B suggest that their complexes will be formed very close to the point of activation— aggregated antibody, in the case of the classical pathway. How such complexes of relatively high-molecular-weight proteins form on, for example, the much smaller Fab sections of the antibody molecules is not apparent.

Though complement fixation has been a routine diagnostic assay for many years, the system itself has been considered an esoteric topic by most biochemists. Recent unraveling of some of the molecular aspects may make it a model system of interest in a much wider context of biochemistry as well as in its own right as a major feature of immunity to infection.

Literature Cited

1. Cooper, N. R., Jensen, F. C., Welsh, R. M., Oldstone, M. B. A. 1976. *J. Exp. Med.* 144:970–84
2. Bartholomew, R. M., Esser, A. F., Müller-Eberhard, H. J. 1978. *J. Exp. Med.* 147:844–53
3. Loos, M., Wellek, B., Thesen, R., Opperkuch, W. 1978. *Infect. Immun.* 22:5–9
4. Clas, F., Loos, M. 1980. *Immunology* 40:547–56
5. Yachnin, S., Ruthenberg, J. M. 1965. *J. Clin. Invest.* 44:518–34
6. Loos, M., Bitter-Suermann, D. 1976. *Immunology* 31:931–34
7. Eisen, V., Loveday, C. 1970. *Br. J. Pharmacol.* 39:831–36
8. Cooper, N. R. 1973. *Contemp. Top. Mol. Immunol.* 2:155–80
9. Bartholomew, R. M., Esser, A. F. 1978. *J. Immunol.* 121:1748–51
10. Sandberg, A. L., Oliveira, B., Osler, A. G. 1971. *J. Immunol.* 106:282–85
11. Reid, K. B. M. 1971. *Immunology* 20:649–58
12. Boackle, R. J., Povitt, K. M., Mestecky, J. 1974. *Immunochemistry* 11:543–48
13. Sandberg, A. L., Osler, A. G., Shiu, H. S., Oliveira, B. 1970. *J. Immunol.* 104:329–34
14. Fujita, T. 1979. *Microbiol. Immunol.* 23:1023–31
15. Hugli, T. E., Müller-Eberhard, H. J. 1978. *Adv. Immunol.* 26:1–53
16. Porter, R. R. 1980. *Proc. R. Soc. London Ser. B* 210:477–98
17. Fothergill, J. E., Anderson, W. H. K. 1978. *Curr. Top. Cell. Regul.* 13:259–311
18. Müller-Eberhard, H. J. 1978. *Molecular Basis of Biological Degradative Processes*, ed. R. D. Berliner, H. Hermann, I. H. Lepow, J. M. Tanzer, pp. 65–112. New York: Academic
19. Porter, R. R. 1979. *Complement.* In *Defense and Recognition IIB. Structural Aspects* (MTP Int. Rev. Sci. Ser.-Biochem.), ed. E. S. Lennox, 23:177–212. Baltimore: Univ. Park Press
20. Porter, R. R., Reid, K. B. M. 1979. *Adv. Protein Chem.* 33:1–71
21. Lachmann, P. J. 1979. *Complement: The Antigens,* ed. M. Sela, 5:283–335. New York: Academic
22. Kolb, W. P., Kolb, L. M., Podack, E. R. 1979. *J. Immunol.* 122: 2103–11
23. Arlaud, G. J., Sim, R. B., Duplaa, A-M., Colomb, M. G. 1979. *Mol. Immunol.* 16:445–50
24. Shinkai, H., Yonemasu, K. 1979. *Biochem. J.* 177:847–52
25. Mizuochi, T., Yonemasu, K., Yamashita, K., Kobata, A. 1979. *J. Biol. Chem.* 253:7404–9
26. Gilmour, S., Randall, J. T., Willan, K. J., Dwek, R. A., Torbet, J. 1980. *Nature* 285:512–14
27. Lepow, I. H., Naff, G. B., Todd, E. W., Pensky, J., Hinz, C. F. 1963. *J. Exp. Med.* 117:983–1008
28. Campbell, D., Booth, N. A., Fothergill, J. E. 1979. *Biochem. J.* 177:531–40
29. Bing, D. H. 1971. *J. Immunol.* 107:1243–44
30. Sledge, C. R., Bing, D. H. 1973. *J. Immunol.* 111:661–66
31. Lin, T-Y., Fletcher, D. S. 1978. *Immunochemistry* 15:107–17
32. Takahashi, K., Nagasawa, S., Koyama, J. 1975. *FEBS Lett.* 50:330–33
33. Chapuis, R. M., Isliker, H., Assimeh, S. N. 1977. *Immunochemistry* 14:313–17
34. Medicus, R. G., Chapuis, R. M. 1980. *J. Immunol.* 125:390–95
35. Taylor, P. A., Fink, S., Bing, D. H., Painter, R. H. 1977. *J. Immunol.* 118:1722–27
36. Sim, R. B., Porter, R. R., Reid, K. B. M., Gigli, I. 1977. *Biochem. J.* 163:219–27
37. Kerr, M. A., Porter, R. R. 1978. *Biochem. J.* 171:99–107
38. Kerr, M. A. 1979. *Biochem. J.* 183:615–22
39. Polley, M. J., Müller-Eberhard, H. J. 1968. *J. Exp. Med.* 128:533–51
40. Cooper, N. R. 1975. *Biochemistry* 14:4245–51
41. Medicus, R. G., Götze, O., Müller-Eberhard, H. J. 1976. *Scand. J. Immunol.* 5:1049–55
42. Christie, D. L., Gagnon, J., Porter, R.

462   REID & PORTER

R. 1980. *Proc. Natl. Acad. Sci. USA*
77:4923–27
43. Schreiber, R. D., Müller-Eberhard, H.
J. 1974. *J. Exp. Med.* 140:1324–35
44. Bolotin, C., Morris, S., Tack, B., Prahl,
J. 1977. *Biochemistry* 16:2008–15
45. Gigli, I., von Zabern, I., Porter, R. R.
1977. *Biochem. J.* 165:439–46
46. Hobart, M. J., Lachmann, P. J. 1976.
Transplant. Rev. 32:26–42
47. Hall, R. E., Colten, H. R. 1977. *Proc.
Natl. Acad. Sci. USA* 74:1707–10
48. Gorski, J. P., Müller-Eberhard, H. J.
1978. *J. Immunol.* 120:1775 (Abstr.)
49. Gigli, I. 1978. *Nature* 272:836–37
50. Abraham, G. N., Goldberger, G. G.,
Colten, H. R., Williams, J. 1980. *Fed.
Proc.* 39:4905 (Abstr.)
51. Dodds, A. W., Sim, R. B., Porter, R. R.,
Kerr, M. A. 1978. *Biochem. J.* 383–90
52. Parker, K. L., Schreffler, D. C., Capra,
J. D. 1980. *Proc. Natl. Acad. Sci. USA.*
77:4275–78
53. Goldberger, G., Colten, H. R. 1980.
Nature 286:514–16
54. Budzko, D. B., Müller-Eberhard, H. J.
1970. *Immunochemistry* 7:227–34
55. Campbell, R. D., Dodds, A. W., Porter,
R. R. 1980. *Biochem. J.* 189:67–80
56. Tack, B. F., Prahl, J. W. 1976. *Biochemistry* 15:4513–21
57. Tack, B. F., Morris, S. C., Prahl, J. W.
1979. *Biochemistry* 18:1490–97
58. Fernandez, H. N., Hugli, T. E. 1976. *J.
Immunol.* 117:1688–94
59. Tack, B. F., Morris, S. C., Prahl, J. W.
1979. *Biochemistry* 18:1497–1503
60. Brade, V., Hall, R. E., Colten, H. R.
1977. *J. Exp. Med.* 146:759–65
61. Haupt, H., Heimburger, N., Kranz, T.,
Schwick, H. G. 1970. *Eur. J. Biochem.*
17:254–61
62. Harpel, P. C., Cooper, N. R. 1975. *J.
Clin. Invest.* 55:593–604
63. Sim, R. B., Arlaud, G. J., Colomb, M.
G. 1979. *Biochem. J.* 179:449–57
64. Ziccardi, R. J., Cooper, N. R. 1979. *J.
Immunol.* 123:788–92
65. Laurell, A. B., MÅrtensson, U., Sjö-
holm, A. G. 1976. *Acta Pathol. Microbiol. Scand. Sect. C* 84:455–64
66. Conradie, J. D., Volanakis, J. E.,
Stroud, R. M. 1975. *Immunochemistry*
12:967–71
67. Scharfstein, J., Ferreira, A., Gigli, I.,
Nussenzweig, V. 1978. *J. Exp. Med.*
148:207–22
68. Fujita, T., Nussenzweig, V. 1979. *J.
Exp. Med.* 150:267–76
69. Shiraishi, S., Stroud, R. M. 1975. *Immunochemistry* 12:935–39

70. Nagasawa, S., Stroud, R. M. 1977. *Immunochemistry* 14:749–57
71. Müller-Eberhard, H. J., Polley, M. J.,
Calcott, M. A. 1967. *J. Exp. Med.*
125:359–80
72. Dodds, A. W., Porter, R. R. 1979. *Mol.
Immunol.* 16:1059–62
73. Polley, M. J., Müller-Eberhard, H. J.
1968. *J. Exp. Med.* 128:533–51
74. Kerr, M. A. 1980. *Biochem. J.*
189:173–81
75. Cooper, N. R., Müller-Eberhard, H. J.
1970. *J. Exp. Med.* 132:775–93
76. Müller-Eberhard, H. J. 1975. *Proteases
and Biological Control.* Presented at
Cold Spring Harbor Conf. Cell Proliferation 2:229
77. Isenman, D. E., Podack, E. R., Cooper,
N. R. 1980. *J. Immunol.* 124:326–31
78. Vogt, W., Schmidt, G., Buttlar, B. V.,
Dieminger, L. 1978. *Immunology*
34:29–40
79. Vogt, W., Schmidt, G., Hinsel, B. 1979.
Immunology 36:139–43
80. von Zabern, I., Molte, R., Vogt, W.
1980. *Scand. J. Immunol.* 9:69–74
81. Strunk, R. C., Giclas, P. C. 1980. *J.
Immunol.* 124:520–26
82. Müller-Eberhard, H. J., Dalmasso, A.
P., Calcott, M. A. 1966. *J. Exp. Med.*
123:33–54
83. Müller-Eberhard, H. J., Lepow, I. H.
1965. *J. Exp. Med.* 121:819–27
84. Willoughby, W. F., Mayer, M. M. 1965.
Science 150:907–8
85. Law, S. K., Levine, R. P. 1977. *Proc.
Natl. Acad. Sci. USA* 74:2701–5
86. Law, S. K., Lichtenberg, N. A., Levine,
R. P. 1979. *J. Immunol.* 123:1388–94
87. Law, S. K., Fearon, D. T., Levine, R. P.
1979. *J. Immunol.* 122:759–65
88. Capel, P. J. A., Groeneboer, O., Grosveld, G., Pondman, K. W. 1978. *J. Immunol.* 121:2566–72
89. Sim, R. B., Twose, T. M., Sim, E., Paterson, D. S. 1981. *Biochem. J.* 193:
115–27
90. Twose, T. M., Sim, R. B., Paterson, D.
S. 1980. *Int. Congr. Immunol., 4th.*
(Abstr.) No. 15, 1.22
91. Law, S. K., Lichtenberg, N. A., Levine,
R. P. 1980. *J. Immunol.* 124:1528
(Abstr.)
92. Goers, J. W. F., Porter, R. R. 1978.
Biochem. J. 175:675–84
93. Folk, J. E., Finlayson, J. S. 1977. *Adv.
Protein. Chem.* 31:1–133
94. Gadd, K. J., Reid, K. B. M. 1981. *Immunology* 42:75–82
95. Linsley, P. S., Blifeld, C., Wrann, M.,
Fox, C. F. 1979. *Nature* 278:745–48

96. Baker, J. B., Simmer, R. L., Glenn, K. C., Cunningham, D. D. 1979. *Nature* 278:743–45

97. Davies, P. J. A., Davies, D. R., Levitzki, A., Maxfield, F. R., Milhaud, P., Willingham, C., Pastan, I. H. 1980. *Nature* 283:162–67

98. Swenson, R. P., Howard, J. B. 1979. *J. Biol. Chem.* 254:4452–56

99. Harpel, P. C., Hayes, M. B. 1979. In *Physiological Inhibitors of Coagulation and Fibrinolysis*, ed. D. Collen, B. Wiman, M. Verstraete, p. 231. Amsterdam: Elsevier

100. Salvesen, G. S., Barrett, A. J. 1980. *Biochem. J.* 187:695–701

101. Janatova, J., Tack, B. F., Prahl, J. W. 1980. *Biochemistry* 19:4479–85

102. Janatova, J., Lorenz, P. E., Schechter, A. N., Prahl, J. W., Tack, B. F. 1980. *Biochemistry* 19:4471–78

103. Sim, R. B., Sim, E. 1980. *Biochem. J.* 193:129–41

104. Harpel, P. C., Hayes, M. B., Hugli, T. E. 1979. *J. Biol. Chem.* 254:8669–78

105. Howard, J. B., Vermeulen, M., Swenson, R. P. 1980. *J. Biol. Chem.* 255:3820–23

106. Tack, B. F., Harrison, R. A., Janatova, J., Thomas, M. L., Prahl, J. W. 1980. *Proc. Natl. Acad. Sci. USA.* 77:5764–68

107. von Zabern, I., Nolte, R., Vogt, W. 1980. *J. Immunol.* 124:1543 (Abstr.)

108. Reboul, A., Thielens, N., Villiers, M. B., Colomb, M. G. 1979. *FEBS Lett.* 103:156–61

109. Volanakis, J. E., Schrohenloher, R. E., Stroud, R. M. 1977. *J. Immunol.* 119:337–42

110. Lesavre, P. H., Müller-Eberhard, H. J. 1978. *J. Exp. Med.* 148:1498–1509

111. Fearon, D. T., Austen, K. F., Ruddy, S. 1974. *J. Exp. Med.* 139:355–66

112. Lesavre, P. H., Hugli, T. E., Esser, A. F., Müller-Eberhard, H. J. 1979. *J. Immunol.* 123:529–34

113. Davies, A. E., Zalut, C., Rosen, F. S., Alper, C. A. 1979. *Biochemistry* 18:5082–87

114. Johnson, M. A., Gagnon, J., Reid, K. B. M. 1980. *Biochem. J.* 187:863–74

115. Volanakis, J. E., Bhown, A. S., Bennett, J. C., Mole, J. E. 1980. *Proc. Natl. Acad. Sci. USA* 77:1116–19

116. Johnson, M. A., Gagnon, J., Reid, K. B. M. 1981. *Method Enzymol.* In press

117. Woodbury, R. G., Katunama, N., Kobayashi, K., Titani, K., Neurath, H. 1978. *Biochemistry* 17:811–19

118. Götze, O., Müller-Eberhard, H. J. 1971. *J. Exp. Med.* 134:90s-108s

119. Lesavre, P., Gaillard, M. H., Halbwachs, L. 1980. *J. Immunol.* 124:1528 (Abstr.)

120. Medicus, R. G., Gotze, O., Müller-Eberhard, H. J. 1976. *Scand. J. Immunol.* 5:1049–55

121. Konno, T., Katsumo, Y., Hirai, H. 1978. *J. Immunol. Methods* 21:325–34

122. Fujita, T., Takiuchi, M., Iida, K., Nagaki, K., Inai, S. 1977. *Immunochemistry* 14:25–30

123. Brade, V., Nicholson, A., Lee, G. D., Mayer, M. M. 1974. *J. Immunol.* 112:1845–54

124. Niemann, M. A., Volanakis, J. E., Mole, J. E. 1980. *Biochemistry* 19:1576–83

125. Müller-Eberhard, H. J., Götze, O. 1972. *J. Exp. Med.* 135:1003–8

126. Vogt, W., Dawes, W., Schmidt, G., Dieminger, L. 1977. *Immunochemistry* 14:201–5

127. Daha, M. R., Fearon, D. T., Austen, K. F. 1976. *J. Immunol.* 116:568–70

128. Day, N. K., Schreiber, R. D., Götze, O., Müller-Eberhard, H. J. 1976. *Scand. J. Immunol.* 5:715–20

129. Brade, V., Bentley, C., Bitter-Suermann, D., Hadding, V. 1977. *Z. Immunitaetsforsch. Exp. Klin. Immunol.* 152:402–14

130. Vogt, W., Dieminger, L., Lynen, R., Schmidt, G. 1974. *Hoppe-Seylers Z. Physiol. Chem.* 355:171–83

131. Cooper, N. R. 1971. *Progress in Immunology*, ed. B. Amos, 1:567–77. New York: Academic

132. Nydegger, U. E., Fearon, D. T., Austen, K. F. 1978. *J. Immunol.* 120:1303–8

133. Lachmann, P. J., Halbwachs, L. 1975. *Clin. Exp. Immunol.* 21:109–14

134. Abramson, N., Alper, C. A., Lachmann, P. J., Rosen, F. S., Jandl, J. H. 1971. *J. Immunol.* 107:19–27

135. Thompson, R. A., Lachmann, P. J. 1977. *Clin. Exp. Immunol.* 27:23–29

136. Whaley, K., Ruddy, S. 1976. *J. Exp. Med.* 144:1147–63

137. Weiler, J. M., Daha, M. R., Austen, K. F., Fearon, D. T. 1976. *Proc. Natl. Acad. Sci. USA* 73:3268–72

138. Ruddy, S., Whaley, K. 1977. *Fed. Proc.* 36:1244 (Abstr.)

139. Pangburn, M. K., Schreiber, R. D., Müller-Eberhard, H. J. 1977. *J. Exp. Med.* 146:257–70

140. Crossley, L. G., Porter, R. R. 1980. *Biochem. J.* 191:173–82

141. Whaley, K., Thompson, R. A. 1976. *Immunology* 35:1045–49

142. Gaither, T. A., Hammer, C. H., Frank,

M. M. 1979. *J. Immunol.* 123:1195–1204

143. Harrison, R. A., Lachmann, P. J. 1980. *Mol. Immunol.* 17:9–20

144. Pangburn, M. K., Müller-Eberhard, H. J. 1978. *Proc. Natl. Acad. Sci. USA* 75:2416–20

145. Pensky, J., Hinz, C. F., Todd, E. W., Wedgewood, R. J., Boyer, J. T., Lepow, I. H. 1968. *J. Immunol.* 100:142–58

146. Minta, J. O., Lepow, I. H. 1974. *Immunochemistry* 11:361–68

147. Medicus, R. G., Esser, A. F., Fernandez, H. N., Müller-Eberhard, H. J. 1980. *J. Immunol.* 124:602–6

148. Reid, K. B. M., Gagnon, J. 1981. *Mol. Immunol.* In press

149. Götze, O., Medicus, R. G., Müller-Eberhard, H. J. 1977. *J. Immunol.* 118:525–28

150. Fearon, D. T., Austen, K. F. 1975. *J. Exp. Med.* 142:856–63

151. Schreiber, R. D., Pangburn, M. K., Lesavre, P. H., Müller-Eberhard, H. J. 1978. *Proc. Natl. Acad. Sci. USA* 75:3948–52

152. Amos, N., Sissons, J. G. P., Girard, J-F., Lachmann, P. J., Peters, D. K. 1976. *Clin. Exp. Immunol.* 24:474–82

153. Parkes, C., Di Scipio, R. D., Kerr, M. K., Prohaska, R. 1981. *Biochem. J.* 193:In press

154. Fearon, D. T., Austen, K. F. 1977. *Proc. Natl. Acad. Sci. USA* 74:1683–87

155. Daha, M., Fearon, D. T., Austen, K. F. 1976. *J. Immunol.* 117:630–34

156. Nicol, P. A. E., Lachmann, P. J. 1973. *Immunology* 24:259–75

157. Pillemer, L., Blum, L., Lepow, I. H., Ross, D. A., Todd, E. W., Wardlaw, A. C. 1954. *Science* 120:279–85

158. Fearon, D. T., Austen, K. F. 1977. *J. Exp. Med.* 146:22–23

159. Marcus, R. L., Shiu, H. S., Mayer, M. M. 1971. *Proc. Natl. Acad. Sci. USA* 68:1351–54

160. Winkelstein, J. A., Tomasz, A. 1978. *J. Immunol.* 120:174–78

161. Schreiber, R. D., Morrison, D. C., Podack, E. R., Müller-Eberhard, H. J. 1979. *J. Exp. Med.* 149:870–82

162. Kierszenbaum, F., Ivanyi, J., Budzko, D. B. 1976. *Immunology* 30:1–6

163. Platts-Mills, T. E., Ishizaka, K. 1974. *J. Immunol.* 113:348–58

164. Fearon, D. T. 1978. *Proc. Natl. Acad. Sci. USA* 75:1971–75

165. Sissons, J. G. P., Cooper, N. R., Oldstone, M. B. A. 1979. *J. Immunol.* 120:2144–49

166. Sissons, J. G. P., Oldstone, M. B. A., Schreiber, R. D. 1980. *Proc. Natl. Acad. Sci. USA* 77:559–62

167. Theophilopoulos, A. N., Bokisch, V. A., Dixon, F. J. 1974. *J. Exp. Med.* 139:696–711

168. Schreiber, R. D., Pangburn, M. K., Medicus, R. G., Müller-Eberhard, H. J. 1980. *Clin. Immunol. Immunopathol.* 15:384–96

169. Loos, M., Raepple, E., Hadding, V., Bitter-Suermann, D. 1974. *Fed. Proc.* 33:775 (Abstr.)

170. König, W., Bitter-Suermann, D., Dierich, M., Limbert, M., Schorlemmer, H.-U., Hadding, V. 1974. *J. Immunol.* 113:501–6

171. Pangburn, M. K., Morrison, D. C., Schreiber, R. D., Müller-Eberhard, H. J. 1980. *J. Immunol.* 124:977–82

172. Kazatchkine, M. D., Fearon, D. T., Austen, K. F. 1979. *J. Immunol.* 122:75–81

173. Austen, K. F. 1978. *J. Immunol.* 121:793–805

174. Nydegger, U. E., Fearon, D. T., Austen, K. F. 1978. *Proc. Natl. Acad. Sci. USA* 75:6078–82

175. Kazatchkine, M. D., Fearon, D. T., Silbert, J. E., Austen, K. F. 1979. *J. Exp. Med.* 150:1202–15

176. Fearon, D. T. 1979. *Proc. Natl. Acad. Sci. USA* 76:5867–71

177. Schreiber, R. D., Götze, O., Müller-Eberhard, H. J. 1976. *J. Exp. Med.* 144:1062–75

178. Schreiber, R. D., Müller-Eberhard, H. J. 1978. *J. Exp. Med.* 148:1722–27

179. Sandberg, A. L., Osler, A. G. 1971. *J. Immunol.* 107:1268–73

180. Frank, M. M., Gaither, T., Adkinson, F., Terry, W. D., May, J. E. 1976. *J. Immunol.* 116:1733 (Abstr.)

181. Ferrone, S., Cooper, N. R., Pellegrino, M. A., Reisfeld, P. A. 1973. *Proc. Natl. Acad. Sci. USA* 70:3665–68

182. Nelson, B., Ruddy, S. 1979. *J. Immunol.* 122:1994–99

183. Press, E. M. 1975. *Biochem. J.* 149:73–82

184. Seegar, G. W., Smith, C. A., Schumaker, V. N. 1979. *Proc. Natl. Acad. Sci. USA* 76:907–11

185. Massoulie, J. 1980. *Trends Biochem. Sci.* 5:160–61

186. Swenson, R. P., Howard, J. B. 1979 *Proc. Natl. Acad. Sci. USA* 76:4313–16

187. Young, C. L., Barker, W. C., Tomaselli, C. M., Dahoff, M. O. 1978. *Atlas of Protein Sequence and Structure,* ed. M. O. Dahoff, 5:Suppl. 3, pp. 73–93. Silver Springs, Md: Natl. Biomed. Res. Found.

Ann. Rev. Biochem. 1981. 50:465–95

GLYCOPROTEIN HORMONES: STRUCTURE AND FUNCTION[1]

♦12085

John G. Pierce and Thomas F. Parsons

Department of Biological Chemistry, University of California School
of Medicine, Center for Health Sciences, Los Angeles, California 90024

CONTENTS

Perspectives and Summary

The term "glycoprotein hormone" usually is applied to gonadotropins, both of pituitary and placental origin, and to pituitary thyroid–stimulating hormone, although some other materials with hormonal activity, e.g. erythropoietin, are also glycosylated. There are two pituitary gonadotropins,

---

[1]Abbreviations used are: b, bovine; h, human; e, equine; p, porcine; o, ovine; PMSG, pregnant mare serum gonadotropin.

465

0066-4154/81/0701-0465$01.00

lutropin [luteinizing hormone (LH) or interstitial cell-stimulating hormone] and follitropin [follicle-stimulating hormone (FSH)]. Their general biological roles are the stimulation of testicular and ovarian functions via the regulation of gametogenesis and steroid hormone synthesis in the gonads (1). Pituitary thyroid–stimulating hormone (thyrotropin or TSH) regulates a wide variety of biochemical and physiological processes in the thyroid that result in the synthesis and secretion of thyroid hormones (2). The most studied glycoprotein hormone of placental origin is human chorionic gonadotropin (hCG); its hormonal effects closely resemble those of pituitary LH. These hormones, with their diverse physiological functions, are closely related in structure and are the most complex protein hormones yet recognized. Each consists of two peptide chains or subunits, designated α and β. Both subunits are glycosylated at specific residues and are highly cross-linked internally by disulfide bonds. These hormones are found in all mammals studied, and hormones with similar properties have also been observed in lower vertebrates including teleost fishes (3–5). Hormonal activity is expressed only after strong and specific noncovalent interactions between an α and a β subunit. Within a species, the α sequence is essentially identical for each hormone; it also is highly conserved from species to species. Both interhormone and interspecies hybrid hormones can be prepared. The hormonal activity of such hybrids is always dictated by the particular β subunit present, though how its specificity is expressed is not clear. The β subunits also show considerable homology in structure; it is most probable that they evolved from a common precursor (6). The hormones elicit their biological responses after interaction with a receptor or receptors on the surface of target cells, presumably by stimulation of the synthesis of cAMP with subsequent effects on the activity of various protein kinases. The identities of specific substrates for the kinases in the case of each hormone are still largely uncertain.

The chemistry of the interaction of these hormones with their receptors is of particular interest because, as far as protein and peptide hormones are concerned, their subunit structure is unique, with the exact role of the common, α subunit unknown. Studies on chemically and enzymatically modified hormones provide considerable evidence that regions of the α subunit, most probably in concert with domains on the β subunit, are involved directly in recognition of receptors. Each subunit is apparently produced by a separate gene, and cell-free translation of enriched mRNAs gives rise to products that can be precipitated with antisera specific to one or the other subunit. As with other secretory proteins, the translated message includes a leader or signal peptide that is removed by processing. Little is known about mechanisms of control in the biosynthesis of individual subunits or how an α and β subunit assemble, in vivo, to yield active hormone. The new techniques of molecular biology will surely answer some questions about the control of synthesis and secretion. This review emphasizes primarily the chemistry and biosynthesis of these hormones, and how modifications affect either the ability of subunits to associate to give an active product or to interact with their receptors. Little structural information is available concerning receptors, though we presume their chemistries are similar inasmuch as receptor structures probably coevolved with hormone structures.

In the past decade, because of the availability of well-characterized hormone preparations, the literature concerning the physiology and immunology of these hormones has expanded enormously and all have become important diagnostic reagents via radioimmunoassays (e.g. 7–9). Research on their mechanisms of action has also expanded. Many details of these areas are beyond the scope of this review and the reader is referred to, for example, the series *Recent Progress in Hormone Research.* Even in the areas covered, the literature has expanded to an extent that the authors were required to be selective [additional reviews are (9–16)]. We attempt to present a balanced view of the subject with most of the important citations.

Occurrence and Isolation

Extracts of the pituitary have been long known to affect many aspects of thyroid and gonadal growth and function. It gradually became clear that there are three separate chemical entities present in the anterior lobe; one a thyroid-stimulating principle, the others gonadotropic materials. Gonad-stimulating material of placental origin was found in human pregnancy urine in 1927 (17), and its presence continues to be the basis of pregnancy tests. It was recognized as differing from pituitary gonadotropins and, as methods for protein purification improved, increasingly potent preparations have been obtained (11, 13.)

The original observations of hCG or hCG-like molecules in the human pituitary were based on immunological comparisons, and recently a partial purification has been reported (18). There is also considerable evidence of hCG (detected by radioimmunoassay) in extracts of many tissues, both normal and tumor, and there is great interest in hCG as a tumor marker (e.g. 19, 20). TSH activity has also been found in chorionic carcinomas (e.g. 21) and in extracts of normal human placenta, but it now appears that chorionic thyrotropin may be artifactual (22) and that HCG has a definite, albeit low, intrinsic thyroid-stimulating activity (23). An hCG-like material originating from a culture of microorganism, named tentatively *Progenitor cryptocides,* has been considerably purified (24), and appears to be a biologically active glycoprotein with physicochemical properties similar to those of hCG. This work, which should be repeated, was preceded by several reports of immunoreactive hCG (e.g. 25) in microbial extracts; because of the nature of the assay, proteases (26) may have caused artifactual results in some instances. The recent finding of hCG-like materials in microorganisms and extraplacental tissues recalls an earlier report of TSH activity in culture media from *Clostridium perfringens* (27).

Materials with the biological properties of LH and FSH are present in relatively high amounts in postmenopausal urine and have been partially purified and characterized (27a).

Considerable characterization of these hormones from lower vertebrates has also been reported (3–6, 28, 29). Early work by Fontaine had suggested that structural similarities between thyrotropins and gonadotropins would be found (30) based on comparisons of composition and the observation that bovine gonadotropins stimulate the thyroids of trout. Only a partial sequence of one nonmammalian gonadotropin, carp α subunit, is available (31); it confirms the conservation of structure (see below) between α subunits of different mammals. Fontaine & Burzawa-Gerard (6) have proposed that all subunits of vertebrate pituitary glycoprotein hormones evolved from a common ancestral molecule that gave rise to two types of molecules, α and β, which then further evolved to give separate hormones. Although two distinct gonadotropins have been identified in most birds, amphibia, and reptiles studied, the question is not settled in fish (5, 28, 32).

In the 1950s the introduction of ion-exchange celluloses for the chromatography of proteins resulted in the isolation of preparations of LH and TSH of high potency. Although the then newly discovered technique of gel electrophoresis showed most preparations to consist of several closely related components, these later were found to have essentially the same amino acid sequence. Bovine TSH (bTSH) was the first thyrotropin to be fully purified; subsequently the hormone has been isolated from human, sheep, whale, and hog pituitaries (see 9) and from mouse tumors. Of the three pituitary hormones, LH is present in the largest amount and is more stable in its biological properties than TSH and FSH. Preparations of ovine LH, which were homogenous by the criteria of the time, were obtained in the late 1950s and early 1960s; subsequently LH has been isolated from many other species (10, 15). FSH has been the most difficult to isolate because of biological instability and the small amounts in pituitaries; human hormone was the first to be fully purified (33, for review see 34) and sequenced. Purification of the glycoprotein hormones has been complicated by polymorphism, which is shown in gel electrophoresis, and because partial inactivation often occurs during purification or storage. In some instances preparations exhibiting only a single electrophoretic component have been described (e.g. 35), but only after the subunits were identified and sequences determined was it generally accepted that fully purified (though microheterogeneous) preparations had been obtained. Based on increased knowledge of chemical properties, including their glycosylated nature, several general protocols for isolation of the glycoprotein hormones in a single run are available (e.g. 35–39).

Primary Structure

The presence of a subunit structure was first suggested by studies of Li and Starman (40) on the apparent dissociation of ovine LH at pH 1.3 (KCl-HCl). The sedimentation rate at this low pH differed from that at neutral pH and indicated a molecular weight for LH of 16,000 rather than ~ 30,000. Evidence that LH contains dissimilar subunits appeared (41), and the subunit nature of LH was firmly established by Papkoff & Samy (42) who developed a countercurrent distribution system that resulted not only in nearly complete dissociation of ovine LH but in excellent separation (see also 43). Evidence for subunits in hCG also appeared (11, 13); the definitive work demonstrating that one subunit was common to at least two of the hormones (LH and TSH) resulted from early peptide mapping and sequence studies (44–47). Additional studies on separation of subunits from different hormones and in different species then proliferated (for review see 10, 45, 48), and it was soon apparent that, within a species, the peptide portion of the α subunit was not only essentially identical between the four hormones but is highly conserved from species to species. Hybrid hormones were made, e.g. bTSH-β plus bLH-α, and the hormonal specificity of a hybrid was always dictated by the β subunit used (e.g. 45). More recent studies show that the binding properties of different LH preparations to heterologous receptors are largely but not exclusively a reflection of the species from which the β subunit is obtained (49, 50).

Figure 1 summarizes the known amino acid sequences of α subunits. The reference sequence is that of the bovine and ovine hormones. Dashes indicate residues identical to those in the reference sequence, and glycosylated residues are indicated by asterisks. As detailed in the legend, residues whose modification leads to significant changes either in ability to recombine with β subunits or in interaction with receptor are also indicated. A most interesting substitution is the interchange of His 87 and Tyr 93 in the equine versus the other species. The carboxyl-terminal region is otherwise highly conserved and appears necessary for interaction with receptors (see below).

A comparison of known β sequences is given in Figure 2 with references to laboratories carrying out sequence determinations. Not all sequences have been rigidly established in terms of all overlaps between, for example, tryptic peptides; and in some cases, amide assignments are not complete. Some discrepancies in data are still present between different groups; the sequences given represent, in our opinion, the best data. Homology between

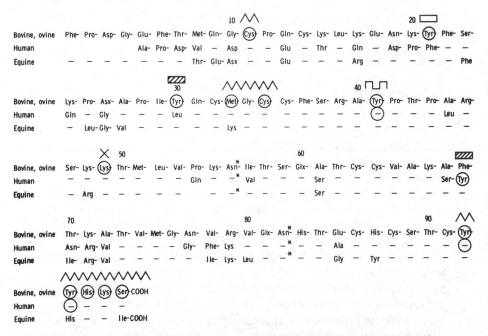

Figure 1 Amino acid sequences of the common, a, subunit of the glycoprotein hormones. The sequence of b,oLH is used as the reference; dashes indicate residues identical to this sequence. Asterisks show the points of attachment of carbohydrate. There are some uncertainties concerning amide groups in some sequences reported. The porcine sequence is six residues shorter than the bovine sequence and contains four replacements (47a). References to these sequences can be found in the legend to Figure 2. Those residues that have been chemically modified are circled, with the degree or effect of modification indicated above each residue: modified in the intact hormone, ▭ ; protected in the intact hormone, ▨ ; involved in an a-β cross-link, X; modification resulting in a significant decrease in receptor interaction, ⋀⋀ ; modification resulting in a significant decrease in subunit interaction,⊓⊔,. [Modified from (48) by permission of Academic Press Inc.]

β sequences becomes immediately apparent when the half cystine residues are placed in register (53), and there are regions of striking conservation, not only for the same hormone between species, but between hormones.

A singularly interesting difference in β sequences is that the two non-pituitary gonadotropins hCG and eCG (PMSG), have unusual COOH-terminal extensions compared to other members of the group. The sequence has been thoroughly worked out with hCG; there are 30 additional residues (relative to hLH) and the segment is rich in proline and serine residues. Several of the latter are O-glycosylated. As points of attachment of oligosaccharides in the rest of the β sequences and in the a sequences are via N-glycosylation, hCG-β and eCG-β represent two of the few instances where both N- and O-glycosylation are found in the same molecule. The function of the extended COOH-terminal domain is unknown; early ideas of an immunological role (69) have not been substantiated; it may confer protection against proteolysis (11).

Carbohydrate Moieties

It has been known since 1940 that carbohydrate is in partially purified preparations of the hormones and that it persists during further purification. Present data concerning the oligosaccharide portions of these hormones and their structures can be summarized as follows. The constituent sugars are D-mannose, D-galactose, L-fucose, N-acetyl neuraminic acid, D-glucosamine, and D-galactosamine (the latter are N-acetylated (70, 71)). Galactosamine is not a usual constituent of N-linked oligosaccharides of extracellular glycoproteins. It was first found in TSH and other soluble glycoproteins of the pituitary (72) and is not present in hCG, except in the O-linked oligosaccharides of the COOH-terminal extension. The percentage of carbohydrate, by weight, ranges from about 16% in bovine LH and TSH to 29–31% in hCG and about 45% in PMSG (13, 15, 68, 73, 74). FSH from all species studied and the chorionic gonadotropins contain neuraminic acid and galactose; however, these two sugars are found in negligible quantities in bovine LH and TSH. All α subunits contain two oligosaccharides of the "complex" type N-linked to asparagines; β subunits contain either one or two similar oligosaccharides, with the additional O-linked moieties in chorionic gonadotropins (see Figure 2). Definitive studies of oligosaccharide structures have been with hCG; the structures, which are those of Kessler et al (71, 75), are in Figure 3. Independently, three N-linked oligosaccharides were separated by Endo et al (76); their basic structures are the same as in Figure 3 with two of the oligosaccharides representing molecules lacking some peripheral sugars. Other structures for hCG and human FSH oligosaccharides were proposed earlier (77, 78); they do not seem as well-substantiated as the above.

While a number of detailed studies of the carbohydrate composition of the pituitary hormones have been made (for review see 15), definitive structural information is now appearing. Bahl et al (79) reported the partial structure of the oligosaccharides of bovine and ovine LH; an unknown acid labile group resides on the N-acetylgalactosamine. In an independent study to determine why bovine LH is resistant to the action of many glycosidases while hCG is much more susceptible (80), we arrived at a similar partial structure (Figure 3). Stability of the LH hexosamines to enzymatic deglycosylation and periodate oxidation, along with Smith degradation data, indicate that one residue of O-sulfated N-acetylgalactosamine and one of O-sulfated N-acetylglucosamine are terminal. The presence of sulfated sugars in protein-carbohydrate complexes was long thought to be limited to sulfated mucopolysaccharides. However, O-sulfated N-acetyl-glucosamines have been reported in rat brain glycoproteins (81), and recently in the glycoproteins of paramyxovirus SV5 (82) and chick embryonic liver and lung (83). We have also shown the presence of acid labile sulfate in bovine TSH-α and human LH, but not in hCG (80). The structures of the complex N-linked oligosaccharides, given in Figure 3, are in full agreement with the considerable body of knowledge concerning the presence of a common core, $(Man)_3GlcNAcGlcNAc \rightarrow Asn$, which is biosynthesized as a larger unit via a dolicholphosphate intermediate, with subsequent processing (84–88). Another structure proposed for the oligosaccharides of the α subunit of the human pituitary hormones shows an inner core with

```
                                      ⋀⋀  10                          ⋀⋀ 20
hTSHβ                                 Phe —  Ile  — Thr-Glx- Tyr(Met, Thr, His, Val,) — Arg-Arg-Glx  —  Ala-
bTSHβ                                 Phe —  Ile  — Thr-Glu- Tyr- Met- Met- His- Val  — Arg-Lys- Glu  —  Ala-
pTSHβ                                  —  —  Ile  — Thr-Glu- Tyr- Met- Met- His- Val  — Arg-Lys- Glu  —  Ala-
b,oLHβ  Ser- Arg-Gly- Pro- Leu-Arg- Pro- Leu-Cys- Gln-Pro- Ile- Asn* Ala- Thr-Leu- Ala- Ala- Glu-(Lys) Glu- Ala-Cys- Pro-
pLHβ     —   —   —    —    —   —    —    —   —    Arg  —   —  —*   —    —    —    —   Asp   —   —   —    —
hLHβ     —   —  Glu   —    —   —   (Trp) —   His   —    —   —   —    Ile  —    —  Val   —    —   —  Gly   —   —
hCGβ     —  Lys- Glu  —    —   —    Arg  —  Arg    —    —   —*  —    —    —  Val    —    —   —  Gly   —   —
hFSHβ                               Asn- Ser  —  Glu- Leu- Thr- —* Ile  —  Ile  —  Ile  —   —  Glu  —  Arg-
pFSHβ                                      —    Glu- Leu- Thr- —  Ile  —  Ile-Thr- Val  — Val- Lys        — Leu-Thr-
eFSHβ                               Asx- Ser  —  Glx- Leu- Thr  —* Ile  —  Ile  —  Val  —   —  Gly  —  Arg-
eCGβ      —   —   —   —    —   —    Arg  —   —*   —    —    —    —    —    —   —  Gln   —   — (Ser,

                                       30                ▨▨           ▨▨               40  ⋀⋀ ·
hTSHβ   Tyr  — Leu  —  Ile- Asn* —  Thr  —   —    —    —    — ( Met, Thr,)Arg- Asx- Ile- Asx- Gly- Lys- Leu- Phe-
bTSHβ   Tyr  — Leu  —  Ile- Asn* —  Thr- Val —    —    —    —   Met- Thr- Arg- Asx- Val- Asx- Gly- Lys- Leu- Phe-
pTSHβ   Tyr  — Leu  —  Ile- Asn* Ser- Thr —   —    —    —    —   Met- Thr- Arg- Asp- Phe- Asp- Gly- Lys- Leu- Phe-
b,oLHβ  Val- Cys- Ile- Thr- Phe- Thr- Thr- Ser- Ile- Cys- Ala- Gly-(Tyr) Cys- Pro- Ser- Met-(Lys) Arg- Val- Leu- Pro- Val- Ile-
pLHβ     —   —   —   —    —    —    —    —    —    —    — (—)  —    —    —   Arg   —    —    —    — Ala- Ala-
hLHβ     —   —   —   —  Val- Asn* —  Thr   —    —    —    —    —   Thr        —    —    —  Gln- Ala- Val-
hCGβ     —   —   —   —  Val- Asn* —  Thr   —    —    —    — (—)  —   Thr  —  Thr   —    —    —  Gln- Gly- Val-
hFSHβ   Phe  — Leu  —  Ile- Asn* —  Thr-(Trp) —    —    —    —   Tyr- Thr- Arg- Asp- Leu   —  Tyr- Lys- Asn- Pro-
pFSHβ   Phe  —  —  Ser- Ile- Asn  —  Thr- Trp  —    —    —    —   Thr- Thr- Gly- Arg- Asx- Leu- Val- Tyr- Lys- Asx-
eFSHβ   Phe  —  —   —  Ile- Asn* —  Trp   —    —    —    —   Tyr- Thr- Arg- Asp- Leu   —  Tyr- Lys- Asp- Pro-
eCGβ    Lys)  — (X  X    X   X)  —  Thr ( —  — )  —    —    —    —    —    —    —  Val   —    —  Thr   — Ala- Ala

          50       ⋀⋀                           60  ▨▨                                   70
hTSHβ    —   —  Lys- Tyr- Ala- Leu- Ser  —  Asx  —    —    —   Arg- Asp- Phe- Ile- Tyr- Arg- Thr   —  Glx- Ile   —
bTSHβ    —   —  Lys- Tyr- Ala- Leu- Ser  —  Asp  —    —    —   Arg- Asp- Phe- Met- Tyr- Lys- Thr- Ala- Glu- Ile   —
pTSHβ    —   —  Lys- Tyr- Ala- Leu- Ser  —  Asp  —    —    —   Arg- Asp- Phe- Met- Tyr- Lys- Thr   —  Glu- Ile   —
b,oLHβ  Leu- Pro- Pro- (Met) Pro  — Gln- Arg- Val- Cys- Thr-(Tyr) His- Glu- Leu- Arg- Phe- Ala- Ser- Val- Arg- Leu- Pro-
pLHβ     —   —   —  Val   —    —   —  Pro   —    —    —    — (—) Arg   —    —  Ile   —    —    —  Ser   —   —   —
hLHβ     —   —   —  Leu   —    —    —    —    —    —   Arg- Asp- Val   —    —  Glu   —  Ile   —    —   —   —
hCGβ     —   —  Ala  Leu   —    —   Val  —  Asn (—) Arg- Asp- Val   —    —  Glu   —  Ile   —    —   —   —
hFSHβ              Ala- Arg- Pro- Lys- Ile  — Lys- Thr   —  Phe- Lys   —    —  Val- Tyr- Glu- Thr   —   —  Val   —
pFSHβ           — Ala- Arg- Pro- Asx- Ile- Glx- Lys- Thr   —    —  Arg- Glx   —  Val- Tyr- Glx- Thr   —  Lys- Val   —
eFSHβ           Ala- Arg- Pro- Lys- Ile- —  Lys- Thr   —  Thr- Phe- Lys   —    —  Val- Tyr- Glu- Thr   —  Lys- Val   —
eCGβ     —   —  Ala  Ile   —    —  Pro   —   —    —    —   Arg   —    —    —    —    —  Ile   —    —   —
```

```
                         80              ▨▨              90              ⊓⊔
hTSHβ   —    —    —   Leu- His  —  (Ala, —, Tyr) Phe  —  Tyr  —    —    —    —    —   Lys  —    —   Lys  —  Asx-
bTSHβ   —    —    —   Arg- His  —   Thr  —  Tyr- Phe  —  Tyr  —    —   Ile  —    —   Lys  —    —   Lys  —  Asx-
pTSHβ   —    —    —   His- His  —   Thr  —  Tyr- Phe  —  Tyr  —    —   Ile  —    —   Lys  —    —   Lys  —  Asp-
b,oLHβ  Gly- Cys- Pro- Pro- Gly- Val- Asp- Pro- Met- Val- Ser- Phe- Pro- Val- Ala- Leu- Ser- Cys- His- Cys- Gly- Pro- (Cys) Arg-
pLHβ    —    —    —    —    —    —    —    —   Thr  —    —    —    —    —    —    —    —    —    —    —    —    —    —    —
hLHβ    —    —    —   Arg  —    —    —    —   Val  —    —    —    —    —    —    —    —   Arg  —    —    —    —    —    —
hCGβ    —    —    —   Arg  —    —   Asn  —   Val  —    —  (Tyr) Ala —    —    —    —    —   Gln  —   Ala- Leu  —    —
hFSHβ   —    —   Ala- His- His- Ala  —   Ser- Leu- Tyr- Thr- Tyr  —    —    —   Thr- Gln  —    —    —    —   Lys  —  Asp-
pFSHβ   —    —   Ala- His- His- Ala- Asx- Ser- Leu- Tyr- Thr- Tyr  —    —    —   Thr- Glu  —    —    —    —   Lys  —  Asx-
eFSHβ   —    —   Ala- His- His- Ala-  —   Ser- Leu- Tyr- Thr- Tyr  —    —    —   Thr- Glx  —    —    —    —   Lys  —  Asx-
eCG β   —    —    —    —    —    —   Asx  —  (Arg)  —   —    —    —   Gly  —    —    —    —   Asx  —    —    —    —  Glx

                        100             ⊓⊔⊓             110              X              ⋀⋁⋀
hTSHβ   Thr- Asx- Tyr- Ser  —    —   Ile- His (Glu, Ala, Ile) Lys- Thr- Asx- Tyr  —   Thr- Lys  —  Glx- Lys- Ser- Tyr- COOH
bTSHβ   Thr- Asx- Tyr- Ser  —    —   Ile- His- Glu- Ala- Ile- Lys- Thr- Asn- Tyr  —   Thr- Lys  —  Gln- Lys- Ser- (Tyr) (Met) COOH
pTSHβ   Thr- Asp- Tyr- Ser  —    —   Ile- His- Glu- Ala- Ile- Lys- Thr- Asn- Tyr  —   Thr- Lys  —  Glu- Lys- Ser- Tyr- COOH
b,oLHβ  Leu- Ser- Ser- Thr- Asp- (Cys) Gly- Pro- Gly- Arg- Thr- Glx- Pro- Leu- Ala- Cys- (Asx) His- Pro- Pro- Leu- Pro- Asp- Ile-
pLHβ    —    —    —   Ser  —    —    —    —    —    —    —   Ala  —    —    —    —    —   Arg  —    —    —    —   Gly- Leu —
hLHβ    Arg  —   Thr- Ser  —    —    —   Gly- Pro- Lys- Asp- His  —    —   Thr  —   Asp- His- Pro- Gln- CONH₂
hCGβ    Arg  —   Thr  —    —    —    —   Gly- Pro- Lys- Asp  His  —    —   Thr  —   Asp- Asp  —   Arg- Phe- Gln  —  Ser-
hFSHβ   Ser- Asp  —    —    —   Thr- Val- Arg- Gly- Leu- Gly  —   Ser- Tyr  —   Ser- Phe- Gly- Glu- Met- Lys- Gln- Tyr-
pFSHβ   Ser- Asx  —    —   Asx  —   Thr- Val- Arg- Gly- Leu- Gly  —   Ser- Tyr  —   Ser- Phe- Gly- Glu-COOH
eFSHβ   Ser- Asp  —    —   Asx  —   Thr- Val- Arg- Gly- Leu- Gly  —   Ser- Tyr  —   Ser- Phe- Gly- Asp- Met- Lys- Glx- Tyr-
eCG β   Ile- Lys- Thr  —    —    —    —    —   Val- Phe  —   Asn- Glu  —    —    —   Asp- Pro  —   Ser- Glx- Thr- Leu- Thr

b, o, pLHβ   Leu-COOH              130                          140
hCGβ    Ser- Ser- Ser- Lys- Ala- Pro- Pro- Pro- Ser*- Leu- Pro- Ser- Pro- Ser*- Arg- Leu- Pro- Gly- Pro- Ser*- Asp- Thr- Pro- Ile-
hFSHβ   Pro- Thr- Ala- Leu- Ser- Tyr-COOH
eFSHβ   Pro- Val- Ala- Leu- Ser- Tyr-COOH
eCG β   Ser (Thr, Ser, Arg) Ala- Pro- Pro- Pro- Glx- Thr- Ser- Pro (Ser, Thr, Pro) Lys (Gly, Pro, Pro, Ser, Arg) (Asx, Ser, Ala,

hCGβ    Leu- Pro- Gln- CONH₂
eCG β   Leu- Pro- Gln) CONH₂
```

Figure 2   Amino acid sequences of the β subunit of the glycoprotein hormones. Asterisks show the points of attachment of carbohydrate. The sequence of b,oLH is used as the reference; dashes indicate residues identical to this sequence; parenthesis indicate regions where sequence is based on composition only; amide groups are also uncertain in several sequences. Deletions are introduced (e.g. in the reference sequence at positions 52 and 55) in order to maximize similarities in the sequences. hTSH (51, 52); bTSH (53); pTSH (54); b,oLH (55–57); pLH (57, 47a); hLH-β (58–61) (the sequence given is that of the latter group); hCG(62, 63); hFSH-β (64, 65) (the sequence given is that of the latter group); pFSH-β (66); eFSH-β (67); eCG (68). Those residues that have been chemically modified are circled, with the degree or effect of modification indicated above each residue: modified in the intact hormone, ▭ ; protected in the intact hormone, ▨ ; involved in an α-β cross-link, X; modification resulting in a significant decrease in receptor interaction, ⋀ ; modification resulting in a significant decrease in subunit interaction, ⊓⊔. [Modified from (48) by permission of Academic Press Inc.]

the sequence Man → GalNAc → GlcNAc → Asn (78). This structure would require a biosynthetic scheme different from that reported for complex N-linked oligosaccharides, and is not consistent with the results obtained for the oligosaccharides of bovine LH.

The biological role of the carbohydrate in most glycoproteins remains unknown (89). The discovery by Ashwell, Morell, and their colleagues (for review see 90) that the exposure of vicinal galactose by desialylation of serum glycoproteins [including hCG (91)] resulted in a drastically reduced half-life in the blood, strongly suggests that sialic acid serves a protective function. This observation [similar ones have been made with FSH (92)] explains the loss of activity of asialo hCG in long term in vivo assays as opposed to in vitro assays, but it is still to be fully demonstrated that desialylation is a physiological mechanism for clearance. Pituitary bovine and ovine LH and TSH do not contain significant amounts of sialic acid,

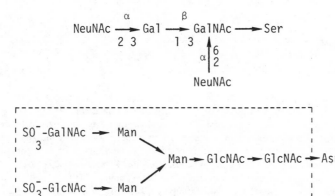

Figure 3 Oligosaccharide structures of the glycoprotein hormones. *Top,* N-linked oligosaccharides of hCG (71, 76); *middle,* O-linked oligosaccharides of hCG (75); *bottom,* N-linked oligosaccharides of bLH (80).

yet are bioactive in numerous in vivo assays, though these are heterologous. The presence of terminal N-acetylglucosamine, N-acetylgalactosamine, or phosphorylated mannose residues (for review see 93) has also been shown to affect cellular uptake of glycoproteins. Our data (80) suggest the possibility that all pituitary glycoprotein hormones contain O-sulfated hexosamines. Thus, it will be of considerable interest to determine whether, in pituitary hormones, the sulfated hexosamines play a role in recognition of receptor, uptake by target cells, or prevention of too rapid clearance from the blood.

As discussed above, exposure of vicinal galactose by desialylation of hCG decreases its in vivo biological activity (91). Further deglycosylation of hCG may be associated with loss of its in vitro biological activity at a post-receptor step (94–96). In a recent report, Sairam and Schiller (97) propose that ovine LH deglycosylated with HF retains receptor binding activity yet has poor steroidogenic activity. Weintraub et al (98), in a study of the de novo biosynthesis and secretion of TSH in dispersed mouse pituitary tumor cells, report that specific glycosylation may be required for α-β subunit combination but not secretion.

Disulfide Bonds

The positioning of the five S-S bridges in the α subunit and the six in the β subunit has proven to be extremely difficult, principally because few points of satisfactory proteolytic cleavage have been found. Several laboratories have tried various combinations of proteolysis, partial acid hydrolysis, and cleavage by cyanogen bromide. Only with the β subunit have any satisfactory results been obtained, with agreement between several laboratories concerning two or three assignments (99–102). With α subunits there is little or no agreement between laboratories. A second approach has been to examine the results of selective reduction of disulfides with subsequent tagging of the opened bonds by derivatization with a radioactive alkylating agent. These bonds are then identified after full reduction, derivatization, and isolation of tryptic peptides. In both approaches disulfide interchange has probably been encountered. Certainly the elegant studies of Creighton (103) on the refolding of pancreatic trypsin inhibitor have shown that extensive interchange can occur during refolding and reoxidation and thus presumably during reduction. However, an apparent specific opening of two of the five bonds in the α subunit has been found; they are between half-cystines 11–35 and 14–36, and these assignments are strengthened by studies on the refolding of 11,35-bis(S-carboxymethyl) LH-α (104). With bLH-β, partial reduction shows two bonds to open (101, 102), which agree with assignments obtained by examining products of proteolysis. Further reduction is accompanied by disulfide interchange, and it should be noted that none of the bonds agreed upon by more than one laboratory were deduced from fragments obtained by partial acid hydrolysis. In the authors' opinion the well-established positions in β subunits are those between half-cystines 95–102 and 26–112 (numbering of Figure 2) with a third bond likely between 23–74. Other proposals have been made (e.g. 105–108) which do not seem to be as well substantiated. When one considers that all β subunits can interact specifically with any α subunit, thus implying very similar conformations to all β subunits, the assumption that the relative positions of the disulfides in different β subunits are the same is logical. The failure to solve fully disulfide placement hampers model building of three-dimensional structures from sequence data; an easy resolution of the problem is not obvious.

At least one and probably all of the disulfide bridges in both subunits are readily accessible to the environment and thus presumably are on the surface, because complete reduction is possible with relatively low concentrations of sulfhydryl reagents in the absence of any denaturing agent (109, 110). Disulfide bonds appear to be a major force in the stabilization of the tertiary structures. For example, in the α subunit, bond 11–35 can be opened without significant change in conformation, but as soon as reduction of a second disulfide (14–36) begins, the native tertiary structure is lost rapidly (104).

Immunological Properties

As radioimmunoassay techniques developed, interest in the immunochemistry of the glycoprotein hormones for studies of their physiological roles, for diagnostic purposes, and for studying structural relations increased (reviewed in 7, 8). Antisera developed against intact hormones often were not specific for a single hormone. After subunits were recognized, interhormone cross-reactivity was recognized to occur not only because of contamination due to incomplete purification but because of the contribution of immunogenic determinants of the α portion of the molecules. Antisera against isolated α or β do not cross-react significantly against the counterpart subunit; the extent of cross-reaction between human α and bovine or rat α subunits is very slight (111) despite extensive homology in sequence. Probably the homologous domains are either inaccessible in intact hormones or are not immunogenic in the rabbit, the species most often used for antiserum production. Non-immunogenicity of the homologous regions would not be surprising because free α subunit is found in the normal circulation (e.g. 112). As an example of how specificity was achieved in RIAs for human TSH, antisera were absorbed with crude hCG, which combines with the antibodies directed against α determinants (113).

There has also been considerable interest in determining which regions of β subunits are immunogenic, particularly in hCG vis à vis hLH (114,

115). Antisera completely specific for hCG-β would have several uses, among them (e.g. 8, 11, 116) the possibility of prevention of pregnancy. Antisera against the unique C-terminal portion of hCG-β have been extensively studied; unfortunately, many either do not cross-react strongly with native hCG (an asialopeptide is often used to generate the antisera) or they do not neutralize the biological activity of hCG [see (11, 117, 118) for more detailed discussions]. Although the other structural differences between hCG-β and hLH-β are minor, occasionally an anti hCG-β antiserum is produced that has low cross-reactivity against hLH-β (119). Partially reduced or chemically modified derivatives have also been used to obtain antisera specific to hCG-β (115). Antisera against completely reduced and derivatized subunits [S-carboxymethyl derivative (120, 121)] do not cross-react significantly with the native structure; however, they do demonstrate increased reactivity between β subunits of different species. Some reports of weak cross-reactivity between intact subunits or hormone and antisera directed against fragments have been reported (e.g. 114), but it must be emphasized that, in general, the antigenic characteristics of these hormones reflect their three-dimensional structures.

Three-Dimensional Structure

Conformation of the glycoprotein hormones has been studied using several probes of three-dimensional structure. As yet, no successful X-ray crystallography has been achieved. That the overall conformations of these hormones are very similar is best shown by measurements of circular dichroism (CD) (122-124). Subsequent studies have taken place in many laboratories and the subject has been well reviewed (125, 126). These molecules appear to contain a low content of α helix (5–8%) and a relatively large amount of β structure (estimated between 25–40%); most of the remainder is aperiodic, though Garnier (125) has estimated about 3% β turn from CD data. A prediction program based on sequence data was utilized also to define regions of secondary structure and shows that all β subunits exhibit very homologous regions of extended (β structure) conformation. This procedure gave the following estimate for β subunits: helix, 1%; β structure 71%; β turns, 27%; and random coil 11%. With the two subunits combined, the amount of β sheet was somewhat less (61%), with a larger amount of β turn predicted. The values are considerably higher than those estimated from CD spectra. That considerable β sheet and β turns are present with a corresponding low α helix content is consonant with the high contents of proline (particularly in b,ö LH-β), the large number of disulfide bonds, and a compact shape; the latter has been deduced from hydrodynamic measurements (see 125). Further evidence of the similarity of conformation between hormones is that, despite differences in the number of tyrosines per hormone (7-16) several are found to have 3-4 residues that titrate abnormally with pKs above 11 (see 125). Similarly, the ease of modification of tyrosines in different hormones is very similar. The three most likely tyrosines to be unreactive and less easily perturbed are $\alpha41$, $\beta37$, and $\beta61$. All tyrosines in both subunits titrate normally after dissociation.

The major characteristic of the CD spectra in the far UV is a minimum at 210 ± 2 nm of -7500 to $-6000°$ cm^2 mol^{-1} which indicates considerable content of β structure. A contribution from carbohydrate becomes evident around 206 nm (e.g. 127, 128). In the near UV, the most obvious features are contributions from oriented cystines and tyrosines. In intact hormones, a maximum often becomes evident at 235 nm that appears to be a tyrosine-cystine interaction; in nonhuman α subunits it probably involves tyrosines 21 and 30. A negative dichroism is also present centering around 280 nm. It also appears in isolated β subunits but is replaced in α subunits by a small positive extremum between 265 nm and 275 nm; this may reflect a different chirality of some of the disulfides in α relative to β subunits. In those hormones containing tryptophan in the β subunit, e.g. hLH and hFSH, additional minima are seen in the region of 285-300 nm. Space does not permit a detailed description of all the spectral changes that occur during dissociation either at low pH or in the presence of urea or guanidine HCl (see 124, 125, 129-131). All studies show that conformational changes occur during dissociation, that are reversed during assembly of the subunits. Though the limitations of CD measurements prevent evaluation of the magnitude or exact location of these changes, examination of spectra after modification is a necessary and useful evaluation of the effects of modification on reassociation of subunits or on biological properties.

Possible Functions of the Subunit Structure

When the subunit nature of these hormones was discovered, it was hoped that new insights would emerge concerning the mechanism of action of polypeptide hormones that stimulate adenylate cyclases. As stated before, significant hormone activity is expressed only after formation of an $\alpha-\beta$ complex, with receptor specificity determined by the information in the β subunit structure. The important question is, why, of all the protein and peptide hormones, has a subunit structure evolved for this particular group; and why, when they have different physiological activities, do they share one subunit with a common structure? Possible functions of the subunit structure include the following: a prohormone (consisting of a single poly-peptide containing α and β sequences) may be required for correct folding; control of the amount of each hormone formed may be dictated by the amount of β subunit biosynthesized; receptor binding domains may be formed on both subunits only after formation of an $\alpha-\beta$ complex; and the α subunit may be internalized, after binding of the hormone to receptor, in order to stimulate adenylate cyclase.

Comparison Between Hormone-Receptor Interaction and Toxin Cell–Surface Interaction

The possibility that the hormone specific β subunit binds to a specific cell surface receptor and that the α subunit is either "internalized" into the cell or "intercalated" within the plasma membrane for stimulation of adenylate cyclase has been suggested on many occasions. This mode of action would explain the function of the common subunit, with the concept of a "binding" and an "active" subunit similar to the proposed actions of some bacterial and plant toxins (132).

Comparisons between the actions of toxins and the glycoprotein hormones were stimulated by the observations of Kohn and his colleagues (for review see 133) who reported that certain gangliosides inhibit the binding of TSH to thyroid membranes (134) and the binding of hCG and bovine LH to membrane components of rat testes (135, 136), thus paralleling the effect of ganglioside G_{M1} on the binding of cholera toxin (137). In addition, a similarity in sequence was found between the binding subunit of cholera toxin and the completely conserved region of residues 34–38 (Figure 2) in the β subunits of the glycoprotein hormones (138). This region is often referred to as the CAGY region (based on the single letter notation for amino acid sequences) and is suggested as a recognition site between either toxin B or hormone β subunits and receptors (138). A more restricted area of similarity is found between glycoprotein hormone α subunits, residues 11–14 (Figure 1), and the toxin A subunit, and is proposed as a possible site for interaction with the adenylate cyclase system (133).

Many contradictions concerning analogous modes of action between these hormones and the toxins can be found. No evidence exists to support the direct interaction of α with either regulatory or catalytic components of a cyclase system. Neither subunit by itself recognizes a target cell receptor (except possibly with an extremely low affinity) as measured by radioligand receptor assays (e.g. 139, 140). One group, however, has proposed that, while isolated β subunit cannot compete with bound [^{125}I]LH, uniodinated β does bind to receptor to a limited extent (141). Recent studies with hCG (142, 143) and bTSH (143a), using more physiological conditions than the earlier assays, show that rat testes gangliosides do not inhibit the binding of hCG to rat testes preparations, and that bovine brain gangliosides do not inhibit the binding of bTSH to human thyroid membrane preparations. Acidic phospholipids, as compared to thyroid gangliosides, were also found to be more potent inhibitors of TSH binding to bovine thyroid membranes (144). Whereas cholera toxin requires NAD for the activation of the cyclase system (145), the requirement for NAD in the stimulation of adenylate cyclase activity by TSH has not been demonstrated (134); and the NAD glycohydrolase activity present in highly purified hCG preparations was found to be a contaminant (146). Finally, the evidence from chemical modification experiments (discussed below) strongly supports the concept that the CAGY region of the β subunits is a region of subunit-subunit contact in the glycoprotein hormones, and therefore is not accessible for interaction with the receptor.

Thus, although it appears that the evidence does not yet support marked similarities in action between the glycoprotein hormones and toxins, the possibility is intriguing. Concerning the precise way that glycoprotein hormones with their subunit structure manifest their specificity and action, much remains to be learned. Considerable data, however, have been obtained concerning the functions of the subunits, both by specific chemical and enzymatic modifications of the hormones and by studies of their biosynthesis in cell-free systems and tumor cell lines.

Identification of Receptor Binding Regions and Contact Regions Between Subunits Via Chemical Modification Studies

The data from modification studies provide considerable evidence that regions of the α subunit may indeed be involved directly in recognition of receptors, most probably in concert with domains on the β subunit (reviewed in 48, 147, 148). Care must be used in attributing the loss of biological activity of a modified hormone to a particular region or residue; significant results are obtained only with reagents that are suitably selective and when the modified residue(s) can be located in the polypeptide chain.

Types of modifications that have been made in the glycoprotein hormones are: acylation of amino groups; nitration and iodination of tyrosine residues; alkylation of methionine and cysteine residues (after specific reduction); oxidation of tryptophan residues; and treatment with carboxypeptidases or CNBr. Two types of results have been obtained: those attributing the loss of biological activity of a modified hormone to a particular region or residue, and those indicating whether residues or regions are exposed or protected in the intact hormone as compared to the free subunits. Figures 1 and 2 summarize results that not only implicate regions involved in binding to receptors and in subunit-subunit interaction, but also locate the modified residue(s). The specific modifications and their respective references are listed in Table 1.

A special problem with these hormones is that a modification may lower the affinity between subunits to a point where the modified hormone dissociates at the low concentrations used for bioassays, receptor binding studies, or radioimmunoassays (1 nM–1 pM), and thus activity is lost because of dissociation rather than directly from modification. Two methods of ruling out dissociation at low concentrations as a consequence of modification have been used in this laboratory (e.g. 104, 149). One is a radioimmunoassay for the β subunit of bovine lutropin. The assay is 40-fold more sensitive to free β subunit than to the complexed β subunit in the intact hormone; thus, if modified hormones produce similar competition curves (displacement of [^{125}I]lutropin-β) as do unmodified controls, dissociation cannot have occurred. The second method is to cross-link covalently the subunits by carbodiimide mediated coupling. Cross-linked lutropin retains 50–100% of its binding and biological activity (149, 154–156), and thus further modifications can be made without subunit dissociation. If a loss in subunit interaction or receptor binding is to be attributed to modification of a particular residue, significant conformational changes should be ruled out, as changes in tertiary structure might involve domains distant from the modified groups. Circular dichroism (102, 104, 149, 172) and proton nuclear magnetic resonance (176) have been utilized in our laboratory to assess conformational changes as a result of a modification (see also 177).

The integrity of the COOH-terminal pentapeptide of the α subunit (Tyr-Tyr-His-Lys-Ser-COOH) apparently is required for receptor-hormone interaction. With α subunits from various sources, iodination of tyrosine residues 92 and 93 results in a modified subunit capable of recombining with native β, yet the recombinant has little biological activity (see Table 1). Removal of this pentapeptide by carboxypeptidase treatment also results in a modified α subunit capable of recombining with native β subunit, but again, biological activity is lost (149, 150). A recombinant with a carboxypeptidase treated α subunit covalently cross-linked to native β (149) also has negligible receptor binding activity, and its CD is not significantly different from that of native hormone. Thus the loss in biological activity resulting from the removal of the COOH-terminal pentapeptide does not appear to be a function of dissociation of the modified α and native β subunit or of significant conformational changes.

The region of the α subunit around the disulfide bond formed between cysteine residues 11 and 35 is also implicated in bLH as a receptor binding domain. The carboxymethylation of methionine-33 (151, 152), or the selective reduction of the disulfide bond formed between residues 11 and 35 and subsequent carboxymethylation of these reduced cysteine residues (104) results in an inactive product. The 11,35-bis(S-carboxymethyl) derivative will recombine with native β, but the recombinant has negligible binding activity. The recombinant formed between native β and a 11,35-bis(S-carbamoylmethyl) derivative, however, binds significantly to gonadotropin

Table 1 Chemical modifications of glycoprotein hormones

Residues modified[a]	Modification	Receptor binding[b]	Interaction of subunits[c]	Modification in intact hormone[d]	References
Bovine LH-α					
Tyr 92–Ser 96	carboxypeptidase	−	+	+	149, 150
Cys 11,35	carboxymethylation	−	+	+	104
Met 33	carboxymethylation	−	+		151, 152
Phe 1–Met 8	CNBr	+	+		153
Lys 49	carbodiimide X-linking	+	+	+	149, 154–156
Tyr 21,92,93	nitration	−	+	+	157
Tyr 30,41	nitration		−	−	157
Ovine LH-α					
Tyr 21,92,93	nitration	−	+	+	158, 159
Tyr 30,41	nitration		−	−	158, 159
Porcine LH-α					
Tyr 21	iodination	+	+	+	160
Tyr 92,93	iodination			+	160
Tyr 30,41	iodination			−	160
Human CG-α					
Tyr 92–Ser 96	carboxypeptidase	−	+	+	161, 162
Tyr 92,93	nitration	−	+	+	163, 164
Tyr 41	nitration		−	−	163
Tyr 69	nitration		+	−	163, 164
Bovine LH-β					
Met 53	carboxymethylation	−	+		151, 152
Lys 42–Met 53	CNBr	+	+		153
Tyr 37	nitration		+	−	157
Asp 113	carbodiimide X-linking	+	+	+	149, 154–156
Cys 95,102	carboxymethylation	−	−	+	Bloomfield, Reeve & Pierce, unpublished experiments
Bovine TSH-β					
Ser 118–Tyr 119	carboxypeptidase	−	+	+	150
Ovine LH-β					
Tyr 37,61	acylation/nitration/ iodination	+	+	−	158, 159, 165, 166
Lys 20,42	Acetylation/acetimi- dinylation/guanidina- tion/carbamylation	+	+		167–169
Lys 20,42	maleylation/citracony- lation	−	+		169, 170
Porcine LH-β					
Tyr 37	dansylation		−		171
Tyr 61	dinitrophenylation	−	+		171
Human LH-β					
Trp 8	N-bromosuccinimide	−	+	+	172

Table 1 *(Continued)*

Residues modified[a]	Modification	Receptor binding[b]	Interaction of subunits[c]	Modification in intact hormone[d]	References
Human FSH-β					
Trp 33	N-bromosuccinimide/2-hydroxy 5-nitrobenzyl bromide			−	172, 173
Human CG-β					
Tyr 37,61,84	nitration/iodination	+	+	−	163, 164, 174
Cys 95,102	carboxymethylation	+	+	+	175

[a] Residue numbers refer to those of the aligned sequences shown in Figures 1 and 2.

[b] −, modification significantly decreases receptor binding; +, modification has little or no effect on receptor binding.

[c] −, modification significantly decreases interaction between subunits; +, modification has little or no effect on interaction between subunits.

[d] +, residues modified in the intact hormone; −, residues protected in the intact hormone.

receptors. The loss of binding activity by the S-carboxymethyl hormone did not occur because of modification-induced dissociation, as shown by the bovine lutropin-β radioimmunoassay system. The CD of the modified recombinant differs very slightly from that of a recombinant of native subunits, thus negative charges at positions 11 and 35 must interfere with binding either by prohibiting the direct interaction of that portion of the α subunit with receptor or possibly by preventing the hormone from attaining an extremely stringent conformation needed for interaction with the receptor.

The effect of lysine modification in the β subunit is described in a series of papers by Ward and his colleagues (see Table 1). If isolated ovine LH-β, which contains two lysines, is modified such that the positive charge is either retained or lost (guanidination or carbamylation), neither recombination with native α nor biological activity of the recombinant is seriously affected. Introduction of a negative charge (maleylation or citraconylation) in the LH-β lysine residues allows recombination in good yield; however, most of the binding activity is lost. The ability of these modified β subunits to recombine with native α is not unexpected in that other β subunits have neutral or negatively charged rather than lysine residues at these positions. The loss of biological activity resulting from the introduction of negative charges could, however, implicate the regions around these lysine residues as possible sites for lutropin receptor recognition. Other residues implicated as possible receptor recognition sites for the β subunits are tryptophan-8 in human LH-β (172), methionine-53 in bovine LH-β (151, 152), and the COOH-terminal dipeptide of bovine TSH-β (150).

Hormonal activity is only expressed after interaction of an α and β subunit, with hormonal specificity dictated by the particular β subunit. Moore et al (178) propose an interesting "determinant loop" region in all the β subunits that may be of primary importance in the expression of hormone specificity. This octapeptide loop is formed by the disulfide bond between the half cystines at positions 95 and 102, and contains a number of nonconservative substitutions (Figure 2) that provide the possibility of net charge transitions on the loop. Their model assumes that a neutral or net positive charge is required on the loop for LH-like activity. A net negative charge is required for FSH/TSH activity, with TSH specificity dictated by a crucial aromatic residue at position 99. An interesting modification implicating the integrity of this disulfide loop required for subunit-subunit interaction is the selective reduction of this disulfide bond and subsequent carboxymethylation or carbamoylmethylation. Neither the 95,102-bis(S-carboxymethyl) nor the 95,102-bis(S-carbamoylmethyl) derivatives of bovine LH-β will recombine with native LH-α subunit (G. A. Bloomfield, J. R. Reeve, and J. G. Pierce, unpublished experiments). However, a contrary result has been reported for hCG (175).

Several chemical modification studies of the α subunit (see Table 1) indicate that tyrosine-41 is protected from modification in all the intact hormones, but nitration of this residue in the free subunit prevents recombination with native β. Tyrosine-30 in the bovine, ovine, and porcine α subunits is not only protected in the intact hormone, but also in the free subunit (157–159). Nitration studies with human α subunits indicate tyrosine residue 69 to be at least partially protected in the intact hormone. It is thus likely that tyrosine-41 as well as the tyrosine/phenylalanine at position 69 is in a contact region with the β subunit.

Soon after the first sequences were determined, conserved β domains were proposed to be, at least in part, regions of subunit-subunit interaction (e.g. 148, 179, 180). Iodination and nitration studies (see Table 1) support the hypothesis that the tyrosines or phenylalanines in the conserved regions of the β subunits (i.e. at positions 37, 61, and 84) are in regions of subunit-subunit interaction. These tyrosines are protected against modification in the intact hormone; however, modification in the free β subunit does not affect either recombination with native α subunit or biological activity of the modified recombinant. The tryptophan residue in position 33 of the human FSH-β subunit is also protected in the intact hormone (172, 173), which strengthens the idea that the highly conserved region containing residues 31–38 is in an area of contact between subunits. The major amide cross-link in carbodiimide mediated cross-linked bovine lutropin is formed by covalent coupling of β-aspartic acid-113 and α-lysine-49 (155), which indicates that these residues are also in regions of subunit-subunit interaction.

Biosynthesis of the Glycoprotein Hormones

While the chemical modifications described above implicate receptor interaction with regions of the "common" α subunit in concert with regions of the "hormone-specific" β subunit, other roles of the α subunit could be in the regulation of assembly and secretion of these hormones. An early suggestion was that a single polypeptide chain containing both the α and β sequence is required for the correct folding of a percursor hormone (181). This role was suggested because of the detection of immunoreactive high-molecular-weight human LH eluting at the void volume of Sephadex G-100 columns. Similar reports of high-molecular-weight species of human FSH (182) and rat TSH (183) have appeared. However, many laboratories have found excess free α subunits present in the normal pituitary (e.g. 112) and in normal placenta (e.g. 7), which supports the concept that synthesis of α and β subunits are independently regulated, with specific β subunit synthesis limiting the production of complete glycoprotein hormones.

Recent reports on the cell-free translation of the mRNAs coding for the subunits of hCG (184), bovine LH (185, 186), and mouse TSH (187–190) demonstrate that α and β subunits are independently translated from separate mRNAs. The nascent α subunit chains ($M_r = 14,000$) include an amino-terminal hydrophobic extension of about 24 residues (189, 191). Amino acid sequences of the glycoprotein hormone "pre-α" and "pre-β" subunits are shown in Figure 4.

The translated pre-α subunits are precipitated with antisera specific for native α; however, an intriguing problem exists with the recognition of translated pre-β subunits. Landefeld & Kepa (186) report that translated bovine "pre-LH-β" is recognized by antiserum specific for native β, in contrast to Daniels-McQueen et al (184) who show that translated "pre-hCG-β" is recognized by antiserum specific for the reduced carboxymethylated β. Giudice & Weintraub (189) report that translated mouse "pre-TSH-β" is only recognized by antiserum specific for the reduced carboxymethylated β; whereas mRNA translated in the presence of microsomal membranes produced a "processed" TSH-β specifically immunoprecipitated by antiserum to native but not denatured TSH-β. This suggests that the conformation of pre-TSH-β differs from that of the subunit processed by microsomal membranes. Giudice & Pierce (194, 195) have shown that isolated α subunits can readily refold in vitro to their native conformation after reduction of their five disulfide bonds. As yet, refolding of reduced β subunits has not been accomplished (J. M. Goverman and J. G. Pierce, unpublished experiments). The necessity for a "pro-β" has not been demonstrated; however, the possibility that the translation product of β mRNA is a "prepro-β" with some residues necessary for folding and formation of the six disulfide bonds cannot be dismissed.

```
              -24              -20                                    -10
hCG-α         Met-Asp-Tyr-Tyr-Arg-Lys-Tyr-Ala-Ala-Ile-Phe-Leu-Val-Thr-Leu-

Bovine pituitary-α Met-                              Leu-        Leu-

Mouse tumor-α      Met-                              Leu-        Met-Leu-

Mouse pituitary-α  Met-                                         Met-

hCG-β                        Met-Glu-Met-Phe-Gln-Gly-Leu-Leu-Leu-Leu-Leu-

                                       -1
hCG-α         Ser-Val-Phe-Leu-His-Val-Leu-His-Ser-

Bovine pituitary-α   Leu-    Leu-        Leu-

Mouse tumor-α        Met-    Leu-        Leu-

Mouse pituitary-α    Met-

hCG-β         Leu-Leu-Ser-Met-Gly-Gly-Thr-Trp-Ala-
```

Figure 4 Amino acid sequences of the prepeptides contained in the α and β subunits of the glycoprotein hormones. Blanks indicate amino acid assignments that have not been made. The hCG-α and -β prepeptide sequences agree with those identified from the recently determined nucleotide sequences of the cDNAs coding for the α (192) and β (193) subunits. hCG-α (191, 192); bovine pituitary, mouse tumor, and mouse pituitary α (188); hCG-β (191, 193) (the sequence given is that of the latter group).

An additional area of current research concerns post-translational processing and secretion of the glycoprotein hormones. Recent studies using cell-free translation systems in the presence of microsomal membranes (189, 190, 196–198) indicated post-translational processing of pre-α (M_r = 14,000) to yield a mannose rich α subunit (M_r = 20,000) containing an oligosaccharide that is sensitive to degradation by mannosidase and endoglycosidase H. Studies using tumor cell lines (199, 200) report three intracellular forms of the α subunit (M_r = 14,000, 18,000, and 21,000) as indicated by immunoprecipitation and SDS gel electrophoresis. The 14,000 form (corresponding to the position of pre-α) is in minor amounts, with the glycosylated 18,000 form converted progressively to the mature 21,000 form.

Less is known about the processing of the β subunit forms, which possibly reflects the difficulties encountered in detecting the products of cell-free translation. The predominant intracellular mouse TSH-β form (M_r = 18,000) is glycosylated and is reported to combine with the excess mature α form (M_r = 21,000) (199). The newly formed TSH and excess free α, but not free β, are secreted into the medium. Similarly, two intracellular hCG-β forms (M_r = 18,000 and 24,000) are found (200), with the 18,000 form having the same apparent molecular weight as pre-hCG-β

(184). The 24,000 form is glycosylated and is converted to a mature β form ($M_r = 34,000$). The secreted hCG contains mature α ($M_r = 22,000$) and mature β ($M_r = 34,000$) subunits.

The intracellular concentrations of α and β subunits could also be regulated at a transcriptional or post-transcriptional level. The recent synthesis and isolation of the cDNA sequences complementary to mRNAs for hCG-α (192), hCG-β (193), and mouse TSH-α, and β subunits (201) should provide powerful tools for quantitating the levels of mRNA and the following mRNA processing. The availability of specific cDNAs should also provide valuable probes for the elucidation of hormonal regulation of glycoprotein hormone gene expression.

In Vitro Dissociation and Reassociation of Subunits

While little is still known about in vivo assembly, the in vitro dissociation and, particularly, reassociation of the subunits of the glycoprotein hormones has been extensively studied. Techniques employed include measurement of the differential absorption, circular dichroism, in vivo biological activity, radioligand receptor assays, immunological probes, and binding of the fluorescent dye, analine napthalene sulfonate (ANS). The interactions between subunits, which are solely noncovalent, are very strong; it has been impossible to detect significant amounts of free subunits under physiological conditions of pH and concentration (10^{-8}–10^{-11} M (202, 203). Thus no direct measurements of an affinity constant have been made. Dissociation is enhanced by lowering pH and/or adding denaturants such as urea or increasing temperature. The kinetics of dissociation at acid pH are found by most workers to be first order, though sometimes some deviations are observed (204, 205). The latter may be due to microheterogeneity of preparations [for example, see (206) in which three preparations of ovine LH of different biological potencies gave qualitatively but not quantitatively the same tyrosine difference spectra upon dissociation]. In acid solution, the relative rates of dissociation of several hormones differ from those found in urea at neutral pH (for review see 207). The data show that despite homologies between β sequences, different groups influence the strength of interaction between subunits.

Little is known about association of subunits in vivo; the rates found in vitro are relatively slow and are pH dependent (e.g. 125). The apparent equilibrium constants for association increase with pH, and Bishop (208) has proposed that a minimum of three proton binding sites are involved in subunit interaction; the associated form is more acidic than the sum of individual subunits. The kinetics of in vitro assembly appear more complex than second order; the rates are low. Data from several laboratories (e.g. 205, 209–211) suggest that formation of an active hormone is at least a two-step process with formation of an initial α-β complex that is followed by at least one slow (and rate-limiting) folding step to yield conformation that can interact with receptor. It should be noted that exact studies of association are hampered by the fact that often full hormonal activity is not recovered due presumably to some "damaged" molecules in preparations; α subunits are particularly susceptible to inactivation in reassociation assays (212). Studies of hydrogen-deuterium exchange of pLH and its subunits by infrared spectroscopy (213) shows LH to be very accessible to solvent. The combination of the subunits gives rise to the shielding of only eight peptide hydrogens that are unshielded in the free subunits. The data indicate that the peptide hydrogens of the α subunit are much more accessible to solvent than those of both intact hormone and β subunits, and that there is considerable flexibility of conformation of both subunits and hormone in solution. The authors conclude that the appearance of biological activity is not due to induction of an active conformation in one subunit by its counterpart, but that residues from both subunits are involved in the activity of the hormone.

Summary of Events Subsequent to Hormone-Receptor Interaction

The glycoprotein hormones elicit their biological responses after interaction with a hormone-specific receptor on the surface of the target cell, and the subsequent stimulation of cAMP production via adenylate cyclase. Although it is beyond the scope of this review, the biochemical properties of the hormone-sensitive adenylate cyclase system have recently been reviewed (214). Most studies of glycoprotein actions have concentrated on the immediate responses that follow the binding of the hormone to its specific receptor; however, much less is known about the possible long-term actions on target-cell growth and differentiation.

It is now clear that receptors for these hormones are individual proteins distinct from adenylate cyclase (for review see 215). The resolution of LH receptors and adenylate cyclase activities (216), and the functional transfer of solubilized ovarian LH receptors to isolated adrenal cells (217), support this concept of distinct proteins. However, the characterization of the solubilized receptors has not yet revealed the mechanism of coupling between receptors and adenylate cyclase. Receptors for LH and hCG (218–221), FSH (222–224), and TSH (225–228) have been partially characterized after detergent solubilization. In most cases, the ligand binding characteristics of the soluble receptor are essentially the same as those of the membrane-bound protein. The possible purification of the LH receptor (220) and the TSH receptor (225), using affinity chromatography, has also been reported.

The binding of polypeptide hormones to specific cell surface receptors was traditionally thought to be a reversible process. It has now become clear that the hormone-receptor interaction is much more complex and cannot be described as a simple bimolecular reaction. The hormone-receptor complex is not only involved with the stimulation of adenylate cyclase, but also functions in the ligand-induced regulation of the receptors and in the internalization and lysosomal degradation of the receptor-bound hormone (reviewed in 215, 229). Desensitization of cellular responses by the gonadotropins (e.g. 215) has been observed, and in some cases has been correlated with a loss in receptor sites (down-regulation). Studies with hCG (for review see 230) suggest not only that hormone degradation occurs in the lysosomes, but also that this degradation is not required for the stimulation of progesterone production in Leydig tumor cells. Thus, other than as a mechanism for terminating hormone action, a biological role of hormone-receptor internalization and degradation remains unknown.

Recently, two derivatives of the glycoprotein hormones that could prove useful in the elucidation of the events subsequent to hormone-receptor interaction have been described. Hormone preparations with the α and β subunits covalently cross-linked (149, 154–156a) or labeled with ^{131}I in the α subunit and ^{125}I in the β subunit (231, 232) have been prepared. Both derivatives retain significant receptor and biological activities. The fact that bovine LH and TSH retain biological activity with their subunits covalently cross-linked indicates that, after binding to their receptors, these hormones need not dissociate to elicit hormonal effects in their target cells. The preparation of doubly labeled bovine LH should prove useful in following the intercellular processing of both subunits simultaneously following interaction of the hormone with the target cell.

Conclusion

While much has been learned in the past ten years, many challenges remain. A particularly interesting one is to utilize the specificity of hormone-receptor interaction in the conveyance of a toxic material to a particular cell type, normal or abnormal. A hybrid between the toxic (A) fragment of the plant lectin, ricin, and hCG-β has been synthesized. It appears to be selectively toxic to a gonadotropin-sensitive cell line, though at concentrations several thousand times greater than the amount of hCG required, in many systems, for hormonal stimulation (233). Further work in this area should be fruitful. Other challenges include the elucidation of three-dimensional structures together with detailed information on how they interact with the almost unknown structures of their receptors. Considerable practical importance could result from the discovery of competitive inhibitors that, while interacting with receptors, would not result in stimulation of further biochemical steps leading to a hormonal response.

ACKNOWLEDGMENTS

Research in this laboratory has been supported by Grant AM 18005 from the National Institutes of Health. J. G. Pierce expresses appreciation to students and associates who have contributed greatly to knowledge of the glycoprotein hormones. On the occasion of Volume 50 of the *Annual Review of Biochemistry*, appreciation is also expressed to Professor J. Murray Luck who along with H. S. Loring introduced the senior author to biochemistry.

Literature Cited

1. Catt, K. J., Pierce, J. G. 1978. In *Reproduction Endocrinology,* ed. S.S.C. Yen, R. B. Jaffe, pp. 34–62. Philadelphia: Saunders
2. Field, J. B. 1978. In *The Thyroid,* ed. S. C. Werner, S. H. Ingbar, pp. 185–95. Hagerstown, Md: Harper & Row. 4th ed.
3. Fontaine, Y. A. 1969. *Acta Endocrinol.* 60: Suppl. 136 p. 154
4. Licht, P., Farmer, S. W., Papkoff, H. 1977. *Recent Prog. Horm. Res.* 33:169–248
5. Farmer, S. W., Papkoff, H. 1979. In *Hormones and Evolution,* ed. E. J. Barrington, 2:325–59. London: Academic
6. Fontaine, Y. A., Burzawa-Gerard, E. 1977. *Gen. Comp. Endocrinol.* 32: 341–47
7. Vaitukaitis, J. L., Ross, G. T., Braunstein, G. D., Rayford, P. L. 1976. *Recent Prog. Horm. Res.* 32:289–31
8. Vaitukaitis, J. L. 1978. In *Structure and Function of the Gonadotropins,* ed. K. W. McKerns, pp. 339–60. New York: Plenum
9. Condliffe, P. G., Weintraub, B. D. 1980. In *Hormones in Blood,* ed. C. H. Gray, A. L. Bacharach, pp. 499–574. New York: Academic. 3rd ed.
10. Liu, W.-K., Ward, D. N. 1975. *Pharm. Therap. Pt. B* 1:545–70
11. Birken, S., Canfield, R. E. 1978. See Ref. 8, pp. 47–80
12. Greep, R. O. 1977. *Frontiers in Reproduction and Fertility Control,* ed. R. O. Greep. Cambridge, Mass: MIT Press. 580 pp.
13. Bahl, O. P. 1977. See Ref. 12, pp. 11–24
14. Bishop, W. H., Nureddin, A., Ryan, R. J. 1976. In *Peptide Hormones,* ed. J. Parsons, pp. 273–98. London: Macmillan
15. Jutisz, M., Tertrin-Clary, C. 1974. *Curr. Top. Exp. Endocrinol.* 2:195–246
16. Catt, K. J., Dufau, M. L. 1977. *Ann. Rev. Physiol.* 39:529–57

17. Asheim, S., Zondek, B. 1927. *Klin. Wochenschr.* 6:1322
18. Matsuura, S., Ohashi, M., Chen, H. C., Skownkeen, R. C., Hartree, A. S., Reichert, L. E. Jr., Stevens, V., Powell, J. G. 1980. *Nature* 286:740–41
19. Braunstein, G. D. 1979. In *Immunodiagnosis of Cancer,* ed. R. B. Haberman, K. R. McIntyre, pp. 383–409. New York: Marcel Dekker
20. Vaitukaitis, J. L. 1979. *N. Engl. J. Med.* 301:324–26
21. Kenimer, J. G., Hershman, J. M., Higgins, H. P. 1969. *J. Clin. Invest.* 48:923–29
22. Harada, A., Hershman, J. M. 1978. *J. Clin. Endocrinol. Metab.* 47:681–85
23. Taliadouros, G. S., Canfield, R. E., Nisula, B. C. 1978. *Endocrinology* 47:855–60
24. Maruo, T., Cohen, H., Segal, S. J., Koide, S. S. 1979. *Proc. Natl. Acad. Sci. USA* 76:6622–26
25. Livingston, V. W. C., Livingston, A. M. 1974. *Trans. NY Acad. Sci.* 36:569–82
26. Maruo, T., Segal, S. J., Koide, S. S. 1979. *Endocrinology* 104:932–37
27. Macchia, V., Bates, R. W., Pastan, I. 1967. *J. Biol. Chem.* 242:3726–30
27a. van Hell, H., Schuurs, A. H. W. M., den Hollander, F. C. 1972. In *Gonadotropins,* ed. B. B. Saxena, C. G. Beling, H. M. Gandy, pp. 185–99. New York: Wiley-Intersci.
28. Fontaine, Y. A., Burzawa-Gerard, E. 1978. See Ref. 8, pp. 361–80
29. Fontaine, Y. A. 1980. *Reprod. Nutr. Dev.* 20:381–418
30. Fontaine, Y. A. 1969. In *Prog. Endocrinol. Int. Congr. Ser. 184,* ed. C. Gual, pp. 453–57. Amsterdam: Excerpta Medica
31. Jolles, J., Burzawa-Gerard, E., Fontaine, Y. A., Jolles, P. 1977. *Biochemic* 59:893–98
32. Ng, T. B., Idler, D. R. 1979. *Gen. Comp. Endocrinol.* 38:410–20

33. Roos, P. 1968. *Acta Endocrinol. Copenhagen* Suppl. 131:1–99
34. Saxena, B. B., Rathnam, P. 1978. See Ref. 8, pp. 183–212
35. Roos, P., Jacobson, G., Wide, L. 1975. *Biochim. Biophys. Acta* 379:247–61
36. Bloomfield, G. A., Faith, M. R., Pierce, J. G. 1978. *Biochim. Biophys. Acta* 533:371–82
37. Hartree, A. S. 1966. *Biochem. J.* 100:754–61
38. Closset, J., Hennen, G. 1978. *Eur. J. Biochem.* 86:105–13
39. Cheng, K.-W. 1976. *Biochem. J.* 159:651–59
40. Li, C.-H., Starman, B. 1964. *Nature* 202:291–92
41. Fujino, M., Lamkin, W. M., Mayfield, J. D., Ward, D. N. 1968. *Gen. Comp. Endocrinol.* 11:444–57
42. Papkoff, H., Samy, T. S. A. 1967. *Biochim. Biophys. Acta* 147:175–77
43. de la Llosa, P., Courte, C., Jutisz, M. 1967. *Biochem. Biophys. Res. Commun.* 26:411–16
44. Liao, T.-H., Pierce, J. G. 1970. *J. Biol. Chem.* 245:3275–81
45. Pierce, J. G. 1971. *Endocrinology* 89:1331–34
46. Pierce, J. G., Liao, T.-H., Carlsen, R. B., Reimo, T. 1971. *J. Biol. Chem.* 246:866–72
47. Ward, D. N., Sweeney, C. M., Holcomb, G. N., Lamkin, W. M., Fujino, M. 1968. See Ref. 30, pp. 385–93
47a. Maghuin-Register, G., Combarnous, Y., Hennen, G. 1973. *Eur. J. Biochem.* 39:255–63
48. Pierce, J. G., Parsons, T. F. 1980. In *The Evolution of Protein and Structure Function,* ed. D. Sigman, M. A. B. Brazier, pp. 99–117. New York: Academic
49. Glenn, S. D., Liu, W.-K., Ward, D. N. 1980. *Endocrinology* 106:Suppl., 143. (Abstr.)
50. Strickland, T. W., Puett, D. 1980. *Endocrinology* 106:Suppl., 77. (Abstr.)
51. Shome, B., Parlow, A. F. 1973. *Endocrinology* 92:Suppl., A-60. (Abstr.)
52. Sairam, M. R., Li, C.-H. 1977. *Can. J. Biochem.* 55:755–60
53. Pierce, J. G., Liao, T.-H., Howard, S. M., Shome, B., Cornell, J. S. 1971. *Recent Prog. Horm. Res.* 27:165–212
54. Maghuin-Register, G., Hennen, G., Closset, J., Kopeyan, C. 1976. *Eur. J. Biochem.* 61:157–63
55. Liu, W.-K., Nahm, H. S., Sweeney, C. M., Holcomb, G. N., Ward, D. N. 1972. *J. Biol. Chem.* 247:4365–81
56. Sairam, M. R., Samy, T. S. A., Papkoff,

57. Maghuin-Register, G., Hennen, G. 1973. *Eur. J. Biochem.* 39:235–53
58. Closset, J., Hennen, G., Lequin, R. M. 1973. *FEBS Lett.* 29:97–100
59. Shome, B., Parlow, A. F. 1973. *J. Clin. Endocrinol. Metab.* 36:618–21
60. Sairam, M. R., Li, C.-H. 1973. *IRCS Res. Commun. System* (Nov.)
61. Keutmann, H. T., Williams, R. M., Ryan, R. J. 1979. *Biochem. Biophys. Res. Commun.* 90:842–48
62. Carlsen, R. B., Bahl, O. P., Swaminathan, N. 1973. *J. Biol. Chem.* 248:6810–27
63. Morgan, F. J., Birken, S., Canfield, R. E. 1975. *J. Biol. Chem.* 250:5247–58
64. Shome, B., Parlow, A. F. 1974. *J. Clin. Endocrinol. Metab.* 39:203–5
65. Saxena, B. B., Rathnam, P. 1976. *J. Biol. Chem.* 251:993–1005
66. Closset, J., Maghuin-Register, G., Hennen, G., Strosberg, A. D. 1978. *Eur. J. Biochem.* 86:115–20
67. Fujiki, Y., Rathnam, P., Saxena, B. B. 1978. *J. Biol. Chem.* 253:5363–68
68. Moore, W. T., Burleigh, B. D., Ward, D. N. 1981. In *Proc. Lake Como Conf. on the Chorionic Gonadotropin Molecule,* ed. S. Siegel. New York: Plenum. In press
69. Adcock, E. W. III, Teasdale, F., August, C. S., Cox, S., Meschia, G., Battaglia, F. C., Naughton, M. A. 1973. *Science* 181:845
70. Ward, D. N., Coffey, J. 1964. *Biochemistry* 3:1575–77
71. Kessler, M., Reddy, M. S., Shah, R. H., Bahl, O. P. 1979. *J. Biol. Chem.* 254:7901–8
72. Wynston, L. K., Free, C. A., Pierce, J. G. 1960. *J. Biol. Chem.* 235:85–90
73. Christakos, S., Bahl, O. P. 1979. *J. Biol. Chem.* 254:4253–61
74. Papkoff, H., Bewley, T. A., Ramachandran, J. 1978. *Biochim. Biophys. Acta* 532:185–94
75. Kessler, M. J., Mise, T., Ghai, R. D., Bahl, O. P. 1979. *J. Biol. Chem.* 254:7909–14
76. Endo, Y., Yamashita, K., Tachibana, Y., Tojo, S., Kobata, A. 1979. *J. Biochem.* 85:669–79
77. Kennedy, J. F., Chaplin, M. F. 1976. *Biochem. J.* 155:303–15
78. Hara, K., Rathnam, P., Saxena, B. B. 1978. *J. Biol. Chem.* 253:1582–91
79. Bahl, O. P., Reddy, M. S., Bedi, G. S. 1980. *Biochem. Biophys. Res. Commun.* 96:1192–99

H., Li, C.-H. 1972. *Arch. Biochem. Biophys.* 153:572–86

80. Parsons, T. F., Pierce, J. G. 1981. *Proc. Natl. Acad. Sci. USA* 77: In press
81. Margolis, R. K., Margolis, R. V. 1970. *Biochemistry* 9:4389–96
82. Prehm, P., Scheid, A., Choppin, P. W. 1979. *J. Biol. Chem.* 254:9669–72
83. Heifetz, A., Kinsey, W. H., Lennarz, W. J. 1980. *J. Biol. Chem.* 255:4528–34
84. Waechter, C. J., Lennarz, W. J. 1976. *Ann. Rev. Biochem.* 45:95–112
85. Li, E., Tabas, I., Kornfeld, S. 1978. *J. Biol. Chem.* 253:7762–70
86. Kornfeld, S., Li, E., Tabas, I. 1978. *J. Biol. Chem.* 253:7771–78
87. Tabas, I., Kornfeld, S. 1978. *J. Biol. Chem.* 253:7779–86
88. Chapman, A., Li, E., Kornfeld, S. 1979. *J. Biol. Chem.* 254:10243–49
89. Kornfeld, S., Kornfeld, R. 1976. *Ann. Rev. Biochem.* 45:217–37
90. Ashwell, G., Morell, A. G. 1974. *Adv. Enzymol.* 4:99–128
91. Van Hall, E. V., Vaitukaitis, J. L., Ross, G. T., Hickman, J. W., Ashwell, G. 1971. *Endocrinology* 88:456–64
92. Amir, S. M., Barker, S. A., Butt, W. R., Crooke, A. C. 1966. *Nature* 209: 1092–93
93. Neufeld, E. P., Ashwell, G. 1980. In *The Biochemistry of Glycoproteins and Proteoglycans,* ed. W. J. Lennarz, pp. 241–66. New York: Plenum
94. Moyle, W. R., Bahl, O. P., Marz, L. 1975. *J. Biol. Chem.* 250:9163–69
95. Channing, C. P., Sakai, C. N., Bahl, O. P. 1978. *Endocrinology* 103:341–47
96. Bahl, O. P., Moyle, W. R. 1978. In *Receptors and Hormone Actions,* ed. L. Birnbaumer, B. W. O'Malley, 3:261–89. New York: Academic
97. Sairam, M. R. 1980. *Arch. Biochem. Biophys.* 204:199–206
98. Weintraub, B. D., Stannard, B. S., Linnekin, D., Marshall, M. 1980. *J. Biol. Chem.* 255:5715–23
99. Chung, D., Sairam, M. R., Li, C.-H. 1975. *Int. J. Peptide Protein Res.* 7:487–93
100. Tsunasawa, S., Liu, W.-K., Burleigh, B. D., Ward, D. N. 1977. *Biochim. Biophys. Acta* 492:340–56
101. Reeve, J. R., Cheng, K.-W., Pierce, J. G. 1975. *Biochem. Biophys. Res. Commun.* 67:149–55
102. Giudice, L. C., Pierce, J. G. 1978. See Ref. 8, pp. 81–110
103. Creighton, T. E. 1978. *Progr. Biophys. Mol. Biol.* 33:231–98
104. Giudice, L. C., Pierce, J. G. 1979. *J. Biol. Chem.* 254:1164–69
105. Chung, D., Sairam, M. R., Li, C.-H.

1973. *Arch. Biochem. Biophys.* 159: 678–82
106. Combarnous, Y., Hennen, G. 1974. *Biochem. Soc. Trans.* 2:135–37
107. Cornell, J. S., Pierce, J. G. 1974. *J. Biol. Chem.* 249:4166–74
108. Fujiki, Y., Rathnam, P., Saxena, B. B. 1980. *Biochim. Biophys. Acta* 624: 428–35
109. Pierce, J. G., Giudice, L. C., Reeve, J. R. 1976. *J. Biol. Chem.* 251:6388–91
110. Holmgren, A., Morgan, F. J. 1976. *Eur. J. Biochem.* 70:377–83
111. Vaitukaitis, J. L., Ross, G. T., Reichert, L. E. Jr. 1973. *Endocrinology* 92: 411–16
112. Kourides, I. A., Weintraub, B. D., Ridgway, E. C., Maloof, F. 1975. *J. Clin. Endocrinol. Metab.* 40:872–85
113. Odell, W. P., Wilber, J. F., Utiger, R. D. 1967. *Recent Prog. Horm. Res.* 16:47–77
114. Swaminathan, N., Braunstein, G. D. 1978. *Biochemistry* 17:5832–38
115. Bahl, O. P., Muralidhar, K., Chanduri, G., Lippes, J. 1980. *Endocrinology* 106:Suppl., 136. (Abstr.)
116. Stevens, V. C. 1976. In *Physiological Effects of Immunity Against Reproductive Hormones,* ed. R. G. Edwards, M. H. Johnson, pp. 249–74. Cambridge: Cambridge Univ. Press
117. Chen, H.-C., Matsuura, S., Hodgen, G. D., Ross, G. T. 1978. In *Novel Aspects of Reproductive Physiology,* ed. C. H. Spilman, J. W. Wilkes, pp. 289–308. New York: Spectrum Publ.
118. Birken, S., Canfield, R., Laner, R., Agosto, G., Gabel, M. 1980. *Endocrinology* 106:1659–64
119. Vaitukaitis, J. L., Brunstein, G. D., Ross, G. T. 1972. *Am. J. Obstet. Gynecol.* 113:751–58
120. Pierce, J. G., Faith, M. R., Donaldson, E. M. 1976. *Gen. Comp. Endocrinol.* 30:47–60
121. Keutmann, H. T., Beitins, I. Z., Johnson, L., McArthur, J. W. 1978. *Endocrinology* 103:2349–56
122. Jirgensons, B., Ward, D. N. 1970. *Tex. Rep. Biol. Med.* 28:553–59
123. Pernollet, J.-C., Garnier, J. 1971. *FEBS Lett.* 18:189–92
124. Bewley, T. A., Sairam, M. R., Li, C.-H. 1972. *Biochemistry* 11:932–36
125. Garnier, J. 1978. See Ref. 8, pp. 381–414
126. Bewley, T. A. 1979. *Recent Prog. Horm. Res.* 35:155–213
127. Holladay, L. A., Puett, D. 1975. *Arch. Biochem. Biophys.* 171:708–20

128. Puett, D., Nureddin, A., Holladay, L. A. 1976. *Int. J. Peptide Protein Res.* 8:183–91
129. Bishop, W. H., Ryan, R. J. 1973. *Biochemistry* 12:3076–84
130. Combarnous, Y., Maghuin-Rogister, G. 1974. *Eur. J. Biochem.* 42:7–12
131. Ingham, K. C., Aloj, S. M., Edelhoch, H. 1974. *Arch. Biochem. Biophys.* 163:589–99
132. Olsnes, S., Pappenheimer, A. M. Jr., Meren, R. 1974. *J. Immunol.* 113:842–47
133. Kohn, L. D. 1978. In *Receptors and Recognition,* ed. P. Cuatrecasas, M. F., Greaves, 5:134–212. London: Chapman & Hall
134. Mullin, B. R., Aloj, S. M., Fishman, P. H., Lee, G., Kohn, L. D., Brady, R. O. 1976. *Proc. Natl. Acad. Sci. USA* 73:1679–83
135. Lee, G., Aloj, S. M., Brady, R. O., Kohn, L. D. 1976. *Biochem. Biophys. Res. Commun.* 73:370–77
136. Lee, G., Aloj, S. M., Kohn, L. D. 1977. *Biochem. Biophys. Res. Commun.* 77:434–41
137. Gill, D. M. 1977. *Adv. Cyclic Nucleotide Res.* 8:85–118
138. Mullin, B. R., Fishman, P. H., Lee, G., Aloj, S. M., Ledley, F. D., Winand, R. J., Kohn, L. D., Brady, R. O. 1976. *Proc. Natl. Acad. Sci. USA* 73:842–46
139. Rayford, P. L., Vaitukaitis, J. L., Ross, G. T., Morgan, F. J., Canfield, R. E. 1972. *Endocrinology* 91:144–46
140. Williams, J. F., Davies, T. F., Catt, K. J., Pierce, J. G. 1980. *Endocrinology* 106:1353–59
141. Muralidhar, K., Mougdal, N. R. 1976. *Biochem. J.* 160:615–19
142. Pacuszka, T., Osborne, J. C. Jr., Brady, R. O., Fishman, P. H. 1978. *Proc. Natl. Acad. Sci. USA* 75:764-68
143. Azhar, S., Menon, K. M. J. 1979. *Eur. J. Biochem.* 94:77–85
143a. Pekonen, F., Weintraub, B. 1980. *J. Biol. Chem.* 255:8121–27
144. Omodeo-Sale, F., Brady, R. O., Fishman, P. H. 1978. *Proc. Natl. Acad. Sci. USA* 75:5301–5
145. Moss, J., Vaughan, M. 1979. *Ann. Rev. Biochem.* 48:581–600
146. Moss, J., Ross, P. S., Agosto, G., Birken, S., Canfield, R. E., Vaughan, M. 1978. *Endocrinology* 102:415–19
147. Pierce, J. G., Faith, M. R., Giudice, L. C., Reeve, J. R. 1976. In *The Peptide Hormones: Molecular and Cellular Aspects, Ciba Found. Symp.,* pp. 225–50. North Holland: Elsevier Excerpta Medica
148. Ward, D. N. 1978. See Ref. 8, pp. 31–45
149. Parsons, T. F., Pierce, J. G. 1979. *J. Biol. Chem.* 254:6010–15
150. Chen, K.-W., Glazer, A. N., Pierce, J. G. 1973. *J. Biol. Chem.* 248:7930–37
151. Cheng, K.-W. 1976. *Biochem. J.* 159:71–77
152. Cheng, K.-W. 1976. *Biochem. J.* 159:79–87
153. de la Llosa, P., Tertrin-Clary, C., Jutisz, M. 1972. In *Structure-Activity Relationships of Protein and Polypeptide Hormones,* ed. M. Margoulies, F. C. Greenwood, pp. 344–47. Amsterdam: Excerpta Medica
154. Weare, J. A., Reichert, L. E. Jr. 1979. *J. Biol. Chem.* 254:6964–71
155. Weare, J. A., Reichert, L. E. Jr. 1979. *J. Biol. Chem.* 254:6972–79
156. Combarnous, Y., Hennen, G. 1974. *FEBS Lett.* 44:224–28
156a. Burleigh, B. D., Liu, W.-K., Ward, D. N. 1978. *J. Biol. Chem.* 253:7179–85
157. Cheng, K.-W., Pierce, J. G. 1972. *J. Biol. Chem.* 247:7163–72
158. Burleigh, B. D., Liu, W.-K., Ward, D. N. 1976. *J. Biol. Chem.* 251:308–15
159. Liu, W.-K., Ward, D. N. 1976. *J. Biol. Chem.* 251:316–19
160. Combarnous, Y., Maghuin-Rogister, G. 1974. *Eur. J. Biochem.* 42:13–19
161. Hum, V. G., Botting, H. G., Mori, K. F. 1977. *Endocrinol. Res. Commun.* 4:205–15
162. Merz, W. E. 1979. *Eur. J. Biochem.* 101:541–53
163. Carlsen, R. B., Bahl, O. P. 1976. *Arch. Biochem. Biophys.* 175:209–20
164. Hum, V. G., Knipfel, J. E., Mori, K. E. 1974. *Biochem.* 13:2359–64
165. Sairam, M. R., Papkoff, H., Li, C.-H. 1972. *Biochim. Biophys. Acta* 278:421–32
166. Yang, K.-P., Ward, D. N. 1972. *Endocrinology* 91:317–20
167. Liu, W.-K., Esfahani, M., Ward, D. N. 1975. *Endocrinol. Res. Commun.* 2:47–63
168. Liu, W.-K., Yang, K.-P., Nakagawa, Y., Ward, D. N. 1974. *J. Biol. Chem.* 249:5544–50
169. Liu, W.-K., Ward, D. N. 1975. *Biochim. Biophys. Acta* 405:522–26
170. Sairam, M. R., Li, C.-H. 1975. *Arch. Biochem. Biophys.* 167:534–39
171. Combarnous, Y., Hennen, G. 1976. *Int. J. Peptide Protein Res.* 8:491–98
172. Giudice, L. C., Pierce, J. G., Cheng, K.-W., Whitley, R., Ryan, R. J. 1978. *Biochem. Biophys. Res. Commun.* 81:725–33

173. Rathnam, P., Saxena, B. B. 1979. *Biochim. Biophys. Acta* 576:81–87
174. Canfield, R. E., Morgan, F. J., Kammerman, S., Ross, G. T. 1972. See Ref. 153, pp. 341–43
175. Mori, K. F., Hum, V. G., Wood, R. J. 1979. *Int. Congr. Biochem., 11th, Toronto,* p. 185. (Abstr.) Natl. Res. Counc. Can.
176. Brown, F. F., Parsons, T. F., Sigman, D. S., Pierce, J. G. 1979. *J. Biol. Chem.* 254:4335–38
177. Maghuin-Rogister, G., Degelaen, J., Roberts, G. C. K. 1979. *Eur. J. Biochem.* 96:59–68
178. Ward, D. N., Moore, W. T. Jr. 1979. In *Animal Models for Research on Contraception and Fertility,* ed. N. J. Alexander, pp. 151–63. New York: Harper & Row
179. Pierce, J. G., Liao, T.-H., Cornell, J. S., Carlsen, R. B. 1971. See Ref. 153, pp. 91–98
180. Stewart, M., Stewart, F. 1977. *J. Mol. Biol.* 116:175–79
181. Prentice, L. G., Ryan, R. J. 1975. *J. Clin. Endocrinol. Metab.* 40:303–12
182. Reichert, L. E. Jr., Ramsey, R. B. 1977. *J. Clin. Endocrin. Metab.* 44:545–52
183. Klug, T. L., Adelman, R. C. 1977. *Biochem. Biophys. Res. Commun.* 77:1431–37
184. Daniels-McQueen, S., McWilliams, D., Birken, S., Canfield, R., Landefeld, T., Boime, I. 1978. *J. Biol. Chem.* 253:7109–14
185. Landefeld, T. D. 1979. *J. Biol. Chem.* 254:3685–88
186. Landefeld, T. D., Kepa, J. 1979. *Biochem. Biophys. Res. Commun.* 90:1111–18
187. Chin, W. W., Habener, J. F., Kieffer, J. D., Maloof, F. 1978. *J. Biol. Chem.* 253:7985–88
188. Giudice, L. C., Waxdal, M. J., Weintraub, B. D. 1979. *Proc. Natl. Acad. Sci. USA* 76:4798–802
189. Giudice, L. C., Weintraub, B. D. 1979. *J. Biol. Chem.* 254:12679–12683
190. Kourides, I. A., Vamvakopoulos, N. C., Maniatis, G. M. 1979. *J. Biol. Chem.* 254:11106–110
191. Birken, S., Featherston, J., Canfield, R. E., Boime, I. 1981. *J. Biol. Chem.* In press
192. Fiddes, J. C., Goodman, H. M. 1979. *Nature* 281:351–56
193. Fiddes, J. C., Goodman, H. M. 1980. *Nature* 286:684–87
194. Giudice, L. C., Pierce, J. G. 1976. *J. Biol. Chem.* 251:6392–99
195. Giudice, L. C., Pierce, J. G. 1978. *Biochim. Biophys. Acta* 533:140–46
196. Bielinska, M., Boime, I. 1978. *Proc. Natl. Acad. Sci. USA* 75:1768–72
197. Bielinska, M., Grant, G. A., Boime, I. 1978. *J. Biol. Chem.* 253:7117–19
198. Bielinska, M., Boime, I. 1979. *Proc. Natl. Acad. Sci. USA* 76:1208–12
199. Weintraub, B. D., Stannard, B. S., Linnekin, D., Marshall, M. 1980. *J. Biol. Chem.* 255:5715–23
200. Ruddon, R. W., Hanson, C. A., Bryan, A. H., Putterman, G. J., White, E. L., Perini, F., Meade, K. S., Aldenderfer, P. H. 1980. *J. Biol. Chem.* 255:1000–7
201. Vamvakopoulos, N. C., Monahan, J. J., Kourides, I. A. 1980. *Proc. Natl. Acad. Sci. USA* 77:3149–53
202. Ryan, R. J., Jiang, H., Hanlon, S. 1970. *Recent Prog. Horm. Res.* 26:105–37
203. Tashjian, A. H. Jr., Weintraub, B. D., Barowsky, N. J., Rabson, A. S., Rosen, S. W. 1973. *Proc. Natl. Acad. Sci. USA* 70:1419–22
204. Reichert, L. E. Jr., Trowbridge, C. G., Bhalla, V. K., Lawson, G. M. 1974. *J. Biol. Chem.* 249:6472–77
205. Salesse, R., Castaing, M., Pernollet, J.-C., Garnier, J. 1975. *J. Mol. Biol.* 95:483–96
206. Garnier, J., Pernollet, J.-C., Tertrin-Clary, C., Salesse, R., Castaing, M., Barnavon, M., de la Llosa, P., Jutisz, M. 1975. *Eur. J. Biochem.* 53:243–54
207. Aloj, S., Ingham, K. 1977. In *Endocrinology,* Int. Congr. Ser. 403, ed. V. H. T. James, 3:108–13. Amsterdam: Excerpta Medica
208. Bishop, W. H. 1978. *J. Mol. Biol.* 122:379–81
209. Reichert, L. E. Jr., Lawson, G. M. Jr., Leidenberger, F. L., Trowbridge, C. G. 1973. *Endocrinology* 93:938–46
210. Reichert, L. E. Jr., Leidenberger, F., Trowbridge, C. G. 1973. *Recent Prog. Horm. Res.* 29:497–532
211. Ingham, K. C., Aloj, S. M., Idelhoch, H. 1967. *Arch. Biochem. Biophys.* 163:589–99
212. Weintraub, B. D., Stannard, B. S., Rosen, S. W. 1977. *Endocrinology* 101:225–35
213. Combarnous, Y., Naberdryk-Viala, E. 1978. *Biochem. Biophys. Res. Commun.* 84:1119–24
214. Ross, E. M., Gilman, A. G. 1980. *Ann. Rev. Biochem.* 49:533–64
215. Catt, K. J., Harwood, J. P., Aguilera, G., Dufau, M. L. 1979. *Nature* 280:109–16
216. Dufau, M. L., Baukal, A., Ryan, D.,

Catt, K. J. 1977. *J. Mol. Cell Endocrinol.* 6:253–69
217. Dufau, M. L., Hayashi, K., Sala, G., Baukal, A., Catt, K. J. 1978. *Proc. Natl. Acad. Sci. USA* 75:4796–73
218. Dufau, M. L., Charreau, E. H., Catt, K. J. 1973. *J. Biol. Chem.* 248:6973–82
219. Charreau, E. H., Dufau, M. L., Catt, K. J. 1974. *J. Biol. Chem.* 249:4189–95
220. Dufau, M. L., Ryan, D. W., Baukal, A. J., Catt, K. J. 1975. *J. Biol. Chem.* 250:4822–24
221. Wimalasena, J., Dufau, M. L. 1980. *Fed. Proc.* 39(6):1796 (Abstr.)
222. Abou-Issa, H., Reichert, L. E. Jr. 1977. *J. Biol. Chem.* 252:4166–74
223. Closset, J., Maghuin-Rogister, G., Ketelslegers, J.-M., Hennen, G. 1977. *Biochem. Biophys. Res. Commun.* 79:372–79
224. Dufau, M. L., Ryan, D. W., Catt, K. J. 1977. *FEBS Lett.* 81:359–62
225. Tate, R. L., Holmes, J. M., Kohn, L., Winand, R. 1975. *J. Biol. Chem.* 250:6527–33
226. Dawes, P. J. D., Petersen, V. B., Rees-Smith, B., Hall, R. 1978. *J. Endocrinol.* 78:103–117
227. Petersen, V. B., Petersen, M. M., Brennan, A., Rees-Smith, B., Hall, R. 1979. *J. Endocrinol.* 83:52P
228. Powell-Jones, C. H. J., Thomas, C. G. Jr., Nayfeh, S. N. 1980. *Fed Proc.* 39(6):-1796
229. Gorden, P., Carpenter, J.-L., Freychet, P., Orci, L. 1980. *Diabetologia.* In press
230. Ascoli, M. 1980. In *Receptor Mediated Binding and Internalization of Toxins and Hormones,* ed. J. Middlebrook. New York: Academic. In press
231. Sharp, S. B., Pierce, J. G. 1980. *Endocrinology* 106:Suppl. 277 (Abstr.)
232. Campbell, K. L., Landefeld, T. D., Midgley, A. R. Jr. 1980. *Proc. Natl. Acad. Sci. USA* 77:4793–97
233. Oeltmann, T. N., Heath, E. C. 1979. *J. Biol. Chem.* 254:1028–32

Ann. Rev. Biochem. 1981. 50:497–532
Copyright © 1981 by Annual Reviews Inc. All rights reserved

PROTEIN FOLDING   ♦12086

Michael G. Rossmann and Patrick Argos

Department of Biological Sciences, Purdue University, West Lafayette,
Indiana 47907

CONTENTS

PERSPECTIVE

Linderstrøm-Lang and his co-workers (1) were the first to recognize structural levels of organization within a protein. They introduced the terms primary, secondary, and tertiary structure. Although a variety of helical secondary structures had been proposed (cf 2, 3), it was Pauling (4, 5) who recognized the α-helix and β-pleated sheet, which provide an acceptable interpretation of Astbury's α- and β-diffraction patterns for fibrous proteins. Nevertheless, details of the α-helix were not seen at high resolution until the advent of the myoglobin structure (6), while the first atomic resolution observation of a β-sheet as a small antiparallel segment in lysozyme was not published until 1965 (7). Since that time well over 100 distinct structures have been determined. This wealth of information has led to a detailed examination of structural hierarchy as displayed by folded polypeptide chains.

The notion was entertained, even in the 1930s, that a protein would spontaneously refold after in vitro denaturation (8, 9). While three-dimensional structures demonstrate the uniqueness of a general fold with respect to a given protein, they do not directly discern the folding pathways. It was not until the 1960s, when the properties of proteins were better understood,

497

0066-4154/81/0701-0497$01.00

that the concept of spontaneous renaturation enjoyed wide acceptance. The pivotal work was that of Anfinsen and his co-workers, who "scrambled" ribonuclease, with its eight sulfhydryl groups, by allowing the reduced protein to reoxidize under denaturing conditions of 8 M urea (10). Removal of the denaturant and addition of mercaptoethanol resulted in a stable, functionally active conformation, though "unscrambling" frequently took hours to complete, an obvious discrepancy with in vivo rates. Denaturation-renaturation investigations have since been performed on a variety of other proteins including myoglobin (11), staphylococcal nuclease (12), lysozyme (13), and pancreatic trypsin inhibitor (14–16). The most detailed work has centered on disulfide proteins, where the covalent formation of S–S bonds can be used to characterize intermediates. In this way Creighton (17–19) was able to draw a folding pathway for bovine pancreatic trypsin inhibitor (Figure 1). It is noteworthy that essential intermediates exhibit some incorrect S–S pairing (17).

With the prompting of crystallographic results that revealed the form of folded proteins and renaturation experiments that demonstrated the spontaneity of refolding, a significant understanding emerged of the physical principles underlying the folding operation. Elementary principles of thermodynamics state that folding in a constant physiological environment

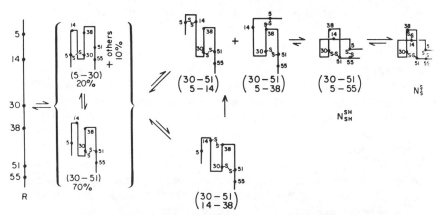

Figure 1 Schematic diagram of the pathway of folding and unfolding of normal bovine pancreatic trypsin inhibitor. The solid line represents the polypeptide backbone, with the positions of the cysteine residues indicated. The configurations of species N $_{SH}^{SH}$ and N $_S^S$ approximate the conformation of the native inhibitor; those of the others are relatively arbitrary except for the relative positions of the cysteine residues involved in disulfide bonds.

The brackets around the single-disulfide intermediates indicate that they are in rapid equilibrium; only the two most predominant species are depicted. The + between intermediates (30–51,5–14) and (30–51,5–38) signifies that both are formed directly from the single-disulfide intermediates, that both are converted directly to N $_{SH}^{SH}$, and that either or both are intermediates in the arrangement of (30–51,14–38) to N $_{SH}^{SH}$. [Reprinted with permission from Creighton (17). Copyright by Academic Press Inc. (London) Ltd.]

must be synonymous with the reduction of Gibbs free energy, though the folded protein may not have attained a "global" minimum (20, 21). Physically this implies burying of hydrophobic groups within the folded molecular core, creation of ion pairs and hydrogen bonds, and reduction of molecular surface. However, such considerations do not provide any information on pathway, since they are concerned with energy. Accordingly, the complex interactions of the polypeptide chain with itself and the environment must be considered. This requires a great simplification of the appearance of the polypeptide (20, 22–24). Alternative simplifications, suggested by light scattering and hydrodynamic measurements, assume nucleation centers, such as helices, around which the polypeptide can condense.

Methods to predict secondary structure from the primary amino acid sequence have been developed to avoid the difficult thermodynamic and statistical calculations. The predictive algorithms are statistical and rely on known protein structures. Perhaps the best known and easiest to apply is the technique of Chou & Fasman (25–27) who rank the amino acids as helix, sheet, and turn formers. They then elaborate on the number and kind of residues required to nucleate and terminate a given structural element. Two international competitions have been held to determine the accuracy of various techniques in the prediction of the adenylate kinase (28) and phage lysozyme (29) secondary structures. It is clear that these methods work better on some proteins than others, but they generally predict with a moderate degree of accuracy (21). Nevertheless, secondary structural predictions have found wide applicability in the analysis of amino acid sequences where the structure is unknown, and are particularly valuable when other functional properties of the structures are known (30). Recently, attempts have been made to extend these methods to predict tertiary structure by analyzing such variables as the packing of α-helices (31, 32) or the frequency of topological arrangements within β-sheets (cf 33).

The relationship between sequences and fold is not rigorous. The code that relates sequence to structure is highly degenerate, and yet remains responsive to the protein solvent. Furthermore there appear to be only a limited number of amino acid sequences that can provide a unique structure in a given environment; all others are nonsense. This may in part be the basis for the apparently small number of folds or architectural classes that have so far been observed. Albeit, with only a few exceptions (34), these observations have been confined to aqueous-soluble proteins.

Enzymes often utilize to their advantage a change of environment to alter conformation, as is implied by the term "induced fit" (35, 36). This is typified by the movement of the loop in lactate dehydrogenase (LDH), which is controlled by NAD binding (37, 38). Huber (39) has drawn attention to the order-disorder phenomenon that can occur in the formation of the trypsin specificity pocket or in the Fc fragment of the immunoglobulins.

Similarly, certain sections of viral coat proteins may fold as α-helices only in the presence of RNA (40, 41). The disordered segments of a polypeptide chain frequently start and end with glycines and contain few, if any, aromatic residues.

The degeneracy of the relationship between sequence and structure is essential in the process of evolution, as it permits an alteration of specific amino acids without destruction of the fold and, hence, without loss of function [see Lesk & Chothia (42) who analyze this degeneracy for the globin structure]. Indeed, the great conservation of residues in the active center of enzymes and the associated conservation of fold clearly demonstrate that function is a controlling aspect in protein evolution [see Doolittle (43) for a recent discussion of protein evolution]. Quaternary interactions can have a regulating effect on the function of each subunit. The presence of functional globin monomers such as lamprey hemoglobin or sperm whale myoglobin is one of many examples that show that the evolution of monomers usually precedes the subsequent evolution of allosteric oligomeric proteins as hemoglobin.

Excellent reviews on protein structure and fold include those of Jane Richardson (44), Schulz & Schirmer (21) and Cantor & Schimmel (45). The valuable book by Dickerson & Geis (46), an updated version of which is soon to be published by Benjamin, has been a standard text for most students and scholars in this area. In addition, there are a long line of reviews on protein structure in the *Annual Review of Biochemistry* (e.g. 47–52) and in other journals (e.g. 53, 54). Mention should also be made of reviews relevant to the dynamics of folding (e.g. 55–64). The present discourse confines itself to the analysis of folds of known protein structures.

SUMMARY

After some general remarks on protein structure, there follows a discussion on primary, secondary, and tertiary organization. The account of primary structure includes a discussion of the conformation of disulfide bonds. Types of helices, sheets, and turns are described in the section on secondary structure, followed by a discussion of super-secondary structure and the effects of metals and prosthetic groups on protein fold.

The crux of the review lies in an examination of tertiary structure, or specifically of domains that are defined, in part, as functional units within a polypeptide chain. An assembly of domains can in turn result in a protein whose function is quite sophisticated. Some consideration of domain recognition is given in the section on taxonomy and in the appendix. The key part of the tertiary structure section concentrates on a taxonomic protein classification dependent not only on structure but also on function. A discussion of the requirements imposed by quaternary structure on a fold are omitted in this review. Finally, no review of this kind can escape a discussion of evolutionary convergence and divergence.

THE ORGANIZATION OF A POLYPEPTIDE

This section concerns the static organization of proteins as found by crystal structure analysis, and is, therefore, limited to those protein domains with unique structures, as opposed to "random coil" conformations. The latter may often be observed for proteins, particularly under solvent denaturing conditions, or for segments of polypeptide chain within a crystal.

The gross structures of proteins are not affected by crystallization. For instance, the structure of trypsin when crystallized on its own (65, 66) or when complexed with inhibitor (67) changes very little. Nor is the detailed structure greatly perturbed, as shown by numerous NMR studies of proteins in solution (e.g. 68) and by the potential for enzyme activity in the crystal (e.g. 69). Furthermore, the fold of many functionally different proteins is often similar, e.g. the structures of triose phosphate isomerase, of the central domain of pyruvate kinase (70), and of bacterial aldolase (71, 72). Sizable forces are usually necessary to upset the unique native structure of a protein. Nevertheless, minor changes do occur between a specific crystalline conformation and the large number of related conformations likely to exist in solution. At times these small structural changes can lead to appreciable differences in the properties of the protein (73, 74). This review focuses primarily on the overall pattern of folding and not on details related to specific side chains.

Primary Structure

The polypeptide backbone consists of a series of planar *trans* linkages with mean dimensions shown in Figure 2 (75–78). The precise bond lengths and angles are derived from single crystal studies of small peptides, where data can be gathered to at least 0.7 Å resolution. It is usually necessary to assume these dimensions in analyzing the conformation of a complete protein. It is only the rare protein that is sufficiently ordered to provide data beyond 2.5 Å resolution. Departure of the peptide unit from planarity has been examined by Ramachandran and co-workers (79, 80). The occurrence of *cis* peptides, discussed by Ramachandran & Mitra (81), has been observed only when proline is one of the residues (82). The nomenclature of the dihedral angles in a peptide linkage is shown in Figure 3. Their definition can be found, for instance, in Dickerson & Geis (46) who give due caution on conventions. The angles ψ and ϕ represent rotations about single bonds and are constrained by steric hindrance. The angle ω is usually assumed to

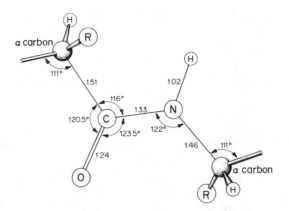

Figure 2 Dimension of planar *trans* peptide linkage.

be zero for a *trans* planar peptide. The backbone angle τ is affected by strain and is somewhat dependent on the type of side chain. Possible combinations of ϕ and ψ are nevertheless limited, and these constraints are best visualized as an orthogonal plot of these variables, known as a Ramachandran diagram (83).

The interrelationship of backbone structure and the sequence of amino acid side chain is clearly vital in the determination of fold. Kendrew et al (84), for instance, pointed out that serine and threonine have a preference

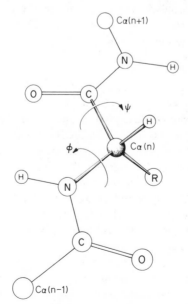

Figure 3 Nomenclature of dihedral angles in peptides.

for hydrogen bonding to backbone NH groups. Preferred orientation of side chain groups in known protein structures has been analyzed by Janin et al (85), Warme & Morgan (86) and Crippen & Kuntz (87). They show a reasonable correlation with conformations predicted from energy calculations (e.g. 62).

An in-depth study of about 60 disulfide bonds observed in proteins has been made by Richardson (44). The most important characteristic is the dihedral angle about the S–S bond, which determines whether the Cys–Cys conformation will be a right-or left-handed spiral. However the dihedral angles on each cysteine side chain also have a limited repertoire because of steric hindrance. Richardson was able to characterize four distinct types of conformations. The common conformations have distances of 6 and 5 Å between the C_α atoms for left- and right-handed bonding systems, respectively.

Secondary Structure

The formation of secondary structure largely relies on hydrogen bonding networks. The existence of a hydrogen bond in a protein structure is usually inferred from the positions of oxygen or nitrogen atoms. In general a hydrogen bond is roughly linear and 2.8–3.0 Å long (88, 89). Thus when the positional error of protein atoms is more than about 0.4 Å, the identification cannot always be made with certainty. In the absence of hydrogen bonds, electrostatic attraction may have some stabilizing effect on secondary structures if charges are closer than about 4 Å.

HELICAL STRUCTURES   Two types of ring systems can be defined within a polypeptide chain:

$$H\text{------------}O$$
$$|\qquad\qquad\qquad |$$
$$-\,N-(CO-CHR-NH)_n-C-\qquad \text{with } m = 3n + 4,\text{ or}$$

$$O\text{------------}H$$
$$|\qquad\qquad\qquad |$$
$$-\,C-(CHR-NH-CO)_n-CHR-N-\qquad \text{with } m = 3n + 5.$$

Any particular structure can then be characterized by three symbols: S, m, and r or l. S denotes the number of residues per turn of the helix; m, the number of atoms in the hydrogen bonded ring; and r or l according to whether the helix is right- or left-handed. A helix is thus designated by the symbol $S_m r$ or $S_m l$. The most stable helix type is the right-handed α-helix, which is denoted $3.6_{13} r$. The only other type of helix that is found with any significant frequency in globular proteins is the $3_{10} r$ helix, which usually appears as single turns ending an α-helix. Variations and intermediates of these conformations have been discussed by Némethy et al (90) who introduced the terms α_I and α_{II} for helices whose carbonyl groups point toward or away from the helix axis. A collagen-like, left-handed helix with proline in every third position has been found in cytochrome c_{551} (91). A left-handed δ-helix ($4.3_{14}l$) has been proposed for the 22 amino-terminal residues of the *lac* repressor (92).

Aggregates of α-helices form left-handed, three-stranded "coiled-coils" in keratin (93). One such short segment has been found in the structure of southern bean mosaic virus (SBMV; 40) where parallel α-helices from three polypeptides twist about threefold axes. A similar left-handed twist can be seen in the packing arrangement of α-helices within the hemoglobin molecule (Figure 4), hemerythrin (94, 95), and cytochrome b_{562} (96).

Figure 4 The hemoglobin molecule, showing the left-handed twist of α-helices. (Reprinted with permission from Max F. Perutz.)

The packing of α-helices was first considered by Crick (97). He found that the side chains can intercollate between each other only when the helix axes cross at an angle of about −70° or +20°. His ideas were modified and extended by Chothia et al (98) whose packing analysis resulted in angles of −82°, −60° and +19° (classes I, II, and III, respectively). These results are reasonably well born out in the analysis of known structures (98). Richards and co-workers (31, 32) have investigated the way α-helices may aggregate and use this to "predict" the structure of myoglobin. Such predictions can be substantially improved when some chemically derived inter-residue distances are known (99). Attempts have been made to show how the four helices in tobacco mosaic virus (TMV) structure (100) and the eight-helical myoglobin structure (101) may self-assemble to produce the known structure. Argos et al (102) have pointed out that four-helical clusters are fairly common and may represent a "super-secondary structure," while Blow et al (103) discuss all possible connectivities between four α-helices formed from a single polypeptide. Whether such structures form due to the selectivity of helix aggregation [convergent evolution to a stable structure (see 104)] or as a result of divergence from a primordial structure is unclear.

SHEET STRUCTURES   Both parallel and antiparallel β-pleated sheet structures, with different characteristic hydrogen bond formations, were proposed by Pauling (105). It has recently been demonstrated that they are also differentiated by their amino acid compositions (106, 107). Chothia (108) was the first to discuss in print the characteristic twist of sheets, whether parallel or antiparallel (Figure 5). Successive strands within a sheet are twisted in a left-handed manner while the succeeding planes of peptide bonds in a given strand are twisted right-handedly. Neighboring strands may rotate from 0° to 30° with respect to each other. Weatherford & Salemme (109) propose that the twist is induced by the deformation of the peptide nitrogen toward a tetrahedral conformation, which causes slight nonplanarity of the peptide and optimizes the hydrogen bond geometry.

Richardson et al (110) investigated anomalies within β-pleated sheets. They found, mostly in antiparallel sheet, one important distortion, a "β-bulge," in which an extra residue is introduced between the closely spaced hydrogen bonds (Figure 6). This disrupts the sheet and can, therefore, only occur on the edge of a β-sheet. Two main types of bulges were explicitly identified: the classic and the G1 type. The values of the ψ and ϕ angles for residues in the bulge represent distinct conformations (110). The G1 bulge is nearly always associated with a glycine in position 1 of the bulge (Figure 6) and is very often associated with a type II tight turn (see section on turns). The latter requires a glycine in its third position, which corresponds to the first position of the G1 β-bulge.

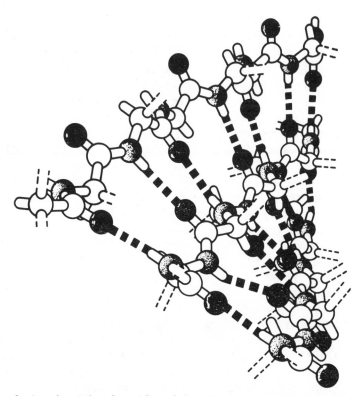

Figure 5   A β-pleated sheet formed from chains with a right-handed twist. The first chain is antiparallel to the second which is parallel to the third. View along the chain illustrates the right-hand twist. [Reprinted with permission from Chothia (108). Copyright by Academic Press Inc. (London) Ltd.]

Both Richardson (111) and Sternberg & Thornton (112, 113) have searched for systematic preferences in the sequential laying-out of strands within parallel, antiparallel, or mixed sheets (Table 1). The principal trend is for strands near each other in the sheet to be sequential along the polypeptide chain. Attempts have been made to predict (115) the sequence of strands within a sheet from these observed natural preferences. However, it would be necessary to predict the occurrence of β-strands within a polypeptide from amino acid sequence data, itself a risky procedure, to further predict tertiary structure (116). Ptitsyn and co-workers (117) have suggested folding pathways, using the principle that all strands must be laid down next to an immediate neighbor. They discuss the formation of both antiparallel (118) and parallel sheets (119).

It has been necessary to establish a nomenclature that permits an easy description of the connectivity within a β-pleated sheet. The procedure of

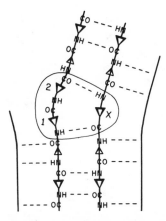

Figure 6 A β-bulge (outlined region) at the edge of an antiparallel β-sheet. Smaller triangles represent side chains that are below the sheet, larger triangles those that are above it. [Reprinted with permission from Richardson et al (110).]

Richardson (114) seems particularly elegant (see Figure 7). Successive segments of the polypeptide chain that form strands of a β-sheet are designated $\pm n$x. The numeral n (Figure 7) designates the number of strands between consecutive portions of the polypeptide chain; the x, if present, designates parallel β-pleated sheet with the required cross-over between successive strands; and the sign can be used to designate whether the next strand is laid down in a positive or negative direction from the current strand.

Table 1 Frequency of antiparallel and parallel strands within a β-pleated sheet

Antiparallel[a]		Parallel[b]			
Turn type	Frequency	Turn type (right)	Frequency	Turn type (left)	Frequency
1	105	1x	44	1x	1
2	9	2x	24	2x	1
3	12	3x	11		
4	2	4x	3		
5	3	5x	—		
6	—	6x	—		
7	1	7x	1		
8	—	8x	1		
		12x	1		

[a]Nomenclature and data are due to Richardson (111). The actual results are strongly affected by the available structural sample, but the trend for strands near each other within the sheet to be sequential along the polypeptide chain is clear.
[b]Nomenclature and data are due to Richardson (114).

(a)

(b)

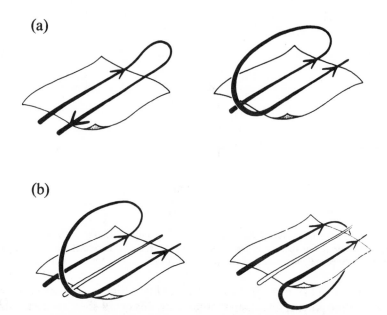

Figure 7 (*a*) Illustration of the two main classes of topological connection in β-sheets: (*left*) "hairpin", "plain", or "same-end" connection; this specific example is type ±1; (*right*) "cross-over", "cross", or "opposite-end" connection; this specific example is type ±1x. (*b*) (*left*) A right-handed ±2x cross-over connection; (*right*) left-handed ±2x cross-over connection. Direction is not indicated for the skipped strand, since it may be either parallel or antiparallel to the others. [Reprinted with permission from Richardson (114).]

TURNS Different secondary structural elements are frequently connected by sharp turns, particularly at the surface of the molecule (120). These turns are called reverse turns, β-turns, β-bends, etc. They generally have distinctly recognizable conformations and are often restricted in their amino acid composition (120a). Crawford et al (121) point out that about one third of all amino acids in globular proteins are in turns. Venkatachalam (122), Crawford et al (121), Lewis et al (123), and Chou & Fasman (124) have all examined the conformation of turns. Their consensus on the definition of a turn is in terms of four residues, connected by three peptide linkages, stabilized by a hydrogen bond between the carbonyl of the first and amide of the fourth group. However, the actual formation of the hydrogen bond is frequently absent, although the distance between the appropriate O and N atoms is reasonably small (less than 4 Å). Some authors (125, 126) have defined turns by using C_a carbons only. This leads to a more general definition that is not as intimately involved with the chemical requirements of hydrogen bonding or steric hindrance.

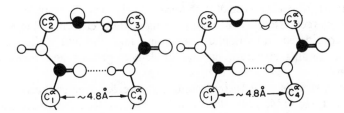

Figure 8   The two types of chain reversal that have hydrogen bonding between the carbonyl of the first and amide of the fourth residue. The four alpha carbons are not meant to appear planar. Type I is on the left, and Type II is on the right. [Reprinted with permission from Crawford et al (121).]

Different types of β-turn conformations were designated I, II, and III and I', II', and III' by Lewis et al (123), a nomenclature that is now often adopted. The primed conformations represent the converse dihedral angles to the unprimed. Conformations I and II are related by a 180° flip of the central peptide (Figure 8). Conformation III cannot readily be distinguished from I, as the dihedral angles differ by less than 30°. Other conformations, IV to VII, were also defined, but they represent special cases (e.g. VI has a *cis* proline at position 3) or encompass unusual situations. Table 2 shows the properties of the four major conformations and the preference for glycine in certain positions because of steric hindrance (Figure 8). This preference had been predicted by Venkatachalam (122), particularly for position 3 in the type II bend. In practice the requirement for glycine does not appear to be as absolute as his predictions would suggest, because of departures from the mean dihedral angles. The possible peptide conformations in antibiotics, where D amino acids are common, have been reviewed by Chandrasekaran & Prasad (127) but lie outside the scope of this review.

SUPER-SECONDARY STRUCTURES   The term super-secondary structure was coined by Rao & Rossmann (128) to describe a recurring fold consisting of a number of secondary structural elements and yet not comprising the complete tertiary structure of a molecule or domain (see section on domain

Table 2   Mean dihedral angles for the two types of chain reversal (123)[a]

	ϕ_2	ψ_2	ϕ_3	ψ_3	Comment
I	−60	−30	−90	0	
II	−60	120	80	0	Glycine likely in position 3
I'	60	30	90	0	Glycine preferred in position 2
II'	60	−120	−80	0	Glycine preferred in position 2

[a] See Figure 8 for definition of subscripts to dihedral angles and turn identifications.

definition). The occurrence of such folds is likely to be the result of pressure to find energetic stability through packing of these elements. Rao & Rossmann considered, in particular, the unique hand and fold of the mononucleotide binding structure present in dehydrogenases and many other proteins.

Two β-strands with their accompanying cross-over connections have been analyzed (114, 129). Both analyses observe that the connection between strands within a parallel β-pleated sheet almost always (Table 1) retains the same hand, which was defined as right-handed (see Figure 7). It is therefore to be concluded that right-handed cross-over structures are stable for reasons about which Richardson (114) and Sternberg & Thornton (129) have speculated. The cross-over connection is often a helix resulting in a β-α-β super-secondary structure.

A series of super-secondary structures can often form a functional domain. For instance, the NAD binding domain in dehydrogenases may be described as $+1x$, $+1x$, $-3x$, $-1x$, $-1x$. Another example is the triose phosphate isomerase structure (130; Figure 9), which occurs also in pyruvate

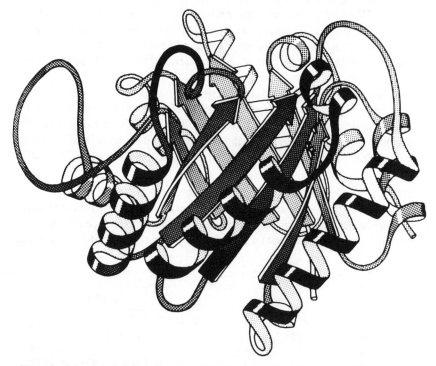

Figure 9 Schematic drawing of a triose phosphate isomerase subunit. Copyright by Jane S. Richardson. [Reprinted with permission from Richardson (71). Copyright by Academic Press, Inc.]

kinase (70, 131) and 2-keto-3-deoxy-6-phosphogluconate aldolase (71, 72), and can be described as $(+1x)_8$. Indeed β-α-β structures form the largest group of observed protein structures. Many of these have been "catalogued" (111, 132–134). The dipoles created by the parallel α-helices on either side of the central parallel sheet produce a significant electrostatic field, which provides suitable orientation and binding forces for the negatively charged phosphates of nucleotides (135).

Schulz (136) introduced the term "β-meander," which represents a $+1$, $+1$ three-stranded antiparallel sheet. This is a common fold (Table 1) and must therefore be classified as a super-secondary structure. A mononucleotide fold plus a β-meander comprises a domain that is twice repeated in the tertiary structure of glutathione reductase (136).

METALS AND PROSTHETIC GROUPS IN PROTEINS   It is not clear whether the presence of metals or prosthetic groups affects the unique fold of a polypeptide. Only minor conformational changes occur in metal-free concanavalin A, relative to the native structure (137). It is possible, in many instances, to remove the prosthetic group without damage (e.g. the heme in globins), whereas in other cases the protein becomes denatured [e.g. the structural Zn in liver alcohol dehydrogenase (LADH) or the heme in catalase]. In any event the fold must adapt itself to the requirement of binding a specific metal. Carp myogen (Table 3) displays three "E–F" hands each of which consists of two sequential helices joined by a right angular turn. Six ligands, situated close to each other within the turn, bind a calcium ion. Kretsinger suggests (244) that the various examples of this fold may all have originated from a common ancestor. In many metal-containing proteins the polypeptide chain makes a tight turn about a metal ion so that most of the liganding residues are situated close to each other along the chain. Examples are carbonic anhydrase, thermolysin, and the Fe–S cage in ferredoxin (48).

Tertiary Structure of Domains

DOMAIN—DEFINITION   The word domain was probably first used to describe the repetition of homologous sequences within the light and heavy chains of an immunoglobulin molecule (245). It was subsequently recognized that similar sequences within a complete polypeptide chain determine similar folds. Such structural domains were often found to be spatially well separated, giving the molecule an appearance of being bi- or multilobal. Later it was observed that differing domains within a single molecule might also have completely different architecture, such as a parallel β-pleated sheet acting as one domain and a "β-barrel" (44) acting as the other. Furthermore, a domain in one molecule might also occur in a quite different

Table 3 Taxonomy of known domains[a] in globular proteins

I. All α structures

    A.   Heme-binding proteins–type A: oxygen and electron carriers (Figure 10)[b]

        1. Globins: α chain of hemoglobin (141, 142); β chain of hemoglobin (141, 142); single-chain hemoglobins in worm (143); single-chain hemoglobins in lamprey (144); single-chain hemoglobins in root nodules of legumes (145); myoglobin (146)

        2. Cytochrome b_5 and related structures such as domains in cytochrome b_2 and sulfite oxidases (147, 148)

        3. Cytochromes c: cytochrome c (149, 150); cytochrome c_2 (151); cytochrome c_{550} (152); cytochrome c_{551} (91); cytochrome c_{555} (153)

    B.   Four-helical protein domains[c]

        1. Oxygen carriers without a heme group, but using Fe: hemerythrin (95); myohemerythrin (154)

        2. Heme-binding proteins–type B, electron carriers: cytochrome b_{562} (96); cytochrome c' (155)

        3. TMV protein (156)

        4. Ferritin (157)

        5. Second domain in Tyr-tRNA synthetase (158)

        6. FB fragment of protein A (159)

        7. First domain of papain (160)

    C.   Calcium-binding proteins[d]: carp myogen (161); troponin C (162)

    D.   Miscellaneous α structures: uteroglobin (163); purple membrane protein[e] (34); pig heart citrate synthetase[f] (165); second domain of thermolysin (166); glucagon[g] (167); 6-phosphogluconate dehydrogenase (168); first and third domains of hemagglutinin (169[h])

II. All β structures

    A.   Single sheets of antiparallel strands: carbonic anhydrase[i] (170); *Streptomyces subtilisin inhibitor*[j] (171); second domain of *p*-hydroxybenzoate hydroxylase (172); second domain of papain[k] (173); bacteriochlorophyll *a*-protein[l] (174)

    B.   Greek key β-barrels with 6–8 strands (Figure 11)

        1. Serine proteases[m]: chymotrypsin and chymotrypsinogen (175, 176); trypsin and trypsinogen (177–179); elastase (180); *Myxobacter 495* α-lytic protease (181); *Streptomyces griseus* protease A and protease B (182)

        2. Acid proteases[m] (183): penicillopepsin (184); *Rhizopus* pepsin (185); *Endothia* pepsin (185, 186); pepsin (187)

        3. "Immunoglobulin domains"

            a. In immunoglobulins[n] (52)

            b. Superoxide dismutase[o] (188)

            c. Plastocyanin[p] (189)

            d. Azurin[p] (190)

Table 3 *(Continued)*

II. All β structures *(continued)*

        4. Miscellaneous barrel structures: soybean trypsin inhibitor (191, 192); prealbumin (193); staphylococcal nuclease (194); second domain of pyruvate kinase (131); DNA-unwinding protein (195); first domain of thermolysin (166)

   C. "Jelly roll" structures (196)

       1. Coat proteins of some spherical viruses

          a. Both domains of tomato bushy stunt virus (197)
          b. SBMV (40)

          c. Middle domain of hemagglutinin (169[h])
       2. Concanavalin A (198, 199)
       3. cAMP receptor protein[q]

III. α/β structures

   A. Domains that bind nucleotides toward the carboxyl end of a predominantly parallel β-pleated sheet

       1. NAD binding domains in dehydrogenases: first large domain of LDH (200); first domain of malate dehydrogenase (201); second domain of LADH (202); first domain of glyceraldehyde-3-phosphate dehydrogenase (203)

       2. Kinases[r]: adenylate kinase (205); two similar domains of phosphoglycerate kinase (206, 207); hexokinase (208); both domains of phosphofructokinase (209)

       3. NADP-dependent dehydrogenases: dihydrofolate reductase (210); first two domains of glutathione reductase (136); NADP binding domain of 6-phosphogluconate dehydrogenase (211[s]); first domain of p-hydroxybenzoate hydroxylase (172)

       4. Pyridoxal phosphate-binding enzymes: first domain of aspartate aminotransferase (212, 213); second domain of phosphorylase (214)

       5. Other nucleotide-binding enzymes: flavodoxin[t] (215); second domain of phosphorylase[u] (214); Tyr-tRNA synthetase[v] (158); catalytic domain of aspartate transcarbamylase[w] (216); Tu elongation factor[v] (217)

       6. Enzymes that bind phosphates: phosphoglycerate mutase (218); glucose 6-phosphate isomerase (219)

   B. Structures based on a single, primarily parallel, β-pleated sheet, with active sites toward the carboxyl end of the sheet, that do not, however, bind nucleotides: carboxypeptidase (220); subtilisin (221, 222); thioredoxin (223); glutathione peroxidase (224)

   C. Structures that contain two domains of primarily parallel β-pleated sheet with the carboxyl ends of the sheet directed toward each other: phosphoglycerate kinase (206); hexokinase (208); phosphofructokinase (209); rhodanese (225); arabinose binding protein (226); phosphorylase (214); catalytic domain of aspartate transcarbamylase (216); glucose 6-phosphate isomerase (219)

   D. Structures with eight successive 1x cross-over turns that form a barrel (Figure 9): triose phosphate isomerase (130); aldolase (71, 72); first domain of pyruvate kinase (131); glycolate oxidase (227); first domain of Taka-amylase A (227a)

Table 3 *(Continued)*

IV. Small $\alpha + \beta$ proteins

    A.   Structures based primarily on β-sheets that frequently contain disulfide bonds. Most of these proteins are extracellular (228)

        1. Neurotoxins and similar structures

           a. Sea snake neurotoxin (229)
           b. Erabutoxin (230)

           c. Wheat germ agglutinin (231)

        2. Structures that can bind polysaccharides (232, 233): hen egg white lysozyme (234); phage lysozyme (235); α-lactalbumin (236)

        3. Miscellaneous structures: ribonuclease (237); papain (173); insulin (238); pancreatic trypsin inhibitor (67); phospholipase A_2 (239)

    B.   Fe-S proteins: rubredoxin (240); both domains of ferredoxin[x] (232, 241); high potential iron protein[y] (242)

    C.   Multiheme cytochrome c_3 (243)

V. Large $\alpha + \beta$ proteins: catalytic domain of LDH (200); catalytic domain of glyceraldehyde 3-phosphate dehydrogenase (203); catalytic domain[z] of LADH (202)

---

[a] Inside bracket indicates strong functional and structural similarity. Outside bracket indicates good structural equivalence. Outside dashed lines indicate some structural similarity.

[b] Structural and functional relationships among these proteins have been discussed by Argos & Rossmann (138). The structurally common part corresponds to a central exon in the globin gene (138–140).

[c] The four α-helices run mostly antiparallel and are connected sequentially (102, 103). The topology of the connectivity is the same in categories 1–4.

[d] These consist of three repeated domains of α-helices, AB, CD, and EF, related by roughly $90°$. Calcium atoms are bound into the elbows of the last two domains. It has been suggested that this domain is a fairly universal calcium-binding domain in muscle proteins (161) and troponin C (162).

[e] This protein is primarily in a lipid membrane environment and thus may have different folding rules to water-soluble proteins. There are seven helices in this structure. Their individual polarities have not been unambiguously established, but the helix axes are essentially parallel and lie perpendicular to the membrane surface (164).

[f] There are at least 12 α-helices in one subunit of this molecule with no β-structure.

[g] A single helix.

[h] D. C. Wiley, personal communication.

[i] Ten antiparallel strands.

[j] 5 Antiparallel strands.

[k] 6 Antiparallel strands.

[l] 15 Antiparallel strands with the ends curled around to make a cavity for seven chlorophyll moieties.

[m] Two similar domains in tandem.

[n] Variable-heavy, variable-light, constant-heavy, constant-light domains in tandem in immunoglobulins, in Bence-Jones protein, and in Fc and Fab fragments of the immunoglobulins. For general review see Amzel & Poljak (52).

[o] Cu-Zn–containing.

[p] Cu–containing.

[q] T. A. Steitz, personal communication.

[r] Most of these are bilobal and slightly change the lobal separation during catalysis (204).

[s] M. J. Adams, personal communication.

[t] Binds FMN.

[u] Binds AMP.

[v] Binds ADP and ATP.

[w] Binds CTP.

[x] Each binds one Fe_4S_4 cluster.

[y] With one Fe_4S_4 cluster.

[z] The catalytic domain consists of two segments; the major portion is at the beginning of the polypeptide and the minor portion at the end, with the NAD binding domain in between.

molecule. For instance, the NAD binding domain has been found to occur in all NAD-linked dehydrogenases whose structures have been determined (246, 247), and, as mentioned earlier, the triose phosphate isomerase structure (Figure 9) occurs in pyruvate kinase (70, 131) and probably in bacterial aldolase (71, 72). This has quite naturally led to the proposal of divergent evolution from a common ancestral structure whose corresponding gene has been copied and fused with a variety of different genes. Domain structures would then be maintained fairly faithfully by the demands of particular functions and structural stability.

The following properties can generally be ascribed to a domain:

1. Similar domain structures or their amino acid sequences can be found either within the same polypeptide or in a different molecule.
2. Domains within a polypeptide are spatially separated from each other, or at least form a compact "glob" or cluster of residues.
3. Domains have a specific function, such as binding a nucleotide or polysaccharides.
4. The active center of a molecule is at the interface between domains, which permits the simple function of each to be brought together to form a more sophisticated molecule (248).

A corollary of these definitions, particularly the spatial separation of compact domains, is that proteolytic cleavage can frequently be used to identify and separate domains. Examples are the separation of the heme and FMN binding domains of cytochrome b_2 (249), the separation of the two elastase domains (250), and the recognition of domains in the β subunit of *E. coli* tryptophan synthetase (251). It is noteworthy that, in the elastase case, the carboxyl-terminal domain retains its ability to bind substrate, but there is no catalytic activity, since the catalytic residues are in part on the amino-terminal domain. Whether excised domains in general retain their fold and function is subject to dispute. The excision may expose large hydrophobic patches, which might destabilize the domain structure in an aqueous environment.

Not all the above listed properties of a domain need be applicable simultaneously. There have been attempts to define domains entirely on the basis of structure using only the second criterion above (252, 253). An exemplary systematic procedure to recognize domains by their spatial separation has been proposed by Janin & Wodak (254). They first devise an algorithm to compute the surface area of any portion of the protein (255). Then they compute the surface area generated by cleaving the polypeptide chain at all possible positions. When no significantly large area is so generated, the globular nature of the two parts would be indicated. Others (132, 253, 256, 257) have assigned domains on a rather more intuitive basis.

Blake (258) and Gilbert (259) have suggested that domain integrity is preserved in eucaryotes by their coding within exons. This is supported by the common helical folding pattern observed in globin, cytochrome c_{551}, and cytochrome b_5 (Figure 10), which has a heme binding function and corresponds to the middle exon of the β globin chain (138). Similarly, conserved specific functional properties can be ascribed to this portion of the globin chain (260). Indeed, Craik et al (261) have been able to excise

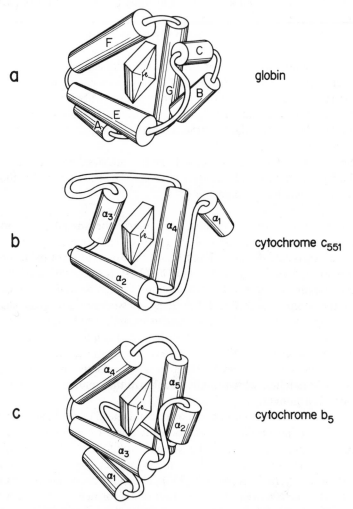

Figure 10 Diagrammatic representation of the similar polypeptide chain topology in (*a*) hemoglobin β chain, (*b*) cytochrome c_{551}, and (*c*) cytochrome b_5. [Reprinted with permission from Argos & Rossmann (138). Copyright by the American Chemical Society.]

this portion of the globin chain and demonstrate its ability to bind heme. Since domains may thus be the result of independent genetic development, they may also represent the basis for protein folding. It is therefore reasonable to assume that the domain will fold independently of other domains in the complete polypeptide and act as a nucleation center. The final tertiary structure will be attained as an assembly of the folded domains. Exons could provide a mode for rapid evolutionary development within the context of energetically favorable folds.

TAXONOMY It had been generally assumed prior to 1959 that every protein structure would be radically different to every other protein structure. The three-dimensional arrangements of secondary structural elements were thought to be almost infinite. Although myoglobin and the α- and β-hemoglobin chains were observed to have similar folds, this could be readily understood in terms of a common precursor for heme-dependent oxygen carriers. In 1971, some 12 years later, it was discovered that malate dehydrogenase and LDH have similar tertiary, if not quaternary, structures (262). This appeared reasonable in that both these enzymes have similar substrates, although their functions are related to different metabolic pathways. Then followed the realization that LDH and flavodoxin share some structural similarity (128) including the approximate orientation and position for coenzyme binding (246). This structural equivalence was a more radical observation, as both molecules show only weak functional relationships and the structural similarity is confined to a portion of the molecule (Table 4). Other surprising similarities, such as lysozyme with α lactalbumin (236) or superoxide dismutase with the immunoglobulin domain (188), provided further enigmas. A debate ensued between proponents of divergence and those supporting convergence to stable super-secondary structures. On the other hand, the concept of convergence in the active center region of molecules had indeed been well illustrated in chymotrypsin and subtilisin (263), which have totally different folds. Nevertheless, it has remained difficult to differentiate between convergence and divergence in the evolution of similar protein folds (see next section).

Once the initial shock of structural similarity in protein structures had passed, it became possible to construct taxonomic classifications. The first comprehensive attempt by Levitt & Chothia (132) used four major classifications: 1. all α proteins with only α-helix secondary structure (e.g. myoglobin); 2. all β proteins with mainly β-sheet secondary structure (e.g. superoxide dismutase or chymotrypsin); 3. α + β proteins with α-helices and β-strands that tend to segregate into all α and all β segments along the peptide chain (e.g. papain or thermolysin); and 4. α/β proteins with mixed or approximately alternating segments of α-helical and β-strand secondary

Table 4   Various criteria for assessing structural equivalence between functional domains in a variety of protein comparisons

Comparison		Number of residues		Number of equivalenced residues	Percentage of equivalenced residues		rms deviation (Å)	MBC/C[a]
Molecule 1	Molecule 2	Molecule 1	Molecule 2		Molecule 1	Molecule 2		
Globins:								
Hemoglobin β	Hemoglobin α	146	141	139	82.7	91.0	1.6	0.70
Heme binding proteins:								
Hemoglobin β	Cytochrome b_5	146	93	48	32.9	51.6	4.1	1.29
Hemoglobin β	Cytochrome c_{551}	146	82	49	33.6	59.8	3.5	1.33
Cytochrome b_5	Cytochrome c_{551}	93	82	41	44.1	50.0	4.9	1.60
Nucleotide binding domains:								
LDH[b]	GAPDH[c]	144	148	83	57.5	56.1	2.9	1.12
LDH	Flavodoxin	144	138	32	22.2	23.2	2.4	1.23
Immunoglobulin folds:								
Superoxide dismutase[d]	Immunoglobulin domain (C_L)[d]	151	110	42	27.8	38.2	1.9	—
"Jelly roll" structure:								
TBSV(S)[e]	TBSV(P)[f]	167	110	69	41.3	62.8	3.8	—
Con A[g]	TBSV(P)	237	110	68	28.7	61.8	3.4	—
Con A	TBSV(S)	237	167	82	34.7	49.2	3.2	—
SBMV-C[h]	TBSV(S)-C[h]	219	200	182	82.7	91.0	2.7	—
Lysozymes:								
T4 phage	Hen egg white	164	129	64	39.1	49.7	4.1	1.53

a Minimum base change per codon.
b Lactate dehydrogenase.
c Glyceraldehyde-3-phosphate dehydrogenase.
d Taken from Richardson et al (188).
e Tomato bushy stunt virus shell domain.
f Tomato bushy stunt virus protruding domain.
g Concanavalin A.
h "C" refers to the C subunit within SBMV and TBSV.

structure (e.g. flavodoxin or kinases). Partial classifications were also concurrently attempted by Sternberg & Thornton (133), and Richardson (111).

Three especially useful categories were introduced by Richardson (111)
for "β-barrel" structures (Figure 11). These were the "Indian basket structures" as in papain or soybean trypsin inhibitor, the "Greek key structures"
as in chymotrypsin or superoxide dismutase, and the "lightning structures"
as in triose phosphate isomerase. The unique hand of all these common

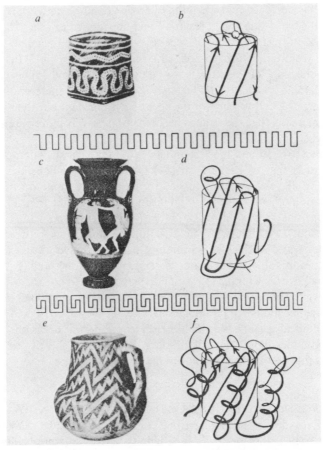

Figure 11   Comparison of geometric motifs common on Greek and American Indian weaving
and pottery with the backbone folding patterns found for cylindrical β-sheet structures in
globular proteins. (*a*) Indian polychrome cane basket from Louisiana: (*b*) Polypeptide backbone of rubredoxin; (*c*) Red-figured Greek amphora showing Cassandra and Ajax (about 450
BC); (*d*) Polypeptide backbone of prealbumin; (*e*) Early Anasazi Indian redware pitcher from
New Mexico. (*f*) Polypeptide backbone of triose phosphate isomerase. [Reprinted with permission from Richardson (111). Copyright by Macmillan Journals Limited.]

structures was noted. Richardson's analogy with art objects is particularly appropriate. The differences in architecture of the various β-barrels is as diverse and recognizable as the different cultures pertaining to the pots and baskets in Figure 11. Just as an art historian can discern the history and function of a particular vase, so the molecular taxonomist should be able to determine the history and function of a polypeptide fold.

A more quantitative approach to the classification of β-barrels has been suggested by McLachlan (264). He defines the shear, S, as the number of residues on a given strand when the hydrogen bonding network is followed around the barrel starting at residue i and finishing at residue $i \pm S$. If the distance along a strand between successive residues is a (usually 3.5 Å), the distance from one hydrogen-bonded strand to the next is b (usually 4.7 Å), and the slope of the strands to the cylindrical barrel axis is a, then $S = [(nb/a)\tan a]$, and $R = \{b/[2\sin(\pi/n)\cos a]\}$ where there are n strands in the barrel. It follows that, if the shear stays constant but the number of strands increase, then the strands become progressively more parallel to the barrel axis.

The most recent and most complete taxonomic classification of protein structures is that of Richardson (44). This is in part dependent on the number of "layers" within a structure. Regrettably, none of the current classification schemes takes any significant account of function. That is, the pendulum has swung far from the early preoccupation with functional relationships at a time when no structural comparisons between diverse molecules was conceivable. An attempt has been made to fuse these two approaches in Table 3. The major classifications follow roughly those suggested by Levitt & Chothia (132). Subdivisions are then created with special sensitivity to the function of domains within the polypeptide chain. Taxonomic classifications can be ambiguous whether they are designed to distinguish botanical specimens or molecules. Hence on some occasions molcules are listed under more than one heading. References generally refer to the most up-to-date information on the molecular structure or to a pertinent review.

CONVERGENCE, DIVERGENCE, GENE DUPLICATION, AND GENE FUSION   Evolutionary studies have had considerable difficulty differentiating between convergence toward a functionally useful structure or divergence from an ancestral structure whose functions have been refined and specialized. A sufficient fossil record usually alleviates this problem. However, in molecular structure there are no fossils; the molecules are those of today. The molecular taxonomist thus faces the perplexing task of reconstructing history from the similarities and variations found in today's specimens (265).

The generally accepted technique for differentiation between convergence and divergence involves a count of the characters that two molecules share or by which they differ. If the majority are the same, it would seem probable that the two molecules (or organisms) have diverged from a common ancestor with a few characters altered. However, such statements are only probabilistic; as the number of similar and dissimilar characters approach each other, it becomes impossible to make a useful differentiation.

The molecular evolutionists have primarily utilized the amino acids within protein sequences for taxonomic characters; but, when divergence has extensively progressed, amino acid sequence comparisons become insensitive. Fortunately, the protein fold is conserved to a far greater extent than the amino acid sequence and thus provides taxonomic properties for examination. However, both amino acid sequence and structural comparisons neglect function as an important characteristic that can delineate evolutionary history. On the other hand, function can be emphasized at the expense of structural information, as exemplified by the work of Dickerson on the evolution of bacterial respiratory processes involving cytochrome c–like molecules (266). For instance, function is not conserved in the divergence of haptoglobin from serine proteases in spite of good homology in sequence (267) and perhaps fold (268).

The first example of the use of structure as a quantitative tool in measuring evolutionary distance was given by Schulz & Schirmer (269) who used the sequence of strands in nucleotide binding folds. On this basis they computed the probability that a particular structure could occur more than once by chance. These values were later corrected (129) by allowing for the apparent preference of right-handed over left-handed cross-over structures. Given such measurements it is possible to construct networks (that can usually be related to phylogenetic trees) by standard techniques (cf 270–272) that permit quantitative testing of all postulated trees. Qualitative assessment of evolutionary relationships (192, 273) must be regarded with suspicion.

Many morphological features (e.g. 232, 270, 274, 275) of molecular structure have been utilized to measure evolutionary distances (see Table 4). A list of such features would include: the minimum base change per codon of residues associated with equivalenced C_α atoms; the rms separation of C_α atoms for spatially superimposed molecules; the similarity of protein folding topology; the number of structural insertions and deletions and their fraction of the structures superimposed; the coincidence of substrate positioning and orientation relative to the compared protein folds; commonality of function; shared active center geometries; axes of symmetry relating two domains; and the like. In projecting convergent or divergent probabilities, it is difficult to assign weights to the various criteria. For example, there could be unequal rates of convergence of different characters toward a functional protein; alternatively, divergence from a common ancestor could have occurred sufficiently early in biological time so that the remaining similarity is obscured. Nonetheless, there are many rich examples of structural comparison (e.g. 276).

Emphasis has been given in the construction of Table 3 to the study of the divergence of polypeptide folds with specific functions and the convergence of fold to energetically favorable structures. A likely example of divergence is found in the structural comparison of the functionally related globins (42). Another example is the relationship among eucaryotic and procaryotic cytochrome c–like molecules (266), where 80% or more of the C_α atoms can be spatially equivalenced with a mean rms separation of approximately 2.0 Å and a minimum base change per codon of around 1.0. The good topological and spatial agreement between the folds of the functionally diverse superoxide dismutase and the immunoglobulin domain (188) provides a likely example of convergence to a stable fold. Here the separation of equivalenced atoms was 1.9 Å with no obvious correlation of amino acid sequences.

Salient enigmas are the β-α-β folds found in various proteins. In dehydrogenases, the six parallel strands and associated helices are well preserved; NAD binding provides the obvious functional link, and the minimum base change per codon value is far from random. However, when manifested in other structures, the number of parallel strands may be less than six, insertions and deletions often occur, antiparallel strands can intrude, and the cofactor is not accurately positioned at the carboxyl end of the sheet (275). The best known example of active center convergence is that of chymotrypsin and subtilisin (263), where 23 catalytic site atoms superimpose to within an rms deviation of 1 Å. A further illustration of particular and perhaps unexpected similarity is the environment of the essential Zn atom in carbonic anhydrase, carboxypeptidase, thermolysin, and LADH (277, 278).

A reasonably common phenomenon is the repeat of a domain along the same polypeptide chain, as for example in the light and heavy chains of immunoglobulins. In some cases (Table 5) these domains are related by a twofold axis (273) as if they were identical subunits in an oligomeric protein with a proper point group (283). The active center may be associated with one of the domains (rhodanese) or both (acid protease). A conceivable mechanism for this phenomenon has been discussed by Tang et al (183) in terms of gene fusion. Obvious examples of fusion of unlike genes is implied by the occurrence of greatly different domains within subunits of NAD-dependent dehydrogenases (284).

Table 5 Gene duplication within the same polypeptide chain with domains related by diad[a]

Protein	Rotation	Translation	Reference
1. Rhodanese	$180° \pm 1°$	< 1 Å	225
2. Calcium-binding protein	$180° \pm 7.5°$	none	279
3. Ferredoxin	$180° \pm 2°24'$	none	232
4. Hexokinase	$180° \pm 17°$	2.1 Å	275
5. Phosphoglycerate kinase	$180° \pm 12°$	16 Å	275
6. Chymotrypsin and other serine proteases	an approximate two-fold screw axis		264
7. Acid proteases	$180° \pm 12°$	0.0 Å	185, 280
8. Myohemerythrin	$180° \pm 1.6°$	3.6 Å	281, 282
9. TMV	$180° \pm 8.8°$	2.9 Å	282
10. Glutathione reductase	$180° \pm 43°$	3.9 Å	136

[a] There are other examples (e.g. arabinose binding protein) where the nature of the relation between the two similar domains has not yet been established quantitatively.

CONCLUSION

With the advent of a relatively large catalogue of structures for water-soluble proteins, order is emerging from diversity. A variety of molecular structural characteristics have been recognized and these have been incorporated in various taxonomic classifications. The "domain," within polypeptides, has generally been recognized as a stable folding unit. Larger and more sophisticated enzymes are built of domains whose evolutionary history precedes that of the enzyme itself. The study of protein folding is thus likely to concentrate on viable domain structures, their cataloguing, and folding pathways.

APPENDIX—DETERMINING STRUCTURAL EQUIVALENCE

In the discussion and classification of folds it is necessary to distinguish between like and unlike structures. It is easy to recognize structures that are very similar or totally different. However in the presence of sizable insertions or deletions, the difficulty of visualizing a three-dimensional object may obscure the underlying similarities. Once equivalent atoms have been selected in two molecules, then quantitative methods are readily available for finding their best relative superpositions and hence the rms value between the superimposed structures (128, 285–288). The problem arises in recognizing significant similarity in structure and determining which atoms are to be assigned as structurally equivalent.

A number of authors have used "diagonal" or "distance" plots for rapid visual recognition of structural domains (see 248, 289, 290). The distance between pairs of atoms, usually only the C_α atoms, are plotted as a symmetrical square matrix. Atoms close together along the polypeptide chain (as in an α-helix) will plot along the diagonal, while parallel and antiparallel strands within a sheet appear as streaks perpendicular and parallel to the diagonal. A complete domain will have a particular pattern and can be readily recognized in such two-dimensional plots.

Further quantitation was not truly attempted until Rossmann & Argos (232) produced a three-dimensional search procedure. Any rigid body can be superimposed onto any other rigid body by a suitable combination of three rotation angles and three translation vectors. Rossmann and Argos searched all orientations of a rigid body with another. (Figure 12 shows the search function whose largest peak corresponds to the superposition seen in Figure 13.) When the proteins have similar folds and orientation, then the vectors between equivalenced atoms will be roughly parallel and of equal length. A search for the orientation with the maximum number of

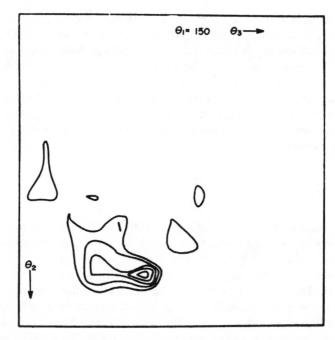

Figure 12 Comparison of phage lysozyme with hen egg white lysozyme showing section $\theta_1=150°$. Contours are drawn at levels 29, 34, 39, ... equivalent amino acid residues. [Reprinted with permission from Rossmann & Argos (232). Copyright by Academic Press Inc. (London) Ltd.]

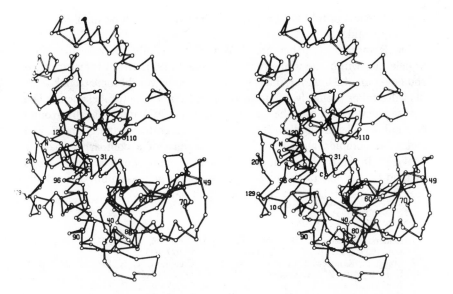

Figure 13 Stereo view of the α-carbon backbone of hen egg white lysozyme (dark lines) superimposed on the phage lysozyme structure (thin double lines). The numbering refers to the hen egg white lysozyme sequence. This superposition corresponds to the large peak in the search function shown in Figure 12. [Reprinted with permission from Rossmann & Argos (232). Copyright by Academic Press Inc. (London) Ltd.]

parallel and equal vectors between atoms selected progressively along the polypeptides provides not only a test for topological equivalence but also identification of insertions and deletions (232, 285). It remained, however, to be seen whether the size and complexity of the equivalenced protein sections were sufficient to permit the description "similar structures" or even "divergence from a common ancestor." A variety of criteria have been proposed by Rossmann & Argos (275) to differentiate these possibilities.

An alternative procedure has been described by Remington & Matthews (232). This method can be rapid if certain computational techniques are adopted (264). It is the three-dimensional equivalent of the Jukes & Cantor (291) method for finding one-dimensional sequence analogies. A section of polypeptide chain of a given number of residues (say 60) is selected in both molecules, and the C_α atoms are then equivalenced sequentially. A least-squares procedure is used to minimize the distance between equivalenced atoms. If the variance is small, then the fit must be good. The procedure is systematically repeated after moving a given number of residues along one chain. All possible starting combinations are tried in turn. The variance can be plotted against the starting residue number for one of the chains. If a good likeness of fold has been found, the variance will be small all along the polypeptide.

The Remington-Matthews method does not readily allow for insertions or deletions within the trial structural segment (292). Thus the test segment should be small. However, if it is too small, the answers become trivial, since fits will confine themselves to individual secondary structural elements. The best segment length must be found empirically (292) and is usually around 40 residues. The method's utility lies in its straightforwardness, speed, and lack of a complex algorithm. The significance of a comparison can be analyzed in both the Remington-Matthews and the Rossmann-Argos method by the size of the respective minimum or maximum value of the search function relative to the background variation.

Literature Cited

1. Linderstrøm-Lang, K. U., Schellman, J. A. 1959. *Enzymes* 1:443–510. 2nd ed.
2. Bragg, L., Kendrew, J. C., Perutz, M. F. 1950. *Proc. R. Soc. London Ser. A* 203:321–57
3. Huggins, M. L. 1943. *Chem. Rev.* 32:195–218
4. Pauling, L., Corey, R. B., Branson, H. R. 1951. *Proc. Natl. Acad. Sci. USA* 37:205–11
5. Pauling, L., Corey, R. B. 1951. *Proc. Natl. Acad. Sci. USA* 37:241–51
6. Kendrew, J. C., Dickerson, R. E., Strandberg, B. E., Hart, R. G., Davies, D. R., Phillips, D. C., Shore, V. C. 1960. *Nature* 185:422–27
7. Blake, C. C. F., Koenig, D. F., Mair, G. A., North, A. C. T., Phillips, D. C., Sarma, V. R. 1965. *Nature* 206:757–61
8. Anson, M. L., Mirsky, A. E. 1934. *J. Gen. Physiol.* 17:399–408
9. Northrop, J. H. 1932. *J. Gen. Physiol.* 16:323–37
10. Anfinsen, C. B., Haber, E., Sela, M., White, F. H. Jr. 1961. *Proc. Natl. Acad. Sci. USA* 47:1309–14
11. Schechter, A. N., Epstein, C. J. 1968. *J. Mol. Biol.* 35:567–89
12. Epstein, H. F., Schechter, A. N., Chen, R. F., Anfinsen, C. B. 1971. *J. Mol. Biol.* 60:499–508
13. Ristow, S. S., Wetlaufer, D. B. 1973. *Biochem. Biophys. Res. Commun.* 50:544–50
14. Creighton, T. E. 1974. *J. Mol. Biol.* 87:563–77
15. Creighton, T. E. 1974. *J. Mol. Biol.* 87:579–602
16. Creighton, T. E. 1974. *J. Mol. Biol.* 87:603–24
17. Creighton, T. E. 1977. *J. Mol. Biol.* 113:275–93
18. Creighton, T. E. 1977. *J. Mol. Biol.* 113:295–312
19. Creighton, T. E. 1977. *J. Mol. Biol.* 113:313–28
20. Némethy, G., Scheraga, H. A. 1977. *Q. Rev. Biophys.* 10:239–352
21. Schulz, G. E., Schirmer, R. H. 1979. *Principles of Protein Structure.* New York: Springer. 314 pp.
22. Levitt, M., Warshel, A. 1975. *Nature* 253:694–98
23. Levitt, M. 1976. *J. Mol. Biol.* 104:59–107
24. Hagler, A. T., Honig, B. 1978. *Proc. Natl. Acad. Sci. USA* 75:554–58
25. Chou, P. Y., Fasman, G. D. 1974. *Biochemistry* 13:211–22
26. Chou, P. Y., Fasman, G. D. 1974. *Biochemistry* 13:222–45
27. Fasman, G. D., Chou, P. Y., Adler, A. J. 1976. *Biophys. J.* 16:1201–38
28. Schulz, G. E., Barry, C. D., Friedman, J., Chou, P. Y., Fasman, G. D., Finkelstein, A. V., Lim, V. I., Ptitsyn, O. B., Kabat, E. A., Wu, T. T., Levitt, M., Robson, B., Nagano, K. 1974. *Nature* 250:140–42
29. Matthews, B. W. 1975. *Biochim. Biophys. Acta* 405:442–51
30. Wootton, J. C. 1974. *Nature* 252:542–46
31. Richmond, T. J., Richards, F. M. 1978. *J. Mol. Biol.* 119:537–55
32. Cohen, F. E., Richmond, T. J., Richards, F. M. 1979. *J. Mol. Biol.* 132:275–88
33. Cohen, F. E., Sternberg, M. J. E., Taylor, W. R. 1980. *Nature* 285:378–82

34. Henderson, R., Unwin, P. N. T. 1975. *Nature* 257:28–32
35. Koshland, D. E. Jr. 1958. *Proc. Natl. Acad. Sci. USA* 44:98–104
36. Koshland, D. E. Jr. 1973. *Sci. Am.* 229(4):52–64
37. Rossmann, M. G., Adams, M. J., Buehner, M., Ford, G. C., Hackert, M. L., Lentz, P. J. Jr., McPherson, A. Jr., Schevitz, R. W., Smiley, I. E. 1972. *Cold Spring Harbor Symp. Quant. Biol.* 36:179–91
38. Musick, W. D. L., Rossmann, M. G. 1979. *J. Biol. Chem.* 254:7611–20
39. Huber, R. 1979. *Trends Biochem. Sci.* 4:271–76
40. Abad-Zapatero, C., Abdel-Meguid, S. S., Johnson, J. E., Leslie, A. G. W., Rayment, I., Rossmann, M. G., Suck, D., Tsukihara, T. 1980. *Nature* 286:33–39
41. Argos, P. 1981. *Proc. Aharon Katzir-Katchalsky Conf. Structural Aspects of Recognition and Assembly in Biological Macromolecules, 7th, Nof Ginossar, Israel, 1979.* In press
42. Lesk, A. M., Chothia, C. 1980. *J. Mol. Biol.* 136:225–70
43. Doolittle, R. F. 1979. In *The Proteins*, ed. H. Neurath, R. L. Hill, 4:1–118. New York: Academic. 679 pp. 3rd ed.
44. Richardson, J. S. 1981. *Adv. Protein Chem.* In press
45. Cantor, C. R., Schimmel, P. R. 1980. *Biophysical Chemistry. Part I: The Conformation of Biological Macromolecules.* San Francisco: Freeman. 365 pp.
46. Dickerson, R. E., Geis, I. 1969. *The Structure and Action of Proteins.* New York: Harper & Row. 120 pp.
47. Jensen, L. H. 1974. *Ann. Rev. Biochem.* 43:461–74
48. Liljas, A., Rossmann, M. G. 1974. *Ann. Rev. Biochem.* 43:475–507
49. Davies, D. R., Padlan, E. A., Segal, D. M. 1975. *Ann. Rev. Biochem.* 44:639–67
50. Kretsinger, R. H. 1976. *Ann. Rev. Biochem.* 45:239–66
51. Salemme, F. R. 1977. *Ann. Rev. Biochem.* 46:299–329
52. Amzel, L. M., Poljak, R. J. 1979. *Ann. Rev. Biochem.* 48:961–97
53. Matthews, B. W. 1977. See Ref. 43, 3:403–590
54. Blake, C. C. F. 1972. *Prog. Biophys. Mol. Biol.* 25:85–130
55. Tanford, C. 1968. *Adv. Protein Chem.* 23:121–282
56. Tanford, C. 1970. *Adv. Protein Chem.* 24:1–95
57. Brandts, J. F. 1969. In *Structure and Stability of Biological Macromolecules*, ed. S. N. Timasheff, G. D. Fasman, pp. 213–90. New York: Dekker. 694 pp.
58. Hermans, J. Jr., Lohr, D., Ferro, D. 1972. *Adv. Polymer Sci.* 9:229–83
59. Wetlaufer, D. B., Ristow, S. 1973. *Ann. Rev. Biochem.* 42:135–58
60. Ptitsyn, O. B., Lim, V. I., Finkelstein, A. V. 1972. *FEBS Fed. Eur. Biochem. Soc. Meet.* 25:421–31
61. Baldwin, R. L. 1975. *Ann. Rev. Biochem.* 44:453–75
62. Anfinsen, C. B., Scheraga, H. A. 1975. *Adv. Protein Chem.* 29:205–300
63. Creighton, T. E. 1978. *Prog. Biophys. Mol. Biol.* 33:231–97
64. Jaenicke, R., ed. 1980. *Protein Folding.* Amsterdam: Elsevier
65. Fehlhammer, H., Bode, W. 1975. *J. Mol. Biol.* 98:683–92
66. Chambers, J. L., Stroud, R. M. 1977. *Acta Crystallogr. Sect. B* 33:1824–37
67. Deisenhofer, J., Steigemann, W. 1975. *Acta Crystallogr. Sect. B* 31:238–50
68. Sloan, D. L., Young, J. M., Mildvan, A. S. 1975. *Biochemistry* 14:1998–2008
69. Fletterick, R. J., Madsen, N. B. 1980. *Ann. Rev. Biochem.* 49:31–61
70. Levine, M., Muirhead, H., Stammers, D. K., Stuart, D. I. 1978. *Nature* 271:626–30
71. Richardson, J. S. 1979. *Biochem. Biophys. Res. Commun.* 90:285–90
72. Mavridis, I. M., Tulinsky, A. 1976. *Biochemistry* 15:4410–17
73. Quiocho, F. A., McMurray, C. H., Lipscomb, W. N. 1972. *Proc. Natl. Acad. Sci. USA* 69:2850–54
74. Johansen, J. T., Vallee, B. L. 1973. *Proc. Natl. Acad. Sci. USA* 70:2006–10
75. Pauling, L., Corey, R. B. 1951. *Proc. Natl. Acad. Sci. USA* 37:235–41
76. Corey, R. B., Pauling, L. 1953. *Proc. R. Soc. London Ser. B* 141:10–20
77. Marsh, R. E., Donohue, J. 1967. *Adv. Protein Chem.* 22:235–56
78. Ramachandran, G. N., Kolaskar, A. S., Ramakrishnan, C., Sasisekharan, V. 1974. *Biochim. Biophys. Acta* 359:298–302
79. Ramachandran, G. N., Lakshminarayanan, A. V., Kolaskar, A. S. 1973. *Biochim. Biophys. Acta* 303:8–13
80. Ramachandran, G. N., Kolaskar, A. S. 1973. *Biochim. Biophys. Acta* 303:385–88
81. Ramachandran, G. N., Mitra, A. K. 1976. *J. Mol. Biol.* 107:85–92
82. Huber, R., Kukla, D., Bode, W., Schwager, P., Bartels, K., Deisenhofer,

J., Steigemann, W. 1974. *J. Mol. Biol.* 89:73–101

83. Ramachandran, G. N., Sasisekharan, V. 1968. *Adv. Protein Chem.* 23:283–438

84. Kendrew, J. C., Watson, H. C., Strandberg, B. E., Dickerson, R. E., Phillips, D. C., Shore, V. C. 1961. *Nature* 190:666–70

85. Janin, J., Wodak, S., Levitt, M., Maigret, B. 1978. *J. Mol. Biol.* 125:357–86

86. Warme, P. K., Morgan, R. S. 1978. *J. Mol. Biol.* 118:289–304

87. Crippen, G. M., Kuntz, I. D. 1978. *Int. J. Peptide Protein Res.* 12:47–56

88. Ramakrishnan, C., Prasad, N. 1971. *Int. J. Protein Res.* 111:209–31

89. Hamilton, W. C., Ibers, J. A. 1968. *Hydrogen Bonding in Solids.* New York: W. A. Benjamin. 284 pp.

90. Némethy, G., Phillips, D. C., Leach, S. J., Scheraga, H. A. 1967. *Nature* 214:363–65

91. Almassy, R. J., Dickerson, R. E. 1978. *Proc. Natl. Acad. Sci. USA* 75:2674–78

92. Chandrasekaran, R., Jardetzky, T. S., Jardetzky, O. 1979. *FEBS Lett.* 101:11–14

93. Crick, F. H. C. 1952. *Nature* 170:882–83

94. Ward, K. B., Hendrickson, W. A., Klippenstein, G. L. 1975. *Nature* 257:818–21

95. Stenkamp, R. E., Sieker, L. C., Jensen, L. H., McQueen, J. E. Jr. 1978. *Biochemistry* 17:2499–504

96. Mathews, F. S., Bethge, P. H., Czerwinski, E. W. 1979. *J. Biol. Chem.* 254:1699–706

97. Crick, F. H. C. 1953. *Acta Crystallogr.* 6:689–97

98. Chothia, C., Levitt, M., Richardson, D. 1977. *Proc. Natl. Acad. Sci. USA* 74:4130–34

99. Cohen, F. E., Sternberg, M. J. E. 1980. *J. Mol. Biol.* 137:9–22

100. Lim, V. I., Efimov, A. V. 1976. *FEBS Lett.* 69:41–44

101. Ptitsyn, O. B., Rashin, A. A. 1975. *Biophys. Chem.* 3:1–20

102. Argos, P., Rossmann, M. G., Johnson, J. E. 1977. *Biochem. Biophys. Res. Commun.* 75:83–86

103. Blow, D. M., Irwin, M. J., Nyborg, J. 1977. *Biochem. Biophys. Res. Commun.* 76:728–34

104. Efimov, A. V. 1979. *J. Mol. Biol.* 134:23–40

105. Pauling, L., Corey, R. B. 1951. *Proc. Natl. Acad. Sci. USA* 37:729–40

106. Lifson, S., Sander, C. 1980. *J. Mol. Biol.* 139:627–39

107. Lifson, S., Sander, C. 1979. *Nature* 282:109–11

108. Chothia, C. 1973. *J. Mol. Biol.* 75:295–302

109. Weatherford, D. W., Salemme, F. R. 1979. *Proc. Natl. Acad. Sci. USA* 76:19–23

110. Richardson, J. S., Getzoff, E. D., Richardson, D. C. 1978. *Proc. Natl. Acad. Sci. USA* 75:2574–78

111. Richardson, J. S. 1977. *Nature* 268:495–500

112. Sternberg, M. J. E., Thornton, J. M. 1977. *J. Mol. Biol.* 110:285–96

113. Sternberg, M. J. E., Thornton, J. M. 1977. *J. Mol. Biol.* 115:1–17

114. Richardson, J. S. 1976. *Proc. Natl. Acad. Sci. USA* 73:2619–23

115. Sternberg, M. J. E., Thornton, J. M. 1977. *J. Mol. Biol.* 113:401–18

116. Sternberg, M. J. E., Thornton, J. M. 1978. *Nature* 271:15–20

117. Ptitsyn, O. B., Finkelstein, A. V., Falk (Bendzko), P. 1979. *FEBS Lett.* 101:1–5

118. Finkelstein, A. V., Ptitsyn, O. B., Bendzko, P. 1979. *Biofizika* 24:21–26

119. Ptitsyn, O. B. Finkelstein, A. V. 1979. *Biofizika* 24:27–31

120. Kuntz, I. D. 1972. *J. Am. Chem. Soc.* 94:4009–12

120a. Ashida, T., Tanaka, I., Yamane, T., Kakudo, M. 1980. In *Biomolecular Structure, Conformation, Function, and Evolution,* ed. R. Srinivasan, 1:607–20. Oxford/New York: Pergamon

121. Crawford, J. L., Lipscomb, W. N., Schellman, C. G. 1973. *Proc. Natl. Acad. Sci. USA* 70:538–42

122. Venkatachalam, C. M. 1968. *Biopolymers* 6:1425–36

123. Lewis, P. N., Momany, F. A., Scheraga, H. A. 1973. *Biochim. Biophys. Acta* 303:211–29

124. Chou, P. Y., Fasman, G. D. 1977. *J. Mol. Biol.* 115:135–75

125. Rose, G. D. 1978. *Nature* 272:586–90

126. Levitt, M., Greer, J. 1977. *J. Mol. Biol.* 114:181–239

127. Chandrasekaran, R., Prasad, B. V. V. 1978. In *CRC Crit. Rev. Biochem.* 5:125–61

128. Rao, S. T., Rossmann, M. G. 1973. *J. Mol. Biol.* 76:241–56

129. Sternberg, M. J. E., Thornton, J. M. 1976. *J. Mol. Biol.* 105:367–82

130. Phillips, D. C., Sternberg, M. J. E., Thornton, J. M., Wilson, I. A. 1978. *J. Mol. Biol.* 119:329–51

131. Stuart, D. I., Levine, M., Muirhead, H., Stammers, D. K. 1979. *J. Mol. Biol.* 134:109–42

132. Levitt, M., Chothia, C. 1976. *Nature* 261:552–58
133. Sternberg, M. J. E., Thornton, J. M. 1977. *J. Mol. Biol.* 110:269–83
134. Janin, J. 1979. *Bull. Inst. Pasteur Paris* 77:337–73
135. Hol, W. G. J., van Duijnen, P. T., Berendsen, H. J. C. 1978. *Nature* 273:443–46
136. Schulz, G. E. 1980. *J. Mol. Biol.* 138:335–47
137. Shoham, M., Yonath, A., Sussman, J. L., Moult, J., Traub, W., Kalb (Gilboa), A. J. 1979. *J. Mol. Biol.* 131:137–55
138. Argos, P., Rossmann, M. G. 1979. *Biochemistry* 18:4951–60
139. Nishioka, Y., Leder, P. 1979. *Cell* 18:875–82
140. Blake, C. C. F. 1978. *Nature* 273:267
141. Perutz, M. F. 1978. *Sci. Am.* 239(6):92–125
142. Perutz, M. F. 1979. *Ann. Rev. Biochem.* 48:327–86
143. Padlan, E. A., Love, W. E. 1974. *J. Biol. Chem.* 249:4067–78
144. Hendrickson, W. A., Love, W. E., Karle, J. 1973. *J. Mol. Biol.* 74:331–61
145. Vainshtein, B. K., Harutyunyan, E. H., Kuranova, I. P., Borisov, V. V., Sosfenov, N. I., Pavlovsky, A. G., Grebenko, A. I., Konareva, N. V. 1975. *Nature* 254:163–64
146. Watson, H. C. 1969. In *Prog. Stereochem.* 4:299–333
147. Mathews, F. S., Levine, M., Argos, P. 1972. *J. Mol. Biol.* 64:449–64
148. Guiard, B., Lederer, F. 1979. *J. Mol. Biol.* 135:639–50
149. Swanson, R., Trus, B. L., Mandel, N., Mandel, G., Kallai, O. B., Dickerson, R. E. 1977. *J. Biol. Chem.* 252:759–75
150. Ashida, T., Tanaka, N., Yamane, T., Tsukihara, T., Kakudo, M. 1973. *J. Biochem. Tokyo* 73:463–65
151. Salemme, F. R., Freer, S. T., Xuong, N. H., Alden, R. A., Kraut, J. 1973. *J. Biol. Chem.* 248:3910–21
152. Timkovich, R., Dickerson, R. E. 1976. *J. Biol. Chem.* 251:4033–46
153. Korszun, Z. R., Salemme, F. R. 1977. *Proc. Natl. Acad. Sci. USA* 74:5244–47
154. Klotz, I. M., Klippenstein, G. L., Hendrickson, W. A. 1976. *Science* 192:335–44
155. Weber, P. C., Bartsch, R. G., Cusanovich, M. A., Hamlin, R. C., Howard, A., Jordan, S. R., Kamen, M. D., Meyer, T. E., Weatherford, D. W., Xuong, N. H., Salemme, F. R. 1980. *Nature* 286:302–4
156. Bloomer, A. C., Champness, J. N., Bricogne, G., Staden, R., Klug, A. 1978. *Nature* 276:362–68
157. Banyard, S. H., Stammers, D. K., Harrison, P. M. 1978. *Nature* 271:282–84
158. Irwin, M. J., Nyborg, J., Reid, B. R., Blow, D. M. 1976. *J. Mol. Biol.* 105:577–86
159. Deisenhofer, J., Jones, T. A., Huber, R., Sjödahl, J., Sjöquist, J. 1978. *Hoppe-Seylers Z. Physiol. Chem.* 359:975–85
160. Drenth, J., Jansonius, J. N., Koekoek, R., Sluyterman, L. A. A., Wolthers, B. G. 1970. *Phil. Trans. R. Soc. London Ser. B* 257:231–36
161. Kretsinger, R. H. 1972. *Nature New Biol.* 240:85–88
162. Kretsinger, R. H. 1981. *CRC Crit. Rev. Biochem.* In press
163. Mornon, J. P., Fridlansky, F., Bally, R., Milgrom, E. 1980. *J. Mol. Biol.* 137:415–29
164. Engelman, D. M., Henderson, R., McLachlan, A. D., Wallace, B. A. 1980. *Proc. Natl. Acad. Sci. USA* 77:2023–27
165. Wiegand, G., Kukla, D., Scholze, H., Jones, T. A., Huber, R. 1979. *Eur. J. Biochem.* 93:41–50
166. Colman, P. M., Jansonius, J. N., Matthews, B. W. 1972. *J. Mol. Biol.* 70:701–24
167. Sasaki, K., Dockerill, S., Adamiak, D. A., Tickle, I. J., Blundell, T. L. 1975. *Nature* 257:751–57
168. Adams, M. J., Helliwell, J. R., Bugg, C. E. 1977. *J. Mol. Biol.* 112:183–97
169. Wiley, D. C., Skehel, J. J. 1977. *J. Mol. Biol.* 112:343–47
170. Kannan, K. K., Notstrand, B., Fridborg, K., Lövgren, S., Ohlsson, A., Petef, M. 1975. *Proc. Natl. Acad. Sci. USA* 72:51–55
171. Mitsui, Y., Satow, Y., Watanabe, Y., Iitaka, Y. 1979. *J. Mol. Biol.* 131:697–724
172. Wierenga, R. K., de Jong, R. J., Kalk, K. H., Hol, W. G. J., Drenth, J. 1979. *J. Mol. Biol.* 131:55–73
173. Drenth, J., Jansonius, J. N., Koekoek, R., Wolthers, B. G. 1971. *Enzymes* 3:485–99 3rd ed.
174. Matthews, B. W., Fenna, R. E., Bolognesi, M. C., Schmid, M. F., Olson, J. M. 1979. *J. Mol. Biol.* 131:259–85
175. Blow, D. M. 1971. *Enzymes* 3:185–212. 3rd ed.
176. Kraut, J. 1971. *Enzymes* 3:165–83. 3rd ed.
177. Bode, W., Schwager, P. 1975. *J. Mol. Biol.* 98:693–717
178. Bode, W., Huber, R. 1978. *FEBS Lett.* 90:265–69

179. Kossiakoff, A. A., Chambers, J. L., Kay, L. M., Stroud, R. M. 1977. *Biochemistry* 16:654–64
180. Hartley, B. S., Shotton, D. M. 1971. *Enzymes* 3:323–73. 3rd ed.
181. Brayer, G. D., Delbaere, L. T. J., James, M. N. G. 1979. *J. Mol. Biol.* 131:743–75
182. James, M. N. G., Delbaere, L. T. J., Brayer, G. D. 1978. *Can. J. Biochem.* 56:396–402
183. Tang, J., James, M. N. G., Hsu, I. N., Jenkins, J. A., Blundell, T. L. 1978. *Nature* 271:618–21
184. James, M. N. G., Hsu, I. N., Delbaere, L. T. J. 1977. *Nature* 267:808–13
185. Subramanian, E., Swan, I. D. A., Liu, M., Davies, D. R., Jenkins, J. A., Tickle, I. J., Blundell, T. L. 1977. *Proc. Natl. Acad. Sci. USA* 74:556–59
186. Wong, C. H., Lee, T. J., Lee, T. Y., Lu, T. H., Yang, I. H. 1979. *Biochemistry* 18:1638–40
187. Andreeva, N. S., Gustchina, A. E. 1979. *Biochem. Biophys. Res. Commun.* 87:32–42
188. Richardson, J. S., Richardson, D. C., Thomas, K. A., Silverton, E. W., Davies, D. R. 1976. *J. Mol. Biol.* 102:221–35
189. Colman, P. M., Freeman, H. C., Guss, J. M., Murata, M., Norris, V. A., Ramshaw, J. A. M., Venkatappa, M. P. 1978. *Nature* 272:319–24
190. Adman, E. T., Stenkamp, R. E., Sieker, L. C., Jensen, L. H. 1978. *J. Mol. Biol.* 123:35–47
191. Blow, D. M., Janin, J., Sweet, R. M. 1974. *Nature* 249:54–57
192. McLachlan, A. D. 1979. *J. Mol. Biol.* 133:557–63
193. Blake, C. C. F., Geisow, M. J., Oatley, S. J., Rérat, B., Rérat, C. 1978. *J. Mol. Biol.* 121:339–56
194. Arnone, A., Bier, C. J., Cotton, F. A., Day, V. W., Hazen, E. E. Jr., Richardson, D. C., Richardson, J. S., Yonath, A. 1971. *J. Biol. Chem.* 246:2302–16
195. McPherson, A., Jurnak, F. A., Wang, A. H. J., Molineux, I., Rich, A. 1979. *J. Mol. Biol.* 134:379–400
196. Argos, P., Tsukihara, T., Rossmann, M. G. 1980. *J. Mol. Evol.* 15:169–79
197. Harrison, S. C., Olson, A. J., Schutt, C. E., Winkler, F. K., Bricogne, G. 1978. *Nature* 276:368–73
198. Reeke, G. N. Jr., Becker, J. W., Edelman, G. M. 1975. *J. Biol. Chem.* 250:1525–47
199. Hardman, K. D., Ainsworth, C. F. 1972. *Biochemistry* 11:4910–19
200. Holbrook, J. J., Liljas, A., Steindel, S.

J., Rossmann, M. G. 1975. *Enzymes* 11:191–292. 3rd ed.
201. Banaszak, L. J., Bradshaw, R. A. 1975. *Enzymes* 11:269–96 3rd ed.
202. Brändén, C. I., Jörnvall, H., Eklund, H., Furugren, B. 1975. *Enzymes* 11:103–90 3rd ed.
203. Moras, D., Olsen, K. W., Sabesan, M. N., Buehner, M., Ford, G. C., Rossmann, M. G. 1975. *J. Biol. Chem.* 250:9137–62
204. Pickover, C. A., McKay, D. B., Engelman, D. M., Steitz, T. A. 1979. *J. Biol. Chem.* 254:11323–29
205. Pai, E. F., Sachsenheimer, W., Schirmer, R. H., Schulz, G. E. 1977. *J. Mol. Biol.* 114:37–45
206. Banks, R. D., Blake, C. C. F., Evans, P. R., Haser, R., Rice, D. W., Hardy, G. W., Merrett, M., Phillips, A. W. 1979. *Nature* 279:773–77
207. Bryant, T. N., Watson, H. C., Wendell, P. L. 1974. *Nature* 247:14–17
208. Steitz, T. A., Anderson, W. F., Fletterick, R. J., Anderson, C. M. 1977. *J. Biol. Chem.* 252:4494–500
209. Evans, P. R., Hudson, P. J. 1979. *Nature* 279:500–4
210. Matthews, D. A., Alden, R. A., Freer, S. T., Xuong, N., Kraut, J. 1979. *J. Biol. Chem.* 254:4144–51
211. Abdallah, M. A., Adams, M. J., Archibald, I. G., Biellmann, J. F., Helliwell, J. R., Jenkins, S. E. 1979. *Eur. J. Biochem.* 98:121–30
212. Eichele, G., Ford, G. C., Glor, M., Jansonius, J. N., Mavrides, C., Christen, P. 1979. *J. Mol. Biol.* 133:161–80
213. Borisov, V. V., Borisova, S. N., Sosfenov, N. I., Vainshtein, B. K. 1980. *Nature* 284:189–90
214. Sprang, S., Fletterick, R. J. 1979. *J. Mol. Biol.* 131:523–51
215. Smith, W. W., Burnett, R. M., Darling, G. D., Ludwig, M. L. 1977. *J. Mol. Biol.* 117:195–225
216. Monaco, H. L., Crawford, J. L., Lipscomb, W. N. 1978. *Proc. Natl. Acad. Sci. USA* 75:5276–80
217. Morikawa, K., la Cour, T. F. M., Nyborg, J., Rasmussen, K. M., Miller, D. L., Clark, B. F. C. 1978. *J. Mol. Biol.* 125:325–38
218. Campbell, J. W., Watson, H. C., Hodgson, G. I. 1974. *Nature* 250:301–3
219. Shaw, P. J., Muirhead, H. 1977. *J. Mol. Biol.* 109:475–85
220. Hartsuck, J. A., Lipscomb, W. N. 1971. *Enzymes* 3:1–56. 3rd ed.
221. Drenth, J., Hol, W. G. J., Jansonius, J. N., Koekoek, R. 1972. *Eur. J. Biochem.* 26:177–81

222. Wright, C. S., Alden, R. A., Kraut, J. 1969. *Nature* 221:235–42

223. Söderberg, B. O., Sjöberg, B. M., Sonnerstam, U., Brändén, C. I. 1978. *Proc. Natl. Acad. Sci. USA* 75:5827–30

224. Ladenstein, R., Epp, O, Bartels, K., Jones, A., Huber, R., Wendel, A. 1979. *J. Mol. Biol.* 134:199–218

225. Ploegman, J. H., Drent, G., Kalk, K. H., Hol, W. G. J. 1978. *J. Mol. Biol.* 123:557–94

226. Newcomer, M. E., Miller, D. M. III, Quiocho, F. A. 1979. *J. Biol. Chem.* 254:7529–33

227. Lindqvist, Y., Brändén, C. I. 1979. *J. Biol. Chem.* 254:7403–4

227a. Matsuura, Y., Kusunoki, M., Harada, W., Tanaka, N., Iga, Y., Yasuoka, N., Toda, H., Narita, K., Kakudo, M. 1980. *J. Biochem. Tokyo* 87:1555–58

228. Drenth, J., Low, B. W., Richardson, J. S., Wright, C. S. 1980. *J. Biol. Chem* 255:2652–55

229. Tsernoglou, D., Petsko, G. A. 1977. *Proc. Natl. Acad. Sci. USA* 74:971–74

230. Kimball, M. R., Sato, A., Richardson, J. S., Rosen, L. S., Low, B. W. 1979. *Biochem. Biophys. Res. Commun.* 88:950–59

231. Wright, C. S. 1977. *J. Mol. Biol.* 111:439–57

232. Rossmann, M. G., Argos, P. 1976. *J. Mol. Biol.* 105:75–95

233. Remington, S. J., Matthews, B. W. 1978. *Proc. Natl. Acad. Sci. USA* 75:2180–84

234. Ford, L. O., Johnson, L. N., Machin, P. A., Phillips, D. C., Tjian, R. 1974. *J. Mol. Biol.* 88:349–71

235. Remington, S. J., Anderson, W. F., Owen, J., Ten Eyck, L. F., Grainger, C. T., Matthews, B. W. 1978. *J. Mol. Biol.* 118:81–98

236. Brew, K., Vanaman, T. C., Hill, R. L. 1967. *J. Biol. Chem.* 242:3747–49

237. Wyckoff, H. W., Tsernoglou, D., Hanson, A. W., Knox, J. R., Lee, B., Richards, F. M. 1970. *J. Biol. Chem.* 245:305–28

238. Cutfield, J. F., Cutfield, S. M., Dodson, E. J., Dodson, G. G., Emdin, S. F., Reynolds, C. D. 1979. *J. Mol. Biol.* 132:85–100

239. Dijkstra, B. W., Drenth, J., Kalk, K. H., Vandermaelen, P. J. 1978. *J. Mol. Biol.* 124:53–60

240. Watenpaugh, K. D., Sieker, L. C., Jensen, L. H. 1979. *J. Mol. Biol.* 131:509–22

241. Adman, E. T., Sieker, L. C., Jensen, L. H. 1976. *J. Biol. Chem.* 251:3801–6

242. Freer, S. T., Alden, R. A., Carter, C. W. Jr., Kraut, J. 1975. *J. Biol. Chem.* 250:46–54

243. Haser, R., Pierrot, M., Frey, M., Payan, F., Astier, J. P., Bruschi, M., Le Gall, J. 1979. *Nature* 282:806–10

244. Kretsinger, R. H. 1975. In *Calcium Transport in Secretion and Contraction,* ed. E. Carafoli, F. Clementi, W. Drabikowski, A. Margreth, pp. 469–78. Amsterdam: North Holland

245. Hill, R. L., Delaney, R., Fellows, R. E. Jr., Lebovitz, H. E. 1966. *Proc. Natl. Acad. Sci. USA* 56:1762–69

246. Rossmann, M. G., Moras, D., Olsen, K. W. 1974. *Nature* 250:194–99

247. Ohlsson, I., Nordström, B., Brändén, C. I. 1974. *J. Mol. Biol.* 89:339–54

248. Rossmann, M. G., Liljas, A. 1974. *J. Mol. Biol.* 85:177–81

249. Pompon, D., Lederer, F. 1976. *Eur. J. Biochem.* 68:415–23

250. Ghelis, C., Tempete-Gaillourdet, M., Yon, J. M. 1978. *Biochem. Biophys. Res. Commun.* 84:31–36

251. Goldberg, M. E., Zetina, C. R. 1980. See Ref. 64, pp. 469–84

252. Rose, G. D. 1979. *J. Mol. Biol.* 134:447–70

253. Crippen, G. M. 1978. *J. Mol. Biol.* 126:315–32

254. Rashin, A. A., Janin, J., Wodak, S. J. 1981. *J. Mol. Biol.* In press

255. Wodak, S. J., Janin, J. 1980. *Proc. Natl. Acad. Sci. USA* 77:1736–40

256. Wetlaufer, D. B. 1973. *Proc. Natl. Acad. Sci. USA* 70:697–701

257. Wetlaufer, D. B., Rose, G. D., Taaffe, L. 1976. Biochemistry 15:5154–57

258. Blake, C. C. F. 1979. *Nature* 277:598

259. Gilbert, W. 1981. In *Introns and Exons: Playground of Evolution, ICN-UCLA Symp., Los Angeles, 1979.* In press

260. Eaton, W. A. 1980. *Nature* 284:183–85

261. Craik, C. S., Buchman, S. R., Beychok, S. 1980. *Proc. Natl. Acad. Sci. USA* 77:1384–88

262. Hill, E., Tsernoglou, D., Webb, L., Banaszak, L. J. 1972. *J. Mol. Biol.* 72:577–91

263. Kraut, J., Robertus, J. D., Birktoft, J. J., Alden, R. A., Wilcox, P. E., Powers, J. C. 1972. *Cold Spring Harbor Symp. Quant. Biol.* 36:117–23

264. McLachlan, A. D. 1979. *J. Mol. Biol.* 128:49–79

265. Sokal, R. R., Sneath, P. H. A. 1963. *Principles of Numerical Taxonomy.* San Francisco: Freeman. 359 pp.

266. Dickerson, R. E. 1980. *Sci. Am.* 242(3):136–53

267. Kurosky, A., Barnett, D. R., Lee, T. H., Touchstone, B., Hay, R. E., Arnott, M. S., Bowman, B. H., Fitch, W. M. 1980. *Proc. Natl. Acad. Sci. USA* 77:3388–92
268. Greer, J. 1980. *Proc. Natl. Acad. Sci. USA* 77:3393–97
269. Schulz, G. E., Schirmer, R. H. 1974. *Nature* 250:142–44
270. Eventoff, W., Rossmann, M. G. 1975. *CRC Crit. Rev. Biochem.* 3:111–40
271. Fitch, W. M., Margoliash, E. 1967. *Science* 155:279–84
272. Farris, J. S., Kluge, A. G., Eckardt, M. J. 1970. *Syst. Zool.* 19:172–89
273. McLachlan, A. D. 1976. In *Taniguchi Symp. Biophys., Mishima, Japan, Oct.,* pp. 208–39
274. Buehner, M., Ford, G. C., Moras, D., Olsen, K. W., Rossmann, M. G. 1973. *Proc. Natl. Acad. Sci. USA* 70:3052–54
275. Rossmann, M. G., Argos, P. 1977. *J. Mol. Biol.* 109:99–129
276. Rossmann, M. G., Argos, P. 1978. *Mol. Cell. Biochem.* 21:161–82
277. Argos, P., Garavito, R. M., Eventoff, W., Rossmann, M. G., Brändén, C. I. 1978. *J. Mol. Biol.* 126:141–58
278. Kester, W. R., Matthews, B. W. 1977. *J. Biol. Chem.* 252:7704–10
279. McLachlan, A. D. 1972. *Nature New Biol.* 240:83–85
280. Blundell, T. L., Sewell, B. T., McLachlan, A. D. 1979. *Biochim. Biophys. Acta* 580:24–31
281. Hendrickson, W. A., Ward, K. B. 1977. *J. Biol. Chem.* 252:3012–18
282. McLachlan, A. D., Bloomer, A. C., Butler, P. J. G. 1980. *J. Mol. Biol.* 136:203–24
283. Monod, J., Wyman, J., Changeux, J. P. 1965. *J. Mol. Biol.* 12:88–118
284. Rossmann, M. G., Liljas, A., Brändén, C. I., Banaszak, L. J. 1975. *The Enzymes* 11:61–102. 3rd ed.
285. Rossmann, M. G., Argos, P. 1975. *J. Biol. Chem.* 250:7525–32
286. Diamond, R. 1976. *Acta Crystallogr. Sect. A* 32:1–10
287. Kabsch, W. 1976. *Acta Crystallogr. Sect. A* 32:922–23
288. Hendrickson, W. A. 1979. *Acta Crystallogr. Sect. A* 35:158–63
289. Nishikawa, K., Ooi, T. 1974. *J. Theor. Biol.* 43:351–74
290. Némethy, G., Scheraga, H. A. 1979. *Proc. Natl. Acad. Sci. USA* 76:6050–54
291. Jukes, T. H., Cantor, C. R. 1969. In *Mammalian Protein Metabolism,* ed. H. N. Munro, 3:21–132. New York: Academic. 571 pp.
292. Remington, S. J., Matthews, B. W. 1980. *J. Mol. Biol.* 140:77–99

Ann. Rev. Biochem. 1981. 50:533–54

CHROMOSOME MEDIATED
GENE TRANSFER

♦12087

L. A. Klobutcher[1] and F. H. Ruddle

Departments of Biology and Human Genetics, Yale University, New Haven,
Connecticut 06511

CONTENTS

Introduction

Somatic cell genetics depends on parasexuality. The primary parasexual events are those of gene transfer and gene loss. A number of schemes have been developed that facilitate the transfer of genetic material from a donor somatic cell to a recipient. Such transfer (1, 2) can be mediated by cell hybridization (CH), microcell mediated gene transfer (MMGT), chromosome mediated gene transfer (CMGT), and DNA mediated gene transfer (DMGT). In each method, the amount of donor genomic material transferred differs. A whole genome is transferred by cell hybridization, intact chromosomes by microcells, subchromosomal segments by chromosome mediated gene transfer, and DNA fragments in the kilobase range in DNA mediated gene transfer. Thus, each method provides a potentially different degree of genetic resolution. In addition, for all cases there is a possibility of subsequent loss from the recipient cell of the transferred donor material, either partial or complete.

---

[1]Present address, Department of Molecular, Cellular, and Developmental Biology, University of Colorado, Boulder, Colorado 80309.

0066-4154/81/0701-0533$01.00

In this article, we discuss the mechanisms of chromosome mediated gene transfer, and point out the potential of chromosome transfer for determining the order and spacing of genes within chromosomes. We believe this system of regional gene mapping has considerable promise for ordering genes at the centiMorgan level of resolution. As such, it represents a gene mapping method of intermediate resolution between that of nucleotide sequencing and restriction analysis on the one hand and chromosome assignment on the other.

Methodology of CMGT

McBride & Ozer (3) were the first to convincingly demonstrate the ability of purified metaphase chromosomes to serve as a vector for the transfer of genetic information to cultured mammalian cells. In their original experiment (Figure 1), they first arrested cultures of wild-type Chinese hamster

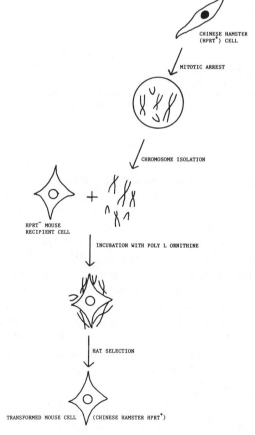

Figure 1   Scheme for chromosome mediated gene transfer.

cells in mitosis. The plasma membranes of these donor cells were then physically disrupted, which released the metaphase, or condensed, chromosomes. The free metaphase chromosomes were then partially purified from whole cells, nuclei, and debris by differential centrifugation. For transfer, the chromosomes were added to recipient mouse cells deficient for the enzyme hypoxanthine phosphoribosyltransferase (HPRT), at a ratio of one cell equivalent of donor chromosomes per recipient cell, and poly-L-ornithine was added to enhance the association of cells and chromosomes. Following a two-hour incubation, the recipient cells were plated at low density in nonselective medium, and three days later HAT selection, which allows only those cells expressing HPRT to grow (4) was applied. Within six weeks HAT resistant colonies appeared at a frequency of about one per 10^7 recipient cells.

The frequency of transformation in this experiment was similar to the spontaneous reversion rate of the HPRT⁻ allele in the recipient cell line. It was therefore necessary to determine the nature of the HPRT activity in the cell lines isolated. Using DEAE-cellulose chromatography and polyacrylamide gel electrophoresis under conditions that separate the donor Chinese hamster and recipient mouse forms of HPRT, many of the cell lines were shown to express the donor form of HPRT and, thus, to be true transformants.

The ability to transfer genes via isolated metaphase chromosomes has been confirmed by other workers using both interspecific and intraspecific differences in enzymes (5–11), ouabain resistance (12), chemically transformed cell characteristics (13), and chromosomally integrated viral sequences (14–16). Moreover, the CMGT method is not limited to mammalian species, since gene transfer between cultured amphibian cells has been demonstrated (17).

In the majority of these experiments, the CMGT method of McBride & Ozer (3) was followed closely, and the frequencies of transfer were near one transformant per 10^7 recipient cell. This low transfer frequency is similar to the reversion rates for most mammalian mutant alleles, which makes it necessary to demonstrate that the transferred phenotype is of donor origin. Moreover, the low transfer frequency makes it difficult to obtain large numbers of transformed cell lines. Therefore, modifications of the CMGT procedure have been sought that enhance the transfer frequency. Miller & Ruddle (18) were able to increase the frequency of transfer by employing reagents normally used for the transformation of mammalian cells by viral DNA (19, 20). By precipitating the donor chromosomes with calcium phosphate and treating the recipient cells first with a mixture of colchicine, Colcemid, and cytochalasin D and then with dimethyl sulfoxide (DMSO), they were able to achieve a 100-fold increase in the frequency of transfer (3.3×10^{-5} transformants/recipient cell). It was subsequently found that the modifications active in increasing the transfer frequency are calcium phosphate precipitation of the donor chromosomes and the DMSO treatment of recipient cells; treatment with colchicine, Colcemid, and cytochalasin D appears to inhibit the transfer frequency (11).

Although these experiments employed mouse L-cells as recipients, a similar increase in transfer frequency has been reported for transfers to human cells using the calcium phosphate and DMSO treatments (12). In the case of human cells, however, higher concentrations of DMSO and shorter treatment times were required for increased transfer frequencies, and the optimal transfer frequency achieved was below that obtained for L-cell recipients. Therefore, it appears that there is some variability among cell lines in their ability to serve as recipients for chromosome transfer, as well as in their sensitivity to frequency enhancing agents.

Wullems et al (21) have reported two additional methods of increasing the transfer frequency. They were able to obtain significant increases in transfer frequency by either employing metaphase cells as recipients or incubating donor chromosomes and recipient cells in the presence of inactivated Sendai virus. It is unfortunate that experiments confirming these results have not yet been reported. These methods may work in combination with the calcium phosphate-DMSO technique to further enhance the frequency of CMGT, and this should certainly be investigated.

Size of the Transferred Genetic Material

The usefulness of the CMGT method as a mapping technique and as a method of genetic analysis depends, in part, on the amount of genetic material transferred to the recipient cell (the term "transgenome" will be used for the donor genetic material in transformants). Current evidence indicates that a range of transgenome sizes is possible. Despite the fact that the donor genetic material consists of full size chromosomes, most transformants receive a subchromosomal transgenome. One indication of the small size of most transgenomes was a failure to detect a change in the chromosome number of transformants relative to the recipient cell (3, 5, 9, 22). Also, in most cases it is not possible to detect chromosomes with donor specific banding patterns in the karyotypes of transformed cells (6, 8, 21). The subchromosomal nature of most transgenomes is also supported by the failure to cotransfer genes that are syntenic with the selected gene. For instance, the gene for HPRT is located on the X-chromosome in man (23) and it is flanked by the genes coding for phosphoglycerate kinase (PGK) and glucose 6-phosphate dehydrogenase (G6PD) (24, 25). Most transformants selected for the transfer of human HPRT have been found not to express the human forms of PGK and G6PD, which indicates that only a portion of the X-chromosome is transferred.

A minority of transformants ($\sim$15%), however, do contain cytologically detectable donor chromosome material (macrotransgenomes). The amount of donor material in these lines is variable, ranging from small fragments of donor chromosomes through whole chromosome arms to apparently complete donor chromosomes (7, 10, 11, 18). As one would expect, lines containing macrotransgenomes display high frequencies of cotransfer for genes syntenic with the selected gene (7, 11, 18).

Although the maximum transgenome size is at least an intact donor chromosome, the minimum transgenome size, or the size of any transgenome that is not detectable by cytological methods (microtransgenome), has not been rigorously determined. Scangos et al (16) have approached this problem in CMGT experiments, transferring a chromosomally integrated copy of the Herpes simplex virus (HSV) thymidine kinase (TK) gene to cultured mouse cells. As a nucleic acid probe exists for the HSV-TK gene, they were able to analyze the size of their microtransgenomes by restriction enzyme and Southern blotting analyses. Using a number of restriction enzymes they found that the donor restriction patterns around the HSV-TK gene were maintained in all but one transformed cell line. Since the largest restriction fragment containing the HSV-TK sequence was 17 kilobases (kb), the microtransgenome must be at least this large.

An estimate for the maximum size of a microtransgenome can be obtained by considering the size of a chromosome fragment just detectable by light microscopy. Double minute chromosomes, which are dot-like chromosomes present in some organisms and cultured cell lines, are just detectable by light microscopy and are estimated to contain about 10^3 kb of DNA (26). As the undetectable microtransgenomes must be smaller than this, they must fall into the size range of 17–1000 kb. Obviously, the extremes of these size estimates are rather far apart, but a more precise sizing of these small transgenomes may require their isolation from transformed cell lines or the development of nucleic acid probes complementary to the DNA of selected chromosomes.

Uptake and Early Fate of Metaphase Chromosomes

The first steps of the CMGT process must involve the uptake of donor chromosomes by recipient cells and the incorporation of transferred material into the host nucleus. The simplest experiments designed to analyze these steps in the process have been to incubate isolated metaphase chromosomes with recipient cells and follow the fate of the chromosomes microscopically (27, 28). In such experiments the chromosomes initially associate with the plasma membranes of the recipient cells. Within two hours after addition, the chromosomes appear to enter intracellular vacuoles, some in close association with the nuclear membrane, where they may persist for days. Most recipient cells take up at least one donor chromosome, and some cells have been observed with as many as twenty.

Additional information has been obtained from experiments employing radioactively labeled chromosomes as donor material and following their fate by cellular autoradiography of the recipient cells. Following uptake, silver grains are first found over cytoplasmic vacuoles in the recipient cell; they reach a peak within 2–6 hr after chromosome addition (21, 29, 30). With time, the labeling pattern shifts to uniformly cover the nucleus of the recipient cell. These results indicate that most donor chromosomes are degraded while in the cytoplasm, and their nucleotide components are reutilized by the recipient cell.

There is evidence, however, that some chromosomes are able to escape degradation while in the cytoplasm. A number of workers have observed labeled donor chromosomes in metaphase spreads of recipient cells (21, 31, 32). Moreover some of these donor chromosomes appear to be functional. When metaphase spreads were examined 2–3 days after addition, some donor chromosomes had only one labeled chromatid (21, 32). This indicates that some donor chromosomes are able to undergo at least two rounds of replication, though such events occur at a low frequency (1×10^{-4} to 1×10^{-7}/recipient cell metaphase).

Although the results of these experiments appear to define the uptake and incorporation of macrotransgenomes, they do not directly address the early steps leading to the formation of a microtransgenome, which is typical of most transformed cell lines. Simmons et al (33) have approached this problem by following the intracellular fate of DNA molecules derived from radioactively labeled donor chromosomes with thin section electron microscopic autoradiography of recipient cells. Following uptake, silver grains were observed over both the nuclear and the cytoplasmic compartments of the recipient cells. The silver grains, however, could not be associated with condensed chromosome-like structures or with cytoplasmic vacuoles. It thus appears that some of the genetic material from donor chromosomes enters the cell via a nonphagocytotic process and is available for transformation either in a subchromosomal form, possibly as naked DNA, or chromatin.

Their second experimental approach followed donor DNA molecules by cellular fractionation and alkaline sucrose gradient sedimentation analysis of the donor DNA obtained from cytoplasmic and nuclear fractions of the recipient cells. On the average, each recipient cell nucleus received about one quarter the amount of DNA present in an average size chromosome. At four hours, the size of the DNA molecules was generally from 1 to 7 $\times 10^6$ daltons. By six hours after chromosome addition, the size of the DNA molecules in the cytoplasm had decreased, which indicates that degradation was occurring. In the nucleus, however, the size of some donor DNA molecules had increased to molecular weights greater than 1×10^7. As the recipient cell's nuclear DNA was much larger than this, and therefore sedimented through the gradient, the increase in donor DNA size could not be attributed to association with recipient cell DNA or to the reutilization of labeled nucleotides from degraded donor DNA. The authors suggest that DNA repair enzymes may be present in the nucleus that restructure the donor DNA molecules.

Although these experiments provide a scenario for the entrance of chromosomes into mammalian cells, the relevance of these steps to the CMGT process is unknown. The results of these experiments indicate that many donor chromosomes enter the cell through phagocytosis or a pathway not involving membrane vesicles, and on the average, a large amount of donor DNA reaches the nucleus of each cell. In contrast to the large amount of donor material in the recipient cell, CMGT occurs at a very low frequency. This apparent discrepancy may be due to the fact that most of the DNA entering the recipient cell is no longer in a form that allows gene expression, perhaps because of breakage. Alternatively, there may be a minor and uncharacterized pathway leading to CMGT that is obscured by more frequent, but nonfunctional modes of chromosome uptake. It is obviously difficult to design experiments to characterize such a minor pathway, or, for that matter, to link an uptake pathway with transformation events.

There is, however, genetic evidence that a restructuring step may be involved early in the CMGT process. Klobutcher & Ruddle (10) screened mouse transformants selected for transfer of the human TK gene, which has been mapped to human chromosome 17 (34), for the expression of 20 human isozyme markers whose genes have been mapped to chromosomes other than human 17. Two of the transformed cell lines expressed normally unlinked human genes. Since this initial observation, additional TK and HPRT transformants have been examined, and about 10% of the lines have been found to express unlinked genes (L. A. Klobutcher, C. L. Miller, and F. H. Ruddle, unpublished data). Moreover, when such cell lines are back-selected for loss of the selected gene, the unlinked cotransferred gene is also lost, indicating that the selected and unlinked cotransferred genes are associated or linked in the transformant.

The cotransfer of unlinked genes might be explained by the abnormal and partially rearranged karyotype of the human cell line (HeLa S3), which was employed as the chromosome donor in these experiments. Rearrangements involving the chromosome carrying the selected gene could have created aritifical linkage groups in the donor cell line, which were subsequently detected in the transfer process. The fact that many different unlinked genes have been cotransferred, in combination with two different selectable genes, argues against such an explanation. Alternatively, the early steps of the CMGT process could be responsible for the cotransfer of genuinely asyntenic genes. Transfer of unlinked markers has been observed in bacterial transformation experiments (35), and is apparently due to the bacterial cell's ability to take up and express more than one piece of donor DNA. As we have mentioned, a mammalian cell is able to ingest more than one donor chromosome. If two donor chromosomes within the same cell are partially degraded and then reconstituted, as postulated above, it is possible to explain both the cotransfer and linkage to the selected gene of asyntenic markers.

Two types of future experiments could distinguish between the two hypotheses presented. First, CMGT experiments should be done using primary cell lines as chromosome donors. Since primary cell lines possess a normal diploid karyotype, the observation of a transformant expressing an asyntenic donor gene would indicate that the CMGT process was responsible. Second, the simultaneous transfer of two selectable genes to a doubly deficient cell line should be attempted. This experiment is ideally done in a trispecific system by mixing the chromosomes of two singly prototrophic cell lines prior to transfer. Again, observation of dual transformation events would indicate that asyntenic gene cotransfer is a result of the CMGT process, and provide additional evidence for the reconstruction of donor chromosomes.

Unstable Expression of the Transferred Phenotype

In their original CMGT experiment, McBride & Ozer (3) observed that transformed cell lines maintained the transferred phenotype in two distinct modes. Upon return to culture in nonselective medium, some transformants (unstable transformants) progressively lost the transferred phenotype. A second class of transformants (stable transformants) maintained the transferred phenotype indefinitely in the absence of selection. These two types of transformants have been observed in all other CMGT experiments where this characteristic has been examined.

The loss of the transferred phenotype in unstable transformants appears to occur at variable rates ranging from 1% loss per cell generation (9) to about 14% per cell generation (22). Upon prolonged culture in selective medium, on the order of months, a number of workers have observed unstable transformants convert to stable expression (10, 22, 36). The reverse process, conversion from stable expression to instability, has not been observed.

A great deal of research has focused on the state of the transgenome in unstable and stable transformants, and on the conversion from unstable to stable expression. The nature of the transgenome in unstable cells is at present poorly understood, and, as such, has been the subject of much speculation. Both McBride & Ozer (3) and Willecke & Ruddle (6) reasoned that as most transgenomes are subchromosomal in size, they probably do not contain centromeres. Therefore, if the unstable transgenome exists as an independent genetic element, one would expect that the transgenome would not be efficiently distributed to daughter cells at mitosis. This would result in some daughter cells with no copies of the transgenome following each division and give the total cell population the characteristic of instability. This model characterizes the unstable transgenome as a genetic element similar to a bacterial plasmid or episome.

An alternative model for instability could be based on a regulatory phenomenon. In this case, epigenetic changes in the transformant would result in a shut-off of expression of the transferred gene. The change in regulation would have to be a random one, occurring at a defined frequency, to explain the constant and progressive loss of the transferred phenotype.

There are no experimental data that rigorously exclude either of these models. An experiment by Willecke & Ruddle (6), however, provides support for the episome model. They obtained subclones from an unstable HPRT transformant that had spontaneously lost the transferred phenotype. When more than 5×10^6 of these cells were replated in HAT-selective medium, no clones could be recovered that reexpressed the donor form of HPRT. This result suggests that the transgenome is physically lost from unstable cells, and supports the episome model.

An additional area of controversy concerning unstable cells is the number of copies of the transgenome per cell. Degnen et al (37) have observed overexpression of the transferred phenotype in unstable transformants. Specifically, they found 7–58 times the donor level of HPRT in a series of unstable intraspecific mouse transformants. Two explanations were proposed to explain this result. The unstable transformants could contain many copies of the transgenome either singly or in tandem, and thus many copies of the selected gene, which would cause the increase in HPRT activity. Alternatively, the selected gene may be under relaxed control while in the unstable state, which would lead to overexpression.

Scangos et al (16) have presented data indicating that multiple copies of the selected gene are present in unstable transformants. In their experiments transferring a chromosomally integrated copy of the HSV-TK gene to mouse cells via CMGT, they observed elevated levels of TK activity in unstable transformants. Using a nucleic acid probe complementary to the HSV-TK gene, they were able to demonstrate that the unstable lines overexpressing TK did have multiple copies of the TK gene.

It should be noted, however, that overexpression of transferred genes in unstable transformants has not been observed in other CMGT transfer experiments. Most workers have observed expression of transferred genes at low levels, consistent with a single copy of the transgenome per unstable cell (3, 36, 38). Moreover, in lines containing cytologically detectable macrotransgenomes there is only one chromosome or fragment present per unstable cell (7, 10, 18).

Some of our own recent work (11) indicates that there are, in fact, two categories of unstable transformants. In transfers of the human TK gene to mouse L cells, we have observed unstable cell populations that either lose the transferred phenotype slowly, at rates of 1.5–2.5% per day when cultured in nonselective medium, or quickly, at rates of 6.5–10% per day. Two differences were noted between these slow loss and fast loss unstable transformants. First, all of the fast loss lines were found to have no cytologically detectable donor material, and thus possess microtransgenomes, while the majority of slow loss lines had cytologically detectable macrotransgenomes. In the cases where a macrotransgenome was present in a slow loss line, it was always an independent chromosome or subchromosomal fragment containing a centromere-like constriction and present at one copy per cell.

Second, slow loss and fast loss unstable lines differed in their levels of TK expression. Slow loss unstable lines had levels of activity consistent with one copy of the TK gene per cell. Fast loss lines, on the other hand, had high levels of TK activity, ranging from 3–10 times higher than that present in slow loss lines. Thus, it is possible to isolate, in the same CMGT experiment, unstable transformants that show the two types of behavior regarding transferred gene expression that had been noted in the literature.

The underlying difference between these two classes of unstable cell lines, which results in different loss rates and levels of gene expression, appears to be in the nature of their transgenomes. As we have mentioned above, the slow loss lines contain independent donor chromosomes or fragments with centromere-like regions. The loss rates for these unstable lines are similar to those observed for the loss of human chromosomes in human x mouse whole cell hybrids (39). The macrotransgenomes may possess normal donor centromeres, or other factors necessary for distribution, that function suboptimally in a heterologous species in a manner analogous to segregating chromosomes in whole cell hybrids. This scheme predicts that one would not find slow loss unstable lines in an intraspecific transfer, as the macrotransgenome in this case would have a centromere homologous to the host. In fact, only unstable lines of the fast loss category have been detected in intraspecific transfers (16, 22).

On the other hand, the small size of the microtransgenome in fast loss unstable transformants suggests that they are acentric and would, therefore, be subject to random distribution among daughter cells at mitosis. As some daughter cells would, by chance, receive no copies of the transgenome, the population as a whole would have the characteristic of instability. Based on this random partitioning model, 3–4 independent copies of the transgenome per cell would be necessary to obtain the loss rates of 6–10% per generation found for fast loss unstable cell lines. If one uses the level of transferred gene expression as an indication of gene copy number, then most fast loss lines fit into this scheme (11, 16). There are, however, instances where gene copy number, as inferred from levels of gene expression, and loss rates are not consistent with this model (37).

Random segregation could also explain the generation of multiple transgenome copies in fast loss lines. The original transformed cell would receive a single copy of the transgenome and replicate it prior to division. Some divisions would result in one daughter cell with two copies of the transgenome. This cell would have a growth advantage over cells with only one transgenome, as more of its daughter cells would survive in selective medium. This argument can be extended to further increase the average gene copy number per cell in the population.

There appears to be some constraint on amplification, however, as most cell lines do not show extreme levels of overexpression. This could be explained by a counterselective pressure, which balances the tendency toward increased gene copy number. Cell lines selected for amplification of the dihydrofolate reductase gene provide a precedent for such a counterselective force (40). Progressive increases in the copy number of dihydrofolate reductase genes has been correlated with an increased cell generation time. Although unproven, it may be that the increased production of gene products coded for by amplified genetic elements is deleterious to cell growth and viability.

In any event, it appears that the phenomenology of unstable expression is becoming well-defined and that this aspect of CMGT will be a major area of interest in the future. The study of experimental systems where selected and/or cotransferred genes are detectable through the use of nucleic acid probes will be particularly important in characterizing the features of unstable expression.

Stable Expression of the Transferred Phenotype

The fate of the transgenome is currently better understood in stable cells. Three types of experimental evidence indicate that the transgenome is associated with chromosomes of the recipient cell in stable transformants. The first indication that the stable transgenome is associated with recipient cell chromosomes came from serial transfer experiments. Chromosomes from a stable transformant were isolated and used as the donor material for a second round of CMGT, again selecting for the transferred gene. Such an experiment has been done with both TK and HPRT transformants (41, 42), and in both cases it was possible to obtain second generation transformants, which indicates that the stable transgenome copurifies with recipient cell chromosomes.

The second experimental approach has relied on whole cell and microcell fusion to essentially "map" the transgenome to recipient cell chromosomes. Fournier (43, 44) produced microcells from a human into mouse stable HPRT transformant and fused them to a third species of cell, HPRT⁻ Chinese hamster cells, selecting for hybrids in HAT medium. In each hybrid cell line isolated they found one mouse chromosome that correlated with the expression of human HPRT. The mouse chromosome involved was never the X-chromosome, which normally codes for mouse HPRT, but did differ between microcell hybrids. Essentially identical results have been reported (45) in whole cell and microcell hybrid experiments with TK transformants. The results of these experiments indicate that there is a strong association, probably covalent, between the stable transgenome and host chromosomes, and that this association is able to withstand the disruptions of the hybridization procedure. Moreover, this association does not involve the homologous genetic site in the recipient cell, and multiple recipient cell chromosomes can be involved.

These first two approaches involved analysis of stable cells that originally had microtransgenomes, as no donor chromosome material was detectable in the transformant analyzed. The third line of evidence for association of the stable transgenome with host chromosomes comes from analysis of lines containing macrotransgenomes. Stable derivatives of an unstable TK transformant containing an independent donor chromosome fragment have been isolated, and the fate of the macrotransgenome followed cytologically (10). In each stable derivative the originally independent chromosome fragment was found to be associated with a recipient cell chromosome in a manner indistinguishable from a translocation. As before, different recipient cell chromosomes were involved, which suggests that host chromosome association was random. In each stable subclone examined, however, the macrotransgenome was associated with a unique recipient cell chromosome.

Thus, stabilization may be viewed as the acquisition of host centromeric activity. The association of the transgenome with a recipient cell chromosome allows for its orderly partitioning to daughter cells at mitosis and gives the cell population the character of stability.

Two additional features of stabilization have been described. First, genetic information is lost from the transgenome following stabilization. Willecke et al (36) obtained a stable derivative of an unstable TK transformant that cotransferred the linked gene, GALK. This stable line no longer expressed the donor form of GALK, which indicates that this gene had been deleted during stabilization. The loss of genetic information is clearly seen in stable derivatives of unstable lines with macrotransgenomes. In this case, there is a visible shortening of the chromosome fragment following association with a host chromosome (10).

Second, only one copy of the transgenome persists in each stable cell. Stable cell lines that were derived from unstable lines that overexpress the selected gene have been found to have low levels of donor gene expression, which is consistent with one copy of the transgenome per cell (16, 37). Moreover, an actual decrease in transferred gene copy number has been observed in stable derivatives of cell lines selected for transfer of the HSV-TK gene (16). Apparently, once one copy of the transgenome undergoes stabilization, the remaining, and no longer necessary, copies of the unstable transgenome are diluted out in subsequent cell divisions.

Currently, it is thought that the stabilization event is rare and occurs in a single cell of the population. This stable cell would then have a growth advantage, relative to the unstable cells in the population, due to its ability to efficiently distribute the transgenome to all its daughter cells. With time, the stable cells would completely overgrow the unstable cells and give rise to a totally stable population. Based on this scheme, Degnen et al (22) calculated that the frequency of the stabilization event was about 1×10^{-5} per cell generation for a series of intraspecific HPRT transformants.

Though this hypothesis remains unproven, circumstantial evidence suggests that unstable cell populations convert to stability in this manner. The hypothesis predicts that it should be possible to find populations that are transiently mixed with regard to stability. Such mixed cell lines would represent the instance where the stable cells are in the process of selectively overgrowing the population. If such a line were grown on nonselective medium, as in a typical stability test, one would expect to initially see a progressive loss of the transferred phenotype, but eventually a fraction of the cells would retain the selected gene indefinitely. Such mixed lines have, in fact, been observed (10, 11). Moreover, in mixed lines containing a macrotransgenome, two subpopulations can be detected by karyotypic analysis. A portion of the cells contain independent macrotransgenomes, which have been associated with unstable expression, and the remainder of the cells have a macrotransgenome translocated to a host chromosome as in the stable state.

Intrachromosomal Mapping Using CMGT

Most genetic mapping methods depend on disruption, or breakage, of the genome to separate linked genes and define their relative locations. There are at least two steps in the CMGT process that result in breakage of the donor genome and that could provide mapping functions. One breakage step occurs during the initial uptake of chromosomes and incorporation of the transgenome in the cell. The breakage in this case is clearly indicated by the observation that most transgenomes are subchromosomal in size. The second breakage event occurs during the conversion from unstable to stable expression, and is evidenced by the loss of cotransferred genes and a physical decrease in macrotransgenome size. As the breakage event in each of these cases disrupts syntenic relationships of genes on a chromosome, CMGT can be used to derive intrachromosomal mapping data. Moreover, it is possible, as we discuss below, to derive both qualitative and quantitative mapping data.

A number of workers have suggested that the cotransfer frequency of nonselected but syntenic genes in primary transformants may be a source of quantitative intrachromosomal mapping data (9, 18, 36, 46). If it is assumed that the breakage and degradation of donor chromosomes during uptake is random, then the cotransfer frequency for a syntenic gene is an inverse measure of its distance from the selected gene. As such, a gene that is far from the selected gene would be transferred infrequently, while a closely linked gene would be cotransferred often.

Though this potential use of CMGT was realized early on, the low frequencies of transfer made it difficult to obtain significant numbers of cell lines for analysis. With the advent of methods for the isolation of large numbers of transformants, such analyses are made possible. For purposes of discussion we consider a series of human into mouse TK transformants that we have recently described (11). Two features of this group of cell lines make them particularly suitable for this discussion. First, the transformed cell lines were chosen by a random clone-picking procedure. This avoids bias and ensures a representative sample of lines with micro- and macrotransgenomes. Second, the lines were examined for expression of two syntenic but unselected genes. As mentioned above, the GALK gene is closely linked to TK on the long arm of human chromosome 17 (47). A third gene, human procollagen type I (PCI), has also been mapped to the long arm of chromosome 17 (48, 49). If only the unstable transformants in this sample are considered, so as to avoid the second breakage step during stabilization, GALK was cotransferred in 19% (5/27) of the cell lines and PCI in 74% (20/27) of the TK-transformed lines. A CMGT map unit will arbitrarily be defined as the genetic distance between two genes that results in a cotransfer frequency of 0.99. It is then possible to convert to CMGT map units by subtracting the cotransfer frequency from 1.00 and multiplying by 100. Using this convention, GALK is 81 map units from TK and PCI is 26 map units from TK (Figure 2). Because of the close linkage between TK and PCI, it is not possible to determine from this sample whether PCI is located between TK and GALK, or distal to this region. In one transformant, however, TK and GALK were transferred without PCI, which suggests that PCI is outside this segment.

The second method of CMGT mapping makes use of transformants with macrotransgenomes to derive qualitative, or cytological, data. In theory,

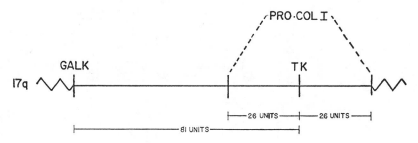

Figure 2 Map of human chromosome 17 based on CMGT cotransfer data. The map and units are described in the text.

this can simply be done by identifying the precise portion of the selected chromosome that is transferred by banding analysis, and correlating this information with the expression of nonselected genes. By considering progressively smaller transgenomes, it would be possible to localize the selected and cotransferred genes to small and defined regions of the chromosome.

Alternatively, it is possible to use the stabilization event to further fragment the transgenome and generate a deletion map. This approach has been used to map the chromosome 17 genes discussed above (10). A CMGT transformant was identified that had a macrotransgenome equivalent to the long arm of human chromosome 17, based on chromosome banding analysis, and expressed all three chromosome 17 genes. Stable subclones were isolated from this line, and in each case the macrotransgenome had associated with a host chromosome and become shortened in the process. Chromosome banding analysis revealed that the centromere-like end of the fragment and various portions of the adjacent region were deleted during stabilization. This process is illustrated diagrammatically in Figure 3. The resulting stable transgenomes essentially generated a deletion map and allowed location of the genes by correlating their expression with the portion of the fragment retained. The genes were thus localized to regions of the long arm of human chromosome 17, and a gene order of centromere-GALK-(TK,PCI) was deduced (Figure 4).

How do these two types of mapping data compare with data obtained by independent methods? McDougall et al (50) have made use of adenovirus induced breaks in human chromosome 17 to localize the TK gene to the region 17q21–22. A similar study by Elsevier et al (47) assigned the GALK gene to this same region. Comparison of this data to the CMGT deletion map data (Figure 4) indicates that the results are compatible. The CMGT deletion map assigns GALK to a region within 17q21–22 and localizes TK to a region that overlaps 17q21–22. In addition, Church et al (49) have used

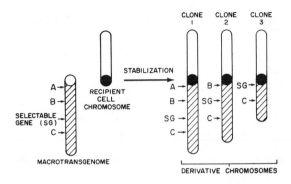

Figure 3 Use of the stabilization event to fragment macrotransgenomes and generate deletion maps. During the stabilization event the transgenome is truncated and becomes attached to a recipient cell chromosome. Genes can be mapped to regions of the chromosome by correlating the portion of the transgenome retained with the expression of donor phenotypes.

adenovirus induced chromosome 17 breaks to determine that PCI is distal to TK and GALK on 17q, which is again consistent with the CMGT map.

The order of chromosome 17 genes deduced by qualitative CMGT mapping [centromere-GALK-(TK,PCI)] is less certain. Church et al (49) obtained one cell line in their study that had lost GALK expression but retained TK, which indicates a gene order of centromere-TK-GALK-PCI.

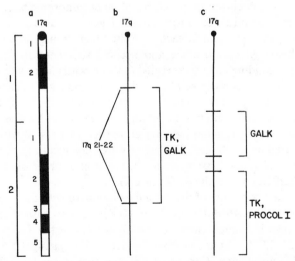

Figure 4 Comparison of human chromosome 17 mapping data. *a,* Schematic of Giemsa-trypsin banded human chromosome 17. *b,* Locations of TK and GALK as determined by McDougall et al (50) and Elsevier et al (47), respectively. *c,* Locations of TK, GALK, and PCI derived from CMGT deletion map.

In both the adenovirus induced deletion and qualitative CMGT mapping methods it is not possible to detect subtle rearrangements, such as a small interstitial deletion, that would influence the ordering of genes. Thus, based on the limited amount of data available, the true order of chromosome 17 genes is uncertain.

A relevant comparison for the CMGT derived quantitative mapping data on chromosome 17 genes, based on cotransfer frequency, does not exist. It is possible, however, to compare contransfer frequency data to human X-linked genes to other mapping data. Miller & Ruddle (18) observed cotransfer frequencies of 0.12 and 0.25 for PGK and G6PD, respectively, in human into mouse HPRT transformants. These values are consistent with the physical distances of PGK and G6PD from HPRT as determined from somatic cell hybridization studies (51) and radiation induced breakage of the X-chromosome (52). Unfortunately, the stability of all transformed cell lines was not assessed in these transfer experiments, nor were random lines chosen for analysis.

A major objection to the use of CMGT for genetic mapping can be raised on the basis of the evidence indicating that rearrangement occurs during the process. Rearrangements undetected by chromosome banding analysis could lead to false gene localizations by the qualitative CMGT mapping procedure. It should be possible to overcome this potential problem by analysis of multiple cell lines. Rearrangements can occur, however, and not affect the validity of quantitative mapping data based on cotransfer frequencies. This method requires only that linked genes be lost or retained during the transfer process in relation to their physical distance from the selected gene. Moreover, it seems unlikely that one could confuse the cotransfer of an asyntenic gene with that of a linked one. For any particular unlinked gene the cotransfer frequency (apparently < 0.01) is much less than that observed for any syntenic gene to date.

A final assessment of these two CMGT mapping methods will require further experimentation. CMGT of genes that are well localized by other methods is highly desirable in this regard. Further investigation may establish CMGT as a mapping method with a level of resolution intermediate between those presently obtainable by somatic cell hybridization and restriction endonuclease analyses. The results discussed here provide an indication of the potential usefulness of CMGT in generating sophisticated mammalian genetic maps.

Conclusion

The data presented in this review can be summarized in a model of the CMGT process (Figure 5). The initial steps in the CMGT process involve

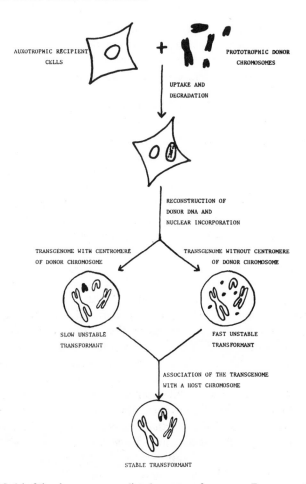

Figure 5 Model of the chromosome mediated gene transfer process. Donor genetic material is shown as solid chromosomes or fragments, while the recipient cell genome is represented by open chromosomes.

entry of donor chromosomes into recipient cells, possibly by phagocytosis. Most donor chromosomes are degraded while in the cytoplasm, but some donor material escapes complete degradation and enters the nuclear compartment of the recipient cell. While in the nucleus, the DNA of the partially degraded chromosome(s) can be reconstructed and ligated to give larger molecules.

Cells that receive and express complementing genes through this process proliferate in selective medium, giving rise to transformed cell lines. Initially, transformed cell lines are unstable and may be grouped into two classes. Transformants that have received a large amount of the complementing donor chromosome, including its centomeric region, lose the transformed phenotype at a slow rate when returned to nonselective medium. Such cell lines have one copy of the transgenome per cell and low levels of transferred gene activity.

The second class of unstable transformants receives a much smaller donor chromosome fragment that is undetectable by cytological methods. This fragment is acentric and subject to random segregation among daughter cells at mitosis. Although such cell lines initially receive one copy of the transgenome, unequal partitioning events at mitosis generate cells with multiple copies. Despite having multiple copies of the transgenome, such cell lines lose the transferred phenotype at high rates due to this random partitioning at division, which generates cells with no copies of the transgenome. Additional characteristics of fast-loss unstable transformants are high levels of transferred gene activity and a low frequency of cotransfer for linked genes.

At a low frequency, the transgenome in each type of unstable transformant associates with a recipient cell chromosome (stabilization). The association process does not involve the homologous genetic site in the recipient cell, but appears to be random. This provides the transgenome with host centromeric activity, which allows for its orderly distribution at mitosis and gives the cell a stable phenotype. Stable cells thus have a growth advantage over the unstable cells in the population and gradually overgrow the population, forming a completely stable cell line.

Two sections of this model are well-documented by experimental data. First, the fate of macrotransgenomes has been well-established since it has been possible to identify them cytologically. Second, a large amount of evidence indicates that in stable lines the transgenome is associated with host chromosomes. Exactly how the stabilization process occurs, particularly whether specific DNA sequences are involved, remains unclear. In regard to lines with microtransgenomes, evidence indicates that they possess multiple copies of the selected gene. It is unclear at present whether such multiple copies represent tandem duplications or independent genetic elements. Moreover, if multiple copies do exist, it will be of interest to determine if their mode of replication, distribution at mitosis, and original mechanisms of amplification concur with the proposed model.

It is clear that the CMGT process shares a number of features with other mammalian genetic systems. The similarities of CMGT to a second transfer process, DNA mediated gene transfer (DMGT), are striking. DMGT uses purified DNA, in the form of a calcium phosphate precipitate, as the vector for transformation (53, 54). Transformants generated by DMGT have been found to be either unstable or stable, in the same manner as CMGT transformants (53; G. A. Scangos, K. M. Huttner, and F. H. Ruddle, unpublished data). Moreover, in stable DMGT transformants, the transferred genetic material is also associated with recipient cell chromosomes (G. A. Scangos and F. H. Ruddle, unpublished data). Finally, it is possible to cotransfer unlinked genes by DMGT. A number of workers have observed that the addition of specific gene sequences, such as the bacterial plasmid pBR322 or mammalian globin genes, to the transforming DNA results in the incorporation of the nonselected sequences in transformed cell lines (55–58).

CMGT transformants also show similarities to mammalian cells selected for amplification of the dihydrofolate reductase gene by growth in the presence of the drug methotrexate. Cells with multiple copies of the dihydrofolate reductase gene have been shown to be either stable or unstable in the absence of selection. In stable cells, the amplified genes are chromosomally associated, apparently in the form of tandem duplications, to form a chromosome with a "homogeneously staining region" (59). In unstable cells, the amplified dihydrofolate reductase genes are located on small acentromeric double minute chromosomes that appear to segregate randomly at mitosis (60). These recurring observations of unstable and stable genetic elements suggest that mammalian cells normally possess mechanisms for both amplifying genes and stably incorporating exogenous genetic information. Exactly what function these events would serve in the life of the cell or organism is obviously open to speculation.

Although we have discussed how CMGT may be used as an intrachromosomal mapping technique, there are at least two other ways in which the CMGT process could be applied to the study of mammalian genetics. First, CMGT can be used to place selectable genes on chromosomes where they are now absent. This can simply be done by isolating stable transformants that have the transgenome associated with a recipient cell chromosome. As transgenome association appears to be random, it is conceivable that every chromosome of a particular recipient cell could be marked by isolating large numbers of stable lines. This would allow each chromosome to be moved at will, through whole cell (61) or microcell hybridization (62), to other cell types. Panels of hybrid cells could be developed, each containing one donor chromosome, and used to quickly and simply determine syntenic relationships.

Secondly, CMGT may aid in the isolation and cloning of specific mammalian genes. If the microtransgenomes are indeed small and independent genetic elements, they may be physically separable from host chromosomes. Double minute chromosomes have been purified on sucrose gradients (26, 60), which suggests that the smaller microtransgenomes may also be purified by this process. Alternatively, the microtransgenome may be separable by flow microfluorometry (63). The DNA obtained from a purified micro-transgenome would represent a significant purification of the selected gene. This DNA could then be further fractionated by cloning in bacterial plasmids, and the selected gene sequence identified by DMGT. As the CMGT process becomes more defined additional uses of the system should become evident.

Literature Cited

1. McKusick, V. A., Ruddle, F. H. 1977. *Science* 196:390–405
2. Deleted in proof
3. McBride, O. W., Ozer, H. 1973. *Proc. Natl. Acad. Sci. USA* 70:1258–62
4. Littlefield, J. 1964. *Science* 145:709–10
5. Burch, J. W., McBride, O. W. 1975. *Proc. Natl. Acad. Sci. USA* 72:1797–1801
6. Willecke, K., Ruddle, F. H. 1975. *Proc. Natl. Acad. Sci. USA* 72:1792–96
7. Wullems, G. J., van der Horst, J., Bootsma, D. 1976. *Somat. Cell Genet.* 2:359–71
8. Wullems, G. J., van der Horst, J., Bootsma, D. 1977. *Somat. Cell Genet.* 3:281–93
9. McBride, O. W., Burch, J. W., Ruddle, F. H. 1978. *Proc. Natl. Acad. Sci. USA* 75:914–18
10. Klobutcher, L. A., Ruddle, F. H. 1979. *Nature* 280:657–60
11. Klobutcher, L. A., Miller, C. L., Ruddle, F. H. 1980. *Proc. Natl. Acad. Sci. USA* 77:3610–14
12. Gross, T. A., Squires, S., Martin, P., Baker, R. M. 1979. *J. Cell Biol.* 83:453a.
13. Shih, C., Shilo, B., Goldfarb, M. P., Dannenberg, A., Weinberg, R. A. 1979. *Proc. Natl. Acad. Sci. USA* 76:5714–18
14. Shani, M., Huberman, E., Aloni, Y., Sachs, L. 1974. *Virology* 61:303–5
15. Cassingena, R., Suarez, H. G., Lavialle, C., Persuy, M. A., Ermonval, M. 1978. *Gene* 4:337–49
16. Scangos, G. A., Huttner, K. M., Silverstein, S., Ruddle, F. H. 1979. *Proc. Natl. Acad. Sci. USA* 76:3987–90
17. Rosenstein, B. S., Ohlsson-Wilson, B. M. 1978. *Somat. Cell Genet.* 4:341–54
18. Miller, C. L., Ruddle, F. H. 1978. *Proc. Natl. Acad. Sci. USA* 75:3346–50
19. Graham, F. L., van der Eb, A. J. 1973. *Virology* 52:456–67
20. Farber, F. E., Eberle, R. 1976. *Exp. Cell Res.* 103:15–22
21. Wullems, G. J., van der Horst, J., Bootsma, D. 1975. *Somat. Cell Genet.* 1:137–52

22. Degnen, G. E., Miller, I. L., Eisenstadt, J. M., Adelberg, E. A. 1976. *Proc. Natl. Acad. Sci. USA* 73:2838–42
23. Nabholz, M., Miggiano, V., Bodmer, W. 1969. *Nature* 223:358–63
24. Meerakhn, P., Westerveld, A., Grzeschik, K. H., Deys, B. F., Garson, O. M., Siniscalco, M. 1971. *Am. J. Hum. Genet.* 23:614–23
25. Ricciuti, F. C., Ruddle, F. H. 1973. *Genetics* 74:661–78
26. Barker, P. E., Stubblefield, E. 1979. *J. Cell Biol.* 83:663–66
27. Yoshida, T. H., Sekiguchi, T. 1968. *Mol. Gen. Genet.* 103:253–57
28. Ittenson, O. L., Hutchison, D. J. 1969. *Exp. Cell Res.* 55:149–54
29. Chorazy, M., Bendich, A., Borenfreund, E., Ittenson, O. L., Hutchinson, D. J. 1963. *J. Cell Biol.* 19:71–77
30. Kato, H., Sekiya, K., Yosida, T. H. 1971. *Exp. Cell Res.* 65:454–62
31. Sekiguchi, T., Sekiguchi, F., Yamada, U. 1973. *Exp. Cell Res.* 80:223–36
32. Ebina, T., Kamo, I., Takahashi, K., Homma, M., Ishida, N. 1970. *Exp. Cell Res.* 62:384–88
33. Simmons, T., Lipman, M., Hodge, L. D. 1978. *Somat. Cell Genet.* 4:55–76
34. Miller, O. J., Allderdice, P. W., Miller, D. A., Breg, W. R., Migeon, B. R. 1971. *Science* 173:244–45
35. Goodgal, S. H. 1961. *J. Gen. Phys.* 45:205–28
36. Willecke, K., Lange, R., Kruger, A., Reber, T. 1976. *Proc. Natl. Acad. Sci. USA* 73:1274–78
37. Degnen, G. E., Miller, I. L., Adelberg, E. A., Eisenstadt, J. E. 1977. *Proc. Natl. Acad. Sci. USA* 74:3956–59
38. Wullems, G. J., van der Horst, J., Bootsma, D. 1976. *Somat. Cell Genet.* 2:155–64
39. Schall, D., Rechsteiner, M. 1978. *Somat. Cell Genet.* 4:661–76
40. Alt, F. W., Kellems, R. E., Bertino, J. R., Schimke, R. T. 1978. *J. Biol. Chem.* 253:1357–70
41. Willecke, K., Davies, P. J., Reber, T. 1976. *Cytogenet. Cell Genet.* 16:405–8

42. Athwal, R. S., McBride, O. W. 1977. *Proc. Natl. Acad. Sci. USA* 74:2943–47
43. Fournier, R. E. K., Ruddle, F. H. 1977. *Proc. Natl. Acad. Sci. USA* 74:3937–41
44. Fournier, R. E. K., Juricek, D. K., Ruddle, F. H. 1979. *Somat. Cell Genet.* 5:1061–77
45. Willecke, K., Mierau, R., Kruger, A., Lange, R. 1978. *Mol. Gen. Genet.* 161:49–57
46. McBride, O. W., Athwal, R. S. 1976. *In Vitro* 12:777–86
47. Elsevier, S. M., Kucherlapati, R. S., Nichols, E. A., Willecke, K., Creagan, R. P., Giles, R. E., McDougall, J. K., Ruddle, F. H. 1974. *Nature* 251:633–35
48. Sundar Raj, C. V., Church, R. L., Klobutcher, L. A., Ruddle, F. H. 1977. *Proc. Natl. Acad. Sci. USA* 74:4444–48
49. Church, R. L., Sundar Raj, N., McDougall, J. K. 1980. *Cytogenet. Cell Genet.* 27:24–30
50. McDougall, J. K., Kucherlapati, R. S., Ruddle, F. H. 1973. *Nature New Biol.* 245:172–75
51. Bergsma, D. 1977. *Int. Workshop Human Gene Mapping, 4th, Winnipeg, Canada.* New York: S. Karger
52. Goss, S. J., Harris, H. 1975. *Nature* 255:680–84
53. Wigler, M., Pellicer, A., Silverstein, S., Axel, R., Urlaub, G., Chasin, L. 1979. *Proc. Natl. Acad. Sci. USA* 76:1373–76
54. Willecke, K., Klomfass, M., Mierau, R., Dohmer, J. 1979. *Mol. Gen. Genet.* 170:179–85
55. Wigler, M., Sweet, R., Sim, G. K., Wold, B., Pellicer, A., Lacy, E., Maniatis, T., Silverstein, S., Axel, R. 1979. *Cell* 16:777–85
56. Huttner, K. M., Scangos, G. A., Ruddle, F. H. 1979. *Proc. Natl. Acad. Sci. USA* 76:5820–24
57. Mantei, N., Boll, W., Weissmann, C. 1979. *Nature* 281:40–46
58. Wigler, M., Perucho, M., Kurtz, D., Dana, S., Pellicer, A., Axel, R., Silverstein, S. 1980. *Proc. Natl. Acad. Sci. USA* 77:3567–70
59. Nunberg, J. H., Kaufman, R. J., Schimke, R. T., Urlaub, G., Chasin, L. A. 1978. *Proc. Natl. Acad. Sci. USA* 75:5553–56
60. Kaufman, R. J., Brown, P. C., Schimke, R. T. 1979. *Proc. Natl. Acad. Sci. USA* 76:5669–73
61. Ruddle, F. H., Creagan, R. P. 1975. *Ann. Rev. Genet.* 9:407–86
62. Fournier, R. E. K., Ruddle, F. H. 1977. *Proc. Natl. Acad. Sci. USA* 74:319–23
63. Gray, J. W., Carrano, A. V., Steimetz, L. L., VanDilla, M. A., Moore, D. H. II, Mayall, B. H., Mendelsohn, M. L. 1975. *Proc. Natl. Acad. Sci. USA* 74:1231–34

Ann. Rev. Biochem. 1981. 50:555–83
Copyright © 1981 by Annual Reviews Inc. All rights reserved

SYNTHESIS AND PROCESSING OF ASPARAGINE-LINKED OLIGOSACCHARIDES[1],[2]

◆12088

S. Catherine Hubbard and Raymond J. Ivatt[3]

Center for Cancer Research, Massachusetts Institute of Technology, Cambridge, Massachusetts 02139

CONTENTS

[1]The other major class of protein-linked glycans, those linked through the hydroxyl group of serine, threonine, or hydroxylysine, is not discussed (see 1). It should be noted that, except in the synthesis of some yeast mannans (2, 3), lipid-linked intermediates have not been implicated in O-linked glycan formation.

[2]Abbreviations used are: VSV, vesicular stomatitis virus; CHO, Chinese hamster ovary; ER, endoplasmic reticulum; endo H, endo-β-N-acetylglucosaminidase H; Man, mannose; Glc, glucose; GlcNAc, N-acetylglucosamine; Fuc, fucose; Gal, galactose; SA, sialic acid.

[3]Present address: Dept. Tumor Biology, M.D. Anderson Hospital, Houston, Tex. 77030

0066-4154/81/0701-0555$01.00

1. INTRODUCTION

The asparagine-linked oligosaccharides of eukaryotic glycoproteins have heterogenous structures that generally fall into two categories: "high-mannose" and "complex" (reviewed in 4, 5). Examples are a high-mannose oligosaccharide from bovine thyroglobulin (6; Figure 1–II) and a complex oligosaccharide from vesicular stomatitis virus (VSV) (7; Figure 1–I). Both classes have an inner core of mannose$_3$N-acetylglucosamine$_2$ (Man$_3$-GlcNAc$_2$) at the reducing terminus; they differ in that high-mannose glycans contain additional α-linked Man residues (usually two to six in vertebrate cells), while complex oligosaccharides carry other external sugars, such as GlcNAc, galactose (Gal), fucose (Fuc), and sialic acid (SA). The oligosaccharide chains of both classes exhibit microheterogeneity; a number of related structures are usually found in a single purified glycopeptide preparation.

Experiments with a number of viral, plasma membrane, and secretory glycoproteins have revealed that high-mannose and complex glycans have a common biosynthetic origin: a large, high-mannose precursor oligosaccharide. This species is not assembled on the protein but is synthesized

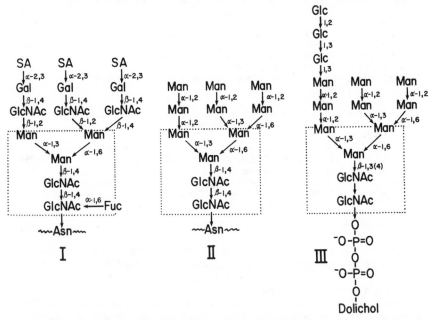

Figure 1 Structures of representative asparagine-linked oligosaccharides from glycoproteins (I, II) and of a dolichol-linked precursor oligosaccharide (III).

while linked to a carrier lipid (dolichol, Dol); completed oligosaccharide chains are then transferred en bloc from lipid to acceptor proteins. In a variety of systems the lipid-linked precursor oligosaccharide has the composition $Glc_3Man_9GlcNAc_2$, and its structure in Chinese hamster ovary (CHO) cells has been determined (8; Figure 1–III).

The strongest evidence that lipid-linked oligosaccharides are the only source of Asn-linked glycans has come from studies with inhibitors and with mutant cell lines. Drugs that inhibit lipid-linked oligosaccharide assembly (Table 1) have invariably been found to block glycosylation of Asn residues as well; both high-mannose and complex glycans are absent in proteins from inhibited cells (reviewed in 9, 10). Similarly, in mutant cells that synthesize truncated lipid-linked oligosaccharides, newly synthesized proteins acquire the abnormal oligosaccharide rather than the normal $Glc_{-3}Man_9GlcNAc_2$ species (11, 12). The case for lipid-linked oligosaccharide involvement has been further strengthened by studies with a temperature-sensitive mutant cell line in which the glycoproteins have a reduced content of N-linked carbohydrate. Under nonpermissive conditions, these cells appear to synthesize a normal lipid-linked oligosaccharide but to transfer it to protein at a greatly reduced rate (13, 14).

This review describes the assembly of the lipid-linked precursor oligosaccharide, the factors that control its transfer to protein, and the sequence of modifications by which it is converted to the high-mannose and complex glycans of mature glycoproteins.

Table 1   Inhibitors of lipid-linked oligosaccharide assembly[a]

Inhibitor	Site of inhibition
Compactin and 25-HC[b]	Suppress synthesis of dolichol (17, 18) by competitively inhibiting hydroxymethylglutaryl-coenzyme A reductase; also suppress synthesis of cholesterol and coenzyme Q
2-Deoxyglucose (2DG)	Is converted to GDP-2DG, which inhibits formation of Glc-P-Dol, Man-P-Dol, and GlcNAc-P-P-Dol by competing for Dol-P to form 2DG-P-Dol (19, 20); 2DG is also incorporated into abnormal lipid-linked oligosaccharides that are not transferred to protein (21)
Glucosamine	Apparently glucosamine itself, rather than a metabolite, inhibits N-glycosylation (22), probably by blocking an early step of lipid-linked oligosaccharide assembly (23); exact mechanism of inhibition unknown
Tunicamycin	Inhibits formation of GlcNAc-P-P-Dol (24, 25), but not its elongation (26–28)

[a] This is an abbreviated presentation of the extensive literature on glycosylation inhibitors. More complete treatments are given in (9, 10, 15, 16).
[b] 25-HC, 25-hydroxycholesterol.

2. STRUCTURE AND ASSEMBLY OF THE LIPID-LINKED PRECURSOR OLIGOSACCHARIDE

Parodi et al (29) reported in 1972 that a glucose-containing lipid-linked oligosaccharide comprising approximately 20 monosaccharide units could be synthesized and transferred to protein in cell-free preparations from rat liver. Subsequently, many laboratories have examined the structures of lipid-linked oligosaccharides isolated from vertebrate tissues or synthesized from radioactive precursors in membrane preparations, tissue slices, or intact cells (reviewed in 9, 10, 30–32). The results of these studies, taken together, have suggested that the oligosaccharide(s) contains 1–3 Glc residues, at least one of which is at a nonreducing terminus; 4–10 α linked Man residues; and a core containing the sequence Manβ1$\rightarrow$4GlcNAcβ-1$\rightarrow$4 GlcNAc. The oligosaccharide is linked via a pyrophosphate group to an α-saturated polyisoprenoid carrier lipid (dolichol), the most abundant isomers of which contain 18–20 isoprene units in vertebrate tissues (33–35) and 15–16 units in yeast (34, 36).

The recent introduction of improved methods for estimating the sizes and compositions of oligosaccharides has made it possible to demonstrate that the major lipid-linked oligosaccharide synthesized from radioactive mannose or glucose in cultures of CHO cells (8, 37), NIL-8 (hamster) fibroblasts (37), chick embryo fibroblasts (37, 38), and yeast (39) has the composition Glc$_3$Man$_9$GlcNAc$_2$. A similar oligosaccharide (reported composition, Glc$_{1-2}$Man$_{10-11}$GlcNAc$_2$) was previously detected in a variety of vertebrate tissues (40, 41). A major contribution has been the recent characterization of the lipid-linked oligosaccharide from CHO cells, reported by Li et al (8). Their findings are summarized in Figure 1–III. The nine Man residues have an organization identical to that originally proposed by Ito et al (6) for a unit A glycopeptide of thyroglobulin (Figure 1–II). The common core region of Asn-linked oligosaccharides (5), Manα1$\rightarrow$3(Manα1$\rightarrow$6)Man$\beta$$\rightarrow$-GlcNAc$_2$, is also present. The arrangement and linkages of the three Glc residues have been confirmed for the NIL-8 lipid-linked oligosaccharide, by Liu et al (42), and Spiro et al (43) have recently demonstrated that the Glc residues of the thyroid compound are in the α-anomeric configuration.

2.1 The Carrier Lipid, Dolichol Phosphate

The lipid dolichol phosphate (Dol-P) acts as the "carrier lipid" in the assembly of pyrophosphate-linked oligosaccharides, and is also the acceptor in the synthesis of the sugar donors Man-P-Dol and Glc-P-Dol from GDP-Man and UDP-Glc, respectively (reviewed in 9, 10, 32). Like the other isoprenoid compounds, cholesterol and coenzyme Q, synthesis of dolichol can be abolished by inhibitors of hydroxymethylglutaryl-coenzyme A reductase (17, 18; Table 1). If cellular reserves of dolichol are specifically depleted by incubation with compactin and exogenous cholesterol and co-enzyme Q, N-linked glycosylation is drastically reduced (17). The availability of Dol-P appears to limit Dol-P-Man or oligosaccharide synthesis in several systems (44–47), and may well prove to be a major site of regulation of the assembly pathway (37, 48).

The concentration of dolichol in rat liver is $\sim$60 $\mu g/g$ wet weight (33); however, most of this material is in the form of free dolichol or is esterified with fatty acid (49) and thus cannot directly participate in oligosaccharide assembly. Furthermore, much of the cellular dolichol is distributed in fractions other than the endoplasmic reticulum (ER) (49, 50), where the enzymes of oligosaccharide synthesis are predominantly located (51). Thus, the pool size of Dol-P actively involved in protein glycosylation is unknown. The problem is further complicated by the possible compartmentation of this pool among oligosaccharide-P-P-Dol, GlcNAc-P-P-Dol, Man-P-Dol, and Glc-P-Dol (52–54).

Dol-P appears to be synthesized predominantly in the outer mitochondrial membrane in rabbit liver (55), and its cell-free synthesis from labeled mevalonate (55) or isopentenyl pyrophosphate (55, 56) has been demonstrated. Keller et al (57) have suggested that de novo Dol-P synthesis is regulated late in the pathway, at the level of Dol-P synthetase. Dol-P can also be synthesized from free dolichol in a CTP-dependent reaction in liver (58, 58a) and brain (59). Finally, enzymes that dephosphorylate Dol-P (60, 61) or convert Dol-P-P to Dol-P (60) have been detected in cell-free preparations. The latter activity may be involved in recycling the Dol-P-P, which is presumably released upon transfer of the fully assembled oligosaccharide from lipid carrier to protein. It is critically important to determine how these reactions are coordinated to regulate the pool of Dol-P involved in protein glycosylation.

2.2 Oligosaccharide Chain Initiation

Initiation of the precursor oligosaccharide entails the formation of GlcNAc-P-P-Dol and its conversion to $GlcNAc_2$-P-P-Dol (reviewed in 9, 32). Both the first and second GlcNAc residues are donated by UDP-GlcNAc (62–64); in the synthesis of GlcNAc-P-P-Dol, GlcNAc-P rather than free GlcNAc is transferred (65). Kean (66) has recently reported that in membranes from several embryonic chick tissues, addition of GDP-Man greatly stimulates formation of $GlcNAc_{1-2}$-P-P-Dol in what may be a physiologically important regulatory interaction. Enzymatic activities capable of synthesizing GlcNAc- and $GlcNAc_2$-P-P-Dol have been solubilized from aorta (67), oviduct (27), *Acanthamoeba castellanii* (63), and yeast (68), and the yeast preparation has been used in examining the carrier lipid chain length specificities of the enzymes that synthesize GlcNAc$_{1-2}$-P-P-Dol and Man-P-Dol (68).

The synthesis of a trisaccharide lipid behaving as Manβ→GlcNAc$_2$-P-P-Dol has been demonstrated in a variety of cell-free systems (reviewed in 9, 10, 32), and the enzyme responsible for addition of the β-Man residue has been solubilized from aorta (69) and *Acanthamoeba castellanii* (63). These solubilized enzymes have an absolute requirement for GlcNAc$_2$-P-P-Dol and utilize GDP-Man, but not Man-P-Dol, as a mannose donor. Exogenous Manβ→GlcNAc$_2$-P-P-Dol can be extended to larger lipid-linked oligosaccharides in hen oviduct microsomes (70).

2.3 Addition of α-Mannosyl Residues

Chapman et al (71) have isolated from cultured CHO cells a series of lipid-linked oligosaccharides from Man$_1$GlcNAc$_2$ to Man$_8$GlcNAc$_2$, the proposed structures of which are consistent with an ordered step-by-step addition of α-Man residues during the assembly of lipid-linked Glc$_3$Man$_9$GlcNAc$_2$ (Figure 2). As yet, kinetic evidence that in intact cells lipid-linked Man$_{1-8}$GlcNAc$_2$ are actually assembly intermediates is avail-

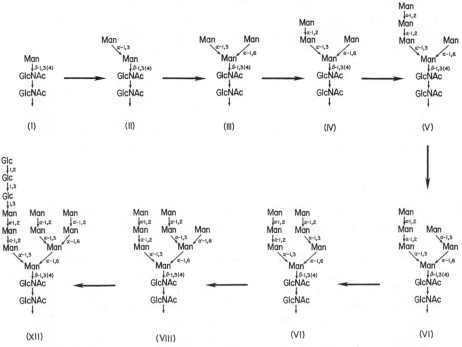

Figure 2 Proposed sequence for the assembly of lipid-linked oligosaccharides in CHO cells (71).

lipid-linked $Man_{1-8}GlcNAc_2$ are actually assembly intermediates is available only for lipid-linked $Man_{5-8}GlcNAc_2$ (37, 72). Furthermore, the isolation of mutant cell lines that accumulate the truncated lipid-linked oligosaccharides $Man_5GlcNAc_2$ (73, 74) or $Man_7GlcNAc_2$ (12) strongly supports the kinetic evidence that these are normally precursors of lipid-linked $Glc_3Man_9GlcNAc_2$.

Early work clearly demonstrated that (β)Man-P-Dol could donate α-Man residues during lipid-linked oligosaccharide assembly in cell-free preparations (reviewed in 9, 10, 30, 32), and until recently this compound was thought to be the only α-Man donor utilized. However, recent biochemical and genetic studies have suggested that some of the α-Man residues can be transferred directly from GDP-(α)Man. Chambers et al (75) and Forsee et al (76) used EDTA to block [^{14}C]Man-P-Dol synthesis from GDP-[^{14}C]-Man in microsomes, and noted that under these conditions radioactive α-linked Man residues were still incorporated into lipid-linked oligosaccharides; both groups were able to solubilize the enzyme(s) responsible for this reaction. Partial structural characterization of the lipid-linked oligosaccharides synthesized in the absence or presence of EDTA suggested that GDP-Man could donate the innermost three or four α-Man residues to generate a heptasaccharide. Elongation to larger species containing at least three additional α-Man residues required Man-P-Dol (75). The results of experiments in which Man-P-Dol formation was inhibited by the antibiotic amphomycin (77, 78) are consistent with this conclusion. Further studies on the lipid-linked heptasaccharide synthesized by solubilized preparations in the absence of Man-P-Dol suggested that it is identical to the $Man_5GlcNAc_2$ species shown in Figure 2, V (79, 80).

These results were dramatically confirmed by Chapman et al (81) who found that a mutant lymphoma cell line that accumulates lipid-linked $Man_5GlcNAc_2$ with the structure shown in Figure 2–V (74) lacks the ability to synthesize Man-P-Dol. Incubation of membrane preparations with GDP-[^{3}H]Man gave rise to labeled lipid-linked oligosaccharides up to $Man_9GlcNAc_2$ for the parent cell line, but $Man_5GlcNAc_2$ was the largest species synthesized by the mutant. In contrast, incubation with [^{3}H]Man-P-Dol gave rise to oligosaccharide-lipids from about $Man_6GlcNAc_2$ to $Man_9GlcNAc_2$ or larger in both cases. Thus, Man-P-Dol is not absolutely required until addition of the sixth Man residue.

2.4 Addition of Glucosyl Residues

At what stage of lipid-linked oligosaccharide assembly are the Glc residues added? Direct enzymatic demonstration is lacking, but indirect evidence suggests that lipid-linked oligosaccharides smaller than $Man_8GlcNAc_2$ are poor substrates for the glucosyltransferase(s) in vivo. The α-mannosyl branch that will ultimately carry the Glc residues is already complete in Man$_5$GlcNAc$_2$ lipid (74; Figure 2). However, glucosylated species are only minor components of the oligosaccharide-lipid fraction from mutant cells that accumulate lipid-linked Man$_5$GlcNAc$_2$ (73, 74) or Man$_7$GlcNAc$_2$ (12). This contrasts with the situation in normal cultured cells, where the fully glucosylated species, Glc$_3$Man$_9$GlcNAc$_2$, is the major lipid-linked oligosaccharide labeled by radioactive sugars (37, 71, 74). Furthermore, while a small amount of Glc$_3$Man$_5$GlcNAc$_2$ lipid has been detected in normal CHO cells (71), kinetic evidence suggests that Man$_{5-8}$GlcNAc$_2$ lipids are intermediates in the assembly of Glc$_3$Man$_9$GlcNAc$_2$ lipid (37, 72). It is not yet known whether the Glc residues are added before or after transfer of the ninth Man residue to lipid-linked Man$_8$GlcNAc$_2$.

The reaction(s) in which Glc residues are transferred to the lipid-linked oligosaccharide have been less thoroughly characterized than those involved in Man or GlcNAc transfer. Fortunately, the synthesis of Glc-P-Dol from UDP-Glc and Dol-P requires divalent cations and is inhibited by EDTA; thus it has been possible to demonstrate that in cell-free preparations from several sources Glc-P-Dol rather than UDP-Glc donates Glc residues during lipid-linked oligosaccharide assembly (reviewed in 32). Recently, Staneloni et al (82) have used this approach to show that all three Glc residues can be donated by Glc-P-Dol in rat liver microsomes.

2.5 Topology, Turnover, and Regulation

Several studies have been directed toward determining whether the enzymes of lipid-linked oligosaccharide assembly are localized on the cytoplasmic or the luminal surface of the ER. This question is of particular interest, since although newly glycosylated proteins are predominantly cisternal (Section 3.2), GDP-Man appears to be cytoplasmic (83), and the polyprenol-linked oligosaccharide probably cannot transverse the membrane unaided (84, 85). From studies involving protease treatments of rat liver microsomes, Snider et al (86) concluded that the enzymes synthesizing Man-P-Dol and Glc-P-Dol, and those transferring GlcNAc and Glc to the lipid-linked oligosaccharide are localized at the cytoplasmic surface of the vesicles, although the possibility could not be excluded that they are transmembrane enzymes with their active sites at the luminal surface. Nilsson et al (87) used a similar approach and also concluded that at least part of the mannose-transferring activity of rat liver smooth and rough ER and Golgi membranes is cytoplasmic. An apparently conflicting result was obtained by Hanover & Lennarz (84), who used a soluble galactosyltransferase as a membrane-impermeant probe to determine the orientation of GlcNAc$_2$-P-P-Dol in hen oviduct microsomes. Formation of Gal-GlcNAc$_2$-P-P-Dol was observed only after disruption of the vesicles, which implies a luminal location for the lipid-linked substrate. Further work is needed to clarify this important question.

Under normal conditions, the available pool of completed lipid-linked oligosaccharide appears to exceed the requirements of protein glycosylation, since during a chase, prelabeled oligosaccharide-lipid does not disappear immediately but falls exponentially with a half-time of about 14 min in cultured hamster lung cells (14), or about 3–5 min in CHO, NIL-8, or chick embryo fibroblasts (37). This rate of turnover is closely tied to the rate at which completed oligosaccharide is transferred to protein; it is greatly decreased when a genetic lesion interferes with oligosaccharide transfer (14) or when protein synthesis is inhibited (37, 88). Since in CHO and chick cells, prelabeled $Glc_3Man_9GlcNAc_2$ lipid is not degraded for at least 20 min in the absence of protein synthesis (37), these turnover rates appear to reflect predominantly transfer of oligosaccharide to protein. Furthermore, the rate at which new lipid-linked oligosaccharide is synthesized is also proportional to the rate at which preassembled chains are transferred to protein (14, 37, 88, 89). This regulation appears to involve oligosaccharide initiation rather than elongation, as no intermediates accumulate during protein synthesis inhibition (37). It has been proposed that under these conditions slowed recycling of the carrier lipid due to impaired transfer of completed oligosaccharides may limit the formation of new lipid-linked oligosaccharides (37).

3. TRANSFER OF OLIGOSACCHARIDE FROM CARRIER LIPID TO PROTEIN

The transfer of the precursor oligosaccharide to an acceptor protein is cotranslational in at least some cases (Section 3.2) and thus involves a complex interplay among a nascent polypeptide chain that is being extruded through a membrane, a polyisoprenoid-linked oligosaccharide, and a membrane-bound oligosaccharide transferring enzyme (dolichyldiphosphoryloligosaccharide: protein oligosaccharyl transferase). Under normal conditions, protein structure appears to be the major determinant of the extent to which a given site is glycosylated in vivo, and is considered first.

3.1 Influence of Protein Structure

It is now well established that a prerequisite for glycosylation is the occurrence of the acceptor Asn in the tripeptide sequence, -Asn-X-Thr(or Ser)-, where X can be any of the 20 amino acids except possibly aspartic acid (reviewed in 10, 90). This structural requirement appears to be rather specific; incorporation of β-hydroxynorvaline (an analogue that differs from threonine by one methylene group) into peptide chains in vitro was found to inhibit cotranslational glycosylation by 30% (91). However, only about ⅓ of the known -Asn-X-Thr/Ser- sites of eukaryotic proteins are glycosylated (10), and of these, some are glycosylated inefficiently. Clearly, other determinants are also involved.

In order to avoid some of the complex variables that affect N-linked glycosylation in vivo, Lennarz and his colleagues approached the problem by utilizing membrane preparations from hen oviduct as an enzyme source, exogenous oligosaccharide-lipid as a donor, and a series of 13 well-characterized proteins containing unglycosylated -Asn-X-Thr/Ser- sites as potential acceptors. Although none of these proteins acted as an acceptor in its native conformation, 6 of the 13 could be glycosylated in vitro after denaturation (10, 92, 93). No correlation other than the known tripeptide was found between amino acid sequence and acceptor activity, and it was concluded that glycosylation requires exposure of carbohydrate attachment sites by peptide chain unfolding. This view was supported by the observation that cyanogen bromide treatment of two of the proteins that had remained inactive despite denaturation, generated oligopeptide fragments that were efficient oligosaccharide acceptors (93). In fact, later studies with membranes from hen oviduct (94, 95) or yeast (96) revealed that a variety of small peptides could be glycosylated in vitro if they met just two criteria: presence of an -Asn-X-Ser- or -Asn-X-Thr- sequence and blockage of the COOH and NH_2 termini by other amino acids or by chemical modification. While other factors such as proximity to charged or hydrophobic groups may exert some effect (96), these results demonstrate that in these two distantly related organisms the predominant structural determinant for exogenous polypeptide glycosylation in defined, cell-free preparations is accessibility of the tripeptide acceptor sequence.

The polypeptide conformation around a carbohydrate attachment site can be roughly estimated from amino acid sequence data by means of predictive procedures based on the frequencies with which particular amino acids occur in configurations such as β-sheets, α-helices, and β-turns in proteins of known three-dimensional structure (97–99). Using this approach, Beeley (100) was able to predict that 30 of 31 N-glycosylated sequences he studied were part of turn or loop structures, which in 22 cases were most probably β-turns. Similarly, Aubert et al (101) estimated that 19 of 28 glycosylated Asn residues examined belonged to a β-turn. These results again emphasize the importance of accessibility, since such turn conformations often occur at the surfaces of globular proteins in exposed regions between sheet or helical structures (102). Indeed, comparison of the predicted secondary structure around the acceptor Asn in rat and bovine α-lactalbumin suggests an explanation for the fact that the rat protein is always glycosylated in vivo (103), while the closely related bovine protein is only 3–5% glycosylated (104, 105): in rat α-lactalbumin, Asn_{45} appears to fall in the exposed portion of a β-turn, while the analogous residue in the bovine protein is less exposed (106). Interestingly, in the bovine protein a second potential acceptor tripeptide at Asn_{74}, which is never glycosylated, was calculated to have a high probability of falling in a β-sheet configuration (106). Although these conformational predictions remain to be confirmed by X-ray crystallographic analysis, it is encouraging that the three-dimensional structure of at least one type of N-glycosylated protein (IgG and a Fc fragment) has been determined (107, 108), and the site is indeed part of a turn (107).

3.2 Time Course and Localization of Oligosaccharide Transfer

Membrane and secretory glycoproteins are believed to be synthesized on membrane-bound polysomes and to penetrate the lumen of the rough ER (109–111), where in many (112–114)—but not all (115–117)—cases a 10–30 amino acid "leader sequence" is rapidly removed by a cisternal peptidase (118, 119). Secretory glycoproteins pass through the bilayer, but membrane glycoproteins remain associated with it; some are anchored in the membrane by a hydrophobic domain at or near their COOH terminus (120–126). Structure develops rapidly in the nascent polypeptide chain; intra- and even some interchain disulfide bonds may be formed before release from the ribosome (127). Is N-glycosylation also cotranslational? It has been demonstrated that ovalbumin (128), a mouse IgG heavy chain (127, 129), and the glycoprotein of VSV (130, 131) can be quantitatively glycosylated before they are released from the ribosome, and some nascent chain glycosylation has also been inferred for several other proteins (132–134). However, there is evidence that oligosaccharides may be transferred after polypeptide chain termination in some cases (127, 129, 135, 136).

Consistent with the close temporal relationship between glycosylation and protein synthesis, subcellular fractionation of hen oviduct indicated that oligosaccharide-transferring activity was associated primarily with the rough ER (51). Although in other systems a considerable amount of activity is also present in smooth membranes (137, 138), the results of kinetic studies in which the intracellular movement of labeled glycoproteins was followed, after a pulse of radioactive carbohydrate or amino acid, suggest that the initial oligosaccharide transfer usually occurs in the rough ER (139–142). As yet, there has been no direct demonstration that oligosaccharide transfer occurs within the lumen of the ER, but several lines of evidence are consistent with this view: (a) Protein-linked carbohydrate is found primarily, if not entirely, on the luminal membrane surface (110, 143, 144); (b) When polypeptides are translated in cell-free systems, glycosylation has not been found to occur in the absence of membrane insertion (130, 137, 145–149); (c) After addition of the acceptor Asn residue to nascent ovalbumin (150) or IgG heavy chain (127), the growing polypeptide is extended by more than ~45 amino acids before glycosylation occurs. This is close to the estimated 60 residues necessary to span the membrane and the interior of the ribosome (112, 151).

3.3 Identification of the Oligosaccharide Transferred in Vivo

The enzymes endo-β-N-acetylglucosaminidase H (endo H) (152, 153) and the closely related endo C_{II} (154), which cleave between the proximal GlcNAc residues of free, peptide-linked (155–157) or protein-linked (158) high-mannose oligosaccharides, have made it possible to directly compare lipid-linked oligosaccharides with protein-linked high-mannose glycans without resorting to harsher procedures, such as hydrazinolysis (159, 160), reductive alkaline cleavage (161), or trifluoroacetolysis (162) to remove residual amino acids. Thus, it was demonstrated that in a variety of cells infected with VSV (158, 163, 164) or in chick embryo fibroblasts infected with Sindbis virus (158), the oligosaccharide initially transferred to the viral glycoproteins was a large high-mannose species similar to the lipid-linked oligosaccharide from the same cells. This was surprising, since the glycoprotein of mature VSV carries only complex oligosaccharides (7, 165), and the two Sindbis virus glycoproteins contain complex and small high-mannose glycans (42, 166–168). Many other glycoproteins have also been found to be synthesized as intracellular precursors that initially carry only endo H–sensitive (i.e. high-mannose) glycans. Examples include histocompatibility antigens (169), immunoglobulins G (163) and M (133), fibronectin (170), and the glycoproteins of Semliki Forest (171), influenza (172), murine leukemia (173), and mammary tumor (174) viruses. The apparent generality of this phenomenon is consistent with the observation that in uninfected fibroblasts labeled with [^{3}H]mannose for less than about 15–20 min, virtually all radioactive protein-linked oligosaccharides are endo H–sensitive and consist primarily of $Man_9GlcNAc_2$ and $Glc_1Man_9GlcNAc_2$ (38, 158).

There is an emerging consensus that the major lipid-linked oligosaccharide in a variety of systems has the composition $Glc_3Man_9GlcNAc_2$ (Section 2). Because glucose residues are rapidly removed after transfer of oligosaccharide to protein (Section 4.1), this species is difficult to detect in vivo. Only one study has demonstrated the presence of protein-linked $Glc_3Man_9GlcNAc_2$ in intact cells (38), although several workers have detected Glc_2- or Glc_1-containing glycans (38, 173, 175, 176). Nevertheless, the kinetics of labeling of protein-linked $Glc_3Man_9GlcNAc_2$ in vivo (38), together with the results of experiments with cell-free systems (see below) permit the conclusion that $Glc_3Man_9GlcNAc_2$ is probably the major oligosaccharide transferred from lipid to protein in many vertebrate cells. Indeed, transfer of an unglucosylated oligosaccharide has yet to be demonstrated in living cells. Even when genetic lesions cause large accumulations of lipid-linked $Man_5GlcNAc_2$ (73, 74, 81) or $Man_7GlcNAc_2$ (12), rather than the normal $Glc_3Man_9GlcNAc_2$, glucosylated forms of these truncated oligosaccharides appear to be preferentially transferred to protein (11, 12).

3.4 Characterization of the Oligosaccharide-Transferring Enzyme in Cell-Free Preparations

Membrane preparations with oligosaccharide-transferring activity have been prepared from a wide variety of eukaryotic sources (for reviews see 10, 32). One of the observations to emerge from this work is that, in apparent contrast to the situation in living cells, unglucosylated oligosaccharides can be transferred to endogenous proteins in vitro. Although direct transfer from GlcNAc-PP-Dol to protein has not yet been observed, $GlcNAc_2$-PP-Dol is an oligosaccharide donor in membranes from hen oviduct (177), yeast (96, 178), and calf brain (179). It is possible that protein glycosylation from lipid-linked $GlcNAc_2$, $Man\beta{\rightarrow}GlcNAc_2$, or $Man_n\alpha{\rightarrow}Man\beta{\rightarrow}GlcNAc_2$ may occur to some extent in vivo; on the other hand, the observed transfer of these small oligosaccharides in vitro may reflect the loss of enzyme specificity in a nonphysiological environment.

When larger lipid-linked oligosaccharides are used as donors, cell-free preparations reflect the specificity for glucosylated species seen in vivo. Turco et al (180) added equimolar amounts of glucosylated and unglucosylated lipid-linked oligosaccharides to microsomes from NIL-8 fibroblasts, and observed that the glucose-containing species was transferred to endogenous acceptors at a four to nine fold higher initial rate. Although the composition of these oligosaccharides was not reported, they were synthesized under conditions later determined to generate nearly homogeneous $Glc_3Man_9GlcNAc_2$ or $Man_9GlcNAc_2$ (42). Similarly, Spiro et al (181) found that removal of Glc residues from thyroid lipid-linked oligosaccharide (reported composition, $Glc_{1-2}Man_{10-11}GlcNAc_2$; see 40) drastically reduced the rate at which it was transferred to endogenous acceptors by a thyroid microsomal preparation. Interestingly, removal of the peripheral mannose residues with jack bean α-mannosidase did not affect donor activity. That the oligosaccharide-transferring enzyme can discriminate between Glc_3- and $Glc_2Man_9GlcNAc_2$ was most clearly demonstrated in an experiment in which a mixture of labeled lipid-linked oligosaccharides containing approximately equal amounts of these two species, along with smaller unglucosylated oligosaccharides, was added to NIL-8 microsomes (182). After a 45 sec incubation, endogenous acceptors carried nearly homogeneous $Glc_3Man_9GlcNAc_2$, with only a trace of $Glc_2Man_9GlcNAc_2$. It was not clear whether the latter oligosaccharide had been transferred intact from the lipid or was a product of glucosidase attack on protein-linked

$Glc_3Man_9GlcNAc_2$. Although it is not yet understood why there should be specificity for donor oligosaccharides containing three glucose residues, the fact that yeast microsomes also synthesize (183, 184) and preferentially transfer $Glc_3Man_9GlcNAc_2$ to proteins (183, 184a) implies that this structure has been conserved through evolution and must therefore have some special significance.

Das & Heath (185) recently reported the successful detergent solubilization of the oligosaccharide-transferring enzyme from hen oviduct. Starting from microsomes, they were able to purify the activity approximately 2000-fold by affinity chromatography on α-lactalbumin–derivatized Sepharose. The purified enzyme, which appeared relatively stable, had an absolute requirement for divalent cation and exogenous lipid-linked oligosaccharide and acceptor. Further characterization of this material should greatly extend current understanding of the oligosaccharide transfer reaction.

4. PROCESSING OF N-LINKED OLIGOSACCHARIDES

Immediately after transfer to acceptor proteins, the initially homogeneous population of precursor oligosaccharides begins to undergo a series of modifications that will eventually produce the diverse N-linked glycans of mature glycoproteins. The following sections deal with the major stages of this processing sequence in roughly chronological order. It should be noted that while its oligosaccharide(s) are being processed, the glycoprotein is transported through several intracellular compartments. This latter aspect is considered in more detail in Section 4.5.

4.1. Removal of the Glucosyl Residues

In VSV-infected CHO cells (175, 186), uninfected chick embryo fibroblasts (38), human diploid fibroblasts (176), BHK21 cells (176), and yeast (187), the first phase of N-linked oligosaccharide processing entails the removal of Glc residues from the precursor oligosaccharide soon after its transfer to protein. Although there appears to be a minor pathway in which one or two Man residues are removed from a small fraction of protein-linked oligosaccharides that still retain one or more Glc residues (38, 175), the majority of the molecules are completely deglucosylated before any Man residues are hydrolyzed. The kinetics of this process in chick embryo fibroblasts have been inferred from a comparison of the rates at which protein-linked Glc_3-, Glc_2-, and $Glc_1Man_9GlcNAc_2$ and $Man_9GlcNAc_2$ acquire label during brief incubations with [^{3}H]mannose (38). The results indicated that almost immediately after transfer of the precursor oligosaccharide ($Glc_3Man_9GlcNAc_2$) to protein, its terminal Glc residue was removed; the apparent half-time for this reaction was less than 2 min. The resulting protein-linked $Glc_2Man_9GlcNAc_2$ was more slowly converted to $Glc_1Man_9GlcNAc_2$ (apparent half-time, about 5 min), and the last Glc residue was removed at a still slower rate. Individual oligosaccharides appeared to be processed at different rates; significant amounts of protein-linked $Man_9GlcNAc_2$ appeared after as little as 5 min of incubation, yet some residual $Glc_1Man_9GlcNAc_2$ persisted after a 2 hr chase. The cause of this inhomogeneity is not known, but it may reflect unequal processing rates for individual glycoproteins (see Section 4.5).

Cell-free preparations capable of removing one or more Glc residues from oligosaccharides released from carrier lipid or protein have been obtained from rat (188–191) and calf (192) liver, calf brain (193), hen oviduct (194, 195), calf thyroid (43), NIL-8 fibroblasts (182), and yeast (184). At least two separate enzymes appear to participate in this reaction (189–192, 195). The first ("glucosidase I") removes only the terminal residue; this activity has been detergent-solubilized from rat liver (189, 191) and hen oviduct, and purified 1600-fold from the latter tissue (195). The second activity ("glucosidase II–III"), which is inactive toward $Glc_3Man_9GlcNAc$ but can convert Glc_2- or $Glc_1Man_9GlcNAc$ to $Man_9GlcNAc$, has been solubilized and partially purified from rat (189, 191) and calf (192) liver. By several criteria, including chromatographic resolution (189, 191), distinct physical properties, and different responses to inhibitors, analogues, and modified substrates (189, 190), rat liver glucosidase I is distinct from glucosidase II–III, and the latter activity probably represents a single enzyme. Removal of the peripheral α-Man residues from the substrate reduced enzyme activity (43, 190, 192), and glucosidase II–III was more sensitive to this substrate modification than glucosidase I (190). The enzymes are active toward lipid-linked as well as toward free and peptide-linked oligosaccharides (43, 190, 195), and it has been proposed (43) that, in addition to its presumed role in protein-linked oligosaccharide processing, glucosidase activity may regulate the level of fully glucosylated lipid-linked oligosaccharide available for protein glycosylation. This does not appear to be the case in cultured chick embryo, NIL-8 fibroblasts, or CHO cells (37), but the hypothesis should be evaluated in other systems.

Rat liver glucosidases I and II–III are integral membrane proteins that appear to be localized on the cisternal surface of the rough and smooth ER (189). This supports the assumption that they participate in the processing of newly synthesized glycoproteins that are sequestered within the lumen of the rough ER (Section 3.2). In hen oviduct, the specific activity of glucosidase I was highest in the rough microsomal fraction (195). These results are quite different from those obtained with the mannose-removing enzymes of the processing pathway, which are located predominantly or exclusively in the Golgi fraction (see below).

4.2 Removal of the First Four Mannosyl Residues

After its Glc residues are removed, the protein-linked oligosaccharide contains only Man and GlcNAc. In a variety of cells, this intermediate has the composition $Man_9GlcNAc_2$ (38, 173, 175, 176, 196, 197); not surprisingly, its structure in VSV-infected CHO cells (175) is identical to that of the CHO lipid-linked oligosaccharide (8; Figure 1–III) without the Glc residues. Only a few mature N-linked oligosaccharides, such as the $Man_9GlcNAc_2$ species from the unit A glycopeptide of calf thyroglobulin (6), retain this structure. More commonly, it is processed further, either to smaller high-mannose oligosaccharides, such as those of ovalbumin (157, 198) or the Sindbis virus glycoproteins (42, 166, 168), or to complex oligosaccharides, which are found in most mature glycoproteins (Section 4.3). In either case, the next stage of the trimming sequence entails hydrolysis of α-Man residues.

In chick embryo fibroblasts, Man residues are not removed from maturing N-linked glycans during the first ten minutes following transfer of the precursor oligosaccharide (38). The delay cannot be entirely due to protection by residual glucose residues, since appreciable amounts of $Man_9GlcNAc_2$ appear at least 5 min earlier (38). Since no rough ER-associated α-mannosidase has been discovered, it appears that release of mannose residues is catalyzed by enzyme(s) localized in a different intracellular compartment, and that the observed lag reflects the transit time for glycoprotein migration between these two compartments. Because newly synthesized glycoproteins are known to pass through the Golgi apparatus en route to the cell surface (142), the α-mannosidases of this organelle are believed to be responsible for this phase of oligosaccharide processing.

Most high-mannose oligosaccharides of mature glycoproteins are thought to arise by removal of a variable number of α-1,2-Man residues from $Man_9GlcNAc_2$ during processing. If all four of these residues are removed, a $Man_5GlcNAc_2$ species containing α-1,3- and α-1,6-Man residues remains. This oligosaccharide, which is still endo H–sensitive, is typically the smallest high-mannose glycan found in appreciable amounts in mature glycoproteins (198–201), and is also an intermediate in the synthesis of complex oligosaccharides (Section 4.3). The second phase of oligosaccharide processing, therefore, can be considered the removal of one to four α-1,2-Man residues from protein-linked $Man_9GlcNAc_2$. In rat liver, at least two Golgi-associated enzymes hydrolyze α-1,2- but not α-1,3-Man residues, and both are active against free or peptide-linked oligosaccharides. The first ("mannosidase I") was solubilized and purified to apparent homogeneity by Touster and his colleagues (202, 203). During its purification, this enzyme was assayed by measuring the hydrolysis of p-nitrophenyl-α-D-mannoside; thus a second Golgi α-mannosidase ("mannosidase II"), which has negligible activity toward this substrate, was not detected. Using labeled high-mannose oligosaccharides as substrates, Tabas & Kornfeld (204) were able to solubilize and partially purify mannosidase II, which can convert free $Man_9GlcNAc$ (an endo H product of protein-linked $Man_9GlcNAc_2$) to the proper $Man_5GlcNAc$ isomer in vitro. Although direct proof is still lacking, it seems probable that these two enzymes participate in N-linked oligosaccharide processing in vivo. Their relative contributions are as yet unknown.

A recent study indicates that during processing of the VSV glycoprotein in CHO cells, the four α-1,2-Man residues (here assigned the symbols A_1, A_2, B, and C as shown in Scheme 1) of each $Man_9GlcNAc_2$ chain are not trimmed randomly but are removed in an ordered sequence. Kornfeld et al (175) found that the high-mannose processing intermediates, Man_9-, Man_8-, and $Man_7GlcNAc_2$ each occurred primarily as a single isomer; comparison of their structures indicated that during processing, the B Man residue was removed first, then the A_1. Although this surprising observation is as yet limited to a single glycoprotein, it acquires broader significance from the results of structural studies on high-mannose oligosaccharides from two glycosylation sites of human myeloma IgM (199–201). If it is assumed that the "families" of high-mannose glycans found at N-glycosylation sites of mature glycoproteins arise by incomplete processing of the $Glc_3Man_9GlcNAc_2$ precursor, then the structures of Man_5- through $Man_9GlcNAc_2$ at Asn_{402} of IgM imply that the four α-1,2-Man residues are removed in the order A_1,C,A_2,B. A different set of isomers was found at the Asn_{563} site; their branching patterns imply a favored trimming sequence of B,A_1,C,A_2. The latter order is identical to what is known of the α-1,2-Man processing sequence of the VSV glycoprotein. Furthermore, the structures of Man_9-, Man_8-, and $Man_6GlcNAc_2$ isolated from bulk CHO cellular glycoproteins (205), as well as those of the known Man_7-, Man_6-, and $Man_5GlcNAc_2$ isomers from ovalbumin (157, 198) are also consistent with a preferred B,A_1,C,A_2 processing sequence. Interestingly, preliminary results obtained by Tabas & Kornfeld (204) suggest that mannosidase II could catalyze the A_1,C,A_2,B processing sequence inferred for the Asn_{402} site of IgM. The specificity of mannosidase I is not yet known.

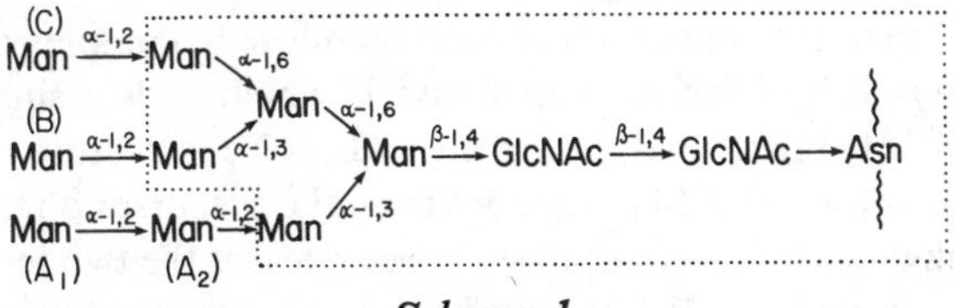

Scheme 1

After removal of the four α-1,2-Man residues, the resulting protein-linked $Man_5GlcNAc_2$ may be converted to endo H–resistant, complex glycans (see below). Before discussion of this next phase in oligosaccharide processing, consideration is given to the question of what determines whether an N-linked oligosaccharide retains a high-mannose structure or is processed to a complex form. Several studies have compared the N-linked oligosaccharides of multiple strains of influenza (206, 207) or murine leukemia virus (173, 208, 209) grown in various host cells, or a single strain of Sindbis (167, 210–213) or VSV (214) grown in several host cell types. The results establish that, although the host cell (i.e. the available processing enzymes) may have some effect on the size of complex glycans, the primary determinant of oligosaccharide structure is the structure of the protein being processed. Thus, a given glycosylation site may carry a complex glycan in one virus strain, while in a distinct but related strain grown in the same cells the analogous site may carry a high-mannose chain or no carbohydrate at all. It is easy to imagine how folded polypeptide chains around a glycosylated Asn residue could sterically hinder processing by, say, a Golgi α-1,2-mannosidase. Since complete removal of α-1,2-Man residues appears to be required for conversion to complex glycans, this inhibition would result in retention of a high-mannose structure. However, there has as yet been no experimental demonstration of this sort of steric interaction between a maturing glycoprotein and a processing enzyme.

4.3. Synthesis of Complex Oligosaccharides

The majority of the protein-linked sugar residues in plasma (215), liver, kidney, and brain (216) occur in complex-type Asn-linked oligosaccharides. Thus, most N-linked oligosaccharides do not retain a high-mannose structure but are converted to complex glycans in what can be considered the third stage of processing. This entails the removal of two Man residues from protein-linked $Man_5GlcNAc_2$ and the addition of "outer sugars:" GlcNAc, Gal, sialic acid, and Fuc (Figure 3). These reactions, which are thought to occur predominantly or solely in the Golgi apparatus, are described only cursorily; the subject is comprehensively treated elsewhere (1, 141, 217). Several of the reactions involved in complex oligosaccharide synthesis were first deduced from studies with mutant cell lines that lack specific glycosyltransferases; this work has also been reviewed (218).

The first step in the conversion of high-mannose to complex oligosaccharides is catalyzed by GlcNAc transferase I, which adds a single GlcNAc residue (219–221; Figure 3–d) to the $Man_5GlcNAc_2$ structure that remains after removal of the α-1,2-Man residues from the precursor oligosaccharide. A second enzyme, "late mannosidase," then releases the two terminal (noncore) Man residues (219, 222; Figure 3–e). The products of reactions d and e in Figure 3 are rapidly processed and have not been identified as interme-

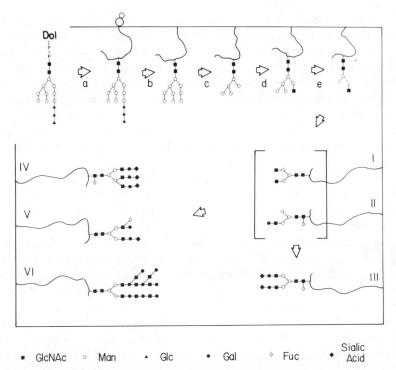

Figure 3 Schematic representation of Asn-linked oligosaccharide processing. Adapted from Kornfeld et al (175).

diates in intact cells. They are, however, the predominant oligosaccharides on mature rhodopsin (223, 224).

The product of GlcNAc transferase I and the late mannosidase, GlcNAc-Man$_3$GlcNAc$_2$, is a substrate for at least three enzymes in vitro: (*a*) GlcNAc transferase II, which can convert it to GlcNAc$_2$Man$_3$GlcNAc$_2$ (220, 221); (*b*) a fucosyltransferase, which can add α-1,6–linked Fuc to the innermost (reducing terminal) GlcNAc residue (225); and (*c*) a galactosyltransferase, which forms Galβ-1,4 linkages to terminal (nonreducing) GlcNAc residues (reviewed in 141). The order in which these reactions occur in vivo is not known; two of the possible products (I and II) are shown in brackets in Figure 3. If both chains are completed with the addition of Gal and sialic acid, the biantennary (two-branch) structure shown in Figure 3–III is obtained. It should be noted that this scheme is an oversimplification that does not take into account the existence of alternative linkages (e.g. α-2,3- vs α-2,6-sialic acid), incomplete processing (i.e. absence of some or all sialic acid, Gal, or Fuc residues), or conversion to other types of complex oligosaccharides (see below).

Many complex oligosaccharides contain more than two external branches. The oligosaccharides of the VSV glycoprotein, for example, have triantennary structures (7, 165; Figure 1–I; shown schematically in Figure 3–IV), and tetraantennary glycans have been detected on several proteins (e.g. 226, 227). Characterization of the reactions and enzymes that initiate these extra branches will be of particular interest, since highly branched complex oligosaccharides appear to be more abundant in transformed than in normal cells (228–230). This phenomenon may be related to other alterations observed in the cell surface carbohydrate composition of transformed cells [recently reviewed in (231)].

Examination of the activities of purified glycosyltransferases toward arrays of potential substrates has suggested that the order in which individual enzymes act can determine the final structure of the oligosaccharide (232; reviewed in 217). For example, Paulson et al (233) prepared various derivatives of asialotransferrin and demonstrated that glycosylation of these substrates by either a purified β-galactoside α-2,6-sialyltransferase (from bovine colostrum) or an N-acetylglucosaminide α-1,3-fucosyltransferase (from human milk) was mutually exclusive. Thus, terminal branches containing both fucose and sialic acid could not be produced, a result shown schematically in Figure 3–V. Studies with other fucosyl and sialyltransferases (232) suggested that this result reflects a fairly general phenomenon; sialylation and fucosylation are often alternative oligosaccharide chain termination reactions. However, the substrate specificities of these and other glycosyltransferases do not completely account for the regulation of glycan synthesis. Since different glycoproteins produced by the same cells carry distinct families of complex oligosaccharides, other factors, such as the structure of the protein being processed, must also direct complex glycan assembly.

4.4 Other Modifications

Many other N-linked glycan structures can also arise during oligosaccharide processing. However, while novel variations are continually being discovered, they usually involve sugar residues or other molecules in linkages exterior to the $Man_3GlcNAc_2$ core. Only a few examples are given.

4.41 HYBRID STRUCTURES   A significant fraction of the family of ovalbumin oligosaccharides combine in a single structure features usually associated exclusively with either high-mannose or complex-type oligosaccharides (156, 234). The α-1,6-Man residue of the inner core carries one or two α-Man residues, a characteristic not found in typical complex oligosaccharides. On the other hand, the α-1,3-Man residue of the inner core carries one or more β-GlcNAc residues, occasionally substituted with a terminal β-Gal residue (234); this arrangement is not found in high-mannose oligosaccharides but is common in tri- or tetraantennary complex glycans. All of the "hybrid" oligosaccharides studied also contain a GlcNAc residue linked β-1,4 to the β-linked (innermost) Man residue. Harpaz & Schachter (222) have recently shown that the late mannosidase (Section 4.3) cannot remove the two noncore α-Man residues from substrates containing this GlcNAc-substituted β-Man residue, an observation that may explain the existence of these usual structures.

4.42 ERYTHROGLYCAN   A large, polymeric oligosaccharide (4,000–13,-000 daltons) has recently been identified in human erythrocyte membranes (235, 236), where it comprises over a third of the stromal carbohydrate (235). This structure, termed erythroglycan (235), contains repeating units of Galβ-1,4-GlcNAc and GlcNAcβ-1,3-Gal, and is therefore structurally related to the keratan sulfates (Section 4.44); a simplified schematic representation is given in Figure 3–VI. Interestingly, the structure of this oligosaccharide undergoes a developmental change. In fetal erythrocytes it is present in a predominantly unbranched form, while the adult material is highly branched; a human variant in which the branching enzyme appears to be lacking has been described (237). Erythroglycan-like structures have been detected on cultured CHO cells (238) and on an established cell line (K-562) with some erythroid characteristics (239, 240).

4.43 PHOSPHORYLATED OLIGOSACCHARIDES   Endo H–sensitive (high-mannose) oligosaccharides of several lysosomal enzymes of human (241–243) or murine (244) origin have been demonstrated to contain phosphate groups. These occupy the 6 position of Man residues (241, 243), which in turn are probably linked α-1,2 to the remainder of the oligosaccharide (245). A recent report (244) suggests that phosphate group(s) may be transferred to the oligosaccharide in a "blocked" form; i.e. in such a way that they are not initially susceptible to alkaline phosphatase. In β-glucuronidase from mouse lymphoma cells, at least 25% of these blocking groups are α-N-acetylglucosamine residues. Thus, the phosphate donor may be UDP-GlcNAc (244). The Man-6-P structure is believed to function primarily as a signal for correct subcellular routing of newly synthesized lysosomal enzymes, but it is also recognized by cell-surface receptors that mediate endocytosis of extracellular enzymes (reviewed in 246, 247). Consistent with this view, fibroblasts from patients with I-cell disease, a condition in which these routing and recognition functions are deranged, synthesize lysosomal enzymes that lack the Man-6-P marker (241).

4.44 SULFATED OLIGOSACCHARIDES   Keratan sulfates, polymeric oligosaccharides containing repeating Galβ-1,4-GlcNAc and GlcNAc-β-1,3-Gal units, occur both in N-linked (Type I, corneal) and O-linked (Type II, skeletal) forms. The sulfate groups occur at position 6 of GlcNAc residues, and to a lesser extent at position 6 of Gal residues (reviewed in 248). Other types of N-linked sulfated oligosaccharides have been detected in sea urchin embryos (249) and in liver and lung from embryonic chicks (250), where they may be involved in embryonic development. The chick lung oligosaccharide has terminal sialic acid and fucose residues, and the sulfate appears to be attached to GlcNAc.

The glycoproteins of a number of enveloped viruses can be labeled with [^{35}S]sulfate (251, 252). In influenza virions, the sulfate is associated with complex-type oligosaccharides, but its attachment site is unknown (252, 253). Sulfated oligosaccharides from Simian virus 5 grown in bovine kidney cells have been more extensively characterized (254). Aside from what is probably a typical Man$_3$GlcNAc$_2$ core, the structure of this material is quite unusual, featuring among other oddities acetaldehyde groups; these do not appear to be unique to the virus since they were also detected in trypsinates of uninfected kidney cells. About 10% of the carbohydrate chains contained sulfate, of which half could be assigned to position 6 of the innermost (reducing terminal) GlcNAc residue.

4.5 Time Course of Processing

The glycoprotein of VSV can be used as an example to summarize what is known of the major stages of N-linked oligosaccharide processing (presented schematically in Figure 4). The indicated times are rough approximations and are intended only to clarify the temporal sequence of processing reactions in vivo. It should be noted that the Golgi localization of the processing enzymes has been rigorously established only in organs such as liver (e.g. 202, 257–259), and not in the tissue culture systems used in studying VSV maturation. Thus, the term Golgi may not be morphologically exact.

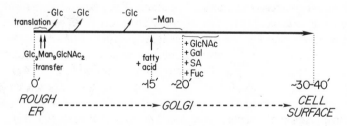

Figure 4 Intracellular migration and processing of the VSV glycoprotein. This glycoprotein also contains covalently bound fatty acid (255); this is attached to the polypeptide chain shortly before the oligosaccharides become resistant to endo H (256).

Precursor oligosaccharides ($Glc_3Man_9GlcNAc_2$; 8) are transferred to both glycosylation sites of the VSV glycoprotein before it is released from the ribosome (130, 131). For approximately 15 min, the protein-linked glycans remain sensitive to endo H or endo C_{II} (158, 163, 256) and consist primarily of $Glc_{1-2}Man_9GlcNAc_2$ and $Man_9GlcNAc_2$ (175). During the next 15 min an increasing fraction of the oligosaccharides becomes resistant to endo H or endo C_{II} (158, 163, 164, 256), which indicates that all the Glc residues and six of the nine Man residues have been removed and addition of the outer sugars GlcNAc, Gal, sialic acid, and Fuc has begun. About 30–40 min after synthesis, viral glycoprotein molecules carrying two triantennary complex oligosaccharides (7, 165) begin appearing at the cell surface (260) and in budding virions (256, 261).

The scheme presented in Figure 4 does not take into account numerous factors that may influence the pathway and time course of processing. Possible variables include:

(a) Growth phase   Evidence suggests that oligosaccharide processing is influenced by the rate of cell growth (168, 262, 263).

(b) Protein structure   Several viral (264–266) or human (267, 268) mutations have been described in which an abnormal glycoprotein is synthesized and acquires precursor oligosaccharides but fails to leave the ER. In each case, oligosaccharide processing does not proceed past the high-mannose stage, although other glycoproteins in the same cells may continue to be processed and transported normally.

(c) Protein-protein interactions   Krangel et al (169) have studied the association of β_2-microglobulin (an unglycosylated protein) with HLA-A and B heavy chains in lymphoblastoid cells. Their results suggest that complexing of a heavy chain with β_2-microglobulin may be required for its efficient oligosaccharide processing and intracellular transport.

(d) Final intracellular localization   The oligosaccharides of lysosomal enzymes have been discussed (Section 4.43). The ER has several intrinsic glycoproteins, and indirect evidence based on lectin binding suggests that they may retain a high-mannose structure (144, 269).

(e) Secreted vs plasma membrane glycoproteins   Strous & Lodish (270) infected rat hepatoma cells with a defective strain of VSV that does not abolish synthesis of host cell proteins, and found that the rate of oligosaccharide processing and intracellular transport was about twice as fast for the VSV glycoprotein as for the secretory glycoprotein transferrin. Further-

more, the ionophore monensin halted processing of transferrin at the high-mannose stage, while the VSV glycoprotein was processed to an endo H–resistant form. These intriguing results suggest that portions of the intracellular processing pathway may differ for secreted and plasma membrane glycoproteins.

5. CONCLUSION

Since the subject was last reviewed in this series (30), a great deal has been learned of the synthesis of asparagine-linked oligosaccharides, from the assembly of the lipid-linked precursor oligosaccharide through its transfer to protein and conversion to the diverse array of final products. Although the main outlines of this pathway are now clear, several areas, including the detailed enzymology of the synthesizing and processing enzymes, the organization of these enzymes in the cell, the regulation of the pathway, and the factors that direct the terminal processing reactions are only beginning to be understood. Future examination of these problems should provide explanations for many of the mysteries that still surround glycoprotein synthesis and function in eukaryotic organisms.

ACKNOWLEDGMENTS

We are grateful to Drs. Phillips W. Robbins, Martin D. Snider, and Clark M. Edson for their helpful criticism during the preparation of this review, and to Ms. Devon Young for typing the manuscript. We would also like to thank all those who have provided us with copies of their unpublished work.

Literature Cited

1. Schachter, H., Roseman, S. 1980. In *The Biochemistry of Glycoproteins and Proteoglycans,* ed. W. Lennarz, pp. 85–160. New York: Plenum
2. Sharma, C. B., Babczinski, P., Lehle, L., Tanner, W. 1974. *Eur. J. Biochem.* 46:35–41
3. Babczinski, P., Haselbeck, A., Tanner, W. 1980. *Eur. J. Biochem.* 105:509–15
4. Kornfeld, R., Kornfeld, S. 1976. *Ann. Rev. Biochem.* 45:217–37
5. Kornfeld, R., Kornfeld, S. 1980. See Ref. 1, pp. 1–34
6. Ito, S., Yamashita, K., Spiro, R. G., Kobata, A. 1977. *J. Biochem. Tokyo* 81:1621–31
7. Reading, C. L., Penhoet, E. E., Ballou, C. E. 1978. *J. Biol. Chem.* 253:5600–12
8. Li, E., Tabas, I., Kornfeld, S. 1978. *J. Biol. Chem.* 253:7762–70
9. Elbein, A. D. 1979. *Ann. Rev. Plant Physiol.* 30:239–72
10. Struck, D. K., Lennarz, W. J. 1980. See Ref. 1, pp. 35–83
11. Kornfeld, S., Gregory, W., Chapman, A. 1979. *J. Biol. Chem.* 254:11649–54
12. Hunt, L. A. 1980. *Cell* 21:407–15
13. Tenner, A. J., Zieg, J., Scheffler, I. E. 1977. *J. Cell Physiol.* 90:145–60
14. Tenner, A. J., Scheffler, I. E. 1979. *J. Cell Physiol.* 98:251–66
15. Schwarz, R. T., Datema, R. 1980. *Trends Biochem. Sci.* 5:65–7
16. Schwarz, R. T., Schmidt, M. F. G., Datema, R. 1979. *Biochem. Soc. Trans.* 7:321–26
17. Carson, D. D., Lennarz, W. J. 1979. *Proc. Natl. Acad. Sci. USA* 76:5709–13
18. Mills, J. T., Adamany, A. M. 1978. *J. Biol. Chem.* 253:5270–73

19. Schwarz, R. T., Schmidt, M. F. G., Lehle, L. 1978. *Eur. J. Biochem.* 85:163–72
20. Datema, R., Pont Lezica, R., Robbins, P. W., Schwarz, R. T. 1981. *Arch. Biochem. Biophys.* 206:65–71
21. Datema, R., Schwarz, R. T. 1978. *Eur. J. Biochem.* 90:505–16
22. Koch, H. U., Schwarz, R. T., Scholtissek, C. 1979. *Eur. J. Biochem.* 94:515–22
23. Datema, R., Schwarz, R. T. 1979. *Biochem. J.* 184:113–23
24. Takatsuki, A., Kohno, K., Tamura, G. 1975. *Agr. Biol. Chem.* 39:2089–91
25. Tkacz, J., Lampen, J. O. 1975. *Biochem. Biophys. Res. Commun.* 65:248–57
26. Lehle, L., Tanner, W. 1976. *FEBS Lett.* 71:167–70
27. Keller, R. K., Boon, D. Y., Crum, F. C. 1979. *Biochemistry* 18:3946–52
28. Heifetz, A., Keenan, R. W., Elbein, A. D. 1979. *Biochemistry* 18:2186–92
29. Parodi, A. J., Behrens, N. H., Leloir, L. F., Carminatti, H. 1972. *Proc. Natl. Acad. Sci. USA* 69:3268–72
30. Waechter, C. J., Lennarz, W. J. 1976. *Ann. Rev. Biochem.* 45:95–112
31. Spiro, R. G., Spiro, M. J. 1979. In *Glycoconjugate Research, Vol. 2,* Proc. Int. Symp. Glycoconjugates, 4th, ed. J. Gregory, R. Jeanloz, pp. 613–35. New York: Academic
32. Parodi, A. J., Leloir, L. F. 1979. *Biochim. Biophys. Acta* 599:1–37
33. Gough, D. P., Hemming, F. W. 1970. *Biochem. J.* 118:163–66
34. Dunphy, P. J., Kerr, J. D., Pennock, J. F., Whittle, K. J., Feeney, J. 1967. *Biochim. Biophys. Acta* 136:136–47
35. Radominska-Pyrek, A., Chojnacki, T., St. Pyrek, J. 1979. *Biochem. Biophys. Res. Commun.* 86:395–401
36. Parodi, A. J. 1977. *Eur. J. Biochem.* 75:171–80
37. Hubbard, S. C., Robbins, P. W. 1980. *J. Biol. Chem.* 255:11782–93
38. Hubbard, S. C., Robbins, P. W. 1979. *J. Biol. Chem.* 254:4568–76
39. Lehle, L., Schulz, I., Tanner, W. 1980. *Arch. Microbiol.* 127:231–37
40. Spiro, R. G., Spiro, M. J., Bhoyroo, V. D. 1976. *J. Biol. Chem.* 251:6409–19
41. Spiro, M. J., Spiro, R. G., Bhoyroo, V. D. 1976. *J. Biol. Chem.* 251:6420–25
42. Liu, T., Stetson, B., Turco, S. J., Hubbard, S. C., Robbins, P. W. 1979. *J. Biol. Chem.* 254:4554–59
43. Spiro, R. G., Spiro, M. J., Bhoyroo, V. D. 1979. *J. Biol. Chem.* 254:7659–67
44. Lucas, J. J., Nevar, C. A. 1978. *Biochim. Biophys. Acta* 528:475–82
45. Lucas, J. J., Levin, E. 1977. *J. Biol. Chem.* 252:4330–36
46. Harford, J. B., Waechter, C. J., Earl, F. L. 1977. *Biochem. Biophys. Res. Commun.* 76:1036–43
47. Harford, J. B., Waechter, C. J. 1980. *Biochem. J.* 188:481–90
48. Hemming, F. 1977. *Biochem. Soc. Trans.* 5:1223–31
49. Butterworth, P. H., Hemming, F. W. 1968. *Arch. Biochem. Biophys.* 128:503–8
50. Martin, H. G., Thorne, K. J. I. 1974. *Biochem. J.* 138:277–80
51. Czichi, U., Lennarz, W. J. 1977. *J. Biol. Chem.* 252:7901–4
52. Ericson, M. C., Gafford, J. T., Elbein, A. D. 1978. *Plant Physiol.* 61:274–77
53. Godelaine, D., Beaufay, H., Wibo, M. 1979. *Eur. J. Biochem.* 96:27–34
54. Kerr, A. K. A., Hemming, F. W. 1978. *Eur. J. Biochem.* 83:581–86
55. Daleo, G. R., Hopp, H. E., Romero, P. A., Pont Lezica, R. 1977. *FEBS Lett.* 81:411–14
56. Grange, D. K., Adair, W. L. Jr. 1977. *Biochem. Biophys. Res. Commun.* 79:734–40
57. Keller, R. K., Adair, W. L. Jr., Ness, G. C. 1979. *J. Biol. Chem.* 254:9966–69
58. Allen, C. M. Jr., Kalin, J. R., Sack, J., Verizzo, D. 1978. *Biochemistry* 17:5020–26
58a. Rip, J. W., Carroll, K. K. 1980. *Can. J. Biochem.* 58;1051–56
59. Burton, W. A., Scher, M. G., Waechter, C. J. 1979. *J. Biol. Chem.* 254:7129–36
60. Wedgwood, J. F., Strominger, J. L. 1980. *J. Biol. Chem.* 255:1120–23
61. Idoyaga-Vargas, V., Belocopitow, E., Mentaberry, A., Carminatti, H. 1980. *FEBS Lett.* 112:63–66
62. Leloir, L. F., Staneloni, R. J., Carminatti, H., Behrens, N. H. 1973. *Biochem. Biophys. Res. Commun.* 52:1285–92
63. Villemez, C., Carlo, P. 1980. *J. Biol. Chem.* 255:8174–78
64. Waechter, C. J., Harford, J. B. 1979. *Arch. Biochem. Biophys.* 192:380–90
65. Molnar, J., Chao, H., Ikehara, Y. 1971. *Biochim. Biophys. Acta* 239:401–10
66. Kean, E. 1980. *J. Biol. Chem.* 255:1921–27
67. Heifetz, A., Elbein, A. D. 1977. *J. Biol. Chem.* 252:3057–63
68. Palamarczyk, G., Lehle, L., Mankowski, T., Chojnacki, T., Tanner, W. 1980. *Eur. J. Biochem.* 105:517–23
69. Heifetz, A., Elbein, A. D. 1977. *Biochem. Biophys. Res. Commun.* 75:20–28

70. Chen, W., Lennarz, W. 1976. *J. Biol. Chem.* 251:7802–9
71. Chapman, A., Li, E., Kornfeld, S. 1979. *J. Biol. Chem.* 254:10243–49
72. Li, E., Kornfeld, S. 1979. *J. Biol. Chem.* 254:2754–58
73. Trowbridge, I. S., Hyman, R. 1979. *Cell* 17:503–8
74. Chapman, A., Trowbridge, I. S., Hyman, R., Kornfeld, S. 1979. *Cell* 17:509–15
75. Chambers, J., Forsee, W. T., Elbein, A. D. 1977. *J. Biol. Chem.* 252:2498–2506
76. Forsee, W. T., Griffin, J. A., Schutzbach, J. S. 1977. *Biochem. Biophys. Res. Commun.* 75:799–805
77. Kang, M. S., Spencer, J. P., Elbein, A. D. 1978. *Biochem. Biophys. Res. Commun.* 82:568–74
78. Kang, M. S., Spencer, J. P., Elbein, A. D. 1978. *J. Biol. Chem.* 253:8860–66
79. Schutzbach, J. S., Springfield, J. D., Jensen, J. W. 1980. *J. Biol. Chem.* 255:4170–75
80. Spencer, J. P., Elbein, A. D. 1980. *Proc. Natl. Acad. Sci. USA* 77:2524–27
81. Chapman, A., Fujimoto, K., Kornfeld, S. 1980. *J. Biol. Chem.* 255:4441–46
82. Staneloni, R. J., Ugalde, R. A., Leloir, L. F. 1980. *Eur. J. Biochem.* 105:275–78
83. Carey, D. J., Sommers, L. W., Hirschberg, C. B. 1980. *Cell* 19:597–605
84. Hanover, J. A., Lennarz, W. J. 1978. *J. Biol. Chem.* 254:9237–46
85. McCloskey, M. A., Troy, F. A. 1980. *Biochemistry* 19:2061–66
86. Snider, M. D., Sultzman, L. A., Robbins, P. W. 1980. *Cell* 21:385–92
87. Nilsson, O. S., DeTomás, M. E., Peterson, E., Bergman, A., Dallner, G., Hemming, F. W. 1978. *Eur. J. Biochem.* 89:619–28
88. Spiro, M. J., Spiro, R. G., Bhoyroo, V. D. 1976. *J. Biol. Chem.* 251:6400–8
89. Schmitt, J. W., Elbein, A. D. 1979. *J. Biol. Chem.* 254:12291–94
90. Marshall, R. D. 1974. *Biochem. Soc. Symp.* 40:17–26
91. Hortin, G., Boime, I. 1980. *J. Biol. Chem.* 255:8007–10
92. Pless, D. D., Lennarz, W. J. 1977. *Proc. Natl. Acad. Sci. USA* 74:134–38
93. Kronquist, K. E., Lennarz, W. J. 1978. *J. Supramol. Struct.* 8:51–65
94. Struck, D. K., Lennarz, W. J., Brew, K. 1978. *J. Biol. Chem.* 253:5786–94
95. Hart, G. W., Brew K., Grant, G. A., Bradshaw, R. A., Lennarz, W. J. 1979. *J. Biol. Chem.* 254:9747–53
96. Bause, E., Lehle, L. 1979. *Eur. J. Biochem.* 101:531–40
97. Chou, P. Y., Fasman, G. D. 1974. *Biochemistry* 13:222–45
98. Chou, P. Y., Fasman, G. D. 1978. *Adv. Enzymol.* 47:45–148
99. Chou, P. Y., Fasman, G. D. 1978. *Ann. Rev. Biochem.* 47:251–76
100. Beeley, J. G. 1977. *Biochem. Biophys. Res. Commun.* 76:1051–55
101. Aubert, J.-P., Biserte, G., Loucheux-Lefebvre, M.-H. 1976. *Arch. Biochem. Biophys.* 175:410–18
102. Kuntz, I. 1972. *J. Am. Chem. Soc.* 94:4009–12
103. Brown, R. C., Fish, W. W., Hudson, B. G., Ebner, K. E. 1977. *Biochim. Biophys. Acta* 441:82–91
104. Barman, T. 1970. *Biochim. Biophys. Acta* 214:242–44
105. Hopper, K. E., McKenzie, H. A. 1973. *Biochim. Biophys. Acta* 295:352–63
106. Prasad, R., Hudson, B. G., Butkowski, R., Hamilton, J. W., Ebner, K. E. 1979. *J. Biol. Chem.* 254:10607–14
107. Huber, R., Deisenhofer, J., Colman, P. M., Matsushima, M., Palm, W. 1976. *Nature* 264:415–20
108. Silverton, E. W., Navia, M. A., Davies, D. R. 1977. *Proc. Natl. Acad. Sci. USA* 74:5140–44
109. Palade, G. 1975. *Science* 189:347–58
110. Rothman, J. E., Lenard, J. 1977. *Science* 195:743–53
111. Davis, B. D., Tai, P.-C. 1980. *Nature* 283:433–38
112. Blobel, G., Dobberstein, B. 1975. *J. Cell Biol.* 67:835–51
113. Blobel, G., Walter, P., Chang, C., Goldman, B., Erickson, A., Lingappa, V. R. 1979. *Symp. Soc. Exp. Biol.* 33:9–36
114. Inouye, M., Halegoua, S. 1980. *CRC Crit. Rev. Biochem.* 7:339–71
115. Palmitter, R. D., Gagnon, J., Walsh, K. A. 1978. *Proc. Natl. Acad. Sci. USA* 75:94–8
116. Bonatti, S., Blobel, G. 1979. *J. Biol. Chem.* 254:12261–64
117. Schechter, I., Burstein, Y., Zemell, R., Ziv, E., Kantor, F., Papermaster, D. S. 1979. *Proc. Natl. Acad. Sci. USA* 76:2654–58
118. Jackson, R., Blobel, G. 1977. *Proc. Natl. Acad. Sci. USA* 74:5598–602
119. Walter, P., Jackson, R. C., Marcus, M. M., Lingappa, V. R., Blobel, G. 1979. *Proc. Natl. Acad. Sci. USA* 76:1795–99
120. Chatis, P. A., Morrison, T. G. 1979. *J. Virol.* 29:957–63
121. Katz, F. N., Lodish, H. F. 1979. *J. Cell Biol.* 80:416–26
122. Rose, J. K., Welch, W. J., Sefton, B. M., Esch, F. S., Ling, N. C. 1980. *Proc. Natl. Acad. Sci. USA* 77:3884–88

123. Garoff, H., Söderlund, H. 1978. *J. Mol. Biol.* 124:535–49
124. Porter, A. G., Barber, C., Carey, N. H., Hallewell, R. A., Threlfall, G., Emtage, J. S. *Nature* 1979. 282:471–77
125. Tomita, M., Furthmayr, H., Marchesi, V. T. 1978. *Biochemistry* 17:4756–70
126. Kaufman, J. F., Strominger, J. L. 1979. *Proc. Natl. Acad. Sci. USA* 76:6304–8
127. Bergman, L. W., Kuehl, W. M. 1979. *J. Supramol. Struct.* 11:9–24
128. Kiely, M., McKnight, G. S., Schimke, R. 1976. *J. Biol. Chem.* 251:5490–95
129. Bergman, L. W., Kuehl, W. M. 1978. *Biochemistry* 17:5174–80
130. Rothman, J. E., Lodish, H. F. 1977. *Nature* 269:775–80
131. Rothman, J. E., Katz, F. N., Lodish, H. F. 1978. *Cell* 15:1447–54
132. Sefton, B. M. 1977. *Cell* 10:659–68
133. Tartakoff, A., Vassalli, P. 1979. *J. Cell Biol.* 83:284–99
134. Bielinska, M., Boime, I. 1978. *Proc. Natl. Acad. Sci. USA* 75:1768–72
135. Bélanger, L., Fleischer, B., Fleischer, S., Guillouzo, A., Lemonnier, M., Chiu, J.-F. 1979. *Biochemistry* 10:1962–68
136. Jamieson, J. C. 1977. *Can. J. Biochem.* 55:408–14
137. Bielinska, M., Rogers, G., Rucinsky, T., Boime, I. 1979. *Proc. Natl. Acad. Sci. USA* 76:6152–56
138. Idoyaga-Vargas, V., Perelmuter, M., Burrone, O., Carminatti, H. 1979. *Mol. Cell Biochem.* 26:123–30
139. Schachter, H. 1974. *Biochem. Soc. Symp.* 40:57–71
140. Carey, D. J., Hirschberg, C. B. 1980. *J. Biol. Chem.* 255:4348–54
141. Schachter, H. 1978. In *The Glycoconjugates, Vol. 2,* ed. M. Horowitz, W. Pigman, pp. 87–181. New York: Academic
142. Morré, D. J., Kartenbeck, J., Franke, W. W. 1979. *Biochim. Biophys. Acta* 599:71–152
143. Hanover, J. A., Lennarz, W. J. 1980. *J. Biol. Chem.* 255:3600–4
144. Rodriguez Boulan, E., Kreibich, G., Sabatini, D. D. 1978. *J. Cell Biol.* 78:874–93
145. Garoff, H., Simons, K., Dobberstein, B. 1978. *J. Mol. Biol.* 124:587–600
146. Dobberstein, B., Garoff, H., Warren, G., Robinson, P. J. 1979. *Cell* 17:759–69
147. Lingappa, V. R., Lingappa, J. R., Prasad, R., Ebner, K. E., Blobel, G. 1978. *Proc. Natl. Acad. Sci. USA* 75:2338–42
148. Toneguzzo, F., Ghosh, H. P. 1978. *Proc. Natl. Acad. Sci. USA* 75:715–19
149. Korman, A. J., Ploegh, H. L., Kauf-man, J. F., Owen, M. J., Strominger, J. L. 1980. *J. Exp. Med.* 152:65s–82s
150. Glabe, C. G., Hanover, J. A., Lennarz, W. J. 1980. *J. Biol. Chem.* 255:9236–42
151. Blobel, G., Sabatini, D. D. 1970. *J. Cell Biol.* 45:130–45
152. Tarentino, A. L., Maley, F. 1974. *J. Biol. Chem.* 249:811–17
153. Tarentino, A. L., Trimble, R. B., Maley, F. 1978. *Methods Enzymol.* 50:574–80
154. Ito, S., Muramatsu, T., Kobata, A. 1975. *Arch. Biochem. Biophys.* 191:78–86
155. Tai, T., Yamashita, K., Kobata, A. 1977. *Biochem. Biophys. Res. Commun.* 78:434–41
156. Tai, T., Yamashita, K., Ito, S., Kobata, A. 1977. *J. Biol. Chem.* 252:6687–94
157. Trimble, R. B., Tarentino, A. L., Plummer, T. H. Jr., Maley, F. 1978. *J. Biol. Chem.* 253:4508–11
158. Robbins, P. W., Hubbard, S. C., Turco, S. J., Wirth, D. F. 1977. *Cell* 12:893–900
159. Yoshizawa, Z., Sato, T., Schmid, K. 1966. *Biochim. Biophys. Acta* 121:417–20
160. Mizuochi, T., Yonemasu, K., Yamashita, K., Kobata, A. 1978. *J. Biol. Chem.* 253:7404–9
161. Austen, B. M., Marshall, R. D. 1971. *Biochem. J.* 124:14p–15p
162. Nilsson, B., Svensson, S. 1979. *Carbohydr. Res.* 72:183–90
163. Tabas, I., Schlesinger, S., Kornfeld, S. 1978. *J. Biol. Chem.* 253:716–22
164. Hunt, L. A., Etchison, J. R., Summers, D. F. 1978. *Proc. Natl. Acad. Sci. USA* 75:754–58
165. Etchison, J. R., Robertson, J. S., Summers, D. F. 1977. *Virology* 78:375–92
166. Sefton, B. M., Keegstra, K. 1974. *J. Virol.* 14:522–30
167. Burke, D., Keegstra, K. 1979. *J. Virol.* 29:546–54
168. Hakimi, J., Atkinson, P. H. 1980. *Biochemistry.* 19:5619–24
169. Krangel, M. S., Orr, H. T., Strominger, J. L. 1979. *Cell* 18:979–91
170. Choi, M. G., Hynes, R. O. 1979. *J. Biol. Chem.* 254:12050–55
171. Rasilo, M., Renkonen, O. 1980. *J. Gen. Virol.* 47:525–28
172. Nakamura, K., Compans, R. W. 1979. *Virology* 93:31–47
173. Rosner, M. R., Grinna, L. S., Robbins, P. W. 1980. *Proc. Natl. Acad. Sci. USA* 77:67–71
174. Dickson, C., Atterwill, M. 1980. *J. Virol.* 35:349–61

175. Kornfeld, S., Li, E., Tabas, I. 1978. *J. Biol. Chem.* 253:7771–78
176. Hunt, L. A. 1979. *J. Supramol. Struct.* 12:209–26
177. Chen, W. W., Lennarz, W. J. 1977. *J. Biol. Chem.* 252:3473–79
178. Lehle, L., Tanner, W. 1978. *Eur. J. Biochem.* 83:563–70
179. Harford, J. B., Waechter, C. J. 1979. *Arch. Biochem. Biophys.* 197:424–35
180. Turco, S. J., Stetson, B., Robbins, P. W. 1977. *Proc. Natl. Acad. Sci. USA* 74:4411–14
181. Spiro, M. J., Spiro, R. G., Bhoyroo, V. D. 1979. *J. Biol. Chem.* 254:7668–74
182. Turco, S. J., Robbins, P. W. 1979. *J. Biol. Chem.* 254:4560–67
183. Trimble, R. B., Maley, F., Tarentino, A. L. 1980. *J. Biol. Chem.* 255:10232–38
184. Lehle, L. 1980. *Eur. J. Biochem.* 109:589–601
184a. Trimble, R. B., Byrd, J. C., Maley, F. 1980. *J. Biol. Chem.* 255:11892–95
185. Das, R. C., Heath, E. C. 1980. *Proc. Natl. Acad. Sci. USA* 77:3811–15
186. Hunt, L. A. 1980. *J. Virol.* 35:362–70
187. Parodi, A. J. 1979. *J. Biol. Chem.* 254:10051–60
188. Ugalde, R. A., Staneloni, R. J., Leloir, L. F. 1978. *FEBS Lett.* 91:209–12
189. Grinna, L. S., Robbins, P. W. 1979. *J. Biol. Chem.* 254:8814–18
190. Grinna, L. S., Robbins, P. W. 1980. *J. Biol. Chem.* 255:2255–58
191. Ugalde, R. A., Staneloni, R. J., Leloir, L. F. 1979. *Biochem. Biophys. Res. Commun.* 91:1174–81
192. Michael, J. M., Kornfeld, S. 1980. *Arch. Biochem. Biophys.* 199:249–58
193. Scher, M. G., Waechter, C. J. 1979. *J. Biol. Chem.* 254:2630–37
194. Chen, W. W., Lennarz, W. J. 1978. *J. Biol. Chem.* 253:5780–85
195. Elting, J. J., Chen, W. W., Lennarz, W. J. 1980. *J. Biol. Chem.* 255:2325–31
196. Koide, N., Muramatsu, T., Kobata, A. 1979. *J. Biochem. Tokyo* 85:149–55
197. Muramatsu, T., Gachelin, G., Jacob, F. 1980. *J. Biochem. Tokyo* 88:685–88
198. Tai, T., Yamashita, K., Ogata-Arakawa, M., Koide, N., Muramatsu, T., Iwashita, S., Inoue, Y., Kobata, A. 1975. *J. Biol. Chem.* 250:8569–75
199. Chapman, A., Kornfeld, R. 1979. *J. Biol. Chem.* 254:816–23
200. Chapman, A., Kornfeld, R. 1979. *J. Biol. Chem.* 254:824–28
201. Cohen, R. E., Ballou, C. E. 1980. *Biochemistry* 19:4345–58
202. Tulsiani, D. R. P., Opheim, D. J.,

Touster, O. 1977. *J. Biol. Chem.* 252:3227–33
203. Opheim, D. J., Touster, O. 1978. *J. Biol. Chem.* 253:1017–23
204. Tabas, I., Kornfeld, S. 1979. *J. Biol. Chem.* 254:11655–63
205. Li, E., Kornfeld, S. 1979. *J. Biol. Chem.* 254:1600–5
206. Nakamura, K., Compans, R. W. 1979. *Virology* 95:8–23
207. Klenk, H.-D., Garten, W., Keil, W., Niemann, H., Schwarz, R. T., Rott, R. 1980. In *Biosynthesis, Modification, and Processing of Cellular and Viral Polyproteins,* ed. G. Koch, D. Richter, pp. 175–84. New York: Academic.
208. Kemp, M. C., Basak, S., Compans, R. W. 1979. *J. Virol.* 31:1–7
209. Kemp, M. C., Famulari, N. G., O'Donnell, P. V., Compans, R. W. 1980. *J. Virol.* 34:154–61
210. Sefton, B. M. 1976. *J. Virol.* 17:85–93
211. Keegstra, K., Sefton, B. M., Burke, D. 1975. *J. Virol.* 16:613–20
212. Weitzman, S., Grennon, M., Keegstra, K. 1979. *J. Biol. Chem.* 254:5377–82
213. Keegstra, K., Burke, D. 1977. *J. Supramol. Struct.* 7:371–79
214. Etchison, J. R., Holland, J. J. 1974. *Proc. Natl. Acad. Sci. USA* 71:4011–14
215. Finne, J., Krusius, T. 1979. *Eur. J. Biochem.* 102:583–88
216. Krusius, T., Finne, J. 1977. *Eur. J. Biochem.* 78:369–79
217. Beyer, T. A., Sadler, J. E., Rearick, J. I., Paulson, J. C., Hill, R. L. 1981. *Methods Enzymol.* In press
218. Stanley, P. 1980. See Ref. 1, pp. 161–89
219. Tabas, I., Kornfeld, S. 1978. *J. Biol. Chem.* 253:7779–86
220. Narasimhan, S., Stanley, P., Schachter, H. 1977. *J. Biol. Chem.* 252:3926–33
221. Harpaz, N., Schachter, H. 1980. *J. Biol. Chem.* 255:4885–93
222. Harpaz, N., Schachter, H. 1980. *J. Biol. Chem.* 255:4894–4902
223. Liang, C.-J., Yamashita, K., Muellenberg, C. G., Shichi, H., Kobata, A. 1979. *J. Biol. Chem.* 254:6414–18
224. Fukuda, M. N., Papermaster, D. S., Hargrave, P. A. 1979. *J. Biol. Chem.* 254:8201–7
225. Wilson, J. R., Williams, D., Schachter, H. 1976. *Biochem. Biophys. Res. Commun.* 72:909–16
226. Mega, T., Lujan, E., Yoshida, A. 1980. *J. Biol. Chem.* 255:4057–61
227. Fournet, B., Montreuil, J., Strecker, G., Dorland, L., Haverkamp, J., Vliegenthart, J. F. G., Binette, J. P., Schmid, K. 1978. *Biochemistry* 17:5206–14

228. Takasaki, S., Ikehira, H., Kobata, A. 1980. *Biochem. Biophys. Res. Commun.* 92:735–42
229. Santer, U. V., Glick, M. C. 1979. *Biochemistry* 18:2533–40
230. Ogata, S.-I., Muramatsu, T., Kobata, A. 1979. *Nature* 259:580–82
231. Atkinson, P. H., Hakimi, J. 1980. See Ref. 1, pp. 191–239
232. Beyer, T. A., Rearick, J. I., Paulson, J. C., Prieels, J.-P., Sadler, J. E., Hill, R. L. 1979. *J. Biol. Chem.* 254:12531–41
233. Paulson, J. C., Prieels, J.-P., Glasgow, L. R., Hill, R. L. 1978. *J. Biol. Chem.* 253:5617–24
234. Yamashita, K., Tachibana, Y., Kobata, A. 1978. *J. Biol. Chem.* 253:3862–69
235. Järnefelt, J., Rush, J., Li, Y.-T., Laine, R. A. 1978. *J. Biol. Chem.* 253:8006–9
236. Krusius, T., Finne, J., Rauvala, H. 1978. *Eur. J. Biochem.* 92:289–300
237. Fukuda, M., Fukuda, M. N., Hakomori, S.-I. 1979. *J. Biol. Chem.* 254:3700–3
238. Li, E., Gibson, R., Kornfeld, S. 1980. *Arch. Biochem. Biophys.* 199:393–99
239. Turco, S. J., Rush, J. S., Laine, R. A. 1980. *J. Biol. Chem.* 255:3266–69
240. Rush, J. S., Turco, S. J., Laine, R. A. 1981. *Biochem. J.* 193:361–65
241. Hasilik, A., Neufeld, E. 1980. *J. Biol. Chem.* 255:4946–50
242. Von Figura, K., Klein, U. 1979. *Eur. J. Biochem.* 94:347–54
243. Natowicz, M. R., Chi, M. M. Y., Lowry, O. H., Sly, W. S. 1979. *Proc. Natl. Acad. Sci. USA* 76:4322–26
244. Tabas, I., Kornfeld, S. 1980. *J. Biol. Chem.* 255:6633–39
245. Distler, J., Hieber, V., Sahagian, G., Schmickel, R., Jourdian, G. W. 1979. *Proc. Natl. Acad. Sci. USA* 76:4235–39
246. Neufeld, E., Ashwell, G. 1980. See Ref. 1, pp. 241–66
247. Hasilik, A. 1980. *Trends Biochem. Sci.* 5:237–40
248. Rodén, L. 1980. See Ref. 1, pp. 267–371

249. Heifetz, A., Lennarz, W. 1979. *J. Biol. Chem.* 254:6119–27
250. Heifetz, A., Kinsey, W., Lennarz, W. 1980. *J. Biol. Chem.* 255:4528–34
251. Pinter, A., Compans, R. W. 1975. *J. Virol.* 16:859–66
252. Nakamura, K., Compans, R. W. 1978. *Virology* 86:432–42
253. Nakamura, K., Compans, R. W. 1977. *Virology* 79:381–92
254. Prehm, P., Scheid, A., Choppin, P. W. 1979. *J. Biol. Chem.* 254:9669–77
255. Schmidt, M. F. G., Schlesinger, M. J. 1979. *Cell* 17:813–19
256. Schmidt, M. F. G., Schlesinger, M. J. 1980. *J. Biol. Chem.* 255:3334–39
257. Fleischer, B. 1978. In *Cell Surface Carbohydrate Chemistry*, ed. R. E. Harmon, pp. 27–47. New York: Academic
258. Bretz, R., Bretz, H., Palade, G. E. 1980. *J. Cell Biol.* 84:87–101
259. Fleischer, B. 1981. *J. Cell Biol.* In press
260. Knipe, D. M., Lodish, H. F., Baltimore, D. 1977. *J. Virol.* 21:1121–27
261. Knipe, D. M., Baltimore, D., Lodish, H. F. 1977. *J. Virol.* 21:1128–39
262. Ceccarini, C., Muramatsu, T., Tsang, J., Atkinson, P. H. 1975. *Proc. Natl. Acad. Sci. USA* 72:3139–43
263. Muramatsu, T., Koide, N., Ceccarini, C., Atkinson, P. H. 1976. *J. Biol. Chem.* 251:4673–79
264. Zilberstein, A., Snider, M. D., Porter, M., Lodish, H. F. 1980. *Cell* 21:417–27
265. Knipe, D., Baltimore, D., Lodish, H. F. 1977. *J. Virol.* 21:1149–58
266. Lohmeyer, J., Klenk, H.-D. 1979. *Virology* 93:134–45
267. Hercz, A., Katona, E., Cutz, E., Wilson, J. R., Barton, M. 1978. *Science* 201:1229–32
268. Hercz, A., Harpaz, N. 1980. *Can. J. Biochem.* 58:644–48
269. Rodriguez Boulan, E., Sabatini, D. D., Pereyra, B. N., Kreibich, G. 1978. *J. Cell Biol.* 78:894–909
270. Strous, G. J. A. M., Lodish, H. F. 1980. *Cell* 22:709–17

Ann. Rev. Biochem. 1981. 50:585–621
Copyright © 1981 by Annual Reviews Inc. All rights reserved

STEROL BIOSYNTHESIS[1] ◆12089

George J. Schroepfer, Jr.

Departments of Biochemistry and Chemistry, Rice University,
Houston, Texas 77001

CONTENTS

Perspectives and Summary

The broad and intense interest in sterols and other isoprenoids is a result of increasing recognition of their roles in cellular structure, function, and replication. It is indeed amazing to consider the broad pattern of involvement of different metabolites of mevalonic acid in the regulation of various metabolic and developmental processes in eukaryotes. Although the biosynthesis of cholesterol and of other metabolites of mevalonic acid has been investigated intensively for some time, understanding of the processes involved in their formation, in their metabolism, and in production of their biological effects is much more limited than commonly assumed. This situation results from the complexity inherent in the chemistry of these compounds, the abundance of closely related compounds in nature, and the extremely low steady-state concentrations of many of the key compounds

[1]A review of the enzymatic conversion of squalene to lanosterol and of sterol intermediates in the biosynthesis of cholesterol, by the same author, will appear in Volume 51 of this series.

585

0066-4154/81/0701-0585$01.00

in many tissues and organisms. In many cases, these problems are compounded by the lack of sophisticated methods that permit complete separation of the individual metabolites. Purification of many of the key enzymes involved in the biosynthesis of these compounds has proved difficult, and has restricted both research into their structure and catalytic mechanisms and the application of modern techniques of immunology, cell biology, and molecular biology to analysis of specific problems. Although investigators now are rapidly defining these complex matters, limitations of existing methodology or of current knowledge have sometimes not been recognized or have been seriously understated in published reports. As a consequence, many conclusions have been unjustified, or represent at best rather simplistic interpretation of experimental data. Critical interpretation and integration of the experimental results in this field are severely hampered by limited descriptions of precise details of experimental design, procedures, and results.

Despite these difficulties, our understanding of the fundamental processes involved in the biosynthesis of cholesterol has advanced remarkably in the last decade. Studies of sterol biosynthesis in cultured cells have provided a virtual explosion of information on the regulation of sterol and isoprenoid biosynthesis in both normal cells and cells derived from individuals with hereditary diseases of high morbidity and mortality. There are new leads concerning the possible roles of sterols and isoprenoids in cellular function and replication and regarding possible mechanisms involved in the regulation of lipoprotein and cholesterol metabolism. Oxygenated derivatives of cholesterol and its sterol precursors, of other sterols, and of vitamin D have been shown to possess extraordinarily high biological activities, and many of these compounds appear to regulate major metabolic, developmental, and cell replication processes in eukaryotes (for example, various ecdysones and oogoniols).

Enzymatic Formation of Mevalonic Acid

The biosynthesis of mevalonic acid has been reviewed recently (1), and there is also a detailed earlier review of various aspects of HMG[2]-CoA reductase and its regulation (2). The formation of mevalonate from acetate involves four enzymes: acetyl-CoA synthetase, cytosolic acetoacetyl-CoA thiolase, HMG-CoA synthase, and HMG-CoA reductase. It is commonly assumed that HMG-CoA reductase is the most important regulatory enzyme in the biosynthesis of both mevalonate and cholesterol. Recent studies indicate that attention should also be directed to other enzymes involved in the biosynthetic pathway. For example, the combined results of White & Rud-

[2]HMG is the abbreviation used for 3-hydroxy-3-methylglutaric acid.

ney (3) and of Lane and co-workers (4–6) strongly suggest that cytosolic acetoacetyl-CoA thiolase and HMG-CoA synthase may also be important in the regulation of hepatic mevalonate synthesis under certain physiological conditions. In addition, Nervi et al (7) have reported a dissociation of hepatic HMG-CoA reductase activity from the overall rate of hepatic sterol synthesis after intravenous administration of intestinal lipoprotein to intact rats.

Studies of sterol synthesis in cells in culture also suggest the possible importance of enzymes other than HMG-CoA reductase in the regulation of mevalonate formation and of sterol synthesis. For example, rapid increases of sterol synthesis and of HMG-CoA reductase activity have been observed upon transfer of cells from media containing serum or lipoproteins to media containing either delipidized serum or no serum (8–11). This increase is also accompanied by increased activity of acetyl-CoA synthetase, cytosolic acetoacetyl-CoA thiolase, HMG-CoA synthase, and mevalonate kinase (9–14). Chang & Limanek (12, 13) recently reported an elevation of cytosolic acetoacetyl-CoA thiolase, HMG-CoA synthase, HMG-CoA reductase, and mevalonate kinase in CHO cells grown in delipidized fetal calf serum. Miller et al (14) reported similar findings in CHO cells with respect to the first three of these enzymes; cholest-8(14)-en-3β-ol-15-one suppressed the increase in these enzymatic activities caused by the use of delipidized newborn calf serum. The increased activities of the four enzymes induced by the use of delipidized serum could also be suppressed by the addition of low density lipoprotein (LDL) or 25-hydroxycholesterol (12, 13). Thus there may be coordinate control of these early enzymes of sterol biosynthesis. However, such coordinate regulation is not invariant (10, 11, 15) and additional studies of the temporal relationships of these changes are needed.

Increases in sterol synthesis in HeLa cells grown in serum-free media could be inhibited by dexamethasone. A parallel suppression of HMG-CoA reductase activity was not observed; instead there was a marked increase in the activity of the reductase relative to serum-free controls, a decrease in HMG-CoA synthase activity, and no effect on the activities of acetyl-CoA synthetase or acetoacetyl-CoA thiolase. Johnston et al (16) subsequently reported that inhibition of sterol synthesis caused by dexamethasone occurred in a wide variety of cell types in culture, but was not consistently associated with an increase in HMG-CoA reductase activity; in some cell types no significant change was observed.

Administration of 4-aminopyrazolopyrimidine, previously reported to increase cholesterol biosynthesis in adrenal tissue (18, 19), caused increases in the activity of both HMG-CoA synthase and HMG-CoA reductase, but not of cytosolic acetoacetyl-CoA thiolase or mevalonate kinase, in this tissue (17). Intravenous administration of LDL caused parallel decreases in the activities of the synthase and the reductase with little or no effect on cytosolic acetoacetyl-CoA thiolase or mevalonate kinase (17). The activity of the enzyme system responsible for the synthesis of squalene from farnesyl pyrophosphate has also been reported to increase upon transfer of fibroblasts to LDL-deficient medium. LDL suppressed this enzyme activity, but only after the suppression of HMG-CoA reductase activity (20). These studies clearly indicate the need to direct attention not only to HMG-CoA reductase, but also to other enzymes in the biosynthesis and metabolism of mevalonic acid.

A great deal of effort has been directed toward the solubilization and purification of microsomal HMG-CoA reductase (21–30). Apparently homogeneous preparations, with specific activities far in excess of previously reported values, have been obtained from livers of rats that had been fed cholestyramine (23, 28–30). It had earlier been suggested (24, 31) that liver enzyme from cholestyramine-fed rats differs from that of normal rats. However, these studies employed enzyme preparations (or antibodies to same) that were of such a low specific activity as to preclude a definitive statement on this matter. More recently, using antibodies prepared against the highly purified reductase of cholestyramine-fed rats and assuming that the antibodies do not differentiate between normal and activated forms of the enzyme, one group has concluded that ~60% of the increase in rat liver HMG-CoA reductase observed upon cholestyramine feeding was due to activation and not to an increase in the amount of the enzyme (28). These results conflict with reports from another laboratory (32, 33) that the increase in activity of HMG-CoA reductase resulted entirely from an increase in the amount of enzyme. Clarification awaits further experimentation. It has also been claimed that cholesterol is associated with purified rat liver HMG-CoA reductase (22, 24) and that the amount associated with the liver reductase from normal rats and from cholestyramine-fed rats differs (24). In view of the low specific activities of the enzymes analyzed and, in the latter case, the use of a new method for cholesterol estimation for which details with regard to validity and precision were not provided, these observations require further investigation. Induction of HMG-CoA reductase in rat liver by compactin (34), Triton WR-1339 (35), and 20, 25-diazacholesterol (36); in cultured cells by compactin (30, 34, 37); and in yeast by glucose (38) has been claimed. Studies utilizing antibodies to highly purified rat liver HMG-CoA reductase indicated that the increased activity of the enzyme in hepatocytes incubated with compactin is due, for the most part, to an increased number of enzyme molecules (30). Several improvements, modifications, or critiques of assays of HMG-CoA reductase in tissues or cell cultures have appeared (23, 26, 27, 39–48).

Inactivation of rat liver microsomal HMG-CoA reductase in the presence of ATP and Mg^{2+}, and its partial reactivation by a factor present in the cytosol (49) suggests a possible mechanism of control of this enzyme. There may also be a more complex system for regulation of the activity of this important enzyme (33, 47, 50–65). The "early" phases of this research have been reviewed by Gibson & Ingebritsen (56). The picture that has evolved from these studies is one of a short-term modulation of the enzyme activity by phosphorylation-dephosphorylation processes. HMG-CoA reductase is inactivated by a microsomal MgATP-dependent reductase kinase that is also present to some extent in liver cytosol, which catalyzes the phosphorylation of the enzyme and is stimulated by cAMP. The inactive, phosphorylated form of the reductase is reactivated by a phosphatase present in the liver cytosol that is inhibited by 50 mM NaF. The reductase kinase also appears to be inactivated by a cAMP-insensitive MgATP-dependent kinase (reductase kinase kinase) and reactivated by a phosphatase. The phosphatase reactions may also be controlled by phosphorylation-dephosphorylation processes. These findings suggest a short-term control mechanism for the regulation of HMG-CoA reductase activity which appears to be very sensitive to hormonal factors (65, 66). Several reports (46, 62, 67), while compatible with the existence of inactive and active forms of HMG-CoA reductase controlled by phosphorylation and dephosphorylation, do not indicate that such a system is important in long-term control of reductase activity and sterol synthesis in liver, brain, or cells in culture, whereas other reports suggest a role for such regulation in long-term control (61, 68). Still others (69, 70) question covalent modification of the reductase by phosphorylation. The apparent modification of the enzyme activity may be artifactual and the reductase kinase may be mevalonate kinase which, in the presence of ATP and Mg^{2+} causes an apparent decrease in mevalonate formation when the reaction is followed by the incorporation of labeled HMG-CoA into mevalonate, but not when the consumption of NADPH is assayed (G. C. Ness, G. A. Benton, S. A. Dieter, personal communication). However, it is possible that preparations of the inactivating protein species (reductase kinase) used by the various investigators are not identical. For example, the native protein species responsible for the inactivation of reductase in the presence of ATP and Mg^{2+} has an estimated M_r 95,000 (G. C. Ness, G. A. Benton, S. A. Dieter, personal communication), over 150,000 (53), 230,000 (33), and 380,000 (60). Further studies to verify the existence and physiological significance of the proposed modulation of reductase activity are clearly indicated.

Conversion of Mevalonic Acid to Squalene and Other Isoprenoids

Mevalonate serves as a precursor of squalene and of sterols derived therefrom, of dolichol, the isoprenoid side chains of ubiquinones, the farnesyl residue of heme a in cytochrome oxidase, the isopentenyl residues of modified adenine moieties in tRNA, and presumably, in the yeast, *Rhodosporidium toruloides*, of the farnesyl residue of a novel undecapeptide, rhodotorucine A, which contains an S-farnesyl cysteine residue (71), and serves as a mating factor in this organism, at concentrations as low as 8 ng per milliliter. In plants, mevalonate is a precursor of sterols, carotenoids, the phytyl side chains of chlorophyll, and a myriad of other metabolites. Some of the latter are extremely potent regulators of various plant processes, and include certain cytokinins, gibberellins, and abscisic acid. In insects, mevalonate and homomevalonate serve as precursors of various juvenile hormones. There are recent reviews on various aspects of polyisoprenoid biosynthesis (1), the biosynthesis of phosphorylated dolichol sugar derivatives and their role in glycosyl transfer reactions (72, 73), the overall biosynthesis of squalene, the enzymatic conversion of farnesyl pyrophosphate to squalene (74), various aspects of isoprenyl transfer reactions, especially the enzymatic formation of farnesyl pyrophosphate (75, 76), allylic pyrophosphate metabolism, with heavy emphasis on the stereochemistry and mechanisms of the concerned reactions (77), and the chemistry of biogenetic-type rearrangements of isoprenoids (78).

In certain tissues (particularly kidney and brain), significant metabolism of mevalonate to n-fatty acids and/or carbon dioxide occurs (79–86), and a "transmethylglutaconate shunt" to account for this metabolism has been proposed (80). This proposal assumes that the branch point leading from mevalonic acid into the shunt, which ultimately yields acetoacetate and acetyl-CoA, is at the level of dimethylallyl pyrophosphate. However, branch points at the level of geranyl pyrophosphate and/or farnesyl pyrophosphate may also exist, and even predominate. Figure 1 shows a possible pathway for the enzymatic degradation of farnesyl pyrophosphate to C_2 and C_4 units based upon analogous reactions described by Seubert and coworkers for the enzymatic degradation of long-chain allylic alcohols in bacteria (87–91). Results presented to date on the transmethylglutaconate shunt in animal tissues do not differentiate between a branch point at the level of dimethylallyl pyrophosphate, geranyl pyrophosphate, or farnesyl pyrophosphate.

Mevalonic acid serves as a precursor of certain cytokinins. These compounds are modified adenine derivatives with Δ^2-isopentenyl residues (or modified isopentenyl residues) attached to the amino group at C–6 of adenine. The term cytokinin was originally applied to substances that promote cell division in plant cells (92). Δ^2-Isopentenyladenosine has biological activity in animal cells that causes a suppression of uridine incorporation into RNA and thymidine into DNA in lectin-treated rat spleen cells (93). These modified adenine derivatives occur in tRNA species of bacteria, yeast, plants, and animals, and are located specifically at the position adja-

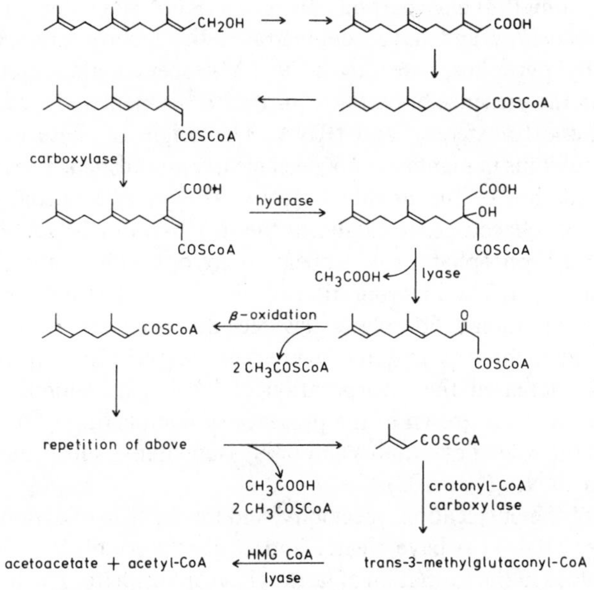

Figure 1 A possible scheme for the enzymatic degradation of farnesyl pyrophosphate [adapted from studies by Seubert et al in bacteria (88–91) and from studies of Edmond & Popjak (80)].

cent to the 3' end of the anticodon (94). This modification possibly leads to an enhancement of the stability of the codon-anticodon complex (95). To date, only Δ^2-isopentenyladenine has been found in tRNA from animal sources. The effects of isopentenyladenine-deficient tRNAs on translation differ in different systems (96). An antisuppressor strain of *Schizosaccharomyces pombe* has been isolated in which the tRNAs appear to lack the modified base N^6-(Δ^2-isopentenyl)adenosine normally present in several tRNAs of wild-type cells (96). Sequence analysis of tyrosine tRNA from this strain demonstrated the presence of an unmodified adenosine adjacent to the anticodon instead of the isopentyl-modified base. A similar situation may exist in tryptophan tRNA and serine tRNA. The growth of the mutant was only slightly reduced relative to that of the wild type. It was postulated (96) that the mutation affected the structural gene for the transfer of the isopentenyl residue to the tRNA rather than an enzyme involved in the overall synthesis of isopentenyl pyrophosphate. Apart from their roles as components of tRNA, these cytokinins appear to have separate, important regulatory activities in plants. Mevalonate serves as the precursor of the isopentenyl residue of isopentenyl adenine of tRNA (97). More recently,

Holtz & Klambt (98) have partially purified a tRNA isopentenyl transferase from *Lactobacillus* and have demonstrated the incorporation of labeled dimethylallyl pyrophosphate into the N^6-(Δ^2-isopentenyl)adenosine moiety of tRNA in the presence of partially purified Δ^2-isopentenylpyrophosphate isomerase, the transferase, and tRNA. The origin of these mevalonate-derived cytokinins in plants has not been clearly established. They may arise from degradation of the modified tRNA. However, a partially purified enzyme from tobacco tissue catalyzed the formation of N^6-(Δ^2-isopentenyl)adenosine-5'-phosphate from isopentenylpyrophosphate and adenosine-5'-phosphate (99). The incorporation of the label of [^{3}H]mevalonolactone into tRNA of human fibroblasts has recently been reported (100); the labeled component was tentatively identified as N^6-(Δ^2-isopentenyl)adenine. LDL increased the incorporation of labeled mevalonolactone into tRNA in fibroblasts grown in the presence of compactin (100). There are reviews on the role of cytokinins and of N^6-isopentenyl-substituted adenine moieties in tRNA (101–103).

Studies of the purification, specificity, and mechanism of action of prenyl transferases (104–113) have clearly and elegantly established the mechanisms involved in the formation of farnesyl pyrophosphate. The latter serves as a precursor of squalene, dolichols, and ubiquinones, as indicated in Figure 2, and can be assumed to contribute the farnesyl residue in the formation of heme *a* of cytochrome *c* oxidase. Incorporation of mevalonate into heme *a* of cytochrome *c* oxidase of yeast has been reported (114) but has not been demonstrated in animal tissues. In addition, earlier studies (115) have demonstrated that farnesyl pyrophosphate can be enzymatically converted to farnesol. Enzymes in liver catalyze the conversion of farnesol to farnesal and of farnesal to farnesoic acid (115). The quantitative importance of these reactions of farnesyl pyrophosphate is not known. Farnesoic acid is metabolized in liver to unidentified products (116). If, as noted before, a branch point in the metabolism of allylic pyrophosphates exists at the level of farnesyl pyrophosphate, this alternate route for its metabolism (Figure 1) could be significant in certain tissues. Substantial metabolism of mevalonate to farnesol has been reported in an ascites tumor (117) and in platelets (118). Farnesoic acid inhibits mevalonate kinase (119) and the overall synthesis of cholesterol from mevalonate (120), and farnesyl pyrophosphate inhibits mevalonate kinase (121). Geranyl pyrophosphate, but not dimethylallyl pyrophosphate or isopentenyl pyrophosphate, is also active in this respect. Phosphomevalonate kinase was unaffected by the allylic pyrophosphates.

The details of the enzymatic formation of dolichols in animal tissues are not well delineated. In animal tissues these compounds are usually composed of 16–22 isoprene residues with 2 internal *trans*-olefinic bonds (the

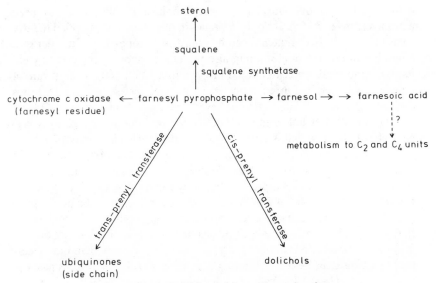

Figure 2   Metabolic fates of farnesyl pyrophosphate.

remainder of the internal olefinic bonds are *cis*-oriented) and with no un-saturation in the α-isoprene unit (122). Considerably more information is available concerning the biosynthesis of polyprenols in bacteria (123–126 and references therein). The incorporation of labeled mevalonate into doli-chol in liver (127–130), into glycosylated phosphate derivatives of dolichol in thyroid tissue (131), and into smooth muscle cells of aorta (132) has been reported. Shorter polyprenols (chiefly 11-isoprene residues with 3 internal *trans*-double bonds) exist in significant quantities (∼16 mg per kilogram of tissue) in liver (133). This material possibly originates from dietary plant sources (133). Daleo et al (134) studied the biosynthesis of dolichol phos-phate from labeled mevalonate and isopentenyl pyrophosphate by subcellu-lar fractions of rabbit and chicken liver. Mitochondria, especially their outer membrane fraction, were especially active in dolichol phosphate syn-thesis, which indicates that the outer membrane of mitochondria must contain isopentenyl pyrophosphate isomerase, a *trans*-prenyl transferase, a *cis*-prenyl transferase, and an enzyme responsible for the reduction of the double bond of the α-isoprene unit. Grange & Adair (135) reported the synthesis of a compound believed to be 2,3-dehydrodolichol phosphate in a particulate fraction of hen oviduct. Such a compound could be considered as the immediate precursor of dolichol phosphate. High speed supernatant fractions of livers of diethylstilbesterol-treated rats catalyze the synthesis of a compound believed to be 2,3-dehydrodolichol pyrophosphate (136). This compound could be considered as the likely immediate precursor of 2,3-

dehydrodolichol monophosphate. The metabolism of dolichol has not been extensively investigated. It exists in tissues in free form and esterified to fatty acids (127, 137). Free dolichol can be esterified to long-chain fatty acids in a number of rat tissues (138), and also undergoes phosphorylation to yield its monophosphate ester in animal preparations (139, 140). A high rate of dolichol synthesis has been reported in normal testes, but not in testes derived from X-irradiated mice or several mutant strains of mice (141).

Synthesis of dolichol in mouse L cells can be inhibited by 25-hydroxy-cholesterol (142), which is known to lower the levels of activity of HMG-CoA reductase in these cells. However, the concentrations of the steroid required to inhibit sterol synthesis from labeled acetate were much lower than those required to suppress dolichol synthesis. HMG-CoA reductase possibly affects the rate of dolichol synthesis by influencing the pool size of one or more sterol pathway intermediates that are common to the biosynthesis of dolichol, and a rate-limiting enzyme unique to the dolichol branch may be saturated at a lower concentration of intermediates than that necessary to saturate the next rate-limiting enzyme in sterol biosynthesis past the branch point (142). The rate of dolichol synthesis in mouse liver is accelerated by cholestyramine feeding and is depressed by fasting and cholesterol feeding (143). Evidence for the existence of a diurnal rhythm for hepatic dolichol synthesis was also presented. James & Kandutsch suggest that in liver, as in mouse L cells in culture, a branch of dolichol biosynthesis is saturated at a lower concentration of isoprenoid intermediates than is required to saturate the branch leading to sterol formation. The conservative nature of the control of dolichol-phosphate function is suggested by the observation that, while dolichol synthesis was depressed after two weeks of cholesterol feeding, the incorporation of labeled mannose into liver and plasma glycoproteins was not changed under these conditions. These findings may indicate that the rate of dolichol synthesis is not rate-limiting for total glycoprotein synthesis under the conditions studied. Dolichol synthesis is of critical importance in embryonic development (144).

Rudney and co-workers have investigated polyprenyl pyrophosphate synthesis and the transfer of the polyprenyl residue to various aromatic acceptors in mitochondria (145–147). The detergent Triton X-100 had marked effects on the nature of the polyprenol pyrophosphates synthesized from isopentenyl pyrophosphate in aged rat and guinea pig liver mitochondria (147). The advantages of isolated rat heart cells in suspension for studies of ubiquinone biosynthesis have been presented (148). Exogenous mevalonate stimulates incorporation of labeled p-hydroxybenzoic acid into ubiquinone in these cells. Starvation, cholesterol feeding, and cholic acid feeding suppressed the formation of ubiquinone from labeled p-hydroxy-benzaldehyde (149). Cholesterol feeding suppressed the incorporation of labeled benzoic acid into ubiquinone by rat liver slices and this inhibition could be overcome by exogenous mevalonate (150). Labeled mevalonolactone is incorporated into ubiquinone in human fibroblasts (151). Cells grown in the presence of lipoprotein-deficient serum and compactin (a competitive inhibitor of HMG-CoA reductase), at levels that have no effect on the concentrations of cellular protein or cholesterol, showed higher incorporation of mevalonolactone into ubiquinone in the presence of added LDL than in its absence. Thin layer chromatography (TLC) of the total lipid extract of fibroblasts has been used to determine the effects of mevalonolactone concentration, time, and LDL on the incorporation of mevalonolactone into material corresponding in mobility to ubiquinone and cholesterol. The results were interpreted as indicating that LDL inhibits cholesterol biosynthesis at one or more points distal to the last intermediate common to the cholesterol and ubiquinone biosynthetic pathways, and that this inhibition allows the synthesis of ubiquinone to proceed in the presence of LDL despite (on the basis of previously published data) a 98% reduction in mevalonate synthesis. Since squalene synthesis from farnesyl pyrophosphate is inhibited in fibroblasts by LDL, the suppression of squalene synthetase activity may allow maintenance of a pool of farnesyl pyrophosphate such that small amounts of mevalonate could be shunted preferentially to the formation of ubiquinone (20).

Rudney and co-workers (152) studied incorporation of labeled substrates into ubiquinone and cholesterol in human fibroblasts by TLC analysis of the total lipid extract. Their results (152) strongly suggest that the major control of ubiquinone synthesis probably is at the level of mevalonate formation, rather than by maintenance of a pool of farnesyl pyrophosphate. 25-Hydroxycholesterol and 7-ketocholesterol inhibited synthesis of ubiquinone from labeled p-hydroxybenzoic acid and lowered the levels of HMG-CoA reductase activity in cells grown in the presence of either whole serum or lipoprotein-deficient serum. LDL also inhibited incorporation of labeled p-hydroxybenzoic acid into ubiquinone and lowered the levels of HMG-CoA reductase activity in cells grown in the presence of lipoprotein-deficient serum. This inhibitory effect of LDL was reversed by exogenous mevalonate. Compactin also inhibited the incorporation of labeled p-hydroxybenzoic acid into ubiquinone in the fibroblasts, and this inhibition was largely reversed by a supply of exogenous mevalonate. The incorporation of ^3H from [^3H]mevalonolactone into ubiquinone and cholesterol in the cells was investigated as a function of the concentration of mevalonolactone in the presence and absence of compactin. In the absence of compactin, incorporation of label into both ubiquinone and cholesterol increased with increasing concentrations of mevalonolactone. Compactin had little or no effect on the incorporation of the label of mevalonolactone into cholesterol, but caused an apparent additional and substantial increase in its incorporation into ubiquinone. When [^{14}C]-p-hydroxybenzoic acid and [^{3}H]-mevalonolactone were added simultaneously as substrates, LDL and compactin reduced the incorporation of the label from mevalonolactone into cholesterol. These observed inhibitions were partially reversed by the addition of mevalonolactone (1 mM). LDL and compactin decreased incorporation of ^{14}C into ubiquinone and increased incorporation of ^{3}H into ubiquinone. Addition of LDL and compactin together markedly decreased incorporation of ^{14}C but increased incorporation of ^{3}H into ubiquinone. Based on these results and the effects of additions of mevalonolactone, it was concluded that when HMG-CoA reductase activity is suppressed, the incorporation of the p-hydroxybenzoic acid into ubiquinone is inhibited and this can be reversed by the addition of unlabeled mevalonolactone. The apparent increases in the incorporation of the labeled mevalonolactone were ascribed primarily to a decreased dilution of the radioactivity of the labeled mevalonolactone by endogenous mevalonate (caused by the compactin and LDL). Preliminary studies indicated that LDL also inhibits the incorporation of labeled acetate into ubiquinone. These results strongly suggest significant control of ubiquinone synthesis at the level of mevalonate formation (152).

Exploration of the regulation of the synthesis of squalene and of the various alternate metabolic fates of mevalonic acid will no doubt represent an expanding area of investigation. The ultimate unraveling of these processes depends not only upon a clearer understanding of the metabolic pathways and the enzymes involved, but also upon the development of sensitive, reliable methods for determination of the steady-state concentrations of the various intermediates. The probable existence of multiple pools of the various intermediates of varying metabolic activity must also be considered.

Oxygenated Sterols as Inhibitors of Sterol Biosynthesis

An important discovery in the area of control of sterol biosynthesis was reported by Kandutsch & Chen in 1973 (153), who used animal cells grown in serum-free, chemically defined media. (Virtually all of the studies from other laboratories cited below have employed culture media supplemented with either serum, delipidated serum, or lipoproteins.) They reported that carefully purified cholesterol did not inhibit sterol biosynthesis from labeled acetate or reduce the levels of HMG-CoA reductase activity in primary cultures of fetal mouse liver cells. Unpurified cholesterol, however, caused significant inhibition of sterol synthesis and a reduction of HMG-CoA reductase activity in these cells. The inhibitory activity of the impure cholesterol was assumed to be due to the presence of autoxidation products of cholesterol. Indeed, highly purified 7-ketocholesterol, 7α-hydroxycholesterol, and 7β-hydroxycholesterol, all known autoxidation products of cholesterol, were all highly inhibitory. These compounds specifically diminished the levels of activity of the reductase without altering the rates of acetate metabolism to carbon dioxide or fatty acids, or the rates of protein or RNA synthesis (153). Several cholesterol derivatives with oxygen functions in the side chain are also potent inhibitors of sterol biosynthesis in mouse L cells and in primary cultures of fetal mouse liver cells (154). 25-Hydroxycholesterol was especially active. As in the cases cited above, these oxygenated sterols reduced HMG-CoA reductase activity in the cells, but had no effect on the rates of acetate incorporation into carbon dioxide or fatty acids, or on the rates of RNA or protein synthesis. In contrast to their effects in intact cells, addition of the inhibitors to cell-free preparations of these cells had no effect on sterol synthesis or on HMG-CoA reductase activity.

In a survey of a large variety of oxygenated sterols and other steroids, it was found (155) that a number of oxygenated derivatives of cholesterol were very active in lowering the levels of HMG-CoA reductase activity in cultured human fibroblasts. These results must be interpreted cautiously, since the purity of only 3 of the 47 compounds tested was evaluated. In our experience, sterols purchased from many commercial sources require extensive purification (see 156). 7-Ketocholesterol, whose purity was assayed prior to use, was effective in reducing the levels of HMG-CoA reductase activity in fibroblasts from both normal and homozygous familial hypercholesterolemic subjects, but did not inhibit HMG-CoA reductase activity when added to a cell-free extract of the normal fibroblasts. Breslow et al (157) reported that both 7-ketocholesterol and 25-hydroxycholesterol, but not purified cholesterol, reduce the levels of HMG-CoA reductase in fibroblasts derived from a normal human. 25-Hydroxycholesterol and 20α-hydroxycholesterol also inhibit sterol synthesis in mouse lymphocytes and abolish the stimulation of sterol synthesis caused by phytohemagglutinin in these cells (158). 25-Hydroxycholecalciferol, 25-hydroxycholesterol, and 25-hydroxy-7-dehydrocholesterol are potent inhibitors of sterol biosynthesis in lymphocytes of normal and leukemic guinea pigs (159). 7-Ketocholesterol causes a reduction in the levels of HMG-CoA reductase activity in lymphoid cell cultures from both normal individuals and individuals with familial hypercholesterolemia. 25-Hydroxycholesterol and/or 7-ketocholesterol have subsequently been reported to be potent inhibitors of sterol synthesis in a wide variety of cells in culture including a strain of HeLa cells, rat hepatoma cells, Chinese hamster ovary and lung cells, neuroblastoma cells, chick embryo fibroblasts, glial and neuronal cells, rat ovary granulosa cells, aortic smooth muscle cells, rat hepatocytes, mouse L cells, Swiss 3T3 cells, and fibroblasts from normal and homozygous familial hypercholesterolemic subjects (132, 161–172). A number of other oxygenated C_{27} sterols also suppress sterol synthesis and/or HMG-CoA reductase activity in cells in culture (173, 174). A synthetic analogue of 7-ketocholesterol, 3β-hydroxy-17β-(4-methylpentyloxy)androst-5-en-7-one, suppressed HMG-CoA reductase activity in fibroblasts from a normal human, but did not reduce serum cholesterol levels of intact animals (175, 176). 25-Hydroxycholesterol oleate, incorporated into LDL, has been reported to suppress HMG-CoA reductase activity in fibroblasts from a normal individual, but not in those from a homozygous familial hypercholesterolemic subject (177). A number of these oxygenated derivatives are known metabolites of cholesterol, and may serve as physiological regulators of HMG-CoA reductase, sterol synthesis, and cell replication in the cells in which they are formed (178).

Another class of oxygenated sterols, Δ^7-14α-hydroxysterols, are potent inhibitors of the synthesis of digitonin-precipitable sterols from acetate in mouse L cells but do not affect the incorporation of acetate into fatty acids. They also reduce HMG-CoA reductase activity (179). Cholest-7-ene-3β, 14α-diol and cholest-7-en-14α-ol-3-one, prepared by chemical synthesis (180), caused 50% inhibition of sterol synthesis at concentrations of 7.0 μM and 5.0 μM, respectively, and 50% reduction of the levels of HMG-CoA reductase activity in the cells at concentrations of 5.0 μM and 8.0 μM, respectively. While 14α-hydroxylation of C_{27} sterols has not been demonstrated in mammalian tissues, the formation of 14α-hydroxylated steroids has been reported in mammalian adrenal and liver preparations (181–183). The finding that Δ^7-14α-hydroxysterols cause a reduction in the levels of HMG-CoA reductase activity may imply a regulatory role of these compounds in other systems. Since the 14α-hydroxy-Δ^7-system is an essential feature of various ecdysones, it is possible that these insect hormones (or 14α-hydroxy-Δ^7-sterol precursors of these hormones) may act by suppressing the level of HMG-CoA reductase in insects and, by virtue of the control of mevalonate (or homomevalonate) formation, thereby control the formation of juvenile hormones and other important products of mevalonate metabolism (179).

15-OXYGENATED STEROLS AS INHIBITORS OF STEROL BIOSYNTHESIS   An observation that certain 15-oxygenated sterols actively suppress sterol synthesis in animal cells in culture (184), coupled with investigations of their possible role in the biosynthesis of cholesterol, stimulated a program of chemical synthesis and structure determination of this class of oxygenated sterols (185–211). These synthetic 15-oxygenated sterols (15-keto, 15α-hydroxy, or 15β-hydroxy) include a wide spectrum of structural features, among which published variations include presence or absence of a nuclear double bond; variations in the number and location of nuclear double bonds; sterols with either a *cis* or *trans* C–D ring junction; a variety of substituents at C–7, C–9, and C–14; variations in the state of oxidation at C–3; variations of orientation of oxygen functions at C–3; and the presence or absence of various esters at this position. The effectiveness of these compounds in inhibiting sterol synthesis and reducing HMG-CoA reductase activity in mouse L cells and, in most cases, primary cultures of fetal mouse liver cells, has been presented (184, 196–198, 202–204, 206, 212). Table 1 presents some results of these studies for four 15-oxygenated sterols (Figure 3) together with results for 7α-hydroxycholesterol, 7-ketocholesterol, 20α-hydroxycholesterol, and 25-hydroxycholesterol in the same cell culture systems. The values were based upon analyses of plots of activity versus concentration of the inhibitory sterol over a minimum of four, and typically five, concentrations. None of these sterols had any significant effect on the incorporation of labeled acetate into fatty acids or carbon dioxide. These results exclude the possibility of an inhibition of sterol synthesis due to an effect of the compounds on the general metabolism of the cells.

A number of the 15-oxygenated sterols have significant hypocholesterolemic action when administered to intact animals (189, 199, 201, 213, 214). For example, cholest-8(14)-en-3β-ol-15-one at a level of 0.1% in the diet caused a decrease in serum cholesterol levels of rats from 86.4 ± 1.2 to 33.4 ± 3.9 mg per 100 ml after 8 days (213). This decrease is particularly noteworthy in view of the relative resistance of normocholesterolemic rats to changes in serum cholesterol. The hypocholesterolemic activity of cholest-8(14)-en-3β-ol-15-one was demonstrated in both male and female rats, and upon long-term administration (201). 9α-Fluoro-cholest-8(14)-en-3β-ol-15-one is also very active in reducing serum cholesterol levels in rats (214). 15-Oxygenated sterols constitute the only oxygenated sterols, active in suppressing sterol synthesis in cells in culture, that have been

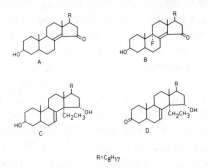

$R=C_8H_{17}$

Figure 3   Four 15-oxygenated sterols of considerable activity in inhibition of cholesterol biosynthesis. *A,* cholest-8(14)-en-3β-ol-15-one; *B,* 9α-fluorocholest-8(14)-en-3β-ol-15-one; *C,* 14α-ethyl-cholest-7-ene-3β,15α-diol; *D,* 14α-ethyl-cholest-7-en-15α-ol-3-one.

Table 1 Effects of selected oxygenated sterols on sterol synthesis and on HMG-CoA reductase activity in mouse L cells and in primary cultures of fetal mouse liver cells

| | Concentrations (μM) required for 50% inhibition | | | |
| | L cell cultures | | Primary cultures of fetal mouse liver cells | |
Inhibitor	Sterol synthesis[a]	HMG-CoA reductase	Sterol synthesis	HMG-CoA reductase
7α-hydroxycholesterol (173)	2.5	10.0	32.0	50.0
7-ketocholesterol (173)	3.0	3.0	21.3	11.3
20α-hydroxycholesterol (173)	3.7	3.5	6.0	5.8
25-hydroxycholesterol (173)	0.07	0.05	1.0	3.0
cholest-8(14)-en-3β-ol-15-one (197)	0.1	0.3	4.0	4.0
9α-fluoro-cholest-8(14)-en-3β-ol-15-one (203)	0.5	0.2	—	—
14α-ethyl-cholest-7-en-3β,15α-diol (196, 212)	0.05	0.2	0.06	2.3
14α-ethyl-cholest-7-en-15α-ol-3-one (198, 212)	0.006	0.05	—	—

[a] Determined as digitonin-precipitable sterols formed from labeled acetate.

reported so far to possess sustained hypocholesterolemic activity in intact animals.

Information concerning the metabolism of the various 15-oxygenated sterols in animal tissues and in cultured cells is limited. 14α-Methyl-cholest-7-ene-3β,15β-diol, but not its 15α-hydroxy epimer, can be converted to cholesterol in rat liver homogenate preparations (185, 215, 216). Both epimers of the corresponding 15-hydroxy derivatives of 24,25-dihydrolanosterol, are converted to cholesterol, are metabolized to more polar material, and undergo diester formation in rat liver homogenate preparations (217). The metabolism of cholest-8(14)-ene-3β,15α-diol and cholest-8(14)-ene-3β,15β-diol in liver homogenate preparations and in rat liver microsomes has been extensively investigated (186, 187, 218, 219). Both sterols are efficiently converted to cholesterol in rat liver homogenate preparations. A scheme to account for the overall conversion of a $\Delta^{8(14)}$-3β,15-dihydroxysterol to cholesterol has been presented (186). Cholest-8(14)-en-3β-ol-15-one is converted to cholesterol in 10,000 X g supernatant fractions of rat liver homogenate preparations from both male and female rats, but more efficiently by the former (220). 14β-Cholest-7-ene-3β,15α-diol, but not its 15β-hydroxy epimer, is efficiently converted to cholesterol in the 10,000 X g supernatant fraction of rat liver homogenate preparations, and a scheme to account for this conversion has been presented (221). As might be expected from its unnatural substituent at C–14, 14α-ethyl-cholest-7-ene-3β,15α-diol, a potent inhibitor of sterol synthesis in cultured cells (196, 222), was not converted to cholesterol in rat liver homogenate preparations (D. L. Raulston and G. J. Schroepfer unpublished). Similarly, 14α-hydroxymethyl-cholest-7-ene-3β,15α-diol, which also inhibits sterol synthesis in animal cells in culture (206), is not converted to cholesterol in rat liver homogenate preparations (R. A. Pascal and G. J. Schroepfer, unpublished). Gibbons et al (222) used TLC to study the distribution of radioactivity after incubation of labeled lanost-8-ene-3β,15α-diol, lanost-8-ene-3β,15β-diol, lanost-7-ene-3β,15α-diol, and lanost-7-ene-3β,15β-diol, with mouse L cells and/or primary cultures of fetal mouse liver cells. Their preliminary results indicate slight conversion to C_{27} sterols, but predominant metabolism to less polar materials, probably esters, and also, in most cases, to more polar materials that were probably more highly oxidized. The results are important in view of the potencies of the 15-oxygenated sterols as inhibitors of sterol synthesis, in reducing HMG-CoA reductase activity in cells in culture, and in reducing serum cholesterol levels in intact animals. Only fragmentary information exists with respect to the metabolism of these compounds in cultured cells. If those that are converted efficiently to cholesterol by liver enzyme preparations are also converted to cholesterol in cells in culture, their actual inhibitory potencies may be considerably greater than the reported values. Moreover, metabolic products of the oxygenated sterols may represent the true inhibitory species, additional inhibitory compounds, or compounds that have been modified so as to be inactive.

The enzymatic formation of 15-oxygenated derivatives of sterols has not been reported in mammalian tissues. However, the formation of 15β-hydroxy and 15α-hydroxy derivatives of a number of steroid hormones has been well documented (223–225 and references therein). Moreover, the presence of a 15-oxygen function (15β-hydroxy or 15-ketone) in various oogoniols is noteworthy. These C_{29} sterols act as sex hormones in the water mold *Achlya,* and cause the formation of oogonial branches at concentrations as low as 50 ng per milliliter (226, 227 and references therein). A 15α-hydroxylated derivative of ergostane has recently been isolated as its acetate and characterized as the major steroidal constituent of *Physalis pubescens* (228). The occurrence of 15-oxygenated sterol derivatives (including digitonin) in plants has been shown previously (229, 230).

OXYGENATED STEROL PRECURSORS OF CHOLESTEROL AS INHIBITORS OF STEROL BIOSYNTHESIS   Various oxygenated cholesterol derivatives may serve as physiological regulators of HMG-CoA reductase, sterol synthesis, and cell replication in the cells in which they are formed (178). Oxygenated sterol precursors of cholesterol may possibly serve a similar function (178, 206, 222, 231). Cholest-8(14)-ene-3β,15α-diol, cholest-8(14)-ene-3β,15β-diol, 14α-methyl-cholest-7-ene-3β,15β-diol, and cholest-8(14)-en-3β-ol-15-one, all of which are converted to cholesterol by rat liver enzyme preparations, are potent inhibitors of sterol synthesis in animal cells in culture (197). The role of these 15-oxygenated sterols as possible intermediates in the biosynthesis of cholesterol is, however, uncertain at this time. The demonstration (206, 222, 231) that sterol synthesis is inhibited in cultured cells and that activity of HMG-CoA reductase in these same cells is reduced by 14α-hydroxymethyl and 14α-formyl sterols, prepared by chemical synthesis (206, 222, 231, 232), is perhaps of more relevance to this matter. Since hydroxylation of the 14α-methyl group and subsequent conversion of the resulting 14α-hydroxymethyl sterol to the corresponding 14α-formyl sterol are commonly considered to be obligatory enzymatic reactions in the removal of this carbon atom in the biosynthesis of cholesterol from lanosterol and other 14α-methyl sterols in all cells capable of de novo synthesis of cholesterol, the possibility that such sterols may represent natural regulators of sterol synthesis and of cell replication in such cells has been considered (206, 222, 231, 232).

ACTIVITIES AND MECHANISMS OF ACTION OF OXYGENATED STEROLS   Although it is premature to conclude that all of the oxygenated sterols that inhibit sterol biosynthesis have a common mode of action, it is clear that most of these compounds reduce the levels of HMG-CoA reductase activity. Where appropriate studies have been conducted, the concentrations of the oxygenated sterols required to inhibit the synthesis of digitonin-precipitable sterols have been similar to those required to reduce HMG-CoA reductase activity (153, 154, 173, 179, 202–204, 206, 212). A few exceptions appear to result from inhibition of sterol synthesis at additional sites distal to the formation of mevalonic acid (see below).

Although the effect of the oxygenated sterols on HMG-CoA reductase has been the major focus of attention, at least two oxygenated sterols (25-hydroxycholesterol and cholest-8(14)-en-3β-ol-15-one) also have significant effects on other enzymes involved in the early steps of cholesterol biosynthesis, i.e. cytosolic acetoacetyl-CoA thiolase and HMG-CoA synthase in CHO-K1 cells (12–14). In addition, 25-hydroxycholesterol affects mevalonate kinase activity in these same cells (12, 13). A few of the oxygenated sterols act at an additional site in the later stages of cholesterol formation. For example, low concentrations (10^{-6} M) of 14α-ethyl-cholest-7-ene-3β,15α-diol inhibit cholesterol formation from both acetate and mevalonate in cell-free preparations of rat liver (156) and in CHO-K1 cells in culture (233); a striking accumulation of lanosterol and lanost-8-en-3β-ol is observed (156). The accumulated lanosterol and lanost-8-en-3β-ol themselves (or a metabolic product derived therefrom) may cause the reduction of the activity of HMG-CoA reductase. However, the same inhibition of the metabolism of lanosterol and lanost-8-en-3β-ol is observed in a mutant of CHO-K1 cells that is resistant to 14α-ethyl-cholest-7-ene-3β,15 α-diol with respect to its effect on HMG-CoA reductase activity (G. J. Schroepfer, F. D. Pinkerton, A. Izumi, L. R. Miller, unpublished). Moreover, even high concentrations of lanosterol or lanost-8-en-3β-ol, had little effect on sterol synthesis from labeled acetate in primary cultures of fetal mouse liver cells (153). The existence of this second site of inhibition of sterol synthesis by 14α-ethyl-cholest-7-ene-3β,15α-diol provides a ready explanation for the discrepancy between the estimates of the potencies of the sterol with respect to its effects on the synthesis of digitonin-precipitable sterols from acetate, and its effects on HMG-CoA reductase activity (Table 1), since lanosterol and lanost-8-en-3β-ol are precipitable with digitonin only to the extent of ~17% under the conditions studied (156). Several other oxygenated sterols also appear to inhibit lanosterol metabolism in cultured cells. A similar, but less striking effect of 7α-hydroxycholesterol, 7β-hydroxycholesterol, and 22(R)-hydroxydesmosterol in HTC cells (174), and of 25-hydroxycholesterol and cholestane-3β,5α,6β-triol (234) in rat hepatocytes has also been reported. Significant accumulation of labeled material with the chromatographic properties of lanosterol was observed in the latter case. However, the concentration of the triol employed was ~50 μM. Characterizations of the methyl sterols that accumulated in these cases were incomplete.

As noted above, the secondary site of action of 14α-ethyl-cholest-7-ene-3β,15α-diol was detected in both cultured cells and in cell-free preparations of rat liver (156). This is a matter of practical importance since it is much simpler to detect such secondary sites and to explore their nature in cell-free preparations of liver than in cells in culture. Only a few other reports have appeared of the inhibition of cholesterol biosynthesis in cell-free preparations by other oxygenated sterols that inhibit sterol synthesis in intact cells in culture, and in these cases, the concentrations of the oxygenated sterols required to cause the former inhibition were very high. Cholestane-3β,5α,6β-triol moderately inhibits synthesis of digitonin-precipitable sterols from labeled acetate and reduces the levels of HMG-CoA reductase activity in mouse L cells and in primary cultures of fetal mouse liver cells (173). This sterol is also effective (50% reduction at ~2.5 μM) in reducing the HMG-CoA reductase activity in rat hepatocytes (234), and inhibits cholesterol biosynthesis from labeled acetate and mevalonate in cell-free preparations of rat liver (235–237). At concentrations of the triol over 25 μM, inhibition is accompanied by the accumulation of C_{30} and C_{29} sterols (237). The inhibition of cholesterol biosynthesis by the triol, at concentrations of 10^{-3} M and 10^{-4} M, is associated with the accumulation of sterols that were assigned structures of 4,4-dimethyl and 4-monomethyl sterols (236). Oxidation of [30,31-^{14}C]-4,4-dimethyl-cholest-7-en-3β-ol by rat liver microsomes (supplemented with a soluble protein fraction) is inhibited about 83% by 50 μM triol (235); similarly, 60 μM cholestane-3β,5α-diol or cholestane-$3\beta,6\beta$-diol inhibit the synthesis of cholesterol from labeled acetate and mevalonate in cell-free preparations of rat liver (237). The former sterol, but not the latter, also caused accumulation of C_{30} and C_{29} sterols, although no experimental details were given. In mouse L cells, both cholestane diols moderately suppressed both the synthesis of digitonin-precipitable sterols from labeled acetate and HMG-CoA reductase activity (173). However, the 3,5-dihydroxysterol was essentially inactive in fetal mouse liver cells (173).

The significance of these inhibitions at later stages (distal to mevalonate formation) of sterol biosynthesis is not known. Not all oxygenated sterols show these apparent secondary sites of inhibition. For example, 5α-cholest-8(14)-en-3β-ol-15-one (at concentrations as high as 10^{-4} M) showed no effects on the distribution of radioactivity in the nonsaponifiable material recovered after incubation of labeled acetate or mevalonate with liver homogenate preparations (L. R. Miller, T. N. Pajewski, G. J. Schroepfer, unpublished).

The mechanism involved in reduction of the levels of HMG-CoA reductase activity in cultured cells by oxygenated sterols has not been determined. It is not due to direct inhibition of the enzyme, since addition of 20α-hydroxycholesterol, 22-ketocholesterol, 25-hydroxycholesterol, 7-ketocholesterol, or 14α-ethyl-cholest-7-ene-$3\beta,15\alpha$-diol to 10,000 X g supernatant fractions of rat liver, rat liver microsomes, mouse liver microsomes, or purified rat liver HMG-CoA reductase all had no effect on the enzyme activity (153, 154, 156, 165, 238).

Nevertheless, a decrease in HMG-CoA reductase activity (in the presence or absence of cycloheximide) does occur over short time periods with cells in culture and with isolated, perfused rat liver (153–155, 157, 158, 161–163, 165, 238). This decrease may be due to an inactivation or degradation of the enzyme. For example, at concentrations $\geq 10\mu g$ per milliliter, 25-hydroxycholesterol caused a prompt and sustained decrease in the activity of HMG-CoA reductase in hepatocytes incubated in lipid-free medium (165). At a concentration of 5 μg per milliliter, reductase activity was significantly decreased after 1 h of incubation, but then increased to near control values after 3 h. The return to near control levels of activity could be blocked by cycloheximide (2 μg per milliliter). 25-Hydroxycholesterol may cause the rapid decrease in reductase activity by effecting an inactivation of the enzyme, while blocking of the restoration of enzyme activity by cycloheximide may result from either inhibition of new enzyme synthesis or of the metabolism of 25-hydroxycholesterol to inactive products. Direct addition of cycloheximide (at concentrations up to 3 mM) has no effect on HMG-CoA reductase activity of rat liver microsomes (239). Kirsten and Watson (240) also reported that cycloheximide had no effect on HMG-CoA reductase activity in cell-free extracts from rat hepatoma cells (240). However, neither the range of concentrations tested nor other details were presented. The effect of 25-hydroxycholesterol in reducing HMG-CoA reductase activity in mouse L cells probably does not involve the reversible conversion of the enzyme to an inactive form via the phosphorylation-dephosphorylation system described previously (46). 7-Ketocholesterol was reported to have no significant effect on the breakdown of cell protein in rat hepatoma cells (161). The conclusion that 25-hydroxycholesterol reduces HMG-CoA reductase activity in rat hepatoma cells by causing "a decrease in the antigenicity of the enzyme rather than a decrease in the number of molecules" (241) appears premature. The authors reported that antiserum, prepared against rat liver HMG-CoA reductase, cross-reacted with solubilized and membrane-bound HMG-CoA reductase from the hepatoma cells. These experiments must be interpreted with extreme caution inasmuch as the enzyme preparation used for the preparation of the antiserum had only ~5% of the specific activity reported for highly purified rat liver HMG-CoA reductase (23, 28–30).

Obviously the combined results cited above are far from definitive, but they suggest that at least part of the action of the oxygenated sterols in suppressing HMG-CoA reductase in intact cells may be via an inactivation or degradation of the reductase. These sterols may also depress synthesis of the reductase, but no direct evidence for this possibility is available. Whether the sterols act by inactivation (or degradation) of the reductase, depression of its synthesis, or a combination of both, the report of two cytosolic protein species from mouse L cells and from fetal mouse liver cells that bind 25-hydroxycholesterol and that appear to be distinct from cytosolic proteins that bind cholesterol (242) is very important. One of the two protein species that binds 25-hydroxycholesterol also appears to bind other oxygenated sterols, and oxygenated sterols that are potent suppressors of HMG-CoA reductase activity compete for the binding of 25-hydroxycholesterol to this species, while those that do not suppress reductase activity do not compete well. Moreover, the competitive activities of various oxygenated sterols are roughly parallel with their reported activities in suppressing HMG-CoA reductase activity.

Cytoplasts (enucleated cells obtained by treatment of CHO-K1 cells with cytochalasin B) show stable levels of sterol synthesis from labeled acetate and of HMG-CoA reductase activity, for periods up to 6h (W. K. Cavenee, H. W. Chen, and A. A. Kandutsch, personal communication). Incubation of the cytoplasts with cycloheximide had little or no effect on HMG-CoA reductase activity, which suggests little degradation of the reductase under the conditions studied. Although compactin (a competitive inhibitor of HMG-CoA reductase) inhibited sterol synthesis and reductase activity in the cytoplasts, 25-hydroxycholesterol had no effect on these parameters, which suggests that the action of the oxygenated sterol on HMG-CoA reductase activity and on sterol synthesis in normal cells is under control of the nucleus.

Since mevalonate serves as a precursor of sterols, dolichols, ubiquinones, the farnesyl residues of heme a, and isopentenyl moieties of modified adenine residues in tRNA, oxygenated sterols that inhibit HMG-CoA reductase could, in principle, affect the synthesis of any of these molecules. Inhibition of the synthesis of dolichols by 25-hydroxycholesterol has been noted (132, 142). Since cholesterol and the other products of mevalonate metabolism are thought to be essential functional entities in eukaryotic cells, it is not surprising that certain oxygenated sterols were found to inhibit the replication of cells in culture (244, 153). These initial observations have been confirmed and extended (158, 166, 245–250) and have led to the isolation of mutant cells resistant to the effects of 25-hydroxycholesterol (168, 169, 251). Two such mutants have shown cross-resistance to a number of other oxygenated sterols (169, 252). This finding, along with the results cited above on the cytosolic binding protein, suggest that the various oxygenated sterol inhibitors of cholesterol synthesis act by mechanisms that share at least some common features.

It has also been reported that 25-hydroxycholesterol, 7-ketocholesterol, and 6-ketocholesterol markedly stimulate the rate at which human fibroblasts esterify their endogenous cholesterol (assayed in most cases by the incorporation of labeled oleate or oleyl-CoA into material with the TLC behavior of cholesteryl oleate) (171, 176, 253, 254). In the presence of LDL, 25-hydroxycholesterol and 7-ketocholesterol appeared to have little if any effect on cholesterol esterification in these cells (176). Kandutsch & Chen (245) studied the concentrations of total and free cholesterol in fetal mouse liver cells grown in the presence and absence of 25-hydroxycholesterol and 7-ketocholesterol. No increase in the cholesteryl ester content of the cells was induced by either sterol. In fact, a decrease in the concentration of both cholesterol and of cholesteryl ester was observed. The ratio of esters to free sterols appeared to be slightly decreased by the oxygenated sterols, but insufficient data was presented to permit definitive evaluation of this point. The concentrations of free and esterified cholesterol in liver microsomes of rats fed 25-hydroxycholesterol did not differ from those of control rats (165). However, increases (small in magnitude but relatively large in terms of percentage) in the concentration of cholesteryl ester in microsomes of rats fed 7-ketocholesterol have been reported (238). Rat hepatoma cells incubated with [^{14}C]-oleate incorporated relatively little of the ^{14}C into cholesterol esters (161). 25-Hydroxycholesterol and 7-ketocholesterol did not stimulate the incorporation of ^{14}C into cholesterol esters. The recent studies of Drevon et al (245a) in cultured rat hepatocytes may explain some of the discrepancies noted above on the effect of oxygenated sterols on sterol esterification. The stimulatory effect of 25-hydroxycholesterol was dependent upon the time that the cells were exposed to the oxygenated sterol; marked stimulation occurred very rapidly (within 15 min) but with subsequent progressive decrease such that at 18 h no effect on cholesterol ester formation was observed. The stimulation of ester formation by 25-hydroxycholesterol was unaffected by cycloheximide (at a concentration that inhibited protein synthesis by 85%). Surprisingly, direct addition of the oxygenated sterol to microsomes, without preincubation, caused a stimulation of acyl-CoA:cholesterol acyltransferase activity. Further studies should clarify not only the mechanism(s) involved but also the physiological significance of the stimulation of sterol ester formation by oxygenated sterols, and its relationship, if any, to the control of sterol synthesis.

25-Hydroxycholesterol and several other oxygenated sterols depress endocytosis in mouse L cells (255). A mixture of cholesterol and 25-hydroxycholesterol suppressed the binding, internalization, and degradation of LDL in cultured cells (256–258). A number of oxygenated sterols, but not cholesterol, are reported to inhibit chemotaxis in human polymorphonuclear leukocytes (259); 25-hydroxycholesterol was inactive in this system.

Nonoxygenated Sterols as Inhibitors of Sterol Biosynthesis

Several new nonoxygenated sterols inhibit sterol synthesis and reduce the levels of activity of HMG-CoA reductase in cultured mammalian cells. A novel dihalogenated sterol, $7\alpha,15\beta$-dichloro-cholest-8(14)-en-3β-ol, was prepared by chemical synthesis and fully characterized by ^{1}H and ^{13}C-nuclear magnetic resonance spectroscopy and X-ray crystallography (260–262). This sterol caused a 50% reduction in the level of HMG-CoA reductase activity in mouse L cells in culture at 6×10^{-7} M, and a 50% inhibition of the synthesis of digitonin-precipitable sterols from labeled acetate at a concentration of 2×10^{-6} M (with no effect on the synthesis of fatty acids from acetate) (260). 25ξ-Azido-25-norcholesterol inhibited synthesis of digitonin-precipitable sterols from acetate, but not from mevalonate, in BHK 21 cells in culture, and reduced the levels of HMG-CoA reductase activity in the same cells (263). The 25-azido-sterol also reduced rat liver microsomal HMG-CoA reductase activity 4 h after intravenous injection of 5 mg per animal (264). A water soluble, polyoxyethylated derivative of cholesterol apparently suppresses sterol synthesis and the level of HMG-CoA reductase activity in fibroblasts derived from normal and homozygous familial hypercholesterolemic subjects (171), but was less active than 25-hydroxycholesterol. Details of the synthesis and characterization of this cholesterol derivative have not been presented. The oxime of 22-ketocholesterol (10^{-5} M) inhibited sterol synthesis from both acetate and mevalonate in a rat liver preparation (265). The site of inhibition, believed to be distal to squalene, was not identified. (22R)-22-Aminocholesterol suppressed sterol synthesis from acetate in liver preparations from rats fed the sterol at a level of 0.3% for 7 days (266). The synthesis of fatty acids was also inhibited and to a greater extent than sterol synthesis.

Other Inhibitors of Sterol Synthesis

In 1976 Endo et al (267) reported the isolation and characterization of three metabolites (designated ML-236A, ML-236B, and ML-236C) of *Penicillium citrinum,* and demonstrated the inhibitory activity of these compounds on sterol synthesis from acetate, but not from mevalonate, in rat liver preparations. ML-236B, the most active of the three compounds, appears to be identical with the antifungal agent compactin isolated from *Penicillium brevicompactum* and fully characterized by Brown et al (268) (Figure 4). This compound is a potent competitive inhibitor of HMG-CoA reductase (33, 36, 269–271) and inhibits sterol synthesis from acetate, but not from mevalonate, in animal and human cells in culture (including those from homozygous familial hypercholesterolemic subjects) (37, 272, 273). Compactin inhibits cellular growth and reduces the concentration of sterols in cultured cells (272–274). Reduction of serum cholesterol levels by dietary compactin in hens, rabbits, dogs, monkeys, and man (275–278) but not in normocholesterolemic or hypercholesterolemic rats (with the exception of rats pretreated with Triton WR-1339) (34, 278–280) has been reported. Increased levels of HMG-CoA reductase have been observed in liver after oral administration of compactin to intact rats (34) and in human fibroblasts and rat hepatocytes grown in the presence of the compound (30, 34, 37), findings that were interpreted as indicating induction of the synthesis of the enzyme by its competitive inhibitor (34, 37). Using antibodies prepared against highly purified rat liver HMG-CoA reductase, Edwards et al (30) found most of the increase in reductase activity to be due to an increased number of enzyme molecules.

Recently, Alberts et al (281), in an impressive paper, reported the isolation and detailed characterization (including a preliminary account of an X-ray crystal analysis) of a methyl-substituted derivative of compactin from *Aspergillus terreus.* In the form of the sodium salt of its hydroxy acid, this

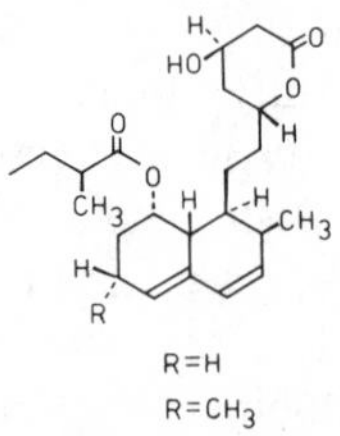

R=H
R=CH₃

Figure 4 Structures of potent competitive inhibitors of HMG-CoA reductase. R = H, 6-demethylmevinolin (Compactin; ML-236B); R = CH₃, Mevinolin (Monacolin K).

compound, named mevinolin (Figure 4), was more active, in the inhibition of HMG-CoA reductase than the corresponding form of compactin. Mevinolin showed hypocholesterolemic activity (~30% reduction after 21 days) upon oral administration (8 mg per kilogram per day) to dogs. The properties of mevinolin appear to correspond to those of a compound isolated from a *Monascus* species (282) and named monacolin K, which was reported to have hypocholesterolemic activity in Triton-WR-1339–treated rats and in rabbits (282).

Colchicine, colcemid, and vinblastin, agents known to disrupt microtubules, have been reported to cause a reduction of HMG-CoA reductase activity in glial cells in culture (283, 284), and the suggestion has been made that microtubules may be involved in the regulation of the reductase and of sterol synthesis in these cells. l-4,4,10β-Trimethyl-trans-decal-3β-ol is a potent inhibitor of sterol synthesis in CHO cells and specifically inhibits the enzymatic cyclization of 2,3-epoxysqualene (285, 286); it had no effect on HMG-CoA reductase activity or on fatty acid synthesis under the conditions studied. Cycloheximide (1 mM), extensively used as an inhibitor of protein synthesis, inhibited sterol biosynthesis from acetate, but not from mevalonate, in cell-free preparations of rat liver (239). This inhibition, shown not to be due to a suppression of protein synthesis under the conditions studied, was associated with a reduction in the levels of acetyl-CoA synthetase and HMG-CoA synthase activities, but not of acetoacetyl-CoA thiolase or HMG-CoA reductase activity. Cycloheximide (50 μM) did not reduce sterol synthesis from acetate. A number of investigators have employed concentrations of cycloheximide considerably in excess of this value to suppress protein synthesis in various cells, tissues, or organisms. Various effects observed upon exposure of cells or tissues to high concentrations of cycloheximide may not be exclusively due to effects of the cycloheximide on the synthesis of protein. Phenylmethylsulfonyl fluoride, commonly used as an inhibitor of proteolysis, causes substantial inhibition of sterol synthesis from acetate, but not from mevalonate, in cell-free preparations of rat liver (D. L. Raulston and G. J. Schroepfer, unpublished observations). This inhibition was associated with a reduction in the levels of activity of acetyl-CoA synthetase, HMG-CoA synthase, and HMG-CoA reductase, but not of cytosolic acetoacetyl-CoA thiolase. These findings obviously complicate, if not preclude, the use of this compound in investigations of the proteolytic degradation of HMG-CoA reductase. More generally, these findings indicate that caution is needed in the interpretation of studies that use this compound to inhibit the enzymatic degradation of proteins in complex systems.

Many other compounds, usually in high concentrations, inhibit sterol synthesis; the precise site of their action is not known. A series of synthetic half acid esters of 1-(4-biphenyl)pentanol and other closely related compounds suppressed HMG-CoA reductase activity in rat liver microsomes (287, 288). 3-Hydroxy-3-methylglutaric acid inhibited in vitro sterol synthesis from acetate and HMG-CoA, but not from mevalonate, in rat liver preparations (289). Clofibrate inhibits sterol synthesis in liver by reducing the level of HMG-CoA reductase activity (290–292). High concentrations ($>10^{-3}$ M) of Clofibrate reduced HMG-CoA reductase activity in rat liver microsomes (292). Kaneko et al (162) have suggested that the observed reduction of HMG-CoA reductase activity in mouse L cells caused by Clofibrate was due to an increased rate of inactivation or degradation of the enzyme, and not to an inhibition of its synthesis. Clofibrate also inhibits synthesis of nonsaponifiable lipids from mevalonate in liver preparations (293). Two sesquiterpene antitumor agents were reported (294) to inhibit sterol synthesis from acetate in ascites tumor cells. A large number of other biochemical parameters were also affected. Racemic 3,5-dihydroxy-3,4,4-trimethylvaleric acid (40 μM) caused a 50% inhibition of sterol synthesis from mevalonate in liver (295), perhaps by inhibiting phosphomevalonate kinase. 5-Hydroxy-4-methyl-valeric acid suppresses sterol synthesis from both acetate and mevalonate (296). 2-Octanone and several substituted cyclooctanones, at rather higher concentrations, inhibit hepatic HMG-CoA reductase activity in vitro (297). A number of 1-alkylimidazoles suppressed sterol synthesis from labeled acetate in liver slices (298). Several local anaesthetics (299) and methylxanthines, such as caffeine and aminophylline (300), inhibited sterol synthesis from acetate in glial cells in culture. Sterol synthesis from both acetate and mevalonate in liver preparations is suppressed by a number of synthetic compounds used as plant growth retardants (301, 302, and references therein). Bloxham (303) reported the inhibition of sterol synthesis from acetate, but not from mevalonate, in rat liver preparations, by a series of alk-3-ynoyl-CoA esters, which inhibit cytosolic acetoacetyl-CoA thiolase. A number of hydroxylated aromatic compounds (2.5 mM) inhibit phosphomevalonate kinase and mevalonate pyrophosphate decarboxylase (304). A series of pyrophosphate farnesyl analogues were shown to inhibit squalene synthetase (305).

Low Density Lipoprotein Receptor Hypothesis for Control of Cholesterol Biosynthesis

In 1974 Brown & Goldstein discovered that binding of LDL to fibroblasts is defective in patients with homozygous familial hypercholesterolemia (306, 307). Coupled with their previous reports (8, 308) that LDL suppressed sterol synthesis and HMG-CoA reductase activity in fibroblasts derived from normal individuals but not those from patients with familial hypercholesterolemia, these new findings led to the LDL receptor hypothesis for the control of cholesterol metabolism, which has almost become dogma and has been frequently reviewed (309–315). Essential features of this hypothesis are: LDL first binds to a specific receptor located in the coated pit regions of the plasma membrane. These receptors are present in fibroblasts from normal individuals, but are absent or deficient in fibroblasts from subjects with familial hypercholesterolemia. The LDL-receptor complex is then internalized by endocytosis. Lysosomal degradation of the protein and its cholesteryl ester components yields amino acids and free cholesterol. The fate and actions of the cholesterol introduced into the cell via these processes are, apart from its utilization for membrane synthesis, of critical importance to this theory, and the following is a quotation from a recent review of this topic (315).

> This cholesterol also regulates three events in cellular cholesterol metabolism. First, it suppresses the activity of 3-hydroxy-3-methylglutaryl coenzyme A reductase (HMG-CoA reductase), the rate-controlling enzyme in cholesterol biosynthesis, thereby suppressing cholesterol synthesis by the cell. Second, it activates an acyl-CoA:cholesterol acyltransferase (ACAT), which re-esterifies some of the incoming cholesterol for storage as cholesteryl ester droplets in the cytoplasm. Third, cholesterol derived from LDL suppresses the synthesis of LDL receptors, allowing the cell to control the rate of entry of LDL by regulating the number of LDL receptors.

This proposal, clearly and forcefully presented by Brown & Goldstein, has stimulated the rapid acquisition of knowledge about the cellular uptake and catabolism of lipoproteins and control of the synthesis of cholesterol and other isoprenoids. However, some of its aspects deserve closer scrutiny. The stated role of LDL-derived cholesterol in controlling the activity of HMG-CoA reductase, and thereby, the biosynthesis of cholesterol and other isoprenoids, deserves particular attention. Oxygenated sterols, produced from LDL-bound cholesterol by enzyme-catalyzed reactions or by nonenzyme-catalyzed autoxidation could account for the reported suppression of HMG-CoA reductase activity and inhibition of sterol synthesis by LDL in normal fibroblasts. This possibility is suggested by the finding that oxygenated sterols are very active in the suppression of sterol synthesis while carefully purified cholesterol had little or no activity in suppressing either sterol synthesis or HMG-CoA reductase activity in mouse L cells or in primary cultures of fetal mouse liver cells (153, 154). Similar results were obtained using fibroblasts from normal human subjects (157). That LDL has no effect in fibroblasts from subjects with familial hypercholesterolemia can be interpreted in terms of the reported defects in binding to the LDL-receptor and/or internalization of the LDL containing the oxygenated sterol(s). This is compatible with suppression of HMG-CoA reductase by a reconstituted LDL containing 25-hydroxycholesteryl-3β-oleate in normal fibroblasts but not in fibroblasts derived from individuals with familial hypercholesterolemia (177). The possibility that oxygenated sterols exist normally in serum (and therefore possibly in LDL) is suggested by the occurrence, in the serum of pregnant mares, of substantial quantities of a sterol assigned the structure 7β-hydroxycholesterol (316); this was subsequently revised to 7α-hydroxycholesterol (see 317). The concentration of the sterol in the unsaponifiable lipid fraction was estimated to be "probably several tenths of 1 per cent" (316). The authors considered the possibility that this material arose by autoxidation of cholesterol, but noted that they exerted considerable effort to avoid autoxidation during the various isolation and purification procedures. The possible natural occurrence of oxygenated sterols in serum and in LDL, and, the possibility of their formation by autoxidation during isolation, purification, and storage of LDL, are important topics that require further investigation. Cholesterol in LDL is present mostly as an ester of linoleic acid. The ease of autoxidation of cholesteryl linoleate, and the possibility of its catalysis by iron-porphyrin components in hemolyzed serum or plasma, present a favorable situation for the autoxidation of cholesterol in LDL. The sensitivity of LDL to autoxidation has been noted by others (318, 319 and references therein). The standard procedure for the isolation of LDL for use in studies in cultured cells (320) involves collection and centrifugation of blood in the presence of 0.1% EDTA, isolation of the LDL by ultracentrifugation for 16–24 h in the presence of KBr, prolonged (36 h) dialysis, and finally sterilization by filtration prior to storage. Apart from the addition of EDTA, no special precautions to suppress autoxidation were noted in the various reports. Lee (319) reported that EDTA alone does not fully protect against autooxidation of LDL. The same concern about autoxidation can be expressed with regard to the reported activities of a reconstituted LDL containing cholesteryl linoleate (321). The critical importance of this matter derives from the extraordinary potency of some oxygenated sterols in suppressing sterol synthesis and HMG-CoA reductase activity: purified cholesterol, at 5×10^{-3} M, had little or no effect on sterol synthesis in primary cultures of fetal mouse liver cells (153), whereas 25-hydroxycholesterol, a known autoxidation product of cholesterol and a compound whose enzymatic formation in liver has also been demonstrated (322, 323), caused a 50% inhibition of sterol synthesis in the same cells at 1×10^{-6} M (159). Binding and internalization of LDL by fibroblasts are clearly important factors in the metabolism of LDL and in the transport of its various components (including cholesterol) into cells. However, the observed effects on HMG-CoA reductase, and on cholesterol and isoprenoid biosynthesis, could easily be due to small amounts of oxygenated sterols in the LDL. Even if this possibility could be excluded, it appears reasonable to consider that, as an excess of cholesterol is delivered into a cell, the enzymatic formation of oxygenated derivatives of cholesterol may be induced, and they then suppress HMG-CoA reductase activity and the synthesis of cholesterol. Reports of significant quantities of oxygenated derivatives of cholesterol in hepatic steryl esters in Wolman's disease (324, 325) are noteworthy in this respect.

Regulation of the "suppression" of the synthesis of LDL receptors by cholesterol should also be viewed in the context of reports of the inhibition of endocytosis in mouse L cells by oxygenated sterols (255) and of the suppression of LDL binding, internalization, and degradation in cultured cells by their exposure to cholesterol and 25-hydroxycholesterol (256-258).

The proposed activation of acyl-CoA: cholesterol acyltransferase by cholesterol derived from LDL also needs further study. LDL causes increased formation of labeled cholesteryl oleate from [^{14}C]oleate in fibroblasts from normal individuals, but not in those from subjects with familial hypercholesterolemia (171, 176, 253, 326, 327). Although these observations could be explained by an activation of the acyl-CoA: cholesterol acyltransferase, as a result of the uptake of LDL by the cells from the normal subjects, the possibility that the apparent accelerated rate of acyl transfer may result from an increase in the intracellular concentration of cholesterol has not been rigorously excluded. If an activation of the transferase is, in fact, induced by cholesterol, exploration of the mechanism of this process would be of considerable interest.

The specificity of the receptor binding site(s) for LDL is not completely resolved in view of the observations that certain HDL species and several other proteins with a net positive charge appear to bind to the same receptor (327–332). Now that the solubilization of the LDL receptor from fibroblasts and bovine adrenal cortex (257) is possible, the specificity of the binding should be clarified soon.

A well-defined working hypothesis can stimulate and guide the design of experiments directed toward an increased understanding of the regulation of cholesterol biosynthesis. However, in spite of the supporting evidence for the LDL-receptor hypothesis for the control of cholesterol biosynthesis, it appears to this reviewer that the basic postulates of this hypothesis are open to alternative interpretation.

Concluding Remarks

Research into the biosynthesis of sterols and other isoprenoids has resulted in major developments which I have attempted to appraise critically. Continued acceleration of the pace and extension of the range of research investigations in this general area can be anticipated. Understanding of fundamental processes in nature is seldom simple, a situation that provides both a challenge and a need for some modicum of conservatism.

ACKNOWLEDGMENTS

The support of the National Institutes of Health (HL-15376 and HL-22532) and the Robert A. Welch Foundation is gratefully acknowledged. The author is especially indebted to Drs. J. L. Vermilion, A. Kisic, J. S. Brabson, and F. D. Pinkerton for careful reading of this manuscript and for valuable assistance in its preparation.

Literature Cited

1. Beytia, E. D., Porter, J. W. 1976. *Ann. Rev. Biochem.* 45:113–42
2. Rodwell, V. W., McNamara, D. J., Shapiro, D. J. 1973. *Adv. Enzymol.* 38:373–412
3. White, L. W., Rudney, H. 1970. *Biochemistry.* 9:2725–31
4. Clinkenbeard, K. D., Sugiyama, T., Moss, J., Reed, W. D., Lane, M. D. 1973. *J. Biol. Chem.* 248:2275–85
5. Clinkenbeard, K. D., Reed, W. D., Mooney, R. A., Lane, M. D. 1975. *J. Biol. Chem.* 250:3108–16
6. Clinkenbeard, K. D., Sugiyama, T., Reed, W. D., Lane, M. D. 1975. *J. Biol. Chem.* 250:3124–35
7. Nervi, F. O., Carrella, M., Dietschy, J. M. 1976. *J. Biol. Chem.* 251:3831–33
8. Brown, M. S., Dana, S. E., Goldstein, J. L. 1973. *Proc. Natl. Acad. Sci. USA* 70:2162–66
9. Howard, B. V., Howard, W. J., Bailey, J. M. 1974. *J. Biol. Chem.* 249:7912–21
10. Melnykovych, G., Matthews, E., Gray, S., Lopez, I. 1976. *Biochem. Biophys. Res. Commun.* 71:506–12
11. Cavenee, W. K., Melnykovych, G. 1977. *J. Biol. Chem.* 252:3272–76
12. Chang, T.-Y. 1980. *Fed. Proc.* 39:1906
13. Chang, T.-Y., Limanek, J. S. 1980. *J. Biol. Chem.* 255:7787–95
14. Miller, L. R., Pinkerton, F. R., Schroepfer, G. J. Jr. 1980. *Biochem. Int.* 1:223–28
15. Ramachandran, C. K., Gray, S. L., Melnykovych, G. 1978. *Arch. Biochem. Biophys.* 189:205–11
16. Johnston, D., Cavenee, W. K., Ramachandran, C. K., Melnykovych, G. 1979. *Biochim. Biophys. Acta* 572: 188–92
17. Balasubramanian, S., Goldstein, J. L., Brown, M. S. 1977. *Proc. Natl. Acad. Sci. USA* 74:1421–25
18. Anderson, J. M., Dietschy, J. M. 1976. *Biochem. Biophys. Res. Commun.* 72: 880–85
19. Balasubramanian, S., Goldstein, J. L., Faust, J. R., Brunschede, G. Y., Brown,

M. S. 1977. *J. Biol. Chem.* 252:1771–79
20. Faust, J. R., Goldstein, J. L., Brown, M. S. 1979. *Proc. Natl. Acad. Sci. USA* 76:5018–22
21. Qureshi, N., Dugan, R. E., Nimmannit, S., Wu, W.-H., Porter, J. W. 1976. *Biochemistry* 15:4185–90
22. Heller, R. A., Shrewsbury, M. A. 1976. *J. Biol. Chem.* 251:3815–22
23. Kleinsek, D. A., Ranganathan, S., Porter, J. W. 1977. *Proc. Natl. Acad. Sci. USA* 74:1431–35
24. Srikantaiah, M. U., Tormanen, C. D., Redd, W. L., Hardgrave, J. E., Scallen, T. J. 1977. *J. Biol. Chem.* 252:6145–50
25. Beg, Z. H., Stonik, J. A., Brewer, H. B. Jr. 1978. *Anal. Biochem.* 86:531–35
26. Beg, Z. H., Stonik, J. A., Brewer, H. B. Jr. 1979. *Biochim. Biophys. Acta* 572: 83–94
27. Edwards, P. A., Lemongello, D., Fogelman, A. M. 1979. *J. Lipid. Res.* 20: 40–46
28. Edwards, P. A., Lemongello, D., Fogelman, A. M. 1976. *Biochim. Biophys. Acta* 574:123–35
29. Kleinsek, D. A., Porter, J. W. 1979. *J. Biol. Chem.* 254:7591–99
30. Edwards, P. A., Lemongello, D., Kane, J., Shecter, I., Fogelman, A. M. 1980. *J. Biol. Chem.* 255:3715–25
31. Higgins, M. J. P., Brady, D., Rudney, H. 1974. *Arch. Biochem. Biophys.* 163: 271–82
32. Kleinsek, D. A., Jabalquinto, A. M., Porter, J. W. 1978. *Fed. Proc.* 37:1427
33. Kleinsek, D. A., Jabalquinto, A. M., Porter, J. W. 1980. *J. Biol. Chem.* 255:3918–23
34. Bench, W. R., Ingebritsen, T. S., Diller, E. R. 1977. *Biochem. Biophys. Acta* 82:247–54.
35. Goldfarb, S. 1978. *J. Lipid Res.* 19:489–94
36. Langdon, R., El-Masry, S., Counsell, R. E. 1977. *J. Lipid Res.* 18:24–30
37. Brown, M. S., Faust, J. R., Goldstein, J.

L., Kaneko, I., Endo, A. 1978. *J. Biol. Chem.* 253:1121–28
38. Berndt, J., Boll, M., Lowel, M., Gaumert, R. 1973. *Biochem. Biophys. Res. Commun.* 51:843–48
39. Goodwin, C. D., Margolis, S. 1976. *J. Lipid Res.* 17:297–303
40. Sanghvi, A., Parikh, B. 1976. *Biochim. Biophys. Acta* 444:727–33
41. Sanghvi, A., Parikh, B. 1978. *Biochim. Biophys. Acta* 531:79–85
42. Murthy, H. R., Moorjani, S., Lupien, P.-J. 1977. *Anal. Biochem.* 81:65–77
43. Langdon, R. B., Counsell, R. E. 1976. *J. Biol. Chem.* 251:5820–23
44. Ackerman, M. E., Redd, W. L., Tormanen, C. D., Hardgrave, J. E., Scallen, T. J. 1977. *J. Lipid. Res.* 18:408–13
45. Ness, G. C., Moffler, M. H. 1978. *Arch. Biochem. Biophys.* 189:221–23
46. Saucier, S. E., Kandutsch, A. A. 1979. *Biochim. Biophys. Acta* 572:541–56
47. Philipp, B. W., Shapiro, D. J. 1979. *J. Lipid Res.* 20:588–93
48. Ness, G. C., Moffler, M. H. 1979. *Biochim. Biophys. Acta* 572:333–44
49. Beg, Z. H., Allman, D. W., Gibson, D. M. 1973. *Biochem. Biophys. Res. Commun.* 54:1362–69
50. Shapiro, D. J., Nordstrom, J. L., Mitschelen, J. J., Rodwell, V. W., Schimke, R. T. 1975. *Biochim. Biophys. Acta* 370:369–77
51. Brown, M. S., Brunschede, G. Y., Goldstein, J. L. 1975. *J. Biol. Chem.* 250:2502–2509
52. Chow, J. C., Higgins, M. J. P., Rudney, H. 1975. *Biochem. Biophys. Res. Commun.* 63:1077–84
53. Nordstrom, J. L., Rodwell, V. W., Mitschelen, J. J. 1977. *J. Biol. Chem.* 252:8924–34
54. Ursini, F., Valente, M., Ferri, L., Gregolin, C. 1977. *FEBS Lett.* 82:97–101
55. Beg, Z. H., Stonik, J. A., Brewer, H. B. Jr. 1978. *Proc. Natl. Acad. Sci. USA* 75:3678–82
56. Gibson, D. M., Ingebritsen, T. S. 1978. *Life Sci.* 23:2649–64
57. Ingebritsen, T. S., Lee, H.-S., Parker, R. A., Gibson, D. M. 1978. *Biochem. Biophys. Res. Commun.* 81:1268–77
58. Bove, J., Hegardt, F. G. 1978. *FEBS Lett.* 90:198–202
59. Keith, M. L., Rodwell, V. W., Rogers, D. H., Rudney, H. 1979. *Biochem. Biophys. Res. Commun.* 90:969–75
60. Beg, Z. H., Stonik, J. A., Brewer, H. B. Jr. 1979. *Proc. Natl. Acad. Sci. USA* 76:4375–79

61. Lin, R. C., Snodgrass, P. J. 1980. *FEBS Lett.* 109:171–74
62. Mitropoulos, K. A., Knight, B. L., Reeves, B. E. A. 1980. *Biochem. J.* 185:435–41
63. Hunter, C. F., Rodwell, V. W. 1980. *J. Lipid Res.* 21:399–405
64. Gil, G., Sitges, M., Bove, J., Hegardt, F. G. 1980. *FEBS Lett.* 110:195–99
65. Beg, Z. H., Stonik, J. A., Brewer, H. B. Jr. 1980. *J. Biol. Chem.* 255:8541–45
66. Ingebritsen, T. S., Geelen, M. J. H., Parker, R. A., Evenson, K. J., Gibson, D. M. 1979. *J. Biol. Chem.* 254:9986–89
67. Brown, M. S., Goldstein, J. L., Dietschy, J. M. 1979. *J. Biol. Chem.* 254:5144–49
68. Maltese, W. A., Volpe, J. J. 1979. *Biochem. J.* 182:367–70
69. Ness, G. C., Spindler, C. D., Benton, G. A. 1980. *Fed. Proc.* 39:1777
70. Ness, G. C., Spindler, C. D., Benton, G. A. 1980. *J. Biol. Chem.* 255:9013–16
71. Kamiya, Y., Sakurai, A., Tamura, S., Takahashi, N., Abe, K., Tsuchiya, E., Fukui, S., Kitada, C., Fujino, M. 1978. *Biochem. Biophys. Res. Commun.* 83:1077–83
72. Lennarz, W. J. 1975. *Science* 188:986–91
73. Waechter, C. J., Lennarz, W. J. 1976. *Ann. Rev. Biochem.* 45:95–112
74. Popjak, G., Agnew, W. S. 1979. *Mol. Cellul. Biochem.* 27:97–116
75. Poulter, C. D., Rilling, H. C. 1978. *Acc. Chem. Res.* 11:307–13
76. Rilling, H. C. 1979. *Pure Appl. Chem.* 51:597–608
77. Cane, D. E. 1980. *Tetrahedron* 36:1109–159
78. Coates, R. M. 1976. *Prog. Chem. Org. Nat. Prod.* 33:73–230
79. Popjak, G. 1970. *Ann. Intern. Med.* 72:106–108
80. Edmond, J., Popjak, G. 1974. *J. Biol. Chem.* 249:2716–21
81. Raskin, P., Siperstein, M. D. 1974. *J. Lipid Res.* 15:20–25
82. Fogelman, A. M., Edmond, J., Popjak, G. 1975. *J. Biol. Chem.* 250:1771–75
83. Edmond, J., Fogelman, A. M., Popjak, G. 1976. *Science* 193:154–56
84. Righetti, M., Wiley, M. H., Murrill, P. A., Siperstein, M. D. 1976. *J. Biol. Chem.* 251:2716–21
85. Linder, J. R., Beitz, D. C. 1978. *J. Lipid Res.* 19:836–40
86. Wiley, M. H., Howton, M. M., Siperstein, M. D. 1977. *J. Biol. Chem.* 252:548–54

87. Seubert, W. 1960. *J. Bacteriol.* 79: 426–34
88. Seubert, W., Fass, E., Remberger, U. 1963. *Biochem. Z.* 338:265–75
89. Seubert, W., Remberger, U. 1963. *Biochem. Z.* 338:245–64
90. Seubert, W., Fass, E. 1964. *Biochem. Z.* 341:23–24
91. Seubert, W., Fass, E. 1964. *Biochem. Z.* 341:35–44
92. Skoog, F., Strong, F. M., Miller, C. O. 1965. *Science* 128:532–33
93. Hacker, B., Feldbush, T. L. 1969. *Biochem. Pharmacol.* 18:847–53
94. McCloskey, J. A., Nishimura, S. 1977. *Acc. Chem. Res.* 10:403–10
95. Grosjean, H. J., deHenau, S., Crothers, D. M. 1978. *Proc. Natl. Acad. Sci. USA* 75:610–14
96. Janner, F., Vogeli, G., Fluri, R. 1980. *J. Mol. Biol.* 139:207–19
97. Fittler, F., Kline, L. K., Hall, R. H. 1968. *Biochemistry* 7:940–44
98. Holtz, J., Klambt, D. 1975. *Hoppe-Seylers Z. Physiol. Chem.* 356:1459–64
99. Chen, C. M., Melitz, D. K. 1979. *FEBS Lett.* 107:15–20
100. Faust, J. R., Brown, M. S., Goldstein, J. L. 1980. *J. Biol. Chem.* 255:6546–48
101. Horgan, R. 1978. *Proc. R. Soc. London Ser. B* 284:439–47
102. Laloue, M. 1978. *Proc. Roy. Soc. London Ser. B* 284:449–57
103. Hall, R. H. 1971. *The Modified Nucleosides in Nucleic Acids.* New York: Columbia Univ. Press. 451 pp.
104. Eberhardt, N. L., Rilling, H. C. 1975. *J. Biol. Chem.* 250:863–66
105. Reed, B. C., Rilling, H. C. 1976. *Biochemistry* 15:3739–45
106. Poulter, C. D., Rilling, H. C. 1976. *Biochemistry* 15:1079–83
107. King, H. L. Jr., Rilling, H. C. 1977. *Biochemistry* 16:3815–19
108. Yeh, L.-S., Rilling, H. C. 1977. *Arch. Biochem. Biophys.* 183:718–25
109. Barnard, G. F., Langton, B., Popjak, G. 1978. *Biochem. Biophys. Res. Commun.* 85:1097–103
110. Brems, D. N., Rilling, H. C. 1979. *Biochemistry* 18:860–64
111. Laskovics, F. M., Krafcik, J. M., Poulter, C. D. 1979. *J. Biol. Chem.* 254:9458–63
112. Saito, A., Rilling, H. C. 1979. *J. Biol. Chem.* 254:8511–15
113. Barnard, G. F., Popjak, G. 1980. *Biochim. Biophys. Acta* 617:169–82
114. Keyhani, J., Keyhani, E. 1978. *FEBS Lett.* 93:271–74
115. Christophe, J., Popjak, G. 1961. *J. Lipid Res.* 2:244–57
116. Popjak, G., Cornforth, J. W. 1960. *Adv. Enzymol.* 22:281–335
117. Schroepfer, G. J. Jr., Gore, I. Y. 1963. *J. Lipid Res.* 4:266–69
118. Dirksen, A., Cohen, P. 1973. *J. Biol. Chem.* 248:7396–7403
119. Levy, H. R., Popjak, G. 1960. *Biochem. J.* 75:417–28
120. Wright, L. D., Cleland, M. 1957. *Proc. Soc. Exp. Biol. Med.* 96:219–24
121. Dorsey, J. K., Porter, J. W. 1968. *J. Biol. Chem.* 243:4667–70
122. Hemming, F. W. 1974. *Med. Tech. Publ. Int. Rev. Sci. Biochem. Ser.* 4:39–98
123. Allen, C. M. Jr., Muth, J. D. 1977. *Biochemistry* 16:2908–15
124. Baba, T., Allen, C. M. 1978. *Biochemistry* 17:5598–5604
125. Baba, T., Allen, C. M. 1980. *Arch. Biochem. Biophys.* 200:474–84
126. Takahashi, I., Ogura, K., Seto, S. 1980. *J. Biol. Chem.* 255:4539–43
127. Butterworth, P. H. W., Hemming, F. W. 1968. *Arch. Biochem. Biophys.* 128:503–8
128. Gough, D. P., Hemming, F. W. 1970. *Biochem. J.* 118:163–66
129. Martin, H. G., Thorne, K. J. I. 1974. *Biochem. J.* 138:277–80
130. James, M. J. Kandutsch, A. A. 1980. *J. Biol. Chem.* 255:16–19
131. Spiro, M. J., Spiro, R. G., Bhoyroo, V. D. 1976. *J. Biol. Chem.* 251:6400–8
132. Mills, J. T., Adamany, A. M. 1978. *J. Biol. Chem.* 253:5270–73
133. Mankowski, T., Jankowski, W., Chojnacki, T., Franke, P. 1976. *Biochemistry* 15:2125–30
134. Daleo, G. R., Hopp, H. E., Roero, P. A., Pont Lezica, R. 1977. *FEBS Lett.* 81:411–14
135. Grange, D. K., Adair, W. L. Jr. 1977. *Biochem. Biophys. Res. Commun.* 79:734–40
136. Wellner, R. B., Lucas, J. J. 1979. *FEBS Lett.* 104:379–83
137. Rupar, C. A., Carroll, K. K. 1978. *Lipids* 13:291–93
138. Keenan, R. W., Kruczek, M. E. 1976. *Biochemistry* 15:1586–90
139. Allen, C. M., Kalin, J. R., Sack, J., Verizzo, D. 1978. *Biochemistry* 17:5020–26
140. Burton, W. A., Scher, M. G., Waechter, C. J. 1979. *J. Biol Chem.* 254:7129–36
141. James, M. J., Kandutsch, A. A. 1980. *J. Biol. Chem.* 255:16–19
142. James, M. J., Kandutsch, A. A. 1979. *J. Biol. Chem.* 254:8442–46
143. James, M. J., Kandutsch, A. A. 1980. *J. Biol. Chem.* 255:8618–22

144. Carson, D. D., Lennarz, W. J. 1979. *Proc. Natl. Acad. Sci. USA* 75:5709–13
145. Alam, S. S., Nambudiri, A. M. D., Rudney, H. 1975. *Arch. Biochem. Biophys.* 171:183–90
146. Nambudiri, A. M. D., Brockman, D., Alam, S. S., Rudney, H. 1977. *Biochem. Biophys. Res. Commun.* 76:282–88
147. Nishino, T., Rudney, H. 1977. *Biochemistry* 16:605–9
148. Ranganathan, S., Nambudiri, A. M. D., Rudney, H. 1979. *Arch. Biochem. Biophys.* 198:506–11
149. Ranganathan, S., Ramasarma, T. 1975. *Biochem. J.* 148:35–39
150. Aiyar, A. S., Olson, R. E. 1972. *Eur. J. Biochem.* 27:60–64
151. Faust, J. R., Goldstein, J. L., Brown, M. S. 1979. *Arch. Biochem. Biophys.* 192:86–99
152. Nambudiri, A. M. D., Ranganathan, S., Rudney, H. 1980. *J. Biol. Chem.* 255:5894–99
153. Kandutsch, A. A., Chen, H. W. 1973. *J. Biol. Chem.* 248:8408–17
154. Kandutsch, A. A., Chen, H. W. 1974. *J. Biol. Chem.* 249:6057–61
155. Brown, M. S., Goldstein, J. L. 1974. *J. Biol. Chem.* 249:7306–14
156. Raulston, D. L., Pajewski, T. N., Miller, L. R., Philip, B. W., Shapiro, D. J., Schroepfer, G. J. Jr. 1980. *Biochem. Int.* 1:113–119
157. Breslow, J. L., Lothrop, D. A., Spaulding, D. R., Kandutsch, A. A. 1975. *Biochim. Biophys. Acta* 398:10–17
158. Chen, H. W., Heiniger, H.-J., Kandutsch, A. A. 1975. *Proc. Natl. Acad. Sci. USA* 72:1950–54
159. Philippot, J. R., Cooper, A. G., Wallach, D. F. H. 1976. *Biochem. Biophys. Res. Commun.* 72:1035–41
160. Kayden, H. J., Hatam, L., Beratis, N. G. 1976. *Biochemistry* 15:521–28
161. Bell, J. J., Sargeant, T. E., Watson, J. A. 1976. *J. Biol. Chem.* 251:1745–58
162. Kaneko, I., Hazama-Shimada, Y., Kuroda, M., Endo, A. 1977. *Biochem. Biophys. Res. Commun.* 76:1207–13
163. Volpe, J. J., Hennessy, S. W. 1977. *Biochim. Biophys. Acta* 486:408–20
164. Cavenee, W. K., Johnston, D., Melnykovych, G. 1978. *Proc. Natl. Acad. Sci. USA* 75:2103–07
165. Erickson, S. K., Matsui, S. M., Shrewsbury, M. A., Cooper, A. D., Gould, R. G. 1978. *J. Biol. Chem.* 253:4159–64
166. Harley, J. B., Goldfine, H. 1979. *Exp. Cell Res.* 118:47–54
167. Schuler, L. A., Scavo, L., Kirsch, T. M., Flickinger, G. L., Straus, J. F. 1979. *J. Biol. Chem.* 254:8662–68
168. Sinensky, M., Duwe, G., Pinkerton, F. 1979. *J. Biol. Chem.* 254:4482–86
169. Chen, H. W., Cavenee, W. K., Kandutsch, A. A. 1979. *J. Biol. Chem.* 254:716–20
170. Maltese, W. A., Volpe, J. J. 1980. *J. Neurochem.* 34:1522–26
171. Fung, C. H., Khachadurian, A. K. 1980. *J. Biol. Chem.* 255:676–80
172. Habenicht, A. J. R., Glomset, J. A., Ross, R. 1980. *J. Biol. Chem.* 255:5134–40
173. Kandutsch, A. A., Chen, H. W. 1978. *Lipids* 13:704–07
174. Ortiz de Montellano, P. R., Beck, J. P., Ourisson, G. 1979. *Biochem. Biophys. Res. Commun.* 90:897–903
175. Dygos, J. H., Desai, B. N. 1979. *J. Org. Chem.* 44:1590–96
176. Goldstein, J. L., Faust, J. R., Dygos, J. H., Chorvat, R. J., Brown, M. S. 1978. *Proc. Natl. Acad. Sci. USA* 75:1877–81
177. Krieger, M., Goldstein, J. L., Brown, M. S. 1978. *Proc. Natl. Acad. Sci. USA* 75:5052–56
178. Kandutsch, A. A., Chen, H. W., Heiniger, H.-J. 1978. *Science* 201:498–501
179. Schroepfer, G. J. Jr., Pascal, R. A. Jr., Kandutsch, A. A. 1980. *Experientia* 36:518
180. Pascal, R. A. Jr., Schroepfer, G. J. Jr. 1980. *J. Lipid Res.* 21:118–22
181. Knuppen, R., Haupt, O., Breuer, H. 1967. *Biochem. J.* 105:971–78
182. Loke, K. H., Chon-Yong, G. 1968. *Steroids* 11:863–75
183. Starka, L., Gustafsson, J.-A., Sjovall, J., Knuppen, R. 1968. *FEBS Lett.* 1:269–71
184. Schroepfer, G. J. Jr., Parish, E. J., Chen, H. W., Kandutsch, A. A. 1976. *Fed. Proc.* 35:1697
185. Spike, T. E., Martin, J. A., Huntoon, S., Wang, A. H.-J., Knapp, F. F. Jr., Schroepfer, G. J. Jr. 1978. *Chem. Phys. Lipids* 21:31–58
186. Huntoon, S., Fourcans, B., Lutsky, B. N., Parish, E. J., Emery, H., Knapp, F. F. Jr., Schroepfer, G. J. Jr. 1978. *J. Biol. Chem.* 253:775–82
187. Phillips, G. N., Quiocho, F. A., Sass, R. L., Werness, P., Emery, H., Knapp, F. F. Jr., Schroepfer, G. J. Jr. 1976. *Bioorg. Chem.* 5:1–10
188. Tsuda, M., Parish, E. J., Schroepfer, G. J. Jr. 1979. *J. Org. Chem.* 44:1282–89
189. Raulston, D. L., Mishaw, C. O., Parish, E. J., Schroepfer, G. J. Jr. 1976. *Biochem. Biophys. Res. Commun.* 71:984–89
190. Parish, E. J., Schroepfer, G. J. Jr. 1976. *Tetrahedron Lett.* 3775–78

191. Parish, E. J., Newcomer, M. E., Gilliland, G. L., Quiocho, F. A., Schroepfer, G. J. Jr. 1976. *Tetrahedron Lett.* 4401–4

192. Parish, E. J., Spike, T. E., Schroepfer, G. J. Jr. 1977. *Chem. Phys. Lipids* 18: 233–39

193. Conner, B. N., Parish, E. J., Schroepfer, G. J. Jr., Quiocho, F. A. 1977. *Chem. Phys. Lipids* 18:240–57

194. Parish, E. J., Schroepfer, G. J. Jr. 1977. *Chem. Phys. Lipids* 18:258–66

195. Parish, E. J., Schroepfer, G. J. Jr. 1977. *Chem. Phys. Lipids* 19:107–13

196. Schroepfer, G. J. Jr., Parish, E. J., Kandutsch, A. A. 1977. *J. Am. Chem. Soc.* 99:5495–96

197. Schroepfer, G. J. Jr., Parish, E. J., Chen, H. W., Kandutsch, A. A. 1977. *J. Biol. Chem.* 252:8975–80

198. Schroepfer, G. J. Jr., Raulston, D. L., Kandutsch, A. A. 1977. *Biochem. Biophys. Res. Commun.* 79:406–10

199. Kisic, A., Monger, D., Parish, E. J., Satterfield, S., Raulston, D. L., Schroepfer, G. J. Jr. 1977. *Artery* 3:421–28

200. Gilliland, G. L., Newcomer, M. E., Parish, E. J., Schroepfer, G. J. Jr. Quiocho, F. A. 1977. *Acta Cryst.* B33: 3117–21

201. Kisic, A., Taylor, A. S., Chamberlain, J. S., Parish, E. J., Schroepfer, G. J. Jr. 1978. *Fed. Proc.* 37:1663

202. Schroepfer, G. J. Jr., Pascal, R. A. Jr., Kandutsch, A. A. 1979. *Biochem. Pharm.* 28:249–52

203. Schroepfer, G. J. Jr., Parish, E. J., Tsuda, M., Kandutsch, A. A. 1979. *Biochem. Biophys. Res. Commun.* 91: 606–13

204. Schroepfer, G. J. Jr., Parish, E. J., Kandutsch, A. A. 1979. *Chem. Phys. Lipids* 25:265–85

205. Monger, D. J., Parish, E. J., Schroepfer, G. J. Jr., Quiocho, F. A. 1980. *Acta Cryst.* B36:1460–66

206. Schroepfer, G. J. Jr., Parish, E. J., Pascal, R. A. Jr., Kandutsch, A. A. 1980. *J. Lipid Res.* 21:571–84

207. Parish, E. J., Tsuda, M., Schroepfer, G. J. Jr. 1979. *Chem. Phys. Lipids.* 24:209–36

208. Parish, E. J., Schroepfer, G. J. Jr. 1979. *Chem. Phys. Lipids.* 25:381–94

209. Parish, E. J., Schroepfer, G. J. Jr. 1980. *Chem. Phys. Lipids.* 26:141–47

210. Parish, E. J., Schroepfer, G. J. Jr. 1980. *Chem. Phys. Lipids.* 27:281–88

211. Parish, E. J., Schroepfer, G. J. Jr. 1980. *J. Org. Chem.* 45:4034–37

212. Schroepfer, G. J. Jr., Parish, E. J., Tsuda, M., Raulston, D. L., Kandutsch, A. A. 1979. *J. Lipid Res.* 20: 994–98

213. Schroepfer, G. J. Jr., Monger, D., Taylor, A. S., Chamberlain, J. S., Parish, E. J., Kisic, A., Kandutsch, A. A. 1977. *Biochem. Biophys. Res. Commun.* 78: 1227–33

214. Schroepfer, G. J. Jr., Walker, V., Parish, E. J., Kisic, A. 1980. *Biochem. Biophys. Res. Commun.* 93:813–18

215. Spike, T. E., Wang, H.-J., Paul, I. C., Schroepfer, G. J. Jr. 1974. *J. Chem. Soc. Chem. Commun.* 477–08

216. Martin, J. A., Huntoon, S., Schroepfer, G. J. Jr. 1970. *Biochem. Biophys. Res. Commun.* 39:1170–74

217. Gibbons, G. F., Mitropoulos, K. A., Pullinger, C. R. 1976. *Biochem. Biophys. Res. Commun.* 69:781–89

218. Schroepfer, G. J. Jr., Lutsky, B. N., Martin, J. A., Huntoon, S., Fourcans, B., Lee, W.-H., Vermilion, J. 1972. *Proc. R Soc. London Ser. B.* 180: 125–46

219. Huntoon, S., Schroepfer, G. J. Jr. 1970. *Biochem. Biophys. Res. Commun.* 40:476–80

220. Monger, D. J., Parish, E. J., Schroepfer, G. J. Jr. 1980. *J. Biol. Chem.* 255: 11122–29

221. Pascal, R. A. Jr., Schroepfer, G. J. Jr. 1980. *J. Biol. Chem.* 255:3565–70

222. Gibbons, G. F., Pullinger, C. R., Chen, H. W., Cavenee, W. K., Kandutsch, A. A. 1980. *J. Biol. Chem.* 255:395–400

223. Eriksson, H., Gustafsson, J.-A. 1971. *Eur. J. Biochem.* 20:231–36

224. Gustafsson, J.-A., Ingelan-Sundberg, M. 1976. *Eur. J. Biochem.* 64:35–43

225. Stanczyk, F. Z., Solomon, S. 1978. *Steroids* 31:627–43

226. McMorris, T. C. 1978. *Proc. R. Soc. London Ser. B.* 284:459–70

227. Wiersig, J. R., Waespe-Sarcevic, N., Djerassi, C. 1979. *J. Org. Chem.* 44: 3374–82

228. Kirson, I., Gottlieb, H. E., Glotter, E. 1980. *J. Chem. Res.(S).* p. 125

229. Djerassi, C., High, L. B., Fried, J., Sabo, E. F. 1955. *J. Am. Chem. Soc.* 77:3673–74

230. Ziegler, R., Tamm, C. 1976. *Helv. Chim. Acta* 59:1997–2011

231. Schroepfer, G. J. Jr., Pascal, R. A. Jr., Shaw, R., Kandutsch, A. A. 1978. *Biochem. Biophys. Res. Commun.* 83: 1024–31

232. Pascal, R. A. Jr., Shaw, R., Schroepfer, G. J. Jr. 1979. *J. Lipid Res.* 20:570–78

233. Miller, L. R., Izumi, A., Pinkerton, F. D. 1980. *Fed. Proc.* 39:1907

234. Havel, C., Hansbury, E., Scallen, T. J., Watson, J. A. 1979. *J. Biol. Chem.* 254:9573–82
235. Gaylor, J. L., Delwiche, C. V. 1976. *J. Biol. Chem.* 251:6638–45
236. Scallen, T. J., Dhar, A. K., Loughran, E. D. 1971. *J. Biol. Chem.* 246:3168–74
237. Witak, D. T., Parker, R. A., Brown, D. R., Dempsey, M. E., Ritter, M. C., Conner, W. E., Brahmankar, D. M. 1971. *J. Med. Chem.* 14:216–22
238. Erickson, S. K., Cooper, A. D., Matsui, S. M., Gould, R. G. 1977. *J. Biol. Chem.* 252:5186–93
239. Raulston, D. L., Miller, L. R., Schroepfer, G. J. Jr. 1980. *J. Biol. Chem.* 255:4706–9
240. Kirsten, E. S., Watson, J. A. 1974. *J. Biol. Chem.* 249:6104–9
241. Beirne, O. R., Heller, R., Watson, J. A. 1977. *J. Biol. Chem.* 252:950–54
242. Kandutsch, A. A., Chen, H. W., Shown, E. P. 1977. *Proc. Natl. Acad. Sci. USA* 74:2500–3
243. Kandutsch, A. A., Thompson, E. B. 1980. *J. Biol. Chem.* 255:10813–26
244. Chen, H. W., Kandutsch, A. A., Waymouth, C. 1974. *Nature* 251:419–21
245. Kandutsch, A. A., Chen, H. W. 1977. *J. Biol. Chem.* 252:409–15
246. Chen, H. W., Heiniger, H.-J., Kandutsch, A. A. 1977. *Exp. Cell Res.* 109:253–62
247. Cheng, K.-P., Nagano, H., Bang, L., Ourisson, G., Beck, J.-P. 1977. *J. Chem. Res.(S).* p. 217
248. Cheng, K.-P., Bang, L., Ourisson, G., Beck, J.-P. 1979. *J. Chem. Res.(S).* pp. 84–85
249. Nagano, H., Poyser, J. P., Cheng, K.-P., Bang, L., Ourisson, G., Beck, J.-P. 1977. *J. Chem. Res.(S)* p. 218
250. Astruc, M., Laporte, M., Tabacik, C., Crastes de Paulet, A. 1978. *Biochem. Biophys. Res. Commun.* 85:691–700
251. Sinensky, M. 1977. *Biochem. Biophys. Res. Commun.* 78:863–67
252. Cavenee, W. K., Gibbons, G. F., Chen, H. W., Kandutsch, A. A. 1979. *Biochim. Biophys. Acta* 575:255–65
253. Brown, M. S., Dana, S. E., Goldstein, J. L. 1975. *J. Biol. Chem.* 250:4025–27
254. Brown, M. S., Goldstein, J. L. 1975. *Cell* 6:307–16
254a. Drevon, C. A., Weinstein, D. B., Steinberg, D. 1980. *J. Biol. Chem.* 255:9128–37
255. Heiniger, H.-J., Kandutsch, A. A., Chen, H. W. 1977. *Nature* 263:515–17
256. Basu, S. K., Goldstein, J. L., Brown, M. S. 1978. *J. Biol. Chem.* 253:3852–56
257. Schneider, W. J., Basu, S. K., McPhaul, M. J., Goldstein, J. L., Brown, M. S. 1979. *Proc. Natl. Acad. Sci. USA* 76:5577–81
258. Goldstein, J. L., Brown, M. S., Krieger, M., Anderson, R. G., Mintz, B. 1979. *Proc. Natl. Acad. Sci. USA* 76:2843–47
259. Gordon, L. I., Bass, J., Yachnin, S. 1980. *Proc. Natl. Acad. Sci. USA* 77:4313–16
260. Schroepfer, G. J. Jr., Parish, E. J., Gilliland, G. L., Newcomer, M. E., Sommerville, L. L., Quiocho, F. A., Kandutsch, A. A. 1978. *Biochem. Biophys. Res. Commun.* 84:823–29
261. Newcomer, M. E., Gilliland, G. L., Parish, E. J., Schroepfer, G. J. Jr., Quiocho, F. A. 1980. *Chem. Phys. Lipids* 26:249–58
262. Parish, E. J., Tsuda, M., Schroepfer, G. J. Jr. 1979. *Chem. Phys. Lipids* 25:111–24
263. Stoffel, W., Klotzbucher, R. 1978. *Hoppe-Seylers Z. Physiol. Chem.* 359:199–209
264. Heller, R. A., Klotzbucher, R., Stoffel, W. 1979. *Proc. Natl. Acad. Sci. USA* 76:1721–25
265. Khuong-Huu, Q., Letourneux, Y., Gut, M., Goutarel, R. 1974. *J. Org. Chem.* 39:1065–68
266. Gut, M., Letourneux, Y., Story, J. A., Tepper, S. A., Kritchevsky, D. 1974. *Experientia* 30:1325–26
267. Endo, A., Kuroda, M., Tsujita, Y. 1976. *J. Antibiot.* 29:1346–48
268. Brown, A. G., Smale, T. C., King, T. J., Hasenkamp, R., Thompson, R. H. 1976. *J. Chem. Soc. Perkin Trans. 1* pp. 1165–73
269. Endo, A., Kuroda, M., Tanzawa, K. 1977. *FEBS Lett.* 72:323–26
270. Endo, A., Tsujita, Y., Kuroda, M., Tanzawa, K. 1977. *Eur. J. Biochem.* 77:31–36
271. Tanzawa, K., Endo, A. 1979. *Eur. J. Biochem.* 98:195–201
272. Doi, O., Endo, A. 1978. *Jpn. J. Med. Sci. Biol.* 31:225–33
273. Kaneko, I., Hazama-Shimada, Y., Endo, A. 1978. *Eur. J. Biochem.* 87:313–21
274. Goldstein, J. L., Helgeson, J. A. S., Brown, M. S. 1979. *J. Biol. Chem.* 254:5403–9
275. Kuroda, M., Tsujita, Y., Tanzawa, K., Endo, A. 1979. *Lipids* 14:555–89
276. Tsujita, Y., Kuroda, M., Tanzawa, K., Kitano, N., Endo, A. 1979. *Atherosclerosis* 32:307–13
277. Yamamoto, A., Sudo, H., Endo, A. 1980. *Atherosclerosis* 35:259–66

278. Endo, A., Tsujita, Y., Kuroda, M., Tanzawa, K. 1979. *Biochim. Biophys. Acta* 575:266–76
279. Kuroda, M., Tanzawa, K., Tsujita, Y., Endo, A. 1977. *Biochim. Biophys. Acta* 489:119–25
280. Fears, R., Richards, D. H., Ferres, H. 1980. *Atherosclerosis* 35:439–49
281. Alberts, A. W., Chen, J., Kuron, G., Hunt, V., Huff, J., Hoffman, C., Rothrock, J., Lopez, M., Joshua, H., Harris, E., Patchett, A., Monaghan, R., Currie, S., Stapley, E., Albers-Schonberg, G., Hensens, O., Hirschfield, J., Hoogsteen, K., Liesch, J., Springer, J. 1980. *Proc. Natl. Acad. Sci. USA* 77:3957–61
282. Endo, A. 1979. *J. Antibiot.* 32:852–54
283. Ottery, F. D., Goldfarb, S. 1976. *FEBS Lett.* 64:346–49
284. Volpe, J. V. 1979. *J. Biol. Chem.* 254:2568–71
285. Chang, T.-Y., Schiavoni, E. S. Jr., McCrae, K. R., Nelson, J. A., Spencer, T. A. 1979. *J. Biol. Chem.* 254:11258–63
286. Nelson, J. A., Czarny, M. R., Spencer, T. A., Limanek, J. S., McCrae, K. R., Chang, T. Y. 1978. *J. Am. Chem. Soc.* 100:4900–2
287. Guyer, K. E., Boots, S. G., Marecki, P. E., Boots, M. R. 1976. *J. Pharm. Sci.* 65:548–552
288. Boots, S. G., Boots, M. R., Guyer, K. E., Marecki, P. E. 1976. *J. Pharm. Sci.* 65:1374–80
289. Beg, Z. H., Lupien, P. J. 1972. *Biochim. Biophys. Acta* 260:439–48
290. White, L. W. 1971. *J. Pharmacol. Exp. Therapeut.* 178:361–70
291. Cohen, B. I., Raicht, R. F., Shefer, S., Mosbach, E. H. 1974. *Biochim. Biophys. Acta* 369:79–85
292. Berndt, J., Gaumert, R., Still, J. 1978. *Atherosclerosis* 30:147–52
293. Ranganathan, S., Ramasarma, T. 1973. *Biochem. J.* 134:737–43
294. Lee, K.-H., Hall, I. H., Mar, E.-C., Starnes, C. O., El Gebaly, S. A., Waddell, T. G., Ruffner, C. G., Weidner, I. 1977. *Science* 196:533–36
295. Hulcher, F. H. 1971. *Arch. Biochem. Biophys.* 146:422–27
296. Diaz-Zagoya, J. C., Hurtado, M. E., Gonzalez, J. 1976. *Experientia* 32:1138–40
297. Hall, I. H., Carlson, G. L. 1976. *J. Med. Chem.* 19:1257–61
298. Baggaley, K. H., Atkin, S. E., English, P. D., Hindley, R. M., Morgan, B., Green, J. 1975. *Biochem. Pharmacol.* 24:1902–3
299. Friedman, S. J., Skehan, P. 1979. *FEBS Lett.* 102:235–40
300. Allan, W. C., Volpe, J. J. 1979. *Pediatr. Res.* 13:1121–24
301. Paley, L., Sabine, J. R. 1971. *Aust. J. Biol. Sci.* 24:1125–30
302. Olson, R. J., Trumble, T. E., Gamble, W. 1974. *Biochem. J.* 142:445–48
303. Bloxham, D. P. 1975. *Biochem. J.* 147:531–39
304. Bhat, C. S., Ramasarma, T. 1979. *Biochem. J.* 181:143–51
305. Ortiz de Montellano, P. R., Wei, J. S., Castillo, R., Hsu, C. K., Boparai, A. 1977. *J. Med. Chem.* 20:243–49
306. Brown, M. S., Goldstein, J. L. 1974. *Proc. Natl. Acad. Sci. USA* 71:788–92
307. Brown, M. S., Goldstein, J. L. 1974. *Science* 185:61–63
308. Goldstein, J. L., Brown, M. S. 1973. *Proc. Natl. Acad. Sci. USA* 70:2804–8
309. Brown, M. S., Goldstein, J. L. 1976. *Science* 191:150–54
310. Brown, M. S., Goldstein, J. L. 1976. *Trends Biochem. Sci.* 1:193–95
311. Goldstein, J. L., Brown, M. S. 1977. *Ann. Rev. Biochem.* 46:897–930
312. Goldstein, J. L., Brown, M. S. 1977. *Metabolism* 26:1257–75
313. Goldstein, J. L., Brown, M. S. 1979. *Ann. Rev. Genet.* 13:259–89
314. Brown, M. S., Goldstein, J. L. 1979. *Proc. Natl. Acad. Sci. USA* 76:3330–37
315. Brown, M. S., Kovanen, P. T., Goldstein, J. L. 1980. *NY Acad. Sci.* 348:48–68
316. Wintersteiner, O., Ritzmann, J. R. 1940. *J. Biol. Chem.* 136:697–707
317. Fieser, L. F., Fieser, M. 1959. *Steroids,* pp. 154–57. New York: Reinhold. 945 pp.
318. Schuh, J., Fairclough, G. F. Jr., Hashemeyer, R. H. 1978. *Proc. Natl. Acad. Sci. USA* 75:3173–77
319. Lee, D. M. 1980. *Biochem. Biophys. Res. Commun.* 95:1163–72
320. Brown, M. S., Dana, S. E., Goldstein, J. L. 1974. *J. Biol. Chem.* 249:789–96
321. Krieger, M., Brown, M. S., Faust, J. R., Goldstein, J. L. 1978. *J. Biol. Chem.* 253:4093–4101
322. Bjorkhem, I., Gustafsson, J. 1974. *J. Biol. Chem.* 249:2528–35
323. Aringer, L., Eneroth, P., Nordstrom, L. 1976. *J. Lipid Res.* 17:263–72
324. Assmann, G., Fredrickson, D. S., Sloan, H. R., Fales, H. M., Highet, R. J. 1975. *J. Lipid Res.* 16:28–38
325. Fredrickson, D. S., Ferrans, V. J. 1978. In *The Metabolic Basis of Inherited Disease,* ed. J. B. Stanbury, J. B.

Wyngaarden, D. S. Fredrickson, 32: 670–87. New York: McGraw-Hill. 1862 pp.

326. Goldstein, J. L., Brunschede, G. Y., Brown, M. S. 1975. *J. Biol. Chem.* 250:7854–62

327. Basu, S. K., Goldstein, J. L., Anderson, R. G. W., Brown, M. S. 1976. *Proc. Natl. Acad. Sci. USA* 73:3178–82

328. Mahley, R. W., Innerarity, T. L., Pitas, R. E., Weisgraber, K. H., Brown, J. H., Gross, E. 1977. *J. Biol. Chem.* 252: 7279–87

329. Bersot, T. P., Mahley, R. W., Brown, M. S., Goldstein, J. L. 1976. *J. Biol. Chem.* 251:2395–98

330. Brown, M. S., Deuel, T. F., Basu, S. K., Goldstein, J. L. 1978. *J. Supramol. Stuct.* 8:223–34

331. Pitas, R. E., Innerarity, T. L., Arnold, K. S., Mahley, R. W. 1979. *Proc. Natl. Acad. Sci. USA* 76:2311–15

332. Mahley, R. W., Weisgraber, K. H., Innerarity, T. L. 1979. *Biochim. Biophys. Acta* 575:81–91

Ann. Rev. Biochem. 1981. 50:623–55
Copyright © 1981 by Annual Reviews Inc. All rights reserved

PROTON-TRANSLOCATING CYTOCHROME COMPLEXES[1]

♦12090

Mårten Wikström, Klaas Krab, and Matti Saraste

Department of Medical Chemistry, University of Helsinki, SF-00170
Helsinki 17, Finland

CONTENTS

[1]The following abbreviations are used: a, a_3, b, b-562, b-566, b_{50}, c, c_1, c_2, the heme group of respective cytochromes, or the cytochrome itself; C side, cytoplasmic side of the mitochondrial membrane; E_h, oxidation-reduction potential relative to the standard hydrogen electrode; ΔE_h, difference in E_h or redox potential span; $E_{m,n}$, midpoint oxidation-reduction potential at pH=n; Fe/S, Rieske's iron-sulfur protein or center; F_o, H^+-conducting membrane segment of H^+-ATPase; H^+/e^-, stoichiometry of H^+ translocation per transferred electron; $\Delta \mu_H$, electrochemical proton gradient (C side minus M side); M side, matrix side of the mitochondrial membrane; $QH_2,QH\cdot,Q$, reduced, semiquinone, and oxidized forms of ubiquinone-10, respectively; Q_{II}, so-called secondary quinone (ubiquinone-10) in photosynthetic bacteria; q^+ /e^-, stoichiometry of translocation of electrical charge (either positive from the M to the C side, or negative from the C to the M side of the membrane) per transferred electron; UHDBT, 5-n-undecyl-6-hydroxy-4,7-dioxobenzothiazole; $ZH_2,ZH\cdot,Z$, reduced, semiquinone, and oxidized forms of a particular protein-bound ubiquinone-10 in bc-type complexes.

623

0066-4154/81/0701-0623$01.00

I. PERSPECTIVES AND SUMMARY

Respiratory and photosynthetic electron transfer chains of mitochondria, bacteria, and chloroplasts include integral membrane protein complexes with heme as a prosthetic group. Of these so-called cytochrome complexes (that also contain other redox centers, *e.g.* nonheme iron and copper), the *bc*- and aa_3-type complexes are the most prominent. Isolation and reconstitution of these complexes into liposome membranes has revealed that redox-linked proton translocation is one of their intrinsic catalytic properties. Energy liberated in electron transfer is therefore transduced into an electrochemical proton gradient across the membrane that subsequently may be utilized to drive either ATP synthesis or ion transport.

Although it is widely agreed that redox energy is transduced in this fashion, the molecular mechanisms of electron transfer and proton translocation, as well as their mutual coupling, are still obscure. Current research is comparable to modern enzymology in general, in which elucidation of protein structure and topography is an essential part. Yet research on the redox-linked energy transducers has specific methodological and conceptual features, in part due to the nature of the redox catalysts and in part to the integration of these enzymes into membranes. The latter poses specific problems in structural studies, but also provides interesting vectorial aspects to the catalytic mechanisms.

Considerable progress has recently been made, particularly in the studies of the aa_3-type cytochrome oxidase complex, but also with the cytochrome bc_1 and bc_2 complexes of mitochondria and photosynthetic bacteria, respectively. In cytochrome oxidase, H+ translocation seems to be specifically linked to oxidoreduction of cytochrome *a,* whereas heme a_3 and copper form a binuclear center of O_2 reduction to water. Kinetic and structural evidence indicates that the energy transduction may be realized in this complex by an anticooperative or reciprocal interaction between cytochromes *a* in an $(aa_3)_2$ dimer. An analogous model has also been proposed for the function of the bc_1 complex, but a more direct redox loop type of coupling between electron transfer and proton translocation, as proposed in the chemiosmotic theory, cannot be excluded in this case.

It is possible that the principle of reciprocal interactions between similar catalytic units is a general one among energy-transducing membrane proteins. Very similar properties have been described for the H$^+$-translocating ATPase (ATP synthase) of mitochondria and chloroplasts, and for the Na$^+$/K$^+$-ATPase. Taken together, these findings indicate that the previously recognized phenomena of anticooperativity, "half-of-the-sites" reactivity, and "alternating site" reactivity may have found a particularly important application in membraneous energy-transducing enzymes.

II. INTRODUCTION

As appropriately stated by Klingenberg (1), ". . . research on chemiosmotic aspects of energy transduction has passed its peak. The main thrust is now to investigate the molecular mechanisms by which the catalysts, *i.e.* membrane proteins, contribute to energy transduction."

General consensus has been reached on the central chemiosmotic dogma, namely the coupling of oxidoreductions to phosphorylation via protonic currents (2–8). Whether the nature of these currents is purely osmotic as proposed by Mitchell (2, 3, 9), more localized to the energy-transducing membrane (10–18), or both (19, 20), the main function of the membraneous cytochrome complexes is nevertheless to generate such currents by redox-linked H$^+$ translocation. Since the molecular mechanisms of generation, transmission, and utilization of these H$^+$ currents are not yet solved (21), it is clear that the mechanism of oxidative and photosynthetic phosphorylation is also not solved on the biochemical/biophysical level of structure and function.

In view of the large number of comprehensive reviews in this area recently and the voluminous literature, it is impossible to attempt a complete account in the limited space available. This paper is therefore restricted to the most recent research. However, we attempt to provide the reader with brief surveys of necessary background information, including frequent citations to other reviews, to facilitate acquisition of a more complete picture.

III. THERMODYNAMIC LIMITS TO MECHANISTIC H$^+$/e$^-$ STOICHIOMETRIES

There has been much work on the stoichiometric aspects of redox-linked proton translocation. The reaction stoichiometry sets limits to possible mechanisms and relates directly to thermodynamic efficiency of the process. Unfortunately these studies have, in part, resulted in unresolved controversies. These are considered in recent reviews (22–26), but only briefly here.

The thermodynamics of redox-linked proton translocation is governed by two main parameters, namely the redox potential difference across which electron transfer occurs (ΔE_h) and the electrochemical proton gradient $(\Delta \mu_{H^+})$ against which the protons are translocated [see (27) for an eloquent review]. The relative magnitude of ΔE_h and $\Delta \mu_{H^+}$ obviously sets an ultimate upper limit of possible H^+/e^- stoichiometries that cannot be reached in any real process. Unfortunately, the range of determined $\Delta \mu_{H^+}$ values by ion distribution and other techniques is wide, varying between 160 and 230 mV (28, 29) in "energized" or State 4 [see (30) for nomenclature] mitochondria. This diminishes the value of the thermodynamic approach. However, it has been less widely recognized that an estimate of $\Delta \mu_{H^+}$ may also be obtained indirectly, but independently, from measured redox equilibria in the respiratory chain under energized conditions. Such data may therefore help to define a minimum $\Delta \mu_{H^+}$ that may be used to put proper thermodynamic limits to H^+/e^- stoichiometries.

Table 1 summarizes the observed stoichiometries of H^+ and charge translocation in three segments of the respiratory chain. The NADH-ubiquinone reductase segment (31) is included for analytical purposes (see below). The table provides standards on which the analysis can be based, namely the generally accepted stoichiometry of one translocated H^+/e^- in the ubiquinol-cytochrome c reductase segment and the maximal stoichiometry of $2H^+/e^-$ in the NADH-ubiquinone segment.

Potentiometric titrations of the b cytochromes in ATP-supplemented mitochondria show that half of the b complement attains an apparent $E_{m,7}$ that is raised by at least 205 mV from the corresponding $E_{m,7}$ in uncoupled mitochondria (14, 44). This effect has been discussed extensively (14, 44–47), but there is little doubt that it is imposed by ATP hydrolysis on the respiratory chain via $\Delta \mu_{H^+}$, i.e. by reversal of the H^+-translocating

Table 1 Stoichiometry of proton translocation in the respiratory chain

Segment	H^+/e^{-a}	q^+/e^{-a}	References
I. NADH-ubiquinone	1	1	31–34
	2	2	35, 36
II. Ubiquinol-cytochrome c	2^b	1	22–26, 32, 35, 37, 38
III. Cytochrome c-O_2	0	1	3, 22, 39–41
	1	2	24–26
	2	3	37, 42, 43

[a] H^+/e^- and q^+/e^- refer to the stoichiometries of protons released on the outside of the mitochondrion and the number of positive electrical charges translocated outward (or negative charge inward), respectively.

[b] Although $2H^+/e^-$ are released from mitochondria, only one H^+/e^- is translocated electrogenically; the other is released on oxidation of ubiquinol by an electron acceptor.

mechanism. Since the H$^+$/e$^-$ stoichiometry is unity in this segment (Table 1), it follows that $\Delta\mu_{H^+}$ must be at least 200 mV under these conditions.

A second estimate may be obtained from data of Klingenberg & Schollmeyer (48). It may be calculated that ΔE_h between cytochrome c and the NADH/NAD couple was 595 mV (assuming pH 8 in the matrix space; see below) and exactly balanced by $\Delta\mu_{H^+}$ in their ATP-supplemented mitochondria. With an H$^+$/e$^-$ quotient of 1 for segment II and a maximal quotient of 2 for segment I (see Table 1), it follows that the minimum $\Delta\mu_{H^+}$ must be 200 mV under these conditions. Obviously, both these estimates are based on the chemiosmotic notion that $\Delta\mu_{H^+}$ is the intermediate form of energy between ATP hydrolysis and the redox reactions of the respiratory chain.

It may be concluded that the minimum $\Delta\mu_{H^+}$ is 200 mV in energized mitochondria, which is roughly the mean of values obtained from more direct measurements (see 28, 29). This number may now be used to define the thermodynamic limits of the H$^+$/e$^-$ stoichiometries in the cytochrome segments of the respiratory chain.

1. The Segment between Succinate and O$_2$

The thermodynamic analysis may be performed with respect to the matrix or M side of the membrane in which the active site of succinate dehydrogenase is located (49, 50). In State 4, M-phase pH may be assumed to be 8 (e.g. 51, 52). Due to the equilibrium constant of 4 for the fumarase reaction, E_h for the succinate/fumarate couple may be estimated to be about -19 mV under these conditions. The corresponding E_h of the ½ O$_2$/H$_2$O couple [air-saturated aqueous medium; 25°C; pH 8 (53, 54)] is then about 740 mV, which yields a ΔE_h of 760 mV for the succinate-O$_2$ span.[2] With the minimum $\Delta\mu_{H^+}$ of 200 mV (see above) it follows that the ultimate upper limit for the H$^+$/e$^-$ ratio is 3.55. This clearly disagrees with a claimed stoichiometry of 4 in this segment (55, 56).

2. The Segment between Cytochrome c and O$_2$

Since cytochrome c is located on the cytoplasmic or C side of the membrane (49), a separate thermodynamic analysis of this segment may be made with respect to this phase. The overall reaction is $1\,e^- + 1\,H_C^+ + ¼\,O_2 \rightarrow H_2O$, by which it is emphasized that the analysis requires the substrate proton

---

[2]As shown by Oshino et al (54a) electron transfer and energy coupling still takes place with [O$_2$] in the submicromolar range. ΔE_h may therefore be much lower.

(H_C^+) to be taken up from the C phase. However, since it is generally believed that this proton is taken from the M phase [irrespective of the particular model used or of whether the oxidase pumps protons or not; (26, 32, 38, 40, 42, 43)], it follows that the reaction, as analyzed, is at least associated with translocation of $1H^+/e^-$ across the membrane. In this thermodynamic treatment, therefore, the H^+/e^- quotient in the different proposals (see Table 1, *III*) equals the mechanistic q^+/e^- ratios.

Ferrocyanide may be used successfully to donate electrons directly to cytochrome *c*, thus initiating electron transfer in the terminal segment of the chain (57, 58). A typical E_h of the ferrocyanide/ferricyanide couple in State 4 may be estimated to be 250 mV or higher [$E_{m,7}$ =355 mV at 25°C and zero ionic strength (59); ferrocyanide/ferricyanide ratio $\simeq$100]. E_h of the ½ O_2/H_2O couple is 800 mV at pH 7 (see above). Hence a typical ΔE_h is <550 mV for this segment.[2] Division with the minimum $\Delta \mu_{H^+}$ derived above shows that the thermodynamic H^+/e^- ratio (or the q^+/e^- ratio; Table 1) cannot be higher than 2.75. This would again exclude the highest proposed stoichiometry for this segment.

However, the H^+/e^- stoichiometries are usually measured under conditions where $\Delta \mu_{H^+}$ is minimal, to minimize backflux of H^+. A priori it is therefore possible that a higher H^+/e^- may be observed than anticipated from thermodynamic limits calculated for State 4. But this would require that the H^+-translocating mechanism can attain variable H^+/e^- stoichiometries depending on the magnitude of $\Delta \mu_{H^+}$. Data of Brand et al (60) suggest, in contrast, that the H^+/e^- ratios are constant over a wide range of electrochemical proton gradients. Possible reasons for the large discrepancies between observed H^+/e^- and q^+/e^- ratios in Table 1 have been recently discussed (24–26).

IV. STRUCTURE OF bc_1 AND aa_3 COMPLEXES

The bc_1 and aa_3 complexes of mitochondria [Complex III and IV, respectively (61–67)] are remarkably similar structurally. These multisubunit enzymes are composed of polypeptides of both mitochondrial and cytoplasmic origin. Both traverse the inner mitochondrial membrane completely and are similar in shape. The question is whether this general structural analogy can be extended to molecular mechanisms.

A more detailed structural account than can be given here [including determinations of heme orientation in the aa_3 complex (68)] may be found in recent reviews of the bc_1 (62–65) and aa_3 (66–68) complexes, respectively.

1. Quaternary Structure

Proposed subunits of the isolated bc_1 complex are frequently numbered I–VIII, but similarly to the case of the isolated aa_3 complex, this does not mean that the preparations contain only the numbered polypeptides (69, 70). Heterogeneities of bands in SDS-polyacrylamide electrophoresis corresponding to the smallest polypeptides have been demonstrated (Table 2; 70–72).

Table 2 summarizes present information on polypeptides of bc_1 and aa_3 complexes. Relative amounts of proposed subunits are controversial for the bc_1 complex, but at least apocytochrome b seems to be present in two copies per monomer, in agreement with spectroscopic data (cf 62). The sum of molecular weights of proposed subunits in appropriate ratios agrees fairly well with the monomer size as calculated on the basis of cytochrome content per protein in the bc_1 complex (Table 3A).

Cytochrome oxidase is proposed to contain seven different subunits in 1 : 1 ratios in both *Bovis* and *Neurospora* mitochondria (Table 2B). However,

Table 2 The proposed subunits of the bc_1 and aa_3 complexes[a]

Subunit	Bovis		Neurospora	
	Apparent mol wt (kD)	Relative stoichiometry	Apparent mol wt (kD)	Relative stoichiometry
A. bc_1 complex				
I. (core protein 1)	46	1	52	1
II. (core protein 2)	43	1	45	2
IV. (cytochrome c_1)	29	1	31	1
III. (cytochrome b)	28	2	30	2
V. (Rieske's Fe/S)	24	1	25	1
VI.	12	2	14	1
VII.	8	2	12	1
VIII. (heterogeneous)	~6	~2	9	1
B. aa_3 complex				
I. (mitochondrial)	35	1	40	1
II. (mitochondrial; Cu?)	26	1	29	1
III. (mitochondrial; H$^+$?)	21	1?	21	1
IV.	17	1	18	1
V.	12	1	14	1
VI.	11	1	12	1
VII. (heterogeneous)	5	>1	9	1

[a] Data for bc_1 complex of *Bovis* and *Neurospora* from (70, 73), respectively. The subunit composition of *Bovis* and *Neurospora* aa_3 complex from (74, 75) and 73, 76), respectively. Relative stoichiometries from (77–79) and (80), respectively. Cu? and H$^+$? refer to proposed involvement in binding of copper and proton translocation, respectively (see text).

the relative amount of subunit III is not yet certain for the mammalian enzyme. Subunit composition of the aa_3 monomer has been subject to some controversy (78, 82–85). This was due to heterogeneity, both with respect to contaminating polypeptides (74, 75, 82) and to more or less extensive loss of prosthetic groups during purification (78, 82). This easily leads to large discrepancies between molecular size as determined from heme/protein ratios, the sum of subunit molecular weights, and hydrodynamic parameters. Until recently, such discrepancies have always been encountered with purified aa_3 preparations. A homogeneous aa_3 preparation was recently described in which these discrepancies were absent (78; Table 3B). The evidence is now strongly in favor of the suggestion that the aa_3 monomer is composed of one copy of each subunit (but with some remaining uncertainty regarding subunit III). There is no doubt that purified homogeneous aa_3 preparations are dimeric in Triton X-100 (Table 3B).

Functional significance of all proposed subunits in bc_1 and aa_3 complexes has by no means been demonstrated. However, in bc_1 such significance is almost certainly linked to subunits III–V, which are associated with redox centers (Table 2A). In aa_3, subunits I–III are also likely to have this status, since they are mitochondrially synthesized polypeptides (86–89). Moreover, the aa_3 complex of *Paracoccus denitrificans,* which in the isolated state appears to be identical to the mitochondrial enzymes with respect to spectra and electron transfer activity, contains only two different subunits (90) that may be closely related to subunits I and II of the mitochondrial enzyme (91, 92). Steffens & Buse (93) have shown that subunit II includes an amino acid

Table 3 Molecular weights of cytochrome bc_1 and aa_3 complexes

Enzyme complex	Molecular weight (kDalton)			
	From subunit composition (see Table 2)	From concentration of prosthetic groups	From hydrodynamic measurements[a]	Remark
A. bc_1 complex				
Bovis[b]	250	240–250 (70)	440–440 (81)	dimeric
Neurospora	290	285 (73)	550 (73)	dimeric
B. aa_3 complex				
Bovis	105[c] (126)	100–110[c] (78)	210[c] (78)	dimeric
Neurospora	145	135–140 (76)	300–340 (73)	dimeric
		220 (73)		

[a] All hydrodynamic molecular weights are corrected for bound detergent (Triton X-100). The oligomeric state of the complex also refers to a solution with this detergent.
[b] The bc_1 preparation from *Bovis* in (81) lacks Rieske's iron-sulfur protein (subunit V).
[c] Values for a homogeneous aa_3 preparation that lacks subunit III (78).

sequence that is closely homologous to certain copper proteins. More direct evidence favoring involvement of subunit III in H^+ translocation by mammalian aa_3 has been obtained by Casey et al (94) on the basis of inhibition and binding studies using dicyclohexylcarbodiimide (DCCD; see Section V.3.).

The heterogeneity of the smallest polypeptides of both bc_1 and aa_3 complexes (see above) could mean that they may include some split-off "addresses" of precursors to cytoplasmically synthesized subunits that may still be associated with the enzyme (see 95).

2. Topography of the Membrane-Bound Complexes

Both the bc_1 and the aa_3 complex form lateral crystals under suitable conditions (62, 96–98). These crystalline preparations have been studied by electron microscopy, and computer-aided image analysis has yielded models of the general topography of the membrane-bound complexes. Both bc_1 (99) and aa_3 (98) apparently form dimers with a symmetry axis perpendicular to the membrane plane in crystal forms obtained by the aid of Triton X-100 (see Table 3). The proteins traverse the membrane completely, dividing the complex into two aqueous and one hydrophobic domain. The shape of both putative bc_1 and aa_3 monomers is similar. Both are elongated structures protruding far into the aqueous phases (bc_1 protrudes on both sides and aa_3 mainly on one side of the membrane). The shape of the proposed aa_3 monomer as seen in crystal structures formed with the aid of deoxycholate has been described as a lopsided "Y" (100). Frey et al (101) showed elegantly by antibody-labeling that the aa_3 complex protrudes far out of the membrane on the C side. The corresponding orientation is not known for bc_1. The overall length of the bc_1 and aa_3 monomers are ~15 (99) and 11 nm (98), respectively.

The topographical distribution of individual subunits in isolated and membrane-bound complexes has been studied by surface labeling techniques that use both hydrophilic and hydrophobic reagents (102–111) and antibodies (112, 113). Even though the picture obtained from these studies is still obscure, some general features are beginning to emerge (see 66, 83, 114).

It has been proposed that aa_3 may be dimeric also in situ (98), as it is in purified preparations dispersed in Triton (Table 3) and in so-called Triton-crystals (98). Evidence for $(aa_3)_2$ being the functional unit has been entirely lacking until recently. Bisson & Capaldi (115) showed that covalent cross-linking of cytochrome c to the enzyme blocks electron transfer maximally at a titer of one c per dimer. Wikström et al (114, 116) have recently suggested a catalytic mechanism (Section VII.3) in which the functional unit would be the $(aa_3)_2$ dimer.

V. MOLECULAR REQUIREMENTS OF REDOX-LINKED PROTON TRANSLOCATION

1. Direct or Indirect Coupling

Distinction has been made between direct and indirect mechanisms of coupling between electron transfer and proton translocation (9). The former is represented by Mitchell's concept of the redox loop (3) and by the more general idea of vectorial metabolism and group translocation (117–119). In such a mechanism H^+ is translocated together with the electron, in that the proton is covalently bonded to the (reduced) redox carrier during translocation. The redox carrier must consequently be of the H-transfer kind (e.g. ubiquinone). This electroneutral step is then supplemented with translocation of the electron back in the opposite direction, catalyzed by classical electron-carriers (e.g. cytochromes) in a way foreshadowed by Lundegårdh (120), but first proposed for mitochondrial electron transport by Robertson (121). However, the electron translocation need not occur over a large distance, but could take place more locally, in which case H^+ transfer to the appropriate redox center would require a channel or a well in the catalytic protein (see 9, 122–125).

Distinction between directly and so-called indirectly coupled mechanisms is not sharp (25). The only difference is that in the latter, H^+ is not directly bound to the (H-transferring) redox center during translocation, but to some other site the function of which is linked to the redox reaction. This site may be far from or close to the redox center, and if it is close, e.g. on heme's axial ligand (26, 126), the situation closely resembles the directly coupled case (25). Therefore, the strong polarization of the field with respect to this distinction may be unnecessary and fruitless. At present it seems that nature may utilize both types of mechanism. Thus, primary light-induced proton translocation in photosynthesis is almost certainly directly coupled (127), while mitochondrial aa_3 (26) and transhydrogenase (128, 129) are almost certainly indirectly coupled systems, using the original definition of these terms.

2. Basic Elements of Redox-Linked Proton Pumps

The basic properties of a redox-linked proton pump that uses a classical electron carrier as redox element have recently been discussed (26, 130). Figure 1 shows a cubic scheme (25) that defines the eight different states that must minimally be considered. The indicated catalytic path is, of course, only one of several possibilities, the correct one of which can be determined only by kinetic data, but not on the basis of bulk thermodynamic properties (cf 131).

This model, though elementary, nicely illustrates some important properties of a proton pump. For instance, two of the four possible translocational or "reorientation" (flip/flop) steps must be forbidden or have low probability in order to prevent short-circuiting and uncoupling. Deviations from an integer mechanistic H$^+$/e$^-$ stoichiometry can occur only when this probability becomes significant. Otherwise the stoichiometry will be independent of changes in pH or electrical potential on either side of the transducer (26, 130). Figure 1 is a direct application of the analysis by De Vault (132), of redox-linked energy transducers, to the special case of a proton pump. Related schemes have been presented by Läuger (133) and by Boyer (134) for redox- and ATP-linked proton pumps, respectively. (See also Section VII.3 for a discussion of electron/H$^+$ coupling.)

The translocational or "reorientation" steps in Figure 1 are essential, and distinguish the pump from mere redox-linked protolytic effects [so-called redox Bohr effects (see 38, 135)], which by themselves cannot account for proton translocation. As shown in Figure 1, a minimal scheme of a redox-linked proton pump is made up of two electronically isolated redox Bohr effects that are interconnected by reorientation steps. The latter need not be extensive in molecular terms, but may involve only subtle changes in position of the H$^+$-carrying residues with respect to proton-conducting channels to the two aqueous phases (see below). Alternatively, the "Bohr residue" may be stationary with respect to the protein frame, and protonic contact is established alternatively to the two sides by controlled opening and closing ("gating") of the proton channels.

Figure 1 may be useful in that it provides several predictions that may be tested experimentally. Some of these have been observed for the aa_3 proton pump [(26, 114, 116, 136–138); see also Sections VII.2 and 3].

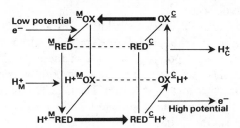

Figure 1 Cubic scheme of eight minimal states of a redox-linked proton pump [adapted from (26)]. M and C indicate protonic contact with the respective sides of the membrane. Thick arrows denote transitions between such M and C states, as well as between redox contacts with electron input and output ends (low and high potential), respectively. Dotted lines denote "forbidden" transitions; arrows and thin lines denote "allowed" transitions. The pathway marked by arrows denotes one possible sequence of catalytic events.

3. Proton Channels

Since several redox centers are not in direct contact with the aqueous phase in the membraneous bc_1 and aa_3 complexes [e.g. heme a_3 (see 139, 140)], proton-conducting pathways in the protein are clearly required whether the H^+/e^- coupling is direct or indirect [see Section V.1 and (26)]. Although such channels may thus be anticipated for theoretical reasons, evidence for their occurrence in redox complexes has been found only recently. Casey et al (94) and more recently Steffens et al (141) have shown that DCCD blocks H^+ translocation by the aa_3 system, probably as a result of specific binding to one carboxylic group of subunit III (cf section IV). This points to inhibition of a H^+-conducting channel, based on the analogy with the established effect of DCCD on the membrane segment (F_o) of the H^+-translocating ATPase (for review see 142). Further support for this conclusion comes from the fact that an aa_3 preparation free of subunit III retains normal spectra of the redox centers and electron transfer activity, but seems to lack the proton-translocating property (78). Possible mechanisms of proton conductance through such a channel may therefore be discussed using the DCCD-binding protein of H^+-ATPase (143) as a model.

It is generally agreed that high proton conductance in a protein may be provided by a chain of hydrogen bonds through which H^+ may move as fast as it moves in ice (see 144–148) by quantum mechanical tunneling. Three kinds of hydrogen-bonded devices have been proposed: (a) a network of protein-bound "crystalline" water (12, 145), (b) hydrogen bonds between peptide linkages of an α-helix (149), and (c) hydrogen bonds between amino acid side chains (hydroxyls) in β-sheets (144) or in α-helices (150, 151). The variety of alternatives suggests that proton conductivity may be the rule rather than the exception in proteins. If so, the problem in H^+-translocating complexes is how to achieve the strict specificity of proton conductance required in these devices.

The F_o domain of ATP synthase is known to induce DCCD-sensitive proton leakage in a phospholipid bilayer (see 152). Although this domain probably contains several polypeptides, one of these, the DCCD-binding protein, possesses this property alone, as shown both by genetic (*e.g.* 153) and reconstitution experiments (154, 155). The DCCD-binding protein may be hexameric in the membrane (156), and binds DCCD to a single amino acid residue that is Glu-59 in the yeast protein (157; cf 141 and above).

The primary structure of the DCCD-binding protein suggests that it may be folded in a hairpin configuration in the membrane (158). This may be visualized as in Figure 2, where the *Saccharomyces* protein has been tentatively arranged as a membrane-traversing hairpin. All charged groups can be arranged on the membrane surfaces except for a single carboxylic residue (Glu-59) which becomes located near the middle of the membrane. This is the DCCD-binding residue, which has been squeezed out of the aqueous milieu by the surrounding hydrophobic segments. Consequently, the pK of the carboxylic acid would tend to increase and may be identical with the monoprotonic site of F_0 that titrates with pK = 6.8 (see 152).

Considering mechanisms a–c above in the system represented by Figure 2, none of them alone seems to be directly applicable in this structure. We are therefore left with the impression that the H⁺ channel may be constructed according to two or all three of these principles simultaneously. If so, the H⁺-conducting property will be uniquely determined by the tertiary structure, which, in turn, may critically depend on the Glu-59 residue. However, analyses of DCCD-binding proteins from mutants have shown that this residue is not the only site at which amino acid substitution affects the H⁺ conductance (156, 158, 160, 161).

Very recently the DNA sequence of subunit III in yeast oxidase was elucidated (162). If the corresponding amino acid sequence is folded across the membrane according to principles similar to those used in producing Figure 2 [i.e. hydrophobic stretches and neutral residues inside the membrane; charged residues in the aqueous phases, as far as possible (see also 163)], subunit III may form seven helical chains traversing the membrane.

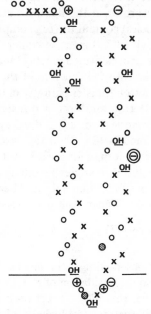

Figure 2   Tentative structure of the DCCD-binding protein of yeast ATPase in the membrane. A proton channel? X, apolar residues; ⊕ and ⊖, basic and acidic residues; O, neutral residues; *OH,* hydroxyl groups;⊖, DCCD-binding glu-59. Data from (157, 159).

Such a model shows four negatively charged residues deep inside the membrane (plus one possible ion pair), one of which is presumably the DCCD-binding site. In contrast to Figure 2, the seven α-helices of the subunit III membrane model carry 38 hydroxylic amino acid residues that are evenly distributed across the membrane thickness. In this case, therefore, the side-chain OH groups may be capable of forming a transmembraneous hydrogen-bonded network (see c above). This requires not only that the bulk of the protein be indeed buried inside the membrane (110, 111), but also that the putative helices be clustered.

VI. ELECTRON TRANSFER AND H+ TRANSLOCATION IN bc-TYPE COMPLEXES

The best known H+-translocating bc-type complexes are the bc_1 and bc_2 complexes of mitochondria and photosynthetic bacteria, respectively. These complexes catalyze oxidation of ubiquinol-10 by a c-type cytochrome. In mitochondria (and aerobic bacteria in the dark), the reduction of ubiquinone is catalyzed by dehydrogenases, utilizing, for example, succinate and α-glycerophosphate as substrate. The entire succinate dehydrogenase/bc_1 complex may be isolated from mitochondria (63) as a functional unit. In photosynthetic bacteria grown in the light, isolated chromatophores (i.e. inverted membrane vesicles) contain the bc_2 complex in functional linkage to the bacteriochlorophyll (BChl) reaction center. The latter donates electrons to ubiquinone in a light-activated reaction and accepts electrons from cytochrome c_2. Together with the function of the bc_2 complex, this then constitutes the cyclic electron transfer pathway of these bacteria (164, 165).

Current models of electron and proton transport may be divided into directly coupled chemiosmotic models [e.g. the Q cycle (166)] and indirectly coupled proton pump models (25, 38, 131). Straightforward distinction between these is more difficult than for the case of the aa_3 complex (Section VII), primarily because the bc complex reactions involve the classical hydrogen carrier, ubiquinone (cf Section V.1), and because the translocational H+/e− stoichiometry is not higher than unity (Section III).

1. Electron and Proton Transfer Models and the Role of Ubiquinone

A crucial experimental observation is the inverse redox coupling between b and c cytochromes, so that oxidation of c_1 induces reduction of b (167; reviewed in 14, 25, 62, 63, 168–170). This effect is apparently enhanced by antimycin. However, this is due to blockade of the otherwise very fast relaxation of this phenomenon, which must itself, therefore, be symptomatic of the electron transfer mechanism in the bc_1 complex. Wikström

& Berden (171) suggested that this phenomenon may be due to branched electron transfer, such that the b and c cytochromes equilibrate with different one-electron couples of ubiquinone, namely $QH_2/QH\cdot$ and $QH\cdot/Q$, so that a sudden change in semiquinone concentration would shift the redox states of the b and c cytochromes in opposite directions. This general model (cf 172) has furnished the basis for several schemes proposed subsequently, notably the Q cycle of Mitchell (166).

As has been pointed out previously (14, 62), there is not sufficient experimental evidence to permit consideration of one proposed mechanism only. The proposals should rather be taken as experimental guides that are in the process of continuous amendment in the light of new data. In this vein Van Ark (173) elegantly tested the credibility of eight models related to those proposed earlier (166, 171), on the basis of six crucial experimental tests. Only one of the considered models, a generalized version of the Q cycle (166; see Figure 3A), survived these tests. Here we cite three experiments, only one of which was used by Van Ark, that also yield the same discrimination.

When anaerobic antimycin-treated mitochondria are pulsed with O_2, the kinetics of formation of ubiquinone from ubiquinol match the kinetics of b reduction, but not the faster kinetics of c_1 oxidation (174). This excludes all schemes in which $QH_2/QH\cdot$ is the reductant of b and $QH\cdot/Q$ is the reductant of c_1 (25, 171, 172, 175). Schemes where these reactants are interchanged survive this test (e.g. 176). The role of Rieske's Fe/S center (177) in reduction of cytochrome b was elegantly elucidated by Trumpower et al (178–180), who showed first, that this center is identical with the "oxidation factor" previously found to be associated with the bc_1 complex (63, 181), and second, that b reduction by substrate becomes fully inhibited by antimycin (in contrast to the fact that, normally, antimycin blocks oxidation of b) after removal of this center. This is related to earlier findings (182, 183) that antimycin prevents reduction of Q by cytochrome b and that the oxidant-induced b reduction is inhibited by reduction of a component in close association with c_1, in the presence of antimycin (184–187). These findings strongly favor cyclic mechanisms such as the Q cycle and exclude linear branched schemes such as the Wikström-Berden model (cf 173).

Finally, the finding by Boveris et al (188) that durohydroquinol reduces b directly and that reduction of Q is inhibited by antimycin (182), tends further to exclude all Q cycle-type schemes in which the antimycin-sensitive step is specifically postulated to be either reduction of Q to QH$\cdot$ or of QH$\cdot$ to QH_2 by cytochrome b. Thus, either dismutation of QH$\cdot$ must occur [in the center i of the Q cycle; see (166) and Figure 3A], or b must be able to reduce both Q and QH$\cdot$ (Figure 3A). The same conclusion was also reached by Van Ark (173), based on other experimental tests.

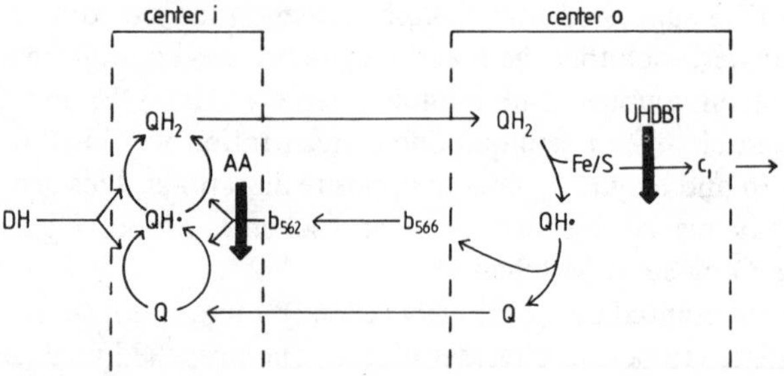

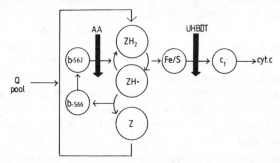

Figure 3A   Model of electron transfer in the bc_1 complex—the Q cycle [modified from (166)]. AA, antimycin; UHDBT, 5-n-undecyl-6-hydroxy-4,7-dioxobenzothiazole.

Wikström & Krab (25) arrived at another model, called the *b* cycle, which is shown in Figure 3B. This working scheme survives the treatment by Van Ark (173) as well as the additional tests described above. It differs from the Q cycle (Figure 3A) in several respects (see 25), of which the most important are (*a*) absence of direct reduction by cytochrome *b* of Q (or QH·) at a center [center i (166); Figure 3A] located on the M side of the membrane (in an electrical sense, not necessarily topographically), (*b*) related unnecessity of electron translocation catalyzed by the *b* cytochromes across the electroosmotic barrier of the membrane. Instead, the bc_1 reactions are suggested to involve only one species of specifically bound ubiquinone [called Z; cf. Figure 3B, to make the analogy to the bacterial bc_2 system; (see 189)]. Z is parted between two electrically insulated domains (ZH₂/ZH· and ZH·/Z) of high and low redox potential, respectively, that are connected only through the cytochrome *b* system (25). Electron transfer

Figure 3B   Model of electron transfer in the bc_1 complex—the *b* cycle [modified from (25)]. This figure is drawn to facilitate visualization of electron transfer. The reader is recommended to draw his own picture with the aid of a quarter (Z) and pennies. Use one quarter to represent Z, with two (ZH₂), one (ZH·), or no (Z) pennies (electrons) on top of it. Electron transfer may be visualized by movement of the pennies between Z and one-electron centers, and of the quarter between the three possible positions. For abbreviations see legend to Figure 3A.

through the latter is proposed to be coupled to H$^+$ translocation much in the same way as seems to be the case in the aa_3 complex (25, 26; Section VII).

A specific catalytic role of ubisemiquinone was earlier considered unlikely (190) due to the expected very fast dismutation. However, more recently, models assuming a fundamentally important role of ubisemiquinone (see above) have gained much support from EPR data (191–195) and demonstration of specific Q-binding proteins (196, 197) in the succinate-cytochrome c reductase segment of the respiratory chain. One of these proteins was recently isolated (197) and shown to confer succinate-ubiquinone reductase activity to purified succinate dehydrogenase after reconstitution. Q bound to this 15,000-dalton protein probably forms the semiquinone radical pair characterized previously (191–193, 195). The other Q radical observed by EPR is probably associated with the bc_1 complex itself (194–197), and may thus be identical with ZH· (Figure 3B) or QH· of center o (Figure 3A). Yu & Yu (197) concluded that the former 15,000-dalton protein is absent from the bc_1 complex as usually isolated [Complex III (198, 199)]. Since Complex III catalyzes both quinol-cytochrome c reductase activity and H$^+$ translocation (200, 201), it may therefore be concluded that the semiquinone radical pair (as bound to the 15,000-dalton protein) cannot be essential for the functioning of the bc_1 complex per se. In the Q cycle scheme (Figure 3A) it can hardly therefore be identical with the ubiquinone at center i as proposed by Mitchell (202). Isolated Complex III contains ⅓–1 ubiquinone per monomer (197, 199), to which the second specific semiquinone EPR signal may then be assigned (194–195), particularly since its properties are consistent with a function such as that of QH· of center o (Figure 3A) or ZH• (Figure 3B). Presently, there seems to be no "room" for the postulated ubiquinone at center i (Figure 3A).

Recent studies with the inhibitory ubiquinone analogue 5-n-undecyl-6-hydroxy-4,7-dioxobenzothiazole (UHDBT) have supported the view that oxidoreduction of ubiquinone may be closely linked to the function of the Fe/S center (203–206). This inhibitor appears to block electron transfer as indicated in Figures 3A and 3B. It is now certain, based on reconstitution experiments, that Fe/S is an obligatory component of electron transfer, at least in the mitochondrial bc_1 segment (180).

2. Problems Associated with the Site of Inhibition by Antimycin

In most current schemes the inhibitory site of antimycin is confined to a single electron transfer step. Although this may be rational, the binding of antimycin to its specific site may have multiple effects on the bc complex

(62, 207). On the other hand, the cyclic electron transfer systems of Figures 3A and 3B presently provide the simplest explanation for a multitude of experimental findings, particularly in the mitochondrial bc_1 system.

However, in the bacterial bc_2 system the situation is more complicated. Even though the oxidant-induced pathway of b reduction has been described here also (208), the second pathway is not blocked by antimycin (see 209). This pathway encompasses light-activated electron transfer from BChl to so-called secondary quinone (Q_{II}) via the quinone/iron complex of the reaction center. $Q_{II}H \cdot$ may then reduce cytochrome b_{50} either directly or via additional intermediates (see below).

Crofts et al (189, 204, 210, 211) prefer a linear scheme of electron transfer [i.e. $Q_{II} \rightarrow b_{50} \rightarrow ZH_2/ZH \cdot \rightarrow Fe/S \rightarrow c_2$; contrast (212) for a scheme related to the Q cycle], with the antimycin block between Z and Fe/S. However, this fails to explain the oxidant-induced b_{50} reduction (208) and is therefore insufficient. Instead, a model may have to be considered in which both routes of b reduction may occur, as proposed by Van den Berg et al (209). Such a model may be plausible given a slight modification of the b cycle (Figure 3B), i.e. if the two b cytochromes are replaced by b_{50}, and the pathway of two-electron reduction of Z is replaced by either a direct reduction of b_{50} by $Q_{II}H \cdot$ or one that takes place via the $ZH \cdot / Z$ couple (cf below).

It seems probable that the bc_2 system functions analogously to the mitochondrial bc_1 complex, and therefore that inhibitory points of action by antimycin are the same. The point suggested by Crofts et al (189, 204, 210, 211; cf above) would however be entirely inconsistent with the models developed for the bc_1 complex (Figure 3; cf above). For this reason we devote some space to a review of the underlying experiments.

Earlier (213) it was thought that re-reduction of c_2 following light-induced oxidation was completely blocked by antimycin in the first turnover and that the requirement of two saturating light flashes to oxidize c_2 completely in the presence of this inhibitor was due to the presence of two hemes c_2 per reaction center. As noted, this would contradict an antimycin block before the $ZH_2/ZH \cdot$ couple, which in that case should be able to re-reduce c_2. However, more recently Bowyer et al (204, 205) showed that c_2 is actually very rapidly, but partially, re-reduced in such conditions; this re-reduction is inhibited by UHDBT (cf Figure 3B). Thus there is only one heme c_2 per reaction center that is proposed to rapidly redox equilibrate with Fe/S in an UHDBT-sensitive step. The antimycin block still appears to be located after the $ZH_2/ZH \cdot$ couple, but before Fe/S (211).

However, Van den Berg et al (209) suggested that the apparent lack of oxidation of ZH_2 in the presence of antimycin may have a thermodynamic explanation such that oxidation may not occur significantly, unless the product $ZH \cdot$ can be oxidized further to Z, which may not occur in these conditions. If so, the antimycin block may still be located as shown in Figure 3B. The finding (204) that the net extent of oxidation of c_2 in the presence of antimycin decreased on decreasing the E_h before the flash from 240 to 0 mV suggests that ZH$_2$ [$E_{m,7}$= 130–150 mV (214, 215)] is indeed partially oxidized by c_2, in agreement with this location. However, this phenomenon was not confirmed in a later study (205), possibly because of different concentrations of redox mediators.

This important problem is clearly unsettled as yet. Rich & Moore (216) showed that cytochrome c is rapidly reduced by substrate in the presence of antimycin before inhibition is attained, which suggests that antimycin indeed may not block oxidation of ZH$_2$ in the plant mitochondria studied. It seems to us that cyclical schemes (Figures 3A and 3B) are still to be preferred over linear ones, but further experimentation is clearly required.

3. Proton Translocation and Charge Separation

Proton translocation is an intrinsic property of the bc_1 complex, as shown by reconstitution experiments (200, 201). Detailed kinetic data are available only for the bc_2 system, however, which offers the unique possibility of studying flash-induced one-turnover events by fast spectroscope techniques. Since the chromatophore membrane is inverted with respect to the mitochondrial or the parent bacterial membrane, H$^+$ translocation is measured as light-driven proton uptake.

2H$^+$/e$^-$ are maximally taken up in the light-driven electron transfer process (e.g. 217). The first proton (H$_I^+$) is linked to the reduction of Q$_{II}$ by the reaction center and therefore does not concern us here further. The second proton (H$_{II}^+$) is taken up by action of the bc_2 complex and is defined as being sensitive to antimycin (217). The kinetics of H$_{II}^+$ uptake are usually synchronous with electron transfer between b_{50} and c_2 (t$_{1/2}$ about 1–2 msec) and with phase III of the electrochromic bandshift of carotenoids, generally taken as a measure of a developed electrical gradient (218).

The central role of Z in the bc_2 system is dramatically demonstrated by initiating the reaction at an E_h high enough to poise Z oxidized before the flash. The rate of re-reduction of c_2 becomes sluggish and quite insensitive to uncoupling agents (215), carotenoid bandshift phase III is lacking, and reduction of b_{50} is markedly slowed (214, 217). Most remarkably, H$_{II}^+$ is still taken up and with a velocity ten times higher than when ZH$_2$ was reduced before the flash (217). Uptake of H$_{II}^+$ may therefore be experimentally dissociated from oxidoreduction of the b and c cytochromes, but it retains sensitivity toward antimycin. The primary effect of antimycin may not be inhibition of electron transfer, or the antibiotic may have several different effects. One could be an inhibition of H$^+$-conducting pathways (section V.3). This could be related to the finding by Artzatbanov & Konstantinov

(219) that the pH-dependence of the E_m values of both b-566 and b-562 in the mitochondrial system is dramatically altered by antimycin. All effects of this antibiotic appear to be directed at the center of the energy-conserving apparatus of the bc-complexes. Thus the H^+-e^- coupling in cytochrome b, as revealed by the pH-dependence of E_m, may indeed be intimately linked with the mechanism of energy transduction (14, 220).

The intriguing finding by Petty et al (217) that the carotenoid bandshift phase III is still observed at a pH high enough to prevent any H^+ uptake from the suspension into the chromatophores, is not easy to reconcile with any simple scheme of proton translocation. This finding must possibly be understood on the basis of local effects in the membraneous proteins that may be discernible when only a few turnovers of the system are under study.

4. The Function of the b-Type Cytochromes

In Q cycle models the b cytochromes merely carry electrons, and therefore electrical charge, vectorially across the membrane (see 189). The weakness of this idea is that the strong H^+-e^- coupling in the b cytochromes (cf above) is neglected (14, 221).

Malviya et al (222) suggested a "Q loop" model in which the cytochrome b dimer (i.e. b-566 and b-562) functions without electron transfer between the hemes. This feature is the main novelty of this model, which becomes very similar, e.g., to the b cycle (Figure 3B), if electron transfer between the b hemes is allowed. The absence of ubisemiquinone (222) may be more a matter of presentation than a true difference, since suggested "concerted" two-electron reduction of b and either c_1 or Fe/S by ubiquinol may well involve radicals.

This model (222) is inconsistent with the kinetic sequence of reduction of b-566 and b-562 as determined by Chance (174), and with the finding that British Antilewisite (BAL; 2,3-dimercaptopropanol) blocks reduction of both b cytochromes in the presence of antimycin (223). In view of this finding and several others relating to the effect of BAL (171, 172, 224, 225) —the inhibition by UHDBT (203–206) and the effect of removal of Rieske's Fe/S center (178–180)—it seems that Fe/S, "X" (172), oxidation factor (181), and the so-called Slater factor (226) may all be the same. The effect of BAL may be similar to that of UHDBT (see Figure 3).

The proposed absence of electron transfer between the cytochrome b hemes (222) was not very well documented. Data by Erecinska & Wilson (207) under similar conditions did not contradict such a reaction. Moreover, the E_m of b-566 may be lowered relative to b-562 in submitochondrial particles (227) so that the former may be highly oxidized and the latter highly reduced at redox equilibrium (cf 222).

Von Jagow & Engel (131) proposed a model in which the cytochrome *b* dimer functions as a proton translocator (cf 14, 25). At this stage this model must be considered highly hypothetical, since experimental backing is scanty. Yet it is very interesting theoretically. The suggested alternation between the "*b*-562" and "*b*-566" identity of the two cytochrome *b* monomers makes this model analogous to models proposed for cytochrome oxidase (114, 116; Section VII.3), H$^+$-ATPase (228), Na$^+$/K$^+$-ATPase (e.g. 229–230a), and for some soluble polymeric enzymes (231). In such models, negative cooperativity (see Section VII) is expected between reactive sites. Although there is no evidence for this so far in the *bc*$_1$ complex, it is also not excluded. For example, the apparent ATP-induced increase in E_m of parts of *b*-566 and *b*-562 (see 14) could be a consequence of a negative redox interaction between the two *b* cytochromes in the complex.

Although considerable progress has been made, it seems to us that central aspects of electron transfer and proton translocation in the *bc*-type complexes still remain to be elucidated. In future research, the one-turnover studies on the photosynthetic *bc*$_2$ complex appear most promising.

VII. ELECTRON TRANSFER AND H$^+$ TRANSLOCATION BY CYTOCHROME *aa*$_3$

Current research on cytochrome oxidase has been covered in several recent reviews (25, 26, 66–68, 126, 130, 232, 233), in a specialized meeting (234), and in a recent monograph (114). After a brief introductory survey with only a few additional references, we discuss here only the most recent developments and their more general implications.

Cytochrome oxidase (or cytochrome *aa*$_3$) contains two hemes (*a* and *a*$_3$) and two coppers (Cu$_A$ and Cu$_B$) per monomer. Heme *a*$_3$ is closely associated with Cu$_B$ both physically and functionally. The two metals form a binuclear center that catalyzes O$_2$ reduction to water. The electrons are donated to the oxidase by cytochrome *c*; heme *a* is the primary acceptor. The second copper (Cu$_A$) appears to be in rapid redox equilibrium with heme *a*. The hemes (but not the coppers) exhibit pH-dependent E_m values, which proves H$^+$-e$^-$ linkage, as with the *b* cytochromes. The hemes also appear to exhibit strong negative redox interactions. Although this has been disputed (68), the basis of the objection is in error as shown by mathematical simulations (114, 233, 235).

Cytochrome oxidase has been found to catalyze H$^+$ translocation linked to electron transfer (cf Section III). Although Mitchell & Moyle (22) have ascribed this and other related findings to ". . . an unfortunate combination of experimental and interpretational difficulties," isolated cytochrome *aa*$_3$

was shown to possess this property after reconstitution into liposome membranes (236, 237). This has subsequently been confirmed and the data extended in several laboratories [(83, 238–240); but see also (241)]. This discovery demonstrates that redox-linked proton translocation need not occur by the redox loop principle championed in Mitchell's chemiosmotic theory (Section V.1).

1. The Mechanism of O_2 Reduction to Water

Evidence from EPR, MCD, and magnetic susceptibility studies (242–246) strongly suggests that heme a_3 and Cu_B are very close to one another; (a figure of less than about 7Å between Fe and Cu has been implicated). Cooperative functioning of the two metals in O_2 reduction is further supported by redox titration data in the presence of CO (247). Since one-electron reduction of O_2 to O_2^- is very unfavourable energetically (248), it is likely, also from a theoretical point of view, that O_2 reduction starts by concerted two-electron transfer, which yields bound peroxide (cf 68).

The low temperature "triple-trapping" method of Chance et al (249) has been instrumental in the elucidation of the mechanism of O_2 reduction. The first intermediate observed by this method is, no doubt, an "oxy" species Fe^{II}-O_2 [Compound A (249–251)]. It is uncertain, however, whether this is indeed a significant catalytic intermediate (68). Recent infrared measurements (252) indicate that photolysis of Fe^{II}-CO (heme a_3) at low temperature may result in rebinding of the CO to Cu_B^I. This would explain the unexpectedly slow rate of recombination of CO with heme iron (249, 253). Since photolysis of Fe^{II}-CO initiates the O_2 reaction in the "triple-trapping" method (249), it is possible that Compound A accumulates only artificially because of inhibition of concerted two-electron transfer by the binding of CO to copper (116). However, Cu^I-CO is unstable above $-130°C$ (252), so that the reaction may proceed by the normal route beyond the Compound A stage in the presence of O_2 as in these experiments.

Identification of subsequent oxygen intermediates has been difficult and even controversial (249–251, 254–258; cf 68). In part this is due to an apparent heterogeneity in the electron-transfer kinetics of cytochrome a (114, 116, 250; Section VII.2). However, the view that bound peroxide may be the first "natural" intermediate (possibly in the μ-peroxo configuration; see no. 1 in Figure 4) has recently gained support. Wikström (258a) demonstrated that O_2 reduction to water may be reversed in ATP-supplemented mitochondria, though only half way, and yield a species with spectral properties identical to one of the low-temperature "oxygen intermediates," Compound C (249, 251, 254–258). This finding excludes the possibility that this "compound" is an artifact due to the presence of CO (256–258), and suggests that it may indeed be a "peroxo" species. This species

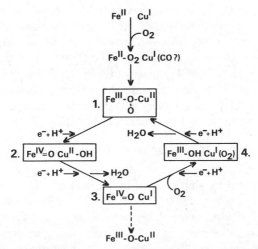

Figure 4 Proposed catalytic cycle of O_2 reduction to water. Fe and Cu denote heme iron (heme a_3) and copper (Cu_B). From (258a). An alternative to the μ-peroxo structure in *1* is peroxide bound to the ferryl (Fe^{IV}) heme iron with copper in the Cu^I state (see 116).

is discernible transitorily in the reaction between fully reduced aa_3 and O_2 (116, 250, 259), but is best observed in the reaction between the "half-reduced" (only heme a_3 and Cu_B reduced) enzyme and O_2 (249, 251, 254, 255, 260).

It was suggested (116, 258a) that the next intermediate is a "ferryl" iron derivative of the a_3/Cu_B center (no. 2 in Figure 4), also observed by ATP-linked reversed electron transfer. Conversion of the oxidized center to this state is associated with a spectral shift that has previously been interpreted as due to formation of an "energized state" of ferricytochrome a_3 (26, 261, 262). However, this species may also be observed as an intermediate in the forward reaction of O_2 and reduced isolated oxidase (263, 264), so that its formation with ATP may best be explained as a simple reversal of the catalytic events at the a_3/Cu_B center (116, 258a).

Figure 4 shows a proposed mechanism of O_2 reduction (116, 258a). In essence, O_2 reduction may occur in two concerted two-electron steps where the ferryl state (no. 2, Figure 4) makes it possible to avoid generation of $HO\cdot$ radicals and to reduce the center by discrete one-electron steps. O_2 may initially be bound to Cu_B^I (68) in accord with the data of Alben et al (252; cf above). Proposed shuttling between Cu_B^I and Cu_B^{II} states is consistent with the recent finding (265) of the EPR signal of the latter under transient conditions. Previously, the existence of Cu_B^{II} during catalysis could be questioned (265a), as long as it remained undetectable by EPR [cf "invisible copper" (e.g. 235)].

Reversal of the O_2 reaction halfway with ATP (258a), but apparently not further (266), suggests that the concerted reduction of bound O_2 to peroxide is highly exergonic (cf 68, 267), and provides the main driving force for the cycle. However, the reversal of steps 3–1 (Figure 4) require very high phosphate potentials as well as a very highly oxidized state of cytochrome c (258a, 261, 268; M. Wikström, unpublished data). Therefore, neither the peroxo nor the ferryl intermediate is expected to accumulate significantly in mitochondria respiring in the controlled State 4.

2. The Role of Cytochrome a

During the past decade interpretation of spectroscopic data has been strongly hampered by ambiguity in the assessment of the contribution of cytochromes a and a_3 to the 605 nm band of the reduced enzyme (see 233). Although this problem may still be matter of some controversy (68), there is by now an overwhelming body of data supporting the original interpretation by Keilin & Hartree (269; see also 270) that ferrous heme a is the main contributor to this band. The evidence for this was recently compiled (138). The importance of this conclusion cannot be overstated as it has far-reaching consequences for the interpretation of nearly all functional data gathered with this enzyme.

There is now little doubt that oxidation of heme a following an O_2 pulse to the reduced enzyme is heterogeneous, and occurs in two kinetically distinct phases with very different temperature dependence (250). Wikström et al (114, 116) rationalized this in terms of the two forms of an energy-transducing element (see Figure 1) that are distinct with respect to redox contact with input and output of the transducer, respectively. They also found much further evidence for this idea in the earlier oxidase literature (cf 259, 271–274), and suggested that cytochrome a is directly involved in the mechanism of proton translocation. Previously [but see (26, 275, 276)] cytochrome a has merely been given an electron-carrying function. Instead, a direct involvement of cytochrome a_3 in energy transduction has been suggested more recently (26, 126, 236, 261, 262, 268, 277, 278), primarily as a result of the studies on the energy-dependent spectral shift in its ferric state. But after the elucidation of the molecular nature of this shift (258a; Section VII.1) these suggestions loose strength. The involved role of cytochrome a_3 in reduction of O_2 (Section VII.1) also makes it difficult to visualize how a_3 could be directly involved in proton translocation as well. However, cytochrome a_3 is probably involved in the uptake of the substrate proton (see Section III.2) from the aqueous M phase (116).

3. A Reciprocating Site Mechanism

Based on the fact that two kinetically distinguishable states of cytochrome a (Section VII.2) always appeared to be equally occupied in different experiments, it was suggested (114, 116) that they may co-exist simultaneously in an $(aa_3)_2$ dimer. The enzyme is indeed dimeric, at least in nonionic detergents and in membraneous vesicles (Section IV). In the model, shown schematically in Figure 5, cytochrome a attains states in which redox contact is established either with cytochrome c or the a_3/Cu_B center in the two companion monomers, respectively (cf analogy to Figure 1). In both states the a_3/Cu_B center would be capable of binding and (partially) reducing O_2. No reasons were found to invoke intermonomeric electron transfer. Due to the reciprocal nature of the monomer/monomer interaction, this model was called the "reciprocating site" mechanism.

In lateral crystals the dimeric enzyme probably has a twofold axis of symmetry (98; Section IV.2). The "head-to-head" arrangement of aa_3 monomers in the dimer (98) makes reciprocal structural interactions between monomers quite feasible for reasons of symmetry, merely by movement of the monomers with respect to each other without the necessity of an asymmetry-inducing ligand such as cytochrome c. Cytochrome c has two binding sites on the enzyme, with different affinity. However, the tight binding-site appears to be associated with each aa_3 monomer (279, 280). Bisson & Capaldi (115) found that covalent binding of a cytochrome c derivative to the high-affinity binding site on subunit II (281, 282) causes 95% inhibition of activity when bound at a stoichiometry of one c per dimer. This most interesting half-of-the-sites effect (cf below) provides strong support to a dimeric function.

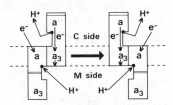

Figure 5 Reciprocating site mechanism of electron transfer and H⁺ translocation by dimeric cytochrome oxidase (from 114, 116). The thick arrow separates two dimeric forms in which aa_3 monomers attain input (downward) or output (upward) states reciprocally. Input state: electron transfer between cytochromes c and a, but not between a and a_3/Cu_B center (double line). An acid/base group (black dot), the pK of which is linked to the redox state of heme a, is in protonic equilibrium with the M phase. Output state: electron transfer between heme a and a_3/Cu_B, but not between c and a (double line). Acid/base group in protonic equilibrium with the C phase. Dotted lines: electroosomotic barrier in the protein separating the two orientations of the acid/base group. After completion of reactions indicated (left) there is a reciprocal switch (thick arrow) by which the monomers in the input and output states are converted into output and input states, respectively. After completion of indicated reactions (right) the dimer is again brought to the state on the left etc. Notes: switch is drawn large for illustrative purposes only; O_2 reduction to water is not shown (see Figure 4); dotted lines do not symbolize the membrane proper.

The notion of symmetric monomers in the fully oxidized state of the enzyme (as isolated), and induction of asymmetry by reduction, is supported by the homogeneity of the EPR resonances of ferric heme a in the "resting" enzyme (283), as opposed to the "split" into two different species of the cytochrome with respect to E_m upon reduction (see 233). This was previously attributed solely to an intramonomeric a/a_3 interaction, but the data do not exclude an intermonomeric a/a interaction. The fact that the anticooperative effect is unaffected by cytochrome c (284) reinforces the impression (cf above) that asymmetry is not imposed by binding of c to the enzyme, but by input of electrons. For the first time there may be a functionally meaningful explanation of the anticooperative properties of this enzyme.

Due to its derivation from Figure 1 (116), the reciprocating site model also accounts for H^+ translocation, but in a more concrete fashion. The specific coupling of the electron input to H^+ uptake from the M phase, and electron output to H^+ release to the C phase (Figures 1 and 5; cf 26) is strongly supported by available data (116, 136–138).

It may be of general interest that the reciprocating site mechanism is closely analogous to the alternating site mechanism proposed for the H^+-translocating ATPase (ATP synthase) by Boyer et al (228, 285–187), to the half-of-the-sites or flip/flop (see 288, 289) mechanism proposed for the Na^+/K^+-ATPase (229–230a), and to the H^+ pump model of cytochrome b of Von Jagow and Engel (131; Section VI.4). Possible advantages of such anticooperative mechanisms in polymeric enzymes have been discussed (288, 289), but mainly for soluble enzymes or membrane receptors. One such advantage is the possibility of energetic coupling between partial reactions occurring simultaneously at the two sets of catalytic centers. Kayalar et al pointed out the specific advantage this would give to the H^+-ATPase (228). Also in cytochrome oxidase this may be important, e.g. to transfer the available free energy change of the O_2 reaction more evenly to the electron input step.

If a mechanism such as that outlined in Figure 5 indeed turns out to be essentially correct and more generally utilized in biological energy transducers, it seems to us that the phenomenon of anticooperativity may be applied in a fashion not previously anticipated. Reciprocal interactions between similar structures [not necessarily monomers of formally dimeric complexes (see 131)] would namely provide all the basic but essential properties that are theoretically required from such a transducer (see 26, 130, 132–134). These may be listed as follows (see Figures 1 and 5; Section V.2): (a) two insulated but readily interconvertible states, of input and output, respectively, for both the exergonic and endergonic elements of the system (e^- and H^+, respectively, in Figure 5), (b) linkage of exergonic and endergonic reactions, (c) a vectorial property of either or both exergonic and endergonic reactions, and (d) necessary control of the interconversion of states (flip/flop) to prevent intrinsic uncoupling.

Finally, we should stress that a more general adaptation of this type of mechanism in biological energy transductions need not mean that other principles, such as the Mitchellian redox loop or the Robertsonian electron translocator, may not also be utilized, as suggested for some bacterial systems (290, 291) and for photosytem II in chloroplasts (e.g. 127, 292).

ACKNOWLEDGMENTS

We are grateful to Ms. Hilkka Vuorenmaa for help with preparation of the manuscript, to the Sigrid Jusélius Foundation for travel support to K. Krab, and to this Foundation and the Finnish Academy for general financial support.

Literature Cited

1. Klingenberg, M. 1980. *Proc. Eur. Bioenerg. Conf., First, Bologna, Italy 1980:* Pàtron Editore. (Foreword)
2. Mitchell, P. 1961. *Nature* 191:144–48
3. Mitchell, P. 1966. *Chemiosmotic Coupling in Oxidative and Photosynthetic Phosphorylation.* Bodmin, UK: Glynn Res. Ltd. 192 pp.
4. Greville, G. D. 1969. *Curr. Top. Bioenerg.* 3:1–72
5. Racker, E. 1976. *A New Look at Mechanisms in Bioenergetics.* New York: Academic. 197 pp.
6. Skulachev, V. P. 1971. *Curr. Top. Bioenerg.* 4:127–90
7. Hauska, G., Trebst, A. 1977. *Curr. Top. Bioenerg.* 6:151–220
8. Harold, F. M. 1977. *Curr. Top. Bioenerg.* 6:83–149
9. Mitchell, P. 1977. *FEBS Lett.* 78:1–20
10. Williams, R. J. P. 1973. *Biochem. Soc. Trans.* 1:1–26
11. Williams, R. J. P. 1974. *Ann. NY Acad. Sci.* 227:98–107
12. Williams, R. J. P. 1975. *Electron Transfer Chains and Oxidative Phosphorylation,* ed. E. Quagliariello et al, pp. 417–22. Amsterdam/New York: North-Holland/Am. Elsevier
13. Wikström, M. K. F. 1971. *Soc. Sci. Fenn. Comment. Biol.* 43:1–42
14. Wikström, M. K. F. 1973. *Biochim. Biophys. Acta* 301:155–93
15. Dilley, R. A., Giaquinta, R. T. 1975. *Curr. Top. Membr. Transp.* 7:49–107
16. Quintanila, A. T., Packer, L. 1977. *FEBS Lett.* 78:161–65

17. Ernster, L. 1977. *Ann. Rev. Biochim.* 46:981–95
18. Kell, D. B. 1979. *Biochim. Biophys. Acta* 549:55–99
19. Padan, E., Rottenberg, H. 1973. *Eur. J. Biochem.* 40:431–37
20. Rottenberg, H. 1978. *FEBS Lett.* 94:295–97
21. Boyer, P. D., Chance, B., Ernster, L., Mitchell, P., Racker, E., Slater, E. C. 1977. *Ann. Rev. Biochem.* 46:955–1026
22. Mitchell, P., Moyle, J. 1979. *Biochem. Soc. Trans.* 7:887–94
23. Brand, M. D. 1977. *Biochem. Soc. Trans.* 5:1615–20
24. Wikström, M. F. K., Krab, K. 1979. *Biochem. Soc. Trans.* 7:880–87
25. Wikström, M. F. K., Krab, K. 1980. *Curr. Top. Bioenerg.* 10:51–101
26. Wikström, M. F. K., Krab, K. 1979. *Biochim. Biophys. Acta* 549:177–222
27. Walz, D. 1979. *Biochim. Biophys. Acta* 505:279–353
28. Rottenberg, H. 1979. *Methods Enzymol.* 55:547–69
29. Wiechmann, A. H. C. A., Beem, E. P., Van Dam, K. 1975. See Ref. 12, pp. 335–42
30. Chance, B., Williams, G. R. 1955. *J. Biol. Chem.* 217:409–27
31. Ragan, C. I. 1976. *Biochim. Biophys. Acta* 456:249–90
32. Mitchell, P., Moyle, J. 1967. *Biochemistry of Mitochondria,* ed. E. C. Slater, et al pp. 53–74. London: Academic
33. Lawford, H. G., Garland, P. B. 1972. *Biochem. J.* 130:1029–44

34. Ragan, C. I., Hinkle, P. C. 1975. *J. Biol. Chem.* 250:8472–76
35. Pozzan, T., Miconi, V., Di Virgilio, F., Azzone, G. F. 1979. *J. Biol. Chem.* 254:10200–5
36. Rottenberg, H., Gutman, M. 1977. *Biochemistry* 16:3220–26
37. Lehninger, A. L., Reynafarje, B., Alexandre, A. 1979. *Cation Flux across Biomembranes,* ed. Y. Mukohata, L. Packer, pp. 343–54. New York: Academic
38. Papa, S. 1976. *Biochim. Biophys. Acta* 456:39–84
39. Moyle, J., Mitchell, P. 1978. *FEBS Lett.* 88:268–72
40. Moyle, J., Mitchell, P. 1978. *FEBS Lett.* 90:361–65
41. Lorusso, M., Capuano, F., Boffoli, D., Stefanelli, R., Papa, S. 1979. *Biochem. J.* 182:133–47
42. Alexandre, A., Lehninger, A. L. 1979. *J. Biol. Chem.* 254:11555–60
43. Azzone, G. F., Pozzan, T., Di Virgilio, F. 1979. *J. Biol. Chem.* 254:10206–12
44. Wilson, D. F., Dutton, P. L. 1970. *Biochem. Biophys. Res. Commun.* 39: 59–64
45. Wilson, D. F., Erecinska, M. 1975. *Arch. Biochem. Biophys.* 167:116–28
46. Lambowitz, A. M., Bonner, W. D. Jr., Wikström, M. K. F. 1974. *Proc. Natl. Acad. Sci. USA* 71:1183–87
47. Wikström, M., Lambowitz, A. M. 1974. *FEBS Lett.* 40:149–53
48. Klingenberg, M., Schollmeyer, P. 1961. *Biochem. Z.* 335:243–62
49. De Pierre, J. W., Ernster, L. 1977. *Ann. Rev. Biochem.* 46:201–62
50. Merli, A., Capaldi, R. A., Ackrell, B. A., Kearney, E. B. 1979. *Biochemistry* 18:1393–1400
51. Mitchell, P., Moyle, J. 1969. *Eur. J. Biochem.* 7:471–84
52. Nicholls, D. G. 1974. *Eur. J. Biochem.* 50:305–15
53. Hodgman, C. D., ed. 1961. *Handbook of Chemistry and Physics,* p. 1706. Cleveland: CRC. 44th ed.
54. George, P. 1965. *Oxidases and Related Redox Systems I,* ed. T. E. King, 1:3–33. New York: Wiley
55. Reynafarje, B., Lehninger, A. L. 1978. *J. Biol. Chem.* 253:6331–34
56. Pozzan, T., Di Virgilio, F., Bragadin, M., Miconi, V., Azzone, G. F. 1979. *Proc. Natl. Acad. Sci. USA* 76:2123–27
57. Jacobs, E. E., Sanadi, D. R. 1960. *Biochim. Biophys. Acta* 38:12–34
58. Jacobs, E. E., Andrews, E. C., Crane, F. L. 1965. See Ref. 54, pp. 784–803
59. Eaton, W. A., George, P., Hanania, G. I. H. 1967. *J. Phys. Chem.* 71:2022–30
60. Brand, M. D., Harper, W. G., Nicholls, D. G., Ingledew, W. J. 1978. *FEBS Lett.* 95:125–29
61. Hatefi, Y., Haavik, A. G., Griffiths, D. E. 1962. *J. Biol. Chem.* 237:1681–85
62. Rieske, J. S. 1976. *Biochim. Biophys. Acta* 456:195–247
63. Trumpower, B. L., Katki, A. G. 1979. *Membrane Proteins in Energy Transduction,* ed. R. A. Capaldi, pp. 89–200. New York: Dekker
64. Weiss, H. 1976. *Biochim. Biophys. Acta* 456:291–313
65. Hatefi, Y. 1976. *The Enzymes of Biological Membranes,* ed A. Martonosi, 4:87–102. New York: Wiley
66. Capaldi, R. A., Briggs, M. 1976. See Ref. 66, pp. 24–29
67. Azzi, A., Casey, R. P. 1979. *Mol. Cell. Biochem.* 28:169–84
68. Erecinska, M., Wilson, D. F. 1978. *Arch. Biochem. Biophys.* 188:1–14
69. Capaldi, R. A., Bell, R. L., Branchek, T. 1977. *Biochem. Biophys. Res. Commun.* 74:425–33
70. Marres, C. A. M., Slater, E. C. 1977. *Biochim. Biophys. Acta* 462:531–48
71. Buse, G., Steffens, G. J. 1978. *Hoppe-Seylers Z. Physiol. Chem.* 359:1005–9
72. Merle, P., Kadenbach, B. 1980. *Eur. J. Biochem.* 105:499–507
73. Weiss, H., Kolb, H. J., 1979. *Eur. J. Biochem.* 99:139–49
74. Briggs, M., Kamp, P. F., Robinson, N. C., Capaldi, R. A. 1975. *Biochemistry* 14:5123–28
75. Downer, N. W., Robinson, N. C., Capaldi, R. A. 1976. *Biochemistry* 15:2930–36
76. Sebald, W., Machleidt, W., Otto, J. 1973. *Eur. J. Biochem.* 38:311–24
77. Steffens, G. J., Buse, G. 1976. *Hoppe-Seylers Z. Physiol. Chem.* 357:1125–37
78. Saraste, M., Penttilä, T., Wikström, M. 1981. *Eur. J. Biochem.* In press
79. Boonman, J. 1979. *The Subunits of Beef Heart Cytochrome c Oxidase,* PhD thesis. Univ. Amsterdam. 128 pp.
80. Weiss, H., Sebald, W. 1978. *Methods Enzymol.* 53:66–73
81. Von Jagow, G., Schägger, H., Riccio, P., Klingenberg, M., Kolb, H. J. 1977. *Biochim. Biophys. Acta* 462:549–58
82. Penttilä, T., Saraste, M., Wikström, M. 1979. *FEBS Lett.* 101:295–300
83. Capaldi, R. A. 1981. *Interaction Between Iron and Proteins in Oxygen and Electron Transport,* ed. C. Ho, W. C. Eaton. New York: Elsevier. In press

84. Saraste, M., Penttilä, T., Wikström, M. 1981. See Ref. 83,
85. Saraste, M., Penttilä, T., Coggins, J. R., Wikström, M. 1980. *FEBS Lett.* 114: 35-38
86. Schatz, G., Mason, T. L. 1974. *Ann. Rev. Biochem.* 43:51-87
87. Borst, P., Grivell, L. A. 1978. *Cell* 15:705-23
88. Bernstein, J. D., Bucher, J. R., Penniall, R. 1978. *J. Bioenerg. Biomembr.* 10: 59-74
89. Hare, J. F., Ching, E., Attardi, G. 1980. *Biochemistry* 19:2023-30
90. Ludwig, B., Schatz, G. 1980. *Proc. Natl. Acad. Sci. USA* 77:196-200
91. Ludwig, B. 1981. See Ref. 83,
92. Ludwig, B. 1980. See Ref. 1, pp. 89-90
93. Steffens, G. J., Buse, G. 1979. *Hoppe-Seylers Z. Physiol. Chem.* 360:613-19
94. Casey, R. P., Thelen, M., Azzi, A. 1980. *J. Biol. Chem.* 255:3994-4000
95. Schatz, G. 1979. *FEBS Lett.* 103: 203-11
96. Wingfield, P., Arad, T., Leonard, K., Weiss, H. 1979. *Nature* 280:696-97
97. Vanderkooi, G. 1974. *Biochim. Biophys. Acta* 344:307-45
98. Henderson, R., Capaldi, R. A., Leigh, J. S. Jr. 1977. *J. Mol. Biol.* 112:631-48
99. Leonard, K. R., Arad, T., Wingfield, P., Weiss, H. 1980. See Ref. 1, pp. 83-84
100. Fuller, S., Capaldi, R. A., Henderson, R. 1979. *J. Mol. Biol.* 134:305-27
101. Frey, T. G., Chan, S. H. P., Schatz, G. 1978. *J. Biol. Chem.* 253:4389-95
102. Briggs, M. M., Capaldi, R. A. 1977. *Biochemistry* 16:73-77
103. Smith, R. J., Capaldi, R. A. 1977. *Biochemistry* 16:2629-33
104. Smith, R. J., Capaldi, R. A, Muchmore, D., Dahlquist, F. 1978. *Biochemistry* 17:3719-23
105. Gellerfors, P., Nelson, B. D. 1977. *Eur. J. Biochem.* 80:275-82
106. Weiss, H., Ziganke, B. 1976. *Genetics and Biogenesis of Chloroplasts and Mitochondria,* ed. Th. Bücher, et al, pp. 259-66. Amsterdam: North-Holland
107. Eytan, G. D., Carroll, R. C., Schatz, G., Racker, E. 1975. *J. Biol. Chem.* 250: 8598-8603
108. Ludwig, B., Downer, N. W., Capaldi, R. A. 1979. *Biochemistry* 18:1401-7
109. Bell, R. L., Sweetland, J., Ludwig, B., Capaldi, R. A. 1979. *Proc. Natl. Acad. Sci. USA* 76:741-45
110. Cerletti, N., Schatz, G. 1979. *J. Biol. Chem.* 254:7746-51
111. Bisson, R., Montecucco, C., Gutweniger, H., Azzi, A. 1979. *J. Biol. Chem.* 254:9962-65

112. Nelson, B. D., Mendel-Hartvig, I. 1977. *Eur. J. Biochem.* 80:267-74
113. Chan, S. H. P., Tracy, R. P. 1978. *Eur. J. Biochem.* 89:595-605
114. Wikström, M., Krab, K., Saraste, M. 1981. *Cytochrome Oxidase—a Synthesis* London: Academic. In press
115. Bisson, R., Capaldi, R. A. 1980. See Ref. 1, pp. 103-4
116. Wikström, M. 1981. *Mitochondria and Microsomes* ed. C. P. Lee, G. Schatz, G. Dallner. Reading, Mass: Addison-Wesley. In press
117. Mitchell, P. 1979. *Eur. J. Biochem.* 95:1-20
118. Mitchell, P. 1976. *Biochem. Soc. Trans.* 4:399-430
119. Mitchell, P. 1979. *Les Prix Nobel en 1978,* pp. 137-72. Stockholm: The Nobel Found.
120. Lundegardh, H. 1945. *Ark. Bot.* A32(2):1-139
121. Robertson, R. N. 1960. *Biol. Rev.* 35:231-64
122. Mitchell, P. 1968. *Chemiosmotic Coupling and Energy Transduction,* Bodmin, UK: Glynn Res. Ltd. 111 pp.
123. Mitchell, P. 1977. *Symp. Soc. Gen. Microbiol.* 27:383-423
124. Wikström, M. K. F., Saari, H. T. 1976. *Mol. Cell. Biochem.* 11:17-33
125. Wikström, M. K. F. 1978. *The Proton and Calcium Pumps,* ed. G. F. Azzone et al, pp. 215-26. Amsterdam: Elsevier/North-Holland Biomed. Press
126. Wikström, M., Krab, K. 1978. *Energy Conservation in Biological Membranes,* ed. G. Schäfer, M. Klingenberg, pp. 128-39. Berlin: Springer
127. Junge, W., McGeer, A. J., Ausländer, W., Kollia, J. 1978. *See Ref. 126, pp. 113-27*
128. Rydström, J. 1977. *Biochim. Biophys. Acta* 463:155-84
129. Rydström, J. 1979. *J. Biol. Chem.* 254:8611-19
130. Wikström, M. 1981. *Curr. Top. Membr. Transp.* In press
131. Von Jagow, G., Engel, W. D. 1980. *FEBS Lett.* 111:1-5
132. De Vault, D. 1971. *Biochim. Biophys. Acta* 226:193-99
133. Läuger, P. 1979. *Biochim. Biophys. Acta* 552:143-61
134. Boyer, P. D. 1975. *FEBS Lett.* 58:1-6
135. Chance, B., Crofts, A. R., Nishimura, M., Price, B. 1970. *Eur. J. Biochem.* 13:364-74
136. Artzatbanov, V. Yu., Konstantinov, A. A., Skulachev, V. P. 1978. *FEBS Lett.* 87:180-85

137. Wikström, M., Krab, K. 1979. See Ref. 37, pp. 321–29
138. Wikström, M. 1981. See Ref. 83,
139. Ohnishi, T., Blum, H., Leigh, J. S. Jr., Salerno, J. C. 1979. *Membrane Bioenergetics*, ed. C. P. Lee et al., pp. 21–30. Reading, Mass: Addison-Wesley
140. Saari, H., Penttilä, T., Wikström, M. 1980. *J. Bioenerg. Biomembr.* 12: 325–38
141. Steffens, G. C. M., Prochaska, L., Capaldi, R. A. 1980. See Ref. 1, pp. 95–96
142. Kozlov, I. A., Skulachev, V. P. 1977. *Biochim. Biophys. Acta* 463:29–89
143. Cattell, K. J., Lindop, C. R., Knight, I. G., Beechey, R. B. 1971. *Biochem. J.* 125:169–77
144. Nagle, J. F., Morowitz, H. J. 1978. *Proc. Natl. Acad. Sci. USA* 75:298–302
145. Morowitz, H. J. 1978. *Am. J. Physiol.* 235:R99–R114
146. Onsager, L. 1973. *Physics and Chemistry of Ice,* ed. E. Whalley, pp. 7–12. Canada: R. Soc.
147. Eigen, M., De Maeyer, L. 1958. *Proc. R. Soc. London Ser. A* 247:505–33
148. Freund, F. 1980. See Ref. 1, pp. 245–46
149. Kayalar, C. 1979. *J. Membr. Biol.* 45:37–42
150. Dunker, A. K., Marvin, D. A. 1978. *J. Theor. Biol.* 72:9–16
151. Dunker, A. K., Zaleske, D. J. 1977. *Biochem. J.* 163:54–57
152. Kagawa, Y. 1978. *Biochim. Biophys. Acta* 505:45–93
153. Sebald, W. 1977. *Biochim. Biophys. Acta* 463:1–27
154. Nelson, N., Eytan, E., Notsani, B., Sigrist, H., Sigrist-Nelson, K., Gitler, C. 1977. *Proc. Natl. Acad. Sci. USA* 74: 2375–78
155. Criddle, R. S., Packer, L., Shieh, P. 1977. *Proc. Natl. Acad. Sci. USA* 74: 4306–10
156. Sebald, W., Graf, T., Lukins, H. B. 1979. *Eur. J. Biochem.* 93:587–99
157. Sebald, W., Machleidt, W., Wachter, E. 1980. *Proc. Natl. Acad. Sci. USA* 77:785–89
158. Sebald, W., Schairer, H. U., Friedl, P., Hoppe, J. 1980. See Ref. 1, pp. 119–20
159. Macino, G., Tzagoloff, A. 1979. *J. Biol. Chem.* 245:4617–23
160. Wachter, E., Schmid, R., Deckers, G., Altendorf, K. 1980. *FEBS Lett.* 113:265–70
161. Hoppe, J., Schairer, H. U., Sebald, W. 1980. *FEBS Lett.* 109:107–11
162. Thalenfeld, B. E., Tzagoloff, A. 1980. *J. Biol. Chem.* 255:6173–80

163. Engelman, D. M., Henderson, R., McLachlan, A. D., Wallace, B. A. 1980. *Proc. Natl. Acad. Sci. USA* 77:2023–27
164. Dutton, P. L., Prince, R. C. 1978. *The Photosynthetic Bacteria,* ed. R. K. Clayton, W. R. Sistrom, pp. 525–70. New York: Plenum
165. Baccarini Melandri, A., Zannoni, D. 1978. *J. Bioenerg. Biomembr.* 10: 109–38
166. Mitchell, P. 1976. *J. Theor. Biol.* 62:327–67
167. Chance, B. 1952. *Int. Biochem. Congr, Paris, Second* p. 32 (Abstr.)
168. Slater, E. C. 1973. *Biochim. Biophys. Acta.* 301:129–54
169. Wilson, D. F., Dutton, P. L., Erecinska, M., Lindsay, J. G., Sato, N. 1972. *Acc. Chem. Res.* 5:234–41
170. Erecinska, M., Wagner, M., Chance, B. 1973. *Curr. Top. Bioenerg.* 5:267–303
171. Wikström, M. K. F., Berden, J. A. 1972. *Biochim. Biophys. Acta* 283: 403–20
172. Baum, H., Rieske, J. S., Silman, H. I., Lipton, S. H. 1967. *Proc. Natl. Acad. Sci. USA* 57:798–805
173. Van Ark, G. 1980 *Electron Transfer through the Ubiquinol: Ferricytochrome c Oxidoreductase Segment of the Mitochondrial Respiratory Chain.* PhD thesis. Univ. Amsterdam, Waarland, The Netherlands. 187 pp.
174. Chance, B. 1974. *Dynamics of Energy-Transducing Membranes* ed. L. Ernster et al., pp. 553–78. Amsterdam: Elsevier
175. Mitchell, P. 1975. *FEBS Lett.* 56:1–6
176. Cadenas, E., Boveris, A., Ragan, C. I., Stoppani, A. O. M. 1977. *Arch. Biochem. Biophys.* 180:248–57
177. Rieske, J. S., Hansen, R. F., Zaugg, W. S. 1964. *J. Biol. Chem.* 239:3017–22
178. Trumpower, B. L. 1976. *Biochem. Biophys. Res. Commun.* 70:73–80
179. Trumpower, B. L., Edwards, C. A. 1979. *J. Biol. Chem.* 254:8697–6
180. Trumpower, B. L., Edwards, C. A., Ohnishi, T. 1980. *J. Biol. Chem.* 255: 7487–93
181. Nishibayashi-Yamashita, H., Cunningham, C., Racker, E. 1972. *J. Biol. Chem.* 247:698–704
182. Boveris, A., Oshino, R., Erecinska, M., Chance, B. 1971. *Biochim. Biophys. Acta* 245:1–16
183. Von Jagow, G., Bohrer, C. 1975. *Biochim. Biophys. Acta* 387:409–24
184. Eisenbach, M., Gutman, M. 1975. *Eur. J. Biochem.* 52:107–16
185. De Kok, J., Slater, E. C., 1975. *Biochem. Biophys. Acta* 376:27–41

186. Lambowitz, A. M., Bonner, W. D. Jr. 1973. *Biochem. Biophys. Res. Commun.* 52:703–11
187. Trumpower, B. L., Katki, A. 1975. *Biochem. Biophys. Res. Commun.* 65: 16–23
188. Boveris, A., Erecinska, M., Wagner, M. 1972. *Biochim. Biophys. Acta* 256: 223–42
189. Crofts, A. R., Crowther, D., Tierney, G. V. 1975. *See Ref. 12, pp. 233–41*
190. Kröger, A. 1976. *FEBS Lett.* 65:278–80
191. Ruzicka, F. J., Beinert, H., Schepler, K. L., Dunham, W. R., Sands, R. H. 1975. *Proc. Natl. Acad. Sci. USA* 72:2886–90
192. Ingledew, W. J., Salerno, J. C., Ohnishi, T. 1975. *Arch. Biochem. Biophys* 177: 176–84
193. Salerno, J. C., Ohnishi, T. 1980. *Biochem. J.* 192:769–81
194. Konstantinov, A. A., Ruuge, E. K. 1977. *FEBS Lett.* 81:137–41
195. Ohnishi, T., Trumpower, B. L. 1980. *J. Biol. Chem.* 255:3278–84
196. Yu, C. A., Nagaoka, S., Yu, L., King. T. E. 1978. *Biochem. Biophys. Res. Commun.* 82:1070–78
197. Yu, C. A., Yu, L. 1980. *Biochim. Biophys. Acta* 591:409–20
198. Rieske, J. S., Zaugg, W. S., Hansen, R. F. 1964. *J. Biol. Chem.* 239:3023–30
199. Rieske, J. S. 1967. *Meth. Enzymol.* 10:239–45
200. Guerrieri, F., Nelson, B. D., 1975. *FEBS Lett.* 54:339–42
201. Leung, K. H., Hinkle, P. C. 1975. *J. Biol. Chem.* 250:8467–71
202. Mitchell, P. 1975. See Ref. 12, pp. 305–16
203. Bowyer, J. R., Trumpower, B. L. 1980. *FEBS Lett.* 115:171–74
204. Bowyer, J. R., Crofts, A. R. 1978. *Frontiers of Biological Energetics*, ed. P. L. Dutton et al, 1:326–33. New York: Academic
205. Bowyer, J. R., Tierney, G. V., Crofts, A. R. 1979. *FEBS Lett.* 101:207–12
206. Trumpower, B. L., Haggerty, J. G. 1980. *J. Bioenerg. Biomembr.* 12: 151–64
207. Erecinska, M., Wilson, D. F. 1976. *Arch. Biochem. Biophys.* 174:143–57
208. Dutton, P. L., Prince, R. C. 1978. *FEBS Lett.* 91:15–20
209. Van den Berg, W. H., Prince, R. C., Bashford, C. L., Takamiya, K., Bonner, W. D., Dutton, P. L. 1979. *J. Biol. Chem.* 254:8594–604
210. Crofts, A. R., Crowther, D., Bowyer, J., Tierney, G. V. 1977. *Structure and Function of Energy-Transducing Membranes*, ed. K. Van Dam and B. F. Van Gelder, pp. 139–55. Amsterdam: Elsevier/North Holland Biomed. Press
211. Crofts, A. R., Bowyer, J. R., Meinhardt, S., Rutherford, A. W. 1980. See Ref. 1, pp. 29–30
212. Petty, K. M., Jackson, J. B., Dutton, P. L. 1977. *FEBS Lett.* 84:299–303
213. Prince, R. C., Baccarini Melandri, A., Hauska, G. A., Melandri, B. A., Crofts, A. R. 1975. *Biochim. Biophys. Acta* 387:212–27
214. Evans, E. H., Crofts, A. R. 1974. *Biochim. Biophys. Acta* 357:89–102
215. Prince, R. C., Dutton, P. L. 1977. *Biochim. Biophys. Acta* 462:731–47
216. Rich, P. R., Moore, A. L. 1976. *FEBS Lett.* 65:339–44
217. Petty, K. M., Jackson, J. B., Dutton, P. L. 1979. *Biochim. Biophys. Acta* 546: 17–42
218. Jackson, J. B., Crofts, A. R. 1969. *FEBS Lett.* 4:185–89
219. Artzatbanov, V. Yu., Konstantinov, A. A. 1980. See Ref. 1, pp. 73–74
220. Wikström, M. K. F. 1972. *Biochemistry and Biophysics of Mitochondrial Membranes*, ed. G. F. Azzone et al, pp. 147–64. New York & London: Academic
221. Petty, K. M., Dutton, P. L. 1976. *Arch. Biochem. Biophys.* 172:346–53
222. Malviya, A. N., Nicholls, P., Elliott, W. B. 1980. *Biochim. Biophys. Acta* 589: 137–49
223. Konstantinov, A. A., Kamensky, Yu., Ksenzenko, M., Surkov, S. 1980. See Ref. 1, pp. 63–64
224. Deul, D. H., Thorn, M. B. 1962. *Biochim. Biophys. Acta* 59:426–36
225. Berden, J. A. 1972. *Site II of the respiratory chain*. PhD thesis. Univ. Amsterdam, Waarland, The Netherlands. 64 pp.
226. Slater, E. C. 1949. *Biochem. J.* 45:14–30
227. Dutton, P. L., Wilson, D. F., Lee, C. P. 1970. *Biochemistry* 9:5077–82
228. Kayalar, C., Rosing, J., Boyer, P. D. 1977. *J. Biol. Chem.* 252:2486–91
229. Repke, K. R. H., Schön, R., Henke, W., Schönfeld, W., Streckenbach, B., Dittrich, F. 1974. *Ann NY Acad. Sci.* 242:203–19
230. Cantley, L. C. Jr., Cantley, L. G., Josephson, L. 1978. *J. Biol. Chem.* 253: 7361–68
230a. Robinson, J. D., Flashner, M. S. 1979. *Biochim. Biophys. Acta* 549:145–76
231. Lazdunski, M., Petitclerc, C., Chappelet, D., Lazdunski, C. 1971. *Eur. J. Biochem.* 20:124–39
232. Caughey, W. S., Wallace, W. J., Volpe, J. A., Yoshikawa, S. 1976. *The Enzymes* 13:299–344 (Pt. C)

233. Wikström, M. K. F., Harmon, H. J., Ingledew, W. J., Chance, B. 1976. *FEBS Lett.* 65:259–77
234. King, T. E., Orii, Y., Chance, B., Okunuki, K., eds. 1979. *Cytochrome Oxidase.* Amsterdam/New York/Oxford: Elsevier/North-Holland Biomed. Press. 426 pp.
235. Malmström, B. G. 1973. *Q. Rev. Biophys.* 6:389–431
236. Wikström, M. K. F., Saari, H. T. 1977. *Biochem. Biophys. Acta* 462:347–61
237. Krab, K., Wikström, M. 1978. *Biochim. Biophys. Acta* 504:200–14
238. Casey, R. P., Chappell, J. B., Azzi, A. 1979. *Biochem. J.* 182:149–56
239. Sigel, E., Carafoli, E. 1979. *J. Biol. Chem.* 254:10572–74
240. Coin, J. T., Hinkle, P. C. 1979. See Ref. 139, pp. 405–12
241. Lorusso, M., Boffoli, D., Capuano, F., Capitanio, N., Pace, V., Papa, S. 1980. See Ref. 1, pp. 99–100
242. Van Gelder, B. F., Beinert, H. 1969. *Biochim. Biophys. Acta* 189:1–24
243. Babcock, G. T., Vickery, L. E., Palmer, G. 1976. *J. Biol. Chem.* 251:7907–19
244. Tweedle, M. F., Wilson, L. J., Garcia-Iniguez, L., Babcock, G. T., Palmer, G. 1978. *J. Biol. Chem.* 253:8065–71
245. Falk, K.-E., Vänngard, T., Ångström, J. 1977. *FEBS Lett.* 75:23–27
246. Stevens, T. H., Brudwig, G. W., Bocian, D. F., Chan, S. I. 1979. *Proc. Natl. Acad. Sci. USA* 76:3320–24
247. Lindsay, J. G., Owen, C. S., Wilson, D. F. 1975. *Arch. Biochem. Biophys.* 169:492–505
248. Wood, P. M. 1974. *FEBS Lett.* 44:22–24
249. Chance, B., Saronio, C., Leigh, J. S. Jr. 1975. *J. Biol. Chem.* 250:9226–37
250. Clore, G. M., Andréasson, L.-E., Karlsson, B., Aasa, R., Malmström, B. G. 1980. *Biochem. J.* 185:139–54
251. Clore, G. M., Andréasson, L.-E., Karlsson, B., Aasa, R., Malmström, B. G. 1980. *Biochem. J.* 185:155–67
252. Alben, R., Altschuld, A., Fiamingo, F., Moh, P. 1980. See Ref. 83, pp.
253. Yonetani, T., Iizuka, T., Yamamoto, H., Chance, B. 1973. *Oxidases and Related Redox Systems II,* ed. T. E. King et al., 1:401–5. Baltimore: Univ. Park Press
254. Chance, B., Saronio, C., Leigh, J. S. Jr. 1979. *Biochem. J.* 177:931–41
255. Denis, M. 1977. *FEBS Lett.* 84:296–98
256. Nicholls, P. 1978. *Biochem. J.* 175:1147–50
257. Nicholls, P. 1979. *Biochemical and Clinical Aspects of Oxygen,* ed W. S.
Caughey, pp. 323–35. New York: Academic
258. Nicholls, P. 1979. *Biochem. J.* 183:519–29
258a. Wikström, M. 1981. *Proc. Natl. Acad. Sci. USA* In press
259. Gibson, Q. H., Greenwood, C. 1965. *J. Biol. Chem.* 240:2694–98
260. Greenwood, C., Wilson, M. T., Brunori, M. 1974. *Biochem. J.* 137:205–15
261. Erecinska, M., Wilson, D. F., Sato, N., Nicholls, P. 1972. *Arch. Biochem. Biophsy.* 151:188–93
262. Wilson, D. F., Dutton, P. L., Wagner, M. 1973. *Curr. Top. Bioenerg.* 5:233–65
263. Orii, Y., King, T. E. 1972. *FEBS Lett.* 21:199–202
264. Shaw, R. W., Hansen, R. E., Beinert, H. 1979. *Biochim. Biophys. Acta* 548:386–96
265. Malmström, B. G., Karlsson, B., Aasa, R., Andréasson, L.-E., Clore, G. M., Vänngard, T. 1980. See Ref. 83, pp.
265a. Seiter, C. H. A., Angelos, S. G. 1980. *Proc. Natl. Acad. Sci. USA* 77:1806–8
266. Bienfait, H. F. 1975. *Reversibility of Site-3 Phosphorylation* PhD thesis. Univ. Amsterdam, Waarland, The Netherlands. 109 pp.
267. Wilson, D. F., Owen, C. S., Holian, A. 1977. *Arch. Biochem. Biophys.* 182:749–62
268. Wilson, D. F., Brocklehurst, E. S. 1973. *Arch. Biochem. Biophys.* 158:200–12
269. Keilin, D., Hartree, E. F. 1939. *Proc. R. Soc. London Ser. B* 127:167–91
270. Lemberg, M. R. 1969. *Physiol. Rev.* 49:48–121
271. Yonetani, T. 1960. *J. Biol. Chem.* 235:3138–43
272. Erecinska, M., Chance, B. 1972. *Arch. Biochem. Biophys.* 151:304–15
273. Andréasson, L.-E. 1975. *Eur. J. Biochem.* 53:591–97
274. Wilson, M. T., Greenwood, C., Brunori, M., Antonini, E. 1975. *Biochem. J.* 147:145–53
275. Wilson, D. F., Chance, B. 1966. *Biochem. Biophys. Res. Commun.* 23:751–56
276. Wilson, D. F., Chance, B. 1967. *Biochem. Biophys. Acta* 131:421–30
277. Wikström, M. K. F. 1977. *Nature* 266:271–73
278. Wikström, M. K. F. 1975. See Ref. 12, pp. 97–103
279. Mochan, E., Nicholls, P. 1972. *Biochim. Biophys. Acta* 267:309–19
280. Ferguson-Miller, S., Brautigan, D. L., Margoliash, E. 1976. *J. Biol. Chem.* 251:1104–15

281. Bisson, R., Azzi, A., Gutweniger, H., Colonna, R., Montecucco, C., Zanotti, A. 1978. *J. Biol. Chem.* 253:1874–80
282. Bisson, R., Jacobs, B., Capaldi, R. A. 1980. *Biochemistry.* 19:4173–78
283. Aasa, R., Albracht, S. P., Falk, K.-E., Lanne, B., Vänngard, T. 1976. *Biochim. Biophys. Acta* 422:260–72
284. Leigh, J. S. Jr., Wilson, D. F., Owen, C. S., King, T. E. 1974. *Arch. Biochem. Biophys.* 160:476–86
285. Hackney, D. D., Rosen, G., Boyer, P. D. 1979. *Proc. Natl. Acad. Sci. USA* 76:3646–50
286. Rosen, G., Gresser, M., Vinkler, C., Boyer, P. D. 1979. *J. Biol. Chem.* 254:10654–61
287. Boyer, P. D. 1980. See Ref. 1, pp. 133–34
288. Lazdunski, M. 1972. *Curr. Top. Cell. Regul.* 6:267–310
289. Levitzki, A., Koshland, D. E. Jr. 1976. *Curr. Top. Cell. Regul.* 10:1–40
290. Jones, R. W., Haddock, B. A., Garland, P. B. 1978. See Ref. 125, pp. 71–80
291. Haddock, B. A., Jones, C. W. 1977. *Bacteriol. Rev.* 41:47–99
292. Junge, W., Ausländer, W. 1978. *Photosynthetic Water Oxidation*, ed. H. Metzner, pp. 213–28. London: Academic

Ann. Rev. Biochem. 1981. 50:657–80

MONOCLONAL ANTIBODIES: A POWERFUL NEW TOOL IN BIOLOGY AND MEDICINE

❖12091

D. E. Yelton and M. D. Scharff

Department of Cell Biology, Albert Einstein College of Medicine, Bronx, New York 10461

CONTENTS

Perspectives and Summary

In 1975 Köhler & Milstein (1) showed that somatic cell hybridization could be used to establish continuous cultures of specific antibody-forming cells. The antibodies produced by such cultures are fundamentally different from antisera obtained from conventionally immunized animals. Each cell line synthesizes a homogeneous, or monoclonal, immunoglobulin that represents one of the many antibodies produced by the immunized animal. Furthermore, the tissue culture cells can be frozen, then recovered later and injected into animals, where they form tumors: large amounts of the homogeneous antibody can be obtained from the tumor-bearing animal. Thus a perpetual supply of the antibody is assured.

0066-4154/81/0701-0657$01.00

Monoclonal antibodies offered such convenience and reliability that many investigators began to explore the possibility of producing them to a diverse array of antigens. The hybridoma technique, as it has come to be called, proved widely applicable. The number of monoclonal antibodies that have been reported is already too vast for them all to be reviewed. We first summarize the development of the hybridoma technology, emphasizing its conceptual rather than technical aspects, and then discuss a few areas in detail to illustrate the advantages and occasional difficulties in using monoclonal antibodies. For investigators who wish to make monoclonal antibodies several descriptions of the technique have been published (2–6).

Monoclonal Immunoglobulins From Myeloma Tumors

The development of the hybridoma technology grew out of immunologists' longstanding interest in antibody structure and the regulation of immunoglobulin synthesis. In fact, hybridomas are neither the first nor only source of monoclonal antibodies. The clonal selection hypothesis advanced by Burnet and his colleagues in the 1950s (7) first established the idea of monoclonality. The hypothesis states that each antibody-forming cell (B-lymphocyte, or its fully differentiated progeny, the plasma cell) is committed to the production of one type of antibody molecule, which has the potential of reacting with one or at most a few structurally similar antigenic determinants. An animal is capable of elaborating millions and perhaps billions of different antibody molecules that react with the many antigens present in nature. Only a small fraction of B-cells recognize any given antigen. During the immune response those B-cells recognizing a particular antigen, in collaboration with other immunoregulatory cells, are stimulated to replicate and differentiate into antibody-secreting cells. Since most macromolecules have a number of distinct antigenic determinants, they stimulate the expansion of many B-cell clones. Furthermore, a single antigenic determinant can stimulate multiple B-cell clones synthesizing antibodies with slightly different specificities. The net result is that immunization with most antigens results in a polyclonal response and the accumulation of many different antibodies in the serum. Conventional antisera are therefore very heterogeneous.

It was soon realized that multiple myeloma is a neoplasm of antibody-forming cells and that each myeloma tumor represents the proliferation of a single clone of plasma cells all secreting the same immunoglobulin. The paraprotein of myeloma patients was an early source of the monoclonal immunoglobulins that were used to elucidate the structural features of the antibody molecule (8). Similar tumors exist in many mammalian species, but human and mouse myeloma proteins have been investigated most extensively. In two inbred mouse strains, BALB/c and NZB, repeated intraperitoneal injections of mineral oil often induce a mouse myeloma tumor. Thousands of such tumors have been induced by Potter (9) and Cohn (10) and their colleagues. These tumors can be passaged indefinitely in syngeneic mice and provide an unlimited supply of homogeneous mouse immunoglobulin for detailed chemical analysis. More recently, these tumors have been used as a source of DNA for direct examination of immunoglobulin gene structure by recombinant DNA technology (11–13).

Myeloma tumors have been useful for studying the biochemistry and cell biology of immunoglobulin production because they can be introduced into tissue culture, where they grow in suspension as continuous, relatively homogeneous cell lines (14–16). These lines can be cloned and manipulated genetically to yield variants in immunoglobulin production and mutants in immunoglobulin structure (17, 18). Since a large proportion of total protein synthesis is devoted to immunoglobulin production, individual variants and mutants can be characterized biochemically (19, 20). Furthermore, the variant cells can be injected back into mice, where hundreds of milligrams of the mutant gene product can be purified from the serum or peritoneal fluid (ascites) of the recipient animals. By introducing drug-resistance markers into cultured myelomas and developing techniques for fusing myeloma cells, it became possible to examine the somatic cell genetics of immunoglobulin production (17, 18, 21). The hybridoma technology is a direct extension of these studies.

In spite of the benefits mentioned, myelomas are not a useful source of monoclonal antibodies to defined antigens. Attempts have been made to produce antigen-binding myelomas by inducing tumors in hyperimmunized animals, but have never succeeded, probably because even in well-immunized animals the proportion of cells making a particular antibody is small. The chance of transforming the right cell is at least as small. One reason the hybridoma technology has succeeded where tumor induction has failed is that several hundred antibody-forming cells are sampled per animal rather than just one. A few myelomas do produce immunoglobulins that react with known antigens (22), most of which are bacterial products. Clearly such an approach cannot provide monoclonal antibodies to the wide variety of interesting and important antigens.

Other Attempts to Obtain Homogeneous Antibodies

Several other approaches to obtain homogeneous antibodies have been tried. Many investigators have attempted to transform B-cells with viruses. SV40 virus was used in one case to obtain a cell line from a rabbit secreting antibody to pneumococcal type III polysaccharide (23). Abelson virus transforms mouse lymphocytes (24), and Epstein-Barr (EB) virus transforms human lymphocytes (25). To our knowledge no antigen-specific cell lines have been obtained from mouse lymphocytes using Abelson virus, though many immunoglobulin-producing lines have been generated (26). Abelson virus appears to preferentially infect immature B-cells that make only small amounts of immunoglobulin (27). EB transformation has recently yielded lymphoblastoid cell lines producing human immunoglobulins specific for diphtheria (28) and tetanus (29) toxoids, the nitrophenyl-hapten (30), and human RBC antigens (31). Most of these lines secrete relatively low amounts of antibody. This technique is limited in the range of antigens to which it can be applied since it is difficult to immunize humans specifically for this purpose. It may be possible, however, to obtain a wider range of lines from human peripheral blood, tonsilar, or spleen cells after in vitro immunization (28).

Homogeneous serum antibodies have been generated using simple antigens with small, repeating structural subunits, such as bacterial polysaccharides (32, 33). However, different clones usually respond in different animals or at different times in the same animal, which limits the amount of homogeneous antibody produced. It was hoped that these restricted responses might serve as models of regulation that would allow immunologists to obtain homogeneous antibodies even with complex antigens. This hope remains unfulfilled and, furthermore, the dissection of antibody responses made possible by the hybridoma technology has called into question how homogeneous such responses actually are.

Very limited amounts of monoclonal antibody can be obtained by in vivo and in vitro techniques that clone normal (untransformed) B-cells stimulated to replicate and secrete immunoglobulin. One such technique is the spleen-focus assay developed by Klinman and his colleagues (34). Small numbers of spleen cells from immunized animals are injected into irradiated recipients to form foci of progeny. These spleens are dissected into small fragments each of which contains only one focus. The fragments are then cultured individually for a few days during which time very small amounts of antibody are secreted. Enough material can be obtained in this way to perform a limited number of very sensitive radioimmune assays. Alternatively, mitogen-stimulated B-cells can be cloned by limiting dilution into carefully defined culture medium that will support growth long enough to obtain small, but usable, amounts of immunoglobulin (35).

The Use of Cell Fusion to Produce Monoclonal Antibodies

It is a historical curiosity that a monoclonal antibody may have been generated by cell fusion as early as 1966. Sinkovics et al (36) report injecting a diploid virus-producing lymphoma into the peritoneal cavity of a mouse and recovering a tetraploid line that secreted a putative virus-neutralizing activity and two types of immunoglobulin. It was suggested that the immunoglobulins were the neutralizing factors and that the cell line arose by an in vivo fusion of a lymphoma cell to an antivirus-producing lymphocyte(s). The experiment was never successfully repeated, nor was it conclusive that cell fusion occurred. In 1973, Schwaber & Cohen (37) fused mouse lymphoma cells with human peripheral blood cells and obtained hybrid lines that secreted both mouse and human immunoglobulin. This showed that fusion of a malignant and a normal lymphoid cell could result in a continuous cell line producing the products of the normal cell. However, neither of these studies led directly to the hybridoma technology.

Experiments involving fusion between different myeloma cell lines provided a number of observations that proved essential to the development of hybridoma technology (17, 18). Initial experiments generated very few hybrids, but examination of these led to the following conclusions: (*a*) Expression of immunoglobulin is codominant, that is, immunoglobulin polypeptides of both parents are expressed. When a nonproducing variant for immunoglobulin production is fused with a producer line, immunoglobulin production is not extinguished, nor is production of immunoglobulin by the variant reactivated. Hybrids produce only those immunoglobulin chains that are being produced by the parental cells at the time of fusion. (*b*) Hybrids from two immunoglobulin-producing parents may assemble their heavy (H) and light (L) chains into all possible combinations of H_2L_2 although the parental chains sometimes associate preferentially. (*c*) Fusion of myeloma cells with other cells not of the B-cell lineage, such as fibroblasts or epithelial cells, results in extinction of immunoglobulin production (38).

Köhler & Milstein obtained their first hybridomas from spleen cells of mice immunized with sheep red blood cells (SRBC), a very potent antigen. In three fusion experiments, 0.2%–10% of the hybrids secreted antibodies that would lyse SRBC in the presence of complement (1). They soon generated antibodies that reacted with the trinitrophenyl hapten (39), and in a short time many laboratories began to generate hybridomas to numerous antigens (40). As the technology has become widely used some technical advances have been made, but the basic procedure is essentially similar to the one used by Köhler & Milstein. Some problems persist that currently make it impractical to generate hybridomas to every antigen of interest.

Assuming an antigen is sufficiently immunogenic, the production of hybridomas is straightforward, although the technique is laborious and requires some skill. Mice or rats are immunized and usually boosted. The optimal immunization regimen almost certainly depends on the individual antigen. Two to four days after the last injection animals are sacrificed and their spleens removed. This corresponds to the time when the greatest numbers of antibody-forming cells can be found in the spleen (41). Lymph node and bone marrow cells also serve, but the spleen is the most convenient source of large numbers of B-cells. After a single cell suspension of lymphocytes is mixed with myeloma cells and treated with polyethylene glycol (PEG), the cells are plated in selective medium.

The fusion protocol results in a mixture of parental cells, hybrids of each parent to itself, and most importantly, hybrids between one parent and the other. It is therefore necessary to grow the cells in conditions that allow only spleen-myeloma hybrids to survive. The classic hypoxanthine-aminopterin-thymidine (HAT) selection system of Littlefield is usually employed (42). The myeloma parent has been engineered with a genetic defect in the enzyme hypoxanthine-guanine phosphoribosyl transferase (HPRT). $HPRT^+$ cells cannot grow in the presence of 6-thioguanine, which provides a convenient way to select for $HPRT^-$ cells. The drug-marked myeloma cells ($HPRT^-$) cannot use exogeneous hypoxanthine to synthesize purines, and die in the presence of aminopterin, which blocks de novo synthesis of purines and pyrimidines. The defect is complemented by the B-cell genome in myeloma-spleen hybrids. Spleen cells, although not killed by aminopterin, will not grow in culture, so that after a few days the only rapidly dividing cells remaining are myeloma-spleen hybrids.

In approximately two to four weeks hybrids grow to sufficient size so that the culture medium can be assayed for antibodies. A number of rapid, simple, and sensitive assays have been developed for screening hybridomas (2). This is necessary because only a small percentage of the hybrids generated produce a desired monoclonal, and it is impractical to propagate any but the useful hybrids beyond the initial small culture vessels.

Once a positive hybrid is identified it should be cloned to avoid overgrowth by other hybrids or by nonproducing variants. These variants frequently arise soon after a hybrid is generated, possibly because of the chromosomal segregation that occurs during the first several weeks of growth in culture (43, 44). After subclones are rescreened to identify those still producing monoclonal antibodies, a few of each are grown to mass culture, frozen for future recovery, and injected into mice to generate high-titer ascites.

We and others have explored a number of technical parameters in an attempt to increase the number of hybrids obtained per fusion and to improve reproducibility (40). These include different immunization regimens, various myeloma partners and ratios of spleen to myeloma cells, different types of PEG, conditions of treatment with PEG, and different methods of selecting the desired hybrids. Many combinations of variables work, and no one procedure has emerged as markedly superior. Attention to using a reliable batch of PEG, good serum, and the like may improve reproducibility, but the barrier to increasing the number of hybrids may be biological rather than technical.

reproducibility, but the barrier to increasing the number of hybrids may be biological rather than technical.

The myeloma cell lines originally used to make hybrids produced their own H and L chains, so the hybrids secreted immunoglobulin molecules that were mixtures of spleen cell and myeloma H and L chains. These mixed molecules not only lowered the amount of the desired monoclonal produced, but they could in principle have an altered antigen specificity and could confound experiments. Köhler & Milstein reported a phenomenon like this in one of their early hybrids (39). This problem was easily overcome since variants in immunoglobulin production occur at a high frequency and are readily isolated (45). As mentioned, defects in myeloma immunoglobulin production do not interfere with synthesis of spleen cell–derived H and L chains in the hybrids. Lines making only the myeloma L chain were soon available (39), and now nonproducing lines suitable for fusion are widely used (46, 47). Table 1 lists the cell lines that are commonly used to generate hybridomas.

Almost all successful fusions have used spleen cells from rats or mice, probably because the myeloma lines in culture that fuse well are from these species. Hybrids between mouse myeloma and rabbit lymphocytes segregate rabbit chromosomes extremely rapidly and stop producing rabbit immunoglobulin. Yarmush et al have generated stable hybrids that secrete either H or L chains of rabbit (50). This stability seems to result from a translocation of a portion of rabbit chromosome onto mouse chromosome. Such events are rare and do not result in production of both H and L chains of rabbit, so the secreted immunoglobulin is not antigen binding. Hybrids formed between mouse myeloma cells and human lymphocytes are somewhat more stable, and a number of hybrids secreting human immunoglobulin have been generated (51). To our knowledge none of these has any antigen

Table 1 Cell lines commonly used to make hybridomas

Cell line	Immunoglobulin produced	Ref.
P3–X63Ag8	γ_1, κ	1
45.6TG1.7	γ_{2b}, κ	48
NSI/1–Ag4.1	κ (not secreted)	39
P3X63Ag8U$_1$	κ (not secreted)	48
Sp2/0–Ag14	none	46
X63–Ag8.653	none	47
Y3–Ag1.2.3 (rat myeloma)	κ (rat)	49

specificity. Recently Olsson & Kaplan (52) reported a human myeloma cell line that they fused to immunized human spleen cells to generate hybridomas that secreted antibodies to the hapten DNP. This is a very significant advance, since human antibodies will be extremely useful for passive immunization and the treatment of disease.

Using Monoclonal Antibodies as Reagents

Hybridomas are attractive reagents since, once generated, they provide a perpetual source of a well-defined antibody. In many ways a monoclonal can be used like a conventional antiserum, but because of the fundamental differences between the two reagents it is unsafe to assume one can automatically be substituted for the other.

Antisera are a complex mixture of many antibodies of different classes with different affinities for various antigenic determinants. Because of the complexity and dynamic nature of the immune response, no one antiserum is exactly like another even if both come from the same animal. These differences are reflected in titer as well as ability to perform certain functions such as fixing complement or precipitating and agglutinating antigens. Antisera contain so many different antibodies that most will perform all these activities to some degree. Any individual monoclonal antibody, however, may not be agglutinating or may not be of a subclass that fixes complement. Precipitation is particularly difficult with monoclonal antibodies since they depend on extensive antigen cross-linking to form insoluble lattices. A monoclonal antibody will bind to only one determinant on a monomeric antigen such as a protein and so not be able to cross-link extensively. This problem can in principle be overcome by mixing monoclonal antibodies that are reactive with different sites on an antigen to produce an artificial but well-defined and reproducible polyclonal antibody reagent.

Because monoclonal antibodies have only a subset of the properties of a conventional antiserum, a number of them may need to be generated to find one suitably tailored for a particular purpose. Cytotoxic assays are most conveniently done with antibodies that efficiently fix complement. Radioimmunoassays require high affinity antibodies that do not release antigen during washing. Affinity purification is probably best done with antibodies of high enough affinity to efficiently bind antigens but low enough to allow elution of the antigen without irreversible denaturation. Rosetting assays for measuring immunoglobulin Fc receptors have been shown to be more sensitive when the red cells can be coated with a large amount of a nonagglutinating monoclonal antibody (53). The list of optimum properties is as long as the number of immunological assays and techniques. It is unlikely that any one monoclonal antibody can fulfill all requirements for all assays.

One of the great conveniences of a monoclonal reagent is the lack of necessity for doing absorptions to render it specific. Absorptions of conventional antisera are often cumbersome and generally result in a loss of titer. It is not only unnecessary to absorb a monoclonal antibody, it is impossible. It is important to bear in mind that monoclonal antibodies are not necessarily monospecific. A given one may react with the antigen of interest and also with other antigens with similar structural features. Such cross-reacting antibodies can be removed from conventional antisera leaving others that are specific for one antigen. Attempts to absorb a monoclonal antibody will simply remove all the immunoglobulin molecules, since all are the same and all will react with the cross-reacting antigen. If a cross-reaction is inconvenient, another monoclonal antibody will have to be generated. On the other hand, a cross-reaction observed with a monoclonal antibody demonstrates shared determinants that are indicative of structural similarities. This ability to examine cross-reactivities that are usually absorbed from conventional antisera is one of the most powerful features of monoclonal antibodies. Such shared specificities may have been highly conserved because of their structural importance, and may reveal functional and evolutionary relationships between macromolecules.

When investigators began to examine the binding of multiple monoclonals to one antigen, a surprising phenomenon emerged (54, 55). In some cases the binding of a monoclonal antibody at one site enhanced or decreased binding of a second at another site, presumably due to a conformational change induced by the first monoclonal antibody. It thus appears that antibodies are able to influence the antigenicity of their targets. Whether this phenomenon has physiological significance is unclear.

Another enormous advantage of hybridomas is the ability to generate specific antibodies even with impure antigens. Because the technique clones one antibody-forming cell away from all others present in the immunized spleen, cells reacting to impurities are discarded during the initial screening. This enormously extends the range of antigens for which serological reagents can be made. Once monoclonal antibodies are obtained, it is relatively easy to purify even difficult antigens (56) . Large amounts (hundreds of milligrams) of monoclonal antibody can be obtained with relative ease from the ascites of a few animals. This antibody can be immobilized on solid supports, such as cyanogen bromide–activated sepharose, to make an affinity column. Even membrane-bound antigens have been purified on such columns by solubilizing the membranes in detergents that do not interfere with antigen-antibody binding (57). Elution can usually be carried out with mild acid or alkaline treatment, which will preserve biological activity.

Although a great boon, hybridomas are not a panacea for all serological woes. In view of the quality, reproducibility, and relative cost effectiveness of monoclonal antibodies, it seems inevitable that they will replace polyclonal antibodies in most large scale routine serology. But in the basic science laboratory, where smaller amounts of antibody are required, conventional antisera may still be preferable for many purposes. It usually takes four to six months to generate a stable hybridoma, if all goes well, whereas producing small amounts of antisera requires less time, energy, and expense. If an antigen does not yield a good antiserum it is often difficult to generate monoclonal antibodies to it. If a very restricted specificity is required, it may be easier to absorb antisera than to find a monoclonal antibody with that specificity. Finally, any given monoclonal antibody may be very sensitive to physical conditions such as pH and temperature, which may change its reactivity and functional activity (58).

Applications of Monoclonal Antibodies

MONOCLONAL ANTIBODIES TO VIRUSES AND OTHER INFECTIOUS AGENTS   Viruses are usually potent immunogens, and conventional antisera from patients and immunized animals have been used extensively in diagnosing viral diseases and in studying their epidemiology. Because of its medical importance and frequent changes in antigenic structure the influenza virus has been studied intensively. Since the advent of the hybridoma technology probably more monoclonal antibodies have been generated against the influenza hemagglutinin than against any other single antigen. These monoclonal antibodies have been used to map antigenic determinants, to isolate and study mutant proteins, and in the diagnosis and epidemiology of the disease.

The large amount of information already available on this system is due in part to the fact that Gerhard & Koprowski and their colleagues were among the earliest workers to employ the hybridoma technique (59). In addition, many antigenic variants of the hemagglutinin occur in nature and are available for study. In fact, the antigenic instability of the influenza hemagglutinin and to a lesser degree of its neuraminidase, are thought to be responsible for the frequent appearance of new strains of the virus to which the world population is not immune. It is known that the pandemic strains differ from each other by a number of amino acid substitutions in the viral hemagglutinin and, in some strains, by changes in the neuraminidase as well (60). Every 10–15 years such major changes occur and are associated with pandemics. In the intrapandemic period, however, minor antigenic changes also occur. It is not clear whether the major antigenic changes associated with pandemics are due to the accumulation of such point mutations or are caused by different genetic or epidemiological mechanisms.

Immunization of a mouse with 10 μg of hemagglutinin results in so vigorous an immune response that most of the hybrids obtained in any fusion make antihemagglutinin antibody (59). These monoclonal antibodies have been used to produce genetic drift in vitro (61–64). A particular parent strain of influenza is propagated for short periods of time in the presence of a single antihemagglutinin monoclonal antibody. The surviving viruses do not react to the selecting monoclonal antibody but do react with monoclonal antibodies against other determinants on the hemagglutinin. Such experiments have been carried out with at least three different pandemic strains of virus and several monoclonal antibodies (61–64). In most cases these in vitro variants are indistinguishable from the parent or from other in vitro variants of the same strain using conventional antisera. These subtle changes in antigenicity suggested a small change in the protein sequence. Chemical studies on a few variants confirmed that a single amino acid substitution had occurred (63–65). Repeated selection of the same strain with the same monoclonal antibody frequently resulted in the identical amino acid substitution (63). Interestingly, the frequency of such spontaneous in vitro mutations differed by as much as 5000-fold when different monoclonals antibodies were used as the selecting agent (61–65). Detailed serological analysis of natural and in vitro antigenic variants with a battery of monoclonal antibodies revealed three to five distinct antigenic areas on the hemagglutinin (61–63). Some in vitro variants that have completely lost the ability to react with the selecting monoclonal antibody react more strongly with another one against a separate antigenic determinant in the same domain, which suggests a conformational change adjoining the substitution. Sequence analysis of variants representing three of these areas confirmed that they are due to amino acid substitutions in different parts of the protein (63–65). The independent nature of these domains and their potential physiological significance was confirmed when the ability to select in vitro variants was compared using one and two monoclonals antibodies. The average frequency of virus particles unreactive to a single monoclonal antibody is 10^{-5} (61–65). When selection with two monoclonals mapping to different areas was attempted no variants could be isolated from 10^8 particles (69).

Although most of the in vitro variants are indistinguishable with conventional antibodies, a few variants did differ in their reactivity with postinfection ferret antiserum and hyperimmune rabbit serum. One such variant had only one detectable amino acid substitution, which was identical to that occurring in a natural antigenic variant (66). This finding suggests that monoclonals can be identified that can select for changes that are important in nature. The epidemiological usefulness of monoclonal antibodies was clearly illustrated by using five of them to examine isolates obtained between pandemic years 1947 and 1957. Some of these isolates could not be separated with ferret antisera, which was the most sensitive assay previously available, but were distinguished by monoclonal antibodies (67). In another study (66), examination of isolates from the 1968 epidemic with monoclonal antibodies revealed previously undetected antigenic differences between the strains circulating in England and those isolated in the United States and Australia.

Thus, monoclonal antibodies have proven to be extremely useful in generating and detecting single amino acid substitutions in the influenza hemagglutinin, in mapping antigenic sites on the protein, in studying the mechanism of antigenic drift, in classifying substrains of the virus, and in studying their epidemiology. The availability of these monoclonal antibodies should make it possible to further refine epidemiological studies and to determine the mechanism of the changes that lead to escape from immunological surveillance and pandemics. In a broader sense, the same approaches used in these studies can be applied to other proteins in order to group mutations and study their origin and effect on protein structure. For example, monoclonal antibodies against bacterial surface proteins and bacteriophage can certainly be used to generate new mutant phenotypes.

In contrast to influenza virus, rabies has appeared to be antigenically stable. Conventional antisera do not usually reveal differences between the laboratory strains of virus and street virus isolated from patients or animals in different parts of the world. This has led to some confidence that the vaccine strains, most of which are derived from the original Pasteur strain, would be generally protective. Wiktor & Koprowski (68, 69) and Flamand et al (70, 71) have shown that rabies virus is, in fact, not antigenically stable in the laboratory or in nature. Wiktor & Koprowski (69) have examined the reactivity of 21 monoclonal antibodies against the viral nucleocapsid and 24 that react with the viral glycoprotein. By propagating the virus in vitro in the presence of antiglycoprotein monoclonal antibodies, they were able to select antigenic variants that preexisted in the virus stock at frequencies of between $10^{-4.3}$ and 10^{-5} per infectious unit. These in vitro variants were similar to those isolated from influenza in that they could no longer be neutralized by the selecting monoclonal antibody, but could be neutralized by others. Furthermore, mice immunized with one variant were killed by other variants. These studies showed that in vitro rabies virus is as antigenically unstable as influenza. Monoclonal antibodies have defined several different serotypes in laboratory strains and street isolates from all over the world (69, 71), and in all likelihood additional serotypes will be defined in the future. Antinucleocapsid antibodies distinguish rabies and rabies-related viruses such as Mokola, Lagos bat, and Duvarhage (70).

The subdivision of rabies virus into distinct serotypes defined by monoclonal antibodies has provided new opportunities to follow the epidemiology of rabies. In one case strong evidence for transmission of the disease via corneal transplant was obtained when serologically identical virus was isolated from the brains of the donor and recipient of the cornea. Preliminary protection assays suggested that serological differences may be important, since the vaccine strains do not offer complete protection against one serotype (69).

The usefulness of monoclonal antibodies as probes of molecular fine structure has led many workers to generate monoclonals against other viruses and infectious agents. The structural proteins of Sinbis (72), Herpes Simplex types 1 and 2 (73), murine leukemia (74), Dengue (75), and measles (76, 77) viruses have been examined to date. Birrer, Bloom & Udem, using an antimeasles hemagglutinin, have selected for an in vitro mutant of measles that is antigenically similar to a natural isolate (77).

Because of the enormous impact of parasitic diseases, the potential of vaccination and the role of immunity is beginning to receive considerable attention. Studies have been difficult because of the complex life cycle of these organisms. Monoclonal antibodies offer a simple and attractive approach. As described by Potocnjak et al (78, 79) for the sporozoite of malaria, monoclonal antibodies against different antigens can be screened to determine which one confers protection, and that specific monoclonal antibody can be used to purify the surface antigen and produce a vaccine. Monoclonal antibodies have been generated against a number of parasites (6, 80) whose biochemistry and mechanism of pathogenesis can now be more quickly explored. When human antibodies of the right specificity become available, it may be possible to intervene in the life cycle of these organisms using passive immunization wherever vaccination is not possible or desirable.

TRANSPLANTATION ANTIGENS   The major transplantation antigens of many mouse and rat strains have been studied extensively with conventional antisera. These surface glycoproteins are highly polymorphic. It has been estimated that between 150 and 200 allelic genes code for the two major transplantation antigens (H2-D and K) in mouse (81), and at least 50 different alleles code for the comparable HLA-A -B, and -C antigens in man (82). Polyclonal antisera that detect these many genetic polymorphisms are best generated by immunizing one member of a species with cells from another (allogeneic immunization) who differs from the recipient in one or a few antigenic determinants. Such alloimmunization is quite possible in mouse or rat where there are a number of inbred and H-2 congenic inbred strains. In man it is obviously much more difficult to obtain typing sera, but such sera are essential for organ transplantation, forensic medicine, and for the diagnosis of certain diseases whose occurrence is linked to HLA (83). Such typing sera are usually obtained from multiparous women, patients who have had multiple transfusions, or occasionally from immunized volunteers. Typing sera from any of these sources usually have low titers and sometimes must be absorbed extensively to provide antibody against single antigenic determinants. Sera against some specificities are very precious. Reference sera must be replenished constantly and their specificity verified at frequent international workshops.

A number of monoclonal antibodies have been generated that react with HLA, and much of the recent work has been reviewed by Brodsky et al (82) and Trucco et al (84). In most studies mice have been immunized with either purified HLA antigens or with whole cells. Earlier studies had suggested that the majority of antibodies produced during xenogeneic immunizations would recognize species-specific antigenic determinants rather than polymorphic determinants (85). It was therefore not surprising that only 20% of the monoclonal anti-HLA antibodies characterized so far recognize polymorphic determinants and can be used as tissue-typing reagents (82). Among the small set of monoclonal antibodies described by Brodsky et al (82), a number reacted with all three of the major human transplantation antigens HLA-A, -B and -C on all individuals, and thus recognized a nonpolymorphic determinant shared by three distinct though structurally related surface glycoproteins. This type of antibody has proved useful in affinity purification of these molecules (86). One monoclonal antibody recognizes a known polymorphic determinant and two distinguish between two specificities that previously had been thought identical (82).

As noted above, xenogeneic immunization should also reveal antigenic specificities not seen with alloimmunization or absorbed heteroantisera. In fact other anti-HLA monoclonal antibodies revealed previously unrecognized cross-reactive antigenic specificities shared among alleles (84). Herzenberg and his colleagues have demonstrated the usefulness of both xenogeneic and allogeneic immunization in generating monoclonal antibodies against mouse surface antigens (87). The benefits of alloimmunization are well illustrated by the work reviewed by Lemke et al (88). Of several monoclonal antibodies against polymorphic determinants, four identified antigens that had previously been defined by conventional antisera, whereas the rest defined new specificities. Similar findings have been reported by Ozato et al (89). Thus, some of these monoclonal antibodies can be used in place of polyclonal antibodies while others further refine tissue-typing and genetic analysis.

Smilek et al used allogeneic immunization to generate monoclonal antibodies reactive with rat transplantation antigens (90). It appeared much easier to generate large numbers of such monoclonal antibodies in rats than in mice. While many fusions had to be done to generate a few H-2-reactive monoclonal antibodies in mouse, from a single fusion Smilek et al obtained

155 monoclonal antibodies (out of 738 hybrids) that reacted with rat transplantation antigens. (This finding was repeatable but remains unexplained.) Of these, 24 have been examined in detail and 11 recognized new antigenic determinants. Two that recognized polymorphic determinants in the rat reacted with all normal mouse and human lymphoid cells, apparently by recognizing a shared antigenic determinant on the major transplantation antigen of these three species. A number distinguish polymorphic determinants on mouse or human transplantation antigens. Thus antibodies made by immunizing one rat strain with another may be useful in studying human transplantation antigens.

In the past, mutations in H-2 antigens have been obtained by screening tens of thousands of mice for the ability to reject skin grafts from their progeny. Approximately 30 have been isolated in 15 years (91). Recently, monoclonal antibodies have been used to select for somatic cell mutants in mouse transplantation antigens (92). This will be an important field of study, since many such mutants can be selected in tissue culture and used to study structure-function relationships and the biological role of transplantation antigens.

TUMOR AND DIFFERENTIATION ANTIGENS   Viral and transplantation antigens are well-characterized molecules and good enough immunogens to have been studied extensively with conventional antisera. Some tumor and polymorphic differentiation antigens have also been identified with conventional antisera. Even in the mouse, however, where allogeneic immunization is feasible, antisera that distinguish between subsets of normal cells or between normal and malignant cells are difficult to generate and are available in limited amounts. Conventional antisera against human differentiation and tumor antigens are especially difficult to raise because xenogeneic immunization must be used and most of these antigens have not been identified.

As mentioned, the hybridoma technology can generate antibodies against impure and even unknown antigens. The strategies that have been used to obtain antibodies against human tumor–associated and differentiation antigens are worth reviewing. Mice are immunized with intact human cells that are thought to contain a distinctive antigen. Such cells are so complex that any given antigen represents only a fraction of its surface components. Hybrids are screened for antibody reactive to immunizing cells but not related cells that do not contain the antigen. Monoclonal antibodies that discriminate between these cell populations are characterized in more detail. This approach has been used to generate monoclonal antibodies that recognize antigens associated with human leukemias (93–95), lymphomas (96), melanoma (97–99), colorectal carcinoma (100), and neuroblastoma

(101). Some of these tumor-associated antigens are also found on fetal cells or on undifferentiated cells, so that it is likely that some monoclonal antibodies that identify differentiation antigens will also recognize tumors, and vice versa. A few studies are described in detail below to illustrate the usefulness of some monoclonal antibodies that have already been reported.

Levy and his colleagues (93) immunized mice with the malignant cells from a child with acute lymphocytic leukemia. They screened 1200 hybridoma cultures. About half of the hybrids produced antibody reactive with the immunizing cells, but only two clearly distinguished between the malignant and normal cells. Further studies using more sensitive assays revealed that even these two monoclonal antibodies reacted strongly with normal lymphocytes found in the cortex of the thymus and very weakly with normal cells present in the lymph nodes, spleen, bone marrow, and peripheral blood. It was possible, however, to set up an assay that identified only malignant cells in the bone marrow and peripheral blood (93). Furthermore, they characterized the surface antigen identified by this antibody and have shown that it is expressed in some but not all patients with acute lymphocytic leukemia. These monoclonal antibodies could now be used by oncologists to monitor the number of malignant cells in the bone marrow and peripheral blood and to determine if the presence of the antigen correlates with prognosis or sensitivity to certain treatments.

Schlossman and his colleagues (95) succeeded in making a monoclonal antibody that recognizes an antigen called CALLA that had previously been defined by conventional antisera. It is present on the cells of many patients with acute lymphoblastic leukemia and some patients with chronic myelogenous leukemia who are in blast crises. Mice were immunized with cells from a patient with acute lymphoblastic leukemia and the hybridomas were screened first with a continuous cell line that expresses the CALLA antigen. Positive hybrids were tested against two additional leukemic cell lines that do not bear the antigen and that had been used to absorb conventional antisera to make it CALLA specific. Based on their reactivity with these three cell lines, twenty-two monoclonal antibodies were thought to recognize CALLA. Those were then screened with a panel of 70 leukemic cells that reacted with the conventional anti-CALLA. Only one of the twenty-two monoclonal antibodies reacted with the whole panel of leukemic cells, and this one precipitated a surface glycoprotein of the same molecular weight as the glycoprotein that reacted with the conventional antisera. The conventional antisera reacts with cells in normal fetal bone marrow and in regenerating adult bone marrow. The monoclonal anti-CALLA shows no reactivity with either, and is therefore more specific for a tumor-associated antigen. Schlossman and his colleagues (96) have also generated a monoclonal antibody that appears to be specific for a subgroup of poorly differentiated lymphomas that are of the B-cell lineage. Again, the hybridomas were screened on a panel of target cells to identify the monoclonal antibody with the most restricted specificity. This monoclonal antibody also subdivides patients who had previously been judged similar by histopathology.

During ontogeny and as cells continue to differentiate in adults they acquire new sets of surface molecules presumably used in cell to cell interactions and in performing their differentiated functions. Antisera have been used to identify and in some cases characterize a number of these differentiation antigens. For example, extensive work has been done to define the surface phenotypes of lymphoid cells that circulate through many tissues and whose functional subpopulations are impossible to distinguish by morphology. Once again the availability of inbred mouse strains permitted investigators to generate alloantisera to some polymorphic differentiation antigens. Such antisera distinguish thymus-derived lymphocytes (T-cells) from B-cells as well as define subpopulations of T-cells that carry out immunoregulatory functions. Generating enough of these antisera has been a major impairment to rapid progress in this field. Already monoclonal antibodies to several known differentiation antigens have been produced, including lymphocyte Thy-1 (87) and Lyt (88) antigens. New lymphocyte antigens have been discovered that were not previously recognized because they are weaker immunogens or not polymorphic (87, 88). Some of these define new subpopulations whose functions may now be investigated. In addition, monoclonal antibodies that identify macrophages (102), neuronal cells (101–104), sperm (105), stage-specific embryonal antigens (106), and the Forssman antigen (107), as well as cell surface antigens not strictly classified as differentiation antigens, have been identified. Monoclonal antibodies have been made to an Fc receptor on mouse macrophages (108) and the nicotinic acetylcholine receptor (109–112).

Antibodies to the acetylcholine receptor are found in 90% of patients with myeasthenia gravis and, when administered passively to animals, they cause a defect in neuromuscular transmission that resembles the disease. Monoclonal antibodies to the receptor were prepared by immunizing mice and rats with purified receptor from *Torpedo californica,* an electric eel (109–111). Because they came from a phylogenetically distant species, the receptors had diverged enough to be immunogenic. Animals making significant amounts of antibody were identified by the presence of muscular weakness (109, 110). Only one of a number of monoclonal antibodies competed with the ligand for binding to the receptor; the others bound to determinants distant from the ligand-binding site (110, 111). However, many of these caused experimental myesthenia gravis when administered to animals (111). These antibodies were also useful in affinity purification of the receptor (112). Similar approaches might be used to identify and purify less well-characterized and poorly immunogenic hormone receptors.

It is even more difficult to make antisera that recognize subsets of human cells (113). Kung et al (114) have reported three monoclonal antibodies that distinguish human T-cell subpopulations analogous to those identified in the mouse (115). These have been used thus far to study multiple sclerosis (MS), systemic lupus erythematosis, juvenile rheumatoid arthritis, infectious mononucleosis, lepromatous leprosy, and immunodeficiency disorders (S. F. Schlossman, personal communication). For example, in a study of 33 patients with MS, Reinherz et al (116) found that 11 of 15 patients with active disease had a selective reduction of suppressor/cytotoxic T-cells in their peripheral blood compared to 1 of 18 MS patients with inactive disease. This finding suggests that changes in immunologic regulation could play a role in the pathogenesis of this disease. These monoclonal antibodies are now commercially available and will certainly become widely used.

IMMUNOGLOBULIN STRUCTURE AND FUNCTION   Although antigen-binding myelomas exist, they are in most cases restricted to the IgA class. Hybridoma proteins provide a source of antigen-binding immunoglobulins of each class and subclass. These reagents are superior in studies of immunoglobulin structure and function because of their homogeneity. When purified from culture supernatants rather than ascites they are free from contamination with other classes of immunoglobulin. The antigen-antibody complexes formed with hybridoma proteins are more physiological than the chemically aggregated immunoglobulins previously used to generate molecules that would execute effector functions.

Bötcher et al (117–118) obtained hybridomas producing antigen-binding mouse IgE. These represent the first tumors secreting this class of mouse immunoglobulin and will undoubtedly facilitate the study of the allergic reactions mediated by IgE.

Using monoclonal antibodies of all IgG subclasses, Diamond and co-workers have confirmed and extended earlier studies on the specificity of receptors for immunoglobulins (Fc receptors) on mouse macrophages (119–121). The potential exists to define the number and specificity of these molecules on different types of lymphocytes as well.

Monoclonal antibodies have been used to examine which classes and subclasses of mouse immunoglobulin mediate cell killing by macrophages (122, 123). Ledbetter & Herzenberg (87) have analyzed the cytotoxic and protein A–binding properties of the subclasses of rat IgG in more detail than had been previously possible with myeloma or serum immunoglobulins (124).

The precise regions of the immunoglobulin molecule that mediate each of its functions has been examined in the past by deriving proteolytic fragments and testing these for biological activity (125) or by examining the activities of mutant immunoglobulins (126). By deriving fragments from hybridoma proteins that retain antigen-binding activity, chemical aggregation can be avoided. Oi & Herzenberg (127) have used such fragments to map IgG_{2a} allotypic determinants to precise immunoglobulin domains, and work is in progress to assess their biological activities. Structural mutants in immunoglobulin can be generated from hybridomas (128, 129) using techniques similar to those used to obtain mutants from myeloma lines (45, 130, 131). These provide an alternative source of antigen-binding material for mapping immunoglobulin effector functions.

Another structural feature of immunoglobulins on which hybridomas are shedding new light is the nature of idiotypes. Idiotype refers to the spectrum of antigenic determinants present on the variable region of individual immunoglobulin molecules. It is believed that if two immunoglobulins share the same idiotype they will usually have the same antigen specificity. The concept of idiotype bears on questions of the extent of the number of variable region genes (the repertoire) and on mechanisms of regulation of B-cell differentiation (132). A finer resolution of the repertoire has been achieved by sequencing different monoclonal antibodies of the same idiotype reactive to a given hapten. What appeared to be a homogeneous serum antibody response is, in fact, a collection of closely related but structurally distinct antibody molecules. One implication of these studies is that the antibody repertoire may be even more extensive than previously thought (133, 134). It has also become possible to determine the structural basis for some individual determinants (135).

FUNCTIONAL CELL LINES   The success of hybridoma technology suggested that cell fusions could establish cell lines that would carry out other immunological functions. It is generally accepted, although not proven, that the phenotype of the malignant cells should match as closely as possible that of the normal cell whose properties one hopes to introduce into culture. With this in mind, many investigators have tried to fuse thymus-derived lymphocytes to normal T-cells to generate functional cell lines. Only one such lymphoma, BW5147, has proved a generally successful partner. A few stable T-cell hybrids have been generated that produce suppressor factors (136–138) or bind specific antigens (139). This one line, however, may not be at an appropriate differentiation state to allow hybrids to express the wide range of T-cell phenotypes.

Diamond and colleagues (119) have reported an apparent hybrid that expresses macrophage-like properties. This cell line has many characteristics of other malignant macrophage-like lines as well as traits heretofore unexpressed in continuous lines. It is possible that functional hybrid lines of B-cells, T-cells, and macrophages may some day be used to reconstruct the immune response in vitro with homogeneous populations of cells. It is also likely that a similar approach will be used to establish homogeneous cell lines producing growth factors, hormones, and other differentiated cell products.

HUMAN HYBRIDOMAS AND CLINICAL APPLICATIONS   Ever since Köhler and Milstein described the hybridoma technology, many investigators have been trying to generate human monoclonal antibodies using the same approach. Two major technical problems had to be overcome. First, a human cell line had to be found that could be used as the malignant partner in cell fusions. Second, it was necessary to obtain adequate numbers of human lymphocytes immunized to the desired antigens. Olsson & Kaplan (52) have described a drug-marked human myeloma cell line that they have successfully fused to immune spleen cells to produce antibody-forming hybridomas. By extending the sources of primary cells to peripheral blood and tonsilar lymphocytes human monoclonals against many antigens will probably become available in the near future.

If human monoclonal antibodies are approved for administration to patients, it is likely that they will be used with radiolabels to identify tumors and their metatases by scintigraphy (140) or for targeting cytotoxic agents to tumors or other tissues. Horse and rabbit antibodies are presently used to treat certain human diseases such as tetanus and transplant rejection. Since such heteroantisera immunize the human recipients, second injections can induce acute and chronic allergic reactions. Human monoclonals are obviously preferable for these purposes as well as for treatment in situations that are less life threatening. Passive immunization with human monoclonal antibodies may become the preferred treatment for drug overdoses and some viral, bacterial, and parasitic infections. Monoclonal antibodies of certain subclasses may prove useful in treating allergies. Since monoclonal antibodies have already been made against neurotransmitters and receptors (109, 141–142), they may prove useful in neurological diseases. In fact, monoclonal antibodies have been reported that induce behavioral changes when introduced into rat brain (143). In essence, the potential for the use of human monoclonal antibodies is enormous.

Conclusion

Antibodies have for years been used to probe complicated biological and biochemical systems and to detect, quantify, and localize small amounts of material in complex mixtures. Immunological assays would have been used even more extensively were it not for the heterogeneity and unpredictability of the immune response. The hybridoma technique has overcome most of the practical, aesthetic, and conceptual objections to the application of immunological techniques to basic and clinical questions. Once a hybridoma producing an appropriate monoclonal antibody has been obtained, large amounts of a homogeneous and reliable reagent is available for as long as it is needed. Monoclonal antibodies have been generated against a great many antigens. We have reviewed only a few of the best studied systems in order to illustrate the power and potential of the technique. Monoclonal antibodies have been made to many antigens already characterized with conventional antisera, and they will no doubt replace their polyclonal counterparts in routine serology in both research and diagnostic laboratories. More importantly, monoclonal antibodies have been made that react with antigens that had not been previously identified, thus providing unique opportunities for understanding complex biological processes and for diagnosing and treating disease.

ACKNOWLEDGMENTS

This work was supported in part by the National Institutes of Health training grant number 5T 32GM7288 from the National Institute of General Medical Sciences, and by grants from the National Institutes of Health (AI 10702 and AI 5231) and the National Science Foundation (PCM75-136090).

Literature Cited

1. Köhler, G., Milstein, C. 1975. *Nature* 256:495–97
2. Kennett, R. H., McKearn, T. J., Bechtol, K. B., eds. 1980. *Monoclonal Antibodies.* New York: Plenum. 423 pp.
3. Kwan, S.-P., Yelton, D. E., Scharff, M. D. 1980. In *Genetic Engineering.* ed. J. K. Setlow, A. Hollaender, 2:31–45. New York: Plenum. 289 pp.
4. McKearn, T. J., Fitch, F. W., Smilek, D. E., Sarmiento, M., Stuart, F. P. 1979. *Immunological Reviews* 47:91–115
5. Herzenberg, Leonard A., Herzenberg, Leonare A., Milstein, C. 1978. In *Handbook of Experimental Immunology,* ed. M. Weir, 2:25.1–25.7. Oxford, London, Edinburgh & Melbourne: Blackwell Sci. Publ.
6. Pearson, T. W., Pinder, M., Roelants, G. E., Kar, S. K., Lundin, L. B., Mayor-Withey, K. S., Hewett, R. S. 1980. *J. Immunol. Methods* 34:141–54
7. Burnet, M. 1959. *The Clonal Selection Theory of Acquired Immunity.* Nashville: Vanderbilt Univ. Press. 209 pp.
8. Edelman, G. M., Cunningham, B. A., Gall, W. E., Gottlieb, P. D., Rustis-Hauser, U., Waxdal, M. J. 1969. *Proc. Natl. Acad. Sci. USA* 63:78–85
9. Potter, M. 1972. *Physiol. Rev.* 52:631–719
10. Cohn, M. 1967. *Cold Spring Harbor Symp. Quant. Biol.* 32:211–21

11. Seidman, J. G., Leder, A., Nau, M., Norman, B., Leder, P. 1978. *Science* 202:11–17
12. Bernard, O., Hozumi, N., Tonegawa, S. 1978. *Cell* 15:1133–44
13. Sakano, H., Rogers, J. H., Huppi, K., Brack, C., Traunecker, A., Maki, R., Wall, R., Tonegawa, S. 1978. *Nature* 277:627–33
14. Pettengill, G. S., Sorensen, G. D. 1967. *Exp. Cell Res.* 47:608–13
15. Horibata, K., Harris, A. S. 1970. *Exp. Cell Res.* 60:61–77
16. Laskov, R., Scharff, M. D. 1970. *J. Exp. Med.* 131:515–41
17. Margulies, D. H., Cieplinski, W., Dharmgrongartama, B., Gefter, M. L., Morrison, S. L., Kelly, T., Scharff, M. D. 1977. *Cold Spring Harbor Symp. Quant. Biol.* 41:781–91
18. Milstein, C., Adetugbo, K., Cowan, J. J., Kohler, G., Secher, D. S., Wilde, C. D. 1977. *Cold Spring Harbor Symp. Quant. Biol.* 41:793–803
19. Adetugbo, K., Milstein, C., Secher, D. S. 1977. *Nature* 265:299–304
20. Francus, T., Birshtein, B. K. 1978. *Biochemistry* 17:4324–31
21. Cotton, R. G. H., Milstein, C. 1973. *Nature* 244:42–43
22. Potter, M. 1978. *Adv. Immunol.* 25: 141–211
23. Strosberg, A. D., Collins, J. J., Black, P. H., Malamud, D., Wilbert, S., Block, K. J., Haber, E. 1974. *Proc. Natl. Acad. Sci. USA* 71:263–64
24. Abelson, J. T., Rabstein, L. S. 1970. *Cancer Res.* 30:2213–22
25. Fahler, J. L., Fingold, I., Rabson, A. S., Manaker, R. A. 1966. *Science* 152: 1259–61
26. Premkumar, E., Potter, M., Singer, P. A., Sklar, M. D. 1975. *Cell* 6:149–59
27. Siegler, R., Zajdel, S., Lane, I. 1972. *J. Natl. Cancer Inst.* 48:189–218
28. Tsuchiya, S., Yokoyama, S., Yoshie, O., Ono, Y. 1980. *J. Immunol.* 124: 1970–76
29. Zurawski, V. R. Jr., Haber, E., Black, P. H. 1978. *Science* 199:1439–41
30. Steinitz, M., Klein, G., Koskimies, S., Makela, O. 1977. *Nature* 269:420–22
31. Koskimies, S. 1980. *Int. Congr. Immunol., 4th.* (Abstr. 19.2.10)
32. Krause, R. M. 1970. *Adv. Immunol.* 12:1–56
33. Haber, E. 1970. *Fed Proc.* 29:66–71
34. Klinman, N. R., Sigal, N. H., Metcalf, E. S., Pierce, S. K., Gearhart, P. J. 1976. *Cold Spring Harbor Symp. Quant. Biol.* 41:165–73

35. Melchers, F., Coutinho, A., Heinrich, G., Andersson, J. 1975. *Scand. J. Immunol.* 4:853–58
36. Sinkovics, J. G., Pienta, R. J., Trujillo, J. M., Ahearn, M. J. 1969. *J. Infect. Dis.* 120:250–54
37. Schwaber, J., Cohen, E. P. 1973. *Nature* 244:444–47
38. Coffino, P., Knowles, M., Nathenson, S., Scharff, M. D. 1971. *Nature New Biol.* 231:87–90
39. Köhler, G., Milstein, C. 1976. *Eur. J. Immunol.* 6:511–19
40. Melchers, F., Potter, M., Warner, N., eds. 1978. *Curr. Top. Microbiol. Immunol.* 81:1–246
41. Benner, R., Meima, F., Van der Meulen, G. M., van Ewijk, W. 1974. *Immunology* 27:747–60
42. Littlefield, J. W. 1964. *Science* 145: 709–10
43. Köhler, G. 1980. *Proc. Natl. Acad. Sci. USA* 77:2197–99
44. Williams, A. F., Galfre, G., Milstein, C. 1977. *Cell* 12:663–73
45. Coffino, P., Baumal, R., Laskov, R., Scharff, M. D. 1972. *J. Cell. Physiol.* 79:429–40
46. Shulman, M., Wilde, C. D., Köhler, G. 1978. *Nature* 276:269–70
47. Kearney, J. F., Radbruch, A., Liesegang, B., Rajewsky, K. 1979. *J. Immunol.* 123:1548–50
48. Yelton, D. E., Diamond, B. A., Kwan, S.-P., Scharff, M. D. 1978. *Curr. Top. Microbiol. Immunol.* 81:1–7
49. Galfre, G., Milstein, C., Wright, B. 1979. *Nature* 277:131–33
50. Yarmush, M. L., Gates, F. T. III, Weisfogel, D. R., Kindt, T. J. 1980. *Proc. Natl. Acad. Sci. USA* 77:2899–903
51. Levy, R., Dilley, J. 1978. *Proc. Natl. Acad. Sci. USA* 75:2411–15
52. Olsson, L., Kaplan, H. S. 1980. *Proc. Natl. Acad. Sci. USA* 77:5429–31
53. Kerbel, R. S. 1980. *J. Immunol. Methods* 34:1–10
54. Howard, J. C., Butcher, G. W., Galfre, G., Milstein, C., Milstein, C. P. 1979. *Immunol. Rev.* 47:139–74
55. Howard, J. C., Butcher, G. W., Galfre, G., Milstein, C. 1978. *Curr. Top. Microbiol. Immunol.* 81:54–60
56. Secher, D. S., Burke, D. C. 1980. *Nature* 285:446–50
57. Parham, P. 1979. *J. Biol. Chem.* 254:8709–12
58. Mosmann, T. R., Gallatin, M., Longenecker, B. M. 1980. *J. Immunol.* 124:1152–56
59. Koprowski, H., Gerhard, W., Croce, C.

M. 1977. *Proc. Natl. Acad. Sci. USA* 74:2985–88

60. Webster, R. G., Laver, W. G. 1975. In *The Influenza Viruses and Influenza*, ed. E. D. Kilbourne, pp. 269–314. New York: Academic

61. Gerhard, W., Webster, R. G. 1978. *J. Exp. Med.* 148:383–92

62. Yewdell, J. W., Webster, R. G., Gerhard, W. 1979. *Nature* 279:246–48

63. Laver, W. G., Air, G. M., Webster, R. G., Gerhard, W., Ward, C. W., Dopheide, T. A. 1979. *Virology* 98:226–37

64. Lubeck, M. D., Schulman, T. L., Palese, P. 1980. *Virology* 102:458–62

65. Laver, W. G., Gerhard, W., Webster, R. G., Frankel, M. E., Air, G. M. 1979. *Proc. Natl. Acad. Sci. USA* 76:1425–29

66. Webster, R. G., Laver, W. G. 1980. *Virology* 104:139–48

67. Webster, R. G., Kendal, A. P., Gerhard, W. 1979. *Virology* 96:258–64

68. Wiktor, T. J., Koprowski, H. 1978. *Proc. Natl. Acad. Sci. USA* 75:3938–42

69. Wiktor, T. J., Koprowski, H. 1980. *J. Exp. Med.* 152:99–112

70. Flamand, A., Wiktor, T. J., Koprowski, H. 1980. *J. Gen. Virol.* 48:97–104

71. Flamand, A., Wiktor, T. J., Koprowski, H. 1980. *J. Gen. Virol.* 48:105–9

72. Roehrig, J. T., Corser, J. A., Schlesinger, M. J. 1980. *Virology* 101:41–49

73. Zweig, M., Heilman, C. J. Jr., Rabin, H., Hopkins, R. F. III, Neubauer, R. H., Hampar, B. 1979. *J. Virol.* 32:676–78

74. Lostrom, M. E., Stone, M. R., Tam, M., Burnett, W. N., Pinter, A., Nowinski, R. C. 1979. *Virology* 30:336–50

75. Dittmar, D., Haines, H. G., Castro, A. 1980. *J. Clin. Microbiol.* 12:74–78

76. McFarlin, D. E., Bellini, W. J., Mingioli, E. S., Behar, T. N., Trudgett, A. 1980. *J. Gen. Virol.* 48:425–29

77. Birrer, M. J., Bloom, B. R., Udem, S. 1981. *Virology.* 108:381–90

78. Potocnjak, P., Yoshida, N., Nussenzweig, R. S., Nussenzweig, V. 1980. *J. Exp. Med.* 151:1504–13

79. Yoshida, N., Nussenzweig, R. S., Potocnjak, P., Nussenzweig, V., Aikawa, M. 1980. *Science* 207:71–73

80. Sethi, K. K., Endo, T., Brandis, H. 1980. *J. Parasitol.* 66:192–96

81. Götze, D., Vollmers, H. P. 1979. *Immunol. Rev.* 47:207–18

82. Brodsky, F. M., Parham, P., Barnstable, C. J., Crumpton, M. J., Bodmer, W. F. 1979. *Immunol. Rev.* 47:3–62

83. Festenstein, H., Demant, P. 1978. *Curr. Top. Immunol. Ser.* 1:1–212

84. Trucco, M. M., Garotta, G., Stocker, J. W., Ceppellini, R. 1979. *Immunol. Rev.* 47:219–52

85. Lampson, L. A., Levy, R., Grumet, F., Mess, D., Pious, D. 1978. *Nature* 271:461–62

86. Parham, P., Barnstable, C. J., Bodmer, W. F. 1979. *J. Immunol.* 123:342–49

87. Ledbetter, J. A., Herzenberg, L. A. 1979. *Immunol. Rev.* 47:63–90

88. Lemke, H., Hämmerling, G., Hämmerling, U. 1979. *Immunol. Rev.* 47: 175–206

89. Ozato, K., Mayer, N., Sachs, D. H. 1980. *J. Immunol.* 124:533–40

90. Smilek, D. E., Boyd, H. C., Wilson, D. B., Zmijewski, C. M., Fitch, F. W., McKearn, T. J. 1980. *J. Exp. Med.* 151:1139–50

91. Kohn, H. I., Klein, J., Melvold, R. W., Nathenson, S. G., Pious, D., Schreffler, D. C. 1978. *Immunogenetics* 7:279–94

92. Rajan, T. V. 1980. *Immunogenetics* 10: 423–31

93. Levy, R., Dilley, J., Fox, R. I., Warnke, R. 1979. *Proc. Natl. Acad. Sci. USA* 76:6552–56

94. McMichael, A. J., Pilch, J. R., Galfre, G., Mason, D. Y., Fabre, J. W., Milstein, C. 1979. *Eur. J. Immunol.* 9:205–10

95. Ritz, J., Pesando, J. M., Notis-McConarty, J., Lazarus, H., Schlossman, S. F. 1980. *Nature* 283:583–85

96. Nadler, L. M., Stashenko, P., Hardy, R., Schlossman, S. F. 1980. *J. Immunol.* 125:570–77

97. Koprowski, H., Steplewski, Z., Herlyn, D., Herlyn, M. 1978. *Proc. Natl. Acad. Sci. USA* 75:3405–9

98. Steplewski, Z., Herlyn, M., Herlyn, D., Clark, W. H., Koprowski, H. 1979. *Eur. J. Immunol.* 9:94–96

99. Yeh, M. Y., Hellstrom, I., Brown, J. P., Warner, G. A., Hansen, J. A., Hellstrom, K. E. 1979. *Proc. Natl. Acad. Sci. USA* 76:2927–31

100. Herlyn, M., Steplewski, Z., Herlyn, D., Koprowski, H. 1979. *Proc. Natl. Acad. Sci. USA* 76:1438–52

101. Kennett, R. H., Gilbert, F. 1979. *Science* 203:1120–21

102. Springer, T., Galfre, G., Secher, D. S., Milstein, C. 1979. *Eur. J. Immunol.* 9:301–6

103. Gottlieb, D., Greve, J. 1978. *Curr. Top. Microbiol. Immunol.* 81:40–44

104. Eisenbarth, G. S., Walsh, F. S., Nirenberg, M. 1979. *Proc. Natl. Acad. Sci. USA* 76:4913–17

105. Bechtol, K. B., Brown, S. C., Kennett,

R. H. 1979. *Proc. Natl. Acad. Sci. USA* 76:363–67
106. Solter, D., Knowles, B. B. 1978. *Proc. Natl. Acad. Sci. USA* 75:5565–69
107. Stern, P. L., Willison, K. R., Lennox, E., Galfre, G., Milstein, C., Secher, D., Ziegler, A. 1978. *Cell* 14:775–83
108. Unkeless, J. C. 1979. *J. Exp. Med.* 150:580–96
109. Moshley-Rosen, D., Fuchs, S., Eshhar, Z. 1979. *FEBS Lett.* 106:389–92
110. Gomez, C. M., Richman, D. P., Berman, P. W., Burres, S. A., Arnason, B. G. W., Fitch, F. W. 1979. *Biochem. Biophys. Res. Commun.* 88:575–82
111. Lennon, V. A., Lambert, E. H. 1980. *Nature* 285:238–40
112. Lennon, V. A., Thompson, M., Chen, J. 1980. *J. Biol. Chem.* 255:4395–98
113. Reinherz, E. L., Kung, P. C., Goldstein, G., Schlossman, S. F. 1980. *J. Immunol.* 124:1301–7
114. Kung, P. C., Goldstein, G., Reinherz, E. L., Schlossman, S. F. 1979. *Science* 206:347–49
115. Reinherz, E. L., Kung, P. C., Goldstein, G., Schlossman, S. F. 1979. *Proc. Natl. Acad. Sci. USA* 76:4061–65
116. Reinherz, E. L., Weiner, H. L., Hauser, S. L., Cohen, J. A., Distaso, J. A., Schlossman, S. F. 1980. *N. Engl. J. Med.* 303:125–29
117. Böttcher, I., Hämmerling, G. 1978. *Nature* 275:761–62
118. Böttcher, I., Ulrich, M., Hirayama, N., Ovary, Z. 1980. *Int. Arch. Allergy Appl. Immunol.* 61:248–50
119. Diamond, B., Bloom, B. R., Scharff, M. D. 1978. *J. Immunol.* 121:1329–33
120. Diamond, B., Scharff, M. D. 1980. *J. Immunol.* 125:631–33
121. Diamond, B., Yelton, D. E. 1981. *J. Exp. Med.* In press
122. Greenberg, A. H., Lydyard, P. M. 1979. *J. Immunol.* 123:861–69
123. Ralph, P., Nakoina, I., Diamond, B., Yelton, D. E. 1980. *J. Immunol.* 125:1885–88
124. Medgyesi, G. A., Fust, G., Hergely, J.,

Bazin, H. 1978. *Immunochemistry* 15:125–29
125. Winkelhake, J. L. 1978. *Immunochemistry* 15:695–714
126. Diamond, B., Birshtein, B. K., Scharff, M. D. 1979. *J. Exp. Med.* 150:721–26
127. Oi, V. T., Herzenberg, L. A. 1979. *Mol. Immunol.* 16:1005–17
128. Kohler, G., Shulman, M. J. 1980. *Eur. J. Immunol.* 10:467–76
129. Yelton, D. E., Cook, W. D., Scharff, M. D. 1980. *Transplant. Proc.* 12:439–42
130. Cotton, R. G. H., Secher, D. S. Milstein, C. 1973. *Eur. J. Immunol.* 3:135–40
131. Liesegang, B., Radbruch, A., Rajewsky, K. 1978. *Proc. Natl. Acad. Sci. USA* 75:3901–5
132. Dennis, K., Kennett, R. H., Klinman, N., Molinaro, C., Sherman, L. 1980. See Ref. 2, pp. 49–59
133. Schilling, J., Clevinger, B., Davie, J. M., Hood, L. 1980. *Nature* 283:35–40
134. Estess, P., Lamoyi, E., Nisonoff, A., Capra, J. D. 1980. *J. Exp. Med.* 151:863–75
135. Clevinger, B., Schilling, J., Griffith, R., Hansburg, D., Hood, L., Davie, J. M. 1980. See Ref. 2, pp. 37–48
136. Taussig, M. J., Holliman, A. 1979. *Nature* 277:308–10
137. Taniguchi, M., Takei, I., Tada, T. 1980. *Nature* 283:227–28
138. Watanabe, T., Kimoto, M., Maruyama, S., Kishimoto, T., Yamamura, Y. 1978. *J. Immunol.* 121:2113–17
139. Goodman, J. W., Lewis, G. K., Primi, D., Hornbeck, P., Ruddle, N. H. 1980. In *Regulatory T lymphocytes,* ed. B. Pernis, H. Vogel. New York: Academic. In press
140. Ballou, B., Levine, G., Hakala, T. R., Solter, D. 1979. *Science* 206:844–46
141. Cuello, A. C., Galfre, G., Milstein, C. 1979. *Proc. Natl. Acad. Sci. USA* 76:3532–36
142. Chan-Palay, V. 1979. *Anat. Embryol.* 156:225–40
143. Williams, C. A., Barna, J., Schupf, N. 1980. *Nature* 283:82–84

Ann. Rev. Biochem. 1981. 50:681–714

THE MECHANISM AND REGULATION OF ATP SYNTHESIS BY F₁-ATPases

♦12092

Richard L. Cross

From the Department of Biochemistry, State University of New York, Upstate Medical Center, Syracuse, New York 13210

CONTENTS

INTRODUCTION

Perspectives and Summary

In oxidative phosphorylation and photophosphorylation, exergonic oxidation-reduction reactions are coupled to the endergonic synthesis of ATP.

0066-4154/81/0701-0681$01.00

The main premise of Mitchell's chemiosmotic hypothesis (1) is that a trans-membrane electrochemical gradient serves as a required intermediate in these energy conversion processes. Since this premise is likely to be correct, the major questions that remain unresolved deal with attempts to obtain a chemical description of how the gradient is formed and how it is used.

Energy coupling occurs at those points along the membrane-embedded electron transport chains of mitochondria, chloroplasts, and aerobic and photosynthetic bacteria where the difference in midpoint reduction potentials of adjacent carriers, A and B, is thermodynamically sufficient to drive proton transport against an electrochemical gradient (Reaction 1). The energy stored in this gradient, $\Delta \mu_{H^+}$ is used by membrane-bound coupling factors, F_0F_1, to drive the net synthesis of ATP (Reaction 2).

$$A_{red} + B_{ox} \rightleftharpoons A_{ox} + B_{red} + \Delta \mu_{H^+} \qquad 1.$$

$$ADP + P_i + \Delta \mu_{H^+} \xrightarrow{F_0F_1} ATP + H_2O \qquad 2.$$

The complex of coupling factors that catalyzes Reaction 2 has two empirically defined components. F_0 is an integral membrane complex of three or four distinct proteins that can be extracted from the membrane by the use of detergents. One component of the complex, a dicyclohexylcarbodiimide-sensitive proteolipid (2), has been shown to play an essential role in the translocation of protons by membrane-bound F_0 (3–6). F_1 can be detached from the membrane as a water soluble complex of five distinct subunits, α to ϵ. The stoichiometry of the subunits is not known, but the main alternatives are $\alpha_3\beta_3\gamma\delta\epsilon$ and $\alpha_2\beta_2\gamma_{1-2}\delta_{1-2}\epsilon_2$ (see 7, 8). The enzyme contains catalytic sites for ATP synthesis by mitochondria (9, 10) chloroplasts (11, 12), and bacteria (13, 14). The reaction mechanism of F_1 is the subject of this review.

When physically separated from the exergonic reactions of the electron transport chain, soluble F_1 is only capable of catalyzing the net hydrolysis of ATP. Hence, it is often referred to as an ATPase. In this article F_1 is used as a general term, while MF_1, CF_1, and BF_1 are used to refer to the corresponding enzymes isolated from mitochondria, chloroplasts, and bacteria. The striking similarities in structure, properties, and function of F_1 from various sources (7) suggest that their catalytic mechanisms are very likely to be similar or identical. In discussing various aspects of the mechanism of ATP synthesis, data is freely drawn from studies utilizing MF_1, CF_1, or BF_1.

The review begins with a discussion of data that have led to the formulation of what this author believes to be the best working model for the mechanism of ATP synthesis by F_1. Alternative suggestions are then considered. This is followed by a review of chemical modification studies that have provided a preliminary view of the environment at the catalytic site of F$_1$. Finally a discussion is presented of the possible modes of regulation of F$_1$ by polypeptides, structural transitions, and ligands.

Previous Reviews

Related reviews have appeared recently dealing with MF$_1$ (8, 15, 15a), CF$_1$ (16, 17), BF$_1$ (18, 19), the structure of F$_0$ and F$_1$ (7, 20, 20a), adenine nucleotide–binding sites on F$_1$ (21, 22), the binding change mechanism (23, 24), the chemiosmotic hypothesis (25–27, see also 28), the mitochondrial inhibitor protein (29), and general aspects of oxidative- and photophosphorylation (30–32).

THE MECHANISM OF ATP FORMATION

Is there a Chemical Precursor?

Nature likes to reuse a good idea, and those who consider biochemical precedent in designing experiments are often rewarded with rapid progress in characterizing "new" enzymes. The idea of a chemical precursor to ATP in oxidative phosphorylation was stimulated by early work on glyceraldehyde-3-phosphate dehydrogenase (33) and clearly emphasized in the chemical coupling hypothesis of Slater (34). However, after years of vigorous pursuit by many laboratories, no covalent intermediates have been identified using radioisotope-labeled P$_i$ or ADP.

Although lack of detection of intermediates can never rule out their existence, several approaches that might have provided indirect evidence for a labile intermediate have failed to do so. If an anhydride precursor to ATP contained an activated ADP, it would be expected to be susceptible to nucleophilic attack by an oxygen of P$_i$. This would result in phosphate providing the bridge oxygen between the β- and γ-phosphoryl groups of ATP. Contrary to this prediction, Boyer has shown that the bridge oxygen is provided by ADP (35). If an anhydride precursor to ATP contained a phosphoryl group, reversal of the water elimination step would result in an exchange of oxygen between water and phosphate. The addition of ADP might decrease the exchange by drawing away an intermediate of the partial reaction. Contrary to this prediction, the P$_i$ $\rightleftarrows$ HOH oxygen exchange catalyzed by submitochondrial particles is stimulated by ADP (36, 37). In fact, when loosely bound nucleotides are removed from the particles the exchange shows an apparent absolute dependence on the addition of ADP (38). The water elimination step may thus require ATP formation. Alternatively, the ADP requirement could result from ordered substrate binding or an allosteric effect.

Recently, the stereochemical course of enzyme-catalyzed phosphoryl transfer reactions has been studied using substrates containing isotopically labeled, chiral phosphoryl groups (see 39). An interesting and consistent pattern has emerged from these studies in that the phosphoryl acceptor appears to approach the phosphorus at an angle of 180° to the leaving group. This "in-line" displacement pathway results in the inversion of configuration of the phosphorus atom:

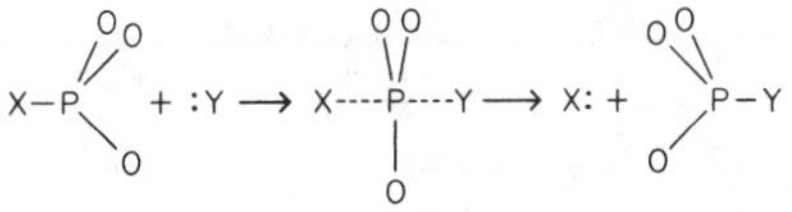

If the overall reaction sequence involves an odd number of such phosphoryl transfers, the product will show inversion, and if an even number of transfers occur, the stereochemistry of the substrate will be retained in the final product. Hence, well-characterized kinases, which are believed to catalyze single displacement reactions, show inversion, while phosphatases, nucleoside diphosphate kinase, and phosphomutases, which are believed to catalyze double displacement reactions via phosphorylated intermediates, show retention of configuration.

Using this approach, Webb et al (40) used beef heart MF_1 to hydrolyze isotopically labeled ATPγS to give [$^{16}O,^{17}O,^{18}O,$] thiophosphate. Analysis of the enantiomers of the inorganic thiophosphate showed that the configuration of the γ phosphorus atom of ATPγS undergoes inversion during hydrolysis. The most likely interpretation of these results is that a phosphoryl group is transferred directly from ATP to water without participation of a phosphorylated intermediate.

In summary, these results indicate a concerted reaction for ATP synthesis by F_1, where bound ATP contains the first covalent derivative of either P_i or ADP.

Interest in the chemical coupling hypothesis was rekindled briefly by reports from D. E. Griffith's and R. S. Criddle's laboratories that thioesters and oleoyl phosphate might participate as intermediates in oxidative phosphorylation. Like other previous proposals, these observations have not withstood close scrutiny (41–45).

Is H_2O Eliminated Before, After, or During Bond Formation?

With the apparent lack of participation of covalent intermediates, the next step in understanding the reaction mechanism would come from determining the sequence of the β-γ bond formation and water elimination steps. If water were eliminated prior to bond formation, a metaphosphate would participate as an intermediate, with nucleophilic attack by a β-phosphoryl oxygen of ADP to give ATP. Such an intermediate would be difficult to detect since it is unstable in aqueous media.

If water were eliminated after bond formation, a pentacoordinate γ-phosphoryl group such as that proposed by Korman & McLick (46) would function as an intermediate. Reversible formation of such an intermediate might allow detection of an oxygen exchange between water and the γ-phosphoryl group of ATP analogues which are not cleaved by F_1. This test has been performed using adenylyl imidodiphosphate (AMPPNP) and adenylyl methylenediphosphonate (AMPPCP) by Holland et al (47). Incorporation of oxygen from water into the ATP analogues was not observed in the presence of submitochondrial particles. Although this is consistent with a requirement for bond cleavage at the water addition step, it is possible that the same properties of the analogues that prevent cleavage also prevent formation of a pentacoordinate intermediate (47).

The third possibility is that bond formation and cleavage occur simultaneously with water elimination and addition, i.e. by a pentacoordinate transition state. An ingenious technique for estimating the relative rates of water addition and bond cleavage was applied to thylakoid membranes by Wimmer & Rose (48). As an indicator of bond cleavage they measured the rate of movement of an ^{18}O-label from a β-γ bridge position to a nonbridge position on the β-phosphoryl group of ATP. This they call bridge to nonbridge scrambling, a process that requires reversible bond cleavage. In a separate experiment, as an indicator of water addition they measured the rate of loss of ^{18}O-label specifically placed in the three nonbridge oxygens of the γ-phosphoryl of ATP. This they refer to as washout (also called medium ATP $\rightleftarrows$ HOH exchange), a process that requires reversible water addition. The rates of scrambling and washout were found to be approximately equal, which suggests that water addition and bond cleavage occur with the same frequency. These results are consistent with the lack of detection of covalent intermediates.

Although these findings should receive serious consideration in formulating models for the mechanism of F_1, they do not provide unequivocal proof for a pentacoordinate transition state, since the rate-limiting steps in the reaction sequences that give rise to the measured scrambling and washout may not be bond cleavage or water addition as assumed. For example, the rate of scrambling might be limited by the rotational freedom of the β-phosphoryl group of enzyme-bound ADP. If the β oxygens of ADP that were in the bridge and nonbridge positions prior to hydrolysis do not exchange by rotation before reformation of ATP, the ^{18}O label may resume the bridge position and the cleavage step will not be detected by measuring scrambling.

In summary, the evidence for a pentacoordinate transition state is strong but not conclusive, and the question may warrant further investigation.

The Energy Requirements of Individual Steps in the Net Synthesis of ATP

Before 1973, all popular models for oxidative phosphorylation and photo-phosphorylation assumed that the point of energy input during net synthesis of ATP was the catalytic step, i.e. the formation of the anhydride bond. This was based upon biochemical precedents and the intuitive feeling that the free energy change for ATP formation at a catalytic site would, as for aqueous solution, be highly endergonic. Evidence that called this assumption into question was obtained from studies of the isotope exchange reactions catalyzed by submitochondrial particles (49). Results from measurements of the exchange reactions have been carefully reviewed (50), and only a brief summary of the rationale and implications of this approach are given here.

In coupled submitochondrial particles all reaction steps involved in ATP synthesis are readily reversible. This results in a number of isotope exchange reactions that can be used to monitor substrate binding and product release steps as well as the catalytic step. Each exchange reaction requires a different combination of binding and release steps, but all require both covalent bond formation and cleavage. If energy input during ATP synthesis were required to drive formation of the anhydride bond (Figure 1A, step 2), then the addition of uncouplers to dissipate the energized state should prevent not only net synthesis of medium ATP but also all exchange reactions. This was found not to be the case (37, 49, 51–53). The intermediate $P_i \rightleftarrows HOH$ oxygen exchange that occurs during net ATP hydrolysis is insensitive to uncoupler. This exchange is most easily measured by following the loss of ^{18}O label placed in the γ-phosphoryl group of ATP during net hydrolysis. Required steps (Figure 1A) include the binding of ATP (step –3), reversible hydrolysis of ATP (steps –2 and 2), and dissociation of ADP and P_i (step –1). The only steps not required by this uncoupler-insensitive exchange are the binding of ADP and P_i (step 1) and the release of ATP (step 3). Since each of the uncoupler-sensitive exchanges requires step 1 or 3 or both, these steps are the most likely candidates for energy input from substrate oxidation. Initially it was proposed that the release of ATP is the main energy-requiring step in oxidative- and photophosphorylation (49, 54). Later studies showed that ADP + P_i binding are also associated with energy input from oxidation (52, 55–57).

Independently, Slater and colleagues (see 58) proposed that the main energy-requiring step in ATP synthesis is the dissociation of ATP. This conclusion was based on their finding that isolated MF_1 (59) and CF_1 (60) contain tightly bound ADP and ATP. Assuming that these nucleotides

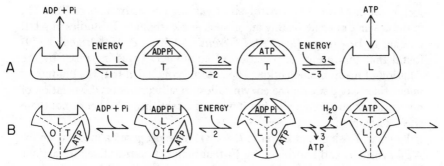

Figure 1 The energy-dependent binding change mechanism. Scheme *A,* an earlier version [see Boyer et al (49, 55)], proposes that upon energization in steps 1 and 3 a catalytic site of F₁ is interconverted between two forms: L, loose binding for ligands and catalytically inactive; T, tight binding for ligands and catalytically active. Scheme *B* is a current version that recognizes catalytic cooperativity [see Kayalar et al (80, 81)] and assumes three interacting catalytic subunits per F₁. Upon energization in step 2 the catalytic sites of F₁ are interconverted between three forms: L and T as defined for scheme *A;* O, open site with very low affinity for ligands and catalytically inactive. Scheme *B* shows one third of the enzyme cycle.

were bound at catalytic sites, they used the ratio of bound ATP to bound ADP to estimate a free energy change for ATP synthesis approximately equal to zero. Although these tightly bound nucleotides have turned out to be of keen interest, they appear not to be directly involved in catalysis (61–69).

The idea that major energy transformations occur during binding changes rather than during synthesis or cleavage of phosphoric acid anhydride bonds has been found to apply not only to oxidative- and photophosphorylation but also to muscle contraction (70) and some membrane transport systems (71–73). This premise is also consistent with the general tendency in enzyme catalysis to have complexes of nearly equal free energy precede and follow the transition state for the catalytic step. A difference in free energy for the overall reaction appears for many enzymes to be accounted for by differences in binding affinities for substrates and products [Table 4 in (39)]. It has been suggested that this "strategy" optimizes the use of substrate-binding energies to promote overall catalysis (74, 75).

Cooperative Interactions Between the Catalytic Subunits of F₁

EARLY INDICATIONS   In studies with 4-chloro-7-nitrobenzofurazan (Nbf-Cl), Ferguson et al (76) found that the modification of one tyrosyl residue on one β subunit per MF₁ causes complete inhibition of ATP hydrolysis. Since the enzyme contains two or three copies of the β subunit, the authors concluded that MF₁ shows ligand-induced asymmetry (76, 77). Ebel & Lardy (78) proposed negative cooperativity between the catalytic subunits of MF₁ as one possible explanation for the nonlinearity of their reciprocal plots for ATP hydrolysis at varying concentrations of MgATP, in the absence of an activating anion such as bicarbonate. In studies of ADP binding to BF_1 from *Alcaligenes faecalis*, Adolfsen & Moudrianakis (79) found that the rate of dissociation of bound ADP was increased by the addition of medium nucleotide. The authors concluded that the binding of nucleotide to one site on the enzyme allosterically promotes dissociation of nucleotide from another site, but it was not established whether or not these sites are catalytic.

Kayalar, Rosing, and Boyer (57, 80, 81) have proposed a mechanism for ATP synthesis and hydrolysis by F_1 that includes cooperative interactions between subunits as well as the energy-dependent binding changes discussed in the previous section. Again, measurements of oxygen exchange reactions provided the necessary information. These exchanges serve not only as an indirect measure of the reversibility of the catalytic step as described, but by measuring the amount of exchange per mole of product formed, they also provide an estimate of how many reversals of the catalytic step take place during a single net turnover of the enzyme (24, 65). Hence, by measuring the effect of substrate concentration on the amount of oxygen exchanged per mole of product formed, it can be determined whether the rate of substrate binding influences the rate of product release. In summary, these experiments provide evidence that the binding of ADP and P_i promotes the release of ATP during net synthesis (81), and that the binding of ATP promotes the release of $ADP + P_i$ during net hydrolysis (82, 83).

The proposed mechanism is shown in Figure 1*B*. As indicated by previous work, the binding of substrates and release of product are associated with energy input from oxidation (Figure 1, compare scheme *A*, steps 1 and 3 with scheme *B*, step 2). Scheme *B* has the obvious advantage of having both of these binding changes occur simultaneously on separate but interacting catalytic sites.

Independently, Kozlov & Skulachev (31) proposed a detailed mechanism for F_1 that involves simultaneous, energy-dependent binding changes for substrates and product. Though complicated by movement of nucleotides between the α and β subunits in an attempt to involve binding sites that have since been shown not to participate directly in catalysis, the basic features of Figure 1*B* are incorporated in their mechanism [Figure 9 in (31)].

OTHER EVIDENCE   In addition to Nbf-Cl (76, 84), other reagents that modify F_1 have recently been shown to exhibit "one-half of the sites" reactivity (or "one-third of the sites" if F_1 contains three catalytic subunits). The ATP hydrolysis activity of BF_1 from *E. coli* is completely inhibited by the covalent incorporation of one mole of dicyclohexylcarbodiimide

(DCCD) per mole of enzyme (85), and inactivation of beef heart MF_1 correlates to the modification of a single reactive residue by phenylglyoxal (86). In addition, one mole of efrapeptin per mole of MF_1 is sufficient to inhibit both ATP synthesis and hydrolysis by submitochondrial particles (87), and ATP hydrolysis is fully inhibited upon binding one mole of the mitochondrial inhibitor protein per mole of MF_1 (88).

Although there are multiple copies of the catalytic subunit, only one high-affinity binding site for either P_i (89) or MgADP (90, 91) can be measured on MF_1. This suggests substrate-induced asymmetry such that only one out of two or three potential tight binding sites may be expressed at one time. The cooperativity (92) and "½ site reactivity" (93, 94, 94a) observed for covalent modification of MF_1 by substrate analogues are also consistent with this interpretation.

One of the clearest means of testing for subunit cooperativity has been to measure the effect of substrate binding on the rate of product release. As summarized above, the oxygen exchange reactions give strong indirect evidence that substrate binding promotes product release. This has been shown directly in binding measurements. The addition of adenine nucleotide accelerates the release of tightly bound P_i (95), and the rate of AMPPNP binding to and release from a very high affinity site on MF_1 is strongly influenced by the binding of adenine nucleotides at additional sites (15, 96, 97).

Alternative explanations for this apparent cooperativity can be found in hysteresis, enzyme heterogeneity, or reversible binding of nucleotides to regulatory sites. These possibilities were carefully considered by Boyer and colleagues, who measured the effect of ATP concentration on distribution of ^{18}O in P_i released during hydrolysis of $[^{18}O\text{-}\gamma]$ATP by submitochondrial particles (95), and the effects of ADP and P_i concentrations on the distribution of ^{18}O in ATP formed during photophosphorylation (98). The results were well predicted by catalytic cooperativity and not by the other possible explanations.

The Overall Synthesis of ATP in Mitochondria, Chloroplasts, and Bacteria

Figure 2 summarizes a working model for the overall process of energy coupling in mitochondrial oxidative phosphorylation. The scheme incorporates mechanistic features suggested by many laboratories.

The stoichiometry of protons transported per electron pair per coupling site has been reported to be at least three (99). These and other results suggest that the transport of protons is at least in part indirectly coupled to the oxidation-reduction reactions of the respiratory chain (Figure 2, step A) by energy-driven H^+-pumps (100–103).

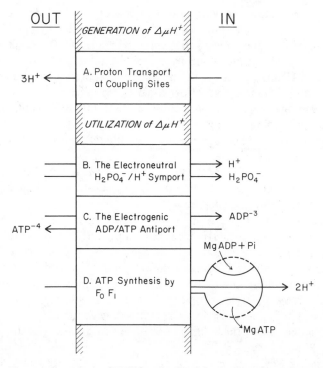

Figure 2 The overall synthesis of ATP in mitochondria. *A,* the passage of two e⁻ through one coupling site pumps three H⁺ outward against an electrochemical gradient. *B,* one phosphate is transported inward against a concentration gradient driven by the inward movement of a proton down a concentration gradient. This is equivalent to and indistinguishable from an $H_2PO_4^-$/OH⁻ antiport. *C,* the transport of one ADP inward and one ATP outward against concentration gradients driven by the outward movement of one negative charge down an electrical gradient. *D,* the inward movement of 2 H⁺ down an electrochemical gradient drives a conformational change in F_1, which promotes ADP + P_i binding at one catalytic site and drives ATP release from a second site. Greater detail for this step is shown in Figure 1*B.*

ATP synthesis in mitochondria takes place on the inside surface of the inner membrane. Hence, ADP and P_i must be transported inward and ATP outward through this membrane. Evidence indicates that these transport processes may be driven by $\Delta\mu_{H^+}$, including the observation by Klingenberg and co-workers (103a) that the phosphorylation potential in the cytosol is higher than in the mitochondrial matrix space. An energy-driven adenine nucleotide translocase has been confirmed by the demonstration that the transport process is electrogenic (104–106). It has been suggested that the uptake of one ADP and one P_i coupled to the efflux of one ATP results in the uptake of one proton (Figure 2, step B) and efflux of one negative charge (Figure 2, step C). Hence, if three protons are transported per coupling site during electron flow down the respiratory chain, then one third of the total energy available is used to move substrates and product against concentration gradients between the cytosol and matrix space (107).

Additional evidence in favor of this view comes from several recent studies. Out et al (108) found that when oxidative phosphorylation by ATP-loaded mitochondria is blocked by oligomycin and aurovertin, the addition of ADP stimulates a short burst of respiration. It was concluded that the transport of ADP into mitochondria in exchange for matrix ATP is energy requiring. Taking the inverse approach, Brand (109) reasoned that the amount of ATP synthesized in the mitochondrial matrix space per atom of oxygen reduced should increase if the energy-requiring transport of adenine nucleotides and P$_i$ were blocked. Under such conditions the ratio of intramitochondrial ATP synthesized per pair of electrons per coupling site was found to be 1.4 in comparison to a normal value of 0.9 for synthesis plus transport. This ratio is consistent with an H$^+$/2e$^-$ ratio of three in step A with one H$^+$ used to drive transport of substrates and product and two H$^+$ used to drive net synthesis of ATP by MF$_0$F$_1$. With the transport systems blocked, three H$^+$ per coupling site would in theory result in synthesis of 1.5 ATP in the matrix space.

If the major energy input step during ATP synthesis occurs during the binding of substrates and release of product, it would seem reasonable to conclude that an energy-dependent conformational change in F$_1$ is required. This is particularly clear when one considers the necessity for disrupting the strong interactions between the protein and ATP that contribute to its tight binding. Boyer (110) has proposed that the F$_0$F$_1$ complex is capable of coupling the energy available when protons or charges move down an electrochemical gradient to an energy-dependent conformational change in F$_1$ that promotes substrate binding and product release (Figure 2, Step D). Chang & Penefsky (111) have measured an energy-dependent fluorescence change in aurovertin complexed to membrane-bound MF$_1$. These studies indicate that upon addition of an oxidizable substrate to submitochondrial particles, MF$_1$ undergoes a conformational change that is sufficiently rapid to be a required step in ATP synthesis. The results also suggest that MgATP dissociates from MF$_1$ during this conformational change (111).

The stoichiometry of 2H$^+$/ATP, indicated by the results discussed above, is supported by direct measurements with submitochondrial particles (112). However, evidence has been presented that suggests that the number of protons transported in step A may actually be four (113, 114). If this is the case, the number of protons expended per ATP synthesized by F$_0$F$_1$ in step D would likely be 3.

In chloroplasts, CF$_1$ faces the stroma, and in bacteria, BF$_1$ faces the cytoplasm, thus eliminating the need to transport adenine nucleotides and phosphate. Hence, schemes for ATP synthesis by chloroplasts and bacteria would not include steps B or C (Figure 2) and for chloroplasts the direction of both steps A and D would be reversed. Although the stoichiometry of step A may be three in chloroplasts (115) as in mitochondria, the H^+/ATP ratio in step D appears to be three in chloroplasts (116).

It should be noted that there is no general consensus in favor of the model described in Figure 2. The reader is referred elsewhere for alternative views on the stoichiometry and energy dependence of steps A, B, and C (117). Three alternatives for step D are described in the next section.

ALTERNATIVE SUGGESTIONS AND VARIATIONS ON THE MECHANISM

The Removal of Phosphate Oxygen by Energized Protons

Mitchell proposes a direct role for energized protons in the synthesis of ATP by oxidative- and photophosphorylation. In his model, two protons flow down an electrochemical gradient through F_0F_1. During their passage the protons are specifically directed to the $H_2PO_4^-$ binding site on F_1 where they are able to combine with and remove an oxygen from phosphate to give H_2O and $H_2PO_3^+$ (118, 119). $MgADP^-$, also bound at the catalytic site, carries out a nucleophilic attack on the phosphorus atom of inorganic phosphate during or following removal of H_2O. A similar proposal has been made by Williams (120, 121) with the exception that intramembrane anhydrous protons provide the driving force for the water elimination step.

Obviously, if the alternative to Mitchell's proposal was that the inward transfer of protons by F_0F_1 drives the formation of ATP at the catalytic site via conformation change, it would be very difficult to distinguish experimentally between the two models. However, there is a clear difference between the proposals of Mitchell and Boyer. In the binding change mechanism the energy available from the electrochemical gradient is used not to drive the synthesis of ATP at the catalytic site but rather to promote substrate binding and product release. In contrast, Mitchell proposes that energized protons promote a shift in the equilibrium for substrates and product bound at the catalytic site, to favor ATP. Since the water elimination step is required for the oxygen exchange reactions, Mitchell's scheme less readily accomodates the fact that the intermediate $P_i \rightleftarrows HOH$ exchange is independent of electrochemical gradients. Perhaps a final choice between these models may require direct measurement of the energy dependence of the equilibrium constant for enzyme-bound substrates and product.

An Adenylate Kinase–Type Reaction Mechanism

Roy & Moudrianakis (122) have proposed that a tightly bound AMP functions as a cofactor in the synthesis of ATP by CF$_1$. According to this scheme, bound AMP and P$_i$ combine in a light-dependent reaction to give ADP. Medium ADP then binds at a second adenine nucleotide site, and a phosphoryl group is transferred directly from the first ADP to give ATP and to regenerate the tightly bound AMP. ATP dissociates and the enzyme is ready for the next cycle.

The current version of this scheme (123) proposes cooperativity between four catalytic sites (two α subunits and two β subunits). An important difference between this mechanism and the mechanism proposed by Boyer and co-workers (49, 81) is that energy is required for the water elimination step. Hence, this mechanism also fails to explain the continuation of the intermediate P$_i$ $\rightleftarrows$ HOH exchange in the absence of energization.

The proposal by Moudrianakis is mainly based on the detection of β-labeled ADP after short incubation of thylakoid membranes with AMP and ^{32}P$_i$. Most other laboratories that have investigated this phenomenon favor the following interpretation: A catalytic amount of endogenous ADP is phosphorylated by ^{32}P$_i$ and released from CF$_1$ in a light-dependent process. An energy-independent adenylate kinase then transfers the ^{32}P-labelled γ phosphoryl to an AMP to give [β–^{32}P]ADP. This product can then bind to CF$_1$, and is detected as membrane-bound, β-labeled ADP. Alternatively, if it binds at a catalytic site on CF$_1$, it may undergo phosphorylation, and appear as soluble or bound β-labeled ATP.

Moudrianakis & Tiefert (124) have countered this argument by demonstrating a transphosphorylation reaction by preparations of CF$_1$ in which the authors believe the conventional adenylate kinase is absent or inactive. However, in recent studies, McCarty (125) has shown that the addition of high levels of hexokinase plus glucose prevents the appearance of membrane-bound or soluble β-labeled ADP. These results suggest that β-labeled ADP arises from medium [γ-^{32}P]ATP rather than functioning as a nondissociable intermediate on CF$_1$. McCarty (125) also finds that a strong inhibitor of adenylate kinase, diadenosine pentaphosphate, has no effect on photophosphorylation, but prevents the appearance of β-labeled ADP.

Additional evidence against the mechanism of Moudrianakis comes from tests of the kinetic competency of β-labeled ADP as an intermediate in ATP synthesis. Regardless of whether an acid-base transition (126, 127) or light (128) is used to drive ATP synthesis, the rate of formation of [β-^{32}P]ADP is much slower on a millisecond time scale than the rate of formation of [γ-^{32}P]ATP, which suggests that ADP and not AMP is the primary acceptor of ^{32}P$_i$. This conclusion is supported by the failure to detect tightly bound AMP in washed chloroplast membranes, although

ATP and ADP are both present (60). Also, if it is assumed that MF_1 and CF_1 have the same catalytic mechanism and that ATP synthesis and hydrolysis by CF_1 occur at the same catalytic sites (see next section) then the double displacement pathway in Moudrianakis' mechanism is difficult to reconcile with inversion of configuration of the γ phosphorus atom during hydrolysis of ATP by MF_1 (40), and with the equal rates of scrambling and washout measured with CF_1 (48).

Separate Catalytic Sites for ATP Synthesis and Hydrolysis

A number of observations have led to the idea that different catalytic sites exist on F_1 for ATP synthesis and hydrolysis. The main type of evidence cited in support of separate pathways concerns the observation that some inhibitors of F_1 have different effects on ATP synthesis and hydrolysis.

In early studies, aurovertin was found to inhibit ATP synthesis to a greater extent than ATP hydrolysis at a given inhibitor to protein ratio (129). However, Linnett et al (130) have shown that this is due to use of a lower protein concentration in ATP hydrolysis assays; hence, less of the added aurovertin is bound under these conditions. When synthesis and hydrolysis of ATP by submitochondrial particles are measured at the same protein concentration as well as the same inhibitor to protein ratio, aurovertin is nearly equally effective in inhibiting both activities.

The most frequent reference to separate pathways comes from the findings that the nonhydrolyzable ATP analogue, AMPPNP, and the mitochondrial inhibitor protein are both potent inhibitors of ATP hydrolysis, but neither has any effect on steady-state rates of ATP synthesis (131–133). Rather than requiring interpretation in terms of separate pathways with different sensitivities to these agents, the results have a very simple explanation. AMPPNP and the inhibitor protein are not bound at their inhibitory sites under conditions for ATP synthesis. This was shown for AMPPNP by Penefsky (15) and Chernyak & Kozlov (96), and for the mitochondrial inhibitor protein by van de Stadt et al (134).

"Tightly-bound" adenine nucleotides may be removed from beef heart MF_1 by gel filtration in the presence of a high ionic strength buffer or by mild trypsin treatment (135). Nucleotide-depleted MF_1 retains the ability to hydrolyze ATP and to bind to MF_1-depleted submitochondrial particles, but it does not reconstitute ATP-dependent reverse electron flow. Although the results have been interpreted as support for separate sites for ATP synthesis and hydrolysis, they are also consistent with a single pathway if it is assumed that removal of the tightly bound nucleotides interferes with energy coupling between MF_1 and the oxidation-reduction reactions of the respiratory chain. Without energy transfer, hydrolysis would remain thermodynamically, and evidently, kinetically favorable. This is essentially similar to the accepted explanation for the fact that uncouplers prevent ATP synthesis but not ATP hydrolysis.

Perhaps the most suggestive evidence for separate pathways comes from recent reconstitution experiments using modified F_1. MF_1-depleted submitochondrial particles, reconstituted with Nbf-Cl-modified MF_1, retain an ATP synthesis rate 35–65% that of native membranes (136). In contrast, the ATP hydrolysis activity is inhibited by about 97%. One potential problem in interpreting this data in terms of separate catalytic sites for ATP synthesis and hydrolysis is that Nbf-Cl does not appear to act at the catalytic site (137). Modification by this reagent may introduce a new rate limitation in catalysis that would give a larger percent inhibition of the rapid ATP hydrolysis reaction (6 μmol/min/mg) than the slower ATP synthesis rate (0.15 μmol/min/mg). Indeed, the rate of ATP hydrolysis by membranes reconstituted with Nbf-Cl-modified F_1 is still twofold greater than the rate of ATP synthesis (138: W. E. Kohlbrenner, personal communication).

In other reconstitution experiments, Brodie and colleagues (139) used BF_1 modified with a 2',3'-dialdehyde derivative of ATP. Membranes reconstituted with the modified enzyme again showed a much larger percent inhibition of ATP hydrolysis than ATP synthesis. Since the analogue did not serve as a substrate for hydrolysis, it is difficult to establish whether regulatory or catalytic sites were modified. Hence, the same limitations to interpretation apply as those discussed for the Nbf-Cl studies. In addition, dialdehyde analogues of adenine nucleotides form reversible Schiff base adducts with amino groups, and it was not reported whether conditions for assaying ATP synthesis caused by reversal of inhibition of ATP hydrolysis activity.

An additional argument made for separate pathways comes from the finding that nucleotide triphosphate (NTP) hydrolysis by membrane-bound F_1 shows a greater specificity for adenine than does the NTP synthesis reaction (17 and references therein). It should be noted, however, that the equilibrium reached under standard assay conditions for synthesis is quite different than that for hydrolysis; i.e. substrate oxidation is used to drive NTP synthesis, whereas uncouplers are used to dissipate the energized state and promote NTP hydrolysis. Consequently, substrate analogues can have very different effects on the maximum velocities of the forward and reverse reactions, and a single Haldane relationship does not apply to both reactions as has been maintained (17).

It has also been noted that the nucleotide specificity for NTP-dependent reverse electron flow is higher than for NTP hydrolysis. Harris et al (140) have provided a reasonable explanation for these results in terms of the binding change mechanism. The higher specificity for NTP-driven reactions may be related to an inability of the base moiety to properly interact with the adenine-binding domain to drive the energy conserving conformation change. Hence a significant portion of the energy may be lost at the cleavage step when nucleotides other than ATP are used.

These experimental observations do not offer convincing evidence in support of separate pathways for ATP synthesis and hydrolysis. In contrast, there are data not easily interpreted in terms of separate pathways. For example, studies of the oxygen exchange reactions catalyzed by submitochondrial particles have shown that ATP synthesis at the catalytic site occurs rapidly and reversibly under conditions for net ATP hydrolysis, and that ATP hydrolysis occurs rapidly and reversibly under conditions for net ATP synthesis (49, 81). Consistent with these results is the finding that respiratory control in mitochondria is determined by the phosphate potential (141). This requires that the catalytic site for ATP synthesis be capable of rapid binding and hydrolysis of medium ATP. It seems reasonable to assume that this activity accounts for the ATP hydrolysis activity of membrane-bound and purified F_1.

In addition to the mechanisms discussed above, Racker (142) has drawn an analogy between F_0F_1 and other cation-translocating ATPases. He suggests that the driving force for ATP synthesis (or the binding changes) may be the binding of Mg^{2+} at a high affinity site. Energized protons then displace the magnesium to complete the enzyme cycle. Although MF_1 has a tight binding site for magnesium (143), this interesting proposal has not been tested experimentally.

In summary, there is no compelling reason to reject any of the proposed mechanisms. However, the general features of the energy-driven binding change mechanism appear unique in the amount of supportive evidence that has been obtained.

THE CATALYTIC SITE OF F_1

The Catalytic Subunit

In modification studies with adenine nucleotide analogues, binding sites on the two largest subunits of F_1 have been identified. However, with both photoaffinity analogues of ATP (144, 145) and alkylating analogues of ATP (146, 147), inactivation of ATP hydrolysis activity best correlates with the modification of the β subunit. In addition, the modification by DCCD of an essential residue, thought to be located at the catalytic site, results in fairly specific labeling of the β subunit of MF_1 (148), BF_1 (85), and CF_1 (149). Although Nbf-Cl (76, 150, 151) and aurovertin (152–154) also inhibit F_1 by interacting with the β subunit, these observations do not provide additional evidence that this subunit contains the active site, since the inhibitors are not likely to act at the catalytic site (137), and their effect could be mediated through subunit interactions.

Reconstitution of F$_1$ from isolated subunits provides additional information. With BF$_1$ from the thermophilic bacterium PS3, only complexes containing β (i.e. $\alpha\beta\delta$ and $\beta\gamma$) are capable of ATP hydrolysis (155).

Although the available data suggest that the β subunits of MF$_1$, CF$_1$, and BF$_1$ contain the catalytic site, it should not be assumed that F$_1$ from all sources will be similar. The α and β subunits are routinely identified on the basis of their electrophoretic mobility. In view of the small differences in molecular weight between α and β ($<$ 10% in most cases), evolutionary change in some organisms may have resulted in a reversal of the relative mobility of the two largest subunits.

Essential Amino Acid Residues

ARGINYL RESIDUES Chemical modification studies with arginine-specific reagents have led Riordan et al (156) to suggest that an arginyl residue is generally present at the active site of enzymes that catalyze phosphoryl transfer or phosphorylation reactions. The guanidinium group is ideally suited to interact with anions. It has a planar structure and can form multiple hydrogen bonds with phosphorylated metabolities. The presence of arginyl residues at anionic substrate-binding sites has been confirmed for several enzymes by X-ray crystallography, but in most cases the presence of an essential arginine has been inferred from the specificity of 2,3-butanedione and phenylglyoxal and from the effect of borate on the formation and stability of the butanedione derivative.

Purified MF$_1$ (157, 158), CF$_1$ (159), and BF$_1$ (160) contain groups essential for catalytic activity that are modified by arginine-specific reagents. While MF$_1$ contains approximately 168 arginyl residues (161), a number of which react slowly with phenylglyoxal, a single fast-reacting essential residue has been observed (86).

LYSYL RESIDUES Pyridoxyl-5'-phosphate (PLP) forms a Schiff-base complex with lysyl residues on a variety of enzymes, including many that do not require it as a cofactor. The complex can be readily converted to a stable form by reduction with sodium borohydride. A common feature of enzymes susceptible to PLP modification is that they contain allosteric or active sites that normally bind sugar phosphates or sugar phosphate–containing cofactors such as adenine nucleotides (162).

In a general survey utilizing group-specific reagents, Ferguson et al (163) found that PLP inhibits MF$_1$. ADP and ATP provided partial protection against inactivation. Godinot et al (164) have recently confirmed this observation and have noted PLP-mediated photoinactivation of MF_1.

Using [^{3}H]PLP, the loss of ATP hydrolysis activity by beef heart MF_1 was shown to be linearly dependent on the level of pyridoxylation, with complete inactivation correlating to 10 mole of reagent incorporated per mole of enzyme (165). This confirms the stoichiometry determined by absorption measurements at a wavelength characteristic of pyridoxamine, and suggests that lysyl residues have been modified (163).

In initial studies, the modification of CF_1 by PLP indicated the presence of a large number of essential lysyl groups (166). Subsequently, conditions were found that allow complete inactivation, with the modification of only one α and one β subunit per CF_1 (167).

TYROSYL RESIDUES   Esch and Allison (146) have modified nucleotide-depleted MF_1 using p-fluorosulfonylbenzoyl-5'-adenosine (FSBA). Although both the α and β subunits are modified, inactivation of MF_1 correlates to incorporation of one nucleotide analogue into each β subunit. A single modified peptide was isolated from digests of purified β subunit. The sequence of this 16 amino acid peptide was determined and a modified tyrosine identified. In contrast, label incorporated into the α subunit is nonspecifically associated with a number of tyrosine and lysine residues (168). FSBA also inactivates CF_1, but the derivative has not been characterized (169).

The earliest indications of an essential tyrosine were obtained by Senior (170). Tetranitromethane and iodoacetate rapidly inactivate MF_1, and inactivation was fully or partially prevented by ATP and ADP. Evidence suggests that the same tyrosyl residues may have been modified as were modifed by FSBA. Although Nbf-Cl also modifies an essential tyrosine, it appears to act at a site different from FSBA (168).

CARBOXYL GROUPS   In addition to reacting with a component of F_0, DCCD inactivates MF_1 (171). Using [^{14}C]DCCD, Vignais and co-workers found that ATP hydrolysis activity is fully inhibited with the incorporation of two moles of reagent per mole of beef heart MF_1 (148) and one mole of reagent per mole of E. coli BF_1 (85). In both cases $MgCl_2$ slowed the rate of inactivation, and the label was found in the β subunit. CF_1 is also susceptible to inhibition by DCCD. Shoshan & Selman (149) found that inactivation correlates to the modification of two β subunits per enzyme and loss of one adenine nucleotide-binding site. $CaCl_2$ provides appreciable protection against modification and inactivation of CF_1 by DCCD (J. L. Arana and R. H. Vallejos, personal communication).

Two other carboxyl reagents that inhibit MF_1, N-cyclohexyl-N'-β-(4-methylmorpholine)-ethylcarbodiimide (31) and N-ethoxycarbonyl-2-ethoxy-1,2-dihydroquinoline (172), do not appear to react with the same carboxyl group, since they do not affect modification by DCCD (148).

POSSIBLE ROLES FOR THE ESSENTIAL RESIDUES   Efrapeptin protects the essential arginyl residue on MF$_1$ from modification by phenylglyoxal (86). Evidence that efrapeptin and phosphate binding are mutually exclusive (173, 174) suggests that an arginyl residue may participate in phosphate binding. Adenine nucleotides, but not efrapeptin or phosphate, protect essential lysyl residues on MF$_1$ from modification by PLP (165). Furthermore, extensive modification of MF$_1$ by PLP decreases the number of adenine nucleotide–binding sites (97), but has no effect on the stoichiometry of phosphate binding (P. G. Koga and R. L. Cross, unpublished data). These results suggest that a lysyl residue may participate in nucleotide binding. The presence of an arginyl and a lysyl residue at the catalytic site would be consistent with the participation of two positively charged groups at a site that simultaneously binds two anionic substrates, H$_2$PO$_4^-$ (173) and MgADP$^-$ (175).

The partial tyrosine spectrum seen in difference spectra of MF$_1$ with and without added ADP or ATP (176) could be due to direct π-π bond interaction between the purine ring of ADP and the tyrosyl residue. Using 3'-O-(5-dimethylaminonaphthoyl-1-)ADP, it was found that the fluorescence of the probe is not enhanced by binding to MF$_1$ (177) even though previous work suggested that the adenine binding site is very hydrophobic. Since the fluorescence of the ADP analogue is quenched by tyrosine in weakly solvating aprotic solvents, the authors proposed an interaction between the probe and an aromatic residue on MF$_1$. Observations of close stacking of the adenine ring with aromatic residues on other enzymes that bind adenine-containing cofactors (178, 179) support such a possibility. Alternatively, the tyrosyl residue may participate as a general acid-base catalyst (146), based on the finding that the pH profile for modification of the essential tyrosyl with FSBA is similar to that for ATP hydrolysis activity.

When magnesium is added in large excess to adenine nucleotide, it inhibits ATP synthesis (175) and hydrolysis (180). Tight-binding sites for magnesium have been identified on MF$_1$ (59, 143) and BF$_1$ (155, 181, 182). The fact that magnesium provides the best protection of MF$_1$ and BF$_1$ against modification by DCCD suggests that the essential carboxyl groups may participate in magnesium binding.

A Model for ATP Synthesis at the Catalytic Site

The modification studies discussed above provide valuable information regarding the topography of the catalytic site of F$_1$. Such information is vital in understanding chemical details not given in Figure 1B. The speculative model for ATP synthesis at the catalytic site of F$_1$ (shown in Figure 1) is based on several assumptions. First, a straightforward interpretation is made of the effects of ligands on the protection of essential residues. If the primary effect of these ligands on modification rates is to influence cooperative interactions between the subunits of F_1 some of these assignments may have to be modified. Second, it is assumed that the mechanisms of phosphoryl transfer by well-characterized enzymes may provide insight into the mechanism of F_1. Studies of kinases, particularly pyruvate kinase, have strongly influenced the model (see below). Finally, Figure 3 is predicated on the belief that net ATP synthesis is indirectly coupled to the electrochemical gradient. Hence, basic amino acid residues at the catalytic site are not proposed to play a role in the overall transfer of protons from one side of the membrane to the other as has been suggested (W. S. Allison, personal communication). Although other valid models can certainly be devised, this one is offered in the hope that it will stimulate interest among investigators in the field.

With these reservations in mind, it is proposed that upon binding MgADP⁻, the adenine ring is stabilized in part by interaction with an aromatic residue, perhaps tyrosine. The Mg^{2+}, which chelates one α- and one β-phosphoryl oxygen of ADP, also coordinates to two carboxyl oxygens of an aspartyl or glutamyl residue. The addition of this negative charge to the magnesium nucleotide complex is compensated for by the interaction of one α- and one β-phosphoryl oxygen of ADP with a positively charged amino acid residue, perhaps lysine. The MgADP⁻ still has a single negative charge, now specifically located on the β oxygen that will carry out a nucleophilic attack on the phosphorus of P_i. The negative charge on the other substrate, $H_2PO_4^-$, is neutralized by interaction with a second positively charged residue, perhaps arginine. Not shown is an adjacent catalytic subunit containing tightly bound ATP that will be released during the next step (see Figure 1B). The enzyme is now in a form that will accept energization (Figure 3A). An energy-dependent conformational change converts the catalytic site from latent to active, and ADP + P_i binding from loose to tight. Concurrent with this conformational change, magnesium breaks its coordination to the carboxyl and α oxygen to form a bridge between one β and one phosphate oxygen. Electron withdrawal from phosphate by magnesium (and the positively charged residue) facilitates elimination of H_2O and formation of the β-γ oxygen bridge in ATP (see arrows, Figure 3A). Only α- and γ-phosphoryl oxygen now coordinate with the two positively charged residues (Figure 3B), and the carboxyl group may be protonated. In the form shown in Figure 3B, the β-γ bond formation and water elimination steps are readily reversible, and H_2O at the catalytic site can exchange with medium water to account for the oxygen exchange reactions.

Magnesium and a charged amino acid residue are proposed to play a role

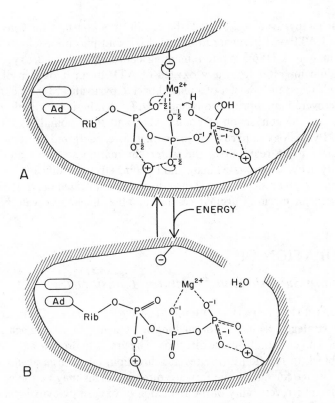

Figure 3 A speculative scheme for ATP synthesis at the catalytic site of F₁.

in withdrawing electrons from phosphate. Unlike Mitchell's scheme (118), where energized protons strip an oxygen from phosphate, Figure 3 depicts a more spontaneous dissociation of water, consistent with the fact that electrochemical gradients are not required for reversible formation of ATP at the catalytic site of soluble F_1 (83).

The proposed movement of magnesium from bridging α and β oxygens on ADP to β and γ oxygens on ATP is in keeping with mechanisms that have been proposed for kinases (183, 184). The idea of a transition state containing a divalent metal ion bridge between a β oxygen of ADP and an oxygen of phosphate receives support from studies with chromium(III)-substituted substrate analogues and beef heart MF_1 (185). These results include the formation of enzyme-bound $P_i \cdot Cr(III)ADP$ from addition of either β-monodentate Cr(III)ADP plus P_i or β,γ-bidentate Cr(III)ATP.

The feasibility of the energetics of this scheme, as well as the role of magnesium, is strengthened by considering the reaction mechanism of pyruvate kinase. Although the overall equilibrium constant for synthesis of phosphoenolpyruvate is 3×10^{-4}, the equilibrium constant for conversion of bound ATP and pyruvate to bound ADP and phosphoenolpyruvate is approximately 1 (186). While remaining coordinated to a β oxygen, the magnesium migrates from a γ oxygen of ATP to an α oxygen of ADP during cleavage and transfer of a phosphoryl to pyruvate (187). In comparison, the overall equilibrium constant for ATP synthesis by soluble F_1 is 4 $\times 10^{-5}$, and the conversion of bound ADP and P_i to bound ATP occurs readily. By analogy to pyruvate kinase, therefore, energized protons would not appear to be required in the water elimination step. An additional similarity between pyruvate kinase and F_1 comes from modification studies. An essential lysyl residue on the kinase has been implicated in ADP binding (188), and an essential arginyl residue in phosphoenolpyruvate binding (189).

REGULATION OF F_1

Inhibitory Subunits and Structural Transitions

GENERAL FUNCTION AND BIOLOGICAL NECESSITY   There are two general strategies available for making a sequence of enzyme-catalyzed reactions run essentially in one direction. The first is to provide a thermodynamic block to reaction in the reverse direction, i.e. a large negative free energy change for flux in the forward direction. This may be the cause of the apparent irreversibility of the reduction of oxygen by cytochrome oxidase in oxidative phosphorylation. The second is to provide a kinetic block by inhibiting an enzyme in the reaction sequence under conditions that would normally favor flux in the reverse direction. In mitochondria, the coupled reactions of electron transport and ATP synthesis are in near equilibrium (190). Hence, the overall free energy change for oxidative phosphorylation at each coupling site approaches zero under steady-state conditions. Considering the capacity for rapid ATP synthesis by oxidative- and photophosphorylation, the reverse reaction, i.e. ATP-driven reverse electron flow, should readily occur under conditions of high phosphate potential and limited supply of oxidizable substrate or light. However, under these conditions, F_1 in mitochondria, chloroplasts, and some bacteria appears to be inactive.

The biological necessity for reversible activation of F_1 is best appreciated by considering photophosphorylation in chloroplasts. In daylight there is an abundance of photons to promote electron flow from water to $NADP^+$. Under these conditions an active form of CF_1 catalyzes net synthesis of ATP. At night, however, if CF_1 were still in an active form, ATP would be rapidly hydrolyzed to drive reverse electron flow to the photocen-

where the energy would be lost as luminescence. Hence, as the sun set, photosynthetic eucaryotes would glow until their stored carbohydrates were exhausted, in an attempt to maintain the phosphate potential and NADH-NADPH levels. Such a catastrophe is avoided, at least in part, by the interconversion of CF_1 between active and inactive forms. Overall, this regulatory process may be somewhat akin to a one-way swinging door that opens under energized conditions to allow net ATP synthesis but closes tightly under deenergized conditions to prevent ATP hydrolysis.

In mitochondria, the reversible activation of MF_1 appears to be mediated by the reversible dissociation of a low-molecular-weight inhibitor protein (191). In chloroplasts the process appears more complex. A chloroplast inhibitor protein (192) has been proposed to play a role similar to that of the mitochondrial inhibitor, except that upon activation of CF_1 the inhibitor is displaced from its inhibitory site but does not dissociate from the enzyme complex. In addition, the interconversion of sulfhydryls and disulfides and the dissociation of inhibitory ligands have been observed upon activation of CF_1. Evidence related to the details of these processes are discussed in the following sections.

PROPERTIES OF THE INHIBITOR PROTEINS   The mitochondrial inhibitor protein is distinct from the five subunits of MF_1 (193); it has an apparent molecular weight less than δ and greater than ϵ. During preparation of submitochondrial particles or MF_1, much of the inhibitor is lost unless ATP is included in the isolation media. In chloroplasts and *E. coli* the smallest subunit of CF_1 (192) and BF_1 (194, 195) may function as an inhibitor. This is based on the comigration of inhibitory activity and the ϵ subunit on one-dimensional sodium dodecyl sulfate polyacrylamide gels. The possibility that this inhibitory activity is due to a contaminating polypeptide has not been ruled out. In contrast to the mitochondrial inhibitor, the inhibitor does not readily dissociate from CF_1 and *E. coli* BF_1 during isolation, and the ATP hydrolysis activity of the soluble enzymes is latent (192, 195). Recently it has been shown that the ϵ subunit of *E. coli* BF_1 does not inhibit ATP hydrolysis by the membrane-bound enzyme (196), although the subunit is required for rebinding BF_1 to depleted membranes (197). The authors suggest that ϵ normally functions in binding BF_1 to F_0, and that its inhibitory activity may be an artifact associated with detachment of the enzyme from the membrane (196).

The inhibitor proteins isolated from mitochondria, chloroplasts, and bacteria share a number of properties. They are low-molecular-weight proteins (7,500–13,000 daltons), heat stable, and very sensitive to trypsin. Comparison of the amino acid compositions of the mitochondrial inhibitor and the ϵ subunits of CF_1 and BF_1 shows that they have a large number of charged residues and few tryptophan or proline (192–194, 198). In most cases they have a basic isoelectric point (pH 7.5–10).

The ϵ subunit can be cross-linked to α, β, and γ in CF_1 (199), and to β and γ in *E. coli* BF_1 (200). Removal of the three smallest subunits of CF_1 or *E. coli* BF_1 by treatment with trypsin results in complexes containing α and β that retain an ATP hydrolysis activity. The inability of isolated ϵ to inhibit ATP hydrolysis by the corresponding α-β complex from CF_1 or BF_1 (150, 201), and the ability of ϵ to inhibit ATP hydrolysis by the α-β-γ complex from BF_1 (202) suggest a role for γ in binding the ϵ subunit. In contrast, the mitochondrial inhibitor has been found to cross-link only to the β subunit of MF_1 (88).

The mitochondrial inhibitor protein blocks phosphate binding to MF_1 (89), and phosphate stimulates uncoupler-dependent ATP hydrolysis in whole mitochondria, presumably by promoting the release of the inhibitor (203). Activation of CF_1 increases the number of AMPPNP binding sites from two to three (204). In addition, thylakoid membranes must be energized in order to obtain inhibition of CF_1 by the phosphate analogues, SO_4^{-2} and MnO_4^{-2} (56). These results were interpreted in terms of a link between substrate binding and energization of CF_1. However, the results are also consistent with a requirement for converting CF_1 from an inactive to an active form prior to phosphate binding.

Aurovertin promotes the dissociation of the mitochondrial inhibitor from MF_1 (205), and oligomycin prevents the binding of inhibitor to inhibitor-depleted submitochondrial particles (206). Tentoxin at high concentrations stimulates ATP hydrolysis by CF_1. This stimulation is reversed by the addition of purified ϵ subunit (207).

The chemical modification of several essential amino acid residues of F_1 does not appear to be influenced by the activity state of the enzyme. The modification of essential tyrosine (169), carboxyl (149), lysine (166), and arginine (208) can be carried out with the latent CF_1. Nbf-Cl modification of BF_1 from *A. faecalis* (209) and *Paranitrococcus denitrificans* (210), and MF_1 from beef heart (211) can also be done with latent enzyme. In each of these studies, however, it was not possible to determine whether or not modification of the enzyme caused a displacement of an inhibitor from its inhibitory site.

REVERSIBLE INHIBITION BY THE INHIBITOR PROTEINS For a long time it was believed that membrane-bound MF_1 and CF_1 could be inactive with respect to ATP hydrolysis, but fully active with respect to ATP synthesis. The demonstration by van de Stadt et al (134) that ATP hydrolysis activity of submitochondrial particles could be activated by centrifuging the particles under conditions for ATP synthesis suggested that the mitochondrial inhibitor protein may dissociate from MF_1 under energized conditions. Consistent with the idea that while bound, the inhibitor protein may inhibit both ATP synthesis and hydrolysis is the long recognized need to preincubate submitochondrial particles with oxidizable substrate, prior to addition of ADP. In the absence of preincubation, a 5–15 sec lag is observed in reaching steady-state ATP synthesis rates. It has recently been shown that this lag may be related to the time required for the energy-dependent dissociation of the mitochondrial inhibitor from its site of inhibition (212, 213).

It also appears that prior to activation of the photosynthetic electron transport chain, membrane-bound CF_1 is incapable of either ATP synthesis or hydrolysis. Using short flashes of light, Harris & Crofts (214) observed the simultaneous appearance of ATP synthesis and hydrolysis activities in spinach chloroplasts. Conditions that increased or decreased the lag in onset of ATP synthesis caused a similar change in the lag preceding ATP hydrolysis. The authors concluded that displacement of an inhibitor protein from its site of inhibition is required for ATP synthesis as well as for hydrolysis. However, these results are also consistent with activation of CF_1 mediated by disulfide-sulfhydryl transitions or by an energy-dependent dissociation of an inhibitory ligand as discussed below.

THE MgATP REQUIREMENT FOR REGULATION BY THE MITOCHONDRIAL INHIBITOR   The inhibition of membrane-bound or soluble MF_1 by the mitochondrial inhibitor protein requires MgATP (215). Several plausible explanations for this requirement appear to have been eliminated. For example, the inhibitor might bind preferentially to the $MF_1 \cdot$ MgATP complex. However, results of experiments in which MgAMPPNP was used to simulate the effect of MgATP binding without the subsequent hydrolysis step make this possibility unlikely. In contrast to MgATP, MgAMPPNP does not promote the binding of ^{14}C-labeled yeast inhibitor to inhibitor-depleted submitochondrial particles (206), and fails to promote binding of ^{125}I-labeled mitochondrial inhibitor to submitochondrial particles (J. W. Power and R. L. Cross, unpublished findings). The products of MgATP hydrolysis, i.e. MgADP, P_i, and the energized state, also do not promote inhibitor binding (134). It has thus been suggested that the mitochondrial inhibitor binds to a transient conformational state of MF_1 that forms during ATP hydrolysis (216).

An additional possibility is that the mitochondrial inhibitor may undergo reversible phosphorylation. The inhibitor can be phosphorylated by a fairly nonspecific cAMP-dependent protein kinase (J. W. Power and R. L. Cross, unpublished findings). A protein kinase activity of unknown function, and unresponsive to cAMP or cGMP, is associated with the matrix side of the inner mitochondrial membrane (217). In recent studies, calcium has been found to affect the rate of ATP hydrolysis by submitochondrial particles, presumably by affecting the binding of the mitochondrial inhibitor (218, 219). These results could reflect the presence in mitochondria of a protein kinase/protein phosphatase system modulated by calcium and specific for inhibitor protein. Such a system might respond to factors other than the energy state of the respiratory chain in modulating the activity of F_1 by the inhibitor protein. For example, conditions may exist in the cell under which mitochondrial calcium transport or other energy-linked functions of the respiratory chain must take precedence over ATP synthesis.

INVOLVEMENT OF CYSTEINYL RESIDUES IN REGULATION OF CF_1
With intact spinach chloroplasts, the ATP hydrolysis activity of membrane-bound CF_1 can be activated by light (see 220). However, following osmotic lysis of chloroplasts, illumination no longer efficiently triggers ATP hydrolysis until dithiothreitol and a mediator of electron transport are added. It has been suggested that during lysis a soluble or loosely bound factor is lost that normally participates in the activation of CF_1 (221). Consistent with this suggestion is the finding that addition of reduced ferredoxin and ferredoxin-thioredoxin reductase restores light-activated ATP hydrolysis in broken chloroplasts (222). A light-dependent inhibition of photophosphorylation by sulfhydryl reagents has also been observed (223) that is due to modification of a group on the γ subunit of membrane-bound CF_1 that is inaccessible to reagent in the dark (224–226).

The latent ATP hydrolysis activity of solubilized CF_1 can be activated under relatively mild conditions by dithiothreitol (227) or by heat (228). CF_1 activated by heat has two additional sulfhydryl groups on the γ subunit that can be modified by N-ethyl maleimide (229). In addition, if N-ethyl maleimide is present during the heat treatment, CF_1 is not activated (230, 231). These results are consistent with the recent demonstration that heat activation of CF_1 is accompanied by the conversion of a disulfide bond on the γ subunit to sulfhydryl groups (232).

These observations suggest a possible role for structural transitions in the regulation of CF_1. The relationship of this process to the mode of action of the inhibitor protein observed in fractions containing the ϵ subunit is not known. The ϵ subunit is more easily removed from active than from latent CF_1, and contains cysteine (233). However, it has not been established whether the inhibitory activity of ϵ has a physiological role.

Role of Ligand-Binding Sites in Regulation

Specific regulatory sites for oxyanions such as bicarbonate have been proposed based on their rate enhancement and modulation of cooperativity in ATP hydrolysis by MF_1 from liver (78, 180, 234). Recktenwald & Hess (235) have reported an anion-dependent displacement of an inhibitory MgATP from a low-affinity regulatory site on yeast MF_1. An anion-dependent displacement of inhibitory magnesium from *A. faecalis* (236) and *Rhodospirillum rubrum* (237) BF_1 has also been proposed. In addition, Penefsky (89) measured an oxyanion stimulation of P_i binding to beef heart MF_1. The presence of significant levels of carbonic anhydrase in liver mitochondria (238), and the particular sensitivity of liver MF_1 to stimulation by bicarbonate (239), suggest an interesting possibility for regulation.

When F_1 is treated to remove all medium nucleotides, endogenous, tightly bound ADP and ATP can still be detected on MF_1 (59, 240–243), CF_1 (60, 244, 245), and BF_1 from several sources (246, 247). Although these tightly bound nucleotides exchange with medium nucleotides when membrane-bound F_1 is energized (60, 248, 249), the rate of exchange does not appear to be rapid enough for these nucleotides to play a direct role in catalysis (61–69). Consequently, a structural or regulatory role has been proposed for these adenine nucleotide sites (22; see 250). The adenine nucleotide binding sites on F_1 have been carefully reviewed (21), and only brief reference to several recent studies is given here.

Measurement of the pre-steady-state kinetics of ATP hydrolysis by MF_1 (251–255), heat-activated CF_1 (256), and BF_1 (236, 257) demonstrate a M^{2+} ATP-dependent activation of F_1 that is independent of the effects of inhibitory peptides or or disulfide-sulfhydryl transitions. The activation process has been attributed to various factors including an ATP-dependent release of inhibitory M^{2+} (236, 251, 253, 256), ADP (255; see 258), or MgADP (252), or to an ATP-dependent conformational change in F_1 (254).

Direct comparisons of the results of these studies is made difficult by the variety of conditions used. Slater and co-workers (255) suggest that the activation that they observe may reflect the dissociation of inhibitory ADP from one of the tight-binding sites. This is consistent with the absence of a lag in ATP hydrolysis seen with adenine nucleotide–depleted MF_1. Recktenwald & Hess (254) present evidence for an activation process independent of the tightly bound nucleotides. The lag they measure is not influenced by conditions that affect the rate of dissociation or exchange of the tightly bound nucleotides. In addition, thermophilic bacterium BF_1, which lacks tightly bound nucleotides, also shows a lag (257). Vinogradov and co-workers (252, 253) interpret a lag in ATP hydrolysis by MF_1 as due to the release of MgADP from a regulatory site. Presumably this is not one of the tight-binding sites, since these sites appear to bind magnesium-free nucleotide (143). They point out that the rate of dissociation of the MgADP is too slow to be a catalytic step in ATP hydrolysis and thus they favor involvement of a regulatory site. However, the slow release of MgADP could reflect a need for bound P_i for rapid release of products during ATP hydrolysis (see 259).

In summary, it is too early to conclude that any of these activation processes play a physiological role in the regulation of F_1. However, it would seem prudent to consider these results when interpreting energy-dependent activation of membrane-bound F_1.

CONCLUSIONS AND FUTURE PROSPECTS

Great strides have been made in understanding the mechanism of enzyme-catalyzed phosphoryl transfer reactions in the last few years due to the development of new probes and techniques that have revealed information regarding previously cryptic steps. These techniques include positional iso-tope exchange, the stereochemical course of phosphoryl transfer, the use of nondissociating metal ion chelates of adenine nucleotides, and substrate modulation of isotope exchange reactions. The latter technique has been applied mainly to F_1, and has provided some of the most convincing evidence yet for subunit cooperativity in enzyme catalysis.

A substantial amount of evidence has accumulated supporting the energy-dependent binding change mechanism (Figure 1B). According to this proposal the most immediate form of energy utilized for the net synthesis of ATP is an energy-dependent conformational change on F_1. Hence this model retains features of conformational coupling first proposed by Boyer (260). However, it also appears that the driving force for this conformational change is a delocalized electrochemical gradient as proposed by Mitchell (1). Thus over the past twenty years those who have suggested that the actual mechanism of ATP synthesis may incorporate features of more than one of the hypotheses may prove to have been not only conciliatory, but prophetic.

A number of exciting questions remain concerning the molecular mechanism and regulation of F_1. Information vital to an understanding of these processes includes: (a) a definition of the environment at the catalytic site and the role of amino acid residues in catalysis; (b) the number, subunit location, and function of the adenine nucleotide binding sites; (c) the mechanism of regulation of F_1 by polypeptides, ligands, and structural transitions; (d) the number of catalytic subunits in F_1; and (e) the mechanism of proton transport by F_0F_1. Research in all of these areas is currently very active, and it seems likely that a clearer picture of oxidative- and photophos-phorylation will emerge as these aspects of the problem are resolved.

ACKNOWLEDGMENTS

I am very grateful to colleagues who provided preprints and helpful suggestions during the preparation of this review. I am particularly indebted to Paul Boyer, William Kohlbrenner, Richard McCarty, Carlo Nalin, and Harvey Penefsky who provided essential criticisms of early drafts of the manuscript. All remaining errors and omissions are mine. Work reported from this laboratory was supported by the United States Public Health Service, grant GM 23152.

Literature Cited

1. Mitchell, P. 1961. *Nature* 191:144–48
2. Cattell, K. J., Lindop, C. R., Knight, I. G., Beechey, R. B. 1971. *Biochem. J.* 125:167–77
3. Shchipakin, V., Chuchlova, E., Evtodienko, Y. 1976. *Biochem. Biophys. Res. Commun.* 69:123–27
3a. Nelson, N., Eytan, E., Notsani, B. E., Sigrist, H., Sigrist-Nelson, K., Gitler, C. 1977. *Proc. Natl. Acad. Sci. USA* 74:2375–78
4. Criddle, R. S., Packer, L., Shieh, P. 1977. *Proc. Natl. Acad. Sci. USA* 74:4306–10
5. Okamoto, H., Sone, N., Hirata, H., Yoshida, M., Kagawa, Y. 1977. *J. Biol. Chem.* 252:6125–31
6. Negrin, R. S., Foster, D. L., Fillingame, R. H. 1980. *J. Biol. Chem.* 255:5643–48
7. Baird, B. A., Hammes, G. G. 1979. *Biochim. Biophys. Acta* 549:31–53
8. Senior, A. E. 1979. In *Membrane Proteins in Energy Transduction,* ed. R. A. Capaldi. pp. 233–78. New York: Dekker
9. Pullman, M. E., Penefsky, H. S., Datta, A., Racker, E. 1960. *J. Biol. Chem.* 235:3322–29
10. Penefsky, H. S., Pullman, M. E., Datta, A., Racker, E. 1960. *J. Biol. Chem.* 235:3330–36
11. Avron, M. 1963. *Biochim. Biophys. Acta* 77:699–702
12. Vambutas, V. K., Racker, E. 1965. *J. Biol. Chem.* 240:2660–67
13. Abrams, A. 1965. *J. Biol. Chem.* 240:3675–81
14. Melandri, B. A., Baccarini-Melandri, A., San Pietro, A., Gest, H. 1970. *Proc. Natl. Acad. Sci. USA* 67:477–84
15. Penefsky, H. S. 1979. *Adv. Enzymol.* 49:223–80
15a. Criddle, R. S., Johnston, R. F., Stack, R. J. 1979. *Curr. Top. Bioenerg.* 9:89–145

16. McCarty, R. E. 1978. *Curr. Top. Bioenerg.* 7:245–78
17. Shavit, N. 1980. *Ann. Rev. Biochem.* 49:111–38
18. Kagawa, Y. 1978. *Adv. Biophys.* 10:209–47
19. Fillingame, R. H. 1981. *Curr. Top. Bioenerg.* 11: In press
20. Fillingame, R. H. 1980. *Ann. Rev. Biochem.* 49:1079–1113
20a. Futai, M., Kanazawa, H. 1980. *Curr. Top. Bioenerg.* 10:181–215
21. Harris, D. A. 1978. *Biochim. Biophys. Acta* 463:245–73
22. Slater, E. C., Kemp, A., van der Kraan, I., Muller, J. L. M., Roveri, O. A., Verschoor, G. J., Wagenvoord, R. J., Wielders, J. P. M. 1979. *FEBS Lett.* 103:7–11
23. Boyer, P. D. 1977. *Trends Biochem. Sci.* 2:38–41
24. Boyer, P. D. 1979. In *Membrane Bioenergetics,* ed. C. P. Lee, G. Schatz, L. Ernster. pp. 461–79. Reading: Addison-Wesley
25. Harold, F. M. 1977. *Curr. Top. Bioenerg.* 6:84–149
26. Mitchell, P. 1979. *Science* 206:1148–59
27. Hinkle, P. C., McCarty, R. E. 1978. *Sci. Am.* 238:104–23
28. Williams, R. J. P. 1978. *Biochim. Biophys. Acta* 505:1–44
29. Ernster, L., Asami, K., Juntti, K., Coleman, J., Nordenbrand, K. 1977. *Struct. Biol. Membr.* 34:135–56. (Nobel Symp.)
30. Boyer, P. D., Chance, B., Ernster, L., Mitchell, P., Racker, E., Slater, E. C. 1977. *Ann. Rev. Biochem.* 46:955–1026
31. Kozlov, I. A., Skulachev, V. P. 1977. *Biochim. Biophys. Acta* 463:29–89
32. Reeves, S. G., Hall, D. O. 1978. *Biochim. Biophys. Acta* 463:275–97
33. Racker, E., Krimsky, I. 1952. *J. Biol. Chem.* 198:731–43

710   CROSS

34. Slater, E. C. 1953. *Nature* 172:975–78
35. Boyer, P. D. 1957. In *Proc. Int. Symp. Enzyme Chem., Tokyo and Kyoto,* pp. 301–7. London: Maruzen
36. Hinkle, P. C., Penefsky, H. S., Racker, E. 1967. *J. Biol. Chem.* 242:1788–92
37. Mitchell, R. A., Hill, R. D., Boyer, P. D. 1967. *J. Biol. Chem.* 242:1793–801
38. Jones, D. H., Boyer, P. D. 1969. *J. Biol. Chem.* 244:5767–72
39. Knowles, J. R. 1980. *Ann. Rev. Biochem.* 49:877–919
40. Webb, M., Grubmeyer, C., Penefsky, H. S., Trentham, D. R. 1980. *J. Biol. Chem.* 255:11637–39
41. Haddock, B. A., Begg, Y. A. 1977. *Biochem. Biophys. Res. Commun.* 79:1150–54
42. Sharma, M., Wang, J. H. 1978. *Biochem. Biophys. Res. Commun.* 84:144–50
43. De Chadarevjan, S., De Santis, A., Melandri, B. A., Melandri, A. B. 1979. *FEBS Lett.* 97:293–95
44. Singh, A. P., Bragg, P. D. 1979. *FEBS Lett.* 98:21–24
45. Kagawa, Y., Sone, N., Hirata, H., Yoshida, M. 1979. *Trends Biochem. Sci.* 4:31–33
46. Korman, E. F., McLick, J. 1970. *Proc. Natl. Acad. Sci. USA* 67:1130–36
47. Holland, P. C., LaBelle, W. C., Lardy, H. A. 1974. *Biochemistry* 13:4549–53
48. Wimmer, M. J., Rose, I. A. 1977. *J. Biol. Chem.* 252:6769–75
49. Boyer, P. D., Cross, R. L., Momsen, W. 1973. *Proc. Natl. Acad. Sci. USA* 70:2837–39
50. Rosing, J., Kayalar, C., Boyer, P. D. 1976. In *The Structural Basis of Membrane Function,* ed. Y. Hatefi, L. Djavadi-Ohaniance. pp. 189–204. New York: Academic
51. Cross, R. L., Boyer, P. D. 1975. *Biochemistry* 14:392–98
52. Rosing, J., Kayalar, C., Boyer, P. D. 1977. *J. Biol. Chem.* 252:2478–85
53. Russo, J. A., Lamos, C. M., Mitchell, R. A. 1978. *Biochemistry* 17:473–80
54. Boyer, P. D. 1974. In *Dynamics of Energy-Transducing Membranes,* ed. L. Ernster, R. Estabrook, E. C. Slater, pp. 289–301. Amsterdam: Elsevier
55. Boyer, P. D., Smith, D. J., Rosing, J., Kayalar, C. 1975. In *Electron Transfer Chains and Oxidative Phosphorylation,* ed. E. Quagliariello, S. Papa, F. Palmieri, E. C. Slater, N. Siliprandi. pp. 361–72. Amsterdam: North Holland
56. Jagendorf, A. T. 1975. *Fed. Proc.* 34:1718–22
57. Kayalar, C., Rosing, J., Boyer, P. D. 1976. *Biochem. Biophys. Res. Commun.* 72:1153–59
58. Slater, E. C. 1974. See Ref. 54, pp. 1–20
59. Harris, D. A., Rosing, J., van de Stadt, R. J., Slater, E. C. 1973. *Biochim. Biophys. Acta* 314:149–53
60. Harris, D. A., Slater, E. C. 1975. *Biochim. Biophys. Acta* 387:335–48
61. Rosing, J., Smith, D. J., Kayalar, C., Boyer, P. D. 1976. *Biochem. Biophys. Res. Commun.* 72:1–8
62. Boyer, P. D., Gresser, M., Vinkler, C., Hackney, D., Choate, G. 1977. In *Structure and Function of Energy-Transducing Membranes,* ed. K. van Dam, B. F. van Gelder pp. 261–74. Amsterdam: Elsevier
63. O'Keefe, D., Dilley, R. A. 1977. *FEBS Lett.* 81:105–10
64. Gresser, M., Cardon, J., Rosen, G., Boyer, P. D. 1979. *J. Biol. Chem.* 254:10649–53
65. Rosen, G., Gresser, M., Vinkler, C., Boyer, P. D. 1979. *J. Biol. Chem.* 254:10654–61
66. McCarty, R. E. 1979. *Trends Biochem. Sci.* 4:28–30
67. Strotmann, H., Bickel-Sandkötter, S., Edelmann, K., Eckstein, F., Schlimme, E., Boos, K. S., Lüstorff, J. 1979. *Biochim. Biophys. Acta* 545:122–30
68. Strotmann, H., Bickel-Sandkötter, S., Shoshan, V. 1979. *FEBS Lett.* 101:316–20
69. Harris, D. A., Baltscheffsky, M. 1979. *Biochem. Biophys. Res. Commun.* 86:1248–55
70. Bagshaw, C. R., Trentham, D. R. 1973. *Biochem. J.* 133:323–28
71. Dahms, A. S., Kanazawa, T., Boyer, P. D. 1973. *J. Biol. Chem.* 248:6592–95
72. Masuda, H., de Meis, L. 1973. *Biochemistry* 12:4581–85
73. Post, R. L., Toda, G., Rogers, F. N. 1975. *J. Biol. Chem.* 250:691–701
74. Jencks, W. P. 1975. *Adv. Enzymol.* 43:219–410
75. Albery, W. J., Knowles, J. R. 1976. *Biochemistry* 15:5631–40
76. Ferguson, S. J., Lloyd, W. J., Radda, G. K. 1975. *Eur. J. Biochem.* 54:127–53
77. Ferguson, S. J. 1977. *Biochem. Soc. Trans.* 5:1281–83
78. Ebel, R. E., Lardy, H. A. 1975. *J. Biol. Chem.* 250:191–96
79. Adolfsen, R., Moudrianakis, E. N. 1976. *Arch. Biochem. Biophys.* 172:425–33
80. Kayalar, C., Rosing, J., Boyer, P. D. 1976. *Fed. Proc.* 35(7):1601

81. Kayalar, C., Rosing, J., Boyer, P. D. 1977. *J. Biol. Chem.* 252:2486–91
82. Hackney, D. D., Boyer, P. D. 1978. *J. Biol. Chem.* 253:3164–70
83. Choate, G. L., Hutton, R. L., Boyer, P. D. 1979. *J. Biol. Chem.* 254:286–90
84. Lunardi, J., Satre, M., Bof, M., Vignais, P. V. 1979. *Biochemistry* 18:5310–16
85. Satre, M., Lunardi, J., Pougeois, R., Vignais, P. V. 1979. *Biochemistry* 18, 3134–40
86. Kohlbrenner, W. E., Cross, R. L. 1978. *J. Biol. Chem.* 253:7609–11
87. Cross, R. L., Kohlbrenner, W. E. 1978. *J. Biol. Chem.* 253:4865–73
88. Klein, G., Satre, M., Dianoux, A.-C., Vignais, P. V. 1980. *Biochemistry* 19:2919–25
89. Penefsky, H. S. 1977. *J. Biol. Chem.* 252:2891–99
90. Catterall, W. A., Pedersen, P. L. 1972. *J. Biol. Chem.* 247:7969–76
91. Hilborn, D. A., Hammes, G. G. 1973. *Biochemistry* 12:983–90
92. Di Pietro, A., Godinot, C., Martin, J.-C., Gautheron, D. C. 1979. *Biochemistry* 18:1738–45
93. Cosson, J. J., Guillory, R. J. 1979. *J. Biol. Chem.* 254:2946–55
94. Drutsa, V. L., Kozlov, I. A., Milgrom, Y. M., Shabarova, Z. A., Sokolova, N. I. 1979. *Biochem. J.* 182:617–19
94a. Lauquin, G., Pougeois, R., Vignais, P. V. 1980. *Biochemistry* 19:4620–26
95. Hutton, R. L., Boyer, P. D. 1979. *J. Biol. Chem.* 254:9990–93
96. Chernyak, B. C., Kozlov, I. A. 1979. *FEBS Lett.* 104:215–19
97. Nalin, C. M., Cross, R. L. 1980. *Fed. Proc.* 39(6):1843
98. Hackney, D. D., Rosen, G., Boyer, P. D. 1979. *Proc. Natl. Acad. Sci. USA* 76:3646–50
99. Brand, M. D., Reynafarje, B., Lehninger, A. L. 1976. *Proc. Natl. Acad. Sci. USA* 73:437–41
100. Chance, B., Crofts, A. R., Nishimura, M., Price, B. 1970. *Eur. J. Biochem.* 13:364–74
101. Massari, S., Azzone, G. F. 1970. *Eur. J. Biochem.* 12:300–9
102. Papa, S., Guerrieri, F., Simone, S., Lorusso, M. 1973. In *Mechanisms in Bioenergetics,* ed. G. F. Azzone, L. Ernster, S. Papa, E. Quagliariello, N. Siliprandi. pp. 451–72. New York: Academic
103. Wikström, M., Krab, K. 1978. In *Energy Conservation in Biological Membranes,* ed. G. Schäfer, M. Klingenberg, pp. 128–39. Berlin: Springer
103a. Heldt, H. W., Klingenberg, M., Milovancev, M. 1972. *Eur. J. Biochem.* 30:434–40
104. Shertzer, H. G., Racker, E. 1976. *J. Biol. Chem.* 251:2446–52
105. La Noue, K., Mizani, S. M., Klingenberg, M. 1978. *J. Biol. Chem.* 253: 191–98
106. Villiers, C., Michejda, J. W., Block, M., Lauquin, G. J. M., Vignais, P. V. 1979. *Biochim. Biophys. Acta* 546:157–70
107. Brand, M. D., Lehninger, A. L. 1977. *Proc. Natl. Acad. Sci. USA* 74:1955–59
108. Out, T. A., Krab, K., Kemp, A., Slater, E. C. 1977. *Biochim. Biophys. Acta* 459:612–16
109. Brand, M. D. 1979. *Biochem. Biophys. Res. Commun.* 91:592–98
110. Boyer, P. D. 1975. *FEBS Lett.* 58:1–6
111. Chang, T.-M., Penefsky, H. S. 1974. *J. Biol. Chem.* 249:1090–98
112. Thayer, W. S., Hinkle, P. C. 1973. *J. Biol. Chem.* 248:5395–5402
113. Alexandre, A., Reynafarje, B., Lehninger, A. L. 1978. *Proc. Natl. Acad. Sci. USA* 75:5296–5300
114. Pozzan, T., Di Virgilio, F., Bragadin, M., Miconi, V., Azzone, G. F. 1979. *Proc. Natl. Acad. Sci. USA* 76:2123–27
115. Junge, W., Rumberg, B., Schroder, H. 1970. *Eur. J. Biochem.* 14:575–81
116. Portis, A. R., McCarty, R. E. 1974. *J. Biol. Chem.* 249:6250–54
117. Mitchell, P. 1979. *Eur. J. Biochem.* 95:1–20
118. Mitchell, P. 1974. *FEBS Lett.* 43:189–94
119. Mitchell, P. 1977. *FEBS Lett.* 78:1–20
120. Williams, R. J. P. 1962. *J. Theoret. Biol.* 3:209–29
121. Williams, R. J. P. 1975. *FEBS Lett.* 53:123–25
122. Roy, H., Moudrianakis, E. N. 1971. *Proc. Natl. Acad. Sci. USA* 68:2720–24
123. Moudrianakis, E. N., Tiefert, M. A. 1979. *J. Biol. Chem.* 254:9509–17
124. Moudrianakis, E. N., Tiefert, M. A. 1976. *J. Biol. Chem.* 251:7796–801
125. McCarty, R. E. 1978. *FEBS Lett.* 95:299–302
126. Yamamoto, T., Tonomura, Y. 1975. *J. Biochem. Tokyo* 77:137–46
127. Smith, D. J., Stokes, B. O., Boyer, P. D. 1976. *J. Biol. Chem.* 251:4165–71
128. Vinkler, C., Rosen, G., Boyer, P. D. 1978. *J. Biol. Chem.* 253:2507–10
129. Lardy, H. A., Connelly, J. L., Johnson, D. 1964. *Biochemistry* 3:1961–68
130. Linnett, P. E., Mitchell, A. D., Beechey, R. B., Baum, H. 1977. *Biochem. Soc. Trans.* 5:1510–11

131. Penefsky, H. S. 1974. *J. Biol. Chem.* 249:3579–85
132. Pedersen, P. L. 1975. *Biochem. Biophys. Res. Commun.* 64:610–16
133. Asami, K., Juntti, K., Ernster, L. 1970. *Biochim. Biophys. Acta* 205:307–11
134. van de Stadt, R. J., de Boer, B. L., van Dam, K. 1973. *Biochim. Biophys. Acta* 292:338–49
135. Leimgruber, R. M., Senior, A. E. 1976. *J. Biol. Chem.* 251:7103–9
136. Steinmeier, R. C., Wang, J. H. 1979. *Biochemistry* 18:11–18
137. Ferguson, S. J., Lloyd, W. J., Lyons, M. H., Radda, G. K. 1975. *Eur. J. Biochem.* 54:117–26
138. Kohlbrenner, W. E., Boyer, P. D. 1980. *Fed. Proc.* 39(6):1979
139. Kumar, G., Kalra, V. K., Brodie, A. F. 1979. *J. Biol. Chem.* 254:1964–71
140. Harris, D. A., Gomez-Fernandez, J. C., Klungsoyr, L., Radda, G. K. 1978. *Biochim. Biophys. Acta* 504:364–83
141. Wilson, D. F., Owen, C., Mela, L., Weiner, L. 1973. *Biochem. Biophys. Res. Commun.* 53:326–33
142. Racker, E. 1977. *Ann. Rev. Biochem.* 46:1006–14
143. Senior, A. E. 1979. *J. Biol. Chem.* 254:11319–22
144. Wagenvoord, R. J., van der Kraan, I., Kemp, A. 1977. *Biochim. Biophys. Acta* 460:17–24
145. Carlier, M.-F., Holowka, D. A., Hammes, G. G. 1979. *Biochemistry* 18:3452–57
146. Esch, F. S., Allison, W. S. 1978. *J. Biol. Chem.* 253:6100–6
147. Hulla, F. W., Höckel, M., Rack, M., Risi, S., Dose, K. 1978. *Biochemistry* 17:823–28
148. Pougeois, R., Satre, M., Vignais, P. V. 1979. *Biochemistry* 18:1408–13
149. Shoshan, V., Selman, B. R. 1980. *J. Biol. Chem.* 255:384–89
150. Deters, D. W., Racker, E., Nelson, N., Nelson, H. 1975. *J. Biol. Chem.* 250:1041–47
151. Nelson, N., Kanner, B. I., Gutnick, D. L. 1974. *Proc. Natl. Acad. Sci. USA* 71:2720–24
152. Verschoor, G. J., van der Sluis, P. R., Slater, E. C. 1977. *Biochim. Biophys. Acta* 462:438–49
153. Douglas, M. G., Koh, Y., Dockter, M. E., Schatz, G. 1977. *J. Biol. Chem.* 252:8333–35
154. Dunn, S. D., Futai, M. 1980. *J. Biol. Chem.* 255:113–18
155. Yoshida, M., Sone, N., Hirata, H., Kagawa, Y. 1977. *J. Biol. Chem.* 252:3480–85
156. Riordan, J. F., McElvany, K. D., Borders, C. L. 1977. *Science* 195:884–86
157. Marcus, F., Schuster, S. M., Lardy, H. A. 1976. *J. Biol. Chem.* 251:1775–80
158. Frigeri, L., Galante, Y. M., Hanstein, W. G., Hatefi, Y. 1977. *J. Biol. Chem.* 252:3147–52
159. Andreo, C. S., Vallejos, R. H. 1977. *FEBS Lett.* 78:207–10
160. Vallejos, R. H., Lescano, W. I. M., Lucero, H. A. 1978. *Arch. Biochem. Biophys.* 190:578–84
161. Knowles, A. F., Penefsky, H. S. 1972. *J. Biol. Chem.* 247:6624–30
162. Colombo, G., Hubert, E., Marcus, F. 1972. *Biochemistry* 11:1798–1803
163. Ferguson, S. J., Lloyd, W. J., Radda, G. K. 1974. *FEBS Lett.* 38:234–36
164. Godinot, C., Penin, F., Gautheron, D. C. 1979. *Arch. Biochem. Biophys.* 192:225–34
165. Cross, R. L., Koga, P. G. 1979. *Abstr. Int. Congr. Biochem., 11th, Toronto*, p. 443, Natl. Res. Counc., Ottawa, Canada
166. Sugiyama, Y., Mukohata, Y. 1978. *FEBS Lett.* 85:211–14
167. Sugiyama, Y., Mukohata, Y. 1979. *FEBS Lett.* 98:276–80
168. Esch, F. S., Allison, W. S. 1979. *J. Biol. Chem.* 254:10740–46
169. DeBenedetti, E., Jagendorf, A. 1979. *Biochem. Biophys. Res. Commun.* 86:440–46
170. Senior, A. E. 1973. *Biochemistry* 12:3622–27
171. Penefsky, H. S. 1967. *J. Biol. Chem.* 242:5789–95
172. Pougeois, R., Satre, M., Vignais, P. V. 1978. *Biochemistry* 17:3018–23
173. Kasahara, M., Penefsky, H. S. 1978. *J. Biol. Chem.* 253:4180–87
174. Kohlbrenner, W. E., Cross, R. L. 1979. *Arch. Biochem. Biophys.* 198:598–607
175. Selwyn, M. J. 1967. *Biochem. J.* 105:279–88
176. Penefsky, H. S., Garrett, N. E., Chang, T. M. 1976. See Ref. 50, pp. 69–79
177. Schäfer, G., Onur, G. 1980. *FEBS Lett.* 109:197–201
178. Vašák, M., Nagayama, K., Wüthrich, K., Mertens, M. L., Kägi, J. H. R. 1979. *Biochemistry* 18:5050–55
179. Rossmann, M. G., Moras, D., Olsen, K. W. 1974. *Nature* 250:194–99
180. Moyle, J., Mitchell, P. 1975. *FEBS Lett.* 56:55–61
181. Abrams, A., Jensen, C., Morris, D. H. 1976. *Biochem. Biophys. Res. Commun.* 69:804–11
182. Senior, A. E., Richardson, L. V., Baker,

K., Wise, J. G. 1980. *J. Biol. Chem.* 255:7211–17
183. Mildvan, A. S. 1979. *Adv. Enzymol.* 49:103–26
184. Dunaway-Mariano, D., Cleland, W. W. 1980. *Biochemistry* 19:1506–15
185. Bossard, M. J., Vik, T. A., Schuster, S. M. 1980. *J. Biol. Chem.* 255:5342–46
186. Nageswara Rao, B. D., Kayne, F. J., Cohn, M. 1979. *J. Biol. Chem.* 254:2689–96
187. Dunaway-Mariano, D., Benovic, J. L., Cleland, W. W., Gupta, R. K., Mildvan, A. S. 1979. *Biochemistry* 18: 4347–54
188. Hollenberg, P. F., Flashner, M., Coon, M. J. 1971. *J. Biol. Chem.* 246:946–53
189. Cardemil, E., Eyzaguirre, J. 1979. *Arch. Biochem. Biophys.* 192:533–38
190. Wilson, D. F., Stubbs, M., Veech, R. L., Ereciñska, M., Krebs, H. A. 1974. *Biochem. J.* 140:57–64
191. Pullman, M. E., Monroy, G. C. 1963. *J. Biol. Chem.* 238:3762–69
192. Nelson, N., Nelson, H., Racker, E. 1972. *J. Biol. Chem.* 247:7657–62
193. Brooks, J. C., Senior, A. E. 1971. *Arch. Biochem. Biophys.* 147:467–70
194. Nieuwenhuis, F. J. R. M., Bakkenist, A. R. J. 1977. *Biochim. Biophys. Acta* 459:596–604
195. Laget, P. P., Smith, J. B. 1979. *Arch. Biochem. Biophys.* 197:83–89
196. Sternweis, P. C., Smith, J. B. 1980. *Biochemistry* 19:526–31
197. Sternweis, P. C. 1978. *J. Biol. Chem.* 253:3123–28
198. Ebner, E., Maier, K. L. 1977. *J. Biol. Chem.* 252:671–76
199. Baird, B. A., Hammes, G. G. 1976. *J. Biol. Chem.* 251:6953–62
200. Bragg, P. D., Hou, C., 1975. *Arch. Biochem. Biophys.* 167:311–21
201. Smith, J. B., Sternweis, P. C., Heppel, L. A. 1975. *J. Supramol. Struct.* 3: 248–55
202. Smith, J. B., Sternweis, P. C. 1977. *Biochemistry* 16:306–11
203. Bertina, R. M., Slater, E. C. 1975. *Biochim. Biophys. Acta* 376:492–504
204. Cantley, L. C., Hammes, G. G. 1975. *Biochemistry* 14:2968–75
205. van de Stadt, R. J., van Dam, K. 1974. *Biochim. Biophys. Acta* 347:240–52
206. Klein, G., Satre, M., Vignais, P. 1977. *FEBS Lett.* 84:129–34
207. Steele, J. A., Uchytil, T. F., Durbin, R. D. 1978. *Biochim. Biophys. Acta* 504: 136–41
208. Vallejos, R. H., Viale, A., Andreo, C. S. 1977. *FEBS Lett.* 84:304–8

209. Adolfsen, R., Moudrianakis, E. N. 1976. *Biochemistry* 15:4163–70
210. Ferguson, S. J., John, P., Lloyd, W. J., Radda, G. K., Whatley, F. R. 1974. *Biochim. Biophys. Acta* 357:457–61
211. Ferguson, S. J., Lloyd, W. J., Radda, G. K., Slater, E. C. 1976. *Biochim. Biophys. Acta* 430:189–93
212. Gómez-Puyou, A., Tuena de Gómez-Puyou, M., Ernster, L. 1979. *Biochim. Biophys. Acta* 547:252–57
213. Harris, D. A., von Tscharner, V., Radda, G. K. 1979. *Biochim. Biophys. Acta* 548:72–84
214. Harris, D. A., Crofts, A. R. 1978. *Biochim. Biophys. Acta* 502:87–102
215. Horstman, L. L., Racker, E. 1970. *J. Biol. Chem.* 245:1336–44
216. Gomez-Fernandez, J. C., Harris, D. A. 1978. *Biochem. J.* 176:967–75
217. Vardanis, A. 1977. *J. Biol. Chem.* 252:807–813
218. Yamada, E. W., Shiffman, F. H., Huzel, N. J. 1980. *J. Biol. Chem.* 255:267–73
219. Tuena de Gómez-Puyou, M., Muller, U., Gavilanes, M., Gómez-Puyou, A. 1980. *Fed. Proc.* 39(6):1845
220. Inoue, Y., Kobayashi, Y., Shibata, K., Heber, U. 1978. *Biochim. Biophys. Acta* 504:142–52
221. Mills, J. D., Hind, G. 1979. *Biochim. Biophys. Acta* 547:455–62
222. Mills, J. D., Mitchell, P., Schurmann, P. 1980. *FEBS Lett.* 112:173–77
223. McCarty, R. E., Pittman, P. R., Tsuchiya, Y. 1972. *J. Biol. Chem.* 247: 3048–51
224. McCarty, R. E., Fagan, J. 1973. *Biochemistry* 12:1503–7
225. Weiss, M. A., McCarty, R. E. 1977. *J. Biol. Chem.* 252:8007–12
226. Moroney, J. V., Andreo, C. S., Vallejos, R. H., McCarty, R. E. 1980. *J. Biol. Chem.* 255:6670–74
227. McCarty, R. E., Racker, E. 1968. *J. Biol. Chem.* 243:129–37
228. Farron, F. 1970. *Biochemistry* 9:3823–28
229. Cantley, L. C., Hammes, G. G. 1976. *Biochemistry* 15:9–14
230. Farron, F., Racker, E. 1970. *Biochemistry* 9:3829–34
231. Vallejos, R. H., Ravizzini, R. A., Andreo, C. S. 1977. *Biochim. Biophys. Acta* 459:20–26
232. Ravizzini, R. A., Andreo, C. S., Vallejos, R. H. 1980. *Biochim. Biophys. Acta* 591:135–41
233. Holowka, D. A., Hammes, G. G. 1977. *Biochemistry* 16:5538–45
234. Schuster, S. M., Ebel, R. E., Lardy, H. A. 1975. *J. Biol. Chem.* 250:7848–53

235. Recktenwald, D., Hess, B. 1977. *FEBS Lett.* 76:25–28
236. Adolfsen, R., Moudrianakis, E. N. 1978. *J. Biol. Chem.* 253:4380–88
237. Webster, G. D., Edwards, P. A., Jackson, J. B. 1977. *FEBS Lett.* 76:29–35
238. Elder, J. A., Lehninger, A. L. 1973. *Biochemistry* 12:976–82
239. Lambeth, D. O., Lardy, H. A. 1971. *Eur. J. Biochem.* 22:355–63
240. Zalkin, H., Pullman, M. E., Racker, E. 1965. *J. Biol. Chem.* 240:4011–16
241. Warshaw, J. B., Lam, K. W., Nagy, B., Sanadi, D. R. 1968. *Arch. Biochem. Biophys.* 123:385–96
242. Cross, R. L., Boyer, P. D. 1973. *Biochem. Biophys. Res. Commun.* 51:59–66
243. Garrett, N. E., Penefsky, H. S. 1975. *J. Biol. Chem.* 250:6640–47
244. Roy, H., Moudrianakis, E. N. 1971. *Proc. Natl. Acad. Sci. USA* 68:464–68
245. Lutz, H. U., Dahl, J. S., Bachofen, R. 1974. *Biochim. Biophys. Acta* 347:359–70
246. Abrams, A., Noland, E. A., Jensen, C., Smith, J. B. 1973. *Biochem. Biophys. Res. Commun.* 55:22–29
247. Maeda, M., Kobayashi, H., Futai, M., Anraku, Y. 1976. *Biochem. Biophys. Res. Commun.* 70:228–34
248. Harris, D. A., Radda, G. K., Slater, E.

C. 1977. *Biochim. Biophys. Acta* 459:560–72
249. Harris, D. A., John, P., Radda, G. K. 1977. *Biochim. Biophys. Acta* 459:546–59
250. Ohta, S., Tsuboi, M., Yoshida, M., Kagawa, Y. 1980. *Biochemistry* 19:2160–65
251. Hackney, D. D. 1979. *Biochem. Biophys. Res. Commun.* 91:233–38
252. Fitin, A. F., Vasilyeva, E. A., Vinogradov, A. D. 1979. *Biochem. Biophys. Res. Commun.* 86:434–39
253. Minkov, I. B., Fitin, A. F., Vasilyeva, E. A., Vinogradov, A. D. 1979. *Biochem. Biophys. Res. Commun.* 89:1300–6
254. Recktenwald, D., Hess, B. 1979. *FEBS Lett.* 108:257–60
255. Roveri, O. A., Muller, J. L. M., Wilms, J., Slater, E. C. 1980. *Biochim. Biophys. Acta* 589:241–55
256. Carmeli, C., Lifshitz, Y., Gutman, M. 1978. *FEBS Lett.* 89:211–14
257. Kagawa, Y. 1978. See Ref. 103, pp. 195–219
258. Shoshan, V., Selman, B. R. 1979. *J. Biol. Chem.* 254:8801–7
259. Schuster, S. M., Reinhart, G. D., Lardy, H. A. 1977. *J. Biol. Chem.* 252:427–32
260. Boyer, P. D. 1965. In *Oxidases and Related Redox Systems,* ed. T. E. King, H. S. Mason, M. Morrison, 2:994–1008. New York: Wiley

Ann. Rev. Biochem. 1981. 50:715–31

MICROBIAL IRON
COMPOUNDS

♦12093

J. B. Neilands[1]

Biochemistry Department, University of California, Berkeley, California 94720

CONTENTS

Perspectives and Summary

The past three decades have witnessed the compilation of a catalogue of microbial iron-containing or iron-binding compounds, most of which can be classified as "siderophores" (Greek for iron bearers). Although the siderophores as chemical entities display considerable structural variation, the majority of them are either hydroxamates or phenolates-catecholates and all exhibit a very strong affinity for Fe(III), the formation constant lying in the range of 10^{30} or higher. The crystal structures of six ferric trihydroxamate type siderophores (ferrichrome A, ferric mycobactin, ferrichrysin, ferrioxamine E, ferrichrome, and ferric N,N',N''-triacetylfusarinine) and of one catechol type siderophore (agrobactin) have been determined. In addi-

---

[1]I wish to thank R. C. Hider for a critical review of the manuscript.

tion, the solution structures of respresentative members of both classes have been worked out by high resolution NMR. This impressive body of knowledge of the chemical and physical properties of the siderophore molecules lays an exceptionally firm foundation for future advances in understanding the mode of receptor-dependent transport of these substances. The mechanism and regulation of transmembrane permeation of iron in the microbial world is a field now ripe for development. It remains to be seen how far the basic features of iron transport in unicellular species may be applied in the domain of higher organisms.

Introduction

The main purpose of this article is to acquaint the reader with a line of nonporphyrin, nonprotein iron-containing or iron-binding compounds from microorganisms, the latter herein defined as bacteria, fungi, and microalgae. For the most part the biological properties of siderophores is not discussed here as that is the subject of a companion review (1). Apart from their role in receptor-dependent, *high affinity* iron transport, siderophores may act as growth or germination factors, and some are potent antibiotics. They are believed to function as virulence factors in animal and plant disease, and the iron-free molecules are employed in chelation therapy for iron overload and actinide poisoning. Although microbial iron compounds as natural products are fascinating in their own right, the fact that they are endowed with unique physiological properties confers upon them a more general interest in the biochemical community.

This is not an encyclopedic review of microbial iron compounds. Such classic substances as pulcherrimin (2) and kojic acid (3), for example, are excluded on the grounds that they do not appear to function as siderophores. The general topic of iron metabolism in microorganisms has been reviewed elsewhere (4). Bacterial and fungal mechanisms of iron assimilation have been discussed extensively by Lankford (5) and Emery (6), respectively, while Raymond & Carrano (7) have surveyed aspects of the coordination chemistry of siderophores.

The nomenclature of microbial iron compounds is confused and it is desirable to recall the historical development of the field. The first substances were obtained some three decades ago from fungi and bacteria and were given such trivial names as ferrichrome, coprogen, and mycobactin (4). The first of these was isolated from the smut fungus *Ustilago sphaerogena* and shown to stimulate the growth of microorganisms requiring chelated iron; the remaining two were obtained as growth factors for specific test organisms. In the middle 1950s it was discovered that the biosynthesis of the ligand moieties of ferrichrome and of other iron-binding compounds could be depressed by culture of the organisms at low iron

(8). The simple expedient of withholding iron from the culture medium has enabled the collection of an array of microbial products giving a reaction with ferric chloride, with the result that several dozen structures have now been elaborated; these form the basis of this review. It is now generally believed that most aerobic and facultative anaerobic microorganisms have the capacity to synthesize such compounds (Table 1).

Two decades ago Swiss workers surveyed the actinomyces and concluded that fully 80% of the strains tested produced iron complexing agents that they called ferrioxamines (9). They suggested the general terms "siderochrome" for all of these red-brown, iron-containing compounds, with the specific designation "sideramine" and "sideromycin" applied to those with growth supporting and antibiotic activity, respectively. The appellations sideramine and sideromycin have not been universally adopted, since the particular response, whether stimulation or inhibition, depends on the test organism. Subsequently, the term siderochrome was extended to all types of small molecular weight, low-iron-inducible microbial iron-binding agents regardless of biological activity (10). "Ferriphore" and siderophore (5) were subsequently proposed and the latter term has become generally accepted.

A siderophore is defined as a low-molecular-weight (500–1000 daltons) virtually ferric-specific legand, the biosynthesis of which is carefully regulated by iron and the function of which is to supply iron to the cell. As iron is involved in several critical stages in metabolism, microbes have evolved

Table 1 Distribution of siderophores among microbial species

Microorganism	Siderophore[a]
Bacteria	
Enteric sp.	Enterobactin, aerobactin
Agrobacterium tumefaciens	Agrobactin
Pseudomonas sp.[b]	Pyochelin, pyoverdine, pseudobactin, ferribactin
Bacillus megaterium	Schizokinen
Anabaena sp.	Schizokinen
Arthrobacter sp.	Arthrobactin
Azotobacter vinelandii	α,ϵ-bis-2,3-dihydroxybenzoyllysine
Actinomyces sp.	Ferrioxamines
Mycobacteria	Mycobactins
Fungi	
Penicillia, Aspergilli, Neurospora, Ustilago	Ferrichromes, coprogen
Rhodotorula sp.	Rhodotorulic acid
Ectomycorrhizal sp.[b]	Hydroxamate type, active for *Arthrobacter flavescens*

[a] Unless otherwise noted, structures appear in the text.
[b] Structures of siderophores not yet known.

multiple systems for its acquisition. The high affinity system is comprised of the siderophores and the matching membrane-associated receptors (1); only the former component of the high affinity system is discussed in this review.

Strictly speaking the term siderophore should be reserved for the metal-free ligand. Thus ferrichrome is properly, if redundantly, described as a "ferric siderophore." The ligand is deferriferrichrome. These terms are awkward and the entire field is in need for a more systematic nomenclature. To amplify the confusion, certain specific siderophores are known by two or three different trivial names. Examples are enterobactin and enterochelin, fusarinine C and fusigen, terregens factor and arthrobactin, nocardamin and deferriferrioxamine E.

In general, siderophores may be classed as either hydroxamates [–CO–N(O⁻)–] or phenolates-catecholates $R=H,OH;X=O,N$.

The former occur widely throughout the display prokaryotic and eukaryotic microbial world, but thus far only bacteria have been shown to produce typical mono- or dihydroxybenzoic acid–bearing siderophores. There is a distinct possibility that substances answering the description of a siderophore will be described that lack either of the functional groups just cited. As we shall see, mycobactin, agrobactin, and parabactin contain an oxazoline N atom as sixth coordination site for iron, and in the various siderophores based on citrate, such as aerobactin, schizokinen, and arthrobactin, the inner carboxylate and hydroxylate functions are presumed to contact the iron. A siderophore lacking the charge transfer bands of ferric hydroxamates or catecholates would be more difficult to detect. There is thus an urgent need for a general test for iron complexing agents that may be present in spent media. By means of such a universal probe some of the organims, such as the common yeast *Saccharomyces cerevisiae*, might be moved from the negative to the positive "column" with respect to siderophore synthesis.

When stressed for iron, microbes often excrete iron-binding substances containing only one or two sets of bidentate chelating groups per mole. Thus *Bacillus subtilis* forms 2,3-dihydroxybenzoyl glycine (11) and *Paracoccus denitrificans* produces 2,3-dihydroxybenzoic acid, N^1,N^8-bis-2,3,-dihydroxybenzoylspermidine (Compound II), and parabactin (12), of which only the latter affords six coordination sites for the ferric ion. Although the formation constants for a tris ferric chelate of a bidentate ligand may approach or even exceed those in which the same brace of functional groups is repeated two or three times within the same molecule, only those ligands with six conveniently disposed coordinating atoms will be effective binding agents for Fe(III) in dilute solution. Thus the less polymerized binders will be less efficient in scavenging iron. In an evolutionary sense, these molecules may represent progression to, or loss of, true siderophore synthesis.

The question arises as to the possible classification of citrate as a sidero-phore. It is certainly an inferior chelating agent when contrasted with the polyhydroxamates or catecholates; a concentration of ligand/Fe(III) of $\sim$ 20/1 is required to force the complex into the mononuclear state (13). On the other hand it is accumulated by certain organisms at low iron, and this process constitutes the basis for the commercial production of the organic acid by *Aspergillus niger*. In addition, *Escherichia coli* has a well-character-ized inducible outer membrane receptor for transport of ferric citrate (14). In sum, citrate and other hydroxy acids (15) could be regarded as sidero-phores of intermediate affinity. These molecules, once embellished with hydroxamate functions as in schizokinen, arthrobactin, and aerobactin (see below) possess greatly increased affinities for Fe(III).

A small number of Fe(II)-binding or- containing substances have been obtained from microorganisms. Of these the best known are pyrimine (16) and ferroverdine (17). There is no evidence that either of these substances play a role in iron metabolism and hence they should not be classed as siderophores.

Biological Rationale for Siderophores

Conspicuous by their absence from Table 1 are the lactic acid bacteria and the strict anaerobes, the former because they may not require iron and the latter, although replete with respect to nonporphyrin iron proteins, may be satisfied with the relatively soluble Fe(II) that is present in their growth environment. Siderophores are viewed as the evolutionary response to the appearance of O_2 in the atmosphere, concomitant oxidation of Fe(II) to Fe(III), and the precipitation of the latter as ferric hydroxide, $K_s = <$ 10^{-38} M (18). This explanation accounts for the general occurrence of siderophores in aerobic and facultative anaerobic microorganisms (19).

Salient Chemical and Physical Properties of Siderophores

DETECTION AND ISOLATION As most, although not all, ferric com-plexes are colored, the mere addition of ferric chloride or ferric perchlorate to the spent medium may serve to detect the presence of a siderophore (20). Tris(hydroxamates) are orange-colored with $a_{mM} \cong 3$ at the maximum of absorbancy, 425–450 nm; for the wine-colored tris(catecholates), the figure is $a_{mM} \cong 5$ at 495 nm. Catechols are most conveniently assayed by the nitrite-molybdate reagent of Arnow (21). This reaction has been investigated by Barnum (22) who showed it to be selective for aromatic vic-diols with the 3 or 4 positions unsubstituted and not sterically hindered. Following nitration, addition of molybdate and alkali affords a complex with absorption maximum at about 500 nm and $a_{mM} \cong 10$. Siderophores containing the 2,3-dihydroxybenzoyl group possess a characteristic electronic absorption spectrum and fluorescence. Hydroxamates may be detected with the Folin reagent (23), with acidified ammonium vanadate (24), or by oxidation with periodate (25) or I_2 (26). The latter, called the Csaky test, is the most popular but, as stressed by Tomlinson et al (27), it is difficult to quantitate.

Alleviation of growth inhibition induced by the presence of one of the iron-binding proteins conalbumin or transferrin (28) has been used widely. The synthetic iron chelating agent ethylenediamine-di-(o-hydroxyphenylacetic acid) has been employed in the same way by Miles & Khimji (29), but the method depends on nonutilization of the reagent by the test organism. Similar erratic responses confound the use of one of the siderophore auxotrophs such as *Salmonella typhimurium enb* or *E. coli* RW193. To repeat, there has yet to be devised a single, comprehensive, sensitive, and reliable test for the siderophore ligand.

Extraction into an organic solvent, ethyl acetate in the case of catechol type siderophores or either benzyl alcohol or 1 : 1 chloroform-phenol for the hydroxamate type, represents an effective purification step for many siderophores, since both salts and macromolecules are thereby excluded. The polystyrene resin, Amberlite XAD, has been employed for isolation of neutral, ferrichrome type siderophores (30), while polyamide resin, long used for the isolation of phenols, has been adapted for chromatographic separation of catechol type siderophores (31). High pressure liquid chromatography has been used for separation of mycobactin (32) and 2,3-dihydroxybenzoic acid (33).

IRON-BINDING PROPERTIES   The single most outstanding feature of the siderophores is their extremely high affinity for the ferric ion (7). Table 2 lists the formation constants of some of the common siderophores and compares the values found with those reported for transferrin and the simple model compound, acethydroxamic acid. The number reported for enterobactin, 10^{52}, is the largest ever found for a ferric ion–binding species (7). Even at neutral pH, where this value must be reduced some 15 powers of 10, owing to proton competition, enterobactin remains fully capable of removing iron from ferrichrome or ferrichrome A (7). At pH values slightly below neutrality, however, there will be cross-over in relative stabilities, since the hydroxamic function is a substantially stronger acid than the proton of *m*-hydroxyl of the typical catechols.

Table 2 Formation constants of deferrisiderophores with ferric ion

Ligand	$\log K_f$	Ref.
Enterobactin	52	7
Agrobactin/parabactin	At neutral pH $\cong$ enterobactin	34
Compound III (parabactin)	Exceeds transferrin	35
Deferriferrioxamine E	32.5	36
Mycobactin	Exceeds deferriferrioxamine B	37
Rhodotorulic acid (dimeric/Fe)	31.2	
Deferriferrioxamine B (Desferal = mesylate)	30.6	36
Deferriferrichrome A	29.6	38
Deferriferrichrome	29.1	36
Acethydroxamic acid	28.3	36
Aerobactin	22.9	7

Ferric siderophores are octahedral complexes in which the coordinated metal ion is d^5, high spin, and rapidly exchangeable. A detailed report has recently appeared on the formation constants, kinetics of formation and aquation, and associated temperature dependencies for a series of six model ferric monohydroxamates (39).

The siderophore ligands, as "hard" bases, show relatively little affinity for the "softer" acid cations, including Fe(II), thus providing a means of release of iron via a reduction step. The aluminum and gallium complexes of ferrichrome and enterobactin have found favor as diamagnetic isostructural analogues suitable for NMR spectroscopy, and the Cr(III) complex of the hydroxamate type siderophores has been exploited as a means of determination of the particular optical coordination isomer favored in the dissolved state (7). None of these ions will compete effectively with Fe(III) for the siderophore ligands under usual conditions in Nature. The Cr(III) is substitution inert; aluminum is abundant but its binding is weaker than that of iron and furthermore a reduction step is probably obligatory in siderophore-mediated iron transport. Gallium, while possessing an ionic radius close to that of Fe(III), is relatively rare in Nature. Thus chemical safeguards have been erected to ensure that the ligands deliver only ferric ion. On the other hand there is a hazard that the siderophores may complex the actinides and move certain of these alpha-emitting, highly toxic ions across biological membranes (40). Binding of the siderophore ligand to ions such as Pu(IV) is a consequence of the similar charge/radius that these ions bear with respect to Fe(III). *Bacillus thuringiensis*, when grown on an iron- and molybdenum-deficient medium, produces phenolate compounds, not yet fully characterized, that coordinate molybdate (41).

In common with other coordination compounds, ferrisiderophores may exhibit both geometrical and optical isomerism.

Siderophores that contain three sets of bidentate ligands with dissimilar coordinating atoms may in theory present these to a centrally located ferric ion in two geometrical orientations, *cis* and *trans*. In siderophores such as enterobactin and ferrichrome, where the ligands are attached to a platform, the *cis* configuration will predominate even though, as in the case of ferrichrome, the *trans* isomer may be constructed in molecular models.

The presence of a single optically active center in the organic moiety suffices to fix in space one ligand set in a potentially preferred spatial orientation. The remaining two ligand sets cluster around the iron in either a left-hand (Λ) or right-hand (Δ) coordination propeller (1).

The X-ray structure of ferrichrome A (42) revealed it to be *cis*, Λ, and subsequent (43) measurements of rotation of polarized light in the region of maximum absorbancy of the chromophore indicated a number of related ferrichromes to have the same optical isomer. That this isomer is Λ, or mainly Λ, in solution was proven by analysis of the chromium complex (7). Other Λ isomers include ferric complexes of mycobactin, aerobactin, agrobactin, and parabactin, while ferric rhodotorulate and ferric enterobactin are Δ. Although not all of these have been characterized in both solid and dissolved states, it is clear that the siderophores of both chemical types may exist in either Λ or Δ forms. Ferric triacetylfusarinine is present in one form in the solid state, and upon solution it assumes the opposite configuration (44).

BIOMIMETIC ANALOGUES OF SIDEROPHORES   The search for an orally effective drug suitable for patients suffering from the iron overload incident to transfusion-induced siderosis has prompted the chemical synthesis of a series of siderophore analogues. Corey & Hurt (45) made the all-*cis* carbocyclic analogue of enterobactin, which proved to be remarkably like the natural product in its capacity to coordinate iron and supply the metal to *E. coli* (46). The same compound actually had been made earlier by Collins et al (47), but only in a fully methylated form. An aromatic analogue based on 1,3,5-aminomethylbenzene has been synthesized in two laboratories (48, 49) and displays similar biological activity.

Iron Storage Compounds

At least two groups have reported the presence of iron storage proteins in bacteria.

Stiefel & Watt (50) concluded that the b-type cytochrome of *Azotobacter vinelandii* is a bacterio-ferritin. From *E. coli* Bauminger et al (51) obtained crystals of an iron protein with Mossbauer spectroscopic properties of ferritin. A biofunction for these iron proteins has yet to be proven by, for example, a genetical analysis.

MAGNETOTATIC BACTERIA   Magnetite (Fe_3O_4) has been identified by Mossbauer spectroscopy as a constituent of a magnetotactic spirillum isolated from freshwater swamp (52). Such bacteria provide one experimental model for study of the basis of geomagnetic orientation of eukaryotic organisms.

Are There Siderophores in Plants and Animals?

Thus far a molecule with the characteristics of a siderophore has not been isolated from the tissue of higher organisms. Citrate is a common constituent of plants where the organic acid is believed to play an important role in iron metabolism (53). Adenochrome, the iron-containing pigment of bronchial heart of *Octupus vulgaris,* has been shown to consist of a mixture of closely related substances derived from catechol and a thiol-substituted histidine (54). Mutants of SV40-transformed BALB/3T3 cells adapted to growth in picolinic acid formed an iron-binding ligand termed "siderophore-like growth factor" (SGF) (55). Although it was demonstrated to stimulate the uptake of Fe(III) the substance is as yet poorly defined in the chemical sense.

Classification of Siderophores

PHENOL-CATECHOL TYPE

Cyclic catechols   **Enterobactin**   The structure of ferric enterobactin is given in Figure 1. The ligand was isolated in 1970 from low-iron cultures of *S. typhimurium* (56) and *E. coli* (57) and named enterobactin and enteroche-

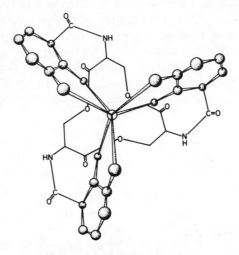

Figure 1   Ferric enterobactin.

lin, respectively. Coordination of the iron to the vic-diols has been demonstrated by a number of techniques, including NMR (59), measurements of electronic spectra (7), and resonance Raman spectroscopy (58), but the complex has thus far defied crystallization. The compound has been synthesized (60), but it is most conveniently isolated from a permease-defective strain of *E. coli.* An enantio-enterobactin has been synthesized from *D* -serine, the iron complex of which is the antipode of that given by the natural substance, which has the Δ configuration illustrated in Figure 1.

Enterobactin-chelated ferric iron is labilized at slightly acidic pH, and according to Mossbauer spectroscopic studies at least some of it may switch over to Fe(II) (61).

Enterobactin, which may be common to all enteric bacteria, is regarded as the prototypical siderophore of the catechol type.

Linear catechols   **Agrobactin**   The compound shown in Figure 2, R=OH, was isolated from the bacterial phytopathogen *Agrobacterium tumefaciens* and named agrobactin (62). The presence of the oxazoline ring, previously detected in mycobactin, was revealed by its ability to form a hydrochloride salt, with concomitant changes in its ultraviolet absorption spectrum. In acidic media the oxazoline opens slowly to yield agrobactin A, the derivative bearing 2,3-dihydroxybenzoylthreonine linked as a tertiary amide to the central N of the N^1,N^8-bis-2,3-dihydroxybenzoylspermidine. Synthesis and examination of model oxazolines showed that the unusual stability of this heterocyclic ring in agrobactin could be attributed to electronic effects arising from the hydroxyl group at the 2-position of the phenyl ring. The duplicated NMR resonances, most easily distinguished in the amino acid moiety, were assigned to *cis-trans* isomerization around the tertiary amide function (63). A number of criteria were involed to propose that in ferric agrobactin the iron is linked to the oxazoline N rather than to the *m*-hydroxyl group (34). Furthermore, bacterial feeding experiments and measurements of circular dichroism spectra suggested the complexes of agrobactin and agrobactin A to be Λ and Δ, respectively. The X-ray structure of agrobactin (64) has been solved, but the metal complexes have not been crystallized.

Figure 2   Agrobactin (R = OH) and parabactin (R = H).

Parabactin Tait (12) isolated a substance from low-iron grown cultures of *Micrococcus* (now *Paracoccus*) *denitrificans*, named "Compound III", and concluded it to be N^1,N^8-bis-2,3-dihydroxybenzoylspermidine ("Compound II") acylated on the secondary amide N with a residue of salicylthreonine. Compound III, like agrobactin, may exist in the oxazoline form (65). This would account for the pH-dependency of the absorption spectrum reported by Tait (12) and provide the sixth coordination site for the ferric ion. The siderophore produced by *P. denitrificans* has been given the trivial name parabactin (Figure 2, R=H), which is apparently what Tait isolated, and the actual structure shown by him has been designated parabactin A (65). These compounds form iron complexes analogous to those of agrobactin (34).

N^1,N^8-**bis-2,3-dihydroxybenzoylspermidine (Compound II)** Compound II is an effective chelating agent (35). It has been synthesized from an N^4-blocked spermidine, a key intermediate for the preparation of siderophores derived from this linear polyamine (66).

α,ϵ-**bis-2,3-dihydroxybenzoyllysine** This substance is of special significance since it was isolated from the nitrogen-fixing species, *Azotobacter vinelandii* (67). As nitrogenase is endowed with a rich complement of iron, the possible role of siderophores in the biosynthesis of this important enzyme is an intriguing but still unanswered question. The spacing of the catechol groups is probably favorable for binding, although binuclear or polymeric species may be formed, and the ligand would be expected to be less affective than agrobactin or parabactin in complexing iron in dilute solution. It has yet to be demonstrated that it serves as a siderophore in *A. vinelandii*.

HYDROXAMIC ACID TYPE

Ferrichromes The core structure of the ferrichrome family of siderophores is illustrated in Figure 3. The fact that the three neutral amino acid residues (R^1,R^2,R^3) of the cyclohexapeptide moiety may be glycine, serine, or alanine, while the acyl substituent (R) on the δ-nitrogen of hydroxylaminoornithine may be any one of a number of small carboxylic acids likely derived from acyl CoA makes for the existence of a large number of variants of the structure shown. The most recently described is malonichrome (68), a ferrichrome-type siderophore carrying 3 mol of malonic acid as the acyl group. It was formed by prolonged culture of *Fusarium roseum*, which in the initial phases of growth produced fusarinine type siderophores.

An X-ray structure of ferrichrome (69) has been elaborated and shows the molecule to contain a single transannular H bond. In general, the conformation is quite similar to that seen in ferrichrome A (42), which has been used as the basis for Figure 3. As the first ferric trihydroxamate identified in Nature, ferrichrome enjoys the status of the prototypical hydroxamate type siderophore.

Measurements of circular dichroism in the ultraviolet suggest that the cyclohexapeptide portion of ferrichrome can bind alkaline earth cations, and it has been proposed that this property facilitates the transport of the siderophore across the cytoplasmic membrane (70).

Ferrichrome has been synthesized by Japanese (71) and Swiss (72) workers; the latter also made enantio-ferrichrome (73). The unnatural compound was relatively ineffective in supplying iron to fungi (74). Ferrichromes occur commonly in fungi and are present in mold-ripened cheese (75). This type of siderophore is utilized efficiently by certain bacteria (76). The structure of the antibiotic albomycin, while not yet known in all details, is closely related to that of ferrichrome (77).

Rhodotorulic acids   The dipeptide shown in Figure 4, rhodotorulic acid, is produced by *Rhodotorula pilimanae* and related yeasts in vast quantities. It has been synthesized chemically by two groups (78, 79). In solution the iron complex may exist as a binuclear species (80). The rhodotorulic acid moiety also occurs in dimerumic acid and in coprogen.

Citrate-hydroxamates   The structures shown in Figure 5 are derivatives of citric acid in which the distal carboxyl groups have been substituted with

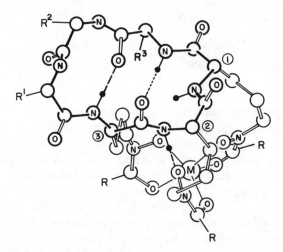

Figure 3   Core structure of the ferrichromes. R^1, R^2, and R^3 are the side chains of small, neutral amino acids such as glycine, serine, or alanine and the acyl substituent R is derived from acyl CoA. For ferrichrome, $R^1 = R^2 = R^3 = H$; R $= CH_3$.

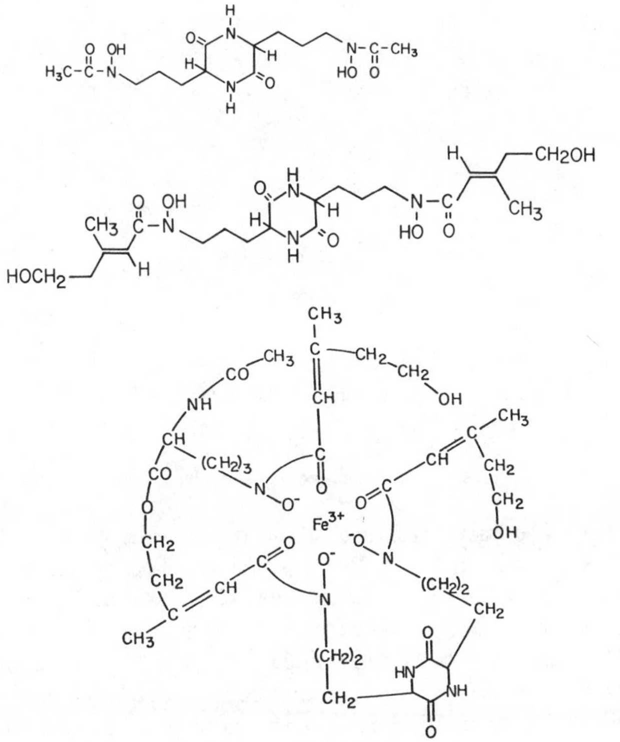

Figure 4 Siderophores derived from rhodotorulic acid. *Upper,* rhodotorulic acid; *center,* dimerumic acid; *lower,* coprogen.

more effective ferric ion complexing units, i.e. hydroxamate groups. All three members of the family known at present, schizokinen, aerobactin, and arthrobactin are derived from bacterial species. Ferric aerobactin has been shown to belong to the Λ isomer series (81). This line of siderophores, as their trivial names imply, arise from various bacterial sources. When grown at low iron, *Aerobacter aerogenes* forms both enterobactin and aerobactin.

Mycobactins   Figure 6 illustrates the general structure of the mycobactins, siderophores of the mycobacteria (37). These compounds display several unique features. Both ligand and iron complex are highly lipid soluble. Also, they may be viewed chemically as hybrids between the hydroxamate and phenolate-catecholate classes of siderophores. They are thought to operate in conjunction with extracellular chelating agents termed exochelins (82), which are presently of unknown chemical constitution.

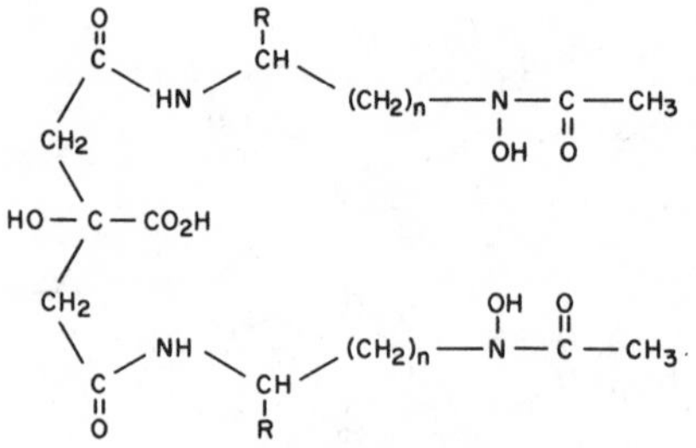

Figure 5 Siderophores based on citric acid.

	R	n
schizokinen —	H	2
aerobactin —	COOH	4
arthrobactin—	H	4

Fusarinines The basic structure of the fungal siderophores, the fusarinines, is given in Figure 7. The functional siderophore is believed to be the cyclic trimer of the unit shown, fusarinine C or fusigen. This substance is stabilized by N-acylation, and the X-ray structure of ferric N,N',N''-triacetylfusarinine has been determined (44). The absolute configuration proved to be Λ,*cis*. However, upon dissolution in aqueous or nonaqueous media, the Δ,*cis* diastereoisomer was shown to predominate. Apparently, the method of crystallization resulted in the separation of the minor, less soluble Λ,*cis* isomer. This study illuminates the necessity of confirming the solution structure, the biologically important state, by circular dichroism spectroscopy. Furthermore, in the case of certain siderophores there may be in solution a facile equilibrium between the V and Δ isomers that could confound attempts to define the biologically active species.

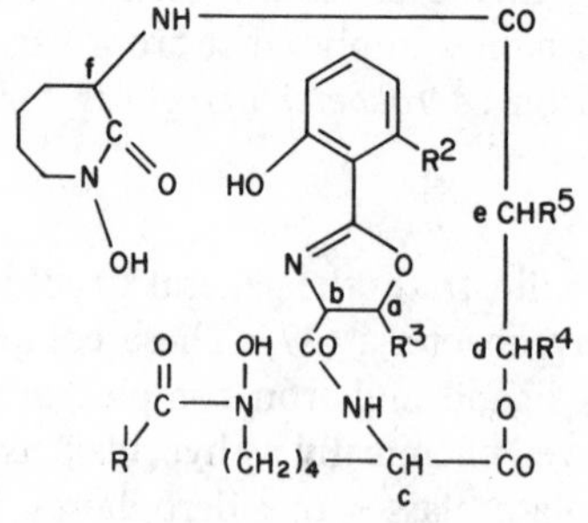

Figure 6 Mycobactins. The R groups and small letters refer to substituents and centers of asymmetry, respectively (37).

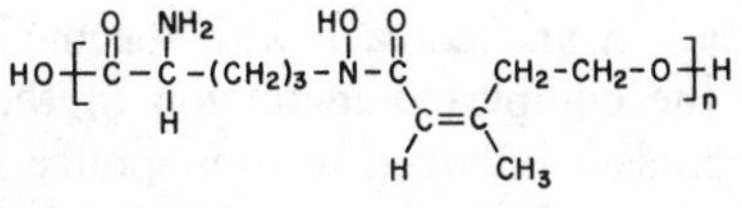

$$HO \left[\begin{array}{c} O \\ \parallel \\ C \end{array} - \begin{array}{c} NH_2 \\ \mid \\ C \\ \mid \\ H \end{array} - (CH_2)_3 - \begin{array}{c} HO \\ \mid \\ N \end{array} - \begin{array}{c} O \\ \parallel \\ C \end{array} \begin{array}{c} \\ C=C \\ \diagup \quad \diagdown \\ H \quad CH_3 \end{array} CH_2 - CH_2 - O \right]_n H$$

Figure 7 Fusarinines. The value of n ranges from 1 to 3. The N-acetyl derivatives, which are more stable, are also known.

Ferrioxamines   As seen in Figure 8, the members of the ferrioxamine family exist in cyclic and open-chain forms. The methane sulfonate salt of deferriferrioxamine B (Figure 8, R=H) is commercially available for Ciba-Geigy as Desferal and is the drug most commonly used for deferration therapy. The ferrioxamines, which include the antibiotic ferrimycin, were found in most actinomycetal strains assayed for their presence (83). Recently ferrioxamine E (nocardamin) has been characterized as a product of *Pseudomonas stutzeri* (84). The X-ray structure (racemic) indicates the molecule to be disc-shaped with a total thickness of 3.6 Å (85).

CONCLUSIONS

Aerobic and facultative anaerobic bacteria, as well as eukaryotic microorganisms, have the general capacity to synthesize Fe(III)-scavenging ligands collectively termed siderophores. This is a testimonial to both the crucial role of iron in microbiology and to the ingenuity of the organism in overcoming the profound insolubility of Fe(III) in an aerobic environment at neutral pH. The siderophore ligands, which are exquisitely designed for complexation of ferric ion, contain multiple sets of bidentate–coordinating groups per mole. The donor atoms are predominantly oxygen and are derived most usually from hydroxamate and catecholate groups. A slightly different mode of binding is noted in mycobactin, agrobactin and parabactin, where the fifth and sixth coordination sites around the ion are filled by the bidentate 2-hydroxyphenyl-2-oxazoline group. Ferric complexes of siderophores in which there is asymmetry in the ligand may exist in two

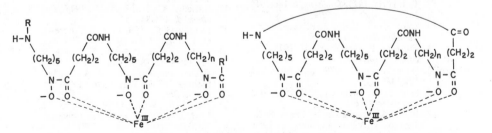

Figure 8   Linear and cyclic ferrioxamines. R = H or –COCH₃; R¹ = CH₃– or HOOC-(CH₂)₂–; n = 4 or 5. For ferrioxamine B, R = H and R¹ = CH₃–. The mesylate salt of deferriferrioxamine B is marketed by Ciba-Geigy as Desferal.

optically isomer species, Λ and Δ, each with specific biological reactivity. Asymmetry around the complexed metal ion in the siderophore series appears to be a mechanism intended to monopolize iron. The ecological advantage in synthesis of a siderophore not utilized by competing species was first pointed out by O'Brien & Gibson (57).

Specific ligands for transiton elements other than iron have not been reported, possibly because the cellular requirements are lower and such metal ions are more soluble than Fe(III). There is no convincing evidence that compounds structurally analogous to the siderophores occur in plants or animals.

Literature cited

1. Neilands, J. B. 1981. *Ann. Rev. Nutr.* 1:27–46
2. Kluyver, A. J., Van der Walt, J. P., Van Triet, A. J. 1953. *Proc. Natl. Acad. Sci. USA* 39:583–93
3. Beélik, A. 1956. *Adv. Carbohydr. Chem.* 11:145–83
4. Neilands, J. B., ed. 1974. *Microbial Iron Metabolism,* New York: Academic. 597 pp.
5. Lankford, C. E. 1973. *Crit. Rev. Microbiol.* 2:273–331
6. Emery, T. 1971. *Adv. Enzymol.* 35:135–85
7. Raymond, K. N., Carrano, C. J. 1979. *Acc. Chem. Res.* 12:183–90
8. Garibaldi, J. A., Neilands, J. B. 1956. *Nature* 177:526–27
9. Keller-Schierlein, W., Prelog, V., Zähner, H. 1964. *Fortschr. Chem. Org. Naturst.* 22:279–333
10. Neilands, J. B. 1973. In *Inorganic Biochemistry,* ed. G. Eichhorn, pp. 167–202. Amsterdam: Elsevier
11. Ito, T., Neilands, J. B. 1958. *J. Am. Chem. Soc.* 80:4645–47
12. Tait, G. H. 1975. *Biochem. J.* 146:191–204
13. Spiro, T. G., Saltman, P. 1969. *Struc. Bonding Berlin* 6:116–56
14. Frost, G. E., Rosenberg, H. 1973. *Biochim. Biophys. Acta* 330:90–101
15. Winkelmann, G. 1979. *Arch. Microbiol.* 121:43–51
16. Shiman, R., Neilands, J. B. 1965. *Biochemistry* 4:2233–36
17. Ballio, A., Barcellona, S., Chain, E. B., Tonolo, A., Vero-Barcellona, L. 1964. *Proc. R. Soc. London Ser. B* 161:384–91
18. Biedermann, G., Schindler, P. 1957. *Acta Chem. Scand.* 11:731–40
19. Neilands, J. B. 1972. *Struct. Bonding Berlin* 11:145–70
20. Atkin, C. L., Neilands, J. B., Phaff, H. J. 1970. *J. Bacteriol.* 103:722–33
21. Arnow, L. E. 1937. *J. Biol. Chem.* 118:531–41
22. Barnum, W. D. 1977. *Anal. Chim. Acta* 89:157–66
23. Subramanian, K. N., Padmanaban, G., Sarma, P. S. 1976. *Anal. Biochem.* 12:106–12
24. Snow, G. A. 1969. *Biochem. J.* 115:199–205
25. Emery, T., Neilands, J. B. 1962. *J. Org. Chem.* 27:1075–76
26. Csaky, T. Z. 1948. *Acta Chem. Scand.* 2:450–54
27. Tomlinson, G., Cruickshank, W. H., Viswanatha, T. 1971. *Anal. Biochem.* 44:670–79
28. Schade, A. L., Caroline, L. 1944. *Science* 100:14–15
29. Miles, A. A., Khimji, P. L. 1975. *J. Med. Microbiol.* 8:477–92
30. Horowitz, N. H., Charlang, G., Horn, G., Williams, N. P. 1976. *J. Bacteriol.* 127:135–40
31. Robinson, A. V. 1979. *Anal. Biochem.* 95:364–70
32. Ratledge, C., Ewing, D. F. 1978. *Biochem. J.* 175:853–57
33. Daumy, G. O., McColl, A. S., Andrews, G. C. 1980. *J. Bacteriol.* 141:293–96
34. Neilands, J. B., Peterson, T., Leong, S. A. 1980. In *Inorganic Chemistry in Biology and Medicine,* ed. A. E. Martell, pp. 263–78 Washington DC: Am. Chem. Soc.
35. Jacobs, A., White, G. P., Tait, G. H. 1977. *Biochem. Biophys. Res. Commun.* 74:1626–30
36. Schwarzenbach, G., Anderegg, G., L'Eplattenier, F. 1963. *Helv. Chim. Acta* 46:1400–9
37. Snow, G. A. 1970. *Bacteriol. Rev.* 34:99–125

38. Wayne, R., Neilands, J. B. 1975. *J. Bacteriol.* 121:497–503
39. Monzyk, B., Crumbliss, A. L. 1979. *J. Am. Chem. Soc.* 101:6203–13
40. Bulman, R. A. 1978. *Struct. Bonding Berlin* 34:39–77
41. Ketchum, P. A., Johnson, D., Taylor, R. C., Young, D. C., Atkinson, A. W. 1977. In *Bioinorganic Chemistry* ed. K. N. Raymond, 2:408–20. Washington DC: Am. Chem. Soc.
42. Zalkin, A., Forrester, J. D., Templeton, D. H. 1966. *J. Am. Chem. Soc.* 88:1810–14
43. Bürer, T., Gulyas, E. 1966. *Proc. Int. Conf. Coordination Chem. 9th,* p. 512
44. Hossain, M. B., Eng-Wilmot, D. L., Loghry, R. A., van der Helm, D. 1980. *J. Am. Chem. Soc.* 102:5766–73
45. Corey, E. J., Hurt, S. D. 1977. *Tetrahedron Lett.* 45:3923–24
46. Hollifield, W. C., Neilands, J. B. 1978. *Biochemistry* 17:1922–28
47. Collins, D. J., Lewis, C., Swan, J. M. 1975. *Aust. J. Chem.* 28:673–79
48. Venuti, M. C., Rastetter, W. H., Neilands, J. B. 1979. *J. Med. Chem.* 22:123–24
49. Weitl, F. L., Raymond, K. N. 1979. *J. Am. Chem. Soc.* 101:2728–31
50. Stiefel, E. I., Watt, G. D. 1979. *Nature* 279:81–83
51. Bauminger, E. R., Cohen, S. G., Dickson, D. P. E., Levy, A., Ofer, S., Yariv, J. 1979. *J. Physiol. Paris* 40:523–25
52. Frankel, R. B., Blakemore, R. P., Wolfe, R. S. 1979. *Science* 203:1355–56
53. Brown, J. C. 1979. In *Iron,* pp. 57–78. Baltimore: Univ. Park Press. (Comm. Med. Biol. Effects Environ. Pollut., Subcomm. Iron, Natl. Res. Counc.)
54. Ito, S., Nardi, G., Palumbo, A., Prota, G. 1979. *J. Chem. Soc. Perkin Trans. 1,* pp. 2617–23
55. Fernandez-Pol, J. A. 1978. *Cell* 14:489–99
56. Pollack, J. R., Neilands, J. B. 1970. *Biochem. Biophys. Res. Commun.* 38:989–92
57. O'Brien, E. G., Gibson, F. 1970. *Biochim. Biophys. Acta* 215:393–402
58. Llinas, M., Wilson, D. M., Neilands, J. B. 1973. *Biochemistry* 12:3836–43
59. Salama, S., Strong, J. D., Neilands, J. B., Spiro, T. G. 1978. *Biochemistry* 17:3781–85
60. Corey, E. J., Bhattacharyya, S. 1977. *Tetrahedron Lett.* 45:3919–22
61. Hider, R. C., Silver, J., Neilands, J. B., Morrison, I. E. G., Rees, L. V. C. 1979. *FEBS Lett.* 102:325–28
62. Ong, S. A., Peterson, T., Neilands, J. B. 1979. *J. Biol. Chem.* 254:1860–65
63. Peterson, T., Falk, K.-E., Leong, S. A., Klein, M. P., Neilands, J. B. 1980. *J. Am. Chem. Soc.* 102:7715–18
64. Eng-Wilmot, D. L., van der Helm, D. 1980. *J. Am. Chem. Soc.* 102:7719–25
65. Peterson, T., Neilands, J. B. 1979. *Tetrahedron Lett.* 50:4805–8
66. Bergeron, R. J., McGovern, K. A., Channing, M. A., Burton, P. S. 1980. *J. Org. Chem.* 45:1589–92
67. Corbin, J. L., Bulen, W. A. 1969. *Biochemistry* 8:757–62
68. Emery, T. 1980. *Biochim. Biophys. Acta* 629:382–90
69. van der Helm, D., Baker, J. R., Eng-Wilmot, D. L., Hossain, M. B., Loghry, R. A. 1980. *J. Am. Chem. Soc.* 102:4224–31
70. Hider, R. C., Drake, A. F., Kuroda, R., Neilands, J. B. 1980. *Naturwissenschaften* 67:136–39
71. Isowa, Y., Ohmori, M., Kurita, H. 1974. *Bull. Chem. Soc. Japan* 47:215–20
72. Keller-Schierlein, W., Maurer, B. 1969. *Helv. Chim. Acta* 52:603–10
73. Naegeli, H.-U., Keller-Schierlein, W. 1978. *Helv. Chim. Acta* 61:2088–95
74. Winkelmann, G. 1979. *FEBS Lett.* 97:43–46
75. Ong, S. A., Neilands, J. B. 1979. *J. Agr. Food Chem.* 27:990–95
76. Luckey, M., Pollack, J. R., Wayne, R., Ames, B. N., Neilands, J. B. 1972. *J. Bacteriol.* 111:731–38
77. Maehr, H., Pitcher, R. G. 1971. *J. Antibiotics* 24:830–34
78. Isowa, Y., Takashima, T., Ohmori, M., Kurita, H., Sato, M., Mori, K. 1972. *Bull. Chem. Soc. Japan* 45:1467–71
79. Fujii, T., Hatanaka, Y. 1973. *Tetrahedron* 29:3825–31
80. Carrano, C. J., Cooper, S. R., Raymond, K. N. 1979. *J. Am. Chem. Soc.* 101:599–604
81. Harris, W. R., Carrano, C. J., Raymond, K. N. 1979. *J. Am. Chem. Soc.* 101:2722–27
82. Stephenson, M. C., Ratledge, C. 1980. *J. Gen. Microbiol.* 116:521–23
83. Keller-Schierlein, W. 1976. In *Development of Iron Chelators for Clinical Use,* eds. W. F. Anderson, M. C. Hiller, pp. 53–82. Bethesda: NIH. (DHEW Publ. No. 76–994.)
84. Meyer, J.-M., Abdallah, M. A. 1980. *J. Gen. Microbiol.* 118:125–29
85. van der Helm, D., Poling, M. 1976. *J. Am. Chem. Soc.* 98:82–86

Ann. Rev. Biochem. 1981. 50:733–64
Copyright © 1981 by Annual Reviews Inc. All rights reserved

GLYCOSPHINGOLIPIDS IN CELLULAR INTERACTION, DIFFERENTIATION, AND ONCOGENESIS[1]

♦12094

Sen-itiroh Hakomori

Division of Biochemical Oncology, Fred Hutchinson Cancer Research Center, and Departments of Pathobiology, Microbiology, and Immunology, University of Washington, Seattle, Washington 98104

---

[1]Abbreviations and nomenclature: GSL = glycosphingolipid(s). Nomenclature of glycolipids is according to the recommendations of the Nomenclature Committee, International Union of The Pure and Applied Chemistry (276). LacCer = Galβ1→4Glc→Cer; GbOs$_3$Cer = Gal α1→4Galβ1→4Glc→Cer; GbOs$_4$Cer = GalNAcβ1→3Galα1→4Galβ1→4Glc→Cer; GbOs$_5$ Cer = GalNAcα1→3GalNAcβ1→3Galα1→4Galβ1→4Glc→Cer; GgOs$_3$Cer = GalNAcβ1→ 4Galβ1→4Glc→Cer; GgOs$_4$Cer =Galβ1→3GalNAcβ1→4Galβ1→4Glc→Cer; LcOs$_3$Cer = GlcNAcβ1→3 Galβ1→4Glc→Cer; LcOs$_4$Cer = Galβ1→3GlcNAcβ1→3Galβ1→4Glc→Cer; LcnOs$_4$Cer = Galβ1→4GlcNAcβ1→3Galβ1→4Glc→Cer; LcOs$_6$Cer = Galβ1→4GlcNAcβ1 →3Galβ1→4GlcNAcβ1→3Galβ1→4Glc→Cer. For others, see Table 2 and its footnote.

Abbreviations of the ganglio series gangliosides are according to the system of Svennerholm (277) and Holmgren et al (264). GM$_3$ = NeuAcα2→3Galβ1→4Glc→Cer; GM$_2$ = GalNAcβ 1→4[NeuAcα2→3]Galβ1→4Glc→Cer; GM$_1$ or GM$_{1a}$ = Galβ1→3GalNAcβ1→4[NeuAcα2→ 3]Galβ1→4Glc→Cer; GM$_{1b}$ = NeuAcα2→3Galβ1→3GalNAcβ1→4Galβ1→4Glc→Cer; GD $_{1a}$ = NeuAcα2→3Galβ1→3GalNAcβ1→4[NeuAcα2→3]Galβ1→4Glc→Cer; GD$_2$ = GalNAc β1→4[NeuAcα2→8NeuAcα2→3]Galβ1→4Glc→Cer; GD$_3$ = NeuAcα2→8NeuAcα2→3Galβ 1→4Glc→Cer; GT$_{1b}$ = NeuAcα2→3Galβ1→3GalNAcβ1→4[NeuAcα2→8NeuAcα2→3] Gal β1→4Glc→Cer; GT$_{1a}$ = NeuAcα2→8NeuAcα2→3Galβ1→3GalNAcβ1→4[NeuAcα2→3] Galβ1→4Glc→Cer; GQ$_{1b}$ = NeuAcα2→8NeuAcα2→3Galβ1→3GalNAc[NeuAcα2→ 8NeuAcα2→3]Galβ1→4Glc→Cer; GQ$_{1c}$ = NeuAcα2→3Galβ1→3GalNAcβ1→4[NeuAcα2 →8NeuAcα2→8NeuAcα2→ 3]Galα1→4Glc→Cer. GP$_{1c}$ = NeuAcα2→8NeuAcα2→3Galβ1 →3GalNAcβ1→4[NeuAcα2→8NeuAcα2→ 8NeuAcα2 →3]Galβ1→4Glc →Cer; GP$_{1b}$ = NeuAcα →8NeuAcα2→8NeuAcα2→3Galβ1→3GalNAcβ1→4[NeuAcα2→8NeuAcα2→3] Galβ1→4 GLC→cer.

733

0066-4154/81/0701-0733$01.00

CONTENTS

PERSPECTIVES AND SUMMARY

Glycosphingolipids are ubiquitous membrane components, and the majority are assumed to be present at the outer leaflet of plasma membranes, although their exact organization in membranes remains unknown. Four classes of glycosphingolipids with basically different carbohydrate core structures have been characterized. These are trivially termed ganglio-, globo-, lacto-, and mucoglycosylceramide series, respectively. Various substitutions added to these core structures result in a large variety of glycolipids, particularly in the lacto series.

The dramatic changes in glycolipid composition and metabolism associated with oncogenic transformation suggest a specific role for membrane glycolipids in regulation of cell growth and cellular interaction. Two general types of changes are 1. deletion of complex glycolipids due to a block in synthesis, which frequently leads to accumulation of precursor structures, and 2. synthesis of new glycolipid due to activation of normally unexpressed glycosyltransferases. Both changes can produce tumor-distinctive glycolipids, some of which may be tumor-associated antigens or markers.

A possible role for glycolipids in cell growth control has been investigated by three approaches: 1. addition of exogenous glycolipid onto cell membranes from culture media, 2. addition of such differentiation modulators as butyric acid and retinoids, and 3. addition of antiglycolipid antibodies and monovalent derivatives. These treatments greatly affect cell growth behavior, morphology, and cell saturation density. Nontransformed cells show increased cellular adhesiveness and reduced saturation density; transformed cells mimic normal cells (reverse transformation) and thereby show significant enhancement of glycolipid synthesis, particularly GM_3.

A remarkable change in glycolipids is associated with cellular interaction and differentiation. Cell contact induces enhanced glycolipid synthesis (contact response), which has been implicated as a basis of contact inhibition of cell growth. The contact response is even more remarkable between heterologous cells (e.g. neuroglial contact). Ontogenesis and differentiation are probably mediated through a continuous change in cell surface structure. Remarkable changes in glycolipid markers at the early embryo stage and in those with differentiation of intestinal epithelia and erythrocytes have been elucidated, and a role for glycolipid in mediating neuroglial, retinotectal, and neuromuscular recognition has been suggested in a series of recent studies.

Another area that has made significant progress in recent years is the characterization of glycolipid antigens and a possible role for glycolipids in regulating immune response. Blood group ABH, P, Ii, Lewis, heterophile Forssman, Hanganutziu-Deicher (H–D) antigens, various stage-specific embryonic antigens, and various tumor antigens have been characterized as glycolipids. Lymphoid cell subsets have been characterized by reactivities with antibodies directed to various glycolipids. Gangliosides and fucolipids may indeed be mediators of interactions between B and T lymphocytes and macrophages.

Glycolipids have been shown to interact with various factors such as bacterial toxins, glycoprotein hormones, interferon, serotonin, etc. However, these interactions may not mean that glycolipids are the real receptors. Some glycolipids may regulate protein receptor function.

INTRODUCTION

The idea that glycosphingolipids (GSL) are ubiquitous plasma membrane components and display cell type–specific patterns may have stemmed from the classical studies of GSL in erythrocyte membranes (1, 2). Subsequently, plasma membranes of various animal cells were successfully isolated and characterized by their much higher content of GSL than intracellular membranes (3–7). It is generally assumed that GSL are present at the outer leaflet of the plasma membrane bilayer, although this assumption is based only on experiments with surface labeling by galactose oxidase–NaB[^{3}H]$_4$ of the intact membranes, the lysed erythrocyte membranes, and the inside-out vesicles (8, 9). Obviously further extensive studies of the organization of GSL in eukaryotic cell membranes are necessary. Interestingly, the majority of neutral GSL with short carbohydrate chains are cryptic at the plasma membranes (see next section under Organization).

A high GSL content at the outer half of the lipid bilayer suggests two possibilities: First, GSL may contribute to the structural rigidity of the surface leaflet (item 1, Table 1). Higher rigidity of sphingolipid liposomes as compared to glyceride liposomes has been demonstrated in studies with NMR and ESR spectroscopy (10, 11) as well as in kinetics of complement-dependent lysis of liposomes by antibodies (12). Ceramides confer greater structural rigidity than glycerides, because ceramides contain both a hydrogen acceptor (amide carbonyl) and a donor (hydroxyl) to form stable hydrogen bonds. In contrast, glycerides have only a hydrogen acceptor (ester carbonyl) (10, 13). Ceramides with sugar may confer even higher stability (13). Second, GSL at the outer leaflet of a bilayer are well suited to interact with exogenous ligands through their carbohydrates or to regulate protein receptor function. Various functional notions of GSL in cellular interactions and differentiation, and their deficiency in transformed cells, can all be viewed from a common standpoint, i.e. GSL as cell surface interactants and transducers, or GSL as regulators of transducers (items 2–6, Table 1).

During the last decade a large number of GSL with new structures (about 25–30) have been discovered. Six or seven GSL from human erythrocyte membranes were known about ten years ago; we now have separated and characterized 15 neutral GSL and 9 monosialosyl gangliosides from the same cell membranes (14–17). Perhaps 10 more neutral GSL and 15 more gangliosides remain to be characterized. Polyglycosylceramides (18, 19) with more than 20 sugar residues are not included in this list. Such complex GSL patterns reinforce the notion that glycolipids are interactants or regu-

Table 1 Functions of glycosphingolipids

1. Constitutive component of outer leaflet of lipid bilayer conferring structural rigidity (10–13).
2. Cell surface markers and antigens (see Table 3).
3. Cell-cell interaction and recognition
   a. cell contact response and contact inhibition (102–105, 149)
   b. neuroglial interaction (106)
   c. retinotectal recognition (107, 108)
   d. neuromuscular recognition (109)
4. Differnetiation marker
   a. early embryo and teratocarcinoma (123–129, 132)
   b. crypt to villus cell differentiation (135, 136)
   c. erythrocyte differentiation (glycolipid branching and i to I) (63–65, 138)
   d. myogenic differentiation (146, 147)
   e. neuroblastoma (144, 145)
   f. lymphoid cell differentiation and mitogenesis (148, 156, 158, 159)
5. Cell growth regulation and oncogenesis (see Table 4)
6. Interaction with bioactive factors: bacterial toxins, glycoprotein hormones, and viruses; implication as receptor (see Table 5)

lators for a large number of specific signals and that they serve as a variety of specific antigens on the cell surface (for review see 20, 21).

This review briefly discusses new developments in four topics: 1. Major new GSL structures and GSL organization in membranes, 2. GSL in cellular interaction and differentiation, 3. GSL in cell growth control and oncogenic transformation, and 4. GSL antigens and immune response.

Since the implication of GSL as cell surface receptors has been reviewed recently (22–25), it is discussed here only briefly. This article does not discuss enzymes involved in the synthesis and degradation of GSL.

STRUCTURE AND ORGANIZATION OF GLYCOLIPIDS

New Glycolipid Structures

Through improved ion exchange chromatography on diethylaminoethyl-Sephadex (26) or Spherosil-diethylaminoethyl dextran (27), and with the use of porous homogeneous silica gel (Iatrobeads), GSL have been separated with great efficiency (28). Thus ganglioside mapping was proposed based on the combined procedure of ion exchange chromatography and silica gel thin layer chromatography (29, 30), through which new gangliosides have been detected and characterized (30). High performance liquid chromatography has been introduced in GSL separation and quantification. GSL have been successfully analyzed as perbenzoylated (31), N-benzoylated and O-acetylated (32), and O-benzoylated (33) derivatives. The last derivatization with benzoylanhydride and 4-methylamino-pyridine is favored since the original GSL could be recovered by base hydrolysis. Three major procedures have been introduced in GSL characterization: 1. Improved methylation analysis (34–36), 2. direct probe mass spectrometry of permethylated or permethylated-reduced compounds (14, 15, 37–40), and 3. enzymatic degradation by exo- (41–43) and endo-β-glycosidase (44).

The basic GSL structures and the structures of new GSL species are shown in Tables 2A and 2B. Space does not allow the listing of all the new structures, most of which have been reviewed recently (45–48). Among the ganglio series, those with a new asialo core are particularly important, i.e. gangliosides with gangliopentaosylceramide (GalNAcβ1→4Galβ1→3Gal-NAcβ1→4Galβ1→4Glc→Cer) (49, 50) and that containing gangliofucopentaosylceramide (Fuca1→2Galβ1→2GalNAcβ1→4Galβ1→4Glc1→1Cer) (40, 51–53). A novel structure, Gala1→3Galβ1→3Gala, linked to GM$_1$ was recently characterized (54). A number of other ganglio series GSL with variations in the number and the location of sialic acid (55–58; for review see 47), or with sialic acid that contains an O-acetyl group (59–61), have been isolated and characterized. Interestingly, some disialogangliosides have both N-glycolyl and N-acetyl neuraminic acid (40, 57, 58).

Two important new additions are the branching structure with GlcNAc $\beta1\rightarrow6$(GlcNAc$\beta1\rightarrow3$)Gal in the lacto series (14, 15, 62–64), or Gal$\beta1\rightarrow$ 3(Gal$\beta1\rightarrow6$)Gal structure in the muco series (67, 68), and a repeating Gal $\beta1\rightarrow4$GlcNAc$\beta1\rightarrow3$Gal structure (repeating N-acetyllactosamine) in the lacto-N-glycosyl series (65, 66). The terminal moiety of these branched or repeating N-acetyllactosamines contains either blood group ABH determinants (14), sialosyl residues (15, 16, 66), or Gal$\alpha1\rightarrow3$Gal, (64, 66, 241), Gal $\alpha1\rightarrow4$Gal (242), or Gal$\beta1\rightarrow3$Gal (69) residues. Thus, many new structures belonging to the lacto-N-glycosyl series have been characterized, and a new lacto series of gangliosides (L series vs G series) has been added (Table 2B). These are all Type II chain series and mainly isolated from human and bovine erythrocyte membranes. Other tissues, particularly gastro-intestinal epithelia and pancreas, contain GSL with Type I chain as well (70–73), and lacto-N-tetraosylceramide was recently isolated and characterized from meconium (73). Some of the lactoglycosyl series contain as many as 30–40 sugar residues and are termed polyglycosylceramide (18, 19), yet their variations and structures remain to be determined. Gastrointestinal mucosa seem to contain GSL with di- to penta-β-galactosyl residues as the carrier of blood group determinants (67, 68), constituting "mucoglycosylceramide series" (Table 2A).

Organization and Dynamic State

CRYPTICITY OF GSL AND ORGANIZATION   The ceramide moiety of GSL is assumed to be inserted into the outer leaflet of the lipid bilayer, and the carbohydrate moieties may extrude towards the outer environment (8, 9). However, much of the GSL is cryptic and its organization is unknown. A striking feature of GSL in the plasma membrane is their crypticity and the change in crypticity associated with a change in cellular functions. Cell surface–labeled activity of GbOs$_4$Cer is greatly enhanced at the G$_1$ phase of the cell cycle (74, 75). The cryptic GSL become highly exposed when cells are transformed (75). Galactose oxidase/NaB[^{3}H]$_4$ labeling, however, does not clearly quantitate the oxidized GSL, i.e. that exposed to the surface. Galactose oxidase treatment followed by reduction with NaBD$_4$ results in deuterated GSL that can be determined by mass spectrometry. Only 10– 15% of GbOs$_4$Cer in human erythrocyte membranes (T. Matsubara and S. Hakomori, unpublished data) and less than 1% of cerebroside at the myelin sheath membrane are oxidized (76). Thus, a large proportion of GSL is masked, probably due to an extensive association with integral membrane components, some of them the yet unknown "co-factor" of membrane enzymes or functional proteins. Sulfatide has been claimed as the K$^+$-selective co-factor site of [Na$^+$, K$^+$]-dependent adenosine triphosphatase

Table 2A Major basic structures of five glycolipid series

Series	Structure	Name of oligosaccharide (abbreviation)
Ganglio series	[a]Galβ1→3GalNAcβ1→4Gal[b]β1→4Glcβ1→1Cer	gangliotetraose (GgOs$_4$)
	GalNAcβ1→4Gal[b]β1→3GalNAcβ1→4Gal[b]β1→4Glcβ1→1Cer	gangliopentaose (GgOs$_5$)
Globo series	[c]GalNAcβ1→3Galα1→4Galβ1→4Glcβ1→1Cer	globotetraose (GbOs$_4$)
	GalNAcβ1→3Galα1→3Galβ1→4Glcβ1→1Cer	globo-*neo*-tetraose (GbnOs$_4$)
Lacto series	[d]Galβ1→4GlcNAcβ1→3Galβ1→4Glcβ1→1Cer	lacto-*neo*-tetraose (LcnOs$_4$)
	Galβ1→3GlcNAcβ1→3Galβ1→4Glcβ1→1Cer	lactotetraose (LcOs$_4$)
	[d]Galβ1→4GlcNAcβ1→3Galβ1→4GlcNAcβ1→3Galβ1→4Glcβ1→1Cer	lacto-*nor*-hexaose (LcOs$_6$)
	[d]Galβ1→4GlcNAcβ1 ⎤⁶ [d]Galβ1→4GlcNAcβ1—3Galβ1→4GlcNAcβ1→3Galβ1→4Glcβ1→1Cer [d]Galβ1→4GlcNAcβ1 ⎦₃	lacto-*iso*-octaose (LciOs$_8$)
Muco series	[e]Galβ1→3Galβ1→4Glcβ1→1Cer	mucotriose (MuOs$_3$)
	[e]Galβ1→3Galβ1→3Gal[f]β1→4Glcβ1→1Cer	mucotetraose (MuOs$_4$)
Gala series	Galα1→4Galβ1→1Cer	galabiose (GalaOs$_2$)

[a]sialosyl 2→3 or fucosyl 1→2 or 1→3; [b]sialosyl 2→3; [c]GalNAcα1→3; [d]sialosyl 2→3, 2→6, Galβ1→3, Galα1→3, Fucα1→2, GalNAcα1→3 [Fucα1→2], Galα1→3 [Fucα1→2]; [e]Fucα1→2, GalNAcα1→3; [f]GalNAcα1→3Galβ1→6.

Table 2B Major new glycolipid structures

Series	Structure	Abbreviation	Ref.
Ganglio series[a]	NeuGlycα2→8NeuAcα2→3Galβ1→4Glc→Cer		58
	NeuGlycα2→3Galβ1→3GalNAcβ1→4[NeuAcα2→3]Galβ1→4Glc→Cer		40, 58
	GalNAcβ1→4[NeuAcα2→3]Galβ1→3GalNAcβ1→4[NeuAcα2→3]Galβ1→4Glc→Cer		49, 50
	Fucα1→2(or 3)Galβ1→3GalNAcβ1→4[NeuAcα2→3]Galβ1→4Glc→Cer	(Fuc-GM$_1$)	51–53
	Galα1→3Galβ1→3Galβ1→3GalNAcβ1→4[NeuAcα2→3]Galβ1→4Glc→Cer		54
Globo series	GalNAcα1→3GalNAcβ1→3Galα1→4Galβ1→4Glc→Cer		36, 236
	NeuAcα2→3Galβ1→GalNAcβ1→3Galα1→4Galβ1→4Glc→Cer		237
Lacto series, neutral GSL[b]	Galβ1→3GlcNAcβ1→3Galβ1→4Glc→Cer		73
	Galβ1→3Galβ1→4GlcNAcβ1→3Galβ1→4Glc→Cer		69
	Galα1→3Galβ1→4GlcNAcβ1→3Galβ1→4Glc→Cer		36, 241
	Galα1→4Galβ1→4GlcNAcβ1→3Galβ1→4Glc→Cer	(P$_1$)	242
	Fucα1→2Galβ1→4GlcNAcβ1→3Galβ1→4Glc→Cer	(H$_1$)	240
	Fucα1→2Galβ1→4GlcNAcβ1→3Galβ1→4GlcNAcβ1→3Galβ1→4Glc→Cer	(H$_2$)	39

Fucα1→2Galβ1→4GlcNAcβ1
$\qquad\qquad\qquad\qquad\searrow^{3}$
$\qquad\qquad\qquad\qquad\quad$ Galβ1→4GlcNAcβ1→3Galβ1→4Glc→Cer   (H$_3$)   39
$\qquad\qquad\qquad\qquad\nearrow^{6}$
Fucα1→2Galβ1→4GlcNAcβ1

Lacto series, gangliosides (LM or LD series ganglioside)

NeuAcα2→3Galβ1→4GlcNAcβ1→3Galβ1→4Glc→Cer   (SPG)c   16

NeuAcα2→6Galβ1→4GlcNAcβ1→3Galβ1→4Glc→Cer   (SPG)   238

NeuAcα2→3Galβ1→4[Fucα1→3]GlcNAcβ1→3Galβ1→4Glc→Cer   239

NeuAcα2→3NeuAcα2→3Galβ1→4GlcNAcβ1→3Galβ1→4Glc→Cer   17

NeuAcα2→3GalNAcβ1→4GlcNAcβ1→3Galβ1→4Glc→Cer   16, 66

NeuAcα2→3Galβ1→4GlcNAcβ1→3Galβ1→4Glc→Cer   17

NeuAcα2→6Galβ1→4GlcNAcβ1→3Galβ1→4Glc→Cer   17

±(Fucα1→2)Galβ1→4GlcNAcβ1
$\qquad\qquad\qquad\qquad\searrow^{6}$
$\qquad\qquad\qquad\qquad\quad$ Galβ1→4GlcNAcβ1→3Galβ1→4Glc→Cer   15, 16
$\qquad\qquad\qquad\qquad\nearrow^{3}$
NeuAcα2→3Galβ1→4GlcNAcβ1

---

aVarious other new structures with variation at sialosyl substitutions (see 30, 47).
bMany other structures with blood group activity (for review see 63).
cSPG = sialylparagloboside.

(77), and gangliosides have been implicated as the Ca^{2+}-binding co-factor in synaptic transmission (78). Gangliosides greatly activate the Mg^{2+}-dependent adenosine triphosphatase but not the Na^+, K^+-dependent enzyme of brain microsomes (79), and they may be a co-factor of membrane adenylate cyclase (80). A type of GSL assumed to be present as annular GSL, surrounding a functional protein, and highly cooperative to the conformation change of such a protein (13), could be in the same category as cryptic GSL.

ASSOCIATION OF MEMBRANE PROTEIN AND GSL   The association of GSL with a specific membrane protein has been suggested not only by the specific requirement of GSL for membrane enzymes described above, but also by cross-linking studies. In a model experiment, the coliphage M13 membrane protein incorporated into the lipid bilayer was preferentially cross-linked with the exogenously added photosensitive glycolipid probe (81). The association of $GbOs_4Cer$ with a membrane protein has been demonstrated by cross-linking studies by heterobifunctional methyl-4-azidobenzoimidate with or without anti-$GbOs_4Cer$ antibodies (82). A low molecular weight amphipathic polypeptide with Paul-Bunnell antigen activity has been isolated from bovine erythrocyte membranes and characterized as having a specific affinity to GM_3 (83). Interestingly, the Paul-Bunnell antigenicity at the cell surface is greatly decreased upon sialidase treatment of intact cells, but the purified polypeptide antigen does not contain sialic acid, and the antigen itself is not sialidase sensitive. The gangliophilic antigen may be associated with gangliosides at the cell surface, and such a complex is sialidase sensitive and displays its maximum antigenic activity. Thus, ganglioside may regulate the antigenicity of polypeptide (83). These results may indicate that membrane protein may be associated with a specific GSL and the function of the membrane protein may be regulated by GSL. Exogenously added GSL in media are taken up quickly by cells (84–86). A kinetic study of GM_1 uptake by lymphocytes indicates the presence of a specific binding site at the cell surface. The binding is Ca^{2+}-dependent and does not compete with LacCer or other gangliosides (87); therefore it may represent a gangliophilic protein as discussed above.

THE DYNAMIC BEHAVIOR OF GLYCOLIPID IN MEMBRANES   The organization and dynamic state of GSL in phospholipid bilayers and in cell membranes were studied by ESR spectrometry with spin-labeled GM_1 (labeled at the sialosyl tail). The results indicate that GM_1 moves rapidly on the lipid bilayer and head group mobility decreases with increasing bilayer concentration of unlabeled GM_1, perhaps as a result of cooperative head group interactions (11, 88). A physiological concentration of Ca^{2+} causes head group cross-linking, leading to decreased mobility in solution as well as on lipid bilayers. Interaction between glycophorin and GM_1 was observed by the spectral change, especially in the presence of Ca^{2+} (89). Spin-labeled GM_1 at a concentration lower than the critical micelle concentration is rapidly incorporated into intact cell membrane and is not released by trypsinization. The mobility of the GM_1 head group increases abruptly upon addition of a very small quantity of wheat germ lectin, which binds to sialic acid, and decreases steadily on further addition of the lectin (90). Perturbation of the entire GM_1 head group is induced when only a small population of GM_1 is bound to wheat germ lectin. Interestingly, the mobility of GM_1 as compared with GM_3, GM_2, GD_{1a}, and GD_{1b} is the most restricted in micelles and in mixed dispersions with phospholipid as measured by fluorescence anisotropy and the nanosecond time-dependent change of the fluorescence using 1,6-diphenyl-1,3,5-hexatriene as the motion probe (91). The cooperativity of phospholipids in liposomes is greatly reduced by GD_{1a} inclusion as shown by broadening of the endotherm curve in differential scanning calorimetry. The endotherm change in phospholipid-GD_{1a} vesicles is much greater in the presence of 10 mM $CaCl_2$. The Ca^{2+} effect is not observed without GD_{1a} (92).

Various physical parameters point to features of ganglioside behavior in lipid bilayers: 1. head group structure may define the intrinsic mobility of each ganglioside; 2. the mobility of one species of ganglioside is greatly reduced by increasing the concentration of the same ganglioside species, perhaps because of aggregation; and 3. ligand-induced mobility increases greatly when a small population is bound to ligands but mobility is reduced upon saturation with ligands.

Cap formation on the choleragen receptor (assumed to be GM_1) has been detected with anticholeragen antibody and the fluorescein-labeled second antibody (93). Dansyl-labeled GM_1 incorporated into lymphocytes is capped upon addition of cholera toxin; the capping is inhibited by treatment with colchicin and trypsin (94). Patching followed by capping of GM_1 in thymocytes is induced by anti-GM_1 antibodies; the capping is inhibited by cytochalasin B (95). Interestingly, anti-GM_1 induces a remarkable mitogenesis (95). There are three possible explanations for the capping of GM_1: 1. capping may be caused by a GM_1-associated transmembrane protein; 2. GSL on membranes are continuously motile according to continuous oriented motion of membranes (the continuous lipid flow model), and cross-linkage by ligands then results in immobilization of GM_1 at the rear "pole" of cells which causes capping (96); 3. the same carbohydrate structure as GM_1 is present on glycoprotein ("ganglioprotein") (97) and the observed capping is in fact a capping of "ganglioprotein" (see next section).

The Common Carbohydrate Chain in GSL and in Glycoproteins

Blood group ABH, Ii determinants, and their carrier carbohydrate chains with a repeating Galβ1→4GlcNAc structure (polylactosamine) are present in simpler glycolipids (39, 62–64), in polyglycosylceramides (19), and in Band 3 and Band 4.5 glycoproteins (98), which are all susceptible to endo-β-galactosidase of *Escherichia freundii* (99). Thus cell surface labeling by galactose oxidase/NaB[^{3}H]$_4$ or by NaIO$_4$/NaB[^{3}H]$_4$ followed by endo-β-galactosidase treatment reveals the presence of the common structure shared between GSL and glycoprotein (99). Another method to detect a common structure between GSL and glycoprotein is the use of anti-glycolipid antibodies. Several methods—cell surface labeling of human erythrocytes, membrane isolation, detergent solubilization, and double immune precipitation with anti-globoside antibodies followed by SDS-polyacrylamide gel electrophoresis—have detected globosides and a few major glycoproteins with the same migration as "PAS 3 and 2" ("globo-protein") (97). The same approach with anti-GM$_1$ antibodies applied to 3T3 cells has detected a few glycoproteins reactive to anti-GM$_1$ ("ganglio-protein") (97). Methylation analysis of brain glycoprotein has detected the sugar residue NeuAc2→8NeuAc2→3Gal, similar to that present in ganglio-side (100). Thus, the presence of a common carbohydrate sequence in GSL and in glycoprotein is increasingly apparent, and many of the same carbohydrate structures as in GSL must be present in the peripheral region of glycoprotein carbohydrates, although the core structure of the carbohy-drates in glycoprotein is entirely different from that of GSL (for review see 101).

The biological significance of the presence of a common carbohydrate chain in GSL and in glycoprotein is not known. However, a two-step reception of ligands from a glycoprotein to GSL is possible.

GLYCOLIPIDS IN CELLULAR INTERACTION AND DIFFERENTIATION

Cellular Interaction

The fact that GSL on cell surfaces may be recognized by other cell surfaces, and vice versa, is an important basis of cellular recognition and differentia-tion. A few examples and the possible mechanism are discussed below.

GANGLIOSIDE CHANGES AND NEUROGLIAL RECOGNITION   Cell contact induces enhanced synthesis of a particular GSL prior to contact inhibition between homologous cells (102–105). A similar GSL response is even more remarkable in heterologous neuroglial cell contact. A pronounced induction of GD_{1b} synthesis has been detected by co-culturing glial and neuronal cell lines, although neither cell line contains GD_{1b} initially. Similarly, GT_{1b} and an unidentified GT appear to be synthesized when two different glial cells, M1 and MT16 lines, are co-cultured, although both cell lines lack GT species initially (106).

RETINOTECTAL RECOGNITION   A GSL marker has been studied that may be involved in the preferential adhesion of chick neuronal retina cells to the surfaces of intact optic tectum. The adhesion of the ventral retina to the dorsal tecta depends on recognition between β-GalNAc residues of GSL on the dorsal tecta and a specific protein located on the ventral retina. A reverse recognition has been observed between a β-GalNAc residue of the dorsal retina and a specific protein at the ventral tecta. The specific GalNAc residue present on the dorsal tecta or dorsal retina is not affected by protease, and liposomes containing GM_2 ($NeuAcGgOs_3Cer$) inhibit these adhesions (107). Consequently, the retinotectal adhesion was found to be inhibited by the affinity purified antibody directed to GM_2, but not by preimmune serum. Furthermore, the oligosaccharides released from GM_2 but not from other GSL affect the adhesive specificity (108).

NEUROMUSCULAR RECOGNITION   The formation of the neuromuscular junction is based on a highly specific recognition process that is now believed to involve GSL. In culture, the formation of the neuromuscular junction is specifically inhibited by $GbOs_4Cer$ at high concentrations (0.-25–0.5 mM) and is stimulated at low concentrations (8–63 μM) of the same GSL (109).

CELL SURFACE RECOGNITION SITES AND MECHANISM FOR CELLULAR INTERACTIONS   All the phenomena described above are explained on the assumption that GSL on one type of cell surface are recognized by the surfaces of the complementary cells involved. Two basically different mechanisms can be considered: 1. The order and number of glycolipids organized with other membrane components at the cell surface is complementary to the same order and number of glycolipids on the counterpart cell surface (Steinberg's self-self interaction model) (110). 2. The molecules that can recognize or interact with GSL, which may be lectins (111) or enzymes (112–114), would be expected to be present. A "lock and key" interaction is possible through these sugar-binding proteins. The presence of a specific binding protein to glycolipid at the cell surface is suggested by the preferential adherence of GSL liposomes containing fluorescein labels (115). The specific binder at the cell surface may be either a lectin or an enzyme. Animal lectins have been isolated from various tissues and cells, but their presence at the cell surface and their involvement in cellular interactions are not clear (for review see 111, 116). A GalNAc transferase has been suggested to be involved in retinotectal recognition (107). Although a classical proposal that a glycosyltransferase occurs at the cell surface (112) has aroused considerable debate (113, 114), the possibility of the presence of glycosylhydrolases at the cell surface has now been suggested (105, 117–119). In fact, specific cell adhesion and spreading can be induced on solid substrata of β-galactosidase or sialidase at physiological pH, under which conditions hydrolytic-catalytic activity is suppressed but binding activity remains (119). Cell surface hydrolase, if present, may well function as a cell surface lectin. A cellular protein released by ethidium bromide may induce growth inhibition (120) and has affinity to GSL, particularly GM_3. One component is adsorbed on a GM_3 column but is difficult to elute (121). A possible role for GSL-associated protein (82, 83) in cellular interaction remains to be studied.

Differentiation Markers

Embryonic or histogenetic differentiation is the orderly process of cellular interaction through a continuous change of cell surface molecules encoded by a genetic program. A remarkable phase-dependent change in GSL composition and synthesis indicates that GSL is involved in cellular recognition during differentiation. Some phase-dependent GSL may represent an "area code molecule" (122), which may define cell-cell interactions and migrations to specific tissues. A few lines of studies are described below.

CARBOHYDRATE MARKERS AT EARLY EMBRYO STAGE AND TERATOCARCINOMA   Changes in cell surface molecules during embryonic development have been detected by immunological methods and such molecules have been described as developmental antigens. These include blood group ABH (123), Forssman (124, 125), Ii (126), F9 (127, 128), TerC (129), and the stage-specific embryonic antigen-1 (SSEA-1) (130). In each case, the antigenic molecules were found to be GSL or the carbohydrate moiety of glycoprotein and were expressed at the defined stage of embryonic development. A stage-specific expression of F9 antigen has been well documented; it expresses most strongly at the morulae and blastocyte stage, and disappears on further development (127). Interestingly, anti-F9 antibody Fab reversibly prevents compaction of mouse morulae (131). The F9 antigen may be a glycoprotein partially susceptible to endo-β-galactosidase (128). A similar surface antigen, SSEA-1, recognized by the monoclonal antibody to F9 cell, appears at the eight-cell stage of embryo, is maximally expressed at the morulae stage, and disappeared at the blastocyte stage

(130). The antigen was recently identified as a type II-chain fucosyl lacto-N-glycosyl GSL with branching and was shown to be susceptible to endo-β-galactosidase. The antigen is therefore closely related if not identical to Ii (132). The carbohydrate structures recognized by monoclonal anti-I (Ma), anti-I (Step), and anti-i (Dench) were shown to be expressed on early postimplantation mouse embryo and teratocarcinoma cells. Undifferentiated stem cells are rich in surface-associated and cytoplasmic I antigen, whereas i antigen appears when cells are differentiated into primary endoderm (126). The Ii antigens are known also as the differentiation marker of later stage erythroid cells (133). The surface antigen, "TerC," of mouse testicular teratoma was recently identified as GSL (129). GSL of teratocarcinoma cell lines are altered by the induction of differentiation. The slowly migrating GSL, corresponding to GM_1 and GD_{1a}, appear upon differentiation (134).

GLYCOLIPIDS IN INTESTINAL EPITHELIUM DIFFERENTIATION
The undifferentiated "crypt" cells of intestinal epithelia are characterized by the presence of LacCer,GlcCer and the absence of GM_3 and GM_3 synthetase (CMP-NeuAc:LacCer-sialyltransferase). They undergo progressive maturation and migration within 12 hr to become villus cells and develop a well-defined brush border with digestive enzymes. The villus cells are characterized by high levels of GM_3 and GM_3 synthetase, and by the low content of GlcCer and LacCer. GM_3 synthesis is therefore correlated with rapid differentiation of intestinal epithelia (135, 136).

ERYTHROCYTE DIFFERENTIATION AND GLYCOLIPID CHANGE   A remarkable change of agglutinability to anti-I and anti-i antibodies after birth (137) is now identified as resulting from the process of branching of the carbohydrate chain (64, 65, 98, 138). The i antigen, which is present in fetal erythrocytes, was identified as a linear repeating N-acetyllactosamine structure (Galβ1→4GlcNAcβ1→3Gal), which is represented by lacto-N-nor-hexaosylceramide (65). I antigen, which is lacking in fetal erythrocytes but is predominantly present in adult erythrocytes, is now identified as a branched structure represented by lacto-N-iso-octaosylceramide (64, 138). Various monoclonal anti-i antibodies recognize various domains within an unbranched linear lacto-N-nor-hexaosylceramide. A linear repeating N-acetyllactosamine structure was found in fetal polyglycosylceramide, while the branched structure was found in adult polyglycosylceramide (19) as well as in glycopeptide fraction derived from Band 3 glycoprotein (98). The branching enzyme that would define a crucial step in erythrocyte differentiation remains to be elucidated. Interestingly, the time for switching synthesis from fetal hemoglobin to adult hemoglobin coincides with the time for i-to-I conversion (139). Endo-β-galactosidase of *E. freundii* (140) hydrolyzes a linear unsubstituted N-acetyllactosamine chain much more readily than branched structure (141), and Ii activity in situ is destroyed (99). The major Ii determinants were identified to be carried by Band 3 and Band 4.5 glycoprotein as well as by lactoglycosyl series GSL (99). The branching process associated with erythrocyte differentiation is a general phenomenon found not only in Ii antigens but also in ABH determinants as well. Cell surface labeling and reaction with anti–branched H (anti-H_3) antibodies showed the branched structures to be much lower in fetal than in adult erythrocytes (142). A weak fetal erythrocyte reaction to the antiglobulin test with anti-A,B IgG was correlated to the lack of the branched, bivalent A and B antigens in fetal erythrocytes (143).

OTHER DIFFERENTIATION SYSTEMS   A slowly migrating ganglioside increases when neuroblastoma cell lines are induced to differentiate. Increased activities of galactosyltransferase and sialyltransferase in an adrenergic clone of neuroblastoma have been observed upon addition of dibutyryl cyclic AMP (144, 145). Enhanced synthesis of GD_{1a} of L6 muscle cells at the aligning stage was observed shortly before cell fusion (146). However, what was described as GD_{1a} has recently been identified as GD_3 (147). In addition, a still unidentified slowly migrating neutral GSL (corresponding to ceramide heptasaccharide) was present in fused mixed culture but not in bromodeoxyuridine-inhibited cultures (147). The presence of asialo GM_1 ($GgOs_4Cer$) in mature T cell population was recently observed (148). A remarkable increase of a specific ganglioside, GM_{1b} was observed when M_1 myeloid cells were induced to differentiate into macrophages (278).

GLYCOLIPIDS IN CELL GROWTH CONTROL AND ONCOGENIC TRANSFORMATION

Growth Regulation

CELL CONTACT RESPONSE AND CONTACT INHIBITION   GSL contact response is closely related to contact inhibition of cell growth, which suggests that GSL may regulate cell growth. Synthesis of $GbOs_3Cer$ in BHK (102), $GbOs_5Cer$ and GM_3 in NIL (103, 104), GD_{1a} in 3T3 (105), C_3H mouse fibroblasts (151), and GD_3 and GM_1 in human fibroblasts 8166 (102) is greatly enhanced prior to density-dependent growth inhibition. The enzymatic basis of this response (149, 150) and the loss of this response have been well correlated to the loss of contact inhibition in oncogenic transformants (102–105). Chemically transformed mouse embryo cells show a very similar GSL pattern to nontransformed cells but lack the density-dependent response of GD_{1a} (151). However, the lack of this response may not be an absolute criterion for tumorigenicity, since there are a few exceptions (152), and GD gangliosides of Ehrlich carcinoma cells also show a density-dependent enhancement (153).

CELL CYCLE AND MITOGENESIS   Changes associated with the cell cycle have also been observed in the exposure of GSL at the cell surface (74, 75) and the metabolic incorporation of precursor sugars (154, 155) into GSL. Lectin-induced mitogenesis of sheep lymphocytes is accompanied by enhanced synthesis of ceramide trihexoside and a change in ganglioside composition (156). Other studies of murine lymphocytes have indicated that synthesis of all GSL is enhanced with a maximum peak 30 hr after adding lectin (157, 158). Moreover, mitogen-activated T and B cells can be distinguished by GSL pattern (159).

To test the possibility that cell surface GSL may control cell proliferation, three approaches have been undertaken: exogenous addition of GSL, application of antibodies directed toward to GSL, and measurement of GSL changes caused by differentiation inducers.

GLYCOLIPID ADDITION IN CELL CULTURE   Incubation of cells with GSL antigen, but not with glycoprotein antigen, causes antigenic conversion of cells (84). Thus, added GSL, but not added glycoprotein, may be incorporated into membrane and function as well as those synthesized in situ. Hamster NIL cells cultured in media containing $GbOs_4Cer$ show more than twofold enhancement of the $GbOs_4Cer$ levels at the plasma membrane. Consequently, cell adhesiveness increases, morphology of NILpy cells resembles that of normal NIL cells, and the prereplicative period, particularly the G_1 phase, becomes twice as long as before treatment (85). Similarly, addition of several gangliosides to culture media reduces both the growth rate and saturation density of SV40-transformed and -untransformed 3T3 cells (86). No apparent cytotoxicity was demonstrated even with the most effective GM_1, GD_{1b}, GT_{1b}; however, free ceramide and GD_{1a} are ineffective. Gangliosides added to the culture media accumulate at the plasma membrane; however, the exogenously supplied gangliosides can be distinguished from those synthesized endogenously by the lability of the former to sialidase (86). While some of the added GSL appear to be removable by trypsin treatment (160), some are certainly oriented with the hydrophobic ceramide moiety embedded in the lipid bilayer (89, 94, 160). This approach is important to establish the function of membrane components as related not only to cellular growth behavior, but also to receptor activity and antigenicity.

MODIFICATION BY ANTIGLYCOLIPID ANTIBODIES   The second approach to studying the function of specific GSL on cell surfaces is to observe the functional changes following the addition of specific anti-GSL antibodies (161). While lectins such as Concanavalin A and their monovalent or divalent derivatives have been used to affect cell growth behavior (162), lectins are cytotoxic and their specificity is too broad to specify the function of a specific carbohydrate. However, affinity-purified anti-GSL antibodies (163) should be suitable for this purpose. Anti-GM$_3$, but not anti-GbOs$_4$ Cer, markedly inhibits the growth of hamster NIL and mouse Balb/c fibroblasts but not their transformed derivatives. In subconfluent cell cultures, the addition of anti-GM$_3$ leads to a 1.5-fold stimulation of [^{14}C]-Gal incorporation specifically into GM$_3$, whereas GSL synthesis in confluent cell cultures is not stimulated. Thus, the interaction with anti-GSL mimics those cellular interactions that stimulate synthesis of density-dependent GSL and may lead to cessation of cell growth (mimic contact inhibition) (161). Significant growth inhibition of astrocytoma cells is induced by anti-GM$_1$ and anti-GM$_3$ antibodies and is accompanied by increased adenylate cyclase and decreased guanylate cyclase activity (106). Incubation of Balb/c 3T3 cells with anti-GM$_1$ antibodies inhibits MSV-dependent expression of transformed phenotype. Rat kidney NRK cells transformed with temperature-sensitive avian sarcoma virus La23 or La25, and preincubated with anti-GM$_3$ antibodies at nonpermissive temperature, completely inhibit such expressions of transformed phenotype as morphology changes and growth in soft agar (164).

GLYCOLIPID CHANGES BY DIFFERENTIATION INDUCER   Lower fatty acids, particularly n-butyric acid, added in culture media have reduced cell saturation density and restored contact inhibition (165). Associated with these changes in cell growth behavior is a greatly increased chemical level of GM$_3$ due to an enhanced GM$_3$ synthetase (CMP-sialic acid : CDH-sialyltransferase (166, 167); the reactivity to cholera toxin is enhanced due to induced synthesis of GM$_1$ (168). Butyrate is now recognized as a potent differentiation inducer for various cell systems, particularly for erythroid cells (169). The tumor promoter (phorbol-12–myristate-13 acetate) for mouse skin was found to induce differentiation of human melanoma cells; it also induces synthesis of melanin and GM$_3$ similar to the effect of butyrate (170).

Other powerful differentiation inducers are the retinoid compounds. Transformed cells tend to regain their normal growth behavior and show density-dependent growth inhibition in vitro upon addition of retinoids (20–100 nmol/ml (171), and chemical and viral carcinogenesis in vivo is prevented by retinoid administration (172). An enhanced synthesis of GM$_3$ in NIL cells and an enhanced cell contact response of ganglioside in 3T3 cells have been observed (173). The mechanism by which butyrate, phorbol esters, and retinoids induce functional changes may in part be due to the change of membrane gangliosides.

Oncogenic Transformation

A NEW TYPE OF GLYCOLIPID CHANGE WITH ONCOGENESIS   Since a GSL change associated with oncogenic transformation was first described (174), many studies have been carried out employing the established fibroblast cell lines or primary cultures transformed by DNA viruses, RNA viruses or chemical carcinogens. The major findings up to 1975–1977 were reviewed previously (175–180). In addition to a classical blocked synthesis of glycolipid (149, 181–185), the synthesis of a new glycolipid that is absent (or present in trace amounts) in progenitor cells has been found in tumor cells and tissues; this change includes the synthesis of Forssman glycolipid antigen (186, 187) in tumors derived from Forssman negative tissue, A-like antigen (188–190) in tumors derived from blood group O or B individuals, and specific types of fucolipid synthesis (191, 192). A large quantity of gangliotriosylceramide (asialo GM_2) has been found in mouse sarcoma (193) and lymphoma (194) that is virtually absent in normal cells and tissues; this change results from an induction of glycolipid synthesis rather than from accumulation due to a blocked synthesis of GM_1 (195).

GSL changes associated with oncogenic transformation can be classified into seven types (shown in Table 3). While transformation from type 1 to 4 in the table represents "incomplete synthesis," transformation from type 5 to 7 represents the induced synthesis of a new carbohydrate residue. An incomplete synthesis sometimes results in accumulation of the precursor. Therefore, either a new synthesis or an incomplete synthesis may result in the appearance of a GSL that is absent (or present in a small amount) in untransformed cells.

GLYCOLIPID CHANGES AS CORRELATED WITH TUMORIGENESIS AND MALIGNANCY   GSL of low tumorigenic mouse L-cell lines (A9 or B82), their high malignant isolates (A9HT, B82HT) and their various hybrids with low or high tumorigenicity have been compared (196). The

Table 3 Types of glycolipid changes associated with oncogenic transformation

Type 1	Decrease or deletion of GM_3, GD_3; CDH or CMH increased at later stage
Type 2	Decrease or deletion of GT_1, GD_{lab}; increase of GM_3, GM_2, or GM_1
Type 3	Accumulation of asialo core ($GgOs_4Cer$, $GgOs_3Cer$, $LcnOs_4Cer$, etc), which is normally absent
Type 4	Decrease or deletion of $GbOs_3Cer$, $GbOs_4Cer$, $GbOs_5Cer$ (Forssman) or longer chain neutral glycolipid
Type 5	Decrease or deletion of GM_3, GM_2; increase of GD, GT
Type 6	New synthesis of $GbOs_5Cer$ (Forssman-negative host)
Type 7	New synthesis of incompatible blood group antigen foreign to the host

level of GM_3 and of long chain neutral GSL reflect tumorigenicity (change from type 4 to 5 in Table 3). The most impressive correlation between the GSL pattern, tumor cell growth behavior, and enzymatic activities was demonstrated with two types of rat ascites hepatomas. Several cell lines with low metastasis and low malignancy form aggregates ("islands") and contain $GbOs_3Cer$, $GbOs_4Cer$, and GM_3. In contrast, several cell lines with high malignancy do not form "islands" do not contain $GbOs_3Cer$, $GbOs_4Cer$, or GM_3, but do contain $GgOs_3Cer$ (asialo GM_2), $GgOs_4Cer$ (asialo GM_1), and fucosyl asialo GM_1 (197–200). The chemical level of total ganglioside in human tumor (201), the loss of glycolipid crypticity in mouse melanoma cells (202), and the level of $GgOs_3Cer$ in mice tumor cell variants transformed by RNA tumor virus (279) have been correlated to their metastatic property.

GLYCOLIPID TUMOR ANTIGEN   The NILpy tumor grown in hamsters causes antibody response directed to paragloboside ($LcnOs_4Cer$); presumably tumorigenic NILpy cells contain this GSL (75). Mouse Balb/c 3T3 cells transformed with KiMSV and grown in Balb/c mice, and L5178Y lymphoma cells grown in DBA mice, contain high levels of $GgOs_3Cer$, although various tissues and organs of these mice do not contain chemically detectable amounts of the same GSL (193, 194). The fact that L5178Y lymphoma growth is completely suppressed by passive immunization of tumor-inoculated DBA mice with the monoclonal anti-$GgOs_3Cer$ may provide a specific marker of this tumor. Vesicular stomatitis virus (VSV) obtained from SV40-transformed hamster cell lines acquires a tumor specific transplantation antigen activity, which causes SV40 tumor rejection. A GSL fraction prepared from VSV incorporated into liposomes was shown to be immunogenic and to suppress tumor growth (205). Antiserum to liposomes containing the polar glycolipids of SV40-transformed hamster tumor cells, after being absorbed on normal hamster tissue, specifically stains the SV40-transformed cells (206). The results imply that a specific GSL acts as an SV40-induced antigen. Forssman antigen ($GbOs_5Cer$) is absent in the majority (70–80%) of human gastrointestinal mucosa, whereas gastrointestinal tumor often contains this GSL (187). Blood group A–like antigen has appeared in tumors of blood group B or O individuals (188, 189); its structure in one case was determined to be a difucosylated heptasaccharide with an A determinant (190). A gastric cancer patient with blood group p contained an incompatible P_1 antigen now identified as a ceramide pentasaccharide (207). One of the distinct antigens found in culture medium of human malignant melanoma cell lines contains GSL antigen that reacts with human melanoma antibodies (208), and a group of monoclonal antibodies directed to human melanoma react to the glycolipid

Table 4   Glycolipid cell surface markers and antigens

1. Blood group antigens
   ABH (14, 62); Le^a, Le^b (84); Ii (64, 65, 138); P, P$_1$, P^k (243).

2. Heterophile antigens
   Forssman (236); Hanganutziu-Deicher (212, 213).

3. Lymphoid subpopulation marker
   GM$_1$, GgOs$_4$Cer for T cell subpopulation (214, 217); LcnOs$_4$Cer for unknown B > T (222); GgOs$_4$Cer for NK (223, 224); Thy-1 ganglioside (221).

4. Differentiation antigen
   Forssman (124, 125); Ii (126); F9 (127, 128); TerC (129); SSEA-1 (132).

5. Glycolipids reacting with "auto-antibodies," or antiglycolipid antibodies causing auto-immune disease
   Ii-active glycolipid (64, 65, 138); sialosylparagloboside (243); "experimental anngloside syndrome" and neuritis (244); sulfatide and cerebroside in experimental encephalomyelitis (245, 246); globoside and Forssman in paroxysmal hemoglobinuria (247).

6. Glycolipid tumor cell surface markers or antigen
   LcnOs$_4$Cer in NILpy tumor (203); GgOs$_3$Cer in KiMSV tumor (193); lymphoma 5178 (194); GgOs$_4$Cer in human acute lymphoblastic leukemia (209); Forssman antigen in human gastrointestinal tumor (187); A-like antigen in human O or B tumor (188–190); P, P$_1$ antigen in human p tumor (207).
   Human melanoma antigen (208, 280); hamster SV40 tumor specific antigen (205, 206).

antigens (280). Acute lymphoblastic leukemia cells were stained with anti-GgOs$_4$Cer antibodies (209). Certainly a search for a GSL-specific marker for tumor cells is important for practical purposes. The knowledge can be directly applied to immunotherapy (204) or targeting of drugs by antibodies (210).

GLYCOLIPIDS IN IMMUNE RECOGNITION

Glycolipid Antigens

During the last decade, many GSL antigens, as listed in Table 4, have been characterized as cell surface antigens. Since this area was extensively reviewed (20, 21, 211) the following major progress has been made: 1. the chemical basis of blood group Ii antigens (64, 65, 138) and the developmental change of blood group A and H antigens in erythrocytes (142) have been established; 2. the heterophile H-D antigen has been identified as gangliosides containing N-glycolyl neuraminic acid (212, 213); 3. some tumor antigens have been identified as GSL (21); 4. some lymphocyte subpopulation markers have been identified as GSL (214); and 5. some data have been found that suggest that immune cell response is mediated by GSL. The last two topics are discussed below; the others are covered in Table 4.

Glycolipid Surface Markers of Lymphoid Cell Subpopulations

The earlier claim that Thy-1 antigen could be GM_1 (215, 216) has not been supported by subsequent studies. Briefly, it has been shown that anti-GM_1 antibodies recognize certain subpopulation independent of Thy-1 type (214), that Thy-1.2 and GM_1 cap independently (217), and that Thy-1 activity is not blocked by prior choleragen treatment (218) and is not found in purified GM_1, GD_{1a}, GD_{1b}, GT_{1b} (219). The active GSL has been separated from GM_1 (220) and its activity abolished by treatment with sialidase or weak acid hydrolysis (221). Thus, Thy-1 may be a minor unidentified ganglioside.

Antibodies to purified GSL have been found to discriminate between lymphoid subpopulations. Anti-GM_1 reacts with peripheral T cells and most thymocytes in A-, AKR-, and Balb/c mice (214). The reactivity is, however, removed by absorption of antisera on a $GgOs_4Cer$ column, which indicates that these cells may contain $GgOs_4Cer$. In C3H and nude mice, anti-GM_1 has identified a population that also expresses surface immuno-globin. Treatment with protease has uncovered reactivities for anti-GM_1 on most cells, which indicates that GM_1 may be cryptic in some cell populations independent from T and B cells (217). Antibodies to purified $GgOs_4Cer$ react with mature T cells in all strains of mice, but do not react with most thymocytes or with pronase-treated B cells (217). Antibodies to $LcnOs_4Cer$ (paragloboside) react with a certain subpopulation of human peripheral lymphocytes (222).

The function of the lymphoid cells recognized by anti-$GgOs_4Cer$ has been recently studied by treating effector cell populations with this antibody and complement (223, 224). These studies revealed that whereas allogenic cytotoxic T killer cells are not affected by this treatment, natural killer (NK) cell activity is abrogated. However, NK activity is not abolished by anti-$GgOs_4Cer$ without complement. This marker may represent an unknown lymphoid cell population that includes an NK cell population. The appearance of $GgOs_4Cer$ in lymphoid cell populations can be well correlated with the differentiation of NK cell activity (148). The suppressor T cell activity selected by peanuts agglutinin is abrogated by anti-$GgOs_4Cer$, anti-$GbOs_4Cer$, and anti-$GbOs_5Cer$ antibodies, respectively, in the presence of complement, whereas the helper T cell activity is abrogated by anti-GM_1 antibodies and complement (225).

Glycolipid as Mediator of Immune Response

MODULATION OF B OR T CELL RESPONSE BY GANGLIOSIDES   When spleen cells are incubated with sheep red blood cells (SRBC) and thymocytes, an anti-SRBC response is stimulated that can be measured by Jerne's hemolytic plaque-forming colony (PFC) assay. GM_1 and GD_{1b}, but not GD $_{1a}$, in lecithin-cholesterol liposome preparations, when added to the above system can inhibit the PFC response, and the inhibition seems to be specific to B cells (220). Furthermore, a factor shed by suppressor T cells delays the PFC by several days. Anti-GM_1 antibodies obviate the inhibitory effect, and a GM_1-like fraction isolated from the medium of suppressor T cells inhibits the PFC response. Thus GM_1 and GD_{1b} either present on the cell surface or released into the medium may modulate the primary immune response of B cells (226). B cell modulation by gangliosides is supported by others (227) who observed that gangliosides in aqueous solution can suppress the lipopolysaccharide-stimulated B cell mitogenesis, but fail to suppress *Phaseolus* lectin-stimulated T cell mitogenesis. A ganglioside peptide complex is, in contrast, mitogenic to B cells. On the other hand, murine T-cell mitogenesis stimulated by Concanavalin A-(Con A) is inhibited by gangliosides, most effectively by GT_1. In this case, A–stimulated DNA synthesis, but not Con A–stimulated lactate production, is inhibited (228).

MEDIATOR OF IMMUNE CELL RECOGNITION When inactivated spleen cells are mixed with allogeneic thymocytes, the thymocytes respond to the spleen cell surface antigens as measured by [^{3}H]thymidine uptake. When syngeneic or autologous thymocytes are mixed, the response is usually not observable; however, when spleen cells are preincubated with brain gangliosides, a remarkable autologous and allogeneic thymocyte response is observed (229). Among gangliosides, GD_{1a} is the most effective. This suggests that properly organized ganglioside(s) on cell surfaces may mediate immune cell response.

GLYCOLIPIDS AS LYMPHOKINE RECEPTOR Antigen- or mitogen-stimulated lymphocytes release a variety of soluble mediators known as lymphokines, which affect the functional properties of macrophages, neutrophiles, etc. One of the most extensively studied lymphokines is the macrophage migration inhibitory factor (MIF). MIF may be related to, if not identical with, the macrophage activation factor (MAF), which renders macrophages cytotoxic for tumor cells (e.g. 230, 231). MIF activity can be eliminated by use of a ganglioside-Sepharose column or by liposomes containing gangliosides. The elimination is not affected by the purified known gangliosides (232). On the other hand, addition of L-fucose or treatment of macrophages with fucosidase eliminates the MIF (232, 233). Of various oligosaccharides, only α-L-fucosyl oligosaccharide inhibits MIF activity (234). Fucose-binding lectins of *Lotus tetragonolobus* and *Ulex europeus* agglutinin I, but not other lectins, interfere with the response of macrophage to MIF (234). Treatment of guinea pig macrophages with liposomes containing GSL from guinea pig macrophages, but not with liposomes containing GSL of neutrophiles and brain tissues, greatly enhances their responsiveness to MIF (232). Recently a new fucolipid with the structure Galα1→3[Fucα1→2]Galβ1→3GalNAcβ1→3Galβ1→4Glc→Cer was isolated from rat granuloma and peritoneal macrophages. This GSL inhibits MIF activity; however, a fucolipid with a lactoglycosyl structure does not inhibit the activity. Therefore, MIF interacts with a fucosylated ganglio series structure (Fuc→Galβ1→3GalNAcβ1→R) but not with the fucosylated lacto series (Fuc→Galβ1→4GlcNAcβ1→R), although it may not distinguish between the various fucosylated linkages, i.e. whether α1→2, 1→3, or 1→6 (235). All these findings suggest that MIF will bind to a specific ganglio series structure that may well be a fucoganglioside or its analogue.

INTERACTION OF GLYCOLIPIDS WITH BIOACTIVE FACTORS

GSL, particularly gangliosides, interact with various bioactive factors, listed in Table 5, by inhibiting or interfering with the physiological effects of these factors on cells. These results have been interpreted as showing that GSL function as receptors; this topic, particularly the subject of gangliosides as receptors for cholera toxin and for glycoprotein hormones, has been extensively reviewed in the last few years (22–25). Therefore, only a few supplemental comments to Table 5 are made here.

Although the interaction of cholera toxin with GM$_1$ is the best characterized (248–250), the interaction per se does not directly indicate that GM$_1$ is the real receptor for cholera toxin for all kinds of cells. The binding of cholera toxin to mouse LY and human KB-3 cell membranes has been shown to be trypsin-sensitive (251). GM$_1$ is absent in fat cells and adrenal cells, but these cells still respond to cholera toxin (252). Glycoproteins of rat microvillus membranes prepared by *Ricinus communis* lectin–Sepharose chromatography were shown to be able to form complexes with iodinated cholera toxin. These glycoproteins were separated by precipitation with anti–cholera toxin antibodies followed by gel electrophoresis. Thus, at least five glycoproteins of microvillus membranes have been shown to react with cholera toxin (253). A similar approach identified the proteins that interact with the toxin in lymphoid cells and Balb/c 3T3 cells (254). GM$_1$ could be the major receptor for some cells, but a ganglioprotein that carries a similar structure as GM$_1$ (97) could also be a receptor.

Table 5 Interaction of glycolipids with biofactors

Biofactor	Major interactant claimed	Ref.
1. Bacterial toxins		
cholera toxin	GM_1	248–250
tetanus toxin	GT_{1b}, GQ_{1b}	255, 256, 275
Botulinus toxin	GT_{1b}	257, 274
Staphylococcus α toxin	SPG	258
Gonococcus pilli protein	GM_1	259
2. Glycoprotein hormones		
thyrotropin	GD_{1b}	260
chorionic gonadotropin	GT_1	261
luteinizing hormone	GT_1	262
3. Sendai virus	GT_{1a}, GQ_{1b}	263, 264
4. Interferon (type 1 only)	GM_2, GT_1	266–268
5. Fibronectin	FT_{1b}	270
6. Lymphokines		
macrophage migration inhibitor	Fucosyl $GgOs_4Cer$	232–235
7. Opiate and morphine	sulfatide	271, 272
8. Serotonin	GD_3	273

The interaction of gangliosides with various glycoprotein hormones (258–260) and with interferon (263–265) has been well established. However, the common gangliosides (GT_1 and GD_{1b}) inhibit the binding of various glycoprotein hormones (thyrotropin, chorionic gonadotropin, and luteinizing hormone) to target tissue membranes (258–260). Various gangliosides (GM_2 and GT) inhibit both mouse and human interferon activity (263, 264), and mouse fibroblast interferon is absorbed on a ganglioside-Sepharose column and eluted with sialyllactose (265). These nonspecific interactions of gangliosides with hormones and interferon strongly conflict with the facts that hormone stimulation is highly specific to target cells and that interferon activity is highly species specific, i.e. human interferon acts on human cells but not on mouse cells, and vice versa. The following possibilities should be considered to explain ganglioside interaction with glycoprotein hormones: 1. The interaction of GT or GD_{1b} with the hormones may be a cross-reaction between a real receptor with a structure similar to that of GT or GD_{1b}. In fact, an unidentified ganglioside was found to be the more potent inhibitor of thyrotropin binding among a large number of thyroid gangliosides (262). 2. A real receptor may be the amphipathic polypeptide associated with ganglioside fraction (77). 3. Gangliosides may be the co-factor for a real receptor and may regulate hormone binding to the receptor. The analogy was recently described for expression of the Paul-Bunnel antigen, which is a gangliophilic polypeptide (77). Similar possibilities should be considered for ganglioside interaction with interferon.

An earlier claim that gangliosides interact with Sendai virus (263) has been recently confirmed (264). The virus binding is most strongly inhibited by GT_{1a}, GQ_{1b}, and GP_{1c}, which have a common terminal sequence: NeuAc $\alpha2{\to}8$NeuAc$\alpha2{\to}3$Gal$\beta1{\to}3$GalNAc${\to}$R. However, these polysialylated gangliosides are absent in various cell lines that are susceptible to Sendai virus-mediated cell fusion. Binding of Sendai virus by HeLa cells was studied by incubation of the virus with various liposomes containing HeLa cell gangliosides or bovine gangliosides, and with various protein fractions extracted with chloroform-methanol. Only HeLa cell protein soluble in chloroform-methanol, but not gangliosides, inhibits virus-induced hemagglutination (265). Thus, glycoproteins, rather than gangliosides having polysialosyl structure, could be the real receptor.

Fibronectin promotes specific cell spreading through its specific cell binding domain (269). It is generally assumed that the specific domain will interact on the cell surface receptor. GT_{1b} and GD_{1b}, but not GM_1 and GM_3, inhibit cell attachment on a fibronectin-collagen layer absorbed on plastic surfaces, which implies that these gangliosides could be the fibronectin receptor (270). However, cell attachment on a collagen or lectin layer is also inhibited by these gangliosides. Higher gangliosides are generally absent in various fibroblasts that have high content of fibronectin (119); therefore, gangliosides may not be the fibronectin receptor.

Thus, some of the interaction of gangliosides with various biofactors as described in this section and listed in Table 5 seem ionic and nonspecific. In such cases gangliosides may not be the receptor. It is possible, however, that gangliosides may regulate or modulate the receptor function, and the susceptibility of membranes to ligands on similar, but uncharacterized mechanisms through which glycolipids play roles for cell growth regulation and immune response.

ACKNOWLEDGMENTS

The author wishes to thank Drs. Y. Barenholz, M. H. Buc, P. H. Fishman, C. W. M. Grant, R. G. Kemp, Y. Kishimoto, L. D. Kohn, R. Langenbach, Y. Nagai, M. Ohashi, and T. Osawa for their kindness in sending their reprints and preprints of their work. The author also wishes to thank his colleagues who read this manuscript, particularly Dr. David Urdal for his suggestions. I am most grateful to Mrs. Charlotte Pagni for arranging the references and typing the manuscript.

Literature Cited

1. Yamakawa, T., Suzuki, S. 1951. *J. Biochem. Tokyo* 38:199–212
2. Klenk, E., Lauenstein, K. 1951. *Z. Physiol. Chem.* 288:220–28
3. Dod, B. J., Gray, G. M. 1968. *Biochim. Biophys. Acta* 150:397–404
4. Weinstein, D. B., Marsh, J. B., Glick, M. C., Warren, L. 1970. *J. Biol. Chem.* 245:3928–37
5. Yogeeswaran, G., Sheinin, R., Wherrett, J. R., Murray, R. K. 1972. *J. Biol. Chem.* 247:5146–58
6. Critchley, D. R., Graham, J. M., Macpherson, I. 1973. *FEBS Lett.* 32:37–40
7. Klenk, H. D., Choppin, P. W. 1979. *Proc. Natl. Acad. Sci. USA* 66:57–64
8. Gahmberg, C. G., Hakomori, S. 1973. *J. Biol. Chem.* 248:4311–17
9. Steck, T. L., Dawson, G. 1974. *J. Biol. Chem.* 249:2135–42
10. Abrahamsson, S., Dahlén, B., Löfgren, H., Pascher, I., Sundell, S. 1977. In *Structure of Biological Membranes,* ed. S. Abrahamsson, I. Pascher, pp. 1–23. New York/London: Plenum
11. Sharom, F. J., Grant, C. W. M. 1977. *J. Supramol. Struct.* 6:249–58
12. Kinsky, S. C. 1972. *Biochim. Biophys. Acta* 265:1–23
13. Yamakawa, T., Nagai, Y. 1978. *Trends Biochem. Sci.* 3:128–31
14. Hakomori, S., Watanabe, K., Laine, R. A. 1977. *Pure Appl. Chem.* 49:1215–27
15. Watanabe, K., Powell, M. E., Hakomori, S. 1978. *J. Biol. Chem.* 253:8962–67
16. Watanabe, K., Powell, M. E., Hakomori, S. 1979. *J. Biol. Chem.* 254:8223–29
17. Watanabe, K., Hakomori, S. 1979. *Biochemistry* 24:5502–4
18. Koscielak, J., Miller-Podraza, H., Krauze, R., Piasek, A. 1976. *Eur. J. Biochem.* 71:9–18
19. Koscielak, J., Zdebska, E., Wilczynska, Z., Miller-Podraza, H., Dzierzkowa-Borodej, W. 1979. *Eur. J. Biochem.* 96:331–37
20. Marcus, D. M., Schwarting, G. A. 1976. *Adv. Immunol.* 23:203–40
21. Hakomori, S., Young, W. W. Jr. 1978. *Scand. J. Immunol.* 6:97–117
22. Fishman, P. H., Brady, R. O. 1976. *Science* 194:906–15
23. Brady, R. O., Fishman, P. H., 1979. *Adv. Enzymol.* 50:303–23
24. Kohn, L. D. 1978. In *Receptors and Recognition, Ser. A,* ed. P. Cuatrecasas, M. F. Greaves, 5:135–212. London: Chapman & Hall
25. Fishman, P. H., Atikkan, E. E. 1980. *J. Membr. Biol.* 54:51–60
26. Momoi, T., Ando, S., Nagai, Y. 1976. *Biochim. Biophys. Acta* 441:488–97
27. Fredman, P., Nilsson, O., Tayot, J.-L., Svennerholm, L. 1980. *Biochim. Biophys. Acta* 618:45–52
28. Ando, S., Isobe, M., Nagai, Y. 1976. *Biochim. Biophys. Acta* 424:98–105
29. Iwamori, M., Nagai, Y. 1978. *Biochim. Biophys. Acta* 528:257–67
30. Nagai, Y., Iwamori, M. 1980. *Mol. Cell. Biochem.* 29:81–90
31. McCluer, R. H., Jungwala, F. B. 1976. *Adv. Exp. Med. Biol.* 68:533–54
32. Suzuki, A., Handa, S., Yamakawa, T. 1977. *J. Biochem. Tokyo* 82:1185–87
33. Gross, S. K., McCluer, R. H. 1980. *Anal. Biochem.* 102:429–33
34. Hakomori, S. 1964. *J. Biochem. Tokyo* 55:205–8
35. Björndal, H., Hellerqvist, C. G., Lindberg, B., Svensson, S. 1970. *Angew. Chem. Int. Ed. Engl.* 9:610–19
36. Stellner, K., Saito, H., Hakomori, S. 1973. *Arch. Biochem. Biophys.* 155:464–72
37. Karlsson, K.-A., Pascher, I., Pimlott, W., Samuelsson, B. E. 1974. *Biomed. Mass Spectrom.* 1:49–56
38. Ledeen, R. W., Kundu, S. K., Price, H. C., Fong, J. W. 1974. *Chem. Phys. Lipids* 13:429–46
39. Watanabe, K., Laine, R. A., Hakomori, S. 1975. *Biochemistry* 14:2725–33
40. Ohashi, M., Yamakawa, T. 1977. *J. Biochem. Tokyo* 81:1675–90
41. Hakomori, S., Siddiqui, B., Li, Y.-T., Li, S.-C., Hellerqvist, C. G. 1971. *J. Biol. Chem.* 246:2271–77
42. Uda, Y., Li, S.-C., Li, Y.-T., McKibbin, J. M. 1977. *J. Biol. Chem.* 252:5194–99
43. Li, Y.-T., King, M. J., Li, S.-C. 1980. *Adv. Exp. Med. Biol.* 125:93–104
44. Fukuda, M. N., Watanabe, K., Hakomori, S. 1978. *J. Biol. Chem.* 253:6814–19
45. Sweeley, C. C., Siddiqui, B. 1977. In *Biochemistry of Mammalian Glycoproteins and Glycolipids,* ed. W. Pigman, M. I. Horowitz, pp. 459–85. New York: Academic
46. Macher, B. A., Sweeley, C. C. 1978. *Methods Enzymol.* 50:236–51
47. Ledeen, R. W. 1978. *J. Supramol. Struct.* 8:1–17
48. McKibbin, J. M. 1978. *J. Lipid Res.* 19:131–47
49. Svennerholm, L., Mansson, J. E., Li, Y.-T. 1973. *J. Biol. Chem.* 248:740–42

50. Iwamori, M., Nagai, Y. 1978. *J. Biochem. Tokyo* 84:1601–8
51. Wiegandt, H. 1973. *Z. Physiol. Chem.* 354:1049–56
52. Macher, B. A., Pacuszka, T., Mullin, B. R., Sweeley, C. C., Brady, R. O., Fishman, P. H. 1979. *Biochim. Biophys. Acta* 588:35–43
53. Suzuki, A., Ishizuka, I., Yamakawa, T. 1975. *J. Biochem. Tokyo* 78:947–54
54. Ohashi, M. 1980. *J. Biochem. Tokyo* 88:583–91
55. Ishizuka, I., Wiegandt, H. 1972. *Biochim. Biophys. Acta* 260:279–89
56. Ando, S., Yu, R. K. 1977. *J. Biol. Chem.* 252:6247–50
57. Ghidoni, R., Sonnino, S., Tettamanti, G., Wiegandt, H., Zambotti, V. 1976. *J. Neurochem.* 27:511–15
58. Iwamori, M., Nagai, Y. 1978. *J. Biol. Chem.* 253:8328–31
59. Hakomori, S., Saito, T. 1969. *Biochemistry* 8:5082–88
60. Veh, R. W., Sander, M., Haverkamp, J., Schauer, R. 1979. In *Glycoconjugate Research, Vol. 1,* ed. J. D. Gregory, R. W. Jeanloz, pp. 557–59
61. Ghidoni, R., Sonnino, S., Tettamanti, G., Baumann, N., Reuter, G., Schauer, R. 1980. *J. Biol. Chem.* 255:6990–95
62. Hakomori, S., Stellner, K., Watanabe, K. 1972. *Biochem. Biophys. Res. Commun.* 49:1061–68
63. Hakomori, S. 1981. *Seminars in Hematology.* New York & London: Grune & Stratton. 18:39–62
64. Watanabe, K., Hakomori, S., Childs, R. A., Feizi, T. 1979. *J. Biol. Chem.* 254:3221–28
65. Niemann, H., Watanabe, K., Hakomori, S., Childs, R. A., Feizi, T. 1978. *Biochem. Biophys. Res. Commun.* 81:1286–93
66. Chien, J.-L., Li, S.-C., Laine, R. A., Li, Y.-T. 1978. *J. Biol. Chem.* 253:4031–35
67. Slomiany, B. L., Slomiany, A., Horowitz, M. 1973. *Biochim. Biophys. Acta* 326:224–31
68. Slomiany, B. L., Slomiany, A. 1977. *Biochim. Biophys. Acta* 486:531–40
69. Stellner, K., Hakomori, S. 1972. *J. Biol. Chem.* 249:1022–25
70. McKibbin, J. M., Smith, E. L., Mansson, J. E., Li, Y.-T. 1977. *Biochemistry* 16:1223–28
71. Wherrett, J. R., Hakomori, S. 1973. *J. Biol. Chem.* 248:3046–51
72. Karlsson, K.-A. 1980. *Proc. IUPAC Meet., 27th, Helsinki,* ed. A. Varmavuori, pp. 171–83. Oxford/New York/Toronto: Pergamon
73. Karlsson, K.-A., Larson, G. 1979. *J. Biol. Chem.* 254:9311–16
74. Gahmberg, C. G., Hakomori, S. 1974. *Biochem. Biophys. Res. Commun.* 59:283–91
75. Gahmberg, C. G., Hakomori, S. 1975. *J. Biol. Chem.* 250:2438–46
76. Kishimoto, Y., Yahara, S., Podulso, J. 1980. In *Biochemistry of Cell Surface Glycolipids,* ed. C. C. Sweeley, pp. 10–23. Am. Chem. Soc. Monogr.
77. Karlsson, K.-A. 1977. In *Structure of Biological Membranes,* ed. S. Abrahamsson, I. Pascher, pp. 245–74. New York: Plenum
78. Svennerholm, L. 1980. *Adv. Exp. Med. Biol.* 125:533–44
79. Caputto, R., Maccioni, A. H. R., Caputto, H. L. 1977. *Biochem. Biophys. Res. Commun.* 74:1046–52
80. Partington, C. R., Daly, J. W. 1979. *Mol. Pharmacol.* 15:484–91
81. Hu, V. W., Wisnieski, B. J. 1979. *Proc. Natl. Acad. Sci. USA* 76:5460–64
82. Lingwood, C. A., Hakomori, S., Ji, T. H. 1980. *FEBS Lett.* 112:265–68
83. Watanabe, K., Hakomori, S., Powell, M. E., Yokota, M. 1980. *Biochem. Biophys. Res. Commun.* 92:638–46
84. Marcus, D. M., Cass, L. 1969. *Science* 164:553–55
85. Laine, R. A., Hakomori, S. 1973. *Biochem. Biophys. Res. Commun.* 54:1039–45
86. Keenan, T. W., Schmid, E., Franke, W. W., Wiegandt, H. 1975. *Exp. Cell Res.* 92:259–70
87. Krishnaraj, R., Saat, Y. A., Kemp, R. G. 1980. *Cancer Res.* 40:2808–13
88. Sharom, F. J., Grant, C. W. M. 1977. *Biochem. Biophys. Res. Commun.* 74:1039–45
89. Sharom, F. J., Grant, C. W. M. 1978. *Biochim. Biophys. Acta* 507:280–93
90. Lee, P. M., Ketis, N. V., Barber, K. R., Grant, C. W. M. 1981. *Biochim. Biophys. Acta.* 601:302–14
91. Uchida, T., Nagai, Y., Kawasaki, Y., Wakayama, M. 1981. *Biochemistry.* 20:162–69
92. Barenholz, Y., Ceastarp, B., Lichtenberg, D., Freire, E., Thompson, T. E., Gott, S. 1980. *Adv. Exp. Med. Biol.* 125:105–35
93. Révèsz, T., Greaves, M. F. 1975. *Nature* 257:103–6
94. Sedlacek, H. H., Stark, J., Seiler, F. R., Ziegler, W., Wiegandt, H. 1976. *FEBS Lett.* 61:272–76
95. Sela, B.-A., Raz, A., Geiger, B. 1978. *Eur. J. Immunol.* 8:268–74
96. Bretscher, J. S. 1976. *Nature* 260:21–25

97. Tonegawa, Y., Hakomori, S. 1977. *Biochem. Biophys. Res. Commun.* 76:9–17
98. Fukuda, M., Fukuda, M. N., Hakomori, S. 1979. *J. Biol. Chem.* 254:3700–3
99. Fukuda, M. N., Fukuda, M., Hakomori, S. 1979. *J. Biol. Chem.* 254: 5458–65
100. Finne, J., Krusius, T., Rauvala, H., Hemminski, K. 1977. *Eur. J. Biochem.* 77:319–23
101. Järnefelt, J., Finne, J., Krusius, T., Rauvala, H. 1978. *Trends Biochem. Sci.* 3:110–14
102. Hakomori, S. 1970. *Proc. Natl. Acad. Sci. USA* 67:1741–47
103. Sakiyama, H., Gross, S. K., Robbins, P. W. 1972. *Proc. Natl. Acad. Sci. USA* 69:872–76
104. Critchley, D. R., Macpherson, I. A. 1973. *Biochim. Biophys. Acta* 296: 145–59
105. Yogeeswaran, G., Hakomori, S. 1975. *Biochemistry* 14:2151–56
106. Mandel, P., Dreyfus, H., Yusufi, A. N. K., Sarlieve, L., Robert, J., Neskovic, N., Harth, S., Rebel, G. 1980. *Adv. Exp. Med. Biol.* 125:515–31
107. Marchase, R. B. 1977. *J. Cell Biol.* 75:237–57
108. Piearce, M. 1980. PhD. thesis. Dep. Biol., Univ. Penn., Philadelphia (mentor, S. Roth)
109. Obata, K., Oide, M., Handa, S. 1977. *Nature* 266:369–71
110. Steinberg, M. S. 1963. *Science* 141: 401–8
111. Barondes, S. H., Rosen, S. D. 1976. In *Neuronal Recognition*, ed. S. H. Barondes, pp. 331–58. New York: Plenum
112. Roseman, S. 1970. *Chem. Phys. Lipids* 5:270–97
113. Marchase, R. B., Vosbeck, K., Roth, S. 1976. *Biochim. Biophys. Acta* 457:385–416
114. Keenan, T. W., Morré, D. J. 1975. *FEBS Lett.* 55:8–43
115. Huang, R. T. C. 1978. *Nature* 276: 624–26
116. Frazier, W., Glaser, L. 1979. *Ann. Rev. Biochem.* 48:491–523
117. Schengrund, C. L., Rosenberg, A., Repman, M. A. 1976. *J. Cell Biol.* 70: 555–61
118. Von Figura, K., Voss, B. 1979. *Exp. Cell Res.* 121:267–76
119. Rauvala, H., Carter, W. G., Hakomori, S. 1980. *J. Cell Biol.* 88:127–37
120. Hakomori, S., Young, W. W. Jr., Patt, L. M., Yoshino, T., Halfpap, L., Lingwood, C. A. 1980. *Adv. Exp. Med. Biol.* 125:257–61
121. Carter, W. G., Fukuda, M., Lingwood, C. A., Hakomori, S. 1978. *Ann. NY Acad. Sci.* 312:160–77
122. Hood, L., Huang, H. V., Dryer, W. J. 1977. *J. Supramol. Struct.* 7:531–59
123. Szulman, A. E. 1960. *J. Exp. Med.* 111:785–800
124. Willison, K. R., Stern, P. L. 1978. *Cell* 14:785–93
125. Stern, P. L., Willison, K. R., Lennox, E., Galfré, G., Milstein, C., Secher, D., Ziegler, A. 1978. *Cell* 14:775–83
126. Kapadia, A., Feizi, T., Evans, M. J. 1981. *Exp. Cell Res.* 131:185–95
127. Artzt, K., Dubois, P., Bennett, D., Condamine, H., Baninet, C., Jacob, F. 1973. *Proc. Natl. Acad. Sci. USA* 70:2988–92
128. Muramatsu, T., Gachelin, G., Damonneville, M., Delarbre, C., Jacob, F. 1979. *Cell* 18:183–91
129. Larraga, V., Edidin, M. 1979. *Proc. Natl. Acad. Sci. USA* 76:2912–16
130. Solter, D., Knowles, B. B. 1978. *Proc. Natl. Acad. Sci. USA* 75:5565–69
131. Kemler, R., Baninet, C., Eisen, H., Jacob, F. 1977. *Proc. Natl. Acad. Sci. USA* 74:4449–52
132. Nudelman, E., Hakomori, S., Knowles, B. B., Solter, D., Nowinski, R. C., Tam, M. R., Young, W. W. Jr. 1980. *Biochim. Biophys. Res. Commun.* 97:443–51
133. Marsh, W. L. 1961. *Br. J. Haemotol.* 7:200–9
134. Coulen-Morelec, M. J., Buc-Caron, M.-H. 1981. *Dev. Biol.* In press
135. Glickman, R. M., Bouhours, J. F. 1975. *Biochim. Biophys. Acta* 424:17–26
136. Bouhours, J. F., Glickman, R. M. 1969. *Biochim. Biophys. Acta* 441:123–33
137. Marsh, W. L. 1961. *Br. J. Haematol.* 7:200–9
138. Feizi, T., Childs, R. A., Watanabe, K., Hakomori, S. 1979. *J. Exp. Med.* 149:975–80
139. Maniatis, A., Papayannopoulou, T., Bertles, J. F. 1979. *Blood* 54:159–68
140. Fukuda, M. N., Matsumura, G. 1976. *J. Biol. Chem.* 251:6218–25
141. Fukuda, M. N., Watanabe, K., Hakomori, S. 1978. *J. Biol. Chem.* 253: 6814–19
142. Watanabe, K., Hakomori, S. 1976. *J. Exp. Med.* 144:644–53
143. Romans, D. G., Tilley, C. A., Dorrington, K. J. 1980. *J. Immunol.* 124: 2807–11
144. Yeung, K.-K., Moskal, J. R., Chien, J.-L., Gardner, D. A., Basu, S. 1974. *Biochem. Biophys. Res. Commun.* 59: 252–60
145. Moskal, J. R., Gardner, D. A., Basu, S.

1974. *Biochem. Biophys. Res. Commun.* 61:751–58

146. Whatley, R., Ng, S. K.-C., Roger, J., McMurray, W. C., Sanwal, B. D. 1976. *Biochem. Biophys. Res. Commun.* 70:180–85

147. McKay, J. 1980. *Glycolipids in myogenesis PhD thesis.* Dep. Biol. Struct., Univ. Wash., Seattle (mentor, D. Nameroff)

148. Schwarting, G., Summers, A. 1980. *J. Immunol.* 124:1691–94

149. Kijimoto, S., Hakomori, S. 1971. *Biochem. Biophys. Res. Commun.* 44:557–63

150. Chandrabose, K. A., Graham, J. M., Macpherson, I. A. 1976. *Biochim. Biophys. Acta* 429:112–22

151. Langenbach, R., Kennedy, S. 1978. *Exp. Cell Res.* 112:361–72

152. Sakiyama, H., Robbins, P. W. 1973. *Fed. Proc.* 32:86–90

153. Prokazova, N. V., Kocharov, S. L., Zvezdina, N. D., Buznikov, G. A., Shaposhnikova, G. I., Bergelson, L. D. 1978. *Biokhimia* 43:1805–8

154. Chatterjee, S., Sweeley, C. C., Velicer, L. F. 1973. *Biochem. Biophys. Res. Commun.* 54:585–92

155. Chatterjee, S., Sweeley, C. C., Velicer, L. F. 1975. *J. Biol. Chem.* 250:61–66

156. Narimsham, R., Hay, J. B., Greaves, M. F., Murray, R. K. 1976. *Biochim. Biophys. Acta* 431:578–91

157. Inouye, Y., Handa, S., Osawa, T. 1974. *J. Biochem. Tokyo* 76:791–99

158. Rosenfelder, G., Van Eijk, R. V. W., Monner, D. A., Mühlradt, P. F. 1978. *Eur. J. Biochem.* 83:571–80

159. Rosenfelder, G., Van Eijk, R. V. W., Mühlradt, P. F. 1979. *Eur. J. Biochem.* 97:229–37

160. Callies, R., Schwarzmann, G., Radsak, K., Siegert, R., Wiegandt, H. 1977. *Eur. J. Biochem.* 80:425–32

161. Lingwood, C. A., Hakomori, S. 1977. *Exp. Cell Res.* 108:385–91

162. Mannino, R. J., Burger, M. M. 1975. *Nature* 256:19–22

163. Laine, R. A., Yogeeswaran, G., Hakomori, S. 1974. *J. Biol. Chem.* 249: 4460–66

164. Lingwood, C. A., Ng, A., Hakomori, S. 1978. *Proc. Natl. Acad. Sci. USA* 75:6049–53

165. Ginsburg, E., Salomon, D., Sreevalson, T., Freese, E. 1973. *Proc. Natl. Acad. Sci. USA* 70:2457–61

166. Fishman, P. H., Simmons, J. L., Brady, R. O., Freese, E. 1974. *Biochem. Biophys. Res. Commun.* 59:292–99

167. Simmons, J. L., Fishman, P. H., Freese, E., Brady, R. O. 1975. *J. Cell Biol.* 66:414–24

168. Fishman, P. H., Atikkan, E. E. 1979. *J. Biol. Chem.* 254:4342–44

169. Leder, A., Leder, P. 1975. *Cell* 5: 319–22

170. Huberman, E., Heckman, C., Langenbach, R. 1979. *Cancer Res.* 39:2618–24

171. Lotan, R., Nicolson, G. L. 1977. *J. Natl. Cancer Inst.* 59:1717–22

172. Sporn, M. B., Dunlop, N. M., Newton, D. L., Smith, J. M. 1976. *Fed. Proc.* 35:1332–38

173. Patt, L. M., Itaya, K., Hakomori, S. 1978. *Nature* 273:379–81

174. Hakomori, S., Murakami, W. T. 1968. *Proc. Natl. Acad. Sci. USA* 59:254–61

175. Hakomori, S. 1973. *Adv. Cancer Res.* 18:265–315

176. Murray, R. K., Yogeeswaran, G., Sheinin, R., Schimmer, B. F. 1973. In *Tumor Lipids,* ed. R. Wood, pp. 285–302. Champaign, Ill: Am. Oil Chem. Soc. Press

177. Hakomori, S. 1975. *Biochim. Biophys. Acta* 417:55–89

178. Brady, R. O., Fishman, P. 1975. *Biochim. Biophys. Acta* 335:121–48

179. Richardson, C. L., Keenan, T. W., Morré, D. J. 1976. *Biochim. Biophys. Acta* 488:88–96

180. Critchley, D. R., Vicker, M. G. 1977. In *Dynamic Aspects of Cell Surface Organization,* ed. G. Poste, G. L. Nicolson, pp. 307–70. Amsterdam: Elsevier/North Holland Biochem.

181. Cumar, L. A., Brady, R. O., Kolodny, R. H., MacFarland, V. W., Mora, P. T. 1970. *Proc. Natl. Acad. Sci. USA* 67:757–64

182. Den, H., Schultz, A. M., Basu, M., Roseman, S. 1971. *J. Biol. Chem.* 246:2721–23

183. Fishman, P. H., Brady, R. O., Bradley, R. M., Aaronson, D. S., Todaro, G. J. 1974. *Proc. Natl. Acad. Sci. USA* 71:298–301

184. Den, H., Sela, B.-A., Roseman, S., Sachs, L. 1974. *J. Biol. Chem.* 249:659–61

185. Keenan, T. W., Morré, D. J. 1973. *Science* 182:935–37

186. Kawanami, J. 1972. *J. Biochem. Tokyo* 72:783–85

187. Hakomori, S., Wang, S.-M., Young, W. W. Jr. 1977. *Proc. Natl. Acad. Sci. USA* 74:3023–27

188. Hakomori, S., Koscielak, J., Bloch, K. J., Jeanloz, R. W. 1967. *J. Immunol.* 98:31–38

189. Häkkinen, I. 1970. *J. Natl. Cancer Inst.* 44:1183–93

190. Breimer, M. E. 1980. *Cancer Res.* 40:897–908

191. Baumann, H., Nudelman, E., Watanabe, K., Hakomori, S. 1979. *Cancer Res.* 39:2637–43

192. Watanabe, K., Matsubara, T., Hakomori, S. 1976. *J. Biol. Chem.* 251:2385–87

193. Rosenfelder, G., Young, W. W. Jr., Hakomori, S. 1977. *Cancer Res.* 37:1333–39

194. Young, W. W. Jr., Durdik, J. M., Urdal, D., Hakomori, S., Henney, C. S. 1981. *J. Immunol.* 126:1–6

195. Lockney, M. W. 1980. *Fed. Proc.* 39:2184

196. Itaya, K., Hakomori, S., Klein, G. 1976. *Proc. Natl. Acad. Sci. USA* 73:1568–71

197. Matsumoto, M., Taki, T. 1975. *Biochem. Biophys. Res. Commun.* 71:472–76

198. Taki, T., Hirabayashi, Y., Suzuki, Y., Matsumoto, M., Kojima, K. 1978. *J. Biochem. Tokyo* 83:1517–20

199. Hirabayashi, Y., Taki, T., Matsumoto, M., Kojima, K. 1978. *Biochim. Biophys. Acta* 529:96–105

200. Taki, T., Hirabayashi, Y., Ishiwata, Y., Matsumoto, M., Kojima, K. 1979. *Biochim. Biophys. Acta* 572:113–20

201. Skipski, V. P., Gitterman, C. O., Prendergast, J. S., Betit-Yen, K., Lee, G., Luell, S., Stock, C. 1980. *J. Natl. Cancer Inst.* 65:249–56

202. Yogeeswaran, G., Stein, B. S., Sebastian, H. 1978. *Cancer Res.* 38:1336–44

203. Sundsmo, J., Hakomori, S. 1976. *Biochem. Biophys. Res. Commun.* 68:799–806

204. Young, W. W. Jr., Hakomori, S. 1981. *Science.* 211:481–89

205. Huet, C., Ansel, S. 1977. *Int. J. Cancer* 20:61–66

206. Ansel, S., Huet, C. 1980. *Int. J. Cancer* 25:797–803

207. Levine, P. 1976. *Ann. NY Acad. Sci.* 277:428–33

208. Gupta, R. K., Irie, R. F., Chee, D. O., Kern, D. H., Morton, D. L. 1979. *J. Natl. Cancer Inst.* 63:347–56

209. Nakahara, K., Ohashi, T., Oda, T., Hirano, T., Kasai, M., Okumura, K., Tada, T. 1980. *N. Engl. J. Med.* 302:674–77

210. Urdal, D., Hakomori, S. 1980. *J. Biol. Chem.* 255:10506–16

211. Rapport, M. M., Graf, L. 1961. *Prog. Allergy* 13:271–331

212. Higashi, H., Naiki, M., Matsuo, S., Ōkouchi, K. 1977. *Biochem Biophys. Res. Commun. Res. Commun.* 79:388–95

213. Merrick, J. M., Zadarlik, K., Milgrom, F. 1978. *Int. Arch. Allergy Appl. Immunol.* 57:477–80

214. Stein-Douglass, K. E., Schwarting, G. A., Naiki, M., Marcus, D. M. 1976. *J. Exp. Med.* 143:822–32

215. Esselman, W. J., Miller, H. C. 1974. *J. Exp. Med.* 139:445–50

216. Miller, H. C., Esselman, W. J. 1975. *J. Immunol.* 115:839–43

217. Stein, K. N., Schwarting, G. A., Marcus, D. M. 1978. *J. Immunol.* 129:767–79

218. Thiele, H.-G., Arndt, R., Stark, R. 1977. *Immunology* 32:767–70

219. Inokuchi, Y., Nagai, Y. 1979. *Mol. Immunol.* 16:791–96

220. Wang, T. J., Freimuth, W. W., Miller, H. C., Esselman, W. J. 1978. *J. Immunol.* 121:1361–65

221. Kato, P. K., Wang, T. J., Esselman, W. J. 1979. *J. Immunol.* 123:1977–84

222. Schwarting, G. A., Marcus, D. M. 1977. *J. Immunol.* 118:1415–19

223. Young, W. W. Jr., Hakomori, S., Durdik, J. M., Henney, C. S. 1980. *J. Immunol.* 124:199–201

224. Kasai, M., Iwamori, M., Nagai, Y., Okumura, K., Tada, T. 1980. *Eur. J. Immunol.* 10:175–80

225. Nakano, T., Imai, Y., Naiki, M., Osawa, T. 1980. *J. Immunol.* 125:1928–32

226. Esselman, W. J., Miller, H. C. 1977. *J. Immunol.* 119:1994–2000

227. Ryan, J. L., Shinitzky, M. 1979. *Eur. J. Immunol.* 9:171–75

228. Lengle, E. E., Krishnaraj, R., Kemp, R. G. 1979. *Cancer Res.* 39:817–22

229. Sela, B.-A. 1980. *Cell. Immunol.* 49:196–201

230. Pick, E. 1977. In *Immunopharmacology*, ed. J. W. Hadden, R. G. Coffey, F. Spreafico, pp. 163–82. New York: Plenum

231. Amsdem, A., Ewan, V., Yoshida, T., Cohen, S. 1978. *J. Immunol.* 120:542–49

232. Higgins, T. J., Sabatino, A. P., Remold, H. G., David, J. R. 1978. *J. Immunol.* 121:880–86

233. Poste, G., Kirsh, R., Fidler, I. J. 1979. *Cell. Immunol.* 44:71–88

234. Poste, G., Allen, H., Matta, K. L. 1979. *Cell. Immunol.* 44:89–98

235. Miura, T., Handa, S., Yamakawa, T. 1979. *J. Biochem. Tokyo* 86:773–76

236. Siddiqui, B., Hakomori, S. 1971. *J. Biol. Chem.* 246:7566–69

237. Chien, J.-L., Hogan, E. L. 1980. *Fed. Proc.* 39:2183 (Abstr. 3040)
238. Rauvala, H. 1976. *J. Biol. Chem.* 251:7517–20
239. Rauvala, H., Krusius, T., Finne, J. 1978. *Biochim. Biophys. Acta* 531:266–74
240. Stellner, K., Watanabe, K., Hakomori, S. 1973. *Biochemistry* 12:656–61
241. Eto, T., Ichikawa, T., Nishimura, K., Ando, S., Yamakawa, T. 1968. *J. Biochem. Tokyo* 64:205–13
242. Naiki, M., Fong, J., Ledeen, R., Marcus, D. M. 1975. *Biochemistry* 14:4831–36
243. Marcus, D. M., Naiki, M., Kundu, S. K. 1976. *Proc. Natl. Acad. Sci. USA* 73:3263–67
244. Nagai, Y., Momoi, T., Saito, M., Mitsuzawa, E., Ohtani, S. 1976. *Neurosci. Lett.* 2:107–11
245. Fry, J. M., Weissbarth, S., Lehrer, G. M., Borenstein, M. B. 1974. *Science* 183:540–42
246. Hurby, S., Alvord, E. C. Jr., Seil, F. J. 1977. *Science* 195:173–75
247. Schwarting, G. A., Kundu, S. K., Marcus, D. M. 1979. *Blood* 53:186–92
248. Van Heyningen, W. E., Carpenter, C. C. L., Pierce, N. F., Greenough, W. H. III. 1971. *J. Infect. Dis.* 124:415–18
249. Van Heyningen, W. E. 1974. *Nature* 249:415–17
250. Holmgren, J., Lönnroth, I., Svennerholm, L. 1973. *Infect. Immun.* 8:208–14
251. Grollman, E. F., Lee, G., Lamos, S., Lazo, P. S., Kabach, H. R., Friedman, R. M., Kohn, L. D. 1978. *Cancer Res.* 38:4172–85
252. Kanfer, J. N., Carter, T. P., Katzan, H. M. 1976. *J. Biol. Chem.* 251:7610–19
253. Morita, A., Tsao, D., Kim, Y. S. 1980. *J. Biol. Chem.* 255:2549–53
254. Critchley, D. R., Ansell, S., Perkins, R., Dilks, S., Ingram, J. 1979. *J. Supramol. Struct.* 12:273–91
255. Van Heyningen, W. E. 1963. *J. Gen. Microbiol.* 31:375–87
256. Ledley, F. D., Lee, G., Kohn, L. D., Habig, W. H., Hardegree, M. C. 1977. *J. Biol. Chem.* 252:4049–55
257. Simpson, L. L., Rapport, M. M. 1971. *J. Neurochem.* 18:1757–59
258. Kato, I., Naiki, M. 1976. *Infect. Immunol.* 13:289–91
259. Buchanan, T. M., Pearce, W.-A., Chen, K. C. S. 1978. In *The Immubobiology of the Nisseria gonorea,* ed. G. Brooks et

al, pp. 242–49. Am. Soc. Microbiol.: Washington DC
260. Mullin, B. R., Fishman, P. H., Lee, G., Aloj, S. M., Ledley, F. D., Winand, R. J., Kohn, L. D., Brady, R. O. 1976. *Proc. Natl. Acad. Sci. USA* 73:842–46
261. Lee, G., Aloj, S. M., Brady, R. O., Kohn, L. D. 1976. *Biochem. Biophys. Res. Commun.* 73:370–77
262. Lee, G., Aloj, S. M., Kohn, L. D. 1977. *Biochem. Biophys. Res. Commun.* 77:434–41
263. Haywood, A. M. 1974. *J. Mol. Biol.* 83:427–36
264. Holmgren, J., Svennerholm, L., Elwing, H., Fredman, P., Strannegård, Ö. 1980. *Proc. Natl. Acad. Sci. USA* 77:1947–50
265. Wu, P.-S., Ledeen, R. W., Uden, S., Isaacson, Y. A. 1980. *J. Virol.* 33:304–10
266. Besancon, F., Ankel, H. 1974. *Nature* 252:478–80
267. Vengris, V. E. Jr., Reynolds, F. H., Hollenberg, M. D., Pitha, P. M. 1976. *Virology* 72:486–89
268. Ankel, H., Krishnamaurti, C., Besancon, F., Stefanos, S., Falcoff, E. 1980. *Proc. Natl. Acad. Sci. USA* 77:2528–32
269. Sekiguchi, K., Hakomori, S. 1980. *Proc. Natl. Acad. Sci. USA* 77:2661–65
270. Kleinman, H. K., Martin, G. R., Fishman, P. H. 1979. *Proc. Natl. Acad. Sci. USA* 76:3367–71
271. Loh, H. H., Cho, T. M., Wu, Y. C., Harris, R. A., Way, E. L. 1975. *Life Sci.* 16:1811–18
272. Craves, F. B., Zalc, B., Leybin, L., Baumann, P., Loh, H. H. 1980. *Science* 207:75
273. Wooley, D. W., Gommi, B. W. 1965. *Proc. Natl. Acad. Sci. USA* 53:959–64
274. Kitamura, M., Iwamori, M., Nagai, Y. 1980. *Biochim. Biophys. Acta* 628:328–35
275. Holmgren, J., Elwing, H., Fredman, P., Strannegård, Ö., Svennerholm, L. 1980. *Adv. Exp. Med. Biol.* 125:453–70
276. IUPAC-IUB Commission on Biochemical Nomenclature. 1977. *Lipids* 12:455–63
277. Svennerholm, L. 1964. *J. Lipid Res.* 5:145–55
278. Saito, M., Nojiri, H., Yamada, M. 1980. *Biochem. Biophys. Res. Commun.* 97:452–62
279. Yogeeswaran, G., Stein, S. 1980. *J. Natl. Cancer Inst.* 65:967–73
280. Dippold, W. G., Lloyd, K. O., Li, L. T. C., Ikeda, H., Oettgen, H. F., Old, L. J. 1980. *Proc. Natl. Acad. Sci. USA* 77:6114–18

Ann. Rev. Biochem. 1981. 50:765–82
Copyright © 1981 by Annual Reviews Inc. All rights reserved

BIOCHEMISTRY OF SENSING AND ADAPTATION IN A SIMPLE BACTERIAL SYSTEM

♦12095

D. E. Koshland, Jr.

Department of Biochemistry, University of California, Berkeley,
California 94720

CONTENTS

Perspectives and Summary

Sensing, adaptation, and regulation are three properties of biological systems that are conceptually distinct and yet interrelated. Sensing involves detection of an external stimulus and processing of the information to modify the behavior of the organism. Adaptation involves a desensitization of a repeated external signal in order to deemphasize that signal in relation to others. Regulation involves a change in the properties of a cell under the influence of external and internal signals in order to adjust the cell's internal biochemistry. Networks that carry out these properties involve allosteric interactions and covalent modifications in the processing system and receptors for the specific signals that control the behavioral output. Much of their biochemistry therefore appears to be similar or at least analogous.

One system that has the properties of sensing and adaptation and operates through a covalent modification processing system is bacterial chemo-

765

taxis. The advantage of this system is not necessarily its simplicity, for it is not apparent that it is less complicated or involves fewer proteins than the analogous regulatory systems. Its advantage is the readily available genetic modification techniques and the possibility of sufficient protein to allow isolation of the components of the system. Because there are a number of recent reviews (1–6), this article attempts to relay the state of the art and recent developments rather than covering all of the past history. Emphasis is placed on those aspects that may relate to other systems involving sensing, adaptation, and regulation.

An overall schematic illustration of the chemotactic system superimposed on a topological version of a bacterium is shown in Figure 1. An outer membrane that is porous to small molecules separates the outside of the bacterium from a periplasmic space that is largely aqueous and contains a number of freely floating proteins that cannot escape through the pores. An inner membrane then separates the cytoplasm from the exterior environment and contains the oxidative machinery of the bacterium, its flagellar motor, its transport systems, etc. The receptors are present both in the periplasmic space and in the inner membrane. Inside the cell are the proteins of the processing system that receive the signal from the receptors. As a result of the processing a signal is finally delivered to the flagella that control smooth swimming versus tumbling.

In this article the functioning of the receptors is considered first, i.e. their molecular mechanisms and the receptor strategy that is deduced from their characteristics. Then the proteins of the processing system are described. Since the bacterial system is subject to genetic analysis the complexity of the bacterial processing has been defined, and consists of a total of nine genes (7), at least six of which code for structural proteins in the cytoplasm (8). Finally, the output of the processing is discussed in terms of the detection of cytoplasmic signaling and the relationships of sensing and adaptation.

The Receptors

There are two types of receptors that have been characterized so far in the chemotactic system. The first are 30,000-dalton soluble proteins in the periplasm. The second are membrane-bound proteins of 60,000 daltons, and appear as integral proteins in the inner membrane.

The three periplasmic receptors that have been identified and purified— the galactose receptor (9), the ribose receptor (10), and the maltose receptor (11)—are proteins with dual functions. In addition to being the receptors for chemotaxis, they are components of the transport systems that carry these sugars from the periplasm to the cytoplasm (12). Although the transport and sensing systems use a common initial receptor, the remaining parts

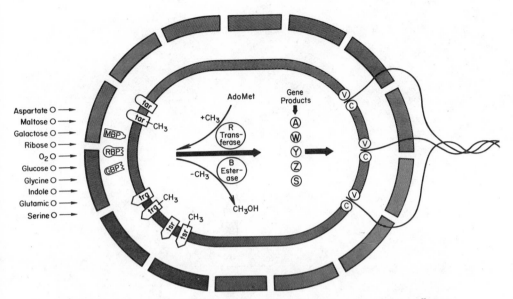

Aspartate O ⟶
Maltose O ⟶
Galactose O ⟶
Ribose O ⟶
O₂ O ⟶
Glucose O ⟶
Glycine O ⟶
Indole O ⟶
Glutamic O ⟶
Serine O ⟶

Figure 1 Schematic summary of the bacterial sensing system. A variety of chemoeffectors (O₂, aspartate, glucose, etc) can pass through pores in the outer membrane and react with receptors in the periplasmic space (GBP, galactose-binding protein; RBP, ribose-binding protein; MBP, maltose-binding protein) and with membrane-bound receptors (*tar* for aspartate, *tsr* for serine). The membrane-bound molecules are methylated by the *cheR* gene product, which utilizes S-adenosyl methionine (AdoMet), and demethylated by the *cheB* gene product. The peptides formed by the A, S, W, Z, and Y genes are cytoplasmic and are known to be essential for the signal processing, but have no specifically identified function as yet. They, with R and B, are believed to control the level of the response regulator X. The *cheC* gene in the flagella is a signal detector that records the level of X and transmits the information to the flagella. The *cheV* gene also contributes to this information transfer in an essential but unknown way.

of their machinery are distinct, as shown by mutation studies that can eliminate transport without eliminating chemotaxis and vice versa (13).

A combination of competition studies and genetic techniques established a mechanism for the action of the periplasmic proteins (14) and this is shown in Figure 2. By fluorescent reporter group techniques, binding of the chemoeffector was shown to induce a conformational change (15) that causes the periplasmic protein to bind to a transmembrane protein of 60,000 daltons. For the ribose and galactose receptors this is the *trg* protein (16). The conformational change induced when the chemoeffector-receptor complex binds to the *trg* protein alters its properties so that the signal is transmitted to the interior of the cell. At the same time the new conformation of the protein exposes glutamic acid residues to methylation by enzymes in the cytoplasm (17, 18).

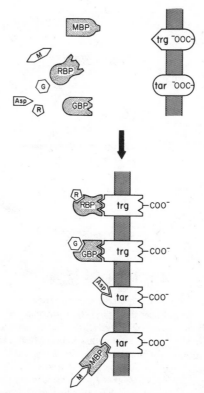

Figure 2 Activation of transmembrane receptors. Ribose (R), galactose (G), and maltose (M) each bind to periplasmic receptors (RBP, GBP, and MBP), which induces conformational changes. The RBP and GBP then bind to the *trg* transmembrane protein, thus inducing a conformational change that transmits a signal in the cytoplasm. The *tar* protein undergoes a similar conformational change induced by the aspartic acid itself or by the maltose-MBP complex. The *tsr* gene product binds serine. The induced conformational changes in the *tar*, *tsr*, and *trg* proteins generate an excitation signal and change the susceptibility of multiple glutamic acid residues to methylation and demethylation; this is shown symbolically by the interior carboxyls before the conformational change and the exterior carboxyls after.

Because the ribose-binding protein and the galactose-binding protein are in excess of the *trg* protein, competition studies were able to show that galactose inhibits ribose taxis and ribose inhibits galactose taxis despite the fact that galactose does not bind to the ribose receptor and vice versa (14). Mutations in the specific binding proteins eliminate responses to galactose and ribose respectively (9, 10), but mutations in the *trg* protein eliminate both galactose and ribose responses (19). This situation, requiring more than one protein to generate a receptor signal, is now familiar in mammalian cells, e.g. the hormone activated adenyl cyclase system (20).

It has been recently shown that the aspartate and serine receptors are also transmembrane proteins (21, 22) of 60,000 daltons, previously identified as the *tar* and *tsr* gene products (28). In each case, the transmembrane protein is capable of transmitting signals for more than one chemoeffector. In the case of the *tar* protein, aspartate, maltose, and divalent metal cation signals are transmitted through it (23). In the case of the *tsr* protein, serine, proton, indole, and possibly temperature signals are transmitted (24–26).

Because of this ability to transmit more than one signal, these "methyl-accepting chemotaxis proteins" (27) were believed to be the second element in the processing system and to act as transmitters from a chemoeffector-receptor complex similar to the *trg* protein (24, 28). This picture now appears to be at least partially in error. The *tar* and *tsr* proteins have been shown to be receptors that bind aspartate and serine directly and transmit their signals to the interior of the cell, as shown in Figure 2 (21, 22). On the other hand the maltose signal operates through a maltose-binding protein similarly to the *trg* case, as indicated by mutation (11) and cross-linking evidence (29). Thus it appears that the *tar* and *tsr* proteins can act as complete recognition proteins with transmembrane signaling for aspartate and serine, and also as part of a receptor complex in the case of maltose. Probably all of the transmembrane proteins were originally receptors designed to receive and transmit the signals from a single chemoeffector. The aspartate receptor, initially designed to transmit information about aspartate gradients, may subsequently, by some chance mutation in evolutionary history, have allowed interaction with the maltose receptor (11), which was present abundantly in the periplasm for transport of maltose. If such an initial mutant could utilize the aspartate receptor to transfer information across the inner membrane this would obviously have given that organism a survival advantage. Hence the element of focusing (14) probably arose through the capacity for a common induced conformational change in molecules that had a receptor function (30).

The finding that these transmembrane proteins are receptors does not diminish the importance of the focusing function. Competition between galactose and ribose for the same transmembrane signaling protein serves a purpose, since both of these compounds act as carbon sources, and an excess of one reduces the need for the other. Competition for serine and repellents is also logical, since the presence of toxic compounds in addition to attractants could be dangerous to the organism. Hence integrative responses would be valuable even at the level of the receptor.

A protein that is methylated and mediates a response to indole and phenol has been discovered (A. T. DeFranco and D. E. Koshland, Jr., manuscript in preparation). Also O_2 chemotaxis has been found to require neither *trg, tsr,* nor *tar* (B. Taylor, private communication), which suggests that its action is transmitted through some other molecule such as cytochrome oxidase, and then through membrane potential (31). Other electron acceptors may operate similarly (31). The role of membrane potential as signal transmitter is discussed extensively elsewhere (1, 32, 32a). Receptors for glucose, mannitol, mannose, and sorbitol were found to be proteins in the PTS system (33). Thus other receptors appear to be present and in need of isolation.

That the chemoeffector-induced conformational change travels a fairly long distance is apparent. The chemoeffector sites that initiate the change are on the outside of the inner membrane, and the signal is released inside. The membrane itself is considered to be from 30–50 Å wide (34). A conformational change extending at least 40 Å has been observed in the ribose receptor (15). Thus a geometry as shown in Figure 2 could require more than a 100 Å transmission. There is precedent for such a long-range alteration, since induced conformational changes in antibodies from binding site to complement fixing site have been demonstrated (35), and immunoglobin X-ray studies indicate that these sites may be as much as 100 Å apart (36).

Receptor theory

"What is the ideal strategy in regard to receptors?" In the first place, an organism will wish to have a repertoire of receptors for those external stimuli that it needs to sense for its survival, and an absence of receptors for signals that would be useless or confusing. In the case of the bacterium, where the individual cell is the whole organism, a wide range of compounds (about 30 in all) are sensed (1, 13, 37). The bacterium needs to swim toward oxygen, toward certain crucial amino acids, toward sugars for carbon sources, toward optimal pH, etc. It could be argued that there are many needed compounds that are not included on this list. However, there is a limited amount of space on the membrane surface, and it may be superfluous to have receptors for all compounds that are useful to the cell. The bacterium generally swims in environments of decaying foodstuffs. Hence any solution that contains a certain amount of serine is likely to contain many of the remaining amino acids. It is unnecessary to synthesize receptors for all 20 amino acids when serine will be a strong indicator of precence of other amino acids. Similarly, toxic conditions can be indicated in many cases by a few compounds. Man is repelled by the odor of putrifying food, but not all the individual components of that decay need be sensed. Thus bacteria seem to have a repertoire of approximately 30 receptors that sample the environment in such a way as to guide their migration to optimal living conditions.

Among the many receptors there is a hierarchy of values (38). A bacterium swims far more effectively to serine and aspartic acid than to any of the sugars. Serine and aspartate can operate as both nitrogen sources and carbon sources. By transamination reactions they immediately contribute to the pyruvate and oxaloacetate pools respectively. Thus the number of receptors and the efficiency of coupling with the next components in the processing system would make a bacterium swim up a serine gradient in preference to glucose if it were faced with a choice.

In a differentiated organism, some cells have a fairly large number of receptors and others only a very limited number. In hormonal cells or neurons, occasionally there may be only a single neurotransmitter or hormone that can act. A higher species with many cells can achieve the same strategy with a few receptors in each of a large number of cells as a bacterium that has all its receptors in a single membrane, but the principles of selection of total receptor repertoire for those conditions in the environment that the organism needs to sense remain the same.

If there are many receptors, the question of efficiency of detection becomes significant. The membrane surface has to provide many functions such as transport, the anchoring of flagella, etc. It cannot afford to have its entire space occupied by receptors and certainly not by a single receptor. What kind of price must be paid to have fewer of each type of receptor and a larger number of types? It has been shown (39) that a small molecule such as serine or glucose that wanders into the neighborhood of the bacterial membrane will spend a reasonable fraction of its time in the vicinity of the membrane before departing. This follows from simple diffusion theory. Hence the probability of a chemoeffector interacting with a receptor during this period is very high even if the surface is not entirely covered with receptors. The calculations indicate that an individual receptor with a 10 Å radius in a cell the size of a bacterium would require 3×10^4 receptors to give a rate that is approximately half of the maximum rate of absorption. Our studies show about 10,000 serine receptors and an equal number of aspartate receptors (21, 22). This equals, with a 15-Å radius per receptor, about 0.1% of the cell surface, and gives an efficiency of about 1/6 that of a surface covered by inner membrane receptors.

The galactose, maltose, and ribose receptors have about 10^4 molecules in the periplasmic space (1), which is more efficient for absorbing the chemoeffector, but the complex must then undergo a second step and bind to the transmembrane protein. The binding constant for that reaction is very high, and therefore there is probably little loss in efficiency of transfer of the signal. Other receptors are in smaller amounts than those listed, so the entire chemotactic detecting system of 30 different types of receptors occupies only a fraction of the bacterial surface and operates with reasonably high efficiency.

The calculated concentration of receptors (40) localized in a portion of a neuron is not too different from the concentration of receptors in the bacterial surface. Even in a specialized tissue that has only one type of receptor, it is not worthwhile to saturate the entire surface with receptors.

The Processing System

The signals received from the receptors must be interpreted by the cellular machinery in order to give a behavioral response. In the bacterial system the concept of a response regulator (38, 41) has been successful in correlating a wide variety of phenomenological observations, even though the identity of the response regulator itself remains to be determined. The further description of a response regulator model is given elsewhere (1, 2), but for this article it suffices to state that the various receptors eventually transmit information through a common parameter termed the response regulator, which may be a molecule or a combination of molecules in the cell. Such a common parameter explains the observation that the bacterial cell is capable of integrating signals from a variety of receptors (42–45). The signals from two attractants, from an attractant and a repellent, an attractant and a membrane potential, etc are processed to give an integrated response. Therefore it is deduced that the receptor signaling changes the level of a parameter common to all signals at some point before the message is delivered to the flagella.

This evidence for a common system gains credibility from the role of methylation in the processing of receptor information. Utilizing mutants and techniques for blocking methionine metabolism, it was possible to show methionine was required for chemotaxis (46), and that blockage of methylation blocked the adaptive response of bacteria (47). The requirement for methionine was shown to be a requirement for S-adenosyl methionine (48, 49), and the latter was found to methylate the 60,000-dalton proteins now identified as the *tsr*, *tar*, and *trg* gene products, respectively (16, 24, 28).

These integral membrane proteins are methylated on a glutamic acid residue (17, 18) catalyzed by a methyltransferase enzyme that is coded for by the *cheR* gene (50). The methyl group is removed by a methylesterase molecule coded for by the *cheB* gene (51). Thus there is a dynamic methylation system that both methylates and demethylates in close analogy to phosphorylation and dephosphorylation in neuronal or hormonal cells.

That this methylation system is intimately tied in with sensory processing and adaptation is shown in several ways. Mutants lacking the transferase enzyme are smooth swimming and do not tumble under normal conditions (50). Mutants in the methylesterase, on the other hand, are constantly tumbling even in the absence of a gradient (51). Thus the normal operation of the sensing system is disrupted when the methylation system is perturbed. Addition of attractant raises the level of methylation, addition of repellent decreases it, and the changes occur during the time period for adaptation (52).

A peculiarity of the methylated proteins is the presence of a banding pattern on sodium dodecyl sulfate chromatographs as revealed by radioactive labeling of these proteins (53). Using a toluenized cell assay that allowed the introduction of S-adenosyl methionine into the cell, it could be shown that these bands change in intensity during the course of the stimulus (54). In general, there was a shift to banding levels of higher mobility on addition of attractant and a shift to banding levels of lower mobility with repellents. Using recombinant DNA techniques (55), peptide mapping (56), and two-dimensional gels (57), the bands were found to be due to multiple methylation of the proteins. Three (57a) or four methyl groups are introduced per 60,000-dalton protein (J. B. Stock, A. Maderis and D. E. Koshland, Jr., manuscript in preparation), with the more highly methylated protein in the higher mobility positions and less methylated protein in the lower mobility region.

Studies in vivo and in broken cell preparations indicate that the attractant (*a*) initially decreases the rate of demethylation (58), (*b*) initially increases the rate of methylation (58, 59), and (*c*) eventually returns to the same rate of methylation and demethylation (58).

An Adaptation Model

Figure 3 shows a mechanism that can explain the existing information on methylation-demethylation and also the adaptive response (60). The transmembrane receptor (R_i) can exist in a number of different methylation states with i designating the state of methylation. This receptor can bind the transferase (Tr) that methylates the carboxyl groups and the esterase (Es) that demethylates them. In this model it is assumed that the transferase and esterase cannot bind simultaneously, a probable but not a necessary requirement for all covalent systems. Since it is known that there is some methylation in the absence of stimuli and that the methylation level changes in proportion to receptor occupancy, two complete pathways are shown, one in the absence and the other in the presence of the chemoeffector (C). Only one symbolic equilibrium is shown for the chemoeffector binding, but it is assumed that C can bind independently to R at each level of methylation from $i=1$ to $i=n$.

Since attractants increase the level of methylation, the pathway labeled +C must have one or more of the following characteristics when C is an attractant: higher rates of methylation ($k_{ic} > k_i$), lower rates of demethylation ($k'_{ic} < k'_i$), higher affinity for transferase ($K_{Tr} > K_{cTr}$), and lower affinity for esterase ($K_{Es} < K_{cEs}$). The studies on the in vitro and in vivo systems indicate that attractant addition causes transient increased methylation and decreased demethylation rates, but precisely which of the above permutations of detailed constants are involved is not certain. Furthermore, the rates need not be the same for each methylation step even though a

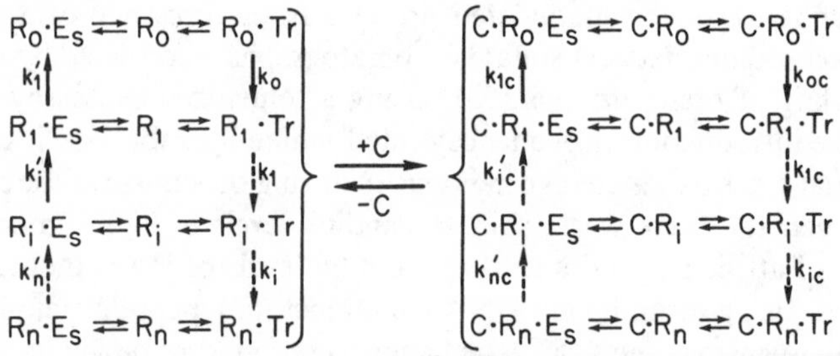

$$
\left.\begin{array}{c}
R_0 \cdot E_S \rightleftharpoons R_0 \rightleftharpoons R_0 \cdot Tr \\
k_1' \uparrow \qquad \downarrow k_o \\
R_1 \cdot E_S \rightleftharpoons R_1 \rightleftharpoons R_1 \cdot Tr \\
k_i' \uparrow \qquad \downarrow k_1 \\
R_i \cdot E_S \rightleftharpoons R_i \rightleftharpoons R_i \cdot Tr \\
k_n' \uparrow \qquad \downarrow k_i \\
R_n \cdot E_S \rightleftharpoons R_n \rightleftharpoons R_n \cdot Tr
\end{array}\right\}
\begin{array}{c} +C \\ \rightleftharpoons \\ -C \end{array}
\left\{\begin{array}{c}
C \cdot R_0 \cdot E_S \rightleftharpoons C \cdot R_0 \rightleftharpoons C \cdot R_0 \cdot Tr \\
k_{1c}' \uparrow \qquad \downarrow k_{oc} \\
C \cdot R_1 \cdot E_S \rightleftharpoons C \cdot R_1 \rightleftharpoons C \cdot R_1 \cdot Tr \\
k_{ic}' \uparrow \qquad \downarrow k_{1c} \\
C \cdot R_i \cdot E_S \rightleftharpoons C \cdot R_i \rightleftharpoons C \cdot R_i \cdot Tr \\
k_{nc}' \uparrow \qquad \downarrow k_{ic} \\
C \cdot R_n \cdot E_S \rightleftharpoons C \cdot R_n \rightleftharpoons C \cdot R_n \cdot Tr
\end{array}\right.
$$

Figure 3 Mechanism for adaptation and control in a multiple covalent modification system. R respresents the receptor, and its subscript $(0, 1, i, n)$ represents the number of covalent groups, e.g. methyl groups. Tr represents the transferase that binds to the receptor, with

binding constants $K_{iTr} = \dfrac{(R_i)\text{Tr.}}{R_{i \cdot Tr}}$

Es stands for the esterase, which removes the methyl groups and binds to the receptor

according to its dissociation constant $[K_{iEs} = \dfrac{(\text{Es}) R_i}{(R_{i \cdot Es})}]$.

The rate constants for interconversion of methylation levels are indicated as k_i for methylation and k'_i for demethylation. When chemoeffector is added to the system it alters the conformation and can change the values of any of the constants, which are listed as K_{icTr}, K_{icEs}, k_{ic}, and k'_{ic}, respectively. The system will come to equilibrium at the level at which the rate of methylation equals the rate of demethylation. This level of methylation at this equilibrium will depend on the specific constants of the system and the concentration of chemoeffector, C. The unliganded receptor (designated as R_i) may include two conformational forms, R'_i and R''_i, the relative amounts of which may vary with the level of methylation.

single transferase and a single esterase are involved in all modifications. The conformation of the substrate as well as the kinetic properties of the enzyme catalyst is a factor in any rate observed, and clearly the receptor conformation is altered. Thus in the absence of a chemoeffector the steady state seems to balance at approximately two methyl groups per receptor. When attractant is added new carboxyl groups are exposed and the transferase/esterase rates change such that the steady state adjusts to a new level with more carboxyl groups modified.

The role of repellents is accommodated in Figure 3 also, but then the pathway labeled +C must have some of the following characteristics: decreased rates of methylation $(k_{ic} < k_i)$, increased rates of demethylation $(k'_{ic}) > k'_i)$, decreased affinity of transperase $(K_{Tr} < K_{cTr})$, and increased affinity of esterase $(K_{Es} > K_{cEs})$. The model explains the finding that repellents and attractants can operate through the same receptors, with the attractants increasing methylation of the receptors and repellents decreasing it.

This model is consistent with a wide variety of data such as the general concordance of adaptation time with changes in level of methylation (52), the lack of adaptation when methylation enzymes are missing (61, 62), and the drastic change in adaptation times when methylation rates are modified or eliminated (49, 50, 65). However, the details that explain level in relation to receptor occupancy quantitatively, the manner in which the rates of methylation vary, and the relation between excitation and adaptation remain to be filled in. Nevertheless, mathematical models based on the scheme in Figure 3 can give good approximations to the experimental findings (A. Goldbeter, J. Stock, B. Rubik, and D. E. Koshland, Jr., manuscript in preparation).

A Requirement for ATP

That the main adaptive system appears to operate through methylation is clear, but the question remains as to whether additional chemical pathways are involved or if there are covalent changes in addition to methylation. Since ATP is required for S-adenosyl methionine synthesis, the behavior of ATP-depleted cells could be explained by a lack of AdoMet (49, 63). However, the finding that AsO_4^{-3} restored smooth swimming to constantly tumbling bacteria, whereas methionine starvation did not, led to the suggestion that an additional role for ATP or its products was indicated (47, 49). Histidine starvation experiments, which deplete ATP levels while maintaining membrane potential (64), and further arsenate experiments on fully methylated or fully demethylated strains (65) strongly support an added role for ATP.

Possible alternatives are a phosphorylation reaction on one of the chemotactic or flagellar components, since regulatory phosphorylation has been found in procaryotes (66) or some compound that requires ATP for synthesis. Among the latter cyclic nucleotides seem obvious, but adenyl cyclase minus mutants chemotax normally. Addition of cGMP to bacteria shows a change in behavior (66a). However, these are only two of the possible cyclic nucleotides and low-molecular-weight compounds that need to be evaluated. In addition, ATP could be needed for synthesis of a protein or small molecule compound in which no chemical component of the ATP remained.

Excitation and Adaptation

In the adapting-sensing system there must be an intimate relationship between excitation and adaptation, but the two processes are conceptually distinct and may also be divided in their molecular mechanisms. The findings that some *cheR* mutants (transferase deficient) can respond but not adapt to repellents (61, 62), and that the *cheB* mutants (esterase deficient) can respond to attractants (47) have now been extended to show that *null* mutants can respond to stimuli, utilizing either attractants or repellents (65). This shows that there must be an excitation response separate from the adaptive response of methylation, a conclusion in excellent agreement with the response regulator model (1).

The question then arises as to the nature of the response regulator that can give this excitation. Suggestions are membrane potential (67), calcium (68), the *tar-tsr* proteins (6), the peptides of the processing system (60, 69), and an ATP derivative (see above). A spike in membrane potential observed on addition of attractants or repellents led to the suggestion that it was the common response signal (67), but it appears that such a spike is an artifact of O_2 depletion in the experimental protocol (32a). Studies on membrane potential indicate that the cell can respond to changes in membrane potential as it does to attractants and repellents (32–32b), but that there is no correlation of membrane potential changes generated by chemoeffector gradients with the behavioral pattern. Alterations of calcium levels and behavior by an ionophore led to the calcium hypothesis (68), but high or very low external levels that alter internal calcium levels did not change behavior (32a, 84), and mutants defective in calcium transport show normal chemotaxis (70). The models involving proteins of the processing system or the receptors are intriguing and cannot be excluded on the basis of present evidence, but no comprehensive and detailed hypothesis has yet emerged that is consistent with all of the data. The response regulator is a parameter that controls behavior, and thus the model can be satisfied by a ratio of molecular species as well as an individual molecule. The experimental as well as the conceptual separation of excitation and adaptation makes the search for the response regulator of primary importance.

Rationale for Adaptation

Not every cell is adaptive, but there is an ubiquitous trend in sensory systems to desensitize repeated stimuli at any level of intensity. Humans can adapt to light, sound, temperature, taste, odor etc, and bacteria show the property of an adaptive sensory system. Since a sensory system records and amplifies a stimulus, adaptation that desensitizes or removes the record would seem to be counterproductive. Why should biological systems show adaptation and sensing within the same cell?

For the bacterial system the answer is apparent from the behavioral responses. If bacteria swam smoothly in a high absolute concentration of serine, they would swim right out of the high serine environment. Moreover, a behavioral response to high absolute levels of serine would make them unable to respond to oxygen or any second attractant when their serine receptors are saturated. Thus desensitization to absolute concentrations is desirable. Moreover the desensitization should be at the level of the individual receptor. Receptor saturation becomes relevant. The difference in background intensity between bright sunlight and pale moonlight is 10^7, and yet we see objects under both conditions. A single receptor usually saturates over two orders of magnitude, so clearly some additional biochemistry must occur. Rhodopsin is multiply phosphorylated (71–73), and the chemotaxis receptors are multiply methylated, which suggests again modification at the receptor level to achieve adaptation.

The logic of covalent modification in adaptation can be explained by the use of the response regulator model described above and is illustrated in Figure 4. The regulator X represents the behavior output of the cell; in bacteria this is the controller of tumbling frequency. When the stimulus (chemoeffector concentration or light) increases, the receptor occupancy increases, thus generating a signaling "pressure" on the cell. If adaptation is to occur the response regulator must return to its previous value. This can occur only by means of some compensatory biochemistry within the cell, which provides a counterpressure to the external stimulus (30).

The level of methylation provides this compensation in the bacteria; phosphorylation provides the compensation in the visual cell. In the absence of effectors, there is an intermediate level of methylation. Addition of attractant raises this level of methylation, and there is a transient increase in signal during the interval of signaling and change in methylation. By the end of the stimulus interval, the level of methylation has reached a new plateau and the response regulator has returned to normal. An outside observer therefore could not deduce the level of a steady background stimulus by observing the behavior of the bacterium, but could do so by measuring the methylation level of the receptor protein.

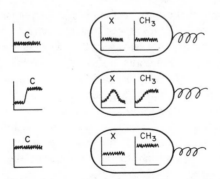

Figure 4 Principles of adaptation. (*a*) Steady-state methylation level is present in the absence of any stimulus. (*b*) When stimulus is increased, level of methylation rises and response regulator X goes through transient change. (*c*) After adaptation with steady outside stimulus, level of methylation remains constant at higher level but response regulators has returned to normal.

The internal biochemistry of the cell has altered the behavioral response so that the same response to a percentage change over background levels is obtained over a wide range of backgrounds. Thus the dilemma posed above is resolved. Adaptation does desensitize the ability to detect absolute levels of stimuli, but only in order to increase sensitivity to relative changes over a wide range of background intensities.

Adaptation to background stimuli was first recognized in a quantitative way in Weber's Law, the beginning of modern psychophysics. Weber stated that the ratio of the just noticeable increment in a stimulus ($\Delta\phi$) divided by the background stimulus (ϕ_B) is a constant over wide ranges of ϕ_B (74). This proved to be correct for a wide variety of stimuli. If Weber's Law is obeyed perfectly a value of 1 is obtained in a power function plot of response versus $k \cdot \ln(\phi)$ (75). Actual numbers for a variety of stimuli in such phenomena as hearing, taste, mechanical pressure, odor, etc show values that are most often in the range 0.8–1.5 (76). Thus the phenomenon is widespread in all sensory systems, and a mechanism of compensatory covalent modification seems a likely general device to achieve adaptation (30).

The bacterial system shows another mechanism that can also be a useful adaptive mechanism. There are at least two serine receptors, a high affinity receptor with a K_d of 4 μM and a low affinity receptor with a K_d of 5 mM (13, 21, 24). Thus when one receptor is saturated a second receptor can be sensitive to signals over a different concentration range.

Desensitization by internalization or by rendering a receptor inactive can diminish a signal but cannot change the range over which the receptor is sensitive. Hence such devices are no substitute for modification of the processing system.

The Detector for the Response Regulator

After the processing of the cytoplasmic signals, the response regulator must interact with the flagella in some way to generate the behavioral response. There is good evidence that at least part of the detector system is the *cheC* gene product (44, 77, 78). Mutants in this gene dramatically alter behavior, even producing a mutant that completely reverses normal behavior (44); it swims toward normal repellents like phenol and away from normal attractants such as serine. Other partial inverse mutants have been obtained by other missense mutations in the same gene (44). All of this peculiar behavior can be explained by assuming that the gene product is the detector of the response regulator (77) and that missense mutations alter the affinity of the protein for the response regulator.

A schematic illustration is shown in Figure 5. If the *cheC* gene product is a part of the flagellar apparatus that undergoes a conformational change on binding the response regulator X, a change from clockwise to counterclockwise rotation can accompany the association or dissociation of X. This change in rotation governs the behavior of the organism, since it has shown that migration depends on control of the tumbling response (41, 79) and tumbling is controlled by reversal of flagella rotation (80, 81). The *cheC* gene has been shown to map in the same place as a flagellar gene, which is consistent with its being a flagellar protein (7, 82).

The above evidence, together with similar explanations for other mutants, makes a strong case for the *cheC* gene, but it may not be the entire detector. Mutations in the *cheV* gene also damage the chemotactic response, and its gene product is also involved in flagellar synthesis (7). The role of this gene, however, is still not understood.

A further understanding of the detector system has been uncovered by two quite different investigations. Membrane potential can change the bias of the detector (83). As the membrane potential decreases a steady progression to more clockwise rotation was observed. Moreover this was found to be an exponential function in which the product of the probabilities for clockwise and counterclockwise rotation were a constant, and each was an exponential function of the membrane protonmotive force. A possible interpretation of these results is that the detector is gated by the membrane potential to alter its conformation so that the switching induced by the response regulator is shifted in scale.

A similar alteration in the responses of the detector was obtained by the use of double mutants (A. T. DeFranco and D. E. Koshland, Jr., manuscript in preparation). Using tetracycline-insertion elements, mutants of the *cheC* gene were inserted into strains with other nonchemotactic mutations to construct double mutants. The *cheC* mutant used (ST120) changes a wild

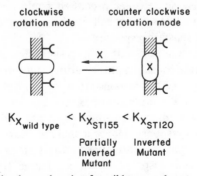

Figure 5 Possible molecular explanation for wild-type and mutant behavior. *CheC* gene product in flagellar protein binds X with dissociation constant K_x in wild type. When it does so the conformation of the flagella are changed from clockwise to counterclockwise rotation. Point mutations in the *cheC* gene produce proteins with different K's and cause aberrant behavior because the detector is responding incorrectly to the normal levels of the response regulator.

type from predominantly counterclockwise rotation to predominantly clockwise rotation. When introduced into a *cheW* mutant that was entirely counterclockwise, the double *cheC-cheW* mutant showed a new, largely clockwise rotation bias. Introduction of the *cheC* into other genes such as the *cheR*, which is also largely counterclockwise and the *cheB*, which is substantially clockwise increased the clockwise ratio in both cases. However, each mutant retained its original properties in regard to response to chemoeffectors. *CheW* mutants do not respond to chemoeffectors but *cheR* mutants do. The double mutants retained the responses to chemoeffectors of the original simple mutation. Thus the mutation has some of the same effects as the protonmotive force, a bias of probabilities of the two modes of rotation.

The Timing of the Response and the Cellular Memory

The finding of a dynamic covalent modification system at the central core of bacterial sensing leads to questions of analogies with mammalian systems in which phosphorylation-dephosphorylation reactions play a role. In some of these systems, the purpose is sensing or the excretion of a compound, as in neuronal or hormonal cells. In others, the covalent system is simply for control of metabolism, as in glycogen synthesis. In each case a perturbation of the system leads to a change in the level of covalent modification. There is a transient period after receipt of the external signal and then return to the steady state in which the covalent modification rate equals the hydrolysis rate. Thus, Figures 3 and 4 can apply to metabolic regulation as well as sensing. It seems tempting to suggest, therefore, that sensing arose from a regulatory system. If one of the components that rises and falls in the transient period alters some other process it could be the signal for the output in the hormonal or sensory cell. Evolution would then use such an output to develop a sensory response.

The timing in the rise and fall of any such signal would then be a consequence of the rate constants of the various steps in the pathway. Each cell would then select for the response time appropriate to it. It has been shown that the bacterial memory time is optional for its survival (1, 84). Bacteria need to respond to immediate gradients and have no need to remember yesterday's response. Hence a short memory is desirable. Some neurons need very rapid response times, others show slower graded responses. Hormonal cells respond with much longer time constants than neuronal cells, except in cases of long-term memory. Yet all of these cells seem to process information by covalent modification in a scheme like that of Figure 3. The level of covalent modification (indicated by n in Figure 3) can vary from 1 to fairly large values in processes like pyruvate dehydrogenase control (85, 86) or rhodopsin adaptation (71–73). Both mono and multiple phosphorylation are found in sensing and in regulatory pathways, which suggests a simple conversion of one type of system to the other. This network also allows one to vary kinetic constants to obtain the maximal sensory signal compatible with an adaptive response. In the case of glycogen metabolism there may be no need for a sensory output signal, and hence the system's response time is optimized for adjustment to metabolic demands. In a neuronal cell some of the phosphorylation-dephosphorylation scheme may have evolved entirely for sensory processing even though its origin was in regulation.

Literature Cited

1. Koshland, D. E. Jr. 1980. *Bacterial Chemotaxis as a Model Behavioral System.* New York: Raven. 193 pp.
2. Koshland, D. E. Jr. 1979. *Physiol. Rev.* 59:812–62
3. Koshland, D. E. Jr. 1980. *Ann. Rev. Neurosci.* 3:43–75
4. Springer, M. S., Goy, M. F., Adler, J. 1979. *Nature* 280:279–84
5. Macnab, R. M. 1979. *Crit. Rev. Biochem.* 5:291–341
6. Taylor, B. L., Lazlo, D. J. 1981. *Perception of Behavioral Chemicals.* New York: Elsevier. In press
7. Warrick, H. M., Taylor, B. L., Koshland, D. E. Jr. 1977. *J. Bacteriol.* 130:223–31
8. Silverman, M., Simon, M. 1977. *J. Bacteriol.* 130:1317–25
9. Hazelbauer, G. L., Adler, J. 1971. *Nature New Biol.* 30:101–4
10. Aksamit, R., Koshland, D. E. Jr. 1974. *Biochemistry* 13:4473–78
11. Hazelbauer, G. L. 1975. *J. Bacteriol.* 122:206–14
12. Heppel, L. A. 1969. *J. Gen. Physiol.* 54:95s–109s
13. Adler, J. 1969. *Science* 166:1588–97
14. Strange, P. G., Koshland, D. E. Jr. 1976. *Proc. Natl. Acad. Sci. USA* 73:762–66
15. Zukin, R. S., Hartig, P. R., Koshland, D. E. Jr. 1979. *Biochemistry* 18:5599–5605
16. Kondoh, H., Ball, C. B., Adler, J. 1979. *Proc. Natl. Acad. Sci. USA* 76:260–64
17. Kleene, S. J., Toews, M. L., Adler, J. 1977. *J. Biol. Chem.* 252:3214–18
18. Van der Werf, P., Koshland, D. E. Jr. 1977. *J. Biol. Chem.* 252:2793–95
19. Ordal, G. W., Adler, J. 1974. *J. Bacteriol.* 117:517–26
20. Ross, E. M., Gilman, A. G. 1980. *Ann. Rev. Biochem.* 49:533–64
21. Clarke, S., Koshland, D. E. Jr. 1979. *J. Biol. Chem.* 254:9695–702
22. Wang, E., Koshland, D. E. Jr. 1980. *Proc. Natl. Acad. Sci. USA.* In press
23. Ordal, G. W., Adler, J. 1974. *J. Bacteriol.* 117:509–16
24. Springer, M. S., Goy, M. F., Adler, J. 1977. *Proc. Natl. Acad. Sci. USA* 74:3312–16
25. Maeda, K., Imae, Y. 1979. *Proc. Natl. Acad. Sci. USA* 76:91–95
26. Reader, R. W., Tso, W., Springer, M. S., Goy, M. F., Adler, J. 1978. *J. Gen. Microbiol.* 111:363–74
27. Kort, E. N., Goy, M. F., Larsen, S. H., Adler, J. 1975. *Proc. Natl. Acad. Sci. USA* 72:3939–42
28. Silverman, M., Simon, M. 1977. *Proc. Natl. Acad. Sci. USA* 74:3317–21
29. Koiwai, O., Hayashi, H. 1979. *J. Biochem. Tokyo* 86:27–34
30. Koshland, D. E. Jr. 1980. *Trends in Biochem. Sci.* 5(11):297–302
31. Taylor, B. L., Miller, J. B., Warrick, H. M., Koshland, D. E. Jr. 1979. *J. Bacteriol.* 140:567–73
32. Miller, J. B., Koshland, D. E. Jr. 1977. *Proc. Natl. Acad. Sci. USA* 74:4752–56
32a. Ordal, G. W., Goldman, D. 1975. *Science* 189:802–4
32b. Snyder, M. A., Stock, J. F., Koshland, D. E. Jr. 1981. *J. Mol. Biol.* In press
33. Adler, J., Epstein, W. 1974. *Proc. Natl. Acad. Sci. USA* 71:2895–99
34. Metzler, D. E. 1977. *Biochemistry,* pp. 252–300. New York: Academic
35. Chiang, H., Koshland, M. E. 1979. *J. Biol. Chem.* 254:2736–41
36. Huber, R., Deisenhofer, J., Colman, P. M., Matsushima, M., Palm, W. 1976. *Nature* 264:415–20
37. Tso, W.-W., Adler, J. 1974. *J. Bacteriol.* 118:560–76
38. Koshland, D. E. Jr. 1977. *Science* 196:1055–63
39. Berg, H. C., Purcell, E. M. 1977. *Biophys. J.* 20:193–219
40. Stevens, C. F. 1979. *Sci. Am.* 241:54–65

41. Macnab, R. M., Koshland, D. E. Jr. 1972. *Proc. Natl. Acad. Sci. USA* 69:2509–12
42. Tsang, N., Macnab, R., Koshland, D. E. Jr. 1973. *Science* 181:60–63
43. Spudich, J. L., Koshland, D. E. Jr. 1975. *Proc. Natl. Acad. Sci. USA* 72:710–13
44. Rubik, B. A., Koshland, D. E. Jr. 1976. *Proc. Natl. Acad. Sci. USA* 75:2820–24
45. Berg, H. C., Tedesco, P. M. 1975. *Proc. Natl. Acad. Sci. USA* 72:3235–39
46. Adler, J., Dahl, M. M. 1967. *J. Gen. Microbiol.* 46:161–73
47. Aswad, D., Koshland, D. E. Jr. 1974. *J. Bacteriol.* 118:640–45
48. Armstrong, J. B. 1972. *Can. J. Microbiol.* 18:1695–701
49. Aswad, D., Koshland, D. E. Jr. 1975. *J. Mol. Biol.* 97:207–23
50. Springer, W. R., Koshland, D. E. Jr. 1977. *Proc. Natl. Acad. Sci. USA* 74:533–37
51. Stock, J. B., Koshland, D. E. Jr. 1978. *Proc. Natl. Acad. Sci. USA* 75:3659–63
52. Goy, M. F., Springer, M. S., Adler, J. 1977. *Proc. Natl. Acad. Sci. USA* 74:4964–69
53. Matsumura, P., Silverman, M., Simon, M. 1977. *J. Bacteriol.* 132:996–1002
54. Paoni, N. F., Koshland, D. E. Jr. 1979. *Proc. Natl. Acad. Sci. USA* 76:3693–97
55. DeFranco, A. T., Koshland, D. E. Jr. 1980. *Proc. Natl. Acad. Sci. USA* 77:2429–33
56. Chelsky, D., Dahlquist, F. R. 1980. *Proc. Natl. Acad. Sci. USA* 77:2434–38
57. Engstrom, P., Hazelbauer, G. L. 1980. *Cell* 20:165–71
57a. Chelsky, D., Dahlquist, F. R. 1981. *Biochemistry.* In press
58. Toews, M. L., Goy, M. F., Springer, M. S., Adler, J. 1979. *Proc. Natl. Acad. Sci. USA* 76:5544–48
59. Clarke, S., Sparrow, K., Koshland, D. E. Jr., Panasenko, S. 1981. *J. Supramol. Biol.* In press
60. Stock, J. B., Koshland, D. E. Jr. 1980. *Curr. Top. Cell. Regul.* In press
61. Goy, M. F., Springer, M. S., Adler, J. 1978. *Cell* 15:1231–40
62. Parkinson, J. S., Revello, P. T. 1978. *Cell* 15:1231–40
63. Larsen, S. H., Adler, J., Gargus, J. J., Hogg, R. W. 1974. *Proc. Natl. Acad. Sci. USA* 71:1239
64. Galloway, R. J., Taylor, B. J. 1981. *J. Bacteriol.* In press
65. Stock, J. B., Yen, L., Yee, A., Koshland, D. E. Jr. 1980. *Fed. Proc.* 39:2629
66. Wang, J. Y. J., Koshland, D. E. Jr. 1978. *J. Biol. Chem.* 253:7605–8
66a. Black, R. A., Hobson, A. C., Adler, J. 1980. *Proc. Natl. Acad. Sci. USA* 77:3879–83
67. Szmelcman, S., Adler, J. 1976. *Proc. Natl. Acad. Sci. USA* 73:4387–91
68. Ordal, G. W., Fields, R. B. 1977. *J. Theor. Biol.* 68:491–500
69. Parkinson, J. S. 1977. *Ann. Rev. Genet.* 11:397–414
70. Brey, R. N., Rosen, B. P. 1979. *J. Bacteriol.* 139:824–34
71. Bownds, D., Dawes, J., Miller, J., Stahlman, M. 1972. *Nature New Biol.* 237:125–27
72. Kuhn, H. 1974. *Nature* 250:588–90
73. Shichi, H., Somers, R. L. 1978. *J. Biol. Chem.* 253:7040–46
74. Marks, L. E. 1974. *Sensory Processes,* pp. 296–99 New York: Academic
75. Stevens, S. S. 1961. *Sensory Communication,* pp. 1–33. Cambridge: MIT Press
76. Thompson, R. F. 1967. *Foundations of Physiological Psychology.* New York: Harper & Row
77. Khan, S., Macnab, R. M., DeFranco, A. L., Koshland, D. E. Jr. 1978. *Proc. Natl. Acad. Sci. USA* 75:4150–54
78. Parkinson, J. S., Parker, S. R. 1979. *Proc. Natl. Acad. Sci. USA* 76:2390–94
79. Berg, H. A., Brown, D. A. 1972. *Nature* 239:500–4
80. Silverman, M., Simon, M. 1974. *Nature* 249:73–74
81. Larsen, S. H., Reader, R. W., Kort, E. N., Tso, W.-W., Adler, J. 1974. *Nature* 249:74–77
82. Silverman, M., Simon, M. 1973. *J. Bacteriol.* 116:114–22
83. Khan, S., Macnab, R. M. 1980. *J. Mol. Biol.* 138:563–97
84. Macnab, R. M., Koshland, D. E. Jr. 1973. *J. Mechanochem. Cell Motility* 2:141–8
85. Reed, L. J. 1980. *Fed. Eur. Biochem. Soc.* 60:47–52
86. Leiter, A., Weinberg, M., Ischashi, S., Utter, M. 1979. *J. Biol. Chem.* 253:2516–23

Ann. Rev. Biochem. 1981. 50:783–814
Copyright © 1981 by Annual Reviews Inc. All rights reserved

IN VIVO CHEMICAL MODIFICATION OF PROTEINS
(Post-Translational Modification)

◆12096

Finn Wold

Department of Biochemistry, University of Minnesota, St. Paul, Minnesota 55108

CONTENTS

0066-4154/81/0701-0783$01.00

PERSPECTIVES AND SUMMARY

The trinucleotide codons ("words") of the genetic code specify exactly 19 α-amino acids and 1 α-imino acid as monomer building blocks in protein synthesis; these are the 20 "common" or "primary" amino acids without which no living organism can exist. The genetic code also specifies a single codon, AUG, as the starting point of synthesis of all proteins; this codon also is the one unique codon for methionine. Because the genetic code by all known criteria is universal, it follows that proteins synthesized by any living cell according to the genetically determined message, should have an amino-terminal methionine and be made up of no more than 20 different amino (imino) acids. An examination of the data available for isolated proteins would not confirm this prediction at all: proteins have very varied amino-terminal residues, and according to the latest count, we can estimate that if all the proteins from a given organism were subject to careful hydrolysis specific for only the peptide bonds, several hundred different amino (imino) acids would be released. Consequently, it can be concluded that the genetically specified polymerization of amino acids on the ribosomes does not constitute all of protein synthesis. Additional covalent changes take place before the finished protein is produced, and those covalent changes are the subject of this review.

   In the broadest sense the final product of protein synthesis should be considered to be the biologically active molecule in its proper compartment of action, and protein synthesis is then best represented by three steps: 1. amino acid activation; 2. polymerization; and 3. processing, as illustrated below.

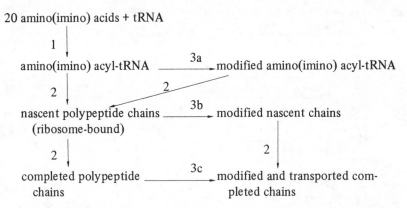

The processing of the genetically determined sequence of primary amino (imino) acid illustrated in this manner by the three horizontal lines in step 3 above thus includes both covalent and noncovalent structural alterations of the linear polymer and, for most proteins, transport from the site of synthesis to the compartment of action. It is probably not useful to visualize a fixed sequence of the individual steps indicated in the above scheme. The sequence of steps for each protein is undoubtedly uniquely determined by its particular amino acid sequence as specified by the appropriate gene, and each individual protein will probably follow its particular biosynthetic pathway in one smoothly coordinated process involving all the steps required to produce that particular end product.

For a review of protein processing it is useful to briefly consider the individual processing steps separately. The covalent modification reactions that are the main topic of this chapter involve the chemical derivatization of all the available functional groups of amino acids, peptides, and protein that take place after the formation of the amino acyl-tRNA in step 1. The functional groups can be either the free amino terminus or carboxyl terminus or the side chains of individual amino acids. The peptide bond itself can be considered a functional part of the polypeptide chain, and proteolytic cleavage of peptide bonds during or after the polymerization process is also considered briefly here. Noncovalent modification, the process by which the growing and/or completed polymer chain acquires its proper three-dimensional structure, or by which individual protomers associate to give rise to multisubunit products, is a universal and essential part of protein processing, but is not included here. Several excellent reviews pertinent to this area have appeared in recent years (1–4). Translocation of proteins, the process by which proteins are transported from the site of synthesis to any one of the proper intra- or extracellular compartments in which they will fulfill the function for which they were synthesized is also given minimal coverage in the present discussion, except as the transport through various compartments relates directly to covalent modification steps known to take place in that compartment. Very little is known at this stage about the type of recognition signals and processes involved in directing the flow of proteins released from free or membrane-bound polysomes to intracellular organelles, plasma membranes, or extracellular space; the total process must be a highly developed and exquisitely specific and precise one, and must be intimately coordinated with the other processing steps. Another chapter of this volume (5) will undoubtedly discuss these aspects as they relate to glycosylation.

In considering at what stage of protein synthesis the various protein processing steps occur, it may be useful to establish a few generalizations. Chemical modifications of amino acid side chains and at least some that involve the amino terminus may occur at any stage during synthesis after the amino acyl-tRNA has been formed. Any modification of free amino acids prior to this step, by virtue of the extreme specificity of the amino acyl-tRNA synthesis, would exclude those amino acids from incorporation into proteins. Step 1 of the simplified scheme above consequently can be considered the commitment step in protein synthesis; once the amino acyl-tRNA is formed from any one of the 19 α-amino acids or proline, that amino acid or its subsequent derivative can enter the polymerization process. Chemical modification of the carboxyl-terminal end obviously cannot take place until the polymerization process has been completed and a carboxyl-terminal end has been formed. Proteolytic cleavage equally obviously requires that a peptide chain exists, but it is clear that proteolytic cleavage occurs equally well with either short, nascent chains or completed proteins. Noncovalent folding undoubtedly requires the completion of a given structural domain before any significant permanent three-dimensional structure is established, and the final folded structure, especially that involving subunit interactions, must await the completion and release of the polypeptide chains. Similarly, the transport step of protein processing requires a completed polymer to be released from the polysomes; however, the commitment to certain transport pathways appears to take place early in the polymerization process.

In preparing this article, it was perceived that one goal should be to compile as complete a listing as possible of known covalent derivatives of proteins, and this has been done primarily by expanding previous reviews (6–9) and updating them. After establishing the nature of the derivatives, an attempt has been made to develop some broad classes of biological functions served uniquely by in vivo protein modifications and some general models for the specificity determinants involved in the modification reactions. Since such classification and model construction inevitably build primarily on the best understood reactions, it was deemed important to include also a brief consideration of reactions that are not so well understood. As in any area of scientific inquiry, the excitement of the field of in vivo protein modification is derived as much from the challenges of the unsolved problems as it is from the impressive volume of new understanding and new concepts that recent research has provided.

INTRODUCTION

Numerous reviews of individual in vivo modification reactions (phosphorylation, glycosylation, methylation, etc) have been written during the last several years, and for a proper in-depth view of the nature and function of each type of reaction, these reviews (which are included in the list of references) are the proper sources. I agreed to write this review because I feel that there may be some merit in "comparing notes" from the many specialty groups, looking for common problem areas, and defining common terminology. Certain aspects of the modification reactions, such as the basis for the specificity markers for the reactions and their relation to sequence, protein folding, and translocation are for example likely to reflect common processing determinants, and as knowledge increases it may become apparent that more and more of the processing steps are extensively interrelated.

Certain assumptions and ground rules need to be stated from the start. In deciding which amino acid derivatives to include as constituents of proteins one inevitably encounters a gray area such as that for short peptides whose biosynthetic origin cannot readily be assessed. On the one hand it is clear from the study of short peptide hormones that these are synthesized as multifunctional large polypeptide precursors whose sequences are encoded by mRNA and assembled by the regular ribosomal synthetic apparatus (9). On the other hand there are short polypeptide antibiotics and cell wall constituents that are assembled in step-by-step amino acid activation and condensation catalyzed by specific enzymes in the absence of genetic information and ribosomes, and these must obviously be excluded from consideration of in vivo modifications of proteins. These latter types of peptides frequently contain rather exotic amino acid derivatives, including D-amino acids; a recently characterized cyclic heptapeptide toxin from the mushroom *Amanita virosa* contains L-Ala, L-Val, D-Ser, D-Thr, 2,3-*trans*-3,4-*trans*-3,4-dihydroxy-L-Pro, and 2'-(methylsulfonyl)-L-Trp, and either γ,δ-dihydroxy-L-Leu or γ,δ,δ' trihydroxy-L-Leu (10). The major criterion used here to decide whether a given peptide with its unique amino acid derivatives is derived from a protein synthesized on the ribosome-mRNA complex or is a product of the action of soluble enzymes, is the presence of D-amino acids. Since the commitment step in proteins synthesis, the formation of the amino acyl-tRNA has an absolute specificity for L-amino (imino) acids, and since we do not know of any case of enzymatically induced inversion of enantiomers in protein (as is discussed below, racemization does occur in proteins to a certain extent), the presence of D-amino acids in stoichiometric amounts in a peptide (protein) is assumed to exclude the biosynthesis of that peptide by the normal protein synthetic apparatus. One guideline in this review has been to use the most recent reference(s) in fields where tens or hundreds of other references would be equally relevant and perhaps more appropriate. Although this is of concern, the implication that this article will contain "all known" amino acid derivatives in proteins has been the source of considerably more consternation. In our first attempt to tabulate the known derivatives (6), we appealed to colleagues for help, and the present compilation includes many very helpful suggestions and additions received at that time. The list of derivatives is more complete now than it was five years ago, but it also is more obvious at this time that the list of omitted derivatives, both known and unknown must be substantial.

INDIVIDUAL REACTIONS

Modifications Involving the Peptide Bonds

Proteolytic cleavage is undoubtedly the most general of all covalent modifications of proteins, and it seems safe to predict that probably all protein products isolated from living cells have been modified by proteolysis from the precursor sequence encoded in the mRNA. The most obvious basis for this prediction is the fact that the universal initiation codon for protein synthesis is the unique methionine codon, AUG. Thus all protein sequences must start with methionine; in prokaryotes, the Met-tRNA$_f$ is formylated to give N-formyl Met-tRNA$_f$, and N-formyl Met consequently is the starting point; since only very few proteins are isolated with the starting amino acid still attached, it is clear that some trimming at the amino-terminal sequence has taken place during and/or after the assembly of the polymer. The total proteolytic cleavage of various proteins represents highly regulated, specific, and complex processes that have been the subject of several recent reviews (11–13). A general scheme for various known cleavages can be represented by the well-established sequence of events in the biosynthesis and secretion of insulin (14): while still attached to the membrane-associated polysomes, two cleavages take place: the removal of the amino-terminal Met, to expose (in the rat) a new amino-terminal Ala, followed by the removal of the 23-amino acids long signal sequence exposing Phe as the new amino terminus (preproinsulin → proinsulin). The second cleavage is presumably associated with initiation of the translocation of proinsulin to the lumen of the endoplasmic reticulum (ER). The next step involves an energy-requiring transport of the completed polypeptide chain to the Golgi, during which the folding and disulfide bond formation (6–SH → 3–S–S–) is likely to occur.

The proinsulin molecule is next packaged in secretory granules and transported to the plasma membrane for release. During this last step, two additional peptide bond cleavages occur at positions 30–31, and 65–66, to release an inactive "propeptide" (35-amino acids long) from the active hormone, which contains two polypeptide chains with Phe (the original) and Gly (res 66 in proinsulin) as amino terminals (proinsulin → insulin). Both hormone and propeptide are finally released to the extracellular compartment. In this case the rat mRNA specified a polypeptide chain of 110 amino acids, from which a single amino acid, a 23 amino acid, and a 35 amino acid peptide are removed before the biologically active product is produced. All these steps are fairly characteristic. The initial trimming of either the formyl or formyl-Met group in prokaryotes or the amino-terminal Met in a variety of systems does not appear to serve any particular function, but must be general, and perhaps relatively nonspecific. The observation that the *Escherichia coli* Trp operon, cloned in a multicopy plasmid, produced a mixture of formylated and normal, deformylated α-subunit is consistent with a model of a fixed level of soluble, deformylating enzyme unable to keep up with the overproduction of new proteins in the modified system (15). While wheat germ ribosomes will produce parathyroid hormone precursor with two amino-terminal Met, an ascites cell system will yield precursors in which both Met residues have been removed (16). The amino-terminal Met residue generally is observed only in short nascent chains (17), which shows that the removal occurs early in the polymerization process.

The next cleavage is the very important segregation step by which proteins are translocated through or inserted into membranes through the specific recognition of sequence information (12, 18). In most instances the information sequence is at the amino-terminal end (the signal or leader sequence) and is removed during the early stages of the polymerization process, but it is also becoming apparent that nature has evolved a number of variations on the general signal peptide theme. In some instances the higher-molecular-weight precursor and its protein product have identical amino termini (19–21), which suggests both that the cleavage of amino-terminal signal-like sequences is not an obligatory step in the segregation, and that in some instances carboxyl-terminal sequences may be cleaved as well. Thus only one of several mitochondrial proteins has been found to have a higher-molecular-weight precursor form and to be cleaved in the process of assembling the active enzyme complexes (22). Although most secreted proteins and proteins to be inserted into membranes are synthesized on ribosomes associated with the appropriate membranes, at least in the case of virus proteins, some can be synthesized by free ribosomes, be transported to the membrane as the fully synthesized soluble precursors, and then be incorporated into the membrane and cleaved (23). In some instances the synthesis cannot take place in association with the proper membrane; in the case of protein components of the outer membrane of *E. coli* the proteins are synthesized and processed in association with the inner membrane, and the finished product is transported through the periplasmic space to its final destination in the outer membrane (24).

Although it seems prudent to be aware of the many variations used in different systems and acknowledge that we do not understand the complete picture, the general model of precursor proteins containing amino-terminal "signal" sequences, which in the early stages of the amino acid assembly on membrane-bound ribosomes are inserted into and through the membrane and cleaved, seems to have broad validity and has been documented in bacteria (25), plants (26), and insects (27), as well as in higher animals (18). It is also possible to discern a certain pattern of signal peptide structures; although the amino acid compositions vary greatly, they all appear to be roughly of the same length (20–30 residues) and to contain a highly hydrophobic sequence in the center. Most importantly, theoretical calculations of structure of the known sequences show that they are conformationally very similar. The chemical synthesis of one signal sequence, that of preproparathyroid hormone, (28) permitted an experimental evaluation of theoretical calculations, and confirmed the conclusion that the highly structured sequence, which favored primarily β-structure in aqueous solution, favored an α-helical conformation in nonaqueous environments (29). This phenomenon of change in conformation in going from an aqueous to a nonaqueous environment, with presumed associated changes in membrane structure, could provide some of the energy needed for the translocation of the peptide, and is consistent with the membrane-trigger model proposed by Wickner (12) as an alternative to the signal hypothesis (18). There is evidence that "signal peptide receptors" exist (30), and since alteration of a single amino acid through mutation (31) or substitution (32) can totally abolish or strongly inhibit protein processing, the information contained in the signal peptide is probably highly specific. It is unlikely that the signal peptide information can be "read" as universally as is the genetic code. Unprocessed precursors are the final products of protein synthesis in many artificial in vitro systems, which suggests that a mismatch of membranes (receptors), ribosomes, and mRNA may make the information inaccessible. Examples of the opposite situation also are known; the normal proteolytic processing can take place with many different combinations of heterologous mixtures of membranes, ribosomes, and mRNAs. Secreted proteins such as ovalbumin (33) and insulin (34), when produced artificially in bacteria, are secreted; in the case of insulin the preproinsulin to proinsulin cleavage was observed to take place in bacteria (35).

The subsequent proteolytic cleavages after the release from the synthetic apparatus are not considered to any significant extent here. These processes, by which inactive precursors are converted to active proteins, have been studied extensively in the past and have been reviewed: proenzyme (zymogen) to enzyme in digestion (36), blood coagulation (37), and complement activation (38), and prohormones to hormones (13, 14). Macromolecular assembly involved in the biosynthesis of complex particles and aggregates such as virus (39) and collagen (40) also involve proteolytic cleavages. One of the fascinating aspects of this area is the recent realization that a given protein molecule contains a large number of physiologically active sequences, and that appropriate cleavages thus may give several distinct active products simultaneously; or different proteolytic cleavages of a given precursor may give different active products. Although this feature is most dramatically illustrated by the peptide hormones (13), it also appears likely that we shall see examples of high-molecular-weight enzyme precursors from which multiple subunits or indeed metabolically related enzymes may be produced by proteolysis. Thus there may be a single chain precursor of the four cytoplasmically produced subunits of cytochrome oxidase in yeast (41); however this conclusion has been seriously questioned (41a).

Modifications Involving the Carboxyl- and Amino-Terminals

DERIVATIVES OF THE α CARBOXYL GROUP   There are three well-characterized derivatives of the carboxyl-terminal end, and at the present time all three types of derivatives appear to be quite specialized and restricted to only a few proteins: *unsubstituted amides* (α-$CONH_2$) involving several different amino acids (Asp, Glu, Gly, His, Met, Phe, Pro, Tyr, Val (7), but limited mostly to short physiologically active peptides (e.g. insect toxins (27, 43) and hormones (42); *ADP-ribosylation* of carboxyl-terminal Lys of histone Hl (44); and the *substituted amide* derivative in tubulin (45) in which a Tyr residue is added in peptide linkage to the α-COOH group in the absence of ribosomes and mRNA.

DERIVATIVES OF THE α-AMINO GROUP   A considerably broader and more general spectrum of derivatives is observed for the amino terminus of proteins. The most common group of derivatives is the result of N^α-acylation, and several different acyl groups have been observed: *formyl* (46), *acetyl* (47), *pyruvoyl* (48), *a-ketobutyryl* (49), *glucuronyl* (50), *a-amino acyl* (51), *pyroglutamate* (52), and *murein* (53). Some of these may be quite rare. The formyl group has only been observed as N^α-formyl-Gly in a fraction of honey bee melittin (46) outside the obligatory N^α-formyl-Met nascent chains and the occasional completed proteins in prokaryotes (15); the murein derivative is the link between *E. coli* peptidoglycan and membrane lipoprotein in which the diaminopimelic acid of the former forms an amide link with the N^α- of the amino terminal Lys; and the glucuronyl-Gly derivative was observed in fungal enzymes. The α-ketoacyl derivatives are produced by deamination of Ser (via dehydroalanine) or Thr (via dehydroaminobutyrate) in the activation of a series of enzymes that appear to substitute the "amino-terminal" carbonyl function for the more common pyridoxal coenzyme in their catalytic apparatus (48, 49, 54–56). The N^α-amino acyl derivatives represent, like the addition of Tyr to the carboxyl-terminal end, an extension of the completed polypeptide chain in the absence of ribosomes and mRNA. The donor is amino acyl-tRNA, and several proteins appear to act as acceptors. In eukaryotes only Arg appears to be transferred in this way, while in *E. coli,* Leu, Phe, and Trp have been found to be involved. The fact that chromosomal proteins can accept Arg in several tissues (57) suggests that the reaction could play an important developmental or regulatory role. Pyroglutamate (pyrrolidone carboxylic acid) is a fairly commonly encountered N^α-acylated derivative in which the γ-carboxyl group of an amino-terminal Glu can be considered as the acylating reagent to form the cyclic amide (52). Although it is clear that this derivative can be formed in enzyme-catalyzed reactions, it also forms readily by spontaneous cyclization of Gln, and its production as an artifact of protein isolation must always be considered as a real possibility. The recent suggestion that amino-terminal pyroglutamate also can arise by spontaneous cleavage of the peptide bond of internal pyroglutamate residues (58) is discussed below. The N^α-acetylation, by far the most common acylation reaction, is discussed in some detail toward the end of this article.

The α-amino group has also been found to be alkylated; again the derivatives appear to be rather specialized. Thus, the nonenzymatic glycosylation of the amino terminus of hemoglobin through an Amadory rearrangement of the initial Schiff base to yield *1-deoxy,1-(N^α-valine)-fructose* (59) must uniquely reflect the high blood glucose concentrations in some species of higher animals, and the best characterized N^α-methylated amino terminals are almost exclusively in ribosomal proteins that contain either *N^α-monomethyl methionine* (60, 61), *N^α-monomethyl alanine* (60), or *N^α-trimethylalanine* (62). *N^α-dimethylproline* has also been identified in a protozoan cytochrome (63).

Modifications Involving Individual Amino Acid Side Chains

ARGININE   Three methylated derivatives of the guanidinium group of arginine have been found in a variety of proteins in both eukaryotes and prokaryotes (64, 65); they are *N^ω-methylarginine, N^ω,N^ω-dimethylarginine,* and *$N^\omega,N^\omega{}'$-dimethylarginine.* Another arginine derivative, *N^ω-ADP-ribosylarginine* (66, 67), is the product of ADP-ribosylation carried out by bacterial enterotoxins (66). The product formed in the reaction is the α-ribosyl anomer that rapidly equilibrates to a mixture of α- and β-anomers. Several proteins can act as acceptors in vitro. Other ADP-ribosyl derivatives are discussed below under Glu, His, and Ser. *N^ω-phosphorylarginine* has been identified in myelin basic protein (67). Apparently, N-phosphoryl derivatives may be much more common than expected, and those of Arg, His and Lys are quite acid labile and are destroyed by the conditions commonly employed in the study of other phosphate derivatives. *Citrulline* appears to be a special derivative found primarily in structural proteins of hair and skin (68, 69). Similarly, *ornithine* is a rare but well-established protein amino acid (70), presumably produced by the action of a protein arginase.

ASPARAGINE This amino acid alone may well account for hundreds of derivatives when all complete structures of asparagine-linked oligosaccharides (*glycosylasparagine*) have been determined. The details of the biosynthesis of these glycoproteins are discussed elsewhere in this volume (5), have been reviewed previously (71), and are not treated here. Once the initial and apparently universal Glc_3-Man_9-$GlcNAc_2$- precursor has been attached to the appropriate Asn residues, the further procession of the oligosaccharide to smaller $Man_{3-7}GlcNAc_2$- (72) or larger $Man_{\sim 300}$-$GlcNAc_2$- such as those in yeast extracellular glycoproteins (73), to complex oligosaccharides containing other sugars, such as Fuc, Gal, sialic acid, Xyl, additional GlcNAc (72), as well as phosphate (73, 74), yields an impressively large number of glycosylated Asn derivatives. Other derivatives of Asn are N^ϵ-(*β-aspartyl)lysine* (75), in which the amide ammonia has been displaced by the ε-amino group of lysine, and the deaminated product, *aspartate*. This latter derivative is interesting in that another primary amino acid is formed as product in a reaction that appears to be spontaneous. The rate with which the deamidation of a given asparagine residue takes place under physiological conditions is determined by the neighboring amino acids in the polypeptide sequence (76), and it seems reasonable to consider the asparagine residues in proteins as built-in clocks determining the life time of the protein according to the rate of the spontaneous conversion of the neutral amide to the charged acid. It appears clear that this conversion indeed does take place in vivo (77) and in fact represents a significant change in long-lived proteins such as α- and β-crystallin from lens (78).

ASPARTATE *β-aspartylphosphate* (O^4-phosphoaspartate) has been identified as the transient phosphorylated derivative in the Na^+,K^+-ATPase of plasma membranes and the Ca^{2+}-ATPase of sarcoplasmic reticulum (79). The indirect evidence that *β-methylaspartate* is a product of protein methylase at least in vitro (64) has recently been substantiated by the demonstration that this derivative is present in erythrocyte proteins after incubation with S-adenosylmethionine (79a). One interesting in vivo modification that probably involves several amino acids, but that has been studied in detail only for Asp, is racemization. This reaction like the deamidation discussed above is apparently not enzyme catalyzed; it is slow and becomes significant only in proteins with long half-lives. Analysis of hard-tissue proteins show that about 0.1% of L-Asp is converted to *D-aspartate* per year (80); in the human lens protein (α-crystallin) the rate is about 0.14% per year in normal individuals, but significantly higher in individuals with cataracts (81).

CYSTEINE   The most obvious and common derivative of Cys is the oxidized disulfide-containing *cystine*. Since this reaction is important in stabilizing the three-dimensional folding of individual domains of whole protomers by intrachain cross-links (1) as well as in cross-linking between protomers in the formation and maturation of multichain structures (69, 82), it is likely that the catalytic agents responsible for the reaction must be active in many compartments of the cell. Other derivatives of Cys include glycosylated forms, *S-galactosylcysteine* (83) and *S-glucosylcysteine* (84), and covalent links to prosthetic groups, the *S-cysteinylheme* of cytochrome *c* (85), the *S-cysteinyl phycocyanobilin* in phycocyanin (85a), *8a-(S-cysteinyl)flavin thiohemiacetal* (86), and *6-(S-cysteinyl)riboflavin-5'-phosphate* (87). Since the thiol group is involved in the catalytic apparatus of several enzymes, several transient covalent Cys derivatives such as thiohemiacetals and thioesters have been demonstrated or deduced, but these are not included here. It is quite possible that derivatives such as the *(peptide)-cysteine-S-S-cysteine* (free) in α-1 antitrypsin (88), the *cysteine-S-phosphate* (89), *cysteine-S-sulfonate* (90), and the *cysteine-S-sulfide* (persulfide) (90) observed in bacterial thioredoxin and in sulfotransferase may also fit in the category of catalytic intermediates. A recent addition to the list is the uniquely reactive Cys analogue present in several enzymes, *seleno-cysteine* (91). One of the major cell envelope proteins of *E. coli* is a lipoprotein containing an amino-terminal Cys uniquely modified with a thioether link to glycerol, *S-(sn-1-glyceryl)cysteine* or diacyl glycerol *S-(sn-1-(2,3-diacyl)-glyceryl)cysteine* (92). Other derivatives such as the thioether *lanthionine, lysinoalanine,* and *dehydroalanine,* which are present in proteins in vitro as a result of alkaline treatment or heat, now also appear to be natural in vivo products. Lanthionine and lysinoalanine have now been found in the outer stratum of epidermal α-keratin (69), and it is highly likely that dehydroalanine is an intermediate in the formation of those derivatives. In addition, it has been clearly established that dehydroalanine is one of the products of thyroxine formation in thyroglobulin (93).

GLUTAMATE   Three derivatives of glutamate are well established as constituents of proteins. *γ-Carboxyglutamate* was originally found in 5 plasma glycoproteins, including prothrombin, and provided the missing link between vitamin K deficiency and faulty blood clotting, when vitamin K was established to be the required cofactor for the carboxylase that introduces the extra carboxyl group (94). γ-Carboxyglutamate has subsequently been found in other tissues, primarily those associated with calcification (95, 96), and most recently it has been reported to be present in ribosomal proteins of both prokaryotes and eukaryotes (97). The function and biosynthesis of this derivative has been reviewed (96). The fact that both the unique Ca^{2+} and Ba^{2+} binding properties of prothrombin and its activity in clotting are lost upon removal of the γ-carboxyl group [by heating *in vacuo* (98)], represents strong evidence that the malonate-like side chain is involved in Ca^{2+} binding as an essential component of biological function. *O^γ-methylglutamate* is another derivative that has received a good deal of recent attention because of its role in the dynamics of bacterial chemotaxis (99) and amoeba movement (100). Also, the presence of alkali-labile methyl groups in proteins of animal cells (64) suggests that the methyl ester of Glu is involved in functions other than chemotaxis. The third derivative is *O^γ-ADP-ribosylglutamate,* which has been identified in histones H1 and H2B (44, 101). ADP-ribosylated histones have been demonstrated in plants (102) as well as in several animal cells (103). A fourth in vivo modification, in which Glu is converted to *glutamine* is another example of a "silent" modification reaction with a primary amino acid as product. In yeast mitochondria (104) (and apparently also in some prokaryotes), glutaminyl-tRNA is formed only from preformed glutamyl-tRNA. No charging enzyme for the normal formation of glutaminyl-tRNA from glutamine, tRNA, and ATP appears to exist.

GLUTAMINE   The reverse of the above reaction, the deamidation of Gln to *glutamate* is also observed in proteins. Like the asparagine to aspartate conversion, it appears to be a spontaneous reaction, the rate of which depends on the sequence environment of the glutamine residue (76). A rather widespread glutamine derivative is the "isopeptide" cross-link formed when the ϵ-NH_2 group of Lys replaces ammonia to yield the substituted amide *N^ϵ-(γ-glutamyl)lysine.* The enzyme responsible for the reaction, transglutaminase, its specificity and occurrence, and the biological significance of the derivative have been reviewed (105). *N^5-methylglutamine* has also been reported in a ribosomal protein (106). In this case it is assumed that the parent primary amino acid is Gln and not Glu or some other precursor. The point is made because it has been demonstrated in for example α_2-macroglobulin that a uniquely reactive residue will incorporate methylamine spontaneously to yield N^5-methylglutamine (107), and it has subsequently been suggested that the uniquely reactive residue is an *internal pyrrolidone carboxylate (pyroglutamate)* residue (58). More recent evidence suggests that the uniquely reactive glutamate represents the thiolester, *S-(γ-glutamyl) cysteine,* involving a neighboring Cys residue (107a). The proposal that a *O^β-(γ-glutamyl)serine* ester provides a unique intrachain cross-link in a clostridium proline reductase (53) represents another possible reactive Glu derivative in proteins. These latter two are derivatives listed under Gln, although their origin could be Glu as well.

HISTIDINE   *$N^{\tau(\pi)}$-phosphorylhistidine* has been demonstrated in myelin basic protein (67) and in nuclear proteins, notably histone H4 (108), N^τ-

(*methyl*)*histidine* in muscle proteins (109), and *4-*(*iodo*)*histidine* in thyroglobulin (110). His is also a site for prosthetic group attachment, and both the N^τ- and the N^π-derivatives of *8a-*(*histidyl*)*flavin* are known (86). Cell-free synthesis of one of these derivatives has recently been reported (111). The most recent addition to the list of His derivatives is compound assigned the structure *2-[3-carboxamido-3-*(*trimethylammonio*)*propyl]histidine;* it is found in yeast elongation factor 2, where it represents the site for the ADP-ribosylation catalyzed by diphtheria toxin. The ribosyl moiety in the ADP-ribosylated derivative is proposed to be linked to the N^τ-(N-1) of the 2-substituted imidazole ring (112) to give *N^τ-ADP-ribosyl,2-[3-carboxamido-3-*(*trimethylammonio*)*propyl]histidine.*

LYSINE   Organic chemists will recognize that the primary ε-amino function of Lys is an excellent candidate for chemical modifications, and in fact a large number of the reagents that have been developed by protein chemists over the years are directed toward the Lys amino group. Nature has done even better in utilizing this functional group, and it is easy to account for about 40 different, naturally occurring Lys derivatives in various proteins. There are in fact two closely related families of derivatives, one family descending from the primary amino acid, Lys itself, and the other descending from the secondary amino acid, *δ-hydroxylysine,* which in turn is derived from Lys. These two families contain many similar derivatives, but will be treated separately in the following. The primary amino group of Lys can be methylated, and all three methylated derivatives, *N^ϵ-methyllysine,* *N^ϵ-dimethyllysine,* and N^ϵ-trimethyllysine (64, 65) are known in a variety of proteins in both prokaryotes and eukaryotes. Enzymes responsible for these methylation reactions (protein-lysine methylases) have been isolated and characterized from a number of sources; some appear capable of using a variety of protein acceptors, while others are as specific as the Neurospora methylase that methylates Lys-72 in cytochrome *c* (113). Neurospora cytochrome *c* contains trimethyllysine in this position. As expected, Lys side chains can be acylated to give rise to *N^ϵ-*(*phosphoryl*)*lysine* (108), *N^ϵ-acetyllysine* (114), the rather unusual derivative *N^ϵ-*(*N^α-methylalanyl*)*lysine,* in which the modified Lys is the amino-terminal residue in *E. coli* ribosomal protein S11 (115), and the *N^ϵ-*(*diaminopimelyl*)*lysine,* which represents a linkage between murein and lipoprotein in gram-negative bacteria (116). In addition, two coenzymes *N^ϵ-biotinyllysine* (117) and *N^ϵ-lipoyllysine* (118), are covalently attached to enzymes through N^ϵ-acylation. Other coenzymes are linked to Lys via their carbonyl functions as Schiff bases such as *N^ϵ-retinallysine aldimine* (119) or *N^ϵ-*(*phosphopyridoxal*)*lysine aldimine* (120). Just as the free amino terminus of hemoglobin can be glycosylated nonenzymatically (59), rat and human serum albumin have been found to contain a certain fraction of *1-deoxy,1-(N^ε-lysine)fructose*, the Amadori rearrangement product of glucosylated Lys (121). The majority of the known Lys derivatives are those found as cross-links in various structural proteins such as collagen and elastin. The parent compound for all of these derivatives is protein *allysine(α-amino adipic acid semialdehyde)*, the product of oxidative deamination of Lys side chains. The oxidase has been well characterized (122, 123) and is found quite widely in nature (124); allysine and some of its derivatives have been demonstrated in bacteria (124, 125) and in animal cells. Allysine, with its reactive carbonyl function, now becomes the initiator of a series of complex chemical reactions leading to a group of cross-links, each involving either two, three, or four amino acid residues. Several of the reactions involved could proceed spontaneously, but it is likely that specific enzymes (e.g. oxidoreductases) are involved as well. The initial two steps in this series of ractions is the reaction of allysine with another allysine or with Lys to yield to aldol condensation product, *allysine aldol* or the aldimine, *dehydrolysinonorleucine*, respectively. The latter compound can be reduced to the stable derivative *lysinonorleucine*, which has been found in both collagen and elastin as a stable two-residue cross-link. The aldol can be dehydrated to the α,β unsaturated analogue that in turn can react with dehydrolysinonorleucine to yield, through a series of precursors, the four-amino acid residue cross-link *desmosine;* or allysine can react with dehydrolysinonorleucine to yield a three-residue cross-link *dehydromero-desmosine*, which in turn can react with another allysine to give another four-residue cross-link *isodesmosine*. Aldol can also react with the imidazole nitrogens of His to yield the three-residue cross-link (*allysinealdol*)-*histidine*. [The actual in vivo reaction pathways leading to these derivatives are not fully established. The reader is referred to excellent reviews for a discussion of these structures and their formation (126–128).]

   δ-Hydroxylysine is apparently not present in elastin, but is a common derivative in all forms of collagen. It is formed from Lys by lysylhydroxylase, which requires α-ketoglutarate, O_2 and peptide-bound Lys as substrates, and Fe^{2+} and a reducing agent such as ascorbate as cofactors (129). It appears that the hydroxylation reaction takes place in the early stages of collagen biosynthesis, perhaps even at the stage of nascent chains. A methylated derivative of hydroxylysine, *N^ε-trimethylhydroxylysine* is known (130); in addition the glycosylated derivatives *O^δ-(glucosyl-galactosyl)* hydroxylysine and *O^δ-(galactosyl) hydroxylysine* are unique derivatives of this hydroxylated secondary amino acid (131). Hydroxylysine can also be oxidized to give the allysine analogue *δ-hydroxyallysine,* which in turn can participate in all the reactions outlined for allysine above to yield another family of two-residue, three-residue, and four-residue cross-links *hydroxyal-*

lysinealdols (syndesines), *(dehydrohydroxylysine)norleucine, hydroxylysino-norleucine, (dehydrohydroxylysine) hydroxynorleucine, (hydroxylysino)hy-droxynorleucine, dehydrohydroxymerodesmosine, (dehydrohistidini)hydro-xymerodesmosine,* and *(hydroxyallysinealdol)histidine* (126, 127, 132). The biosynthesis and "maturation" (the formation of cross-links, mostly intra-chain in elastin and interchain in collagen) of fibrous proteins is under active investigation, and several recent studies (e.g. 133–135) suggest that both the temporal sequence of events and the detailed chemistry will soon be fully elucidated.

PHENYLALANINE   *O^β-glucosyl-β-hydroxyphenylalanine* has been identi-fied as a naturally occurring amino acid derivative in the fungal enzyme cutinase (136). It is assumed that the precursor of this derivative is free *β-hydroxyphenylalanine,* which could be produced in a reaction similar to that demostrated for the β-hydroxylation of Trp (see below). The non-glycosylated precursor has not been observed.

PROLINE   The main derivatives are the different hydroxyprolines: the major derivative in proteins, *4-hydroxyproline* (129), as well as the minor derivative, *3-hydroxyproline* (129) and the rather rare *3,4-dehydroxyproline* (137). Monohydroxyprolines, like the δ-hydroxylysine, are essential com-ponents of collagen, and the hydroxylation is catalyzed by enzymes very similar to the Lys hydroxylase, which require Fe^{2+}, reducing agents, α-ketoglutarate, and O_2 for their action. The major proline hydroxylase activity has been shown to exist almost exclusively in the lumen of the ER; it has been proposed that it is specific for 4-hydroxylation, and that a separate enzyme is required for the 3-hydroxylation (129). A hydroxylase preparation from earth worm cuticle (which contains relatively high levels of 3-hydroxyproline) may hydroxylate both positions (137). As in the case of hydroxylysine, hydroxyproline represents a new potential site for further modification, and glycosylated hydroxyprolines have been reported in plant cell wall material. Two such derivatives are *O^4-arabinosylhydroxyproline* (138) and *O^4-galactosylhydroxyproline* (139); both undoubtedly contain other sugars attached to the peptide-linked monosaccharide.

SERINE   The side chain derivatives of Ser have received a good deal of attention in recent years because they include such widely distributed and important modifications. *O^β-phosphoserine* (140) is the main phos-phorylated derivative in phosphoproteins, both those with stable "perma-nent" phosphate residues and those in which phosphorylation-dephos-phorylation represents an important regulatory on-off or off-on switch for biological activity. The specificity and regulation of protein kinases responsible for the transfer of phosphate from ATP to the acceptor protein have been studied extensively, and excellent reviews are available (140, 141). An interesting new protein phosphate ester has recently been reported for some flavoproteins. Based on the NMR spectra at different pH, it is concluded to be a phosphate diester, and although it has not been definitely established to be associated with Ser, it could represent a phosphate diester cross-link O^β-(O^β-phosphoserine)serine between two Ser residues (142). In another phosphate diester derivative of Ser, O^β-(4'-phosphopantetheine)serine (143), the phosphate diester provides the covalent linkage between the prosthetic group and the protein in the acyl carrier proteins of bacteria, yeast, and higher plants. In yet others, the serine-phosphate is linked to the reducing end of sugars through the acid-labile glycosylphosphate bond: O^β-(ADP-ribosyl-phospho)serine in histones (144), and O^β-(N-acetylglucosaminyl-phospho)serine (145) in a slime mold. Ser is also a major glycosylation site in proteins (72); in terms of the sugar linked directly to Ser by the glycoside bond there are at least four distinct families of glycosylated Ser derivatives, derived from O^β-(N-acetylgalactosaminyl)serine (146), O^β-mannosylserine (147), O^β-galactosylserine (148), and O^β-xylosylserine (149). Two additional derivatives should be mentioned although they have not been definitely identified as products of Ser modifications. A small amount of covalently bound palmitate is found in Sindbis membrane glycoprotein (150), the chemical properties of which are consistent with those of a simple ester, and Ser is thus a possible participant in a putative O^β-palmitylserine derivative. Similarly, the recognition of alkali- and acid-stable O-methyl derivatives in seminal plasma proteins (151) strongly suggests the possible presence of O^β-methylserine. Several enzymes have uniquely reactive Ser residues that participate directly in the catalytic function of the enzyme, frequently by participating in the formation of transient covalent intermediates of the reaction. These derivatives are not included in this review.

THREONINE Many of the enzymes responsible for derivatization of Ser residues in proteins apparently can use Thr equally well as the site of modification, and many of the derivatives discussed above for Ser, also exist for Thr: O^β-phosphothreonine (140), O^β-(N-acetylgalactosaminyl)-threonine (146), O^β-mannosylthreonine (147), and O^β-galactosylthreonine (148). A unique threonine-linked oligosaccharide, O^β-(glucosyl-β-1,3-fucosyl)threonine, has been found as a glycopeptide in human urine (152). The putative phosphate diester between two Ser residues can equally well involve one or two Thr (142), and the base labile palmityl ester (150) as well as the proposed O-methyl ethers (151) listed above for Ser could also be Thr derivatives.

TRYPTOPHAN   Apparently no naturally occurring derivative of Trp has been observed in proteins. It is clear, however, that certain bacterial oxidases will act on peptide-bound Trp in vitro to yield peptide bound a,β dehydrotryptophan (153). Using model compounds, it has also been possible to demonstrate that β-hydroxy- and β-ketotryptophan derivatives would form if the reaction was carried out at low pH (154). Although this enzyme may well be primarily involved in Trp catabolism, the capability of producing a derivative analogous to the reported β-(glycosyloxy)phenylalanine (above) and β-(glycosyloxy)tyrosine (below) makes it interesting to consider a role in protein modification for this enzyme.

TYROSINE   The phenol ring of Tyr runs a close second to the Lys amino group as a site for a large number of in vivo chemical modifications, a fact that again probably should not surprise the organic chemist. A variety of halogenated derivatives are known, *3-chlorotyrosine* (155–157), *3,5-dichlorotyrosine* (157), *3-bromotyrosine* (156, 157), *3,5-dibromotyrosine* (156, 157), *5-bromo-3-chlorotyrosine* (157), *3-iodotyrosine* (110), and *3,5-diiodotyrosine* (110). The latter two are key intermediates in the biosynthesis of the thyroxin family of hormones. This reaction is now believed to involve a β-elimination of mono- and diiodityrosines in thyroglobulin, which leaves a protein-bound dehydroalanine (93) and provides the reactive mono- or diiodophenol radical that can react with the phenolic hydroxyl of another residue of diiodotyrosine to give the two most common forms of the hormone, *3,5,3'-triiodothyronine* (158) and *3,5,3',5'-tetraiodothyronine* (thyroxin) (158), as residues in the thyroglobulin polypeptide chain. Ring substitution by hydroxyl is possibly a Tyr reaction, since the enzyme that converts Tyr to dihydroxylphenylalanine (DOPA) is broadly distributed in nature and capable of oxidizing Tyr in proteins in vitro (159). Apparently DOPA has not been found as a natural constituent of proteins, however. Several derivatives of the phenolic hydroxyl group are known: simple monoesters such as *tyrosine O^4-sulfate* (160), *tyrosine O^4-phosphate* (161), the diesters of the covalently bound regulatory effectors that appear to be unique to glutamine synthetase of gram-negative bacteria, *O^4-adenylyltyrosine* (162), *O^4-uridylyltyrosine* (163), and the phosphate diesters through which protein-Tyr is linked to the 5' phosphate of RNA (*5'-(O^4-tyrosylphospho)RNA*) in polio virus (164) or to the 5' phosphate of DNA (*5'-(O^4-tyrosylphospho)-DNA* in *E. coli* DNA topoisomerase I and in *Micrococcus luteus* DNA gyrase (165). The latter derivative, as a putative intermediate in the reaction catalyzed by the gyrase, should perhaps not be included in a list of true in vivo modifications. A recent addition to the list of covalently linked flavin coenzymes in bacterial flavoproteins involves a derivative of Tyr, *8a-(O^4-tyrosyl)flavin* (166). Oxidative phenolic coupling of Tyr rings to yield *3,3'-bityrosine* (167) and *3,3';5'3'' tertyrosine* (167) cross-links appears to be a natural occurrence in several systems, such as the fertilization membrane of sea urchin eggs (168), adhesives substances of sea mussels (169), and human lens protein (170). Although it is clear that subjecting a variety of proteins to the action of peroxidase at alkaline pH in vitro will lead to the artificial production of these cross-links (171), the evidence available favors this reaction as a natural protein processing step. The final derivative of Tyr is analogous to that discussed above for Phe, O^β-*glycosyl-β-hydroxytyrosine* (136). Again it is assumed that this derivative requires the formation of *β-hydroxytyrosine* as a precursor, although that compound has not been observed in proteins.

OTHER AMINO ACIDS   No derivatives of the side chains of alanine, glycine, isoleucine, leucine, methionine, and valine have been identified to date.

GENERAL CONSIDERATIONS

Modification Reaction Types

As an alternative to listing the derivatives as above, amino acid by amino acid, there may be some merit in following the more common pattern of looking at the different reactions in terms of the new functional groups or properties they introduce into proteins. In this way, a rather long and tedious list of individual derivatives can be classified in a few broad categories. Glycosylation, the introduction of monosaccharides, oligosaccharides, and polysaccharides as covalent components of proteins, would undoubtedly become the largest group of derivatives, including all the glycosyl derivatives of the primary amino acids, Asn, Ser, Thr, Cys, as well as of the secondary ones, hydroxylysine, hydroxyproline, β-hydroxyphenylalanine, and β-hydroxytyrosine. Phosphorylation would encompass the monoesters of Ser, Thr, and Tyr, the phosphoramides of Lys, His, and Arg and perhaps include some of the phosphate diesters by which other functional groups are linked through the phosphate to Ser and Tyr. Methylation would combine the rather stable N-methylated derivatives of Arg, His, Lys, Gln, and free α-amino groups and the putative O-methyl ethers of Ser and Thr with the labile methyl esters of Glu and presumably also Asp. Acylations (other than phosphorylations) would include a large group of primarily acetyl but also the other acylated derivatives of α- and ε-amino groups and Ser (Thr) esters; ADP-ribosylations the mono- and polyADP-ribosyl derivatives of Ser, His, Glu, Arg, and the α-carboxyl group; and halogenations the Cl, Br, and I derivatives of Tyr and His.

We can use this classification to single out a variety of different reactions that appear to proceed spontaneously in proteins as nonenzymatic modifications. These would include the glucosylations of N^α and N^ϵ groups in compartments containing high concentrations of glucose, the conversion of Asn and Gln to Asp and Glu, the formation of a number of the collagen, elastin cross-links, and the slow but inevitable racemization of protein L-amino acids and hydrolysis of peptide bonds. It has been argued that these are not true postsynthetic modification reactions, but apparently any property inherent in a given genetically determined amino acid sequence, which uniquely determines the function and half-life of that sequence whether as a result of enzyme catalysis or spontaneous change, should be considered an in vivo modification. With a final large and heterogeneous category of "miscellaneous reactions" all the derivatives could thus be accounted for. This classification emphasizes Nature's apparent need for additional functional groups in proteins beyond those specified by the genetic code, but deemphasizes the variety both in chemistry and potentially in biological functions derived from the different properties of individual derivatives within each group.

If emphasis were instead placed on either the real or presumed functional attributes of different derivatives, a rather different classification would result. One could, for example, consider one group of derivatives specifically designed to introduce cross-links to stabilize the three-dimensional structure of the products or to provide the proper elasticity or rigidity of structural proteins. Such cross-links would include a broad spectrum of chemical derivatives, from the relatively ubiquitous disulfide bond, the N^ϵ-(γ-glutamyl)-lysine amide, and probably phosphate diesters in all types of proteins, to the complex allysine- and hydroxyallysine- derived cross-links in collagen and elastin, lanthionine and lysinoalanine in keratins, and bis- and tertyrosine in other structural proteins. One could consider a class of derivatives evolved to affix a given coenzyme or prosthetic group permanently to the appropriate enzyme, and again realize that this desirable end has been achieved through a broad spectrum of chemical derivatives including amides, aldimides, ethers, thioethers, phosphate-diesters, and N-alkyl derivatives. One could visualize the desirability for readily reversible modification reactions evolved as on-off switches in sophisticated regulatory mechanisms of biological activity, and the most obvious examples in this category would be most, but not all, O- and N-phosphates and N-acetates, the methyl esters of Glu, the adenylyl- and uridylyl-O-Tyr derivatives, and probably also the (poly)ADP-ribosyl derivatives of nuclear proteins. In view of the recent realization that transient (74) and perhaps also reversible (172) changes occur in the nonreducing terminals of glycoproteins, it seems reasonable to include glycosylated derivatives in this category of reversible, dynamic derivatives. Glycosylations are difficult to assign to a single functional classification, as it appears that many different biological functions can be ascribed to the sugar moieties. In some cases, establishing structural integrity, protease resistance, increased solubility, and high viscosity may be the key roles of polysaccharide chains on proteins, while others much more dynamic functions may be served. Based on the recent explosion of information on lectins, apparently ubiquitous proteins that recognize and bind oligosaccharides and play an important role in a wide spectrum of biological processes, it seems that the glycosylated derivatives are primarily involved in biological communication. After identifying these broad categories based on presumed knowledge of functional properties, there are several derivatives left, some of which can be assigned specific functions (γ-carboxy glutamate in Ca^{2+} binding, di- and triiodo Tyr as essential precursors of thyroxin, etc); but we can still only speculate about the purpose of most of them.

Specificity Determinants

In terms of mechanisms of formation and functions of different derivatives, there are few common features among the known modification reactions, but in terms of the specificity of the different processes, the basis on which specific residues in specific proteins are selected for modification must be quite similar for all the reactions, and a brief discussion of this apsect of in vivo protein modification is included in the following.

Referring back to the scheme in the first section, any modification at the level of amino acyl-tRNA appears clear. The specificity determinant must reside with the tRNA and not with the amino acyl moiety. This has been clearly established in the case of the formation of N-formyl-Met-tRNA in *E. coli;* two distinct t-RNAs are charged with Met to yield Met-tRNA^{Met} and Met-tRNA^{MetF}, but only the latter is a substrate for the transferase that transfers the formyl group from formyl tetrahydrofolate to the amino group of Met (173). The selection of the formylated derivative as the starting amino acid in all proteins apparently depends on the specificity of initiation factor IF-2 together with the initiator codon AUG (174). For internal, modified residues formed at the level of amino acyl-tRNA, the discrimination between modified and unmodified amino acyl-tRNAs would presumably have to involve different codons to match specifically with the two distinct anticodons. Since all amino acids other than Met are specified by more than one codon, this would presumably be feasible.

Modifications at the level of nascent and completed polypeptide chains are considerably more complicated, and any integrated model of specificity determinant for protein processing must include a variety of different components. The integrated model must first of all account for a number of observations made in different systems for different modification reactions:

1. Several proteins that are not modified in vivo are excellent substrates in vitro when incubated with cell-free extracts or purified enzymes (64, 71, 141).
2. Some substrates that have been modified in vivo can be further modified in vitro after denaturation (141).
3. Some substrates are not affected by partial proteolysis, but are destroyed after extensive proteolysis (64).
4. Probably every modification process is specific for a given amino residue (or a set of similar residues like Ser and Thr) but in no known case are all copies of that residue in a given protein modified.
5. In many cases a given process requires that the residue to be modified is located in a specific sequence. [y-Asn-x-Ser(Thr)-z is the required sequence for N-glycosylation (71), Arg(Lys)-Arg-x-y-Ser-z for Ser-phosphorylation by cAMP-dependent kinase (141), while other protein kinases apparently can use other sequences such as N-acetylseryl groups (175), and N^{α}-acetylating activities in a given system have been reported to be specific for a single amino acid in the amino-terminal position of oligopeptides containing about ten residues (176)];
6. In the case of O-glycosylation in yeast, for example, no sequence specificity appears to exist (177).
7. Interference with an early protein processing step may exclude all subsequent steps (30).
8. Based on recent results from in vitro protein synthesis, a given protein may be processed differently in different systems. (Different processed products were obtained when immunoglobulin light chain mRNA were translated in a reticulocyte-dog pancreatic microsome and in an oviduct system.) (178);
9. Enzymes involved in in vivo protein modification are generally highly compartmentalized.

Superimposed on these individual observations are the two general facts that virtually all proteins are transported from the site of synthesis to a compartment of action different from the synthetic site, and that for a given protein undergoing several modification reactions, there appears to be an obligatory sequence of the steps. This latter point becomes obvious in the cases where the secondary steps involve directly the products of the preceding one: N^{α}-acetylation must follow the proteolytical cleavage in which the acetylation site was produced, and the exquisitely specific steps (74, 179) involved in producing the major mammalian N-glycosylated derivatives must follow the incorporation of the precursor oligosaccharide Glc_3Man_9

GlcNAc$_2$- into the protein. To allow for all of these properties, it appears that at least three completely distinct, but interacting types of specificity determinants must be considered; first, the structural determinants of the substrate (the protein to be modified), second, the compartment location of the enzyme catalyzing the modification reaction, and third, the temporal changes in the substrate structure as they relate to the growth of the nascent chain and to the movement of the completed chain through the various compartments. The first of these, the structural features, have many facets of their own. The specificity determinant may, as stated above, be either the single residue or a short primary sequence, but it is clear that the conformational restrictions imposed on this modification site by the rest of the polypeptide chains at the time of reaction are all-important as well. In the extreme, one can visualize an all or none difference between a totally exposed residue (sequence) at the surface of a polypeptide domain and the same residue (sequence) totally buried in the interior of that domain, but more subtle differences could also be expressed. Thus, the residue (sequence) in a short segment of random coil structure may have sufficient flexibility to assume the necessary conformation to combine with the enzyme's binding site, while the same residue (sequence) in a highly structural region may have a much higher activation energy barrier to form the same enzyme-substrate complex.

It seems reasonable to propose that these structural specificity determinants are the principal ones affecting modification reactions that occur after the finished protein molecule has been delivered to its compartment of action. All the regulatory modifications for example are probably adequately determined in this way. Modifications during synthesis and transport require more complex considerations however. As already pointed out, several reactions may have to take place in a tightly coupled sequence or even in concerted fashion, so that what may be considered as a simple lock-key model for common enzyme-catalyzed reactions become more like a highly complex combination lock system in protein processing. The second specificity component, the compartment-location of the enzymes involved in processing may at first sight look trivial as more and more processing enzymes are determined to reside in specific compartments such as RER, Golgi, secretory granules, mitochondria, nuclei, plasma membranes, cytosol, and extracellular fluids. If a given protein bypasses any of these compartments it is inevitably excluded from the modification reactions of that compartment whether it has the necessary structural specificity determinants or not. The main reason that the compartment specificity is so complex and puzzling is that it must involve a whole system of additional recognition signals specifically involved in "directing traffic," i.e. determining which processing path a given protein must take. The first decision that

must be made is whether a given mRNA is to be translated by free ribosomes, with release of the finished product to the cytosol, or by membrane-
bound ribosomes, with release of the product to either the membrane itself
or to the opposite side (to the lumen of the ER or to the extracellular side
of plasma membrane), and that decision presumably could be made on the
basis of ribosome and mRNA structure before amino acid polymerization
even starts to provide the additional signals of the amino-terminal end of
the peptide.

Once a protein has been committed to a given path at this first step, there
must be a large number of similar decisions to be made throughout its
processing. Take for example a protein that is committed to the lumen of
the ER and acquires the universal precursor N-glycosyl group (Glc_3
$Man_9GlcNAc_2$) early in the biosynthetic process. By what process is it
decided whether the modified protein is to become a "high mannose"-type
glycoprotein by removing 3 Glc and 3 Man, or a "complex type" glycoprotein in which additional GlcNAc, Gal, Fuc, and sialic acid are added
to the trimmed core oligosaccharide, and whether the processed protein is
to be secreted, delivered to the lysosomes, or become a membrane component? The third specificity determinant is added to emphasize what has been
implied in the last few paragraphs, that substrate structure is continuously
changing as biosynthesis and processing proceeds. When conformational
changes and changing environments are included in the considerations of
the dynamics of the polypeptide chain elongation, the proteolytic cleavages,
and the covalent modifications that accompany the biosynthesis and transport of each protein, its combined appearance to the enzymes and pathway-
directing agents in the compartments along the way must be totally
kaleidoscopic. A given structural determinant may exist only transiently,
and if its appearance does not coincide with the presence of the appropriate
enzyme, no modification can take place.

Since conformation is also a result of amino acid sequence, the final
conclusion of these considerations is that all the specificity of all protein
processing steps ultimately is determined by the amino acid sequence of
each individual protein. The staggering amount of sequence information
involved is simply interpreted in many different ways: as the relative rates
of folding and refolding of the polypeptide chain, as the rate of transfer
through and the nature of different compartments of residence, and as the
specific exposed sequence sites that the appropriate enzyme can modify at
the appropriate time and place.

Some Enigmas

THE ROLE OF THE GLYCOSYL MOIETY OF GLYCOPROTEINS   In considering the role of the sugar moieties in glycoproteins above, it was

proposed that in general they are involved in biological communications. The basic idea was that the sugars directly participate in fundamental biological processes involving cell-cell and cell-molecule interactions, through specific interactions of receptors (e.g. lectins), which recognize and bind sugars in specific oligosaccharide sequences. There are numerous examples that suggest that sequence recognition is the key to the specificity of these processes: Several lectins require specific oligosaccharide sequences for tight binding (182); it appears that the immune system can distinguish between Manα1-2Man-X, Manα1-3Man-X, and Manα1-6Man-X (183), and the activity of many glycosyltransferases and glycosidases is exquisitely sensitive to the total composition and sequence of their substrates (179). Thus the case can be made that oligosaccharide recognition and binding are based on detailed sequence information, not only in terms of which sugars are involved in the sequence, but also in terms of anomeric configuration of the glycosidic linkage, the linkage position, and the branching ("valency") as additional specificity determinants. If this is indeed the way the oligosaccharide structures are "read" in biological communication, it should be obvious that an incredibly varied and complex "language" can be derived from a relatively simple alphabet consisting only of the common monosaccharide constituents of glycoproteins and glycolipids.

Having an additional biological language does not necessarily mean that it is always easy to discern how the communication proceeds, and one minor puzzle in this area is briefly discussed here, namely the role of the carbohydrate groups in pancreatic enzymes. There appears to be a fairly consistent picture from several species, of the processing of the ten common enzymes secreted in pancreatic juice of animals (trypsin, chymotrypsin, elastase, carboxypeptidase A and B, lipase, phospholipase A, RNase, DNase, and α-amylase). Although the data are incomplete, they reasonably support the generalization that all ten enzymes undergo proteolytic cleavage and disulfide formation in the course of their processing, but that only three of them, RNase, DNase, and α-amylase may be glycosylated to any appreciable extent. The main problem is that the extent of glycosylation of these proteins appears to be a virtually random process.

Very careful phylogenetic comparisons have been made for pancreatic RNases (180) and are used to illustrate the problem. Based on the amino acid sequence determined for RNases from 30 species, a total of 4 x-Asn-y-Ser(Thr)- sequences can be recognized in positions Asn-21, Asn-34, Asn-62, and Asn-76; any one of the 30 ribonucleases has either 1, 2, or 3 of these glycosylation sites, and each of the 4 sites has been found to be glycosylated in at least 1 species. In spite of these facts, 12 of the 30 species produce no glycosylated enzyme, 7 species glycosylate all their RNase, and the remaining 11 species produce a mixture of glycosylated and nonglycosylated

RNases. As far as can be deduced from the data, there is no difference in the amino acid sequence or in the enzymatic activity of the two forms, and furthermore, there is no obvious phylogenetic basis for the differences in glycosylation. The enigma here is best represented by the 11 species that produce the mixture of glycosylated and nonglycosylated, enzymatically active RNases. What additional specificity signals are involved in discriminating between two pools of a single protein? Since all the RNase presumably is synthesized with a leader sequence (181), which should be removed only by membrane-associated proteases, it does not seem reasonable to presume that the nonglycosylated forms are synthesized on free ribosomes away from the membranes. Is it possible that the rate of proteolysis, folding, and passage through the glycosylating compartment relative to the transferase activity permits only a fraction of the molecules to react? This almost trivial explanation inevitably leads to the next questions: Is the fractional glycosylation itself trivial, or has it been preserved as what appears to be a constant property of a given species because it does serve an important function? What might that function be? Since the enzymatic activity of the RNase is not affected by glycosylation, since nonglycosylated and glycosylated forms are being secreted with equal facility and since closely related species apparently get along equally well with or without glycosylated RNase, it seems that the function served by the glycosylation must be very highly specialized. One such specialized function has been proposed: The oligosaccharide may permit the glycosylated enzyme to be transported to and take extended residence in a specific location in the alimentary canal, but no evidence is available on this point.

THE CRYPTIC N^α-ACETYL GROUP   Since the first N^α-acetylated protein was characterized almost a quarter of a century ago (184), a large number of similarly "blocked" proteins have been identified, mostly from animals and plants (and their viruses) but also occasionally from bacteria and fungi (47, 185). It has been suggested that the majority (80%) of the soluble proteins in Ehrlich ascites cells are N^α-acetylated (186). A survey of some 40 acetylated proteins (47, 176, 185, 187–189) shows that the N^α-acetylation is not random (Ala or Ser are acetylated in the majority of known cases, Met, Gly, and Asp are found in a few cases, Asn, Ile, Thr, and Val each has been observed once, and none of the other 11 amino acids has been found acetylated so far) and that protein N^α-acetylation seems to follow a certain phylogenetic pattern. Thus, cytochrome c is always N^α-acetylated in vertebrates (N^α-AcGly) and in higher plants (N^α-AcAla), but never in yeast and bacteria. A similar pattern appears to hold for the glycolytic enzyme enolase, which is "N^α-blocked" in vertebrates (N^α-AcAla in rabbit muscle) but not in yeast (190); and in the case of the 10 pancreatic enzymes discussed above, α-amylase appears to be unique among them in having a "blocked" amino terminal in all species studied. The blocking group has been shown to be acetate in pig (191) and cow amylase (R. Uy and F. Wold, unpublished results). Both the nonrandomness of N^α-acetylations and the phylogenetic consistency suggest that that acetylation may serve an important function, and one of the enigmas of this reaction is to deduce what that function may be. The observation that a minor, nonacetylated form of actin in slime mold turns over much more rapidly than does the major, N^α-acetylated actin (192) suggests that the acetylation imparts increased stability, but no such stability enhancement was observed when the turnover of acetylated and nonacetylated forms of cat hemoglobin were compared (193). Similar contradictory observations tend to confound other suggested possible functions in biological activity, secretion, or processing, and the puzzle remains unsolved.

Another enigma in N^α-acetylations is associated with its place in the sequence of processing steps for different proteins. As a basis for this point it should be established that although all protein acetyl transferases catalyze the transfer of acetyl from acetyl CoA to proteins, the N^ϵ-acetylases are distinct from the N^α-acetylases. The former are located either in the nucleus or in the cytosol, and are specific for the ε-amino group of lysine (194), while the latter seem to be associated with ribosomes, from which they can be released by high salt, and are specific for α-amino groups (195, 196). In fact, based on the nonrandomness of N^α-acetylation as well as on specific cases studied, a given N^α-acetyl transferase must be quite specific for a single amino acid terminal of a peptide chain as the acetyl acceptor (176). With this information, it is easy to visualize the biosynthesis of an acetylated soluble protein. In the early stages of polymerization, the initiator Met is removed, and if the proper amino-terminal amino acid is now exposed, the ribosomal transferase can catalyze the acetylation, and the finished acetylated product can ultimately be released to the cytoplasm. However, in the case of secreted or membrane-transported acetylated proteins, there is an obvious dilemma. If the first step of secretion is insertion of the amino-terminal end of the nascent chain into the lumen of the ER, with the subsequent removal of the signal peptide in the lumen, the new amino terminal is on one side of the membrane and the acetyl transferase on the other side. In order to resolve this dilemma, one well-characterized secreted acetyl protein, ovalbumin (47), was investigated, and the following sequence of processing steps was established (197): the amino-terminal Met was removed when the nascent chain was about 20 residues long, and the new amino-terminal Gly was acetylated when the chain was 44 residues long. In other words, ovalbumin could perhaps be acetylated before it is delivered into the ER, and is then secreted without any subsequent peptide bond cleavage. Although this suggests that an alternate processing mechanism might be used for acetylated secreted proteins, the problem still remains, since, as already mentioned, pancreatic α-amylase is synthesized as a high-molecular-weight precursor (198, 199), which presumably must have its signal peptide removed (in the lumen of the ER) before it can be acetylated. It seems that the best way out of the present dilemma is to propose the existence of other (highly specific) N^α-acetylating enzymes in the ER or in storage granules. Such enzymes could incorporate the acetyl group any time after the removal of the signal peptide, and could also be responsible for N^α-acetylation of some of the low-molecular-weight peptide hormones that are produced via extensive proteolysis of high-molecular-weight precursors.

ACKNOWLEDGMENTS

I thank Dr. Rosa Uy for her many contributions to this review. The stimulating interest of a large number of colleagues who provided new information, both published and unpublished, is also gratefully acknowledged. The research background for parts of the review and the preparation of the review itself were supported in part by a grant (GM 15053) from the National Institutes of Health.

Literature Cited

1. Baldwin, R. L. 1975. *Ann. Rev. Biochem.* 44:453-75
2. Chou, P. Y., Fasman, G. D. 1978. *Ann. Rev. Biochem.* 47:251-76
3. Friedman, F. K., Beychok, S. 1979. *Ann. Rev. Biochem.* 48:217-50
4. Timasheff, S. N., Grisham, L. M. 1980. *Ann. Rev. Biochem.* 49:565-91
5. Hubbard, S. C., Ivatt, R. J. 1981. *Ann. Rev. Biochem.* 50:555-83
6. Whitaker, J. R. 1977. In *Food Proteins, Improvement through Chemical and Enzymatic Modifications,* ed. R. E. Feeney, J. R. Whitaker, pp. 95-155. Washington DC: Am. Chem. Soc.
7. Uy, R., Wold, F. 1977. *Science* 198: 890-96
8. Uy, R., Wold, F. 1980. In *Chemical Deterioration of Proteins,* ed. J. R. Whitaker, M. Fujimaki, pp. 49-62, Washington DC: Am. Chem. Soc.
9. Wold, F. 1980. In *Novel ADP-Ribosylations of Regulatory Enzymes and Proteins,* ed. M. Smulson, T. Sugimura, pp. 325-32. New York: Elsevier/N. Holland
10. Faulstich, H., Baku, A., Bodenmüller, H., Wieland, T. 1980. *Biochemistry* 19: 3334-42
11. Holzer, H., Heinrich, P. C. 1980. *Ann. Rev. Biochem.* 49:63-91
12. Wickner, W. 1979. *Ann. Rev. Biochem.* 48:23-45
13. Krieger, D. T., Ganong, W. F., eds. 1977. *ATCH and Related Peptides: Structure, Regulation and Action.* Ann. NY Acad. Sci. 297: 664
14. Chan, S. J., Patzelt, C., Duguid, J. R., Quinn, P., Labrecque, A., Noyes, B., Keim, P., Heinrikson, R. L., Steiner, D. F. 1979. In *From Gene to Protein: Information Transfer in Normal and Abnormal Cells,* ed. T. R. Russell, K. Brew, H. Faber, J. Schultz, pp. 361-77. New York: Academic
15. Sugino, Y., Tsunasawa, S., Yutani, K., Ogasahara, K., Suzuki, M. 1980. *J. Biochem. Tokyo* 87:351-55
16. Habener, J. F., Kemper, B., Potts, J. T. Jr., Rich, A. 1975. *Biochem. Biophys. Res. Commun.* 67:1114-21
17. Wilson, D. B., Dintzis, H. D. 1970. *Proc. Natl. Acad. Sci. USA* 66:1282-89
18. Blobel, G. 1979. See Ref. 14, pp. 347-58
19. Bonatti, S., Blobel, G. 1979. *J. Biol. Chem.* 254:12261-64
20. Bar-Nun, S., Kreibich, G., Adesnik, M., Alterman, L., Negishi, M.,

Sabatini, D. D. 1980. *Proc. Natl. Acad. Sci. USA.* 77:965–69

21. Parker, K. L., Shreffler, D. C., Capra, J. D. 1980. *Proc. Natl. Acad. Sci. USA* 77:4275–78

22. Sewarino, K. A., Poyton, R. O. 1980. *Proc. Natl. Acad. Sci. USA* 77:142–46

23. Ito, K., Mandel, G., Wickner, W. 1979. *Proc. Natl. Acad. Sci. USA* 76:1199–1203

24. Lin, J. J. C., Giam, C. Z., Wu, H. C. 1980. *J. Biol. Chem.* 255:807–11

25. Movva, N. R., Nakamura, K., Inouye, M. 1980. *J. Biol. Chem.* 255:27–29

26. Nelson, C. E., Ryan, C. A. 1980. *Proc. Natl. Acad. Sci. USA* 77:1975–79

27. Suchanek, G., Kreil, G., Hermodson, M. A. 1980. *Proc. Natl. Acad. Sci. USA* 75:701–4

28. Rosenblatt, M., Habener, J. F., Tyler, G. A., Shepard, G. L., Potts, J. T. Jr. 1979. *J. Biol. Chem.* 254:1414–21

29. Rosenblatt, M., Beaudette, N. V., Fasman, G. D. 1980. *Proc. Natl. Acad. Sci. USA* 77:3983–87

30. Prehn, S., Tsamaloukas, A., Rapoport, T. A. 1980. *Eur. J. Biochem.* 107:185–95

31. Lin, J. J., Kamazawa, H., Wu, H. C. 1980. *J. Biol. Chem.* 255:1160–63

32. Hortin, G., Boime, I. 1980. *Proc. Natl. Acad. Sci. USA* 77:1356–60

33. Fraser, T., Bruce, B. J. 1978. *Proc. Natl. Acad. Sci. USA* 75:5936–40

34. Talmadge, K., Stahl, S., Gilbert, W. 1980. *Proc. Natl. Acad. Sci. USA* 77:3369–73

35. Talmadge, K., Kaufman, J., Gilbert, W. 1980. *Proc. Natl. Acad. Sci. USA* 77:3988–92

36. Neurath, H., Walsh, K. A. 1976. *Proc. Natl. Acad. Sci. USA* 73:3825–32

37. Jackson, C. M., Nemerson, Y. 1980. *Ann. Rev. Biochem.* 49:765–811

38. Reid, K. B. M., Porter, R. R. 1981. *Ann Rev. Biochem.* 50:433–64

39. Hershko, A., Fry, M. 1975. *Ann. Rev. Biochem.* 44:775–97

40. Fessler, J. H., Fessler, L. I. 1978. *Ann. Rev. Biochem.* 47:129–62

41. Poyton, R. O., McKemmie, E. 1979. *J. Biol. Chem.* 254:6763–80

41a. Lewin, A. S., Gregor, I., Mason, T. L., Nelson, N., Schatz, O. 1980. *Proc. Natl. Acad. Sci. USA* 77:3998–4002

42. Lowry, P. J., Chadwick, A. 1970. *Nature* 226:219–24

43. Kreil, G. 1973. *FEBS Lett.* 33:241–44

44. Ogata, N., Ueda, K., Kagamiyama, H. O. 1980. *J. Biol. Chem.* 255:7616–20

45. Nath, J., Flavin, M. 1979. *J. Biol. Chem.* 254:11505–10

46. Kreil, G., Kreil-Kiss, G. 1967. *Biochem. Biophys. Res. Commun.* 27:275–80

47. Narita, K. 1975. in *Protein Structure and Function,* ed. M. Fanatsu, K. Hiromi, K. Imahori, T. Murachi, K. Narita, 2:227–59. New York: Wiley

48. Snell, E. E. 1977. *Trends Biochem. Sci.* 2:131–35

49. Kapke, G., Davis, L. 1975. *Biochemistry* 14:4273–76

50. Lin, T. S., Kolattukudy, P. E. 1980. *Eur. J. Biochem.* 106:341–51

51. Kaji, A., Kaji, H., Novelli, G. D. 1965. *J. Biol. Chem.* 240:1185–97

52. Orlowski, M., Meister, A. 1971. *Enzymes* 4:123–51

53. Braun, V., Sieglin, U. 1970. *Eur. J. Biochem.* 13:336–46

54. Seto, B. 1978. *J. Biol. Chem.* 253:4525–29

55. Satre, M., Kennedy, E. P. 1978. *J. Biol. Chem.* 253:479–83

56. Demetriou, A. A., Cohn, M. S., Tabor, C. W., Tabor, H. 1978. *J. Biol. Chem.* 253:1684–86

57. Kaji, H. 1976. *Biochemistry* 15:5121–25

58. Howard, J. B., Vermeulen, M., Swenson, R. P. 1980. *J. Biol. Chem.* 255:3820–23

59. Bunn, H. F., Haney, D. N., Gabbay, K. H., Gallop, P. M. 1975. *Biochem. Biophys. Res. Commun.* 67:103–9

60. Chen, R., Brosius, J., Wittmann-Liebold, B., Schafer, W. 1977. *J. Mol. Biol.* 111:173–81

61. Brauer, D., Wittmann-Liebold, B. 1977. *FEBS Lett.* 79:269–75

62. Lederer, P., Alix, J. H., Hayes, D. 1977. *Biochem. Biophys. Res. Commun.* 77:470–80

63. Pettigrew, G. W., Smith, G. M. 1977. *Nature* 265:661–62

64. Paik, W. K., Kim, S. 1975. *Adv. Enzymol.* 42:227–86

65. Klagsbrun, M., Furano, A. V. 1975. *Arch. Biochem. Biophys.* 169:529–39

66. Watkin, P. A., Moss, J., Vaughan, M. 1980. *J. Biol. Chem.* 255:3959–63

67. Smith, L. S., Kern, C. W., Halpern, R. M., Smith, R. A. 1976. *Biochem. Biophys. Res. Commun.* 71:459–65

68. Rogers, G. E., Harding, H. W. J., Llewellyn-Smith, I. J. 1977. *Biochim. Biophys. Acta* 495:159–75

69. Steinert, P. M., Idler, W. W. 1979. *Biochemistry* 18:5664–69

70. Allen, A. K., Neuberger, A. 1973. *Biochem. J.* 135:307–14

71. Lennarz, W. J. 1979. See Ref. 14, pp. 407–23

72. Kornfeld, R., Kornfeld, S. 1976. *Ann. Rev. Biochem.* 45:217–37
73. Cohen, R. E., Ballou, L., Ballou, C. E. 1980. *J. Biol. Chem.* 255:7700–7
74. Tabas, I., Kornfeld, S. 1980. *J. Biol. Chem.* 255:6633–39
75. Klostermeyer, H., Rabbel, K., Reimerdes, E. H. 1976. *Hoppe-Seylers Z. Physiol. Chem.* 357:1197–1202
76. Robinson, A. B., Scotchler, J. W., McKerrow, J. H. 1973. *J. Am. Chem. Soc.* 95:8156–59
77. Midelfort, C. F., Mehler, A. H. 1972. *Proc. Natl. Acad. Sci. USA* 69:1816–19
78. Van Kleef, F. S. M., deJong, W. W., Hoenders, H. J. 1975. *Nature* 258:264–66
79. Degani, C., Boyer, P. D. 1973. *J. Biol. Chem.* 248:8222–26
79a. Janson, C. A., Clarke, S. 1980. *J. Biol. Chem.* 255:11640–43
80. Helfman, P. M., Bada, J. L. 1976. *Nature* 262:279–81
81. Masters, P. M., Bada, J. L., Zigler, J. S. Jr. 1977. *Nature* 268:71–73
82. Lukens, L. N. 1976. *J. Biol. Chem.* 251:3530–38
83. Lote, C. J., Weiss, J. B. 1971. *FEBS Lett.* 16:81–85
84. Weiss, J. B., Lote, C. J., Bobinski, H. 1971. *Nature New Biol.* 234:25–26
85. Margoliash, E., Schejter, A. 1966. *Adv. Protein Chem.* 21:113–286
85a. Williams, V. P., Glazer, A. N. 1978. *J. Biol. Chem.* 253:202–11
86. Edmondson, D. E., Singer, T. P. 1976. *FEBS Lett.* 64:255–65
87. Ghisla, S., Kenney, W. C., Knapper, W. R., McIntire, W., Singer, T. P. 1980. *Biochemistry* 19:2537–44
88. Glaser, C. B., Karic, L. 1976. *Fed. Proc.* 35:1465
89. Pigiet, V., Conley, R. P. 1978. *J. Biol. Chem.* 253:1910–20
90. Tsang, M. L.-S., Schiff, J. A. 1976. *J. Bacteriol.* 125:923–33
91. Stadtman, T. C. 1980. *Ann. Rev. Biochem.* 49:93–110
92. Hantke, K., Braun, V. 1973. *Eur. J. Biochem.* 34:284–96
93. Gavaret, J.-M., Cahnmann, H. J., Nunez, J. 1979. *J. Biol. Chem.* 254:11218–22
94. Carlisle, T. L., Suttie, J. W. 1980. *Biochemistry* 19:1161–67
95. Hauschka, P. V., Lian, J. B., Gallop, P. M. 1975. *Proc. Natl. Acad. Sci. USA* 72:3925–29
96. Stenflo, J., Suttie, J. W. 1977. *Ann. Rev. Biochem.* 46:157–72
97. Van Buskirk, J. J., Kirsch, W. M. 1978. *Biochem. Biophys. Res. Commun.* 82:1329–31
98. Tuhy, P. M., Bloom, J. W., Mann, K. G. 1979. *Biochemistry* 18:5842–48
99. Koshland, D. E. Jr. 1981. *Ann. Rev. Biochem.* 50:765–82
100. Mato, J. M., Marin-Cao, D. 1979. *Proc. Natl. Acad. Sci. USA* 76:6106–9
101. Ogata, N., Ueda, K., Hayaishi, O. 1980. *J. Biol. Chem.* 255:7610–15
102. Whitby, A. J., STone, P. R., Whish, W. J. D. 1979. *Biochem. Biophys. Res. Commun.* 90:1295–1304
103. Hayaishi, O., Ueda, K. 1977. *Ann. Rev. Biochem.* 46:95–116
104. Martin, N. C., Rabinowitz, M., Fukuhara, H. 1977. *Biochemistry* 16:4672–77
105. Folk, J. E. 1979. *Ann. Rev. Biochem.* 49:517–31
106. Lhoest, J., Colson, C. 1977. *Mol. Gen. Genet.* 154:175–80
107. Swenson, R. P., Howard, J. B. 1979. *Proc. Natl. Acad. Sci. USA* 76:4313–16
107a. Tack, B. F., Harrison, R. A., Janatova, J., Thomas, M. L., Prahl, J. W. 1980. *Proc. Natl. Acad. Sci. USA* 77:5764–68
108. Chen, C. C., Bruegger, B. B., Kern, C. W., Lin, Y. C., Halpern, R. M., Smith, R. A. 1977. *Biochemistry* 16:4852–55
109. Young, V. R., Munro, H. N. 1978. *Fed. Proc.* 37:2291–300
110. Wolff, J., Covelli, I. 1969. *Eur. J. Biochem.* 9:371–77
111. Hamm, H. H., Decker, K. 1980. *Eur. J. Biochem.* 104:391–95
112. Van Ness, B. G., Howard, J. B., Bodley, J. W. 1980. *J. Biol. Chem.* 255:10710–20
113. Durban, E., Nochumson, S., Kim, S., Paik, W. K., Chan, S.-K. 1978. *J. Biol. Chem.* 253:1427–35
114. Sterner, R., Vidali, G., Allfrey, V. G. 1979. *J. Biol. Chem.* 254:11577–83
115. Chen, R., Chen-Schmeisser, U. 1977. *Proc. Natl. Acad. Sci. USA* 74:4905–8
116. Braun, V., Rehn, K., Wolff, H. 1970. *Biochemistry* 9:5041–49
117. Maloy, W. L., Bowien, B. U., Zwolinski, G. K., Kumar, K. G., Wood, H. G., Ericsson, L. H., Walsh, K. A. 1979. *J. Biol. Chem.* 254:11615–22
118. Hale, G., Perham, R. N. 1980. *Biochem. J.* 187:905–8
119. Oesterhelt, D., Stoeckenius, W. 1971. *Nature* 233:149–52
120. Tanase, S., Kojima, H., Morino, Y. 1979. *Biochemistry* 18:3002–7
121. Day, J. F., Thornburg, R. W., Thorp, S. R., Baynes, J. W. 1979. *J. Biol. Chem.* 254:9394–9400

122. Siegel, R. C., Fu, J. C. C. 1976. *J. Biol. Chem.* 251:5779–85
123. Siegel, R. C. 1976. *J. Biol. Chem.* 251:5786–92
124. Mirelman, D., Siegel, R. C. 1979. *J. Biol. Chem.* 254:571–74
125. Diedrich, D. L., Schnaitman, C. A. 1978. *Proc. Natl. Acad. Sci. USA* 75:3708–12
126. Gallop, P. M., Blumenfeld, O. O., Seifter, S. 1972. *Ann. Rev. Biochem.* 41:617–72
127. Gallop, P. M., Paz, M. A. 1975. *Physiol. Rev.* 55:418–72
128. Guay, M., Lamy, F. 1979. *Trends Biochem. Sci.* 4:160–64
129. Bornstein, P. 1974. *Ann. Rev. Biochem.* 43:567–603
130. Nakajima, T., Volcani, B. 1970. *Biochem. Biophys. Res. Commun.* 39:28–33
131. Levine, M. J., Spiro, R. G. 1979. *J. Biol. Chem.* 254:8121–24
132. Housley, T., Tanzer, M. L., Henson, E., Gallop, P. M. 1975. *Biochem. Biophys. Res. Commun.* 67:824–29
133. Fukae, M., Mechanic, G. L. 1980. *J. Biol. Chem.* 255:6511–18
134. Fujimoto, D., Moriguchi, T., Ishida, T., Hayaishi, H. 1978. *Biochem. Biophys. Res. Commun.* 84:52–57
135. Narayanan, A. S., Page, R. C. 1976. *J. Biol. Chem.* 251:1125–30
136. Lin, T.-S., Kolattukudy, P. E. 1979. *Arch. Biochem. Biophys.* 196:255–64
137. Nordwig, A., Pfab, F. K. 1969. *Biochim. Biophys. Acta* 181:52–58
138. Lamport, D. T. A. 1969. *Biochemistry* 8:1155–63
139. Miller, D. H., Lamport, D. T. A., Miller, M. 1972. *Science* 176:918–20
140. Taborsky, G. 1974. *Adv. Protein Chem.* 28:1–210
141. Krebs, E. G., Beavo, J. A. 1979. *Ann. Rev. Biochem.* 48:923–59
142. Edmondson, D. E., James, T. L. 1979. *Proc. Natl. Acad. Sci. USA* 76:3786–89
143. Vagelos, P. R. 1973. *Enzymes* 8:155–199
144. Ord, M. G., Stocken, L. A. 1977. *Biochem. J.* 161:583–92
145. Gustafson, G. L., Milner, L. A. 1980. *J. Biol. Chem.* 255:7208–10
146. Lombart, C. G., Winzler, R. J. 1974. *Eur. J. Biochem.* 49:77–86
147. Nakajima, T., Ballou, C. E. 1974. *J. Biol. Chem.* 249:7685–94
148. Muir, L., Lee, Y. C. 1969. *J. Biol. Chem.* 244:2343–49
149. Lindahl, B., Roden, L. 1972. In *Glycoproteins, Their Composition, Structure and Function,* ed. A. Gottschalk, Pt. A, pp. 491–517. New York: Elsevier
150. Schmidt, M. F. G., Schlesinger, M. J. 1980. *J. Biol. Chem.* 255:3334–39
151. Sheid, B., Pedrinan, L. 1975. *Biochemistry* 14:4357–61
152. Hallgren, P., Lundblad, A., Svensson, S. 1975. *J. Biol. Chem.* 250:5312–14
153. Noda, Y., Takai, K., Tokuyama, T., Narumiya, S., Ushiro, H., Hayaishi, O. 1977. *J. Biol. Chem.* 252:4413–15
154. Noda, Y., Takai, K., Tokuyama, T., Narumiya, S., Ushiro, H., Hayaishi, O. 1978. *J. Biol. Chem.* 253:4819–22
155. Hunt, S., Breuer, S. W. 1971. *Biochim. Biophys. Acta* 252:401–4
156. Hunt, S. 1972. *FEBS Lett.* 24:109–12
157. Wellinder, B. S. 1972. *Biochim. Biophys. Acta* 279:491–97
158. McQuillan, M. T., Trikojus, V. M. 1972. See Ref. 149, Pt. B, pp. 926–63
159. Cory, J. G., Frieden, E. 1967. *Biochemistry* 6:116–20
160. Bettelheim, F. R. 1954. *J. Am. Chem. Soc.* 76:2838–39
161. Ushiro, H., Cohen, S. 1980. *J. Biol. Chem.* 255:8363–65
162. Shapiro, B. M., Stadtman, E. R. 1968. *J. Biol. Chem.* 243:3769–71
163. Adler, S. P., Purich, D., Stadtman, E. R. 1975. *J. Biol. Chem.* 250:6264–72
164. Ambros, V., Baltimore, D. 1978. *J. Biol. Chem.* 253:5263–66
165. Tse, Y.-C., Kirkegaard, K., Wang, J. C. 1980. *J. Biol. Chem.* 235:5560–65
166. McIntire, W., Edmondson, D. E., Singer, T. P., Hopper, D. J. 1980. *J. Biol. Chem.* 255:6553–55
167. Anderson, S. O. 1963. *Biochim. Biophys. Acta* 69:249–62
168. Foerder, C. A., Shapiro, B. M. 1977. *Proc. Natl. Acad. Sci. USA* 74:4214–18
169. DeVore, D. P., Gruebel, R. J. 1978. *Biochem. Biophys. Res. Commun.* 80:993–99
170. Garcia-Castineiras, S., Dillon, J., Spector, A. 1978. *Science* 199:897–99
171. Aeschbach, R., Amado, R., Neukom, H. 1976. *Biochim. Biophys. Acta* 439:292–301
172. Paulson, J. C., Hill, R. L., Tanabe, T., Ashwell, G. 1977. *J. Biol. Chem.* 252:8624–28
173. Marcker, K. A. 1965. *J. Mol. Biol.* 14:63–70
174. van der Laken, K., Bakker-Steenevelt, H., Berkhout, B., van Knippenberg, P. H. 1980. *Eur. J. Biochem.* 104:19–24
175. Williams, R. E. 1976. *Science* 192:473–74
176. Woodford, T. A., Dixon, J. E. 1979. *J. Biol. Chem.* 254:4993–99

177. Bause, E., Lehle, L. 1979. *Eur. J. Biochem.* 101:531–40
178. Das, R. C., Brinkley, S. A., Heath, E. C. 1980. *J. Biol. Chem.* 255:7933–40
179. Beyer, T. A., Rearick, J. I., Paulson, J. C., Prieels, J.-P., Sadler, J. E., Hill, R. L. 1979. *J. Biol. Chem.* 254:12531–41
180. Beintema, J. J., Gaastra, W., Scheffer, A. J., Welling, G. W. 1976. *Eur. J. Biochem.* 63:441–48
181. Haugen, T. H., Heath, E. C. 1979. *Proc. Natl. Acad. Sci. USA* 76:2689–93
182. Goldstein, I. J., Hayes, C. E. 1978. *Adv. Carbohydr. Chem. Biochem.* 35:127–340
183. Ballou, C. E. 1970. *J. Biol. Chem.* 245:1197–1203
184. Narita, K. 1958. *Biochim. Biophys. Acta* 30:352–59
185. Dayhoff, M. O. 1976. *Atlas of Protein Sequence and Structure,* Vol. 5, Suppl. 2. Washington DC: Natl. Biomed. Res. Found.
186. Brown, J. L., Roberts, W. K. 1976. *J. Biol. Chem.* 254:525–29
187. Watterson, D. M., Scharief, F., Vanaman, T. C. 1980. *J. Biol. Chem.* 255:962–75
188. Maruyama, T., Watt, K. W. K., Riggs, A. 1980. *J. Biol. Chem.* 255:3285–3301
189. Korri, K. K., Chippel, D., Chauvin, M. M., Tirpak, A., Scrimgeour, K. G. 1977. *Can. J. Biochem.* 55:1145–52
190. Wold, F. 1971. *Enzymes* 5:499–538
191. Romano, A. T., Strumeyer, D. H. 1976. *Biochim. Biophys. Acta* 446:19–29
192. Rubenstein, P., Deuchler, J. 1979. *J. Biol. Chem.* 254:1142–47
193. Mank, M. R., Putz, G. R., Taketa, F. 1976. *Biochem. Biophys. Res. Commun.* 71:768–75
194. Sterner, R., Vidali, G., Allfrey, V. G. 1979. *J. Biol. Chem.* 254:11577–83
195. Pestana, A., Pitot, H. C. 1975. *Biochemistry* 14:1404–12
196. Pestana, A., Pitot, H. C. 1975. *Biochemistry* 14:1397–1403
197. Palmiter, R. D., Gagnon, J., Welsh, K. A. 1978. *Proc. Natl. Acad. Sci. USA* 75:94–98
198. MacDonald, R. J., Przybyla, A. E., Rutter, W. J. 1977. *J. Biol. Chem.* 252:5522–28
199. Gorecki, M., Zeelon, E. P. 1979. *J. Biol. Chem.* 254:525–29

Ann. Rev. Biochem. 1981. 50:815–43
Copyright © 1981 by Annual Reviews Inc. All rights reserved

MOLECULAR APPROACHES TO THE STUDY OF FERTILIZATION

♦12097

*Bennett M. Shapiro, Robert W. Schackmann,
and Christopher A. Gabel*

Department of Biochemistry, University of Washington, Seattle,
Washington 98195

CONTENTS

PERSPECTIVES AND SUMMARY

The union of sperm and egg is a dramatic moment for the species as well as for the biochemist, who uses it to decipher the mysteries of cell biology. Gametes are highly specialized cells with a myriad of activities waiting to be ignited by the spark of fertilization. Although we are only beginning to achieve a molecular understanding of this remarkable process, certain generalizations are emerging. Sperm and egg are quiescent until they participate in fertilization, when their metabolism is rapidly and specifically activated. Some critical effectors of metabolic activation are ions taken up or released by the cells. For example, in both gametes, alterations in in-

815

0066-4154/81/0701-0815$01.00

tracellular free calcium and pH modulate the activation response. The gametes recognize one another by receptor-ligand interactions of high specificity, and provide a very well-characterized system for the study of cell-cell interactions. Gamete membrane fusion is followed by marked alterations in the egg surface and results in the striking persistence of sperm surface components in a localized region of the embryo. After fertilization, the egg secretes a coordinated biochemical system for modification of its surface coat, both to protect the embryo and to prevent the entry of additional sperm. Biochemical mechanisms employed in this process include limited proteolytic cleavage of surface components, lectin-mediated extracellular assembly of a macromolecular complex, and covalent cross-linking by formation of dityrosine residues catalyzed by a complex peroxidative system. Thus, in just a few minutes, a coordinated system of biochemical reactions is initiated in the zygote. Studies of these events are providing new insights not only into fertilization but also into the chemistry of the regulation of complex functions in differentiated cells.

INTRODUCTION

Fertilization is a life saving event. The union of sperm and egg in rescuing these two terminal cells acts as an evolutionary springboard to launch a new individual into the future. The gametes carrying the haploid genomes that are recombined at fertilization are differentiated; their forms reflect their functions. The sperm is generally streamlined and motile, with little cytoplasm, and contains a nucleus, mitochondria, tail, and centriole; also a secretory vesicle, the acrosome, to mediate sperm-egg interaction. The large, spherical egg contains proteins, mRNA, and nutrients to carry the zygote through early development; in addition, it is armed with a battery of defensive weaponry to ensure that only one sperm fertilizes it. From a biochemist's point of view, the moments just before and after union of these gametes provide remarkably rich opportunities to study many fundamental problems in cell biology. In appropriate animals, fertilization occurs rapidly and synchronously, with copious material for study. In many cases, notably in marine invertebrates, eggs are shed in quantities equivalent to high density bacterial cultures. Additionally, the early events of fertilization occur without the need for transcription or translation, which focuses the problem on mechanisms of regulation of protein activity, rather than of protein synthesis.

BIOLOGY OF GAMETES AT FERTILIZATION

As ejaculated, many types of sperm cannot fertilize eggs, an ability that develops after removal of seminal fluid components and the occurrence of poorly characterized molecular alterations in the sperm surface. A defined change in the sperm required for fertilization follows. This is the acrosome reaction, which consists of exocytosis of the acrosomal vesicle, the release of lytic enzymes to penetrate the egg coats, and the exposure of binding components for sperm-egg interaction. During the acrosome reaction of many marine invertebrate sperm, an actin-containing filament elongates from the head of the sperm to contact the egg; the length of the filament depends on the thickness of the egg coat. After sperm-egg binding, membrane fusion occurs; in invertebrates the egg plasma membrane fuses with the tip of the acrosomal process of the fertilizing sperm, and in mammals (where no acrosomal filament is extended) sperm-egg fusion begins at the postacrosomal region of the sperm. Following membrane fusion, sperm components are incorporated into the zygote and the egg responds rapidly with an exocytosis of some thousands of cortical vesicles releasing components that alter its surface to decrease its receptivity to sperm. Activation of egg metabolism follows this cortical reaction; now the dormant egg awakens to begin a complex developmental pathway that will lead to a mature reproductive individual. Thus, the brief time window around the moment of fertilization includes several membrane fusions and metabolic activations, with many associated biochemical changes occurring in both gametes. The complexity of fertilization provides multiple sites for regulatory controls to ensure that fertilization occurs between homologous gametes with high efficiency, and with only one sperm for each egg. This carefully orchestrated sequence is the agent of both the continuation and diversity of the species.

This review defines some biochemical aspects of this complex cellular interaction by concentrating on three areas where recent insights have greatly expanded our knowledge of molecular mechanisms of sperm-egg interaction. The first area concerns the role of the ionic regulation of sperm activation prior to fertilization, and of egg activation following it. Several important events seem to be controlled by altered intracellular ionic compositions, and the evidence for this regulation is explored. The next section deals with the nature of sperm-egg interaction, specifically with the mechanisms by which gametes bind to one another and effect membrane fusion, as well as the attendant alterations in the egg plasma membrane and cytoskeleton that follow sperm entry. The final topic concerns the modifications in the egg surface that occur after fertilization, when a complex structure is assembled outside of the zygote from preformed elements of the egg, by employing discretely activated catalytic functions. Thus, our principal concern is with areas where understanding of biochemical mechanisms that are close to our research interests are beginning to emerge. Other recent reviews deal with more general aspects of fertilization (1–4) or provide specific insights into the mechanism in invertebrates (5–7) and mammals (8, 9).

Additionally, we do not discuss the nature and role of lytic enzymes that are released by the sperm during the acrosome reaction, as this subject has been extensively reviewed (10, 11).

IONIC MECHANISMS IN GAMETE ACTIVATION

Triggering of the Acrosome Reaction

INDUCING AGENTS The components of the female that elicit the acrosome reaction are beginning to be defined. In echinoderms the reactions occur when sperm reach the egg surface (12, 13); the best candidate for the triggering agent in sea urchins is a fucose-sulfate polymer of the hydrated egg surface "jelly" (14–16). Although the purified material requires Ca^{2+} in excess of physiological levels for in vitro triggering, the sulfated polymer retains the species specificity expected for the reaction (15), which suggests that this relatively simple sugar polymer has multiple structural isomers. When sperm encounter egg jelly, both exocytosis and acrosomal filament polymerization occur within seconds (17, 18). The mammalian acrosome reaction consists only of an exocytosis and requires several hours for completion in a population of sperm (19, 20). The site in the female tract where the reaction occurs is not defined: ultrastructural studies in the mouse suggest that sperm undergo the reaction after binding to the zona pellucida (ZP) (21), but other studies (22) have found acrosome-reacted sperm distributed throughout the female tract. In part this difficulty may relate to the problem of distinguishing a "true" mammalian acrosome reaction (23). In addition, the physiological trigger of the mammalian acrosome reaction is not defined, although albumin (24, 25) and catecholamines (20, 26, 27) are among the candidates, for they are effective in vitro.

Ca^{2+} REQUIREMENT In her pioneering studies of the mechanism of the sea urchin acrosome reaction, Dan showed that depletion of Ca^{2+} (< 2 mM) from seawater prevented the reaction (17, 28) and suggested that Ca^{2+} influx was required. It has subsequently been demonstrated that an absolute requirement for external Ca^{2+} for induction of the reaction exists for both echinoderms (18, 29–34) and mammalian (19, 34–36) sperm. Several drugs that inhibit Ca^{2+} binding or transport prevent the egg jelly–induced acrosome reaction of sea urchin sperm. Thus, local anesthetics like xylocaine and procaine (31), and Ca^{2+} channel inhibitors like verapamil and D600 (16, 29), prevent the reaction and act competitively with Ca^{2+}. Several other observations suggest that Ca^{2+} influx is needed. First, the acrosome reaction is induced by the divalent cation ionophore A23187, only in the presence of extracellular Ca^{2+} (31, 32). Second, increased Ca^{2+} uptake (as shown by $^{45}Ca^{2+}$ accumulation) is seen in sea urchin sperm after triggering of the acrosome reaction (16, 18, 29). Much of this Ca^{2+} uptake is into the mitochondria (M. Cantino and R. W. Schackmann, unpublished data), and most occurs after the reaction is completed, but some $^{45}Ca^{2+}$ uptake is found even when mitochondrial function is inhibited. In guinea pig sperm, $^{45}Ca^{2+}$ uptake occurs during capacitation, prior to induction of the acrosome reaction (35). Removal of Ca^{2+} seems to prevent the primary interaction of the egg jelly with sperm (18), and to block subsequent events in the reaction. Taken together, the data suggest that Ca^{2+} acts at several sites in the pathway of acrosome-reaction triggering (18, 37).

pH CONTROLS   Dan (38) first demonstrated that increased seawater pH induced the echinoderm acrosome reaction in the absence of egg jelly; further studies showed that at low pH ($<$ 7.0) the acrosome reaction is prevented, while at elevated pH ($>$ 8.5) the reaction occurs spontaneously (12, 18, 29–33, 38). Extracellular pH alterations may affect the intracellular pH (R. Christen, unpublished data) or the membrane permeability to Ca^{2+} (31, 32). Other data suggest that intracellular pH variations affect the acrosome reaction. Ionophores that catalyze the exchange of metal ions for H^+ (X537A, A23187, nigericin) cause actin polymerization and H^+ efflux in starfish sperm (30): Ca^{2+} is required for exocytosis, but not for H^+ efflux or actin polymerization. Similarly, acid is released upon triggering of the acrosome reaction in several echinoderms by egg jelly (18, 29–32). That this acid efflux is accompanied by an increased intracellular pH is suggested by a nigericin-induced release of 9-aminoacridine (9AA) from sea urchin sperm, even when the overall reaction is inhibited by Ca^{2+} removal (R. W. Schackmann, unpublished result). Nigericin also induces 9AA release from mammalian sperm (39). As an unprotonated amine, 9AA passes through membranes and accumulates according to a pH gradient (40, 41); hence, some intracellular compartment of the sperm seems to become more basic when the reaction is triggered. By fluorescence microscopy, 9AA fluorescence is localized around the acrosome vesicle of sea urchin, starfish (R. W. Schackmann, unpublished results), and hamster (39) sperm.

Treatment of sea urchin sperm with egg jelly also induces $^{22}Na^+$ uptake; the kinetics and stoichiometry suggest a link between Na^+ uptake and H^+ efflux (18). Both $^{22}Na^+$ uptake and H^+ efflux occur at the same time as extension of the acrosomal filament, whereas Ca^{2+} accumulation is slower (28, 29). That Na^+ uptake is required for H^+ efflux is further suggested by the discovery that neither induction of the acrosome reaction nor H^+ efflux occur with addition of egg jelly to sperm suspended with choline instead of Na^+, despite the fact that $^{45}Ca^{2+}$ accumulates. The acrosome reaction is then spontaneously triggered (without egg jelly) when 20 mM Na^+ is added back (18).

CYCLIC NUCLEOTIDES   Cyclic nucleotides appear to function in the acrosome reaction, although their roles have not been clearly defined. cAMP levels increase in sea urchin sperm after they are exposed to egg jelly (42). The cAMP analogues dibutyryl cAMP and 8-bromo-cAMP (8BrcAMP) and several phosphodiesterase inhibitors either stimulate (43, 44) or inhibit (45) the acrosome reaction in mammalian sperm, and the reasons for the different results are not clear. Dibutyryl cAMP plus theophylline inhibit $^{45}Ca^{2+}$ uptake by both boar and human sperm (46). Additionally, dibutryl cGMP and 8BrcGMP also elicit the acrosome reaction in guinea pig sperm (47), but cGMP is not found in these cells at appreciable levels (44).

There are at least two factors in sea urchin egg jelly that affect sperm cyclic nucleotide levels (16, 48–50). One is the fucose-sulfate polymer responsible for triggering the acrosome reaction. It elicits a Ca^{2+}-dependent increase in intracellular cAMP and increased cAMP-dependent protein kinase activity (16, 48), but has no effect on cGMP levels or on respiration. Another pronase-sensitive low-molecular-weight factor from egg jelly stimulates both cAMP and cGMP levels in a Ca^{2+}-*independent* fashion (49), leads to enhanced sperm respiration, and stimulates the oxidation of long- and medium-chain fatty acids (50). As is clear from the above, the definition of the biochemical responses of sperm to specific components is at an early stage, for complex interactions occur between sperm metabolism, Ca^{2+} fluxes, and cyclic nucleotide levels. This may refect the fact that the activation of the sperm is a complex process, not only because of the acrosome reaction, but also because of increased motility [for a discussion of the latter see (51)].

Activation of the Egg

The ionic events in egg activation can be subdivided into three major areas: (a) The fertilization or activation potential, which is a change in the egg membrane potential following fertilization or parthenogenic activation. This typically includes a rapid (~ one second) depolarization of the (initially) negative membrane potential to a positive value, followed by a gradual (within a few minutes) repolarization or hyperpolarization; (b) The transient increase in intracellular Ca^{2+} that precedes cortical granule exocytosis with the concomitant release of cortical granule contents to change the egg surface and block the entry of additional sperm; and (c) The increase in intracellular pH, which is associated with the activation of protein and DNA synthesis and the chromosome cycle.

ACTIVATION POTENTIAL AND OTHER ELECTRICAL CHANGES   The first physiological changes after fertilization are electrical ones. A rapid depolarization of the egg membrane potential is found in many species, including sea urchins (52–56), the echiuroid worm, *Urechis caupo* (57, 58), amphibians (59–62), teleost fish (63, 64), and starfish (65, 66). Although the magnitude and the ionic bases for the potential changes vary from one species to another, in the best studied cases, the echinoids (sea urchins and starfish) and echiuroid worm, the depolarization is largely dependent upon an increased membrane permeability to Na^+ and Ca^{2+} (55, 57, 58, 65, 66). Although in *Urechis* the membrane potential remains positive for 10 min instead of 2 min as in echinoids, in both cases there is a gradual return, lasting a few minutes, to a negative membrane potential. In sea urchins this is largely the result of increased K^+ conductance (56, 57, 68). In unfertilized sea urchin eggs with an initial membrane potential of –60 mV, fertilization takes the egg from –60 mV to +10 mV and then back to –60 mV; the final level in the fertilized egg is characterized by an approximately eightfold increase in the membrane conductance for K^+ (55, 56, 67, 68). In starfish (65) and in *Urechis* (57) the increases in K^+ conductance are smaller and the repolarization presumably results from closure of Na^+ and Ca^{2+} channels.

There remains substantial disagreement about the resting potentials in sea urchin eggs prior to fertilization. Aside from technical problems in accurate measurement, the discrepancies are in part accounted for by there being essentially two populations of eggs with different resting potentials in a number of species (52–56, 67, 68). One population shows resting potentials in the unfertilized egg of –10 to –20 mV; the other has resting potentials in the range of –60 mV. Some workers (67, 68) believe that the correct value is about –60 mV for all eggs, and that –10 to –20 mV values reflect electrical leaks around the electrodes. While this is reasonable, only 1% of freshly isolated *Paracentrotus* eggs show the value of –60 mV (53, 54), and this fraction increases as the eggs age, so the difference may have physiological significance. In any event, both classes of eggs depolarize upon fertilization, to values of +10 to +20 mV.

In amphibians, fertilization also causes a rapid depolarization to a positive membrane potential (59, 62), apparently due to an increase in Cl^- permeability (59). The return to a negative membrane potential occurs over 20–30 min. In the teleost fish, medaka, the depolarization is quite small ($\sim$ 4mV), followed by a hyperpolarization that is partially dependent upon external K^+ (63).

In echinoid, echiuran, and amphibian eggs, the rapid depolarization seems to be involved in a rapid block to polyspermy (52, 58, 59, 65, 66), for eggs that have been voltage clamped at a depolarized value by passage of current, or by manipulation of the external ionic medium, fail to fertilize. However, when the voltage clamp is terminated and the membrane potential becomes more negative, the eggs can be fertilized. Moreover, in starfish eggs the ability to achieve monospermic fertilization correlates with the magnitude of depolarization obtained: immature and overripe eggs, which become polyspermic at fertilization, do not give sufficiently positive or rapid depolarizations (66). However, in medaka, fertilization occurs over a wide range of membrane potentials (−80 mV to +40 mV), and there does not appear to be a rapid electrical block to polyspermy (64). Since fertilization occurs at only one point on the medaka egg surface, there may not be the same requirement for the rapid polyspermy block as found with eggs that have sperm penetration over the entire sphere. Despite many observations on the correlation between the depolarization at fertilization and the early block to polyspermy in *Paracentrotus* and *Psammechinis* species of sea urchins, no primary depolarizations larger than 1–2 mV were found (53, 54), and these were thought to be due directly to sperm penetration. The true activation potential (the depolarization phase) followed only after a substantial period of time, as long as 13 sec. While these data were interpreted (53) to mean that no rapid polyspermy block is established immediately following sperm penetration, no attempt was made to see whether positive membrane potentials prevented fertilization in these eggs.

Ca^{2+} AND THE CORTICAL REACTION   Over 40 years ago, experiments of Moser (69) and Mazia (70) suggested that Ca^{2+} was involved in activation of the egg, a hypothesis that has recently been vigorously pursued. Several lines of evidence implicate a causal role for free intracellular Ca^{2+} as a mediator of the exocytosis of the cortical vesicles. First, the divalent ionophore, A23187, elicits the cortical reaction in sea urchins (71, 72), amphibians, and mammals (73) even in the absence of extracellular Ca^{2+}. Second, isolated preparations of cortical vesicles attached to plasma membranes of sea urchin eggs lyse after Ca^{2+} addition, and local anesthetics inhibit exocytosis (74, 75). The clearest evidence for the involvement of Ca^{2+} comes from experiments using the Ca^{2+}-sensitive photoprotein, aequorin, which emits light in the presence of free Ca^{2+}. In fertilized eggs of medaka, that had been injected with aequorin, a flash of light propagated as a wave starting from the micropyle, the site of sperm penetration, and advanced around the spherical egg in a wave that preceded cortical vesicle exocytosis (76). The luminescence decreased relatively rapidly, which suggests that fertilization results in a transient increase in free Ca^{2+} after which Ca^{2+} is resequestered. A transient emission of light is also found in aequorin-loaded eggs of the sea urchin *L. pictus* after fertilization or activation with A23187, in the presence or absence of external Ca^{2+} (77). Additionally, injection of

EGTA into sea urchin eggs prevents cortical granule exocytosis, in agreement with the idea that a rise in free Ca^{2+} is associated with the exocytosis of the cortical granules (78). Attempts to measure the Ca^{2+} levels necessary for exocytosis in isolated cortices yielded values of 9–18 μM, approximately four times that resulting from fertilization; this suggests that Ca^{2+} release may be confined to a localized area of the egg, presumably in the cortex, where the vesicles lie (77). In a separate study, eggs made transiently permeable to Ca^{2+} gave vesicle exocytosis at 1 μM free Ca^{2+}, which is in reasonable agreement with the quantitative aequorin estimates (79–81).

Many treatments that lead to parthenogenetic activation of eggs (urea, hypertonic seawater, and A23187) cause release of Ca^{2+} from intracellular sites, whereas activation by increasing the intracellular pH (see below) does not (82). With isotopic tracers, increases in both uptake and release of Ca^{2+} were found at fertilization (83–86), with net efflux predominating in sea urchins (84). The intracellular store that releases Ca^{2+} has not been identified, but it seems to require ATP to be filled (79). Since Ca^{2+} accumulation has been found in a microsomal fraction of sea urchin eggs (79, 87), a structure analogous to sarcoplasmic reticulum might be operative in Ca^{2+} uptake and release in eggs, but no direct evidence for this has been obtained. Both eggs (88) and sperm (89) have high concentrations of calmodulin, so that a Ca^{2+}-calmodulin complex may mediate some of the Ca^{2+}-dependent events in the cortical reaction. Once sperm have undergone the acrosome reaction in media containing Ca^{2+}, they can fertilize at much lower Ca^{2+} concentrations (91, 92). An attractive hypothesis is that Ca^{2+} accumulated in the sperm during the acrosome reaction (29) may be released into the egg at fertilization to trigger the cortical reaction, via Ca^{2+}-stimulated Ca^{2+} release, as is found in sarcoplasmic reticulum (90). Ionophoretic injection of Ca^{2+} into mouse eggs results in both cortical vesicle exocytosis and cell division (93). In sea urchins, topical application of A23187 results in only a limited cortical vesicle exocytosis (94) that does not propagate, which indicates that localized Ca^{2+} influx may not be sufficient to continue the reaction. Clearly, more needs to be known about the regulation of the propagation of the Ca^{2+}-dependent exocytosis.

INTRACELLULAR pH INCREASE   The "fertilization acid" released after insemination (95) was suggested to be due to the release of protons rather than of weak acids (96, 97), and is associated with $^{22}Na^+$ uptake and an increase in the pH of egg homogenates from 6.5 to 6.7 (98). Based upon the stoichiometry of Na^+ influx to H^+ efflux and the ability of amiloride to inhibit both ion movements (98), a $Na^+:H^+$ exchange through amiloride-sensitive Na^+ channels was implicated in the mechanism. However, subsequent studies failed to find inhibition by amiloride (99, 100), although at high concentrations it retarded the rate of H^+ efflux. The cell homogenization techniques for estimating cytoplasmic pH have several serious difficulties (101), but surprisingly the pH estimates thus obtained are in reasonable agreement with those found with intracellular pH electrodes (102): a pH shift, from 6.8 to 7.2, occurs following fertilization. When eggs are activated with A23187 in the absence of Na^+, no intracellular pH shift occurs until Na^+ is replaced (100), which further implicates a role for extracellular Na^+ in the intracellular pH increase. As initially defined by Chambers (103), Na^+ is essential only during a short period (< 10 min) following fertilization, and only low concentrations (5 mM) are needed to support the change in intracellular pH (100), which suggests that Na^+ is required to activate a system that regulates the intracellular pH, rather than to participate in a $Na^+ : H^+$ exchange (100).

Evidence that increased intracellular pH activates the egg has also come from the use of weak bases (NH_4, procaine, etc). These agents activate protein synthesis without inducing the cortical reaction, and are thought to increase intracellular pH (104–106) via the transmembrane diffusion of the uncharged amine. Just as increasing the intracellular pH stimulates protein synthesis, decreasing intracellular pH inhibits it (107). Similar responses of protein synthesis have been found in lysates from sea urchin eggs, where pH shifts from 6.8 to 7.3 increased the rate of protein synthesis in vitro (108), but this may only reflect the pH-activity curve for some component in the system, and not the physiological control mechanism.

Other metabolic activations that respond to the pH increase are increased K^+ conductance (109, 110), changes in membrane permeability to Cl^- and urea (111, 112), altered protein phosphorylation patterns in egg homogenates (113), and increased DNA synthesis and chromosome cycling (114). Regulation by changes in intracellular pH has been recently suggested to occur in other systems as diverse as the cellular response to insulin and the control of bacterial growth (115–117).

SPERM-EGG INTERACTION

The Binding of Sperm to Egg

The fertilizing sperm encounters the egg's jelly coat, then an extracellular glycoprotein coat, the vitelline layer (VL), before reaching the plasma membrane (reviewed in 118–124). Although many sperm bind to each egg (119, 120, 125), only one fuses with the plasma membrane (PM) to initiate development. Eggs rapidly fixed after insemination contain many sperm adhering to their VL (119–123) via an extracellular coat that covers the acrosomal filament in echinoids (17, 124) and molluscs (126). The interaction of acrosome-reacted sperm with the VL of an egg is referred to as primary gamete binding (123), and is species specific, which shows greater selectivity than the triggering of the acrosome reaction (123, 127, 128). For example, eggs of the sea urchin *Lytechinus veriegatus* cause homologous sperm as well as those from three other species to undergo normal acrosomal reactions, but only the homologous sperm are able to bind to the VL. Primary gamete binding accounts for much of the species-specificity of fertilization, which suggests that a receptor exists in the VL for a specific sperm determinant.

Aketa et al (129) isolated an acid-soluble fraction from sea urchin sperm that reduced the fertilizability of homologous, but not heterologous, eggs by preventing sperm binding. When analyzed by cellulose acetate electrophoresis (129), the purified factor contained a single component with 5% protein and 95% carbohydrate. The location of this glycoprotein on the sperm is unknown. Using a different approach, Vacquier & Moy extracted sperm with Triton X-100, then isolated granular material from the acrosomal vesicle to examine its egg binding ability. SDS-gel electrophoresis of the isolated, purified material showed it to be a single, carbohydrate-free protein, "bindin," with an apparent molecular weight of 30,500 (130). That bindin might play a role in the binding of sperm to eggs was suggested by its species-specific agglutination of unfertilized eggs (131, 132). The subcellular distribution of bindin was determined by immunoperoxidase electron microscopy with rabbit antibindin serum (133). Bindin was present over the acrosomal filament of acrosome-reacted sperm, a location compatible with the protein's proposed role.

Since the VL of unfertilized echinoid eggs has been implicated as the site of sperm binding, it should contain structures complementary to the sperm factors described above. Aketa et al (134) found that an antiserum prepared against a 1 M urea extract of unfertilized sea urchin eggs (a treatment known to disrupt the VL) prevented sperm from fertilizing homologous eggs. The responsible component of the 1 M urea extract was purified to homogeneity (as assessed by cellulose acetate electrophoresis) and shown to be a high-molecular-weight, amido black positive component (135). The urea-solubilized factor neither prevented sperm motility nor induced the acrosome reaction; it did, however, inhibit fertilization in a species-specific fashion (136), which suggests that the solubilized factor bound to the sperm surface to prevent subsequent interaction with the egg. Sepharose beads covalently attached to the factor bound homologous but not heterologous sperm (137), which further suggests that the factor was a sperm receptor. Unlike its soluble counterpart, the factor attached to Sepharose induced the acrosome reaction in the bound sperm (137). The factor was localized to the surface of unfertilized eggs by indirect immunofluorescence microscopy and found to be trypsin sensitive (138), consistent with its localization in the VL, which is destroyed by limited trypsin digestion (118, 139). The isolated egg factor interacted with the sperm factor, and the inhibitory effects on fertilization were neutralized only if factors from the same species were employed (129), which suggests that the two are complementary agents in gamete interaction.

Another approach to identifying and isolating a sperm receptor from the egg surface has utilized membrane fractions of unfertilized sea urchin eggs (140), which specifically inhibit fertilization by homologous gametes, but not by heterologous ones (140). Trypsin treatment of eggs prior to preparation of the membrane fraction destroyed this inhibitory activity, and a solubilized form of the membrane receptor for sperm bound to concanavalin A-Sepharose. More recently, a surface complex was isolated from unfertilized sea urchin eggs that contained the VL, the PM, and the attached cortical granules (141), and this complex bound only homologous sperm. Attempts at purifying the receptor have not yet succeeded.

An egg receptor for sea urchin bindin has also been purified (142). The isolated receptor is trypsin sensitive, contains neutral sugars (mannose and galactose), and has a molecular weight of $> 5 \times 10^6$. The bindin receptor was released from eggs that were artificially activated with the ionophore A23187 in the presence of soybean trypsin inhibitor, which was included to inhibit the trypsin-like enzyme released from the egg at fertilization (143). In the absence of soybean trypsin inhibitor, no bindin receptor activity was recovered. Although the purified material associates with bindin, it has not been shown to either bind to sperm or have any effect on fertilization. Thus, the classes of molecules that mediate sperm-egg interaction in invertebrates are beginning to be defined, and some of their properties known, but the interaction is a complex one and is probably mediated at several sites.

Mammalian eggs are contained within a glycoprotein envelope, the zona pellucida (ZP), which is analogous to the VL of invertebrates (144, 145). Although little is known of the chemistry of mammalian sperm-egg binding, sperm interact with the ZP in a species-specific fashion (146–148). Heterologous sperm neither bind to the ZP nor fertilize intact eggs (147, 149), but after removal of the ZP, penetration by heterologous sperm occurs (150–153). Solubilized ZP preparations inhibit the binding of capacitated hamster sperm to eggs and block fertilization in vitro (8, 154), which suggests that they contain a sperm receptor; recently a specific component has been implicated (273; see below).

The Fusion of Sperm and Egg

After binding, the echinoid sperm penetrates the VL, apparently with the aid of a protease that is activated at the time of the acrosome reaction (155,

156). Penetration of the egg's outer vestment is mediated by the acrosomal filament, an impressive example of which is found in the sea cucumber, where sperm attach to an outer layer of the egg 90 μm away from the plasma membrane (PM), which is then approached by the elongating acrosomal filament (157). A similar situation exists in other invertebrates, where the distance between the sperm binding site and egg PM is not so great (158).

The juxtaposition of sperm and egg membranes, initiated by contact of the acrosomal filament tip with the egg PM, affords the opportunity for cell-cell fusion. That fusion of the gamete membranes actually occurs was shown by electron microscopy of fertilization in annelids (159), enteropneusts (160), sea urchins (161, 162), and molluscs (24, 162), where the gamete membranes were shown to be continuous at the point of contact between the acrosomal filament and the egg's PM. Sperm of the domestic fowl, *Gallus gallus,* fuse with the egg in an analogous fashion (163). Sperm and eggs of mammalian species also undergo fusion of their membranes, but fusion is not initiated at the apex of the sperm head. Rather, after the acrosomal cap has been lost, the sperm head interacts tangentially with the egg, and continuity of the gamete membranes first is seen at the postacrosomal or equatorial segment of the sperm (164–171). In all cases, sperm-egg fusion occurs between membranes that were either exposed or altered during the acrosome reaction (172, 173). The cryptic localization of the membrane fusion sites prior to gamete contact may be necessary if these surfaces have an inherent "fusibility" (6), in order to prevent unwanted fusions between gametes and other cells, and this may be why the acrosome reaction occurs proximate to the egg. This notion is supported by observations that acrosome-reacted sea urchin sperm fuse with each other (174) and that acrosome-reacted mouse sperm can spontaneously fuse with mouse fibroblasts (175).

NATURE OF THE GAMETE MEMBRANES   Very little information exists on the mechanism of sperm-egg fusion, but similarities exist between gamete fusion and fusion of other cell membranes, so some effort has been made to define the biochemical characteristics of gamete surfaces prior to fusion. Sperm are asymmetric cells with regional heterogeneity in lectin (176–183) and fluorochrome (184) binding, charge distribution (185), antigen placement (180, 186–189), and in the presence and distribution of intramembraneous particles (190–192). This topographic heterogeneity of the sperm plasma membrane suggests that there are constraints on the movement of membrane components, which is supported by the limited lateral mobility found in studies of lectin- and antibody-induced clustering of sperm membrane components (181, 187). Exceptions to this mobility restriction are few (183, 193).

Fusion occurs between lipid bilayers only above the phase transition temperature of the component phospholipids (194, 195), and a similar temperature effect is seen in viral-induced cell fusion (196) and myoblast fusion (197, 198). When the temperature dependence of membrane fusion was studied in hamster gametes, sperm bound to the PM of eggs at 4°C, but fusion did not occur until the temperature was elevated to 37°C (199). This is a crude assessment, complicated by the many temperature-sensitive processes in gametes (e.g. sperm mobility, respiration), but it is consistent with the notion that appropriate membrane fluidity is required. Surfaces devoid of intramembranous particles have been found in fusing membranes, as for example during exocytosis (201–204) or myoblast fusion (205). Similar regions are found in the fusion site of guinea pig (173) but not hamster (200) sperm only after induction of the acrosome reaction, but the significance of these sites in all cases is not clear.

The unfertilized egg has a relatively symmetric distribution of microvilli (206, 207), intramembranous particles (208), and lectin binding sites (176, 209, 210). Estimates of mouse egg plasma membrane mobility by fluorescence photobleaching recovery (FPR) for a fluorescent lipid, dioctadecylindocarbocyanine iodine (diI-C18), gives a diffusion coefficient of 1.91×10^{-8} cm^2 s^{-1} (211), a value similar to that found for other mammalian cells (212). The mobility of lipid probes in unfertilized sea urchin eggs measured by FPR varied according to the length of the acyl side chain (213), from 2 to 6×10^{-9} cm^2 s^{-1}, and the fraction of the probe that was diffusible ranged from 50–80%, which suggests that sea urchin egg PM may possess distinct domains of differing lipid states (213).

In both echinoid (92) and mammalian (214) fertilization, Ca^{2+} is needed not only for the induction of the acrosome reaction but also for sperm-egg fusion. A Ca^{2+} requirement for membrane fusion is found in myoblasts (215) and in viral-induced cell fusion (216). From studies with phospholipid vesicles (217, 218), Ca^{2+} seems to induce crystalline domains of acidic phospholipids that may destabilize the bilayer, thereby generating sites at which membrane fusion could occur (219).

FUSION AS A REQUISITE FOR ACTIVATION   Some inhibitors that have been used in an attempt to define the sequence of steps in sperm-egg interaction appear to affect the process of gamete binding and fusion. For example, cytochalasins B (CB) or D seem to inhibit fertilization just after membrane fusion.

Pretreatment of eggs from several invertebrates with CB inhibits sperm pronuclear incorporation (220–222) but does not affect either metabolic activation or the cortical or acrosomal reactions. When fluorescent sperm were used to fertilize sea urchin eggs, (thus providing a convenient and sensitive assay for the presence of sperm components after fertilization), 50% of the eggs that elevated a fertilization membrane possessed a fluorescent sperm associated with their surfaces, even though the normal pathway of pronuclear decondensation did not occur (223). When cytochalasin B was washed away, the pronucleus migrated to an internal site and became decondensed. This suggests that the cytoskeleton plays a role, after sperm-egg fusion, in internalization of sperm components. Cytochalasins have similar inhibitory effects in other systems in not blocking fusion itself [e.g. of Sendai virus (224) with the host plasma membrane, or of phospholipid vesicles with cultured cells (225)], but rather in inhibiting the actin-dependent microfilament system (226).

Another inhibitor that acts during the gamete interaction phase is erythrosin B. Sea urchin eggs fertilized in the presence of erythrosin B bind sperm but do not become activated (227). The fluorochrome does not affect sperm motility, respiration, or sperm agglutination, but the effect on acrosomal filament formation was not determined. Erythrosin-treated eggs neither elevate fertilization membranes nor contain sperm pronuclei, which suggests that gamete fusion is blocked and that sperm binding is not sufficient to activate unfertilized sea urchin eggs.

Postfertilization Alterations of the Gamete Surfaces

FATE OF SPERM SURFACE COMPONENTS IN THE ZYGOTE   After membrane fusion, sperm cytoplasmic components enter the egg. The male pronucleus decondenses and interacts with the egg pronucleus to form the zygote nucleus (228, 229). The centrioles participate in the formation of the sperm aster (230, 231), and sperm mitochondria and axomenes are found in the zygote cytoplasm during the early cleavages (159, 161, 231–234). However, neither the molecular fates of these structures nor the mechanisms by which those of the sperm interact with those of the egg are defined.

Based on the behavior of membranes following fusion of other cell types (235), one might expect sperm surface components to diffuse laterally within the plane of the zygote membrane to form a uniform mosaic surface. The zygote surface, however, undergoes a rigidification after fertilization (211) that could limit the lateral diffusion of its membrane components, thereby resulting in a heterogeneous distribution of sperm and egg components. Several approaches have been utilized to determine the distribution of sperm surface components. Yanagimachi et al (236), using a colloidal iron hydroxide labeling technique, reported that sperm and egg membrane components intermix following fusion; the data, however, only show the egg membrane components surrounding the sperm and do not localize the sperm components themselves. Early embryos have been treated with antibodies prepared against specific sperm determinants to localize sperm surface components immunologically. Despite several attempts with different antisera (237–239), there has been no successful localization of sperm antigens immediately after fertilization. In many instances, the sperm antigens are expressed later during development, possibly due to the de novo synthesis of the antigens by the zygote (240).

Another approach for tracing the fate of the sperm surface after fertilization involves the covalent modification of sperm surface components prior to fertilization, and localization of the labeled surface components after fusion by following the attached probe. Fluorescein isothiocyanate (FITC) and its radioactive congener, [^{125}I]diiodofluorescein isothiocyanate (^{125}IFC), covalently label surface components of the sperm midpiece (184). Following fertilization of mouse or sea urchin eggs by fluorescently labeled sperm, the labeled components are transferred to the zygote and persist as a distinct patch (241). However, autoradiographic analysis of sections of embryos fertilized with ^{125}IFC sperm revealed that the labeled components are localized intracellularly following fertilization (223) at a cortical site. This suggests that components of the sperm surface may be internalized following initial gamete fusion; such an internalization would also explain the inaccessibility of many sperm surface antigens following fertilization.

ALTERATIONS OF THE EGG CORTEX   The most dramatic change in the egg cortex following fertilization is the exocytosis of the cortical granules and, in many invertebrates, the elevation of the fertilization membrane; the details of this process are described below. The insertion of the cortical granule membrane into the PM results in a doubling in the number of concanavalin-A receptors (242) and of apparent surface area (206). As the diameter of the zygote is the same as the unfertilized egg, the increase in surface area is accomplished by an increase in the length of the surface microvilli (206, 207). Proteolytic activity increases at the egg surface after fertilization, following the release of trypsin-like proteases from the dehiscing cortical granules (146, 243, 244). The proteases act in the removal of sperm receptors (143, 244) and in the limited proteolysis (245) of some egg surface components.

Mobility measurements of components of the egg PM indicate that a change in membrane fluidity occurs postfertilization. By incorporation of a spin label fatty acid, 5-doxylstearate, into sea urchin eggs, it was shown by electron spin resonance (ESR) that an increase in membrane fluidity occurs following fertilization (246, 247). However, the ESR probe partitioned into lipids throughout the cell, and the increase in fluidity reflected changes in the total zygote and not just in the plasma membrane. This increase in bulk membrane fluidity was independent of the cortical reaction, as similar changes were observed upon treatment of eggs with ammonia, a parthenogenetic activator that does not elicit cortical granule exocytosis (247).

In contrast, measurements of the mobility of plasma membrane components suggest that a *decrease* in fluidity occurs following fertilization. Incorporation of 5-doxylstearate into cortices isolated from both unfertilized and fertilized sea urchin eggs indicated that a decrease in membrane mobility occurs with fertilization (248). Addition of Ca^{2+} to unfertilized egg cortices induced a similar reduction in mobility. No consistent change occurred in the diffusion constants of fluorescent lipid probes (derivatives of 3,3'-diacylindocarbocyanine iodides) at fertilization of sea urchin eggs (213); however, a consistent decrease in the percentage of freely diffusible molecules was found. The situation in the mouse zygote is more pronounced, for a significant reduction in the fraction of mobile label occurred after fertilization for both a lipid and protein probe (211), and there was a 10–100-fold reduction in the measured diffusion coefficients for both probes. Thus, the surface of the mouse zygote undergoes a tremendous reduction in fluidity following fertilization.

The significance and mechanisms by which the reduction in membrane fluidity occur are unclear. The mobility of PM components in other cell types is, to some extent, controlled by underlying cytoskeletal elements (microfilaments, microtubules, and, in the red blood cell, spectrin) (249–252). The reduction in egg membrane mobility after fertilization is accompanied by a general increase in the rigidity or stiffness of the egg cortex (253, 254). In addition, following fertilization of sea urchin eggs, there is a dramatic rearrangement of cytoskeletal components; the distribution and amount of actin associated with the zygote cortex changes (255–260), there is an aggregation of cortical fibers (261), and the microtubules undergo a profound reorganization (262). Therefore, the reduction in membrane fluidity after fertilization and increased cortical rigidity may result from a reorganization of the zygote cytoskeleton.

MODIFICATION OF THE EGG SURFACE COAT

The secretion products released during the cortical reaction alter the properties of the egg coat both to protect the embryo and to prevent the penetration of additional sperm. In general, the egg surface coats are multimeric assemblies of heterogeneous glycoproteins. An interesting exception is found in the keyhole limpet, *Megathura crenulata;* in its egg the vitelline envelope (VE) contains polypeptide chains consisting principally of threonine residues (61 mol percent) almost all of which are attached to carbohydrate (263). In this structure, 20 mol percent of the residues were proline and hydroxyproline, which further suggests that a repeating unit exists. Since the VE was solubilized in dithiothreitol (264), it is probably crosslinked by disulfide bonds. High-threonine glycoproteins are also found in the VE of *Tegula pfefferi* (265) and the jelly coat of *Bufo vulgaris* (266), but these seem to be specialized classes of molecules not generally found in vitelline or fertilization envelopes.

Modifications in egg coat components at fertilization are general phenomena, and seem to exhibit certain similarities in molecular mechanisms between different animals. For example, a decreased solubility of the surface coats occurs after fertilization of mammals (267, 268), amphibians (269), and echinoderms (5). The change in solubility of the mammalian zona pellucida at fertilization appears to be associated with limited proteolytic cleavage of zona pellucida proteins (270), perhaps caused by a trypsin-like protease released from the cortical granules (271). Although it is difficult to obtain much material for analysis, the specific molecular composition of the ZP of several mammalian eggs is being studied. The mouse ZP accounts for 17% of the oocyte protein and is composed of only three major components of 200, 120, and 83 K_d (272), all of which are synthesized by the maturing oocyte (273). The 83-K_d protein appears to be the sperm receptor, and loses sperm binding properties after fertilization (273). The porcine ZP also has three to four major classes of proteins by size, but exhibits marked microheterogeneity in composition (274).

Limited Proteolysis

A cortical vesicle protease of sea urchin eggs (275–277) is implicated in limited proteolytic cleavage of the VE (245) during the time that it is being converted into the modified, hardened fertilization membrane (FE). Likewise, a component of the VE of the amphibian egg is modified at fertilization in a fashion consistent with its being cleaved by limited proteolysis (278). Thus, cortical vesicle proteases may play several roles after fertilization, in removing sperm binding sites (see above) as well as in modifying the glycocalyx. In fact, distinct proteases were implicated in these two functions after a partial purification of sea urchin egg secretion product (243, 279). A more highly purified preparation of the active enzyme exists as a single serine protease (280) of 22,500 daltons.

Lectin-Mediated Fertilization Envelope Assembly

The molecular mechanism of the interaction of released cortical vesicle components with molecules of the egg surface has been most clearly demonstrated at fertilization of the amphibian, *Xenopus laevis* (209, 281–284). The *Xenopus* egg is surrounded by a vitelline layer adjacent to the plasma membrane and a jelly coat of three layers (J_1–J_3) each with a distinct glycoprotein composition (21). After fertilization, a sulfated glycoprotein of J_1 interacts with a component released from the cortical vesicles (282, 283) to form a fertilization layer (F layer) at the interface of the VE and J_1, which blocks the entry of additional sperm. Isolated cortical vesicle and J_1 layer components agglutinate in a reaction involving α-galactoside residues (assessed by competition) and Ca^{2+} (282) that is chemically and phylogenetically specific. Thus, a cortical vesicle lectin is implicated in the interaction with a sulfated ligand of the J_1 layer (284) in a complex reaction with the VE that leads to FE production and a concomitant block to sperm entry. The FE contained the same antigenic and electrophoretic components as the VE, plus those from the cortical vesicles and J_1 (278), but FE prepared in the absence of Ca^{2+} contained only VE components, which implicates Ca^+ in the assembly process.

The FE of the sea urchin embryo assembles from molecules released from the cortical vesicle and VE components in a defined sequence of events that are susceptible to inhibition at different steps (209). Assembly includes the deposition of components on the top and bottom of the existing VE fibrillar network and degradation of preexisting VE fibers (285). The biochemical events occurring during assembly of the sea urchin FE have been refractory to study until recently, for the structure is extensively cross-linked by peroxidase-dependent dityrosine formation (286–289), as discussed below. Some attempts have been made to analyze the composition of the cross-linked FE to determine the number of components involved in the structure. Under optimal conditions (290) (100°C, pH 10 in 6 M urea: 1.5 M mercaptoethanol) up to 70% of the protein is extracted, but no unequivocal evidence was provided that peptide bonds were not cleaved during extraction, aside from the lack of marked heterogeneity in FE composition. The hardened cross-linked FE contained three glycoproteins (of 91.6, 71.2, and 53 kilodaltons) and two other proteins (of 32.6, and 18.2 kilodaltons) as determined by SDS-gel electrophoresis. By inhibiting the peroxidase responsible for cross-linking the FE with aminotriazole, as discussed below, "soft," uncross-linked fertilization membranes can be isolated and their properties examined (291; E. Kay, E. Turner, P. Weidman, and B. M. Shapiro, unpublished data). Seven major proteins are seen by SDS-gel electrophoresis of disaggregated soft fertilization membrane, of approximately 150, 120, 96, 70, 51, 31, and 29.5 kilodaltons and some minor components. Preliminary evidence suggests that at least the 50- and 120-kilodalton components are secretion products of the cortical vesicles; the latter appears to be cross-linked by dityrosine residues in the FE (B. M. Shapiro, unpublished data), and the former to be the ovoperoxidase (see below).

These soft FE are stabilized by divalent cations and ionic strength (291), so that whereas in seawater they retain their structure, in distilled water they rapidly disaggregate ($t_{1/2} \cong 10$ min). In this disaggregation they become filmy and wraith-like, an alteration that can be followed by a decrease in light scattering, or by the release of protein complexes from the soft FE. This disaggregation, termed "wraithing," is blocked by Ca^{2+} and Mg^{2+} at concentrations below those in seawater. The effect of Ca^{2+} is of interest because a Ca^{2+}-dependent paracrystalline component has been described in the cortical vesicle secretion product (292, 293). Thus, although less well-characterized, sea urchin FE assembly is similar to that in the amphibian egg. In both, Ca^{2+} modulates the interactions of specific components of the cortical vesicle secretion product and the egg cell coat, although no evidence of a lectin-ligand type of interaction has yet been obtained in the sea urchin.

Dityrosine Cross-Linking

Although it had been known for over 70 years that the FE of the sea urchin was stable to many disruptive agents (reviewed in 5, 7, 209; 294) the mechanism of this resistance has only recently been demonstrated to be caused by dityrosine cross-links that "harden" the FE (286, 287). These cross-links are synthesized after fertilization in a reaction catalyzed by an ovoperoxidase that is released from the cortical granules and inserted into the assembling FE (286, 287, 289). One dityrosine residue is found for each 50,000–100,000 daltons of hardened FE protein. All inhibitors of the ovoperoxidase block the hardening reaction, and the classically studied inhibitors of hardening are peroxidase inhibitors at equivalent concentrations (286, 287). The peroxidase is localized in the cortical granules before fertilization, and in the fertilization membrane afterwards (289). The oxidizing substrate, hydrogen peroxide, is synthesized by the egg (288) in a burst after fertilization. This burst of hydrogen peroxide synthesis accounts for at least two thirds of the oxygen uptake first described by Warburg (295) as the archetype of metabolic activation of eggs.

The occurrence of dityrosine as the cross-linking residue in hard fertilization membranes is not the first description of this amino acid. Dityrosine can be synthesized from free tyrosine (296, 297), or from surface tyrosyl residues of many globular proteins and collagen (298, 299) in similar peroxidase-catalyzed reactions. Dityrosine is also produced when tyrosine or tyrosyl-containing polypeptides or proteins are subjected to ultraviolet or ionizing radiation (300, 301). When galactose oxidase is treated with horseradish peroxidase, its activity increases with the production of dityrosine residues (302), which was suggested to occur at lower (and variable) levels as an intramolecular cross-link even in the native enzyme.

Dityrosine is naturally found in many types of structural proteins, including the insect elastic protein, resilin (303), the adhesive disc of mussels (304), *Bacillus subtilis* spore coats (305), *Tussah* silk fibroin and keratin (306), and human lens proteins (307, 308). It is often seen in various classes of mammalian connective tissue proteins (309–313), although in these cases the residue is probably not present in sufficient quantities to be physiologically important as a cross-link. It is also found in *Drosophila* (314) and dragon fly (315) egg envelopes.

Peroxidative System of Fertilization

Although dityrosine has been found in other proteins, including those from egg coats, in none of these cases is the biosynthetic pathway as clearly defined as in the sea urchin fertilization envelope, where a complex peroxidative system exists to catalyze its synthesis (286–289). The hypothetical pathway for synthesis of dityrosine cross-links is shown in Equations 1 and 2:

$$NAD(P)H + H^+ + O_2 \rightarrow NAD(P)^+ + H_2O_2 \qquad 1.$$

$$H_2O_2 + \text{peptide A}(tyr_{An}) + \text{peptide B } (tyr_{Bn}) \rightarrow$$

$$\text{Peptide A}(try_{An} - tyr_{Bn})\text{peptide B} + 2 H_2O \qquad 2.$$

Equation 2 is supported by data given above (286, 287), but reaction 1 is speculative, and is advanced for several reasons. Hydrogen peroxide is made by fertilized sea urchin eggs (288). The reason for suggesting that NAD(P)H is the reductant is inferential, based upon several similarities between the peroxidative system of the sea urchin egg and that of the polymorphonuclear leucocyte (PMN) when it begins phagocytosis (289). The PMN responds to foreign stimuli with a phagocytic reaction using activated oxygen species and a myeloperoxidase (reviewed in 316, 317). In both the egg and PMN, there is an increase in oxygen consumption (289, 295, 318) and concomitant production of H_2O_2 (288, 319). Both cells contain peroxidases in cytoplasmic granules, the myeloperoxidase of the PMN, and the ovoperoxidase of the sea urchin egg (289, 320), and both enzymes are released in response to specific stimuli. With both peroxidative systems, there are chemilluminescence (288, 320), increased hexose monophosphate shunt activity (318, 321), and other physiological similarities (289). In the PMN, the hexose monophosphate shunt may produce the substrate for a transiently active NAD(P)H oxidase (316, 317, 322–324) that is present in the particulate fraction of the PMN and catalyzes the synthesis of superoxide anion. A similar enzyme system may exist for the sea urchin egg: Equation 1 suggests the possibility of a membrane-bound NAD(P)H oxidase that would be transiently active after fertilization to produce active oxygen intermediates.

The myeloperoxidase of the PMN (325) along with other peroxidase reactions (326) effect potent bacteriocidal processes and are spermicidal in model systems (327). Additionally, H_2O_2 itself is toxic for both echinoderm (328) and mammalian (329) sperm. Thus, the peroxidative system of the egg may play a role in addition to hardening the fertilization membrane, that of killing supernumerary sperm and acting as another block to polyspermy. This is made more likely in view of the fact that mammalian sperm are rich in polyunsaturated long-chain fatty acids (330) that can be oxidized to spermicidal fatty acid hydroperoxides (331–334). There are suggestions that similar peroxidative systems may exist in brine shrimp oocytes (335) and mammalian eggs (336). Thus, activated oxygen species provide potent intermediates for biochemical redox reactions (337–339), and both PMN's and sea urchin eggs have evolved a complex system for generating such intermediates outside of the cell. By the production of such activated redox intermediates and the use of appropriate coupling reactions, these cells are able to carry on complex molecular transformations in an appropriately defined environment. The regulation of the systems is a most interesting subject for study, since the oxidases are transiently activated, and the peroxidase reactions [dityrosine synthesis, bacterial (and sperm?) killing] are well aimed in order to accomplish these potent reactions in an appropriate place. One factor in the targeting of the sea urchin egg reaction is the faithful placement of the ovoperoxidase in the fertilization membrane (289; B. M. Shapiro and E. Kay, unpublished data) by a coordinated assembly reaction, but the other regulatory aspects of this process remain to be explored.

CONCLUSIONS

The impression we take from the limited molecular studies on the biochemistry of fertilization is that we are in a position equivalent to the study of metabolism almost a century ago. Certain of the biochemical reactions in the overall process are clear, and from the properties of the system we can infer some aspects of others. Yet we are still in no position to design an overall scheme that is analogous to the pathway of glycolysis or of the tricarboxylic acid cycle. Nonetheless, such a possibility is clearly anticipated, for the events in the fertilization sequence seem carefully orchestrated, with defined regulation and systematic programing, as in well-regulated metabolic pathways. Behind the apparent chaos of interspecies differences and multiple rapid and unidirectional reactions, we can begin to discover the outline of a pattern. Ultimately, we should be able to write the entire biochemical pathway for this complex process, so central to the propagation and evolution of metazoan species.

Literature Cited

1. Austin, C. R. 1975. *J. Reprod. Fertil.* 44:155–66
2. Longo, F. J. 1973. *Biol. Reprod.* 9:149–215
3. Moscona, A. A., Monroy, A., eds. 1978. *Curr. Top. Dev. Biol.* 12:1–254
4. Shapiro, B. M., Eddy, E. M. 1980. *Int. Rev. Cytol.* 66:257–303
5. Giudice, G. 1973. *Developmental Biology of the Sea Urchin Embryo.* New York: Academic. 469 pp.
6. Epel, D., Vacquier, V. D. 1978. *Cell Surface Rev.* 5:1–63
7. Czihak, G., ed. 1975. *The Sea Urchin Embryo: Biochemistry and Morphogenesis,* Berlin & New York: Springer. 700 pp.
8. Gwatkin, R. B. L. 1977. *Fertilization Mechanisms in Man and Mammals.* New York: Plenum. 161 pp.
9. Bedford, J. M., Cooper, G. W. 1978. See Ref. 6, pp. 65–125
10. McRorie, R. A., Williams, W. L. 1974. *Ann. Rev. Biochem.* 43:777–803
11. Stambaugh, R. 1978. *Gamete Res.* 1:65–85
12. Dan, J. C. 1967. In *Fertilization: Comparative Morphology, Biochemistry and Immunology,* ed. C. B. Metz, A. Monroy, 1:237–93. New York: Academic. 489 pp.
13. Afzelius, B. A., Murray, A. 1957. *Exp. Cell Res.* 12:325–37
14. Hotta, K., Hamazaki, H., Kurokawa, M. 1970. *J. Biol. Chem.* 245:5434–40
15. Segall, G. K., Lennarz, W. J. 1979. *Dev. Biol.* 71:33–48
16. Kopf, G. S., Garbers, D. L. 1980. *Biol. Reprod.* 22:1118–26
17. Dan, J. C., Ohori, Y., Kushida, H. 1964. *J. Ultrastruct. Res.* 11:508–24
18. Schackmann, R. W., Shapiro, B. M. 1981. *Dev. Biol.* 81:145–54

19. Yanagimachi, R., Usui, N. 1974. *Exp. Cell Res.* 89:161–74
20. Cornett, L. E., Meizel, S. 1978. *Proc. Natl. Acad. Sci. USA* 75:4954–58
21. Saling, P. M., Sowinski, J., Storey, B. T. 1979. *J. Exp. Zool.* 209:229–38
22. Overstreet, J. W., Cooper, G. W. 1979. *J. Exp. Zool.* 209:97–103
23. Bedford, J. M. 1969. *Adv. Biosci.* 4:35–50
24. Lui, C. W., Meizel, S. 1977. *Differentiation* 9:59–66
25. Lui, C. W., Cornett, L. E., Meizel, S. 1977. *Biol. Reprod.* 17:34–41
26. Meizel, S., Working, P. K. 1980. *Biol. Reprod.* 22:211–16
27. Cornett, L. E., Bavister, B. D., Meizel, S. 1979. *Biol. Reprod.* 20:925–29
28. Dan, J. C. 1954. *Biol. Bull. Woods Hole, Mass.* 107:335–49
29. Schackmann, R. W., Eddy, E. M., Shapiro, B. M. 1978. *Dev. Biol.* 65:483–95
30. Tilney, L. G., Kiehart, D. P., Sardet, C., Tilney, M. 1978. *J. Cell Biol.* 77:536–50
31. Collins, F., Epel, D. 1977. *Exp. Cell Res.* 106:211–22
32. Decker, G. L., Joseph, D. B., Lennarz, W. J. 1976. *Dev. Biol.* 53:115–25
33. Gregg, K. W., Metz, C. B. 1976. *Biol. Reprod.* 14:405–11
34. Talbot, P., Summers, R. G., Hylander, B. L., Keough, E. M., Franklin, L. E. 1976. *J. Exp. Zool.* 198:383–92
35. Singh, J. P., Babcock, D. F., Lardy, H. A. 1978. *Biochem. J.* 172:549–56
36. Russell, L., Peterson, R. N., Freund, M. 1979. *Fertil. Steril.* 32:87–92
37. Gregg, K. W. 1979. *Biol. Reprod.* 20:338–45
38. Dan, J. C. 1952. *Biol. Bull. Woods Hole, Mass.* 103:54–66
39. Meizel, S., Deamer, D. W. 1978. *J. Histochem. Cytochem.* 26:98–105

40. Schuldiner, S., Rottenberg, H., Avron, M. 1972. *Eur. J. Biochem.* 25:64–70
41. Deamer, D. W., Prince, R. C., Crofts, A. R. 1972. *Biochim. Biophys. Acta* 274:323–35
42. Garbers, D. L., Hardman, J. G. 1975. *Nature* 257:677–78
43. Mrsny, R. J., Meizel, S. 1980. *J. Exp. Zool.* 211:153–57
44. Hyne, R. V., Garbers, D. L. 1979. *Proc. Natl. Acad. Sci. USA* 76:5699–703
45. Rogers, B. J., Garcia, L. 1979. *Biol. Reprod.* 21:365–72
46. Peterson, R. N., Seyler, D., Bundman, D., Freund, M. 1979. *J. Reprod. Fertil.* 55:385–90
47. Santos-Sacchi, J., Gordon, M. 1980. *J. Cell Biol.* 85:798–803
48. Garbers, D. L., Tubb, D. J., Kopf, G. S. 1980. *Biol. Reprod.* 22:526–32
49. Kopf, G. S., Tubb, D. J., Garbers, D. L. 1979. *J. Biol. Chem.* 254:8554–60
50. Hansbrough, J. R., Kopf, G. S., Garbers, D. L. 1980. *Biochim. Biophys. Acta* 630:82–91
51. Hoskins, D. D., Casillas, E. R. 1975. In *Handbook of Physiology*, ed. S. R. Geiger, Sect. 7, 5:453–60. Baltimore: Waverly Press. 519 pp.
52. Jaffe, L. A. 1976. *Nature* 261:68–71
53. DeFelice, L. J., Dale, B. 1979. *Dev. Biol.* 72:327–41
54. Taglietti, V. 1979. *Exp. Cell Res.* 120:448–51
55. Chambers, E. L., deArmendi, J. 1979. *Exp. Cell Res.* 122:203–18
56. Steinhardt, R. A., Mazia, D. 1973. *Nature* 241:400–1
57. Jaffe, L. A., Gould-Somero, M., Holland, L. 1979. *J. Gen. Physiol.* 73:469–92
58. Gould-Somero, M., Jaffe, L. A., Holland, L. Z. 1979. *J. Cell Biol.* 82:426–40
59. Cross, N. L., Elinson, R. P. 1980. *Dev. Biol.* 75:187–98
60. Elinson, R. P. 1975. *Dev. Biol.* 47:257–68
61. Ito, S. 1972. *Dev. Growth Differ.* 14:217–27
62. Maeno, T. 1959. *J. Gen. Physiol.* 43:139–57
63. Nuccitelli, R. 1980. *Dev. Biol.* 76:483–98
64. Nuccitelli, R. 1980. *Dev. Biol.* 76:499–504
65. Miyazaki, S. 1979. *Dev. Biol.* 70:341–54
66. Miyazaki, S., Hirai, S. 1979. *Dev. Biol.* 70:327–40
67. Hagiwara, S., Jaffe, L. A. 1979. *Ann. Rev. Biophys. Bioeng.* 8:385–416
68. Jaffe, L. A., Robinson, K. R. 1978. *Dev. Biol.* 62:215–28
69. Moser, F. 1939. *J. Exp. Zool.* 80:448–71
70. Mazia, D. 1937. *J. Cell Comp. Physiol.* 10:291–304
71. Chambers, E. L., Pressman, B. C., Rose, B. 1974. *Biochem. Biophys. Res. Commun.* 60:126–32
72. Steinhardt, R. A., Epel, D. 1974. *Proc. Natl. Acad. Sci. USA* 71:1915–19
73. Steinhardt, R. A., Epel, D., Carroll, E. J., Yanagimachi, R. 1974. *Nature* 252:41–3
74. Vacquier, V. D. 1975. *Dev. Biol.* 43:62–74
75. Detering, N. K., Decker, G. L., Schmell, E. D., Lennarz, W. J. 1977. *J. Cell Biol.* 75:899–914
76. Ridgway, E. B., Gilkey, J. C., Jaffe, L. F. 1977. *Proc. Natl. Acad. Sci. USA* 74:623–27
77. Steinhardt, R., Zucker, R., Schatten, G. 1977. *Dev. Biol.* 58:185–96
78. Zucker, R. S., Steinhardt, R. A. 1978. *Biochim. Biophys. Acta* 541:459–66
79. Baker, P. F., Whitaker, M. J. 1978. *Nature* 276:513–15
80. Baker, P. F., Whitaker, M. J. 1979. *Nature* 279:820–21
81. Zucker, R. S., Steinhardt, R. A. 1979. *Nature* 279:820
82. Zucker, R. S., Steinhardt, R. A., Winkler, M. M. 1978. *Dev. Biol.* 65:285–95
83. Nakazawa, T., Asami, K., Shoger, R., Fujiwara, A., Yasumasu, I. 1970. *Exp. Cell Res.* 63:143–46
84. Arzania, R., Chambers, E. L. 1976. *J. Exp. Zool.* 198:65–78
85. Paul, M., Johnston, R. N. 1978. *J. Exp. Zool.* 203:143–49
86. Johnston, R. N., Paul, M. 1977. *Dev. Biol.* 57:364–74
87. Nakamura, M., Yasumasu, I. 1974. *J. Gen. Physiol.* 63:374–88
88. Head, J. F., Mader, S., Kaminer, B. 1979. *J. Cell Biol.* 80:211–18
89. Jones, H. P., Bradford, M. M., McRorie, R. A., Cormier, M. J. 1978. *Biochem. Biophys. Res. Commun.* 82:1264–72
90. Katz, A. M., Repke, D. I., Dunnett, J., Hasselbach, W. 1977. *J. Biol. Chem.* 252:1950–56
91. Takahashi, Y. M., Sugiyama, M. 1973. *Dev. Growth Differ.* 15:261–67
92. Sano, K., Kanatani, H. 1980. *Dev. Biol.* 78:242–46
93. Fulton, B. P., Whittingham, D. G. 1978. *Nature* 273:149–51
94. Chambers, E. L., Hinkley, R. E. 1979. *Exp. Cell Res.* 124:441–46
95. Runnström, J. 1933. *Biochem. Z.* 258:257–79

96. Rothschild, L. 1956. *Fertilization* pp. 65–66. London: Methuen. 170 pp.
97. Mehl, J. W., Swann, M. M. 1961. *Exp. Cell Res.* 22:233–45
98. Johnson, J. D., Epel, D., Paul, M. 1976. *Nature* 262:661–64
99. Cuthbert, A., Cuthbert, A. W. 1978. *Exp. Cell Res.* 114:409–15
100. Shen, S. S., Steinhardt, R. A. 1979. *Nature* 282:87–89
101. Needham, J. 1931. *Chemical Embryology* 2:839–55. New York: Macmillan. 1253 pp.
102. Shen, S. S., Steinhardt, R. A. 1978. *Nature* 272:253–54
103. Chambers, E. L. 1976. *J. Exp. Zool.* 197:149–54
104. Winkler, M. M., Grainger, J. L. 1978. *Nature* 273:536–38
105. Boron, W. F., Roos, A., DeWeer, P. 1979. *Nature* 274:190
106. Hutchens, J. O., Krahl, M. E., Clowes, G. H. A. 1939. *J. Cell. Comp. Physiol.* 14:313–25
107. Grainger, J. L., Winkler, M. M., Shen, S. S., Steinhardt, R. A. 1979. *Dev. Biol.* 68:396–406
108. Winkler, M. M., Steinhardt, R. A. 1978. *J. Cell Biol.* 79(2)Pt.2:171a (Abstr.)
109. Shen, S. S., Steinhardt, R. A. 1980. *Exp. Cell Res.* 125:55–61
110. Tupper, J. T. 1974. *Dev. Biol.* 38:332–45
111. Christen, R., Sardet, C., Lallier, R. 1979. *Cell Biol. Int. Rep.* 3:121–28
112. Christen, R., Sardet, C. 1980. *J. Physiol.* 305:1–11
113. Keller, C. H., Gundersen, G. G., Shapiro, B. M. 1980. *Dev. Biol.* 74:86–100
114. Epel, D., Steinhardt, R., Humphreys, T., Mazia, D. 1974. *Dev. Biol.* 40:245–55
115. Fidelman, M. L., Moore, R. D., Seeholzer, S. H. 1980. *Fed. Proc. Fed. Am. Soc. Exp. Biol.* 39(6):1925 (Abstr.)
116. Moore, R. D. 1979. *Biochem. Biophys. Res. Commun.* 91:900–4
117. Plack, R. H., Rosen, B. P. 1980. *J. Biol. Chem.* 255:3824–25
118. Runnström, J. 1966. *Adv. Morphog.* 5:221–325
119. Vacquier, V. D., Payne, J. E. 1973. *Exp. Cell Res.* 82:227–35
120. Vacquier, V. D., Tegner, M. J., Epel, D. 1972. *Nature* 240:352–53
121. Tegner, M. J., Epel, D. 1973. *Science* 179:685–88
122. Schatten, G., Mazia, D. 1976. *J. Supramol. Struct.* 5:343–69
123. Summers, R. G., Hylander, B. L. 1976. *Exp. Cell Res.* 100:190–94
124. Summers, R. G., Hylander, B. L. 1974. *Cell Tissue Res.* 150:343–68
125. Tegner, M. J., Epel, D. 1976. *J. Exp. Zool.* 197:31–58
126. Hylander, B. L., Summers, R. G. 1977. *Cell Tissue Res.* 182:469–89
127. Summers, R. G., Hylander, B. L. 1975. *Exp. Cell Res.* 96:63–68
128. Metz, C. B. 1978. *Curr. Top. Dev. Biol.* 12:107–47
129. Aketa, K., Miyazaki, S., Yoshida, M., Tsuzuki, H. 1978. *Biochem. Biophys. Res. Commun.* 80:917–22
130. Vacquier, V. D., Moy, G. W. 1977. *Proc. Natl. Acad. Sci. USA* 74:2456–60
131. Glabe, C. G., Vacquier, V. D. 1977. *Nature* 267:836–38
132. Glabe, C. G., Lennarz, W. J. 1979. *J. Cell Biol.* 83:595–604
133. Moy, G. W., Vacquier, V. D. 1979. *Curr. Top. Dev. Biol.* 13:31–44
134. Aketa, K., Onitake, K. 1969. *Exp. Cell Res.* 56:84–86
135. Tsuzuki, H., Yoshida, M., Onitake, K., Aketa, K. 1977. *Biochem. Biophys. Res. Commun.* 76:502–11
136. Aketa, K. 1973. *Exp. Cell Res.* 80:439–41
137. Aketa, K., Yoshida, M., Miyazaki, S., Ohta, T. 1979. *Exp. Cell Res.* 123:281–84
138. Yoshida, M., Aketa, K. 1978. *Acta Embryol. Exp.* 3:269–78
139. Epel, D. 1970. *Exp. Cell Res.* 61:69–70
140. Schmell, E., Earles, B. J., Breaux, C., Lennarz, W. J. 1977. *J. Cell Biol.* 72:35–46
141. Decker, G. L., Lennarz, W. J. 1979. *J. Cell Biol.* 81:92–103
142. Glabe, C. G., Vacquier, V. D. 1978. *Proc. Natl. Acad. Sci. USA* 75:881–85
143. Vacquier, V. D., Tegner, M. J., Epel, D. 1973. *Exp. Cell Res.* 80:111–19
144. Austin, C. R. 1961. *The Mammalian Egg.* Oxford: Blackwell Sci. Publ. 183 pp.
145. Piko, L. 1969. In *Fertilization: Comparative Morphology, Biochemistry and Immunology,* ed. C. B. Metz, A. Monroy. 2:325–403. New York: Academic. 553 pp.
146. Franklin, L. E., Barros, C., Fussell, E. N. 1970. *Biol. Reprod.* 3:180–200
147. Hartmann, J. F., Gwatkin, R. B. L., Hutchison, C. F. 1972. *Proc. Natl. Acad. Sci. USA* 69:2767–69
148. Hartmann, J. F., Gwatkin, R. B. L. 1971. *Nature* 234:479–81
149. Yanagimachi, R. 1977. *Clin. Exp. Immunoreprod.* 4:255–89

150. Yanagimachi, R., Yanagimachi, H., Rogers, B. J. 1976. *Biol. Reprod.* 15:471–76
151. Barros, C., Berrios, M., Herrera, E. 1973. *J. Reprod. Fertil.* 34:547–49
152. Yanagimachi, R. 1972. *J. Reprod. Fertil.* 28:477–80
153. Hanada, A., Chang, M. C. 1972. *Biol. Reprod.* 6:300–9
154. Gwatkin, R. B. L., Williams, D. T. 1976. *J. Reprod. Fertil.* 49:55–9
155. Levine, A. E., Walsh, K. A., Fodor, E. J. B. 1978. *Dev. Biol.* 63:299–306
156. Levine, A. E., Walsh, K. A. 1979. *Dev. Biol.* 72:126–37
157. Colwin, L. H., Colwin, A. L. 1956. *Biol. Bull. Woods Hole, Mass.* 110:243–57
158. Colwin, L. H., Colwin, A. L. 1967. See Ref. 12, pp. 295–367
159. Colwin, A. L., Colwin, L. H. 1961. *J. Biophys. Biochem. Cytol.* 10:231–74
160. Colwin, L. H., Colwin, A. L. 1963. *J. Cell Biol.* 19:501–18
161. Franklin, L. E. 1965. *J. Cell Biol.* 25:81–100
162. Pasteels, J. J. 1965. *Arch. Biol. (Liège)* 76:463–509
163. Okamura, F., Nishiyama, H. 1978. *Cell Tissue Res.* 190:89–98
164. Shalgi, R., Phillips, D. 1980. *J. Ultrastruct. Res.* 71:154–61
165. Stefanini, M., Oura, C., Zamboni, L. 1969. *J. Submicrosc. Cytol.* 1:1–23
166. Barros, C., Herrera, E. 1977. *J. Reprod. Fertil.* 49:47–50
167. Barros, C., Franklin, L. E. 1968. *J. Cell Biol.* 37:c13–c18
168. Szollosi, D. G., Ris, H. 1961. *J. Biophys. Biochem. Cytol.* 10:275–83
169. Yanagimachi, R., Noda, Y. D. 1970. *J. Ultrastruct. Res.* 31:486–93
170. Yanagimachi, R., Noda, Y. D. 1970. *Am. J. Anat.* 128:429–62
171. Bedford, J. M. 1972. *Am. J. Anat.* 133:213–53
172. Koehler, J. K. 1976. *Biol. Reprod.* 15:444–56
173. Friend, D. S., Orci, L., Perrelet, A., Yanagimachi, R. 1977. *J. Cell Biol.* 74:561–77
174. Collins, F. 1976. *Dev. Biol.* 49:381–94
175. Siroky, J., Spurna, V., Kopecny, V., Tkadlecek, L. 1979. *J. Exp. Zool.* 208:245–54
176. Aketa, K. 1975. *Exp. Cell Res.* 90:56–62
177. Edelman, G. M., Millette, C. F. 1971. *Proc. Natl. Acad. Sci. USA* 68:2436–40
178. Gall, W. E., Millette, C. F., Edelman, G. M. 1974. In *Physiology and Genetics of Reproduction*, ed. E. M. Coutinho, F.

179. Fuchs, A:241–57. New York: Plenum. 417 pp.
179. Kinsey, W. H., Koehler, J. K. 1976. *J. Supramol. Struct.* 5:185–98
180. Koehler, J. K. 1978. *Int. Rev. Cytol.* 54:73–108
181. Millette, C. F. 1977. *Clin. Exp. Immunoreprod.* 4:51–71
182. Nicolson, G. L., Usui, N., Yanagimachi, R., Yanagimachi, H., Smith, J. R. 1977. *J. Cell Biol.* 74:950–62
183. Nicolson, G. L., Yanagimachi, R. 1974. *Science* 184:1294–96
184. Gabel, C. A., Eddy, E. M., Shapiro, B. M. 1979. *J. Cell Biol.* 82:742–54
185. Yanagimachi, R., Noda, Y. D., Fujimoto, M., Nicolson, G. L. 1972. *Am. J. Anat.* 135:497–520
186. Fellous, M., Gachelin, G., Buc-Caron, M. H., Dubois, P., Jacob, F. 1975. *Dev. Biol.* 41:331–37
187. Koehler, J. K. 1975. *J. Cell Biol.* 67:647–59
188. Koo, G. C., Stackpole, C. W., Boyse, E. A., Hammerling, U., Lardis, M. P. 1973. *Proc. Natl. Acad. Sci. USA* 70:1502–5
189. Metz, C. B. 1967. See Ref. 12, pp. 163–236
190. Friend, D. S. 1977. *Clin. Exp. Immunoreprod.* 4:5–30
191. Friend, D. S., Fawcett, D. W. 1974. *J. Cell Biol.* 63:641–64
192. Koehler, J. K., Gaddum-Rosse, P. 1975. *J. Ultrastruct. Res.* 51:106–18
193. O'Rand, M. G. 1977. *Dev. Biol.* 55:260–70
194. Papahadjopoulos, D., Poste, G., Schaeffer, B. E. 1973. *Biochim. Biophys. Acta* 323:23–42
195. Papahadjopoulos, D., Poste, G., Schaeffer, B. E., Vail, W. J. 1974. *Biochim. Biophys. Acta* 352:10–28
196. Bächi, T., Aguet, M., Howe, C. 1973. *J. Virol.* 11:1004–12
197. Van Der Bosch, J., Schudt, C., Pette, D. 1973. *Exp. Cell Res.* 82:433–38
198. Prives, J., Shinitzky, M. 1977. *Nature* 268:761–63
199. Hirao, Y., Yanagimachi, R. 1978. *J. Exp. Zool.* 205:433–38
200. Kinsey, W. H., Koehler, J. K. 1978. *J. Ultrastruct. Res.* 64:1–13
201. Lawson, D., Raff, M. C., Gomperts, B., Fewtrell, C., Gilula, N. B. 1977. *J. Cell Biol.* 72:242–59
202. Neutra, M. R., Schaeffer, S. F. 1977. *J. Cell Biol.* 74:983–91
203. Peixota de Menezes, A., Pinto da Silva, P. 1978. *J. Cell Biol.* 76:767–78
204. Pinto da Silva, P., Nogueira, M. L. 1977. *J. Cell Biol.* 73:161–81

205. Kalderon, N., Gilula, N. B. 1979. *J. Cell Biol.* 81:411–25
206. Eddy, E. M., Shapiro, B. M. 1976. *J. Cell Biol.* 71:35–48
207. Schroeder, T. E. 1978. *Dev. Biol.* 64:342–46
208. Chandler, D. E., Heuser, J. 1979. *J. Cell Biol.* 83:91–108
209. Veron, M., Foerder, C., Eddy, E. M., Shapiro, B. M. 1977. *Cell* 10:321–28
210. Nicolson, G. L., Yanagimachi, R., Yanagimachi, H. 1975. *J. Cell Biol.* 66:263–74
211. Johnson, M., Edidin, M. 1978. *Nature* 272:448–50
212. Edidin, M. 1974. *Ann. Rev. Biophys. Bioeng.* 3:179–201
213. Wolf, D. E., Kinsey, W., Lennarz, W., Edidin, M. 1980. *Dev. Biol.* In press
214. Yanagimachi, R. 1978. *Biol. Reprod.* 19:949–58
215. Shainberg, A., Yagil, G., Yaffe, D. 1969. *Exp. Cell Res.* 58:163–67
216. Okada, Y. 1969. *Curr. Top. Microbiol. Immunol.* 48:102–28
217. Papahadjopoulos, D., Vail, W., Newton, C., Nir, S., Jacobson, K., Poste, G., Lazo, R. 1977. *Biochim. Biophys. Acta* 465:579–98
218. Ohnishi, S., Ito, T. 1974. *Biochemistry* 13:881–87
219. Papahadjopoulos, D. 1978. *Cell Surface Rev.* 5:765–90
220. Gould-Somero, M., Holland, L., Paul, M. 1977. *Dev. Biol.* 58:11–22
221. Longo, F. J. 1978. *Dev. Biol.* 67:249–65
222. Byrd, W., Perry, G. 1980. *Exp. Cell Res.* 126:333–42
223. Gundersen, G. G., Gabel, C. A., Shapiro, B. M. 1980. *J. Cell Biol.* 143a (2) Pt.2(Abstr.)
224. Miyake, Y., Kim, J., Okada, Y. 1978. *Exp. Cell Res.* 116:167–78
225. Poste, G., Papahadjopoulos, D. 1976. *Proc. Natl. Acad. Sci. USA* 73:1603–7
226. Wessells, N. K., Spooner, B. S., Ash, J. F., Bradley, M. O., Luduena, M. A., Taylor, E. L., Wrenn, J. T., Yamada, K. M. 1971. *Science* 171:135–43
227. Carroll, E. J., Levitan, H. 1978. *Dev. Biol.* 63:432–40
228. Longo, F. J., Anderson, E. 1969. *J. Exp. Zool.* 172:97–120
229. Longo, F. J., Kunkle, M. 1978. *Curr. Top. Dev. Biol.* 12:149–84
230. Longo, F. J. 1976. *J. Cell Biol.* 69:539–47
231. Szollosi, D. 1965. *J. Exp. Zool.* 159:367–78
232. Anderson, W. A. 1968. *J. Ultrastruct. Res.* 24:311–21
233. Anderson, W. A., Perotti, M. E. 1975. *J. Cell Biol.* 66:367–76
234. Blandau, R. J., Odor, D. L. 1952. *Fertil. Steril.* 3:13–26
235. Frye, L. D., Edidin, M. 1970. *J. Cell Sci.* 7:319–35
236. Yanagimachi, R., Nicolson, G. L., Noda, Y. D., Fujimoto, M. 1973. *J. Ultrastruct. Res.* 43:344–53
237. Menge, A. C., Fleming, C. H. 1978. *Dev. Biol.* 63:111–17
238. Artzt, K., Dubois, P., Bennett, D., Condamine, H., Babinet, C., Jacob, F. 1973. *Proc. Natl. Acad. Sci. USA* 70:2988–92
239. Zimmerman, A., Vadeboncoeur, M., Press, J. L. 1979. *Dev. Biol.* 72:138–54
240. O'Rand, M. G. 1977. *J. Exp. Zool.* 202:267–73
241. Gabel, C. A., Eddy, E. M., Shapiro, B. M. 1979. *Cell* 18:207–15
242. Veron, M., Shapiro, B. M. 1977. *J. Biol. Chem.* 252:1286–92
243. Carroll, E. J., Epel, D. 1975. *Dev. Biol.* 44:22–32
244. Schuel, H., Longo, F. J., Wilson, W. L., Troll, W. 1976. *Dev. Biol.* 49:178–84
245. Shapiro, B. M. 1975. *Dev. Biol.* 46:88–102
246. Campisi, J., Scandella, C. J. 1978. *Science* 199:1336–37
247. Campisi, J., Scandella, C. J. 1980. *J. Biol. Chem.* 255:5411–19
248. Campisi, J., Scandella, C. J. 1980. *Nature* 286:185–86
249. Edelman, G. M. 1976. *Science* 192:218–26
250. Elgsaeter, A., Branton, D. 1974. *J. Cell Biol.* 63:1018–30
251. Elgsaeter, A., Shotton, D. M., Branton, D. 1976. *Biochim. Biophys. Acta* 426:101–22
252. Golan, D. E., Veatch, W. 1980. *Proc. Natl. Acad. Sci. USA* 77:2537–41
253. Mitchison, J. M., Swann, M. M. 1955. *J. Exp. Biol.* 32:734–50
254. Hiramoto, Y. 1974. *Exp. Cell Res.* 89:320–26
255. Burgess, D. R., Schroeder, T. E. 1977. *J. Cell Biol.* 74:1032–37
256. Begg, D. A., Rodewald, R., Rebhun, L. I. 1978. *J. Cell Biol.* 79:846–52
257. Begg, D. A., Rebhun, L. I. 1979. *J. Cell Biol.* 83:241–48
258. Wang, Y., Taylor, L. 1979. *J. Cell Biol.* 81:672–79
259. Spudich, A., Spudich, J. A. 1979. *J. Cell Biol.* 82:212–26
260. Vacquier, V. D., Moy, G. W. 1980. *Dev. Biol.* 77:178–90
261. Harris, P. 1979. *Dev. Biol.* 68:525–32
262. Harris, P., Osborn, M., Weber, K. 1980. *J. Cell Biol.* 84:668–79

263. Heller, E., Raftery, M. A. 1976. *Biochemistry* 15:1194–98
264. Heller, E., Raftery, M. A. 1976. *Biochemistry* 15:1199–1203
265. Haino, K., Kigawa, M. 1966. *Exp. Cell Res.* 42:625–37
266. Kawai, Y., Anno, K. 1975. *Biochim. Biophys. Acta* 381:195–202
267. Barros, C., Yanagimachi, R. 1971. *Nature* 233:268–69
268. Inoue, M., Wolf, D. P. 1974. *Biol. Reprod.* 11:558–65
269. Wolf, D. P. 1974. *Dev. Biol.* 38:14–29
270. Repin, V. S., Akimova, I. M. 1976. *Biokhimiia* 41:50–58
271. Gwatkin, R. B. L., Williams, D. J., Hartmann, J. F., Kniazuk, M. 1973. *J. Reprod. Fertil.* 32:259–65
272. Bleil, J. D., Wassarman, P. M. 1980. *Dev. Biol.* 76:185–202
273. Bleil, J. D., Wassarman, P. M. 1980. *Proc. Natl. Acad. Sci. USA* 77:1029–33; *Cell* 20:873–83
274. Dunbar, B. S., Wardrip, N., Hedrick, J. L. 1980. *Biochemistry* 19:356–65
275. Vacquier, V. D., Epel, D., Douglas, L. A. 1972. *Nature* 237:34–36
276. Grossman, A., Levy, M., Troll, W., Weissman, G. 1973. *Nature New Biol.* 243:277–78
277. Schuel, H., Wilson, W. L., Chen, K., Lorand, L. 1973. *Dev. Biol.* 34:175–86
278. Wolf, D. P., Nishihara, T., West, D. M., Wyrick, R. E., Hedrick, J. L. 1976. *Biochemistry* 15:3671–78
279. Carroll, E. J., Baginski, R. M. 1978. *J. Cell Biol.* 79(2)Pt.2:162a (Abstr.)
280. Fodor, E. J. B., Ako, H., Walsh, K. A. 1975. *Biochemistry* 14:4923–27
281. Yurewicz, E. C., Oliphant, G., Hedrick, J. L. 1975. *Biochemistry* 14:3101–7
282. Wyrick, R. E., Nishihara, T., Hedrick, J. L. 1974. *Proc. Natl. Acad. Sci. USA* 71:2067–71
283. Grey, R. D., Wolf, D. P., Hedrick, J. L. 1974. *Dev. Biol.* 36:44–61
284. Birr, C., Hedrick, J. L. 1979. *Fed. Proc.* 38(3):466 (Abstr.)
285. Chandler, D. E., Heuser, J. 1980. *J. Cell Biol.* 84:618–33
286. Foerder, C. A., Shapiro, B. M. 1977. *Proc. Natl. Acad. Sci. USA* 74:4214–18
287. Hall, H. G. 1978. *Cell* 15:343–55
288. Foerder, C. A., Klebanoff, S. J., Shapiro, B. M. 1978. *Proc. Natl. Acad. Sci. USA* 75:3183–87
289. Klebanoff, S. J., Foerder, C. A., Eddy, E. M., Shapiro, B. M. 1979. *J. Exp. Med.* 149:938–53
290. Carroll, E. J., Baginski, R. M. 1978. *Biochemistry* 17:2605–12
291. Kay, E., Shapiro, B. M. 1980. *Fed. Proc.* 39(6):2210 (Abstr.)
292. Bryan, J. 1970. *J. Cell Biol.* 44:635–44
293. Bryan, J. 1970. *J. Cell Biol.* 45:606–14
294. Harvey, E. N. 1910. *J. Exp. Zool.* 8:355–76
295. Warburg, O. 1908. *Z. Physiol. Chem.* 57:1–16
296. Gross, A. J., Sizer, I. W. 1959. *J. Biol. Chem.* 234:1611–14
297. Malanik, V., Ledvina, M. 1979. *Prep. Biochem.* 9:273–80
298. Aeschbach, R., Amadò, R., Neukom, H. 1976. *Biochim. Biophys. Acta* 439:292–301
299. Labella, F., Waykole, P., Queen, G. 1968. *Biochem. Biophys. Res. Commun.* 30:333–38
300. Lehrer, S. S., Fasman, G. D. 1967. *Biochemistry* 6:757–67
301. Majewska, M. R., Dancewicz, A. M. 1976. *Acta Biochim. Pol.* 23:353–55
302. Tressel, P., Kosman, D. J. 1980. *Biochem. Biophys. Res. Commun.* 92:781–86
303. Anderson, S. O. 1966. *Acta Physiol. Scand.* 66: Suppl. 28, pp. 1–81
304. DeVore, D. P., Gruebel, R. J. 1977. *Fed. Proc. Fed. Am. Soc. Exp. Biol.* 36(3):679 (Abstr.)
305. Pandey, N. K., Aronson, A. I. 1979. *J. Bacteriol.* 137:1208–18
306. Raven, P. J., Earland, C., Little, M. 1971. *Biochim. Biophys. Acta* 251:96–99
307. Garcia-Castineiras, S., Dillon, J., Spector, A. 1978. *Exp. Eye Res.* 26:461–76
308. Garcia-Castineiras, S., Dillon, J., Spector, A. 1978. *Science* 199:897–99
309. Labella, F., Keeley, F., Vivian, S., Thornhill, D. 1967. *Biochem. Biophys. Res. Commun.* 26:748–53
310. Keeley, F. W., Labella, F., Queen, G. 1969. *Biochem. Biophys. Res. Commun.* 34:156–61
311. Keeley, F. W., Labella, F. S. 1972. *Biochim. Biophys. Acta* 263:52–59
312. Waykole, P., Heidemann, E. 1976. *Connect. Tissue Res.* 4:219–22
313. Malanik, V., Ledvina, M., 1979. *Connect. Tissue Res.* 6:235–40
314. Petri, W. H., Wyman, A. R., Kafatos, F. C. 1976. *Dev. Biol.* 49:185–99
315. Kawasaki, H., Sato, H., Suzuki, M., 1974. *Insect Biochem.* 4:99–111
316. Babior, B. M. 1978. *N. Engl. J. Med.* 298:659–68
317. Babior, B. M. 1978. *N. Engl. J. Med.* 298:721–25
318. Sbarra, A. J., Karnovsky, M. L. 1959. *J. Biol. Chem.* 234:1355–62
319. Iyer, G. Y. N., Islam, D. M. F., Quostel, J. H. 1961. *Nature* 192:535–41

320. Katsura, S., Tominaga, A. 1974. *Dev. Biol.* 40:292–97
321. Isono, N., Yasumasu, I. 1968. *Exp. Cell Res.* 50:616–26
322. Jandl, R. C., Andre-Schwartz, J., Borges-DuBois, L., Kipnes, R. S., McMurrioh, B. J., Babior, B. M. 1978. *J. Clin. Invest.* 61:1176–85
323. Gabig, T. G., Babior, B. M. 1979. *J. Biol. Chem.* 254:9070–74
324. Tauber, A. I., Goetzl, E. J. 1979. *Biochemistry* 18:5576–84
325. Klebanoff, S. J., Rosen, H. 1979. *Ciba Found. Symp.* 65:263–84
326. Thomas, E. L., Aune, T. M. 1978. *Infect. Immunol.* 20:456–63
327. Smith, D. C., Klebanoff, S. J. 1970. *Biol. Reprod.* 3:229–35
328. Evans, T. C. 1947. *Biol. Bull. Woods Hole, Mass.* 92:99–109
329. Wales, R. G., White, I. G., Lamond, D.

R. 1959. *J. Endocrinol.* 18:236–44
330. Jones, R., Mann, T. 1976. *Proc. R. Soc. London Ser. B* 193:317–33
331. Buege, J. A., Aust, S. D. 1976. *Biochem. Biophys. Acta* 444:192–201
332. Kumar, K. S., Walls, R., Hochstein, P. 1977. *Arch. Biochem. Biophys.* 180:514–21
333. Jones, R., Mann, T. 1973. *Proc. R. Soc. London Ser. B* 184:103–7
334. Jones, R., Mann, T. 1977. *J. Reprod. Fertil.* 50:261–68
335. Roels, F. 1971. *Exp. Cell Res.* 69:452–56
336. Schmell, E. D., Gulyas, B. J. 1979. *J. Cell Biol.* 83(2)Pt.2:203a (Abstr.)
337. Fridovich, I. 1978. *Science* 201:875–80
338. Hill, H. A. 1979. *Exp. Cell Res.* 69:5–17
339. Slater, T. F. 1979. See Ref. 325, pp. 143–76.

Ann. Rev. Biochem. 1981. 50:845–77

BIOCHEMISTRY OF DISEASES ♦12098
OF IMMUNODEVELOPMENT

David W. Martin, Jr.

Howard Hughes Medical Institute Laboratory and Department of Medicine and Department of Biochemistry and Biophysics, University of California, San Francisco, California 94143

Erwin W. Gelfand

Division of Immunology, Research Institute, Hospital for Sick Children, Toronto, Ontario M5G 1X8 Canada

CONTENTS

PERSPECTIVES AND SUMMARY

The discovery in 1972 that an inherited disorder of adenine nucleoside metabolism was associated with a specific immunodeficiency disease in

0066-4154/81/0701-0845$01.00

humans prompted a resurgence of interest in the metabolism and effects of purine nucleosides and led to the timely review, *The Role of Adenosine and 2'-Deoxyadenosine in Mammalian Cells,* in Volume 47 of this *Review* (1). Since that time, investigators in many disciplines have focused attention on the basic pathogenesis of the immunodeficiency diseases associated with the inherited absence of adenosine deaminase or of purine nucleoside phosphorylase, and on the implications of these disorders for immunology, oncology, and the study of purine metabolism. In this review we describe the current understanding of the biochemical ontogeny of the human immunity system, the metabolic derangements in immunodeficiency diseases, and the effects of the latter on the former. We hope to offer biochemical explanations for the immunologic disorders, with particular emphasis on the biochemical basis for the remarkable tissue-specific and function-specific pathologic effects of these inborn errors of metabolism.

At present the understanding of pathogenesis and tissue specificity is incomplete. Using model systems, many laboratories have clearly defined metabolic and even molecular mechanisms of cellular toxicity resulting from the loss of the specific gene products. However it is not yet possible to conclude which if any of these defined mechanisms is responsible for the in vivo pathogenesis of these specific human immunodeficiency diseases.

In neither the deficiency of adenosine deaminase or purine nucleoside phosphorylase does a deficiency of a product or products of the missing enzyme seem to be detrimental to the immunity system. Thus attention has been directed toward the potentially toxic effects of accumulated substrates of the missing enzymes. Of the four substrates of purine nucleoside phosphorylase, all of which accumulate in extracellular fluids of patients deficient in that enzyme, only deoxyguanosine is particularly toxic when added to the medium of cultured mammalian cells or lectin-stimulated lymphocytes. For at least two decades deoxyguanosine has been recognized as an inhibitor of DNA synthesis, and recent genetic evidence has demonstrated that this effect is mediated by deoxyGTP inhibition of the enzyme, ribonucleotide reductase, which causes an intracellular depletion of the DNA substrate, deoxyCTP. Of the purine nucleoside phosphorylase substrates, only deoxyguanosine can be phosphorylated; in the absence of the phosphorylase the other three substrates, inosine, deoxyinosine, and guanosine are simply excreted in urine. Thus, at the present time, it appears that the depletion of intracellular deoxyCTP may provide the best explanation for the pathogenesis of the immunodeficiency disease associated with the loss of purine nucleoside phosphorylase.

In the absence of adenosine deaminase, its accumulated substrates, adenosine and 2'-deoxyadenosine, have numerous alternative pathways of metabolism, most of which have been recently reviewed (1). From studies in model systems the two pathways most relevant to the immunodeficiency disease involves: (a) the inhibition of ribonucleotide reductase by the phosphorylated metabolite of deoxyadenosine, deoxyATP; and (b) the accumulation of S-adenosylhomocysteine, a potent inhibitor of methylation reactions mediated by S-adenosylmethionine. The effects of the inhibition of ribonucleotide reductase by deoxyATP are closely related to those caused by deoxyGTP in purine nucleoside phosphorylase deficient model systems. The accumulation of S-adenosylhomocysteine may occur by either of two mechanisms, one an effect of adenosine and the other of deoxyadenosine.

The deficiency of transcobalamin II, a cobalamin- or vitamin B_{12}-binding protein present in blood plasma, is also associated with an immunodeficiency disease. A disturbance in the cobalamin-dependent transmethylation of homocysteine to methionine may be responsible for the pathogenesis of this disorder, and thus may be closely related to the disturbance of methylation reactions proposed to exist in adenosine deaminase deficiency.

From recent studies in cultured human cells, it is clear that those that are thymus-dependent, or of T-cell lineage, have a much greater capacity to trap deoxyguanosine and deoxyadenosine as intracellular deoxyribonucleoside triphosphates. This enhanced trapping probably results from two phenomena. The T cells and their precursors, thymocytes, contain greater deoxyadenosine and deoxyguanosine phosphorylating activity and significantly less purine deoxyribonucleotide dephosphorylating activity than do either thymus-independent lymphocytes or other human tissues. The accumulated purine deoxyribonucleoside triphosphates, by inhibiting ribonucleotide reductase, are capable of effecting an inhibition of DNA synthesis. Thus, one would expect that T cells and/or their immunologic functions that are proliferative-dependent, such as population expansions or T cell–mediated immunosuppression, would be particularly vulnerable to the presence of deoxyguanosine or deoxyadenosine. Hence, from the data currently available it is possible to construct a feasible explanation for the tissue-specific and function-specific effects of the loss of adenosine deaminase or of purine nucleoside phosphorylase. The loss of B-cell function in adenosine deaminase deficiency and in transcobalamin II deficiency might be related to a dependence of those cell types on S-adenosylhomocysteine hydrolase, which functions to destroy the methylation inhibitor, S-adenosylhomocysteine. However in the case of adenosine deaminase deficiency, the severe disturbance of thymus and T-cell function may be sufficient to preclude normal B-cell ontogeny and/or function.

Observations from these few, rare inherited diseases have provided important insights concerning the biochemical differentiation of the cells of the immunity system and the hope that one may eventually be able to control specifically thymus-dependent immunity and its malignancies in humans by mimicking pharmacologically what nature has done genetically.

OVERVIEW OF NUCLEOSIDE AND NUCLEOTIDE METABOLISM

The processes of de novo synthesis of purines and pyrimidines generate inosine monophosphate and uridine monophosphate, respectively, as the first compounds with completed heterocyclic rings. The presence of the charged phosphates of the nucleotides greatly hinders their transport across cell membranes. However, mammalian cells possess intracellular (2, 3) and/or extracellular nucleotidases (4–6) to remove the α-phosphates, thereby generating nucleosides that transverse cell membranes both by passive diffusion (7) and facilitated transport (7–10). Seemingly in order to prevent the loss of these heterocyclic compounds synthesized at considerable energy and substrate expense, mammalian cells have evolved with mechanisms to salvage these purine and pyrimidine metabolites, both within the same cell in which they are produced and from their extracellular environment. Nevertheless, some of these purine and pyrimidine nucleotide degradation products are potentially toxic to the cell if salvaged in excessive quantities. Accordingly, mammalian cells possess a means of detoxifying excessive quantities of these useful but potentially harmful purine and pyrimidine metabolites.

Crucial to the well-being of the cell is a proper balance between the production of dephosphorylated purine and pyrimidine metabolites, their detoxification by further degradation, and their salvage by metabolism back to the nucleotide level. The recently discovered immunodeficiency diseases in humans associated with disorders of purine metabolism seem to result from an imbalance in these pathways.

Adenosine and deoxyadenosine generated by the action of nucleotidases on the corresponding nucleotides are substrates of adenosine deaminase (11, 12), a ubiquitous mammalian enzyme (13). Although the deamination of adenosine and deoxyadenosine is probably a necessary detoxification process, it is in humans a quantitatively minor contributor to the overall degradation of purines to the usual excreted form, uric acid. Those patients with adenosine deaminase deficiency have normal levels of plasma and urinary urate (14, 15), and only a few percent of their normal total purine excretion occurs in the form of adenosine deaminase substrates (14–16).

The in vivo site of generation of adenosine and deoxyadenosine is not known, but the fact that relatively small quantities are generated (14–18) (compared with purine nucleoside phosphorylase substrates, see below) is consistent with the suggestion of Chan that reticuloendothelial cells release adenosine and deoxyadenosine during the process of digesting nucleic acids of phagocytized cells (19).

Both adenosine and deoxyadenosine can be phosphorylated by the enzyme adenosine kinase (20–22), and deoxyadenosine is also a substrate for deoxycytidine kinase (1, 22–26).

The K_m value of adenosine kinase for adenosine is significantly lower than that of adenosine deaminase for adenosine (1, 11, 12, 21, 22, 26). Thus the more active metabolic route for adenosine is rephosphorylation to AMP. However the situation is different for deoxyadenosine. The K_m of adenosine deaminase for deoxyadenosine (11, 12) is significantly lower than the K_m value of either deoxycytidine kinase (23, 24) or adenosine kinase (1, 21, 22, 26) for deoxyadenosine. Thus, under normal conditions, deoxyadenosine follows the route to deamination rather than rephosphorylation; presumably this is a selective advantage to the organism and suggests that the phosphorylation of deoxyadenosine may be more detrimental than that of adenosine. There is controversy concerning whether adenosine kinase or deoxycytidine kinase in human tissues is responsible for the phosphorylation of low concentrations of deoxyadenosine to deoxyAMP (1, 26–29).

Adenosine is also a substrate for the enzyme S-adenosylhomocysteine hydrolase (30, 31), which generates S-adenosylhomocysteine from the homocysteine normally present in body fluids. Since the equilibrium for the reaction of adenosine and homocysteine strongly favors S-adenosyl-homocysteine (30), the latter acts as a sink for adenosine when it has not been deaminated by the deaminase. As discussed in more detail below, S-adenosylhomocysteine is a potent competitive inhibitor of all transmethylation reactions mediated by S-adenosylmethionine (32–35).

The products of adenosine deaminase, inosine, and deoxyinosine, along with guanosine and 2'-deoxyguanosine are all naturally occurring substrates for the enzyme purine nucleoside phosphorylase (36, 37). This ubiquitous phosphorylase generates the corresponding purine bases, hypoxanthine or guanine, and the corresponding sugar phosphate, ribose-1-phosphate or 2-deoxyribose-1-phosphate. The flux of substrates through purine nucleoside phosphorylase is much greater than that through adenosine deaminase as evidenced by the fact that patients with a complete deficiency of the phosphorylase excrete minimal uric acid but massive quantities of phosphorylase substrates, quantities severalfold greater than the total urates excreted in a normal person (38, 39). Since the generation of inosine and deoxyinosine by way of adenosine deaminase appears to be minimal, the vast majority of purine nucleoside phosphorylase substrates normally must be generated by the action of nucleotidases on the hypoxanthine and guanine ribo- and deoxyribonucleotides.

From observations in patients with the Lesch-Nyhan syndrome, it is apparent that most of the hypoxanthine and guanine produced by purine nucleoside phosphorylase, are normally salvaged by phosphoribosylation via hypoxanthine-guanine phosphoribosyl transferase (HGPRTase), a PPriboseP consuming reaction (40). While the HGPRTase deficient Lesch-Nyhan patients cannot salvage hypoxanthine or guanine due to the deficiency of the catalyst, those patients with purine nucleoside phosphorylase deficiency cannot salvage hypoxanthine and guanine due to a deficiency of these substrates for the salvage reaction. Accordingly, both Lesch-Nyhan patients and the purine nucleoside phosphorylase deficient patients "spare" PPriboseP, which in turn drives the de novo purine synthetic pathway to overproduction (38–40). The latter seems to account for the massive total purine overproduction in the patients with purine nucleoside phosphylase deficiency and those with the HGPRTase deficiency.

The liver in mammals has been thought to be a major supplier of purines and purine nucleosides to peripheral tissues (41, 42), and it seems likely to be a major contributor to the massive quantities of purine nucleoside phosphorylase substrates generated in patients deficient in that enzyme.

Of the purine nucleoside phosphorylase substrates that accumulate in excess in the enzyme deficient, immunodeficient patients, only one, deoxyguanosine, can be salvaged directly by phosphorylation to the corresponding nucleotide, 2'-deoxyguanosine monophosphate (22, 24, 43). For practical purposes, the other three substrates have but one metabolic path to follow and that is phosphorylysis by purine nucleoside phosphorylase. The enzyme deficient patients excrete in their urine millimolar quantities of the four substrates, of which deoxyguanosine is present at the lowest concentration (38). Deoxyguanosine is phosphorylated exclusively by deoxycytidine kinase (22, 24, 44, 45), an enzyme that is also capable of phosphorylating deoxyadenosine. None of the purine nucleoside phosphorylase substrates seems to accumulate in vivo as purine nucleosidyl homocysteine, even though S-inosylhomocysteine can be generated from inosine and homocysteine by S-adenosylhomocysteine hydrolase in vitro and will accumulate in phosphorylase deficient mouse T-lymphosarcoma cells exposed to high concentrations of 200 μM) of inosine and homocysteine (46).

The other area of metabolism of interest here is the role of cobalamin, or vitamin B_{12}, in purine and pyrimidine metabolism. Methylcobalamin is a cofactor necessary for the transmethylation of homocysteine to methionine by homocysteine transmethylase, a reaction that utilizes as a substrate, N^5-methyltetrahydrofolate (47). The latter normally transfers its methyl group to cobalamin during the reaction process, but in the absence of the methyl acceptor, cobalamin, N^5-methyltetrahydrofolate accumulates and acts as a sink for tetrahydrofolate (48). The latter is a required carrier for the methylene and methenyl moieties necessary in the de novo synthesis of thymidylate and purine nucleitides (49, 50). Patients with cobalamin deficiency accumulate homocysteine in their body fluids and are, in effect, deficient in available tetrahydrofolate for reactions other than the trans-methylation of homocysteine to methionine.

CAUSE AND EFFECT RELATIONSHIPS

The discoveries of the deficiencies of adenosine deaminase and of purine nucleoside phosphorylase among children with severe immunodeficiency diseases were clearly serendipitous and have been reviewed (51–53). The discovery of adenosine deaminase deficiency was related to the fact that erythrocyte adenosine deaminase activity is polymorphic in the human population (54–56), and the discovery of purine nucleoside phosphorylase deficiency resulted from the assay for erythrocyte adenosine deaminase activity in Dr. Giblett's laboratory being dependent upon endogenous purine nucleoside phosphorylase of the hemolysate being studied. The discovery in 1972 of a patient with an agammaglobulinemia and a deficiency of the cobalamin transport protein, transcobalamin II, stemmed from the observations of Hitzig et al (57, 58) that the child in question developed a hematologic disorder indistinguishable from pernicious anemia but not responding to therapeutic doses of vitamin B_{12}. Although transcobalamin II (TCII) has no recognized catalytic activity, it is included in this review because it is a gene product with a known function (59) and its deficiency is accompanied by a recognized biochemical disorder and an impairment of immune function.

The diseases associated with the complete deficiencies of adenosine deaminase, purine nucleoside phosphorylase, and TCII are inherited in an autosomal recessive manner. Consanguinity has been described in pedigrees of patients affected with each of these diseases (52, 57, 60), and in general the obligate heterozygotes (parents) have half-normal levels of the biochemical function in question but do not exhibit detectable immunologic abnormalities.

Evidence that the detectable missing genetic functions have causal roles in the immunologic diseases comes first from the genetic information mentioned above, second from the descriptions of feasible pathogenic mechanisms, and third from observations of responses of patients or of their cells in vitro to specific manipulations.

In the case of TCII deficiency, therapeutic responses of patients to pharmacologic doses of parenteral cobalamin and folinic acid leaves little doubt that this deficiency causes the observed agammaglobulinemia (57, 58).

The first biochemical evidence that deficiency of adenosine deaminase was responsible for the immunologic dysfunction observed in enzyme deficient patients was provided by in vitro studies of Polmar et al (61) on peripheral blood lymphocytes from an adenosine deaminase deficient patient. They observed that while the lymphocytes did not respond to lectin stimulation by increasing the incorporation of thymidine into DNA, the introduction of purified calf thymus adenosine deaminase to the in vitro mitogenic assay permitted a significant DNA synthetic response. In another study, the addition of adenosine deaminase to enzyme deficient peripheral blood lymphocytes promoted E rosetting, a mature T-cell function, among the population of cells previously without that in vitro function (62).

Polmar et al (63) and subsequently others (64) have demonstrated that the infusion of irradiated normal erythrocytes containing adenosine deaminase allows the transient reconstitution of the immunity system in some adenosine deaminase deficient patients. Furthermore, children from three different families have been described with adenosine deaminase deficiency in their erythrocytes but with normal immunity systems (65–67). In each case, significant (10–50%) residual adenosine deaminase activity was present in the lymphocytes and fibroblasts, and as described below, they did not exhibit other metabolic abnormalities characteristic of adenosine deaminase deficient patients (66). Interestingly, all three of those families are black.

Although there has been some evidence that the infusion of erythrocytes containing purine nucleoside phosphorylase into two purine nucleoside phosphorylase deficient, immunodeficient children had some beneficial effects on the immunity system, the data are less convincing than are those from the adenosine deaminase deficient patients (68, 69). However, two brothers with incomplete deficiencies of purine nucleoside phosphorylase do exhibit less severe abnormalities of purine metabolism and less severe immunologic dysfunction than those children with the complete deficiency of the enzyme (70, 71). Thus, for all three immunodeficiency diseases there is good evidence that the immunologic disorders result from the observed loss of function of the defective gene products.

The importance of these pathways of purine metabolism is further stressed by the selectivity of the enzyme deficiencies on the lymphoid system. The precise mechanism(s) by which damage is done, the selectivity for lymphocytic cells, and the reasons why the characteristic syndromes result from the individual enzyme deficiencies are slowly being unraveled. To understand these phenomena, one must consider the differentiation, expansion, and functional maturation of the interacting lymphocyte populations, since the purine enzyme deficiencies probably express themselves by negative rather than by positive selection. That is, the immunodeficiencies likely result from the death of specific precursor cells through the accumulation of toxic nucleosides or their metabolites rather than by differential promotion of the proliferation of other cell types.

DIFFERENTIATION OF THE LYMPHOID SYSTEM

The basis of the immune response resides in the ability of subpopulations of lymphocytes to recognize, react, and interact in a specific manner to the wide variety of encountered stimuli. These responses can be roughly separated into those mediated by thymus-dependent T lymphocytes (cell-mediated immunity) and those mediated by antibody or thymus-independent B lymphocytes (humoral immunity). This two-component concept is supported by observed differences in differentiation pathways, organization of lymphoid tissues, antigenic markers, function and, as is stressed in this review, nucleoside and nucleotide metabolism.

Differentiation of the lymphoid precursor cells may be divided into two stages. In the first stage, pluripotent hematopoietic stem cells differentiate into erythrocytes, granulocytes, megakaryocytes, monocytes, or lymphocytes, depending on the microenvironment. These stem cells originate in the yolk sac during the first trimester of fetal life and are subsequently found in fetal liver and then fetal bone marrow. In the second stage, cells committed to the development of the lymphoid system mature along one of the two recognized pathways.

T Lymphocytes

Although the thymus has long been recognized as the organ that controls the development of T lymphocytes, there are many gaps in our knowledge concerning the mechanisms by which the thymus exerts these controls, regulates T-lymphocyte subset distribution, and influences B-cell functional maturation.

A combination of clinical and experimental evidence has established two major roles for the thymus and T-cell differentiation. The first involves the proliferation and differentiation of prethymic precursor T cells under the influence of the thymic microenvironment into the more mature post-thymic precursor T cells and intrathymic T cells (reviewed in 72, 73). The prethymic precursors enter the thymus from the circulation and undergo a series of maturation events. The initial stage(s) of T-cell differentiation require intimate contact between the prethymic precursor T cells, identified in fetal liver, peripheral blood, and marrow, and the thymic epithelium. Following this event the precursor T cells, now identified as post-thymic precursors, become susceptible to a variety of thymic-derived humoral or hormonal-like influences. It is these latter factors that are probably essential for the second role of the thymus in T-cell differentiation, namely the maturation, modulation, and expansion of the population of post-thymic T lymphocytes into immunocompetent T cells.

Although this concept of T-cell maturation from the prethymic precursor cell → post thymic precursor cell → intrathymic T cell → post-thymic T cell is based on experimental data, it has been difficult to establish what proportion of cells "make it through," at what stage immunocompetence is achieved, and when T-cell heterogeneity arises, which results in the functional diversification of post-thymic T-cell subclasses identified in peripheral blood and lymphoid tissues. The thymus exports to the periphery a post-thymic T cell with limited immunocompetence, but in the thymus the rate of generation of new thymocytes far exceeds that necessary to maintain the post-thymic T-cell pool (74).

T-CELL HETEROGENEITY   In recent years it has become increasingly apparent that human post-thymic T lymphocytes are heterogenous with regard to specific immunologic functions and to their constellation or phenotype of cell surface receptors. Most T lymphocytes are defined as such by their ability to bind sheep erythrocytes (E) using a simple rosetting technique (75). E receptors appear initially at the intrathymic stage of maturation: more than 95% of thymocytes bind E. The thymocytes form large, firm E rosettes equally well at both 37°C and at 4°C. At the post-thymic stage the E receptor can be used to advantage to identify T-cell heterogeneity. Most post-thymic T cells bind E optimally at 4°C with few rosettes forming at 37°C. Moreover, a proportion of these cells bind E less avidly than others, thus inferring heterogeneity for the E receptor. A number of drugs can interfere with the T cell's capacity to form rosettes. Thus, reagents that increase intracellular levels of cAMP, e.g. adenosine or histamine, result in varying degrees of inhibition of E-rosette formation. Not all peripheral or post-thymic T cells are susceptible to such inhibition using these reagents. In extensive studies with theophylline, an inhibitor of cAMP phosphodiesterase, two populations of E-rosetting T cells could be delineated: a thoephylline-sensitive subset that fails to bind E in the presence of the drug, and a theophylline-resistant subset (76). Two major subsets of peripheral T cells have also been defined using heteroantisera (77) autoantibodies, or monoclonal antibodies directed at stable cell surface antigens.

During the process of differentiation and maturation, post-thymic T cells develop receptors for the Fc portion of immunoglobulin. Even though thymocytes do not, the majority of post-thymic T cells express a receptor for the Fc portion of IgM (Tμ), and a minority express receptors for the Fc portion of IgG (Tγ) (78).

FUNCTIONAL HETEROGENEITY   The heterogeneity of T cells revealed by their surface receptors and antigens is paralleled by a degree of functional heterogeneity. This is most evident in the mature, post-thymic T-cell compartment where the distinct subsets exert a regulatory influence on the differentiation of B lymphocytes to antibody-secreting cells or plasma cells.

On the basis of surface receptors for the Fc portion of immunoglobulin, theophylline sensitivity of E receptors, or a number of specific antisera, a population of mature T cells with the capacity to promote (help) B-cell differentiation can be distinguished from another population that prevents (suppresses) this process (77, 79–82).

B Lymphocytes

Our understanding of the ontogeny of B lymphocytes in humans is far less extensive (reviewed in 83). There are significant gaps in our knowledge of the cellular and molecular events underlying the stepwise acquisition of B-cell immunocompetence and the generation of the diversity of the mature humoral antibody response. In birds, differentiation along B-cell lines is initiated in the Bursa of Fabricius; in mammals, the analogous inductive environment may reside in fetal liver and bone marrow.

The B lymphocytes have been identified by their expression of surface immunoglobulin and, as noted earlier, can differentiate into mature immunoglobulin-secreting plasma cells under the influence of "helper" T cells. The earliest recognizable cells in the B-cell differentiation pathway are called pre-B cells. Dividing and nondividing pre-B cells synthesize the heavy chain (μ chain) of IgM molecules in small amounts that are detectable in their cytoplasm by immunofluorescent techniques. The differentiation of pre-B cells to immature B cells (which possess surface IgM) appears to be both T-cell independent and antigen independent. As the immature B cells mature and memory B cells are generated surface immunoglobulin isotype diversity appears, a process that is antigen dependent but proliferative independent. At the B-cell stage, in addition to exhibiting surface immunoglobulin, the mature B cells demonstrate receptors for the Fc portion of immunoglobulin, the third component of complement, Epstein-Barr virus, and Ia-like determinants. These receptors and antigens are lost for the most part as the cells further mature to the immunoglobulin-secreting or plasma cell stage. This final process in the ontogeny of the humoral arm of immunity is proliferative dependent (82).

Enzyme Expression In Lymphocyte Ontogeny

In addition to the presence of receptors or cell surface antigens, which may be used to identify distinct stages of lymphocyte maturation, a number of enzymes important in nucleotide/nucleoside metabolism have a restricted distribution in different lymphocyte subpopulations.

TERMINAL DEOXYNUCLEOTIDYL TRANSFERASE Terminal deoxynucleotidyl transferase has the capacity to add deoxymononucleotides to any 3'-OH-terminated segment of DNA without template direction (84).

There is increasing evidence that terminal deoxynucleotidyl transferase is restricted in distribution to immature lymphocytes and that many terminal deoxynucleotidyl transferase positive bone marrow cells are prethymic precursor T cells (85). This is supported by the findings that terminal deoxynucleotidyl transferase activity can be significantly increased in athymic animals by in vivo and in vitro treatment with thymic hormone (86). The majority of intra-thymic T lymphocytes (thymocytes) also contain a high level of terminal deoxynucleotidyl transferase, whereas post-thymic T cells contain very little.

ADENOSINE DEAMINASE   Adenosine deaminase activity is highest in thymus, with decreasing levels in spleen, lymph nodes, and other lymphoid organs (87, 88). As T-cell maturation proceeds, adenosine deaminase activity falls (89). Indeed, in rats, highest activity was localized in cortical thymocytes with three- to tenfold lower levels in medullary thymocytes and peripheral T cells (89). B lymphocytes also contain adenosine deaminase. The adenosine deaminase levels in normal B cells have not been determined, but levels in B-cell lines are roughly the same as those in peripheral T cells (90). It has been recently demonstrated that the human thymus/leukemia associated antigen (HThy-L), identified initially in thymus and T-cell leukemias (91), is a thymic isozyme of the adenosine deaminase (92).

PURINE NUCLEOSIDE PHOSPHORYLASE   Purine nucleoside phosphorylase (PNP) seems to bear a reciprocal relationship with adenosine deaminase in cells of T lineage at various stages of differentiation (89). Thus PNP activity appears to increase as T cells mature from the intrathymic stage to the post-thymic peripheral T-cell stage. Levels of purine nucleoside phosphorylase activity in B lymphocytes are not documented but are similar to peripheral blood T cells when assayed in cultured B-cell lines (90).

DEOXYCYTIDINE KINASE   Several studies have shown that the thymus is highly enriched in deoxycytidine kinase (78, 93). As T cells mature there is a decrease in the activity of this enzyme; B lymphocytes have slightly higher activities than peripheral blood T lymphocytes (93).

ECTO-5'-NUCLEOTIDASE   This ectoenzyme can also serve as a marker of T-cell maturation. Intrathymic T cells have very low activities when compared to the more mature post-thymic T cells in peripheral blood or lymphoid tissue (93, 94). Thymic-derived factors, known to promote T-cell differentiation, can increase the expression of 5'-nucleotidase on human thymocytes two- to threefold (95). These data suggest a role for thymic humoral factors in the expression of the enzyme activity during T-cell maturation. The levels of enzyme activity appear to be highest in mature B lymphocytes (93, 94). However, chronic lymphatic leukemia cells, which may represent the clonal expansion of an immature B-cell population, have low activity (95a).

PURINE NUCLEOSIDE PHOSPHORYLASE DEFICIENCY

Natural History of the Disease

The natural history of the immunodeficiency associated with absence of purine nucleoside phosphorylase in humans has been reviewed (96). Briefly, most children with the disease first show signs and symptoms of immunodeficiency between 3 and 18 months of age. However, one of the brothers with an incomplete deficiency of purine nucleoside phosphorylase was without symptoms until six years of age (70). Many of the affected children were immunized during infancy and early childhood with live virus vaccines without unusual responses, which strongly suggests that at least in early life T-cell function may be relatively intact. Subsequently, immune attrition likely occurs with loss of T-cell function (96). The children become increasingly susceptible to viral infections, particularly varicella, and often have succumbed to viral illness.

At autopsy the thymus has been seen to be severely hypocellular but to contain Hassall's corpuscles (97), which indicates the one-time existence of a functioning organ and suggests that involution rather than congenital absence or hypoplasia of the thymus is responsible for the dysfunction.

In contrast to the involvement of cell-mediated immunity, humoral immune function was universally preserved and often exaggerated. Thus, immunoglobulin levels and antibody responses to specific challenges were normal and often higher than in controls. Indeed, several of the patients have had excessive antibody formation, which results in autoimmune disease.

This dichotomy between T- and B-cell immunity in purine nucleoside phosphorylase deficient patients clearly distinguishes them from adenosine deaminase deficient individuals (see below). The pathogenic mechanisms for the dichotomy require consideration of T-cell differentiation and the role of T cells in B-cell immunity. On the basis of the laboratory data, B cells, B memory cells, and plasma cell differentiation appear perfectly intact in purine nucleoside phosphorylase patients (71). In addition, since antibody responses are normal, it may be assumed that T-helper cells are present in adequate numbers. The presence of auto-antibodies and excessive immunoglobulin production implies some form of regulatory abnormality or imbalance between helper cells and suppressor cell events.

These questions have been addressed in the two brothers with less than complete deficiency of purine nucleoside phosphorylase (71). Although the course and immune deficiency were milder than the others described, laboratory analysis of their lymphocyte function has provided some clues as to the relationship of immune function to the biochemical abnormality. Lymphopenia was severe in these patients, but T cells with E receptors were detected. Further, T-cell precursors, responsive to thymic inductive influences, were present, which suggests that the initial stages of T-cell differentiation and expansion could occur, at least to some extent. However, post-thymic mature T cells were deficient in the peripheral blood, and a lymph node biopsy revealed a paucity of these cells in the T-cell or paracortical region. Although the patients are lymphopenic, all E rosetting T cells were theophylline resistant. The cells that are present may be somewhat immature and nonproliferating, but are capable of mediating help (71). Theophylline-resistant cells are known to provide helper cell activity without requiring proliferation themselves. In contrast, the functions of theophylline-sensitive cells, capable of mediating suppression, are proliferative dependent. Thus, it appears that these patients lack the subpopulation of proliferative-dependent post-thymic suppressor cells but do possess T cells capable of providing help.

The substrates of the missing purine nucleoside phosphorylase, inosine, guanosine, deoxyinosine, and deoxyguanosine are present at extraordinarily high concentrations in the plasma and urine of affected patients (38, 39), and were detected in the umbilical cord blood plasma of two affected siblings (98). Small but abnormal levels of orotate have also been demonstrated in the urine of two unrelated patients (99, 100), and in one of them the urinary orotate disappeared after the oral administration of uridine (99).

The seemingly most significant metabolic abnormality yet found in purine nucleoside phosphorylase deficient patients is the elevated deoxyGTP in their erythrocytes (101, 102), a metabolite that seemed to decrease when one of the patients was treated with infusions of irradiated erythrocytes containing normal levels of the phosphorylase (68). Because the patients are severely lymphopenic, sufficient numbers of peripheral blood lymphocytes have not been available for determinations of deoxyGTP in those cell types.

The Purine Nucleoside Phosphorylase Enzyme

The structural gene for purine nucleoside phosphorylase in humans has been assigned to the q13 band of chromosome 14 (103, 104) and codes for a 30,000-molecular weight subunit that exists natively as a trimeric molecule (105–108). The enzyme is ubiquitous in human tissues, but the peripheral blood lymphocytes, granulocytes, and erythrocytes and the kidney contain the highest specific catalytic activities (88). Histochemical data suggest that peripheral blood T lymphocytes contain greater quantities than the immunoglobulin-producing B lymphocytes (109). Studies of the enzyme from several homozygous deficient patients have revealed altered ratios of catalytic to immunoreactive activity (110, 111), which suggests structural defects in the enzyme molecule. Peptide mapping of the abnormal subunit from one of the six reported families with the deficiency has demonstrated abnormalities within the subunit, which confirms that the mutation in that family is in the structural gene for purine nucleoside phosphorylase (112, 113).

Specific Metabolic Abnormalities and Effects In Model Systems

In general for inborn errors of metabolism one expects to observe relative deficiencies of the products of the missing function and an accumulation of its substrate(s) or their secondary metabolites. Accordingly, the detrimental effects of the metabolic abnormalities may be secondary to either the deficiency of the product or toxic effects of the accumulated substrate(s) or secondary metabolites. Most investigators studying the pathogenesis of the enzyme deficient immunodeficiency diseases have utilized in vitro model systems and the tools of somatic cell genetics to control the cellular environment. The model systems have included mouse T-cell lymphosarcomas in continuous culture, peripheral blood lymphocytes from humans and other mammals, and human T- and B lymphoblasts in continuous culture.

Deoxyguanosine has long been recognized as a potent cytotoxic agent to cultured mammalian cells (114, 120), and it has been known that deoxycytidine partially reverses or protects from the cytotoxic effects (115–120).

Ullman et al isolated a clone of purine nucleoside phosphorylase deficient T-lymphosarcoma (S49) cells and demonstrated that although they accumulate all four substrates of the phosphorylase, intracellularly they grow at a normal rate even in culture medium containing dialyzed serum (121). Thus it was clear that in the S49 model system the purine nucleoside phosphorylase deficiency per se was not intrinsically lethal to the T cell. Ullman et al went on to demonstrate that of the four phosphorylase PNP substrates, only deoxyguanosine significantly inhibited the growth of the purine nucleoside phosphorylase deficient cell line ($EC_{50}=20$ μM). By remutagenizing the phosphorylase deficient clone of cells and selecting for mutants resistant to the cytotoxic deoxyguanosine, three classes of mutant cells were isolated (121). One class was unable to transport deoxyguanosine or any other purine or pyrimidine nucleoside or deoxyribonucleoside into the cell (10). The second class of mutants was unable to phosphorylate deoxyguanosine to deoxyGMP and was deficient in the enzyme, deoxycytidine kinase, known to be responsible for the phosphorylation of deoxyguanosine (22, 25). The third class of mutants could transport deoxy-guanosine into the cell and accumulated it as deoxyGTP but was still resistant to the growth inhibitory effects of deoxyguanosine up to an EC_{50} of 200 μM. This latter class of mutants is heterozygous for an altered M1 subunit of ribonucleotide reductase, which is feedback-resistant to deoxy-GTP and TTP, but normally sensitive to feedback-inhibition by deoxyATP (122). From these studies in the S49 T-lymphosarcoma model the authors concluded that of the four purine nucleoside phosphorylase substrates, all of which accumulate in the body fluids of purine nucleoside phosphorylase deficient patients, only deoxyguanosine is toxic at low concentrations. Furthermore, the genetics studies demonstrated that in order to be toxic, deoxy-guanosine must be transported into the cell, be phosphorylated, and accumulate as deoxyGTP, which in turn acts allosterically on the target enzyme, ribonucleotide reductase, to inhibit CDP and GDP reduction, which depletes the cell of deoxyCTP, a necessary precursor for DNA synthesis.

The S49 cell mutants, with a feedback-resistant ribonucleotide reductase, still exhibit sensitivity to deoxyguanosine at very high concentrations, even though at these high concentrations of deoxyguanosine they still do not deplete their pools of deoxyCTP. Thus there is likely to be a secondary mechanism of deoxyguanosine toxicity above and beyond that mediated by ribonucleotide reductase. However, this secondary mechanism does require the transport of deoxyguanosine into the cell and its phosphorylation by deoxycytidine kinase, since mutants deficient in either of these two func-tions exhibit much greater resistance to deoxyguanosine than do the feed-back-resistant reductase mutants (10, 121). The mechanism reponsible for this apparent secondary toxicity of deoxyguanosine is not known but may well simply be the inhibition of de novo purine synthesis and depletion of adenine nucleotides.

The phosphorylation by deoxycytidine kinase and the phosphorolysis by purine nucleoside phosphorylase are the only two pathways known to metabolize deoxyguanosine. Deoxyguanosine can be efficiently excreted in urine (28) and transported across the placenta to the maternal circulation (123).

Deoxyguanosine has also been demonstrated to inhibit DNA synthesis in peripheral blood T lymphocytes from normal humans (124, 125). Gelfand et al further demonstrated that in a plaque-forming cell assay of human cells specific for IgM production against ovalbumin, T-helper cell function was insensitive, but the T-suppressor function was sensitive to micromolar quantities of deoxyguanosine present during the five days of the assay incubation (125). These data confirm the observations that the very early steps in T-cell differentiation and T-helper cell activity are intact in at least some purine nucleoside phosphorylase patients (71, 126). As noted above, the T-suppressor function is proliferative-dependent, while helper cell function of peripheral T lymphocytes is not.

To determine whether these in vitro observations could have in vivo relevance, guanine nucleosides were administered intraperitoneally to mice daily for five days and the generation of suppressor cells monitored. Deoxyguanosine, but not guanosine, was shown to inhibit selectively the development of murine T-suppressor cells (127).

Numerous investigators have compared the sensitivities of human T-lymphoblast lines with those of B-lymphoblast lines to exogenous deoxyguanosine. With the exception of a B-lymphoblast line derived from a purine nucleoside phosphorylase deficient patient (128), the phosphorylase was intact in these model systems. In the presence of nucleoside phosphorylase, deoxyguanosine is metabolized to guanine and deoxyribosephosphate; in the presence of HGPRTase, guanine is toxic to cells, an effect that can be prevented by the simultaneous presence of micromolar quantities of adenine in the medium (44; D. W. Martin, unpublished observations). Thus the toxicity of deoxyguanosine in cells possessing purine nucleoside phosphorylase is at least partially due to the toxicity of guanine generated from deoxyguanosine. Nevertheless, numerous studies have demonstrated that T-lymphoblastoid human cells are more sensitive than B-lymphoblastoid cells to the cytotoxic effects of deoxyguanosine (125, 129–131). In some of these studies deoxycytidine has been demonstrated to antagonize the deoxyguanosine and deoxyadensoine toxicity (125, 131). However, it is known that deoxycytidine competes with both and deoxyguanosine and deoxyadenosine not only for phosphorylation by deoxycytidine kinase but also for a common facilitated transport system that has a K_t value between 10 and 50 μM nucleoside (7–10).

In mammalian tissues the specific catalytic activity of deoxycytidine kinase, which phosphorylates deoxyguanosine, is highest in the thymus (23). Carson et al (88) have also demonstrated that of ten human tissues examined, the thymus possesses the greatest deoxyguanosine and deoxyadenosine phosphorylating activity, at least threefold greater than any other tissue.

Carson has also recently examined a cytoplasmic deoxynucleotidase activity in human T- and B-cell lines (132). This activity, which may be the same as that previously described (2, 3), is from ten- to twentyfold more effective in removing the phosphate from deoxyAMP and deoxyIMP than from the corresponding ribonucleotides, a substrate specificity that is the opposite to that seen for the 5'-ectonucleotidase (4–6). He has also demonstrated that human T-lymphoblastoid lines contain only about 30% as much deoxynucleotidase activity as do B cells (132). Thus it appears that

T-lymphoblastoid lines contain more deoxyguanosine phosphorylating activity and less deoxyguanylate hydrolyzing activity than do B-lymphoblastoid lines.

In a recent series of experiments Cohen et al have correlated a number of these enzymes with the stages of T-cell differentiation. Thymocytes, compared to peripheral T cells and B cells, have the highest deoxycytidine kinase activity, lowest ecto-5'-nucleotidase, most rapid accumulation of dGTP from deoxyguanosine, and the greatest sensitivity to the cytotoxicity of exogenous deoxyguanosine. As T cells mature they fail to accumulate dGTP and lose their sensitivity to deoxyguanosine (127).

ADENOSINE DEAMINASE DEFICIENCY

Natural History of the Disease

The clinical and immunologic phenotypes of adenosine deaminase deficient patients have been reviewed (133). Approximately one third of the patients have significantly decreased numbers of lymphocytes in peripheral blood at birth, and the majority of other infants with the disease will develop its signs and symptoms by three months of age (133). Even within a sibship there may be significant differences in the age of onset of the immunodeficiency (133). Patients with adenosine deaminase deficiency are subject to recurrent chronic infection with viral, fungal, protozoal, and bacterial organisms from which they are likely to succumb (60, 133). The children with adenosine deaminase deficiency usually fail to thrive and may have associated bony abnormalities, hair growth abnormalities and occasionally nonspecific neurologic disorders (60, 133, 134).

Adenosine deaminase deficiency has been diagnosed prenatally (135, 136), and in at least one affected individual who was immunodeficient at birth, the cord blood contained adenine metabolites at abnormal concentrations, comparable to those seen in older affected children (18). The majority of the affected children have a severe combined immunodeficiency (133).

Many have no detectable lymphocytes in peripheral blood or bone marrow. Cell-mediated immunity is nonexistent; T cells, defined by E rosettes, are most often absent; and lymphocyte proliferative responses are not demonstrable. Agammaglobulinemia is the rule (some residual maternal IgG may be detected), and specific antibody responses are absent.

In 10% of the patients with adenosine deaminase deficiency, the presentation and disease manifestations are less-clear-cut. In some, the cellular immune deficit may not be as profound, although most are severely affected. More impressive is the fact that the patients in this group have normal or near-normal immunoglobulin levels and some possess the capacity to mount an antibody response to some antigens but not to others (51). The natural history in this group is that of immunologic attrition with loss of residual cellular and humoral immune function over time.

At the present time, it is somewhat difficult to explain the heterogeneity of clinical manifestations in adenosine deaminase deficiency. Both environmental and genetic factors may contribute to this phenotypic diversity. Hirschhorn has recently reviewed a number of these infants to analyze the influence of environment and heredity on the clinical and immunological state. She presented evidence to suggest similarities between siblings but environmentally determined differences in phenotypic expression (133).

The most obvious pitfall is to attempt to correlate red cell enzyme activity with the specific abnormalities of lymphocyte function. The extreme lymphopenia in these infants precludes studies of enzyme activity in the target cells themselves, the T or B lymphocytes. Studies of fibroblasts or B-cell lines may not adequately reflect the conditions within defined lymphocyte subpopulations.

Enzyme deficient children with the immunodeficiency accumulate abnormally high concentrations of adenosine and deoxyadenosine in their plasma and urine but excrete normal quantities of total purines (14–17); the adenosine and deoxyadenosine in absolute quantitative terms are a relatively minor component of total purines excreted. The erythrocytes, lymphocytes, and bone marrow cells from adenosine deaminase deficient patients exhibit markedly elevated concentrations of deoxyATP (16, 137–140) and, where it has been looked for, deoxyADP (138, 140). In several cases the concentration of deoxyATP in erythrocytes has been equal to or greater than that of ATP, practically millimolar concentrations.

Infusions of irradiated erythrocytes containing normal levels of adenosine deaminase activity have been shown to be efficacious for at least partial reconstitution of the immunity system in some adenosine deaminase deficient patients (63, 64). In essentially all of the patients who have received infusion of irradiated erythrocytes, the erythrocytic dATP concentrations have fallen with a half-life of 3 to 5 days (137), only to gradually reappear at the markedly elevated levels some several weeks after the infusion (137; E. W. Gelfand and D. W. Martin,).

Four children from three different families have been noted to lack erythrocyte adenosine deaminase activity but possess normal immunity (65–67). In children from at least two of these families it has been demonstrated that their culture fibroblasts or lymphoblasts contain 25–50% of the normal level of adenosine deaminase activity (66), whereas the comparable cells from the immunodeficient adenosine deaminase deficient patients usually contain less than 1 percent of and certainly less than 10% of the normal adenosine deaminase activity (141–143). The erythrocytes and lymphocytes from these children do not contain elevated deoxyATP (137), and the children do not accumulate adenosine or deoxyadenosine in their plasma or urine at concentrations anywhere near those occuring in the comparable fluids of immunodeficient, adenosine deaminase deficient patients (66). Interestingly all three of these families are black, two from Africa and one from the United States.

The Adenosine Deaminase Enzyme

Adenosine deaminase is a ubiquitous enzyme (13), the structural gene for which is located on chromosome 20 in humans (144). However the physical state of the enzyme differs in different human tissues (13, 145–147), where the 38,000-dalton catalytic subunit is associated to variable extents with another high-molecular-weight protein referred to as complexing protein or binding protein (148–150), the expression of which is controlled by genes on human chromosomes 6 and 2 (151). In addition, it has been recently shown that the human thymus leukemia associated antigen (HThy-L) on normal and malignant thymocytes is identical to the 38,000-dalton catalytic subunit of adenosine deaminase (92).

By immunologic techniques the ratio of catalytic activity to immunoreactive activity of adenosine deaminase is altered in a number of patients deficient for that enzyme (141–143, 147), thus providing evidence that they harbor a mutation in the structural gene for adenosine deaminase. In addition, those children with absent erythrocyte deaminase activity but immunocompetency have a high level of residual enzyme activity in nucleated cells, and the residual activities exhibit unusual heat lability, which suggests that those children also harbor mutations in the structural gene of that enzyme, perhaps even the same allele. All enzyme deficient patients examined have an adenosine deaminase complexing protein that is normal, qualitatively and quantitatively (151a).

The specific catalytic activity of adenosine deaminase is highest in normal lymphoid tissues, particularly the thymus and spleen (13, 88, 152–154), and in the lymphoid cells it is highest in T lymphocytes (155). The distribution of the complexing protein is highest in kidney but present in many other tissues including human serum (149, 150). There is an interesting inverse correlation between the level of adenosine deaminase and that of purine nucleoside phosphorylase in rat lymphocyte populations (89).

Adenosine deaminase deaminates adenosine and deoxyadenosine, and its kinetic parameters have been recently reviewed (1). Adenosine deaminase is also capable of demethylating O^6-methyl deoxyguanosine (156), a metabolite that may occur naturally in individuals exposed to agents that methylate DNA (157).

Specific Metabolic Abnormalities and Effects in Model Systems

Because both the thymus-dependent and thymus-independent or B-cell lineage of the immunity system are affected in adenosine deaminase deficiency, both T-cell and B-cell model systems have been deployed to study the metabolic abnormalities in adenosine deaminase deficiency. The availability of pharmacologic inhibitors of adenosine deaminase, particularly erythro-9-(2-hydroxy-3-nonyl) adenine (158) and deoxycoformycin (159) have greatly aided these studies, but not without the usual risks inherent in using pharmacologic agents to mimic genetic diseases.

In the presence of an adenosine deaminase inhibitor, cultured mouse T-lymphosarcoma (S49) cells are sensitive to the presence of micromolar adenosine in the culture medium. (160). It has been demonstrated that the toxicity of low concentrations of adenosine is dependent upon the transport of adenosine into the cell and phosphorylation by the enzyme adenosine kinase (10, 161) and is associated with depletion of PPriboseP and thereby inhibition of de novo purine and de novo pyrimidine biosynthesis (1, 20, 161–163). The toxicity of these low concentrations of adenosine in the S49 cells can be prevented by the simultaneous presence of micromolar quantities of uridine. Mutants of S49 cells that are unresponsive to cAMP are normally sensitive to adenosine toxicity (14). At higher adenosine concentrations, in the absence of adenosine kinase activity or in the presence of uridine, adenosine is toxic by a second mechanism that involves the accumulation of S-adenosylhomocysteine from adenosine and homocysteine in a reaction catalyzed by the enzyme S-adenosylhomocysteine hydrolase (30, 164). S-adenosylhomocysteine is a potent competitive inhibitor of S-adenosylmethionine-mediated methylation reactions, including the methylation of DNA and ribosomal RNA precursors (32–35). Adenosine has a specific effect localized to the nucleolus in cultured cells (165); when S-adenosylhomocysteine is increased in cultured cells, the 45S ribosomal RNA precursor accumulates due to a block in its methylation-dependent processing (166). S-adenosylmethionine-mediated methylation is also necessary for chemotaxis of human leukocytes (167).

In the presence or absence of an inhibitor of adenosine deaminase, 2'-deoxyadenosine is also a potent inhibitor of cell growth (168–176). In the mouse T-lymphosarcoma (S49) cell system with erythro-9(2-hydroxy-3-nonyl) adenine (EHNA) or deoxycoformycin (described above) it has been demonstrated with the use of somatic cell mutants that in order for deoxyadenosine to exert its growth inhibitory effects it must be transported into the cytoplasm and phosphorylated by the enzymes adenosine kinase or deoxycytidine kinase to accumulate as deoxyATP (25). A mutant S49 cell deficient in adenosine kinase exhibits partial resistance to the toxicity of deoxyadenosine, but the deficiency of deoxycytidine kinase exhibits no decrease in sensitivity to deoxyadenosine toxicity (25). However, a mutant deficient in both adenosine kinase and deoxycytidine kinase exhibits markedly diminished sensitivity to deoxyadenosine. Thus in the cultured mouse T-lymphosarcoma cells, adenosine kinase is primarily responsible for the phosphorylation of adenosine and deoxyadenosine, while deoxycytidine kinase can phosphorylate deoxyadenosine but only at higher concentrations (25).

It has recently been demonstrated that deoxyadenosine is a potent suicide inhibitor of the enzyme S-adenosylhomocysteine hydrolase in vitro (176). Hershfield et al have also demonstrated that adenosine deaminase deficient patients have severely decreased S-adenosylhomocysteine hydrolase activity in their erythrocytes before treatment (177, 178). Thus one would anticipate that the inhibition of S-adenosylhomocysteine hydrolase, which is responsible for the removal of S-adenosylhomocysteine, may well lead to the accumulation of this product of the S-adenosylmethionine-mediated methylation reactions, and eventually to an inhibition of the latter reactions.

Two other types of T-cell model systems have been employed to study the effects of adenosine deaminase deficiency primary cells and established human cell lines. Thymocytes or peripheral blood lymphocytes have been used to study the toxicity of adenosine and deoxyadenosine in the presence of an adenosine deaminase inhibitor.

Deoxyadenosine in the presence of deoxycoformycin or EHNA will markedly inhibit human lymphocyte activation by phytohemoglutinin (109, 124, 125, 130, 179, 180). However, it seems that the inhibition of stimulation is expressed only when the deoxyadenosine or, for that matter, adenosine are present with deoxycoformycin during the first 24 hr of exposure to lectin (181), which suggests that the mechanism may not involve ribonucleotide reductase inhibition, even though deoxycytidine and thymidine reverse deoxyadenosine toxicity (181). Adenosine has also been demonstrated to inhibit S-adenosylmethionine-mediated methylation reactions in lectin-stimulated peripheral blood lymphocytes (182).

Established lymphoblast lines from humans have been used as model systems in which to study the toxicity of adenosine and deoxyadenosine. In the presence of an adenosine deaminase inhibitor, B-lymphoblast lines are sensitive to the toxic effects of adenosine (183), even when their adenosine kinase is defective (184). The majority of this adenosine-mediated toxicity in B lymphoblast appears to be secondary to an accumulation of S-adenosylhomocysteine as described for the S49 cell (164, 185). The toxicity of deoxyadenosine in the B lymphoblasts requires significantly higher concentrations than those effective in S49 cells (29). In human B lymphoblasts, deoxyadenosine can be phosphorylated by either adenosine kinase or deoxycytidine kinase, but mutants in either one or both of these phosphorylating enzymes behave in a manner analogous to the comparable S49 cell mutants described above (29). That is, a deficiency of adenosine kinase causes decreased sensitivity to deoxyadenosine, deoxycytidine kinase deficiency does not affect the sensitivity of the cell to deoxyadenosine, but a deficiency of both adenosine kinase and deoxycytidine kinase results in a cell that is insensitive to deoxyadenosine toxicity (29). In addition, deoxycytidine conveys some protection from deoxyadenosine toxicity in wild-type B lymphoblasts (29), which suggests that depletion of deoxyCTP may mediate some of the deoxyadenosine toxicity even in the B lymphoblasts.

Deoxyadenosine clearly is capable of inactivating the S-adenosylhomocysteine hydrolase of the B-lymphoblasts lines; and, in fact, adenosine kinase deficiency conveys some protection against deoxyadenosine toxicity as a result of the accumulation of adenosine, which protects against the suicide inactivation of the hydrolase by deoxyadenosine (186).

Some of the most informative experiments have been comparisons of the sensitivities of human B lymphoblasts and human T lymphoblasts in continuous culture exposed to deoxyadenosine in the presence of a deaminase inhibitor. Using different T- and B-lymphoblast lines several groups have demonstrated independently the greater sensitivity of the T-lymphoblast lines to growth inhibition by deoxyadenosine (129–131, 187, 188). The concentrations of exogenous deoxyadenosine necessary to inhibit growth of T lymphoblasts by 50% have been an order of magnitude lower than those necessary to inhibit the growth of B lymphoblasts. In addition, Mitchell et al (129) and Carson et al (188) have demonstrated that T-cell lines exposed to EHNA and any given level of exogenous deoxyadenosine accumulate much higher concentrations of dATP than do similarly exposed B lymphoblasts. This increased "trapping" of deoxynucleosides in the form of deoxynucleotides in the T lymphoblasts seems to result from two phenomena. First, the T lymphoblasts contain significantly greater deoxyadenosine and deoxyguanosine phosphorylating activities (128, 188), and second, the deoxynucleotides are degraded more slowly in the T lymphoblasts than in the B lymphoblasts (188, 189).

It is clear that in both T and B lymphoblasts deoxycytidine kinase is responsible for the phosphorylation of deoxyguanosine; and T lymphoblasts contain significantly higher levels of deoxycytidine kinase than do B cells or B lymphoblasts or, for that matter, other cells in the human body (see above). However, the responsibility for the phosphorylation of deoxyadenosine is not so clearly defined. In cell-free extracts, deoxycytidine kinase is more active than is adenosine kinase in the phosphorylation of deoxyadenosine, the latter having higher K_m and lower V for deoxyadenosine than the former (1, 26). However, deoxycytidine kinase deficient human B lymphoblasts exhibit normal sensitivity to deoxyadenosine and normal accumulation of deoxyATP from deoxyadenosine (29). Adenosine kinase deficient B lymphoblasts, on the other hand, exhibit decreased sensitivity to deoxyadenosine and decreased accumulation of deoxyATP from deoxyadenosine (29). As in the mouse T-lymphosarcoma cell system, the human B-lymphoblast mutants deficient in both adenosine kinase and deoxycytidine kinase accumulate no deoxyATP from deoxyadenosine and exhibit a much greater insensitivity to deoxyadenosine than do the adenosine kinase deficient mutants (29). Thus, it seems clear that in human B lymphoblasts adenosine kinase is responsible for the phosphorylation of deoxyadenosine, but in B lymphoblasts the concentration of deoxyadenosine necessary to exert a cytotoxic effect and to allow the detection of accumulated deoxyATP is considerably greater than is required for similar effects in T lymphoblasts (129–131). It is possible that at the lower deoxyadenosine levels that affect the function of a T lymphoblast, the relative roles of these two phosphorylating enzymes in the trapping of deoxyATP are probably different. The appropriate adenosine kinase-, deoxycitidine kinase-, and the double enzyme-deficient mutants have recently been isolated and appear to confirm the suggestion that at low deoxyadenosine concentrations ($<2 \mu$M) the toxicity is mediateo primarily by deoxycytidine kinase (B. Ullman and D. W. Martin, unpublished observations; M. S. Hershfield, personal communication; D. A. Carson, personal communication).

Most human cells contain an ecto-5'-nucleotidase that is capable of dephosphorylating deoxyAMP or deoxyGMP, but only from the outside of the cell (4–6). B lymphocytes from the peripheral blood and B lymphoblasts appear to contain more of the ectonucleotidase than do their T-cell counterparts. However, it is in cytoplasm that the deoxynucleoside triphosphates are more stable in T cells than in B cells (129–131), and therefore it seems that there should be a difference in a cytoplasmic nucleotidase to account for the different sensitivities of T cells and B cells. Carson has recently studied a cytoplasmic deoxynucleotidase in T lymphoblasts and B lymphoblasts (132). This deoxynucleotidase is probably one described previously (2, 3); it has a much lower K_m and higher V for deoxynucleotides, particularly deoxyAMP and deoxyIMP, than for the corresponding ribonucleotides, properties quite contrary to those of the ecto-5'-nucleotidase. The human B-lymphoblast lines studied by Carson contain three- to fivefold higher concentrations of this deoxynucleotidase activity than do human T-lymphoblast lines (132). It remains to be seen whether this specific deoxynucleotidase is responsible for the differential labilities of deoxynucleoside triphosphates in T cells versus B cells.

In many of the above-mentioned studies of lymphoblast lines and peripheral blood lymphocytes, it has been noted that deoxycytidine provides protection against deoxyguanosine and less frequently to deoxyadenosine toxicity. However, the enzyme responsible for the phosphorylation of deoxyguanosine and to some extent, perhaps, for the phosphorylation of deoxyadenosine is deoxycytidine kinase, which of course phosphorylates deoxycytidine. Thus the protection of a cell from deoxynucleoside toxicity by the presence of deoxycytidine in many cases will be due to competition for the phosphorylating activity, which results in diminished accumulation of the toxic purine deoxynucleotide. Only from experiments in which deoxynucleoside-resistant mutants do not become depleted of deoxyCTP when accumulating large quantities of deoxyGTP or deoxyATP can one conclude that the mechanism of toxicity results from deoxyCTP depletion (121). No experiments comparable to those in S49 cells have been reported for human T-lymphoblast cells.

ANIMAL MODELS OF ADENOSINE DEAMINASE AND PURINE NUCLEOSIDE PHOSPHORYLASE DEFICIENCY

Mice, dogs, and humans have been administered the adenosine deaminase inhibitor deoxycoformycin with variable responses. Mice given daily injections of deoxycoformycin (0.2 or 1 mg/kilogram) exhibit greatly increased levels of dATP in erythrocytes and thymus but not in liver, spleen, intestinal mucosa, bone marrow, skin, or peripheral lymphocytes (190). Above 0.3 mg/kilogram deoxycoformycin was toxic to newborn animals (190). The administration of deoxycoformycin causes a severe lymphopenia in about three days (191).

Similar observations have been made in terminally ill humans given parenteral deoxycoformycin (192). The use of deoxycoformycin as a chemotherapeutic agent in the treatment of a T-cell leukemia has recently been reported (193). The pharmacologic simulation of adenosine deaminase deficiency in humans seems to be accompanied by both the cellular and the biochemical mimicry of adenosine deaminase deficiency. This mimicry includes a diminished S-adenosylhomocysteine hydrolase in the erythrocytes of deoxycoformycin-treated patients (166) and, interestingly, an increase in S-adenosylhomocystine concentration in cells (166).

Allopurinol riboside is an inhibitor of mouse purine nucleoside phosphorylase, although it has a high K_i value (300 μM). Mice administered high doses of allopurinol riboside exhibit a diminished cellular response but normal humoral response to immunization with sheep erythrocytes (194). No metabolic studies have accompanied these interesting immunologic studies, and the validity of the model for the human disease remains to be shown.

TRANSCOBALAMIN II DEFICIENCY

The first child with transcobalamin II deficiency was initially recognized to have a macromegalocytic anemia, leukopenia, granulocytopenia, and thrombocytopenia (57). He required massive doses of vitamin B_{12} to promote reticulocytosis and leukocytosis and when further studied was shown to have a complete absence of cobalamin binding by transcobalamin II (57, 58). Further studies revealed that his plasma contained 15% of the normal level of immunoreactive transcobalamin II (58). The child had recurrent respiratory infections including bronchopneumonia before treatment with cobalamin, and was hypogammaglobulinemic in spite of normal numbers of B lymphocytes in his peripheral blood. After the commencement of cobalamin therapy, circulating specific antibodies began to appear without new antigenic stimulation and, in fact, booster injections of diphtheria toxoid induced an IgG class anamnestic response (58). Thus, the B lymphocytes had recognized the specific antigen injected earlier, had been primed to become memory cells, but had been unable to progress to the stage of antibody secretion, a process that requires clonal expansion. The child also exhibited abnormal killing of staphylococci by his granulocytes, but the addition of folinic acid (N^5-formyltetrahydrofolate) to the vitamin B_{12} supplement normalized his leukocyte killing function (58).

Interestingly these patients do not exhibit a methylmalonic aciduria or homocysteinuria, both of which usually accompany congenital or acquired vitamin B_{12} deficiencies (195).

There have been no published observations in model systems relevant to this vitamin-responsive immunodeficiency disease, but, as mentioned above, the involvement of cobalamin in homocysteine methylation suggests an intriguing interface between the hypogammaglobulinemia seen with adenosine deaminase deficiency and that present in transcobalamin II deficiency. It is certainly possible that S-adenosylmethionine-mediated methylation reactions are necessary for normal B-cell function and that in adenosine deaminase deficiency those reactions are inhibited by the accumulation of S-adenosylhomocysteine, whereas in the transcobalamin II defect there is a relative deficiency of methionine due to the blocked homocysteine transmethylation reaction.

CAN THE BIOCHEMISTRY EXPLAIN THE IMMUNODEFICIENCY?

From immunologic studies of immunodeficiency these questions await explanation: 1. Why does a deficiency of adenosine deaminase or of purine nucleoside phosphorylase, both ubiquitous enzymes, lead to dysfunction, exclusively or predominantly, of the immunity system? 2. Why is only the thymus-dependent arm of the immunity system defective in the purine nucleoside phosphorylase deficiency, whereas a severe combined immunodeficiency results from loss of adenosine deaminase activity? 3. Why in purine nucleoside phosphorylase deficiency is there a compromise of suppressor cell activity while T-helper function is unimpaired? 4. Is the pathogenesis of the B-cell dysfunction in transcobalamin II deficiency related pathogenically to the dysfunction of the humoral immunity in adenosine deaminase deficiency? 5. Are other related biochemical abnormalities responsible for other specific inherited immunodeficiency diseases in humans.

The prethymic and intrathymic T-precursor cells appear to be uniquely endowed with the ability to salvage deoxyguanosine and deoxyadenosine to deoxyGTP and deoxyATP, respectively. This unique endowment seems to be related directly to the relatively high specific catalytic activity of deoxycytidine kinase, an enzyme that is one of two capable of phosphorylating deoxyadenosine but the only enzyme capable of phosphorylating deoxyguanosine. In addition, preliminary evidence suggests that compared to other cells in the body and particularly other cells of the immunity system, these immature T cells may be relatively deficient in a cytoplasmic deoxyribonucleotidase activity capable of enhancing the turnover of deoxyribonucleoside triphosphates. Why immature T-cells exhibit such properties is not clear; perhaps because of the need for extensive expansion of the T-lymphocyte precursor population during ontogeny, they acquire a significant advantage from being able to salvage deoxyribonucleosides as DNA precursors, thereby conserving energy and material. The presence of prethymic and post-thymic precursor T cells in purine nucleoside phosphorylase deficiency, evidence for some maturation of these cells to the intrathymic or early post-thymic stage of maturation (capable of providing help but not suppression), and absence of post-thymic cells suggest that these intermediate stages of T lineage are most susceptible, at least to deoxyguanosine.

It is not clear why thymus-dependent lymphocyte function is exclusively affected in purine nucleoside phosphorylase deficiency whereas adenosine deaminase deficiency is nearly always a combined T- and B-cell deficiency disease. There are several possible but as yet unproven explanations. Since all studies of B cell function have been carried out only in cultured B cell lines, it is conceivable that the earliest stages of B-cell differentiation are as uniquely susceptible as immature T cells to the accumulation of deoxyadenosine. Alternatively, the B-cell dysfunction may be due to another metabolic abnormality associated only with adenosine deaminase deficiency. As described above, deoxyATP, which does accumulate in lymphocytes of adenosine deaminase deficient patients, is a potent inhibitor of the reduction of all four ribonucleoside diphosphates by ribonucleotide reductase. Hence the effect of the accumulated deoxyATP may have a more profound effect on DNA synthesis than would the accumulation of deoxyGTP, since the latter inhibits the reduction of the pyrimidine nucleoside diphosphates.

As for other potentially detrimental metabolic abnormalities occurring exclusively in adenosine deaminase deficiency, it is possible that the B-cell dysfunction results from an inhibition of S-adenosylmethionine-mediated methylation reactions on which the B cells may be uniquely dependent. Although there is evidence that S-adenosylhomocysteine hydrolase is somewhat inhibited in erythrocytes of purine nucleoside phosphorylase deficient patients, there is no evidence of accumulation of a methylation inhibitor analogous to S-adenosylhomocysteine in cells of those patients. If S-adenosylhomocysteine is playing a significant role in the pathogenesis of the immunodeficiency diseases associated with the loss of these two enzyme activities, it is likely to be at the level of B-cell differentiation and function.

There is strong evidence from purine nucleoside phosphorylase deficient children in several families that T-suppressor function is more severely affected than is T-helper function. As noted above, the intrathymic expansion of precursor T cells is of course proliferative dependent; thus inhibition of DNA synthesis due to precursor pool depletion will affect all subsets of T lymphocytes. However, in the event that the block of proliferation is not absolute and some T-lymphocyte precursor cells mature beyond the intrathymic stage, one might expect that depletion of intracellular deoxynucleoside triphosphate pools would more severely affect the proliferative functions than the nonproliferative functions of those peripheral T lymphocytes. T-cell suppressor functions have been shown by numerous laboratories and various criteria to be proliferative dependent, while T-helper functions are not. Thus in purine nucleoside phosphorylase deficiency, there may exist a trickle of T-lymphocyte maturation beyond the intrathymic stage, and those that do escape the potentially detrimental effects of accumulated deoxyGTP on their proliferative expansion may be prevented from participating in any further proliferative-dependent mature functions.

If as suggested above, B-cell dysfunction in adenosine deaminase deficiency is secondary to loss of S-adenosylmethionine-mediated methylation functions, a deficiency of intracellular cobalamin due to defective transcobalamin II would greatly hinder the formation of methionine from homocysteine and have the potential of diminishing the availability of S-adenosylmethionine.

In many hypotheses for the pathogenesis of these immunodeficiency diseases, the allosteric regulation of ribonucleotide reductase has been given a major role. To determine whether this mechanism is actually responsible and which of the many possible explanations described in this review lead to the diseases, it will be necessary to combine more extensive clinical studies with more sophisticated animal model systems. The ability to isolate mutant cells containing ribonucleotide reductase molecules that are genetically unresponsive to feedback inhibition by deoxynucleoside triphosphates may assist in the resolution of these interesting questions and allow one to determine whether the hypotheses derived from in vitro cell culture model systems are valid in vivo.

The interdisciplinary approaches used over the last six to eight years to study the molecular pathogenesis of these inborn errors of metabolism have provided a surge of enthusiasm for the use of biochemical and genetic approaches to understanding both the physiologic and pathophysiologic aspects of the immunity system. Thus it seems that once the precise biochemical pathogenesis of the immunodeficiency diseases associated with the loss of purine nucleoside phosphorylase or of adenosine deaminase have been defined, one should be able to predict with reasonable confidence which specific inborn error of metabolism leads to which specific dysfunction of the immunity system. Of course, a more thorough "biochemical anatomy" of the ontogeny and function of the numerous cell types of the immunity system would no doubt accelerate this process. The expansion of this knowledge of the normal and abnormal biochemistry of the immunity system should provide information that will be of use in the pharmacologic manipulation of normal immune responsiveness and the elimination of abnormal or malignant cells.

Literature Cited

1. Fox, I. H., Kelley, W. N. 1978. *Ann. Rev. Biochem.* 47:655–86
2. Magnusson, G. 1971. *Eur. J. Biochem.* 20:225–30
3. Fritzson, P. 1978. *Adv. Enzyme Regul.* 16:43–61
4. DePierre, J. W., Karnovsky, M. L. 1974. *Science* 183:109
5. Uusitalo, R. J., Karnovsky, M. J. 1977. *J. Histochem. Cytochem.* 25:87
6. Fleit, H., Conklyn, M., Stebbins, R. D., Silber, R. 1975. *J. Biol. Chem.* 250:8889
7. Plagemann, P. G. W. 1971. *J. Cell Physiol.* 77:213–40
8. Plagemann, P. G. W., Erbe, J. 1974. *J. Cell Physiol.* 83:337–43
9. Oliver, J. M., Paterson, A. R. P. 1971. *Can. J. Biochem.* 49:262–70
10. Cohen, A., Ullman, B., Martin, D. W., Jr. 1979. *J. Biol. Chem.* 254:112–16
11. Frederiksen, S. 1966. *Arch. Biochem. Biophys.* 113:383–88
12. Agarwal, R. P., Sagar, S. M., Parks, R. E. 1975. *Biochem. Pharmacol.* 24:693–701
13. Van der Weyden, M., Kelley, W. N. 1976. *J. Biol. Chem.* 251:5448–56
14. Simmonds, H. A., Sahota, A., Potter, C. F., Cameron, J. S. 1978. *Clin. Sci. Mol. Med.* 54:579–84
15. Mills, G. C., Goldblum, R. M., Schmalstieg, F., Newkirk, K. F. 1978. *Biochem. Med.* 20:180–99
16. Donofrio, J., Coleman, M. S., Hutton, J. J. 1978. *J. Clin. Invest.* 62:884–87
17. Mills, G. C., Schmalstieg, F. C., Trimmer, K. B., Goldman, A. S., Goldblum, R. M. 1976. *Proc. Natl. Acad. Sci. USA* 73:2867–71
18. Hirschhorn, R., Roegner, V., Rubinstein, A., Papageorgiou, P. 1980. *J. Clin. Invest.* 65:768–71
19. Chan, T.-S. 1979. *Proc. Natl. Acad. Sci. USA* 76:925–29

20. Ishii, K., Green, H. 1973. *J. Cell. Sci.* 13:429
21. Lindberg, B., Klenow, H., Hansen, K. 1967. *J. Biol. Chem.* 242:350–56
22. Anderson, E. P. 1973. *Enzymes* 9:49–96
23. Durham, J. P., Ives, D. H. 1969. *Mol. Pharmacol.* 5:358–75
24. Krygier, V., Momparler, R. L. 1971. *J. Biol. Chem.* 246:2745–51
25. Ullman, B., Gudas, L. J., Cohen, A., Martin, D. W. Jr. 1978. *Cell* 14:365–75
26. Carson, D. A., Kaye, J., Seegmiller, J. E. 1979. In *Inborn Errors of Specific Immunity,* ed. B. Pollara, R. J. Pickering, H. J. Meuwissen, I. H. Porter, pp. 221–33. New York: Academic. 469 pp.
27. Fox, I. H., 1979. See Ref. 25, pp. 93–114
28. Carson, D. A., Kaye, J. 1978. *Ciba Found. Symp.* 68:115–33
29. Ullman, B., Levinson, B. B., Hershfield, M. S., Martin, D. W. Jr. 1981. *J. Biol. Chem.* 256:848–52
30. De la Haba, G., Cantoni, G. L. 1959. *J. Biol. Chem.* 234:603–8
31. Hershfield, M. S., Kredich, N. M. 1978. *Science* 202:757–60
32. Zappia, V., Zydek-Cwick, C. R., Schlenk, F. 1969. *J. Biol. Chem.* 244:4499–4509
33. Pegg, A. E. 1971. *FEBS Lett.* 16:13–16
34. Kerr, S. J. 1972. *J. Biol. Chem.* 247:4248–52
35. Glick, J. M., Ross, S., Leboy, P. S. 1975. *Nucleic Acids Res.* 2:1639–51
36. Krenitsky, T. A. 1967. *Mol. Pharmacol.* 3:526–36
37. Parks, R. E., Agarwal, R. P. 1973. *Enzymes* 9:483–514
38. Cohen, A., Doyle, D., Martin, D. W. Jr., Ammann, A. J. 1976. *N. Engl. J. Med.* 295:1449–54
39. Siegenbeek van Heukelom, L. H., Akkerman, J. W. N., Staal, G. E. J., DeBruyn, C. H. M. M., Stoop, J. W., Zegers, B. J. M., DeBree, P. K., Wadman, S. K. 1977. *Clin. Chim. Acta* 74:271–79
40. Rosenbloom, F. M., Henderson, J. F., Caldwell, I. C., Kelley, W. N., Seegmiller, J. E. 1968. *J. Biol. Chem.* 243:1166–73
41. Lajtha, L. G., Vane, J. R. 1958. *Nature* 182:191
42. Mager, J., Hershko, A., Zeitlin-Beck, R., Shoshami, A. Razin, A. 1967. *Biochim. Biophys. Acta* 149:50
43. Friedman, T., Seegmiller, J. E., Subak-Sharpe, J. H., 1969. *Exp. Cell Res.* 56:425–29
44. Gudas, L. J., Ullman, B., Cohen, A., Martin, D. W. Jr. 1978. *Cell* 14:531–8
45. Krenitsky, T. A., Tuttle, J. V., Koszalka, G. W., Chen, I. S., Beachman, L.,

Rideout, J. L., Elion, G. B. 1976. *J. Biol. Chem.* 51:4055–61
46. Hershfield, M. S. 1981. *J. Clin. Invest.* 67:696
47. Taylor, R. T., Weissbach, H. 1972. *Enzymes* 9:121
48. Larrabee, A. R., Rosenthal, S., Cathrow, R. E., Buchanan, J. M. 1963. *J. Biol. Chem.* 238:1025
49. Warren, L., Flaks, J. G., Buchanan, J. M. 1957. *J. Biol. Chem.* 229:627
50. Frearson, P. M., Kit, S., Dubbs, D. R. 1965. *Cancer Res.* 25:737
51. Giblett, E. R., Anderson, J. E., Cohen, F., Pollara, B., Meuwissen, H. J. 1972. *Lancet* ii:1067–69
52. Giblett, E. R., Ammann, A. J., Wara, D. W., Sandman, R., Diamond, L. K. 1975. *Lancet* i:1010–13
53. Giblett, E. R. 1978. *Ciba Found. Symp.* 68:318
54. Spencer, N., Hopkinson, D. A., Harris, H. 1968. *Ann. Hum. Genet.* 32:9–14
55. Hopkinson, D. A., Cook, P. J., Harris, H. 1969. *Ann. Hum. Genet.* 32:361–67
56. Detter, J. C., Stamatoyannopoulos, G., Giblett, E. R., Motulsky, A. G. 1970. *J. Med. Genet.* 71:356–57
57. Hitzig, W. H., Dohrmann, U., Pluss, H. J., Vischer, D. 1974. *J. Ped.* 85:622–28
58. Hitzig, W. H., Frater-Schroder, M., Seger, R. 1978. *Ciba Found. Symp.* 68:77–88
59. Hall, C. A., Finkler, A. E. 1965. *J. Lab. Clin. Med.* 65:459–68
60. Meuwissen, H. J., Pollara, B., Pickering, R. J. 1975. *J. Pediatr.* 86:169–81
61. Polmar, S. H., Wetzler, E. M., Stern, R. C. 1975. *Lancet* ii:743–46
62. Rubinstein, A., Hirschhorn, R., Sicklick, M., Murphy, R. A. 1979. *N. Engl. J. Med.* 300:387–92
63. Polmar, S. H., Stern, R. C., Schwartz, A. L., Wetzler, E. M., Chase, P. A., Hirschhorn, R. 1976. *N. Engl. J. Med.* 295:1337–43
64. Polmar, S. H. 1978. *Ciba Found. Symp.* 68:213–30
65. Jenkins, T., Robson, A. R., Nurse, G. T., Lane, A. B., Hopkinson, D. A. 1976. *J. Pediatr.* 89:732–36
66. Hirschhorn, R., Roegner, V., Jenkins, T., Seaman, C., Piomelli, S., Borkowsky, W. 1979. *J. Clin. Invest.* 64:1130–39
67. Kelley, W. N. 1978. *Ciba Found. Symp.* 68:50
68. Staal, G. E. J., Stoop, J. W., Zegers, B. J. M., Siegenbeek van Heukelom, L. H., van der Vlist, M. J. M., Wadman, S. K., Martin, D. W. Jr. 1980. *J. Clin. Invest.* 65:103–8

69. Rich, K. C., Arnold, W. J., Pallela, T., Fox, I. H. 1979. See Ref 26, pp. 31–41
70. Biggar, W. D., Giblett, E. R., Ozere, R. L., Grover, B. D. 1978. *J. Pediatr.* 92:354–57
71. Gelfand, E. W., Dosch, H. M., Biggar, W. D., Fox, I. H. 1978. *J. Clin. Invest.* 61:1071–80
72. Gelfand, E. W., Dosch, H. M., Shore, A. 1978. In *Hematopoietic Cell Differentiation,* ed. D. W. Golde, M. J. Cline, D. Metcalfe, C. F. Fox, pp. 277–93, New York: Academic
73. Gelfand, E. W., Dosch, H. M., Shore, A., Limatibul, S., Lee, J. W. W. 1980. In *The Biological Basis for Immunodeficiency Disease,* ed. E. W. Gelfand, H. M. Dosch, pp. 39–56, New York: Raven
74. Scollay, R. G., Butcher, E. C., Weissman, I. L. 1980. *Eur. J. Immunol.* 10:210–18
75. Gelfand, E. W., Shore, A., Green, B., Lin, M. T., Dosch, H. M. 1979. *Clin. Immunol. Immunopath.* 12:119–23
76. Limatibul, S., Shore, A. H., Dosch, H. M., Gelfand, E. W. 1978. *Clin. Exp. Immunol.* 33:503–13
77. Reinherz, E. L., Strelkauskas, A. J., O'Brien, C., Schlossman, S. F. 1979. *J. Immunol.* 123:83–86
78. Moretta, L., Webb, S. R., Gross, C. E., Lydyard, P. M., Cooper, M. D. 1977. *J. Exp. Med.* 146:184
79. Strelkauskas, A. J., Callery, R. T., McDowell, J., Borel, Y., Schlossman, S. F. 1978. *Proc. Natl. Acad. Sci. USA* 75:5150–54
80. Reinherz, E. L., Schlossman, S. F. 1980. *Cell* 19:821–27
81. Shore, A., Dosch, H. M., Gelfand, E. W. 1978. *Nature* 274:586–87
82. Dosch, H. M., Gelfand, E. W. 1979. *Immunol. Rev.* 45:243–74
83. Cooper, M. D., Burrows, P. D., Lawton, A. R. 1980. See Ref. 73, pp. 57–70
84. Bollum, F. J. 1978. *Adv. Enzymol.* 47:347
85. Gregoire, K. E., Goldschneider, I., Barton, R. W., Bollum, F. J. 1979. *J. Immunol.* 123:1347–52
86. Pazimo, N. H., Ihle, J. N., Goldstein, A. L. 1978. *J. Exp. Med.* 147:708
87. Adams, A., Harkness, R. A. 1976. *Clin. Exp. Immunol.* 26:647–49
88. Carson, D. A., Kaye, J., Seegmiller, J. E. *Proc. Natl. Acad. Sci. USA* 74:5677–81
89. Barton, R., Martinuik, F., Hirschhorn, R., Goldschneider, I. 1980. *Cell Immunol.* 49:208–14

90. Tritsch, G. L., Minowada, J. 1978. *J. Natl. Cancer Inst.* 60:1301–4
91. Chechik, B. E., Pyke, K. W., Gelfand, E. W. 1976. *Int. J. Cancer* 18:551–56
92. Chechik, B. E., Schrader, W. P., Dadonna, P. E. 1980. *J. Natl. Cancer Inst.* 64:1077–83
93. Cohen, A., Lee, J. W. W., Dosch, H. M., Gelfand, E. W. 1980. *J. Immunol.* 125:1578–82
94. Edwards, N. L., Gelfand, E. W., Burk, L., Dosch, H. M., Fox, I. H., 1979. *Proc. Natl. Acad. Sci. USA* 76:3474–76
95. Cohen, A., Dosch, H. M., Gelfand, E. W. 1981. *Clin. Immunol. Immunopath.* In press
95a. Lopes, J., Zucker-Franklin, D., Silber, R. J. 1973. *J. Clin. Invest.* 52:1297
96. Ammann, A. J. 1978. *Ciba Found. Symp.* 68:55–69
97. Ammann, A. J., Wara, D. W., Allen, T. 1978. *Clin. Immunol. Immunopathol.* 10:262–69
98. Stoop, J. W., Zegers, B. J. M., Hendricks, G. F. M., Siegenbeek van Heukelom, L. H., Staal, G. E. J., DeBree, P. K., Wadman, S. K., Ballieux, R. E. 1977. *N. Engl. J. Med.* 296:651
99. Cohen, A., Staal, G. E. J., Ammann, A. J., Martin, D. W. Jr. 1977. *J. Clin. Invest.* 60:491–94
100. Van Gennip, A. H., Grift, J., DeBree, P. K., Zegers, B. J. M., Stoop, J. W., Wadman, S. K. 1979. *Clin. Chim. Acta* 93:419–28
101. Cohen, A., Gudas, L. J., Ammann, A. J., Staal, G. E. J., Martin, D. W. Jr. 1978. *J. Clin. Invest.* 61:1405–9
102. Rich, K. C., Mejias, E., Fox, I. H. 1980. *N. Engl. J. Med.* 303:973–77
103. Aitken, D. A., Ferguson-Smith, M. A. 1978. *Cytogenet. Cell Genet.* 22:490–92
104. Dallapiccola, B. Magnani, M., Dach, A. M., Giorgi, P. L. 1979. *Human Genet.* 50:341–43
105. Kim, B. K., Cha, S., Parks, R. E. 1968. *J. Biol. Chem.* 243:1763–70
106. Agarwal, K. C., Agarwal, R. P., Parks, R. E. 1973. *Fed. Proc.* 32:581
107. Edwards, Y. H., Edwards, P. A., Hopkinson, D. A. 1973. *FEBS Lett.* 32:235–37
108. Zannis, V. I., Doyle, D., Martin, D. W. Jr. 1978. *J. Biol. Chem.* 253:504–10
109. Borgers, M., Thone, F. 1978. *Histochem. J.* 10:721–30
110. Fox, I. H., Andres, C. M., Kaminska, J., Wortmann, R. L. 1978. *Ciba Found. Symp.* 68:193–205
111. Osborne, W. R., Chen, S. H., Giblett, E. R., Biggar, W. D., Ammann, A. J.,

Scott, C. R. 1977. *J. Clin. Invest.* 60:741–46

112. Gudas, L. J., Zannis, V. I., Clift, S. M., Ammann, A. J., Staal, G. E. J., Martin, D. W. Jr. 1978. *J. Biol. Chem.* 253:8916–24

113. McRoberts, J. A., Martin, D. W. Jr. 1980. *J. Biol. Chem.* 255:5605–15

114. Xeros, N. 1962. *Nature* 194:682–83

115. Galavazi, G., Schenk, H., Bootsma, D. 1966. *Exp. Cell Res.* 41:428–37

116. Morris, N. R., Fischer, G. A. 1961. *Fed. Proc.* 20:358

117. Schachtschabel, D. O., Cunze, P. 1970. *Humangenetik* 10:127–37

118. Thomson, M. J., Garland, M. R., Richards, J. F. 1971. *J. Cell Physiol.* 77:17–30

119. Theiss, J. C., Morris, N. R., Fischer, G. A. 1976. *Cancer Biochem. Biophys.* 1:211–14

120. Henderson, J. F., Scott, F. W., Lowe, J. K. 1980. *Pharmacol. Ther.* 8:573–604

121. Ullman, B., Gudas, L. J., Clift, S. M., Martin, D. W. Jr. 1979. *Proc. Natl. Acad. Sci. USA* 76:1074–78

122. Eriksson, S., Gudas, L. J., Ullman, B., Clift, S. M., Martin, D. W. Jr. 1981. *J. Biol. Chem.* In press

123. Hayashi, T. T., Garvey, B. I. 1968. *Am. J. Obstet. Gynecol.* 68:1154–61

124. Tattersall, M. H. B., Ganeshaguru, K., Hoffbrand, A. V. 1975. *Plant Physiol.* 42:907–10

125. Gelfand, E. W., Lee, J. J., Dosch, H. M. 1979. *Proc. Natl. Acad. Sci. USA* 76:1998–2002

126. Wortmann, R. L., Andres, C., Kaminska, J., Mejias, E., Gelfand, E., Arnold, W., Rich, K., Fox, I. H. 1979. *Arthritis Rheum.* 22:524–31

127. Dosch, H. M., Mansour, A., Cohen, A., Shore, A., Gelfand, E. W. 1980. *Nature* 285:494–96

128. Ochs, U. H., Chen, S. H., Ochs, H. D., Osborne, W. R., Scott, C. R. 1979. *J. Immunol.* 22:2424–29

129. Mitchell, B. S., Mejias, E., Daddona, P. E., Kelley, W. N. 1978. *Proc. Natl. Acad. Sci. USA* 75:5011–14

130. Hirschhorn, R., Bajaj, S., Borkowsky, W., Kowalski, A., Hong, R., Rubinstein, A., Papageorgiou, P. 1979. *Cell. Immunol.* 42:418–23

131. Ochs, U. H., Chen, S. H., Ochs, H. D., Scott, C. R., Wedgewood, R. 1979. *Fed. Proc.* 38:1222

132. Carson, D. A. 1981. *J. Immunol.*, In press

133. Hirschhorn, R. 1978. *Ciba Found. Symp.* 68:35–54

134. Hirschhorn, R., Papageorgiou, P. S., Kesarwala, H. H., Taft, L. T. 1980. *N. Engl. J. Med.* 303:377–80

135. Hirschhorn, R., Beratis, N., Rosen, F. S., Parkman, R., Stern, R., Polmar, S. 1975. *Lancet* i:73–75

136. Hirschhorn, R. 1979. In *Inborn Errors of Immunity and Phagocytosis,* ed. F. Guttler, pp. 121–8. Lancaster, England: MTP Press

137. Cohen, A., Hirschhorn, R., Horowitz, S. D., Rubinstein, A., Polmar, S. H., Hong, R., Martin, D. W. Jr. 1978. *Proc. Natl. Acad. Sci. USA* 75:472

138. Coleman, M. S., Donofrio, J., Hutton, J. J., Hahn, L., Daoud, A., Lampkin, B., Dyminski, J. 1978. *J. Biol. Chem.* 253:1619–26

139. Cohen, A., Gudas, L. J., Ullman, B., Martin, D. W. Jr. 1978. *Ciba Found. Symp.* 68:1386–89

140. Chen, S. H., Ochs, H. D., Scott, C. R., Giblett, E. R., Tingle, A. J. 1978. *J. Clin. Invest.* 62:1386–89

141. Chen, S. H., Scott, C. R., Swedberg, K. R. 1975. *Am. J. Hum. Genet.* 27:46–53

142. Hirschhorn, R., Beratis, N., Rosen, F. S. 1976. *Proc. Natl. Acad. Sci. USA* 73:213–17

143. Carson, D. A., Goldblum, R., Seegmiller, J. E. 1977. *J. Immunol.* 118:270–73

144. Mohandes, T., Sparkes, R. S., Passage, M. B., Sparkes, M. C., Miles, J. H., Kaback, M. M. 1980. *Cytogenet. Cell Genet.* 26:28–35

145. Hirschhorn, R. 1975. *J. Clin. Invest.* 55:661–67

146. Daddona, P. E., Kelley, W. N. 1977. *J. Biol. Chem.* 252:110–15

147. Daddona, P. E., Kelley, W. N. 1980. *Mol. Cell. Biochem.* 29:91–101

148. Daddona, P. E., Kelley, W. N. 1970. *J. Biol. Chem.* 253:4617–23

149. Schrader, W. P., Stacey, A. R. 1979. *J. Biol. Chem.* 254:11958–63

150. Schrader, W. P., Woodward, F. J., Pollara, B. 1979. *J. Biol. Chem.* 254:11964–68

151. Koch, G., Shows, T. B. 1980. *Proc. Natl. Acad. Sci. USA* 77: 4211–15

151a. Daddona, P. E., Kelley, W. N. 1978. *Ciba Found. Symp.* 68:178

152. Hirschhorn, R., Martiniuk, F., Rosen, F. S. 1978. *Clin. Immunol. Immunopath.* 9:287

153. Adams, A., Harkness, R. A. 1976. *Clin. Exp. Immunol.* 26:647

154. Brady, T. G., O'Donovan, C. I. 1965. *Comp. Biochem. Physiol.* 14:101

155. Barton, R., Martiniuk, F., Hirschhorn,

R., Goldschneider, I. 1979. *J. Immunol.* 122:216–20

156. Pegg, A. E., Swann, P. F. 1979. *Biochim. Biophys. Acta* 565:241–52
157. Pegg, A. E. 1977. *Adv. Cancer Res.* 25:195–269
158. Schaeffer, H. J., Schwender, C. F. 1974. *J. Med. Chem.* 17:6
159. Agarwal, R. P., Spector, T., Parks, R. E. 1977. *Biochem. Pharmacol.* 26: 359–67
160. Ullman, B., Cohen, A., Martin, D. W. Jr. 1976. *Cell* 9:205–11
161. Gudas, L. J., Cohen, A., Ullman, B., Martin, D. W. Jr. 1978. *Somatic Cell Genet.* 1:201–19
162. Green, H., Chan, T. S. 1973. *Science* 182:836
163. Planet, G., Fox, I. H. 1976. *J. Biol. Chem.* 251:5839
164. Kredich, N. M., Martin, D. W. Jr. 1977. *Cell* 12:931–38
165. Hughes, A. 1952. *Exp. Cell Res.* 3: 108–20
166. Kredich, N. M. 1981. *Clin. Res.* In press
167. Pike, M. C., Kredich, N. M., Snyderman, R. 1978. *Proc. Natl. Acad. Sci. USA* 75:3928–32
168. Hakala, M. T., Taylor, E. 1959. *J. Biol. Chem.* 234:126–28
169. Langer, L., Klenow, H. 1960. *Biochim. Biophys. Acta* 37:33–37
170. Morris, N. R., Fischer, G. A. 1960. *Biochim. Biophys. Acta* 42:183–84
171. Meuth, M., Green, H. 1974. *Cell* 3:367–74
172. Wanka, F. 1974. *Exp. Cell. Res.* 85:409–14
173. Lowe, J. K., Gowans, B., Brox, L. 1977. *Cancer Res.* 37:3013–17
174. Kim, J. H., Kim, S. H., Eidinoff, M. L. 1965. *Biochem. Pharmacol.* 14:1821–29
175. Young, C. W., Hodas, S. 1965. *Biochem. Pharmacol.* 14:204–14
176. Lapi, L., Cohen, S. S. 1977. *Biochem. Pharmacol.* 26:71–76

177. Hershfield, M. S. 1979. *J. Biol. Chem.* 254:22–25
178. Hershfield, M. S., Kredich, N. M., Ownby, D. R., Ownby, H., Buckley, R. 1979. *J. Clin. Invest.* 63:807–11
179. Simmonds, H. A., Panayi, G. S., Corrigall, V. 1978. *Lancet* i:60
180. Blustein, H. G., Seegmiller, J. E. 1978. *Fed. Proc.* 37:1465
181. Uberti, J., Lightbody, J. J., Johnson, R. M. 1979. *J. Immunol.* 123:189–93
182. Johnston, J. M., Kredich, N. M. 1979. *J. Immunol.* 123:97–103
183. Snyder, F. F., Hershfield, M. S., Seegmiller, J. E. 1978. *Cancer Res.* 38: 2357–62
184. Hershfield, M. S., Snyder, F. F., Seegmiller, J. E. 1977. *Science* 197:1284
185. Kredich, N. M., Hershfield, M. S. 1979. *Proc. Natl. Acad. Sci. USA* 76:2450–54
186. Hershfield, M. S., Kredich, N. M. 1980. *Proc. Natl. Acad. Sci. USA* 77:4292–96
187. Carson, D. A., Kaye, J., Seegmiller, J. E. 1978. *J. Immunol.* 121:1726–31
188. Carson, D. A., Kaye, J., Matsumoto, S., Seegmiller, J. E., Thompson, L. 1979. *Proc. Natl. Acad. Sci. USA* 76: 2430–33
189. Wortmann, R. L., Mitchell, B. S., Edwards, N. L., Fox, I. H. 1979. *Proc. Natl. Acad. Sci. USA* 76:2434–37
190. Nelson, D. J., LaFon, S., Lambe, C. U. 1979. See Ref. 26, p. 327
191. Smyth, J. F., Young, R. C., Young, D. M. 1978. *Cancer Chemother. Pharmacol.* 1:49–51
192. Smyth, J. F. 1978. *Ciba Found. Symp.* 68:263–72
193. Prentice, H. G., Smyth, J. F., Ganeshaguru, K., Wonke, B., Bradstock, K. F., Janossy, G., Goldstone, A. H., Hoffbrand, A. V. 1980. *Lancet* ii:170–72
194. Nishida, Y., Kamatani, N., Tanimoto, K., Akaoka, I. 1979. *Agents and Actions* 9:549–52
195. Rosenberg, L. E. 1978. In *The Metabolic Basis of Inherited Disease*, ed. J. B. Stanbury, J. B. Wyngaarden, D. S. Fredrickson, p. 423. New York: McGraw-Hill

Ann. Rev. Biochem. 1981. 50:879–910

DNA TOPOISOMERASES ♦12099

Martin Gellert[1]

Laboratory of Molecular Biology, National Institute of Arthritis, Metabolism, and Digestive Diseases, National Institutes of Health, Bethesda, Maryland 20205

CONTENTS

PROSPECTUS AND SUMMARY

The topological state of DNA, and the enzymes that control this state, play a crucial role in determining the function of DNA in cells. In eukaryotes, DNA supercoiling is an essential step in forming the DNA-histone complexes of chromatin, and thus in folding and organizing the chromosome. In prokaryotic cells, DNA supercoiling results in torsional strain, and the DNA can be considered to be in an energetically activated state. This activation is an important element of the processes of DNA replication, transcription, and genetic recombination.

[1]The US Government has the right to retain a nonexclusive royalty-free license in and to any copyright covering this paper.

DNA supercoiling is controlled by a class of enzymes called topoisomerases. Some of these, originally named DNA relaxing enzymes, swivelases, untwisting enzymes, or nicking-closing enzymes carry out the relaxation of superhelical DNA. These enzymes are found in all cell types that have been examined. Another group, the DNA gyrases, carry out the reverse reaction of converting relaxed closed-circular DNA to a superhelical form, in a reaction coupled to the hydrolysis of ATP. DNA gyrases have been found so far only in bacteria. The level of DNA supercoiling in bacteria is presumably set by a balance of DNA supercoiling and DNA relaxing activities.

Topoisomerases can also catalyze the interconversion of other topological isomers of DNA. Various enzymes of this class carry out the formation of knotted structures in single-stranded DNA or the formation and resolution of knots and catenanes in duplex DNA. Of these reactions, the separation of catenated circles may be particularly significant biologically, because catenanes are known to be produced in some processes of replication and genetic recombination, and the products must be separated to allow segregation of the DNA.

All topological interconversions of DNA require the transient breakage and rejoining of DNA strands. It is therefore not surprising to find close relations between topoisomerases and other proteins whose activities involve the reversible breakage of DNA. Topoisomerase activity has been found to be associated with the phage ϕX174 cistron A protein (and the related gene 2 protein of phage fd) which breaks and attaches to the replicative form of the DNA at a specific site, and with the Int protein of phage λ, which catalyzes an intermolecular strand transfer during the integrative recombination reaction. Conversely, one can expect that proteins first characterized as topoisomerases will prove to have other activities requiring transient DNA strand breakage.

Several recent articles have reviewed various aspects of work on DNA supercoiling and on topoisomerases (1–4). In view of the rapid development of the field a new summary of research in the area is appropriate. This review focuses on recent advances in understanding the modes of action of topoisomerases, and on new insights into the influence of topoisomerases and DNA supercoiling on biological functions. It is now evident that topoisomerases belong to one of two basic types, depending on whether the reaction proceeds by a transient single- or double-strand break in DNA. These types are discussed separately below.

TOPOLOGICAL CONSIDERATIONS AND THE ENERGETICS OF SUPERCOILING

The topology of DNA supercoiling has been extensively discussed (5–7) and for our purposes only a summary is necessary. We adopt the notation of Liu & Wang (8). Supercoiling of a circular duplex DNA is defined in terms of a quantity called the linking number (α), which specifies the number of times the two strands are intertwined. The linking number is necessarily an integer. A convenient way to arrive at this number is to consider a projection of the molecule in a plane, and to count the excess of right-handed over left-handed crossings of one strand over the other. Because DNA is a right-handed helix, the linking number is normally positive. DNA in solution takes up a conformation close to the B form, with a pitch that has been measured as 10.4 base pairs (bp) per helical turn (9). A closed-circular duplex DNA with this pitch is under no torsional strain and is termed relaxed. Its linking number, α°, will be distributed over a narrow range of integer values (10, 11) centered around 1/10.4 times the number of base pairs. DNA species with a mean linking number smaller than this are termed negatively supercoiled, or underwound; DNA with a larger linking number is positively supercoiled, or overwound. The linking difference (deviation of the linking number from its "relaxed" value) $\Delta\alpha = (\alpha-\alpha^\circ)$, can be partitioned between *twist* (altered pitch of the helix) and *writhe* (macroscopic contortion of the helix axis): $\Delta\alpha = Tw + Wr$ (5). It is the writhe that corresponds to the intuitive concept of supercoiling as a spiraling of the axis of the double helix in space. However, for a molecule in solution, there is no method devised so far of partitioning $\Delta\alpha$ into fixed proportions of twist and writhe; the intuitive (or electron microscopic) picture of supercoiling cannot be simply correlated with the thermodynamic or hydrodynamic properties of the DNA species. The linking number serves as a suitable invariant to describe a particular covalently circular molecule; it cannot be altered except by breaking and rejoining covalent bonds. A convenient measure of degree of supercoiling that is independent of molecular length is given by the specific linking difference $\Delta\alpha/\alpha^\circ$ [a closely related quantity was originally termed the superhelix density (12)].

Among methods for measuring $\Delta\alpha$, the most widely used have been velocity sedimentation (12,13) or equilibrium sedimentation (14), both in the presence of an intercalating dye, and electrophoresis in agarose gel (15). For DNA with molecular weight less than 10^7, the last method has the advantages of simplicity, capacity to handle many samples, and high resolving power. In a suitable range of $\Delta\alpha/\alpha^\circ$, DNA isomers differing by 1 in linking number form separate bands in the gel. This permits detailed information about the distribution of topological isomers to be collected. For rapid assay of state of supercoiling, measurement of the fluorescence of a bound dye (16,17) has also been found useful.

The excess free energy of a highly supercoiled DNA is quite appreciable. Two methods have been used to investigate the energetics: one is based on the preferential binding of the intercalating dye ethidium bromide to negatively supercoiled DNA (18,19), the other on the Boltzmann distribution of linking numbers in a relaxed closed-circular DNA (10,11). By both methods, the free energy of supercoiling is found to be closely proportional to the square of the linking difference. The free energy per unit length of DNA can be expressed as $\Delta G/a° = k(\Delta a/a°)^2$, where k is a constant independent of length. Between them, the two methods have explored the range of $\Delta a/a°$ from +0.06 to –0.1; this formulation is found to be valid over at least that range. For DNA with $\Delta a/a° = -0.06$, a typical value for natural DNA species, an increase of one unit in a (toward the relaxed state) is favored by about 9 kcal/mol. Correspondingly, the unwinding of one turn of duplex DNA (e.g. by a bound protein) would also be favored by 9 kcal/mol relative to the unwinding of a nonsupercoiled DNA. Supercoiling will greatly influence the equilibrium state of such reactions. Negative supercoiling also makes DNA more accessible to agents that recognize single-stranded regions (20) and favors the looping out of palindromic DNA sequences into cruciform structures (21,21a). It has also been suggested that conversion of alternating purine-pyrimidine sequence tracts to the left-handed Z-DNA structure (22) may be favored by negative super-coiling (23). Supercoiling can thus serve in many ways to modify the recognition properties of DNA.

Nomenclature of Enzymes

The term DNA topoisomerase (3) for enzymes that interconvert topological isomers of DNA is replacing older names for this class of activities. It has been further suggested (24) that enzymes that act by means of a transient double-strand break be named type II DNA topoisomerases, while type I enzymes are those that make a transient single-strand break. This nomenclature is adopted here. A previous terminology that distinguished type I and type II enzymes on the basis of their ability to relax positive as well as negative supercoiling (2) is not in wide use. There is unfortunately some additional confusion possible with the serial naming of topoisomerases in order of discovery in particular cell types (one could find, for instance, that DNA topoisomerase II in a given organism was a type I enzyme in the new nomenclature).

TYPE I TOPOISOMERASES

Topoisomerases were first identified as activities capable of relaxing negatively supercoiled DNA in *Escherichia coli* [ω protein or *E. coli* topoisomerase I; (25)] and in mouse embryo cells (26). Enzymes later detected in many cell types have broad functional similarities to one or the other of these two. The sources include *Micrococcus luteus* (27), *Bacillus megaterium* (28), yeast (29), sea urchin eggs (30), *Drosophila* eggs (31), *Xenopus* eggs (32,33) duck and chicken erythrocytes (34,35), mouse cells (36,37), human KB cells (15) and lymphocytes (38), monkey cells (39), HeLa cells (36), calf thymus (40,41), salmon testis (42), and rat liver (43). A distinct mitochondrial topoisomerase has been found in rat liver (44). In addition, topoisomerase activities have been found associated with vaccinia virions (45) and in *E. coli* infected with λ phage [Int protein; (46)] and phages φX174 [cistron A protein; (47)] and fd [gene 2 protein; (48)].

In several cases (15,27,43) it has been shown that these proteins relax DNA catalytically; all DNA molecules become relaxed even in molar excess of DNA over enzyme. The basic mechanism of action must involve a change in the linking number of the DNA and this can be obtained only by breaking and rejoining the DNA chain. Because the enzymes can break and rejoin DNA repeatedly without any added cofactor to supply the energy for rejoining, one can presume that the reaction intermediate conserves the energy of the DNA phosphodiester bond. The most plausible scheme (25) employs a transient covalent bond between the protein and one broken end of DNA, followed by resealing of DNA and release of protein after the linking number has changed by one or more units. Experiments supporting this reaction scheme are described in the following sections.

The prokaryotic and eukaryotic enzymes listed above appear to operate by a mechanism involving single-strand breaks. This is shown by an experiment first described by Pulleyblank et al (11). When the mouse cell topoisomerase is presented with a single purified topological isomer of relaxed DNA, it reequilibrates the linking number distribution to give the same population of species with neighboring linking numbers as is produced by sealing the nicked DNA with DNA ligase. The topoisomerization thus causes unit changes in linking number. Relaxation by *E. coli* topoisomerase I also produces unit changes in linking number (49). In contrast, a reaction proceeding by transient double-strand breaks will cause the linking number to change in steps of two (see the discussion of type II topoisomerases below).

Prokaryotic Type I Topoisomerases

The following discussion centers on *E. coli* topoisomerase I, the most intensively studied enzyme of this class. This enzyme has been purified to a nearly homogeneous state (50). Its molecular weight is 110,000 as determined by denaturing gel electrophoresis, and the native protein appears to exist in solution as a monomer (51). An *E. coli* topoisomerase preparation with similar enzymatic properties but a different molecular constitution has been reported; it has two subunits with molecular weights of 56,000 and 31,000 (51a). It is possible that this method of preparation yields proteolysis products of the larger protein.

Relaxation of DNA by *E. coli* DNA topoisomerase I exhibits a distinct substrate preference. Highly negatively supercoiled DNA is initially relaxed with high efficiency, but the rate becomes very slow as the linking number approaches that of fully relaxed DNA. Positively supercoiled DNA is essentially unaffected by the enzyme (25). This preference for a negatively supercoiled substrate, and the enzyme's sensitivity to inhibition by single-stranded DNA, have led to the suggestion that the topoisomerase may require single-stranded regions of DNA for activity, the transient existence of such regions being favored by negative but not positive supercoiling (25).

Several reactions of the enzyme with single-stranded DNA have been demonstrated. The interconversion of single-stranded circles of phage fd DNA with knotted forms of varying complexity has been found (52). Formation of knots is favored at low temperature in high salt; conditions of high temperature and low salt promote the resolution of knotted species to the simple circular form. This result suggests that knot formation depends on the presence of regions of local base pairing (52). Topoisomerase I also catalyzes the formation of duplex circular DNA from complementary single-stranded rings under conditions where the duplex DNA is more stable (53). This reaction, first demonstrated with the rat liver nuclear topoisomerase (54), changes the linking number from $a = 0$ for the two separate rings to

$$a \simeq \frac{N}{10.4}$$

(where N is the number of base pairs) for the duplex circular product.

Topoisomerase I forms several distinct complexes with single- and double-stranded DNA. These complexes have yielded information about likely intermediates in the relaxation reaction. All the catalytic activities of topoisomerase I require Mg^{2+}. When incubated with single-stranded DNA in the absence of added Mg^{2+}, the enzyme forms a complex that is stable even in concentrated salt solutions, and can be identified by equilibrium centrifugation in CsCl or Cs_2SO_4 (50). Although no added Mg^{2+} is needed, formation of the complex is inhibited by EDTA, which implies a requirement for at least a low concentration of metal ion presumably introduced with the enzyme preparation. Addition of Mg^{2+} to the complex leads to release of the enzyme in active form, and the DNA is recovered unbroken. If, however, the complex is treated with alkali, DNA breakage occurs with covalent attachment of the protein to the DNA. The resulting DNA fragments are susceptible to digestion by *E. coli* exonuclease I or the exonuclease activity of T4 DNA polymerase, and therefore have free 3'-hydroxyl ends. When single-stranded DNA labeled at its 5' end with ^{32}P is broken, protein does not adhere to the ^{32}P-labeled fragment; therefore protein is attached to the 5' end of the newly formed break (50). The chemical linkage has now been identified as an O^4-phosphotyrosine bond (55). Though the cleavage reaction shows no absolute site specificity, the distribution of sites on the

DNA has been found to be nonrandom. Similar sites of cleavage are produced by *M. luteus* topoisomerase I (55).

With duplex DNA of at least moderate negative supercoiling (specific linking difference $<$ –0.03), the enzyme forms a very similar complex to that described for single-strand DNA, i.e. the complex is dissociated by Mg^{2+} and the DNA is cleaved in alkali. With nonsupercoiled duplex DNA, a different type of complex is found; it is stable in concentrated salt, but neither dissociated by Mg^{2+} nor cleaved by alkali (8). It was suggested that unwinding of the DNA double helix, favored by the negative supercoiling, is needed to form the alkali-cleavable complex. Together with the strong preference of *E. coli* topoisomerase I for a highly negatively supercoiled substrate DNA, this suggests that the alkali-cleavable complex is an intermediate in DNA topoisomerization.

Another reaction catalyzed by this enzyme has recently come to light. Double-stranded circular DNA can become catenated, or catenanes can be separated into simple circles, by either the *E. coli* or *M. luteus* topoisomerase I, provided at least one circle contains a single-stranded break (56). Apparently eukaryotic type I topoisomerases can perform the same reaction (57). In contrast, type II topoisomerases can catenate or unlink catenanes of covalently circular duplex DNA, as is discussed below.

It has been pointed out that this catenation reaction is mechanistically different from the other reactions promoted by type I topoisomerases (57). The other reactions—relaxation of a duplex DNA, knotting of single-stranded rings, and intertwining of complementary single-stranded rings—are all satisfactorily accounted for by rotating a transiently broken strand around an unbroken one (or, equivalently, passing one single strand through a transient break in another without rotation). To form a catenane between one nicked and one closed-circular duplex DNA, it is necessary either to make a transient break opposite the nick and pass a *double* strand through it, or to make a transient break in the covalent circle, admit the nicked circle to make a fused-ring intermediate, and then transform this to the catenated product by means of a second transient break. No available evidence implicates one of these two pathways. Apparently the two circular partners do not have to possess any extensive homology, since dimeric catenanes can be formed between phage PM2 DNA and plasmid pBR322 DNA (56). Thus the passing of a duplex DNA segment through a transient break in another may occur without extensive base pairing of complementary regions.

Eukaryotic Type I Topoisomerases

Highly purified topoisomerases have been obtained from rat liver (43), human KB cells (15), mouse LA9 cells (36), and calf thymus (41). These proteins are found associated with the nuclear chromatin, and have a denatured molecular weight in the range of 60,000–75,000. The enzymes differ in several aspects from the prokaryotic type I topoisomerases. The eukaryotic enzymes are active in the absence of Mg^{2+} and in the presence of chelators, though activity can be obtained in the presence of Mg^{2+} (15,36). The eukaryotic enzymes can relax both positively and negatively supercoiled DNA (15,26,36).

If the reaction of rat liver topoisomerase or KB cell topoisomerase with duplex DNA is arrested with alkali, single-strand breaks are produced and protein remains attached to the broken strands (58,59). The attachment of rat liver topoisomerase to duplex DNA shows some sequence preference, but no absolute site specificity (60). The protein-DNA bond is covalent as judged by its stability under a variety of denaturing conditions. The broken strands are refractory to digestion by *E. coli* exonuclease III but are readily phosphorylated by polynucleotide kinase; thus the protein is attached to the 3' end at the break (58,61), unlike *E. coli* topoisomerase I. A similar linkage has been found with a calf thymus topoisomerase (41). The chemical linkage between protein and DNA has been shown to be a 3'-phosphotyrosine (62). It has also been possible to isolate the protein-DNA complex in active form by incubating enzyme with single-stranded DNA. The DNA in the isolated complex can be recircularized under suitable conditions, or, alternatively, the complexed end can be joined to a second DNA strand having a free 5'-OH end (J. J. Champoux and M. D. Been, personal communication). This is the first direct evidence that the covalent protein-DNA complex is a functional intermediate in the breakage-rejoining reaction.

Although the earlier isolation procedures of eukaryotic topoisomerases mentioned above produced homogeneous enzyme in good yield, it has recently been suggested that these preparations contain proteolytic fragments of larger proteins (57,63). Greater precautions against proteolysis have resulted in preparations of type I topoisomerases with denatured molecular weights of 120,000 from HeLa cells (63) and 110,000 from *Drosophila melanogaster* (57). A considerable stimulation by Mg^{2+} has been reported (63).

Other Proteins with Associated Topoisomerase Activity

The Int protein coded by phage λ promotes the DNA strand interchange reaction of λ integrative recombination. The Int protein possesses a topoisomerase activity (46). The enzyme changes linking numbers in increments of one, as is characteristic of type I topoisomerases (56,64). Surprisingly, although the integrative recombination reaction is site-specific, and although Int protein binds preferentially to the phage attachment site (65), the topoisomerization reaction has no preference for DNA containing the phage attachment site over DNA without it.

The cistron A protein of phage ϕX174 and the gene II protein of phage fd are known to break and attach to the viral (+) strand of the replicative form (RFI) of the phage DNA at a unique site (47,48,66,67). This step can initiate a round of (+)-strand replication that finishes when the protein again breaks at the origin site and the displaced strand is joined to form a circle (48,67). These proteins are also able to relax negatively supercoiled DNA (47,48). The proteins appear to be capable of either breaking DNA and rejoining it promptly, which leads to topoisomerization, or of separating the breakage and rejoining steps by a round of replication, which leads to release of single-strand circles. Here, as in a number of other contexts, topoisomerization can be viewed as one of the possible outcomes of a process involving covalent protein-DNA intermediates.

TYPE II TOPOISOMERASES

DNA Gyrase

DNA gyrase is distinct from other topoisomerases in being able to catalyze the conversion of relaxed duplex DNA to a superhelical form. As the supercoiled product has a higher free energy than the starting DNA, the reaction needs energy, which is supplied by the hydrolysis of ATP. DNA gyrase was first isolated from E. coli (68) and has also been detected in M. luteus (69), Pseudomonus aeruginosa (70), and B. subtilis (71). The enzyme carries out a number of reactions:

1. ATP-dependent negative supercoiling (reduction of linking number) of closed circular duplex DNA;
2. Relaxation of superhelical DNA, in the absence of ATP;
3. Oxolinic acid–promoted introduction of double-strand breaks at specific sites in DNA;
4. DNA-dependent hydrolysis of ATP;
5. Formation and resolution of catenated and knotted duplex DNA.

These reactions are discussed in more detail below.

DNA gyrase from E. coli, M. luteus, or B. subtilis is specifically inhibited by two classes of antibiotics, one of which includes coumermycin A_1 and novobiocin, and the other nalidixic acid and oxolinic acid (69,71–74). E. coli mutants resistant to one or the other class of drugs are known and the map locations of the genetic loci responsible have been determined (75–77); DNA gyrase isolated from a resistant mutant is also drug resistant. Drug resistance of the cells and of the enzyme are cotransduced by phage P1. Thus DNA gyrase appears to be the principal target of these antibiotics. Because of this identification, the genetic loci in question have been renamed (78). The locus for resistance to nalidixic and oxolinic acid at 48 min on the standard E. coli map, formerly nalA, is now denoted by gyrA. The locus for resistance to coumermycin A_1 and novobiocin [and clorobiocin (79)] at 82 min on the map (formerly *cou*) is now designated *gyr*B. Both groups of drugs are known to be inhibitors of DNA replication (75,76,80), which immediately suggests that DNA gyrase is involved in that process.

The first preparations of DNA gyrase (68,81) were only partly purified. More extensive purification has shown that the enzyme consists of two different proteins, each the target of one family of antibiotics. These proteins have been purified to near homogeneity either separately (74,82) or as a complex (83). The complex contains equimolar amounts of the two proteins (83), and probably exists in solution as a tetramer, as is observed with the *M. luteus* enzyme (see below). The product of the *gyr*A gene, here called the gyrase A protein, has a denatured molecular weight of 100,000–105,000 and exists in solution as a dimer (74). The gyrase B protein has a denatured molecular weight of 90,000–95,000 (78,82,83). The full activity of the enzyme is reconstituted by mixing the subunits; neither subunit alone has any of the activities of DNA gyrase (82). A report that the gyrase A protein alone is capable of relaxing supercoiled DNA (74) was later shown to be incorrect (82).

The DNA gyrase of *M. luteus* has also been reconstituted from two highly purified proteins, α and β, of denatured molecular weights 115,000 and 97,000 respectively (69). Interspecies complementation experiments have shown that α can complement *E. coli* gyrase B protein and β can complement gyrase A protein for the activities of DNA gyrase, thus identifying α and β as the functional equivalents of the A and B proteins respectively (84). The *M. luteus* gyrase subunits have therefore been renamed A and B, consistent with the *E. coli* proteins (55). The *M. luteus* enzyme binds to DNA as an A_2B_2 complex, and at least some of the enzyme in free solution was found by cross-linking experiments to exist in the same form (85).

SUPERCOILING BY DNA GYRASE   ATP-driven supercoiling catalyzed by DNA gyrase is always in the direction of reduced linking number (negative supercoiling) (68); this corresponds to the sense of supercoiling found intracellularly. There is a limit to the extent of supercoiling that can be achieved. With high concentrations of the *E. coli* enzyme, the specific linking difference reaches values of approximately –0.10 (68). The smaller extent of supercoiling normally found in DNA isolated from *E. coli* ($\Delta\alpha/\alpha \simeq -0.06$) may reflect the presence of competing topoisomerases in vivo. The reaction is catalytic; the linking number of all DNA molecules in the solution is changed by many units even when DNA is in considerable molar excess over enzyme (82,83). In addition to ATP, the reaction requires a divalent cation, usually Mg^{2+}, and under some reaction conditions is stimulated severalfold by spermidine (68). Deoxy ATP can substitute for ATP, though with a much higher K_m; other nucleoside triphosphates are ineffective (68).

The supercoiling reaction is inhibited by novobiocin and coumermycin (72), which inhibit competitively with ATP (86). An inhibition constant, K_i, of 4 X 10^{-9} M for coumermycin and 10^{-8} M for novobiocin has been determined (86). Inhibition is also obtained with oxolinic acid (73,74); the half-inhibitory concentration is 4 X 10^{-5} M. Nalidixic acid is a sufficiently weaker inhibitor that it has not been widely used for in vitro experiments.

The ATP analogue (β,γ-imido) adenosine triphosphate also inhibits the ATP-driven supercoiling reaction. However, when a high level of DNA gyrase is incubated with DNA and (β,γ-imido) ATP, limited negative supercoiling is induced, stoichiometric with the amount of enzyme added (86). The linking number is reduced by about 1.4 per enzyme tetramer added; this value is presumably a lower bound, because of the possibility that some enzyme is inactive. As the analogue is not noticeably hydrolyzed by the enzyme, its binding alone evidently permits one cycle of supercoiling. Hydrolysis of ATP appears to return the enzyme to the starting state for another round of supercoiling (86). The decrease of the linking number is thus coupled to binding of the triphosphate rather than its hydrolysis.

RELAXATION OF DNA BY DNA GYRASE   In the absence of ATP, DNA gyrase relaxes negatively supercoiled DNA (73,74). The relaxing activity is much less efficient than the supercoiling activity; about 20–40 times as much enzyme is required to give a comparable rate (82, 87). Relaxation of negatively supercoiled DNA by DNA gyrase is inhibited by oxolinic acid but not by novobiocin or coumermycin (73,74). The required concentration of oxolinic acid is comparable to that needed to block supercoiling. Thus there is a strong presumption that oxolinic acid interferes with the breakage-rejoining step of the overall supercoiling reaction, and that the A protein, being sensitive to oxolinic acid, is involved primarily in that step.

In the absence of nucleotide, DNA gyrase can neither reequilibrate the linking number of a relaxed DNA (88) nor relax positively supercoiled DNA (84). [Earlier reports to the contrary (73,74) appear to have been based on the contaminating presence of a fragment of gyrase B protein that complements the A protein to yield such an activity (see below).] Interestingly, DNA gyrase acquires the ability to relax positively supercoiled DNA when (β,γ-imido) ATP is present (88), an additional reaction to the single cycle of supercoiling induced by the analogue. This result argues for the presence of a different functional state, and possibly a different conformation, of the enzyme in the presence of the nucleotide.

OXOLINIC ACID–DEPENDENT BREAKAGE OF DNA   In the presence of oxolinic acid, an additional reaction of DNA gyrase is found. Incubation of *E. coli* DNA gyrase with DNA and oxolinic acid, followed by treatment with sodium dodecyl sulfate, results in the appearance of double-strand breaks in the DNA (73,74). A similar reaction is seen with *M. luteus* DNA gyrase, but it is less efficient and occurs more readily in this case if the complex is treated with alkali instead of detergent (55). The DNA is broken into defined fragments, but not all cleavage sites are equally sensitive (73,74). Subsequent work has shown that the broken ends have a uniform chemical structure: the two strands are always broken four bases apart, with the 5' ends protruding and blocked by protein (88–90). The bound moiety is the A protein, and the linkage, as in the case of type I topoisomerases, is a phosphotyrosine bond (55).

So far, determination of nucleotide sequences around the cleavage sites has failed to show any extensive homologies. A study of sites cleaved by *E. coli* DNA gyrase in SV40, ϕX174-RFI, and ColE1 DNA indicated that cleavage on one strand commonly occurred within a TpG doublet, and it was proposed that this dinucleotide was a general feature of gyrase cleavage sites (89). However, another study of sites cleaved by *E. coli* DNA gyrase in the plasmids pBR322 and pNT1 (a ColE1 derivative) failed to confirm even this limited homology (88); cleavage within a TpG dinucleotide was found in only one out of five sites. Similarly, an analysis of sites cleaved by *M. luteus* DNA gyrase revealed one TpG cleavage among three sites (90). Strong cleavage sites occur relatively rarely in DNA, less frequently than once in every thousand base pairs. This implies that a nucleotide sequence of considerable extent is recognized. However, gyrase binds to a region of DNA about 140-bp long (see below), and the recognition sequence could be dispersed anywhere within that length. It is likely that many more site sequences will have to be determined before the recognition rule becomes apparent. The pattern of fragments produced in a given DNA by DNA gyrase from *E. coli, M. luteus,* and *B. subtilis* is very similar, which raises the possibility of a common set of recognition sequences for all these enzymes (71,84). However, no analysis at the nucleotide level has yet been done to confirm the matching of sites.

The overall frequency of cleavage is enhanced by ATP or (β,γ-imido) ATP, and the pattern of preference among cleavage sites is altered (86,91). Originally it was proposed that ATP or its analogue caused the enzyme to move and cleave uniformly 400 bp away from its previous site (86,91). However, a reinvestigation has shown that no movement occurs; gyrase remains firmly bound and the nucleotides act rather to enhance cleavage at certain sites (92).

ATP HYDROLYSIS AND ITS INHIBITION BY COUMERMYCIN AND NOVOBIOCIN   Hydrolysis of ATP to ADP and phosphate by *E. coli* DNA gyrase is strongly stimulated by duplex DNA (83,86,93). Single-stranded DNA is a relatively poor cofactor. Since some synthetic duplex DNA homopolymers and alternating copolymers are good cofactors, recognition of specific sites in DNA is probably not important for this reaction (93). The tertiary structure of DNA clearly does have an effect on the ATPase activity. Linear, nicked-circular, and relaxed closed-circular duplex DNA stimulate the ATPase strongly, while highly negatively supercoiled DNA is less effective (83,93). This is evidence for coupling between ATP hydrolysis and the supercoiling reaction. A degree of coupling is also suggested by the rough stoichiometry between ATP-hydrolyzed and superhelical turns introduced in the early stages of reaction with a relaxed DNA (93).

Hydrolysis of ATP is inhibited by novobiocin and coumermycin, but not appreciably by oxolinic acid (83,86). Even the basal ATPase activity found in the absence of DNA is found to be largely sensitive to novobiocin, which implies that it is also a reaction mediated by DNA gyrase rather than a contaminant (83). Inhibition is competitive with ATP binding (86) even though coumermycin and novobiocin have no obvious structural similarity to ATP. Two other kinds of experiments also indicate that the drugs interfere with the binding of ATP to the enzyme. First, they block the shift in oxolinic acid–induced cleavage pattern caused by ATP or (β,γ-imido) ATP (86). Since this shift does not require hydrolysis of the triphosphate, it is plausible that the drugs block its binding. More directly, novobiocin prevents the covalent attachment of the ATP analogue oATP (94) to the gyrase B protein (83). Coumermycin and novobiocin are thus seen to interfere with the energy coupling of the supercoiling reaction, while oxolinic acid and nalidixic acid block the breakage and rejoining of DNA.

BINDING OF DNA GYRASE TO DNA: WRAPPING OF DNA AROUND THE ENZYME   Several results important to an understanding of the action of DNA gyrase have come from studies of the binding of enzyme to DNA. Liu & Wang showed that *M. luteus* gyrase protects a region of about 140 bp from digestion by staphylococcal nuclease (95). They further showed that binding of gyrase to a nicked-circular DNA, followed by sealing of the DNA with DNA ligase and deproteinization of the DNA leads to an increase in the linking number of the DNA (i.e. positive supercoiling) by about one unit for each gyrase tetramer bound (69). The length of the protected region implies that the DNA may be wrapped around the enzyme; the introduction of supercoiling implies that this wrapping is asymmetric. More evidence for wrapping of DNA around the enzyme comes from digestion of the gyrase-DNA complex with pancreatic DNase: this generates a series of fragments whose lengths differ by 10 ± 1 bases (95), as has been found in nucleosomes where the DNA is known to be bound externally (96).

More detailed studies have shown that specific binding of either *E. coli* or *M. luteus* gyrase to DNA fragments known to contain a site for oxolinic acid–dependent cleavage can often be demonstrated (88,90,92). By "foot-printing" methods (97) it has been shown that the sequence protected by DNA gyrase both in the absence and presence of oxolinic acid in such cases is roughly centered on the cleavage site, with strongest protection found near that site (88,90). Flanking regions, though still protected, have enhanced bands (increased DNase I sensitivity) at 10–11 bp intervals. The DNase cutting sites on complementary strands are staggered by about 2 bp, again reminiscent of the action of DNase I on nucleosomal DNA (98). The overall range of protection is 100–140 bp. In one case, *M. luteus* gyrase protected a region in which cleavage did not occur (90); thus some caution is necessary in identifying sites of binding, and possibly of gyrase action, uniformly with known cleavage sites.

FORMATION AND RESOLUTION OF KNOTTED AND CATENATED DNA   Recent developments in the study of DNA gyrase have been guided by the finding that the ATP-dependent T4 DNA topoisomerase, described in more detail below, is capable of introducing and removing knots in covalently closed duplex DNA (24,99). Such reactions can only take place by a coordinated breakage and rejoining of both strands of DNA. DNA gyrase can also remove knots from DNA (24,100), form catenanes (101), and separate catenated rings (100,101). Unknotting and separation of catenanes occur under normal gyrase reaction conditions. To form catenanes, it is necessary to use high spermidine and low salt concentrations, which presumably serve to aggregate the DNA and increase the local concentration of chain segments (101). These reactions require ATP (100,101) and are inhibited by both novobiocin and oxolinic acid (88,101); their requirements are thus more akin to the supercoiling than the relaxation reaction of DNA gyrase. Nicked DNA as well as covalently closed DNA is an effective substrate for the resolution of knots and catenanes (100). All these reactions are severalfold slower than DNA supercoiling, possibly because they require juxtaposition of two distant segments of DNA; this point is mentioned again below.

TRANSIENT DOUBLE-STRAND BREAKAGE OF DNA DURING THE SUPERCOILING REACTION   While the mere existence of the topological interconversions described in the last section is sufficient to establish that double-strand breakage is involved in those processes, a more subtle test is required to show the participation of double-strand breakage in the super-coiling reaction. It was pointed out by Fuller (6) that passage of one bihelical segment of DNA through another, without rotation, would change the linking number by two units. An experimental test requires the isolation of circular DNA with a unique linking number, most conveniently done by purifying DNA present in one band following gel electrophoresis of a closed-circular DNA (11). When such a test is carried out, it is found that DNA gyrase does indeed change linking numbers in steps of two, both during supercoiling and during relaxation (49,100). Thus all interconversions of DNA topoisomers by DNA gyrase are seen to involve double-strand breakage.

MECHANISTIC MODELS OF DNA GYRASE   Earlier models of DNA gyrase action involved translocation of DNA past the enzyme (69,83,86), and usually invoked breakage of one strand of the DNA. These models have now been abandoned in the light of evidence for transient double-strand breakage and the lack of evidence for translocation when ATP effects on cleavage patterns are examined (90,92). The current models (24,49,57, 100,102) all involve passage of a DNA segment through a double-strand break and generally make use of site-specific binding of the enzyme. They differ in the mechanics of strand transfer and, in particular, in the means used to ensure that the supercoiling reaction has the correct polarity. One model is briefly described below (100) and its differences from others are pointed out.

In this scheme (Figure 1) the specific sites of oxolinic acid–promoted cleavage are the sites of enzyme binding and transient double-strand breakage in the supercoiling reaction. It is suggested that the DNA segment that is to be transported through the break lies within or very near the same protected region (of about 140 bp) as the site, and becomes wrapped over the enzyme (Figure 1,B) with the positive supercoiling indicated by the experiments of Liu & Wang (69). It is this wrapping that provides the directionality of the supercoiling reaction. Binding of ATP then causes a conformational change in the enzyme, which leads to a coupled opening of the double-strand break and transport of the wrapped DNA segment through it, followed by resealing of the break. This process decreases the linking number of the DNA by two and introduces the observed negative supercoiling (Figure 1,C). Because the nonhydrolyzed ATP analogue (β, γ-imido) ATP drives one cycle of supercoiling but then blocks further reaction (86), one can plausibly assume that release of the translocated segment is coupled to hydrolysis of ATP (Figure 1,D). The enzyme is then ready to undergo another cycle of reaction (Figure 1, E, F, and G), which results in a further reduction of DNA linking number by two units. To retain the supercoiling during the reaction, the transiently broken DNA ends must not rotate relative to each other. Presumably the ends are held by the enzyme in a fixed orientation, with the 5' ends attached by reversible covalent linkage to the gyrase A protein; this is the bond observed after cleavage in the presence of oxolinic acid (55). DNA gyrase binds to DNA as a tetrameric complex containing two molecules each of A and B protein (85); it is thus possible that the enzyme-DNA complex opens on one side to admit the transported DNA segment and then in a coupled process closes on that side and opens on the other to expel the segment. With such a two-step process, it is easier to envisage how a rotation that would frustrate the supercoiling reaction can be avoided.

Relaxation of negatively supercoiled DNA, in this model, requires that the DNA segment to be transported must be bound in the opposite sense from that used for the supercoiling reaction, i.e. crossing over the enzyme with a negative superhelical sense (Figure 1,*H–J*). The higher efficiency of supercoiling compared to relaxation is thought to be partly due to the ability of ATP to speed up the operation of the gating mechanism. Reactions involving knots and catenanes would require the transport of a more distant DNA segment, and would thus also be expected to be less efficient, because

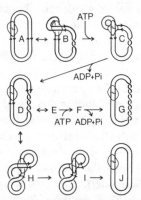

Figure 1   A model for DNA gyrase–induced DNA supercoiling by means of transient double-strand breaks. The enzyme binds preferentially to certain sites on DNA and induces a left-handed (positive superhelical) wrapping of a local DNA region (B). ATP binding then leads to transport of the upper double helix through the lower, via a transient double-strand break (C), with an accompanying conformational change in the enzyme. This reaction decreases the linking number of the DNA by two. Subsequent hydrolysis of ATP and release of the transported DNA segment prepares the system for another cycle of supercoiling (D–G). During relaxation of negatively supercoiled DNA (bottom series of drawings), the superhelical coiling causes a loop of DNA to fold over the enzyme with the opposite (right) handedness to that used in the supercoiling reaction (H). Transport through a transient double-strand break causes an increase of linking number (relaxation) by two units (I–J). [Reprinted from (100)].

binding of a distant segment has to compete with the preferred local wrapping.

In a model termed "sign inversion" (49), the enzyme is envisaged as binding two segments of DNA so as to produce a node of positive superhelical sense (sign). Binding of ATP again causes passage of one duplex through a break in the other, with a reversal of the local sense of supercoiling (sign) and a net decrease of linking number by two units. The sign of the node is thought to be determined either by local wrapping, in the sense of Liu & Wang (69), or by binding of a more remote DNA segment with a polarity dependent on a sequence-specific orientation.

Another model (57) proposes that the positive superhelical wrapping of DNA around the enzyme does not change during the reaction cycle, and that the ATP-driven reaction moves a second DNA loop from outside to inside this wrapped complex through a transient double-strand break generated in the wrapped DNA, again with a linking number change of two units. Hydrolysis of ATP then allows release of the inside DNA segment and repetition of the cycle. In this model, the orientation of the strand crossings that produce net negative supercoiling is left somewhat unspecified, but could presumably be accommodated by a requirement for binding within a small loop with a preferred sense.

TOPOISOMERASE ASSOCIATED WITH A FRAGMENT OF GYRASE B PROTEIN   Two groups have isolated from *E. coli* a protein of molecular weight 50,000 that complements the gyrase A protein to produce a DNA relaxing activity (84,87). The protein has been shown by partial proteolytic digestion to be most probably a fragment of the gyrase B protein (87). The name topoisomerase II' has been proposed for the complex of the gyrase B fragment with gyrase A protein (84). The complex relaxes negatively superhelical DNA (84,87) and, unlike DNA gyrase, is also capable of relaxing positively superhelical DNA in the absence of ATP analogues (84). Like DNA gyrase, this activity alters linking numbers in steps of two (M. Gellert, unpublished results), and can thus be classified as a type II topoisomerase. The activity is inhibited by oxolinic acid but not by novobiocin; oxolinic acid–dependent cleavage of DNA is found at a spectrum of sites that is the same as observed with DNA gyrase (84,87). The pattern of cleavage is not altered by ATP, unlike the situation with DNA gyrase. Furthermore, the complex has no ATP-dependent supercoiling activity, nor any DNA-dependent ATPase. It thus seems that the fragment has lost a portion of the B protein structure required for energy coupling.

The mechanism by which the fragment is generated has not yet been clarified; neither has its genetic origin. There appears to be several times more B fragment than B protein in extracts. If the proportion reflects the intracellular situation, the B fragment could in principle make an appreciable contribution to the topoisomerase activity in the cells.

Other Type II Topoisomerases

Type II topoisomerases other than DNA gyrase have been isolated from both prokaryotic and eukaryotic sources. These enzymes have generally been detected as ATP-dependent DNA-relaxing activities; all such enzymes tested have subsequently been shown to function by a type II mechanism.

The first enzyme of this class was purified from *E. coli* infected with phage T4 (99,103). The activity is not present in cells infected with phages mutant in genes 39 or 52 (or possibly gene 60, though here the genetic situation is more complicated). These genes are of the "DNA-delay" class, whose products are required for substantial replication of T4 DNA at early times of infection (104). The purified enzyme has protein components of molecular weights 63,000 and 52,000, identified as the gene 39 and gene 52 products, respectively (99,103). Some preparations contain in addition a 16,000-molecular weight protein that may be the product of gene 60 (99). The T4 DNA topoisomerase can relax both positively and negatively supercoiled DNA, and one enzyme molecule can relax many DNA molecules (99). The enzyme requires ATP hydrolysis to relax DNA catalytically (99,103) and has a DNA-dependent ATPase activity (99); it has been estimated that the linking number of a DNA molecule changes by one to two units for each ATP molecule hydrolyzed (99). No inhibition by novobiocin, and little or no inhibition by oxolinic acid, has been found (99,103).

When large amounts of T4 DNA topoisomerase are incubated with circular duplex DNA in the absence of ATP, knotted molecules of varying complexity are formed (24). The knotted molecules can be distinguished from other possible complex structures by: (*a*) anomalous gel mobility even when nicked; (*b*) conversion to a simple linear structure by a single restriction nuclease cut; and (*c*) appearance in the electron microscope (24). These knotted circles can be efficiently restored to simple circular form by much lower (catalytic) amounts of the enzyme in the presence of ATP (24). Both reactions occur with covalently closed DNA, which implies that an intermediate with a double-strand break must be involved. The relaxation of DNA by T4 DNA topoisomerase also proceeds through transient double-strand breaks, since the linking number is changed by multiples of two (24).

Enzymes with similar properties have recently been isolated from a number of eukaryotic sources. An ATP-dependent topoisomerase has been purified from *Drosophila* embryos and shown to relax DNA by a double-strand break mechanism (105). It can relax either negatively or positively supercoiled DNA. This enzyme can separate catenated rings and resolve knotted molecules (24,105); in the presence of a second protein component it can also form large catenated networks from circular DNA (105). No homology between DNA species is required for catenation. The second protein has not been purified so far, but can be replaced by histone H1 (105); the latter is known to have the ability to condense DNA (106), and so may function by raising the local strand concentration to the point where catenation is favored.

Relaxation by the *Drosophila* enzyme is inhibited by novobiocin, though a much higher concentration (about 100 μg/ml) is needed than required to inhibit DNA gyrase (105). The inhibition appears to be competitive with ATP, and thus may operate by the same mechanism as with DNA gyrase. Nalidixic acid is not inhibitory.

An ATP-dependent topoisomerase from *Xenopus laevis* germinal vesicles has been partly purified and shown to form or resolve catenanes of super-coiled DNA (107). Catenation is stimulated by spermidine, and decatena-tion is favored by high ionic strength, as with DNA gyrase (101). Catenation is inhibited by coumermycin A_1 at concentrations above 100 μg/ml (107).

Extracts of rat liver mitochondria have also been reported to contain an ATP-dependent topoisomerase capable of forming catenanes (108). This activity is insensitive to the inhibitor Berenil, and is thus distinct from the previously described mitochondrial topoisomerase (44).

Type II topoisomerases have also been partly purified from HeLa cells and from calf thymus (63); they appear to be widely distributed among eukaryotes.

The role of ATP in the activity of type II topoisomerases has not been entirely clarified. Liu et al (99) found that T4 DNA topoisomerase could produce a very limited relaxation of DNA in the presence of ATP-γS[adenosine 5'-O-(3-thiotriphosphate)]. They suggested that this re-action might be analogous to the single cycle of DNA supercoiling pro-duced by DNA gyrase in the presence of $(\beta,\gamma$-imido) ATP (86). One could envisage an ATP-stimulated gating mechanism such as that described above for DNA gyrase. It has been suggested that the reactions of other type II enzymes may be very similar to that of DNA gyrases, with the exception that the other type II enzymes lack the ability to wrap DNA with a positive twist, and thus are unable to supercoil DNA (105). Without this orienting factor, DNA segments should more frequently pass through a transient break with a sense determined by the DNA's supercoiling, thus relaxing the DNA.

Although the eukaryotic and T4 type II DNA topoisomerases require ATP for their catalytic activities, this is not necessarily true of all type II activities. Relaxation by DNA gyrase and the related topoisomerase II' is an example of a type II reaction proceeding without ATP. It is not safe to assume that any ATP-independent topoisomerase will work by a type I mechanism.

BIOLOGICAL ROLE OF TOPOISOMERASES

Topoisomerases can participate in cellular processes in two ways: either indirectly, through their effect in controlling the superhelical state of DNA, as well as its state of catenation and knotting; or directly, by their participation in processes requiring the coupled breakage and rejoining of DNA strands (as, for example, in recombination reactions). Evidence for both functional roles of topoisomerases has been obtained in the past few years by combining studies of topoisomerase mutants with information gained by the use of specific inhibitors. We discuss first the present state of knowledge about supercoiling of DNA in cells and its functional implications.

Intracellular Supercoiling of DNA

Closed-circular duplex DNA isolated from cells is negatively supercoiled, with a specific linking difference ranging from –0.03 to –0.09, depending on the DNA species and the source (reviewed in 2). Evidence indicates that in eukaryotic cells most or all supercoiling can be accounted for by histone binding combined with the relaxing activity of a topoisomerase. In prokaryotes, supercoiling appears to be mainly due to the action of DNA gyrases. The physiological consequences are very different in the two cases.

Binding of histone octamers to DNA, thus forming nucleosome core particles, results in a negative superhelical wrapping of DNA on the histone complex that is equivalent to a change in writhing number of about 1 ¼ per nucleosome. If it were not for the activity of topoisomerases, the parts of the DNA outside the nucleosome cores would remain positively supercoiled. Relaxation of the DNA by a topoisomerase leads to a corresponding reduction in linking number (109; for review see 110). This mechanism should lead to an absence of torsional stress on eukaryotic nuclear DNA. In agreement with this idea, the rate of binding of trimethylpsoralen to intracellular DNA in *Drosophila* and HeLa cells is characteristic of unstressed rather than stressed DNA (111). The change in linking number found in SV40 DNA isolated from infected cells agrees at least semiquantitatively with the change expected from the nucleosomal structure alone (109).

On the other hand, there is evidence that in *E. coli* (and by inference in other prokaryotic cells), DNA is actively supercoiled and remains under torsional stress. The first indication was provided by superinfection of λ-lysogenic *E. coli* with phage λ. Under these conditions, the infecting DNA becomes circularized and in the normal case is isolated in supercoiled form (112). When λ superinfection is carried out in the presence of the

DNA gyrase inhibitors coumermycin or oxolinic acid, however, the closed-circular λ DNA is isolated with very little supercoiling (72,73). It was estimated that at most 15% of normal supercoiling is introduced by mechanisms independent of DNA gyrase (72).

The chromosomal DNA of *E. coli* cells treated with coumermycin also loses a large proportion of its supercoiling. The chromosome isolated as a folded nucleoid structure normally contains about 50 negatively supercoiled loops (113,114). Upon incubation of cells with coumermycin, the supercoiling decreases three to four fold within 10–20 min (115). DNA replication is arrested with a similar time course. When a coumermycin-resistant (*gyr*B) strain of *E. coli* is used, no decrease in supercoiling is found, which implies that both effects of coumermycin are due solely to inhibition of DNA gyrase.

The results described above argue that: (*a*) DNA gyrase is needed to supercoil DNA in *E. coli* and is distributed on the chromosome in a way that makes this possible. (*b*) Cellular DNA will be relaxed eventually if supercoiling by DNA gyrase is blocked. However, they do not examine the instantaneous state of the DNA. Is the DNA torsionally stressed, or is it restrained by binding proteins so that the torsional force is not expressed? In a first attempt to answer this question (116), cells containing the F plasmid were γ-irradiated to introduce single-strand breaks, and coumermycin was added during repair of the breaks to prevent any additional supercoiling after repair. When repaired F DNA was isolated, it retained 50–60% of its original supercoiling, which indicates that this fraction of the supercoiling could not be quickly relaxed even in a nicked molecule (the half-time of repair was about 3 min in these experiments). However, examination of the instantaneous state of intracellular DNA gives evidence of its being under torsional strain. The technique measures the rate of covalent photobinding of trimethylpsoralen; this rate is enhanced when the DNA is under the torsional strain of negative supercoiling. Chromosomal DNA in *E. coli* cells is found to react like purified superhelical DNA with a specific linking difference of –0.05 (111). This value is close to that found for isolated nucleoids (113,114), which suggests that most of the supercoiling is not restrained in the cell.

Topoisomerases in DNA Replication

In principle, topoisomerases could play several roles in replication of closed-circular duplex DNA.

1. At initiation, supercoiling could assist the binding of factors required to start replication. This function would be specific for DNA gyrases.

2. During chain elongation, a topoisomerase capable of relaxing positive superhelical turns could function as a swivel (117) to remove the positive supercoiling that would otherwise accumulate in the unreplicated portion of the molecule. A DNA gyrase, in addition to satisfying this requirement, could maintain the DNA under negative superhelical strain and thus facilitate unwinding at the replication fork (68).

3. At the end of a round of replication, a topoisomerase could unlink catenated products, by means of either a single- or a double-strand break.

We discuss below the evidence implicating topoisomerase function in replication, and identifying the step involved where possible.

DNA replication in *E. coli* is inhibited by novobiocin (118) and by coumermycin (75,115); replication in coumermycin-resistant (*gyr* B) mutants is unaffected. Coumermycin acts preferentially on semiconservative replicative DNA synthesis in both whole and toluene-permeabilized *E. coli* cells, while repair synthesis is less affected (75,119). Similarly, novobiocin blocks semiconservative DNA replication in toluenized cells, but leaves DNA repair synthesis unaffected (120).

Replication of circular duplex DNA species in vitro shows a similar sensitivity. Replication of Col E1 DNA in a cell-free system is blocked by novobiocin (72,121). Synthesis can be restored by adding novobiocin-resistant DNA gyrase to the system (72). Supercoiling of DNA is required in this system for initiation of DNA synthesis (72,73); however, elongation of strands can also be stimulated by DNA gyrase action (122). Replication of ϕX174 RFI DNA is also inhibited by novobiocin (123), although a tenfold higher concentration is required than for ColE1. In the ϕX174 system, supercoiling of DNA is required for attack by the ϕX174 *cis*A protein, which is the first step in replication of RFI DNA (81). Relaxed RFI DNA can be replicated if the system is supplemented with DNA gyrase.

E. coli chromosomal replication has also been studied with the aid of a temperature-sensitive lethal mutation at the *gyr*B locus (124). At the restrictive temperature, initiation is blocked, but the rate of chain elongation is essentially unchanged. The mutant thus responds differently than do cells exposed to coumermycin, which shuts off DNA replication rapidly (75, 115), and so must block elongation. Either coumermycin has some gyrase-mediated effect (so far unknown) in addition to blocking supercoiling, or else the temperature-sensitive mutation is sufficiently leaky to permit one but not the other step in replication. The decrease in DNA gyrase activity in this mutant has not yet been measured. A role for DNA gyrase in initiation has also been proposed as an explanation of the hypersensitivity of *dna*A mutants to gyrase inhibitors (125).

Novobiocin and coumermycin also cause the selective elimination of some plasmids from *E. coli* (126,127). It is possible that DNA replication of these plasmids is more sensitive to inhibition of DNA gyrase than replication of the host genome.

Although attention has been focused mainly on topochemical effects in the replication of circular DNA, replication of linear T7 DNA in vivo is also blocked by coumermycin (128,129). It is possible that axial rotation even of linear DNA is sufficiently restricted in the cell so that torsional stress can be maintained and can be functionally significant.

Inhibitors of the *gyr*A protein, oxolinic and nalidixic acid, also interfere with DNA replication. Replication is inhibited in whole bacteria (76, 130,131), in toluenized cells (132), and in cell-free systems for the synthesis of ColE1 DNA (133) and ϕX174 RFI DNA (123).

There is evidence that this family of drugs does not simply interfere with supercoiling. First, prolonged incubation of *E. coli* with nalidixic acid leads to some breakdown of DNA (75,134), an effect not seen with coumermycin A_1 or novobiocin. Second, even brief exposure of cells to oxolinic acid leads to formation of a protein DNA complex that, upon exposure of the isolated nucleoid to sodium dodecyl sulfate, breaks the DNA at about 45 widely dispersed sites per chromosome (135). At low concentrations of oxolinic acid (< 50 ng/ml) mainly single-strand breaks are produced; at higher concentrations double-strand breaks predominate. Protein remains tightly bound to the cleavage products. No breakage is found in an oxolinic acid–resistant (*gyr*A) mutant. This reaction is reminiscent of the cleavage of DNA by purified gyrase and oxolinic acid. It is suggested that there may be one preferred gyrase binding site in each loop of the folded chromosome (135). Third, the concentration of oxolinic acid required to block intracellular supercoiling of λ DNA (73) is at least 100-fold higher than that needed to block growth of the cells or to produce breakage of the chromosome (135). Fourth, inactivation of the enzyme by a conditional lethal *gyr*A temperature-sensitive mutation has some quite different consequences from use of nalidixic acid (136). The mutant cells produce a temperature-sensitive DNA gyrase and stop DNA replication at high temperature. However, while nalidixic acid blocks growth and DNA replication of phage T7 (137), the mutation does not, and, in addition, the mutation renders the phage-infected cells insensitive to nalidixic acid at high temperature. These various lines of evidence suggest that nalidixic and oxolinic acid interfere with replication by forming a gyrase-DNA complex that is on a path leading to chromosome breakage, and that inhibition of supercoiling is of secondary importance in the physiological effect of these drugs. In this connection, it has been suggested that DNA gyrase may function directly at replication forks in addition to its role in maintaining general supercoiling of the chromosome (138). Inhibition by nalidixic or oxolinic acid is seen in this model as causing a physical block to fork movement, for example by forming a very stable DNA-enzyme complex.

INVOLVEMENT OF T4 DNA TOPOISOMERASE IN REPLICATION   Because cells infected with T4 phages mutant in genes 39, 52, or 60 eventually produce a burst of phage, it has been proposed that the missing gene function is partly replaced by a host function (139). Growth of these phage mutants is much more sensitive to coumermycin and novobiocin than growth of wild-type T4 (140); the complementing function might thus plausibly be DNA gyrase. Liu et al (99) have extended this line of argument by suggesting that T4 DNA topoisomerase, though so far known only to relax DNA, may have a gyrase-like supercoiling activity when bound at specific DNA sites forming a loop around the T4 replication origin, and so may help to melt out the DNA sequence at the initiation point.

MUTATIONAL STUDY OF *E. COLI* DNA TOPOISOMERASE I   Mutants of *E. coli* deficient in topoisomerase I have recently been found (141). Two strains have been characterized, one with 10% residual activity, the other with 1–5%. The mutations map close to the *trp* locus, near 27 min on the *E. coli* genetic map. Neither strain has any obvious physiological defect in growth, recombination, or repair of UV damage. In addition, a strain with topoisomerase I entirely inactivated by a deletion is still able to grow (R. Sternglanz, personal communication). Thus topoisomerase I activity is dispensable. The physiological role of the enzyme remains to be clarified.

POSSIBLE ROLE OF EUKARYOTIC TOPOISOMERASES IN REPLICATION   High concentrations of novobiocin inhibit initiation of DNA synthesis in a mammalian cell system (142) and in adenovirus-infected cells (143); novobiocin also inhibits UV repair in HeLa cells (144). These results might implicate an enzyme such as the *Drosophila* type II topoisomerase, which is known to be inhibited by novobiocin (105), or a (so far undiscovered) DNA gyrase. However, a direct inhibition by novobiocin of monkey cell DNA polymerase α has been found (145), and may be sufficient to explain at least some of the effects on DNA replication.

The role of topoisomerases in mitochondrial DNA replication in animal cells is a particularly interesting question. The mitrochondrial DNA is supercoiled but not complexed with histones; the cause of its supercoiling is unknown. Newly replicated mouse cell mitochondrial DNA is in relaxed closed-circular form (146) and only becomes supercoiled after a lag of approximately one hour (147). Since relaxed DNA is also found transiently after replication of some prokaryotic plasmids (148,149), whose later supercoiling is effected by DNA gyrase, and since coumermycin A_1 and novobiocin are known to inhibit DNA synthesis in rat liver mitochondria (150), these analogies have led to suggestions that a DNA gyrase may exist in mitrochondria (147,150). Until now, no direct evidence for such an activity has been found.

The cellular function of eukaryotic type I topoisomerases is also unknown. Rat liver topoisomerase I, in addition to its DNA-relaxing activity, helps assemble the four core histones into nucleosomal particles on DNA (151). Since replicating DNA has to attach new histones, such a reaction could be important in DNA replication. However, another protein with similar activity is known (152), and the cellular role of both reactions is yet to be assessed.

CATENATED PRODUCTS OF DNA REPLICATION   Catenanes have been found as a minor fraction of Col E1 DNA replicated in a cell-free system (153). These catenanes are stable once formed. However, catenanes found after replication in vivo of SV40 DNA (154) or bacterial plasmids (155,156) are chased into monomeric closed circles. [Catenanes are also found in mouse mitochondrial DNA (157) but there is evidence that they do not result from replication (146).] While several schemes have been suggested that would allow segregation of circular products at the end of a replication cycle and avoid the formation of catenanes (158–160), the newly found reactions of type I and type II topoisomerases that unlink catenanes now make it possible, at least in principle, to avoid a special segregation step. Catenanes could be a normal product of replication and could be unlinked later, in nicked-circular form by a type I topoisomerase, and in nicked- or closed-circular form by a type II enzyme. The very complex catenated DNA networks found in the kinetoplasts of trypanosomes can be unlinked by DNA gyrase or T4 DNA topoisomerase (161).

Effects of DNA Supercoiling on Transcription

It has been known for some time that transcription of negatively supercoiled DNA with purified *E. coli* RNA polymerase is generally enhanced over that of relaxed, nicked, or linear DNA (20,162–166). A study of initiation by *E. coli* RNA polymerase on PM2 DNA showed that the number of binding sites, the rate of binding of enzyme, and the stability of the polymerase-DNA complex were all enhanced when a superhelical PM2 DNA was used instead of nicked or relaxed DNA (167). These results are thermodynamically plausible in light of the fact that binding of *E. coli* RNA polymerase unwinds DNA by almost one turn (168). Binding to negatively supercoiled DNA is thus aided by a favorable free energy change possibly as large as

–9 kcal/mol (for DNA with $\Delta\alpha/\alpha° = -0.06$), which corresponds to an additional factor of more than 10^6 in equilibrium constant.

Inhibitors of DNA gyrase are known to depress transcription in vivo (75,118,169) and in vitro (77). An interesting result obtained more recently is that different transcripts are affected very differently. Sanzey (170) found that the expression in *E. coli* cells of the maltose and lactose operons is reduced five to ten fold by nalidixic acid, while that of the threonine, tryptophan, and tryptophanase loci is almost unaffected. Expression from the lactose repressor (*lac*I^q) and the *lac* UV5 mutant promoters, which is insensitive to catabolite repression, is even increased by nalidixic acid. Novobiocin has similar, though smaller, effects. It was suggested that catabolite-repressible (CAP-protein dependent) promoters are particularly sensitive to the inhibitors. Differential sensitivity has also been found in expression of the *trp* genes contained in a ϕ80 p*trp* phage; transcription from the phage P_L promoter is more sensitive to gyrase inhibitors than transcription from the *trp* promoter (171,172). Differential sensitivity has also been found for various other *E. coli* genes (173) and among T7 promoters (129), this last case showing that, as in replication (128), circularity of the DNA is not required for DNA gyrase action to be expressed. A specific inhibition by novobiocin of the outgrowth of *B. subtilis* spores has also been noted (174). In all these situations, inhibition is abolished by the use of *gyr*A or *gyr*B drug-resistant mutants, thus linking the inhibitory effect to DNA gyrase function.

With phage N4, a particularly close connection is seen between transcription and DNA conformation. Purified N4 RNA polymerase transcribes only single-stranded DNA. Intracellular transcription of N4 is abolished by gyrase inhibitors or a *gyr*A temperature-sensitive mutation (175). Apparently the unwinding of DNA that is facilitated by negative supercoiling is essential to allow N4 RNA polymerase to function.

Another possible level of interaction between transcription and supercoiling is suggested by the work of Baas et al (176), who find that ϕX174-RFI DNA isolated from cells often contains displacement loops (R-loops) of nascent RNA. The RNA is removed by treatments that disrupt the supercoiled DNA structure (e.g. DNase I nicking). As such an R-loop would remove torsional strain from the DNA molecule, the authors propose that RNA initiation at one promoter could block synthesis from other promoters in the same DNA molecule.

The influence of supercoiling on transcription is seen most directly in a cell-free coupled transcription-translation system (177). For example, with relaxed ColE1 DNA as template, expression of the colicin gene is strongly depressed by gyrase inhibitors, while other proteins coded by the plasmid are less affected. However, there is no inhibition if the DNA is added in supercoiled form, which shows that the need for gyrase function can be bypassed if the DNA is already supercoiled. By using rifampicin to block later initiation events, it was also possible to show that the effect of gyrase inhibition is on chain initiation; growth of previously initiated RNA chains is unaffected (177).

In eukaryotic systems, the possible relation between supercoiling and transcription has not been extensively studied. It has been reported that transcription by RNA polymerase II from calf thymus (177a) or from wheat germ (178) is greatly stimulated by supercoiling of the template DNA.

Role of Supercoiling and of Topoisomerases in DNA Recombination and Repair

DNA gyrase was first detected as a factor needed to activate DNA for the phage λ integrative recombination reaction in vitro (68). The requirement for gyrase activity can be bypassed by using a supercoiled DNA substrate (179,180). While reaction conditions can be modified to avoid an absolute requirement for supercoiling, recombination is always much more rapid with a supercoiled substrate (181). Intracellular λ integrative recombination also appears to require supercoiling, since it is inhibited by coumermycin (182). It has been suggested that the strand rotations involved in exchanging DNA partners during this reciprocal recombination would be favored by supercoiling (64). However, the requirement for supercoiling is not shared equally between the parental DNA species. Since only the molecule containing attP, not that containing attB, needs to be supercoiled (183), there is presumably also a requirement of supercoiling for protein recognition.

The Int protein that catalyzes this reaction has type I topoisomerase activity [(46); see above]. Nash et al (64) have suggested that strand exchanges occur after cleavage of pairs of strands in a four-strand synapse, with the Int protein bound to strand ends of the intermediates in analogy to a topoisomerization reaction. Two such pairwise exchanges would generate a double-strand crossover between DNA duplexes.

Supercoiling also stimulates repair of DNA damage in UV-irradiated cells, as well as some types of general recombination (184, 185); both processes are inhibited by coumermycin A_1. Supercoiling aids in vitro uptake of single strands into duplex DNA (reviewed in 186); whether its enhancement of recombination in vivo is due to this effect is not yet known. Another gyrase-mediated effect on DNA repair is known. Nalidixic acid treatment of *E. coli* induces the so-called SOS repair system (reviewed in 187), as do many other DNA-damaging and replication-inhibiting agents. Coumermycin A_1 is not an inducer, which suggests that mere inhibition of supercoiling is not the inducing event. Possibly the DNA breakage resulting from nalidixic acid treatment (discussed above) is responsible.

A more direct participation of DNA gyrase in one type of recombination has been found. Illegitimate recombination (i.e. between nonhomologous sequences) in a cell-free *E. coli* system is stimulated by *inhibition* of DNA gyrase by oxolinic acid (188). The stimulation is abolished by coumermycin A_1, and is not found in extracts of nalidixic acid-resistant mutants. The authors propose that strand ends cleaved by the gyrase-oxolinic acid combination may cross over by an exchange of gyrase subunits followed by resealing. This model suggests a preference for crossovers at gyrase cleavage sites.

CONCLUSIONS

The advances described above open the possibility of arriving at a detailed understanding of the chemistry and biology of topoisomerases. Now that the outlines of topoisomerization reactions are known, one can ask questions about the mechanics of DNA strand movement. Particularly for the ATP-dependent enzymes, where a protein conformation change is an almost essential feature of the reaction, knowledge about the coupling of protein motion to DNA transport should lead to new insights about these very simple mechanochemical systems.

While the cellular functions of DNA gyrase and T4 DNA topoisomerase are partly understood, much more information is needed about the multiple possible effects of these enzymes on DNA replication and, in the case of DNA gyrase, on transcription. The reasons for the coexistence of multiple topoisomerases in a single cell need to be examined; do they have competing, overlapping, or separate functions? In eukaryotes, there is even less information available. No eukaryotic topoisomerase has yet been identified with a known cellular function. Work in this area would be greatly advanced by the isolation of suitable mutants and specific enzyme inhibitors.

ACKNOWLEDGMENTS

I thank G. Felsenfeld, L. M. Fisher, K. Mizuuchi, and G. Selzer for their advice, and B. Leis for skilful secretarial help. I also thank the many colleagues who sent preprints of unpublished work.

Literature Cited

1. Champoux, J. J. 1978. *Ann. Rev. Biochem.* 47:449–79
2. Bauer, W. R. 1978. *Ann. Rev. Biophys. Bioeng.* 7:287–313
3. Wang, J. C., Liu, L. F. 1979. In *Molecular Genetics,* Taylor, J. H., ed., Pt 3, pp. 65–88. New York: Academic
4. Cozzarelli, N. R. 1980. *Science* 207:953–60
5. Fuller, F. B. 1971. *Proc. Natl. Acad. Sci. USA* 68:815–19
6. Fuller, F. B. 1978. *Proc. Natl. Acad. Sci. USA* 75:3557–61
7. Crick, F. H. C. 1976. *Proc. Natl. Acad. Sci. USA* 73:2639–43
8. Liu, L. F., Wang, J. C. 1979. *J. Biol. Chem.* 254:11082–88
9. Wang, J. C. 1979. *Proc. Natl. Acad. Sci. USA* 76:200–3
10. Depew, R. E., Wang, J. C. 1975. *Proc. Natl. Acad. Sci. USA* 72:4275–79
11. Pulleyblank, D. E., Shure, M., Tang, D., Vinograd, J., Vosberg, H.-P. 1975. *Proc. Natl. Acad. Sci. USA* 72:4280–84
12. Bauer, W., Vinograd, J. 1968. *J. Mol. Biol.* 33:141–72
13. Crawford, L. V., Waring, M. J. 1967. *J. Mol. Biol.* 25:23–30
14. Radloff, R., Bauer, W., Vinograd, J. 1967. *Proc. Natl. Acad. Sci. USA* 57:1514–21
15. Keller, W. 1975. *Proc. Natl. Acad. Sci. USA* 72:4876–80
16. Morgan, A. R., Lee, J. S., Pulleyblank, D. E., Murray, N. L., Evans, D. H. 1979. *Nucleic Acids Res.* 7:547–69
17. Morgan, A. R., Evans, D. H., Lee, J. S., Pulleyblank, D. E. 1979. *Nucleic Acids Res.* 7:571–94
18. Bauer, W., Vinograd, J. 1970. *J. Mol. Biol.* 47:419–35
19. Hsieh, T.-S., Wang, J. C. 1975. *Biochemistry* 14:527–35
20. Wang, J. C. 1974. *J. Mol. Biol.* 87:797–816
21. Gellert, M., Mizuuchi, K., O'Dea, M. H., Ohmori, H., Tomizawa, J. 1978. *Cold Spring Harbor Symp. Quant. Biol.* 43:35–40
21a. Vologodskii, A. V., Lukashin, A. V., Anshelevich, V. V., Frank-Kamenetskii, M. D. 1979. *Nucleic Acids Res.* 6:967–82
22. Wang, A. H.-J., Quigley, G. J., Kolpak, F. J., Crawford, J. L., van Boom, J. H., van der Marel, G., Rich, A. 1979. *Nature* 282:680–86
23. Davies, D. R., Zimmerman, S. 1980. *Nature* 283:11–12
24. Liu, L. F., Liu, C.-C., Alberts, B. M. 1980. *Cell* 19:697–708
25. Wang, J. C. 1971. *J. Mol. Biol.* 55:523–33
26. Champoux, J. J., Dulbecco, R. 1972. *Proc. Natl. Acad. Sci. USA* 69:143–46
27. Kung, V., Wang, J. C. 1977. *J. Biol. Chem.* 252:5398–402
28. Burrington, M. G., Morgan, A. R. 1976. *Can. J. Biochem.* 56:123–28
29. Durnford, J. M., Champoux, J. J. 1978. *J. Biol. Chem.* 253:1086–89
30. Poccia, D. L., LeVine, D., Wang, J. C. 1978. *Dev. Biol.* 64:273–83
31. Baase, W. A., Wang, J. C. 1974. *Biochemistry* 13:4299–303
32. Mattoccia, E., Attardi, D. G., Tocchini-Valentini, G. P. 1976. *Proc. Natl. Acad. Sci. USA* 73:4551–54
33. Laskey, R. A., Mills, A. D., Morris, N. R. 1977. *Cell* 10:237–43
34. Camerini-Otero, R. D., Felsenfeld, G. 1977. *Nucleic Acids Res.* 4:1159–81
35. Bina-Stein, M., Vogel, T., Singer, D. S., Singer, M. F. 1976. *J. Biol. Chem.* 251:7363–66
36. Vosberg, H.-P., Grossman, L. I., Vinograd, J. 1975. *Eur. J. Biochem.* 55:79–93
37. Vosberg, H.-P., Vinograd, J. 1976. *Biochem. Biophys. Res. Commun.* 68:456–64
38. Yoshida, S., Ungers, G., Rosenberg, B. H. 1977. *Nucleic Acids Res.* 4:223–28
39. DeLeys, R. J., Jackson, D. A. 1976. *Nucleic Acids Res.* 3:641–52
40. Pulleyblank, D. E., Morgan, A. R. 1975. *Biochemistry* 14:5205–9
41. Prell, B., Vosberg, H.-P. 1980. *Eur. J. Biochem.* 108:389–98
42. Eskin, B., Morgan, A. R. 1978. *Can. J. Biochem.* 56:89–91
43. Champoux, J. J., McConaughy, B. L. 1976. *Biochemistry* 15:4638–42
44. Fairfield, F. R., Bauer, W. R., Simpson, M. V. 1979. *J. Biol. Chem.* 254:9352–54
45. Bauer, W. R., Ressner, E. C., Kates, J., Patzke, J. V. 1977. *Proc. Natl. Acad. Sci. USA* 74:1841–45
46. Kikuchi, Y., Nash, H. A. 1979. *Proc. Natl. Acad. Sci. USA* 76:3760–64
47. Ikeda, J., Yudelevich, A., Hurwitz, J. 1976. *Proc. Natl. Acad. Sci.* 73:2669–73
48. Meyer, T. F., Geider, K. 1979. *J. Biol. Chem.* 254:12642–46
49. Brown, P. O., Cozzarelli, N. R. 1979. *Science* 206:1081–83
50. Depew, R. E., Liu, L. F., Wang, J. C. 1978. *J. Biol. Chem.* 253:511–18
51. Wang, J. C. 1973. In *DNA Synthesis in Vitro,* ed. R. B. Inman, R. D. Wells, pp. 163–74. Baltimore, Md: Univ. Park Press

51a. Burrington, M. G., Morgan, A. R. 1976. *Can. J. Biochem.* 54:301–6
52. Liu, L. F., Depew, R. E., Wang, J. C. 1976. *J. Mol. Biol.* 106:439–52
53. Kirkegaard, K., Wang, J. C. 1978. *Nucleic Acids Res.* 5:3811–20
54. Champoux, J. J. 1977. *Proc. Natl. Acad. Sci. USA* 74:5328–32
55. Tse, Y.-C., Kirkegaard, K., Wang, J. C. 1980. *J. Biol. Chem.* 255:5560–65
56. Tse, Y.-C., Wang, J. C. 1980. *Cell.* 22:269–76
57. Wang, J. C., Gumport, R. I., Javaherian, K., Kirkegaard, K., Klevan, L., Kotewicz, M. L., Tse, Y.-C. 1981. In *Mechanistic Studies of DNA Replication and Genetic Recombination,* ed. B. M. Alberts, C. F. Fox. New York: Academic. In press
58. Champoux, J. J. 1977. *Proc. Natl. Acad. Sci. USA* 74:3800–4
59. Keller, W., Muller, U., Eicken, I., Wendel, I., Zentgraf, H. 1977. *Cold Spring Harbor Symp. Quant. Biol.* 42:227–44
60. Champoux, J. J., Young, L. S., Been, M. D. 1978. *Cold Spring Harbor Symp. Quant. Biol.* 43:53–58
61. Champoux, J. J. 1978. *J. Mol. Biol.* 118:441–46
62. Champoux, J. J., Been, M. D. 1981. See Ref. 57. In press
63. Liu, L. F. 1981. See Ref. 57. In press
64. Nash, H.A., Mizuuchi, K., Enquist, L. W., Weisberg, R. A. 1981. *Cold Spring Harbor Symp. Quant. Biol.* 45:In press
65. Kikuchi, Y., Nash, H. A. 1978. *J. Biol. Chem.* 253:7149–57
66. Henry, T. J., Knippers, R. 1974. *Proc. Natl. Acad. Sci. USA* 71:1549–53
67. Eisenberg, S., Griffith, J., Kornberg, A. 1977. *Proc. Natl. Acad. Sci. USA* 74:3198–3202
68. Gellert, M., Mizuuchi, K., O'Dea, M. H., Nash, H. A. 1976. *Proc. Natl. Acad. Sci. USA* 73:3872–76
69. Liu, L. F., Wang, J. C. 1978. *Proc. Natl. Acad. Sci. USA* 75:2098–2102
70. Scurlock, T. R., Miller, R. V. 1981. *J. Bact.* In press
71. Sugino, A., Bott, K. F. 1980. *J. Bacteriol.* 141:1331–39
72. Gellert, M., O'Dea, M. H., Itoh, T., Tomizawa, J. 1976. *Proc. Natl. Acad. Sci. USA* 73:4474–78
73. Gellert, M., Mizuuchi, K., O'Dea, M. H., Itoh, T., Tomizawa, J. 1977. *Proc. Natl. Acad. Sci. USA* 74:4772–76
74. Sugino, A., Peebles, C. L., Kreuzer, K. N., Cozzarelli, N. R. 1977. *Proc. Natl. Acad. Sci. USA* 74:4767–71
75. Ryan, M. J. 1976. *Biochemistry* 15:3769–77
76. Goss, W. A., Deitz, W. H., Cook, T. M. 1965. *J. Bacteriol.* 89:1068–74
77. Staudenbauer, W. L. 1976. *Eur. J. Biochem.* 62:491–97
78. Hansen, F. G., von Meyenburg, K. 1979. *Mol. Gen. Genet.* 175:135–44
79. Fairweather, N. F., Orr, E., Holland, I. B. 1980. *J. Bacteriol.* 142:153–61
80. Bourguignon, G. J., Levitt, M., Sternglanz, R. 1973. *Antimicrob. Agents Chemother.* 4:479–86
81. Marians, K. J., Ikeda, J., Schlagman, S., Hurwitz, J. 1977. *Proc. Natl. Acad. Sci. USA* 74:1965–69
82. Higgins, N. P., Peebles, C. L., Sugino, A., Cozzarelli, N. R. 1978. *Proc. Natl. Acad. Sci. USA* 75:1773–77
83. Mizuuchi, K., O'Dea, M. H., Gellert, M. 1978. *Proc. Natl. Acad. Sci. USA* 75:5960–63
84. Brown, P. O., Peebles, C. L., Cozzarelli, N. R. 1979. *Proc. Natl. Acad. Sci. USA* 76:6110–14
85. Klevan, L., Wang, J. C. 1980. *Biochemistry.* 19:5229–34
86. Sugino, A., Higgins, N. P., Brown, P. O., Peebles, C. L., Cozzarelli, N. R. 1978. *Proc. Natl. Acad. Sci. USA* 75:4838–42
87. Gellert, M., Fisher, L. M., O'Dea, M. H. 1979. *Proc. Natl. Acad. Sci. USA* 76:6289–93
88. Gellert, M., Fisher, L. M., Ohmori, H., O'Dea, M. H., Mizuuchi, K. 1981. *Cold Spring Harbor Symp. Quant. Biol.* 45:In press
89. Morrison, A., Cozzarelli, N. R. 1979. *Cell* 17:175–84
90. Kirkegaard, K., Wang, J. C. 1981. *Cell.* In press
91. Peebles, C. L., Higgins, N. P., Kreuzer, K. N., Morrison, A., Brown, P. O., Sugino, A., Cozzarelli, N. R. 1978. *Cold Spring Harbor Symp. Quant. Biol.* 43:41–52
92. Morrison, A., Higgins, N. P., Cozzarelli, N. R. 1980. *J. Biol. Chem.* 255:2211–19
93. Sugino, A., Cozzarelli, N. R. 1980. *J. Biol. Chem.* 255:6299–306
94. Easterbrook-Smith, S. B., Wallace, J. C., Keech, D. B. 1976. *Eur. J. Biochem.* 62:125–30
95. Liu, L. F., Wang, J. C. 1978. *Cell* 15:979–84
96. McGhee, J. D., Felsenfeld, G. 1980. *Ann. Rev. Biochem.* 49:1115–56
97. Galas, D. J., Schmitz, A. 1978. *Nucleic Acids Res.* 5:3157–70
98. Sollner-Webb, B., Felsenfeld, G. 1977. *Cell* 10:537–47

99. Liu, L. F., Liu, C.-C., Alberts, B. M. 1979. *Nature* 281:456–61

100. Mizuuchi, K., Fisher, L. M., O'Dea, M. H., Gellert, M. 1980. *Proc. Natl. Acad. Sci. USA* 77:1847–51

101. Kreuzer, K. N., Cozzarelli, N. R. 1980. *Cell* 20:245–54

102. Forterre, P. 1980. *J. Theor. Biol.* 82:255–69

103. Stetler, G. L., King, G. J., Huang, W. M. 1979. *Proc. Natl. Acad. Sci. USA* 76:3737–41

104. Epstein, R. H., Bolle, A., Steinberg, C. H., Kellenberger, E., Boy de la Tour, E., Chevalley, R., Edgar, R. S., Susman, M., Denhardt, G., Lielausis, A. 1963. *Cold Spring Harbor Symp. Quant. Biol.* 28:375–94

105. Hsieh, T.-S., Brutlag, D. 1980. *Cell* 21:115–25

106. Hsiang, M. W., Cole, R. D. 1977. *Proc. Natl. Acad. Sci. USA* 74:4852–56

107. Baldi, M. I., Benedetti, P., Mattoccia, E., Tocchini-Valentini, G. P. 1980. *Cell* 20:461–67

108. Castora, F. J., Brown, G. G., Simpson, M. V. 1981. In *The Organization and Expression of the Mitochondrial Genome,* ed. A. M. Kroon, C. Saccone. Amsterdam: Elsevier. In press

109. Germond, J. E., Hirt, B., Oudet, P., Gross-Bellard, M., Chambon, P. 1975. *Proc. Natl. Acad. Sci. USA* 72:1843–47

110. Felsenfeld, G. 1978. *Nature* 271:115–22

111. Sinden, R. R., Carlson, J. O., Pettijohn, D. E. 1980. *Cell.* 21:773–83

112. Young, E. T. II, Sinsheimer, R. L. 1964. *J. Mol. Biol.* 10:562–64

113. Worcel, A., Burgi, E. 1972. *J. Mol. Biol.* 71:127–47

114. Pettijohn, D. E., Hecht, R. 1973. *Cold Spring Harbor Symp. Quant. Biol.* 38:31–41

115. Drlica, K., Snyder, M. 1978. *J. Mol. Biol.* 120:145–54

116. Pettijohn, D. E., Pfenninger, O. 1980. *Proc. Natl. Acad. Sci. USA* 77:1331–35

117. Cairns, J. 1963. *J. Mol. Biol.* 6:208–13

118. Smith, D. H., Davis, B. D. 1967. *J. Bacteriol.* 93:71–79

119. Ryan, M. J. 1976. *Biochemistry* 15:3778–82

120. Staudenbauer, W. L. 1975. *J. Mol. Biol.* 96:201–5

121. Staudenbauer, W. L. 1976. *Mol. Gen. Genet.* 145:273–80

122. Itoh, T., Tomizawa, J. 1978. *Cold Spring Harbor Symp. Quant. Biol.* 43:409–17

123. Sumida-Yasumoto, C., Yudelevich, A., Hurwitz, J. 1976. *Proc. Natl. Acad. Sci. USA* 73:1887–91

124. Orr, E., Fairweather, N. F., Holland, I. B., Pritchard, R. H. 1979. *Mol. Gen. Genet.* 177:103–12

125. Filutowicz, M. 1980. *Mol. Gen. Genet.* 177:301–9

126. McHugh, G. L., Swartz, M. N. 1977. *Antimicrob. Agents Chemother.* 12:423–26

127. Taylor, D. E., Levine, J. G. 1979. *Mol. Gen. Genet.* 174:127–33

128. Itoh, T., Tomizawa, J. 1977. *Nature* 270:78–79

129. De Wyngaert, M. A., Hinkle, D. C. 1979. *J. Virol.* 29:529–35

130. Cook, T. M., Brown, K. G., Boyle, J. V., Goss, W. A. 1966. *J. Bacteriol.* 92:1510–14

131. Dermody, J. J., Bourguignon, G. J., Foglesong, P. D., Sternglanz, R. 1974. *Biochem. Biophys. Res. Commun.* 61:1340–47

132. Pedrini, A. M., Geroldi, D., Siccardi, A., Falaschi, A. 1972. *Eur. J. Biochem.* 25:359–65

133. Sakakibara, Y., Tomizawa, J. 1974. *Proc. Natl. Acad. Sci. USA* 71:802–6

134. Pisetsky, D., Berkower, I., Wickner, R., Hurwitz, J. 1972. *J. Mol. Biol.* 71:557–71

135. Snyder, M., Drlica, K. 1979. *J. Mol. Biol.* 131:287–302

136. Kreuzer, K. N., Cozzarelli, N. R. 1979. *J. Bacteriol.* 140:424–35

137. Baird, J. P., Bourguignon, G. J., Sternglanz, R. 1972. *J. Virol.* 9:17–21

138. Drlica, K., Engle, E., Manes, S. H. 1980. *Proc. Natl. Acad. Sci. USA* 77:6879–83

139. Mufti, S., Bernstein, H. 1974. *J. Virol.* 14:860–71

140. McCarthy, D. 1979. *J. Mol. Biol.* 127:265–83

141. Sternglanz, R., Di Nardo, S., Wang, J. C., Nishimura, Y., Hirota, Y. 1981. See Ref. 57. In press

142. Mattern, M. R., Painter, R. B. 1979. *Biochim. Biophys. Acta* 563:306–12

143. D'Halluin, J.-C., Milleville, M., Boulanger, P. 1980. *Nucleic Acids Res.* 8:1625–41

144. Collins, A., Johnson, R. 1979. *Nucleic Acids Res.* 7:1311–20

145. Edenberg, H. J. 1980. *Nature* 286:529–31

146. Berk, A. J., Clayton, D. A. 1976. *J. Mol. Biol.* 100:85–102

147. Bogenhagen, D., Clayton, D. A. 1978. *J. Mol. Biol.* 119:69–81

148. Timmis, K., Cabello, C., Cohen, S. N. 1976. *Nature* 261:512–16

149. Crosa, J. H., Luttropp, L. K., Falkow, S. 1976. *Nature* 261:516–19

150. Castora, F. J., Simpson, M. V. 1979. *J. Biol. Chem.* 254:11193–95
151. Germond, J. E., Rouviere-Yaniv, J., Yaniv, M., Brutlag, D. L. 1979. *Proc. Natl. Acad. Sci. USA* 76:3779–83
152. Laskey, R. A., Honda, B. M., Mills, A. D., Finch, J. T. 1978. *Nature* 275:416–20
153. Sakakibara, Y., Suzuki, K., Tomizawa, J. 1976. *J. Mol. Biol.* 108:569–82
154. Jaenisch, R., Levine, A. J. 1973. *J. Mol. Biol.* 73:199–212
155. Kupersztoch, Y. M., Helinski, D. R. 1973. *Biochem. Biophys. Res. Commun.* 54:1451–59
156. Novick, R. P., Smith, K., Sheehy, R. J., Murphy, E. 1973. *Biochem. Biophys. Res. Commun.* 54:1460–69
157. Flory, P. J. Jr., Vinograd, J. 1973. *J. Mol. Biol.* 74:81–94
158. Meinke, W., Goldstein, D. A. 1971. *J. Mol. Biol.* 61:543–63
159. Gefter, M. L. 1975. *Ann. Rev. Biochem.* 44:45–78
160. Tomizawa, J. 1978. In *DNA Synthesis: Present and Future,* ed. M. Kohiyama, I. Molineux, pp. 797–826. New York: Plenum
161. Marini, J. C., Miller, K. G., Englund, P. T. 1980. *J. Biol. Chem.* 255:4976–79
162. Hayashi, Y., Hayashi, M. 1971. *Biochemistry* 10:4212–18
163. Botchan, P., Wang, J. C., Echols, H. 1973. *Proc. Natl. Acad. Sci. USA* 70:3077–81
164. Botchan, P. 1976. *J. Mol. Biol.* 105:161–76
165. Seeburg, P. H., Nusslein, C., Schaller, H. 1977. *Eur. J. Biochem.* 74:107–13
166. Levine, A., Rupp, W. D. 1978. In *Microbiology,* ed. D. Schlessinger, pp. 163–65. Washington DC: Am. Soc. Microbiol.
167. Richardson, J. P. 1975. *J. Mol. Biol.* 91:477–87
168. Saucier, J.-M., Wang, J. C. 1972. *Nature New Biol.* 239:167–70
169. Puga, A., Tessman, I. 1973. *J. Mol. Biol.* 75:99–108
170. Sanzey, B. 1979. *J. Bacteriol.* 138:40–47
171. Smith, C. L., Kubo, M., Imamoto, F. 1978. *Nature* 275:420–23
172. Kubo, M., Kano, Y., Nakamura, H., Nagata, A., Imamoto, F. 1979. *Gene* 7:153–71
173. Shuman, H., Schwartz, M. 1975. *Biochem. Biophys. Res. Commun.* 64:204–9
174. Gottfried, M., Orrego, C., Keynan, A., Halvorson, H. O. 1979. *J. Bacteriol.* 138:314–19
175. Falco, S. C., Zivin, R., Rothman-Denes, L. B. 1978. *Proc. Natl. Acad. Sci. USA* 75:3220–24
176. Baas, P. D., Keegstra, W., Teertstra, W. R., Jansz, H. S. 1978. *J. Mol. Biol.* 125:187–205
177. Yang, H.-L., Heller, K., Gellert, M., Zubay, G. 1979. *Proc. Natl. Acad. Sci. USA* 76:3304–8
177a. Lescure, B., Chestier, A., Yaniv, M. 1978. *J. Mol. Biol.* 124:73–85
178. Akrigg, A., Cook, P. R. 1980. *Nucleic Acids Res.* 8:845–53
179. Nash, H. A., Mizuuchi, K., Weisberg, R. A., Gellert, M. 1977. In *DNA Insertion Elements, Plasmids, and Episomes,* ed. S. L. Adhya A. I. Bukhari, J. A. Shapiro, pp. 363–73. Cold Spring Harbor, New York: Cold Spring Harbor Lab.
180. Mizuuchi, K., Gellert, M., Nash, H. A. 1978. *J. Mol. Biol.* 121:375–92
181. Pollock, T. J., Abremski, K. 1979. *J. Mol. Biol.* 131:651–54
182. Kikuchi, Y., Nash, H. A. 1978. *Cold Spring Harbor Symp. Quant. Biol.* 43:1099–109
183. Mizuuchi, K., Mizuuchi, M. 1978. *Cold Spring Harbor Symp. Quant. Biol.* 43:1111–14
184. Hays, J. B., Boehmer, S. 1978. *Proc. Natl. Acad. Sci. USA* 75:4125–29
185. Raina, J. L., Ravin, A. W. 1979. *Mol. Gen. Genet.* 176:171–81
186. Radding, C. M. 1978. *Ann. Rev. Biochem.* 47:847–80
187. Witkin, E. M., Wermundsen, I. E. 1978. *Cold Spring Harbor Symp. Quant. Biol.* 43:881–86
188. Ikeda, H., Moriya, K., Matsumoto, T. 1981. *Cold Spring Harbor Symp. Quant. Biol.* 45: In press

Ann. Rev. Biochem. 1981. 50:911–68

A SURVEY OF INBORN ERRORS OF AMINO ACID METABOLISM AND TRANSPORT IN MAN

♦12100

Daniel Wellner and Alton Meister

Department of Biochemistry, Cornell University Medical College, New York, New York 10021

CONTENTS

0066-4154/81/0701-0911$01.00

Perspectives and Summary

Studies on the inborn errors of metabolism and transport of amino acids may be considered to have begun with the finding in 1810 by Wollaston (20e) of cystine, the second natural amino acid to be isolated, in a urinary stone (most likely from a patient with cystinuria). About a century later, Garrod (7a) introduced the term "inborn error of metabolism" and accurately defined the biochemical and genetic concepts involved, but his work was not fully appreciated at the time. Beadle and Tatum were able to effectively develop the "one gene-one enzyme" concept using an organism (Neurospora) that was more readily available for laboratory study. Later work, especially on mutants of *E. coli*, further defined many pathways of amino acid metabolism. The use of mutant organisms is now well established in biochemistry. However, the occurrence of genetically induced deficiencies in man clearly justifies special consideration. There is some tendency to regard human inborn errors of metabolism as rare and unusual types of human disease; however, it is now generally appreciated that both these and more common human diseases can be understood in molecular terms. The study of inborn errors may illuminate some of the more common human disorders; for example, the development of premature atherosclerosis in homocystinuria, the frequent occurrence of neoplasms in certain forms of tyrosinemia, and the early development of arthritis in alcaptonuria offer potentially important clues to the pathogenesis of other human diseases. The development of practical treatments for some inborn errors has thus not only benefited the patients, but has also increased the general body of medical information and experience. It is likely that the ultimate therapy for inborn errors will involve application of genetic engineering technology; there is therefore a need to carefully define the exact nature of the biochemical and genetic defects involved.

Information about amino acid metabolism in man has been achieved through studies on man, experimental animals, and microorganisms. However, study of inborn errors has been of particular value, and has revealed how incomplete and inaccurate some of our ideas have been. Not only is our knowledge of the various pathways of metabolism incomplete and in some respects defective, but we do not know important facts about the relationships between metabolic events, physiological findings, and the pathogenesis of disease. Study of inborn errors has often revealed unusual metabolites and quantitatively minor pathways that have turned out to be of physiological importance. Such investigations have revealed the existence of new enzymes and new pathways; for example, work on hypervalinemia and on β-aminoisobutyric aciduria led to finding of new enzymes, and studies on nonketotic hyperglycinemia showed that the major pathway for the degradation of glycine is mediated by the glycine cleavage system. Studies on phenylketonuria have clarified the function of a cofactor that functions in three hydroxylation reactions in different pathways. Studies on 5-oxoprolinuria have elucidated the feed-back control mechanism of glutathione biosynthesis. Examination of the aminoacidurias has led to information not previously available, about individual amino acid transport systems. Such developments have greatly enlarged our knowledge of amino acid metabolism, and the associated enzymology and transport phenomena.

The more than one hundred diseases surveyed here include some conditions that represent heterogeneous groups of closely related disorders. Such heterogeneity arises from a number of causes. Thus, a different enzyme may be affected in different patients. For example, hyperphenylalaninemia, albinism, homocystinuria, hyperglycinemia, hyperammonemia, and methylmalonic acidemia are not truly single disease entities; rather, they consist of a number of different diseases each associated with a different enzyme deficiency. However, even a disease produced by a single enzyme deficiency may present a spectrum of heterogeneity. In some instances this is due to alleles of the same gene that code for forms of the enzyme that vary in activity or degree of stability. In other cases, heterogeneity may be caused by the interaction of a mutant gene with different total genetic endowments, which may provide different degrees of compensation for the primary defect.

Several general points emerge from consideration of the large mass of data surveyed here. Most of the inborn errors are recessive, rather than dominant. The term "recessive" implies that the affected individual can escape symptoms if he has about half of the normal amount of the enzyme involved. It is apparent that the recessiveness or dominance of a trait might well be affected by the environment; e.g. a substantial increase in the dietary intake of phenylalanine by an individual heterozygous for the phenylketonuria trait might lead to phenylketonuria. The morbidity associated with inborn errors of metabolism may exceed that expected on the basis of purely genetic considerations. Thus, a fetus heterozygous for a recessive trait or genetically normal may not develop normally in the uterus of a heterozygous mother.

Amino acids, although essential for the synthesis of proteins and certain other compounds, can act as poisons. Indeed, the amino acid toxicities of the type seen in certain inborn errors are much more severe than many of the environmental ones that are currently under scrutiny. One may conclude that the major function of phenylalanine hydroxylase is to destroy phenylalanine; without this enzyme, phenylalanine would accumulate and produce severe toxicity inconsistent with the development of normal intelligence. It seems less reasonable to argue that the major function of this enzyme is to produce tyrosine, which is almost always readily available in the diet.

The metabolic consequences of blocks due to inborn errors may lead to deficiency or accumulation of a precursor, development of alternative metabolic pathways, lack of synthesis of a product, or abnormal control phenomena. The product that accumulates may have effects on other pathways and this may obscure the basic defect. Reliable observations on the biochemical changes that accompany inborn errors should thus lead to greater understanding of the physiological consequences of chemical events. Some metabolic defects are accompanied by little or no physiological effect, which illustrates an aspect of the variations that occur among normal individuals. Finally, a significant number of inborn errors are, to some extent, correctable by administration of particular vitamins that serve as precursors of necessary coenzymes. Whereas these observations of themselves do not necessarily support all that has been claimed for "megavitamin" therapy, the thoughtful physician might well consider the possibility of a trial of vitamin therapy (usually an innocuous procedure except for vitamins A and D) in certain cases involving inborn errors of metabolism and perhaps in other diseases as well.

Introduction

This survey of the inborn errors of amino acid metabolism and transport is in the form of a relatively brief text, a rather large table, a few diagrams of metabolic pathways, and some references to the literature. Table 1, which has 122 entries, was compiled after an extensive examination of the literature.

We have probably erred by including some disorders that are very poorly described and some have undoubtedly been overlooked. The disorders listed in Table 1 involve derangements in the formation, degradation, or transport of amino acids. Some are associated with deficiency of an enzyme directly concerned with the metabolism of an amino acid, while others affect amino acid metabolism indirectly. Disorders associated with deficiencies of transport of amino acids in kidney, intestine, and other types of cells are included. Table 1 gives the most prominant biochemical and clinical findings for each disease, the probable primary defect, if known, and references selected to provide the reader with an introduction to the literature[1]. We have selected references to recent reviews, reports of the initial discovery of new diseases, and relatively recent findings. *The reference numbers correspond to the disease numbers given in Table 1 and to the enzymatic steps given in the figures.* In certain respects, the nomenclature of the metabolic diseases serves as a considerable source of confusion. No official nomenclature has been established for inborn errors, and we feel that recourse to the officially recommended nomenclature for the enzymes would lead to even further obfuscation. A number of the diseases listed have synonyms and occasionally the same name has been used for different diseases. To illustrate the problem, a condition in which the level of glycine is elevated in both blood plasma and urine may be variously referred to as glycinemia, hyperglycinemia, glycinuria, or hyperglycinuria; clearly none of these terms defines the disease. The term tyrosinemia may refer to conditions in which there is a deficiency of fumarylacetoacetase or of tyrosine transaminase. Tyrosinemia is an inappropriate term in any event, since tyrosine is a normal constituent of blood plasma. We have decided to resist the temptation to invent new and logical names for these disorders, but have simply used the terms that have already been used in the literature together with cross references to their synonyms.

Several excellent reviews of this field have appeared previously; these are cited by roman numerals in the text and the list of references.

Disorders of Aromatic Amino Acid Metabolism

INTRODUCTION Phenylalanine, a dietary essential amino acid, is converted to tyrosine, the first step in its main degradative pathway (see Figure 1). Tyrosine is not essential in the diet, unless its formation from phenylala-

---

[1]The authors deeply regret that, due to space limitations, not only the text, but also the list of references had to be greatly restricted in length; a more detailed monograph covering this extensive field is being prepared.

Table 1 Inherited disorders of amino acid metabolism and transport in man

Name[a] (probable primary defect)	Major chemical and clinical findings
Albinism (oculocutaneous) (No. 1-6) [At least 6 types are known, all with defective melanin formation, nystagmus, photophobia, and decreased visual acuity].	
1 Tyrosinase-negative albinism [Tyrosinase]	No detectable tyrosinase or melanin.
2 Tyrosinase-positive albinism [?]	Tyrosinase present, some melanin formed.
3 Chediak-Higashi syndrome [?]	Tyrosinase present, abnormal melanin, neutropenia, susceptibility to infection, giant granules in granulocytes.
4 Hermansky-Pudlak Syndrome [?]	Tyrosinase present, abnormal platelets, accumulation of lipids in reticuloendothelial and other cells.
5 Yellow mutant albinism [?]	Melanocytes present with abnormal, yellowish pigment.
6 Cross syndrome [?]	Microphthalmia, oligophrenia, athetosis. Some tyrosinase activity.
7 Alcaptonuria [Homogentisate dioxygenase]	Urinary excretion of homogentisic acid, ochronosis, arthritis, dark pigmentation of cerumen.
8 α-Aminoadipic aciduria [?]	Urinary excretion of α-aminoadipate. Asymptomatic. (High urinary excretion of α-aminoadipate may also occur in Reye's syndrome).
9 β-Aminoisobutyric aciduria [D-β-aminoisobutyrate-pyruvate transaminase (liver)]	High urinary excretion of D-β-aminoisobutyrate. Asymptomatic.
Argininemia	(see hyperargininemia).
10 Argininosuccinic aciduria [Argininosuccinase]	Accumulation of argininosuccinate in blood, urine, and cerebrospinal fluid. Convulsions, hepatomegaly, mental retardation, hair defect.
11 Aspartylglycosaminuria [N-aspartyl-β-glucosaminidase]	Mental retardation, high urinary excretion of 2-acetamido-1-N-(4-L-aspartyl)-2-deoxy-β-D-glucopyranosylamine and other glycoasparagines. A lysosomal storage disease.
12 Blue diaper syndrome [Intestinal tryptophan transport]	Hypercalcemia, nephrocalcinosis, high urinary excretion of indican, indoleacetic acid, indolelactic acid, indoleacetamide, and indoleacetylglutamine. Soiled diapers turn blue on standing due to formation of indigotin.
13 Branched-chain ketoaciduria (intermittent) [Branched-chain α-keto acid decarboxylase] Busby syndrome	In several milder variants of maple syrup urine disease (q.v.), there is residual activity of the deficient enzyme. Some patients respond to thiamin. (See Rowley-Rosenberg syndrome).

[a] The numbers assigned to the diseases in the table correspond to the reference numbers (given in the text and in the literature cited section and also to the enzymatic steps numbered in the figures.

Table 1 *(Continued)*

Name[a] (probable primary defect)	Major chemical and clinical findings
14 Carnosinemia [Carnosinase (plasma)]	High plasma levels and urinary excretion of carnosine. Since some, but not all patients have mental retardation and other neurological abnormalities, these signs may be unrelated to the enzyme defect.
15 Citrullinemia [Argininosuccinate synthetase]	Elevated plasma and urinary citrulline. Other plasma amino acids and ammonia may also be high. Clinical signs vary in severity. They may include vomiting, convulsions and mental retardation, or, in the acute neonatal form, coma and death soon after birth.
16 Cutis laxa [Lysyl oxidase (skin)]	Loose, sagging skin. X-linked.
17 Cystathioninuria [γ-Cystathionase]	High urinary excretion of cystathionine. Cystathionine may also be elevated in plasma and tissues. Large doses of vitamin B_6 results in decreased cystathionine excretion in some patients. Probably asymptomatic.
18 Cysteine peptiduria [?]	One patient reported, with mental retardation and urinary excretion of a peptide containing cysteine and glycine in a ratio of 2:1.
19 Cystinosis [Transport or reduction of cystine]	Intracellular accumulation of cystine, often resulting in crystallization within lysosomes. There is a severe nephropathic form, a benign form, and an intermediate form. Growth impairment, rickets, generalized aminoaciduria (Fanconi syndrome), progressive deterioration of renal function. Cystine crystal deposition in cornea, conjunctiva, bone marrow, and leukocytes.
20 Cystinuria [Transport of cystine, lysine, ornithine, and arginine (renal and intestinal)]	High urinary excretion of cystine, lysine, ornithine, and arginine. Cystine stones often form in the urinary tract. Occurs in three genetically distinct forms. (See also glutathionuria, hypercystinuria).
21 Dicarboxylic aminoaciduria [Transport of aspartic and glutamic acids (renal)]	High urinary excretion of aspartic and glutamic acids. Asymptomatic.
22 Dicarboxylic aminoaciduria [Transport of aspartic and glutamic acids (renal and intestinal)]	High urinary excretion of aspartic and glutamic acids. Also moderate hyperprolinemia and hypoglycemia.
Ehlers-Danlos syndrome	A group of diseases related to collagen synthesis, which vary in severity and in mode of inheritance. Symptoms include hyperextensibility of joints, fragile skin, easy bruising, and poor wound healing.
23 Type I [?]	Autosomal dominant. Severe form.
24 Type II [?]	Autosomal dominant. Mild form.
25 Type III [?]	Autosomal dominant. Mild form, hyperextensibility of joints.

Table 1 *(Continued)*

Name[a] (probable primary defect)	Major chemical and clinical findings
26 Type IV [Decrease in the synthesis of Type III collagen]	Autosomal recessive. Tendency for arterial rupture.
27 Type V [Lysyl oxidase]	X-linked recessive. Mild form.
28 Type VI [Lysyl hydroxylase]	Autosomal recessive. Kyphoscoliosis, arachnodactily.
29 Type VII [Abnormal pro α2 chain]	Autosomal recessive. Short stature, joint dislocations. Defective conversion of procollagen to collagen.
30 Ethanolaminosis [Ethanolamine kinase (liver)]	Ethanolamine elevated in urine and liver. Vomiting, physical and mental retardation, cardiomegaly, hypotonia, cerebral dysfunction, death in infancy. A lysosomal storage disease.
31 Fanconi syndrome [Renal tubular reabsorption of amino acids, as well as glucose, phosphate, bicarbonate, water, sodium, and potassium]	May result from inborn errors of metabolism such as cystinosis, Lowe's syndrome, tyrosinemia, galactosemia, fructose intolerance, or Wilson's disease. Also seen in poisoning by lead, mercury, or maleic acid and related compounds and in other non-hereditary conditions.
32 Formiminoglutamic aciduria [Glutamate formiminotransferase (liver)]	High urinary excretion of formiminoglutamate, increased by histidine loading. Some elevation of urinary histidine, carnosine, and anserine, and plasma histidine. Abnormally low plasma methionine. Excretion of formiminoglutamate decreased by administration of methionine. Hyperkinesia and retardation of speech maturation.
Friedreich's ataxia	(See hyperalaninemia (dihydrolipoyl dehydrogenase deficiency)).
33 Fructose-1,6-diphosphatase deficiency [Fructose-1,6-diphosphatase]	Lactic acidosis, hepatomegaly, hypoglycemia, seizures, coma. Elevation of many amino acids in plasma, lactic aciduria, and generalized aminoaciduria.
34 Fructose intolerance (hereditary) [Fructose-1-phosphate aldolase (liver aldolase)]	Hypoglycemia and vomiting following fructose ingestion. Hepatomegaly, jaundice, albuminuria, aminoaciduria. Elevated plasma tyrosine and methionine. Strong aversion for foods containing fructose or sucrose.
35 Galactosemia [Galactose-1-phosphate-uridyl transferase]	Galactose elevated in blood and urine, acidosis, albuminuria, aminoaciduria. Failure to thrive, vomiting, hepatomegaly, cataracts, mental retardation.
36 Glucoglycinuria [Renal reabsorption of glucose and glycine]	Asymptomatic. Autosomal dominant. High urinary excretion of glucose and glycine.
37 γ-Glutamylcysteine synthetase deficiency [γ-Glutamylcysteine synthetase]	Hemolytic anemia, spinocerebellar degeneration, peripheral neuropathy, myopathy, glutathione deficiency, and general aminoaciduria.

Table 1 *(Continued)*

Name[a] (probable primary defect)	Major chemical and clinical findings
γ-Glutamyl cyst(e)inuria	(See glutathionuria)
γ-Glutamyl transpeptidase deficiency	(See glutathionuria)
38 Glutaric aciduria, Type I [Glutaryl CoA dehydrogenase]	Progressive dystonic cerebral palsy and choreo-athetosis. Urinary excretion of glutarate, β-hydroxyglutarate, and glutaconate. Glutaric acidemia. Impaired ability of leukocytes to metabolize glutaryl CoA.
39 Glutaric aciduria, Type II [?]	Fatal neonatal metabolic acidosis, hypoglycemia. High excretion of glutaric, lactic, isobutyric, isovaleric, α-methylbutyric, and other organic acids. Hyperlysinemia and hypervalinemia. Strong sweaty-feet odor.
40 Glutathione peroxidase deficiency [Glutathione peroxidase (erythrocytes)]	Compensated hemolytic anemia, neonatal jaundice, drug-induced hemolysis. (Enzyme activity in leukocytes was not reported).
41 Glutathione peroxidase deficiency [Glutathione peroxidase (erythrocytes and leukocytes)]	Hemolytic anemia, reticulocytosis. May be the same as No. 40.
42 Glutathione peroxidase deficiency [Glutathione peroxidase (leukocytes)]	Chronic granulomatous disease. Apparent autosomal recessive inheritance. Enzyme activity normal in erythrocytes.
43 Glutathione reductase deficiency [Glutathione reductase (erythrocytes)]	Hemolytic crisis after ingestion of fava beans. No glutathione reductase detectable in erythrocytes, low values in leukocytes.
44 Glutathione synthetase deficiency [Glutathione synthetase (erythrocytes)]	Hemolytic anemia, low levels of glutathione and glutathione synthetase in erythrocytes, higher levels of both in leukocytes and fibroblasts. Absence of, or moderate oxoprolinuria. No clinical signs of acidosis. (Cf. 5-oxoprolinuria).
Glutathione synthetase deficiency (generalized)	(See 5-oxoprolinuria).
45 Glutathionuria [γ-Glutamyl transpeptidase]	Glutathionemia and glutathionuria. Substantial urinary excretion of cysteine and γ-glutamylcysteine moieties. Mental retardation.
46 D-Glyceric acidemia [?]	High plasma and urine concentrations of D-glycerate. No hyperoxaluria (cf. L-glyceric aciduria). One patient also had non-ketotic hyperglycinemia (q.v.) but no signs of acidosis. Another had chronic metabolic acidosis but no hyperglycinemia.
L-Glyceric aciduria	(See hyperoxaluria, primary Type II).
Glycolic aciduria	(See hyperoxaluria, primary Type I).
47 Glycylprolinuria [Prolidase]	Otitis, sinusitis, dermatitis, splenomegaly. High urinary excretion of dipeptides with carboxyl-terminal proline or hydroxyproline. Prolidase deficient in erythrocytes, leukocytes, and fibroblasts.
48 Gyrate atrophy of the choroid and retina [Ornithine transaminase]	Chorioretinal degeneration, night blindness, loss of peripheral vision. High urinary excretion of ornithine, high concentrations of ornithine and low concentrations of lysine in plasma.
49 Hartnup's disease [Renal and intestinal transport of a number of neutral amino acids]	Variable and intermittent clinical signs which may include rashes and cerebellar ataxia. Some patients are asymptomatic. Amino acids excreted in the urine in

Table 1 *(Continued)*

Name[a] (probable primary defect)	Major chemical and clinical findings
	excess include alanine, serine, threonine, valine, leucine, isoleucine, asparagine, glutamine, tyrosine, phenylalanine, tryptophan, and histidine. Indican and other indole derivatives may also be excreted in the urine.
Hepatolenticular degeneration	(See Wilson's disease).
50 Histidinemia [Histidase]	Histidine elevated in blood and urine. High urinary excretion of imidazolepyruvate, imidazoleacetate. Variable degree of mental retardation and speech defects. Some patients are asymptomatic.
51 Histidinuria [Renal and intestinal histidine transport]	High urinary excretion, normal blood levels of histidine. Mild mental retardation.
52 Homocarnosinosis [?]	Progressive spastic paraplegia, progressive mental deterioration, and retinal pigmentation. Elevated homocarnosine in cerebrospinal fluid.
Homocitrullinuria	(See hyperornithinemia, periodic hyperlysinemia).
53 Homocystinuria [Cystathionine-β-synthase]	Mental retardation, ectopia lentis, skeletal defects resembling those of Marfan's syndrome. Atherosclerosis-like pathology resulting in arterial and venous thromboembolism. High plasma homocystine, methionine, low plasma cystine, high urinary homocystine. Some patients respond to high vitamin B_6 administration with decreased biochemical abnormalities.
54 Homocystinuria [N^5, N^{10}-methylenetetrahydrofolate reductase]	Mental retardation. High plasma homocystine, normal or low plasma methionine, high urinary homocystine. Variation in symptoms suggests genetic heterogeneity. Biochemical and clinical response to folate therapy was observed in one case.
55 Homocystinuria [Vitamin B_{12} metabolism]	Low functional N^5-methyltetrahydrofolate-homocysteine transmethylase activity due to inability to make its methylcobalamin cofactor. High plasma and urinary homocystine and cystathionine, low plasma methionine, high urinary methylmalonate.
56 Homocystinuria (Imerslund syndrome) [Vitamin B_{12} transport]	High urinary homocystine and methylmalonic acid. Pernicious anemia and proteinuria. Defective intestinal B_{12} absorption with normal intrinsic factor. Responds to parenteral B_{12} administration.
α-Hydroxybutyric aciduria	(See methionine malabsorption syndrome).
β-Hydroxyisovaleric aciduria	(See β-methylcrotonylglycinuria).
57 Hydroxykynureninuria [Kynureninase]	High urinary excretion of hydroxykynurenine, kynurenine, and xanthurenic acid. Symptoms may include mental retardation, ulceration of the mouth. In some cases, excretion of tryptophan metabolites decreased after administration of vitamin B_6. Different patterns of urinary metabolites suggest that genetic heterogeneity exists or that a different enzyme is affected in some cases.

Table 1 *(Continued)*

Name[a] (probable primary defect)	Major chemical and clinical findings
58 Hydroxylysinemia [Degradation of hydroxylysine (?)]	High urinary excretion of free hydroxylysine and its monoacetylated derivatives. High blood hydroxylysine. May be associated with mental retardation.
59 Hydroxylysinuria [Renal transport of hydroxylysine (?)]	High urinary excretion of free hydroxylysine. No hydroxylysine detected in fasting blood. May be associated with mental retardation.
60 β-Hydroxy-β-methylglutaric aciduria [β-Hydroxy-β-methylglutaryl-CoA lyase]	Severe metabolic acidosis, urinary excretion of β-hydroxy-β-methylglutarate, β-methylglutaconate, β-hydroxy-β-methylbutyrate, and β-methylglutarate.
61 Hydroxyprolinemia [Hydroxyproline oxidase]	High plasma and urinary hydroxyproline. Probably asymptomatic.
62 Hyperalaninemia [Dihydrolipoyl dehydrogenase]	Progressive spinocerebellar degeneration (Friedreich's ataxia) or mental retardation and seizures. Metabolic acidosis with high blood lactate, pyruvate, and alanine. Sometimes also high blood glutamate and proline. Low activity of both pyruvate and α-ketoglutarate dehydrogenase complexes, which have dihydrolipoyl dehydrogenase as a common subunit. In one case, there was elevation of blood pyruvate, lactate, α-ketoglutarate and branched-chain amino acids, with low levels of branched-chain keto acid dehydrogenase in addition to the other two complexes. (See maple syrup urine disease).

63 Hyperalaninemia
     [Pyruvate dehydrogenase complex
     (decarboxylase subunit)]
	Metabolic acidosis, high blood lactate, alanine, and serine. High urinary excretion of lactate and pyruvate, but not α-ketoglutarate. Only one subunit (the decarboxylase) of the pyruvate dehydrogenase complex was deficient in fibroblasts.

64 Hyperalaninemia [Pyruvate dehydrogenase phosphatase]	Metabolic acidosis with high blood lactate, pyruvate and free fatty acids. Hypotonicity, coma. Defect in reactivation of pyruvate dehydrogenase in the presence of Ca^{++} and Mg^{++} was found in muscle and liver, but not in brain. In another patient, defect was found in brain.
65 Hyperalaninemia [Pyruvate carboxylase (liver)]	Subacute necrotizing encephalomyelopathy (Leigh's syndrome), metabolic acidosis, high lactate, pyruvate, and alanine in blood and urine.
Hyperalaninemia	(See No. 121).

66 Hyper-β-alaninemia
     [β-Alanine-α-ketoglutarate trans-
     aminase (?)]
	β-Alanine and γ-aminobutyrate elevated in plasma and cerebrospinal fluid. High urinary excretion of β-alanine, taurine, and γ-aminobutyrate.

67 Hyperammonemia [Ornithine transcarbamylase]	Sex-linked dominant. More severe in males than in females. Vomiting, headaches, convulsions, coma. High blood ammonia. Orotate in urine.
68 Hyperammonemia [Carbamyl phosphate synthetase]	Protein intolerance, vomiting, lethargy, coma. High blood ammonia.
Hyperammonemia	(See periodic hyperlysinemia, lysinuric protein intolerance, methylmalonic aciduria, citrullinemia, hyperornithinemia).

Table 1 *(Continued)*

Name[a] (probable primary defect)	Major chemical and clinical findings
69 Hyperargininemia [Arginase]	Mental retardation, convulsions, spastic diplegia, high arginine levels in blood, urine, and cerebrospinal fluid.
Hypercarnosinemia	(See carnosinemia).
70 Hypercystinuria [Renal tubular cystine transport]	High urinary excretion of cystine not accompanied by increased excretion of lysine, ornithine, or arginine.
71 Hyperdibasic aminoaciduria [Renal and intestinal transport of lysine, ornithine and arginine]	Dominant, probably asymptomatic. High urinary excretion of lysine, ornithine, and arginine. Normal plasma concentration of these amino acids.
Hyperglycinemia	(See ketotic hyperglycinemia, non-ketotic hyperglycinemia).
72 Hyperglycinuria [Renal transport of glycine]	Elevated urinary excretion of glycine with normal plasma concentration of glycine. May be associated with calcium oxalate nephrolithiasis. Hyperglycinuria is also observed in heterozygotes for iminoglycinuria (q.v.).
73 Hyperlysinemia (periodic) [Lysine dehydrogenase (?)]	Periodic vomiting, spasticity, convulsions, and coma. Increase in blood lysine, arginine, and ammonia concentrations after protein ingestion. Urinary excretion of homocitrulline reported in one patient.
74 Hyperlysinemia (persistent) [Lysine-α-ketoglutarate reductase, saccharopine dehydrogenase, and saccharopine oxidoreductase]	High plasma concentration and urinary excretion of lysine. May be asymptomatic. No hyperammonia. Persistent hyperlysinemia and hyperlysinuria also occur in saccharopinuria (q.v.) and in other syndromes with unknown primary defects.
Hyperlysinuria	(See lysine malabsorption syndrome, lysinuric protein intolerance, hyperdibasic aminoaciduria, cystinuria).
75 Hypermethioninemia [Methionine adenosyltransferase (liver)]	High concentrations of methionine in blood. May be asymptomatic. (Hypermethioninemia also occurs in other hereditary diseases, such as cystathionine synthase deficiency, tyrosinemia, and fructose intolerance).
Hyperornithinemia	(See gyrate atrophy of the choroid and retina).
76 Hyperornithinemia [Mitochondrial transport of ornithine (?)]	Mental impairment, seizures. High blood levels of ornithine and ammonia. Low levels of carbamyl phosphate synthetase I in liver and leukocytes. Low level of ornithine decarboxylase in skin fibroblasts. Abnormal liver mitochondria. Homocitrullinuria.
77 Hyperoxaluria (primary, Type I) [α-Ketoglutarate-glyoxylate carboligase]	Calcium oxalate nephrolithiasis, nephrocalcinosis, oxalosis. High urinary excretion of oxalate, glycolate, and glyoxylate.
78 Hyperoxaluria (primary, Type II) [D-glycerate dehydrogenase]	Calcium oxalate nephrolithiasis, nephrocalcinosis, oxalosis. High urinary excretion of oxalate and L-glycerate.
Hyperphenylalaninemia	(See phenylketonuria).
Hyperpipecolatemia	(See pipecolic acidemia).
79 Hyperprolinemia (Type I) [Proline oxidase]	High concentration of proline in blood. High urinary excretion of proline, hydroxyproline, and glycine. Probably asymptomatic.

Table 1 *(Continued)*

Name[a] (probable primary defect)	Major chemical and clinical findings
80 Hyperprolinemia (Type II) [Δ^1-Pyrroline-5-carboxylate dehydrogenase]	High concentrations of proline in blood. High urinary excretion of proline, hydroxyproline, glycine, Δ^1-pyrroline-5-carboxylate, and Δ^1-pyrroline-3-hydroxy-5-carboxylate. Probably asymptomatic.
81 Hypersarcosinemia [Sarcosine dehydrogenase]	Elevated plasma concentrations and urinary excretion of sarcosine. Symptoms variable. Signs and symptoms may improve on prolonged administration of folate.
82 Hyperthreoninemia [?]	Growth retardation, convulsions. Elevated blood threonine concentration and increased urinary excretion of threonine.
Hypertyrosinemia	(See tyrosinemia).
83 Hypervalinemia [Valine transaminase]	Mental and physical retardation, frequent vomiting. High blood valine concentrations.
84 Iminoglycinuria [Renal transport of glycine, proline and hydroxyproline]	High urinary excretion of glycine, proline, and hydroxyproline. Probably asymptomatic. Autosomal recessive. Some heterozygotes have hyperglycinuria, others do not, indicating genetic heterogeneity.
Iminopeptiduria	(See glycylprolinuria).
Indicanuria	(See blue diaper syndrome).
85 Iodotyrosine deiodinase deficiency [Iodotyrosine deiodinase]	Goiter, hypothyroidism, mental retardation. Elevated concentrations of monoiodotyrosine and diiodotyrosine in plasma and urine.
86 Isovaleric acidemia [Isovaleryl-CoA dehydrogenase]	Acute attacks of vomiting, acidosis, dehydration, convulsions, coma. Also anemia, leukopenia, thrombocytopenia. Characteristic odor of sweaty feet. High blood concentrations of isovalerate, high urinary excretion of isovalerate and isovalerylglycine. Signs and symptoms are ameliorated by low leucine diets and administration of glycine, which helps prevent the accumulation of isovalerate by forming isovalerylglycine, a readily excreted compound.
87 α-Ketoadipic aciduria [?]	Psychomotor retardation, high urinary excretion of α-ketoadipate, α-hydroxyadipate, α-aminoadipate, 1,2-butenedicarboxylate, high plasma α-aminoadipate.
Ketotic hyperglycinemia	(See propionic acidemia, methylmalonic aciduria, α-methyl-β-hydroxybutyric aciduria).
Lactic acidosis	(See hyperalaninemia).
Leigh's syndrome	(See hyperalaninemia).
Lowe's syndrome	(See oculocerebrorenal syndrome).
88 Luder-Sheldon syndrome [Renal transport of glucose, amino acids, and phosphate]	Autosomal dominant, (cf. Fanconi syndrome).
Lysine intolerance	(See hyperlysinemia (periodic)).
89 Lysine malabsorption syndrome [Renal and intestinal transport of lysine]	Physical and mental retardation. Increased urinary excretion of lysine, low lysine concentration in plasma. Other amino acids normal.
90 Lysinuric protein intolerance [Transport of lysine, arginine, and	Recessive. Diarrhea, vomiting, hyperammonemia following protein meal. Aversion for protein-containing

Table 1 *(Continued)*

Name[a] (probable primary defect)	Major chemical and clinical findings
ornithine in kidney, intestines, and into hepatocytes]	foods. Growth failure, hepatomegaly, osteoporosis. Lysine, arginine, and ornithine high in urine, low in plasma. Impaired urea formation. Genetic heterogeneity seems probable.
91 Maple syrup urine disease [Branched-chain α-keto acid decarboxylase]	Poor feeding and neurological disturbances appearing during first week of life. Urine has odor of maple syrup. High concentrations of leucine, isoleucine, alloisoleucine, valine, and the corresponding α-keto acids in blood and urine. Milder variants of the disease have been reported [see branched-chain ketoaciduria (intermittent)].
92 Marfan's syndrome [?]	Autosomal dominant. Long extremities, arachnodactily, skeletal deformities, ectopia lentis, cardiovascular degeneration.
93 β-Mercaptolactate-cysteine disulfiduria [β-Mercaptopyruvate sulfur transferase]	May be asymptomatic. High urinary excretion of the mixed disulfide of cystine and β-mercaptolactate.
94 Methionine malabsorption syndrome [Methionine transport (?)]	Excretion of α-hydroxybutyrate in urine and stools. Peculiar body odor ressembling that of an oasthouse.
α-Methylacetoacetic aciduria	(See α-methyl-β-hydroxybutyric aciduria).
95 β-Methylcrotonylglycinuria [β-Methylcrotonyl-CoA carboxylase]	Muscular hypotonia and atrophy. High urinary excretion of β-methylcrotonylglycine and β-hydroxyisovalerate. Some cases are responsive to biotin treatment.
96 β-Methylcrotonylglycinuria [Holocarboxylase synthetase]	Ketoacidosis, urinary excretion of β-methylcrotonylglycine, β-hydroxyisovalerate, tiglylglycine, β-hydroxypropionate and methylcitrate. In cultured fibroblasts, β-Methylcrotonyl-CoA carboxylase propionyl CoA carboxylase, and pyruvate carboxylase activities were deficient, unless the medium was supplemented with biotin. Symptoms may respond to large doses of biotin.
97 α-Methyl-β-hydroxybutyric aciduria [β-Ketothiolase]	Ketoacidosis, vomiting. In some cases, convulsions, hyperglycinuria, hyperglycinemia. High urinary excretion of α-methyl-β-hydroxybutyrate, α-methylacetoacetate, butanone. In some patients, tiglate and tiglylglycine were also found in urine.
β-Methylglutaconic aciduria	(See No. 120).
98 Methylmalonic aciduria [Methylmalonyl-CoA mutase]	Severe ketoacidosis, failure to thrive, high blood and urine methylmalonate, varying degrees propionic aciduria, hyperglycinuria, hyperglycinemia, and hyperammonemia.
99 Methylmalonic aciduria [Methylmalonyl-CoA racemase]	As for the mutase deficiency (see above).
100 Methylmalonic aciduria [Cob(III)alamin reductase (?)]	As for the mutase deficiency (see above). Methylmalonyl-CoA mutase activity of fibroblast extracts deficient unless adenosylcobalamin is added. Adenosylcobalamin synthesis by intact cells deficient; normal in extracts in the presence of thiols. In some patients, administration of large amounts of vitamin B$_{12}$ results in decreased excretion of methyl malonate.

Table 1 *(Continued)*

Name[a] (probable primary defect)	Major chemical and clinical findings
101 Methylmalonic aciduria [Cobalamin-deoxyadenosyl transferase (?)]	As for the mutase deficiency (see above). Impaired synthesis of adenosylcobalamin in cell-free extracts of fibroblasts. In some patients, administration of large amounts of vitamin B_{12} results in decreased excretion of methylmalonate.
Methylmalonic aciduria [Vitamin B_{12} metabolism]	Inability to form either adenosylcobalamin or methylcobalamin. (See homocystinuria).
Methylmalonic aciduria [Vitamin B_{12} transport]	(See homocystinuria).
102 Non-ketotic hyperglycinemia [Glycine cleavage enzyme]	Hypotonia, lethargy, myoclonic convulsions, respiratory failure, mental retardation. High glycine concentrations in blood, urine and cerebrospinal fluid. One patient was described who also had D-glyceric acidemia (q.v.). Milder forms of the disease have also been described.
Oasthouse urine disease	(See methionine malabsorption syndrome).
103 Oculocerebrorenal syndrome (Lowe's syndrome) [Renal transport defect (?)]	Cataracts, glaucoma, mental and physical retardation, generalized aminoaciduria, metabolic acidosis, proteinuria, hyperphosphaturia. Sex linked.
104 Osteogenesis imperfecta [Collagen synthesis (?)]	Fragile bones, skeletal deformities, abnormal teeth, blue sclera. Genetically heterogeneous.
Oxalosis	(See hyperoxaluria).
105 5-Oxoprolinuria [Glutathione synthetase (generalized deficiency)]	Metabolic acidosis, hemolysis, high concentrations of 5-oxoproline in blood, urine, and cerebrospinal fluid. Low glutathione concentrations in erythrocytes, leukocytes, and fibroblasts. Mental deficiency.
106 Phenylketonuria [Phenylalanine hydroxylase (liver)]	Mental deficiency, convulsions, eczema, muscular hypertonicity, hyperactivity. High blood phenylalanine concentration. Increased urinary excretion of phenylalanine, phenylpyruvate, phenyllactate, phenylacetate, phenylacetylglutamine, o-hydroxyphenylacetate. Symptoms may be ameliorated by low-phenylalanine diet started in infancy. Genetic heterogeneity. Variants with hyperphenylalanemia, with or without phenylketonuria, and with milder symptoms have been reported.
107 Phenylketonuria [Dihydropteridine reductase (liver, brain, fibroblasts)]	Mental retardation, convulsions, elevated concentrations of phenylalanine in blood. Low phenylalanine diet reduces blood phenylalanine concentration but does not ameliorate symptoms. Neurological symptoms respond to administration of dopa, 5-hydroxytryptophan, and tetrahydrobiopterin.
108 Phenylketonuria [Dihydrobiopterin synthetase]	Findings are similar to those in dihydropteridine reductase deficiency. Serum phenylalanine concentrations may be lowered to normal by oral administration of tetrahydrobiopterin, or its immediate precursors, dihydrobiopterin or sepiapterin. Administration of dopa and 5-hydroxytryptophan may ameliorate the neurological symptoms.
109 Pipecolic acidemia [?]	Elevated plasma pipecolate, mental retardation, hepatomegaly, visual defects, progressive neurological

Table 1 *(Continued)*

Name[a] (probable primary defect)	Major chemical and clinical findings
	deterioration. Elevated pipecolate levels in blood and urine have also been observed in the cerebro-hepatorenal syndrome of Zellweger.
Prolidase deficiency	(See glycylprolinemia).
110 Propionic acidemia [[Propionyl-CoA carboxylase]	Ketoacidosis, vomiting, lethargy, neutropenia, thrombocytopenia, osteoporosis, protein intolerance. High plasma and urinary glycine and propionate. Also found in urine: 2-methylcitrate, 3-hydroxy-propionate, tiglate, tiglylglycine, and propionyl-glycine. Some cases respond to treatment with biotin.
Protein intolerance (familial)	(See lysinuric protein intolerance).
111 Pyridoxine dependency with seizures [Glutamate decarboxylase]	Severe convulsions beginning around the time of birth. Symptoms do not respond to the usual anti-convulsants, but disappear on administration of vitamin B6. In vitro, the activity of the affected enzyme from a kidney biopsy was undetectable, but was brought to control values by addition of pyridoxal phosphate.
Pyroglutamic aciduria	(See 5-oxoprolinuria).
Pyruvic acidemia	(See hyperalaninemia).
Richner-Hanhart syndrome	(See tyrosinemia).
112 Rowley-Rosenberg syndrome [Renal amino acid transport (?)]	Retarded growth, poor muscular development, scanty adipose tissue, cardiac hypertrophy, partially collapsed lungs. Generalized aminoaciduria with normal plasma amino acid concentrations.
113 Saccharopinuria [Saccharopine dehydrogenase]	Elevated lysine and saccharopine concentrations in plasma and urine. Mental retardation.
Sarcosinemia	(See hypersarcosinemia).
114 Sulfituria (sulfocysteinuria) [Sulfite oxidase]	Neurological abnormalities, dislocated lenses. High urinary excretion of sulfite, thiosulfate, and S-sulfocysteine. Abnormally low excretion of inorganic sulfate. Sulfite and S-sulfocysteine also present in plasma.
Threoninemia	(See hyperthreoninemia).
115 Tryptophanuria [?]	Pellagra-like skin rash, mental retardation, dwarfism. Higher than normal urinary excretion of tryptophan, lower than normal excretion of kynurenine. Elevated plasma concentrations of tryptophan.
116 Tyrosinemia (hepatorenal) [Fumarylacetoacetase (?)]	Severe liver failure, hepatic cirrhosis, hepatoma. Generalized aminoaciduria, glycosuria, phosphaturia, rickets. May occur acutely in early infancy or be chronic. High plasma concentrations of tyrosine, me-thionine, and other amino acids. High urinary excretion of p-hydroxyphenyllactate, p-hydroxyphenyl-pyruvate, p-hydroxyphenylacetate, δ-aminolevulinic acid, succinylacetone, succinylacetoacetate. Several enzyme activities may be low or absent, including p-hydroxyphenylpyruvate dioxygenase, porphobilinogen

Table 1 *(Continued)*

Name[a] (probable primary defect)	Major chemical and clinical findings
	synthetase, and fumarylacetoacetase. Symptoms are ameliorated by treatment with diet low in phenylalanine, tyrosine, and methionine. (Similar symptoms are seen in other hereditary or acquired diseases such as fructose intolerance, fructose-1,6-diphosphatase deficiency, galactosemia, hepatitis. Transient neonatal tyrosinemia due to delayed maturation of liver enzymes may also occur).
117 Tyrosinemia (Richner-Hanhart syndrome) [Cytosolic tyrosine transaminase]	Mental retardation, hyperkeratosis of palms and soles, corneal opacities. High blood tyrosine concentrations, high urinary excretion of tyrosine, p-hydroxyphenyllactate, p-hydroxyphenylpyruvate, p-hydroxyphenylacetate, N-acetyltyrosine, p-tyramine. Normal activities of p-hydroxyphenylpyruvate dioxygenase and mitochondrial tyrosine transaminase. Normal liver and kidney function. Symptoms may respond to diet low in phenylalanine and tyrosine.
118 Tyrosinosis [p-Hydroxyphenylpyruvate dioxygenase (?)]	One patient, with myasthenia gravis (probably unrelated to tyrosinosis) had high urinary excretion of p-hydroxyphenylpyruvate. After ingestion of tyrosine or a high-protein diet, tyrosine, p-hydroxyphenyllactate, and dopa were also excreted in urine. Recently, several members of a family were found to have elevated plasma and urinary p-hydroxyphenylpyruvate and undetectable liver p-hydroxyphenylpyruvate dioxygenase activity. They were clinically normal.
Urocanic aciduria	(See No. 122).
119 Wilson's disease (Hepatolenticular degeneration) [Copper transport]	Brain degeneration, cirrhosis of the liver, hemolytic anemia. Deposition of excess copper in liver, brain, kidneys, cornea, and other tissues. Low serum levels of copper and ceruloplasmin. High urinary excretion of amino acids, glucose, inorganic phosphate. May be effectively treated by administration of D-penicillamine.
Xanthurenic aciduria	(See hydroxykynureninuria).
Zellweger syndrome	(See pipecolic acidemia).
120 β-Methylglutaconic aciduria [β-Methylglutaconyl-CoA hydratase (?)]	Progressive neurological deterioration. Urinary excretion of β-methylglutaconate and β-methylglutarate.
121 Hyperalaninemia [?]	Mental retardation, dwarfism, diabetes mellitus, enamel hypoplasia. Alanine levels elevated in blood and urine. High blood lactate and pyruvate.
122 Urocanic aciduria [Urocanase]	Mental retardation, urocanate in urine, delayed utilization of histidine.

nine is blocked. Tyrosine is converted in melanocytes to melanin by way of dopa and dopaquinone. Blocks in this pathway result in albinism. Tyrosine is a precursor of the neurotransmitters dopamine and norepinephrine; neurological disturbances may occur when this pathway is inhibited. The major degradative pathway of tyrosine is by transamination to *p*-hydroxy-phenylpyruvate, which is, in turn, oxidized to homogentisate, and finally broken down to fumarate and acetoacetate.

Hyperphenylalaninemia (106a–108k)

Recent advances in the understanding of phenylketonuria and related forms of hyperphenylalaninemia have shown this disease to be far more complex and variable than was previously thought. In addition to "classical phenylketonuria" (106a–h), two additional diseases (107a–108k), are now known to be associated with elevated blood levels of phenylalanine. These involve deficiencies in enzymes required for synthesis of the cofactor for phenylalanine hydroxylase, L-*erythro*-5,6,7,8-tetrahydrobiopterin. Since this cofactor is required also for hydroxylation of tyrosine and tryptophan, these metabolic blocks result in deficiencies in the neurotransmitters dopamine, norepinephrine, and serotonin as well as phenylalanine accumulation.

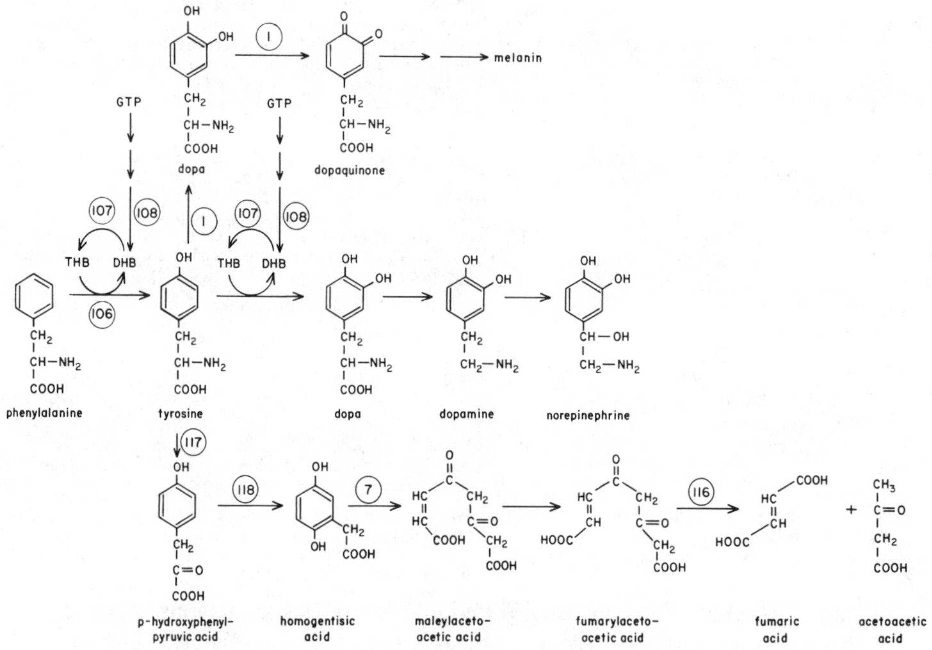

Figure 1 Inborn errors of aromatic amino acid metabolism. The numbers in this figure correspond to those given to the diseases listed in Table 1.

Such blocks are associated with severe neurological symptoms that are not reversed or prevented by lowering the blood levels of phenylalanine.

Phenylketonuria due to deficiency of phenylalanine hydroxylase has been reviewed recently (106a); there is much clinical variability, presumably due to genetic heterogeneity. Patients with "classical" phenylketonuria do not necessarily have a complete absence of enzyme activity (106d). A cross-reacting protein has been detected immunologically (106c) that suggests the presence of a modified enzyme or subunits of the enzyme. Two or three isozymes of the hydroxylase occur in the liver of normal individuals, whereas patients with phenylketonuria had one or none (106d). A surprising finding was that the activity of the enzyme in liver is not proportional to gene dosage; parents of patients with hyperphenylalaninemia (heterozygotes) averaged 10% of the normal activity, rather than the 50% expected (106b). This could be explained in terms of a multimeric structure of the enzyme with overproduction of a defective subunit (106b). The human enzyme has been isolated; like the rat enzyme, it has subunits of about 50,000 mol wt (106h).

Many abnormal metabolites, including 2-phenylethylamine (106e), 3',4'-deoxynorlaudanosoline carboxylic acid (106f), and γ-glutamylphenylalanine (106g) are found in the urine of phenylketonurics; their relationship to the clinical findings is not clear. Reports of patients with hyperphenylalaninemia and progressive neurological disturbances that did not respond to dietary therapy appeared about seven years ago (107a,b,d); some of these had near-normal liver phenylalanine hydroxylase activity. A new molecular defect, deficiency in dihydropteridine reductase was found; this enzyme catalyzes the NADH-dependent regeneration of tetrahydrobiopterin from the quinonoid form of dihydrobiopterin formed in hydroxylation reactions. No dihydropteridine reductase activity was detected in the liver, brain, or fibroblasts of a patient (107d), while the parents had about 50% or less of control activity in their fibroblasts (107e). Improved methods for the diagnosis of this form of hyperphenylalaninemia and its differentiation from classical phenylketonuria involve analysis of biopterin derivatives in urine (107f,g, 108k). Methylenetetrahydrofolate reductase, an FAD-containing enzyme involved in the biosynthesis of methionine from homocysteine, also catalyzes the formation of tetrahydrobiopterin from quinonoid dihydrobiopterin in vitro (107h). Thus this enzyme may provide a means for some hydroxylation of phenylalanine to occur in patients with dihydropteridine reductase deficiency. A third molecular defect responsible for hyperphenylalaninemia and neurological disease is a block in the biosynthesis of dihydrobiopterin from GTP (108a,b,e–g); the block is at the level of dihydrobiopterin synthetase (108f). Treatment of these patients and those with dihydropteridine reductase deficiency with dopa, carbidopa, 5-hydroxytryptophan, and tetrahydrobiopterin together with low phenylalanine diets has led to significant improvement (107c, 108c,d,i,j).

TYROSINEMIA AND TYROSINOSIS (116a–118e)   Diseases in which tyrosine and its metabolites accumulate in body fluids have been reviewed recently (116a). They include nonhereditary conditions such as cirrhosis of the liver and hepatitis, as well as transient neonatal tyrosinemia that may be due to vitamin C deficiency or delayed maturation of enzymes of tyrosine metabolism in the liver. Transient tyrosinemia is frequently seen in premature infants. Hereditary diseases of tyrosine metabolism include: hepatorenal tyrosinemia (116a–j), a severe childhood disease associated with liver and kidney dysfunction; the Richner-Hanhart syndrome (117a–h), associated with mental retardation, palmar and plantar hyperkeratosis, and the appearance of tyrosine crystals in the cornea; and a rare condition known as tyrosinosis (118a–e), which is probably asymptomatic. Clinical and biochemical findings similar to those seen in hepatorenal tyrosinemia are also seen in some hereditary diseases of carbohydrate metabolism such as fructose intolerance (34a,b), fructose-1,6-diphosphatase deficiency (33a,b), and galactosemia (35a).

Hepatorenal tyrosinemia (116a–j) involves several metabolic pathways. Tyrosine utilization is impaired, which results in its accumulation in blood and urine. Tyrosine metabolites such as p-hydroxyphenyllactate, p-hydroxyphenylpyruvate, and p-hydroxyphenylacetate are also excreted in large amounts. Methionine metabolism is disturbed, which results in hypermethioninemia. The activity of a number of enzymes is decreased including p-hydroxyphenylpyruvate dioxygenase (116a), methionine activating enzyme (116c), cystathionine synthetase (116c), porphobilinogen synthetase (116e), and fumarylacetoacetase (116j). Succinylacetone and succinylacetoacetic acid were found in the urine of these patients (116e). Succinylacetone inhibits porphobilinogen synthetase, which explains excretion of δ-aminolevulinate (116e). It was suggested that the deficiency of fumarylacetoacetase is the primary defect, and that succinylacetone and succinylacetoacetate are formed from fumarylacetoacetate or maleylacetoacetate that accumulate behind the enzymatic block (116e,j).[2] Many of the symptoms are alleviated by diet low in phenylalanine, tyrosine, and methionine (116b). Many of these patients develop hepatomas.

Another abnormality in which plasma tyrosine levels are very high and

[2]A patient (of Dr. Maria New) with hepatorenal tyrosinemia was recently treated at the New York Hospital-Cornell Medical Center and was found to have marked deficiency of fumarylacetoacetase, low glutathione levels in liver, low glutathione levels in erythrocytes, and low activity of porphobilinogen synthetase.

tyrosine metabolites are excreted in the urine is associated with deficiency of cytosolic liver tyrosine transaminase (117c). One patient (117c) had multiple congenital anomalies, including severe mental retardation. His urine contained large amounts of p-hydroxyphenylpyruvate, p-hydroxyphenylacetate, p-hydroxyphenyllactate, N-acetyltyrosine, and p-tyramine. Paradoxically, hepatic p-hydroxyphenylpyruvate dioxygenase activity was normal, as was hepatic mitochondrial tyrosine transaminase. The problem remains of why the α-keto acid analogue of tyrosine is excreted in large amounts in the presence of normal amounts of the enzyme that catalyzes the next reaction in its degradative pathway, and in the absence of one of the enzymes that would be expected to participate in its formation. Since p-hydroxyphenylpyruvate dioxygenase is a cytosolic enzyme and is not found in mitochondria, compartmentation probably plays an important role in the catabolism of tyrosine in liver (117b). The different tissue distribution of the enzymes involved may also be important in this regard (117h).

Several additional patients with Richner-Hanhart syndrome have subsequently been found to have elevated plasma tyrosine levels and high urinary excretion of tyrosine metabolites (117d–g), but enzyme studies were not reported for most of these. Both biochemical and clinical findings were significantly improved by a low phenylalanine, low tyrosine diet (for a review see 117a; 117d–g).

Tyrosinosis, a very rare condition, was described in 1932 (118a). Although this patient had myasthenia gravis and creatinuria, these seem unrelated to the disturbances in tyrosine metabolism, which was expressed as urinary excretion of p-hydroxyphenylpyruvate. It is curious that appreciable amounts of p-hydroxyphenyllactate were not excreted, except after tyrosine loading. It was concluded that the primary defect was a complete block in the oxidation of p-hydroxyphenylpyruvate (118a), but there are doubts about this conclusion (118b). Very recently, six clinically normal members of a family were reported with high plasma tyrosine and p-hydroxyphenylpyruvate, crystalline deposits of tyrosine in the bone marrow, and high urinary excretion of p-hydroxyphenylpyruvate (118c). Enzyme studies on liver biopsies revealed undetectable p-hydroxyphenylpyruvate dioxygenase and normal levels of tyrosine-α-ketoglutarate transaminase.

A different, but possibly related metabolic error was found in a child with prolonged transient tyrosinemia combined with a metabolic acidosis (118d). Both the baby and her mother excreted large amounts of a new amino acid, named hawkinsin, (2-L-cystein-S-yl-1,4-dihydroxycyclohex-5-en-1-yl)acetic acid, believed to be formed by addition of cysteine to an intermediate in the oxidation of p-hydroxyphenylpyruvate (118e).

OTHERS   Other diseases of tyrosine metabolism include albinism and alcaptonuria, both of which were discussed by Garrod (7a). Albinism, a
hereditary inability to synthesize melanin, consists of a group of distinct
metabolic disorders (1a). Alcaptonuria (7b,d) is characterized by a deficiency of homogentisate dioxygenase. Homogentisate therefore accumulates and is excreted in the urine. Clinical findings include arthritis and a
dark pigmentation (ochronosis) of the skin, bone, cartilage, cerumen (7c)
and elsewhere. When renal failure occurs in conjunction with alcaptonuria,
ochronosis increases rapidly and becomes much more extensive, probably
because of increased retention of homogentisate (7e).

Tyrosine is a precursor of the thyroid hormones thyroxine and triiodo-
thyronine. There are a number of conditions involving derangements in the
metabolism of these hormones, which are grouped together as "familial
goiter" (85a). One involves inability to remove iodine from monoidotyro-
sine and diiodotyrosine; these amino acids are excreted in the urine, and the
consequent depletion of iodine results in cretinism and goiter (85a). A new
variant of the disease was recently described in which there was no mental
retardation (85b). The patients had some ability to deiodinate monoiodotyrosine, but not diiodotyrosine.

Disorders of Branched-Chain Amino Acid Metabolism

INTRODUCTION   The pathways of metabolism of the branched-chain
amino acids—leucine, isoleucine, and valine—have a number of common
features (Figure 2). The first step is a transamination; the second, oxidative
decarboxylation to yield the acyl-CoA derivative with one less carbon atom;
the third, the formation, by dehydrogenation, of an α,β-unsaturated acyl-
CoA derivative which is subsequently hydrated to yield a β-hydroxyacyl-
CoA intermediate. In the case of leucine, a biotin-dependent carboxylation
precedes the hydration reaction. A number of diseases are known involving
enzyme deficiencies in these pathways. Others are known involving deficiencies in the breakdown of propionyl-CoA and methylmalonyl-CoA (see
section on sulfur-containing amino acids).

HYPERVALINEMIA (83a–g)   This rare disease was first reported in 1963
(83a). Valine concentrations were elevated in the blood and urine, but there
was no elevation in the levels of leucine or isoleucine. Valine tolerance
curves were abnormal in the patient, while leucine, isoleucine, and methionine were metabolized normally (83b). The enzymatic defect was shown to
involve a specific valine transaminase (83c). Leukocytes from the patient
had no detectable valine transaminase activity, while transamination of
leucine. isoleucine, methionine, and phenylalanine were normal. This is of

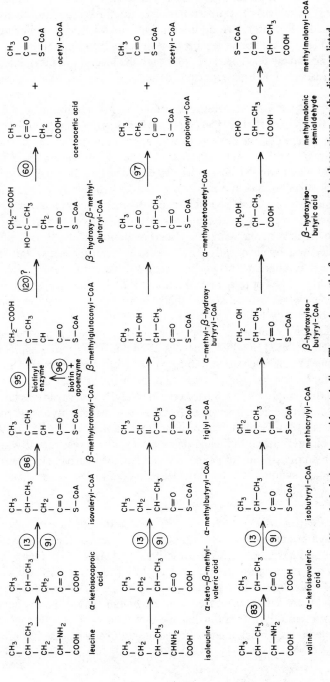

Figure 2 Inborn errors of branched-chain amino acid metabolism. The numbers in this figure correspond to those given to the diseases listed in Table 1.

interest in view of evidence that a single transaminase from animal tissues (pig heart) catalyzes the transamination of the three branched-chain amino acids (83d). Recently, two additional patients with hypervalinemia were reported (83e). A block in leucine or isoleucine transamination has not yet been conclusively demonstrated, although evidence for a partial deficiency in this reaction has been reported (83g).

MAPLE SYRUP URINE DISEASE (91a–e)   This syndrome was first described by Menkes et al in 1954 (91a,b). Although the infants appear normal at birth, severe neurological symptoms appear after a few days. The urine has a characteristic odor resembling that of maple syrup. All three branched-chain amino acids, as well as the corresponding α-keto acids, are elevated in blood and urine. Because d-α-keto-β-methylvaleric acid (the α-keto analogue of L-isoleucine) enolizes and thus racemizes (91e), L-alloisoleucine may be formed and found in the blood and urine. The enzymatic block is at the oxidative decarboxylation of the three keto acids. Clinically and biochemically, the disease is heterogeneous, and there is evidence that different defects may occur in different patients (91c). In one patient, high doses of thiamine ameliorated the biochemical abnormalities, and evidence was presented that this was mediated by a stabilizing effect on an abnormally labile enzyme (91d).

In addition to "classical" maple-syrup urine disease, a number of cases have been reported in which episodes of convulsions, coma, and elevation of plasma branched-chain amino acids and keto acids occurred intermittently (for review see 13a). In addition, both the severity and the time of onset of the symptoms have varied considerably. These variations have been correlated with different levels of residual enzyme activity in the patients' fibroblasts (13a). Some cases respond to thiamine (13b,c, 91d). A new variant of intermittent branched-chain keto aciduria was recently reported in which administration of valine, but not leucine or isoleucine, resulted in severe toxic symptoms, including lethargy, vomiting, ataxia, and hypoglycemia (13d).

ISOVALERIC ACIDURIA (86a–i)   This severe disease of early childhood, whose symptoms include vomiting, acidosis, convulsions, pancytopenia, and mental retardation was first described by Tanaka et al (86a). Patients were noted to exude a peculiar odor; thus the name "sweaty-feet syndrome." This disease is associated with a block in leucine catabolism at isovaleryl-CoA dehydrogenease (86b). Although the symptoms have been treated successfully with a low-leucine diet (86b,g), a new therapeutic approach based on a biochemical understanding of the disease has proved to be remarkably effective. This consists of giving glycine (86c,e,f,h). Reaction between glycine and isovaleryl-CoA, mediated by glycine-N-acylase, results in formation of the nontoxic product, isovaleryl-glycine, which is rapidly excreted; this helps to remove isovaleryl-CoA, which accumulates behind the enzymatic block, and decreases its hydrolysis to the toxic isovaleric acid.

β-METHYLCROTONYLGLYCINURIA   Advances in our understanding of this disease, first reported in 1970 (95a,b), have been made recently. The first patient was a 4½ month-old girl who excreted large amounts of two unusual metabolites, β-hydroxyisovaleric acid and β-methylcrotonylglycine. The authors concluded that the metabolic block was at the biotin-dependent carboxylation of β-methylcrotonyl-CoA. However, administration of large doses of biotin had no effect. Other cases of β-methylcrotonylglycinuria, on the other hand, responded to biotin (95c, 96a, 96b). Several reports have appeared of patients with combined deficiency of β-methylcrotonyl-CoA carboxylase and propionyl-CoA carboxylase (96c,e,f). When biotin was added to the growth medium of fibroblasts from such a patient, the activity of both enzymes returned to normal (96c,f). It was suggested that the defect might be in the enzyme holocarboxylase synthetase, which activates the carboxyl group of biotin and attaches it to the ε-amino group of a lysine residue in the apocarboxylase. More recently, fibroblasts from patients with biotin-responsive organic acidemia were found to be deficient in three carboxylase activities: propionyl-CoA carboxylase, β-methylcrotonyl-CoA carboxylase, and pyruvate carboxylase. All three activities were restored to normal after biotin supplementation of the medium (96g,i). Another patient with holocarboxylase synthetase deficiency responded dramatically to biotin (96h).

OTHERS   In the leucine degradative pathway, a block in β-methylglutaconyl-CoA hydrase has been postulated to account for the urinary excretion of β-methylglutaconate and β-methylglutarate (120a,b). β-Hydroxy-β-methylglutarate and other organic acids have been found in the urine of a baby with severe metabolic acidosis (60b,c). Skin fibroblasts from the patient were deficient in β-hydroxy-β-methylglutaryl-CoA lyase (60d), and both parents had reduced levels of the enzyme in leukocytes (60e). β-Hydroxy-β-methylglutaric aciduria may also occur in Reye's syndrome (60a,f,g). An additional block observed in isoleucine degradation occurs at β-ketothiolase, the enzyme that converts α-methylacetoacetyl-CoA to propionyl-CoA and acetyl-CoA. This results in the excretion of α-methylacetoacetate and α-methylhydroxybutyrate (7a, 97b–d).

Disorders of Sulfur-Containing Amino Acid Metabolism

INTRODUCTION   These diseases include the homocystinurias (53a–56e) and methylmalonic acidurias (98a–101c) (which are related to each other and may occur together), propionic aciduria (110a–j), cystinosis (19a–i), cystathioninuria (17a–c), hypermethioninemia (75a–c), sulfituria (114a–c), and β-mercaptolactate-cysteine disulfiduria (93a,b) (Figure 3). Propionic aciduria and methylmalonic aciduria may also be classified as diseases of threonine and of branched-chain amino acid metabolism.

The sulfur atom of methionine, a dietary essential amino acid, is transferred to the carbon chain of serine by the transsulfuration pathway, which involves intermediate formation of cystathionine. In this process, the carbon chain of methionine is converted to α-ketobutyrate which, through the intermediary formation of propionyl-CoA and methylmalonyl-CoA, enters the citric acid cycle as succinyl-CoA. Homocysteine represents a branch point in this pathway. It can either go to cystathionine and contribute its sulfur atom to the synthesis of cysteine, as described above, or it can be

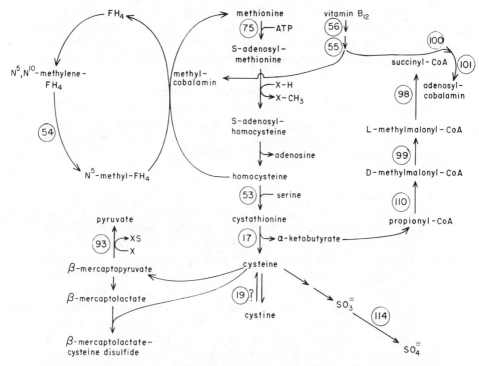

Figure 3   Inborn errors of sulfur-containing amino acid metabolism. The numbers in this figure correspond to those given to the diseases listed in Table 1.

reconverted to methionine by accepting a methyl group from N^5-methylte-trahydrofolic acid in a reaction that requires methylcobalamin as a cofactor. Homocysteine accumulates when either of these reactions is blocked. However, an important difference between these two blocks is that blood methionine levels are elevated in homocystinuria due to cystathionine synthase deficiency, whereas they are low or normal in homocystinuria due to deficiencies in the remethylation pathway.

Cysteine can undergo many transformations, only three of which are indicated in Figure 3: It can be oxidized to cystine; its sulfur atom can be oxidized to sulfite, then to sulfate, which is excreted in the urine; or it can undergo transamination to β-mercaptopyruvate, which is converted to pyruvate.

HOMOCYSTINURIA (53a–56e)   There are at least four known familial disorders in which high levels of homocystine are formed in blood and urine. The most common of these is due to a defect in cystathionine-β-synthase (53a). In this form of homocystinuria, blood methionine levels are elevated and blood cystine levels are low. A number of unusual sulfur-containing amino acids are also found in urine (53b). There is probably considerable genetic heterogeneity as evidenced by the variation in severity of the symptoms and responsiveness to vitamin B_6 therapy (53b–e). Three classes of mutants were revealed in a study of cultured fibroblasts from 14 patients (53i). Human liver cystathionine-β-synthase has been purified to homogeneity and was shown to have a mol wt of 94,000 and to consist of two identical subunits (53j). One patient had a structurally altered, immunologically cross-reacting protein, differing from the normal enzyme in its isoelectric point (53h).

A different type of homocystinuria was found to be due to a deficiency in N^5,N^{10}-methylenetetrahydrofolate reductase (54a,b). Genetic heterogeneity was suggested by the finding of a more severe, infantile form of the disease (54e) and by the fact that some symptoms responded to folate therapy in one case (54c) but not in another (54f). Furthermore, the enzyme in extracts of fibroblasts from different patients varied in their heat stability (54g). In some patients, the activity of cystathionine-β-synthase was deficient in addition to that of the reductase (54d,e). Since the reductase is an FAD-containing protein, it may be of interest to test the effect of large doses of riboflavin on these patients. Two additional forms of homocystinuria (55a–56e) are due to defects in the metabolism or absorption of vitamin B_{12}. A cofactor derived from vitamin B_{12} is known to be required in two enzymatic reactions in the human body. These are N^5-methyltetrahydrofo-late-homocysteine transmethylase, which requires methylcobalamin as a cofactor, and methylmalonyl-CoA mutase, which requires adenosylcobalamin. Since neither of these coenzymes is apparently synthesized in these forms of homocystinuria, methylmalonic aciduria is also observed in these patients (55a,b). Defects in vitamin B_{12} metabolism have been studied extensively and reveal both biochemical and genetic heterogeneity (55c–f). In Imerslund's syndrome, also characterized by homocystinuria and methylmalonic aciduria (56a), there is a defect in the absorption of vitamin B_{12} in the presence of intrinsic factor (56b,c). Recently, a combination of homocystinuria, methylmalonic aciduria, and megaloblastic anemia was observed in the baby of a strictly vegetarian mother, as a result of a vitamin B_{12} deficiency (56e).

METHYLMALONIC ACIDURIA (98a–101c)   Six inborn errors of metabolism are known to be characterized by methylmalonic acidemia and aciduria. In addition to the two discussed above in connection with homocystinuria (55a–56e), two others involve defects in the formation of adenosylcobalamin, but not methylcobalamin, from vitamin B_{12} (100a–101c). One, reported in a single patient, involves a deficiency in methylmalonyl-CoA racemase (99a). The sixth results from a deficiency in methylmalonyl-CoA mutase apoenzyme (98a–h). The clinical findings resemble those of propionic aciduria (propionyl-CoA carboxylase deficiency) (110a–j), and include ketotic hyperglycinemia and hyperammonemia (98g,h). One case was responsive to vitamin B_{12} therapy and was shown to involve a structural defect in the apoenzyme, which caused a decrease in its affinity for the adenosylcobalamin coenzyme (98f).

PROPIONIC ACIDURIA (110a–j)   This severe childhood disease is characterized clinically by ketoacidosis, vomiting, neutropenia, and thrombocytopenia (110a,c). Propionate, abnormal products of propionate metabolism such as 3-hydroxypropionate and 2-methylcitrate, and other organic acids are excreted in the urine (110a,i). Other characteristics of this "ketotic hyperglycinemia" syndrome, which is similar to that seen in methylmalonic aciduria (98a–101c), include hyperglycinemia and hyperammonemia (110a,d,j). The deficient enzyme is propionyl-CoA carboxylase (110a), a biotin-dependent enzyme. Clinical, biochemical, and genetic heterogeneity have been demonstrated. Some cases are biotin-responsive (110 e,f). Genetic heterogeneity has been demonstrated in cultured fibroblasts (110b). Purification of the enzyme from fibroblasts of patients and comparison with that from controls have provided evidence for mutations in the structural gene, which results in an altered enzyme protein (110h). The cause of the hyperglycinemia characteristic of this and other diseases of the branched-chain amino acids is still obscure. One possibility is that the glycine cleavage enzyme system may be inhibited by products that accumulate in these disorders (110g).

CYSTINOSIS (19a–i)   Cystine is extremely insoluble. In nephropathic cystinosis, it accumulates intracellularly to such an extent that cystine crystals form in the bone marrow, leukocytes, conjunctiva, cornea, and other tissues (for review see 19a). The crystals appear to be intralysosomal. It is not known whether cystine deposition is due to an inability to reduce cystine to the much more soluble cysteine, an inability to transport cystine across the lysosomal membrane, or some other cause. In the more severe, nephropathic form of the disease, end-stage renal failure usually occurs by the end of the first decade. Milder forms of the disease also occur. In addition to the Fanconi syndrome (31a), growth retardation, and photophobia, a frequent finding is hypothyroidism (19e). Cultured fibroblasts of patients with cystinosis accumulate more than 100 times as much cystine as control fibroblasts (19a). Although the level of glutathione was about the same in cystinotic cells as in controls, the level of glutathione disulfide was less than half that in normal cells (19c). The primary defect in this disorder is still unknown.

Several compounds that reduce cystine intracellularly when added to the medium of cultured cells have been tried clinically. One of these, ascorbic acid, is clinically ineffective (19f). This may be because of its inability to significantly lower intracellular cystine concentrations in vivo (19i). Other compounds being tried include cysteamine (19d,g), phosphocysteamine (19h), and dithiothreitol (19b).

OTHERS   The primary defect in cystathioninuria is a block in the conversion of cystathionine to cysteine and α-ketobutyrate (17a,b). It appears to by asymptomatic. The urinary excretion of cystathionine usually decreases after administration of vitamin B_6. Genetic heterogeneity has been demonstrated (17c). Hypermethioninemia due to methionine adenosyl transferase activity has been reported (75a–c). It is apparently asymptomatic. Hypermethioninemia due to cystathionine synthase deficiency (53a–j), and to hepatic diseases such as hepatorenal tyrosinemia (116a–j) are mentioned elsewhere in this review. Sulfite oxidase deficiency is characterized by neurological abnormalities, dislocated lenses, and high urinary excretion of sulfite, thiosulfate, and S-sulfocysteine (114a–c). The absence of immunologically cross-reacting material in a patient's liver (114b) suggests that the enzyme protein may not be made.

Disorders of the Urea Cycle

INTRODUCTION   There are five enzymes involved in the urea cycle, and five diseases are known, one for each enzyme (Figure 4). Diseases of the urea cycle are characterized by mental retardation and hyperammonemia.

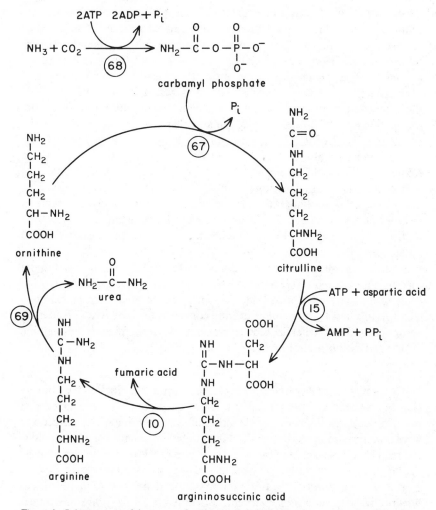

Figure 4   Inborn errors of the urea cycle. The numbers in this figure correspond to those given to the diseases listed in Table 1.

HYPERAMMONEMIA (CARBAMYL PHOSPHATE SYNTHETASE DEFI-CIENCY) (68a–f)   The first patient described (68a,b) had metabolic acidosis and hyperglycinemia in addition to hyperammonemia. The findings in other patients have varied, which suggests genetic heterogeneity or a different primary defect in some cases (68c). Three cases of the neonatal form were recently described (68d,e) in which liver carbamyl phosphate synthetase activity was undetectable, while the other urea cycle enzymes were normal. There was severe hyperammonemia and all three patients died within five days after birth. Patients with less severe hyperammonemia may improve when placed on a low-protein diet (68c). A means of decreasing blood ammonia in carbamyl phosphate synthetase deficiency and in other disorders of the urea cycle consists of treating patients with benzoate or phenylacetate (68f). These are excreted in the urine as hippurate and phenylacetylglutamine, respectively, thus helping to eliminate nitrogen.

HYPERAMMONEMIA (ORNITHINE TRANSCARBAMYLASE DEFICIENCY) (67a–f)   This condition was first described in 1962 (67a); the trait is sex-linked dominant. In males, the disease is almost invariably lethal, whereas the severity varies in females, depending in part on the extent of inactivation of the X-chromosome that carries the mutant gene, according to the Lyon hypothesis. A characteristic finding, in addition to hyperammonemia, is orotic aciduria due to the diversion of carbamyl phosphate to the pyrimidine biosynthetic pathway. (for review see 67b). One case was found to be associated with propionic acidemia and hyperlysinemia (67c). Even asymptomatic female carriers have significantly decreased IQ scores (67e).

CITRULLINEMIA (15a–e)   In this disorder, the deficient enzyme is argininosuccinate synthetase. The clinical symptoms are variable, which suggests genetic heterogeneity (15b,c). The abnormal enzyme from one patient was found to exhibit sigmoid kinetics with citrulline, whereas the normal enzyme gave a hyperbolic curve (15d).

ARGININOSUCCINIC ACIDURIA (10a–c)   This condition is characterized by high concentrations of argininosuccinic acid in blood, urine, and cerebrospinal fluid; the symptoms vary widely in their severity (10b). The deficient enzyme is argininosuccinase. Supplementation of the diet with arginine was beneficial (10c), possibly because replenishing the arginine (or ornithine) used in argininosuccinic acid synthesis permits more of the latter to be formed and excreted, thus promoting ammonia elimination.

HYPERARGININEMIA (69a–g)   This rare disorder is due to arginase deficiency (69a). Because of competition for a common transport system [see cystinuria (20a–e)], there is elevated excretion of cystine, ornithine, and lysine (69b). Also found in the urine is the α-keto acid analogue of arginine and its cyclized form (69f). Several new cases have been reported recently (69c–e,g).

Disorders of Lysine Metabolism

INTRODUCTION   The metabolic transformations of lysine are not as well understood as those of most other amino acids. The main degradative pathway is through saccharopine and α-aminoadipic-δ-semialdehyde (in equilibrium with its cyclized form, Δ^1-piperideine-6-carboxylate) to α-aminoadipate (see Figure 5). There may also be alternative pathways. One of these is through pipecolate, as indicated by the existence of a metabolic block in which pipecolate accumulates. α-Aminoadipate is converted through α-ketoadipate, glutaryl-CoA, and glutaconyl-CoA, to crotonyl-CoA and, eventually, acetyl-CoA. A number of diseases involve enzymes in this series. There are also several hereditary diseases characterized by lysine intolerance or excessive lysine excretion that are probably due to transport defects.

HYPERLYSINEMIA (PERSISTENT) (74a–j)   Hyperlysinemia was first described in 1964 (74a). The patient had convulsions, mental retardation, and high concentrations of lysine in plasma and urine. Three additional patients with hyperlysinemia were also mentally retarded (74b,d). However, the association of mental retardation with hyperlysinemia may have been fortuitous, since a sister and a cousin of the first patient had equally high plasma lysine levels but were clinically normal (74c). The deficient enzyme was found (74e) to be lysine-α-keto-glutarate reductase, which catalyzes formation of saccharopine. Two additional enzyme activities were also absent from the liver and fibroblasts of these patients (74h): saccharopine dehydrogenase, which catalyzes the conversion of saccharopine to glutamate and α-aminoadipic-δ-semialdehyde, and saccharopine oxidoreductase, which catalyzes the oxidation of saccharopine to α-ketoglutarate and lysine. These three reactions were believed to be catalyzed by separate

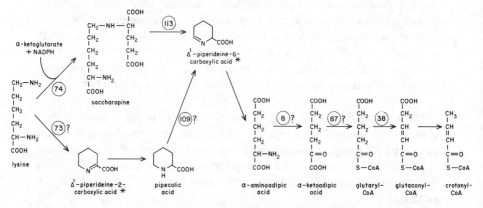

Figure 5 Inborn errors of lysine metabolism. Δ^1-piperideine-2-carboxylic acid and Δ^1-piperideine-6-carboxylic acid are in equilibrium with their open-chain forms, α-keto-ε-aminocaproic acid and α-aminoadipic-δ-semialdehyde, respectively. The numbers in this figure correspond to those given to the diseases listed in Table 1.

enzymes. However, all three activities purified together and exhibited similar heat stability (74j), which suggests that they are associated with a single protein. However, in saccharopinuria (see below), only saccharopine dehydrogenase appears to be deficient. Among the urinary metabolites of lysine found in these patients were pipecolate, N-α-acetyllysine, N-ε-acetyllysine, homoarginine, and homocitrulline (74f). Hypusine, a lysine derivative found in normal blood and urine, was also greatly elevated in the urine (74g). In a recently reported mentally retarded patient, a low-lysine diet was instituted at 8.5 yr. Although lysine levels in blood and urine became normal, there was no significant effect on mental development (74i).

HYPERLYSINEMIA (PERIODIC) (73a–d)   The first patient, a 3-month old girl (73a,b) had episodes of coma accompanied by hyperammonemia. A mentally retarded 18 yr-old man with periodic hyperammonemia, was reported later 74c,d). Both had elevated plasma lysine, but some of the other findings were different.

SACCHAROPINURIA (113a–d)   The first patient was mentally retarded, and excreted large amounts of saccharopine, lysine and citrulline (113a). A second patient (113b) had similar findings but had normal blood and urinary citrulline levels. Saccharopine dehydrogenease was completely absent in the first patient (113d) and partially deficient in the second (113c). The first patient also had a partial deficiency of lysine-α-ketoglutarate reductase (113d), which is of great interest in light of recent evidence that it is an activity of the same enzyme (74j). The high citrulline level remains unexplained.

GLUTARIC ACIDURIA (38a–g, 20a–c)   Two cases of glutaric aciduria (now designated as Type I) were reported in 1975 (38a). There was progressive neurological degeneration and high urinary excretion of glutarate, which was increased by oral administration of lysine. Three additional cases were recently found (38b,c). Leukocytes and fibroblasts showed a severe but not complete deficiency of glutaryl-CoA dehydrogenase in some patients, while the activity was undetectable in the others (38d). One patient also had periodic episodes of ketosis during which he excreted a number of organic acids including branched-chain and dicarboxylic acids (38e). Small improvements were seen in patients treated with a low-protein diet, riboflavin, and a γ-aminobutyrate analogue (38g).

A different syndrome (glutaric aciduria, Type II) is characterized by hypoglycemia and fatal neonatal acidosis (39a). Glutaryl-CoA dehydrogenease activity in fibroblasts is normal in these patients (39b). The findings suggest a defect in several acyl-CoA dehydrogenases, perhaps due to a deficiency in a component required by all (39b,c).

OTHERS   Several patients with α-aminoadipic aciduria have been reported (8a,c); the defect has not yet been defined. α-Aminoadipate has also been found in the urine of patients with Reye's syndrome (8b). A patient with α-ketoadipic aciduria together with α-aminoadipic aciduria and α-aminoadipic acidemia has been described (87b). Another disease of lysine metabolism has been reported in two patients (109a,b) who have high plasma levels of pipecolate. Elevated levels of pipecolate in blood and urine have also been observed in Zellweger's syndrome (109c).

Disorders of Glycine and Serine Metabolism

NONKETOTIC HYPERGLYCINEMIA (102a–n)   This severe neonatal disease affects the brain primarily or exclusively (102g). Glycine levels are elevated not only in blood and urine, but also in cerebrospinal fluid and brain (102e). The symptoms include severe mental retardation in patients who survive the neonatal period, convulsions, hypotonia, and respiratory failure (for review see 102a). Some patients have a milder disease (102d,f,h). Interest in this disease drew attention to the physiological importance of the glycine cleavage system, which is deficient in nonketotic hyperglycinemia, and catalyzes the following reaction:

$$\text{Glycine} + \text{FH}_4 + \text{NAD}^+ \rightleftharpoons N^5,N^{10}\text{-methylene-FH}_4 + \text{NADH} + \text{CO}_2 + \text{NH}_4^+$$

This mitochondrial enzyme system uses four coenzymes: NAD, FAD, FH_4, and pyridoxal 5'-phosphate, and consists of four proteins: 1. P-protein, which contains bound pyridoxal 5'-phosphate, 2. H-protein, a small, acidic, heat-stable protein with a reactive disulfide group, 3. T-protein, which interacts with tetrahydrofolic acid (FH_4), and 4. L-protein, an FAD-containing protein with lipoyl dehydrogenase activity (for review see 102b). Glycine cleavage catalyzed by this enzyme is the major catabolic pathway in mammals for both glycine and serine, which is converted to glycine by serine hydroxymethyltransferase.

Although low enzyme activity was present in the liver of patients (102e,k), none was found in brain (102e). By reconstitution experiments with enzyme components purified from bacteria, it was shown that the H-protein was absent or nonfunctional in the brain of a patient (102e). Other components may have been absent as well.

Treatment with sodium benzoate may be successful in lowering glycine levels in blood and cerebrospinal fluid by conjugation and excretion as hippurate. However, such treatment, started from birth in one patient, did not prevent mental retardation (102c). Another treatment consists of giving strychnine, a glycine antagonist with high affinity for glycine receptors in the central nervous system; although some report improvement in symptoms (102i,j,m,n), others found no beneficial effects (102k,l).

"Ketotic hyperglycinemia," a syndrome characterized by severe ketoacidosis and hyperammonemia in addition to hyperglycinemia and hyperglycinuria, is associated with deficiencies in other enzymes, such as propionyl-CoA carboxylase or methylmalonyl-CoA mutase. In these diseases, the cause of the hyperglycinemia is not known, but it may be associated with a secondary inhibition of the glycine cleavage system by products that accumulate (110g).

HYPEROXALURIA (77a, 78a)   Two enzyme deficiencies are known to lead to hyperoxaluria (for review see 77a). In both, there is increased synthesis and urinary excretion of oxalate, and deposition in tissues of the highly insoluble salt, calcium oxalate. The accumulation of calcium oxalate in the kidney leads to early death from uremia.

Oxalate is a product of glyoxylate oxidation. Glyoxylate may originate from glycine by transamination or from hydroxyproline through γ-hydroxyglutamate. In addition to being oxidized to oxalate, glyoxylate may be reduced to glycolic acid or it may combine with α-ketoglutarate to yield α-hydroxy-β-ketoadipate and CO_2. The latter reaction is catalyzed by α-ketoglutarate:glyoxylate carboligase, a thiamin pyrophosphate requiring enzyme.

The two types of hyperoxaluria are clinically similar, but they may be differentiated by urine analysis and by enzymatic tests. In primary hyperoxaluria type I, high urinary excretion of glycolate, glyoxylate, and oxalate is observed. A deficiency in cytosolic α-ketoglutarate:glyoxylate carboligase in the liver, kidney, and spleen was demonstrated in five patients, but mitochondrial enzyme activity was normal (77a).

In primary hyperoxaluria, type II, the urine contains high levels of oxalate and L-glycerate. The leukocytes from four patients have been found to be deficient in D-glycerate dehydrogenase. This enzyme catalyzes the formation of D-glycerate from hydroxypyruvate, a product of serine transamination. Although the relationship between this enzyme defect and hyperoxaluria is not yet clear, a plausible hypothesis has been advanced (78a), based on the fact that both hydroxypyruvate and glyoxylate are substrates of lactate dehydrogenase. It is proposed that, in the absence of D-glycerate dehydrogenase, the hydroxypyruvate the accumulates is reduced by lactate dehydrogenase to L-glycerate, which accounts for the high excretion of this compound, and that this process takes place at the expense of glyoxylate, which is oxidized to oxalate in a coupled reaction. Two patients were recently reported, however, whose metabolic disturbances are not easily explained by this hypothesis. Both had D-glyceric acidemia (46a–c) but no oxaluria. One of the patients, a severely mentally retarded boy, also had nonketotic hyperglycinemia (46a,b). There was no acidosis, and assays of the patient's leukocytes demonstrated a level of D-glycerate dehydrogenase that was about 20–30% of control values (46a). The other had severe metabolic acidosis, but no hyperglycinemia (46c).

Disorders of Proline and Hydroxyproline Metabolism

INTRODUCTION   Proline, ornithine, and glutamate are metabolically interconvertible (see Figure 6). Ornithine can undergo reversible transamination of the δ-amino group to yield glutamic-γ-semialdehyde (in equilibrium with its cyclic form, Δ^1-pyrroline-5-carboxylate). This compound can be

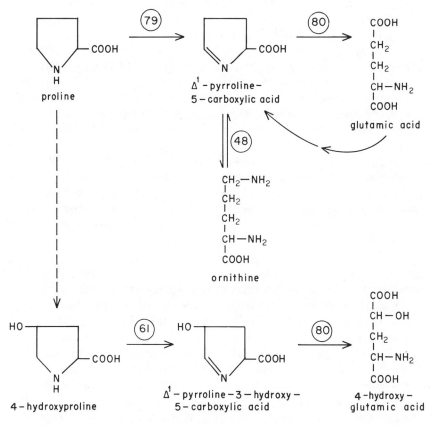

Figure 6   Inborn errors of proline and hydroxyproline metabolism. The numbers in this figure correspond to those given to the diseases listed in Table 1.

oxidized to glutamate or reduced to proline. It can also be formed from either of these amino acids. Hydroxyproline is released after proteolytic degradation of collagen; it may be oxidized to Δ^1-pyrroline-3-hydroxy-5-carboxylate, which may be oxidized further to 4-hydroxyglutamate.

HYPERPROLINEMIA (79a–80f)   Two defects are associated with elevated proline levels in blood and urine (for review see 79a). In both, urine contains excess glycine and hydroxyproline in addition to proline because of competition for a common transport system [see iminoglycinuria (84a–d)]. In view of the variety of symptoms observed, and the fact that patients with Type I and Type II hyperprolinemia are known who are asymptomatic (79a, 80d), it appears that both types are benign. Type I hyperprolinemia is associated with a deficiency of proline oxidase (79b,c). Type II hyperprolinemia is associated with complete absence of the second enzyme in proline degradation, Δ^1-pyrroline-5-carboxylate dehydrogenase (80d,e). It can be differentiated biochemically from Type I by the presence in urine of a compound that gives a yellow color with o-aminobenzaldehyde. Although this compound was originally assumed to be Δ^1-pyrroline-5-carboxylic acid, it was recently shown to consist primarily of Δ^1-pyrroline-3-hydroxy-5-carboxylic acid (80c), which indicates a block in hydroxyproline degradation. Subsequently, enzyme studies have demonstrated that patients with Type II hyperprolinemia lack not only Δ^1-pyrroline-5-carboxylate dehydrogenase activity, but also Δ^1-pyrroline-3-hydroxy-5-carboxylate dehydrogenase activity, (80f). In addition, there was an excellent correlation, spanning a wide range, between the two activities in the leukocytes of 25 individuals (80f). It therefore appears that both activities are properties of the same enzyme.

HYDROXYPROLINEMIA (61a–d)   A patient with elevated blood and urine hydroxyproline levels was reported in 1962 (61a). Although the patient was mentally retarded, other individuals with this disorder are normal (61c). Since Δ^1-pyrroline-3-hydroxy-5-carboxylate was not excreted in the urine following an oral load of hydroxyproline, in contrast to increased excretion of this compound in controls (61b), it was concluded that the deficient enzyme is hydroxyproline oxidase, which is distinct from proline oxidase. New cases of hydroxyprolinemia were reported recently (61d).

GYRATE ATROPHY OF THE CHOROID AND RETINA WITH HYPEROR-NITHINEMIA (48a–g)   This syndrome was first described in 1973 (48a). Ornithine levels are elevated in plasma and urine, while plasma lysine levels are low. The enzyme defect was shown to be an almost complete absence of ornithine-α-keto acid transaminase (48b–f). Heterozygotes had about half of the activity of controls. A study of amino acid loading in such patients was recently published (48g). A different type of hyperornithinemia accompanied by hyperammonemia and homocitrullinuria (76a–d) is a different disease in which the primary defect is unknown.

Disorders of Glutathione Metabolism

INTRODUCTION   The metabolism of glutathione, a tripeptide that occurs in substantial concentrations in virtually all mammalian cells, involves cycles of interconversion between glutathione and glutathione disulfide as well as the reactions of the γ-glutamyl cycle, which accounts for the degradation of glutathione to its constituent amino acids and the resynthesis of this tripeptide (37c). Glutathione is synthesized intracellularly from glutamate, cysteine, and glycine by the combined actions of γ-glutamylcysteine and glutathione synthetases. γ-Glutamyl transpeptidase, a membrane-bound enzyme especially concentrated at anatomic sites that are involved in transport processes, such as renal and intestinal brush-borders, is extensively distributed on the external surface of cells such as those of the proximal renal tubule, but the enzyme is found elsewhere as well. This enzyme catalyzes transpeptidation reactions between glutathione and amino acids that lead to the formation of γ-glutamyl amino acids. The enzyme also catalyzes hydrolysis of glutathione, other γ-glutamyl compounds, and S-substituted glutathione derivatives (formed endogenously, e.g. in reactions involving estrogens, prostaglandins, and leukotrienes, and also after administration of foreign compounds). Intracellular glutathione S-transferases mediate the formation of S-substituted glutathione derivatives, which are converted to the corresponding S-substituted cysteinylglycine derivatives. These become mercapturic acids after cleavage of the glycine moiety and intracellular N-acetylation. Translocation of glutathione across cell membranes is a discrete step in the γ-glutamyl cycle. Glutathione is translocated chiefly in the form of GSH, some of which is nonenzymatically oxidized to GSSG extracellularly. Glutathione synthesized in kidney cells is translocated and used extracellularly or in the cell membrane by transpeptidase, and the products of such metabolism, which include γ-glutamyl amino acids, glutamate, and cysteinylglycine, are taken up by the kidney, thus constituting an intraorgan cycle of glutathione metabolism. The γ-glutamyl amino acids that are transported into the cell are converted to the corresponding amino acids and 5-oxoproline by γ-glutamyl cyclotransferase. 5-Oxoproline is converted to glutamate in an ATP-dependent reaction catalyzed by 5-oxoprolinase. Tissues such as liver and muscle, which have low levels of γ-glutamyl transpeptidase, translocate glutathione to the blood plasma; glutahione is removed from the plasma by the kidney and other tissues that have transpeptidase activity, thus constituting an interorgan cycle of glutathione metabolism.

Glutathione is synthesized in the form of GSH, which may be converted to GSSG by a number of pathways involving the activity of various transhydrogenases and also that of glutathione peroxidase, which catalyzes the interaction of glutathione and peroxides to form GSSG. The reduction of GSSG to GSH is catalyzed by the NADPH-dependent flavoprotein, glutathione reductase. Glutathione is also a coenzyme for a number of enzymatic reactions.

Glutathione therefore functions in processes that include the reduction of key compounds and protection of cell membranes and other cellular components, and in transport. The symptoms and signs found in patients with errors of glutathione metabolism include increased tendency to erythrocyte breakdown, phenomena referable to impaired central nervous system function, and defects of amino acid transport.

DEFECTS OF GLUTATHIONE SYNTHESIS (37a–c, 44a–d, 105a–g)   Two patients (siblings) who have a deficiency of γ-glutamylcysteine synthetase, exhibit a syndrome of hemolytic anemia, spinocerebellar degeneration, peripheral neuropathy, myopathy, and aminoaciduria associated with a generalized deficiency of glutathione. The concentrations of glutathione in their erythrocytes, leukocytes, and muscle are, respectively, less than 3%, about 50%, and about 25% of normal (37a,b).

Two general types of glutathione synthetase deficiency are known. In one, the defect seems restricted to the erythrocyte, whereas in the other, there is a more severe and generalized deficiency. In the latter type, known as 5-oxoprolinuria (pryoglutamic aciduria), there is massive urinary excretion of 5-oxoproline and elevated levels of 5-oxoproline in the blood plasma and cerebrospinal fluid (105a–g). Such patients have severe metabolic acidosis that requires continuous therapy with sodium bicarbonate. Most have a tendency to hemolysis, and a number are mentally retarded. The mechanism that accounts for the extensive 5-oxoprolinuria exhibited by these patients is closely associated with their deficiency of glutathione (105a,b). Glutathione normally regulates its own synthesis by inhibiting γ-glutamylcysteine synthetase (105g). Thus, the severe decrease in glutathione levels leads to increased formation of γ-glutamylcysteine, which is readily converted to 5-oxoproline by γ-glutamyl cyclotransferase. The extensive overproduction of 5-oxoproline exceeds the capacity of 5-oxoprolinase to convert this substrate to glutamate, and some of the 5-oxoproline (about 30% of that formed) is excreted in the urine. The metabolic defect in this disease results in a modified γ-glutamyl cycle in which there is a futile synthesis of γ-glutamylcysteine followed by conversion of this dipeptide to

5-oxoproline and cysteine by action of γ-glutamyl cyclotransferase (105a).

Patients who have the mild form of glutathione synthetase deficiency have reduced levels of glutathione in their erythrocytes and exhibit well-compensated hemolytic disease (44a–d). Such patients do not excrete abnormally large quantities of 5-oxoproline. In this form, the genetic lesion appears to be associated with synthesis of an unstable glutathione synthetase molecule. Presumably, the defective but active enzyme can be synthesized in most tissues at a sufficient rate to compensate for the defect. However, such compensation is impossible in the erythrocyte, in which protein synthesis does not occur; therefore, the disease appears to be restricted to the erythrocyte.

γ-GLUTAMYL TRANSPEPTIDASE DEFICIENCY (45a–d) Two patients with marked deficiency of γ-glutamyl transpeptidase have been described. The first patient, a mildly retarded adult male, exhibited substantial glutathionuria amounting to about 850 mg of glutathione per day (45a,b). This patient also has elevated blood plasma levels of glutathione and his cultured skin fibroblasts show marked reduction in transpeptidase activity. The second patient, a mentally retarded young woman with severe behavioral problems, was also reported to have glutathionuria, glutathionemia, and decreased transpeptidase activity in her cultured fibroblasts (45c). Although standard amino acid analyses of the urine of these patients and calculations of the patient's renal reabsorption of amino acids were interpreted to be "within or very close to the normal range under normal dietary conditions" (45b) and to be "normal" (45c), later consideration of these findings suggests that the reabsorption of several amino acids and of cystine (45d) may be impaired. Recent studies on the urine obtained from the second patient indicate that there is an abnormally large excretion of γ-glutamylcysteine and cysteine (as well as glutathione); these compounds were found in disulfide form (45d). The occurrence of γ-glutamylcyst(e)ine in the urine of this patient, and other considerations, indicate that this dipeptide is formed as an extracellular metabolite of glutathione. γ-Glutamylcyst(e)ine may be formed by a pathway involving transpeptidation or by cleavage of the CYS-GLY bond of glutathione. A similar pattern of excretion [glutathione, cyst(e)ine, γ-glutamylcyst(e)ine] was found in experimental animals given γ-glutamyl transpeptidase inhibitors (45d). γ-Glutamylcyst(e)ine is probably produced by the action of the residual transpeptidase (in both patients and experimental animals) on glutathione and cyst(e)ine. The excretion of γ-glutamylcyst(e)ine and cyst(e)ine in γ-glutamyl transpeptidase deficiency indicates a functional role of the transpeptidase in the transport or metabolism (or both) of sulfur-containing compounds.

DEFICIENCIES OF GLUTATHIONE REDUCTASE AND PEROXIDASE
(40a–43c) Maintenance of the normally high level of GSH in erythrocytes
and other cells depends upon the activity of glutathione reductase, which
utilizes NADPH for reduction of GSSG to GSH. In the erythrocyte,
NADP is reduced to NADPH by the actions of glucose 6-phosphate dehydrogenase and 6-phosphogluconate dehydrogenase. Glucose 6-phosphate
dehydrogenase deficiency is probably the commonest enzyme deficiency of
man associated with disease, and results from a variety of mutations (43b).
Patients with these disorders exhibit hemolytic disease, especially when
exposed to certain drugs (e.g. primaquine, certain sulfonamides, phenace-
tin) and certain other materials such as fava beans. It is generally thought
that in the erythrocyte, glutathione is continually oxidized by the action of
glutathione peroxidase on peroxide normally formed by the action of superoxide dismutase on superoxide produced from oxyhemoglobin. Presumably, drugs such as primaquine increase the formation of peroxide and thus
the utilization of GSH. Glutathione is therefore considered essential for the
protection of hemoglobin, other components of the erythrocyte, and the
erythrocyte membrane. Such protection requires not only GSH, but also the
activity of glutathione reductase, the production of NADPH, and the activity of glutathione peroxidase. A number of reports have appeared describing
glutathione reductase deficiency and also deficiency of glutathione peroxidase. Although such deficiencies might well be expected to be associated
with hemolytic disease, additional studies on patients with such apparent
deficiencies are needed (see 43c).

Disorders of Amino Acid Transport

INTRODUCTION Amino acids are transported from the intestines into
the blood; from the blood into other body fluids, tissues, and cells; from one
cell compartment to another; and from the glomerular filtrate into the
blood. It is generally believed that amino acids are transported by "systems," that have usually been characterized by studies of the uptake of
amino acids in the presence and absence of other amino acids that may or
may not compete for uptake with the amino acid being examined. Other
evidence for such systems has come from studies of inherited defects of
transport. It must be emphasized that although the term systems is widely
applied in this area of research, virtually nothing is known about the
molecular details of such transport systems. There are various complexities
that need to be considered. For example, there is evidence (from competition studies and genetic observations) that a given amino acid may be
transported by more than one system. Since amino acid transport is a

crucially important cellular function, it is understandable that multiple ("back-up") systems have developed through evolution. It follows, therefore, that the loss of a specific system may not significantly affect the overall transport of a particular amino acid into the cell. Thus, patients with Hartnup disease (in which there is defective transport of certain neutral amino acids) and patients with cystinuria (in which there is defective transport of cystine, lysine, ornithine, and arginine) do not become amino acid deficient, because the amino acids are transported from the gut in the form of peptides. It is now known that a substantial fraction of the amino acids absorbed in the gut is normally taken up as peptides, and there is good experimental evidence that certain peptides, especially dipeptides, are transported more efficiently than mixtures of the corresponding free amino acids. It may be noted in this connection that transport of amino acids by the γ-glutamyl cycle (37c) involved dipeptides; thus the amino acids are transported in this pathway as γ-glutamyl amino acids.

As noted above, the existence of inherited diseases of amino acid transport has led to some interesting conclusions. For example, the evidence indicates that there are groups of amino acids that use a common component for transport. One such group consists of cystine, lysine, arginine, and ornithine. Another includes glycine, proline, and hydroxyproline. In certain diseases in which a common component appears to be defective, the transport of all members of the group is affected. Surprisingly, however, other diseases exist in which one, or a few members of the group are selectively affected. This suggests the possibility that, in addition to some common components, each member of the group may use a part of the system that is specifically required for its own transport. Alternatively, a single amino acid may be transported by more than one system, one of which is shared with the group and the other specific for itself. Another conclusion that has arisen from study of these diseases is that some components of the intestinal amino acid transport system may be very similar or identical to a corresponding component of the renal system; thus they appear to be coded for by the same gene. There are some inherited diseases in which both the intestinal and renal transport systems for the same group of amino acids are deficient. It should be noted, however, that there are other diseases in which only one of the tissue systems is deficient. No explanation in molecular terms is yet available for these apparent paradoxes.

Although amino acid transport is often considered and studied as an isolated cellular function, separate from metabolism, it is well known that individual cells may take up certain amino acids and release others, and that some cells are extensively involved in amino acid metabolism, whereas others function chiefly in protein synthesis. There are undoubtedly close connections between transport of amino acids into cells and the utilization of amino acids intracellularly. It seems probable that some metabolic reactions have been modified through evolution to function in transport. Similarities between enzymatic and transport phenomena have often been noted, as indicated, for example, by the term permease. Although all amino acid transport probably does not require energy, the fact that amino acids can be transported against a concentration gradient indicates that a means of coupling amino acid translocation with expenditure of metabolic energy must exist. These considerations emphasize again both the complexities and our great ignorance of the chemical mechanisms involved in the transport of the 20 or more amino acids into and out of various types of cells and subcellular compartments. Inborn defects of amino acid transport provide some potentially valuable clues about the mechanisms involved. At least 20 disorders of amino acid transport are known; some of these are briefly discussed below.

DIBASIC AMINO ACIDS   Although cystine is a neutral amino acid, it shares with lysine and ornithine the property of possessing two amino groups. It is perhaps because of this common chemical feature that cystine shares with lysine and ornithine, as well as with arginine—which has amino and guanidino moieties—the component of renal and intestinal transport systems that is defective in cystinuria. An excellent review of this disease is available (20a). There is evidence that there are separate transport systems for cystine and for lysine, ornithine and arginine (20a). This is supported by the finding of (a) two siblings with hypercystinuria not accompanied by elevated excretion of lysine, ornithine, or arginine (70a), and (b) patients with hyperdibasic aminoaciduria who do not have increased excretion of cystine (71a,b). In the latter disorder, impaired intestinal transport of lysine, but not cystine, was also observed. A study of lysine and arginine excretion in cystinurics, after intravenous infusion of these amino acids, showed normal tubular reabsorption at very high amino acid loads (20c). This was interpreted to indicate the existence of two separate transport systems for dibasic amino acids: a low-affinity, high-capacity system that is normal in cystinurics, and a high-affinity, low-capacity system that is defective.

Cystinuria is genetically heterogeneous; three types have been distinguished by analysis of intestinal amino acid transport and examination of heterozygote phenotypes (20a). The complexities involved are indicated by the finding of a genetic variant who appears to be intermediate between cystinuria and "hypercystinuria" (20b), and evidence suggesting that the same cystinuria gene may be expressed in a highly variable manner in different individuals as well as in the same individual at different times (20d). The finding of greatly increased cystine excretion in a patient with γ-glutamyl transpeptidase deficiency (45d) is consistent with the idea that the γ-glutamyl cycle mechanism functions in cystine transport.

Lysine malabsorption syndrome is another interesting transport defect in which there is increased urinary excretion and impaired intestinal absorption of lysine, but not of arginine, ornithine, or cystine (89a). This finding suggests the existence of a specific lysine transport system.

Another disorder involving the renal and intestinal transport of the dibasic amino acids was first reported from Finland and given the name "lysinuric protein intolerance," among many others (90a,b,g). It is characterized by aversion for proteins, diarrhea, vomiting, growth failure, hepatomegaly, hyperammonemia, deficient urea formation, and low plasma levels of lysine, arginine, and ornithine. Additional cases and several variants have been described (90c–g,k,l). In addition to the intestinal and renal transport defects, there is a deficiency of dibasic amino acid transport into hepatocytes (90h), so that the hyperammonemia and deficient urea synthesis might be explained in terms of low hepatic levels of urea cycle intermediates. Dietary supplements of arginine, citrulline, and lysine were found to be beneficial (90i).

Although the primary defect in hyperornithinemia with hyperammonemia and homocitrullinuria is unknown (76a,b,d,e), ultra-structural abnormalities of the mitochondria suggest the possibility of defective transport of ornithine into these organelles (76b,d). Deficiencies of ornithine decarboxylase in fibroblasts (76c) and carbamyl phosphate synthetase I in liver and leukocytes have been reported in these patients.

Hydroxylysinuria, with and without hydroxylysinemia, have been reported, but it is not certain whether these are defects in the transport or in the metabolism of hydroxylysine (58a–c, 59a).

OTHER AMINO ACIDS   The existence of a common transport system for glycine, proline, and hydroxyproline is revealed by the discovery of the familial defect termed iminoglycinuria (for review see 84a). Many individuals with this disorder are asymptomatic, but it sometimes occurs in association with mental retardation (94a,b). The condition is transmitted as an autosomal recessive, but there is considerable genetic heterogeneity. In some patients, the renal transport defect is associated with a defect in intestinal absorption, while in others it is not. Some heterozygotes have a normal urinary amino acid pattern, whereas others excrete an excess of glycine only. Thus, a form of hyperglycinuria that is apparently inherited as a dominant trait (72a) would be difficult to distinguish from the heterozygotic form of iminoglycinuria. Another type of heterozygote has been described who excretes large amounts of proline and normal amounts of glycine (84c). Although a homozygote for this gene has not been found, a "genetic compound" or double heterozygote for this gene and that associated with hyperglycinuria has a urinary amino acid pattern characteristic of iminoglycinuria (84c). Studies of additional families have revealed even greater complexity of these transport systems (72b, 84d). An unusual renal transport disorder found in 14 members of a family and inherited as an asymptomatic autosomal dominant, "glucoglycinuria," consists of a combined defect in the reabsorption of glucose and glycine (36a).

Other amino acids or groups of amino acids appear to have their own specific transport systems in kidney and gut, as revealed by hereditary aminoacidurias that are not associated with elevated plasma levels of amino acids and defects in intestinal absorption. One of these is Hartnup disease (49a–e), which involves intestinal and renal defects in the transport of many neutral amino acids, including tryptophan, phenylalanine, tyrosine, valine, leucine, isoleucine, alanine, glutamine, asparagine, serine, threonine, and histidine. Specific defects in the transport of some of these amino acids include: histidinuria (51a) (renal and intestinal transport of histidine); methionine malabsorption syndrome (94a–d) (intestinal defect); and "blue diaper syndrome" (12a) (intestinal defect in tryptophan absorption). A common system appears to be involved in the transport of aspartic and glutamic acids, as evidenced by two forms of dicarboxylic aminoaciduria; one of these appears to involve both renal and intestinal systems (22a), while the other affects only the kidney (21a).

GENERALIZED AMINOACIDURIA   There are several inherited disorders that result in kidney damage. These may result in massive excretion of substances that are normally efficiently reabsorbed, such as amino acids, glucose, water, phosphate, bicarbonate, sodium, and potassium. All of these substances are excreted in excessive amounts in the Fanconi syndrome (31a). Inherited forms of the Fanconi syndrome may be "idiopathic", i.e. of unknown origin, or they may be secondary to other metabolic or transport abnormalities such as cystinosis (19a–i), hepatorenal tyrosinemia (116a–j), hereditary fructose intolerance (34a,b), galactosemia (35a), or Wilson's disease (119a,b). Other inherited forms of generalized aminoaciduria include the Luder-Sheldon syndrome (99a), the Rowley-Rosenberg syndrome (112a,b), and the oculocerebrorenal syndrome of Lowe (103a–d).

Other Inborn Errors of Amino Acid Metabolism

COLLAGEN DISORDERS   There are many inherited diseases involving collagen synthesis, e.g. the different types of Ehlers-Danlos syndrome, the

Marfan syndrome, cutis laxa, and osteogenesis imperfecta. These defects are increasingly being understood at the molecular level, and have been reviewed (23a–d). In several, the primary defect is known. Thus both cutis laxa (16b) and the Ehlers-Danlos syndrome, Type V (27b) appear to result from a deficiency in the copper-dependent enzyme, lysyl oxidase. Ehlers-Danlos syndrome, Type VI is due to a deficiency in lysyl hydroxylase (28b,c). Ehlers-Danlos syndrome, Type IV appears to be due to a decrease in the synthesis of Type III collagen (26b), while Ehlers-Danlos syndrome, Type VII has a defect in the structure the pro $\alpha 2$ chain that apparently prevents the normal removal of the amino-terminal propeptide (29b,c).

HYPERALANINEMIA   Hereditary blocks in the metabolism of pyruvate result in the accumulation of this compound and its metabolites. Thus, these disorders are usually associated with lactic acidosis, hyperalaninemia, and hyperalaninuria. Five such diseases have been described. One of these, Friedreich's ataxia (62a–f), which is characterized by progressive spinocerebellar degeneration, is due to a defect in dihydrolipoyl dehydrogenase. Since this enzyme forms a part of several different dehydrogenase complexes that catalyze the oxidative decarboxylation of pyruvate, α-ketoglutarate, and the branched-chain α-keto acids, respectively, there is a derangement in the metabolism of all these substances. Another form of hyperalaninemia is due to a deficiency in the decarboxylase subunit of the pyruvate dehydrogenase complex (63a–c). The activity of the pyruvate dehydrogenase complex is controlled by a phosphorylation-dephosphorylation mechanism. Defects in the phosphatase that catalyzes the reactivation of the phosphorylated enzyme have been found. In one case, the defect was found in the brain (64b), while in another, it was found in muscle and liver, but not in brain. In Leigh's syndrome (subacute necrotizing encephalomyelopathy), a defect in liver pyruvate carboxylase was found (65a–e). The enzyme defect in the fifth type of hyperalanemia, which was associated with microcephaly, dwarfism, enamel hypoplasia, and diabetes mellitus is unknown (121a).

DISORDERS OF HISTIDINE METABOLISM   Several hereditary blocks in the metabolism of histidine are known (see Figure 7). In histidinemia (50a), the block is in the conversion of histidine to urocanic acid (50b). Histidase activity, which can be measured in the skin, is undetectable in most patients. Urocanase, which converts urocanic acid to imidazolonepropionate, was deficient in the liver of a mentally retarded patient with urocanic aciduria (122a). Another individual who excreted large amounts of urocanate and urocanylgylcine after histidine loading was clinically normal (122b). Five

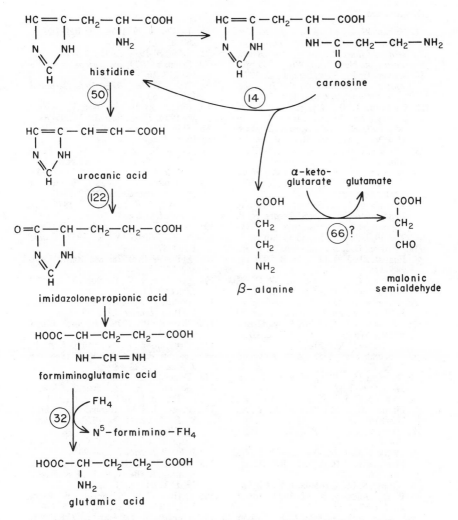

Figure 7 Inborn errors of histidine metabolism. The numbers in this figure correspond to those given to the diseases listed in Table 1.

patients with a deficiency of liver formimino transferase were described by Arakawa (32b). Formiminoglutamate was excreted in the urine after histidine loading. In another patient, this disorder was associated with low plasma methionine concentration (32a). Administration of methionine resulted in a decrease in the excretion of formiminoglutamate as well as clinical improvement.

Literature Cited

I. Nyhan, W. L., ed. 1974. *Heritable Disorders of Amino Acid Metabolism: Patterns of Clinical Expression and Genetic Variation.* New York: Wiley. 765 pp.

II. Scriver, C. R., Rosenberg, L. E. 1973. *Amino Acid Metabolism and Its Disorders.* Philadelphia: Saunders. 491 pp.

III. Stanbury, J. B., Wyngaarden, J. B., Fredrickson, D. S., eds. 1978. *The Metabolic Basis of Inherited Diseases.* New York: McGraw-Hill. 1862 pp.

IV. Wacker, W. E. C., Coombs, T. L. 1969. *Ann. Rev. Biochem.* 38:539–68

V. Meister, A. 1965. *Biochemistry of the Amino Acids. Vol. 2,* pp. 1021–84. New York: Academic

VI. Rosenberg, L. E., Scriver, C. R. 1980. In *Metabolic Control and Disease,* ed. P. K. Bondy, L. E. Rosenberg, pp. 583–776. Philadelphia: Saunders. 8th ed.

1a. Witkop, C. J. Jr., Quevedo, W. C. Jr., Fitzpatrick, T. B. 1978. See Ref. III, pp. 283–316

2a. Witkop, C. J. Jr., Quevedo, W. C. Jr., Fitzpatrick, T. B. 1978. See Ref. 1a

2b. King, R. A., Olds, D. P., Witkop, C. J. 1978. *J. Invest. Dermatol.* 71:136–39

2c. O'Donnell, F. E., King, R. A., Green, W. R., Witkop, C. J. Jr. 1978. *Arch. Ophthalmol.* 96:1621–25

3a. Chédiak, M. 1952. *Rev. Hematol.* 7:362–67

3b. Higashi, O. 1954. *Tohoku J. Exp. Med.* 59:315–32

3c. Witkop, C. J. Jr., Quevedo, W. C. Jr., Fitzpatrick, T. B. 1978. See Ref. III, pp. 283–316

4a. Hermansky, F., Pudlak, P. 1959. *Blood* 14:162–69

4b. Witkop, C. J. Jr., Quevedo, W. C. Jr., Fitzpatrick, T. B. 1978. See Ref. III, pp. 283–316

4c. Garay, S. M., Gardella, J. E., Fazzini, E. P., Goldring, R. M. 1979. *Am. J. Med.* 66:737–47

5a. Nance, W. E., Jackson, C. E., Witkop, C. J. Jr. 1970. *Am. J. Hum. Genet.* 22:579–86

5b. Witkop, C. J. Jr., Quevedo, W. C. Jr., Fitzpatrick, T. B. 1978. See Ref. III, pp. 283–316

6a. Cross, H. E., McKusick, V. A., Breen, W. 1967. *J. Pediatr.* 70:398–406

6b. Witkop, C. J. Jr., Quevedo, W. C. Jr., Fitzpatrick, T. B. 1978. See Ref. III, pp. 283–316

6c. Deleted in proof

7a. Garrod, A. E. 1923. *Inborn Errors of Metabolism.* London: Frowde, Hodder, Stoughton. 2nd ed.

7b. La Du, B. N. 1978. See Ref. III, pp. 268–82

7c. Sršen, Š. 1978. *Lancet* 2:577

7d. Sršen, Š. 1979. *Johns Hopkins Med. J.* 145:217–26

7e. Wyre, H. W. 1979. *Arch. Dermatol.* 115:461–63

8a. Fischer, M. H., Gerritsen, T., Optiz, J. M. 1974. *Humangenetik* 24:265–70

8b. Shih, V. E., Glick, T. H., Bercu, B. B. 1974. *Lancet* 2:163–64

8c. Casey, R. E., Zaleski, W. A., Philp, M., Mendelson, I. S., MacKenzie, S. L. 1978. *J. Inher. Metabl. Dis.* 1:129–35

9a. Kakimoto, Y., Taniguchi, K., Sano, I. 1969. *J. Biol. Chem.* 244:335–40

9b. Evered, D. F., Barley, J. F. 1978. *Clin. Chim. Acta* 84:339–46

10a. Allan, J. D., Cusworth, D. C., Dent, C. E., Wilson, V. K. 1958. *Lancet* 1:182–87

10b. Shih, V. E. 1978. See Ref. III, pp. 362–86

10c. Brusilow, S. W., Batshaw, M. L. 1979. *Lancet* 1:124–27

11a. Pollitt, R. J., Jenner, F. A., Merskey, H. 1968. *Lancet* 2:253–55

11b. Maury, P. 1979. *J. Biol. Chem.* 254:1513–15

12a. Drummond, K. N., Michael, A. F., Ulstrom, R. A., Good, R. A. 1964. *Am. J. Med.* 37:928–48

13a. Dancis, J. 1974. See Ref. I, pp. 32–36

13b. Scriver, C. R., Mackenzie, S., Clow, C. L., Delvin, E. 1971. *Lancet* 1:310–312

13c. Pueschel, S. M., Bresnan, M. J., Shih, V. E., Levy, H. L. 1979. *J. Pediatr.* 94:628–31

13d. Zipf, W. B., Hieber, V. C., Allen, R. J. 1979. *Pediatrics* 63:286–94

14a. Perry, T. L., Hansen, S., Tischler, B., Bunting, R., Berry, K. 1967. *N. Engl. J. Med.* 277:1219–27

14b. Scriver, C. R., Nutzenadel, W., Perry, T. L. 1978. See Ref. III, pp. 528–42

14c. Fleisher, L. D., Rassin, D. K., Wisniewski, K., Salwen, H. R. 1980. *Pediatr. Res.* 14:269–71

15a. McMurray, W. C., Mohyruddin, F., Rossiter, R. J., Rathburn, J. C., Valentine, G. H., Koegler, S. J., Zarfas, D. E. 1962. *Lancet* 1:138

15b. Shih, V. E. 1978. See Ref. III, pp. 362–86

15c. Burgess, E. A., Oberholzer, V. G., Semmens, J. M., Stern, J. 1978. *Arch. Dis. Child.* 53:179–82

15d. Matsuda, Y., Tsuji, A., Katunama, N., Hayashi, M., Takahashi, Y. 1979. *J. Biochem.* 85:191–95

15e. Saheki, T., Tsuda, M., Takada, S., Kusumi, K., Katsunuma, T. 1980. *Adv. Enz. Regul.* 18:221–38

16a. Pinnell, S. R. 1978. See Ref. III, pp. 1366–94

16b. Byers, P. H., Narayanan, A. S., Bornstein, P., Hall, J. G. 1976. *Birth Defects, Orig. Artic. Ser.* 12:293–98

17a. Harris, H., Penrose, L. S., Thomas, D. H. H. 1959. *Ann. Hum. Genet.* 23:442–53

17b. Frimpter, G. W. 1974. See Ref. I, pp. 452–66

17c. Pascal, T. A., Gaull, G. E., Beratis, N. G., Gillam, B. M., Tallan, H. H. 1978. *Pediatr. Res.* 12:125–33

18a. Ben-Ami, E., Burstein, I., Cohen, B. E., Szeinberg, A. 1973. *Clin. Chim. Acta* 45:335–39

19a. Schneider, J. A., Schulman, J. D., Seegmiller, J. E. 1978. See Ref. III, pp. 1660–82

19b. Depape-Brigger, D., Goldman, H., Scriver, C. R., Delvin, E., Mamer, O. 1977. *Pediatr. Res.* 11:124–31

19c. States, B., Scardigli, K., Segal, S. 1978. *Life Sci.* 22:31–37

19d. Roy, L. P., Pollard, A. C. 1978. *Lancet* 2:729–30

19e. Burke, J. R., El-Bishti, M. M., Chantler, C. 1978. *Arch. Dis. Child.* 53:947–51

19f. Schneider, J. A., Schlesselman, J. J., Mendoza, S. A., Orloff, S., Thoene, J. G., Kroll, W. A., Godfrey, A. D., Schulman, J. D. 1979. *N. Engl. J. Med.* 300:756–59

19g. Girardin, E. P., DeWolfe, M. S., Crocker, J. F. S. 1979. *J. Pediatr.* 94:838–40

19h. Thoene, J. G., Lemons, R. 1980. *J. Pediatr.* 96:1043–44

19i. Oberfield, S. E., Levine, L. S., Wellner, D., Novogroder, M., Laino, P., New, M. I. 1980. *Dev. Pharmacol. Ther.* 2:80–90

20a. Thier, S. O., Segal, S. 1978. See Ref. III, pp. 1578–92

20b. Stephens, A. D., Perrett, D. 1976. *Clin. Sci. Mol. Med.* 51:27–32

20c. Kato, T. 1977. *Clin. Sci. Mol. Med.* 53:9–15

20d. Caldwell, R. J., Townsend, J. I., Smith, M. J. V. 1978. *J. Urol.* 119:531–33

20e. Wollaston, W. H. 1810. *Ann. Chim. Paris* 76:21

21a. Melançon, S. B., Dallaire, L., Lemieux, B., Robitaille, P., Potier, M. 1977. *J. Pediatr.* 91:422–27

22a. Teijema, H. L., van Gelderan, H. H., Giesberts, M. A. H., Laurent de Angulo, M. S. L. 1974. *Metabolism* 23:115–23

23a. Pinnell, S. R. 1978. See Ref. III, pp. 1366–94

23b. Prockop, D. J., Kivirikko, K. I., Tuderman, L., Guzman, N. A. 1979. *N. Engl. J. Med.* 301:13–23

23c. Prockop, D. J., Kivirikko, K. I., Tuderman, L., Guzman, N. A. 1979. *N. Engl. J. Med.* 301:77–85

23d. Bornstein, P., Traub, W. 1979. In Proteins, Third Edition. ed. H. Neurath, R. L. Hill, Vol. 4, pp. 412–632. New York: Academic

24a. Pinnell, S. R. 1978. See Ref. III, pp. 1366–94

25a. Pinnell, S. R. 1978. See Ref. III, pp. 1366–94

26a. Pinnell, S. R. 1978. See Ref. III, pp. 1366–94

26b. Pope, F. M., Martin, G. R., Lichtenstein, J. R., Penttinen, R., Gerson, B., Rowe, D. W., McKusick, V. A. 1975. *Proc. Natl. Acad. Sci. USA* 72:1314–16

27a. Pinnell, S. R. 1978. See Ref. III, pp. 1366–94

27b. DiFerrante, N., Leachman, R. D., Donnelly, P. V., Francis, G., Almazan, A. 1975. *Connect. Tissue Res.* 3:49–53

28a. Pinnell, S. R. 1978. See Ref. III, pp. 1366–94

28b. Krane, S. M., Pinnell, S. R., Erbe, R. W. 1972. *Proc. Natl. Acad. Sci. USA* 69:2899–2903

28c. Quinn, R. S., Krane, S. M. 1976. *J. Clin. Invest.* 57:83–93

29a. Pinnell, S. R. 1978. See Ref. III, pp. 1366–94

29b. Steinmann, B., Tuderman, L., Martin, G. R., Prockop, D. J. 1980. *Pediatr. Res.* 14:176

29c. Lichtenstein, J. R., Martin, G. R., Kohn, L. D., Byers, P. H., McKusick, V. A. 1973. *Science* 182:298–300

30a. Vietor, K. W., Havsteen, B., Harms, D., Busse, H., Heyne, K. 1977. *Eur. J. Pediatr.* 126:61–75

31a. Schneider, J. A., Schulman, J. D., Seegmiller, J. E. 1978. See Ref. III, pp. 1660–82

32a. Russell, A., Statter, M., Abzug-Horowitz, S. 1978. *Monogr. Hum. Genet.* 9:65–74

32b. Arakawa, T. 1974. *Clin. Endocrinol. and Metab.* 3:17–35

33a. Baker, L., Winegrad, A. I. 1970. *Lancet* 2:13–16

33b. Bakker, H. D., De Bree, P. K., Ketting, D., Van Sprang, F. J., Wadman, S. K. 1974. *Clin. Chim. Acta* 55:41–47

34a. Froesch, E. R. 1978. See Ref. III, pp. 121–36

34b. Lindemann, R., Gjessing, L. R., Merton, B., Löken, A. C., Halvorsen, S. 1970. *Acta Pediatr. Scand.* 59:141–47
35a. Segal, S. 1978. See Ref. III, pp. 160–81
36a. Käser, H., Cottier, P., Antener, I. 1962. *J. Pediatr.* 61:386–94
37a. Konrad, P. N., Richards, F., Valentine, W. N., Paglia, D. E. 1972. *N. Engl. J. Med.* 286:557–61
37b. Richards, F., Cooper, M. R., Pearce, L. A., Cowan, R. J., Spurr, C. L. 1974. *Arch. Intern. Med.* 134:534–37
37c. Meister, A., Tate, S. S. 1976. *Ann. Rev. Biochem.* 45:559–604
38a. Goodman, S. I., Markey, S. P., Moe, P. G., Miles, B. S., Teng, C. C. 1975. *Biochem. Med.* 12:12–21
38b. Gregersen, N., Brandt, N. J., Christensen, E., Grøn, I., Rasmussen, K., Brandt, S. 1977. *J. Pediatr.* 90:740–45
38c. Brandt, N. J., Brandt, S., Christensen, E., Gregersen, N., Rasmussen, K. 1978. *Clin. Genet.* 13:77–80
38d. Christensen, E., Brandt, N. J. 1978. *Clin. Chim. Acta* 88:267–76
38e. Gregersen, N., Brandt, N. J. 1979. *Pediatr. Res.* 13:977–81
38f. Whelan, D. T., Hill, R., Ryan, E. D., Spate, M. 1979. *Pediatrics* 63:88–93
38g. Brandt, N. J., Gregersen, N., Christensen, E., Grøn, I. H., Rasmussen, K. 1979. *Pediatrics* 94:669–73
39a. Przyrembel, H., Wendel, U., Becker, K., Bremer, H. J., Bruinvis, L., Ketting, D., Wadman, S. K. 1976. *Clin. Chim. Acta* 66:227–39
39b. Goodman, S. I., McCabe, E. R. B., Fennessey, P. V., Mace, J. W. 1979. *Pediatr. Res.* 13:419
39c. Sweetman, L., Nyhan, W. L., Trauner, D. A., Merritt, T. A. 1980. *J. Pediatr.* 96:1020–26
40a. Necheles, T. F., Steinberg, M. H., Cameron, D. 1970. *Br. J. Haematol.* 19:605–12
40b. Steinberg, M. H., Necheles, T. F. 1971. *Am. J. Med.* 50:542–46
41a. Nishimura, Y., Chida, N., Hayashi, T., Arakawa, T. 1972. *Tohoku J. Exp. Med.* 108:207–17
42a. Holmes, B., Park, B. H., Malawista, S. E., Quie, P. G., Nelson, D. L., Good, R. A. 1970. *N. Engl. J. Med.* 283:217–21
42b. Matsuda, I., Oka, Y., Taniguchi, N., Furuyama, M., Kodama, S., Arashima, S., Mitsuyama, T. 1976. *J. Pediatr.* 88:581–83
43a. Loos, H., Roos, D., Weening, R., Houwerzijl, J. 1976. *Blood* 48:53–62
43b. Beutler, E. 1978. See Ref. III, pp. 1430–51

43c. Valentine, W. N., Tanaka, K. R. 1978. See Ref. III, pp. 1410–29
44a. Boivin, P., Galand, C. 1966. *Nouv. Rev. Fr. Hematol.* 6:859–66
44b. Mohler, D. N., Majerus, P. W., Minnich, V., Hess, C. E., Garrick, M. D. 1970. *New Engl. J. Med.* 283:1253–57
44c. Boivin, P., Galand, C., Schaison, G. 1978. *Nouv. Presse Med.* 7:1531–1535
44d. Spielberg, S. P., Garrick, M. D., Corash, L. M., Butler, J. D., Tietze, F., Rogers, L., Schulman, J. D. 1978. *J. Clin. Invest.* 61:1417–20
45a. Goodman, S. I., Mace, J. W., Pollak, S. 1971. *Lancet* 1:234–35
45b. Schulman, J. D., Goodman, S. I., Mace, J. W., Patrick, A. D., Tietze, F., Butler, E. J. 1975. *Biochem. Biophys. Res. Commun.* 65:68–74
45c. Wright, E. C., Stern, J., Ersser, R., Patrick, A. D. 1979. *J. Inher. Metab. Dis.* 2:3–7
45d. Griffith, O. W., Meister, A. 1980. *Proc. Natl. Acad. Sci. USA* 77:3384–87
46a. Kølvraa, S., Rasmussen, K., Brandt, N. J. 1976. *Pediatr. Res.* 10:825–30
46b. Brandt, N. J., Rasmussen, K., Brandt, S., Kølvraa, S., Schønheyder, F. 1976. *Acta Paediatr. Scand.* 65:17–22
46c. Wadman, S. K., Duran, M., Ketting, D., Bruinvis, L., De Bree, P. K., Kamerling, J. P., Gerwig, G. J., Vliegenthart, J. F. G., Przyrembel, H., Becker, K., Bremer, H. J. 1976. *Clin. Chim. Acta* 71:477–84
47a. Powell, G. F., Rasco, M. A., Maniscalco, R. M. 1974. *Metabolism* 23:505–13
47b. Powell, G. F., Kurosky, A., Maniscalco, R. M. 1977. *J. Pediatr.* 91:242–46
47c. Sheffield, L. J., Schlesinger, P., Faull, K., Halpern, B. J., Schier, G. M., Cotton, R. G. H., Hammond, J., Danks, D. M. 1977. *J. Pediatr.* 91:578–83
47d. Isemura, M., Hanyu, T., Gejyo, F., Nakazawa, R., Igarashi, R., Matsuo, S., Ikeda, K., Sato, Y. 1979. *Clin. Chim. Acta.* 93:401–7
48a. Simmell, O., Takki, K. 1973. *Lancet* 1:1031–33
48b. Kennaway, N. G., Weleber, R. G., Buist, N. R. M. 1977. *New Engl. J. Med.* 297:1180
48c. Trijbels, J. M. F., Sengers, R. C. A., Bakkeren, J. A. J. M., de Kort, A. F. M., Deutman, A. F. 1977. *Clin. Chim. Acta* 79:371–77
48d. Valle, D., Kaiser-Kupfer, M. I., Del Valle, L. A. 1977. *Proc. Natl. Acad. Sci. USA* 74:5159–61
48e. O'Donnell, J. J., Sandman, R. P., Martin, S. R. 1978. *Science* 200:200–1

48f. Shih, V. E., Berson, E. L., Mandell, R., Schmidt, S. Y. 1978. *Am. J. Hum. Genet.* 30:174–79

48g. Yatziv, S., Statter, M., Merin, S. 1979. *J. Lab. Clin. Med.* 93:749–57

49a. Baron, D. N., Dent, C. E., Harris, H., Hart, E. W., Jepson, J. B. 1956. *Lancet* 2:421–28

49b. Jepson, J. B. 1978. See Ref. III, pp. 1563–77

49c. Wilcken, B., Yu, J. S., Brown, D. A. 1977. *Arch. Dis. Child.* 52:38–40

49d. Tahmoush, A. J., Alpers, D. H., Feigin, R. D., Armbrustmacher, V., Prensky, A. L. 1976. *Arch Neurol.* 33:797–807

49e. Prensky, A. L., Nelson, J. S., Tahmoush, A. J. 1978. *Adv. Neurol.* 21:339–43

50a. Ghadimi, H., Partington, M. W., Hunter, A. 1961. *N. Engl. J. Med.* 265:221–24

50b. La Du, B. N. 1978. See Ref. III, pp. 317–27

51a. Sabater, J., Ferré, C., Puliol, M., Maya, A. 1976. *Clin. Genet.* 9:117–24

52a. Sjaastad, O., Berstad, J., Gjesdahl, P., Gjessing, L. 1976. *Acta Neurol. Scand.* 53:275–90

53a. Mudd, S. H., Finkelstein, J. D., Irreverre, F., Laster, L. 1964. *Science* 143:1443–45

53b. Perry, T. L. 1974. See Ref. I, pp. 395–428

53c. Rassin, D. K., Longhi, R. C., Sternowsky, H. J., Sturman, J. A., Gaull, G. E. 1977. *Clin. Chim. Acta* 79:197–210

53d. Rassin, D. K., Longhi, R. C., Gaull, G. E. 1977. *J. Pediatr.* 91:574–77

53e. Longhi, R. C., Fleisher, L. D., Tallan, H. H., Gaull, G. E. 1977. *Pediatr. Res.* 11:100–3

53f. Uhlemann, E. R., Ten Pas, J. H., Lucky, A. W., Schulman, J. D., Mudd, S. H., Shulman, N. R. 1976. *N. Engl. J. Med.* 295:1283–86

53g. Harker, L. A., Scott, C. R. 1977. *N. Engl. J. Med.* 296:818

53h. Griffiths, R. 1978. *Monogr. Hum. Genet.* 9:135–39

53i. Fowler, B., Kraus, J., Packman, S., Rosenberg, L. E. 1978. *J. Clin. Invest.* 61:645–53

53j. Kraus, J., Packman, S., Fowler, B., Rosenberg, L. E. 1978. *J. Biol. Chem.* 253:6523–28

54a. Mudd, S. H., Uhlendorf, B. W., Freeman, J. M., Shih, V. E. 1972. *Biochem. Biophys. Res. Commun.* 46:905–12

54b. Mudd, S. H. 1974. See Ref. I, pp. 429–51

54c. Freeman, J. M., Finkelstein, J. D., Mudd, S. H. 1975. *N. Engl. J. Med.* 292:491–96

54d. Kanwar, Y. S., Manaligod, J. R., Wong, P. W. K. 1976. *Pediatr. Res.* 10:598–609

54e. Narisawa, K., Wada, Y., Saito, T., Suzuki, H., Kudo, M., Arakawa, T., Katsushima, N., Tsuboi, R. 1977. *Tohoku J. Exp. Med.* 121:185–94

54f. Wong, P. W. K., Justice, P., Hruby, M., Weiss, E. B., Diamond, E. 1977. *Pediatrics* 59:749–56

54g. Rosenblatt, D. S., Erbe, R. W. 1977. *Pediatr Res.* 11:1141–43

55a. Mudd, S. H., Levy, H. L., Abeles, R. H. 1969. *Biochem. Biophys. Res. Commun.* 35:121–26

55b. Mudd, S. H. 1974. See Ref. I, pp. 429–51

55c. Mahoney, M. J., Hart, A. C., Steen, V. D., Rosenberg, L. E. 1975. *Proc. Natl. Acad. Sci. USA* 72:2799–2803

55d. Gravel, R. A., Mahoney, M. J., Ruddle, F. H., Rosenberg, L. E. 1975. *Proc. Natl. Acad. Sci. USA* 72:3181–85

55e. Fenton, W. A., Rosenberg, L. E. 1978. *Ann. Rev. Genet.* 12:223–48

55f. Baumgartner, E. R., Wick, H., Linnell, J. C., Gaull, G. E., Bachmann, C., Steinmann, B. 1979. *Helv. Paediatr. Acta* 34:483–96

56a. Hollowell, J. G. Jr., Hall, W. K., Coryell, M. E., McPherson, J. Jr., Hahn, D. A. 1969. *Lancet* 2:1428

56b. Imerslund, O. 1960. *Acta Paediatr. Scand. Suppl.* 119:1–115

56c. Gräsbeck, R., Gordin, R., Kantero, I., Kuhlbäck, B. 1960. *Acta Med. Scand.* 167:289–96

56d. Mudd, S. H. 1974. See Ref. I, pp. 429–51

56e. Higginbottom, M. C., Sweetman, L., Nyhan, W. L. 1978. *N. Engl. J. Med.* 299:317–23

57a. Knapp, A. 1960. *Clin. Chim. Acta* 5:6–13

57b. O'Brien, D., Jensen, C. B. 1963. *Clin. Sci.* 24:179–86

57c. Komrower, G. M., Wilson, V., Clamp, J. R., Westall, R. G. 1964. *Arch. Dis. Child.* 39:250–56

57d. Komrower, G. M., Westall, R. 1967. *Am. J. Dis. Child.* 113:77–80

57e. Price, J. M., Yess, N., Brown, R. R., Johnson, S. A. M. 1967. *Arch. Dermatol.* 95:462–72

57f. Tada, K., Yokoyama, Y., Nakagawa, H., Yoshida, T., Arakawa, T. 1967. *Tohoku J. Exp. Med.* 93:115–24

57g. Leklem, J. E., Brown, R. R. 1976. *J. Natl. Cancer Inst.* 56:1101–4

57h. Reddi, O. S., Reddy, M. V. R., Reddy, K. R. S. 1978. *Hum. Hered.* 28:238–40

58a. Parker, C. E., Shaw, K. N. F., Jacobs, E. E., Gutenstein, M. 1970. *Lancet* 1:1119–20

58b. Hoefnagel, D., Pomeroy, J. 1970. *Lancet.* 1:1342–43

58c. Goodman, S. I., Browder, J. A., Hiles, R. A., Miles, B. S. 1972. *Biochem. Med.* 6:344–54

59a. Benson, P. F., Swift, P. N., Young, V. K. 1969. *Arch. Dis. Child.* 44:134–35

60a. Faull, K., Bolton, P., Halpern, B., Hammond, J., Danks, D. M., Hähnel, R., Wilkinson, S. P., Wysocki, S. J., Masters, P. L. 1976. *N. Engl. J. Med.* 294:1013

60b. Faull, K. F., Bolton, P. D., Halpern, B., Hammond, J., Danks, D. M. 1976. *Clin. Chim. Acta* 73:553–59

60c. Wysocki, S. J., Wilkinson, S. P., Hähnel, R., Wong, C. Y. B., Panegyres, P. K. 1976. *Clin. Chim. Acta* 70:399–406

60d. Wysocki, S. J., Hähnel, R. 1976. *Clin. Chim. Acta* 71:349–51

60e. Wysocki, S. J., Hähnel, R. 1976. *Clin. Chim. Acta* 73:373–75

60f. Leonard, J. V., Seakins, J. W. T., Griffin, N. K. 1979. *Lancet* 1:680

60g. Leonard, J. V., Seakins, J. W. T., Griffin, N. K., Marshall, W. C. 1979. *Lancet* 1:1147

61a. Efron, M. L., Bixby, E. M., Palattao, L. G., Pryles, C. V. 1962. *N. Engl. J. Med.* 267:1193–94

61b. Efron, M. L., Bixby, E. M., Pryles, C. V. 1965. *N. Engl. J. Med.* 272:1299–1309

61c. Scriver, C. R. 1978. See Ref. III, pp. 336–61

61d. Roesel, R. A., Blankenship, P. R., Lynch, W. R., Coryell, M. E., Thevaos, T. G., Hall, W. K. 1979. *Hum. Hered.* 29:364–70

62a. Blass, J. P., Kark, R. A. P., Menon, N. K. 1976. *N. Engl. J. Med.* 295:62–67

62b. Haworth, J. C., Perry, T. L., Blass, J. P., Hansen, S., Urquhart, N. 1976. *Pediatrics* 58:564–72

62c. Robinson, B. H., Taylor, J., Sherwood, W. G. 1977. *Pediatr. Res.* 11:1198–1202

62d. Taylor, J., Robinson, B. H., Sherwood, W. G. 1978. *Pediatr. Res.* 12:60–62

62e. Kark, R. A. P., Rodriguez-Budelli, M., Blass, J. P. 1978. *Adv. Neurol.* 21:163–80

62f. Barbeau, A., Melançon, S., Butterworth, R. F., Filla, A., Izumi, K., Ngo, T. T. 1978. *Adv. Neurol.* 21:203–17

63a. Blass, J. P., Avigan, J., Uhlendorf, B. W. 1970. *J. Clin. Invest.* 49:423–32

63b. Strömme, J. H., Borud, O., Moe, P. J. 1976. *Pediatr. Res.* 10:62–66

63c. Blass, J. P., Cederbaum, S. D., Kark, R. A. P., Rodiguez-Budelli, M. 1978. *Monogr. Hum. Genet.* 9:12–15

64a. Robinson, B. H., Sherwood, W. G. 1975. *Pediatr. Res.* 9:935–39

64b. Koster, J. F., Slee, R. G., Fernandes, J. 1978. *Monogr. Hum. Genet.* 9:7–11

65a. Lonsdale, D., Faulkner, W. R., Price, J. W., Smeby, R. R. 1979. *Pediatrics* 43:1025–34

65b. Tada, K., Yoshida, T., Konno, T., Wada, Y., Yokoyama, Y., Arakawa, T. 1969. *Tohoku J. Exp. Med.* 97:99–100

65c. Maesaka, H., Komiya, K., Misugi, K., Tada, K. 1976. *Eur. J. Pediatr.* 122:159–68

65d. Saudubray, J. M., Marsac, C., Charpentier, C., Cathelineau, L., Besson Leaud, M., Leroux, J. P. 1976. *Acta Paediatr. Scand.* 65:717–24

65e. Atkin, B. M., Utter, M. F., Weinberg, M. B. 1979. *Pediatr. Res.* 13:38–43

66a. Scriver, C. R., Pueschel, S., Davies, E. 1966. *N. Engl. J. Med.* 274:635–43

67a. Russell, A., Levin, B., Oberholzer, V. G., Sinclair, L. 1962. *Lancet* 2:699–700

67b. Shih, V. E. 1978. See Ref. III, pp. 362–86

67c. Krieger, I., Bachmann, C., Gronemeyer, W. H., Cejka, J. 1976. *J. Clin. Endocrinol. Metab.* 43:796–802

67d. Gray, R. G. F., Black, J. A., Lyons, V. H., Pollitt, R. J. 1976. *Pediatr. Res.* 10:918–23

67e. Batshaw, M. L., Roan, Y., Jung, A. L., Rosenberg, L. A., Brusilow, S. W. 1980. *N. Engl. J. Med.* 302:482–85

67f. Mori, M., Uchiyama, C., Miura, S., Tatibana, M., Nagayama, E. 1980. *Clin. Chim. Acta* 104:291–99

68a. Freeman, J. M., Nicholson, J. F., Masland, W. S., Rowland, L. P., Carter, S. 1964. *J. Pediatr.* 65:1039–40

68b. Freeman, J. M., Nicholson, J. F., Schimke, R. T., Rowland, L. P., Carter, S. 1970. *Arch. Neurol.* 23:430–37

68c. Shih, V. E. 1978. See Ref. III, pp. 362–86

68d. Farriaux, J. P., Ponte, C., Pollitt, R. J., Lequien, P., Formstecher, P., Dhondt, J. L. 1977. *Acta Paediatr. Scand.* 66:529–34

68e. Mantagos, S., Tsagaraki, S., Burgess, E. A., Oberholzer, V., Palmer, T., Sacks, J., Baibas, S., Valaes, T. 1978. *Arch. Dis. Child.* 53:230–34

68f. Brusilow, S., Tinker, J., Batshaw, M. L. 1980. *Science* 207:659–61

69a. Terheggen, H. G., Schwenk, A., Lowenthal, A., van Sande, M., Colombo, J. P. 1969. *Lancet* 2:748–49

69b. Terheggen, H. G., Lowenthal, A., Lavinha, F., Colombo, J. P. 1975. *Arch. Dis. Child.* 50:57–62

69c. Snyderman, S. E., Sansaricq, C., Chen, W. J., Norton, P. M., Phansalkar, S. V. 1977. *J. Pediatr.* 90:563–68

69d. Cederbaum, S. D., Shaw, K. N. F., Valente, M. 1977. *J. Pediatr.* 90:569–73

69e. Michels, V. V., Beaudet, A. L. 1978. *Clin. Genet.* 13:61–67

69f. Marescau, B., Pintens, J., Lowenthal, A., Terheggen, H. G. 1979. *Clin. Chim. Acta* 98:35–38

69g. Cederbaum, S. D., Shaw, K. N. F., Spector, E. B., Verity, M. A., Snodgrass, P. J., Sugarman, G. I. 1979. *Pediatr. Res.* 13:827–33

70a. Brodehl, J., Gellissen, K., Kowalewski, S. 1967. *Klin. Wochenschr.* 45:38–40

71a. Whelan, D. T., Scriver, C. R. 1968. *Pediatr. Res.* 2:525–534

71b. Endres, W., Zoulek, G., Schaub, J. 1979. *Eur. J. Pediatr.* 131:33–41

72a. de Vries, A., Kochwa, S., Lazebnik, J., Frank, M., Djaldetti, M. 1957. *Am. J. Med.* 23:408–15

72b. Greene, M. L., Lietman, P. S., Rosenberg, L. E., Seegmiller, J. E. 1973. *Am. J. Med.* 54:265–71

73a. Colombo, J. P., Vassella, F., Humbel, R., Buergi, W. 1967. *Am. J. Dis. Child.* 113:138–41

73b. Colombo, J. P., Bürgi, W., Richterich, R., Rossi, E. 1967. *Metabolism* 16:910–25

73c. Oyanagi, K., Sogawa, H., Sato, S., Orii, T., Nakao, T., Fujita, S. 1976. *Tohoku J. Exp. Med.* 120:105–12

73d. Sogawa, H., Oyanagi, K., Nakao, T. 1977. *Pediatr. Res.* 11:949–53

74a. Woody, N. C. 1964. *Am. J. Dis. Child.* 108:543–53

74b. Ghadimi, H., Binnington, V. I., Pecora, P. 1965. *N. Engl. J. Med.* 273:723–29

74c. Woody, N. C., Hutzler, J., Dancis, J. 1966. *Am. J. Dis. Child.* 112:577–80

74d. Armstrong, M. D., Robinow, M. 1967. *Pediatrics* 39:546–54

74e. Dancis, J., Hutzler, J., Cox, R. P., Woody, N. C. 1969. *J. Clin. Invest.* 48:1447–52

74f. Woody, N. C., Pupene, M. B. 1971. *Pediatr. Res.* 5:511–13

74g. Woody, N. C., Pupene, M. B. 1973. *Pediatr. Res.* 7:994–95

74h. Dancis, J., Hutzler, J., Woody, N. C., Cox, R. P. 1976. *Pediatr. Res.* 10:686–91

74i. van der Heiden, C., Brink, M., de Bree, P. K., van Sprang, F. J., Wadman, S. K., de Pater, J. M., van Biervliet, J. P. G. M. 1978. *J. Inher. Metab. Dis.* 1:89–94

74j. Dancis, J., Hutzler, J., Cox, R. P. 1979. *Am. J. Hum. Genet.* 31:290–99

75a. Gaull, G. E., Tallan, H. H. 1974. *Science* 186:59–60

75b. Gaull, G. E. 1975. In *Normal and Pathological Development of Energy Metabolism*, F. A. Hommes, C. J. Van den Berg, pp. 11–23. New York: Academic

75c. Finkelstein, J. D., Kyle, W. E., Martin, J. J. 1975. *Biochem. Biophys. Res. Commun.* 66:1491–97

76a. Shih, V. E., Efron, M. L., Moser, H. W. 1969. *Am. J. Dis. Child.* 117:83–92

76b. Fell, V., Pollitt, R. J., Sampson, G. A., Wright, T. 1974. *Am. J. Dis. Child.* 127:752–56

76c. Shih, V. E., Mandell, R. 1974. *Lancet* 2:1522–23

76d. Gatfield, P. D., Taller, E., Wolfe, D. M., Haust, M. D. 1975. *Pediatr. Res.* 9:488–97

77a. Williams, H. E. Smith, L. H. Jr. 1978. See Ref. III, pp. 182–204

78a. Williams, H. E., Smith, L. H. Jr. 1978. See Ref. III, pp. 182–204

79a. Scriver, C. R. 1978. See Ref. III, pp. 336–61

79b. Schafer, I. A., Scriver, C. R., Efron, M. L. 1962. *N. Engl. J. Med.* 267:51–60

79c. Efron, M. L. 1965. *N. Engl. J. Med.* 272:1243–54

80a. Scriver, C. R. 1978. See Ref. III, pp. 336–61

80b. Valle, D. L., Phang, J. M. 1974. *Science* 185:1053–54

80c. Goodman, S. I., Mace, J. W., Miles, B. S., Teng, C. C., Brown, S. B. 1974. *Biochem. Med.* 10:329–36

80d. Pavone, L., Mollica, F., Levy, H. L. 1975. *Arch. Dis. Child.* 50:637–41

80e. Valle, D. L., Goodman, S. I., Applegarth, D. A., Shih, V. E., Phang, J. M. 1976. *J. Clin. Invest.* 58:598–603

80f. Valle, D., Goodman, S. I., Harris, S. C., Phang, J. M. 1979. *J. Clin. Invest.* 64:1365–70

81a. Gerritsen, T., Waisman, H. A. 1978. See Ref. III, pp. 514–17

81b. Tippett, P., Danks, D. M. 1974. *Helv. Paediatr. Acta* 29:261–67

81c. Blom, W., Fernandes, J. 1979. *Clin. Chim. Acta* 91:117–25

82a. Reddi, O. S. 1978. *J. Pediatr.* 93:814–16

83a. Wada, Y., Tada, K., Minagawa, A., Yoshida, T., Morikawa, T., Okamura,

T. 1963. *Tohoku J. Exp. Med.* 81:46-55

83b. Wada, Y. 1965. *Tohoku J. Exp. Med.* 87:322-31

83c. Dancis, J., Hutzler, J., Tada, K., Wada, Y., Morikawa, T., Arakawa, T. 1967. *Pediatrics* 39:813-17

83d. Tada, K., Wada, Y., Arakawa, T. 1967. *Am. J. Dis. Child.* 113:64-67

83e. Reddi, O. S., Reddy, S. V., Reddy, K. R. 1977. *Hum. Genet.* 39:139-42

83f. Taylor, R. T., Jenkins, W. T. 1966. *J. Biol. Chem.* 241:4396-4405

83g. Jeune, M., Collombel, C., Michel, M., David, M., Guibaud, P., Guerrier, G., Albert, J. 1970. *Ann. Pediatr. Paris* 17:85-99

84a. Scriver, C. R. 1978. See Ref. III, pp. 1593-1606

84b. Statter, M., Ben-Zvi, A., Shina, A., Schein, R., Russell, A. 1976. *Helv Paediatr. Acta* 31:173-82

84c. Law, E. A., Sardharwalla, I. B. 1978. *Monogr. Hum. Genet.* 9:152-54

84d. Lasley, L., Scriver, C. R. 1979. *Pediatr. Res.* 13:65-70

85a. Stanbury, J. B. 1978. See Ref. III, pp. 206-39

85b. Ismail-Beigi, F., Rahimifar, M. 1977. *J. Clin. Endocrinol. Metab.* 44:499-506

86a. Tanaka, K., Budd, M. A., Efron, M. L., Isselbacher, K. J. 1966. *Proc. Natl. Acad. Sci. USA* 56:236-42

86b. Levy, H. L., Erickson, A. M. 1974. See Ref. I, pp. 81-97

86c. Krieger, I., Tanaka, K. 1976. *Pediatr. Res.* 10:25-29

86d. Malan, C., Neethling, A. C., Shanley, B. C., Gompertz, D., Bartlett, K., Schraader, E. B. 1977. *S. Afr. Med. J.* 51:980-83

86e. Yudkoff, M., Cohn, R. M., Puschak, R., Rothman, R., Segal, S. 1978. *J. Pediatr.* 92:813-17

86f. Cohn, R. M., Yudkoff, M., Rothman, R., Segal, S. 1978. *N. Engl. J. Med.* 299:996-99

86g. Winokur, P. A., Vashistha, K., Seshamani, R. 1978. *Pediatrics* 61:902-3

86h. Velazquez, A., Prieto, E. C. 1980. *Lancet* 1:313-14

86i. Kelleher, J. F. Jr., Yudkoff, M., Hutchinson, R., August, C. S., Cohn, R. M. 1980. *Pediatrics* 65:1023-27

87a. Przyrembel, H., Bachmann, D., Lombeck, I., Becker, K., Wendel, U., Wadman, S. K., Bremer, H. J. 1975. *Clin. Chim. Acta* 58:257-69

87b. Wendel, U., Rüdiger, H. W., Przyrembel, H., Bremer, H. J. 1975. *Clin. Chim. Acta* 58:271-76

88a. Luder, J., Sheldon, W. 1955. *Arch. Dis. Child.* 30:160-64

88b. Sheldon, W., Luder, J., Webb, B. 1961. *Arch. Dis. Child.* 36:90-95

89a. Omura, K., Yamanka, N., Higami, S., Matsuoka, O., Fujimoto, A., Issiki, G., Tada, K. 1976. *Pediatrics* 57:102-5

90a. Perheentupa, J., Visakorpi, J. K. 1965. *Lancet* 2:813-16

90b. Kekomäki, M., Visakorpi, J. K., Perheentupa, J., Saxén, L. 1967. *Acta Paediatr. Scand.* 56:617-30

90c. Oyanagi, K., Miura, R., Yamanouchi, T. 1970. *J. Pediatr.* 77:259-66

90d. Malmquist, J., Jagenburg, R., Lindstedt, G. 1971. *N. Engl. J. Med.* 284:997-1002

90e. Brown, J. H., Fabre, L. F. Jr., Farrell, G. L., Adams, E. D. 1972. *Am. J. Dis. Child.* 124:127-32

90f. Kihara, H., Valente, M., Porter, M. T., Fluharty, A. L. 1973. *Pediatrics* 51:223-29

90g. Norio, R., Perheentupa, J., Kekomäki, M., Visakorpi, J. K. 1971. *Clin. Genet.* 2:214-22

90h. Simell, O., Perheentupa, J., Rapola, J., Visakorpi, J. K., Eskelin, L. 1975. *Am. J. Med.* 59:229-40

90i. Awrich, A. E., Stackhouse, W. J., Cantrell, J. E., Patterson, J. H., Rudman, D. 1975. *J. Pediatr.* 87:731-38

90j. Kato, T., Tanaka, E., Horisawa, S. 1976. *Am. J. Dis. Child.* 130:1340-44

90k. Chan, H., Billmeier, G. J. Jr., Molinary, S. V., Tucker, H. N., Shin, B. C., Schaffer, A., Cavallo, K. 1977. *J. Pediatr.* 91:79-81

90l. Oyanagi, K., Sogawa, H., Minami, R., Nakao, T., Chiba, T. 1979. *J. Pediatr.* 94:255-57

91a. Menkes, J. H., Hurst, P. L., Craig, J. M. 1954. *Pediatrics* 14:462-66

91b. Snyderman, S. E. 1974. See Ref. I, pp. 17-31

91c. Singh, S., Willers, I., Goedde, H. W. 1977. *Clin. Genet.* 11:277-84

91d. Danner, D. J., Wheeler, F. B., Lemmon, S. K., Elsas, L. J. II, 1978. *Pediatr. Res.* 12:235-38

91e. Meister, A. 1952. *J. Biol. Chem.* 195:813-26

92a. Pinnell, S. R. 1978. See Ref. III, pp. 1366-94

93a. Crawhall, J. C. 1978. See Ref. III, pp. 504-13

93b. Shih, V. E., Carney, M. M., Fitzgerald, L., Monedjikova, V. 1977. *Pediatr. Res.* 11:464

94a. Smith, A. J., Strang, L. B. 1958. *Arch. Dis. Child.* 33:109-13

94b. Jepson, J. B., Smith, A. J., Strang, L. B. 1958. *Lancet* 2:1334-35

94c. Hooft, C., Timmermans, J., Snoeck, J., Antener, I., Oyaert, W., van den Hende, C. 1965. *Ann. Paediatr.* 205:73–104

94d. Hooft, C., Carton, D., Snoeck, J., Timmermans, J., Antener, I., van den Hende, C., Oyaert, W. 1968. *Helv. Paediatr. Acta* 23:334–49

95a. Eldjarn, L., Jellum, E., Stokke, O., Pande, H., Waaler, P. E. 1970. *Lancet* 2:521–22

95b. Stokke, O., Eldjarn, L., Jellum, E., Pande, H., Waaler, P. E. 1972. *Pediatrics* 49:726–35

95c. Lehnert, W., Niederhoff, H., Junker, A., Saule, H., Frasch, W. 1979. *Eur. J. Pediatr.* 132:107–14

96a. Gompertz, D., Draffan, G. H., Watts, J. L., Hull, D. 1971. *Lancet* 2:22–24

96b. Gompertz, D., Bartlett, K., Blair, D., Stern, C. M. M. 1973. *Arch. Dis. Child.* 48:975–77

96c. Bartlett, K., Gompertz, D. 1976. *Lancet* 2:804

96d. Roth, K., Cohn, R., Yandrasitz, J., Preti, G., Dodd, P., Segal, S. 1976. *J. Pediatr.* 88:229–35

96e. Sweetman, L., Bates, S. P., Hull, D., Nyhan, W. L. 1977. *Pediatr. Res.* 11:1144–47

96f. Weyler, W., Sweetman, L., Maggio, D. C., Nyhan, W. L. 1977. *Clin. Chim. Acta* 76:324–28

96g. Saunders, M., Sweetman, L., Robinson, B., Roth, K., Cohn, R., Gravel, R. A. 1979. *J. Clin. Invest.* 64:1695–1702

96h. Roth, K. S., Yang, W., Foreman, J. W., Rothman, R., Segal, S. 1980. *J. Pediatr.* 96:845–49

96i. Bartlett, K., Ng, H., Leonard, J. V. 1980. *Clin. Chim. Acta* 100:183–86

97a. Daum, R. S., Scriver, C. R., Mamer, O. A., Delvin, E., Lamm, P., Goldman, H. 1973. *Pediatr. Res.* 7:149–60

97b. Hillman, R. E., Keating, J. P. 1974. *Pediatrics* 53:221–25

97c. Gompertz, D., Saudubray, J. M., Charpentier, C., Bartlett, K., Goodey, P. A., Draffan, G. H. 1974. *Clin. Chim. Acta* 57:269–81

97d. Halvorsen, S., Stokke, O., Jellum, E. 1979. *Acta Paediatr. Scand.* 68:123–28

98a. Oberholzer, V. G., Levin, B., Burgess, E. A., Young, W. F. 1967. *Arch. Dis. Child.* 42:492–504

98b. Stokke, O., Eldjarn, L., Norum, K. R., Steen-Johnsen, J., Halvorsen, S. 1967. *Scand. J. Clin. Lab. Invest.* 20:313–28

98c. Morrow, G., Barness, L. A., Cardinale, G. J., Abeles, R. H., Flaks, J. G. 1969. *Proc. Natl. Acad. Sci. USA* 63:191–97

98d. Giorgio, A. J., Trowbridge, M., Boone, A. W., Patten, R. S. 1976. *N. Engl. J. Med.* 295:310–13

98e. Wilcken, B., Kilham, H. A., Faull, K. 1977. *J. Pediatr.* 91:428–30

98f. Morrow, G. III, Revsin, B., Clark, R., Lebowitz, J., Whelan, D. T. 1978. *Clin. Chim. Acta* 85:67–72

98g. Shapiro, L. J., Bocian, M. E., Raijman, L., Cederbaum, S. D., Shaw, K. N. F. 1978. *J. Pediatr.* 93:986–88

98h. Packman, S., Mahoney, M. J., Tanaka, K., Hsia, Y. E. 1978. *J. Pediatr.* 92:769–71

99a. Kang, E. S., Snodgrass, P. J., Gerald, P. S. 1972. *Pediatr. Res.* 6:875–79

100a. Mahoney, M. J., Hart, A. C., Steen, V. D., Rosenberg, L. E. 1975. *Proc. Natl. Acad. Sci. USA* 72:2799–2803

100b. Gravel, R. A., Mahoney, M. J., Ruddle, F. H., Rosenberg, L. E. 1975. *Proc. Natl. Acad. Sci. USA* 72:3181–85

100c. Fenton, W. A., Rosenberg, L. E. 1978. *Ann. Rev. Genet.* 12:223–48

101a. Mahoney, M. J. 1975. See Ref. 55c, pp. 2799–2803

101b. Gravel, R. A. 1975. See Ref. 55d, pp. 3181–85

101c. Fenton, W. A. 1978. See Ref. 55e, pp. 223–48

102a. Nyhan, W. L. 1978. See Ref. III, pp. 518–27

102b. Kikuchi, G. 1973. *Mol. Cell. Biochem.* 1:169–87

102c. Krieger, I., Winbaum, E. S., Eisenbrey, A. B. 1977. *Metabolism* 26:517–24

102d. Holmgren, G., Blomquist, H. K. 1977. *Neuropaediatrie* 8:67–72

102e. Perry, T. L., Urquhart, N., Hansen, S. 1977. *Pediatr. Res.* 11:1192–97

102f. Bank, W. J., Pizer, L., Pfendner, W. 1978. *Adv. Neurol.* 21:267–78

102g. de Groot, C. J., Boelleverts, V., Touwen, B. C. L., Hommes, F. A. 1978. *Prog. Brain Res.* 48:199–207

102h. Frazier, D. M., Summer, G. K., Chamberlin, H. R. 1978. *Am. J. Dis. Child.* 132:777–81

102i. Gitzelmann, R., Steinmann, B., Cuénod, M. 1978. *N. Engl. J. Med.* 298:1424

102j. Arneson, D., Ch'ien, L. T., Chance, P., Wilroy, R. S. 1979. *Pediatrics* 63:369–73

102k. Kølvraa, S., Brandt, N. J., Christensen, E. 1979. *Acta Pediatr. Scand.* 68:629–34

102l. von Wendt, L., Similä, S., Saukkonen, A.-L., Koivisto, M. 1980. *Pediatrics* 65:1166–69

102m. MacDermot, K. D., Nelson, W., Reichert, C. M., Schulman, J. D. 1980. *Pediatrics* 65:61–64

102n. Warburton, D., Boyle, R. J., Keats, J. P., Vohr, B., Peuschel, S., Oh, W. 1980. *Am. J. Dis. Child.* 134:273–75

103a. Lowe, C. U., Terrey, M., MacLachlan, E. A. 1952. *Am. J. Dis. Child.* 83:164–84

103b. Chutorian, A., Rowland, L. P. 1966. *Neurology* 16:115–22

103c. Matsuda, I., Takeda, T., Sugai, M., Matsuura, N. 1969. *Am. J. Child.* 117:205–12

103d. Delleman, J. W., Bleeker-Wagemakers, E. M., van Veelen, A. W. C. 1977. *J. Pediatr. Ophthalmol.* 14:205–12

104a. Pinnell, S. R. 1978. See Ref. III, pp. 1366–94

104b. Penttinen, R. P., Lichtenstein, J. R., Martin, G. R., McKusick, V. A. 1975. *Proc. Natl. Acad. Sci. USA* 72:586–89

105a. Meister, A. 1978. See Ref. III, pp. 328–35

105b. Wellner, V. P. Sekura, R., Meister, A., Larsson, A. 1974. *Proc. Natl. Acad. Sci. USA* 71:2505–9

105c. Larsson, A., Zetterström, R., Hörnell, H., Porath, U. 1976. *Clin. Chim. Acta* 73:19–23

105d. Spielberg, S. P., Boxer, L. A., Oliver, J. M., Butler, E. J., Schulman, J. D. 1978. *Monogr. Hum. Genet.* 9:90–94

105e. Eldjarn, L., Jellum, E., Stokke, O. 1973. *Inborn Errors of Metabolism.* ed. F. A. Hommes, C. J. Van den Berg, pp. 255–60. New York: Academic

105f. Eldjarn, L., Stokke, O., Jellum, E. 1972. In *Organic Acidurias,* J. Stern, C. Toothill, eds, pp. 113–120. Baltimore: Williams & Wilkins

105g. Richman, P., Meister, A. 1975. *J. Biol. Chem.* 250:1422–26

106a. Tourian, A. Y., Sidbury, J. B. 1978. See Ref. III, pp. 240–55

106b. Kaufman, S., Max, E. E., Kang, E. S. 1975. *Pediat. Res.* 9:632–34

106c. Bartholomé, K., Ertel, E. 1976. *Lancet* 2:862–63

106d. Parker, C. E., Barranger, J., Newhouse, R., Bessman, S. 1977. *Biochem. Med.* 17:8–12

106e. Reynolds, G. P., Seakins, J. W. T., Gray, D. O. 1978. *Clin. Chim. Acta* 83:33–39

106f. Lasala, J. M., Coscia, C. J. 1979. *Science* 203:283–84

106g. Peck, H., Pollitt, R. J. 1979. *Clin. Chim. Acta* 94:237–40

106h. Choo, K. H., Cotton, R. G. H., Danks, D. M., Jennings, I. G. 1979. *Biochem. J.* 181:285–94

107a. Bartholomé, K. 1974. *Lancet* 2:1580

107b. Smith, I., Clayton, B. E., Wolff, O. H. 1975. *Lancet* 1:1108–11

107c. Bartholomé, K., Byrd, D. J. 1975. *Lancet* 2:1042–43

107d. Kaufman, S., Holtzman, N. A., Milstien, S., Butler, I. J., Krumholz, A. 1975. *N. Engl. J. Med.* 293:785–90

107e. Milstien, S., Holtzman, N. A., O'-Flynn, M. E., Thomas, G. H., Butler, I. J., Kaufman, S. 1976. *J. Pediatr.* 89:763–66

107f. Schlesinger, P., Watson, B. M., Cotton, R. G. H., Danks, D. M. 1979. *Clin. Chim. Acta* 92:187–95

107g. Milstien, S., Kaufman, S., Summer, G. K. 1980. *Pediatrics* 65:806–10

107h. Matthews, R. G., Kaufman, S. 1980. *J. Biol. Chem.* 255:6014–17

108a. Leeming, R. J., Blair, J. A., Rey, F. 1976. *Lancet* 1:99–100

108b. Rey, F., Blandin-Savoja, F., Rey, J. 1976. *N. Engl. J. Med.* 295:1138–39

108c. Bartholomé, K., Byrd, D. J., Kaufman, S., Milstien, S. 1977. *Pediatrics* 59:757–61

108d. Schaub, J., Däumling, S., Curtius, H.-C., Niederwieser, A., Bartholomé, K., Viscontini, M., Schircks, B., Bieri, J. H. 1978. *Arch. Dis. Child.* 53:674–76

108e. Kaufman, S., Berlow, S., Summer, G. K., Milstien, S., Schulman, J. D., Orloff, S., Spielberg, S., Pueschel, S. 1978. *N. Engl. J. Med.* 299:673–79

108f. Niederwieser, A., Curtius, H.-C., Bettoni, O., Bieri, J., Schircks, B., Viscontini, M., Schaub, J. 1979. *Lancet* 1:131–33

108g. Leeming, R. J., Smith, I., Schaub, J., Niederwieser, A., Curtius, H.-C. 1979. *Arch. Dis. Child.* 54:166–67

108h. Gál, E. M., Sherman, A. D. 1979. *Lancet* 1:448

108i. Niederwieser, A., Curtius, H.-C., Viscontini, M., Schaub, J., Schmidt, H. 1979. *Lancet* 1:550

108j. Curtius, H.-C., Niederwieser, A., Viscontini, M., Otten, A., Schaub, J., Scheibenreiter, S., Schmidt, H. 1979. *Clin. Chim. Acta* 93:251–62

108k. Kaufman, S. 1980. *Pediatrics* 65: 840–42

109a. Gatfield, P. D., Taller, E., Hinton, G. G., Wallace, A. C., Abdelnour, G. M., Haust, M. D. 1968. *Can. Med. Assoc. J.* 99:1215–33

109b. Thomas, G. H., Haslam, R. H. A., Batshaw, M. L., Capute, A. J., Neidengard, L., Ransom, J. L. 1975. *Clin. Genet.* 8:376–82

109c. Danks, D. M., Tippett, P., Adams, C.,

Campbell, P. 1975. *J. Pediatr.* 86: 382–87

110a. Ando, T., Nyhan, W. L. 1974. See Ref. I, pp. 37–60

110b. Gravel, R. A., Lam, K.-F., Scully, K. J., Hsia, Y. E. 1977. *Am. J. Hum. Genet* 29:378–88

110c. Branski, D., Gale, R., Gross-Kieselstein, E., Abrahamov, A. 1977. *Am. J. Dis. Child.* 131:1379–81

110d. Shafai, T., Sweetman, L., Weyler, W., Goodman, S. I., Fennessey, P. V., Nyhan, W. L. 1978. *J. Pediatr.* 92:84–86

110e. Hillman, R. E., Keating, J. P., Williams, J. C. 1978. *J. Pediatr.* 92:439–41

110f. de Cespedes, C., Loria, A. R., Estrada, Y., Sweetman, L., Nyhan, W. L. 1978. *Monogr. Hum. Genet.* 9:80–83

110g. O'Brien, W. E. 1978. *Arch. Biochem. Biophys.* 189:291–97

110h. Hsia, Y. E., Scully, K. J., Rosenberg, L. E. 1979. *Pediatr. Res.* 13:746–51

110i. Przyrembel, H., Bremer, H. J., Duran, M., Bruinvis, L., Ketting, D., Wadman, S. K., Baumgartner, R., Irle, U., Bachmann, C. 1979. *Eur. J. Pediatr.* 130:1–14

110j. Harris, D. J., Yang, B. I. Y., Wolf, B., Snodgrass, P. J. 1980. *Pediatrics* 65:107–10

111a. Hunt, A. D. Jr., Stokes, J. Jr., McCrory, W. W., Stroud, H. H. 1954. *Pediatrics* 13:140–45

111b. Scriver, C. R., Whelan, D. T. 1969. *Ann. NY Acad. Sci.* 166:83–96

111c. Yoshida, T., Tada, K., Arakawa, T. 1971. *Tohoku J. Exp. Med.* 104:195–98

112a. Rowley, P. T., Mueller, P. S., Watkin, D. M., Rosenberg, L. E. 1961. *Am. J. Med.* 31:187–204

112b. Rosenberg, L. E., Mueller, P. S., Watkin, D. M. 1961. *Am. J. Med.* 31:205–15

113a. Carson, N. A. J., Scally, B. G., Neil, D. W., Carré, I. J. 1968. *Nature* 218:679

113b. Simell, O., Visakorpi, J. K., Donner, M. 1972. *Arch. Dis. Child.* 47:52–55

113c. Simell, O., Johansson, T., Aula, P. 1973. *J. Pediatr.* 82:54–57

113d. Fellows, F. C. I., Carson, N. A. J. 1974. *Pediatr. Res.* 8:42–49

114a. Mudd, S. H., Irreverre, F., Laster, L. 1967. *Science* 156:1599–602

114b. Johnson, J. L., Rajagopalan, K. V. 1976. *J. Clin. Invest.* 58:551–56

114c. Shih, V. E., Abroms, I. F., Johnson, J. L., Carney, M., Mandell, R., Robb, R. M., Cloherty, J. P., Rajagopalan, K. V. 1977. *N. Engl. J. Med.* 297:1022–28

115a. Tada, K., Ito, H., Wada, Y., Arakawa, T. 1963. *Tohoku J. Exp. Med.* 80:118–34

116a. La Du, B. N., Gjessing, L. R. 1978. See Ref. III, pp. 256–67

116b. Scriver, C. R., Larochelle, J., Silverberg, M. 1967. *Am. J. Dis. Child.* 113:41–46

116c. Gaull, G. E., Rassin, D. K., Solomon, G. E., Harris, R. C., Sturman, J. A. 1970. *Pediatr. Res.* 4:337–44

116d. Weinberg, A. G., Mize, C. E., Worthen, H. G. 1976. *J. Pediat.* 88:434–38

116e. Lindblad, B., Lindstedt, S., Steen, G. 1977. *Proc. Natl. Acad. Sci. USA* 74:4641–45

116f. Tomer, K. B., Rothman, R., Yudkoff, M., Segal, S. 1977. *Clin. Chim. Acta* 81:109–17

116g. Fisch, R. O., McCabe, E. R. B., Doeden, D., Koep, L. J., Kohlhoff, J. G., Silverman, A., Starzl, T. E. 1978. *J. Pediatr.* 93:592–96

116h. Michals, K., Matalon, R., Wong, P. W. K. 1978. *J. Am. Diet. Assoc.* 73:507–14

116i. Shinohara, Y., Hasegawa, R., Ito, N., Kanayama, M., Kato, T., Wada, Y. 1979. *Acta Path. Jp.* 29:615–22

116j. Fällstrom, S.-P., Lindblad, B., Lindstedt, S., Steen, G. 1979. *Pediatr. Res.* 13:78

117a. Goldsmith, L. A. 1978. *Exp. Cell. Biol.* 46:96–113

117b. Fellman, J. H., Vanbellinghen, P. J., Jones, R. T., Koler, R. D. 1969. *Biochemistry* 8:615–22

117c. Kenneway, N. G., Buist, N. R. M. 1971. *Pediatr. Res.* 5:287–97

117d. Goldsmith, L. A., Kang, E., Bienfang, D. C., Jimbow, K., Gerald, P., Baden, H. P. 1973. *J. Pediatr.* 83:798–805

117e. Bienfang, D. C., Kuwabara, T., Pueschel, S. M. 1976. *Arch. Ophthalmol.* 94:1133–37

117f. Garibaldi, L. R., Siliato, F., DeMartini, I., Scarsi, M. R., Romano, C. 1977. *Helv. Paediatr. Acta* 32:173–80

117g. Faull, K. F., Gan, I., Halpern, B., Hammond, J., Im, S., Cotton, R. G. H., Danks, D. M., Freeman, R. 1977. *Pediatr. Res.* 11:631–37

117h. Fellman, J. H., Fujita, T. S., Roth, E. S. 1972. *Biochim. Biophys. Acta* 284:90–100

118a. Medes, G. 1932 *Biochem. J.* 26:917–40

118b. La Du, B. N., Gjessing, L. R. 1978. See Ref. III, pp. 256–67

118c. Jaiswal, R. B., Bhai, I., Daginawala, H. F., Nath, N., Nath, M. C. 1978. *Indian Pediatr.* 15:893–99

118d. Danks, D. M., Tippett, P., Rogers, J. 1975. *Acta Pediatr. Scand.* 64:209–14

118e. Niederwieser, A., Matasovic, A., Tippett, P., Danks, D. M. 1977. *Clin. Chim. Acta* 76:345–56

119a. Sass-Kortsak, A., Bearn, A. G. 1978. See Ref. III, pp. 1098–1126

119b. Chan, W. Y., Cushing, W., Coffman, M. A., Rennert, O. M. 1980. *Science* 208:299–300

120a. Robinson, B. H., Sherwood, W. G., Lampty, M., Lowden, J. A. 1976. *Pediatr. Res.* 10:371

120b. Greter, J., Hagberg, B., Steen, G., Soderhjelm, U. 1978. *Eur. J. Pediatr.* 129:231–38

121a. Stimmler, L., Jensen, N., Toseland, P. 1970. *Arch. Dis. Child.* 45:682–85

122a. Yoshida, T., Tada, K., Honda, Y., Arakawa, T. 1971. *Tohoku J. Exp. Med.* 104:305–12

122b. Ayalon, D., Chayen, R., Herzberg, M., Menachem, H., Neufeld, E., Toaff, R. 1970. *Isr. J. Med. Sci.* 6:488–93

Ann. Rev. Biochem. 1981. 50:969–96

NMR STUDIES ON RNA STRUCTURE AND DYNAMICS[1]

♦12101

Brian R. Reid

Department of Chemistry, University of Washington and Department of Biochemistry, University of Washington Medical School, Seattle, Washington 98195

CONTENTS

Perspectives and Summary

Over the last five to ten years high-resolution nuclear magnetic resonance (NMR) has increasingly been used to study the solution properties of relatively small cellular RNA molecules such as tRNA and, to a lesser extent, ribosomal 5S RNA. The two natural isotopes in RNA that are

---

[1]The following abbreviations are used: FTNMR, Fourier transform nuclear magnetic resonance; D or DHU, dihydrouridine; T or rT, ribothymidine; NOE, nuclear Overhauser effect; FID, free induction decay: CW, continuous wave.

0066-4154/81/0701-0969$01.00

amenable to NMR studies are ^{1}H and ^{31}P, and the vast majority of NMR studies on RNA to date involve proton NMR. Isotope enrichment methods are currently being developed in order to study other NMR nuclei e.g. ^{13}C, ^{15}N, and ^{19}F.

RNA molecules of biochemical interest range in size from a few dozen nucleotides to ribosomal and viral RNAs with molecular weights in the millions, but not all of these are amenable to structural investigation by NMR. Each nucleotide in RNA contains 10–12 protons, and proton NMR linewidths increase with polymer molecular weight; the large number of resonances and their broad linewidths lead to overlapping unresolved NMR spectra for molecules larger than 100–200 nucleotides. Spectral resolution is increased by the high magnetic field strengths of modern very-high-resolution NMR spectrometers operating around 500 MHz, but even with the resolution afforded by the highest available field strengths (11–12 Tesla) only selected regions at the extremities of the spectrum are sufficiently resolved to attempt resonance assignments and begin structural interpretation.

The most intensively studied regions of tRNA NMR spectra are the extreme low field region (–15 to –11 ppm), which contains ring NH or imino protons that are hydrogen bonded between complementary base pairs, and the high field region (–4 to 0 ppm), which contains methyl protons from modified bases. The methyl resonance positions are often shifted by ring-current effects from neighboring stacked bases in the native structure and can be used as local reporter groups, since they move back to their characteristic unshifted resonance positions when that region of the molecule unfolds. The low field spectrum is potentially more informative, since it contains a reporter resonance from each base pair; several resonances, including some from tertiary folding, have been assigned to specific base pairs by a variety of methods, but some assignments remain controversial.

A particularly undesirable aspect of low field spectroscopy of exchangeable protons is the enormous proton resonance of the solvent, since such studies must be carried out in H_2O (rather than D_2O). Recently, Redfield has introduced pulsed FTNMR methods that greatly suppress the H_2O resonance and allow the application of double-resonance and time-resolved techniques. Double-resonance NOE measurements have unambiguously identified guanine-uracil (GU) base pair resonances and have also assigned other low field protons. Time-resolved FTNMR techniques have been used to measure the helix-coil breathing rates of individual base pairs in isolated helices and in intact tRNA. Short helices are not fully cooperative; they are more stable in the middle and relatively labile at the ends. Time-resolved studies on selected tertiary base pairs in intact tRNA are beginning to reveal the overall dynamics of the folded structure. These fundamental investigations on assignments and dynamics are opening the way to future experiments in which the effects of other molecules such as codons, cations, drugs, enzymes etc on tRNA structure can be studied.

Background Introduction

NMR spectroscopy involves inherently weak signals; compared to other forms of spectroscopy it requires relatively high sample concentrations ($\sim$ 1 mM or higher). Also, the sensitivity to detection by NMR methods is not the same for all atomic nuclei. By far the most easily detected nucleus is ^{1}H, which is why most biochemical NMR investigations of proteins and nucleic acids involve proton NMR, although ^{31}P NMR has been used in some nucleic acid studies. The natural isotopes of carbon and oxygen do not possess nuclear spin, but prior substitution of the sample with ^{13}C (and ^{19}F via fluorouracil) has been used recently in RNA structural analysis by NMR and is discussed briefly at the end of this review.

In proton NMR, the resonance linewidth is governed by interproton dipolar relaxation, the rate of which usually depends on the rotational correlation time, or tumbling time, of the entire molecule. In general, this tumbling time is directly related to the molecular weight, thus the linewidth is a linear function of polymer size. (This relationship breaks down for elongated rod-like molecules or for very flexible polymers with rapid local motion that is independent of the global tumbling.) There are several general textbooks on biochemical applications of NMR that discuss linewidths and relaxation (82, 83). Hence for high-molecular-weight samples the proton resonances are generally too broad to be resolved. The tumbling time and linewidth can be reduced by heating the sample, but biochemists like to work close to physiological temperature where the molecule carries out its biological function. The spectral resolution can also be increased by higher magnetic field strength; however, even at very-high-resolution (300–500 MHz) the maximal tractable polymer size is $\sim$ 40,000 daltons where linewidths of carbon protons are around 20 Hz at 37°C. The major RNA molecules of biological interest in this size range are tRNA and 5S RNA, and RNA NMR studies have focused principally on these molecules, especially the former.

The field of RNA proton NMR is approximately ten years old and was the topic of several reviews, now somewhat outdated, in the mid-1970s (1–4). In addition, several more general articles on nucleic acids have been written that include sections on NMR of RNA (5–10). There are some relatively recent research articles and reviews in the specific area of NMR studies on transfer RNA (11–13), and a recent review by Schweizer (14) updates the general area of RNA NMR, including studies on oligonucleotides. Many of the above articles are directed more toward a spectroscopic audience than a biochemical one; in contrast this review is aimed more at the biochemist already familiar with many aspects of tRNA, but perhaps not so familiar with the techniques and advantages of modern high-resolution NMR, and it will of necessity be somewhat restricted. The most significant development in the last two to three years has been the ability to carry out double resonance and time-resolved FTNMR experiments in H_2O solvents, and the second half of this review focuses on the application of these more recent techniques and the results obtained from them.

Types of Protons and Resonance Positions

An average tRNA or 5S RNA molecule contains approximately 1000 protons including ribose CH, ribose OH, aromatic base CH, base exocyclic NH_2, base ring NH, and CH_3 groups from methylated nucleosides. The range of chemical shifts for these protons covers the spectral region from −15 to 0 ppm, but they are far from evenly distributed. The low field region from −15 to −9 ppm contains the base ring NHs of G and U, which generally resonate below −11 ppm when they are hydrogen bonded to other bases (there are typically about 30 protons in this region in a tRNA spectrum). The high field region from −4 to 0 ppm contains methyl and aliphatic resonances from modified nucleosides and usually has an intensity corresponding to 6–30 protons. The midfield region in between contains the remaining several hundred protons of which the ribose resonances are up around −5 ppm. The spectrum can be simplified somewhat by working in D_2O, thus substituting all the exchangeable NH_2, OH, and ring NH protons for deuterons (15, 16); this eliminates the low field spectrum, reduces the midfield spectrum, and leaves the high field "methyl" region unaffected. Even in D_2O the midfield region is too crowded with an excessive number of overlapping nonexchangeable protons to be amenable to detailed interpretation, although Schmidt & Kastrup have succeeded in resolving 15 single protons out of the 89 aromatic CH resonances (−9 to −7 ppm) in *Escherichia coli* valine tRNA (17).

THE HIGH FIELD METHYL SPECTRUM (−4 TO 0 PPM)   The alkyl and methyl protons of modified bases are relatively straightforward to assign since their resonance positions in the random coil state are the same (or very similar) as the chemical shifts in the model methylated nucleosides, which are usually known. Hence a common strategy has been to monitor the high field spectrum of tRNA at several temperatures up to ∼ 90°C; the assigned peaks are then traced back to their native chemical shifts by analyzing the spectra in reverse order. Once assigned, these resonances can be used as reporters to monitor intermediate structures and local unfolding by plotting resonance position (which changes due to loss of ring-current effects during unstacking) as a function of temperature. Kastrup & Schmidt have used such an approach to monitor conformational transitions in the DHU loop (D17 methylene), the anticodon loop (m⁶A37 methyl), and the ribothymidine loop (rT54 methyl) of *E. coli* valine tRNA (18, 19). The former two signals are from unpaired single-stranded residues in loop regions, whereas the latter residue is involved in a tertiary base pair in the crystal structure of yeast phenylalanine tRNA (20, 37, 38) and probably all other tRNAs as well (21, 22). Temperature-dependent chemical shift plots carried out in low salt (0.045 M free Na⁺) revealed transitions centered at 55° (D17), 58° (m⁶A37) and 67°, (rT54), which were interpreted to reflect unfolding of the DHU stem loop, anticodon stem loop, and rT stem loop, respectively (18). However, these chemical shift changes actually monitor events only in the loop region and reveal no direct information about the helical stem. The rT resonance and the DHU resonance indicated a slow exchange conformational equilibrium (long lifetime in each state compared to the chemical shift difference, which results in distinct resonances from each state), whereas the anticodon loop signal reflected a fast exchange two-state transition. Furthermore, the rT signal exhibited interesting multiple states at 1.9, 1.8, and 1.25 ppm; an additional peak at 1.0 ppm was initially attributed to contaminant (18), but subsequently proved to be a fourth state (actually the native state) of rT54 (19). The presence of multiple states at physiological temperature was attributed to "non-native" conformers peculiar to the low salt conditions, because they largely disappeared above 0.2 M Na⁺, under which conditions a single rT54 methyl resonance was found at −1.0 ppm (19). At 0.2 M Na⁺ the alternate conformations can be made to reappear, but only at higher temperature; however, they can be trapped by slowly lowering the temperature back to room temperature, whereas rapid quenching in the presence of magnesium largely abolishes these alternate states (19). This hysteresis emphasizes the importance of the past history of the sample and starting conformation(s) in analyzing unfolding.

The above approach to conformational analysis of tRNA suffers somewhat from not having enough "handles" reporting events from the many regions of interest in tRNA; this is especially true for tRNA species containing only one or two methylated nucleosides. Furthermore, some methyl resonances are not useful reporters in that they do not shift during the unfolding process or are buried downfield among the ribose protons. In general, eukaryotic tRNAs contain a much higher level of methylation and other modifications than prokaryotic tRNAs, thus they present more high field spectroscopic handles with which to monitor conformational events. For instance, yeast phenylalanine tRNA contains resolved aliphatic resonances from modified nucleosides in the D stem, D loop, anticodon stem, anticodon loop, extra loop, rT stem, and rT helix. The thermal unfolding of this molecule has been investigated via the chemical shifts of these methyl groups (23–25). The added complexity of eukaryotic tRNA high field spectra, while permitting observation of more regions of the molecule, also increases the possibility of misassigning transitions. For instance the D16,D17 resonance (C5H) and the m_2^2G26 resonance coalesce at -2.6 ppm at intermediate temperature and separate out at lower temperature. The component that moves upfield to -2.4 ppm was attributed to D16,D17 by Kan et al (23, 24), and the transition was therefore interpreted to reflect events occurring in the DHU loop. However, by means of decoupling the D16,D17 C6H, Robillard et al (25) were able to show that the D16,D17 resonance hardly moves from -2.6 ppm at all and that it is the $_2$G26 resonance that crosses over to -2.4 ppm; hence the observed transition is reporting events at the DHU stem-anticodon stem junction rather than in the DHU loop. The NMR transitions observed for the nonexchangeable methyl protons can be directly related to independently determined thermodynamic parameters derived from optical studies in order to cross-check the assignments and interpretations. Kan et al (23, 24) carried out parallel UV hyperchromicity studies in the same buffer as their NMR studies, whereas Robillard et al (25) adjusted their buffers to the same conditions used in the optical studies of Romer et al (26, 27) so that these T_m values and temperature-jump relaxation times could be directly related to the NMR results. In the absence of magnesium, the first unfolding events reflect changes in the acceptor helix, tertiary structure, and anticodon helix (20–40°C), followed by the rT helix transition (45–55°C); the most pronounced transitions are those of m_2^2G26 and D16,D17 at both ends of the DHU helix, which reflect the unfolding of this helix at the surprising temperature of 60–65°C (25). In the presence of magnesium the unfolding of secondary and tertiary structure becomes unresolved in a cooperative transition at $\sim$ 74°C (24). Although much important information is beginning to emerge from high field NMR spectroscopy this spectral region contains only a few signals that directly report events in the base paired helical regions.

RESONANCES FROM BASE-PAIR HYDROGEN BONDS   The hydrogen bonded protons of complementary base pairs are, of course, solvent exchangeable and cannot be observed in D_2O where most "aqueous" NMR is carried out. Hence one must work in H_2O to detect these resonances, and this causes special problems in detection and instrumentation that are discussed later. Since solvent exchange from the exposed coil state is usually rapid, a further restriction is that these base pairing resonances will only be observed when their helix lifetimes are long compared to the chemical shift difference between the bonded and nonbonded states (for modern high-resolution spectrometers this corresponds to base pair lifetimes greater than 2–3 msec). However, compensating advantages are that the ring NH hydrogen bonds resonate in the uncrowded extreme low field region of the spectrum between –11 and –15 ppm and there is only one such signal per base pair (U N3H in an AU pair and G N1H in a GC pair). Furthermore, these low field resonances are derived from the double helical parts of the RNA and can, under appropriate conditions, reveal dynamic lifetime information about these helices.

The hydrogen bonded ring NH resonances of tRNA were first observed in the early 1970s by Kearns, Shulman, and collaborators (28, 29), and the early results have been reviewed by Kearns & Shulman (1) and by Kearns (2). The majority of tRNA species studied by NMR (class I) contain 20 secondary base pairs, and the integrated low field intensity was initially interpreted to contain 20 protons; however these interpretations are undoubtedly wrong. The first evidence that the spectra actually indicated the presence of approximately seven additional resonances from tertiary base pairs was presented by Reid et al (30), Daniel & Cohn (31), and Reid & Robillard (32). Although this point remained the subject of some controversy (33–35), the analysis of extremely pure tRNA samples at higher field strengths, combined with computer analysis and simulation of the resolved spectra, has established beyond doubt the existence of low field resonances between –15 and –11 ppm from approximately seven extra base pairs in several class I tRNAs (36). As examples Figure 1 shows the low field spectra of E. coli tRNA$_1$^{Val} and tRNA^{Ile}, both of which contain 20 base pairs in their secondary cloverleaf sequence. The isoleucine tRNA spectrum contains 23 partially resolved peaks between –15 and –11 ppm, but the peaks at –11.8 and –13.2 ppm obviously contain 2 protons, and the peak at –12.8 ppm contains at least 3 protons. Both spectra contain 4 peaks between –11 and –9 ppm including the narrow aromatic resonance from C8H of m^{7}G46 at ∼–9.1 ppm (see below). Perhaps the best evidence for the existence of the controversial tertiary resonances is that several of them have recently been assigned (a spectacular feat if they do not exist!).

ORIGIN OF LOW FIELD RESONANCES   Among the ∼ 27 low field resonances observed in most class I tRNA spectra the first 20 or so are obviously derived from the hydrogen bonded ring NHs of the 19–21 Watson-Crick base pairs in the cloverleaf secondary structure of such tRNAs. The origin of the extra seven resonances between –15 and –11 ppm is difficult to determine by NMR methods alone, and progress in this area has been greatly facilitated by X-ray crystallographic structure determination. The three-dimensional structure of yeast tRNA^{Phe} has been determined in two crystal forms by four laboratories, and the structural aspects of tertiary base pairing in this molecule have been reviewed recently (20, 37, 38). Although

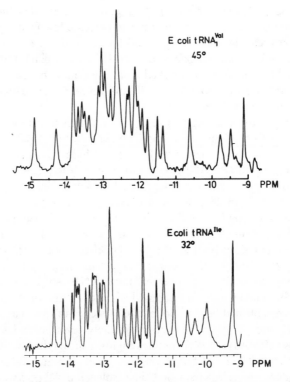

Figure 1   The 360-MHz low field NMR spectra of *E. coli* tRNA$^{Val}_1$ at 45°C (*upper*) and *E. coli* tRNA^{Ile} at 32°C (*lower*). The region between −11 and −15 ppm contains hydrogen bonded ring NHs; each proton resonance is from a different base pair in the tRNA except for the two resonances from the GU pair (−11.35 and −11.95 ppm in tRNA^{Val}; −10.9 and −11.8 ppm in tRNA^{Ile}). Both samples contained 6 mg of tRNA in 0.2 ml of buffer. The upper spectrum was obtained using rapid sweep correlation spectroscopy (2500 Hz in 0.8 sec) after 17 min of signal-averaging; the lower spectrum was obtained using Redfield 214 FTNMR with a 0.37 msec pulse and 10 min of signal averaging. The lower spectrum has been resolution-enhanced by Gaussian multiplication. The tRNA$^{Val}_1$ sample was fully thiolated at uridine 8 (note the full proton at −14.9 ppm) whereas tRNA^{Ile} contains no s^{4}U.

the structures of other tRNA species have been solved at lower resolution, the yeast tRNA^{Phe} molecule remains the only refined high-resolution structure available and hence assumes the role of a "reference tRNA structure." Klug et al (21) and Kim et al (22) have pointed out that most other class I tRNAs contain either identical base pairs or coordinated base pair changes at the strategic tertiary base pairing positions, which indicates that the crystal structure is probably a good generalized model for other tRNA species. Only the tertiary base pairs involving a hydrogen bonded ring NH are expected to generate low field NMR resonances. These hydrogen bonds involve:

1. 8–14 (reverse Hoogsteen pair involving U8 in eukaryotic tRNAs and s⁴U8 in most bacterial tRNAs);
2. 54–58 (reverse Hoogsteen pair involving the unique T54 common to most tRNAs);
3. 46–22 (GGC triple connecting the variable loop to the DHU stem via m⁷G46);
4. 19–56 (a Watson-Crick GC pair connecting the DHU loop to the rT loop);
5. 15–48 (a reverse Watson-Crick GC connecting the DHU loop and variable loop);
6. 26–44 (a twisted AG "base pair" at the junction of the anticodon helix).

Additional tertiary interactions that might possibly generate extra low field resonances are Ψ55–P58 (a ring NH-phosphate hydrogen bond) and possibly G18–Ψ55 (20, 37, 38). Furthermore, most tRNAs contain a GU pair in their secondary structure; in wobble geometry this might be expected to generate two low field resonances. Thus there is no dearth of potential candidates for the extra resonances observed in well-resolved low field spectra.

ASSIGNMENT OF SECONDARY RESONANCES The complexity of intact tRNA spectra can be simplified by analysis of isolated helical hairpin fragments produced by controlled chemical or enzymatic cleavage, and this technique has been applied to several tRNA species (39–41). Ring current shifts from neighboring bases are undoubtedly responsible for the spectral resolution of resonances in fragments and intact tRNA, but the resulting resonance position depends on two unknowns, namely the starting inherent resonance position of an AU or GC pair and the net upfield shift from stacked neighbors. Thus there are several solutions for the two unknowns in this equation each of which can empirically rationalize the position of a given base pair resonance, but each rationalization gives rise to a different set of values for the neighboring ring current shift (2, 43, 44). Ring current shifts are both distance- and angle-dependent, and Reid et al (41) have attempted to lower the ambiguity by *assuming* that helical fragments and intact tRNA in solution maintain the 11-fold RNA geometry observed in the crystal structure. With this assumption the net shifts predicted for 11-fold geometry by Arter & Schmidt (42), using Pullman ring current values, leads to starting resonance positions of −14.35 ppm for AU pairs and −13.45 ppm for GC pairs, and accounts for the observed fragment spectra reasonably well (41). Analysis of the intact tRNA spectrum involves the assumption that it can be approximated by the sum of the component helical parts; this has been disputed by P. D. Johnston and A. G. Redfield (data submitted for publication) and by Sanchez et al (73) who have claimed that, in at least one case involving base pair 6, tertiary folding in intact tRNA causes variations in donor-acceptor geometry, which modify the starting resonance position of this base pair. The discrepancy between different sets of theoretically predicted assignments (see next paragraph) and between theoretical and experimentally calibrated ring current shift "rules" may be as great as 0.3 ppm; if the error level is this high it is, unfortunately, large enough to transpose base pair assignments between adjacent multiple-proton peaks, especially in the more crowded region between −14 and −12 ppm.

In a completely different approach Robillard et al (45) have carried out a computer calculation of the ring current effects from all nucleotides on each ring NH, using the phenylalanine tRNA X-ray coordinates. A good fit to the experimental spectrum is obtained only when unshifted AU and GC offsets of −14.35 and −13.54 ppm were used, and the resulting computer spectrum indicated no secondary base pairs below −14 ppm (45). Geerdes & Hilbers (46) and Kan & Ts'o (47) have used the same approach, but their assignments differ from those of Robillard et al (45). However, impressive support was given to the Robillard calculations by the fact that they generated a remarkably close facsimile of the observed spectrum of E. coli tRNA^{Val} when this sequence was substituted into the program (48). Nevertheless, there is obviously room for new experimental approaches to assigning secondary base pair resonances.

ASSIGNMENT OF TERTIARY RESONANCES   The tertiary folding of tRNA involves many nonstandard interactions e.g. base triples, reverse Hoogsteen, and reverse Watson-Crick pairs etc (20, 37, 38). This creates special problems in calculating their resonance positions; e.g. the two protons below −14 ppm in yeast tRNA^{Phe} were calculated to be AU6 and AU12 by Kan & Ts'o (47), whereas the Robillard calculations assign them to the tertiary Hoogsteen pairs 8–14 and 54–58 (45). The assignments based on calculated spectra from three different laboratories have been compared and discussed in detail (49). Experimental assignment of tertiary resonances has been attempted by several laboratories and agreement has been reached in some cases.

The resonance at −14.8 to −14.9 ppm has been confidently assigned to the s^{4}U8–A14 reverse Hoogsteen pair in species that are thiolated at residue 8. This assignment is based on the observation that this resonance moves upfield to ∼ −14.3 ppm upon dethiolating s^{4}U8 (compare the valine tRNA spectra in Figures 1 and 2), and also leads to the assignment of the 8–14

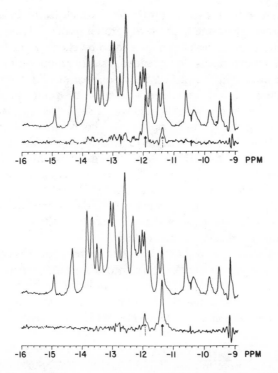

Figure 2 Identification of the resonances from the GU base pair in *E. coli* tRNA$^{Val}_1$ via the ring NH-ring NH NOE. The upper spectra are controls that were preirradiated in the valley at −11.6 ppm. The difference spectrum below each control was obtained by subtracting the spectrum after preirradiating at −11.95 ppm (*upper*) or −11.35 ppm (*lower*) from the control spectrum. Preirradiation was carried out for 0.1 sec with a 1 msec delay before the 0.37 msec observation pulse. The tRNA$^{Val}_1$ sample was only 50% thiolated at residue 8 (note the half-proton resonance at −14.9 ppm with the other half-resonance appearing as a shoulder at −14.3 ppm). [Figure taken from (55).]

resonance at −14.3 ppm in species containing unmodified uridine at position 8 (30, 31, 50, 51). This is perhaps the most definitive and unanimous tertiary assignment.

Hurd & Reid used chemical removal of the unique m^{7}G residue to assign the m^{7}G46–G22 tertiary interaction at −13.35 ppm in yeast tRNA^{Phe} and *E. coli* tRNA^{Val} (55). In the course of these studies they also established that the narrow aromatic resonance at ∼ −9.1 ppm is the C8H of m^{7}G46, and thus created a built-in measure of the extent of m^{7}G removal, which was corroborated by chemical analysis. This assignment was disputed by Salemink et al who used a similar approach but claim the m^{7}G46 imino resonance is at −12.5 ppm in yeast tRNA^{Phe} (56). However their difference spectra appear to be mis-scaled by 10–15%, which led to an apparent one proton difference spectrum in the seven-proton peak at –12.5 ppm; also the extent of m^7G removal was not documented by chemical analysis, and the spectra were unfortunately not extended far enough upfield to monitor the extent of removal via the m^7G C8H resonance at –9.1 ppm. In *E. coli* tRNA^{fMet} the m^7G is in a markedly different environment, and Hurd & Reid showed that in this case the removal of m^7G was correlated with the loss of the resonance at –14.55 ppm (55). This is in contrast to the –13.4 ppm assignment of Daniel & Cohn (57) based on s^4U8 spin labeling and comparison with fMet-3 tRNA (which contains no m^7G); however the fMet-3 sample used was not purified by the authors themselves and was not characterized with respect to m^7G content (57). Bolton & Kearns (2, 35) claim, on very weak evidence, that the m^7G tertiary resonance is not present in the spectrum. In the light of the above conflicts the assignment of this tertiary resonance remains controversial.

The G15–C48 reverse Watson-Crick tertiary base pair has been problematical and is not yet satisfactorily assigned. From the relaxation effects of paramagnetic ions including Co^{2+}, which binds to G15 in the crystal structure (79), Hurd et al conclude that the resonance at –12.25 ppm is the G15–C48 imino proton in *E. coli* tRNA^{Val} (59). This assignment has neither been disputed nor corroborated.

The G19–C56 Watson-Crick tertiary resonance is generally assigned at, or close to, –12.9 ppm in most tRNAs, and has not been disputed despite (or perhaps because of) the lack of direct evidence. This is the only tertiary interaction involving a normal Watson-Crick pair, and its deductive assignment is based on the calculated upfield shift of ~ 0.6 ppm (45, 49) combined with the standard GC offset of $\sim –13.5$ ppm, and the observation that an extra (nonsecondary) resonance is always observed at this position (41). Some support has been lent by thermal unfolding studies (25) and by exchange dynamics (52, 70, 71). Although highly reasonable, this assignment cannot be regarded as definitive in the absence of direct evidence.

The T54–A58 tertiary interaction is a reverse Hoogsteen pair and its assignment is quite controversial. The immediate environment surrounding T54–A58 is identical in yeast tRNA^{Phe} and *E. coli* tRNA^{Val}, so that its resonance position should be the same in both spectra. Reid et al (41) have argued that the five AU pairs in *E. coli* tRNA^{Val} should have similar resonance positions around –13.7 ppm; since the –14.9 ppm resonance has already been assigned to the 8–14 Hoogsteen, and the lone resonance at –14.3 ppm is too low field to be a GC pair, the only remaining candidate is T54–A58, since there are only two UA-type tertiary interactions in the crystal structure. The early loss of the –14.3 ppm resonance (and the 8–14 Hoogsteen) as the temperature is raised (41, 51), and its rapid exchange dynamics in both yeast tRNA^{Phe} (52, 70, 71) and *E. coli* tRNA^{Val} (R. E. Hurd and B. R. Reid, unpublished observations; P. D. Johnston and A. G. Redfield, unpublished observations) are consistent with this assignment; the Robillard calculations of the total ring current shift on this proton place it at −14.4 ppm when the reverse Hoogsteen unshifted offset of −14.9 ppm is used (45, 48, 49). In addition, Hurd & Reid (51) have examined the fragment encompassing residues 47–76, which contains T54 and A58 in an intact rT loop and stem; the spectrum contains a resonance at −14.35 ppm that cannot be assigned to any secondary base pairs in the stem (which is devoid of AU pairs).

However, there is evidence against the −14.3 ppm assignment for T54–A58. Kearns & Bolton (53) have observed that ethidium bromide abolishes the −14.3 ppm resonance in *E. coli* tRNA^{Val} without shifting the T54 methyl resonance at −1 ppm, and Reid has observed that the *E. coli* tRNA species Ala-1B and Phe (which contain the same rT loop as tRNA^{Val} and yeast tRNA^{Phe}) contain no resonances near −14.3 ppm (unpublished observations). The ethidium result might be due to breaking the T54–A58 hydrogen bond without directly stacking on T54, but the comparative data are difficult to rationalize. Very recently Sanchez et al (73) and P. D. Johnston and A. G. Redfield (manuscript submitted for publication) have presented additional evidence on this point; they have used double-resonance NOE techniques (described in the next section on FTNMR) that indicate that the −14.3 ppm resonance is a secondary Watson-Crick AU pair rather than a reverse Hoogsteen tertiary pair. Thus the T54–A58 assignment remains controversial and further work is needed to resolve this problem.

The best estimates of the total proton intensity between −11 and −15 ppm are 27 or 28 for *E. coli* tRNA^{Val} and 26 or 27 for yeast tRNA^{Phe} (36, 41). Both tRNAs contain a secondary GU pair, and there is now incontrovertible NOE evidence that the −11 to −15 ppm region contains two protons from the tRNA^{Val} GU pair (55, 71) and one proton from the yeast tRNA^{Phe} GU pair (70, 71). In combination with the 20 secondary resonances and the 5 tertiary resonances discussed above it now appears that, at most, only 1 other proton may be present in this spectral region. Of the three remaining crystallographic candidates the 24–44 "base pair" has been experimentally displaced upfield to the −9 to −11 ppm region (P. D. Johnston and A. G. Redfield, data submitted for publication), and this spectral region probably also contains the Ψ55 ring NH hydrogen bonded to P58 that may even have exchanged out of the spectrum as a result of its accessibility to solvent; if there is one more resonance to be accounted for in the −11 to −15 ppm region it is probably G18-Ψ55, and almost all tRNAs contain an unassigned resonance at ∼ −11.5 ppm, which is a likely position for such a "GU-type" interaction. From the foregoing it is apparent that reasonable progress in assignment has been made by extensive use of a variety of chemical and biochemical approaches, but that ambiguities still remain. Future progress in this area will probably involve spectroscopic methods using FTNMR techniques that have recently been developed for the special case of H_2O solutions.

FTNMR in H_2O Solution

Fourier transform NMR, with its advantages in sensitivity, time resolution, and double-resonance capabilities has not, until recently, been applicable to exchangeable proton studies in H_2O because of computer-digitizer dynamic range limitations. In FTNMR all resonances are excited simultaneously by a short radio frequency pulse encompassing all proton frequencies in the sample. The resulting time domain signal is a free induction decay (FID) containing the mixed frequencies of the sample resonances, which is Fourier transformed in a computer to reveal each resonance in the more normal frequency domain (60, 61). For dilute samples around 1 mM the pulse-FID sequence is repeated hundreds of times to generate reasonable signal-to-noise ratios in the final spectrum (signal averaging). The signal must first be digitized to carry out these computer manipulations, and the H_2O proton signal (110 M) obviously dominates the FID. In fact a normal 12-bit digitizer (4096 points in the vertical axis), when filled with a 110 M signal, will not register signals much below about 25 mM (i.e. 1 mM solute signals are not digitized and are lost). Even with a 16-bit digitizer, a computer with a word size of 20 bits will overflow memory after only 16 (2^4) pulses, which is insufficient signal averaging. For these reasons, until about 1977, spectra in H_2O were usually collected by CW frequency-sweep or rapid scan correlation spectroscopy methods (62), which avoided sweeping through the H_2O resonance. Such methods do not easily lend themselves to double-resonance techniques by which assignments can often be made, or to time-resolved experiments in which the dynamic aspects of the molecule can be studied.

This problem has been tackled and elegantly solved by the introduction of a new phase-shifted pulse (the 214 pulse) by Redfield and colleagues (63–67). The Redfield 214 pulse is a selective excitation pulse the most important aspect of which is that its frequency transform falls to zero amplitude at the H_2O frequency thus exciting the water resonance $< 1\%$ of a normal FT pulse. More recent "magic knob" modifications of the 214 pulse (67) lead to reductions of over 1000-fold for the H_2O amplitude in the FID, so that the effective size of the solvent signal can be reduced to ~ 100 mM, thus placing 1 mM solute signals well within the dynamic range of 12-bit digitizers and allowing at least 256 (2^8) pulses to be signal averaged in a 20-bit word computer. The length of the 214 pulse is tailored to fit the offset between the pulse carrier frequency and the water frequency, and is usually around 0.4 msec, which permits time resolution of processes in the millisecond range.

Assignments from Double-Resonance: The NOE

The nuclear Overhauser effect is a particularly useful double-resonance technique for biological polymers in which the magnetic saturation of a selectively irradiated proton resonance is transferred, via cross-relaxation, to other resonances from protons in the immediate neighborhood (68, 69). The marked distance dependence (r^6) limits the first-order NOE (in the absence of further spin diffusion) to distances in the 1–4 Å range, and it is mainly used to relate pairs of resonances that may be greatly separated in the spectrum but are very close to each other in space.

GU ASSIGNMENT VIA NOE   The vertical distance of 3.4 Å combined with the lateral displacement of a normal RNA helical pitch is such that the ring NHs of adjacent base pairs are separated by more than 4 Å and, as expected, do not exhibit ring NH-ring NH NOEs. Johnston & Redfield (70) noted that two low field protons in yeast tRNA^{Phe}, located at −11.8 ppm and −10.4 ppm, nevertheless did show mutual NOEs, and concluded that both ring NHs must be in the same base pair, i.e. the GU pair at position 4. *E. coli* tRNA^{Val} also contains a secondary GU pair (at position 50), which also has been studied by Johnston & Redfield (71) and by Hurd & Reid (55); in this case the NOE-related pair of imino protons are located at −11.95 ppm and −11.35 ppm. Figure 2 shows the results from such an experiment presented as difference spectra in which the preirradiated spectrum is subtracted from the control so that the selectively saturated line appears as a positive full proton resonance. No other low field ring NHs show mutual NOEs, but it is obvious that the −11.95 ppm and −11.35 ppm resonances exhibit partial cross-relaxation (NOE).

Further proof that the two imino resonances related by mutual NOE are in fact derived from the wobble GU pair comes from the observation that two tRNAs of different sequence and different low field spectra, but with their GU pair in identical nearest neighbor environments, exhibit NOE-paired resonances at identical spectral positions (K. A. Jones and B. R. Reid, unpublished observations). Although their presence in the spectrum was not even acknowledged a few years ago (2, 4), the recent use of the NOE has now turned the GU resonances into perhaps the most reliable base pair assignment. Despite the ease of experimentally identifying GU resonances via the NOE, Geerdes & Hilbers have published a set of ring current shift rules for theoretically predicting GU resonance positions based on nearest neighbor sequence (72). However in cases such as *E. coli* tRNA^{Phe} and tRNA $^{Ala}_{1B}$ the observed GU resonances differ by 0.5–0.7 ppm from their predicted positions (K. A. Jones and B. R. Reid, unpublished observations); Geerdes & Hilbers assumed that the helical distortion required to form a wobble pair is the same for all GC pairs regardless of nearest neighbors, but this is apparently not the case (i.e. either the G or the U, or perhaps both, can move laterally to assume wobble geometry depending perhaps on the particular stacking sequence).

SECONDARY RESONANCE ASSIGNMENT VIA NOE   Although the ring NHs of adjacent base pairs are normally too far apart to exhibit mutual NOEs, other protons located vertically above (3.4Å) and methyl protons on the same nucleoside are within NOE range of some ring NHs (see Figure 3). P. D. Johnston and A. G. Redfield (manuscript submitted for publication) have used the NOE from the m^2G10 methyl resonance at –2.75 ppm to assign the ring NH of base pair 10 at –12.63 ppm in yeast tRNA^{Phe}. The fact that Robillard et al (45) previously calculated from the crystal coordinates that this resonance should be at –12.55 ppm lends credence to the theoretical calculation approach to assignments when carried out properly. Also, the experimental approach using a fragment containing intact anticodon and DHU helices as well as fragment-calibrated ring current shifts assigned base pair 10 at –12.7 ppm in yeast tRNA^{Phe} and at –12.6 ppm in E. coli tRNA^{Val} (41).

Sanchez et al (73) have used the fact that the NOE acceptor from an AU ring NH donor is the adenine C2H (narrow line), whereas the acceptor from a GC ring NH donor is the guanosine amino group (broad line), to discriminate between AU and GC resonances.

TERTIARY RESONANCE ASSIGNMENT VIA NOE   As mentioned earlier, the reverse Hoogsteen T54–A58 ring NH and the m^7G46–G22 ring NH have been assigned by Hurd & Reid (51, 55) at –14.3 ppm and –13.35 ppm, respectively, in both E. coli tRNA^{Val} and yeast tRNA^{Phe}, but these assignments have been disputed (53, 56). Sanchez et al have used double-resonance methods to address the 54–58 assignment (73). The observed sharp NOE at –7.8 ppm upon irradiating yeast tRNA^{Phe} at –14.3 ppm should be the C8H of m^1A58 if the –14.3 ppm resonance is in fact the 54–58 reverse Hoogsteen (see Figure 3). However substitution of purine C8 protons with deuterium by heating at 90°C in D_2O (or more recently by growth of an auxotroph on C8 deuteroadenine), and subsequent reisolation in H_2O, did not abolish the –14.3 to –7.8 ppm NOE as expected (73). The most reasonable conclusion is that the –7.8 ppm NOE is actually an adenine C2 proton, which indicates that the –14.3 ppm resonance is a standard Watson-Crick AU pair rather than the 54–58 Hoogsteen pair. Sanchez et al (73) thus

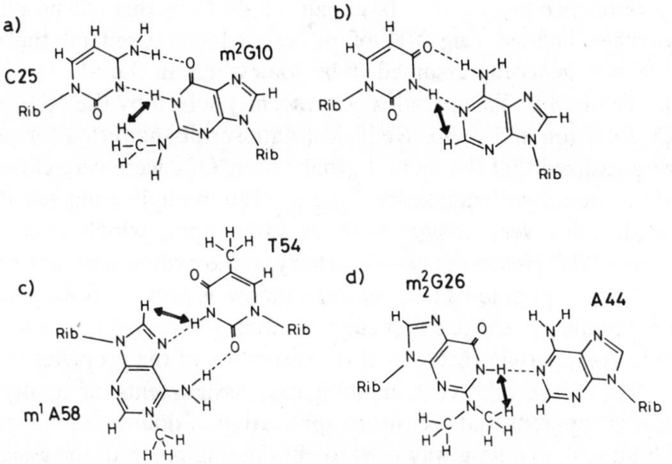

Figure 3 Diagrammatic representation of secondary and tertiary base pairs observed in the crystal structure of yeast phenylalanine tRNA: (*a*) m²G10-C25, (*b*) standard Watson-Crick AU pair, (*c*) T54-m¹A58, and (*d*) m²₂G26-A44. In each case the proximity of the hydrogen bonded ring NH to adjacent aromatic or methyl protons is shown by a double-headed arrow. (See text concerning assignment via the NOE.)

tentatively conclude that the -14.3 ppm resonance must be UA6 with the 54–58 tertiary resonance further upfield in the group of protons around -13.8 ppm, where it was assigned by Kearns (53). However, they have not been able to observe a demonstrable NOE to C8H from the low field spectrum, and it is unfortunate that yeast tRNA^{Phe} contains m¹A at position 58 instead of A; m¹A has peculiar chemical properties and may exhibit anomalous exchange behavior (54). Although the evidence of Sanchez et al points against assigning T54–A58 at -14.3 ppm it must remain tentative until it is found elsewhere by H/D elimination of the NOE, and analysis of the final sample proves that m¹A58–C8H has in fact been replaced by deuterium and retained the isotope. The NOE results on m⁷G by P. D. Johnston and A. G. Redfield (data submitted for publication) are consistent with, but do not prove, the m⁷G46–G22 assignment of Hurd & Reid (55).

The NOE approach has been very informative concerning the m²₂G26–A44 tertiary interaction, which appears, from the crystal structure, to involve a ring NH-ring N hydrogen bond between propeller-twisted purines (20, 37, 38; see Figure 3). Using subtractive inference and comparative spectra of different tRNAs, the extra resonance observed near -12.3 ppm was initially deduced by Reid & Hurd to be base pair 26–44 (3). However it was subsequently displaced from this position by later evidence from Hurd el al, using paramagnetic ions, which indicated that the -12.3 ppm tertiary resonance belonged to base pair 15–48 (59); this left no place for the hydrogen bonded ring NH of m_2^2G26 which, based on the crystal structure, was generally assumed to be somewhere in the −11 to −15 ppm spectral region (49). This paradox was recently solved by the NOE studies of P. D. Johnston and A. G. Redfield (manuscript submitted for publication) who realized that the methyl groups of m_2^2G26 were very close to the ring NH in question; irradiation of the m_2^2G26 methyl resonance at −2.45 ppm produced a very strong NOE at −10.4 ppm, which must be the m_2^2G26 ring NH. Hence the 26–44 tertiary imino proton does not resonate in the −11 to −15 ppm low field spectral window as previously assumed, and its unexpectedly upfield position suggests that it is probably not hydrogen bonded to A44, perhaps because of the distortion of the propeller twist. By making use of the relatively unambiguous assignments of methyl resonances, it is apparent that the future application of double-resonance NOE methods should go a long way toward eliminating much of the assignment ambiguity in the potentially more informative ring NH spectral region; the use of these methods in H_2O was made possible by the introduction of the Redfield 214 selective observation pulse (67).

RNA Dynamics Using FTNMR

Helix-coil dynamics can, in principle, be studied using steady-state CW-sweep NMR by analysis of the exchange contribution to the ring NH linewidth (74, 75). However such studies are only applicable in the restricted temperature ranges where broadening can be observed, and they are inherently less accurate than time-resolved FTNMR methods. The ring NH protons in a RNA double helix are not directly accessible to water, and double resonance FTNMR methods are beginning to reveal a great deal about helix-coil kinetics.

The helix-coil equilibrium can be described by the two rate constants k_{open} and k_{close} (the latter is obviously the greater if the structure is in the native state). The helix-coil transition is obligatory before exchange with water can occur; coil-water exchange is buffer catalyzed and can be made very rapid ($\sim 10^6$ sec^{-1}) by addition of small amounts (10–100 mM) of buffer (76, 77). If the coil lifetime is long compared to exchange from the coil state ($k_{close} \ll k_{cw} = k$ [cat] = 10^5–10^6 sec^{-1} in 10 mM buffer) then every opening event results in exchange, and the observed helix-water exchange rate is limited by, and equal to, k_{open} [For a more detailed discussion of exchange see Teitelbaum & Englander (80), Kallenbach & Berman (81), and Hilbers (75).]

Double-resonance FTNMR methods are particularly well-suited to monitoring exchange processes in the 1–1000 sec^{-1} rate range, and hence offer an exciting new approach to directly measuring k_{open} for individual base pairs with resolved ring NH resonances. After selective preirradiation of a single resonance to saturation (destruction of net magnetization) that proton will recover relatively slowly via interaction with fluctuating megnetic fields (T_1 or spin-lattice relaxation processes) in its environment (82, 83). However if, during the recovery period, that proton population is being exchanged with magnetized water protons (which are not perturbed magnetically by the 214 observation pulse or the preirradiation), recovery will be greatly accelerated; the increase directly reflects the rate of exchange with water and can be equated with k_{open} if the conditions described in the previous paragraph are met. Such experiments have been carried out on the acceptor helix base pairs of *E. coli* tRNA^{Phe} by Hurd & Reid (78), and an example is shown in Figure 4. After saturation of GC5 at −12.4 ppm the spectrum is collected at various time intervals during the recovery period, using a 0.3-msec 214 observation pulse, with the results shown. The recovery rate was observed to be 28 sec^{-1} at 62°C of which 5–6 sec^{-1} is the (temperature-independent) magnetic recovery rate, which leads to a value of 22 sec^{-1} for the helix-water exchange rate. The lack of any stimulation of this rate by increasing the buffer concentration tenfold proved that exchange was open limited, and hence k_{open} is therefore 22 sec^{-1} (78). This experiment can be extended to all six base pairs in the helix and such studies reveal a positional dependence on the opening rate; for instance CG4 opens at only 5 sec^{-1} at 58°C, whereas GC6 opens at 67 sec^{-1} (78). The whole cycle of experiments can be repeated at various temperatures, thus revealing, at least in theory, the activation energy for opening each base pair; in practice such a goal for even a small six-base pair helix would require ∼ 150 NMR spectra, each signal-averaged for about 15 min, and would yield only a moderately accurate result (four or five points on an Arrhenius plot). Nevertheless semiquantitative interpretation of the very large activation energies in the helix interior (∼ 70 kcal mol^{-1}) are consistent with no independent internal opening i.e. exchange only via sequential peeling back from the termini (78).

Saturation recovery determination of k_{open} would obviously be extremely useful in probing base pair dynamics in various regions of intact tRNA under physiological conditions. Unfortunately the interpretations of such analyses are not straightforward due to the previously discussed ambiguities in assigning the individual ring NH resonances. Undaunted by this, Redfield's group has carried out a detailed exchange analysis of all peaks in the low field spectrum of yeast tRNA^{Phe} (52, 70, 71) and, more recently, the more resolved *E. coli* tRNA^{Val} (P. D. Johnston et al, manuscript in preparation).

Interestingly, the technique can be used as a two-edged sword in that the results can be turned around and used to address the problem of identifying,

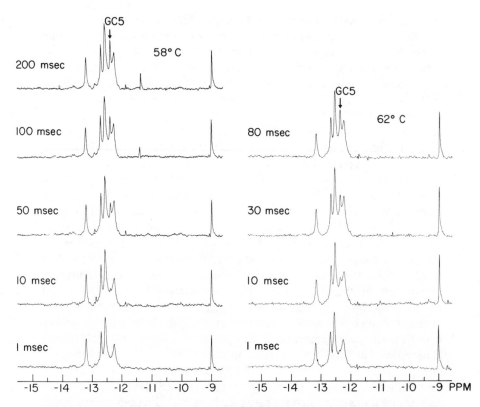

Figure 4 Measurement of k_{open} via saturation recovery for GC5 in the acceptor helix of *E. coli* tRNA[Phe]. The intact tRNA was partially unfolded in the absence of magnesium at intermediate temperature leaving only the six GC resonances of the acceptor stem, as described previously (3). The GC5 resonance was then saturated by selective preirradiation at −12.45 ppm for 0.1 sec, after which the extent of recovery was monitored at various time intervals using a 0.37 msec Redfield 214 observation pulse. The exchange rate (after correction for magnetic recovery) was found to be 6 sec⁻¹ at 58°C and 22 sec⁻¹ at 62°C; these values can be directly equated with k_{open} since exchange is open limited under these conditions (78).

without necessarily assigning, tertiary resonances based on their atypical dynamics. In the presence of magnesium, exchange was too slow to be measured accurately, but at moderately low ionic strength in the absence of magnesium the saturation recovery rates were exchange-dominated at around 40°C which allows the individual opening rates of some resonances to be measured (52, 70, 71). In the case of yeast tRNA[Phe] the large peaks at −13.25 and −12.5 ppm contain five base pairs and seven base pairs, respectively, and are too complex to extract rates for single base pairs from their multiphasic recovery behavior. However, the single resonances at −14.4 and −14.2 ppm both exhibit anomalously rapid (>60 sec⁻¹) water exchange at 42°C; the latter is the 8–14 tertiary pair, but the former, although originally assigned to the tertiary 54–58 pair (49, 51, 71) has been reassigned to UA6 by Sanchez et al (73), based on its NOE behavior. In fully thiolated *E. coli* tRNA^{Val} there is only one resonance between –14.5 and –14.0 ppm (see Figure 1), and it also exhibits rapid solvent exchange (R. E. Hurd and B. R. Reid, unpublished observations; P. D. Johnston et al, manuscript in preparation): however its assignment to T54–A58 (49, 51) has been weakened by the yeast tRNA^{Phe} NOE results of Sanchez et al (73). Interestingly, *E. coli* tRNA^{Phe} only contains one proton between –15 and –14 ppm, and this is the s^{4}U8–A14 tertiary resonance at –14.9 ppm (3). In samples that are only 50% thiolated at residue 8, this peak is reduced to half-intensity, and a new half-proton resonance appears at –14.3 ppm from the population of molecules containing U8–A14; this latter peak exhibits more rapid water exchange than the half proton at –14.9 ppm, which indicates that s^{4}U8–A14 is a more stable tertiary base pair than U8–A14, and perhaps suggests a functional reason for the common thiol modification at residue 8 in bacterial tRNAs (R. E. Hurd and B. R. Reid, unpublished observations). It is obvious that time-resolved helix-coil rate studies will reveal a great deal about RNA dynamics in the future. Although there are several unambiguous ring NH assignments, dynamic studies are presently somewhat limited by the absence of a complete set of thoroughly proven and incontrovertible assignments, but the information from dynamic studies is helping to bring this goal closer.

Interactions of tRNA

With reliable assignments for the high field methyl resonances, and the emergence of assignments for at least some resonances in the more complicated low field spectrum, the interaction of tRNA with cations, drugs, enzymes etc can now begin to be studied and interpreted by NMR. From the pronounced shifting of some low field resonances it is apparent that the solution structure of tRNA is different in the presence and absence of magnesium (25, 41, 70), although the spectral changes have not been interpreted in terms of detailed structural changes. From the most up-to-date assignments the changes appear to involve the D-loop interactions (15–48, 14–8, and 13–22), and perhaps the acceptor stem-rT stem junction (base pairs 7 and 49). The binding of Mn^{2+} and Co^{2+} has been studied via their specific paramagnetic relaxation effects. Chao & Kearns (84) have interpreted the selective broadening by Mn^{2+} on unfractionated tRNA and yeast tRNA^{Phe} to reflect binding at several sites in the order 8–14, U33, 54–58 and 19–56. Hurd et al (59) investigated several tRNAs at slightly lower Mn^{2+} levels and found a single site close to tertiary base pair 8–14 consistent with the P8–P9 crystallographic Mg^{2+} site. Divalent cobalt also relaxed the

8–14 resonance but occupied a different site, since it relaxed several additional resonances not relaxed by Mn^{2+} (59); the Co^{2+} effects were consistent with occupation of the crystallographic cobalt site on G15 (79).

Although all of the foregoing studies assumed the existence of discrete cation sites, a rigorous ESR and ^{31}P NMR study of tRNA-Mn^{2+} complexes by Gueron & Leroy indicates very similar Mn^{2+} lifetimes on *all* phosphates with no specific sites (85). It might be possible to explain the selective effects of Mn^{2+} on the basis of greater access of solvent ions to ring NHs in the region of the D-stem "backbone turn," but further work is needed to rationalize this point. Several tRNAs contain a single binding site for ethidium bromide, and this interaction has been studied by Kearns and colleagues (11, 86). They report that binding involves intercalation at the sixth base pair in the acceptor stem; however the interpretation is based upon the effect on the −14.3 ppm resonance, the assignment of which is controversial as discussed earlier. X-ray studies on ethidium bromide–soaked crystals by Liebman et al (87) indicate binding at the "D turn" without intercalation, but crystals of the solution complex have not been obtained.

The effect of binding the complementary codon to yeast tRNA^{Phe} has been monitored via the high field methyl resonances, by Davanloo et al (88) and also via the ring NH resonances, by Geerdes et al (89); in neither case could any conformational change be detected.

One of the ultimate goals of NMR is the analysis of tRNA recognition by specific enzymes of protein biosynthesis, although serious broadening of resonances in these high-molecular-weight complexes can be expected. Shulman and co-workers have investigated the effects on *E. coli* tRNA^{Glu} of binding glutamyl-tRNA synthetase and the elongation factor EFTu (90). These studies could only be performed at lower temperature because of the lability of the enzymes, and very broad line unresolved spectra were obtained; the authors concluded that no tRNA base pairs were broken in the complexes. More recently, advantage has been taken of thermophilic synthetase enzymes that form functional complexes with tRNA up to 65°C, and the rapid tumbling at these elevated temperatures leads to greatly reduced linewidths. In the cognate complex between *E. coli* tRNA$_1$^{Val} and *B. stearothermophilus* valyl-tRNA synthetase, at least two base pairs in the tRNA are broken (R. E. Hurd, M. LaBelle and B. R. Reid, unpublished observations); however the ring NH intensity losses are in multiple proton peaks, and cannot, as yet, be assigned to specific base pairs. The development of higher resolving spectrometers (500–600 MHz) combined with recent resolution enhancement software should make this an exciting area for further study.

NMR Studies on 5S RNA

Investigation of 5S RNA structure by NMR is in a much more primitive stage of development than the corresponding tRNA studies. There are no methyl resonances to study in the high field spectrum and as yet there is no uniform agreement on the intensity and number of base pairs in the low field spectrum; no assignments have been reported. Kearns & Wong (91) studied *E. coli* 5S RNA at temperatures between 40 and 63°C and interpreted the intensity of the spectrum at 63°C to reflect about 19 base pairs from three helices in a thermally resistant core. At lower temperatures the intensity increased by about 50% which led these authors to claim 28 ± 3 base pairs at 40°C; using -13.2 ppm as a somewhat arbitrary dividing line for AU resonances they deduced the existence of 4 AU pairs and 24 GC pairs and proposed a model (4, 91). Burns et al (92), taking care to prevent conversion to the "A form," studied the "B form" of *E. coli* 5S RNA and found evidence for 33 base pairs in the low field spectrum with resonance positions that were consistent with their structural data from Raman spectra (93). Smith & Marshall (94) used ^{19}F-^{1}H NOE methods to show that *all* of the fluorouracil (FU) residues in FU-substituted *E. coli* 5S RNA had long rotational correlation times ($>10^{-8}$ sec), which indicates a highly rigid molecular framework for the entire molecule in solution. Luoma et al (95) investigated yeast 5S RNA by NMR, UV, and CD methods and concluded that it contains at least 40 base pairs of which 40% were AU pairs. Salemink et al (96) carried out similar studies on *B. licheniformis* 5S RNA and found $\sim$ 36 base pairs, which, from their chemical shifts, were largely GC pairs. In general, 5S RNA proton spectra are not as well resolved as tRNA spectra, perhaps due to slower molecular tumbling because of their larger size or extended shape, or both; the spectra may well prove to be very difficult to assign and very little progress has been made to date.

NMR Studies with Other Nuclei

^{31}PHOSPHORUS   As mentioned in the introduction the natural isotope ^{31}P has spin ½ and is amenable to NMR, although the sensitivity to detection is only one fifteenth that of protons. In RNA we should expect one resonance from each nucleotide, and a few studies on tRNA have been reported using this nucleus. Gueron & Shulman (97) investigated *E. coli* tRNA^{Glu} and yeast tRNA^{Phe} at 40 MHz and 109 MHz (the phosphorus resonance frequency is about 40% that of protons in a given magnetic field) and observed much broader lines at the higher frequency, perhaps indicating relaxation via chemical shift anisotropy. The total range of chemical shifts encompassed almost 8 ppm, but the spectra contained a large central peak of about 65 unresolved resonances with 5 or 6 peaks on the low field side and 2 or 3 peaks at higher field. The extreme low field resonance was assigned to the 5'-terminal phosphate, since it was the only peak whose position titrated with the secondary pK of a phosphomonoester (97); and several peaks were affected by Mg^{2+}, which suggests they may be from nucleotides involved in tertiary folding. Salemink et al (98) have extended these findings in a more detailed study of yeast tRNA^{Phe} using chemical and biochemical modification; using selective cleavage at U33 with pancreatic RNase they assigned the resolved resonances c (lowfield) and j_2 (the highest field peak) to the anticodon loop. Although a few assignments are beginning to emerge, the tRNA ^{31}P NMR spectrum has not been particularly informative; its disadvantages are poor resolution and low sensitivity ($\sim$ 20 hr of signal averaging for 1 mM samples).

^{13}CARBON   The wide range of chemical shifts makes ^{13}C NMR a promising approach to studying RNA molecules, especially their complexes with enzymes. Although Komoroski & Allerhand (99, 100) succeeded in obtaining natural abundance ^{13}C spectra of unfractionated tRNA, the prohibitively long signal-averaging times make this impractical for most biochemical applications. Agris and co-workers (101, 102) have used the E. coli rel met cys mutant together with ^{13}C-methyl methionine to produce bulk tRNA with ^{13}C-enriched methyl groups. Using mononucleoside reference resonance positions and biochemical correlation of the extent of incorporation they assigned the ^{13}C-methyl resonances of rT, m^{6}A, and m^{7}G at 12 ppm, 29.1 ppm, and 36.6 ppm, respectively (101); more recently the signals from ms^{2}i^{6}A, m^{2}A, m^{1}G, mam^{5}s^{2}U, mo^{5}U, and Gm have been assigned (103). Temperature-dependent studies in unbuffered water revealed changes in the resonance positions of rT (tertiary folding) as well as m^{2}A and m^{1}G (anticodon loop) between 9 and 30°C (103). However, all the above studies were carried out with unfractionated tRNA containing $\sim$ 60 molecular species, and it is likely that individual tRNAs will show different thermal unfolding behavior. Along similar lines Hamill et al (104, 105) used an auxotroph to incorporate ^{13}C-4 uracil into tRNA, and Tompson & Agris (106) used controlled growth conditions to produce ^{13}C-2 adenine-substituted tRNA and ^{13}C-2 U/C-substituted tRNA. NMR analysis of the ^{13}C-2-substituted tRNA indicated that the broader than expected linewidth was due to chemical shift nonequivalence for the various environments of the ^{13}C in the unfractionated tRNA (107). Very recently investigations have been carried out on purified single species of tRNA that have been selectively labeled with ^{13}C. Agris & Schmidt (108) have studied E. coli tRNA

[Cys], tRNA[Tyr], and tRNA[Phe] after ^{13}C-methyl enrichment. The much simplified spectra contain resolved peaks from rT, ms^{2}i^{6}A, and m^{7}G (present only in tRNA[Phe]). In the presence of magnesium the m^{7}G shows intermediate exchange between environments, whereas in the absence of magnesium the rT of tRNA[Phe] shows separate resonances (slow exchange) corresponding to at least two conformations, as was shown earlier for the proton resonances of the rT methyl group (18, 19) and the low field resonances of tRNA[Val] (41). Schweizer et al (109) have examined the spectrum of purified *E. coli* tRNA$_1$[Val] substituted with ^{13}C-4 uracil. The spectrum revealed 11 peaks for the 14 uracils (or uracil derivatives) in this tRNA, several of which were assigned and observed to undergo differential shifts reporting local unfolding as the temperature was raised. Although the isotope substitution is somewhat laborious, and the lower sensitivity of ^{13}C necessitates a few hours of signal averaging even with enriched samples, there are several biochemical applications in which ^{13}C NMR, with its approximately ten times wider range of chemical shifts, offers distinct advantages over proton NMR, and this is obviously a fruitful avenue for further research.

^{19}FLUORINE Although fluorine does not naturally occur in RNA, the high sensitivity and reasonably large chemical shift range combined with the incorporation of fluoro-uracil (5FU) for U in bacterial systems (110) has led to RNA NMR studies using this nucleus. In all biochemical assays, fully substituted FU-tRNA has been found to be functional (100–112). Horowitz et al have examined the ^{19}F spectrum of FU-substituted *E. coli* tRNA$_1$[Val] (113) and succeeded in resolving 13 peaks for the 14 FU residues in the molecule; however individual assignments have not been reported. The spectrum of FU-substituted 5S RNA is disappointingly unresolved (114), but Smith & Marshall have used ^{19}F-^{1}H NOE (94) to show that there are no flexible FU residues in the rigid 5S RNA framework.

ACKNOWLEDGMENTS

This review was written in mid-1980 while on sabbatical leave at the University of Oxford. Thanks are due to the Inorganic Chemistry Laboratory and Professor R. J. P. Williams for hospitality at Oxford, and to the Fogarty Center of the NIH for the award of a Senior International Fellowship. I would like to thank all those who sent their unpublished observations and manuscripts, and also my own students, especially Ralph Hurd, Ed Azhderian, and Kathy Jones, for the use of their research data.

Literature Cited

1. Kearns, D. R., Shulman, R. G. 1974. *Acc. Chem. Res.* 7:33–39
2. Kearns, D. R. 1976. Prog. *Nucleic Acids Res. Mol. Biol.* 18:91–149
3. Reid, B. R., Hurd, R. E. 1977. *Acc. Chem. Res.* 10:396–402
4. Kearns, D. R. 1977. *Ann. Rev. Biophys. Bioeng.* 6:477–523
5. Kallenbach, N. R., Berman, H. M. 1977. *Q. Rev. Biophys.* 10:138–236
6. Crothers, D. M., Cole, P. E. 1978. In *Transfer RNA,* ed. S. Altman, pp. 196–247. Cambridge, Mass: MIT Press. 356 pp
7. Opella, S. J., Lu, P., eds. 1979. *NMR and Biochemistry.* New York: Dekker. 434 pp
8. Shulman, R. G., ed. 1979. *Biological Applications of Magnetic Resonance.* New York: Academic. 595 pp
9. Abelson, J., Schimmel, P. R., Soll, D. eds. 1979. *Transfer RNA: Structure, Properties and Recognition,* Cold Spring Harbor New York: Cold Spring Harbor Labs.
10. Schimmel, P. R., Redfield, A. G. 1980. *Ann. Rev. Biophys. Bioeng.* 9:181–221
11. Kearns, D. R., Bolton, P. H. 1978. In *Biomolecular Structure and Function,* ed. P. F. Agris, pp. 493–516. New York: Academic
12. Patel, D. J. 1978. *Ann. Rev. Phys. Chem.* 29:337–62
13. Reid, B. R. 1980. In *Topics in Nucleic Acid Structure,* ed. S. Neidle, pp. 113–39. London:Macmillan
14. Schweizer, M. 1980. In *Magnetic Resonance in Biology,* ed. J. S. Cohen. New York: Wiley. In press
15. MacDonald, C. C., Phillips, W. D., Penman, S. 1964. *Science* 144:1234–38
16. Smith, I. C. P., Yamane, T., Shulman, R. G. 1969. *Can. J. Biochem.* 47:480–84
17. Schmidt, P. G., Kastrup, R. V. 1978. In *Biomolecular Structure and Function,* ed. P. F. Agris, pp. 517–25. New York: Academic
18. Kastrup, R. V., Schmidt, P. G. 1975. *Biochemistry* 14: 3612–18
19. Kastrup, R. V., Schmidt, P. G. 1978. *Nucleic Acids Res.* 5:257–64
20. Rich, A., RajBhandary, U. L. 1976. *Ann. Rev. Biochem.* 45:805–60.
21. Klug, A., Ladner, J., Robertus, J. D. 1974. *J. Mol. Biol.* 89:511–20
22. Kim, S. H., Sussman, J. L., Suddath, F. L., Quigley, G. J., McPherson, A., Wang, A., Seeman, N. C., Rich, A. 1974. *Proc. Natl. Acad. Sci. USA* 71: 4970–75
23. Kan, L. S., Ts'o, P. O. P., von der Haar, F., Sprinzl, M., Cramer, F. 1974. *Biochem. Biophys. Res. Commun.* 59: 22–29
24. Kan, L. S., Ts'o, P. O. P., Sprinzl, M., von der Haar, F., Cramer, F. 1977. *Biochemistry* 16: 3143–52
25. Robillard, G. T., Tarr, C. E., Vosman, F., Reid, B. R. 1977. *Biochemistry* 16:5261–73
26. Romer, R., Riesner, D., Maass, G., Wintermeyer, W., Thiebe, R., Zachau, H. G. 1969. *FEBS Lett.* 5:15–19
27. Romer, R., Riesner, D., Maass, G. 1970. *FEBS Lett.* 10:352–57
28. Kearns, D. R., Patel, D., Shulman, R. G. 1971. *Nature* 229:338–39
29. Kearns, D. R., Patel, D., Shulman, R. G., Yamane, T. 1971. *J. Mol. Biol.* 61:265–72
30. Reid, B. R., Ribeiro, N. S., Gould, G., Robillard, G., Hilbers, C. W., Shulman, R. G. 1975. *Proc. Natl. Acad. Sci. USA* 72: 2049–55
31. Daniel, W. E., Cohn, M. 1975. *Proc. Natl. Acad. Sci. USA* 72:2582–88
32. Reid, B. R., Robillard, G. T. 1975. *Nature* 257: 287–91
33. Bolton, P. H., Kearns, D. R. 1976. *Nature* 262:423–24
34. Reid, B. R. 1976. *Nature* 262:424
35. Bolton, P. H., Jones, C. R., Bastedo-Lerner, D., Wong, K. L. Kearns, D. R. 1976. *Biochemistry* 15:4370–76
36. Reid, B. R., Ribeiro, N. S., McCollum, L., Abbate, J., Hurd, R. E. 1977. *Biochemistry* 16:2086–94
37. Kim, S. H. 1976. *Prog. Nucleic Acid Res. Mol. Biol.* 17: 181–216
38. Kim, S. H. 1978. In *Transfer RNA,* ed. S. Altman, pp. 248–93. Cambridge, Mass: MIT Press. 356 pp
39. Lightfoot, D. R., Wong, K. L., Kearns, D. R., Reid, B. R., Shulman, R. G. 1973. *J. Mol. Biol.* 78:71–82
40. Wong, K. L., Wong, Y. P., Kearns, D. R. 1975. *Biopolymers* 14:749–757
41. Reid, B. R., McCollum, L., Ribeiro, N. S., Abbate, J., Hurd, R. E. 1979. *Biochemistry* 18:3996–4005
42. Arter, D. B., Schmidt, P. G. 1976. *Nucleic Acids Res.* 3:1437–46
43. Shulman, R. G., Hilbers, C. W., Kearns, D. R., Reid, B. R., Wong, Y. P. 1973. *J. Mol. Biol.* 78:57–69
44. Reid, B. R., Azhderian, E., Hurd, R. E. 1979. See Ref. 7, pp. 91–115
45. Robillard, G. T., Tarr, C. E., Vosman, F., Berendsen, H. J. C. 1976. *Nature* 262:363–68

46. Geerdes, H. A. M., Hilbers, C. W. 1977. *Nucleic Acids Res.* 4:207–12
47. Kan, L. S., Ts'o, P. O. P. 1977. *Nucleic Acids Res.* 4:1633–38
48. Robillard, G. T., Tarr, C. E., Vosman, F., Sussman, J. L. 1977. *Biophys. Chem.* 6:291–97
49. Robillard, G. T., Reid, B. R. 1979. See Ref. 8, pp. 45–112
50. Wong, K. L., Bolton, P. H., Kearns, D. R. 1975. *Biochim. Biophys. Acta* 382: 446–51
51. Hurd, R. E., Reid, B. R. 1979. *Biochemistry* 18:4005–11
52. Johnston, P. D., Redfield, A. G. 1977. *Nucleic Acids Res.* 4:3599–15
53. Kearns, D. R., Bolton, P. H. 1978. In *Biomolecular Structure and Function,* ed. P. F. Agris, pp. 493–516. New York: Academic
54. Macon, J. B., Wolfenden, R. 1968. *Biochemistry* 7:3453–58
55. Hurd, R. E., Reid, B. R. 1979. *Biochemistry* 18:4017–24
56. Salemink, P. J. M., Yamane, T., Hilbers, C. W. 1977. *Nucleic Acids Res.* 4:3737–38
57. Daniel, W. E., Cohn, M. 1976. *Biochemistry* 15:3917–24
58. Romer, R., Varadi, V. 1977. *Proc. Natl. Acad. Sci. USA* 74:1561–66
59. Hurd, R. E., Azhderian, E., Reid, B. R. 1979. *Biochemistry* 18:4012–17
60. Farrar, T. C., Becker, E. D. 1971. *Pulse and Fourier Transform NMR.* New York: Academic. 115 pp.
61. Mullen, K., Pregosin, P. S. 1976. *Fourier Transform NMR Techniques.* New York: Academic. 149 pp.
62. Dadok, J., Sprecher, R. F. 1974. *J. Magn. Reson.* 13:243–53
63. Redfield, A. G. 1976. In *NMR: Basic Principles and Progress,* eds. P. Diehl, E. Fluck, R. Kosfeld, 13:137–52. Berling: Springer-Verlag
64. Redfield, A. G. 1978. *Methods Enzymol.* 49:253–70
65. Redfield, A. G., Kunz, S. D. 1975. *J. Magn. Reson.* 19:250–54
66. Redfield, A. G., Kunz, S. D., Ralph, E. K. 1975. *J. Magn. Reson.* 19:114–17
67. Redfield, A. G., Kinz, S. D. 1979. See Ref. 7, pp. 225–39
68. Noggle, J. H., Schirmer, R. E. 1971. *The Nuclear Overhauser Effect: Chemical Applications.* New York: Academic
69. Bothner-By, A. A. 1979. See Ref. 8, pp. 177–219
70. Johnston, P. D., Redfield, A. G. 1978. *Nucleic Acids Res.* 5:3913–27
71. Johnston, P. D., Redfield, A. G. 1979. See Ref. 9, pp. 191–206
72. Geerdes, H. A. M., Hilbers, C. W. 1979. *FEBS Lett.* 107:125–31
73. Sanchez, V., Redfield, A. G., Johnston, P. D., Tropp, J. 1980. *Proc. Natl. Acad. Sci. USA.* 77:5659–62
74. Crothers, D. M., Cole, P. E., Hilbers, C. W., Shulman, R. G. 1974. *J. Mol. Biol.* 87:63–88
75. Hilbers, C. W. 1979. See Ref. 8, pp. 1–43
76. Eigen, M. 1964. *Angew. Chem. Int. Ed.* 13:1–19
77. Kallenbach, N. R., Daniel, W. E., Kaminker, M. A. 1976. *Biochemistry* 15:1218–28
78. Hurd, R. E., Reid, B. R. 1980. *J. Mol. Biol..* In press
79. Jack, A., Ladner, J. E., Rhodes, D., Brown, R. S., Klug, A. 1977. *J. Mol. Biol.* 111:315–28
80. Teitelbaum, H., Englander, S. W. 1975. *J. Mol. Biol.* 92:79–92
81. Kallenbach, N. R., Berman, H. M. 1977. *Q. Rev. Biophys.* 10:138–236
82. Dwek, R. A. 1973. *Nuclear Magnetic Resonance in Biochemistry.* London: Oxford Univ. Press. 395 pp
83. James, T. L. 1975. *Nuclear Magnetic Resonance in Biochemistry.* New York: Academic. 413 pp
84. Chao, Y. H., Kearns, D. R. 1977. *Biochim. Biophys. Acta* 477:20–27
85. Gueron, M., Leroy, J. L. 1979. In *ESR and NMR of Paramagnetic Species in Biological and Related Systems,* ed. I. Bertini, R. S. Drago, pp. 327–67. Dordrecht, Netherlands: Reidel
86. Jones, C. R., Kearns, D. R. 1975. *Biochemistry* 14:2660–65
87. Liebman, M., Rubin, J., Sundaralingam, M. 1977. *Proc. Natl. Acad. Sci. USA* 74:4821–25
88. Davanloo, P., Sprinzl, M., Cramer, F. 1979. *Biochemistry* 15:3189–99
89. Geerdes, H. A. M., Van Boom, J. H., Hilbers, C. W. 1978. *FEBS Lett.* 88:27–32
90. Shulman, R. G., Hilbers, C. W., Miller, D. L., Yang, S. K., Soll, D. 1974. In *Structure and Conformation of Nucleic Acids and Protein-Nucleic Acid Interactions,* ed. M. Sundaralingam, S. T. Rao, pp. 149–70. Baltimore: Univ. Park Press
91. Kearns, D. R., Wong, Y. P. 1974 *J. Mol. Biol.* 87:755–74
92. Burns, P. D., Luoma, G. A., Marshall, A. G. 1980. *Nucleic Acids Res.* In press
93. Luoma, G. A., Marshall, A. G. 1978. *Proc. Natl. Acad. Sci. USA* 75:4901–4
94. Smith, J. L., Marshall, A. G. 1980. *Biochemistry.* In press

95. Luoma, G. A., Burns, P. D., Bruce, R. E., Marshall, A. G. 1980. *Biochemistry.* In press
96. Salemink, P. J. M., Raue, H. A., Heerschap, A., Planta, R. J., Hilbers, C. W. 1980. *Biochemistry.* In press
97. Gueron, M. and Shulman, R. G. 1975. *Proc. Natl. Acad. Sci. USA* 72:3482–85
98. Salemink, P. J. M., Swarthof, T., Hilbers, C. W. 1979. *Biochemistry* 18:3477–85
99. Komoroski, R. A., Allerhand, A. 1972. *Proc. Natl. Acad. Sci. USA* 69:1804–8
100. Komoroski, R. A., Allerhand, A. 1974. *Biochemistry* 13:369–72
101. Agris, P. F., Fujiwara, F. G., Schmidt, C. F., Loeppky, R. N. 1975. *Nucleic Acids Res.* 2:1503–12
102. Fujiwara, F. G., Tompson, J., Loeppky, R. N., Agris, P. F. 1978. In *Biomolecular Structure and Function,* ed. P. F. Agris, pp. 527–33. New York: Academic
103. Tompson, J. G., Hayashi, F., Paukstelis, J. V., Loeppky, R. N., Agris, P. F. 1979. *Biochemistry* 10:2079–85
104. Hamill, W. D., Grant, D. M., Horton, W. J., Lundquist, R., Dickman, S. 1976. *J. Am. Chem. Soc.* 98:1276–78
105. Hamill, W. D., Horton, W. J., Grant, D. M. 1980. *J. Am. Chem. Soc.* In press
106. Tompson, J. G., Agris, P. F. 1979. *Nucleic Acids Res.* 7:765–79
107. Schmidt, P. G., Tompson, J. G., Agris, P. F. 1980. *Nucleic Acids Res.* 8:643–56
108. Agris, P. F., Schmidt, P. G. 1980. *Nucleic Acids Res.* In press
109. Schweizer, M. P., Hamill, W. D., Walkiw, I. J., Horton, W. J., Grant, D. M. 1980. *Nucleic Acids Res.* In press
110. Kaiser, I. I. 1969. *Biochemistry* 8:231–38
111. Horowitz, J., Ou, C. N., Ishaq, M., Ofengand, J., Bierbaum, J. 1974. *J. Mol. Biol.* 88:301–12
112. Ofengand, J., Bierbaum, J., Horowitz, J., Ou, C. N., Ishaq, M. 1974. *J. Mol. Biol.* 88:313–25
113. Horowitz, J., Ofengand, J., Daniel, W. E., Cohn, M. 1977. *J. Biol. Chem.* 252:4418–20
114. Marshall, A. G., Smith, J. L. 1977. *J. Am. Chem. Soc.* 99:635–36

Ann. Rev. Biochem. 1981. 50:997–1024

DOUBLE HELICAL DNA: CONFORMATIONS, PHYSICAL PROPERTIES, AND INTERACTIONS WITH LIGANDS[1]

♦12102

M. T. Record, Jr., S. J. Mazur, P. Melançon, J.-H. Roe, S. L. Shaner, and L. Unger

Department of Chemistry, University of Wisconsin, Madison, Wisconsin 53706

CONTENTS

Perspectives and Summary

Substantial progress has been made recently in understanding the unique physical features of double helical DNA and their implications for its interactions with ligands and for packaging DNA in vivo. These unique features of DNA include: 1. its lateral and torsional stiffness; 2. its high axial

[1]Abbreviations used are: bp, base pair; deg, angular degree; e.u., entropy units (cal mol^{-1} K^{-1}); mbp, mole base pairs.

997

0066-4154/81/0701-0997$01.00

charge density (two phosphate groups per 3.4 Å) and moderate surface charge density (two phosphate groups per 210 Å²); and 3. its superposition of two binding lattices, one a relatively homogeneous array of phosphate charges that will interact electrostatically and nonspecifically with cationic ligands, the other a heterogeneous array of stacked base pairs that will interact with ligands nonelectrostatically and specifically. We examine the molecular basis of these physical features and discuss their implications for a variety of thermodynamic properties of DNA, including the secondary and higher order structure of the molecule as a function of solution conditions, the interaction of DNA with ligands, and the ligand or solvent induced collapse of DNA into a folded state. Our intent is to provide a basic quantitative as well as qualitative description and a critical analysis of these topics. We have attempted to be comprehensible, if not comprehensive. Fortunately a number of general (1, 2) and specific (3–8) reviews of topics in DNA physical chemistry have recently appeared.

DNA Conformation

The double helical nature of DNA has now been convincingly established (9), and except for minor details the experimental evidence confirms the canonical B-DNA structure: an antiparallel, right-handed double stranded helix with internal Watson-Crick base pairing. The refined structure was shown by model building to fit the X-ray diffraction pattern of DNA fibers at high humidity. This approach and the results have been discussed extensively elsewhere (1–3, 10, 11); we only list here, in Table 1, some useful molecular parameters. Of particular interest for modeling DNA as a polyelectrolyte or as a lattice of binding sites are the actual helical charge distribution and geometric quantities such as groove depth and width. (To date, it has been difficult to incorporate real geometry into polyelectrolyte theories, and with a few exceptions (12–14; LeBret, M., personal communication), DNA is usually modeled as a line charge or cylinder.) The major portion of this review is concerned with the effect of environmental factors on the thermodynamic properties of DNA. It therefore is important to

Table 1   Structural parameters of DNA duplex helices (10)

Sample	Humidity (%)	Winding angle (deg)	Rise per residue (Å)	Width of grooves (Å)		Depth of grooves (Å)	
				minor	major	minor	major
DNA·A,Na⁺	75	32.7	2.56	11.0	2.7	2.8	13.5
DNA·B,Na⁺	92	36	3.46	—	—	—	—
DNA·B,Li⁺	66	36	3.37	5.7	11.7	7.5	8.5
DNA·C,Li⁺	66	38.6	3.31	4.8	10.5	7.9	7.5

understand the extent to which these factors affect DNA conformation. The present discussion focuses on what the preferred conformation of DNA in solution is, as a function of environmental variables, and how much the structure can vary about that average conformation.

SOLUTION CONFORMATION   There are several reasons to suspect that DNA conformation in solution differs somewhat from the structure in the fiber. First, the conformation in the fiber (see Table 1) and in solution (see below) depends on the environment. Second, it has been suggested that intermolecular packing forces will affect the observed structural parameters (15, 16), although an increase in the distance between molecules through extreme hydration does not affect the values obtained from fibers (17). The lack of precise information on DNA conformation in solution has motivated much research on the subject. Some controversy exists with regard to the exact number of base pairs per turn (helical repeat), the tilt of the base pairs relative to the helix axis, and the sequence dependence of the structure. Information on the helical repeat of DNA under physiological conditions has come mainly from two different approaches. Using the bandshift method, which exploits the topological properties of closed circular DNA, Wang measured an average helical repeat of 10.6 ± 0.1 (18; J. Wang, personal communication). The interpretation of the precise size distribution of DNA fragments produced by nuclease digestion of DNA immobilized on different surfaces gave a similar nonintegral value for the helical repeat of 10.6 ± 0.1 (19). Related results from the nuclease digestion of nucleosome cores (20, 21) can only yield model-dependent values (22). Neither the bandshift nor the nuclease digestion techniques can distinguish between the reported smaller helical winding and a combination of right-handed helical winding and left-handed intrinsic writhe of the helix axis (23).

Electric dichroism measurements on DNA fragments indicated that the tilt of the bases was not 83 deg as in the B-DNA structure but closer to 73 deg (16, 24), a value similar to that predicted by the conformational energy minimization technique of Levitt (25). However a tilt of 83 deg was observed if measurements were done under conditions at which aggregation of DNA fragments occurs, which suggests that intermolecular interactions may affect structure (16).

Investigation of the sequence dependence of the conformation with the bandshift method has so far yielded preliminary results that indicate that nonalternating AT has 10.1 ± 0.1 bp per turn whereas nonalternating GC gives 10.7 ± 0.1 bp per turn (Wang, J. C., personal communication). Also, the very recent determination of the atomic coordinates for the dodecamer d(CpGpCpGpApApTpTpGpCpGpC) allows a direct observation of variation of the winding angle along the sequence (23). An extreme case of a sequence dependent effect is given by the alternating B-structure proposed for poly(dA-dT) (27) and by the left-handed helices observed at high salt for alternating GC oligomers (28, 29) and polymers (30, 31).

ENVIRONMENTAL EFFECTS   As shown in Table 1, DNA takes on different conformations in different environments in the fiber. For example, depending on the type and amount of salt or the amount of hydration, a B, C, or A structure is observed. This polymorphism has been recently discussed (32). Circular dichroism (CD) spectra of isotropic DNA films prepared under conditions similar to those of the fibers have been measured (33). Correlation between optical properties and geometrical structure suggests that transitions of one form into another can be observed directly. Changes in the near UV CD spectrum of DNA induced by variations in salt concentration, type of salt, solvent composition, temperature, or in the DNA itself (superhelicity, collapse) have been reported and interpreted by many authors (34, 35). Although useful as a characterization of the different transitions, the approach is limited by the fact that interpretation in terms of defined conformational changes is rather difficult and controversial (34, 36, 36a, 37). [An example is the decrease in the 275 nm CD band observed at high salt, and attributed to a B-form to C-form transition. This interpretation would require a change in the winding angle of up to 2.6 deg (see Table 1), whereas a direct determination of that change by ethidium bromide sedimentation velocity titration gives a maximum value of 0.8 deg (36).]

A more precise estimate of the effect of salt and temperature on linear DNA conformation has come from studies that take advantage of the topological properties of closed circular DNA molecules. The average change in the winding of DNA upon transfer from an incubation solution to a fixed electrophoresis condition has been measured as a function of such parameters as temperature, type of salt, and salt concentration. The winding angle shows a significant temperature dependence equivalent to $-13 \pm 2 \times 10^{-3}$ deg bp^{-1} K^{-1}, which implies that the helical structure unwinds continuously as the temperature is increased (38). The winding angle also varies with counterion type at constant salt concentration, and increases in the order $Na^+ < K^+ < Li^+ < Rb^+ < Cs^+ < NH_4^+$. In all cases the winding angle increases linearly with the logarithm of the salt concentration over the range studied (39). No effect of anion substitution was observed. The slopes are such that at 20°C, one observes a decrease in winding angle of about 0.1 deg per base pair in going from 0.2 M NaCl to 0.05 M NaCl, or 0.2 deg per base pair in changing the salt from NH_4Cl to NaCl at 0.2 M. It is plausible to interpret a decrease in the winding angle as corresponding to a decrease in the axial charge density of the helix (10). Consequently, the observed linear dependence of the winding angle on the logarithm of the salt concentration can be understood as a general polyelectrolyte effect. A decrease in axial charge density reduces the requirement for the local accumulation of counterions near the polyion [thermodynamic counterion binding (40)]. Consequently counterions are released to the bulk solution as the axial charge density is reduced; the entropic contribution from counterion release (a free energy of dilution) varies as the logarithm of the bulk salt concentration. Manning is developing a more quantitative description of these observations (G. S. Manning, personal communication).

The effect of cation type on the winding angle could be related to a more specific interaction of the counterion with the DNA. It is interesting to note that there is a good correlation between the ordering of counterions derived from their effect on DNA conformation (41–45) and their relative affinity for DNA as determined by NMR from their ability to displace ^{23}Na from the surface region (46). The origin of these cation-specific effects is unclear. Site-binding of cations to either phosphates or bases, driven by the high local ion concentration, is a possible explanation. However, the formation of inner sphere complexes, as observed (47, 48) in crystals of the dinucleotides ApU and GpC (under low hydration conditions), is unlikely in solution. Neither field jump kinetic measurements on the interaction of Na$^+$ with poly (dA) (49), nor ^{23}Na NMR relaxation measurements on DNA (46, 50, 51), detected the degree of dehydration that accompanies inner sphere site binding. In addition, differential equilibrium dialysis experiments failed to show any base-specific interactions of these counterions with DNA (52).

Thermodynamic Analysis of DNA Stability

HELIX-COIL EQUILIBRIA   The stability of native DNA as a function of base composition and solution conditions is characterized by the denaturation temperature T_m. From a calorimetric determination of the enthalpy of denaturation ($\Delta H^0{}_{T_m}$), the entropy of denaturation ($\Delta S^0{}_{T_m}$) is obtained as $\Delta S^0{}_{T_m} = \Delta H^0{}_{T_m}/T_m$. Since in general the heat capacity difference between denatured and native states ($\Delta C^0{}_p$) will be nonzero, $\Delta H^0{}_{T_m}$ and $\Delta S^0{}_{T_m}$ will be functions of absolute temperature as well as of the variables used to affect T_m. Consequently it is important to obtain accurate values of $\Delta C^0{}_p$ and use them to extrapolate thermodynamic functions to a standard reference temperature (e.g. 298 K). Such standard enthalpies, entropies, and free energies, determined as a function of base composition and solution variables, will provide information about the contributions of various noncovalent interactions to helix stability. (Stability is of course a relative term, and the extent of residual noncovalent interactions in the denatured state must be known in such an analysis.) The development of high precision differential adiabatic scanning microcalorimeters has allowed the determination of ΔC_p^0 and $\Delta H^0_{T_m}$ with sufficient accuracy to perform the above analysis [see e.g. (53, 54)]. To date, only the transition of poly(rA)·poly(rU) has been investigated (55, 56). Interpretation of the calorimetric results is complicated by the existence of residual base stacking in rA above the transition temperature of poly(rA)·poly(rU) (57, 58). Though different thermodynamic parameters for poly(rA) stacking have been obtained by different groups, and therefore different corrections of the poly(rA)· poly(rU) calorimetric data to obtain thermodynamic functions for the transition to unstructured single strands have been applied, the potential of the method is such that we cover one plausible treatment of the data here.

Filimonov & Privalov (56) measured ΔC_p^0 and $\Delta H^0_{T_m}$ for the denaturation of poly(rA)·poly(rU) as a function of [NaCl] (0.01 – 0.1 M). An apparent ΔC_p^0 of approximately 80 cal K^{-1} mol^{-1} bp^{-1} was obtained; calorimetric enthalpies ranged from $\sim$6.8 kcal mbp^{-1} (0.01 M NaCl; $T_m = 310.5$ K) to 8.6 kcal mbp^{-1} (0.1 M NaCl; $T_m = 331$ K). Single stranded poly(rA) shows a broad, salt independent, noncooperative transition (from a stacked to an unstacked state) centered at 313 K, for which a calorimetric enthalpy of 3.0 kcal per mole base and entropy of 9.8 cal K^{-1} per mole base were obtained. Correction for the contribution of residual stacking in poly(rA) to the observed ΔC_p^0 gave a heat capacity difference between unstructured strands and native helix of 30 ± 4 cal K^{-1} mol^{-1} bp^{-1}, and increased the enthalpies of denaturation to 8.5 kcal mol^{-1} bp^{-1} (at 0.01 M NaCl) and to 9.2 kcal mbp^{-1} (at 0.1 M NaCl). Filimonov & Privalov observed that the variation of the corrected $\Delta H^0_{T_m}$ with T_m is exactly that predicted from the value of ΔC_p^0 for the effect of temperature itself on the enthalpy change. Consequently, upon converting the thermodynamic functions to a standard temperature of 298 K, the authors found that the enthalpy difference between denatured and native states of poly(rA)·poly(rU) was independent of [NaCl]; the salt effect on helix stability has an entropic origin. Values of ΔS^0 at 298 K range from 26.3 cal K^{-1} mol^{-1} bp^{-1} at 0.01 M NaCl to 24.3 cal K^{-1} mol^{-1} bp^{-1} at 0.1 M NaCl. The corresponding free energy differences are 0.3 kcal mol^{-1} bp^{-1} and 0.8 kcal mol^{-1} bp^{-1}, respectively. Earlier calorimetric measurements of the enthalpy of denaturation of T2 phage DNA (59) and calf thymus DNA (60) as a function of NaCl concentration showed a similar dependence of $\Delta H^0_{T_m}$ on T_m to that observed for poly(rA)· poly(rU), which suggests that in general the effect of electrolyte concentration on the stability of the double helix is primarily entropic.

A molecular basis for this result is provided by polyelectrolyte theory (7, 8, 40). The high structural charge density on a polyion such as DNA generates locally steep gradients in the concentrations of counterions and coions surrounding it. (Even at low concentrations of added salt, the local counterion concentration at the DNA surface is predicted to be in the molar range; the concentration of coions near the surface is essentially zero.) This effect is analogous to, but much more dramatic than, the Debye-Hückel screening effect in an ordinary electrolyte solution. Because the local counterion concentration is so high, a significant amount of binding of counterions to the polyion charged groups will occur if there is any chemical (as distinct from electrostatic) affinity between these species. The fraction of a counterion physically bound per polyion monomer is denoted $1 - \alpha$. Both of the above effects (ion gradients, counterion binding) contribute to a reduction in the chemical potential of the electrolyte component as a result of the presence of the polyelectrolyte, and may be considered together to define the extent of thermodynamic binding of counterions. The fraction of a counterion thermodynamically bound per polyion monomer is denoted $1 - i$ or ψ (40, 61, 62). Such thermodynamic quantities as the Donnan coefficient, osmotic coefficient, and electrolyte activity coefficient are directly related to the extent of thermodynamic binding of counterions (40); effects of electrolyte concentration on conformational equilibria or ligand binding reactions of nucleic acids result from differences in the extent of thermodynamic binding of counterions between product and reactant states of the nucleic acid (8, 40). At low salt concentrations, ψ is entirely determined by the structural charge density of the polyion; in the experimental range of salt concentrations, ψ is *reduced* from its low salt value to an extent that depends on the amount of physical counterion binding and the salt concentration (40). ^{23}Na NMR experiments are consistent with an extent of physical binding of Na^+ ions to helical DNA that is relatively independent of salt, and in the range $0.25 \leq 1 - \alpha \leq 0.75$ (41, 51).

The denaturation of a nucleic acid reduces its structural charge density, which reduces the steepness of the local ion concentration gradients and therefore reduces ψ. Consequently, denaturation is accompanied by the release of $\psi_h - \psi_c$ thermodynamically bound counterions per nucleotide transformed from the helix (h) to the coil (c) state; $\psi_h - \psi_c$ appears from both theory and experiment to be relatively independent of salt concentration (7, 8, 40, 59, 63). As the salt concentration (m_3) is reduced, the contribution from the release of these ions to the observed free energy difference (at a specified temperature) between denatured and native states of a base pair (ΔG^0_{obs}) becomes more important:

$$\frac{d\Delta G^0_{obs}}{d \ln m_3} = 2RT (\psi_h - \psi_c)$$

To the extent that ψ is independent of temperature, this effect is entropic in origin (a free energy of dilution); that is

$$\frac{d\Delta S^{\circ}_{obs}}{d \ln m_3} \cong 2R(\psi_h - \psi_c) = \frac{\Delta H^{\circ}_{T_m}}{T^2_m} \frac{dT_m}{d \ln m_3}$$

and

$$\frac{d\Delta H^{\circ}_{obs}}{d \ln m_3} \cong 0$$

where $\Delta H^0_{T_m}$ is the enthalpy of denaturation at T_m, and ΔS^0_{obs} and ΔH^0_{obs} are the entropy and enthalpy differences between denatured and native states of a base pair at a fixed reference temperature. From either the calorimetric or T_m data on poly(rA)·poly(rU) denaturation to unstacked coils, one finds that 0.19 ions are released (in the thermodynamic sense) per nucleotide denatured ($\psi_h - \psi_c = 0.19$).

Values of ΔC^0_p as a function of base composition and solution conditions for natural DNA molecules are not available, and only limited calorimetric data is available on the variation of the denaturation enthalpy with T_m at constant electrolyte concentration and variable base composition (64). It is interesting to note that the dependence of $\Delta H^0_{T_m}$ on T_m is in close agreement with that predicted from the ΔC^0_p of poly(rA)·poly(rU) denaturation (56, 60, 64), which suggests (as in the case where T_m was varied by changing the salt concentration) that much of the apparent variation of $\Delta H^0_{T_m}$ with base composition may be a heat capacity effect and not directly an effect of base composition. This would imply that the additional thermodynamic stability of the G·C base pair as compared to the A·T base pair may have a primarily entropic origin. One effect that may contribute to a variation in ΔS^0_{obs} with base composition is the apparent variation in the thermodynamic extent of ion release ($\psi_h - \psi_c$) with base composition. Evidence has accumulated to indicate that the dependence of T_m on the fraction of G·C base pairs (X_{GC}) is a function of salt concentration. At low salt concentrations dT_m/X_{GC} is larger than the canonical Marmur-Doty value ($dT_m/dX_{GC} = 41$ deg) (65) applicable at 0.2 M salt; dT_m/dX_{GC} decreases dramatically in the molar salt concentration range (66–68). If dT_m/dX_{GC} is a function of salt concentration (m_3), then $dT_m/d \ln m_3$ must be a function of X_{GC}. An empirical equation that represents one set of denaturation data in NaCl is T_m (°C) = 176 − (2.6 − X_{GC}) (36 − 3.06 ln m_3) (68), from which one obtains $\psi_h − \psi_c$ = 0.16 − 0.06 X_{GC}. A similar result has recently been derived by Blake & Haydock (69) from an elegant study of the salt dependence of the subtransitions of 34 cooperatively melting regions of phage λ DNA, using a high resolution differential melting technique. As these authors point out, ~40% more ions are released (in the thermodynamic sense) per nucleotide denatured in an AT-rich region than in a GC-rich region. Consequently AT-rich regions are predicted to be preferentially destabilized by a reduction in salt concentration; the effect is predicted to be primarily entropic. Small changes in the structural charge density or in the chemical affinity for counterions in either the denatured and/or the native form as a function of G·C content could account for this result. If the G·C dependence of thermodynamic counterion binding resides in part in a greater extent of thermodynamic binding of electrolyte ions to AT-rich regions in the helix (which would be of great importance physiologically), this effect must not be ion-specific, since difference equilibrium dialysis experiments using alkali metal cations did not detect any ion-specific effects of base composition (52).

CYCLIZATION   There has been renewed interest in the cyclization and oligomerization reactions of DNA molecules or fragments with complementary single-stranded termini, both as a result of the ability to generate such fragments by digestion of DNA with a variety of restriction endonucleases, and because of the importance of such end-joining reactions in molecular cloning procedures. The definitive work on the kinetics and equilibria of the end-joining reactions is that of Wang & Davidson (70–72) with λ DNA and sheared half-molecules thereof. The process is adequately represented as a two-state equilibrium between cohered and free ends, which proceeds through a transient intermediate in which the cohesive ends are aligned but not yet base paired. Comparison of the van't Hoff enthalpy obtained from the temperature dependence of the end-joining equilibrium constant with the calorimetric enthalpy of base pair formation provided an accurate estimate of the number of nucleotides in the single-stranded ends. Recently, the dependence of the end-joining equilibrium constant on salt concentration (72) has been analyzed by the method described in the previous section to obtain the total increase in thermodynamic binding of counterions that accompanies the reaction (73). From this, the number of nucleotides in the single-stranded ends could be determined; again agreement with the chemically determined value is excellent.

Equilibrium constants (and also association rate constants) for cyclization and linear dimer formation are related by the Jacobson-Stockmayer factor $j = (3/2\pi Lb_e)^{3/2}$, where L is the contour length and b_e is the statistical segment length or Kuhn length of the DNA under the ionic conditions of the experiment (see below); j is the concentration of one end of a randomly coiling polymer chain in the vicinity of the other (70). In predicting the distribution of cyclic and linear products from an end-joining reaction involving molecules with $L \gg b_e$ so that Gaussian chain statistics apply, the significant quantity is the ratio j/i, where i is the bulk concentration of cohesive ends (75). For large DNA molecules, j will accurately represent the concentration of proximate ends. As the fragment contour length decreases, the applicability of Gaussian statistics breaks down and j will increasingly overestimate the concentration of ends sufficiently close to cyclize. The ratio j/i can still be used as a semiempirical measure of concentrations that promote cyclization. The extent of cyclization may be maximized for short fragments by selecting incubation conditions (salt concentrations and/or temperature) that stabilize the closed form (74, 75). Recent computations of the radial distribution functions of short, ideal DNA chains predict that fragments as small as 256 bp may cyclize, although those with only 128 bp should not (75a).

Flexibility of DNA

LATERAL RIGIDITY   Unstructured single-stranded polynucleotides possess a limited extent of stiffness (statistical segment length of ~40 Å) (76). The statistical segment length does increase (by a factor of 2–3) under conditions favoring a large amount of base stacking, but the single-stranded molecules fail to achieve the rigidity displayed by double helical DNA.

Stiff macromolecules such as DNA are conveniently modeled as wormlike coils (2). This representation treats the polymer as a continuously curving chain for which the direction of curvature at any point along the chain is random. In this model, the stiffness of the molecule is characterized by its persistence length a which represents, in the limit of infinite contour length, the average projection of its end-to-end distance vector along a z-axis defined by the orientation of the first segment of the chain. Larger values of a correspond to greater local rigidity. For contour lengths of the order of a, the chain approximates rodlike behavior. For very long contour lengths, the chain approaches the behavior of a Gaussian coil. The persistence length and the statistical segment length b_e (used in the Gaussian chain model of the polymer as a series of segments connected by universal joints) are simply related: $a = b_e/2$.

Long-range excluded volume effects in addition to short-range stiffness perturb the dimensions and properties of real polymer chains from those of the ideal flexible random coil. Excluded volume effects arise both because the segments occupy a finite volume and because of differences in segment-

segment and segment-solvent interactions. Obtaining a rigorous theoretical description of excluded volume effects on chain statistics has been a difficult problem (77). One commonly used method for introducing these effects is by introducing a parameter ϵ into the equation for the mean square end-to-end distance: $\langle h^2 \rangle = N_e^{1+\epsilon} b_e^2$ where N_e is the number of statistical segments in the total contour length.

Most of the determinations of the persistence length of DNA are at 0.2 M NaCl, neutral pH, and 25°C. Within the last decade, a substantial degree of consensus on the value of a has finally appeared. When excluded volume has been neglected, the persistence length has been found to be 600 ± 100 Å (78–84). Consideration of excluded volume in determining a has produced values of 400 ± 100 Å with $\epsilon \approx 0.1 \pm 0.02$ (85–87a).

Whether excluded volume need be considered has been much debated (88–91). The sedimentation coefficient and intrinsic viscosity of native DNA vary with powers of the molecular weight (0.445 and 0.665, respectively), which differ from the 0.5 power dependence expected for an unperturbed random coil. This deviation was postulated to be due to excluded volume. However, it has also been argued, based on measured values of the second virial coefficient, that the effects of excluded volume on chain dimensions are small at 0.2 M salt (a 10% increase in the radius of gyration over its unperturbed value for molecules with $M = 30 \times 10^6$) (88). Recent measurements of the second virial coefficient as a function of salt indicates that it increases from 2.5×10^{-4} mL mol g^{-2} to 5.0×10^{-4} mL mol g^{-2} between 4.0 and 0.2 M NaCl. A reduction of the salt concentration to 0.005 M increases the second virial coefficient to 22.2×10^{-4} mL mol g^{-2} (86). This suggests that excluded volume effects are more significant at lower salt, where long-range polyelectrolyte effects are important.

Several studies of the salt dependence of the persistence length exist. The results are summarized in Table 2. All of the available data agree that as the salt concentration is lowered, a increases. A large dependence of the persistence length upon salt concentration was obtained from a statistical analysis of local curvature of T2 DNA fragments examined by electron microscopy (93). Fragments were adsorbed on to a cytochrome c coated grid from solutions containing ammonium acetate. Evaluation of these results is impeded by lack of knowledge of the effect of salt upon the cytochrome c-DNA interaction. Moreover, the effects of excluded volume on the conformation of these fragments, stated to be a few microns in length and therefore possibly subject to such constraints in solution, were not taken into account (93a). Harrington (91) combined flow birefringence and intrinsic viscosity data on T2 DNA to obtain a steep dependence of a on salt when the experimental data was calibrated with $a = 660$ Å at 0.2 M NaCl. A recent light-scattering study on the linear form of the ColE1 plasmid ($M = 4.4 \times 10^6$) (86) determined a with and without an excluded volume correction (see Table 2). When excluded volume is considered, $a = 370$ Å at 0.2 M NaCl, and the increase in a with decreasing salt is reduced. Recalibration of Harrington's data (R. E. Harrington, personal communication) using this value of a results in a less dramatic salt dependence that is in reasonable agreement with the light-scattering data, as well as with that obtained in an earlier analysis (87) of salt dependent hydrodynamic data.

Analysis of the rotational relaxation times of the transient electric birefringence of short (587 bp) blunt-ended restriction fragments indicates that the persistence length is 500 ± 50 Å at or above 1 mM NaCl and that electrostatic contributions to the persistence length are negligible for these fragments above 1 mM NaCl. Similar experiments in $MgCl_2$ show that the asymptotic value (530 ± 50 Å) is obtained at concentrations above 0.1 mM (93a).

Two similar theoretical treatments have derived an expression for the electrostatic component of the persistence length (94–96). The polyelectrolyte is modeled as a structureless space curve and electrostatic interactions are assumed to occur via screened Coulombic potentials. The electrostatic contribution is predicted to be negligible at 1 M NaCl and to increase to 65 Å at 0.005 M NaCl (97). This predicted dependence is not in quantitative agreement with any of the available experimental data.

The total free energy of bending has been considered for a model equivalent to the wormlike coil (2, 98). It has been found that $\Delta G^\circ = \beta\theta^2/2L$, where β, the bending force constant of the chain of contour length L, is related to the persistence length ($\beta = akT$) and θ is the angle between the tangent vectors of the origin and the end of the chain. A similar result has been obtained by considering the discrete analogue of this model (99). Thus by studying the variation of the persistence length as a function of temperature, the thermodynamic parameters for bending may be obtained.

Table 2 Salt dependence of the persistence length of DNA

[NaCl] M	Hearst et al (87) a ($\epsilon = 0.072$) Å	Frontali et al (93)[a] a ($\epsilon = 0$) Å	Borochov et al (86) a ($\epsilon = 0$) Å	Borochov et al (86) a (Å)	ϵ	Harrington (91) a^b Å	a^c Å	ϵ
0.005	660	2100	910	530	0.200	1540	530	—
0.1	400	800	550	400	0.118	740	370	0.102
0.2	400	560	500	370	0.100	660	370	0.085
1.0	330	540	360	290	0.090	460	280	0.070

[a] Values interpolated from the available salt data.
[b] Values obtained using $a = 660$ Å at 0.2 M salt as calibration.
[c] Values obtained using $a = 370$ Å at 0.2 M salt as calibration (R. E. Harrington, personal communication).

Only two such studies exist. Gray & Hearst (100) obtained the temperature dependence (5–49°C) of a from a sedimentation study on several phage DNAs at 0.2 M salt. The enthalpy and entropy of bending were determined to be 6.9 ± 2.2 cal bp mol^{-1} deg^{-2} and $(-1.5 \pm .8) \times 10^{-2}$ e.u. bp deg^{-2}. A recent flow birefringence study by Harrington (101) on T2 DNA in 2.0 M NaCl determined that $\Delta H = 12.4 \pm .9$ cal bp mol^{-1} deg^{-2} and $\Delta S = (1.8 \pm 3.2) \times 10^{-3}$ e.u. bp deg^{-2}. These values for the enthalpy and entropy were obtained by assuming a value of 660 Å for the persistence length at 25° C. If the flow birefringence data is calibrated with a value of 370 Å for a at 25°C, the values for the enthalpy (6.7 ± .5 cal bp mol^{-1} deg^{-2}) and entropy (0.002 ± 0.004 e.u. bp deg^{-2}) are in agreement with the results of Gray & Hearst.

TORSIONAL RIGIDITY   Several studies concerned with the torsional rigidity of DNA have appeared recently. These studies arrive at the common conclusion that DNA is much less resistant to twisting than was previously supposed (102–106). The torsional rigidity determines the energy required to change the angle between base pairs by a given amount and is related to the extent to which the conformation of DNA may be perturbed by solution conditions or deformed by the binding of a protein.

The fluorescence depolarization of intercalated ethidium bromide demonstrated the existence of twisting and bending motions of the double helix occurring on a time scale of 10^{-8} seconds (107). The dynamics of a bead-spring model (102, 103) and an elastic continuum model (102) have been developed in order to gain an understanding of the motions that contribute to the depolarization of fluorescence. For an ideal elastic cylinder, the energy required for deformation by twisting is given by $U = \frac{1}{2} C \theta^2$, where C is the torsional force constant and θ is the angular displacement per unit length. The torsional force constants found in these studies were in the range $C = 23$–53 cal bp mol^{-1} deg^{-2}. Recently, the fluorescence depolarization has been measured with increased accuracy and subnanosecond time resolution (103a, 104). The observed polarization of the fluorescence decays as an exponential in (time)$^{1/2}$. This is the time dependence predicted for twisting motions by the elastic continuum model and also by the discrete model after an initial period. Thus the torsional motions of the DNA helix occurring in a time range of 10^{-8} to 10^{-7} seconds can be described by simple models and are characterized by a torsional force constant $C = 17$ cal bp mol^{-1} deg^{-2}. In addition, these results have been interpreted as evidence against the existence of widely spaced torsion joints in an otherwise stiff molecule (103a).

A similar model has also been developed to describe the contributions of twisting and bending motions to the electron spin resonance (ESR) correlation time of an intercalated spin labeled propidium derivative (105). The ESR measurements can be described by a model in which the intercalated compound undergoes an average torsional motion of 4 deg (at 20°C). This average amplitude corresponds to a lower limit of $C = 40$ cal bp mol^{-1} deg^{-2}. This model approximately predicts the observed correlation times and their dependence on temperature, segment length, and magnetic relaxation rates.

Another estimate of the torsional force constant has been obtained from a combination of experimental and theoretical work on supercoiled DNA (106, 106a). The resulting torsional force constant is $C = 21$ cal bp mol^{-1} deg^{-2}.

For an isolated torsion spring, the root-mean-squared amplitude of a fluctuation in the displacement is given by $\theta_{rms} = (RT/C)^{1/2}$. The force constants given here correspond to a base pair motion of $\theta_{rms} = 3$–6 deg at 25°C, which is substantially larger than expected.

Thermodynamic and Molecular Description of Supercoiling

Circular, double-stranded DNA isolated from a variety of sources has been found to have a convoluted tertiary structure and altered chemical and biochemical reactivities. DNA in this state, known as supercoiled or superhelical DNA, has been the subject of a great deal of research and several reviews (4, 108). The emphasis in this section is on the recent advances toward a thermodynamic and molecular description of DNA in the supercoiled state.

The physical constraint that causes superhelicity is that the number of times one strand encircles the other remains invariant as long as the strands are not broken. This number is called the linking number or the topological winding number and is designated by Lk, L, or α (4, 109, 110). The linking number is the sum of the twisting of one strand about the other in the double helix, which is referred to as the twist or Tw, and the three dimensional disposition of the helix axis, which is called the writhe or Wr. This is described by the equation $Lk = Tw + Wr$. For a closed, double-stranded molecule, a change in the twist results in a compensating change in writhe. A useful quantity is the number of titratable superhelical turns, τ, which may be thought of as the writhe of a molecule of a given linking number when Tw is fixed at its value for a linear molecule. The form of τ that is independent of molecular weight is called the superhelix density and is defined by $\sigma = 10\tau/N$, where N is the number of base pairs. The properties of these quantities and methods for the measurement of τ are discussed in detail in the references cited above and in several texts (1, 2).

The evidence from sedimentation (91, 111–113), intrinsic viscosity (113), light scattering (114, 115), and electron microscopy studies (112, 116) suggests that the shape of a closed-circular DNA changes as the superhelix density is increased. At low superhelix densities ($|\sigma| \gtrsim 0.05$), the DNA remains a random coil that becomes more compact as the superhelix density increases. Molecules with higher superhelix densities ($0.06 < |\sigma| < 0.19$) have been characterized as interwound or branched interwound superhelices. This behavior is approximately symmetric for positive and negative superhelix densities (112, 117). The details of molecular structure in these limiting regions, and the transition between them are influenced by temperature and ionic environment (111, 112) as well as by the origin of the DNA (115). In addition to these physical studies that characterize superhelical DNA by shape, and therefore indirectly by writhe, there are spectroscopic studies that bear indirectly on twist. The circular dichroism of superhelical DNA differs significantly from that of linear or nicked DNA (43, 118, 119). There is evidence that this effect may either decrease (120) or increase (120–122) with increasing salt concentration, depending on the nature of the DNA. The present lack of understanding of the effect of tertiary structure on CD (37) forestalls its use in a quantitative description of superhelical DNA.

The thermodynamic description of supercoiling is comparatively simple. In terms of intensive quantities, the free energy of supercoiling is given by $\Delta G^0 = uRT\sigma^2$, where u is a reduced force constant. From buoyant density sedimentation (123) and gel electrophoresis experiments (38, 124, 125), it was found that $u = 10.2 \pm 0.6$. Little variation in u was observed for different salt types and concentrations (0.002 M $MgCl_2$, 0.2 M NaCl, 5.8 M CsCl), temperatures (2–28°C), superhelix densities ($|\sigma| < 0.06$), and for DNA from various sources. For higher superhelix densities, a cubic term is included in the free energy expression (123). The value of u found by spectrophotometric titration with ethidium bromide was significantly lower: $u = 5.4 \pm 0.4$ (3M CsCl, 2.5–20°C) (126). These authors also report a slight temperature dependence above 20°C. The insensitivity of u to temperature suggests that the free energy of supercoiling is largely entropic. The precision of the available data, however, does not permit a more quantitative statement (126).

Theoretical descriptions of highly supercoiled molecules have developed from an elastic model of DNA. The molecule is described as an isotropic rod characterized by a force constant for bending, A, and a force constant for twisting, C (see section on flexibility). The elastic model is expected to be an appropriate model for highly supercoiled molecules (127). The equilibrium shapes of an elastic rod subject to torsional stress have been obtained by solution of the equations of linear elasticity (128, 129). These results have not been usefully applied to supercoiled, circular molecules because the constraint of ring closure is not explicitly included. The general solutions of the equilibrium shapes of twisted rings are apparently not known at present. Another approach has been to assume that a highly supercoiled DNA takes the shape of an interwound superhelix (130). Subject to the constraint of a distance of closest approach of the arms of the superhelix, r_o, the free energy of twisting and bending is minimized with respect to the number of physical superhelical turns. By specifying either a torsional force constant or a value for r_o, these authors have explained the observed insensitivity of the sedimentation coefficient to the superhelix density in the range $0.06 < |\sigma| < 0.09$. Introduction of recent estimates of the torsional force constant (102–106) results in a solution only when r_o is large.

Molecules with low superhelix density have been the subject of recent theoretical developments based on ideal models of polymer chains. In this formalism, supercoiling is a constraint that reduces the configurational entropy of the closed random coil. By calculating the writhes of many computer generated random walks, it was found that a population of circular molecules equilibrated by strand scission and closure will have a Gaussian distribution of writhes (106, 106a). The variance of the writhe is proportional to the number of statistical segments in the random walk (see section on flexibility). When combined with the experimental observation that the equilibrium distribution in linking number is Gaussian (38, 124), this result predicts that the distribution in twist is also Gaussian, and gives an independent estimate of the torsional force constant. When the statistical segment length is taken as 1150 Å, the resulting theoretical variance in the writhe accounts for about half of the observed variance in the linking number. In this case, a change in the linking number of 1 causes a change in the writhe of ½ and a change in the twist of ½. If the more recently devalued statistical segment lengths are used instead (see section on flexibility), the variance in the writhe is increased. The resulting, larger torsional force constant is in better agreement with other estimates (see section on torsional rigidity).

The investigation of an interesting form of DNA, produced by annealing complementary closed circular single strands, suggests that there are limits to the variance in writhe (119). Analysis by spectroscopy and electron microscopy show that this DNA is 60–90% double stranded. The linking number of this structure must be zero, but the sedimentation and gel electrophoresis characterization is not consistent with the compensation of all of the right-handed double-helix turns by left-handed writhing. The authors propose that a large fraction of the compensating turns takes the form of left-handed duplex turns.

The expression of the free energy of supercoiling in terms of fluctuations in secondary structure or altered chemical and biochemical reactivity has been recently reviewed (4). Briefly, DNA with a negative superhelix density is characterized by an enhanced reactivity toward ligands or proteins that unwind the double helix and toward reagents or enzymes that react specifically with single-stranded DNA. This is generally attributed to a superhelix density dependent increase in either transient fluctuations in secondary structure or stable alternate structures such as cruciforms and melted regions. Some sequences in negatively supercoiled DNA are reactive with single-strand specific reagents (131). Recent work identifies sites of nuclease sensitivity as inverted-repeated sequences and argues for the existence of cruciform structures (131a). A closely related phenomenon is the uptake of homologous single strands by supercoiled DNA (132–134). This reaction is driven by the free energy of supercoiling. A careful study analysis of the kinetics of this reaction has resulted in a model in which the rate limiting step is the opening or unstacking of a small number of bases in the double helix (133). This is taken as evidence against the existence of stable denatured regions.

Helix-coil transition theories have been applied to the problem of local melting induced by the free energy of supercoiling (135–139). The theories predict a substantial destabilization of the double helix at high superhelix densities, although there is some disagreement about the probability of stable melted regions and cruciform structures. An interesting prediction is that significant fluctuations disrupting base pairing will appear at positive ($\sigma > 0.2$) as well as negative superhelix densities ($\sigma < -0.02$) (136). Recent experimental work has shown that PM2 DNA becomes sensitive to *Alteromonas* nuclease at positive and negative superhelix densities very close to those predicted (117).

Thermodynamic Analysis of Ligand Binding

Thermodynamic studies complement structural studies by revealing how much each type of interaction contributes to the free energy of complex formation under specified conditions, and the extent to which each contribution is enthalpic or entropic. Moreover, a molecular interpretation of thermodynamic data can be used to obtain estimates of the number of each type of interaction (see section on cyclization). Here we review some elements of the thermodynamic analysis of the noncovalent interactions of proteins and other large cationic ligands with double helical DNA. No attempt is made to review the totality of thermodynamic information available on ligand-DNA interactions. Comprehensive reviews of the interactions of DNA with cationic ligands (7, 8), oligopeptides (140), intercalators (141), and various proteins (3, 5, 6, 142–144) are available.

THE DNA DOUBLE HELIX FROM A LIGAND'S PERSPECTIVE   Double helical DNA may be viewed as the superposition of two lattices: the regular exterior array of phosphate charges, and the interior array of stacked base pairs, accessible to ligand binding from either the major or minor groove.

To an approaching ligand the phosphate lattice appears relatively homogeneous, affording sites for nonspecific electrostatic interactions. The lattice(s)
of base pairs, which exhibit both sequence heterogeneity and, consequently,
local conformational heterogeneity, may bind ligands either through intercalation or by specific interactions with accessible functional groups on the
bases. Both lattices are involved in the formation of high affinity, site-
specific complexes between proteins such as *lac* repressor or RNA polymerase and DNA (145–147). These proteins, like any ligand having positively
charged groups, also interact in a weaker, nonspecific electrostatic mode
with the lattice of phosphates (148–152).

ANALYSIS OF NONSPECIFIC BINDING EQUILIBRIA
McGhee & von Hippel (153) have derived a modified version of the Scatch-
ard binding isotherm that has proved to be widely applicable in the analysis
of both electrostatic and intercalative types of binding to nucleic acid lattices [see e.g. (141, 148)]. In a relatively simple and powerful way their
probabilistic derivation incorporates the consequences of nearest-neighbor
cooperativity and overlap. The latter effect is characterized by n, the number of contiguous sites (nucleotides) rendered inaccessible to further binding
by the binding of one ligand. A potential ambiguity in the analysis of
ligand-nucleic acid binding equilibria arises in the physical interpretation
of n: does it represent a number of nucleotides or half that number of
nucleotide pairs? For example, the value of n deduced for the nonspecific
binding of *lac* repressor is approximately 24 nucleotides or (equivalently)
12 base pairs. The latter alternative suggests that the ligand occupies (sterically blocks) all radial access to the helix for ~1.2 helical turns. The former
number indicates that over a span of ~2.4 helical turns only half of the
cylindrical surface is inaccessible to further binding; this interpretation
appears to be favored by evidence from electron microscopy on the nonspecific DNA-repressor complexes (154). Thus, a second repressor molecule could bind nonspecifically to the helical region opposite to a bound
ligand. Electron microscopy (155) and chemical modification experiments
(156) both indicate that the specific binding of *lac* repressor and RNA
polymerase involves contacts on only one side of the DNA helix.

DEPENDENCE OF K_{obs} ON SOLUTION CONDITIONS   A unique and often
dramatic feature of the interaction of cationic ligands (or proteins with a
cationic binding site) with nucleic acids is the dependence of the binding
constant K_{obs} on salt concentration (m_3) in the solution. The reaction of
a protein with a nucleic acid in solution is a multiple equilibrium, involving
a variety of different microscopic states of binding of small ions (e.g. protons
or electrolyte ions) or other solvent components to both protein, nucleic

acid, and complex. These small ions are consequently net participants in the equilibrium, and the observed equilibrium constant becomes a function of their concentrations. (For example, in the nonspecific interaction of *lac* repressor with DNA, $-d\ln K_{obs}/d\ln m_3 = 11 \pm 1$ (148, 149, 151).) This effect of small ions is *not* simply an ionic strength effect but rather is ion specific. Ionic strength (I) is not a useful quantity for interpreting competitive binding equilibria, such as the effects of Mg^{2+} and Na^+ on the nonspecific binding constant of lac repressor to double stranded DNA (145, 148). The value of K_{obs} is essentially the same ($\sim 3 \times 10^5$ M^{-1}) in 0.12 M NaCl ($I = 0.12$), in 0.09 M NaCl, 0.003 M $MgCl_2$ ($I = 0.099$), and in 0.01 M NaCl, 0.01 M $MgCl_2$ ($I = 0.04$). Since, in the absence of $MgCl_2$, the nonspecific binding affinity of *lac* repressor increases dramatically as the NaCl concentration is reduced, the Mg^{2+} ion clearly functions as a direct competitor with the protein for DNA sites; this effect of Mg^{2+} on K_{obs} is much larger than that which might be expected from the contribution of $MgCl_2$ to the ionic strength.

A second source of the dependence of K_{obs} on electrolyte concentration is the extreme deviation from uniformity of electrolyte concentrations near the DNA polyanion. As a result of this screening effect, the activity coefficient of the DNA is a strong function of salt concentration. The neutralization of phosphate charges by the ligand may be expected to reduce the local nonuniformity of ion concentrations. An approximate theory that interprets salt effects on K_{obs} in terms of the release of thermodynamically bound (by binding and screening interactions) counterions from the DNA upon the neutralization of phosphate charges by cationic groups on the ligand has been developed by Record and co-workers (8, 40, 61). Based originally on Manning's limiting-law (low salt) polyelectrolyte theory (61, 157), the binding theory has recently been extended to higher (physiologically relevant) salt concentrations using thermodynamic results obtained from the cylindrical Poisson-Boltzmann model under conditions of excess salt (40, 158). Alternative molecular thermodynamic models to describe the effects of salt on K_{obs} have been developed by Manning (7) and by Bloomfield and co-workers (159).

Since the thermodynamic ion binding theory neglects end effects in the region of the bound ligand and cannot treat the interactions of electrolyte ions with the (unknown) distribution of charges on the oligocationic binding site of the ligand, it is not expected to be as successful in this context as it has proven to be in the analysis of helix-coil equilibria of nucleic acids (see section on helix-coil equilibria), where both reactant and product species are polyelectrolytes, and where the binding of anions need not be considered. Use of the thermodynamic ion binding parameter ψ gives the result that

$$-\frac{d \ln K_{obs}}{d \ln m_3} = Z\psi + k$$

where Z is the net cationic valence of the binding site of the ligand (or equivalently the number of DNA phosphates neutralized by the ligand), ψ is the fraction of an electrolyte ion thermodynamically bound per DNA phosphate (and therefore released from each phosphate neutralized), and k is the unknown net amount of ion release from the ligand upon complex formation (8, 40). At low electrolyte concentrations, ψ approaches its limiting-law value (0.88 for double helical DNA); at higher salt concentrations ψ is reduced by an amount that depends on the salt concentration and the amount of physical counterion binding. At 0.2 M NaCl, the Poisson-Boltzmann thermodynamic theory (40) predicts a range of values of ψ from 0.53 in the absence of any physical counterion binding, to 0.80 if the extent of counterion binding is as great as that hypothesized by Manning (157). In the absence of information about the anion release term k, the logarithmic derivative $-d \ln K_{obs}/d \ln m_3$ may be considered to provide a maximum estimate of the amount of thermodynamic ion release from the nucleic acid $(Z\psi)_{max}$, from which a maximum value of Z can be obtained.

The dependence of K_{obs} on m_3 has been measured for a large number of oligocationic ligands of known charge, including Mg^{2+}, oligopeptides, and various intercalators (see e.g. 7, 8, 160–164). Without exception, $\ln K_{obs}$ is found to be a linear function of $\ln m_3$, in the absence of competition by other cationic species; moreover, values of Z estimated from the slopes of such plots are in good agreement ($\pm$ 10–20%) with the charge on the ligand, using the limiting-law value of ψ ($\psi = 0.88$) and neglecting possible effects of anion release from the ligand. These studies may be considered to calibrate the binding theory, and may further indicate a substantial amount of physical counterion binding to the DNA, since this is required by the Poisson-Boltzmann analysis in order for ψ to remain in the vicinity of the low salt limiting value at higher electrolyte concentrations. The above results can also be interpreted using the molecular thermodynamic model of Manning (7) (which assumes that the ligand replaces counterions in the delocalized surface (condensed) layer surrounding the polyion, and predicts that $-d \ln K_{obs}/d \ln m_3 = Z + k$) or by the numerical Poisson-Boltzmann calculations of Bloomfield and co-workers (159) (which evaluate the actual amount of counterion release from a cylindrical shell around the DNA upon introducing an oligocationic ligand). Each theory has its own limitations and approximations. At present it is difficult to distinguish between these approaches since all provide a reasonable fit to experimental results.

The strong dependences of K_{obs} on solution conditions make it possible to obtain a wide range of binding constants for the interaction of a cationic ligand with DNA by proper choice of pH and ion concentration. Consequently it is useful to define a standard condition for comparison of binding constants. Since the thermodynamic equivalent of the binding reaction is

$$\text{ligand} + \text{DNA} \rightleftharpoons \text{complex} + (Z\psi + k) \text{ electrolyte ions,}$$

an approximate standard condition is at an electrolyte concentration of 1 M. [The questions of whether the standard state should be at unit activity or unit concentration, and the choice of concentration scale, are as yet unresolved (8, 40).] Such standard state binding constants for the interactions of oligolysines with nucleic acid helices are in the range $0.1–1 M^{-1}$, and provide an estimate of the standard free energy of formation of a single lysine-phosphate interaction at 1 M NaCl: $\Delta G° \simeq +0.2 \pm 0.1$ kcal. The large increase in K_{obs} obtained at lower electrolyte concentration results from the favorable free energy of dilution of the $(Z\psi + k)$ ions from the standard state to the salt concentration of the experiment. In favorable cases, where protonation and anion effects on the binding constant of a ligand-nucleic acid interaction have been quantified, then extrapolation of K_{obs} to the 1 M salt standard state provides an estimate of the nonelectrostatic contribution to the binding free energy, by comparison with the binding free energy expected for an oligolysine with the same valence as that of the binding site on the ligand (61).

For *lac* repressor, assuming no anion contribution to the salt dependence of K_{obs}, analysis of nonspecific binding data indicated that $Z = 12 \pm 2$ (148, 149, 151) and that the nonspecific complex probably involved only electrostatic interactions. Estimates of Z ranging from 5 to 11 have been obtained for the specific interaction of repressor with the operator site on λ p*lac* DNA (145, 147, 165); there is direct evidence that the amount of ion release accompanying formation of the specific complex is less than for the nonspecific complex (166, 167). A probable value of Z for the specific interaction is 8 ± 1 (145); this is in general agreement with the number of phosphates in the operator region that cannot be chemically modified by an ethylating agent without blocking specific binding (5). [Curiously, the amount of ion release (~ 1.5) accompanying the interaction of repressor with a synthetic operator fragment is much less than that observed with λ p*lac* DNA (165); a further decrease can be caused by 5-bromouracil substitution in the fragment (168).] Estimates of the nonelectrostatic component of the binding free energy of the repressor-operator interaction from the extrapolated binding constant at 1 M salt are in the range 9–12 kcal/mol (145, 147); these values agree with that obtained from binding studies (165) using a modified (core) repressor lacking the 59-residue NH_2 termini that contribute the ionic interactions with operator. Under "physiological" ionic conditions (0.2 M NaCl, 0.003 M MgCl$_2$), approximately 40% of the binding free energy is contributed by electrostatic interactions and counterion release (145). Further work should lead to a detailed molecular and thermodynamic picture of the specific and nonspecific interactions.

Investigation of the kinetics of ligand-DNA interactions in vitro provides information about the reaction mechanism. The kinetics of these binding processes exhibit unusual features that are traceable to the polymeric and polyelectrolyte characteristics of DNA (147, 169, 170). Thus in addition to defining the time scale of the reaction (and therefore indicating what time dependent conformational events might be involved), kinetic studies provide insights into mechanisms not encountered in nonpolymeric systems. Although the polymeric and polyelectrolyte character of DNA introduces mechanistic complexities, it also suggests logical variables (chain length and ion concentrations) to use in distinguishing between alternative mechanisms.

Intramolecular Folding of DNA

The in vitro monomolecular collapse (condensation) of DNA into highly compacted structures may be caused by the addition of neutral or anionic (171, 172) polymers, or a variety of cationic species [polyamines (173–175), histones (176), poly-L-lysine (177) etc] to very dilute solutions of DNA. The structure of the DNA in the collapsed state is not well characterized (178–181). The experimental evidence available suggests that the structure the compacted form assumes may depend upon the species inducing collapse, and/or upon the solution conditions.

Intramolecular segment-segment interactions are normally highly unfavorable for DNA because of the strong repulsive forces between the charges of the backbone. To produce a compacted molecule one must find a means of either increasing the favorability of the segment-segment interactions and/or of making segment-solvent interactions even more unfavorable than the intramolecular interactions. The cation induced collapse occurs because neutralization of a large fraction [$\gtrsim$90% has been observed experimentally (182)] of the phosphate charge by the cationic species provides the necessary reduction in the electrostatic free energy of the compacted form (183). The collapse caused by addition of neutral or anionic polymers and salt is an example of the alternate strategy. These polymers exclude DNA from the solution volume. Collapse of the DNA reduces this entropically unfavorable effect on free energy. [To obtain collapse in the presence of a polymer, the salt concentration must be sufficiently high ($>$.3 M) to stabilize the collapsed DNA.] In both ways of inducing compaction, monomolecular collapse occurs rather than aggregation and/or precipitation simply because the DNA is at high dilution, and intramolecular interactions are favored over intermolecular ones (171, 174, 180, 184, 185).

A theoretical description of the monomolecular collapse of DNA has appeared (184). The expression for the free energy of mixing a polymer and solvent, obtained from the lattice theory of Flory and Huggins, was extended to include third virial coefficient effects. Calculation of the free energy of mixing as a function of polymer chain expansion indicated that when segment-segment interactions are sufficiently favorable, the preferred polymer configuration is highly compacted. For comparatively stiff polymers such as DNA, the transition from the extended coil to the collapsed form is predicted to occur abruptly as segment-solvent interactions become increasingly unfavorable compared to segment-segment interactions. This theoretical prediction is in agreement with the experimental observation that collapse occurs over a narrow concentration range of the species inducing the effect (173–175, 180, 182, 185).

It was suggested by Manning (7) that DNA may spontaneously fold upon the neutralization of some critical fraction of the backbone charge. Wilson & Bloomfield (182) used light scattering to study the effects of cation concentration and valence on DNA collapse, and then applied Manning's theory (7) for a mixed ion system to calculate that about 90% of the phosphate charge was neutralized under the solution conditions at which collapse occurred. According to Manning's theory, the divalent ions putrescine and Mg^{2+} neutralize only 88% of the charge on DNA (7) and consequently are unable to produce collapse. A study of several inert trivalent metal ion complexes (predicted by Manning to neutralize 92% of the phosphate charge) showed that each was capable of inducing collapse in aqueous solution (175).

The finding that a high-degree of charge neutralization is required for collapse is not surprising in view of the analysis of Bloomfield and co-workers (183, 186) of the energetic factors for the packaging of T4 DNA into its phage head. Order of magnitude estimates for the various possible sources of unfavorable contributions to the free energy of packaging were obtained. Polyelectrolyte repulsions are found to be the major thermodynamic obstacle to the process of packaging T4 viral DNA (10^5 kcal of free energy per mole of virus), although the free energy of bending (10^3 kcal per mole of virus) and that due to the loss of configurational entropy upon collapse (10^2 kcal per mole of virus) were also found to be nonnegligible. Similar rankings for T5 and λ have been calculated (186). Riemer & Bloomfield (183) estimated that polyamine interactions with DNA could be sufficiently favorable to cancel the unfavorable electrostatic repulsions. Another calculation (186), based on a Poisson-Boltzmann analysis of the repulsive potential generated in one model for the collapsed state, indicates that with the reduction in surface charge density produced by multivalent counterions, dispersion forces may be sufficient to drive compaction. Ion-dipole interactions (187), and the cationic cross-bridging of DNA segments (175) have also been proposed as possible attractive forces that could drive collapse.

ACKNOWLEDGMENTS

We acknowledge with thanks the preprints and discussions with many colleagues that contributed to this review. Dr. Charles F. Anderson provided helpful comments on the manuscript. Work from this laboratory was supported by NSF grant PCM79-04607 and NIH grant GM 23467. The assistance of Mary Ehren in preparing the manuscript is gratefully acknowledged.

Literature Cited

1. Cantor, C. R., Schimmel, P. R. 1980. *Biophysical Chemistry.* San Francisco: Freeman. 1371 pp.
2. Bloomfield, V. A., Crothers, D. M., Tinoco, I. Jr. 1974. *Physical Chemistry of Nucleic Acids.* New York: Harper & Row. 517 pp.
3. Wells, R. D., Goodman, T. C., Hillen, W., Horn, G. T., Klein, R. D., Larson, J. E., Müller, U. R., Neuendorf, S. K., Panayotatos, N., Stirdivant, S. M. 1980. *Prog. Nucl. Acids Res. Mol. Biol.* 24:167–267
4. Bauer, W. R. 1978. *Ann. Rev. Biophys. Bioeng.* 7:287–313
5. von Hippel, P. H. 1979. In *Biological Regulation and Development,* Vol. 1, ed. R. F. Goldberger, pp. 279–347. New York: Plenum
6. von Hippel, P. H., McGhee, J. D. 1972. *Ann. Rev. Biochem.* 41:231–300
7. Manning, G. S. 1978. *Q. Rev. Biophys.* 11:179–246
8. Record, M. T. Jr., Anderson, C. F., Lohman, T. M. 1978. *Q. Rev. Biophys.* 11:103–78
9. Crick, F. H. C., Wang, J. C., Bauer, W. R. 1979. *J. Mol. Biol.* 129:449–61
10. Arnott, S. 1977. In Proc. *Cleveland Symp. Macromolecules 1st,* ed. A. G. Walton, pp. 87–104. Amsterdam: Elsevier
11. Davies, D. R. 1967. *Ann. Rev. Biochem.* 36:321–64
12. Deleted in proof
13. Skolnick, J. 1979. *Macromolecules* 12:515–21
14. Soumpasis, D. 1978. *J. Chem. Phys.* 69:3190–96
15. Dover, S. D. 1977. *J. Mol. Biol.* 110:699–700
16. Mandelkern, M., Dattagupta, N., Crothers, D. M. 1981. *Nature.* In press
17. Zimmerman, S. B., Pheiffer, B. H. 1979. *Proc. Natl. Acad. Sci. USA* 76:2703–7
18. Wang, J. C. 1979. *Proc. Natl. Acad. Sci. USA* 76:200–3
19. Rhodes, D., Klug, A. 1980. *Nature* 286:573–78
20. Trifonov, E. N., Bettecken, T. 1979. *Biochemistry* 18:454–56
21. Prunnel, A., Kornberg, R. D., Lutter, L., Klug, A., Levitt, M., Crick, F. H. C. 1979. *Science* 204:855–58
22. McGhee, J. D., Felsenfeld, G. 1980. *Ann. Rev. Biochem.* 49:1115–56
23. Drew, H. R., Wing, R. M., Takano, T., Broka, C., Tanaka, S., Itakura, K., Dickerson, R. E. 1981. *Proc. Natl. Acad. Sci. USA.* In press
24. Hogan, M., Dattagupta, N., Crothers, D. M. 1978. *Proc. Natl. Acad. Sci. USA* 75:195–99
25. Levitt, M. 1978. *Proc. Natl. Acad. Sci. USA* 75:640–44
26. Deleted in proof
27. Klug, A., Jack, A., Viswamitra, M. A., Kennard, O., Shakked, Z., Steitz, T. A. 1979. *J. Mol. Biol.* 131:669–80
28. Wang, A. H. J., Quigley, G. J., Kolpak, F. J., Crawford, J. L., van Boom, J. H., Marel, G., Rich, A. 1979. *Nature* 282:680–86
29. Drew, H. R., Takano, T., Tanaka, S., Itakura, K., Dickerson, R. E. 1980. *Nature* 286:567–73

30. Pohl, F. M., Jovin, T. M. 1972. *J. Mol. Biol.* 67:375–96
31. Arnott, S., Chandrasekaran, R., Birdsall, D. L., Wheslie, A. G., Ratliff, R. L. 1980. *Nature* 283:743–45
32. Arnott, S. 1980. *Trends Biochem. Sci.* 5:231–34
33. Tunis-Schneider, M. J., Maestre, M. F. 1970. *J. Mol. Biol.* 52:521–41
34. Tinoco, I. Jr., Bustamante, C., Maestre, M. F. 1980. *Ann. Rev. Biophys. Bioeng.* 9:107–41
35. Woody, R. W. 1977. *J. Poly. Sci. Macromol. Rev.* 12:181–321
36. Baase, W. A., Johnson, W. C. Jr. 1979. *Nucleic Acids Res.* 6:797–814
36a. Zimmerman, S. B., Pheiffer, B. H. 1980. *J. Mol. Biol.* 142:315–30
37. Parthasarathy, N., Schmitz, K. 1980. *Biopolymers* 19:1137–1151
38. Depew, R. E., Wang, J. C. 1975. *Proc. Natl. Acad. Sci. USA* 72:4275–79
39. Anderson, P., Bauer, W. 1978. *Biochemistry* 17:594–601
40. Klein, B. J. 1980. *Thermodynamic studies of the polyelectrolyte behavior of nucleic acids.* PhD thesis. Univ. Wisconsin-Madison. 198 pp.
41. Anderson, C. F., Record, M. T. Jr., Hart, P. A. 1978. *Biophys. Chem.* 7:301–16
42. Ivanov, V. I., Minchenkow, L. E., Schyolkina, A. K., Poletayev, A. I. 1973. *Biopolymers* 12:89–110
43. Zimmer, C., Luck, G. 1973. *Biochem. Biophys. Acta* 312:215–27
44. Hanlon, S., Brudno, S., Wu, T. T., Wolf, B. 1975. *Biochemistry* 14:1648–60
45. Chan, A., Kilkuskie, R., Hanlon, S. 1979. *Biochemistry* 18:84–91
46. Bleam, M. L., Anderson, C. F., Record, M. T. Jr. 1980. *Proc. Natl. Acad. Sci. USA* 77:3085–89
47. Seeman, N. C., Rosenberg, J. M., Suddath, F. L., Parkkim, J. J., Rich, A. 1976. *J. Mol. Biol.* 104:105–44
48. Rosenberg, J. M., Seeman, N. C., Day, R. O., Rich, A. 1976. *J. Mol. Biol.* 104:145–67
49. Pörschke, D. 1976. *Biophys. Chem.* 4:383–94
50. Bleam, M. L. 1980. *NMR Studies of the interactions of small cations with DNA.* PhD thesis. Univ. Wisconsin-Madison. 211 pp.
51. Reuben, J., Shporer, M., Gabbay, E. 1975. *Proc. Natl. Acad. Sci. USA* 72:245–47
52. Shapiro, J. T., Stannard, B. S., Felsenfeld, G. 1969. *Biochemistry* 8:3232–41
53. Privalov, P. L., Khechinashvili, N. N. 1974. *J. Mol. Biol.* 86:665–84
54. Privalov, P. L. 1979. *Adv. Protein Chem.* 33:167–241
55. Suurkuusk, J., Alvarez, J., Freire, E., Biltonen, R. 1977. *Biopolymers* 16:2641–52
56. Filimonov, V. V., Privalov, P. L. 1978. *J. Mol. Biol.* 122:465–70
57. Breslauer, K. J., Sturtevant, J. M. 1977. *Biophys. Chem.* 7:205–9
58. Dewey, T. G., Turner, D. H. 1979. *Biochemistry* 18:5757–62
59. Privalov, P. L., Ptitsyn, O. B., Birshtein, T. M. 1969. *Biopolymers* 8:559–71
60. Shiao, D. D. F., Sturtevant, J. M. 1973. *Biopolymers* 12:1829–36
61. Record, M. T. Jr., Lohman, T. M., deHaseth, P. L. 1976. *J. Mol. Biol.* 107:145–58
62. Gross, L. M., Strauss, U. P. 1966. In *Chemical Physics of Ionic Solutions,* ed. B. E., Conway, R. G. Baradas. pp. 361–89. New York: Wiley
63. Krakauer, H., Sturtevant, J. M. 1968. *Biopolymers* 6:491–512
64. Klump, H., Ackermann, T. 1971. *Biopolymers* 10:513–22
65. Marmur, J., Doty, P. 1962. *J. Mol. Biol.* 5:109–31
66. Owen, R. J., Hill, L. R., LaPage, S. P. 1969. *Biopolymers* 7:503–16
67. Gruenwedel, D. W., Han, C. H. 1969. *Biopolymers* 7:557–70
68. Frank-Kamenetskii, M. D. 1971. *Biopolymers* 10:2623–24
69. Blake, R. D., Haydock, P. V. 1979. *Biopolymers* 18:3089–3109
70. Wang, J. C., Davidson, N. 1966. *J. Mol. Biol.* 15:111–23
71. Wang, J. C., Davidson, N. 1966. *J. Mol. Biol.* 19:469–82
72. Wang, J. C., Davidson, N. 1968. *Cold Spring Harbor Symp. Quant. Biol.* 33:409–15
73. Record, M. T. Jr., Lohman, T. M. 1978. *Biopolymers* 17:159–66
74. Mertz, J. E., Davis, R. W. 1972. *Proc. Natl. Acad. Sci. USA* 69:3370–74
75. Dugaiczyk, A., Boyer, H. W., Goodman, H. M. 1975. *J. Mol. Biol.* 96:174–84
75a. Olson, W. K. 1979. *Biopolymers* 18:1213–33
76. Inners, L. D., Felsenfeld, G. 1970. *J. Mol. Biol.* 50:373–89
77. Yamakawa, H. 1971. *Modern Theory of Polymer Solutions.* New York: Harper & Row. 419 pp.
78. Godfrey, J. E., Eisenberg, H. 1976. *Biophys. Chem.* 5:301–18

79. Jolly, D., Eisenberg, H. 1976. *Biopolymers* 15:61–95
80. Voordouw, G., Kam, Z., Borochov, N., Eisenberg, H. 1978. *Biophys. Chem.* 8:171–89
81. Yamakawa, H., Fujii, M. 1974. *Macromolecules* 7:128–35
82. Yamakawa, H., Fujii, M. 1974. *Macromolecules* 7:649–54
83. Kovacic, R. T., van Holde, K. E. 1977. *Biochemistry* 16:1490–98
84. Record, M. T. Jr., Woodbury, C. P., Inman, R. B. 1975. *Biopolymers* 14:393–408
85. Sharp, P., Bloomfield, V. A. 1968. *Biopolymers* 6:1201–11
86. Borochov, N., Eisenberg, H., Kam, Z. 1981. *Biopolymers* In press
87. Hearst, J. E., Schmid, C. W., Rinehart, F. P. 1968. *Macromolecules* 1:491–94
87a. Harpst, J. A. 1980. *Biophys. Chem.* 11:295–302
88. Hays, J. B., Magar, M. E., Zimm, B. H. 1969. *Biopolymers* 8:531–36
89. Sharp, P., Bloomfield, V. A. 1968. *J. Chem. Phys.* 48:2149–55
90. Schmid, C. W., Rinehart, F. P., Hearst, J. E. 1971. *Biopolymers* 10:883–93
91. Harrington, R. E. 1978. *Biopolymers* 17:919–36
92. Crothers, D. M., Zimm, B. H. 1965. *J. Mol. Biol.* 12:525–36
93. Frontali, C., Dore, E., Ferrauto, A., Gratton, E., Bettini, A., Pozzan, M. R., Valdevit, E. 1979. *Biopolymers* 18:1353–73
93a. Hagerman, P. J. 1981. *Biopolymers.* In press
94. Skolnick, J., Fixman, M. 1977. *Macromolecules* 10:944–48
95. Odijk, T. 1977. *J. Polymer Sci. Polymer Phys. Ed.* 15:477–83
96. Odijk, T., Houwaart, A. C. 1978. *J. Polymer Sci. Polymer Phys. Ed.* 16:627–39
97. Odijk, T. 1979. *Biopolymers* 18:3111–13
98. Landau, L., Lifshitz, E. 1958. *Statistical Physics,* pp. 478–82. London: Pergamon
99. Schellman, J. A. 1974. *Biopolymers* 13:217–26
100. Gray, H. B. Jr., Hearst, J. E. 1968. *J. Mol. Biol.* 35:111–29
101. Harrington, R. E. 1977. *Nucleic Acids Res.* 4:3519–35
102. Barkley, M. D., Zimm, B. H. 1979. *J. Chem. Phys.* 70:2991–3007
103. Allison, S. A., Schurr, J. M. 1979. *Chem. Phys.* 41:35–59
103a. Thomas, J. C., Allison, S. A., Appel-lof, C. J., Schurr, J. M. 1980. *Biophys. Chem.* 12:177–88
104. Millar, D. P., Robbins, R. J., Zewail, A. H. 1980. *Proc. Natl. Acad. Sci. USA* 77:5593–97
105. Robinson, B. H., Lerman, L. S., Beth, A. H., Frisch, H. L., Dalton, L. R., Aver, C. 1980. *J. Mol. Biol.* 139:19–44
106. Vologodskii, A. V., Anshelevich, V. V., Lukashin, A. V., Frank-Kamenetskii, M. D. 1979. *Nature* 280:294–98
106a. LeBret, M. 1980. *Biopolymers* 19:619–37
107. Wahl, Ph., Paoletti, J., LePecq, J. B. 1970. *Proc. Natl. Acad. Sci. USA* 65:417–21
108. Bauer, W., Vinograd, J. 1974. In *Basic Principles in Nucleic Acid Chemistry,* ed. P. O. P. Ts'o, 2:262–303. New York: Academic
109. Crick, F. H. C. 1976. *Proc. Natl. Acad. Sci. USA* 73:2639–43
110. Fuller, F. B. 1978. *Proc. Natl. Acad. Sci. USA* 75:3557–61
111. Wang, J. C. 1969. *J. Mol. Biol.* 43:25–39
112. Upholt, W. B., Gray, H. B. Jr., Vinograd, J. 1971. *J. Mol. Biol.* 62:21–38
113. Ostrander, D. A., Gray, H. B. Jr. 1973. *Biopolymers* 12:1387–1419
114. Campbell, A. M., Jolly, D. J. 1973. *Biochem J.* 133:209–66
115. Campbell, A. M., Eason, R. 1975. *FEBS Lett.* 55:212
116. Campbell, A. M. 1976. *Biochem. J.* 155:101–5
117. Lau, P. P., Gray, H. B. Jr. 1979. *Nucleic Acids Res.* 6:331–57
118. Bram, S. 1971. *J. Mol. Biol.* 58:277–88
119. Stettler, U. H., Weber, H., Koller, T., Weissman, C. 1979. *J. Mol. Biol.* 131:21–40
120. Maestre, M. F., Wang, J. C. 1971. *Biopolymers* 10:1021–30
121. Campbell, A. M., Lochhead, D. S. 1971. *Biochem. J.* 123:661–63
122. Belintsev, B. N., Gagua, A. V., Nedospasov, S. A. 1979. *Nucleic Acids Res.* 6:983–92
123. Bauer, W., Vinograd, J. 1970. *J. Mol. Biol.* 47:419–35
124. Pulleyblank, D. E., Shure, M., Tang, D., Vinograd, J., Vosberg, H. P. 1975. *Proc. Natl. Acad. Sci. USA* 72:4280–84
125. Shure, M., Pulleyblank, D. E., Vinograd, J. 1977. *Nucleic Acids Res.* 4:1183–1205
126. Hsieh, T. S., Wang, J. C. 1975. *Biochemistry* 14:527–35
127. Fuller, F. B. 1971. *Proc. Natl. Acad. Sci. USA* 68:815–19

128. Benham, C. J. 1977. *Proc. Natl. Acad. Sci. USA* 74:2397–41
129. Benham, C. J. 1979. *Biopolymers* 18:609–23
130. Camerini-Otero, R. D., Felsenfeld, G. 1978. *Proc. Natl. Acad. Sci. USA* 75:1708–12
131. Hale, P., Woodward, R. S., Lebowitz, J. 1980. *Nature* 284:640–44
131a. Panayotatos, N., Wells, R. D. 1981. *Nature* 289:466–70
132. Holloman, W. K., Wiegand, R., Hoessli, C., Radding, C. M. 1975. *Proc. Natl. Acad. Sci. USA* 72:2394–98
133. Beattie, K. L., Wiegand, R. C., Radding, C. M. 1977. *J. Mol. Biol.* 116:783–803
134. Wiegand, R. C., Beattie, K. L., Holloman, W. K., Radding, C. M. 1977. *J. Mol. Biol.* 116:805–24
135. Laiken, N. 1973. *Biopolymers* 12:11–26
136. Benham, C. J. 1979. *Proc. Natl. Acad. Sci. USA* 76:3870–74
137. Benham, C. J. 1980. *J. Chem. Phys.* 77:3633–39
138. Anshelevich, V. V., Vologodskii, A. V., Lukashin, A. V., Frank-Kamenetskii, M. D. 1979. *Biopolymers* 18:2733–44
139. Vologodskii, A. V., Lukashin, A. V., Anshelevich, V. V., Frank-Kamenetskii, M. D. 1979. *Nucleic Acids Res.* 6:967–82
140. Helene, C. 1981. *Crit. Rev. Biochem.* In press
141. Wilson, W. D., Jones, R. L. 1981. In *Intercalation Chemistry*, ed. M. S. Whittingham, A. J. Jacobson. New York: Academic: In press
142. Barkley, M. D., Bourgeois, S. 1980. In *The Operon*, ed. W. S. Reznikoff, J. H. Miller, pp. 171–220. Cold Spring Harbor, New York: Cold Spring Harbor Lab.
143. Chamberlin, M. J. 1976. In *RNA Polymerase*, ed. R. Losick, M. Chamberlin. pp. 159–92. Cold Spring Harbor, New York: Cold Spring Harbor Lab.
144. Jovin, T. M. 1976. *Ann. Rev. Biochem.* 45:889–920
145. Record, M. T. Jr., deHaseth, P. L., Lohman, T. M. 1977. *Biochemistry* 16:4791–95
146. Strauss, H. S., Burgess, R. R., Record, M. T. Jr. 1980. *Biochemistry* 19:3504–15
147. Barkley, M. D., Lewis, P. A., Sullivan, G. E. 1981. *Biochemistry.* In press
148. Revzin, A., von Hippel, P. H. 1977. *Biochemistry* 16:4769–76
149. deHaseth, P. L., Lohman, T. M., Record, M. T. Jr. 1977. *Biochemistry* 16:4783–90
150. deHaseth, P. L., Lohman, T. M., Burgess, R. R., Record, M. T. Jr. 1978. *Biochemistry* 17:1612–22
151. Lohman, T. M., Wensley, C. G., Cina, J., Burgess, R. R., Record, M. T. Jr. 1980. *Biochemistry* 18:3516–22
152. Revzin, A., Woychik, R. P. 1981. *Biochemistry* 30:251–55
153. McGhee, J. D., von Hippel, P. H. 1974. *J. Mol. Biol.* 86:469–89
154. Zingsheim, H. P., Geisler, N., Weber, K., Mayer, F. 1977. *J. Mol. Biol.* 115:565–70
155. Hirsh, J., Schleif, R. 1976. *J. Mol. Biol.* 108:471–90
156. Goeddel, D. V., Yansura, D. G., Caruthers, M. H. 1978. *Proc. Natl. Acad. Sci. USA* 75:3578–82
157. Manning, G. S. 1969. *J. Chem. Phys.* 51:924–33
158. Anderson, C. F., Record, M. T. Jr. 1980. *Biophys. Chem.* 11:353–60
159. Wilson, R. W., Rau, D. C., Bloomfield, V. A. 1980. *Biophys. J.* 30:317–26
160. Lohman, T. M., deHaseth, P. L., Record, M. T. Jr. 1980. *Biochemistry* 19:3522–30
161a. Howe-Grant, M., Lippard, S. J. 1979. *Biochemistry* 18:5762–69
161b. Becker, M. M., Dervan, P. B. 1979. *J. Am. Chem. Soc.* 101:3664–66
162. Saucier, J. M. 1977. *Biochemistry* 16:5879–89
163. Capelle, N., Barbet, J., Dessen, P., Blanquet, S., Roques, B. P., LePecq, J. B. 1979. *Biochemistry* 18:3354–62
164. Wilson, W. D., Lopp, I. G. 1979. *Biopolymers* 18:3025–41
165. O'Gorman, R. B., Dunaway, M., Matthews, K. S. 1980. *J. Biol. Chem.* 255:10100–10106
166. Riggs, A. D., Suzuki, H., Bourgeois, S. 1970. *J. Mol. Biol.* 48:67–83
167. Lin, S.-Y., Riggs, A. D. 1975. *Cell* 4:107–11
168. Goeddel, D. V., Yansura, D. G., Winston, C., Caruthers, M. H. 1978. *J. Mol. Biol.* 123:661–87
169. Lohman, T. M., deHaseth, P. L., Record, M. T. Jr. 1978. *Biophys. Chem.* 8:281–94
170. Belintsev, B. N., Zauriev, S. K., Shemyakin, M. F. 1980. *Nucleic Acids Res.* 8:1391–404
171. Lerman, L. 1971. *Proc. Natl. Acad. Sci. USA* 68:1886–90
172. Evdokimov, Y. M., Platonov, A. L., Tikhonenko, A. S., Varshavsky, Ya. M. 1972. *FEBS Lett.* 23:180–84
173. Gosule, L. C., Schellman, J. A. 1976. *Nature* 259:333–35

174. Gosule, L. C., Schellman, J. A. 1978. *J. Mol. Biol.* 121:311–26
175. Widom, J., Baldwin, R. L. 1980. *J. Mol. Biol.* 144:431–53
176. Olins, D. E., Olins, A. L. 1971. *J. Mol. Biol.* 57:437–55
177. Haynes, M., Garrett, R. A., Gratzer, W. B. 1970. *Biochemistry* 9:4410–16
178. Chattoraj, D. K., Gosule, L. C., Schellman, J. A. 1978. *J. Mol. Biol.* 121:327–37
179. Laemmli, U. K. 1975. *Proc. Natl. Acad. Sci. USA* 72:4288–92
180. Jordan, C. F., Lerman, L. S., Venable, J. H. Jr. 1972. *Nature New Biol.* 236:67–70
181. Maniatis, T., Venable, J. H. Jr., Lerman, L. S. 1974. *J. Mol. Biol.* 84:37–64
182. Wilson, R. W., Bloomfield, V. A. 1979. *Biochemistry* 18:2192–96
183. Riemer, S. C., Bloomfield, V. A. 1978. *Biopolymers* 17:785–94
184. Post, C. B., Zimm, B. 1979. *Biopolymers* 18:1487–501
185. Dore, E., Frontali, C., Gratton, E. 1972. *Biopolymers* 11:443–59
186. Bloomfield, V. A., Wilson, R. W., Rau, D. C. 1980. *Biophys. Chem.* 11:1339–43
187. Oosawa, F. 1971. *Polyelectrolytes.* New York: Dekker

Ann. Rev. Biochem. 1981. 50:1025–52

PRIMARY STRUCTURAL ANALYSIS OF THE TRANSPLANTATION ANTIGENS OF THE MURINE H-2 MAJOR HISTOCOMPATIBILITY COMPLEX[1]

◆12103

Stanley G. Nathenson, Hiroshi Uehara, and Bruce M. Ewenstein

Department of Microbiology and Immunology and Department of Cell Biology, Albert Einstein College of Medicine, Bronx, New York 10461

Thomas J. Kindt and John E. Coligan

Laboratory of Immunogenetics, Building 8, Room 100, National Institutes of Health, Bethesda, Maryland 20205

CONTENTS

[1]Abbreviations used are: MHC, major histocompatibility complex; β_2m, β_2-microglobulin; K^b, D^d, D^b D^d, product of the respective H-2 gene, (e.g. H-2K^b or K^b is the product of the H-2K^b gene).

0066-4154/81/0701-1025$01.00

PERSPECTIVES AND SUMMARY

The murine H-2 Class I histocompatibility alloantigens consist of two non-covalently associated polypeptide chains: a highly polymorphic 45,000-mol wt glycoprotein (also referred to as the H-2 heavy, or 45K chain) coded for by the major histocompatibility complex (MHC), and a 11,600-mol wt non-MHC coded protein of limited variability termed β_2-microglobulin (β_2M) (also referred to as the light or 12K chain). These antigens are integral components of the cell surface membrane of all cells, but are more densely concentrated on lymphocytes (1–5).

Recent advances in the technology for determining the amino acid sequence of molecules that are available only in small amounts have permitted detailing the primary structure of one of the H-2 antigens (6; H. Uehara, J. E. Coligan, S. G. Nathenson, submitted for publication) as well as the partial sequence of several others (7–11). Concomitant advances in preparation and purification of large amounts of MHC Class I HLA alloantigens from human cell lines have permitted similar studies to proceed in the human system using conventional protein sequencing techniques (12–14). Data from these two sources, mouse and human, have provided information on the intramolecular organization of these Class I histocompatibility molecules as well as insights into their genetic interrelationships.

The H-2 Class I K^b glycoprotein consists of 346 amino acid residues and two carbohydrate moieties. On the basis of its postulated structure in the cell surface membrane, the molecule can be divided into three functional regions. The extracellular region consists of approximately the amino terminal 283 residues and includes two linearly arranged intrachain disulfide loops and two carbohydrate moieties. This region carries the alloantigenic and T cell–recognized determinants (15). Proteolytic action by papain cleaves this moiety from the cell surface membrane as a fragment soluble in aqueous solution (16). The carboxyl-terminal 63 residues contain two functional regions: a highly hydrophobic membrane binding region (approximately 24-residues long) followed by a hydrophilic intracellular region at the extreme carboxyl-terminus (H. Uehara, J. E. Coligan, S. G. Nathenson, manuscript submitted for publication).

Sequence comparisons among four H-2 molecules and two HLA molecules for the N-terminal 100 residues have shown that H-2 and HLA heavy chains are highly homologous molecules (approximately 70% homology). A similar degree of sequence homology was also observed by sequence comparison of the entire papain fragments H-2K^b and HLA-B7. Homology between four H-2 molecules (two K and two D antigens) for the N-terminal 100 residues ranges from 74 to 87%. Most amino acid differences occur in discrete regions distributed throughout the entire length; these are interspersed with regions of near identity. One extraordinarily diverse region among all molecules examined so far occurs from positions 61–83. This may be important with regard to the phenomenon of associative recognition in which immune cytotoxic T lymphocytes recognize foreign cell surface antigens in the context of a specific Class I glycoprotein (17–19).

Complexity at three levels characterizes the Class I H-2 MHC system. First, an extraordinarily large number of alleles (i.e. polymorphism), have been found to be present in the mouse population, second, an extensive degree of diversity exists among the allelic products, and third, there are at least three closely linked genes, K, D, and L, that coexpress homologous Class I products on the cell surface. These unique properties raise interesting questions as to the generation and regulation of the complexity and diversity of this system, questions that can be approached by direct analysis of the MHC genetic structure at the DNA level using gene cloning techniques. New directions must also include three-dimensional structure analysis of these antigens in order to localize functional sites.

This article reviews the present state of knowledge on the primary structure of the murine H-2 alloantigens. A speculative model of the antigen on the membrane is presented and interpretation of the results of primary structure comparisons are used to probe possible answers to the questions of the generation of the polymorphism and genetic interrelationships. References are made where pertinent to the homologous human products, the HLA antigens. Details of the structural and biological properties of HLA antigens are presented in current comprehensive reviews by Strominger and colleagues (20, 21).

INTRODUCTION

Originally discovered by Gorer (22) as a mouse blood group system associated with histoincompatibility between mouse strains, the H-2 antigen system has been the object of increasing interest for the past 35 years. Discovery that genetic control of tissue graft rejection is controlled by a number of loci (e.g. K, D and L) (1, 2, 3) and that tightly linked to these are loci controlling antibody immune responsiveness (I region) to foreign antigens (23–25), and also a locus controlling a component of complement (S region) (26, 27) has led to an unparalleled development of immunogenetic research on the MHC in the past few years. Continuing reevaluation of the fine genetic structure of this highly complex region has produced a genetic map as shown in Figure 1. This "super gene" (2, 28) is now called the H-2 major histocompatibility complex (MHC) and is about 1.5 centimorgans in length, approximately 17–20 centimorgans from the centromere on chromosome 17. Similar MHC complexes exist in other species (3, 5).

The classical transplantation antigens are the K and D region gene products (K, D, and L), and they are generally referred to as Class I MHC products (1, 2). In situ, these integral cell membrane antigens are a noncovalent complex of two components of which only the heavy chain (the 45-K chain) is determined by the MHC, whereas the light chain, the β_2m, is coded elsewhere (Figure 1).

While the functional properties of the Class I, K, D, and L antigens are not yet unequivocably defined, it is already clear that they provide the specificity for certain cell-cell interactions. Although they are targets of tissue graft rejection, present evidence suggests that allograft rejection is a specific instance of their more general role in the cell mediated immune response to viral and other infections (17). In such cases, foreign antigens are recognized by the immune system in the context of the H-2 Class I glycoproteins. This phenomenon has been termed associative recognition because of the dual requirement for the "self" Class I molecule and the foreign antigen (17–19, 30).

The I region of the H-2 MHC contains at least four subregions that appear to control a number of different phases of the humoral immune response, as well as the expression of another set of cell surface glyco-

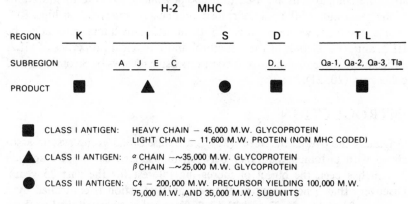

Figure 1 Abbreviated genetic structure of the H-2 major histocompatibility complex, and overall biochemical properties of three classes of antigenic products.

proteins, the Ia antigens (1, 23–25). These are referred to as the Class II antigens (Figure 1). The Ia antigens are not associated with $\beta_2 m$ and are found in the membrane as noncovalently associated complexes of glyco-protein chains (α chain of 35,000 mol wt and β chain of 28,000 mol wt). Recent evidence has suggested that there is also a third invariant (31) molecule associated with the α and β chains. The S region of the MHC codes for another class of molecules that comprise the fourth component of complement (4, 26, 27).

The TL (32) and Qa (33, 34) antigens that are encoded by genes that map to the right of the D region appear to have biochemical properties similar to the K/D/L glycoproteins, and therefore are also referred to as MHC Class I molecules; however, their products differ notably from K, D, and L in a number of ways including restricted tissue distribution, and, for TL, anomolous appearance in tumors. Relatively little is known about their structure.

DETERMINATION OF PRIMARY STRUCTURE BY RADIOCHEMICAL METHODOLOGY

General Methodology

When biochemical studies of biologically important membrane bound molecules such as the H-2 alloantigens were begun nearly 20 years ago available technology was not adequate for isolation and characterization of such moieties. One particular problem was their aqueous insolubility due to the presence of hydrophobic regions in the molecules, a property that prevented the use of conventional biochemical purification techniques. A second problem was the extremely minute amount of materials available on cells.

These problems were resolved for the H-2 system by the use first of proteolytic cleavage, and later of nonionic detergents for solubilization. Along with the utilization of radiochemical microsequence methodology for the analysis of biosynthetically labeled materials, such techniques have greatly facilitated the studies described below.

In 1966 (35), it was shown that papain or ficin proteolytic digestion of spleen or tumor cell membranes, or of whole cells, would release an H-2 antigenically active fragment that was soluble in aqueous solution. Analyses of material, purified by classical techniques, provided the first proof that the H-2 antigens were glycoproteins (16). The use of nonionic detergents, such as Triton X-100 and NP-40 to extract H-2 (36), as well as other cell membrane components was introduced somewhat later. This technique has been utilized extensively because of the advantage that H-2 molecules are isolated with considerably higher yield and in the intact, native form. Purification of relatively large amounts of H-2 antigens from spleens or livers by solubilization with papain (37), and from tumor cells by solubilization with Triton X-100 or deoxycholate (38–40) has been moderately successful, but it has not been widely used because of the large quantities of starting materials required. Recent studies on nonradioactive H-2 antigens purified by affinity chromatography with homogeneous antibodies show promise of circumventing previous purification problems (39), especially if one can use cell culture facilities to prepare large amounts of starting material (40).

For most studies on the isolation and characterization of the H-2 molecules, radiochemical micromethodology has been used.

The use of biosynthetically labeled material is advantageous for three reasons. First, the sensitivity of detection of radioactivity enables the analysis of very small amounts of material. Second, antibodies can be used to purify the H-2 molecule by either immune precipitation or by immunoadsorbant chromatography since one does not have to be concerned about contamination of the radiolabeled H-2 molecules with unlabeled proteins (in fact it is helpful to have additional protein in order to prevent losses). Third, many different molecules can be isolated for study by their sequential immunoprecipitation from the cell extract, since all cell proteins are radiolabeled by the biosynthetic incorporation of radioactive amino acids.

Isolation of Radiolabeled Class I Antigens

The methods used here for isolation of MHC membrane glycoproteins have general applicability for isolation of any membrane molecules that can be precipitated with a specific antibody. The procedure used by our laboratories for the isolation of the H-2 45K-glycoprotein is shown in Figure 2.

Radiolabeled antigens are usually obtained from cultured tumor lines. Tumor lines are easily cultivated and express large amounts of the antigen in question. Spleen cells, however, can also be used as the source of H-2 antigen for structural studies of H-2 molecules from mutant strains of mice, although the efficiency of radioactive incorporation is lower (R. Nairn, S. Schumacher, S. G. Nathenson, manuscript in preparation). Nonionic detergent NP-40 is usually used to solubilize radiolabeled H-2 antigens (36) although other detergents [e.g. deoxycholate (47), Brij (48, 49)] have been used to solubilize HLA molecules.

After *Lens culinaris* lectin affinity chromatography is used for partial fractionation of the glycoprotein fraction (50), the H-2 antigen is isolated by immunoprecipitation. Separation of the 45K-glycoprotein from the non-covalently associated β_2m is achieved by gel filtration of the acid solubilized immunoprecipitate in 1 M formic acid. A number of different molecules can be isolated from the same radiolabeled cell lysate. Thus, the H-2D^b (9) molecule has been recovered from the glycoprotein fraction supernatant

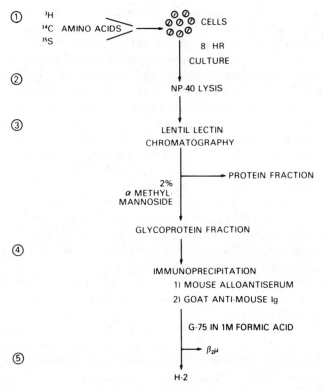

Figure 2   General methodology for isolation of labeled H-2 alloantigens.

from EL4 cell lysates after removal of the H-2K^b molecules. Similarly, H-2K^d, H-2D^d, and H-2L^d have been isolated successfully from the same cell lysate of C-14 cells (H-2^d) (8, 10).

Radiochemical Sequence Determination

As mentioned previously, the radiochemical approach to sequence determination is most appropriate in cases for which materials are too difficult to obtain in large amounts and in sufficient purity for conventional sequencing methods (51). The approach was first applied for the determination of partial amino acid sequences of small proteins (52, 53), hormones (54), and viral peptides (51). The early studies on H-2 molecules consisted of the determination of partial amino terminal sequences (reviewed in 4, 29). However, the applicability of this methodology for the primary structural analysis of larger proteins has been demonstrated by the successful sequence analysis of the H-2K^b glycoprotein, a molecule of 346 amino acids (6,

41–46; H. Uehara, J. E. Coligan, S. G. Nathenson, submitted for publication).

Radiochemical sequencing differs from the conventional approaches only in the sense that certain modifications are necessary due to the specific requirements for analyses of radiolabeled materials (42, 43). Of particular importance is which radiolabeled amino acids to use together for most efficient utilization of time and materials. In general, it is impractical to label with more than seven or eight amino acids. Such multilabelings are best restricted to certain amino acids, since each amino acid is incorporated into cells at different levels, and the relative specific activities for two amino acids can differ by as much as 20-fold. This leads to difficulties in identifying the low specific activity PTH amino acid residues when they are adjacent in sequence position to high specific activity ones. Amino acids that are incorporated less efficiently are generally used in single label preparations. Since aspartic acid is incorporated poorly, Asp residues are usually determined indirectly (45, 46). A more complete analysis of the details of these methods is discussed elsewhere (42, 43).

In sequence determination of the H-2 molecules carried out with the radioactive methods, three major criteria were used as guidelines for positive assignment of the amino acid residues at a given position (44, 45). First, for analysis of the peptides labeled with a single amino acid, assignments were made at positions showing significant radioactivity in the butyl chloride extract obtained from the automatic sequencer. These assignments were usually verified by high pressure liquid chromatographic analysis of the phenylthiohydantoin amino acid derivatives. Second, when the same amino acid appeared at more than one step in the sequencer runs, identification was considered valid only when the radioactivity at these steps fitted the expected repetitive yield line (45). Recoveries of radioactivity for other residues in multilabel preparations were also required to be related to this curve as determined by the ratio of specific activities of the amino acids involved. Third, when possible, assignments were made only after an amino acid was observed at a given position in more than one analysis. This requirement was fulfilled by the isolation and sequence analysis of peptides from different digests of the same region of the molecule or by duplicate analysis of separate samples of the same peptide.

PRIMARY STRUCTURAL STUDIES

The Primary Structure of H-2K^b

The complete primary structure of the H-2K^b molecule, (Figure 3) has been determined by radiochemical microsequencing methodology. The strategies for generation and purification of peptides employed in this study were basically the same as those that have been used for primary structure analyses of nonradioactive proteins. Initial fragmentation of the molecule was performed by CNBr cleavage at Met residues (55), which results in the generation of five major fragments derived from the NH$_2$-terminal 284 residues and several smaller fragments derived from the COOH-terminal portion of the molecule. These CNBr fragments were purified by gel filtration and ion exchange column chromatography (42). After NH$_2$-terminal sequence analysis of each CNBr fragment established its unique identity, the total sequence of each fragment and its alignment were determined. The sequence of large CNBr fragments was determined by analysis of smaller peptides obtained from these fragments by proteolytic cleavage (6, 45, 46). The hydrophobic portion of the molecule was not isolated as a CNBr fragment due to its insolubility. Therefore, it was isolated from the intact molecule as a tryptic fragment.

The molecule consists of 346 amino acid residues to which are attached two carbohydrate chains vis Asn residues (Figure 3). Two linearly arranged intrachain disulfide bridges are formed between Cys 101 and Cys 164 and between Cys 203 and Cys 259 (J. Martinko, J. Adlersberg, and S. G. Nathenson, manuscript submitted for publication). Following the second

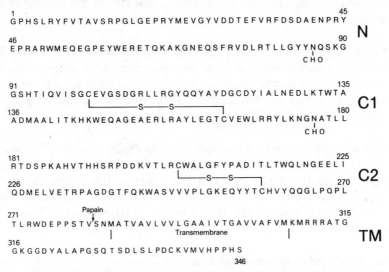

Figure 3 Complete amino acid sequence of H-2K^b molecule. Amino acid residues are indicated in the single letter amino acid code. (IUPAC-IUB Commission on Biochemical Nomenclature 1968). CHO, carbohydrate moiety; S-S, cysteine disulfide bridge. The prominent letters on the right side indicate: N,N-terminal 90 amino acid residues; C1, first disulfide loop region; C2, second disulfide loop region; TM, hydrophobic region and COOH-terminal hydrophilic region. [Data taken from (45) and (H. Uehara, J. E. Coligan, and S. G. Nathenson, manuscript submitted for publication).

disulfide bridge, a large stretch containing a high percentage of uncharged and hydrophobic amino acid residues is found between position 277 and 307, which is followed by a COOH-terminal hydrophilic region.

The molecule can be functionally divided into three distinct portions on the basis of its postulated orientation with respect to the cell membrane: (*a*) an NH_2-terminal region of about 280 residues, which is situated external to the cell membrane; (*b*) a hydrophobic region (approximately 25 residues), which appears to interact with the cell membrane; and (*c*) a COOH-terminal hydrophilic region, which is thought to be situated intracellularly.

THE EXTRACELLULAR PORTION   A water soluble fragment of H-2 (37,000 mol wt) can be solubilized from the cell membrane by proteolysis with papain (16, 35). This fragment is derived from the NH_2-terminus of the intact molecule, as demonstrated by sequence analysis of the fragment (41), and retains the serological and antigenic activities of the intact molecule (15, 16, 56). In situ this NH_2-terminal portion of the molecule is presumed to be situated outside of the cell membrane, since it can be removed by proteolysis without destroying the integrity of the cells. This extracellular region can be further subdivided into three regions based on the two intrachain disulfide loops arrangement and on the location of the carbohydrate moieties.

NH_2-terminal 90 residues (referred to as N, or α1 for the HLA chain)   No intrachain disulfide bridge is found in this region. An attachment site for the carbohydrate moiety is assigned as Asn 86. The sequence including this residue is Asn-Glu-Ser, which fulfills the recognition sequence for glycosylation at Asn residues (57).

First disulfide loop region (residues 91–108) (referred to as C1 or α2 in HLA)   One of the two disulfide bridges in the molecule is formed between Cys 101 and Cys 164, roughly in the center of this region. A second carbohydrate moiety is attached to Asn 176. The existence of this carbohydrate chain may be specific for H-2 molecules: in HLA-B7 and A2, only one carbohydrate moiety attached at Asn 86 has been reported (58, 59).

Second disulfide loop region (referred to as C2, or α3 in HLA)   This region extends from residue 181 to the COOH-terminus of the papain fragment, and includes a second disulfide bridge formed between Cys 203 and Cys 259. A papain cleavage site of H-2K^b was determined as Val 281, whereas the COOH-terminus of the papain fragment of HLA-B7, and the mixture of HLA-A, B, and C proteins were reported as Thr 271 and Arg 273, respectively. (13, 14, 59).

HYDROPHOBIC REGION (APPROXIMATELY POSITIONS 284–307)
Studies using lactoperoxidase catalyzed iodination of Tyr residues in intact lymphocytes, and inside-out membrane vesicles have suggested that HLA antigens are transmembrane proteins (60). Although no such studies have been carried out for the H-2 molecules so far, it is likely that H-2 has a similar cell membrane orientation. An extremely hydrophobic segment of H-2K^b was found from position 284–307 that consists of a continuous stretch of uncharged amino acid residues. It is localized between two hydrophilic regions. Since the papain derived H-2K^b molecule lacks this segment and also has lost its membrane binding properties, this extremely hydrophobic region appears to be a transmembrane region. It is notable that a peptide of this length, if in a α helix conformation, would span a distance approximately equal to that of the hydrocarbon core of a lipid bilayer (61). It is pertinent that a similar 23 uncharged amino acid stretch has been found in glycophorin near the COOH-terminus (62, 63). Glycophorin is a major sialic acid–containing glycoprotein of the human erythrocyte membrane (64, 65), and is believed to be a transmembrane protein (66, 67).

COOH-TERMINAL HYDROPHILIC REGION (POSITIONS 308–346)   This region consists of 39 amino acid residues with a high content of polar amino acids (50%), and is localized to the COOH-terminus of the hydrophobic region. A cluster of basic residues, –ARG–ARG–ARG–, is found near the NH$_2$ terminus of this region. A similar cluster of basic residues at the beginning of the COOH-terminal hydrophobic region is also found in HLA-B7, HLA-2, (68), and glycophorin (63, 69). These basic residues may interact with the phosphate moiety of phospholipids and may function as a free-energy barrier preventing the COOH terminus from entering the hydrophobic lipid bilayer (20). A free Cys residue is assigned at position 337. This residue may be involved in the postulated dimerization of 45-K molecules that has been reported for detergent solubilized histocompatibility antigens (70–72). Two Cys residues are also assigned in the COOH-terminus hydrophilic region of HLA-B7, although the positions of these residues are different from those of H-2K^b.

In the HLA molecule, phosphorylation of one or more Ser residues in the COOH-terminus region has been observed (73). The phosphorylation of H-2K^b molecule has also been suggested from preliminary experiments testing for incorporation of ^{32}P-phosphate, although the site of phosphorylation has not yet been identified (J. E. Coligan, H. Uehara, and S. G. Nathenson, unpublished observation). Recently, the phosphorylation of Ser residues at the COOH-terminal region of the H-2K^k molecule has also been observed (11). Since phosphorylation of Ser or Thr residues plays a regulatory role important for enzyme function (74) and in contractile or cyto-

Table 1 Seguence of β_2-microglobulin from mouse and other species[a]

		10	20	30	40	50
Mouse β₂M		I Q K T P Q I Q V Y	S R H P P E N G K P	N I L N C Y V T Q F	H P P H I E I Q M L	K N G K K I P K V E
Guinea pig β₂M	V L H A – R V ——————— A ——— Q – F I ——— S G ——— Q — V E L ——————— D N —
Rabbit β₂M	V – R A – N V ——————— A ——————— F ——— S G ——— Q – D – E L ——— V — E N —
Human β₂-M	— R — K ——————————— A ——— S – F ——— S G ——— S D — V D L — D – E R – E ——

		60	70	80	90	99
Mouse β₂-M		M S D M S F S K D W	S F Y I L A []H T E F	T P T E T D T Y A C	R V K H $^{A}_{D}$ S M A E P	K T V Y W D R D M
Guinea pig β₂M	—— L ——————— T — L – V []– A A ——— N D S – E – S ——— S – I T L S ———— I – K — P N K
Rabbit β₂M	Q — L — N ——————— L – V [] ———— N N K N E – S ——————— V T L K — M — K ——————— Y
Human β₂M	H — L ——————————— L – Y S Y ——————— K – E ——————— N – V T L S Q — I – K ————

[a] Sequence is expressed in one letter code (IUPAC–IUB Commission on Biochemical Nomenclature). Lines indicate homology to mouse sequence. Data are taken from the following references: mouse (87); guinea pig (85); rabbit (86); and human (84).

sketal systems (75–77), the phosphorylation reaction might be involved in the interactions of MHC antigens with the cytoskeletal components.

Radiosequence Determination of the H-2 Associated Chain, β_2-Microglobulin

Although β_2m is not genetically determined by the MHC, its association as the smaller subunit of the H-2 and HLA antigens justifies consideration of its primary structure. β_2m (mol wt 11,800) was first discovered in human urine (78) and later was found to be noncovalently associated with the HLA glycoproteins (79, 80) or H-2 glycoproteins (81, 82). Despite its attachment to various cell surface proteins, the function remains unclear, although there is some evidence that the molecule may stabilize the tertiary structure of associated proteins (83).

The complete amino acid sequence has been determined for the human (84), guinea pig (85) and rabbit molecules (86). Most recently the complete sequence of murine β_2m has been determined by radiosequence methodology (87). The murine β_2m was isolated as a by-product during the purification of H-2^d and H-2^b alloantigens. As shown in Table 1, murine β_2m and its homologues from the other species examined differ in sequence by 30–40%. Previous amino acid sequence determinations had not detected any polymorphism in β_2m within the species, unlike the extraordinary intra species diversity for the H-2 45-K chain. However, the primary structural analysis (87) of murine β_2m isolated from the tumor cell lines E14.BU (H-2^b) and C-14 (H-2^d) have revealed the presence of an amino acid difference at position 85 of this molecule (see Table 1). The β_2m isolated from these cell lines also showed differences in electrophoretic mobility that correspond to similar differences noted for material obtained from H-2^b and H-2^d mouse lines (88).

Of considerable interest is the finding that the amino acid sequence of mouse β_2m shows a significant homology to the sequence of the second disulfide loop region of the K^b 45-K chain, just as has been observed in the comparison of human β_2m with HLA-B7 (89). Although only 20 residues of murine β_2m are homologous to amino acids in positions 179–278 of H-2K^b, 16 of these substitutions are conserved in the human, rabbit, and guinea pig sequence of β_2m. In combination with a large number of conservative amino acid interchanges, this limited homology suggests that β_2m and the region of the second disulfide loop of histocompatibility antigens may have a common evolutionary origin. The functional significance of this homology must await an analysis of the biological role and three-dimensional structure of the histocompatibility antigens.

Overall Structural Properties of H-2 and HLA 45-K Glycoproteins

Examination of the overall properties of six H-2 and one HLA 45-K chains reveals the striking similarity in structure of these transplantation antigens. As shown in Figure 4, a number of major features are shared by the H-2 molecules. First, there are two intrachain disulfide bridges defined in H-2K^b (J. M. Martinko et al, submitted for publication) that appear to be found in all other H-2 molecules that have been analyzed for this characteristic. The size and location of the disulfide loops in H-2 [K^d (8), D^b (9), D^d (7),

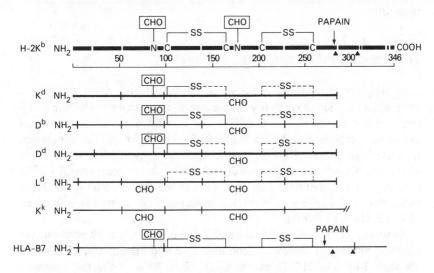

Figure 4 Schematic outlines of the six H-2 molecules and one HLA molecule. Information taken from the following sources: H-2K^b (6, 42); H-2K^d (8); H-2D^b (9); H-2D^d (7); H-2L^d (10); H-2K^k (11); HLA-B7 (12, 13). CHO, carbohydrate moiety; CHO in box denotes those moieties whose attachment site is identified; N, Asn; C, Cys; ▲, indicates approximate position of hydrophobic segment.

and L^d (10)] seem to be very similar, if not identical, with those of H-2K^b. Second, all molecules contain two carbohydrate moieties, and in K^b, K^d, D^b, and D^d one of the carbohydrate attachment sites has been identified as an Asn residue at position 86 (44). The attachment of the second carbohydrate chain in K^b has been determined to be an Asn residue at position 176, and in the other molecules the carbohydrate moiety has been localized to the homologous region and is presumably present at an identical position. Third, the Met residues where cleavage by CNBr occurs are highly conserved. This localization of Met residues permits application of similar strategies for studies of primary structure of the H-2 molecules.

Similar structural features to those found in H-2 molecules, are also found in the HLA molecule. HLA-B7 (59) has two intrachain disulfide bridges in identical positions with H-2K^b, and although it has not yet been shown, a similar structure is expected for HLA-A2. Further, as in the H-2 molecules, HLA-B7 contains a carbohydrate moiety attached at the Asn residue at position 86 (58).

Certain prominent differences also occur between H-2 and HLA molecules. First, while all six H-2 molecules seem to contain at least two carbohydrate moieties, HLA-B7 and probably HLA-A2 have only one (13). Second, a free Cys residue at position 121 so far found in K^b and D^d is lacking in B7 and A2. Whether these differences reflect species differences must await forthcoming information from studies on other H-2 and HLA molecules.

Amino Acid Sequence Comparison Between H-2 and Other Histocompatibility Antigens

THE NH$_2$-TERMINAL 100 RESIDUES OF FOUR H-2 AND TWO HLA MOLECULES   The availability of the amino acid sequences of the NH$_2$-terminal 100 residues from four H-2 and two HLA molecules allows an extensive comparison to be made (Table 2). The most obvious conclusion is that a high degree of primary structure homology exists among the six molecules. Thus, when sequences of each molecule are compared, 75–87% homology is observed within the H-2 series as well as between the two HLA molecules, and 67–73% homology is observed from comparison between H-2 and HLA (Table 3).

As scored in the bottom two lines of Table 2, residues at 56 positions are identical among six molecules (common H-2 and HLA), and 65 residues are shared by four H-2 (common H-2). Thus, 85% of the H-2 common residues (56/65) are also common to HLA, a finding that underscores the remarkably close relationship between the H-2 and the HLA molecules.

The sequence comparisons, however, also demonstrate the existence of

Table 2 Sequence comparison at positions 1–100 among mouse and human MHC molecules

```
              1                   2                   3                   4                   5
Protein  1 2 3 4 5 6 7 8 9 0 1 2 3 4 5 6 7 8 9 0 1 2 3 4 5 6 7 8 9 0 1 2 3 4 5 6 7 8 9 0 1 2 3 4 5 6 7 8 9 0

Kᵇ       G P H S L R Y F V T A V S R P G L G E P R Y M E V G Y V D D T E F V R F D S D A E N P R Y E P R A R

Kᵈ       ——————————————————————————————— F I A ——————————— Q ——————————————— D — A — F ——————— P

Dᵇ       ——————— M ——————— E ——————————————— E ——————— I S ——————— N K ——————————————————————————— P

Dᵈ       — S ——————————————————————————— F ——————————————— N ———————————————————————————————————

B7       — S ——— M ——————— Y — S ——————————— R ——————— F I S ——————— Q ——————————————— A S ——— E ——————— P

A2       — S ——— M ——————— F — S ——————————— R ——————— F I A ——————— Q ——————————————— A S Z — M ——————— P

Common   G   H S   R Y F   T A V S R P G     E P R       V G Y V D       F V R F D S D A   N   R   E P R A
 H-2

Common   G   H S   R Y F   T   V S R P G     E P R       V G Y V D       F V R F D S D A       R   E P R A
 H-2/HLA

                   *                   *                   Δ                           *   *   *
```

```
              6                   7                   8                   9                  10
Protein  1 2 3 4 5 6 7 8 9 0 1 2 3 4 5 6 7 8 9 0 1 2 3 4 5 6 7 8 9 0 1 2 3 4 5 6 7 8 9 0 1 2 3 4 5 6 7 8 9 0

Kᵇ       W M E Q E G P E Y W E R E T Q K A K G N E Q S F R V D L R T L L G Y Y N Q S K G G S H T I Q V I S G

Kᵈ       ——————————————————— E Q ——— R V — S D ——— W ——— S T ——— A Q R ——————————————— F — R M F

Dᵇ       ——————————————————————————— Q — W ——— S — N ——————————— A ——————————— L — Q M

Dᵈ       — I ——————————————————— R R ——————————————— A — R ——————— A ——— • ——— L — W M A —

B7       — I ——————————— D — N — I Y — A Q A — T D — E S ——— N — R ——————— E A ——— L — S M Y —

A2       I ——————————— D ——— — V — A H A H T V ——————— G — R ——————— E A ——— L — R M Y —

Common   W   E Q E G P E Y W E     T     K     E Q   F R V   R         Y Y N Q S   G G S H T   Q         G
 H-2

Common   W   E Q E G P E Y W       T     K             R                 Y Y N Q S     G S H T   Q         G
 H-2/HLA

                           *   * *   * Δ   *                       *                 *         Δ   Δ   Δ
```

Notes: Sequence is expressed in one letter code. Lines indicate homology to H-2Kᵇ sequence. Data are taken from: Kᵇ (45); Kᵈ (8); Dᵇ (9); Dᵈ (7; R. Nairn, J. E. Coligan, and S. G. Nathenson, submitted for publication); B7 (13); A2 (59).

Δ, indicates position where more than two different amino acids are assigned among the four H-2 molecules.

*, indicates position where more than two different amino acids are assigned among the six molecules.

"species specific residues." Thus at positions 11, 42, 61, 71, and 90, the four H-2 molecules and two HLA molecules each have residues in common within the species that are not shared by molecules of the other species. The species specific residues distribute throughout the NH₂-terminal 100 residues.

Comparison among the H-2 molecules reveals an intriguing and unexpected relationship between H-2Kᵈ and other H-2 molecules. The homology between Kᵈ and the other H-2 molecules is significantly lower

Table 3 Percent sequence homologies among six molecules

	K^b	K^d	D^b	D^d	B7	A2	
H-2K^b		48	83	83	43	50	P O S I T I O N S 61 – 83
K^d	75		57	57	30	25	
D^b	85	76		65	57	40	
D^d	87	75	81		30	50	
HLA-B7	68	68	73	67		55	
A2	69	69	68	67	85		

P O S I T I O N S   1   –   100

than when K^b, D^b, or D^d are compared to each other. Thus, K^d compared with K^b, D^b, or D^d yields 74–76% homology, while comparisons among K^b, D^b, and D^d gives 80–87%. However, the homology between K^d and two HLA molecules (e.g., 67–68%) is the same as that seen when the other three H-2 molecules are compared to HLA (68–73%). This finding for K^d suggests that the relationship between the K and D gene families may not be a simple one. Clearly, complete analyses of these molecules is needed to extend these comparisons.

With regard to the analysis of the specific differences among all the H-2 molecules, there are 35 positions where at least two different amino acid residues have occurred in the sequence. The majority of these positions are nonrandomly distributed, which result in discrete clusters of sequence differences among the four H-2 molecules. The largest cluster is between residues 61–83. This stretch of amino acid sequence heterogeneity both between the species (H-2 and HLA) and within species (HLA) has been pointed out previously (7–10, 59). The distinctive features of this region as judged from the amino acid sequence comparisons of six molecules can be summarized as follows:

1. Only 3 of the 23 compared residues are common between the 6 molecules (i.e. 13%), and only 9 residues are shared among the 4 H-2 (i.e. 39%). This degree of homology is strikingly lower than that observed in the remainder of the residues compared.
2. At 7 positions in this region, more than 3 different amino acid assignments are made, whereas only 17 positions of such diversity are found throughout the entire NH$_2$-terminal 100 residues.

3. Of 15 positions in the NH_2-terminal 100 residues that require 2 base changes at the nucleic acid level for interconversion, 12 positions are found in residues 61–83.

The extent of diversity in this region is further exemplified when the amino acid sequences at residues 61–83 are compared for each molecule (Table 2). The percentage of homology between any given H-2 and any given HLA molecule ranges from 25–57%. Homologies among four H-2 molecules range from 48–83%, and the homology between the HLA molecules is 55%. These numbers are significantly lower than the number scored in Table 3 for the overall homology (positions 1–100) between the corresponding two molecules, with two exceptions.

These two exceptions are obtained by comparison between H-2K^b and D^b and between H-2K^b and D^d. Both comparisons show homology of 83%, and this is only slightly lower than that observed by comparison of the entire NH_2-terminal 100 residues for these pairs. However, when residues 61–83 of D^b and D^d are compared, the homology is only 64%, and this degree of homology is significantly lower than the overall homology between these molecules. Therefore, while these observations do not change our overall conclusions, the exact degree of amino acid sequence homology in positions 61–83 between the H-2 molecules varies considerably for different combinations.

Among the four H-2 sequences, the lower homologies between K^d and the other molecules for the NH_2-terminal 100 residues are also found for positions 61–83 (Table 3).

A second highly variable region among the H-2 molecules is seen in the stretch from positions 95–99. The four H-2 molecules do not share identical residues at four positions in this region. Furthermore, more than three different assignments were made at two of these four positions. Of special note is the finding of a different amino acid residue at position 97 for each H-2 molecule. This is the only position throughout the NH_2-terminal 100 residues where four different amino acids have been assigned. Position 99 is potentially as diverse as 97. However, when the sequences of this region of the two HLA molecules are compared, 80% homology is obtained. Therefore, an extensive degree of heterogeneity in this region is only observed within H-2 molecules and not within HLA, which suggests the possibility that positions 95–99 constitute a diverse region within the H-2 series only, whereas 61–83 shows diversity for products of both species. There are relatively long stretches of conserved sequences near these highly variable stretches. Thus, within positions 46–60, 13 of 15 residues are identical, and within positions 83–94 10 out of 11 are identical among all the histocompatibility antigens.

Although the regions described above constitute major differences in the NH$_2$-terminal 100 residues, other clusters of differences, such as positions 22–24, positions 30–32, and positions 41–45 are present. For example, three different assignments are made at position 24 among the four H-2 molecules, and extensive amino acid differences between H-2 and HLA are found at positions 41–45.

One of the major concerns of studies on primary structure of histocompatibility antigens is the possible localization of the functional sites that are recognized as determinants for alloantibodies and for activated lymphocytes. It is possible that regions such as positions 61–83, in which a high degree of amino acid sequence heterogeneity is found, are involved in antigenic determinant(s). However, as shown in Table 2, sequence differences are localized in at least several discrete regions. Since it is known from the studies on the structure of the H-2K^b mutants (90) that one or two amino acid differences are sufficient to cause profound biological changes, it is especially difficult to define a precise function-structure correlation in these molecules.

Certain structural relationships have emerged from the sequence comparisons of the six MHC molecules that are summarized schematically in Figure 5. Although based only on comparisons from the amino-terminal 100 residues, the data suggest: (*a*) that all four H-2 molecules have a similar degree of homology with HLA-B7 and A2; (*b*) that K^d is less related to the other three H-2 molecules than they are to each other; and (*c*) K^b and D^b, and K^b and D^d are more homologous than other H-2 pairs.

THE PAPAIN FRAGMENTS FROM H-2K^b AND HLA-B7.   The extensive sequence studies carried out on the K^b and B7 molecules allow sequence comparisons to be made for the entire length of their papain fragments (Figure 6) i.e. the antigenically active region of the molecule located outside the cell surface membrane.

In addition to the features described in the section on overall structural

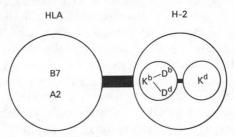

Figure 5   Schematic representation of the structural relationships between four H-2 and two HLA molecules.

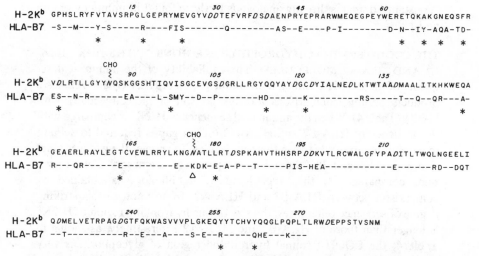

Figure 6 Comparison of the amino acid sequence of H-2K^b and HLA-B7. Identical residues are indicated by solid line. The single letter code for amino acids is used. CHO, carbohydrate moiety; *, represent differences of an amino acid residue between these two molecules that require more than a single base change; Δ, indicates lack of glycosylation in HLA-B7. A papain cleavage site of H-2K^b was identified as Val 281 (6). The COOH terminus of HLA-B7 papain fragment was reported to be Thr (12). [Figure reprinted from (6) with permission of Am. Chem. Soc.]

properties of H-2 and HLA 45-K glycoproteins, several additional points are noteworthy from the sequence comparison of 271 residues.

First, the sequence homology between the two molecules is 71% for 271 residues, as compared to 68% for the NH_2-terminal 100 residues. Second, in addition to positions 61–83 and positions 95–99, a high degree of heterogeneity is found at positions 173–199. In this region, only 13 residues are identical between H-2K^b and HLA-B7, and the sequence homology is only 48%. However, only one of the amino acid substitutions in this region requires more than a single nucleotide base pair change, while five of the differences observed for positions 61–83 need two changes. Of 80 differences between the two molecules in the NH_2-terminal 271 residues, 16 require a minimum of 2 or 3 base pair changes. Third, the position of the two intrachain disulfide bridges are identical. Fourth, the second carbohydrate chain, attached to the Asn residue at position 176 in the H-2K^b, has not been detected in HLA-B7. This is one of the major differences found between H-2K^b and HLA-B7.

In general, clusters of amino acid differences, as were observed in the NH_2-terminal 100 residues, are also found throughout the entire papain fragment. There is no evidence to suggest the existence of "variable" or "constant" domains directly analogous to those found in immunoglobulin molecules.

THE COOH-TERMINAL HYDROPHILIC REGIONS OF THE H-2K^b, HLA-B7 AND HLA-A2 MOLECULES   The availability of the amino acid sequence of the COOH-terminal hydrophilic region of HLA-B7 and HLA-A2 products permits comparisons for this portion of these molecules with H-2K^b (Table 4). When the amino acid sequence of H-2K^b is compared with the sequence of HLA-B7 and HLA-A2 (68), a gap is required to achieve maximum homology between these molecules. In spite of the gap introduction, only 34% and 28% homology, respectively, are observed at 32 positions compared. On the other hand 87% homology is obtained by comparison between HLA-B7 and HLA-A2. In addition to this striking degree of sequence differences, the size of this hydrophilic region of H-2K^b is longer than those of HLA-B7 and HLA-A2 Interestingly, as shown in Table 4, the COOH-terminal hydrophilic region of glycophorin is very similar in size to that of the H-2K^b molecule, although little homology is observed between these two sequences. Furthermore, the position of a free Cys residue in H-2K^b is different from the positions of two Cys residues in HLA-B7 assigned in the COOH-terminal hydrophilic region. (In HLA-A2 no Cys residue was assigned.) Thus there appear to be major differences between these two HLA and one H-2 molecule in this region.

HISTOCOMPATIBILITY ANTIGENS FROM OTHER SPECIES   Although extensive amino acid sequence data are only available for mouse (6–11) and human MHC antigens (12–14, 91, 92) partial sequence data have been obtained from MHC antigens of rabbit (93), rat (94), guinea pig (95), and chicken (96, 97). Comparison of presently available sequences are shown in

Table 4 Amino acid sequence comparison of COOH-terminal hydrophilic region

	310	320	330	340
H–2K^b	K M R R R A T G G K G G D Y A L A P G S Q T S D L S L P D C K V M V H P P H S			
HLA–B7	* C — K S S ———— S — S Q — A C — D S A Q G — D V S L T A			
A2	* W — K S S D R ——— S — S Q — A S — D S A Z G — B V S L T A			
Glycophorin	— L I K K S P S D V K P — P S P D T D V P —— S V E I E N P E T S D Q			

Notes: Sequence is expressed in one letter code Lines indicate homology to H–2K^b sequence. Data are taken from the following sources: H–2K^b, (H. Uehara, J. E. Coligan, and S. G. Nathenson, data submitted for publication); HLA–B7 and A2 (67); Glycophorin (62).
  *, indicates that a gap is required to obtain maximum sequence homology between H–2K^b and HLA–B7 or HLA–A2. The NH$_2$-terminal Cys (C) residue of HLA–B7 and the Trp (W) residue of HLA–A2 is position 308. The sequence of Glycophorin is aligned based on the cluster of basic residues corresponding to K^b position 310 and 311.

Table 5. Clearly, a high degree of sequence homology can be seen throughout all species, although considerable differences also exist. Residues at six positions (3, 6, 7, 15, 27, and 28) are identical among MHC antigens from all species studied, and six other positions (1, 4, 14, 16, 20, and 26) have identical residues for those species with amino acid assignments available for comparison. Thus, as many as 40% of NH_2-terminal 28 residues are completely conserved among MHC antigens from six species.

Description of the evolutionary and immunological relationships among these molecules awaits further structural and genetic information.

SPECULATIONS AND CONCLUDING REMARKS

An overall picture of the Class I histocompatibility antigens is emerging from a number of different types of studies. As described in this review, studies on the primary structure of the H-2 and HLA histocompatibility antigens are sufficiently far advanced as to give new insight into their molecular organization and how their structure might be related to their biological function(s).

The H-2 and HLA molecules are transmembrane proteins, as inferred from the amino acid sequence data and other biochemical studies. The H-2K^b molecule, which is a glycoprotein of 346 amino acid residues with two carbohydrate moieties, appears to consist of three functional regions. The extracellular region extends from the amino terminus to the papain cleavage site and has the alloreactive determinants. The hydrophobic mem-

Table 5 Homology among Class I MHC molecules from different species[a]

Species	1	2	3	4	5	6	7	8	9	10	11	12	13	14	15	16	17	18	19	20	21	22	23	24	25	26	27	28
Mouse H-2	G	P	H	S	L	R	Y	F	V	T	A	V	S	R	P	G	L	G	E	P	R	Y	M	E	V	G	Y	V
		S		M					E							F	E	K			F	I	A					
				H																				S				
Human HLA	G	S	H	D	M	R	Y	F	Y	T	S	V	S	R	P	G	R	G	E	P	E	F	I	S	V	G	Y	V
				F	S	A	M	A																A				
Rabbit RLA	G	S	H	S	M	R	Y	F	Y		S	V	S	R	P	G	L	G		P	R	F			V	G	Y	V
Rat AGB4					L	R	Y	F	Y		A	V									F	I	A	V			Y	
					I																							
Guinea Pig GPLA			H		L	R	Y	F	Y	I	A	V			P							F			V		Y	
					I		V																					
Chicken B		L	H		L	R	Y	I	F		A	M			P		L				F	V		V		Y	V	
								F	R													Y						

[a] The single letter code for amino acids is used. Data are compiled from the following sources: mouse H-2 (4, 6–10, 11, 71, 116); human HLA (13, 14, 59, 91, 92); rabbit RLA (93); Rat AGB4 (94); guinea pig GPLA (95); and chicken B (96, 97).

brane segment anchors the molecule in the membrane and is followed by the COOH-terminal intracellular region that may interact with cytoplasmic components.

The extracellular hydrophilic region may be organized into three independent structural domains (shown schematically in Figure 7). The predicted domains of H-2 (and HLA) (20, 98) correspond to the three subregions described in the section on the extracellular portion of H-2K^b. Thus, each domain consists of about 90 amino acid residues, and the second and third domains contain a disulfide loop. The domain-like organization of the histocompatibility antigens is postulated both from the structural relationship between these molecules and immunoglobulins and from proteolytic cleavage studies of the molecule. Thus, immunoglobulins are known to be organized in a domain structure (99–101) as are many other proteins, e.g. globin (102), bacterial ferodoxin (103), and serum albumin (104). A structural relationship between H-2 or HLA and immunoglobulin has been suggested by: (a) a modest but statistically significant amino acid

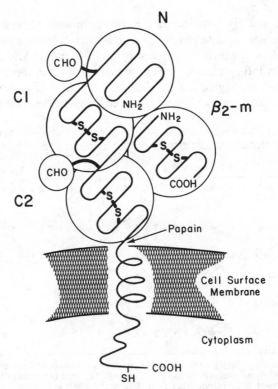

Figure 7 A schematic representation of the H-2K^b molecule. See legend to Figure 3.

sequence homology between the second disulfide loop of H-2K^b (H. Uehara, unpublished observations), or the HLA molecules (13, 89) and immunoglobulin constant region domains; (b) the association with β_2m, a protein with an immunoglobulin-like domain (105, 106); and (c) a high content of β-pleated sheet structure in the HLA-B7 papain fragment as predicted by far ultraviolet-circular dichroism spectra (83, 98).

Further biochemical evidence for the postulated domain structure has also come from studies on the proteolytic cleavage sites in H-2 and HLA molecules (16, 70). Such sites were compatible with cleavage of the molecules between the predicted domains. In addition, marginal amino acid sequence homology between the NH$_2$-terminal 90 residues and the first disulfide loop region was observed in HLA-B7 (13), and in a mixture of HLA-A, -B, and -C molecules (14). Thus, although no direct data are available that support the hypothesis, a schematic model of the H-2 molecule incorporating domain-like regions is proposed in Figure 7.

Recent studies of the structure of the K^b gene product from a series of histogenic mutants (for review see 90) are also pertinent with regard to the molecular organization of the molecule. In approximately nine well-characterized K^b mutant molecules, limited amino acid changes of one or two amino acid residues are correlated with profound alterations in biological reactivity properties. All the substitutions occurred in the NH$_2$-terminal two thirds of the extracellular portion of the molecule, which is consistent with the postulation that this region carries T-cell recognition properties. The positions where amino acid alterations are observed do not cluster in one particular linear segment of the sequence, but this does not exclude the possibility that these amino acid interchanges occur in a region of the folded molecule as part of a "site" that is important in the process of T-cell recognition of H-2 molecules.

It is beyond the scope of this review to discuss the function of the MHC encoded molecules, and for this data the interested reader is referred to other reviews (17, 18, 23, 30, 108). At this time it is not possible to correlate structure and biological properties of the Class I molecules, but the continuing studies with the MHC mutant strains offer a promising approach toward understanding structure-function relationships in the future.

As we discuss above, a very notable feature of the MHC is the extent of genetic complexity and polymorphism. These features are documented by the occurrence of an unusually large number of alleles in the mouse and human populations, by the expression of multiple Class I products on each cell surface (i.e. K, D, and L) (2, 3, 109), and by the extensive degree of amino acid sequence differences observed between molecules from different allelic products (see above; 107).

These properties raise questions of how the diversity and complexity in the MHC system were generated and what evolutionary pressures maintain them. Present thinking would suggest that the evolutionary pressures to generate and maintain such polymorphism are the necessity for biological "uniqueness" and the advantages this may confer upon the organism with regards to disease resistance (2, 3, 108, 109). However, additional experimentation is needed to further evaluate this concept.

While there are no direct answers to the question of the mechanism for generation of the diversity and complexity, the sequence data reviewed here offer the opportunity for some speculations on certain aspects of this process. Thus, the presence of a number of genes (e.g. K, D, and L) that encode histocompatibility antigens of very similar structure, and with supposedly similar function(s), implies that these genes may have arisen by gene duplication of a primordial gene. If the diversity of the histocompatibility genes was generated by the accumulation of independent mutations following the gene duplication, then there should be "K-ness," "D-ness," or "L-ness" reflected in the primary structure of these molecules. However, the amino acid sequence comparison of the NH_2-terminal 100 residues of molecules encoded by each of these genes does not support this view. Thus, as shown in Table 2 and illustrated in Figure 5, the sequence of K^b seems to be closer to the sequence of D^b and D^d than to the sequence of its allele, K^d.

In fact this relationship becomes clearer when an evolutionary tree (110, 111) is constructed for the four H-2 molecules based on the amino acid sequence of the NH_2-terminal 100 residues. The tree that best fits the observed amino acid sequence similarity between the four molecules is shown in Figure 8A. It reveals a relationship inconsistent with the idea that the four H-2 antigens arose by the accumulation of mutations following the tandem duplication of a primordial gene. If this were the case, then the tree shown in Figure 8B would be the best fit.

The availability of small but increasing amounts of sequence information

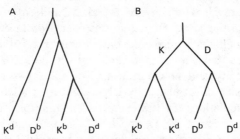

Figure 8 Possible evolutionary relationships among four H-2 molecules. *A*, evolutionary tree obtained by analysis of NH_2-terminal 100 residues of four H-2 molecules. *B*, evolutionary tree constructed by extrapolating from the classical hypothesis of gene duplication and accumulation of mutations.

about the Class I molecules raises intriguing questions about the genetic organization of the MHC genes. Possibly some of the mechanisms involved in the generation of diversity of antibody genes (113–116) may apply to the MHC also, although certainly new mechanisms may also be anticipated.

The last five years or so has seen the transition from the phase of biochemical characterization of the MHC gene products to the stage of primary structure determination. Even now, as these data are being generated and the first exciting insights into the evolution of the MHC are being made, we must anticipate the next transition to the level of the gene itself where many of the answers to our questions about diversity and complexity are presumably to be found.

ACKNOWLEDGMENTS

We gratefully thank our colleagues Rod Nairn, John Martinko, Tosiki Nisizawa, Lee Maloy, and Ed Kimball for their participation in some of the structural studies reported here and for making available certain unpublished material. We further thank Drs. Larry Pease and Kazushige Yokoyama, in addition to the others mentioned above, for their helpful discussions in the preparation of this manuscript, and Catherine Whelan for her secretarial assistance. Some of the studies reported in this review were supported by grants AI-07289 and AI-10702 from the National Institutes of Health and grant IM-236 (SGN) from the American Cancer Society.

Literature Cited

1. Klein, J. 1975. *Biology of the Mouse Histocompatibility-2 Complex.* New York: Springer. 620 pp.
2. Snell, G. D., Dausset, J., Nathenson, S. G. 1976. *Histocompatibility.* New York: Academic. 401 pp.
3. Festenstein, H., Demant, P. 1979. *HLA and H-2 Basic Immunogenetics, Biology and Clinical Relevance.* Great Britain: Edward Arnold.
4. Vitetta, E. S., Capra, J. D. 1979. *Adv. Immunol.* 26:147–93
5. Goetze, D., ed. 1977. In *The Major Histocompatibility System in Man and Animal,* Berlin/New York: Springer
6. Martinko, J. M., Uehara, H., Ewenstein, B. M., Kindt, T. J., Coligan, J. E., Nathenson, S. G. 1981. *Biochemistry.* In press
7. Nairn, R., Nathenson, S. G., Coligan, J. E. 1980. *Eur. J. Immunol.* 10:495–503
8. Kimball, E. S., Nathenson, S. G., Coligan, J. E. 1981. *Biochemistry.* In press
9. Maloy, W. L., Nathenson, S. G., Coligan, J. E. 1981. *J. Biol. Chem.* In press
10. Coligan, J. E., Kindt, T. J., Nairn, R., Nathenson, S. G., Sachs, D. H., Hansen, T. H. 1980. *Proc. Natl. Acad. Sci. USA* 77:1134–38
11. Rothbard, J. B., Hopp, T. P., Edelman, G. M., Cunningham, B. A. 1980. *Proc. Natl Acad. Sci. USA* 77:4239–43
12. Lopez de Castro, J. A., Orr, H. T., Robb, R. J., Kostyk, T. G., Mann, D. L., Strominger, J. L. 1979. *Biochemistry* 18:5704–10
13. Orr, H. T., Lopez de Castro, J. A., Lancet, D., Strominger, J. L. 1979. *Biochemistry* 18:5711–20
14. Trägärdh, L., Rask, L., Wiman, K., Fohlman, J., Peterson, P. A. 1980. *Proc. Natl. Acad. Sci. USA* 77:1129–33
15. Nathenson, S. G., Cullen, S. E. 1974. *Biochim. Biophys. Acta* 344:1–25
16. Shimada, A., Nathenson, S. G. 1969. *Biochemistry* 8:4048–62
17. Zinkernagel, R., Doherty, P. C. 1979. *Adv. Immunol.* 27:51–177
18. Shearer, G. M., Schmitt-Verhulst, A. 1977. *Adv. Immunol.* 25:55–91

19. Forman, J. 1975. *J. Exp. Med.* 142:403–18
20. Strominger, J. L., Engelhard, V. H., Fuks, A., Guild, B. C., Hyafil, F., Kaufman, J. F., Korman, A. J., Kostyk, T. G., Krangel, M. S., Lancet, D., Lopez de Castro, J. A., Mann, D. L., Orr, H. T., Parham, P. R., Parker, K. C., Ploegh, H. L., Pober, J. S., Robb, R. J., Shakelford, D. A. 1980. In *The Role of the Major Histocompatibility Complex in Immunobiology.* New York:Garland Press. In press
21. Krangel, M. S., Orr, H. T., Strominger, J. L. 1980. *Scand. J. Immunol.* 11:567–71
22. Gorer, P. A. 1936. *Br. J. Exp. Pathol.* 17:42–50
23. McDevitt, H. O., ed. 1978. *Ir Genes and Ia Antigens.* New York: Academic. 613 pp.
24. Shreffler, D. C., David, C. S. 1978. *Adv. Immunol.* 20:125–95
25. Moller, G. 1976. *Transpl. Rev.* 30:1–322
26. Roos, M. H., Atkinson, J. P., Shreffler, D. C. 1978. *J. Immunol.* 121:1106–15
27. Ferreira, A., Nussenzweig, V., Gigli, I. 1978. *J. Exp. Med.* 148:1186–97
28. Bodmer, W. F. 1972. *Nature* 237:139–45
29. Silver, J., Hood, L. E. 1976. *Contemp. Top. Mol. Immunol.* 5:35–68
30. Paul, W., Benacerraf, B. 1977. *Science* 195:1293–1300
31. Jones, P., Murphy, D. G., Hewgill, D., McDevitt, H. O. 1978. *Mol. Immunol.* 16:51–60
32. Old, L. J., Stockert, E. 1977. *Ann. Rev. Genet.* 11:127–60
33. Stanton, T. H., Boyse, E. A. 1976. *Immunogenetics* 3:525–31
34. Flaherty, L. 1976. *Immunogenetics* 3:533–39
35. Nathenson, S. G., Davies, D. A. L. 1966. *Proc. Natl. Acad. Sci. USA* 56:467–83
36. Schwartz, B. D., Kato, K., Cullen, S. E., Nathenson, S. G. 1973. *Biochemistry* 12:2157–64
37. Henrikson, O., Robinson, E. A., Appella, E. 1978. *Proc. Natl. Acad. Sci. USA* 75:3322–26
38. Freed, J. H., Sears, D. W., Brown, J. L., Nathenson, S. G. 1979. *Molec. Immunol.* 16:9–21
39. Hermann, S. H., Mescher, M. F. 1979. *J. Biol. Chem.* 254:8713–16
40. Mole, J. E., Hunter, F., Paslag, J. W., Brown, A. S., Bennett, J. C. 1981. *Mol. Immunol.* In press
41. Ewenstein, B. M., Freed, J. H., Mole, L. E., Nathenson, S. G. 1976. *Proc. Natl. Acad. Sci. USA* 73:915–18
42. Ewenstein, B. M., Nisizawa, T., Uehara, H., Nathenson, S. G., Coligan, J. E., Kindt, T. J. 1978. *Proc. Natl. Acad. Sci. USA* 75:2909–13
43. Coligan, J. E., Kindt, T. J., Ewenstein, B. M., Uehara, H., Nisizawa, T., Nathenson, S. G. 1978. *Proc. Natl. Acad. Sci. USA* 75:3390–94
44. Coligan, J. E., Kindt, T. J., Ewenstein, B. M., Uehara, H., Martinko, J. M., Nathenson, S. G. 1979. *Molec. Immunol.* 16:3–8
45. Uehara, H., Ewenstein, B. M., Martinko, J. M., Nathenson, S. G., Coligan, J. E., Kindt, T. J. 1980. *Biochemistry* 19:306–15
46. Uehara, H., Ewenstein, B. M., Martinko, J. M., Nathenson, S. G., Kindt, T. J., Coligan, J. E. 1980. *Biochemistry* 19:6182–88
47. Snary, D., Goodfellow, P., Hayman, M. J., Bodmer, W. F., Crumpton, M. J. 1974. *Nature* 247:457–61
48. Springer, T. A., Strominger, J. L., Mann, D. 1974. *Proc. Natl. Acad. Sci. USA* 71:1539–43
49. Springer, T. A., Mann, D. L., DeFrance, A. L., Strominger, J. L. 1976. *J. Biol. Chem.* 252:4682–93
50. Hayman, M. J., Crumpton, M. S. 1972. *Biochem. Biophys. Res. Comm.* 47:923–30
51. McKean, D. J., Peters, E. H., Waldby, J. I., Smithies, O. 1974. *Biochemistry* 13:3048–51
52. Devillers-Thiery, A., Kindt, T., Schelle, G., Blobel, G. 1975. *Proc. Natl. Acad. Sci. USA* 72:5016–20
53. Palmiter, R. D., Gagnon, J., Ericsson, L. H., Walsh, K. A. 1977. *J. Biol. Chem.* 252:6386–93
54. Kemper, B., Habener, J. F., Ernst, M. D., Potts, J. T. Jr., Rich, A. 1976. *Biochemistry* 15:15–19
55. Gross, E., Witkop, B. 1961. *J. Am. Chem. Soc.* 83:1510–1512
56. Graff, R., Mann, D. L., Nathenson, S. G. 1970. *Transplantation* 10:59–65
57. Marshall, R. D. 1972. *Ann. Rev. Biochem.* 41:673–702
58. Parham, P., Alpert, B. N., Orr, H. T., Strominger, J. L. 1977. *J. Biol. Chem.* 252:7555–67
59. Orr, H. T., Lopez de Castro, J. A., Parham, P., Ploegh, H. L., Strominger, J. L. 1979. *Proc. Natl. Acad. Sci. USA* 76:4395–99
60. Walsh, F. S., Crumpton, M. J. 1977. *Nature* 269:307–11

61. Guidotti, G. 1977. *J. Supramol. Struct.* 7:489–97
62. Furthmayr, H., Galardy, R. E., Tomita, M., Marchesi, V. T. 1978. *Arch. Biochem. Biophys.* 185:21–29
63. Tomita, M., Furthmayr, H., Marchesi, V. T. 1978. *Biochemistry* 17:4756–70
64. Winzler, R. J. 1969. In *Red Cell Membrane, Structure and Function,* ed. G. A. Jamieson, T. J. Greenwalt, pp. 157–71. Philadelphia, Penn: Lippincott
65. Marchesi, V. T., Tillack, T. W., Jackson, R. L., Segrest, J. P., Scott, R. E. 1972. *Proc. Natl. Acad. Sci. USA* 69:1445–49
66. Bretscher, M. D. 1971. *Nature New Biol.* 231:229–32
67. Cotmore, S. F., Furthmayr, H., Marchesi, V. T. 1977. *J. Mol. Biol.* 113:539–53
68. Robb, R. J., Terhorst, C., Strominger, J. L. 1978. *J. Biol. Chem.* 253:5319–24
69. Tomita, M., Marchesi, V. T. 1975. *Proc. Natl. Acad. Sci. USA* 72:2964–68
70. Peterson, P. A., Rask, L., Sege, K., Klareskog, L., Anundi, H., Ostberg, L. 1975. *Proc. Natl. Acad. Sci. USA* 72:1612–16
71. Henning, R., Milner, R. J., Reske, K., Cunningham, B. A., Edelman, G. M. 1976. *Proc. Natl. Acad. Sci. USA* 73:118–22
72. Springer, T. A., Robb, R. J., Terhorst, C., Strominger, J. L. 1977. *J. Biol. Chem.* 252:4694–700
73. Pober, J. S., Guild, B. C., Strominger, J. L. 1978. *Proc. Natl. Acad. Sci. USA* 75:6002–6
74. Robin, C. S., Rosen, O. M. 1975. *Ann. Rev. Biochem.* 44:831–87
75. Barany, K., Barany, M. 1977. *J. Biol. Chem.* 252:4752–54
76. Edelman, G. M. 1976. *Science* 192:218–26
77. Adelstein, R. S. 1978. *Trends Biol. Sci.* 3:27–30
78. Berggard, I., Bearn, A. G. 1968. *J. Biol. Chem.* 243:4095–4103
79. Nakamuro, K., Tanigaki, N., Pressman, D. 1973. *Proc. Natl. Acad. Sci. USA* 70:2863–65
80. Cresswell, P., Turner, M. J., Strominger, J. L. 1973. *Proc. Natl. Acad. Sci. USA* 70:1603–7
81. Rask, L., Lindblom, J. B., Peterson, P. A. 1974. *Nature* 249:833–34
82. Silver, J., Hood, L. 1974. *Nature* 249:764–65
83. Lancet, D., Parham, P., Strominger, J. L. 1979. *Proc. Natl. Acad. Sci. USA* 76:3844–48
84. Cunningham, B. A., Wang, J. L., Berggard, I., Peterson, P. A. 1973. *Biochemistry* 12:4811–21
85. Wolfe, P. B., Cebra, J. J. 1980. *Mol. Immunol.* In press
86. Gates, F. T. III, Coligan, J. E., Kindt, T. J. 1979. *Biochemistry* 18:2267–72
87. Gates, F. T. III, Coligan, J. E., Kindt, T. J. 1981. *Proc. Natl. Acad. Sci. USA* 78:554–58
88. Michaelson, J., Rothenberg, E., Boyse, E. A. 1980. *Immunogenetics* 11:93–96
89. Orr, H. T., Lancet, D., Robb, R. J., Lopez de Castro, J. A., Strominger, J. L. 1979. *Nature* 282:266–70
90. Nairn, R., Yamaga, K., Nathenson, S. G. 1980. *Ann. Rev. Genet.* 14:241–77
91. Bridgen, J., Snary, D., Crumpton, M. J., Barnstable, C., Goodfellow, P., Bodmer, W. F. 1976. *Nature* 261:200–5
92. Ballou, B., McKean, D. J., Freedlender, E. F., Smithies, O. 1976. *Proc. Natl. Acad. Sci. USA* 73:4487–91
93. Kimball, E. S., Coligan, J. E., Kindt, T. J. 1979. *Immunogenetics* 8:201–11
94. Blackenhorn, E. P., Cecka, J. M., Goetze, D., Hood, L. 1979. *Nature* 274:90–92
95. Schwartz, B. D., McMillan, M., Shevach, E., Hahn, Y., Rose, S. M., Hood, L. 1980. *J. Immunol.* 125:1055–59
96. Huser, H. A., Zeigler, A., Knecht, R., Pink, J. R. L. 1978. *Immunogenetics* 6:301–7
97. Vitetta, E. S., Uhr, J. W., Klein, J., Pazderka, F., Moticka, E. J., Ruth, R. F., Capra, J. D. 1977. *Nature* 270:535–36
98. Lancet, D., Strominger, J. L. 1979. In *Molecular Mechanisms of Biological Recognition.* ed. M. Balaban, pp. 289–98 Amsterdam: Elsevier/North Holland Biomed. Press
99. Nisonoff, A., Hopper, J. E., Spring, S. B. 1975. *The Antibody Molecule.* New York: Academic. 542 pp.
100. Beale, D., Feinstein, A. 1976. *Q. Rev. Biophys.* 9:135–80
101. Amzel, L. M., Poljak, R. J. 1979. *Ann. Rev. Biochem.* 48:961–97
102. Dayhoff, M. O., ed. 1972. *Atlas of Protein Sequence and Structure,* Vol. 5. Natl. Washington, DC: Biomed. Res. Found. Georgetown Univ. Med. Center
103. Adman, E. T., Sieker, L. C., Jensen, L. H. 1973. *J. Biol. Chem.* 248:3987–96
104. McLachlan, A. D., Walker, J. E. 1977. *J. Mol. Biol.* 112:543–58
105. Smithies, O., Poulik, M. D. 1972. *Science* 175:187–89
106. Peterson, P. A., Cunningham, B. A., Berggard, I., Edelman, G. M. 1972.

Proc. Natl. Acad. Sci. USA 69:1697–1701

107. Brown, J. L., Kato, K., Silver, J., Nathenson, S. G. 1974. *Biochemistry* 13:3174–78

108. Klein, J. 1979. *Science* 203:516–21

109. Klein, J. 1981. In *The Mouse in Biomedical Research,* ed. H. L. Foster, J. D. Small, J. G. Fox. New York: Academic. In press

110. Fitch, W. M., Margolish, E. 1967. *Science* 155:279–84

111. Smith, G. P., Hood, L., Fitch, W. M. 1971. *Ann. Rev. Biochem.* 40:969–1012

112. Seidman, J. G., Leder, A., Nau, M., Norman, B., Leder, P. 1978. *Science* 202:11–17

113. Davis, M. M., Calame, K., Early, P. W., Livant, D. L., Joho, R., Weissman, I. L., Hood, L. 1980. *Nature* 283:733–39

114. Sakano, H., Huppi, K., Heinrich, G., Tonegawa, S. 1979. *Nature* 280:288–94

115. Gottlieb, P. 1980. *Molec. Immunol.* 17:1423–35

116. Cook, R. G., Vitetta, E. S., Uhr, J. W., Klein, J., Wilde, C. E., Capra, J. D. 1978. *J. Immunol.* 121:1015–22

Ann. Rev. Biochem. 1981. 50:1053–86

TRANSMEMBRANE TRANSPORT OF COBALAMIN IN PROKARYOTIC AND EUKARYOTIC CELLS

♦12104

Cary Sennett and Leon E. Rosenberg

Department of Human Genetics, Yale University School of Medicine, New Haven, Connecticut 06510

Ira S. Mellman

Laboratory of Cellular Physiology and Immunology, The Rockefeller University, New York, New York 10021

CONTENTS

PERSPECTIVES AND SUMMARY

All eukaryotic animal cells and most prokaryotic ones require cobalamin (vitamin B_{12}) coenzymes to sustain life. Two such coenzymes, methylcobalamin and adenosylcobalamin, are known. Methylcobalamin is formed

1053

0066-4154/81/0701-1053$01.00

and used by a ubiquitous methyltransferase enzyme that "shuttles" methyl groups from N^5-methyltetrahydrofolate to homocysteine, thereby forming tetrahydrofolate and methionine. Adenosylcobalamin participates in the enzymatic reduction of ribonucleotides, and in a number of enzyme-catalyzed iosmerization reactions in prokaryotes, only one of which (methylmalonyl CoA mutase) exists in mammalian cells. Each cobalamin coenzyme is formed by a unique and complex reaction sequence that results in the formation of a covalent carbon-cobalt bond between the central cobalt nucleus of the cobalamin vitamer and the ligand, either a methyl group or 5'-deoxy-5'-adenosyl moiety, which confers coenzyme specificity.

Whereas most cobalamin-requiring prokaryotes and simple eukaryotes can synthesize the entire cobalamin entity de novo, some microorganisms and all higher animals require that it be supplied either from dietary sources (as is the case in man) or from intestinal flora (as in ruminants). This implies that mechanisms must exist in these organisms for the cellular accumulation and uptake of exogenous cobalamins.

Escherichia coli is the only prokaryotic species in which cobalamin transport has been studied in detail. Although exogenous cobalamins are not essential for growth of wild-type *E. coli,* a process exists that facilitates uptake of this group of compounds. Neither the mechanisms of this process nor the cellular components that mediate it are understood completely, but current evidence favors the following scheme. First, cobalamin binds reversibly to a multifunctional protein receptor on the external surface of the outer membrane of the bacterial envelope. A conformational change in this receptor consequent to cobalamin binding effects the movement of cobalamin across the outer membrane bilayer. Next, cobalamin (or the cobalamin-receptor complex) is transferred to a specific, cobalamin-binding periplasmic protein that facilitates movement across the periplasmic space. Finally, the cobalamin (or the cobalamin-periplasmic protein complex) binds to another receptor on the inner bacterial membrane, traverses this barrier from periplasmic to cytoplasmic surface, and is released into the cytosol.

The above scheme posits the existence of three proteins in *E. coli* that act sequentially to facilitate the movement of cobalamin across the two discrete membranes of the cell envelope. In man, on the other hand, ingested cobalamin must be transported within the gut lumen, across the intestinal mucosa, within the blood plasma, and across the plasma membrane of body cells before the vitamin gains access to the intracellular space. Each of these steps is mediated by discrete proteins whose structural and functional properties are well known (see Table 1): gastric intrinsic factor (IF), a glycoprotein that binds ingested cobalamins and carries them to the ileum; an ileal mucosal receptor that binds the IF-cobalamin complex and,

Table 1 Properties of proteins mediating transport of cobalamins in mammals

Protein	Estimated monomer molecular weight (daltons)	Glycosylated	Estimated affinity for ligand (K_a)	Distribution
TC II	38,000	no	1×10^{11} M^{-1} (for Cb1)	plasma; ileal enterocyte; other peripheral tissues
TC II receptor	50,000	yes	5.6×10^{10} M^{-1} (for TC II-Cb1) 2.8×10^{10} M^{-1} (for TC II)	probably all peripheral tissues
TC III (or asialo-TC I)	58,000	yes	1×10^{11} M^{-1} (for Cb1)	plasma; granulocytes
Asialoglycoprotein receptor	40,000 and 48,000	yes		hepatocytes
IF	44,200 (human) 54,500 (hog)	yes	1×10^{10} M^{-1} (for Cb1)	gastric juice; gastric parietal cells
IF receptor	70,000 and 110,000 (hog) 90,000 and 140,000 (human)	yes	2.5×10^{10} M^{-1} (for IF–Cb1)	ileal enterocyte

in some as yet poorly understood way, facilitates dissociation of the complex and liberation of the cobalamin on the serosal side; transcobalamin II (TC II), a plasma protein that binds newly absorbed cobalamin and carries the vitamer to tissue cells; and a TC II receptor on tissue cell surfaces that binds the TC II-cobalamin complex. This binding initiates a process of adsorptive endocytosis, and the TC II-cobalamin complex is internalized, first into pinocytotic vesicles, and then into secondary lysosomes. There the TC II is hydrolyzed by proteases, and the cobalamin moiety moves, probably by a mediated process, into the cytosol. Finally, some of the intracellular cobalamin vitamer traverses both the outer and inner mitochondrial membranes to serve as substrate for the enzymes that catalyze adenosylcobalamin synthesis.

Although the TC II–facilitated process just described is the most widely distributed and, probably, the major physiological mechanism for cobalamin accumulation by mammalian cells, it is not the exclusive one. Hepatocytes contain an asialoglycoprotein receptor that binds a different protein-cobalamin complex—namely that between cobalamins and other specific plasma-binding proteins, transcobalamin I, and transcobalamin III. These complexes, too, are internalized via adsorptive endocytosis. Yet a third transmembrane transport system for cobalamin exists, this one for free, rather than bound, cobalamins. The latter system, of no apparent consequence under physiological circumstances, becomes of importance if TC II is deficient.

In those steps that ordinarily move cobalamins across the intestinal cell and into tissue cells, it is the binding protein (IF, TC II, TC I) that is recognized by a cell surface receptor, the cobalamin moiety travelling "piggy back." It now appears, however, that free cobalamin is the ligand for those intracellular transport processes that move the vitamer across the lysosomal and mitochondrial membranes. Much still remains to be learned about the transmembrane transport of cobalamin. It is already clear, however, that this sytem, with its many human and bacterial mutants, is a useful model for comparing processes in man with those in microbes, for relating function to structure, and for understanding health and disease.

INTRODUCTION

Cobalamin (Cbl, vitamin B_{12}) is a large (1356 mol wt), complex, water-soluble molecule essential for the health of some microorganisms and all higher animals. Structurally, Cbls are tripartite corrinoid molecules comprising (Figure 1): a cobalt atom in the 3^{+} oxidation state coordinated to the four nitrogens of a planar tetrapyrrole corrin ring, substituted at the R and R' positions with acetamide (CH_2CONH_2) and propionamide (CH_2CH_2-$CONH_2$) residues, respectively; a 1-α-D-ribofuranosyl-5,6-dimethylbenzimidazolyl-3-phosphate, α to (below) the ring, in phosphodiester linkage to a 1-amino-2-propanol substituent on the (f) propionamide, and coordinated to the central cobalt through one of its nitrogens; and a β substituent (denoted X) coordinated to the remaining cobalt ligand position above the ring. It is the β substituent that confers biological specificity on the molecule, and after which the cobalamin is named. Related compounds, lacking the nucleotide, are called cobinamides (Cbi). The chemistry, structure, and nomenclature of Cbls and related compounds have been reviewed extensively elsewhere (1–3).

Intracellular cobalamin exists primarily in cofactor form, as methylcobalamin (MeCbl) in the cytoplasm, and as adenosylcobalamin (AdoCbl) in the mitochondria of eukaryotic cells. In many prokaryotes, and all higher eukaryotes, MeCbl is a cofactor for the catalytic transfer of a methyl group from N^5-methyltetrahydrofolate to homocysteine, a reaction in which methionine is synthesized and tetrahydrofolate is generated. This is the only known MeCbl-dependent reaction in higher animals. In many species of the anaerobic bacterium Clostridium, MeCbl is involved as well in the biosynthesis of methane and acetate. AdoCbl is a cofactor for a diverse group of mechanistically related rearrangement reactions. The only one of these of known importance to animal cells is the AdoCbl-dependent rearrangement of methylmalonyl CoA to succinyl CoA, a reaction catalyzed by the mitochondrial enzyme, methylmalonyl CoA mutase. In some prokaryotes

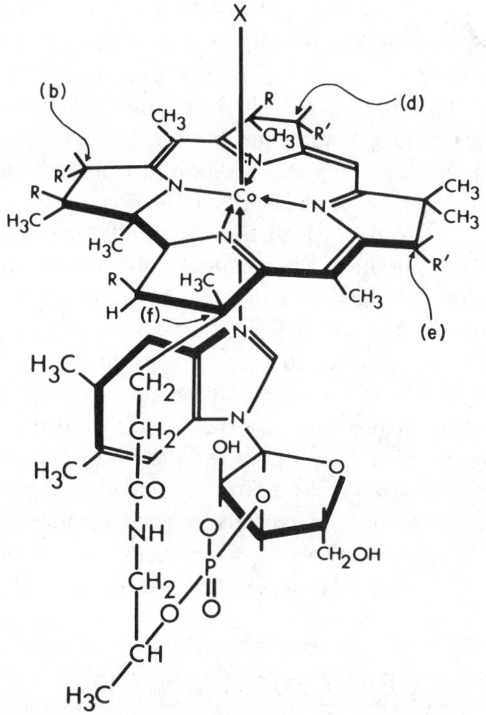

Figure 1 Cobalamin structure. X denotes the β substituent (CN, OH, CH$_3$, 5'-deoxy-5'-adenosyl) that coordinates to the central cobalt nucleus. R and R' refer to the acetamide and propionamide side chains located on the periphery of the corrin ring. Substitutions at the (*b*), (*d*), (*e*), and (*f*) propionamide side chains are discussed further in the text.

(*Lactobacilli*) and some unicellular eukaryotes (*Euglena*), there is also an AdoCbl-dependent ribonucleotide reductase that catalyzes the conversion of ribonucleotides to 2'-deoxyribonucleotides. For a further discussion of the Cbl-dependent reactions, the reader is referred to reviews (2, 4–7).

Many microorganisms contain another Cbl-independent methyltransferase system, and therefore do not require Cbl. However, in those microorganisms that have no such alternate pathway, and in animal cells, MeCbl deficiency is manifested as an ultimately lethal disorder with prominent hematologic and neurologic manifestations. The absence of AdoCbl-dependent rearrangements is rarely a major problem among unicellular organisms. In man, however, and in other higher eukaryotes, the inability to metabolize methylmalonate ultimately leads to its accumulation in cells and body fluids, a derangement that often has lethal consequences. Cbl deficiency states have been reviewed elsewhere (8).

Most Cbl-requiring microorganisms are able to synthesize Cbl de novo.

In those that are not, however, and in all higher eukaryotic cells, mechanisms must exist by which extracellular, preformed Cbl can be acquired. Several major problems face the cell, or organism, attempting to utilize environmental Cbl. First, Cbls are found naturally in minute quantities. The concentration of Cbl in sea water has been estimated to be between 10^{-13} and 10^{-12} M (9), whereas the concentration of Cbl in the prokaryotic cell has been estimated to be at least 3×10^{-7} M (10), and that in human serum to be between 1 and 3×10^{-10} M (8). Second, at least in mammals, Cbl must be transported from the site of environmental contact (the gut) for distribution throughout the organism. Finally, Cbl must be moved across the one, or several, membrane barriers that limit access to the intracellular sites of Cbl utilization. For a molecule to diffuse freely through the membranes of the bacterial cell envelope, its molecular weight can be no larger than 800 (11); for eukaryotic membranes, estimates are even smaller (12, 13). As noted, Cbl is appreciably larger than even the largest of these estimates. Moreover, the Cbl molecule has a large, extended tetrapyrrole plane, and it is hydrophilic, properties not expected to permit much diffusion through a lipid bilayer. The mechanisms by which cells and organisms deal with these problems, Cbl accumulation and translocation, are the subject of this review.

UPTAKE OF COBALAMIN BY E. COLI

Cbl uptake has been observed in a variety of bacterial species (14) but has been studied in detail only in *E. coli*. Early work on the kinetics of uptake of CN-[^{60}Co]Cbl by *E. coli* established that uptake was a biphasic process comprising a rapid, early phase essentially complete within one minute, and a slower, secondary phase, requiring 30–60 min for completion (14–16). Subsequent work has led to the proposal of a relatively simple model for uptake of Cbls by coliforms. According to this model, shown schematically in Figure 2, Cbl uptake proceeds first by the rapid binding of Cbl to an outer membrane receptor. Subsequently, there is a slower release of Cbl into the interior of the cell, an event probably requiring two additional steps. The evidence for, and details of, this model are considered in the following sections.

Binding to Outer Membrane

INTERACTION WITH MEMBRANE RECEPTOR   The initial association of Cbl with *E. coli* is rapid and saturable (15). This interaction obeys Michaelis-Menten kinetics, with a K_m that varies with Cbl structure. Binding occurs at 0°C (16, 17) and is relatively insensitive to temperature over a range from 15°C–35°C (Q_{10} calculated to be 1.2) (15). Furthermore,

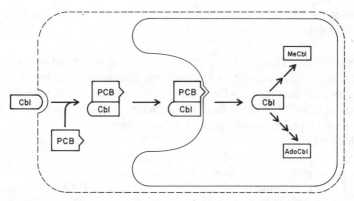

Figure 2 Proposed pathway of uptake of Cbl by *E. coli.* The outer membrane of the bacterial envelope is shown by the dashed line; the inner membrane by the solid line; and the periplasmic space by the space between the two membranes, expanded at left. PCB, periplasmic cobalamin binding protein; AdoCbl, adenosylcobalamin; MeCbl, methylcobalamin. See text for details.

binding has been shown to occur even with heat-killed bacteria (14). The rate of binding is sensitive to changes in pH, occurring maximally near pH 6.0 (15). This suggests that there is an ionizable group, with a pK at or near pH 6, either on the Cbl receptor, or on Cbl itself, that is of major import in allowing the association of one with the other.

EFFECTS OF INHIBITORS   Cbl binding to *E. coli* is not affected by 2,4-dinitrophenol (14, 15), sodium arsenite (15), potassium cyanide (15), sodium fluoride (15), or chloroamphenicol (17). Binding is inhibited by the sulfhydryl reagent mercuric chloride (at 0.1 mM), but not by *N*-ethylmaleimide (at 1 mM) or chloromercuribenzoate (at 1 mM) (18). This suggests that, if a sulfhydryl bond is crucial to the activity of the Cbl receptor, it is a bond not exposed on the surface of the molecule. Binding of Cbl is also decreased by EDTA or EGTA (18). This inhibition is completely reversed by the addition of divalent cations, most effectively by Ca^{2+} (18), which suggests that the physiological association of Cbl with the Cbl receptor is a calcium-dependent event.

CORRINOID SPECIFICITY   The affinity of the Cbl receptor for a Cbl (or related corrinoid) depends upon the specific structure of that compound. The high affinity of receptor for CN-Cbl is reflected by a K_s (dissociation constant) for CN-Cbl binding of between 0.3 and 0.5 X 10^{-9} M (19, 20). The affinity of related corrinoids for the Cbl receptor has usually been assayed by determining the K_i (inhibitor constant) of these compounds. In this way, all of the major structural components of the Cbl molecule have been examined to determine those that contribute most to the interaction with the Cbl receptor.

The nucleotide   There is some controversy as to whether the nucleotide portion of Cbl is essential to Cbl-receptor interaction. Taylor et al (17) report that dicyanocobinamide [$(CN)_2$-Cbi] and MeCbi are ineffective inhibitors of CN-Cbl binding. In contrast, White et al (18) found CN-Cbi to be a potent competitive inhibitor of CN-Cbl binding.

Substitutions of the ribose portion of the nucleotide chain appear to reduce the affinity of a Cbl for the Cbl receptor significantly. White et al (18) were unable to demonstrate any competition by CN-Cbl-5'P. Similarly, Toraya et al (21) found no competition by 2'-*O*-succinyl Cbl, 5'-*O*-succinyl Cbl, and 2'-*O*,5'-*O*-disuccinyl Cbl.

The cobalt atom and "X"   Elimination of the cobalt atom from CN-Cbl or from CN-Cbi increases the K_i of these compounds by approximately a factor of 10 (20). Substitution of the X group has no appreciable effect on the ability of Cbls to bind to the receptor. White et al (18) reported K_i's for MeCbl, AdoCbl, and OH-Cbl of 1.2×10^{-9} M, 4.2×10^{-9} M, and 0.3×10^{-9} M, respectively, compared to a K_i value of 2.8×10^{-9} M for CN-Cbl.

The corrin nucleus   Additional information on the corrinoid specificity of the Cbl receptor comes from the work of Kenley et al (19, 20). Their results may be summarized as follows: 1. The hydrolysis of either the b- or d-propionamide side chain (see Figure 1) of a Cbl to the corresponding monocarboxylic acid increases the K_i by no more than a factor of 10. 2. Addition of a bulky substituent to the d-propionamide of a Cbl increases the K_i by about the same magnitude. 3. Addition of a bulky substituent to the b-propionamide of a Cbl increases the K_i by a factor of 100–500. 4. Hydrolysis of the e-propionamide of a Cbl or Cbi, or substitution there of even a small aliphatic group increases the K_i by a factor of 100–500. From these results, they concluded that multiple small interactions occur between the corrinoid nucleus and the Cbl receptor, each of which contributes slightly to the overall energy of association of the complex. Associations at or near the e-propionamide seem to be especially important. The fact that the e-carboxylate is an exceedingly poor substrate for the receptor suggests that the presence of an anionic terminal group on the e-propionamide is extremely disruptive to Cbl-receptor interaction. The marked effect of substitution at the e-propionamide amide suggests that a change in the nature of the amide is equally disruptive to binding. While this may be simply a steric effect, it is seen both with small substituent groups (e.g. methyl) and for binding of the much less hindered Cbis. It has been proposed (20) that anionic nucleophilic groups (e.g. COO⁻) on the Cbl receptor form critical hydrogen bonds with the e-propionamide amide, and that these provide much of the energy for the Cbl-receptor association. The existence of such critical hydrogen bonding explains the observed reduction in affinity of CN-Cbl for receptor at low pH. Hydrogen bonds in other regions of the corrin nucleus (e.g. the b- and d-propionamides, and conceivably corrin acetamides) may also provide associative energy elsewhere in the complex. Moreover, the fact that a bulky substituent (6-amino-n-hexylamide) at the b-propionamide is tolerated much less well than hydrolysis of that amide makes it clear that other sorts of interactions, more sensitive to steric than to electrostatic interference, are also of major importance.

Properties of Outer Membrane Cobalamin Receptor

STRUCTURE It has been estimated on the basis of equilibrium binding that there are approximately 200 binding sites for Cbl on the haploid cell surface (18). This ability to bind free Cbl to the outer membrane resides in a protein component (18, 22). Bradbeer's group has solubilized this cobalamin binder from outer membrane preparations using Triton X-100 with EDTA. The binding characteristics of the soluble Cbl receptor are similar to those of intact cells (18, 22). They have also purified the receptor to about 95% homogeneity using Cbl-Sepharose affinity chromatography (19). The purified receptor has a distinctly reduced affinity for CN-Cbl, which suggests that part of it (or perhaps an essential cofactor) has been lost in the purification process. This receptor has wild-type affinity for colicin E3, though less than wild-type affinity for colicin E1 (see next section for consideration of binding properties of the Cbl receptor).

Although there are no further published studies of the Cbl receptor per se, the colicin E3 receptor (see below), purified by means of ion exchange chromatography, has been characterized further. It is a 60,000-mol wt glycoprotein, containing glucose, galactose, rhamnose, uronic acid, and amino sugars (23). The relevance of this structure to that of the Cbl receptor is uncertain (see below), but it would be of interest to examine the Cbl-binding properties of this purified colicin receptor.

BIOLOGICAL SPECIFICITY There is considerable evidence that the Cbl receptor has a remarkably broad range of biological functions. DiMasi et al (24, 25) reported that all colicins[1] tested reduced the overall accumulation of Cbl by *E. coli*. Surprisingly, there was, in addition, a concentration-

---

[1]Colicins are strain-specific colicidal agents synthesized by *E. coli* and known to inhibit some energy-coupled transport systems in sensitive strains (26).

dependent decrease in the magnitude of energy-independent (initial phase) Cbl accumulation with colicins E1 and E3, but not with colicin K. Reverse competition was seen as well; Cbl provided protection against the cidal effects of colicins E1 and E3 (but not K) in a manner kinetically consistent with competition for a single binding site. Moverover, mutants putatively deficient in the Cbl receptor (*btu,* discussed below) showed a decreased sensitivity to the E colicins, but wild-type sensitivity to K. In these "receptorless" strains, Cbl no longer protected against the colicins. Finally, recent work (27) suggests the coordinate loss of Cbl binding activity and E colicin sensitivity after growth in media containing high concentrations of Cbl (discussed further below). The simplest explanation for these data is that the receptors for the E colicins and for Cbl share an essential component. [2]

A similar line of evidence suggests yet a third function. Cbl significantly inhibits the binding of the bacteriophage BF23 to *E. coli* (28) and, in receptor-deficient strains, there is a strong positive correlation between the rate of BF23 absorption and the estimated number of Cbl receptors per cell.

These data provide strong, though indirect, evidence that the Cbl receptor and the binding sites for two other classes of biologically important macromolecules include a common element. Importantly, other data suggest that these sites are nonidentical. Bassford & Kadner (29) describe a class of mutants (BtuIV—see section on genetics below) that display drastically reduced ability to bind Cbl, but wild-type sensitivity to both the E colicins and phage BF23. We may then imagine the Cbl receptor (or a component of it) as a molecule with (at least) two distinct but related binding sites, one for Cbl and (at least) one for the E colicins and BF23. Occupation of either site by an appropriate substrate prevents binding to the second, either by physically blocking the second binding site, or by inducing some sort of conformational change in the molecule such that the affinity of the second site is changed.

GENETIC CONTROL   Kadner and his co-workers (29–32) have succeeded in generating a series of mutants with greatly reduced ability to transport Cbl. Most of these map to a single locus on the *E. coli* chromosome. This locus, now named *btuB,* appears to encode the Cbl receptor. Mutations at this locus may have either of two phenotypes with regard to Cbl binding. One class [Class I (31), or BtuA (32) in some publications; most recently BtuI (29)] retains the ability to bind Cbl at the cell surface, but is deficient in the subsequent energy-dependent phase of transport. The other [Class II

---

[2]Interestingly, the number of cell surface colicin receptors has been estimated to be 220 per cell (23), a value similar to the estimated number of Cbl receptors.

(31), BtuB (32), or BtuII-IV(29)] lacks the ability to bind Cbl. These phenotypes are noncomplementing and map together; thus they appear to be
manifestations of alterations in the same cistron. If this is so, the existence
of the Class I phenotype argues for the presence on the receptor molecule
of a region uninvolved with Cbl binding, but essential for its subsequent
internalization.

The Class II phenotype appears to reflect diminished representation on
the cell surface of the high affinity Cbl binding site. This could result either
from a defect in the Cbl binding site itself or from a defect elsewhere in the
molecule (or a precursor) such that its insertion into the membrane is
impaired. In fact, the Class II phenotype is heterogeneous and both types
of defects may exist. Bassford & Kadner's (29) BtuIV phenotype does not
bind Cbl, but retains wild-type sensitivity to colicin E3 and bacteriophage
BF23. In the light of the data presented in the previous section, this phenotype would appear to reflect the selective loss of the Cbl binding site from
the polyfunctional membrane receptor. However, the BtuII phenotype has
lost all binding functions; it is resistant to colicin E3 and at least partially
resistant to BF23, as well as unable to bind Cbl. In addition, one BtuII
strain studied elsewhere (KBT069) (28) has Cbl receptors of normal affinity
though fewer than 0.5% of wild-type number.

MODULATION   An additional feature of the Cbl receptor is its reported
regulation by intracellular Cbl. Kadner (27) reported that, in the presence
of CN-Cbl, Cbl binding activity fell exponentially with time, as would be
expected for dilution of receptors with bacterial growth. In addition, on gel
electrophoresis of cells grown in the presence of Cbl, there is loss of an outer
membrane protein band, a band absent from *btu* mutants and, hence,
presumably the membrane receptor. It remains to be seen whether subsequent work will bear out this interesting suggestion of receptor regulation
by cell Cbl content.

Internalization

In those *btuB* mutants in which Cbl binding is not significantly impaired
(Class I phenotype), 95% of cell-associated Cbl can be recovered as CN-
Cbl, compared with $\sim$ 15% in wild type (31). Clearly, the utilization of Cbl
is not achieved merely by its association with outer membrane receptor. We
may conclude that an additional mechanism must exist to facilitate the
internalization of this receptor bound Cbl.

Some of the properties of this internalization mechanism already have
been noted in brief. The rate of internalization of Cbl is slow compared to
that of binding, and is nearly constant for between 30 and 60 min (14, 15).
Internalization, unlike binding, is markedly temperature dependent (Q_{10}

estimated to be 2.7 between 15°C and 35°C), and notably susceptible to inhibitors of energy metabolism (potassium cyanide, 2,4-dinitrophenol, sodium azide) (18, 33).

Cbl internalization does not appear to require, however, the immediate participation of ATP. Bradbeer & Woodrow (33) report that *E. coli* strain AN120, a mutant deficient in the Ca^{2+}, Mg^{2+}-stimulated ATPase known to be coupled to several active transport systems, is able to internalize CN-Cbl if provided with glucose or lactate and oxygen. The implication of this result is that Cbl transport falls into the class of energy-dependent transport systems in *E. coli* that do not require ATP but rather are able to couple changes in the energy state of the bacterial inner membrane (consequent to respiratory electron transport) directly to substrate movement. Like other such active transport processes, Cbl transport is able to utilize ATP if respiratory electron flow is blocked with cyanide or anaerobiosis (33); under such conditions ATP presumably energizes the inner membrane, the reverse of the reaction by which aerobic ATP synthesis occurs.

The Cbl specificity for the internalization mechanism is difficult to study, given that access to it is limited by the specificities of the Cbl receptor. Nonetheless, the rates of internalization for a variety of corrinoids have been estimated (34); these studies suggest that the rate of internalization for a corrinoid is not predictable from its affinity for the outer membrane Cbl receptor. This datum is compatible with the notion that a carrier protein, distinct from the Cbl receptor and with Cbl binding properties different from it, participates in the internalization process. In accord with this idea is the observation that Cbl internalization is reduced distinctly following osmotic shock (15–18). This sensitivity has been shown, in other transport systems, to reflect the loss of a required periplasmic binding protein. In fact, such a binding protein for Cbl is known to exist (16, 34) and its properties are considered in greater detail subsequently. It should be emphasized, as discussed by Kenley et al (34), that none of these observations proves that such a protein is a required component of the Cbl uptake mechanism. Cbl binding and transport could be kinetically distinct functions of a single receptor protein whose association and dissociation constants for different corrinoid species differ markedly. It is also possible that osmotic shock does not inhibit Cbl uptake by releasing some needed component but rather by disrupting essential structural elements of the bacterial envelope.

Finally, there are genetic data that add another level of complexity to the Cbl internalization process. In *tonB* mutants, a class of *E. coli* variants that have lost the ability to transport a wide range of molecules for which outer membrane receptors exist, Cbl internalization is deficient (35). The receptors for such molecules, including Cbl, function normally (35). Since the nature of the *tonB* gene product is unknown, and since it may even be consumed as it is used (36), there is presently no easy way to fit these observations into a scheme based on simple transfer of Cbl from outer membrane receptor through the inner membrane via a periplasmic binder.

MOVEMENT ACROSS OUTER MEMBRANE   Cbl must traverse the outer membrane either as free vitamin (subsequent to its release from the receptor), or complexed with its receptor. Previous arguments have emphasized the improbability that free Cbl simply diffuses through the lipid membrane. Because release of Cbl from receptor is much slower in whole cells ($t_{1/2} \sim$ 220 sec) than in isolated membrane particles ($t_{1/2} \sim 50$ sec), Bradbeer et al (34) proposed that there is, in the intact cell, a conformational change in the Cbl-receptor complex prior to release of Cbl such that substrate is reoriented toward the periplasmic surface of the outer membrane. This reversible reorientation would lead to a decrease, at any instant, in the amount of Cbl-receptor complex able to liberate Cbl into the incubation medium; therefore, it would lead to an apparent increase in the half-life of the Cbl-receptor complex. Such a model is attractive, reasonable, and testable.

RELEASE FROM OUTER MEMBRANE RECEPTOR   Release of receptor-bound Cbl into the incubation medium can be effected by the addition of chelating agents (EDTA/EGTA) (18) and by Sephadex G-25 filtration in phosphate buffer (17). These properties establish the reversibility in vitro of the association of Cbl and its outer membrane receptor, but do not speak directly to the question of how this dissociation is accomplished in the intact bacterium. Bradbeer's group (34) argue that the dissociation is not passive. They calculated a theoretical maximal rate for Cbl release into the cell, based upon values derived for the maximal rate of dissociation of the Cbl-receptor complex and the maximal number of receptors occupied per cell. This calculated rate was, in fact, 1.3–1.8 times less than they observed. They conclude that a catalyst exists in situ that accelerates the release of Cbl from this complex. No other evidence exists that bears on the presence of such a putative catalyst.

MOVEMENT ACROSS PERIPLASMIC SPACE   The fate of Cbl after its release from the outer membrane receptor is not known. It seems improbable that Cbl remains free in the periplasm until it is subsequently moved across the inner membrane, because the inner membrane appears to be unable to bind free Cbl (18). It may be that the inner membrane recognizes Cbl bound to the outer membrane receptor, and that release from the latter occurs simultaneous with its transfer to an inner membrane carrier. Alternatively, Cbl may be released to a periplasmic binding protein that would serve to carry Cbl across the intermembranous space to an inner membrane receptor recognizing some unique structural aspect either of the complex, or of the periplasmic protein component itself. As mentioned earlier, a 22,000-mol wt protein first described by Taylor (16) is postulated to be such a carrier. This protein is present in the periplasmic space in only trace amounts (1–3 binding sites per cell) (18, 34) and is unaffected by mutations at *btuB* (18). Its involvement in Cbl transport is presumed by virtue of its high affinity for Cbl (K_s for CN-Cbl of 0.3–0.8 $\times$ 10^{-9} M (34)), by the correlation between the affinity of various corrinoids for this protein and the overall rate at which these corrinoids are transported (19), and by the concomitant decrease in the rate of Cbl transport and the release of this periplasmic binder after osmotic shock (17, 18).

The precise functions of this binder, should it prove to facilitate Cbl transport, are not known. It seems unreasonable that it is the *tonB* gene product, because the pleiotropic *tonB* mutation cannot be explained by alteration in a specific Cbl-binding protein. It may serve simply to carry Cbl from outer membrane receptor to inner membrane, but it could equally well have other catalytic or structural properties.

MOVEMENT ACROSS THE INNER MEMBRANE   As already stated, the means by which, and form in which, Cbl reaches the inner membrane are not known. The details of the mechanism by which penetration of this barrier is achieved are unknown as well. Simple diffusion, here as before, seems unlikely. Evidence for the existence of an inner membrane carrier comes from the work of Bassford & Kadner (29). They describe a mutation, *btuC,* which in *tonB⁺/btuB⁺* strains of *E. coli* marginally increases the requirement for supplementary Cbl, but which, in *tonB* or *btuB* mutant strains, increases this requirement by more than four orders of mangitude. This exaggerated requirement cannot be overcome by adding agents that increase membrane permeability, which suggests that the *btuC* gene product is not an outer membrane factor. *btuC* strains (on *tonb⁺/btuB⁺* background) were shown to accumulate Cbl almost normally, but, in efflux experiments with cobalt-labeled Cbl, were seen to have a marked inability to retain it intracellularly. In contrast with *btuB* strains, outer membrane receptor binding of Cbl appears normal in *btuC* strains. Furthermore, *btuC* maps approximately halfway around the *E. coli* chromosome from *btuB*.

These data establish the existence of a locus, *btuC,* that encodes a gene product distinct from those of the *tonB* and *btuB* loci. A reasonable explanation for these data is that the *btuC* gene product is an inner membrane carrier protein or "permease" that moves periplasmic Cbl into the bacterial cytoplasm. In the absence of the *btuB* or *tonB* gene products, insufficient Cbl is accumulated in the periplasm to provide substrate for the *btuC* gene product; the external concentration of Cbl must be increased to overcome this block, and the magnitude of increase necessary is a measure of the dependence of the system on the products of the *btuB* and *tonB* gene products. In the absence of normal *btuC* gene product, Cbl accumulates in the periplasm because its further movement through the inner membrane is blocked. It therefore leaks out of the cell, or is shuttled out by a reverse of the *tonB/btuB*-dependent reactions by which it was accumulated. Finally, if *btuC* is deficient in a strain lacking *tonB* or *btuB* function, virtually no external concentration of Cbl is sufficient to produce a periplasmic concentration of substrate high enough to permit adequate function of the aberrant carrier, and growth is impossible.

UPTAKE OF COBALAMIN BY LOWER EUKARYOTES

Protozoans

Interest in the mechanism by which protozoans accumulate Cbls has been stimulated by their usefulness in bioassays for Cbl content (reviewed in 2). In particular these mechanisms have been studied in the protozoans, *Ochromonas malhamensis* and *Euglena gracilis*. The limited data do not permit the same kind of mechanistic analysis permitted by the work on *E. coli;* nonetheless, a comparison of these data with that from *E. coli* is instructive.

Like *E. coli*, both *Ochromonas* (37) and *Euglena* (38) accumulate free Cbls. Like *E. coli*, the kinetics of cobalamin accumulation by both protozoans are biphasic, with rapid, initial uptake phases, and slower, secondary phases (37, 39). As in *E. coli*, the initial phases of uptake in these organisms are saturable, rapid, independent of temperature, and insensitive to metabolic inhibitors (37, 39, 40). As in coliforms, too, the secondary phases are slower, temperature-dependent, and sensitive to a variety of metabolic inhibitors (37, 39, 40). There are, however, notable differences, as in the pH sensitivities (37, 40). Even so, the data speak for an underlying similarity in the overall process in protozoans and bacteria.

A suggestion of major difference comes from work attempting to identify the site of initial interaction of Cbl with the *Euglena* cell. Varma et al (41) conclude that this initial interaction (binding) is at the cell wall membrane, whereas Sarhan et al (39) argue that it is at the chloroplast. Both studies depend upon subcellular fractionation techniques for the localization of newly accumulated Cbl; both studies then attempt to show that the in vitro association of cobalamin with the appropriate subcellular fraction mirrors the in vivo association. Varma's group studied Cbl localization after two hours of uptake, an unfortunately long incubation period if one is to focus on the early events of uptake. Furthermore, in their in vitro studies, in which they demonstrate rapid, high affinity binding of Cbl to cell wall membrane material, they fail to establish whether this binding is saturable or specific. Its relevance to the situation in vivo is, therefore, uncertain. Unfortunately, Sarhan et al do not specify the time of uptake in their subcellular fractionation experiments. They do, however, establish that the interaction of Cbl with purified chloroplasts is kinetically indistinguishable from initial phase cellular accumulation. The data are taken to indicate that initial phase uptake in *Euglena* represents the binding of Cbl to a site on the chloroplast membrane. An obvious implication is that the cell wall membrane of *Euglena* does not substantially retard the movement of Cbl into the protozoan cell. Additional experiments are needed to resolve these disparate findings.

Metazoans

A Cbl-deficiency syndrome in man can be a prominent manifestation of tapeworm infection. It is now clear that this results from uptake of dietary Cbls by the intestinal parasite (42, 43). The mechanisms by which this uptake is effected have not been well studied; at present there are only isolated reports of Cbl uptake by a variety of metazoan species (42–46). On the basis of these, it seems clear that it is free Cbl that is accumulated (apparently liberated from the host intrinsic factor-Cbl complex by a specific parasite function (47)). From the limited data available, it seems that metazoan Cbl uptake is unidirectional (44, 46), kinetically monophasic (46, 48), and (in the larvae of the tapeworm, *Spirometra mansonoides*) specific, saturable, temperature-dependent, and able to distingiush between a variety of structurally related Cbls (46). It seems probable, then, that cobalamin accumulation by the metazoan species is a facilitated process of some sort.

UPTAKE OF COBALAMIN BY MAMMALS

The mechanisms by which mammals normally sequester and utilize Cbl depend largely on the vitamin's association with specific proteins in extracellular fluids. At least three distinct receptor-mediated systems are known to effect the transmembrane movement of protein-bound Cbl in a variety of cell types. In the gut, dietary Cbl is complexed by the gastric glycoprotein IF, thus facilitating the transport of Cbl to the circulation via a specific receptor for IF on the surface of the ileal enterocyte. In the blood, newly absorbed (or parenterally administered) Cbl is bound by the serum protein transcobalamin II (TC II), which greatly enhances Cbl uptake by peripheral tissues due to the existence of a cell surface receptor for TC II. Serum Cbl is also found in association with one or more other Cbl-binding proteins, the so-called R binders, TC I and TC III. These glycoproteins, when desialylated, have been shown to mediate the hepatic uptake of Cbl via the asialoglycoprotein receptor. The structural features of these extracellular and cellular Cbl-binding proteins are summarized in Table 1. We now review in turn the mechanisms by which Cbl moves between the three major compartments in question: the gut, the blood, and the tissues.

Transport Within and Across the Gut

STRUCTURE AND FUNCTION OF GASTRIC IF   IF is a glycoprotein that is present in the gastric secretions of many, if not all, mammals (reviewed in 49–51). IF has been extensively purified from human gastric juice and hog pyloric mucosa in several laboratories (52–57). When analyzed by SDS-polyacrylamide gel electrophoresis or by gel filtration, molecular weight estimates for IF have ranged between 50,000 and 70,000. In the presence of Cbl, however, IF has a tendency to aggregate, forming dimers (mol wt ~ 115,000) and perhaps higher order oligomers (52–59). Each IF monomer binds one molecule of Cbl. Allen and co-workers have performed detailed analyses of both human (52–57) and hog (53) IF, using material purified by the technique of affinity chromatography employing Sepharose-Cbl columns (60). More accurate molecular weight determinations were inferred from equilibrium sedimentation ultracentrifugation and from direct analysis of amino acid and carbohydrate composition. For human IF, these values were ~ 46,500 daltons and 44,200 daltons, respectively. Approximately 15% of the mass of IF was found to be due to carbohydrate. Hog IF was slightly larger (~ 54,500 daltons) and somewhat higher in carbohydrate content (17.5%). It is not clear whether differences between human and hog IF's reflect species variation, or the fact that the two proteins were isolated from different sources (gastric juice versus mucosa).

In humans, IF appears to be formed uniquely by the stomach (49, 51). In particular, using immunofluorescence (61, 62) and autoradiography (with [57Co]Cbl) (63), IF has been localized to the HCl-secreting parietal cells of the cardiac and fundic regions. More recently, evidence has begun to accumulate that the gastric mucosa, and presumably the parietal cell, is the site of IF biosynthesis. In a preliminary report (64), evidence is presented that radiolabeled IF can be isolated from cultures of rabbit gastric mucosal biopsies incubated in either [35S]methionine or [3H]leucine. In addition, using electron microscopic immunocytochemistry, Allen and co-workers (65) have obtained data that suggest that IF is synthesized in the endoplasmic reticulum of the parietal cell—as is the case for all proteins destined for secretion—and is then transported to the tubular vesicular system of these cells (as opposed to typical secretion granules) prior to release.

Cbl is bound very tightly by IF, with an association constant of between 10^9 and 10^{10} M^{-1} (52, 53, 56, 66, 67). The nature of the substituent on the β-coordination position of the Cbl molecule is apparently unimportant in the formation of the IF-Cbl complex, since all of the commonly occurring Cbl derivatives (OH-Cbl, CN-Cbl, MeCbl, AdoCbl) bind with equal affinity. In contrast, Cbl analogues in which there has been some alteration either of the corrin ring or of the nucleotide portion of the molecule exhibit little affinity for IF (66–69).

The binding of Cbl to IF has been found to occur over a wide range of pH values, from pH 3 to pH 9 (67, 70). Moreover, the binding is quite rapid, being virtually complete within 5 min at 4°C. Once formed, the IF-Cbl complex is exceedingly stable, and exhibits a low rate constant for dissociation at neutral pH (66, 68, 70). Cbl can be removed by treatment of the complex with 5–7.5 M guanidine hydrochloride (52, 53, 59) or with buffer at pH $\geq$ 12.6 (71). Interestingly, the binding of Cbl by IF also seems to confer resistance to the effects of various proteases (72). It is not known whether this effect is a direct result of a conformational change induced by the Cbl itself or by the dimerization that follows the formation of the complex.

It is probably significant that IF exhibits a markedly reduced affinity for Cbl at pH $<$ 3. Under these conditions, it has been demonstrated in vitro that Cbl is bound preferentially to another protein—"nonintrinsic factor" or R protein (see below)—which is present in saliva and gastric mucosa (73). Accordingly, Allen et al have proposed that dietary Cbl may first become associated with R protein and subsequently be transferred to IF following degradation of the R protein in the intestine by pancreatic proteases (74, 75). This concept is supported by several lines of evidence. First, patients with pancreatic insufficiency have long been known to be deficient in Cbl absorption (76), a condition that is often reversed by the ingestion of pancreatic extract or trypsin (for review see 51). Second, duodenal Cbl in untreated patients is associated predominantly with R protein, while in healthy individuals most Cbl is bound to IF. Finally, when the binding of Cbl to R protein is inhibited by prior administration of Cbi, which binds to R protein but not to IF, Cbl absorption in these patients is markedly improved (75). In contrast, pancreatic proteases have no demonstrable effect on IF or on the IF-Cbl complex nor are they necessary for the attachment of IF to its ileal receptor (74) (see below).

INTERACTION OF THE IF-CBL COMPLEX WITH THE ILEAL MUCOSAL RECEPTOR   It has been well-established that the intestinal absorption of physiological amounts of Cbl is dependent on the presence of a receptor for the IF-Cbl complex on the surface of the ileal enterocyte (41, 49). The specificity of this interaction has been demonstrated in vitro by the fact that the binding of IF-[^{57}Co]Cbl to everted sacs or ileal brush borders is saturable, specific, and inhibited by preincubation with excess unlabelled IF-Cbl but not free Cbl (65, 77, 78). Biologically inert IF (see below) has been shown not to bind to ileal membranes in vitro (79, 80). Furthermore, species specificity in the binding of IF to hamster brush borders has been observed such that only IF from hamsters, rabbits, and rats is efficacious (77). Attachment of the IF-Cbl complex to its receptor is calcium dependent, does not require metabolic energy, and occurs preferentially between pH 6.0 and 8.0 (66, 77, 78). At pH values below 5.6 the IF-Cbl complex will detach from its receptor. The dissociation constant at neutral pH for the binding of the complex to microvillus membranes has been estimated to be as low as 2.5×10^{-10} M (66).

Interestingly, the binding of the ileal receptor and the binding of Cbl appear to be mediated by different domains of the IF glycoprotein. In many cases of pernicious anemia, anti-IF autoantibodies are produced that block either the initial formation of the IF-Cbl complex ("blocking antibody") or the binding of the complex to its receptor ("binding antibody") (for review see 75). Moreover, the form of Cbl bound by IF has little influence on the affinity of the complex for the receptor (66).

Several investigators have reported the solubilization and purification of the IF receptor from human and hog ileal mucosa (77, 80–84). Solubilized receptor preparations have been shown to retain functional activity and, moreover, to exhibit many of the same characteristics (Ca^{2+} dependence, pH range, specificity, affinity) as when associated with microvillus membranes (80–84). Receptor activity has also been reported to be variably sensitive to low pH, 56°C, chymotrypsin, dithiothreitol, pronase, and subtilisin while resistant to the effects of neuraminidase, trypsin, and phospholipase A (79, 80). Thus, it appears that the IF receptor is, at least in part, composed of some protein component. That this component is an integral protein of the enterocyte plasma membrane, however, is not entirely clear. Although receptor activity is most effectively solubilized by nonionic detergent (80), some investigators have reported that significant amounts of IF receptor can be stripped from mucosal membranes by homogenization in low salt buffer without added detergent (80–84).

Gräsbeck and co-workers (83, 84) have employed several cycles of affinity chromatography to purify to apparent homogeneity the IF receptor from hog ileal mucosa. Their results suggest that the receptor exists as a large complex ($\sim 3.5 \times 10^6$ daltons) in solution that can be resolved by SDS-polyacrylamide gel electrophoresis into two polypeptides with molecular weights of 110,000 and 70,000. Similar data were obtained regarding the structure of the human receptor, which was found to consist of 140,000- and

90,000-dalton subunits. Both IF receptors were found to contain carbohydrate. The human receptor was also iodinated in vitro, and the 90,000 polypeptide was shown to cross-react with an anti-IF antiserum. However, it was not determined whether this polypeptide was simply IF that copurified with the receptor or a distinct but structurally related component of the receptor itself.

One reason it has, thus far, been difficult to purify quantities of IF receptor sufficient for detailed analysis is that only small amounts of receptor are present on the enterocyte plasma membrane. In the hamster it has been estimated that there is no more than one binding site per microvillus on average (85). This tiny number of receptors, however, can account quantitatively for the capacity of the ileum to absorb Cbl in vivo.

TRANSINTESTINAL TRANSPORT   Surprisingly little definitive information exists regarding the mechanism by which IF-bound Cbl is transported across the ileal enterocyte to the portal circulation. It is clear, however, that this is a slow ($>$ 3 hr in vivo), energy-dependent process (86–89). It seems reasonable to hypothesize that the transport of IF-Cbl occurs via some type of adsorptive endocytosis, by analogy with the transepithelial transport of immunoglobulin in the neonatal rat intestine (90). However, little data has been obtained to support such a mechanism. Subcellular fractionation studies have suggested a transient association of Cbl with a poorly characterized "mitochondrial" fraction, perhaps contaminated with secondary lysosomes (91–93). However, there is disagreement over whether significant amounts of this Cbl are associated with IF (87, 91–93). An added difficulty in interpreting such experiments is the possibility that free Cbl present in homogenates may partition artifactually among various organelles. This is of particular concern in the case of mitochondria, which have been shown to efficiently take up, concentrate, and metabolize certain chemical species of Cbl in vitro (94, 95).

Consistent with an endocytic mechanism of IF-Cbl uptake are recent preliminary data of Donaldson and co-workers (96, 97). Using [^{35}S] methionine-labeled IF, they have found that, at 37°C, IF-[^{57}Co]Cbl complexes become irreversibly associated with everted ileal sacs, while at 4°C, both [^{35}S]IF and [^{57}Co]Cbl can be eluted with EDTA or pH $<$ 5.5. The "internalized" [^{57}Co]Cbl remains associated with IF. As would be expected for a pinocytic process, IF-Cbl "internalization" requires energy, is rapid, and is resistant to treatment with cycloheximide. Clearly, much additional biochemical and morphological work is required to elucidate this process. Unfortunately, the small number of IF receptors per cell is likely to render such studies difficult.

Following its transport across the enterocyte, Cbl appears in the portal blood bound to the plasma Cbl transport protein, TC II. The metabolic fate of IF or its ileal receptor as a consequence of this transmembrane movement of Cbl is unknown. While it was initially presumed that newly absorbed Cbl became bound to unsaturated (apo-) TC II in the blood, evidence is accumulating that indicates this is not the case. Prior saturation of circulating TC II with a large excess of Cbl does not prevent newly absorbed Cbl from appearing in the bloodstream bound to TC II (89). A likely explanation for this is provided by the immunofluorescent identification of an apparently large preformed pool of TC II in the ileal enterocyte (98). The transfer of Cbl from IF to TC II may be a complex process, however, since it evidently requires the structural integrity of the mucosa; free Cbl added to ileal homogenates does not become associated with TC II (98).

GENETIC CONTROL OF IF-MEDIATED CBL TRANSPORT   The study of human Cbl metabolism has been aided significantly by the existence of at least eleven distinct mutations that affect various aspects of Cbl uptake and utilization (for review see 99). The importance of IF and its receptor in mediating the transmembrane movement of Cbl across the ileal enterocyte is emphasized by the phenotypes caused by inherited disorders in this pathway. Affected individuals exhibit a decreased ability to absorb dietary Cbl and usually develop typical pernicious anemia during childhood (for review see 76). Although as yet poorly characterized, two distinct classes of mutation are known to affect the IF glycoprotein itself. One results in the apparent failure to produce and secrete immunologically demonstrable IF (100, 101), while the other causes the production of a mutant protein (79, 102). This altered IF binds Cbl normally and exhibits physicochemical characteristics indistinguishable from the normal protein. However, the mutant IF-Cbl complex has a greatly reduced affinity for the IF receptor. This defect is of additional interest because it corroborates genetically the immunological evidence discussed above that suggests that the Cbl binding and receptor binding properties of IF are due to different functional domains of the molecule.

A third and particularly fascinating defect has also been described that apparently affects the IF receptor itself (103). Three brothers were studied who exhibited selective Cbl malabsorption, adequate concentrations of biologically active IF and transcobalamin II, an absence of anti-IF autoantibodies, and histologically normal intestinal mucosa. Moreover, homogenates of ileal mucosal biopsies were found to bind IF-Cbl normally. thus, it would appear that this defect specifically affects the transport of Cbl across the enterocyte. If IF-Cbl uptake in fact occurs via adsorptive endocytosis (see above), then this phenotype might be analogous to the defect in low density lipoprotein uptake that results in the failure of receptor-bound lipoprotein to be internalized (104).

Transport in Blood Plasma

TRANSCOBALAMIN II (TC II)   TC II is one of three Cbl-binding proteins normally found in human plasma. The other two are TC I and TC III, the so-called Cbl R binders (for review see 49, 105–107). Virtually all Cbl in the circulation ($\sim$ 300 pg/ml plasma) is protein bound; most is associated with TC I (108). However, TC II accounts for most of the unsaturated Cbl-binding activity found in human plasma ($\sim$ 1 ng/ml) (108). In addition, TC II has long been known to be responsible for facilitating the uptake of Cbl by a variety of cell types and peripheral tissues (109, 110). Mutations affecting TC II have been shown to result in severe tissue Cbl insufficiency —in spite of normal plasma Cbl concentrations (111–113)—while the congenital deficiency of the Cbl R binders is clinically harmless despite abnormally low concentrations of Cbl in plasma (114, 115) (see below). Parenterally administered or newly absorbed Cbl appears in the circulation bound to TC II, which is then rapidly cleared. Estimates of the rate of clearance have yielded $t_{1/2}$'s ranging from 5 min to 1.5 hr, depending on the species examined (116, 117). The quantitative aspects of the plasma transport of TC II–(and R binder–) bound Cbl have been reviewed recently by Allen (105).

TC II is a small, nonglycosylated protein with an apparent molecular weight of 38,000 (73). it has been extensively studied by Allen and his co-workers who have purified to homogeneity the trace amounts of TC II present in rabbit and human plasma ($<$ 100 μg/L) by affinity chromatography (108, 118). TC II is immunologically distinct from all other Cbl-binding proteins and, unlike IF, will bind a wide range of Cbl species and other corrinoids. TC II exhibits a high affinity for Cbl ($K_a \sim 10^{11}$ M^{-1}) and is active over a wide pH range from 4–11 (67). TC II–bound Cbl is virtually nonexchangeable and can be removed only after denaturation (e.g. in 7.5 M guanidine-HCl). One mole of Cbl is bound per mole of TC II.

In vivo, TC II is known to be synthesized by the liver (119–122), although it is apparent that other organs contribute as well (119, 123). Moreover, TC II production by several types of cultured cells has been observed: isolated rat liver parenchymal cells (124); mouse L- cells (125); mouse thioglyco-late–elicited peritoneal macrophages (126); and human diploid fibroblasts (127).

THE R BINDERS: TRANSCOBALAMIN I (TC I) AND TRANSCOBALAMIN III (TC III)   TC I and TC III are glycoproteins that exhibit apparent molecular weights of $\sim$ 120,000 when determined by gel filtration or SDS-polyacrylamide gel electrophoresis (105). They can be distinguished, however, by DEAE-cellulose chromatography. They have been purified to homogeneity (some several million fold) from human plasma using affinity chromatography on Cbl-Sepharose (108, 128). Based on their amino acid and carbohydrate compositions and on results of ultracentrifugation studies, a more accurate molecular weight estimate of ∼ 58,000 has been obtained. TC I and TC III are immunologically indistinguishable and probably differ only in their carbohydrate compositions; in particular, TC I is richer in sialic acid while TC III contains proportionately more fucose. TC I is the primary R binder found in the circulation; TC III is largely derived from granulocytes and is released artificially in vitro following blood collection (105, 129, 130). Although most of the endogenous Cbl content of human plasma is associated with TC I and TC III, their function is not known. Individuals expressing a congenital deficiency of R binders are clinically asymptomatic (114, 115).

Transport into Tissue Cells

Having been absorbed from the intestine and transported in blood plasma, Cbls must finally be moved into cells throughout the body. For most, perhaps all, cells this appears to be accomplished by an intricate process initiated by the binding of the TC II-Cbl complex to specific receptors on the cell surface. Such TC II–mediated uptake, however, is not the exclusive means by which all tissues can obtain Cbls. Hepatocytes, for instance, contain a surface receptor for asialoglycoproteins, and this receptor interacts with TC I-Cbl (and perhaps with TC III-Cbl) complexes, thereby providing a second potential means by which this particular tissue obtains Cbls. Finally, there is growing evidence that at least some tissues are capable of taking up free (i.e. unbound) Cbl if the concentration of unbound vitamer is raised to concentrations large enough to exceed the binding capacity of the transcobalamins, thereby permitting interaction between the free vitamin and another cell surface carrier system. These three systems, schematically illustrated in Figure 3, ultimately deliver Cbls to the cytosol and, after traversing the final membrane barrier posed by the outer and inner mitochondrial membranes, to the mitochondrial matrix.

BINDING OF TC II-CBL TO PLASMA MEMBRANE RECEPTOR   As mentioned earlier, TC II has long been known to facilitate the uptake of Cbl by a variety of cell types both in vivo and in cell culture. Evidence for TC II–mediated Cbl uptake has been obtained in liver (116, 131–134), kidney (116, 123, 134), heart (116), spleen (116), small intestine (116), lung (116), as well as cultured human fibroblasts (135, 136), HeLa cells (110), Chinese hamster ovary cells (I. S. Mellman, and L. E., Rosenberg, unpublished data), mouse L cells (125), phytohemagglutinin-stimulated lymphocytes (137), and L1210 lymphoma cells (138). Accordingly, it has been presumed

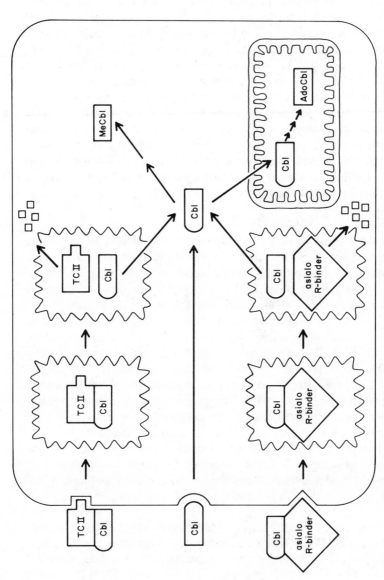

Figure 3 Pathways of uptake and subcellular compartmentation of Cbl by mammalian cells. TC II, transcobalamin II; asialo, asialoglycoprotein; R binder, transcobalamins I and III; MeCbl, methylcobalamin; AdoCbl, adenosylcobalamin. The nature and role of the different membrane receptors for protein-bound and free cobalamin are discussed further in the text, as are the details of translysosomal and transmitochondrial transport.

that the TC II-Cbl complex is recognized by a specific receptor expressed on the plasma membranes of these cells. Only recently, however, has direct biochemical evidence been obtained that proves the existence of a cell surface receptor for TC II.

Using ^{125}I-labeled TC II-Cbl complexes, Youngdahl-Turner et al were able to demonstrate the specific binding of rabbit and human TC II to cultured human fibroblasts (136). At 4°C, binding obeyed saturation kinetics and plateaued at a ^{125}I-TC II concentration of $\sim 0.13 \times 10^{-9}$ M. Binding was inhibited almost completely by a large excess of unlabeled TC II-Cbl, slightly less well by high concentrations of apo-TC II (devoid of Cbl), and not at all by free Cbl or TC I. The attachment of ^{125}I-TC II to its receptor was also found to be calcium dependent. The apparent association constant for specific binding was calculated to be 2×10^{10} M^{-1}, and between 3000 and 4000 high affinity sites per cell were found.

TC II receptor activity has also been identified in crude membrane preparations. Using human placental membranes, Friedman et al (139) demonstrated specific binding of TC II-[^{57}Co]Cbl which was trypsin sensitive, inhibited by EDTA, and resistant to the effects of phospholipase C and neuraminidase. Receptor activity was solubilized with Triton X-100. In a manner similar to TC II binding to intact fibroblasts, apo-TC II appeared to bind slightly less well than holo-TC II (estimated K_a's were 2×10^9 M^{-1} and 7.2×10^9 M^{-1}, respectively). TC I-Cbl, IF-Cbl, and bovine TC II-Cbl did not bind. Specific binding of TC II to a partially purified preparation of rat liver plasma membranes has also been observed (140). While basically in agreement, these latter data are somewhat more difficult to evaluate since crude rat serum was used as the TC II source, and the binding assay employed yielded high background values.

More recently, Seligman & Allen (141) employed affinity chromatography on TC II-Sepharose to purify the TC II receptor from Triton X-100-solubilized human placenta. The receptor was shown to be a glycoprotein ($\sim \frac{1}{3}$ carbohydrate) with a nominal molecular weight of 50,000. The purified material retained receptor activity in vitro, and exhibited an association constant for human holo-TC II (5.6×10^{10} M^{-1}) twice that for apo-TC II (2.3×10^{10} M^{-1}). Free Cbl, IF-Cbl, and TC III-Cbl did not bind to the isolated receptor. Of additional interest is their observation that the amino acid compositions of TC II and its receptor are quite similar, which leads to the speculation that both receptor and ligand may have evolved from a single homologous ancestral gene.

Taken together, these data provide strong support for the existence of a specific receptor for TC II on the plasma membranes of a variety of mammalian cell types. Interestingly, this receptor appears to exhibit a somewhat higher affinity for the TC II-Cbl complex than for apo-TC II, consistent with its role in facilitating the transmembrane movement of Cbl.

ADSORPTIVE ENDOCYTOSIS OF THE TC II-CBL COMPLEX   Following its binding to the cell surface, it is now apparent that the TC II-Cbl complex is internalized intact via pinocytosis. Incoming pinocytic vesicles then fuse with preexisting lysosomes, which results in the proteolytic degradation of the TC II moiety and concomitant release of free Cbl. The subsequent movement of free Cbl across the lysosomal membrane to the cytosol occurs by an, as yet, uncharacterized process (see below).

Early evidence in support of this mechanism of Cbl uptake came from subcellular fractionation studies. Within 1 hr following an intracardiac injection of [^{57}Co]Cbl in rats, both Pletsch and Coffey (131, 132) and Newmark et al (134) found a transient localization of radioactivity with partially purified lysosomal fractions of rat liver and kidney, respectively. Moreover, the [^{57}Co]Cbl observed in this fraction initially appeared to be associated with a protein similar in size to TC II.

Subsequent experimentation has been concerned with analyzing the uptake and metabolic fate of TC II-Cbl in cultured human fibroblasts (136). When cell monolayers were exposed to ^{125}I-TC II at 4°C, radioactivity bound relatively slowly and reached a plateau in 4–6 hr. Virtually all of the ^{125}I-TC II was released by brief trypsinization, which confirms that the association of ^{125}I-TC II with fibroblasts under conditions that inhibit pinocytosis reflects only cell surface binding. At 37°C, however, ^{125}I-TC II bound much more rapidly (maximal within 1–2 hr) and became increasingly resistant to trypsin treatment. Moreover, after a 1–2 hr lag period, TCA-soluble ^{125}I (as monoiodotyrosine) steadily accumulated in the culture medium, which indicates the degradation of TC II. That ^{125}I-TC II was being degraded in lysosomes was suggested by the fact that low concentrations of chloroquine (50 μM) completely but reversibly inhibited the appearance of TCA-soluble radioactivity without affecting the accumulation of ^{125}I-TC II by the cells.

That Cbl was being internalized concomitantly with TC II was shown by experiments using doubly-labeled ^{125}I-TC II-[^{57}Co]Cbl complexes (142). When these complexes were incubated with monolayers at 37°C, the percentage of trypsin-releasable ^{125}I and ^{57}Co decreased coordinately. Similarly, the partial inhibition of pinocytosis by sodium fluoride or sodium azide inhibited equally the accumulation of both ^{125}I and ^{57}Co. Finally, chloroquine-treated cells were shown to be incapable either of degrading ^{125}I-TC II or of releasing [^{57}Co]Cbl from the TC II-Cbl complex. These results, implying that the TC II-Cbl complex is internalized intact, are further supported by in vivo experiments that show that ^{125}I-TC II and [^{57}Co]Cbl are cleared coordinately from the circulation following intravenous injection of ^{125}I-TC II-[^{57}Co]Cbl (116).

Unfortunately, little is known regarding the molecular events involved in the receptor-mediated uptake of TC II-Cbl. Since this pathway has yet to be examined morphologically, it is unknown whether TC II is interiorized in specialized structures such as coated vesicles, as seems to be the case for low density lipoprotein (104). The TC II receptor does not appear to be subject to metabolic regulation, however; its expression has been shown to be unaffected by Cbl or TC II concentrations in the growth medium (142). Given that the major function of this system appears to be the delivery of the trace amounts of available Cbl to the two Cbl-dependent apoenzymes in mammalian cells (133, 143), it is not surprising that TC II–mediated Cbl uptake might be a constitutive process.

BINDING AND UPTAKE OF TC I-CBL AND TC III-CBL VIA THE HEPATOCYTE ASIALOGLYCOPROTEIN RECEPTOR   TC I and TC III are each rapidly cleared from the circulation ($t_{1/2} = 3$ min) by virtue of their interaction with the hepatocyte cell surface receptor for asialoglycoproteins originally described by Ashwell & Morrell (108, 144–146). TC I, however, must first be desialylated by neuraminidase treatment to expose the penultimate galactose residues recognized by the receptor (144). Native TC I, in contrast to the asialo compound, exhibits a half-life in the circulation of ~ 10 days in rabbits. Injections of ^{125}I-TC III-[^{57}Co]Cbl or ^{125}I-asialo-TC I-[^{57}Co]Cbl result in the simultaneous and rapid clearance of both ^{125}I and ^{57}Co. This effect is highly specific and can be inhibited by a prior injection of a large dose of unlabeled asialofetuin. Cbl delivered to the liver by this route appears to be handled in much the same way as Cbl bound to TC II; in both instances, the protein moieties are degraded and [^{57}Co]Cbl becomes associated with hepatic Cbl-dependent apoenzymes (145). However, within 1 hr following injection, most of the [^{57}Co]Cbl reappears in the plasma now bound to endogenous TC II (89). This TC II-bound Cbl is subsequently taken up by a variety of tissues. In addition, approximately 15% of the injected ^{125}I-TC III-[^{57}Co]Cbl appears in biliary secretions within 10–80 min. Thus, it is clear that R binders can facilitate the transmembrane movement of Cbl, although the physiological significance of this mechanism is not immediately clear (see below).

The asialoglycoprotein receptor has been the subject of extensive study during the last few years. Using electron microscope autoradiography, Hubbard et al (147, 148) have established that the uptake of ^{125}I-asialofetuin (and ^{125}I-asialo-orosomucoid) occurred uniquely at the blood sinusoidal front of the rat liver hepatocyte. Little or no ligand was taken up by endothelial cells or Kupffer cells. In addition, these asialoglycoproteins were apparently taken up by pinocytosis and eventually delivered to the hepatocyte lysosomal compartment within 15–60 min. During this time period, the iodinated proteins were also found to be degraded, releasing monoiodotyrosine. These data are in general agreement with the in vivo data of Allen et al (144) regarding the metabolic fate of ^{125}I-TC III.

The asialoglycoprotein receptor from rabbit liver has been purified by affinity chromatography in Ashwell's laboratory (149, 150). The receptor is comprised of two nonidentical subunits with molecular weights of 40,000 and 48,000, both of which are glycosylated. In solution, the receptor apparently exists as a complex ($>$ 500,000 daltons) that has been shown to bind asialoglycoproteins in a calcium-dependent fashion. Recently, an antiserum has been produced against the isolated protein that blocks receptor activity (151).

A novel approach to the functional study of this receptor has been suggested by experiments of Doyle and co-workers (152). In this work, membrane vesicles prepared from rat liver were fused with cultured mouse L cells using polyethylene glycol. Interestingly, functional asialoglycoprotein receptor activity became stably associated with the L-cell membranes, which ordinarily do not express this receptor. ^{125}I-asialoorosomucoid was specifically bound, internalized, and degraded by these hybrids, which demonstrates the epigenetic transfer of a complex process (normally limited to the hepatocyte) to a heterologous cell type.

GENETIC CONTROL OF CELLULAR COBALAMIN TRANSPORT As mentioned earlier, neither the function of R protein–bound Cbl nor that of the TC III–mediated transport of Cbl into the hepatocyte is known. In particular, the significance of these phenomena is questioned by the fact that two brothers found to be congenitally deficient in functional R protein did not exhibit any symptoms normally associated with a defect in Cbl metabolism (114, 115). In contrast, TC II–deficient children exhibit severe pernicious anemia (111–113), which indicates the importance of TC II–mediated Cbl transport. However, it is interesting to note that these children, although anemic, do not show either of two other chemical abnormalities characteristic of Cbl insufficiency, increased plasma concentrations of methylmalonic acid and homocystine (153). Because the liver is a rich source of both Cbl-dependent enzymes, it may be that, in TC II–deficient subjects, Cbl is delivered to hepatocytes by an alternative route, perhaps bound to TC III. In fact, as pointed out by Allen et al (144), 3–9 μg of Cbl per day are excreted in human bile, thereby indicating a significant clearance of R protein–bound Cbl by the liver. Furthermore, it has been estimated that sufficient TC III is produced daily by granulocytes to bind between 100 and 150 μg Cbl.

UPTAKE OF FREE COBALAMIN While it is clear that the most, and perhaps the only, physiologically important mechanism for the uptake of Cbl by the cells of higher eukaryotes requires the mediation of a Cbl-binding protein, there is both in vivo and in vitro evidence to suggest that free Cbl is able, under some circumstances, to traverse the plasma membrane. Circumstantial evidence for the ability of cells to utilize free Cbl comes from the study of children congenitally and totally deficient in TC II, in whom parenteral administration of very large amounts of free Cbl results in striking remission of the clinical and chemical signs of intracellular Cbl deficiency (111, 154). Insofar as no other protein is able to substitute for TC II as a protein mediator of cellular uptake, such observations are compelling. They are supported by recent studies of the uptake of free vitamin in HeLa cells (154) and human fibroblasts (127). In HeLa cells, uptake of free Cbl is between 1 and 2% of that seen for TC II–bound Cbl; with human fibroblasts, free Cbl accumulation in a two-hour interval amounts to about 20% of that noted with TC II-bound vitamin. These results cannot be explained by contamination with endogenous TC II. HeLa cells do not make apo-TC II (154). Uptake in fibroblasts cannot be dependent on cellular apo-TC II because cells from a child with congenital TC II deficiency take up free Cbl as well as control cells do (127). Free vitamin taken up in these experiments can be recovered as intracellular Cbl cofactors (AdoCbl and MeCbl) which establishes that the accumulated vitamin is being internalized and metabolized in a physiological way.

This free vitamin uptake system in human fibroblasts has been studied in some detail by Berliner & Rosenberg (127). Uptake of free CN-[^{57}Co]Cbl is biphasic. The initial component is rapid, saturable, specifically inhibited by excess unlabeled CN-Cbl and OH-Cbl, and complete within 30 min. The second component is slower, linear with time, not inhibited by excess unlabeled Cbl, and does not plateau even after 8 hr. The latter component thus has the properties of a nonspecific process, and has not been examined further. The initial component has properties of a mediated process: it has been shown to be independent of cellular metabolism (uninhibited by potassium cyanide or sodium fluoride); it is sensitive to the sulfhydryl reagent, N-ethylmaleimide; and it is inhibited markedly by cycloheximide (50 μM). These properties are consistent with the presence of a protein-mediated, facilitated diffusion system. The details of this mechanism, the properties of the carrier, and its raison d'être remain to be clarified.

TRANSLYSOSOMAL TRANSPORT OF FREE COBALAMIN   According to the current model of cellular uptake of Cbls (Figure 3) the vitamin enters the lysosomal compartment bound to TC II or the R binders and leaves this compartment in the free state after the transcobalamin-Cbl complexes are hydrolyzed by lysosomal proteases. Little can be said at this time about this important exit process.

The lysosomal membrane has been shown to be freely permeable only to molecules of about 400 mol wt or less (12, 13). Thus, it is reasonable to assume the existence of some process that facilitates Cbl movement across this membrane. To date, the only published studies that bear on this question are those of Pletsch & Coffey (131, 132), and Newmark and co-workers (134, 155). In these studies, the association of Cbl with lysosomal fractions, following in vivo administration of radiolabeled CN-Cbl, was established in rat liver (131, 132) and kidney (134, 155). In each of these systems, there was a time-dependent increase in the amount of lysosomal fraction–associated Cbl over relatively short periods (0–4 hr) after Cbl administration (131, 134). These workers also observed time-dependent conversion of TC II-Cbl to free Cbl in lysosomal fractions (132, 134). In studies of the egress of label from these lysosomal fractions in vitro, however, no distinction could be made between the rate of loss of Cbl and the rate of loss of intralysosomal enzymes (155). In recent work from our laboratory using rat liver lysosomal fractions of substantially higher purity, quantitative distinctions between these rates have been made, but conditions have not been found under which Cbl release occurs independent of hydrolase release (C. Sennett, and L. E. Rosenberg, unpublished data). Because hydrolase release marks lysosomal membrane disruption and because the rates of escape for small molecules (e.g. Cbl) and large molecules (e.g. marker hydrolases) through damaged membrane are expected to vary, these data leave unanswered questions concerning the means by which Cbl moves out of the lysosome.

TRANSMITOCHONDRIAL TRANSPORT OF COBALAMINS   If the model that we have presented in Figure 3 is correct, it is free Cbl that comes in contact with its final membrane barriers, the outer and inner mitochondrial membranes. Assuming this to be the case, Fenton et al (94) conducted a detailed study in vitro of the movement of free OH-[^{57}Co]Cbl into a highly purified preparation of rat liver mitochondria. In unswollen mitochondria, little uptake was observed. When conditions were employed that allowed mitochondrial swelling, however, uptake was rapid, saturable, essentially undirectional, and capable of accumulating Cbl against a large concentration gradient. Other features of this uptake process included: substrate specificity, in that CN-Cbl was taken up much less well than OH-Cbl; a lack of sensitivity to inhibitors of electron transport or respiration; and quantitative binding of the accumulated Cbl to an intramitochondrial protein (of mol wt $\sim$ 120,000 on Sephadex G-150). These data are consistent with the thesis that free Cbl diffuses passively through swollen mitochondrial membranes and then binds to a matrix protein (probably methylmalonyl CoA mutase). The relevance of this model to the situation in vivo is conjectural, particularly since it is not clear whether mitochondria swell, and, if they do, by how much and for what reason.

Additional uncertainty derives from a very different set of results obtained by Gams and co-workers (156–158). This group reported that uptake of TC II-bound Cbl by mitochondria in vitro was 10–30-fold greater than that of free Cbl. Such TC II-Cbl uptake was Ca^{2+}-dependent, inhibited by barbital and cyanide, and unaffected by atractylate and 2,4-dinitrophenol (157, 158). From these and other data, these workers suggest that Cbl enters mitochondria bound to TC II and then is liberated intramitochondrially. These results are very difficult to reconcile with those of Fenton et al (94), who noted that uptake of protein-bound Cbl was much less than that of free Cbl. Moreover, they are difficult to reconcile with a model that places the site of TC II hydrolysis and, *peri passu,* "liberation" of Cbl from its bound state in the lysosomal compartment.

CONCLUDING REMARKS

From all of the foregoing, it should be apparent that, despite nearly 150 years of biological history relevant to Cbl deficiency—the last 50 of which have seen intensive scientific experimentation—much still remains to be learned about the mechanisms that mediate the movement of this unique vitamer. This review has, we hope, underscored some of the unanswered questions regarding the biochemical components, the physiological events, the genetic control systems, and the cell biological phenomena whose understanding will complete the picture of a substance germane to the life cycles of prokaryotes and eukaryotes alike.

ACKNOWLEDGMENTS

We are grateful to Wayne A. Fenton for timely assistance with the revision of this manuscript, and to Marilyn Feldman for expert and patient secretarial work throughout its preparation.

Literature Cited

1. Smith, E. L. 1955. In *The Biochemistry of Vitamin B_{12},* ed. R. J. Williams, pp. 3–14. Cambridge: Cambridge Univ. Press. 123 pp.
2. Smith, E. L. 1960. *Vitamin B_{12}.* London: Methuen. 196 pp.
3. Hogenkamp, H. P. C. 1975. In *Cobalamin: Biochemistry and Pathophysiology,* ed. B. M. Babior, pp. 21–73. New York: Wiley. 477 pp.
4. Arnstein, H. R. V. 1955. See Ref. 1, pp. 92–108
5. Lascelles, J.,Cross, M. J. 1955. See Ref. 1, pp. 109–23
6. Poston, J. M., Stadtman, T. C. 1975. See Ref. 3, pp. 111–40
7. Babior, B. M. 1975. See Ref. 3, pp. 141–212
8. Beck, W. S. 1975. See Ref. 3, pp. 403–50
9. Droop, M. R. 1957. *Nature* 180:1041–42
10. Bradbeer, C., Woodrow, M. L. 1976. *J. Bacteriol.* 128:99–104
11. Payne, J. W. Gilvarg, C. 1968. *J. Biol. Chem.* 243:6291–99
12. Cohn, Z. A., Ehrenreich, B. A. 1969. *J. Exp. Med.* 129:201–25
13. Ehrenreich, B. A., Cohn, Z. A. 1969. *J. Exp. Med.* 129:227–45
14. Giannella, R. A., Broitman, S. A., Zamcheck, N. 1969. *Clin. Res.* 17:594 (Abstr.)
15. DiGirolamo, P. M.,Bradbeer, C. 1971. *J. Bacteriol.* 106:745–50
16. Taylor, R. T., Norrell, S. A., Hanna, M. L. 1972. *Arch Biochem. Biophys.* 148:366–81
17. Taylor, R. T., Nevins, M. P., Hanna, M.

L. 1972. *Arch. Biochem. Biophys.*
149:232–43
18. White, J. C., DiGirolamo, D. R., Fu,
M. L., Preston, Y. A., Bradbeer, C.
1973. *J. Biol. Chem.* 248:3978–86
19. Kenley, J. S., White, J. C., DiMasi, D.
R., Bradbeer, C. 1976. *Fed. Proc.*
35:1700 (Abstr.)
20. Kenley, J. S., Leighton, M., Bradbeer,
C. 1978. *J. Biol. Chem.* 253:1341–46
21. Toraya, T., Kazumoto, D., Ueno, H.,
Fukui, S. 1975. *Bioinorg. Chem.*
4:245–55
22. White, J. C., Fu, M. L., Bradbeer, C.
1972. *Ann. Meet. Am. Soc. Microbiol.*
72:158 (Abstr.)
23. Sabet, S. F., Schanitman, C. A. 1973. *J.
Biol. Chem.* 248:1797–1806
24. DiMasi, D. R., White, J. C., Bradbeer,
C. 1973. *Fed. Proc.* 32:600 (Abstr.)
25. DiMasi, D. R., White, J. C., Schnaitman, C. A., Bradbeer, C. 1973. *J. Bacteriol.* 115:506–13
26. Fields, K. L., Luria, S. E. 1969. *J. Bacteriol.* 97:57–63
27. Kadner, R. J. 1978. *J. Bacteriol.*
136:1050–57
28. Bradbeer, C., Woodrow, M. L., Khali-
fah, L. I. 1976. *J. Bacteriol.* 125:
1032–39
29. Bassford, P. J., Jr., Kadner, R. J. 1977.
J. Bacteriol. 132:796–805
30. Liggins, G. L., Kadner, R. J. 1972. *Ann.
Meet. Am. Soc. Microbiol.* 72:158
(Abstr.)
31. DiGirolamo, P. M., Kadner, R. J.,
Bradbeer, C. 1971. *J. Bacteriol.*
106:751–57
32. Kadner, R. J., Liggins, G. L. 1973. *J.
Bacteriol.* 115:514–21
33. Bradbeer, C., Woodrow, M. L. 1976. *J.
Bacteriol.* 128:99–104
34. Bradbeer, C., Kenley, J. S., DiMasi, D.
R., Leighton, M. 1978. *J. Biol. Chem.*
253:1347–52
35. Bassford, P. J., Jr., Bradbeer, C., Kadner, R. J., Schnaitman, C. A. 1976. *J.
Bacteriol.* 128:242–47
36. Kadner, R. J., McElhaney, G. 1978. *J.
Bacteriol.* 134:1020–29
37. Reeves, R. B., Fay, F. S. 1966. *Am. J.
Physiol.* 210:1273–78
38. Ross, G. I. M. 1952. *J. Clin. Pathol.*
5:250–56
39. Sarhan, F., Houde, M., Cheneval, J. P.
1980. *J. Protozool.* 27:235–38
40. Bradbeer, C. 1971. *Arch. Biochem. Biophys.* 144:184–92
41. Varma, T. N. S., Abraham, A., Hansen,
I. A. 1961. *J. Protozool.* 8:212–16
42. Brante, G., Enberg, T. 1957. *Scand. J.
Clin. Lab. Invest.* 9:313–14

43. Scudamore, H. H., Thompson, J. H.,
Jr., Owen, C. A. 1961. *J. Lab. Clin.
Med.* 57:240–46
44. Weinstein, P. P., Mueller, J. F. 1970. *J.
Parasitol.* 56:363 (Abstr.)
45. Zam, S. G., Martin, W. E., Thomas, L.
J. Jr. 1963. *J. Parasitol.* 49:190–96
46. Tkachuck, R. D., Weinstein, P. P.,
Mueller, J. F. 1976. *J. Parasitol* 62:94–
101
47. Nyberg, W., Saarni, M., Gothoni, G.,
Järventie, G. 1961. *Acta Med. Scand.*
170:257–62
48. Nyberg, W. 1958. *Exp. Parasitol.*
7:178–90
49. Ellenbogen, L. 1979. In *Biochemistry of
Nutrition I.,* ed. A. Neuberger, T. H.
Jukes, pp. 45–96. Baltimore: Univ. Park
Press. 331 pp.
50. Ellenbogen, L. 1975. See Ref. 3., pp.
215–86
51. Donaldson, R. M. 1981. In *Physiology
of the Digestive Tract,* ed. L. R. Johnson. New York: Raven. In press
52. Allen, R. H., Mehlman, C. S. 1973. *J.
Biol. Chem.* 248:3660–69
53. Allen, R. H., Mehlman, C. S. 1973. *J.
Biol. Chem.* 248:3670–80
54. Ellenbogen, L., Highley, D. R. 1967. *J.
Biol. Chem.* 242:1004–9
55. Visuri, K., Gräsbeck, R. 1973. *Biochim.
Biophys. Acta* 310:508–17
56. Christensen, J. M., Hippe, E., Olesen,
H., Rye, M., Haber, E., Lee, L.,
Thomsen, J. 1973. *Biochim. Biophys.
Acta* 303:319–32
57. Katz, M., Mehlman, C. S., Allen, R. H.
1974. *J. Clin. Invest.* 53:1274–83
58. Gräsbeck, R., Simons, K., Sinkkonen, I.
1966. *Biochim. Biophys. Acta* 127:47–58
59. Highley, D. R., Davies, M. C., Ellenbogen, L. 1967. *J. Biol. Chem.*
242:1010–15
60. Allen, R. H., Majerus, P. W. 1972. *J.
Biol. Chem.* 247:7695–7708
61. Jacob, E., Glass, G. B. J. 1971. *Clin.
Exp. Immunol.* 8:517–27
62. Jacob, E., Glass, G. B. J. 1971. *Proc.
Soc. Exp. Biol. Med.* 137:243–48
63. Hoedemaeker, P. J., Abels, J., Watch-
ers, J. J., Arends, A. H., Niewig, H. O.
1964. *Lab Invest.* 13:1394–99
64. Serfillipi, D., Donaldson, R. M., 1979.
Gastroenterology 76:1241
65. Levine, J. S., Nakane, P. K., Allen, R.
H. 1981. *Gastroenterology.* In press
66. Mathan, V. I., Babior, B. M., Donaldson, R. M. 1974. *J. Clin. Invest.* 54:598–
608
67. Hippe, E., Oleson, H. 1971. *Biochim.
Biophys. Acta* 243:83–89

68. Hippe, E., Haber, E., Oleson, H. 1971. *Biochim. Biophys. Acta* 243:75–82
69. Gottlieb, C. W., Retief. F. P., Herbert, V. 1967. *Biochim. Biophys. Acta* 143:560–72
70. Wagstaff, M., Broughton, A., Jones, F. R. 1973. *Biochim. Biophys. Acta* 320:406–15
71. Gräsbeck, R., Stenman, U. H., Puutula, L., Visuri, K. 1968. *Biochim. Biophys. Acta* 158:292–95
72. Gräsbeck, R., Kantero, I., Slurala, M. 1959. *Lancet* 1:234
73. Allen, R. H. 1975. *Prog. Hematol.* 9:57–84
74. Allen, R. H., Seetharam, B., Podell, E., Alpers, D. H. 1978. *J. Clin. Invest.* 61:47–54
75. Allen, R. H., Allen, N. C., Podell, E. R., Alpers, D. H. 1978. *J. Clin. Invest.* 61:1628–34
76. Donaldson, R. M. Jr. 1975. See Ref. 3, pp. 335–68
77. Donaldson, R. M. Jr., Mackenzie, I. L., Trier, J. S. 1967. *J. Clin. Invest.* 46:1215–28
78. Herbert, V., Castle, W. B. 1961. *J. Clin. Invest.* 40:1978–83
79. Katz, M., Mehlman, C. S., Allen, R. H. 1974. *J. Clin. Invest.* 53:1274–83
80. Katz, M., Cooper, B. A. 1974. *J. Clin. Invest.* 54:733–39
81. Cotter, R., Rothenberg, S. P. 1976. *Brit. J. Haematol.* 34:477–87
82. Cotter, R., Rothenberg, S. P., Weiss, J. P. 1977. *Biochim. Biophys. Acta* 490:19–26
83. Marcoullis, G., Gräsbeck, R. 1977. *Biochim. Biophys. Acta* 499:309–14
84. Kouvonen, I., Gräsbeck, R. 1979. *Biophys. Biochem. Res. Commun.* 86:358–64
85. Donaldson, R. M., Small, D. M., Robins, S., Mathan, V. I. 1973. *Biochim Biophys. Acta* 311:477–81
86. Cooper, B. A., Castle, W. B. 1960. *J. Clin. Invest.* 39:199–214
87. Hines, J. D., Rosenberg, A., Harris, J. W. 1968. *Proc. Soc. Exp. Biol. Med.* 129:653–58
88. Strauss, E. W., Wilson, T. H. 1960. *Am. J. Physiol.* 198:103–7
89. Chanarin, I., Muir, M., Hughes, A., Hoffbrand, A. V. 1978. *Br. Med. J.* 1:1453–55
90. Rodewald, R. 1980. *J. Cell Biol.* 85:18–32
91. Peters, T. J., Hoffbrand, A. V. 1970. *Br. J. Haematol.* 19:369–82
92. Peters, T. J., Quinlan, A., Hoffbrand, A. V. 1971. *Br. J. Haematol.* 20:123–29
93. Rothenberg, S. P., Weisberg, H., Ficarra, A. 1972. *J. Lab. Clin. Med.* 79:587–97
94. Fenton, W. A., Ambani, L. A., Rosenberg, L. E. 1976. *J. Biol. Chem.* 251:6616–23
95. Fenton, W. A., Rosenberg, L. E. 1978. *Arch. Biochem. Biophys.* 189:441–47
96. Kapadia, C. R., Serfillipi, D., Donaldson, R. M. Jr. 1979. *Clin. Res.* 27:455A (Abstr.)
97. Kapadia, C. R., Donaldson, R. M. Jr. 1979. *Gastroenterology* 76:1163
98. Rothenberg, S. P., Weiss, J. P., Cotter, R. 1978. *Br. J. Haematol.* 40:401–14
99. Fenton, W. A., Rosenberg, L. E. 1978. *Ann. Rev. Genet.* 12:223–48
100. McIntyre, O. R., Sullivan, L. W., Jeffries, G. H., Silver, R. H. 1965. *N. Engl. J. Med.* 272:981–86
101. Spurling, C. L., Sacks, M. S., Jiji, R. M. 1964. *N. Engl. J. Med.* 271:995–1003
102. Katz, M., Lee, S. K., Cooper, B. A. 1972. *N. Engl. J. Med.* 287:425–29
103. Mackenzie, I. L., Donaldson, R. M., Jr., Trier, J. S. Mathan, V. I. 1972. *N. Engl. J. Med.* 286:1021–25
104. Anderson, R. G. W., Goldstein, J. L., Brown, M. S. 1977. *Nature* 270:695–99
105. Allen, R. H. 1976. *Br. J. Haematol.* 33:161–71
106. Hall, C. A. 1975. *J. Clin. Invest.* 56:1125–31
107. Mahoney, M. J., Rosenberg, L. E. 1975. See Ref. 3, pp. 369–402
108. Burger, R. L., Mehlman, C. S., Allen, R. H. 1975. *J. Biol. Chem.* 250:7700–6
109. Hall, C. A., Finkler, A. E. 1963. *Biochim. Biophys. Acta* 78:234–36
110. Finkler, A. E., Hall, C. A. 1967. *Arch. Biochem. Biophys.* 120:79–85
111. Gimpert, E., Jakob, M., Hitzig, W. H. 1975. *Blood* 45:71–82
112. Hakami, N., Neiman, P. E., Canellos, G. P., Lazerson, J. 1971. *N. Engl. J. Med.* 285:1163–70
113. Hitzig, W. H., Dohmann, V., Pluss, H. J., Vischer, D. 1974. *J. Pediatr.* 85:622–28
114. Carmel, R., Herbert, V. 1969. *Blood* 33:1–12
115. Hall, C. A., Begley, J. A. 1977. *Am. J. Hum. Genet.* 29:619–26
116. Schneider, R. J., Burger, R. L., Mehlman, C. S., Allen, R. H. 1976. *J. Clin. Invest.* 57:27–38
117. Hom, B. L., Olesen, H. A. 1969. *Scand. J. Clin. Lab. Invest.* 23:201–11
118. Allen, R. H., Majerus, P. W. 1972. *J. Biol. Chem.* 247:7709–17
119. Rapazzo, M. E., Hall, C. A. 1972. *J. Clin. Invest.* 51:1915–18

120. England, J. M., Tavill, A. S., Chanarin, I. 1973. *Clin. Sci. Mol. Med.* 45:479–83
121. Cooksley, W. G. E., England, J. M., Louis, L., Down, M. C., Tavill, A. S. 1974. *Clin. Sci. Mol. Med.* 47:531
122. Tan, C. H., Hansen, H. J. 1968. *Proc. Soc. Exp. Biol. Med.* 127:740–44
123. Sonneborn, D. W., Abouna, G., Mendez-Pican, G. 1972. *Biochim. Biophys. Acta* 273:283–86
124. Savage, C. R. Jr., Green, P. D. 1976. *Arch. Biochem. Biophys.* 173:691–702
125. Green, P. D., Savage, C. R. Jr., Hall, C. A. 1976. *Arch. Biochem. Biophys.* 176:683–89
126. Rachmilewitz, B., Rachmilewitz, M., Chaouat, M., Schlesinger, M. 1978. *Blood* 52:1089–98
127. Berliner, N., Rosenberg, L. E. 1981. *Metabolism.* 30:230–36
128. Allen, R. H., Majerus, P. W. 1972. *J. Biol. Chem.* 247:7702–8
129. Scott, J. M., Bloomfield, F. J., Stebbins, R., Herbert, V. 1974. *J. Clin. Invest.* 53:228–39
130. Carmel, R., Herbert, V. 1972. *Blood* 40:542–49
131. Pletsch, Q. A., Coffey, J. W. 1971. *J. Biol. Chem.* 246:4619–29
132. Pletsch, Q. A., Coffey, J. W. 1972. *Arch. Biochem. Biophys.* 151:157–67
133. Mellman, I. S., Youngdahl-Turner, P., Willard, H. F., Rosenberg, L. E. 1977. *Proc. Natl. Acad. Sci. USA* 74:916–20
134. Newmark, P. A. 1972. *Biochim. Biophys. Acta* 261:85–93
135. Rosenberg, L. E., Patel, L., Lilljeqvist, A.-C. 1975. *Proc. Natl. Acad. Sci. USA* 72:4617–21
136. Youngdahl-Turner, P., Rosenberg, L. E., Allen, R. H. 1978. *J. Clin. Invest.* 61:133–41
137. Hoffbrand, A. V., Tripp, E., Das, K. C. 1973. *Br. J. Haematol.* 24:147–56
138. DiGirolamo, P. M., Huennekens, F. M. 1975. *Arch. Biochem. Biophys.* 168:386–93

139. Friedman, P. A., Shia, M. A., Wallace, J. K. 1977. *J. Clin. Invest.* 59:51–58
140. Fiedler-Nagy, C., Rowley, G. R., Coffey, J. W., Miller, O. N. 1975. *Br. J. Haematol.* 31:311–21
141. Seligman, P. A., Allen, R. H. 1978. *J. Biol. Chem.* 253:1766–72
142. Youngdahl-Turner, P., Mellman, I. S., Allen, R. H., Rosenberg, L. E. 1979. *Exp. Cell Res.* 118:127–34
143. Mellman, I., Willard, H. F., Rosenberg, L. E. 1978. *J. Clin. Invest.* 62:952–60
144. Burger, R. L., Schneider, R. J., Mehlman, C. S., Allen, R. H. 1975. *J. Biol. Chem.* 250:7707–13
145. Kolhouse, J. F., Allen, R. H. 1977. *Proc. Natl. Acad. Sci. USA* 74:921–25
146. Ashwell, G., Morell, A. G. 1974. *Adv. Enzymol.* 44:99–128
147. Hubbard, A. L., Wilson, G., Ashwell, G., Stukenbrok, H. 1979. *J. Cell Biol.* 83:47–64
148. Hubbard, A. L., Stukenbrok, H. 1979. *J. Cell Biol.* 83:65–81
149. Hudgin, R. L., Pricer, W. E. Jr., Ashwell, G., Stockert, R. J., Morell, A. G. 1974. *J. Biol. Chem.* 249:5536–43
150. Kawasaki, T., Ashwell, G. 1976. *J. Biol. Chem.* 251:1296–1302
151. Tanabe, T., Pricer, W. E. Jr., Ashwell, G. 1979. *J. Biol. Chem.* 254:1038–43
152. Doyle, D., Hou, E., Warren, R. 1979. *J. Biol. Chem.* 254:6853–56
153. Scott, C. R., Hakami, N., Teng, C. C., Sagerson, R. N. 1972. *J. Pediat.* 81:1106–11
154. Hall, C. A., Hitzig, W. H., Green, P. D., Begley, J. A. 1979. *Blood* 53:251–63
155. Newmark, P., Newman, G. E., O'Brien, J. R. P. 1970. *Arch. Biochem. Biophys.* 141:121–30
156. Gams, R. A., Ostroy, F. 1976. *Clin. Res.* 24:308A (Abstr.)
157. Gams, R. A. 1975. *Clin. Res.* 23:273A (Abstr.)
158. Gams, R. A., Ryel, E. M., Ostroy, F. 1976. *Blood* 47:923–30

AUTHOR INDEX

(Names appearing in capital letters indicate authors of chapters in this volume.)

SUBJECT INDEX

CUMULATIVE INDEXES

CONTRIBUTING AUTHORS, VOLUMES 46–50

1169

CHAPTER TITLES, VOLUMES 46–50

ORDER FORM   ANNUAL REVIEWS INC.

Please list on the order blank on the reverse side the volumes you wish to order and whether you wish a standing order (the latest volume sent to you automatically each year). Volumes not yet published will be shipped in month and year indicated. Prices subject to change without notice. Out of print volumes subject to special order.

NEW TITLES FOR 1981

ANNUAL REVIEW OF NUTRITION   ISSN 0199-9885
    Vol. 1 (avail. July 1981): $20.00 (USA), $21.00 (elsewhere) per copy

INTELLIGENCE AND AFFECTIVITY: Their Relationship During Child Development
    A monograph, translated from a course of lectures by Jean Piaget   ISBN 0-8243-2901-5
    Avail. Feb. 1981   Hard cover: $8.00 (USA), 9.00 (elsewhere) per copy

SPECIAL PUBLICATIONS

ANNUAL REVIEWS REPRINTS: CELL MEMBRANES, 1975–1977   ISBN 0-8243-2501-X
    A collection of articles reprinted from recent *Annual Review* series
    Published 1978   Soft cover: $12.00 (USA), $12.50 (elsewhere) per copy

ANNUAL REVIEWS REPRINTS: IMMUNOLOGY, 1977–1979   ISBN 0-8243-2502-8
    A collection of articles reprinted from recent *Annual Review* series
    Published 1980   Soft cover: $12.00 (USA), $12.50 (elsewhere) per copy

THE EXCITEMENT AND FASCINATION OF SCIENCE, VOLUME 1   ISBN 0-8243-1602-9
    A collection of autobiographical and philosophical articles by leading scientists
    Published 1965   Clothbound: $6.50 (USA), $7.00 (elsewhere) per copy

THE EXCITEMENT AND FASCINATION OF SCIENCE, VOLUME 2: Reflections by Eminent Scientists
    Published 1978   Hard cover: $12.00 (USA), $12.50 (elsewhere) per copy   ISBN 0-8243-2601-6
                         Soft cover: $10.00 (USA), $10.50 (elsewhere) per copy   ISBN 0-8243-2602-4

THE HISTORY OF ENTOMOLOGY   ISBN 0-8243-2101-7
    A special supplement to the *Annual Review of Entomology* series
    Published 1973   Clothbound: $10.00 (USA), $10.50 (elsewhere) per copy

ANNUAL REVIEW SERIES

Annual Review of ANTHROPOLOGY   ISSN 0084-6570
    Vols. 1–8 (1972–79): $17.00 (USA), $17.50 (elsewhere) per copy
    Vol. 9 (1980): $20.00 (USA), $21.00 (elsewhere) per copy
    Vol. 10 (avail. Oct. 1981): $20.00 (USA), $21.00 (elsewhere) per copy

Annual Review of ASTRONOMY AND ASTROPHYSICS   ISSN 0066-4146
    Vols. 1–17 (1963–79): $17.00 (USA), $17.50 (elsewhere) per copy
    Vol. 18 (1980): $20.00 (USA), $21.00 (elsewhere) per copy
    Vol. 19 (avail. Sept. 1981): $20.00 (USA), $21.00 (elsewhere) per copy

Annual Review of BIOCHEMISTRY   ISSN 0066-4154
    Vols. 28–48 (1959–79): $18.00 (USA), $18.50 (elsewhere) per copy
    Vol. 49 (1980): $21.00 (USA), $22.00 (elsewhere) per copy
    Vol. 50 (avail. July 1981): $21.00 (USA), $22.00 (elsewhere) per copy

Annual Review of BIOPHYSICS AND BIOENGINEERING   ISSN 0084-6589
    Vols. 1–9 (1972–80): $17.00 (USA), $17.50 (elsewhere) per copy
    Vol. 10 (avail. June 1981): $20.00 (USA), $21.00 (elsewhere) per copy

Annual Review of EARTH AND PLANETARY SCIENCES   ISSN 0084-6597
    Vols. 1–8 (1973–80): $17.00 (USA), $17.50 (elsewhere) per copy
    Vol. 9 (avail. May 1981): $20.00 (USA), $21.00 (elsewhere) per copy

Annual Review of ECOLOGY AND SYSTEMATICS   ISSN 0066-4162
    Vols. 1–10 (1970–79): $17.00 (USA), $17.50 (elsewhere) per copy
    Vol. 11 (1980): $20.00 (USA), $21.00 (elsewhere) per copy
    Vol. 12 (avail. Nov. 1981): $20.00 (USA), $21.00 (elsewhere) per copy

Annual Review of ENERGY   ISSN 0362-1626
    Vols. 1–4 (1976–79): $17.00 (USA), $17.50 (elsewhere) per copy
    Vol. 5 (1980): $20.00 (USA), $21.00 (elsewhere) per copy
    Vol. 6 (avail. Oct. 1981): $20.00 (USA), $21.00 (elsewhere) per copy

Annual Review of ENTOMOLOGY   ISSN 0066-4170
    Vols. 7–25 (1962–80): $17.00 (USA), $17.50 (elsewhere) per copy
    Vol. 26 (avail. Jan. 1981): $20.00 (USA), $21.00 (elsewhere) per copy

Annual Review of FLUID MECHANICS   ISSN 0066-4189
    Vols. 1–12 (1969–80): $17.00 (USA), $17.50 (elsewhere) per copy
    Vol. 13 (avail. Jan. 1981): $20.00 (USA), $21.00 (elsewhere) per copy

Annual Review of GENETICS   ISSN 0066-4197
    Vols. 1–13 (1967–79): $17.00 (USA), $17.50 (elsewhere) per copy
    Vol. 14 (1980): $20.00 (USA), $21.00 (elsewhere) per copy
    Vol. 15 (avail. Dec. 1981): $20.00 (USA), $21.00 (elsewhere) per copy

(continued on reverse)

Annual Review of MATERIALS SCIENCE   ISSN 0084-6600
     Vols. 1–9 (1971–79): $17.00 (USA), $17.50 (elsewhere) per copy
     Vol. 10 (1980): $20.00 (USA), $21.00 (elsewhere) per copy
     Vol. 11 (avail. Aug. 1981): $20.00 (USA), $21.00 (elsewhere) per copy

Annual Review of MEDICINE: Selected Topics in the Clinical Sciences   ISSN 0066-4219
     Vols. 1–3, 5–15, 17–31 (1950–52, 1954–64, 1966–80): $17.00 (USA), $17.50 (elsewhere) per copy
     Vol. 32 (avail. Apr. 1981): $20.00 (USA), $21.00 (elsewhere) per copy

Annual Review of MICROBIOLOGY   ISSN 0066-4227
     Vols. 15–33 (1961–79): $17.00 (USA), $17.50 (elsewhere) per copy
     Vol. 34 (1980): $20.00 (USA), $21.00 (elsewhere) per copy
     Vol. 35 (avail. Oct. 1981): $20.00 (USA), $21.00 (elsewhere) per copy

Annual Review of NEUROSCIENCE   ISSN 0147-006X
     Vols. 1–3 (1978–80): $17.00 (USA), $17.50 (elsewhere) per copy
     Vol. 4 (avail. Mar. 1981): $20.00 (USA), $21.00 (elsewhere) per copy

Annual Review of NUCLEAR AND PARTICLE SCIENCE   ISSN 0066-4243
     Vols. 10–29 (1960–79): $19.50 (USA), $20.00 (elsewhere) per copy
     Vol. 30 (1980): $22.50 (USA), $23.50 (elsewhere) per copy
     Vol. 31 (avail. Dec. 1981): $22.50 (USA), $23.50 (elsewhere) per copy

Annual Review of PHARMACOLOGY AND TOXICOLOGY   ISSN 0362-1642
     Vols. 1–3, 5–20 (1961–63, 1965–80): $17.00 (USA), $17.50 (elsewhere) per copy
     Vol. 21 (avail. Apr. 1981): $20.00 (USA), $21.00 (elsewhere) per copy

Annual Review of PHYSICAL CHEMISTRY   ISSN 0066-426X
     Vols. 10–21, 23–30 (1959–70, 1972–79): $17.00 (USA), $17.50 (elsewhere) per copy
     Vol. 31 (1980): $20.00 (USA), $21.00 (elsewhere) per copy
     Vol. 32 (avail. Nov. 1981): $20.00 (USA), $21.00 (elsewhere) per copy

Annual Review of PHYSIOLOGY   ISSN 0066-4278
     Vols. 18–42 (1956–80): $17.00 (USA), $17.50 (elsewhere) per copy
     Vol. 43 (avail. Mar. 1981): $20.00 (USA), $21.00 (elsewhere) per copy

Annual Review of PHYTOPATHOLOGY   ISSN 0066-4286
     Vols. 1–17 (1963–79): $17.00 (USA), $17.50 (elsewhere) per copy
     Vol. 18 (1980): $20.00 (USA), $21.00 (elsewhere) per copy
     Vol. 19 (avail. Sept. 1981): $20.00 (USA), $21.00 (elsewhere) per copy

Annual Review of PLANT PHYSIOLOGY   ISSN 0066-4294
     Vols. 10–31 (1959–80): $17.00 (USA), $17.50 (elsewhere) per copy
     Vol. 32 (avail. June 1981): $20.00 (USA), $21.00 (elsewhere) per copy

Annual Review of PSYCHOLOGY   ISSN 0066-4308
     Vols. 4, 5, 8, 10–31 (1953, 1954, 1957, 1959–80): $17.00 (USA), $17.50 (elsewhere) per copy
     Vol. 32 (avail. Feb. 1981): $20.00 (USA), $21.00 (elsewhere) per copy

Annual Review of PUBLIC HEALTH   ISSN 0163-7525
     Vol. 1 (1980): $17.00 (USA), $17.50 (elsewhere) per copy
     Vol. 2 (avail. May 1981): $20.00 (USA), $21.00 (elsewhere) per copy

Annual Review of SOCIOLOGY   ISSN 0360-0572
     Vols. 1–5 (1975–79): $17.00 (USA), $17.50 (elsewhere) per copy
     Vol. 6 (1980): $20.00 (USA), $21.00 (elsewhere) per copy
     Vol. 7 (avail. Aug. 1981): $20.00 (USA), $21.00 (elsewhere) per copy

To ANNUAL REVIEWS INC., 4139 El Camino Way, Palo Alto, CA 94306 USA   (Tel. 415-493-4400)
Please enter my order for the following publications:
(Standing orders: indicate which volume you wish order to begin with)

_____, Vol(s). _____ Standing order ☐

_____, Vol(s). _____ Standing order ☐

_____, Vol(s). _____ Standing order ☐

_____, Vol(s). _____ Standing order ☐

Amount of remittance enclosed $_____   California residents please add applicable sales tax.
Please bill me ☐   Prices subject to change without notice.

SHIP TO (include institutional purchase order if billing address is different)

Name _____

Address _____

_____

_____ Zip Code _____

Signed _____ Date _____

☐ Please add my name to your mailing list to receive a free copy of the current Prospectus each year.
☐ Send free brochure listing contents of recent back volumes for *Annual Review(s)* of _____